# Peptide and Protein Drug Analysis

# DRUGS AND THE PHARMACEUTICAL SCIENCES

**James Swarbrick, Executive Editor**
*AAI, Inc.*
*Wilmington, North Carolina*

## Advisory Board

# DRUGS AND THE PHARMACEUTICAL SCIENCES

A Series of Textbooks and Monographs

*edited by*

**James Swarbrick**
*AAI, Inc.*
*Wilmington, North Carolina*

1. Pharmacokinetics, *Milo Gibaldi and Donald Perrier*
2. Good Manufacturing Practices for Pharmaceuticals: A Plan for Total Quality Control, *Sidney H. Willig, Murray M. Tuckerman, and William S. Hitchings IV*
3. Microencapsulation, *edited by J. R. Nixon*
4. Drug Metabolism: Chemical and Biochemical Aspects, *Bernard Testa and Peter Jenner*
5. New Drugs: Discovery and Development, *edited by Alan A. Rubin*
6. Sustained and Controlled Release Drug Delivery Systems, *edited by Joseph R. Robinson*
7. Modern Pharmaceutics, *edited by Gilbert S. Banker and Christopher T. Rhodes*
8. Prescription Drugs in Short Supply: Case Histories, *Michael A. Schwartz*
9. Activated Charcoal: Antidotal and Other Medical Uses, *David O. Cooney*
10. Concepts in Drug Metabolism (in two parts), *edited by Peter Jenner and Bernard Testa*
11. Pharmaceutical Analysis: Modern Methods (in two parts), *edited by James W. Munson*
12. Techniques of Solubilization of Drugs, *edited by Samuel H. Yalkowsky*
13. Orphan Drugs, *edited by Fred E. Karch*
14. Novel Drug Delivery Systems: Fundamentals, Developmental Concepts, Biomedical Assessments, *Yie W. Chien*
15. Pharmacokinetics: Second Edition, Revised and Expanded, *Milo Gibaldi and Donald Perrier*
16. Good Manufacturing Practices for Pharmaceuticals: A Plan for Total Quality Control, Second Edition, Revised and Expanded, *Sidney H. Willig, Murray M. Tuckerman, and William S. Hitchings IV*
17. Formulation of Veterinary Dosage Forms, *edited by Jack Blodinger*
18. Dermatological Formulations: Percutaneous Absorption, *Brian W. Barry*
19. The Clinical Research Process in the Pharmaceutical Industry, *edited by Gary M. Matoren*
20. Microencapsulation and Related Drug Processes, *Patrick B. Deasy*
21. Drugs and Nutrients: The Interactive Effects, *edited by Daphne A. Roe and T. Colin Campbell*
22. Biotechnology of Industrial Antibiotics, *Erick J. Vandamme*
23. Pharmaceutical Process Validation, *edited by Bernard T. Loftus and Robert A. Nash*

# Peptide and Protein Drug Analysis

edited by

## Ronald E. Reid
*University of British Columbia*
*Vancouver, British Columbia, Canada*

**CRC Press**
Taylor & Francis Group
Boca Raton  London  New York

CRC Press is an imprint of the
Taylor & Francis Group, an **informa** business
A TAYLOR & FRANCIS BOOK

CRC Press
Taylor & Francis Group
6000 Broken Sound Parkway NW, Suite 300
Boca Raton, FL 33487-2742

First issued in paperback 2019

© 2000 by Taylor & Francis Group, LLC
CRC Press is an imprint of Taylor & Francis Group, an Informa business

No claim to original U.S. Government works

ISBN-13: 978-0-8247-7859-0 (hbk)
ISBN-13: 978-0-367-39926-9 (pbk)

**Visit the Taylor & Francis Web site at**
**http://www.taylorandfrancis.com**

**and the CRC Press Web site at**
**http://www.crcpress.com**

# Preface

Peptides and proteins are biological workhorses performing many structural, functional, and regulatory roles in the cell and its surrounding environment. Movement, proliferation, differentiation, and communication are just a few of the complicated cellular functions involving peptides and proteins. The potency and specificity of polypeptide function are the standards set by nature and are frequently sought after by the pharmaceutical scientist. In order to harness this potency and specificity to therapeutic use, the structural, chemical and functional properties of proteins and peptides must be examined in detail. The major obstacle to accurate chemical, structural, and functional analysis of proteins and peptides is the complexity of the molecules, in terms of both conformational flexibility and amino acid sequence variability. Also, the frequent association of polypeptides with lipids and carbohydrates further complicates the analysis of polypeptides as potential therapeutic agents.

Strategies for polypeptide analysis have been aided by the invention of solid-phase peptide synthesis and the development of recombinant DNA technology that provide the means by which peptides and proteins can be produced in large quantities with specific changes in sequence. These technological advances have fostered the development of protein and peptide structure–function relationships. The recent application of solid-phase technology to single-batch multiple synthesis of peptides and expression of proteins on the surface of bacteriophage has led to the development of combinatorial chemical and biological techniques that allow the chemist to prepare a large number of peptide analogs in one synthesis and to rapidly screen them for pharmacological activity, thus providing a ready source of lead molecules for further study. The development of novel in vitro translation systems for the production of proteins has opened up the area of protein chemistry to the introduction of new, nonproteinogenic amino acids into proteins as a tool for the production of modified and novel activities. The evolution of the desktop computer into a powerful computing tool and the design of algorithms for the theoretical analysis of amino acid sequences have added versatility to the prediction of protein and peptide structure and the interpretation of experimental structure–function studies. These tools and techniques expose the biochemical machinery of the cell to functional analysis with the aim of developing a detailed understanding of the biochemical mechanisms of disease and producing novel therapeutic agents.

Microheterogeneity of proteins and peptides prepared chemically or biologically places tremendous importance on the analysis of chemical purity which relies mainly on the strengths of high pressure liquid chromatography (HPLC) and gel electrophoresis. The basic normal and reversed phase HPLC techniques have been augmented with a number of variations that have strengthened HPLC as an analytical and preparative tool. The recent development of capillary electrophoresis and isoelectric focusing has taken the analytical capabilities of HPLC and gel electrophoresis to yet higher resolution. A major advance in the analysis of chemical purity of polypeptides has come with the use of mass spectrometry as a method of molecular weight determination and amino acid sequence analysis.

The functional analysis of peptides and proteins as therapeutic agents is problematic. These agents differ from the conventional chemical therapeutic agent in that polypeptides are normal constituents of the cell and distinguishing between the levels of therapeutic and endogenous agent is extremely difficult, if not impossible. Pharmacokinetics and pharmacodynamics of the therapeutic agent differ very little, if at all, from the endogenous compound, and a better understanding of these areas will lead to a more detailed understanding of the distribution, elimination, and function of polypeptide therapeutic agents. Immunoassays have been the mainstay of the investigation of functional analysis of polypeptides. Recent attempts have been successful in developing bioassays that are more in tune with the molecular mechanism of the polypeptide under investigation.

The tremendous advances in molecular biology over the past 20 years have afforded new techniques to aid in the analysis of therapeutic agents. Pharmacogenomics and bioinformatics have arisen in response to the research implicating an individual's genetic makeup with effective drug therapy and the development of toxicities. These sciences will grow in importance as more and more information becomes available with development of the Human Genome Project.

Structural analysis of polypeptides has also been assisted by the chemical and biochemical advances in peptide and protein synthesis. Large quantities of pure polypeptide are needed both for the preparation of crystals suitable for x-ray analysis and for the determination of solution structure using nuclear magnetic resonance spectroscopy. Pulse excitation followed by Fourier transformation, multinuclear flexibility, increasingly powerful high field superconducting magnets, and major advances in computer technology for data handling have advanced nuclear magnetic resonance as a major source of polypeptide structural information.

A unique opportunity exists for pharmaceutical sciences to fill the expertise gap between the chemistry of polypeptides and the biological activity of these molecules as they can be manipulated for therapeutic use. The aim of this book is to provide a broad, basic understanding of the strategy and techniques behind the structure, function, and chemical analysis of polypeptides as therapeutic agents upon which chemists, biochemists, and pharmaceutical scientists alike can build a knowledge base that will provide a strong background in drug design. To this end, the book is divided into four parts. Part I, Chapters 1 through 10, provides a background to the strategies of peptide and protein design and synthesis. Part II, Chapters 11 through 16, presents the basics of chemical purity analysis through HPLC; gel and capillary electrophoresis; and carbohydrate, amino acid sequence, and mass spectrometric analysis. Part III, Chapters 17 through 21, develops the basic analytical techniques in determining the biological and pharmacological activity of polypeptide therapeutics. Finally, in Part IV, Chapters 22 through 27 detail some of the techniques of structural analysis of proteins and peptides.

This book will be a valuable aid for the student of pharmaceutical sciences, both graduate and advanced undergraduate, and scientists working in the academic or industrial environment of drug research. It is intended that *Peptide and Protein Drug Analysis* will complement Vincent Lee's book, *Peptide and Protein Drug Delivery* (Marcel Dekker, 1990), and that the two books together will provide complete coverage of the biotechnology of polypeptide drug design.

This book is dedicated to Sheena, Cara, and Aly.

*Ronald E. Reid*

# Contents

## II. Chemical Analysis

## III. Functional Analysis

## IV. Structural Analysis

# Contributors

**Marie-Isabel Aguilar, Ph.D.**   Department of Biochemistry and Molecular Biology, Monash University, Clayton, Victoria, Australia

**G. M. Anantharamaiah, Ph.D.**   Department of Medicine, Atherosclerosis Research Unit, University of Alabama at Birmingham, Birmingham, Alabama

**Graeme J. Anderson, B.Sc., Ph.D.**   Department of Chemistry, Manchester Metropolitan University, Manchester, United Kingdom

**Rene Braeckman, Ph.D.\***   Department of Pharmacokinetics and Pharmacodynamics, Chiron Corporation, Emeryville, California

**Lorraine D. Buckberry**   Biomolecular Science Laboratory Science and Engineering Research Center, De Montfort University, Leicester, United Kingdom

**Walter J. Chazin, Ph.D.**   Department of Molecular Biology, The Scripps Research Institute, La Jolla, California

**Anthony B. Chen, Ph.D.**   Quality Control Department, Genentech Inc., South San Francisco, California

**Herbert C. Cheung, Ph.D.**   Department of Biochemistry and Molecular Genetics, University of Alabama at Birmingham, Birmingham, Alabama

**Graham J. Cotton, Ph.D.**   Laboratory of Synthetic Protein Chemistry, The Rockefeller University, New York, New York

---

\* *Current affiliation*: Vice President of Pharmaceutical Development, Ceptyr, Inc., Bothell, Washington

**Ann K. Daly, Ph.D.**   Department of Pharmacological Sciences, University of Newcastle upon Tyne, Newcastle upon Tyne, United Kingdom

**Geeta Datta, Ph.D.**   Department of Medicine, Atherosclerosis Research Unit, University of Alabama at Birmingham, Birmingham, Alabama

**Nancy D. Denslow, Ph.D.**   Department of Biochemistry and Molecular Biology, University of Florida, Gainesville, Florida

**Michael J. Dunn, Ph.D.**   National Heart and Lung Institute, Imperial College School of Medicine, London, United Kingdom

**Mark R. Ehrhardt, Ph.D.**   Department of Biochemistry and Biophysics, University of Pennsylvania, Philadelphia, Pennsylvania

**Henryk Eisenberg, Ph.D.***   National Institutes of Health, Bethesda, Maryland

**Gregg B. Fields, Ph.D.**   Department of Chemistry and Biochemistry, Florida Atlantic University, Boca Raton, Florida

**Michael C. Fitzgerald, Ph.D.**   Department of Chemistry, Duke University, Durham, North Carolina

**Patrick L. Franchini**   Faculty of Pharmaceutical Sciences, University of British Columbia, Vancouver, British Columbia, Canada

**Vic Garner, Ph.D.**   Department of Medicine, University of Manchester, Manchester, United Kingdom

**Jane Grove**   Department of Pharmacological Sciences, University of Newcastle upon Tyne, Newcastle upon Tyne, United Kingdom

**Andreas Hühmer, Ph.D.**[†]   Department of Chemistry, University of Pittsburgh, Pittsburgh, Pennsylvania

**Q. Khai Huynh, Ph.D.**   Searle Research and Development, Monsanto/Searle Company, St. Louis, Missouri

**Nigel Jenkins, Ph.D.**   Eli Lilly & Co., Indianapolis, Indiana

**James P. Landers, Ph.D.**   Department of Chemistry, University of Pittsburgh, Pittsburgh, Pennsylvania

---

*Current affiliation*: Structural Biology Department, Weizmann Institute of Science, Rehovot, Israel
[†] *Current affiliation*: Texas Instruments Incorporated, Dallas, Texas

**Andrew L. Lee, Ph.D.**  Department of Biochemistry and Biophysics, University of Pennsylvania, Philadelphia, Pennsylvania

**Scott A. Lesley, Ph.D.**  Novartis Institute for Functional Genomics, La Jolla, California

**Jin Li, Ph.D.**  Department of Computational Chemistry, Proteus Molecular Design Ltd., Macclesfield, Cheshire, United Kingdom

**Ming Li, M.D.**  Department of Orthopaedics and Traumatology, Prince of Wales Hospital, The Chinese University of Hong Kong, Hong Kong

**Hsieng S. Lu, Ph.D.**  Department of Protein Structure, Amgen Inc., Thousand Oaks, California

**Derek Marsh, D.Phil.**  Max-Planck-Institut für biophyskalische Chemie, Gottingen, Germany

**Lee Anne Merewether**  Department of Protein Structure, Amgen Inc., Thousand Oaks, California

**Vinod K. Mishra, Ph.D.**  Department of Medicine, Atherosclerosis Research Unit, University of Alabama at Birmingham, Birmingham, Alabama

**Joseph B. Monahan, Ph.D.**  Department of Biochemistry and Molecular Biology, Monsanto/Searle Company, St. Louis, Missouri

**Tom W. Muir, Ph.D.**  Laboratory of Synthetic Protein Chemistry, The Rockefeller University, New York, New York

**Nicole J. Munro**  Department of Chemistry, University of Pittsburgh, Pittsburgh, Pennsylvania

**Robin E. J. Munro\***  Division of Mathematical Biology, National Institute for Medical Research, London, United Kingdom

**Maria C. Pietanza**  Laboratory of Synthetic Protein Chemistry, The Rockefeller University, New York, New York

**Ronald E. Reid, Ph.D.**  Faculty of Pharmaceutical Sciences, University of British Columbia, Vancouver, British Columbia, Canada

**Henriette A. Remmer, Ph.D.**  University of Illinois, Urbana, Illinois

**Barry Robson**  Computational Biology Center, IBM Thomas J. Watson Research Center, Hawthorne, New York

---

\* *Current affiliation*: LION Bioscience AG, Heidelberg, Germany

**Michael F. Rohde**   Department of Protein Structure, Amgen Inc., Thousand Oaks, California

**Tomi K. Sawyer, Ph.D.**   Department of Drug Discovery, ARIAD Pharmaceuticals, Cambridge, Massachusetts

**Pir M. Shah**   Biomolecular Science Laboratory Science and Engineering Research Center, De Montfort University, Leicester, United Kingdom

**Ned R. Siegel, B.S., M.S.**   Department of Biochemistry and Molecular Biology, Searle Discovery Research and Development, Monsanto/Searle Company, St. Louis, Missouri

**Nicholas J. Skelton, Ph.D.**   Department of Protein Engineering, Genentech Inc., South San Francisco, California

**Elizabeth Strickland**   Program in Molecular Biophysics, The University of Texas Southwestern Medical Center at Dallas, Dallas, Texas

**William R. Taylor**   Division of Mathematical Biology, National Institute for Medical Research, London, United Kingdom

**Ian R. Tebbett, Ph.D.**   Center for Environmental and Human Toxicology, College of Veterinary Medicine, University of Florida, Gainesville, Florida

**Philip J. Thomas, Ph.D.**   Department of Physiology, The University of Texas Southwestern Medical Center at Dallas, Dallas, Texas

**Jeffrey L. Urbauer, Ph.D.**   Department of Biochemistry and Biophysics, University of Pennsylvania, Philadelphia, Pennsylvania

**Bernard N. Violand, Ph.D.**   Department of Biochemistry and Molecular Biology, Searle Discovery Research and Development, Monsanto/Searle Company, St. Louis, Missouri

**A. Joshua Wand, Ph.D.**   Department of Biochemistry and Biophysics, University of Pennsylvania, Philadelphia, Pennsylvania

**Rachel L. Winston**   The Scripps Research Institute, La Jolla, California

**David C. Wood, Ph.D.**   Department of Biochemistry and Molecular Biology, Monsanto/Searle Company, St. Louis, Missouri

# 1
# Practical Principles of Protein Design

**Ronald E. Reid and Patrick L. Franchini**
*University of British Columbia, Vancouver, British Columbia, Canada*

## I.  INTRODUCTION

Molecular biology, a term with a multitude of meanings, brought together the classical basic sciences of physics and chemistry in an intellectual and empirical fusion that began with the description of the molecular structure of deoxyribonucleic acid in 1953 (1). A mere 45 years later, the science of molecular biology is on the verge of mapping the entire human genome (2). As a result of these efforts, molecular genetics has developed into a power-house of diagnostic science which has related a large number of gene sequences to such relevant diseases as muscular dystrophy and cystic fibrosis (3). Diagnosis and prevention are powerful tools in health care, but disease therapy is the most dramatic. Molecular medicine is in its infancy but in the guise of "gene therapy" has grabbed the headlines and public interest not only for its revolutionary approach to disease therapy but also for the social, ethical, economic, and legal implications following in its wake.

The movement of biological information from DNA to RNA to protein is well understood at the molecular level. However, this chain of events has yet another step that involves the biological function of the final product—the protein. Determination of the relationship between the protein and its biological function is necessary to an understanding of the genetic basis of life and disease. Anfinsen's studies on protein folding and his subsequent "Thermodynamic Hypothesis" were primarily responsible for the evolution of the concept that a protein's function is a consequence of it's conformation, and conformation is specified by the amino acid sequence or primary structure (4). The nature of the precise relationship between conformation and sequence has been a major topic of empirical and theoretical research since Anfinsen's seminal studies on ribonuclease A and has come to be known as "the protein folding problem" (5–10). The other half of the problem, the relationship of conformation to function, is another aspect of the protein folding problem – perhaps "the functional protein problem." The idea that protein function involves movement in loops or domains may require a certain amount of instability in the folded protein structure (11–14). It could be that a balance must be struck between protein conformational stability, and protein functional flexibility to obtain a truly native functional protein.

### A.  Protein Structural Hierarchy

The concept of structural hierarchy in proteins has greatly simplified the conceptual development of protein function from amino acid sequence. The dipeptide, -NH-HC$_{\alpha 1}$ (R$_1$)-CO-NH-HC$_{\alpha 2}$ (R$_2$)-CO-, is the fundamental structural unit of proteins and peptides. The peptide bond coupling the two amino acids in the dipeptide structure is repeated throughout the length of the polypeptide chain and will thus make major contributions to the structure, and possibly the function of the polypeptide. The partial double-bonded character of this bond is due to the resonance structures shown in Fig. 1a. As a result of its double bond character, rotation about the CO—NH bond is restricted and the atoms comprising the peptide bond tend to be coplanar. Two configurations about the CO—NH bond are possible, one in which the C$^{\alpha}$ atoms are *trans* to one another, and the other in which they are *cis* (Fig. 1b). With the exception of the Pro-X peptide bond, the *trans* isomer should have the least interaction between non-bonded atoms in the peptide bond, and thus is the more energetically favored configuration. The second major implication of the resonance structures of the peptide bond is that there is a redistribution of electrons in the structure, such that the carbonyl oxygen bears a 0.42 negative charge and the amide hydrogen bears a 0.2 positive charge resulting in a dipole moment of 3.5 Debye (Fig. 1c) (15).

A complete description of the conformational relationships of the atoms composing the dipeptide backbone can be made using the torsion angles denoted by $\phi, \psi, \omega$. If we have four atoms A, B, C and D in the arrangement shown in Fig. 2 and the plane normal to bond B—C, the angle between the projection of A—B and the projection of C—D is the torsion angle. This angle is considered positive ($+\theta$) or negative ($-\theta$) according as, when the system is viewed along the central bond in the direction of B → C (or C → B), the bond to the front atom A (or D) requires to be rotated to the right or to the left,

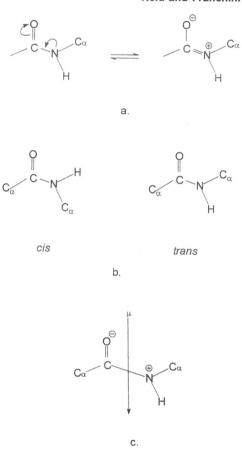

**Fig. 1** Schematic representation of the peptide bond: (a) resonance structures; (b) *cis* and *trans* isomers; (c) dipole moment.

respectively, in order that it may eclipse the bond to the rear of atom D (or A) (Fig. 2) (16). For the eclipsed conformation in which the projections of A—B and C—D coincide, $\theta$ is given the value 0° (*synplanar* conformation or *cis*). The bonds between the main chain atoms are denoted by the symbols of the two atoms terminating them, $N_i$—$C_i^{\alpha}$, $C_i^{\alpha}$—$C_i'$, $C_i'$—$N_{i+1}$. There are two torsion angles that describe the backbone conformation of each amino acid residue in the protein. The $\phi$ angle is the projection of $C_{i-1}'$—$N_i$ and $C_i^{\alpha}$—$C_i'$ (rotation about the $N_i$—$C_i^{\alpha}$ bond) and the $\psi$ torsion angle is the projection of $N_iC_i^{\alpha}$ and $C_i'$—$N_{i+1}$ (rotation about $C_i^{\alpha}$—$C_i'$

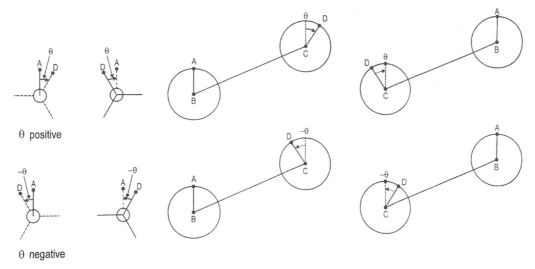

θ positive

θ negative

**Fig. 2** Newman and perspective projections illustrating positive and negative torsion angles. Note that a right-handed turn of the bond to the front atom about the central bond gives a positive value of $\theta$ from whichever end of the system is viewed. (Taken from Ref. 16 with permission of the publisher Academic Press).

bond) (Fig. 3). The $\omega$ torsion angle describes the conformational relationship of the $C^\alpha$ atoms which are the projections of $C_i^\alpha$—$C_i'$ and $N_{i+1}$—$C_{i+1}^\alpha$ (rotation about the $C_i'$—$N_{i+1}$ bond) and normally assume a value of 180° (*trans-planar*) because of the partial double bond charac-

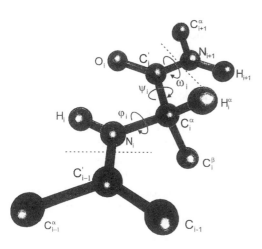

**Fig. 3** Schematic drawing of the dipeptide structural unit of a polypeptide showing the $\phi$, $\psi$, and $\omega$ angles.

ter of the amide bond. The conformation of amino acid side chains can be treated in the same way and the torsion angles are denoted by $\chi_i$. As an example, the projection of $N_1$—$C_1^\alpha$ and $C_1^\beta$—$C_1^\gamma$ (rotation about the $C_1^\alpha$—$C_1^\beta$ bond) is defined as $\chi_1$ and the projection of $C_1^\alpha$—$C_1^\beta$ on $C_1^\gamma$—$C_1^\delta$ (rotation about the $C_1^\beta$—$C_1^\gamma$ bond) is $\chi_2$, etc.

Assuming the $\omega$ torsion angle is roughly 180°, the linear amino acid sequence (primary structure) will fold into the protein's secondary structure that is described by the $\phi$ and $\psi$ torsion angles of every amino acid component of the protein or peptide. A plot of the $\phi$ and $\psi$ angles (Ramachandran map) of a large number of known protein structures (glycine residues omitted due to high conformational flexibility and proline residues omitted due to conformational constraints) indicates three areas in which the values for the possible $\phi/\psi$ combinations are primarily distributed (Fig. 4) (17). These conformational restrictions are due mainly to backbone interactions with adjacent residues and steric clashes with

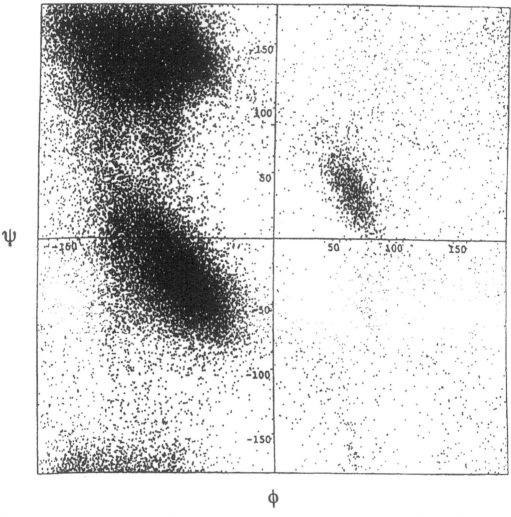

**Fig. 4** Ramachandran maps of known protein structures showing the $\phi$, $\psi$ distribution for all residues, except Gly and Pro, taken from 310 proteins in the Brookhaven Protein Databank.

the side chain $C_\beta$ atoms. The three areas describe the $\phi/\psi$ angles of folded protein secondary structure—the right handed $\alpha$-helix ($\alpha_R$) ($\phi/\psi \approx -60/-40$), the $\beta$-strand ($\beta$) ($\phi/\psi \approx -135/135$), and the left handed $\alpha$-helix ($\alpha_L$) ($\phi/\psi \approx 60/40$). For simplicity, the secondary structure of proteins is categorized into two regular structures, $\alpha$-helix and $\beta$-sheet, and two non-regular structures, loops and turns. Similarly, the side chain $\chi_1$ angles are assumed to adopt one of three preferred conformers: $g^-$, $t$ or $g^+$ (Fig. 5) (18).

Simple combinations of a few secondary elements with a specific geometric arrangement have been found to occur frequently in protein structures. These units have been called either supersecondary structures or motifs (19, 20). Some motifs have been associated with biological function, however, motifs usually have no specific biological function on their own, but are part of larger structural and functional assemblies (20). Several motifs usually combine to form compact globular structures called domains. For example,

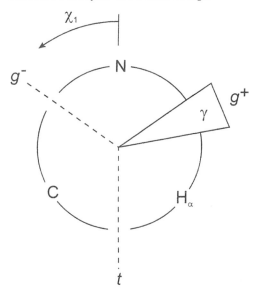

**Fig. 5** The geometry of rotation around the $C_\alpha$—$C_\beta$ bond. The side-chain is represented above the plane of the $C_\alpha$ substituents in Newman projection down the $C_\beta$—$C_\alpha$ bond. The $g^-$ position is defined as the position trans to the $H_\alpha$ atom; the $g^+$ position is defined as the position trans to the carbonyl group; and the t position is defined as the position trans to the nitrogen. The idealized $g^-$ (+60°), $t$ (180°), $g^-$ (+60°) are indicated. (Reprinted from J Janin, S Wodak, M Levitt, B Maigret. Journal of Molecular Biology 125: 357–386, 1978). Conformation of amino acid side-chains in proteins. (By permission of the publisher Academic Press).

the Rossman fold or nucleotide binding domain is composed of a number of $\beta\alpha\beta$ motifs (21). Protein tertiary structure is descriptive of both motifs arranged into domain structures, and a single polypeptide chain folded into one or several domains. The fundamental unit of tertiary structure is the domain and it is defined as a polypeptide chain or part of a polypeptide chain consisting of combinations of secondary elements and motifs that can independently fold into a stable tertiary structure (20). Frequently, domains are associated with a biological function and it is possible that the domain could be con-

sidered the basic functional unit of proteins, and may be important in the eventual description of the relationship between protein conformation and biological function. There are obviously possibilities for "grey areas" in this definition where a single independently folding functional motif within a protein will be called a domain. For our purposes we will equate domain with other terms, module and fold, appearing in the literature in a variety of contexts (22).

The single chain, tertiary structural units of some proteins associate into multiple units of quaternary structure. The subunits so formed can function independently or cooperatively, can be different polypeptide chains (heterologous subunits) or identical (homologous subunits), and may have different functions within the protein quaternary structure.

### B. Protein Folding

The design and engineering or re-engineering of therapeutically active proteins and peptides should be considered an exercise in controlled protein folding in the absence of chaperones (special proteins that assist protein folding in a living cell). With recognition that helices and sheets are the fundamental structural units of a protein, the folding problem becomes one of describing how these fundamental structural units are formed and arranged in conformational space. The "Core Hypothesis" states that, in general, non-polar side chains typically pack tightly in the interior of a protein forming the solvent-inaccessible protein core while the polar residue side chains are located on the protein surface (23–25). The hydrophobic effect, opposed by the loss of conformational entropy, is considered the driving force of protein folding, and this efficient search of conformational space for the unique, stable, and biologically active conformation is currently described by two mechanisms (26). The "Frame-

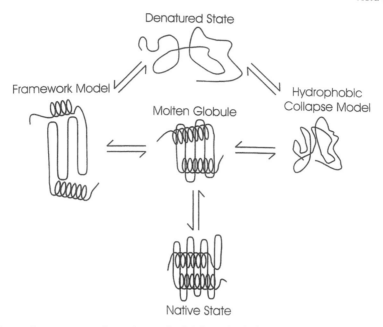

**Fig. 6** Schematic representation of protein folding depicting the Framework and Hydrophobic Collapse models of protein folding. The molten globule is shown as the common feature of both models.

work Model" of protein folding postulates the rapid, local formation of native-like secondary structure which then either coalesces through diffusion and collision to form the tertiary structure or acts as a nucleation site for initiation and propagation of structure to form the tertiary structure (Fig. 6) (7, 27–32). The "Hydrophobic Collapse Model" of protein folding postulates a general hydrophobic collapse of the protein followed by structural rearrangement leading to formation of secondary structure (Fig. 6) (33, 34). It has been suggested that small single domain protein molecules may fold via the "Hydrophobic Collapse" model and larger multidomain proteins may fold domain by domain via the "Framework Model" (26). However, the "new view" of "folding funnels" proposes multiple folding pathways (10).

The hierarchic organization of the "Framework Model" utilizes the fact that the hydrophobic core of a protein is com-posed of $\beta$-sheets and $\alpha$-helices connected by loops and turns located on the surface of the protein (7). The polar backbone atoms of the helices and sheets are neutralized through hydrogen bond formation prior to condensing to form the core of the protein. These secondary structural elements coalesce through hydrophobic interactions to form the super-secondary structural motifs and domains. The interesting feature of this aspect of the 'Framework Model" is that each level of structure formation during folding utilizes cooperativity between new "quasi-independent" stabilizing factors which enhance those of earlier steps, with no energetic compromises required at any level (7). Thus the selection of secondary structural elements by an arbitrary chain segment is the key feature of Rose and Wolfenden's "Hierarchic Framework Model" of protein folding and the authors propose a redundant stereochemical code embracing the interplay between shape and polarity of residue side chains and secondary structure conformation (7). The pattern of hydrophobic and hydrophilic residues occurring in the primary structure may contribute to the folding pattern through both the polar interactions of secondary structure formation and the hydrophobic interactions of super-secondary and tertiary structure formation (35). This model anticipates the fact that many proteins with non-homologous primary structures have similar tertiary structure (36) and predicts that conformational information is distributed throughout the primary structure and will be fairly stable to alteration or deletion of a few amino acid residues in the sequence (37). The model also anticipates four important goals that protein design must achieve (38):

1. Initiate and terminate elements of secondary structure.
2. Create a hydrophobic core.
3. Supply sequences that encode loops and hairpin turns in the protein backbone, to connect sequential, secondary structural elements
4. Create specific tertiary links, that is, energetically favorable interactions between amino acids that are distant in the sequence.

The molten globule (Fig. 6) was predicted to be a kinetic intermediate in protein folding formed by the collapse of preexisting secondary structural units with the arrangement of side chains in the molten globule satisfying the criterion of maximal screening of hydrophobic side groups from contacts with water (39). This predicted intermediate has compact, native-like secondary structure and a native-like tertiary fold, but differs from the native state by a lack of specific tertiary interactions. The molten globule corresponds to either the nonspecific collapsed intermediate of the "Hydrophobic Collapse Model" or an expanded native-like protein in the "Framework Model" (Fig.

6). The molten globule has been detected in a number of protein mutants and under non-native conditions as well as being a characteristic feature of a number of proteins designed *ab initio*. This structure has drastically changed properties when compared to native proteins. These changed properties include the following (40):

1. Loss of activity.
2. Do not bind antibodies to the native protein
3. Do not display cooperative thermal or denaturant induced unfolding and are marginally stable.
4. Nuclear magnetic resonance signals are broad and less dispersed than those of the native protein. However, many resonances are well shifted from those characteristic of random coil polypeptides.
5. Many amides show protection from H/D exchange with the "protection factors" intermediate between a random polypeptide and native protein.
6. Retain much of the native far UV absorption and CD ellipticity but lose aromatic CD.
7. Have increased hydrophobic surface demonstrated by increased anilinonaphthalene sulfonamide (ANS) binding and increased quenching of tryptophan fluorescence by acrylamide.
8. Have a radius of gyration less than that of a random coil polymer and also slightly bigger than the native protein.

## C. Protein Structure and Function

A corollary to the protein folding problem is determining the relationship between protein conformation and protein function. This relationship is critical for those designing proteins for a particular biological function or therapeutic activity. If function is a consequence of conformation, then the hierarchy describing protein structure must be a component of protein functional description. We will assume that the concept of "function" is a complex of the measurable entities of binding, specificity, and activity. These components can be related to the arrangement of protein side chain functional groups in space for interaction with a particular conformation of a ligand or substrate. The arrangement of the histidine, serine, and aspartic acid catalytic triad in serine proteases is a classical example of the "activity" category in the concept of function. Features of the fundamental structural unit (dipeptide) such as planarity, polarity, hydrophobicity, and hydrophilicity could be translated into function via the formation of secondary, super-secondary and tertiary structural units. The dipole moment of the $\alpha$-helix has been associated with ionic binding sites in proteins (41), and super-secondary structural elements have been associated with ligand binding to proteins (11). Enzyme substrate and cofactor binding sites have been found in crevices formed at the carboxy end of parallel strands of $\beta$-sheet which are a result of the right-handedness of the $\beta\alpha\beta$ motif (see Fig. 17b) and can be predicted from the strand order (the order in which the $\beta$-strands appear in the linear amino acid sequence compared to the order of their appearance in the $\beta$-sheet) of the sheet (11). Differences in strand order give rise to differences in the position and number of crevices.

The domain is considered a unit of protein structure that plays a key role in protein folding; however, domains in some proteins have a specific effect on function (11, 12, 42). It would appear that the domain is the subglobular unit that links protein conformation to protein function and is one of the general features of protein structure that is defined by function (42). Movement of domains within enzyme structures as part of the catalytic mechanism has been documented (12, 14). Multidomain proteins generally have a cleft between domains, and protein active sites occur at these interfaces. In proteins with more than one ligand, the different binding sites usually occur on different domains, and proteins with multiple functions usually have the different functions residing in different domains.

The principles guiding the design of biologically or therapeutically active proteins will create a hierarchy of structural elements building functional domains within proteins. This chapter attempts to describe this hierarchy.

## II. ENGINEERING STRUCTURE

### A. Secondary Structure

#### 1. $\alpha$-Helix

##### a. Structure

The $\alpha$-helix is a secondary structural element in which the polypeptide backbone adopts a coiled arrangement that is stabilized by a system of hydrogen bonds between backbone amide hydrogens and carbonyl oxygens four residues apart (see Fig. 7). Existence of the $\alpha$-helix was predicted by Linus Pauling in 1951 to explain x-ray diffraction patterns observed in crystalline proteins (43). This prediction was confirmed with the solution of the x-ray structure of myoglobin (44). Approximately 30% of all residues in proteins exist in the $\alpha$-helical conformation (45) making this the most common secondary structural type. The right-handed $\alpha$-helix which predominates in natural proteins, can be as short as five residues, and as long as 40 but has an average length of 10 residues (46). The residues that make up an $\alpha$-helix display characteristic $\phi$, $\psi$ angles of approximately $-60°$ and $-40°$, respectively. Three parameters; the pitch, the radius, and the number of residues per turn characterize the geometry of the $\alpha$-helix. The ideal $\alpha$-helix has a pitch of

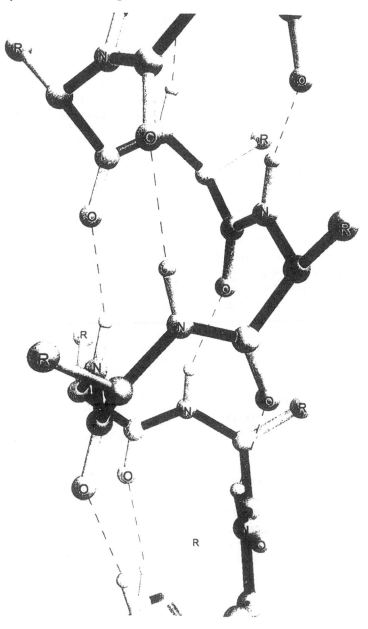

**Fig. 7** Structural model of the classic α-helix. The amino acid side chains are designated "R". The hydrogen bonds are formed between the C=O of the $i$th residue and the NH of the $(i + 4)$th residue.

1.5 Å, a radius of 2.3 Å and 3.6 residues per turn. Another designation for the α-helix is the $3.6_{13}$ helix, reflecting that the loop formed by the hydrogen bond between the NH and C=O contains 13 atoms.

In addition to the classic α-helix, two other types of helix, designated the $3_{10}$ helix and the π helix, are observed in proteins. The $3_{10}$ helix has three residues per turn and is stabilized by backbone hydrogen bonds that occur between residues three residues apart. As indicated by the lower case "10" in the designation "$3_{10}$" helix, the loop formed by the hydrogen bond encompasses 10 atoms. The $3_{10}$ helix constitutes 3–4% of all residues in proteins and generally consists of three or four residues (45). $3_{10}$ helices are commonly observed as extensions at either the N or C termini of α-helices in proteins, with 32% existing as extensions at the N-terminus and 34% at the C-terminus (47). In π-helices the hydrogen bond occurs between the NH and the CO five residues along the helix. The π-helix is rare, having been observed in catalase (48) and cytochrome p450 (49). Of the three types of helices, the α-helix has been the most studied, and the factors that influence its stability are the best understood.

To engineer a polypeptide sequence that will fold into an α-helix, two basic considerations are reqired. First, the amino acid sequence chosen must favor α-helix formation. The 20 different amino acid side chains have been shown experimentally to influence helical stability. In addition, residue side chains can affect helical stability through interactions with helix dipoles, capping interactions or H-bond and salt bridge formation. The second consideration is the desired characteristics of the helix. For example, the choice of the residues can determine if the helix is hydophilic, hydrophobic or amphipathic. The choice of residues can also determine the shape of the helix, i.e. whether it is kinked or curved. A more detailed discussion of both considerations follows.

### b.   α-Helical Stability

There are two fundamental factors that influence α-helical stability: the intrinsic helical propensity of the residues involved and specific interactions between these residues in the helix. N and C capping, interaction of charged residues with the

helix dipole, salt-bridge formation and hydrogen bonding are among the relevant structural factors.

Helical Propensity.   Helical propensity can be described as the ability of the side chain of an amino acid to influence the polypeptide backbone to adopt an α-helical conformation, i.e. the transition from random coil to helix. Two closely related helix-coil transition theories have been developed to describe the process of α-helix formation: the Zimm–Bragg theory (50), and the Lifson–Roig theory (51). Both theories describe two basic steps, nucleation and propagation of the α-helix. Nucleation is the spontaneous formation of a helical segment anywhere along the unfolded polypeptide chain whereas propagation involves the addition of a helical residue to an existing helical segment. In both theories, each residue has both nucleation and propagation parameters which are $\sigma$ and $s$ in the Zimm-Bragg theory, and $v$ and $w$ in the Lifson-Roig theory. The relationship between the parameters described by these two theories has been derived by Qian and Schellman (52). A number of attempts have been made to quantify the nucleation and propagation parameters (i.e. the helical propensity) of the various amino acids as described below.

In 1974 Chou and Fasman proposed that the frequencies of amino acids in middle positions of protein helices should quantitatively reflect the helical propensity (53). This proposal resulted in the development of a helical propensity scale based on helical residue frequency statistics. Since then a number of model helical systems have been studied to derive scales for α-helical propensity using the *guest host system* i.e. an amino acid in the host helix is substituted at one position by the guest amino acids, and the change in helix content is measured. Helical propensity studies have been carried out using monomeric helices (54–61), coiléd-coils (62), and whole protein models (63–65). Some of the helix propensity scales, standardized to glycine, are

**Table 1** Measured Free Energy Change on Helix Formation[a]

| Residue | AK/AQ[b] | E$_4$K$_4$[c] | Coiled coil[d] | T4 lysozyme[e] | Barnase[f] |
|---|---|---|---|---|---|
| Ala | −1.88 | −0.74 | −0.77 | −0.96 | −0.91 |
| Arg$^+$ | −1.67 | — | −0.68 | −0.77 | −0.77 |
| Asn | −0.99 | −0.14 | −0.07 | −0.39 | −0.25 |
| Asp$^-$ | −1.00 | — | −0.15 | −0.42 | −0.20 |
| Asp$^0$ | −1.00 | — | — | — | — |
| Cys | −1.06 | — | −0.23 | −0.42 | −0.09 |
| Gln | −1.31 | −0.41 | −0.33 | −0.80 | −0.43 |
| Glu$^-$ | −1.20 | — | −0.27 | −0.53 | −0.36 |
| Glu$^0$ | −1.40 | — | — | — | — |
| Gly | −0.00 | −0.00 | −0.00 | −0.00 | −0.00 |
| His$^0$ | −1.07 | — | −0.06 | −0.57 | −0.13 |
| His$^+$ | −0.10 | — | — | — | — |
| Ile | −1.18 | −0.32 | −0.23 | −0.84 | −0.10 |
| Leu | −1.60 | −0.55 | −0.62 | −0.92 | −0.56 |
| Lys$^+$ | −1.52 | — | −0.65 | −0.73 | −0.72 |
| Met | −1.37 | −0.50 | −0.50 | −0.86 | −0.60 |
| Phe | −0.95 | — | −0.41 | −0.59 | −0.22 |
| Pro | $\gtrless$3.00 | — | $\gtrless$3.00 | +2.50 | — |
| Ser | −1.10 | −0.23 | −0.35 | −0.53 | −0.50 |
| Thr | −0.56 | −0.18 | −0.11 | −0.54 | −0.12 |
| Trp | −1.1 to −1.97 | — | −0.45 | −0.58 | −0.07 |
| Tyr | −1.3 to −1.1 | — | −0.17 | −0.72 | −0.09 |
| Val | −0.83 | −0.27 | −0.14 | −0.63 | −0.03 |

Correlation coefficient[g]

| | | | | | |
|---|---|---|---|---|---|
| AK/AQ | 1.00 | (0.93) | 0.81 (0.89) | 0.88 (0.91) | 0.77 (0.85) |
| E$_4$K$_4$ | — | 1.00 | (0.95) | (0.86) | (0.85) |
| Coiled coil | — | — | 1.00 | 0.70 (0.79) | 0.76 (0.93) |
| T4 lysozyme | — | — | — | 1.00 | 0.59 (0.66) |
| Barnase | — | — | — | — | 1.00 |

[a] Data in kcal/mol. Values relative to glycine ($\Delta\Delta G°$) from different peptide, coiled-coil and protein systems. (Table reprinted from A Chakrabartty, RL Baldwin. (1995) Advances in Protein Chemistry 46: 141–176 with permission of Academic Press).

[b] From (58).

[c] From (59).

[d] From (62).

[e] From (63).

[f] From (64).

[g] Correlation coefficients in parentheses are from analysis of uncharged nonaromatic residues.

summarized in Table 1. In all the scales save one, the alanine residue demonstrates the highest helical propensity. This result can be explained by the shortness of the alanine side chain, which does not lose entropy on helix formation (66) and by its non-polar nature, which allows alanine to participate in hydrophobic interactions with the backbone (63). Amino acids with long non $\beta$-branched side chains (Arg, Leu, Lys, Gln, Glu, Met) follow alanine in helical propensity. $\beta$-branched amino acids (Val, Ile, Thr) have a restricted $\chi_1$ angle in the $\alpha$-helix (18, 67), and undergo greater loss of side chain entropy on $\alpha$-helix formation which is reflected in lower helical propensity. The aromatic residues (Phe, Trp, Tyr) generally display lower helical propensity than non-aromatic residues. This trend has been attributed to a greater relative loss of side chain entropy by the aromatic residues (64, 66). The rest of the residues have intermediate to low helical propensities with Gly and Pro having the lowest helical propensity in all scales. The low helix propensity of Pro can be explained by the fact that it lacks an amide proton that is part of helical H-bonding in the backbone (68). The low helix propensity of Gly can be explained by the greater flexibility of the backbone due to the lack of a side-chain.

All the scales correlate positively with each other (average correlation = 0.76), with the greatest correlation occurring between scales based on monomeric peptides (correlation = 0.93) (69).

Helix Capping.   The residues located at the N and C-termini of an $\alpha$-helix are termed the Ncap and Ccap residues respectively (70). The nomenclature for the termini of helices is as follows:

N-terminus
　　$\cdots$N″-N′-Ncap-N1-N2-N3-$\cdots$

C-terminus
　　$\cdots$C3-C2-C1-Ccap-C′-C″-$\cdots$

where N1 to N3 and C1 to C3 designate helical residues and the primed residues designate non-helical residues flanking the helix at the N and C termini. The first (Ncap to N3), and last four residues (C3 to Ccap) of an $\alpha$-helix have only one of the two backbone hydrogen bonds that characterize this secondary structural type. This characteristic makes the terminal regions of an $\alpha$-helix less stable and prone to "fraying." Helix capping can stabilize the terminus of an $\alpha$-helix through short range interactions between the Ncap or Ccap residue side chains, and the backbone and solvent or through interactions

of the backbone in the vicinity of the Ncap and Ccap with the solvent.

Helix capping interactions have been shown to stabilize both peptide (71–73) and protein (74–77) helices. The relative Ncap propensities of the amino acids determined in a variety of systems are listed in Table 2. The Ncap propensity of an amino acid is independent of the $\alpha$-helical propensity and has been given the designation $n$ in the modified Lifson–Roig theory (78). Residues with short polar side chains (Asn, Ser, Thr) have the highest N-capping propensity because of their ability to form H bonds with the backbone which stabilize the helix (70, 79). The N-capping residues have been shown to be involved in a structure called the "N-capping box" (Fig. 8). The N-terminal capping box was first proposed by Presta and Rose and can be described as a system of reciprocal hydrogen bonds between the side chains and backbone amide hydrogens of the Ncap residue and the N3 helical residue (see Fig. 8) (79). The residues that are preferred in the Ncap and N3 positions are those that can form side chain hydrogen bonds such as Ser, Glu, Asp, Thr, Asn, His and Gln (80). A number of studies involving both peptide and protein helical models have demonstrated that Thr, Ser, Asn and Asp provide the greatest stability at the Ncap position (71, 76, 81–84). It has been observed that the Glu residue predominates at the N3 position (7, 47, 70). Work on the protein barnase has shown that N-capping can stabilize an $\alpha$-helix by 2.5 kcal mol$^{-1}$ (74). Stabilization provided by the H-bond from the Ncap side chain can be twice that from the H-bond provided by the N3 residue (85). The Ncap residue exhibits $\phi$ and $\psi$ angles of $94 \pm 15°$ and $167 \pm 5°$, respectively (86). In addition to the Ncap and N3 residues which are directly involved in the hydrogen bonding, it has been suggested that the capping box also includes a hydrophobic interaction that occurs between residues on either side of the

**Table 2** Measured Free Energy Change on N-Cap Formation[a]

| Residue | AK peptide[b] | T4 Lysozyme[c] | Barnase[d] |
|---------|---------------|----------------|------------|
| Ala | −0.00 | −0.00 | −0.00 |
| Asn | −1.41 | −2.20 | −0.86 |
| Asp[−] | — | — | −2.02 |
| Asp[0] | — | −1.90 | — |
| Gln | +0.93 | — | −0.42 |
| Glu[−] | — | — | −0.25 |
| Gly | −1.08 | −0.60 | -0.69 |
| Ile | −0.58 | — | -0.16 |
| Leu | −0.71 | — | — |
| Lys[+] | +0.50 | — | — |
| Met | −0.39 | — | — |
| Pro | −0.33 | — | +0.87 |
| Ser | −1.12 | −1.90 | −1.64 |
| Thr | −0.64 | −2.80 | −2.05 |
| Val | −0.10 | −0.00 | +0.15 |

[a] Data in kcal/mol. Values relative to alanine ($\Delta\Delta G°$) from different peptide and protein systems. (Table reprinted from A Chakrabartty, RL Baldwin. (1995) Advances in Protein Chemistry 46: 141–176 with permission of Academic Press).
[b] Values from applying Lifson–Roig theory modified to include helix capping (78) to Ala-Lys peptides (58).
[c] Values from substitutions at Ncap in T4 lysozyme (site 59) at pH 2.0 (76).
[d] Values from substitutions at two Ncap positions in barnase (sites 6 and 26) (75, 77).

capping box (87). In addition, electrostatic and capping interactions have been reported between the Ncap and the N4 residue (88).

It has been suggested that the N-terminal capping box acts as a stop signal for $\alpha$-helix formation at the N-terminus (80). However, sequences constituting the capping box have been identified in the middle of helices, thereby demonstrating that they are not necessarily stop signals (89). The capping box does appear to initiate the helical structure even if there are no residues N-terminal to the capping box (90). It has been reported that the N-cap residue can influence the type of helix. For example, Ser promotes $\alpha$-helix and discourages $3_{10}$-helix formation (91).

Approximately one-third of helical Ccap residues are glycine (70). Due to the lack of a side-chain, glycine can hydrogen bond to two successive backbone C=O groups while turning the chain to prevent helix continuation (70). Two H-bonding motifs have been identified which can be used to classify helices with a Ccap Gly (See Fig. 9) (92). The Shellman motif is characterized by a double hydrogen bonding pattern between the N—H at C' and C=O of C4 and between the N—H of the Ccap Gly and C=O of C3 (92). The $\alpha_L$ motif has only one H-bond between the N-H of the Ccap Gly and the C=O at C4 (92). Glycine followed by an apolar residue at C' will result in a Shellman motif provided that C3, C4 or C5 are also apolar (92). A polar residue following the Ccap Gly will in most cases result in an $\alpha_L$ motif (92). Although Gly predominates at the Ccap, attempts have been made to quantify the relative preferences for all amino acids at the Ccap position. In the protein barnase the Ccap preferences are Gly $\gg$

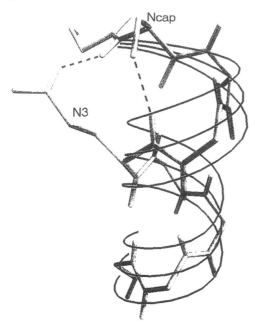

**Fig. 8** N-capping box from lamprey hemoglobin (PDB code 2LHB) formed between Ser12 and Glu15. Reciprocal H-bonds are formed between the side chains of the Ncap residue and the backbone amide hydrogen of the N3 residue and the side chain of the N3 residue and the backbone amide hydrogen of the Ncap residue.

His > Asn > Arg > Lys > Ala $\approx$ Ser > Asp (75). However, in a peptide model system little variation in preferences at the Ccap position was observed (83). As with the N-cap residue, C-capping boxes have been described. A possible C-capping box H-bond has been reported to form between a Ccap lysine and the C=O of a N4 leucine (93). However, this H-bonding is not reciprocal as in the N-Capping box. Another proposed capping motif involves a C′ Pro which stabilizes the C-terminus by restricting the Ccap $\phi$ and $\psi$ dihedral angles (94).

Dipole. The electric dipole arises from the parallel alignment of the individual peptide dipole moments along the helix axis. An accepted value for this dipole moment is 3.5 Debye which is equivalent to 0.7 e Å or $1.2 \times 10^{-29}$ C m (41). The dipole moment can be approximated by a 0.5–0.7 positive and negative unit charge near the N-terminus and C-terminus of the helix, respectively (41, 95, 96). The helix dipole has been demonstrated to have a number of roles in proteins from binding phosphate or sulphate groups at the positively charged N-termini of helices (41) to increasing the pKa of a histidine residue at the C-terminus of an $\alpha$-helix (97, 98) and decreasing it at the N-terminus (98). It has been suggested that modification of side chain pKa's by the helix dipole is essential for catalytic activity of a number of proteins (15).

The stabilizing effect of the helix dipole has been supported by the observation that in an anti-parallel arrangement of adjacent helices there is a stabilization of 5–7 kcal/mol whereas a parallel arrangement is relatively destabilized by 20 kcal/mol (96). In addition, it has been shown that the removal of the charge from the N-terminal amine and C-terminal carboxylate of a model peptide by pH titration stabilizes an $\alpha$-helix (99, 100). Altering the nature of the side chains along the length of the helix can also affect helix stability through interaction with the helix dipole. Negatively charged residues such as aspartate or glutamate increase helix stability at the N-terminus (101–104) whereas positively charged residues such as histidine or lysine (104, 105) decrease stability with neutral residues in between (98, 104). Positively charged residues increase helix stability at the C-terminus of an $\alpha$-helix (97, 98, 101, 105), and neutral residues have a greater stabilizing effect when located at the C-terminus than do negatively charged residues (106). The effect of a charged residue on helix stability is not localized to the N or C-terminus but affects stability throughout the helical region (101).

Salt Bridges, H-bonding and Hydrophobic Interactions. A salt bridge is an ionic interaction between positively and negatively charged amino acid side chains as well as between the $\alpha$-amino and carboxyl

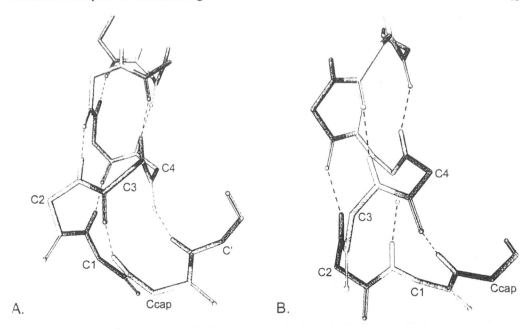

**Fig. 9** Examples of the Shellman (A) and αL (B) glycine based C-capping motifs. The Shellman motif is from the protein Triose Phosphate Isomerase (1TIM) and is characterized by a two backbone H-bonds from the Ccap (Gly120) and C3(Leu117) residues ($i$, $i$-3) and the C′ (Leu121) and C4 (Ala116) ($i$, $i$-5) residues. The αL motif is from the protein Thermolysin (8TLN) and is characterized by a single H-bond between the Ccap (Gly247) and C4 (Leu243) residues ($i$, $i$-4).

groups at the N and C-termini of peptides and proteins. A number of salt bridges in both model peptides and whole protein models have been observed to stabilize α-helices. Examples of salt bridges observed to stabilize α-helices in model peptides include Glu$^-$ ··· His$^+$ ($i$, $i+3$) (107–109), Glu$^-$ Arg$^+$ ($i$, $i+8$) (108), and Glu$^-$ ··· Lys$^+$ ($i$, $i+4$) (110). In whole proteins Asp$^-$ Arg$^+$ ($i$, $i+4$) (111), and Asp$^-$ ··· Arg$^+$ ($i$, $i+3$), (112) salt bridges have been observed. A salt bridge oriented so that it positively interacts with the helix dipole is more stable than if it does not (113, 114).

In addition to salt bridges between side chains in a helix, hydrogen bonding and hydrophobic interactions have also been observed. A hydrogen bond is observed between the side chains of glutamine and aspartate ($i$, $i+4$) (101). Studies of His-Asp side chain hydrogen bonding in an alanine based peptide demonstrate

that H-bonding only occurs in the His-Asp orientation and only occurs in the ($i$, $i+3$), ($i$, $i+4$) but not the ($i$, $i+5$) spacing (115). On average there are 18 sidechain hydrogen bonds for every 160 residues (116). A Gln ··· Asp ($i$, $i+4$) sidechain H-bonding interaction stabilizes peptide helices by −1.0 kcal/mol (117). Hydrophobic interactions have been observed between a residue in the N′ position and the residue in the N + 4 position (118).

*c.  Macroscopic Helical Characteristics*

The hydrophilicity or hydrophobicity of an α-helix can be controlled by including polar or non-polar residues in the sequence. Generation of an amphipathic α-helix involves designing a sequence that on folding forms an α-helix with opposing polar and non-polar faces oriented along the long axis of the helix. The amphipathic nature of α-helices was first described by Segrest et al. (119) and is characteristic of α-helices in lipid associated proteins (120–122). Segrest and co-workers have successfully created synthetic peptides that have the potential to form amphipathic α-helices (123, 124). In addition to its polar and non-polar sides, an amphipathic α-helix displays curvature through distortion of the H-bonding (125). This curvature results from the fact that carbonyl H-bonds are longer and less linear on the hydrophilic side than are H-bonds from carbonyls on the hydrophobic side (126). Another way to influence the shape of an α-helix is to introduce a kink into it with a proline residue. Approximately 3% of helices in globular proteins have been observed to have a proline residue in the middle of the helix that introduces a kink without terminating the helix (45). The kink results from the lack of a free amide proton, $N_i$ of proline at location $i$, which causes disruption of the $N_i \ldots O_{i-4}$ and $N_{i+1} \ldots O_{i-3}$ H-bonds. The unbonded amide hydrogen and carbonyl oxygens H-bond either to the solvent or to other groups in the protein (127).

## 2. β-Sheet

### a. Structure

Unlike the α-helix, the β-strand is not stable as an isolated secondary structural unit, and the mechanism of folding the β-strand to form a β-sheet has not been as well characterized as that for α-helices (128). However, a considerable understanding of the forces involved in secondary structure formation and their role in the formation of β-sheet structure is developing rapidly. The main complication in designing β-sheet structure is that this motif depends upon interactions between neighboring strands that are not necessarily but frequently separated in the linear sequence of the protein. This feature is in direct contrast to the $i \rightarrow i+4$ hydrogen bonding of a contiguous linear

amino acid sequence that is characteristic of α-helix formation. β-Sheet stability is dependent on intra- and inter-strand hydrogen bonding and hydrophobic interactions as well as the association of inter-sheet hydrophobic surfaces (129).

β-Sheets can associate to form tertiary or quaternary structure, and it is, therefore, probably more appropriate to consider the β-strand as secondary structure and the β-sheet as tertiary or quaternary structure. Strand order, number, and direction will combine to characterize a β-sheet and the β-sheet orientation, twist, and number will define the tertiary or quaternary structure formed by β-sheets. Pauling and Corey first described the parallel and antiparallel arrangements of β-strands to form the simple two stranded β-sheet (130). Both the antiparallel and parallel β-sheets are "pleated" with the α-carbon atoms alternating along the strands slightly above and below the plane of the sheet (Fig. 10) β-strands usually combine in numbers from 2 to 6 to form pure antiparallel sheets, pure parallel sheets, or mixed sheets containing both parallel and antiparallel strands. In fact there is a strong bias against the mixed sheets which may be due to the fact that the two different orientations of the strands require slightly different peptide bond orientation to permit the hydrogen bonding to occur. The number of residues in a β-strand varies from 2 to 17, but the majority of strands contain 3 to 10 residues with a preference for an odd number (131).

The antiparallel β-sheet consists of strands that have extended conformations with $\phi/\psi$ angles in the region of $-139°/+135°$ and alternate in chain direction or orientation (Fig. 11). The identity distance (the distance between repeating backbone structure) of the antiparallel sheet is 6.68 Å which corresponds to a translation of 3.34 Å/residue, and the inter-strand distance is approximately 5 Å. The inter-strand hydrogen

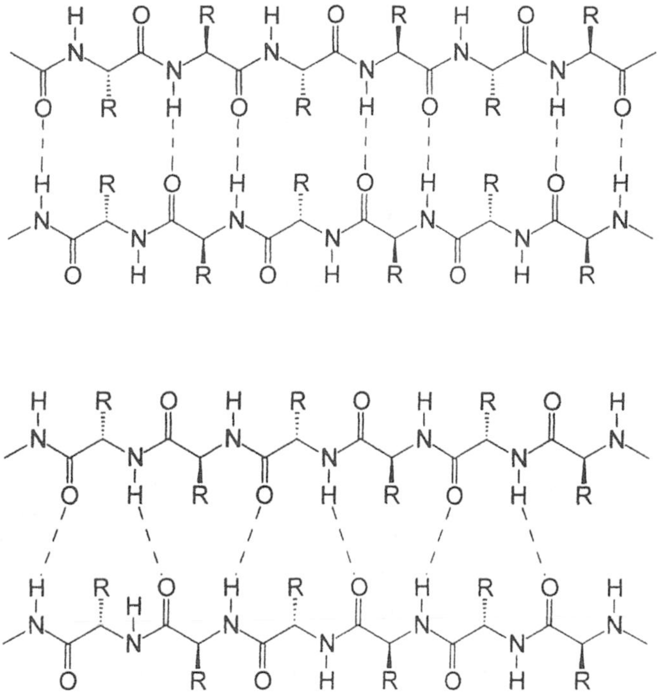

**Fig. 10** Schematic representation of parallel and antiparallel β-sheets.

bonds that are formed between the amide carbonyl oxygen and amino hydrogen of one chain with the similar groups of the adjacent chain are perpendicular to the strands and form alternating 10- and 14-member rings. The side chains of adjacent residues on the same strand protrude on opposite sides of the sheet while the side chains of adjacent residues on neighboring chains are located on the same side of the sheet and form residue pairs. The antiparallel β-sheet has two distinct types of residue pairs (132–134) (Fig. 11):

1. A pair for which the backbone atoms of the residues are hydrogen bonded to each other (H-bonded site—forms a 10 membered-bonded ring)
2. A pair for which the backbone atoms of the residues are not hydrogen bonded to each other (non-H-bonded site—forms a 14 membered H-bonded ring)

These two different pairs exhibit different $C_\alpha$ atom separation distances. The H-bonded 10 membered ring $C_\alpha$ atoms are separated by 5.5 Å while the non-H-bonded 14 membered ring $C_\alpha$ atoms are separated by 4.5 Å (134). For β-sheets with more than two strands there are one or more center strands in which all the amide oxygens and hydrogens are involved in hydrogen bonds with neighboring strands and two edge strands in which half of the amide oxygens and hydrogens are involved in interstrand hydrogen bonds. Antiparallel sheets generally have one face exposed to solvent and the other face buried in the protein hydrophobic core. As a result, the antiparallel strands often show an alternation of side chain hydrophobicity in the linear amino acid sequence. In general, the hydrophobicity of the central strand of a β-sheet is greater than that of the edge strand (135).

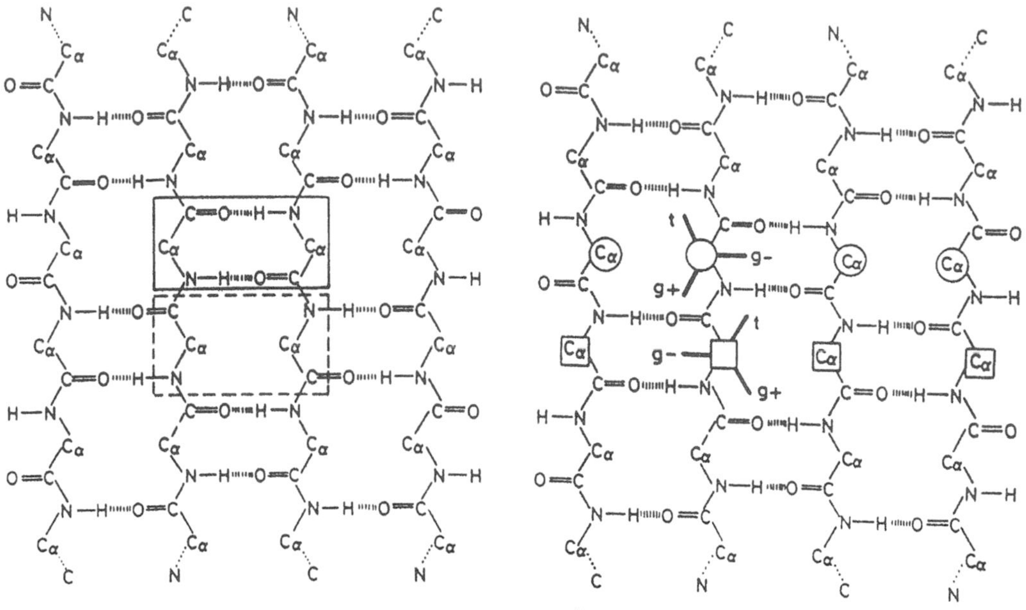

**a.**                                                          **b.**

**Fig. 11** (a) Antiparallel β-sheet showing the H-bonded (unbroken box) and non-H-bonded (dashed) sites. Each residue within a site is equivalent due to the 2-fold symmetry. Side chain rotamers for pairs of residues in the two sites are different; (b) Rotamers of non-β-branched residues are shown for one member of the pair in the H-bonded (○) and non-H-bonded (□) sites. The rotamers are indicated by g−, gauche −; g+, gauche +; and t, trans. Side chains of circled $C_{\alpha}$s point out of the page while those of boxed $C_{\alpha}$s point into the page. Taken from MA Wouters, PMG Curmi. Proteins: Structure, Function, and Genetics 22: 119–131, 1995. An analysis of side chain interactions and pair correlations with antiparallel β-sheets: Differences between backbone hydrogen-bonded and non-hydrogen bonded residue pairs. (Reprinted by permission of Wiley-Liss, Inc., a subsidiary of John Wiley and Sons, Inc.).

The β-strands of a parallel β-sheet have a generally less extended conformation than the anti-parallel β-strand with $\phi/\psi$ angles around −119°/113° which results in more pronounced pleating of the strand. The parallel strand has a slightly shorter identity distance (6.5 Å) than the antiparallel strand, and the inter-strand hydrogen bonding produces uniform 12-membered rings (Fig. 10). Parallel sheets rarely occur with fewer than five strands while antiparallel sheets often have two strands arranged in a twisted ribbon. The parallel sheet and parallel portions of mixed sheets are generally buried in a hydrophobic region of the protein with other main chain conformations protecting both faces of the sheet from exposure to the solvent. These facts suggest that the isolated two stranded parallel β-sheet is less stable than the isolated two stranded antiparallel β-sheet (136).

Some features of the β-sheet sound like a 60's dance craze, and include structures such as the bulge, bend, twist, and a variety of connectivities. The β-bulge consists of a region between two consecutive residues on one β-strand (the bulge strand, residues 1 and 2 in Fig. 12) which are hydrogen bonded to a single residue on the opposite β-strand (residue X in Fig. 12) (137–139). The bulge has been subdivided into three predominant types:

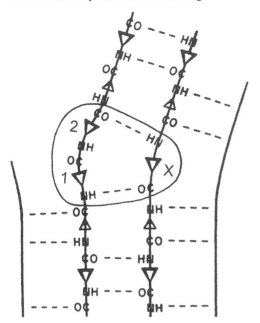

**Fig. 12** A β bulge (outlined region) at the edge of an antiparallel β-sheet. Smaller triangles represent side chains that are below the sheet, larger triangles those that are above it. (Reprinted from JS Richardson, ED Getzoff, DC Richardson. Proceedings of the National Academy of Science USA 75: 2574–2578, 1978. The β-bulge: A common small unit of non-repetitive protein structure. By permission of JS Richardson.)

classic, G1, and wide; and two less common types: bent and special (138).

The bulge appears to have important structural and functional implications for β-structure in a protein. Because the presence of a bulge disrupts the regular hydrogen bonding of the β-sheet, it is frequently seen in edge strands or at the end of a β-sheet. It is possible that this struc-ture could be used to prevent the uncontrolled growth of β-sheet but it has not yet been demonstrated that the β-bulge can be engineered *ab initio*\*. The bulge can be seen to line up the active site residues in enzymes as well as accentuate the twist in a β-sheet and accommodate insertions or deletions in β-strands while preserving the hydrogen-bonding pattern in the remainder of the strand (138).

β-strands may also exhibit various degrees of bending as initially described by Chothia for the orthogonal packing of β-pleated sheets in which a β-strand that is part of one β-sheet turns through a right-handed bend of 90° to become part of a second β-sheet (Fig. 13) (140). These β-bends are characterized by a very strong right-handed twist that occurs over three peptide units in the middle β-strand (strand II in Fig. 13). The middle strand may also change direction through the insertion of a residue in an α-conformation producing a β-bulge.

Large β-bends of greater than 25° have been detected, and the majority of these occur in antiparallel strands (141). These bends are highly distorted globally and occur in strands of more than five residues in length that are located in large sheet structures. These bends are uniform and not related to the presence of a bulge. The type of residues that occur most frequently near strand bends are not those inclined to be in turn conformations, are rarely preferred in extended substructures, and are usually charged (Asp, Lys, Arg, His) or polar (Ser, Thr) residues or the important loop residues Gly and Pro. It would appear that the formation of salt bridges and hydrogen bonds is important to the stabilization of the bend structure. The most frequent pairs of residues associated in side chain/side chain interactions include Glu-Lys, Asp-Arg, Asp-His, Asn-Tyr, Asp-Lys, Tyr-Asp, and Thr-Thr. Thus, although interactions between hydrophobic residues must also be significant for bend stabilization, the relevance

---

\* Notation for the process of engineering proteins from first principles can be found in the literature as *ab initio* as well as *de novo*. We prefer to use *ab initio* which is the Latin equivalent of "from the start" whereas *de novo* is an expression used in describing a fresh start ignoring past experience.

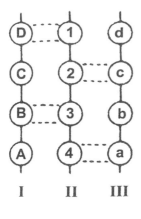

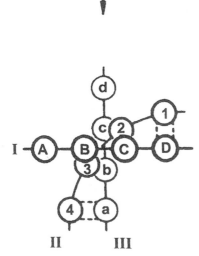

## Model β-bend

**Fig. 13** Schematic representation of a model β-bend with three β-strands which form an antiparallel β-sheet (top) to which a strong right-handed twist is applied (bottom) such that strands I and III now form a right angle. H-bonds are maintained. The side chains of residues labeled b, 3 and B now pack against each other. (Reprinted with permission from C Chothia, J Janin. Biochemistry 21: 3955–3965. Copyright 1982 American Chemical Society).

of charged and polar residues at bend sites points to their special importance to bend stability (141).

Polypeptide chains in the β-sheet conformation seldom exhibit the classical flat arrangement originally described by Pauling and Cory (130). Unlike α-helices, the $\phi/\psi$ angles for β-sheets are not localized but are spread over most of the upper left quadrant of the Ramachandran Plot (Fig. 14) (17). The $\phi/\psi$ angles of the classical parallel and anti-parallel β-sheets describe two-fold helical chains and lie on the N = 2 line (Fig. 14). The majority of

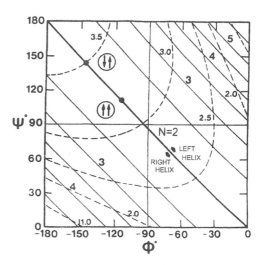

**Fig. 14** Graphical representation of how various geometrical parameters of structures composed of conformationally identical residues vary over the range of $\phi = -180$ to $0°$, and $\psi = 0$ to $180°$. A given point on the plot defines a helical structure characterized by the number of residues per turn (solid lines – N value) and the rise along the helix axis per residue (dashed lines). The classical flat β-sheets are composed of two-fold helical chains lying on the N = 2 line (indicated by the bold dot and adjacent symbols). Observed sheets have slightly left-twisted chains whose conformations lie to the right of the N = 2 line on the plot. (Reprinted from FR Salemme. Progress in Biophysics and Molecular Biology 42: 95–133. Structural properties of protein β-sheets. Copyright 1983, with permission from Elsevier Science).

$\phi/\psi$ angles of the peptide chains composing the $\beta$-strands found in the $\beta$-pleated sheet lie to the right of the N = 2 line indicating a left-handedness or helical sense to the $\beta$-strand (Fig. 15a) (142). However, the interactions between left-handed strands composing a $\beta$-sheet occur for alternating residues which define a $\beta$-sheet with a right-twist between 0°–30° (Fig. 15a and 15b). Conversely, a left-twisted sheet would be composed of strands that are right-handed or have a right helical sense (Fig. 15c). Further confusion arises in the literature concerning the handedness of the sheet twist. Some authors define twist in terms of the angle at which neighboring $\beta$-strands cross each other, a definition that results in a twist which is opposite to that seen when one defines the sheet twist in terms of the rotation viewed along the longitudinal sheet axis (Fig. 15b and 15c) (143). The tendency for peptide chains to assume left-handed helical conformations results from the presence of L-amino acids (142). Twist is a reflection of the compensatory interactions of the minimization of conformational energy due to the local intra-chain interactions and the preservation of inter-chain hydrogen bonds (142). The degree of twist decreases with increasing length and strand number in the sheet to permit hydrogen bonding at the ends of the strands (129, 143). There is considerable evidence for the participation of the amino acid side chains in sheet twist, and hydrophobic packing may be the main reason for sheet twist (144–146). Studies

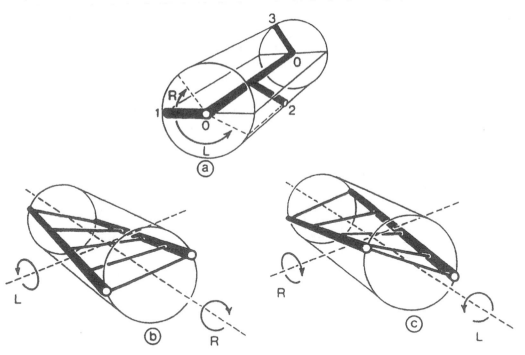

**Fig. 15** Schematic representation of twist sense in $\beta$-sheets. Chains forming twisted $\beta$-sheets have conformations lying to the right of the $N = 2$ line on the $\phi,\psi$ plot (Fig. 14), and so are left-handed helices (a). However, interactions between sheet chains occur for alternating residues which define a right-hand helix. Extended sheets then appear to have a net right-handed twist when viewed along the chain axis direction (b). Conversely, a left-twisted sheet (c) would be composed of locally right-handed helices. (Reprinted from FR Salemme. Progress in Biophysics and Molecular Biology 42: 95–133. Structural properties of protein $\beta$-sheets. Copyright 1983, with permission from Elsevier Science).

of homopolypeptides and polydipeptides indicate that antiparallel twisted sheets are more stable when the sheets are composed of small, unbranched residues like Gly, Ala, and Leu. The twisted parallel sheet is more stable if the amino acids are $\beta$-branched (Val, Ile, Thr), polar (Ser, Thr), aromatic (Phe, Tyr), or contain a long side chain (Lys) (146).

Nishikawa and Scheraga argued that Chothia's "crossing model;" of the right-twisted $\beta$-sheet structure was an adequate treatment of the geometry of short $\beta$-structures but that longer structures should be treated as coiled coils (147). Such modeled structures were shown to occur naturally in double stranded antiparallel sheets and not parallel sheets because the antiparallel sheets can have greater conformational diversity than parallel sheets (133, 148). Chothia established that for strongly twisted $\beta$-sheets to coil in the right-handed direction the $\phi/\psi$ values must have the following relations (149):

$$\psi_i \approx -\phi_{i+1}, \psi_{i+1}, > -\phi_{i+2}, \psi_{i+2}$$
$$\approx -\phi_{i+3} \psi_{i+3} > -\phi_{i+4} \cdots$$

An interesting structural feature of the double stranded antiparallel sheet is that the two surfaces are asymmetric, one surface is populated by residues situated in the 10-member hydrogen bonded rings, and the other by residues in the 14-member hydrogen bonded rings (Fig. 11) (142). Of the two alternative right-handed double helical arrangements, only the structure with the small ring residues on the coil interior is observed. Salemme suggests that coiling the sheet involves the cooperative superposition of both components of sheet twist and bend that result from geometrical constraints imposed by the interchain hydrogen bonds (Fig. 16) (142).

Two $\beta$-strands composing a $\beta$-sheet may be connected in one of two ways: $\beta$-hairpin connections occur when the peptide backbone reenters the same end of the $\beta$-sheet that it left; and crossover connections occur if the peptide backbone loops around to reenter the sheet on the opposite end (Fig. 17a) (143, 150, 151). The crossover connections have a handedness due to the fact that these structures form a helical turn from one strand,

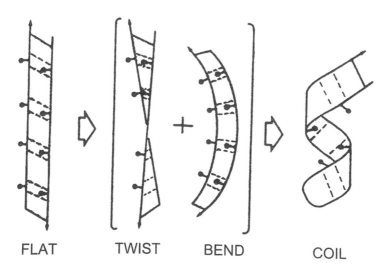

FLAT          TWIST          BEND          COIL

**Fig. 16** Schematic representation of the coupling between sheet twist and bend, both of which are assymetric. These effects act in concert to produce the observed chiral supercoil geometry. (Reprinted from FR Salemme. Progress in Biophysics and Molecular Biology 42: 95–133. Structural properties of protein $\beta$-sheets. Copyright 1983, with permission from Elsevier Science).

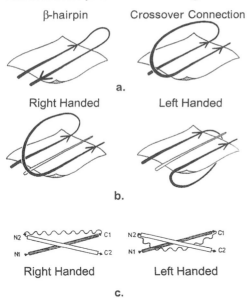

β-hairpin    Crossover Connection

a.

Right Handed    Left Handed

b.

Right Handed    Left Handed

c.

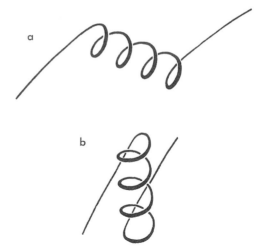

**Fig. 18** A Possible folding pathway which forms right-handed crossover loops from a right-handed α-helix with a β-strand at each end of it. (Reprinted from JS Richardson. Proceedings of the National Academy of Science USA 73: 2619–2623, 1976. Handedness of crossover in β-sheets. With permission of JS Richardson).

**Fig. 17** (a) Illustration of the two main classes of topological connection in β-sheets. The hairpin, plain or same-end connection and the crossover, cross, or opposite-end connection; (b) Illustrations of right- and left-handed crossover connections. (Reprinted from JS Richardson. Proceedings of the National Academy of Science USA 73: 2619–2623, 1976. Handedness of crossover in β-sheets. With permission of JS Richarson.) (c) Schematic representation of the differences of length between the right- and left-handed crossover connections. The strands are viewed normal to the direction of the strands and have the usual right-handed twist. (Reprinted from MJE Sternberg, JM Thornton. Journal of Molecular Biology 105: 367–382, 1976. On the conformation of proteins: The handedness of the β-strand-α-helix-β-strand unit. By permission of the publisher Academic Press).

up or down and around with the loop, and back into the next strand (Fig.17b). Essentially all the crossover connections in the known protein structures to date are right-handed. The handedness has been explained by the fact that the β-sheet twist makes the right-handed connection shorter and more compact (Fig. 17c) (152). The most common crossover connection is an α-helix producing the very common

βαβ structural motif. The right-handedness of these units is proposed to arise from folding in which a right-handed helix that is flanked by extended peptide chains has its axis roughly perpendicular to the direction of the extended strands. The two extended regions move towards each other to form the β-structure, and the result is always a right-handed unit (Fig. 18).

*b. Folding and Stability*

Designing a particular folding pattern for a linear sequence of amino acids involves logical use of the fundamental forces that stabilize a unique, defined protein structure. These forces involved in protein folding include hydrophobic interactions, hydrogen bonding, Van der Waals forces, and ion pairing (7, 9, 23, 153–155). The intrinsic propensity of amino acid residues and peptide fragments (motifs) to adopt specific conformations can also be used in the design of an amino acid sequence that will fold into a predetermined three dimensional structure (19, 156).

While none of the above forces should be considered as unimportant, hydrophobicity is considered the dominant driving force for folding that produces the extraordinarily compact structure of the protein with a core of non-polar residues and a complex internal architecture (23, 155, 157). The folding forces consist of local and non-local interactions between connected and topological neighbors (9). The folding pattern is dominated by non-local hydrophobic interactions, but local interactions like helix and turn propensities as well as hydrogen bonds also contribute to the nucleating conformations (158). This hydrophobic nucleation process is a critical feature of the "Hydrophobic Collapse Model" of protein folding, and is exemplified by the hydrophobic zipper (HZ) (158) and the hydrophobic cluster (159–164). Many authors have demonstrated the existence of hydrophobic clusters of aromatic and aliphatic amino acid side chains in proteins under denaturing conditions which appear to serve as nuclei for the subsequent folding of the protein (163–166).

The fact that very few biophysical studies have been carried out on the folding of $\beta$-structure is primarily due to the difficulties in creating a model system in which to examine the forces involved in $\beta$-sheet formation. As a result, there is inherent difficulty in predicting a $\beta$-sheet fold as opposed to an $\alpha$-helical fold and such difficulty has its impact in the specific design of $\beta$-structure. Despite these shortcomings, a few generalizations can be made to assist in the design of $\beta$-structure in proteins designed *ab initio*.

The mechanism of $\beta$-structure formation probably differs from that of $\alpha$ or $\alpha/\beta$-structure because the $\beta$-sheets are primarily stabilized by long-range or non-local interactions rather than the local interactions that predominate in the for-

mation of $\alpha$-structure. This characteristic is similar to the non-local hydrophobic interactions that are assumed to be the driving force in formation of the compact hydrophobic core (23, 158). The fact that the folding of $\beta$-sheet structure is generally slower than that of an $\alpha$-helix implies that the folding mechanisms of the two structures may differ (167–171). There is the implication that the formation of tertiary structure may be critically important to the formation of $\beta$-structure during protein folding and that the initiation of $\beta$-structure formation may depend upon the protection of the structure from aqueous solvent through the formation of tertiary structure (129). This point is made in dramatic fashion by Liu's study on the cellular retinoic acid binding protein (167). Liu found that this protein contains a substantially larger amount of $\alpha$-helix when denatured below pH 2.6 which, upon the addition of small amounts of sodium sulfate, undergoes a conformational change to predominantly $\beta$-sheet structure that is more in line with the native protein.

"Intrinsic propensity" is a term frequently used to describe the tendency of a particular di or tripeptide sequence to form particular protein structures as a result of local interactions (23). The helix-coil transition has been studied extensively to define the intrinsic helix forming propensity of specific amino acid sequences. The $\alpha$-helix is a good model for the study of intrinsic propensities because helices form from predominately local interactions, The $\beta$-sheet, on the other hand, is principally derived from forces that are intrinsically non-local and, therefore, intrinsic propensity should not be expected to be a good indicator of $\beta$-sheet structure, However, a number of studies have proposed thermodynamic scales that define $\beta$-sheet-forming propensities of amino acids and these scales have been compared with the Chou and Fasman statistical probabilities for $\beta$-sheet formation ($P_\beta$) (Table 3)

**Table 3** Experimentally Determined $\beta$-Sheet Propensities

| Guest residue | $P_\beta$ (Chou and Fasman) | $\Delta\Delta G$ (kcal/mol) (Kim and Berg edge strand) | $\Delta\Delta G$ (kcal/mol) (Smith and Regan central strand) | $\Delta\Delta G$ (kcal/mol) (Minor and Kim central strand) | $\Delta\Delta G$ (kcal/mol) (Minor and Kim edge strand) |
|---|---|---|---|---|---|
| Val | 1.64 | −0.53 | −0.94 | 0.82 | 0.17 |
| Ile | 1.57 | −0.56 | −1.25 | 1.0 | 0.02 |
| Thr | 1.33 | −0.48 | −1.36 | 1.1 | 0.83 |
| Tyr | 1.31 | −0.50 | −1.63 | 0.96 | 0.11 |
| Trp | 1.24 | −0.48 | −1.04 | 0.54 | −0.17 |
| Phe | 1.23 | −0.55 | −1.08 | 0.86 | 0.16 |
| Leu | 1.17 | −0.48 | −0.45 | 0.51 | −0.24 |
| Cys | 1.07 | −0.47 | −0.78 | 0.52 | 0.08 |
| Met | 1.01 | −0.46 | −0.90 | 0.72 | 0.02 |
| Gln | 1.00 | −0.40 | −0.38 | 0.23 | 0.04 |
| Ser | 0.94 | −0.39 | −0.87 | 0.70 | 0.63 |
| Arg | 0.94 | −0.44 | −0.40 | 0.45 | −0.43 |
| Gly | 0.87 | 0 | 1.21 | −1.2 | −0.85 |
| His | 0.83 | −0.46 | −0.37 | −0.20 | −0.01 |
| Ala | 0.79 | −0.35 | 0 | 0 | 0 |
| Lys | 0.73 | −0.41 | −0.35 | 0.27 | −0.40 |
| Asp | 0.66 | −0.41 | 0.85 | −0.94 | −0.10 |
| Asn | 0.66 | −0.38 | −0.52 | −0.08 | −0.24 |
| Pro | 0.62 | 0.23 | ND | <−3 | <−4 |
| Glu | 0.52 | −0.41 | −0.23 | 0.01 | 0.31 |

Note: Due to differences in the methods in which the $\Delta\Delta G$s were determined in the different studies, the reported $\Delta\Delta G$ values for the Minor and Kim studies are opposite in sign to the Kim and Berg and Smith and Regan studies. In the latter studies, negative $\Delta\Delta G$s indicate an increase in stability while in the Kim and Minor study, positive $\Delta\Delta G$s indicate an increase in stability.
(Reprinted from CL Nesloney, JW Kelly. Bioorganic and Medicinal Chemistry 4: 739–766, 1995 Progress toward understanding $\beta$-sheet structure. Copyright 1995 with permission from Elsevier Science).

(172–178). The main conclusion that can be drawn from such work is that the $\beta$-sheet propensities have minimal predictive value in terms of the ranking or range of values obtained. Moreover, such propensities are context sensitive in terms of local conformational preferences and non-local factors. In some cases these factors can be entirely non-local (176). As $\beta$-sheets are formed through non-local topological interactions, the residues surrounding a $\beta$-sheet amino acid residue are likely to play a large role in determining the $\beta$-sheet propensities.

Experimental studies have determined that there are intrinsic differences in the propensity of individual amino acids to adopt helical or sheet structure, and these intrinsic propensities correlate fairly well with the amino acid rankings obtained through statistical probability studies. However, the physico-chemical and structural basis for these differences is not clear, and it is evident that the intrinsic propensities are context sensitive, especially in the case of $\beta$-sheet formation. Contexts that may be relevant to $\beta$-sheet stability include whether the amino acids are located in parallel or antiparallel strands and whether the residues are central to the sheet or located in an edge strand. If the amino acid residue is located in an

**Table 4**   Residue Abundances in Antiparallel β-Sheet Compared to the Global Distribution*

|        | Global (%)       | Antiparallel (%)  | $\beta_A$ |
|--------|------------------|-------------------|-----------|
| Ala    | $8.44 \pm 0.12$  | $5.97 \pm 0.28$   | 0.71      |
| Arg    | $4.53 \pm 0.09$  | $4.51 \pm 0.24$   | 1.00      |
| Asn    | $4.63 \pm 0.09$  | $2.78 \pm 0.19$   | 0.60      |
| Asp    | $5.93 \pm 0.01$  | $2.92 \pm 0.20$   | 0.49      |
| Cys    | $1.05 \pm 0.04$  | $1.11 \pm 0.12$   | 1.06      |
| SS-cys | $0.78 \pm 0.04$  | $1.49 \pm 0.14$   | 1.92      |
| Phe    | $4.06 \pm 0.09$  | $6.17 \pm 0.28$   | 1.52      |
| Gln    | $3.58 \pm 0.08$  | $3.15 \pm 0.21$   | 0.88      |
| Glu    | $6.15 \pm 0.01$  | $4.41 \pm 0.24$   | 0.72      |
| Gly    | $7.98 \pm 0.12$  | $5.13 \pm 0.26$   | 0.64      |
| His    | $2.26 \pm 0.06$  | $2.19 \pm 0.17$   | 0.97      |
| Ile    | $5.38 \pm 0.01$  | $7.38 \pm 0.32$   | 1.45      |
| Leu    | $8.21 \pm 0.12$  | $9.11 \pm 0.34$   | 1.11      |
| Lys    | $6.12 \pm 0.01$  | $5.31 \pm 0.26$   | 0.87      |
| Met    | $2.13 \pm 0.06$  | $2.32 \pm 0.18$   | 1.09      |
| Pro    | $4.70 \pm 0.09$  | $2.27 \pm 0.18$   | 0.48      |
| Ser    | $6.10 \pm 0.01$  | $5.82 \pm 0.28$   | 0.95      |
| Thr    | $6.03 \pm 0.01$  | $8.17 \pm 0.32$   | 1.36      |
| Trp    | $1.49 \pm 0.05$  | $2.12 \pm 0.17$   | 1.42      |
| Tyr    | $3.72 \pm 0.08$  | $5.91 \pm 0.28$   | 1.59      |
| Val    | $6.72 \pm 0.11$  | $11.31 \pm 0.37$  | 1.68      |
| Σ      | 52357            | 7231              |           |

* A comparison of the global abundance of the amino acids with that found in antiparallel sheet. The last column is the propensity for antiparallel β-sheet. It is calculated by dividing the antiparallel frequency for a residue $i$ by the global frequency. Residues which are more abundant in antiparallel-sheet than globally are the β-branched and aromatic residues as well as disulfide-bonded cystines. The elevated level of the latter is principally due to their abundance in small proteins with minimal hydrophobic cores. Residues which do not favor antiparallel β-sheet include proline and aspartate, followed by asparagine, glycine , alanine, and glutamate. Taken from Ref. 134.

antiparallel strand, its location in an H-bonded or non-H-bonded site will have major implications on its contribution to the stability of the antiparallel sheet (134).

Some practical aspects of β-sheet design result from the facts that sheet formation is driven by hydrophobic clustering, the sheets are primarily buried in the hydrophobic center of the protein, and the polar amide atoms are shielded from water. Amino acids most commonly found in β-sheets are the hydrophobic, β-branched residues that would likely be most efficient in shielding the peptide backbone from solvent. Wouters and Curmi (134) analyzed the amino acid composition of antiparallel β-sheets and found

that cystines, aromatic, and β-branched amino acid residues are relatively abundant in such structures (Table 4). The interior strands of an antiparallel β-sheet have more hydrophobic and aromatic residues than charged or polar residues (Table 5).

Cysteine, aspartate, asparagine, glycine and proline appear most frequently in bulges in the antiparallel β-sheet, and therefore, would be most useful in the edge strands or located at the ends of the sheet strand. All the residues in the interior of the β-sheet are involved in both an H-bonded (10-member ring) and non-H-bonded (14-member ring) interaction. In contrast, the edge residues can occupy

**Table 5** The Location of Specific Residues in Antiparallel $\beta$-Sheet*

| | Residue location | | | Edge residue distribution | | | |
|---|---|---|---|---|---|---|---|
| | Interior (%) | Bulge (%) | Edge (%) | H-bonded (%) | Non-H-bonded (%) | Total counts | $\chi^2$ |
| Ala | 31.2 ± 2.1 | 9.1 ± 1.3 | 59.8 ± 2.2 | 55.3 ± 3.0 | 44.7 ± 3.0 | 284 | 3.17 |
| Arg | 22.1 ± 2.2 | 6.6 ± 1.3 | 71.3 ± 2.4 | 56.2 ± 3.1 | 43.8 ± 3.1 | 249 | 3.86 |
| Asn | 17.2 ± 2.4 | 21.5 ± 2.6 | 61.3 ± 3.0 | 51.6 ± 4.0 | 48.4 ± 4.0 | 157 | 0.16 |
| Asp | 14.7 ± 2.2 | 20.7 ± 2.5 | 64.7 ± 2.9 | 48.8 ± 3.8 | 51.2 ± 3.8 | 172 | 0.09 |
| Cys | 32.7 ± 4.7 | 18.4 ± 3.9 | 49.0 ± 5.0 | 41.7 ± 7.1 | 58.3 ± 7.1 | 48 | 1.33 |
| SS-cys | 22.1 ± 3.8 | 11.5 ± 2.9 | 66.4 ± 4.3 | 39.5 ± 5.4 | 60.5 ± 5.4 | 81 | 3.57 |
| Phe | 34.3 ± 2.2 | 4.9 ± 1.0 | 60.8 ± 2.3 | 51.2 ± 3.0 | 48.8 ± 3.0 | 285 | 0.17 |
| Gln | 24.4 ± 2.7 | 8.8 ± 1.8 | 66.8 ± 3.0 | 49.1 ± 3.9 | 50.9 ± 3.9 | 167 | 0.05 |
| Glu | 20.7 ± 2.1 | 11.9 ± 1.7 | 67.4 ± 2.5 | 52.9 ± 3.2 | 47.1 ± 3.2 | 244 | 0.80 |
| Gly | 22.9 ± 2.0 | 18.3 ± 1.8 | 58.8 ± 2.3 | 68.9 ± 2.8 | 31.1 ± 2.8 | 267 | 38.21 |
| His | 20.5 ± 3.0 | 14.6 ± 2.6 | 64.9 ± 3.5 | 58.3 ± 4.5 | 41.7 ± 4.5 | 120 | 3.33 |
| Ile | 31.3 ± 1.9 | 9.6 ± 1.2 | 59.1 ± 2.0 | 51.6 ± 2.6 | 48.4 ± 2.6 | 370 | 0.39 |
| Leu | 35.1 ± 1.7 | 12.5 ± 1.2 | 52.5 ± 1.8 | 47.8 ± 2.5 | 52.2 ± 2.5 | 395 | 0.73 |
| Lys | 24.8 ± 2.1 | 6.8 ± 1.2 | 68.4 ± 2.3 | 60.6 ± 2.9 | 39.4 ± 2.9 | 282 | 12.77 |
| Met | 33.2 ± 3.4 | 10.2 ± 2.2 | 56.7 ± 3.6 | 51.9 ± 4.9 | 48.1 ± 4.9 | 106 | 0.15 |
| Pro | 0.0 ± 0.0 | 19.2 ± 2.8 | 80.8 ± 2.8 | 0.0 ± 0.0 | 100.0 ± 0.0 | 164 | 164.00 |
| Ser | 22.0 ± 1.9 | 11.0 ± 1.4 | 67.0 ± 2.2 | 47.3 ± 2.8 | 52.7 ± 2.8 | 317 | 0.91 |
| Thr | 25.1 ± 1.7 | 7.8 ± 1.1 | 67.1 ± 1.9 | 46.0 ± 2.4 | 54.0 ± 2.4 | 430 | 2.69 |
| Trp | 33.5 ± 3.8 | 3.2 ± 1.4 | 63.3 ± 3.8 | 47.0 ± 5.0 | 53.0 ± 5.0 | 100 | 0.36 |
| Tyr | 38.5 ± 2.3 | 4.9 ± 1.0 | 56.6 ± 2.3 | 55.5 ± 3.1 | 44.5 ± 3.1 | 254 | 3.09 |
| Val | 31.3 ± 1.6 | 8.2 ± 0.9 | 60.5 ± 1.6 | 51.0 ± 2.2 | 49.0 ± 2.2 | 539 | 0.22 |

*The left-hand side of the table shows the location of each residue within antiparallel $\beta$-sheet. Edge residues are those which only have a $\beta$-sheet partner on one side. Interior residues have partners on both sides. Bulge residues are those classified by DSSP as not having $(\phi, \psi)$ angles in $\beta$ conformation. They are always located on edges but the distribution of residues found in bulges is different to the edge residue distribution. Standard deviations quoted were calculated using $(p(1-p)/N)^{\frac{1}{2}}$ where $p$ is the fraction of each residue $i$ found in a particular location and $N$ is the total number of residue $i$ in all locations. They are a measure of the accuracy of the percentages based on the number of counts. They do not reflect variations between proteins. The right-hand side of the table shows how the edge residues are distributed between the H-bonded and non-H-bonded sites. The results of a $\chi^2$ test to determine whether the partitioning is nonrandom are shown in the last column. For a 50 : 50 partitioning of residues the expectation value would be half of the number of counts shown in the second last column. A $\chi^2$ value greater than 3.84 indicates a nonrandom partitioning at the 95% confidence level. Proline, glycine, lysine, and arginine show a nonrandom partitioning at a high confidence level. Taken from Ref. 134.

only one of these sites, and of the edge residues, Pro is restricted to the non-H-bonded site while Gly, Lys, and Arg favor the H-bonded site. Residue pairing in the H-bonded and non-H-bonded sites also shows some interesting features. Electrostatic compatibility is the strongest determinant of residue pairing. Pairs that occur in both the non- and H-bonded sites are Glu-Arg, Glu-Lys, Gln-Arg, and Gln-Lys. Ser and Asn form H-bonds with Ser in the $t$ rotamer and Asn in the $g^+$ rotamer. Lys-Asp and Ser-Ser pairs do not interact with each other. In general, polar and charged residues are on the surface of the proteins and interact with the solvent and other surface polar residues. Of strongly correlated pairs, specific intra-pair interaction patterns are not observed. Cysteine has an extraordinary preference to be its own partner in both the H-bonded and non-H-bonded sites. Six membered aromatic ring side chains favor the H-bonded site when pairing with themselves. Phe has a strong preference for the $g^-$ rotamer in the $\beta$-sheet and when paired with itself will interact through a $\pi/\pi$ interaction. The less favorable interactions between Tyr side chains may result from the OH group interfering with $\pi/\pi$ interaction. The residues Val, Ile, Thr, and Leu favor the non-H-bonded site in the antiparallel sheet. The $\beta$-branched side chains favor the rotamer that staggers the side chain with respect to the backbone which would be the $t$ rotamer for Val and the $g^+$ rotamer for Ile and Thr. In the H-bonded site, these rotamers cause the side chains to point away from each other, and in the non-H-bonded site, these rotamers facilitate inter-side chain interaction.

Theoretical considerations concerning the driving forces behind protein folding tend to support the concept of the hydrophobically driven model of protein folding. Hecht has demonstrated that it is possible for secondary structure to correlate with the polar and nonpolar period-icity of a protein or peptide amino acid sequence (179–181). Therefore, alternating polar and nonpolar residues in the sequence may produce $\beta$-structure while a 3.6 periodicity of polar and nonpolar residues may result in an $\alpha$-helical structure. This correlation is one possible explanation for the fact that a given fold may be generated by many different amino acid sequences (182, 183).

## 3. Non-Repetitive Structure

Early studies of proteins classified structure into $\alpha$-helix, $\beta$-sheet, and random coil. However, it was soon realized that there is nothing random about random coils. The non-repeating $\phi/\psi$ angles of the coil structure result in conformationally variable, aperiodic structures of varying length. These random coils became known as loops and are defined as regions of polypeptide chain connecting two segments of regular secondary structure. Loops vary in length from three to approximately 20 residues with the smaller loops characterized as turns, the three residue $\gamma$-turns (184), the four residue $\beta$-turns (143), or three, four and five residue $\beta$-hairpin turns (185). The longer loop regions are classified according to their linearity and flatness as in the strap, omega ($\Omega$) and zeta ($\zeta$) loops (186, 187). As loops are usually found at the surface of proteins and by definition are the spaces between the regular secondary structure which composes the protein core, these structures likely provide more than simple links between secondary structural elements and could be involved in the functional aspects of protein structure. Hence, any attempt to simplify the random coil conceptually into structural motifs could lead to a better understanding of protein function.

### a. Turns

The smaller three to five residue loops are usually found at sites on the surface of a

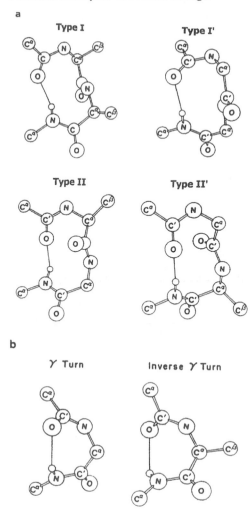

a

Type I    Type I'

Type II    Type II'

b

γ Turn    Inverse γ Turn

**Fig. 19** Schematic representation of Type I, II, I', and II' β-turns and γ and inverse γ-turns. (Reprinted from GD Rose, LM Gierasch, JA Smith. Advances in Protein Chemistry 37: 1–109. Copyright 1985, by permission of Academic Press).

protein where the polypeptide chain reverses its overall direction, and as a result, are called "turns." Turns are generally characterized by the distance between the $C_\alpha$ atoms of the first and last residue, the backbone hydrogen bonding within the turn, and the $\phi/\psi$ angles of the residues between the first and last residue of the turn.

The three residue γ-turn is defined by the existence of a hydrogen bond between the carbonyl oxygen of the $i$-residue and the backbone amide hydrogen of the residue in the $i+2$ position (Fig. 19b). Due to the relatively strained nature of the seven member ring structure formed by the H-bonded system, the stability of the γ-turn is dependent upon the absence of competing hydrogen bonding interactions with solvent or other atoms. The γ-turn is divided into two classes called the inverse turn and the classic turn. The $\phi/\psi$ angles are inverse to each other in these turns so that the backbone conformations of inverse and classic turns are mirror images of one another (Table 6). An L configuration of the residue in the $i+1$ position favors the inverse γ-turn, so inverse γ-turns are more prevalent in natural proteins. Inverse turns occur at the N- and C-termini of α-helices and β-sheets as well as in the active sites of serine and thiol proteases (184).

The four residue β-turn is characterized by a $C_i^\alpha$ to $C_{i+1}^\alpha$ distance of $<17$ Å and a hydrogen bond between the carbonyl oxygen of the $i$ residue and the amide hydrogen of the $i+3$ residue (Fig. 19a). This class is subdivided into various types of β-turn on the basis of the $\phi/\psi$ angles of the residues in the and $i+1$ and $i+2$ positions and the presence of a hydrogen bond (Table 6) (188–191). Types I, II, and VIII are the major types of β-turn accounting for over 50% of such structures (192). In these types, the chain folds back on itself so that the $i$ and $i+3$ $C^\alpha$ atoms are approximately 5 Å apart. Type I turns have $\phi_{(i+1)}/\psi_{(i+1)}$ and $\phi_{(i+2)}/\psi_{(i+2)}$ values of approximately $-60°/-30°$ and $-90°/0°$, respectively. Type II turns have $\phi_{(i+1)}/\psi_{(i+1)}$ and $\phi_{(i+2)}/\psi_{(i+2)}$ values of approximately $-60°/120°$ and $80°/0°$, respectively. Type I' and II' are mirror images of Type I and II with the inverse of the $\phi/\psi$ angles shown above. The $\phi_{(i+1)}/\psi_{(i+1)}$ values of the Type VIII turn are identical to Type I whereas the

**Table 6** Frequency and Mean Dihedral Angles for Standard $\beta$- and $\gamma$-Turn Types

| Turn type | Ramachandran nomenclature[a] | Mean dihedral angles[b] | | | |
|---|---|---|---|---|---|
| | | $\phi(i+1)$ | $\psi(i+1)$ | $\phi(i+2)$ | $\psi(i+2)$ |
| I | $\alpha_R\alpha_R$ | −64 (−60) | −27 (−30) | −90 (−90) | −7 (0) |
| II | $\beta\gamma_L$ | −60 (60) | 131 (120) | 84 (80) | 1 (0) |
| VIII | $\alpha_R\beta$ | −72 (60) | −33 (−30) | −123 (−120) | 121 (120) |
| I′ | $\alpha_L\gamma_L$ | 55 (60) | 38 (30) | 78 (90) | 6 (0) |
| II′ | $\varepsilon\alpha_R$ | 60 (60) | −126 (−120) | −91 (−80) | 1 (0) |
| VIa1 | $\beta\alpha_R$ | −64 (60) | 142 (120) | −93 (−90) | 5 (0) |
| VIa2 | $\beta\alpha_R$ | −132 (120) | 139 (120) | −80 (−60) | −10 (0) |
| VIb | $\beta\beta$ | −135 (135) | 131 (135) | −76 (−75) | 157 (160) |
| IV | | −61 | 10 | −53 | 17 |
| $\gamma$[c] | | 70 to 85 | −60 to −70 | | |
| Inverse $\gamma$[c] | | −70 to −85 | 60 to 70 | | |

[a] Ramachandran nomenclature for turn type as in reference 191. The nomenclature describes the regions of the Ramachandran plot occupied by residues $i+1$ and $i+2$ of the turn.
[b] The idealized $\phi,\psi$ values as determined by Ref. 190 are given in parentheses after the averaged values determined from the Ref. 192. The values for the type VI turns are taken from Ref. 143. Types VIa and VIa2 are the two subclasses of type VIa turns identified by Richardson (Ref. 143). Adapted from Ref. 192.
[c] Taken from Ref. 188.

$\phi_{(i+2)}/\psi_{(i+2)}$ values are −120°/120°. Type VI and its subgroups have a *cis* proline residue in the $i+2$ position of the turn and are rare in known protein structures. Finally the Type IV category contains the large number of turns which do not fit into any of the categories derived from the stringent restrictions on the $\phi/\psi$ angles set out by Lewis, et al. (190).

It would appear that solvation of backbone atoms that are not H-bonded imposes a greater restriction on the conformational flexibility of the tetrapeptide structure than does the sequence. In a turn there are usually four unsatisfied main chain atoms, $NH_{i+1}$, $CO_{i+1}$, $NH_{i+2}$, and $CO_{i+2}$. These groups must either be solvated or H-bonded with other protein atoms. Although there are a few buried turns in which protein atoms provide such an environment, the majority of turns are at the surface of the protein surrounded by water molecules. The conformation of $\beta$-turns is not extremely dependent on the sequence of the tetrapeptide; however, a recent study by Hutchinson and Thornton

has redefined the residue preferences in the major turn types (192).

Position $i$ of Type I turns is most frequently occupied by residues with side chains that are hydrogen bond acceptors, Asn, Asp, Cys, Ser, and His. Proline is also preferred at this position in Type I turns. Proline is the most common residue in the $i+1$ position probably because the ˜angle is restricted to −60°. Glu and Ser in this second position may also stabilize the $\phi$ angle through side chain H-bonding with the main chain amide, so they are frequently observed in the $i+1$ position of Type I turns. Asp, Asn, Ser, and Thr are the residues most likely to adopt the correct conformation for the $i+2$ position and are thus favored. Finally, Gly is frequently observed in the $i+3$ position of the Type I turn possibly because it is a useful residue for facilitating the return of the polypeptide chain to run antiparallel to its original direction.

Tyr and Pro are favored for the $i$ position in Type II turns. The $i+1$ position is favored by Pro although Lys is also common. The $i + 2$ position is overwhelmingly favored by Gly, and to a lesser extent, Arg. Both residues readily adopt the $\phi/\psi$ angles represented by the $\alpha_L$ region in the Ramachandran plot which are representative of residues in this position in the Type II turn. The $i + 3$ position is favored by Ser, Lys, and Cys. The side chains of Ser and Lys tend to hydrogen bond to other parts of the turn region.

The $i$ position of Type VIII turns has a preference for Pro or Gly. The $\phi/\psi$ angles of the $i + 1$ residue are identical to Type I turns, Pro being favored due to the $\phi/\psi$ angle restrictions. Asp is also favored in this position due to the side chain H-bond to the main chain amide which stabilizes the $\phi$ angle. The side chains of Asn and Asp in the $i + 2$ position of the Type VIII turn hydrogen bond to the main chain amide of the $i + 4$ residue resulting in formation of the classic Asx turn (193) and stabilization of the $\phi/\psi$ angles of this turn. The $i + 2$ residue also has a preference for Val, Phe, and Ile which is very unusual for a turn region, however, these residues have a conformational preference for the $\beta$ region of the Ramachandran plot and that structure is preferred in this position of the Type VIII turn. Finally, Pro dominates the $i + 3$ position of this turn, and this residue has a tendency to dictate the Type VIII structure as opposed to the Type I by restricting the conformation of the $i + 2$ residue.

Type I' and II'' turns are frequently incorporated into $\beta$-hairpin turns. Since in the $\beta$-hairpin structure the $i$ position is the last residue in a $\beta$-sheet, tyrosine, a $\beta$-sheet preferring residue, is frequently found in the first position of both turns. Position $i + 1$ of the Type I' turn is dominated by Asn, Asp, and Gly while Gly is preferred in $i + 2$. Gly is the preferred residue in the $i + 1$ position of the Type II' turn with Asp, Asn, and Ser preferred at the $i + 2$ position. Charged polar residues are preferred in the $i + 3$ position of the Type I' turn with Thr and

Gly being preferred in this position of the Type II'' turn.

The quasi-ring like structure of the $\beta$-turn conformation places the side chains of the $i + 1$ and $i + 2$ residues in the equatorial and axial orientations, respectively (Fig. 20) (188). The restricted orientation of the side chains of residues in these two middle positions of the turn coupled with the solvent exposed nature of the turn itself and the predominance of amino acids with functional groups in the side chain leads to a potential for this structure to be involved in structurally specific biological interactions (11, 188, 194). Many small biologically active peptides including somatostatin, luteinizing hormone-releasing hormone (LHRH), $\alpha$-melanocyte stimulating hormone ($\alpha$MSH), bradykinin, and enkephalin are predicted to contain a turn region which may be responsible for much of the biological activity of these peptide hormones. Many antagonists have been designed to these hormones through conformational restricted analogs which mimic the turn structure (188, 194). The surface location of turn regions lend them to function as potential antigenic sites, as well as sites of phosphorylation, glycosylation, and other types of posttranslational modification. The requirement of a stable folded globular structure for biological function in a protein leads to the observation that many of the variable residues in the sequences of related proteins are found on the surface of the protein while the sequence of the core scaffolding is relatively invariant. As the turn structures are at the protein surface, it stands to reason that variations in protein function could be attained through modification of turn residues (195).

*b. Loops*

While there is no universally accepted definition distinguishing the turn from the loop, the shorter length of the turn—three

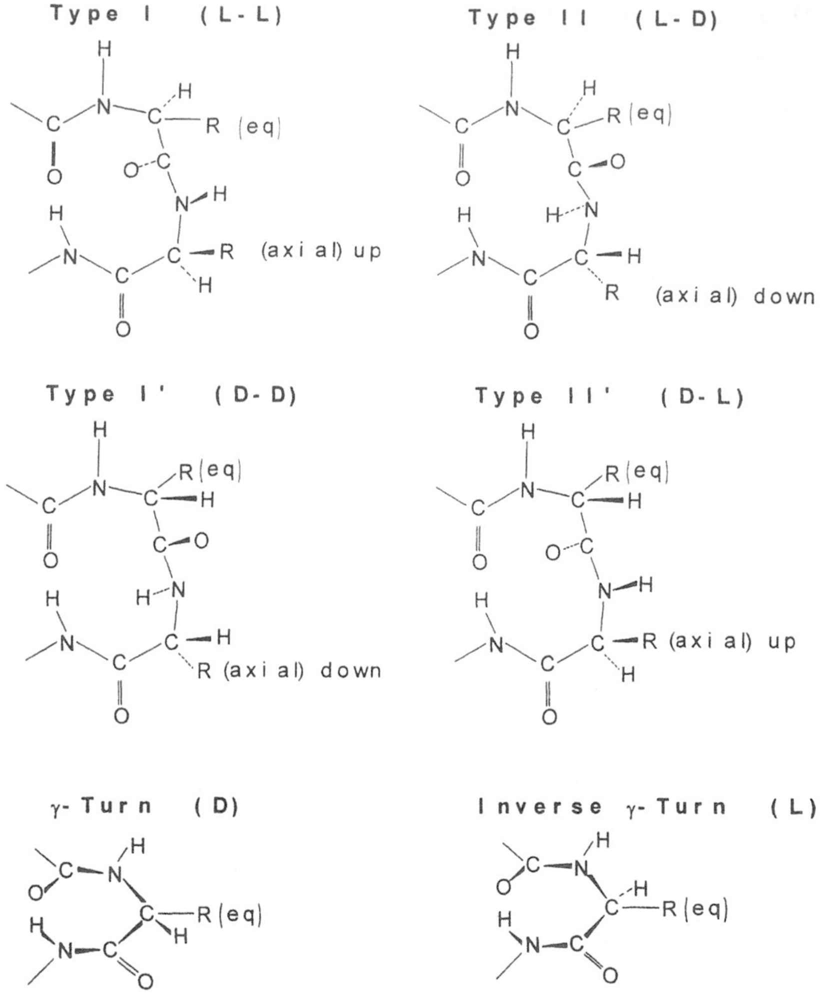

**Fig. 20** Diagrammatic representation of the orientation of side chains in various hydrogen-bonded turns. The configurations of the residues shown are listed on the figure: they correspond to the favored configurations for the standard turn types. Note that the residue in positon $i + 1$ of standard $\beta$-turns always has its side chain oriented equatorially, and that in position $i + 2$ always has its side chain oriented axially (up or down). (Reprinted from GD Rose, LM Gierasch, JA Smith. Advances in Protein Chemistry 37: 1–109. Copyright 1985, by permission of Academic Press).

to five residues—has made these structures more amenable to study and characterization. For our purposes, we will define the loop as an aperiodic polypeptide fragment of six or more residues connecting two secondary structures. Loops are divided into two groups (Fig. 21). Simple loops contain no subsets that are also loops while compound loops contain at least one smaller embedded loop (196).

Simple loops can be classified according to their linearity and flatness into three classes: linear loops (straps), non-linear and planar loops (omegas), and non-linear and non-planar loops (zetas) (Fig. 22) (186).

The $\Omega$-loop, the best characterized of loop structure, was originally named for its structural resemblance to the Greek letter in that the 6–16 residue protein back-

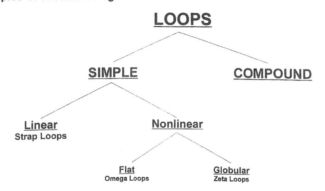

# LOOPS

SIMPLE COMPOUND

<u>Linear</u>
Strap Loops

<u>Nonlinear</u>

<u>Flat</u>
Omega Loops

<u>Globular</u>
Zeta Loops

**Fig. 21** Loop taxonomy based on the linearity and flatness of the geometric parameters. (Reprinted from CS Ring, DG Kneller, R Langridge, FE Cohen. Journal of Molecular Biology 224: 685–699. Copyright 1992, by permission of Academic Press).

bone follows a loop or lariate-shaped course the ends of which are within 10 Å of one another (187). A study of nearly 1000 examples of this structure has permitted an estimation of the frequency of appearance of the amino acids in such a structure (Table 7). The functional nature of $\Omega$-loops has been demonstrated through studies of the hypervariable loops of immunoglobulin (197) and the P-loop in ATP and GTP binding proteins (198). $\Omega$-loops have been demonstrated as sites of tyrosine sulfation (199) and prohormone cleavage (200). Loops have

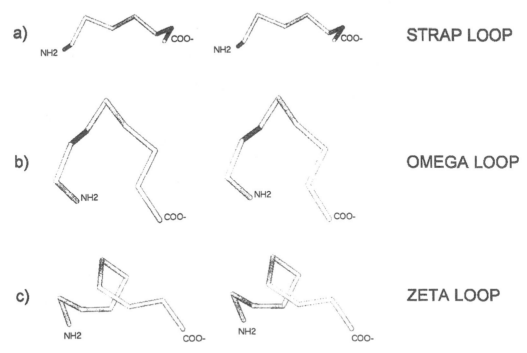

**Fig. 22** Schematic representation of the $C_\alpha$ backbone of the three main loop classes. (Reprinted from CS Ring, DG Kneller, R Langridge, FE Cohen. Journal of Molecular Biology 224: 685-699. Copyright 1992, by permission of Academic Press).

**Table 7**   Amino Acid Frequencies Found in 979 Ω-Loops

| | | | | | |
|---|---|---|---|---|---|
| Asn | 1.48[a] | Tyr | 1.06 | Trp | 0.90 |
| ½Cys[b] | 1.35 | Lys | 1.02 | Phe | 0.85 |
| Gly | 1.34 | His | 1.01 | Ala | 0.80 |
| Pro | 1.33 | Thr | 1.0 | Leu | 0.75 |
| Asp | 1.29 | Arg | 1.00 | Val | 0.69 |
| Cys[b] | 1.16 | Glu | 0.93 | Ile | 0.68 |
| Ser | 1.14 | Gln | 0.90 | Met | 0.57 |

[a] Amino acid frequency, $f$, was calculated as the ration of the fraction of the specific amino acid residue found in Ω-loops to the fraction of the number of total amino acids found in Ω-loops.

[b] Cys was calculated for the free cysteine residues in the database and ½Cys was calculated for the cysteines involved in disulfide bonds. (Reproduced with permission from JS Fetrow. FASEB Journal 9: 708–717, 1995. Omega loops: Non-regular secondary structures significant in protein function and stability).

been demonstrated as functional entities in protein activities, and the specificity of a given protein function has been altered through deletion or replacement of surface loops (201–203). The Ω-loop has also been implicated in the active site of a number of enzymes as well as in protein structure stabilization and folding (187).

### B. Super-Secondary Structure

*Ab initio* design of a protein tests the principles of protein folding and stability. Designing protein structure can be simplified by starting at the secondary structural elements and building complex globular proteins by packing helices and sheets to form a core structure to which functional entities may be added. Helix bundles are probably easier to design than sheet structures owing to the modularity of the α-helix and to solubility problems inherent in the β-sheet design.

### 1. α-Structure: Bundles and Coils

*Ab initio* design of supersecondary α-helical structure must take into consideration factors governing helix formation, termination, stability, and packing as well as loop formation between secondary helical structure. The packing of α-helices has been variously described as "knobs into holes" by Crick and as "ridges into grooves" by Chothia (Fig. 23) (204, 205).

The "ridges into grooves" model divides the ridges and grooves of α-helices into three main types that are formed by:

1. Residues one apart in the helical sequence (±1).
2. Residues three apart in the helical sequence (±3).

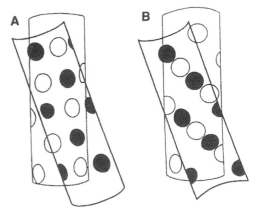

**Fig. 23** A and B illustrate "ridges into grooves" and "knobs into holes" side chain-side chain packing arrangements, respectively. The helices are depicted as cylinders of approximately 10 Å radius, and the positions of the side chains are denoted as open and closed circles for the bottom and top helix, respectively. The upper part of the top helix has been cut away to allow inspection of the packing. (Reprinted with permission from WF DeGrado, ZA Wasserman, JD Lear. Science 243: 622–643. Copyright 1989 American Association for the Advancement of Science).

3. Residues four apart in the helical sequence (±4).

When both packed helices use the ±4 ridges into ±4 grooves, the helix angle is approximately −50° while packing of ±4 ridges into ±3 grooves produces a helix angle of approximately 20°.

The "knobs into holes" model of helix packing will also result in a 20° angle between the helix axes of the two packed helices. The side chain "knob" of one α-helix is inserted into a "hole" formed by residues $i$, $i+3$, $i+4$, and $i+7$ of a neighboring helix. This helical overlap will eventually result in the helices diverging at the ends, unless they wrap around one another with a left handed sense producing a 3.5 residue repeat giving rise to the heptad repeat pattern of residues which is characteristic of coiled-coils (206).

The αα motif is the most elementary of protein α-structure and is divided into two classes: coiled-coils and helix-bundles (206). The major difference between the two structures is the number of residues/turn of helix. The helical overlap and "knobs into holes" interaction between helices in the coiled-coil causes the helices to wrap around one another with a left handed sense producing a 3.5 residue repeat and proximity of the helix ends. The helix-bundle has a 3.6 residue repeat, and the helix ends tend to splay apart as the bundle gets longer. The coiled-coil displays a heptad repeat (207) which is descriptive of a "knobs into holes" packing of hydrophobic residues in the coil's core whereas hydrophobic packing in the helix-bundle is most easily described by the "ridges into grooves" model.

The importance of helix formation, termination, stability, and packing have been addressed in the design and synthesis of a number of model four helix-bundle proteins (208–214). The prototypical helix-bundle is composed of four amphipathic α-helices, each with a minimum of four turns or sixteen residues. The

sequence is designed such that the helical wheel display (see Fig. 27 for a helical wheel display) of the sequence shows a distinct division between the hydrophobic and hydrophilic residues. The "minimalist" approach to design of helix bundles (215) uses a single helical sequence with leucine as the apolar residue and glutamic acid and lysine as the polar hydrophilic residues, all three chosen for their high intrinsic helical propensity (208, 209). The helix is stabilized by positioning the glutamate and lysine residues at the $i$ and $i+4$ positions of the helical sequence so that ionic interactions may occur between them. A negatively charged glutamate residue is positioned near the N-terminus of the helix, and a positively charged lysine residue is positioned near the C-terminus to stabilize the helical macro-dipole. Acetylation of the N-terminal amino function and synthesis of a C-terminal carboxamide are also designed to eliminate the charges that are unfavorable to helix dipole stability. Considerable attention is paid to stabilization of this dipole since it may be instrumental in stabilizing the antiparallel packing of the helix bundle. However, there is some indication that the helix dipole may not play a major role in the helix-bundle packing (216). Glycine is placed at the beginning and end of the helical segments as a helix breaker in an attempt to prevent formation of long helical segments when the identical helices are covalently linked. This continuation of the helix through the linker can also be deterred by altering the sequence to offset the hydrophobic stripes by one-third to one-half helical turn (211). Helix stabilization may also be increased through judicial choice of N and C terminal helical residues to maximize N- and C-capping of the helices (211, 70).

Design of helix-bundle packing topology should consider the number of helices involved, the parallel/antiparallel relationship of the helices, and the handedness of the bundle (216). Of the four

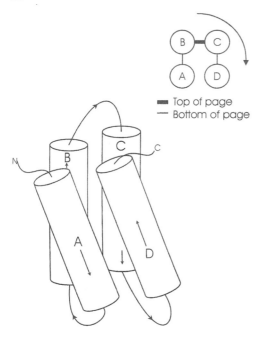

Fig. 24 Schematic representation of the four helix bundle. Taken from C Cohen, DAD Parry. Proteins: Structure, Function, and Genetics 7: 1–15, 1990. α-Helical coiled coils and bundles: How to design an α-helical protein. (Reprinted by permission of Wiley-Liss, Inc., a subsidiary of John Wiley and Sons, Inc.).

possible helical arrangements of the four helix bundle (all up or all down, down-up-down-up, down-down-up-up, or down-up-up-down) the down-up-down-up is the most common. In this case each of the four helices is oriented antiparallel to its two nearest neighbors and parallel to its single, more distant, diagonal neighbor (Fig. 24). The bundle can either be right- or left-handed depending upon the spatial relationship of the seqeuntial helices (Fig. 25).

The rules dictating the number of helices involved in bundle formation are not as yet clearly defined. Studies involving single unconnected identical helices resulted in a concentration dependent formation of a four-helix bundle (208). The four-helix bundle formed through dimerization of a peptide with two identical helices connected by a short Pro-Arg-Arg loop sequence resulted in a less stable bundle compared to that formed by four unlinked helices (209). The importance of the loop in helix packing is also demonstrated in the same study where two helices linked by a single Pro residue formed a

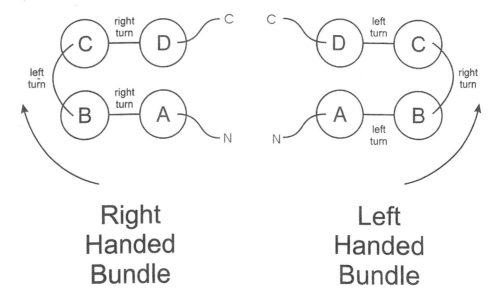

Fig. 25 Schematic representation of the end on view of right-handed and left-handed four helix bundles.

trimer instead of a tetramer. The bundle formed by a single protein designed with four identical helices connected by the PRR loop region was more stable than the single helix tetramer or two helix dimer (210). Dimerization and tetramerization of two helices linked by a four residue loop or single helices, respectively, to form four-helix-bundles was confirmed using an *ab initio* design of an antigenic determinant (211).

An αα motif designed using similar principles described above resulted in a stable monomer of helix-turn-helix structure. It is not entirely clear why this design failed to dimerize into a four-helix-bundle although the authors did attempt to design against dimerization or nonspecific aggregation by arranging charged amino acids in the helices to flank the hydrophobic interface (217). These authors also describe their design as similar to that for "coiled-coils" without the heptad repeat. Thus, there is a variety of hydrophobic residues present in the non-identical helices ranging from alanine to leucine and the β-branched valine and isoleucine. Perhaps, the packing of these hydrophobic residues resembles the "knobs into holes" mode of the

coiled-coils more than the "ridges into grooves" of the helix-bundles thereby producing a much closer packed hydrophobic core. Clearly more studies are required to clarify the reasons for dimerization in one case and not in the other.

The up and down topology of the helix bundle does not appear to be affected by the nature of the connecting loop region (216, 218). However, short linking segments of less than 11 residues will lead to a regular antiparallel arrangement of the four α-helices. Longer linkers will allow other packing modes and connectivities (206, 219). The down-down-up-up arrangement will require two loops which form crossovers as seen in the hematopoietic cytokines and the down-up-up-down arrangement will require one long interconnecting loop which traverses the length of the bundle as seen in the ferritin super-family (Fig. 26) (216). Both left and right handedness of the helix bundle occurs with equal frequency (220). The left handed up and down bundle has a sequence of left, right, left turns between the helices where as the right handed up and down bundle has a sequence of right, left, right turns between helices (Fig. 25).

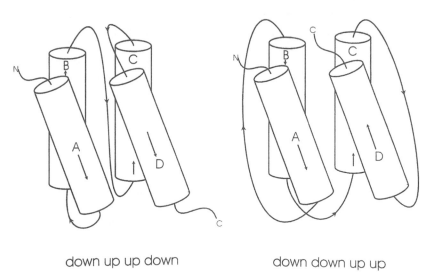

down up up down                down down up up

**Fig. 26**  Arrangement of four helix bundles requiring long crossover links between helices.

Since left turning connections can be longer than right turning connections, it is possible to engineer the helix connections to influence the handedness of the helix bundles (212, 221). The handedness of a bundle can also be engineered using strategically placed cysteine residues which would oxidize to form an intramolecular disulfide bridge in only the bundle with the desired handedness leaving the unwanted fold to polymerize through formation of intermolecular disulfide bonds during oxidation of the cysteine residues (212).

The packing of the amphipathic helices should be accomplished by a correct balance of hydrophobic and hydrophilic interactions. A shortage of hydrophobic residues may lead to poor stability of the bundle while too many may lead to highly stable but dynamic structures (128). This problem has been approached in the synthesis of the 4-helix-bundle by producing a peptide with a moderate hydrophobic moment through a sequence with an equal number of charged residues, uncharged hydrophilic residues, large hydrophobic residues, and alanine residues. Every effort was made to position the charged residues so that they formed salt links between neighboring helices or between the $i$ and $i + 4$ residues within a helix (212). A major tool used in designing proper packing of the hydrophobic residues was computer based molecular modeling and model building.

The prototypes for the coiled-coil class of α-helical supersecondary structure are the rod shaped α-fibrous proteins such as tropomyosin, intermediate filament protein, lamin, M-protein, paramyosin, myosin, and fibrinogen, and the leucine zipper found in numerous globular DNA binding proteins. These are two- and three-stranded homo or hetero di-and trimeric structures of parallel amphipathic helices wound around one another in axial register (Fig. 27) (206). Such structures

are characterized by the repetition of a seven residue sequence called the heptad with amino acid positions in the heptad labeled a through g (Fig. 27) (222). The heptad is characterized by a 3,4 hydrophobic repeat where hydrophobic residues, frequently leucine, are found in positions a and d. The a and d positions are completely buried in the coiled-coil core and the a and d positions of one helix pack against the a and d positions of the parallel partner helix with interactions being in register (a-a and d-d interactions). With large residues like leucine, isoleucine, and valine in the a and d positions, the "knobs into holes" packing in the hydrophobic core ensures that there are interactions between the a and d residues of each helix as well. The b, c, and f positions which are most exposed to the environment are occupied by charged or polar residues. Positions e and g are located at the edges of the hydrophobic/hydrophilic sites in the amphipathic helices and the placement of glutamic acid and lysine, respectively, in these positions results in ionic interactions between helix partners which shield the hydrophobic core from solvent and stabilize the parallel arrangement of helices and destabilize the antiparallel arrangement (Fig. 27) (222). However, this salt bridge stabilization of coiled-coils is currently under debate (223–225).

A stable coiled-coil of the rod structure of fibrous proteins requires at least five heptad repeats although it is possible to have smaller coiled-coil structures with other stabilizing features in combination with the arrangement of hydrophobic residues. While most of the studies on the stability of coiled coils has been carried out on the dimer, it is possible to engineer tri- and tetramers of the rod structure by manipulating the nature of the hydrophobic residues in the a and d positions of the heptad repeat (226–228). Model peptides with Val or Leu at positon a and Leu at position d exist in a

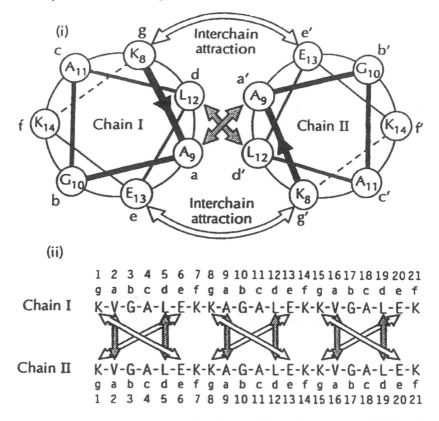

**Fig. 27** Interactions stabilizing the parallel coiled-coil homodimer of a model 21 residue peptide. Helical wheel representations (i) show only the central heptad (labeled a through g) whereas sequences (ii) show all 21 residues in the one-letter amino acid code. Shaded arrows indicate hydrophobic interactions while unshaded arrows depict interchain ionic interactions. (Reprinted with permission from JG Adamson, NE Zhou, RS Hodges. Current Opinion in Biotechnology 4: 428–437. Copyright 1993 Current Biology Ltd.).

monomer-dimer-trimer equilibrium. A Leu residue in position d and a preponderance of $\beta$-branched residues at position a favors dimer formation as does Asn in position a. Leu in position a with the $\beta$-branched Ile residue in positon d favors tetramer formation while the $\beta$-branched Ile in both a and d positions facilitates trimer formation. Residues in the other positions of the heptad repeat will also help in dictating the aggregation state. Peptides with Leu at the a and d positions will form dimers, trimers, tetramers, pentamers, or hexamers depending upon the hydrophobicity and steric properties

of the residues in positions e and g (123, and references therein).

The coiled-coil is also found in globular proteins other than the leucine zipper transcription factors, and the number of heptad repeats is considerably reduced in comparison to the $\alpha$-fibrous proteins coiled-coils. The sequences are usually unique depending upon the protein, and the arrangement of neighboring helices is generally antiparallel (206). *Ab initio* design of the coiled-coil globular proteins is modeled on the ColE1 Rop (repressor of primer) protein which contains a very regular 4-$\alpha$-helical coiled-coil

**Fig. 28**  Ribbon diagram illustration of Rop.

bundle composed of two 63 residue monomers (Fig. 28) (229). When the
hydrophobic core of wild type Rop was
altered such that all the a heptad positions
are Ala and the d positions are Leu, the
resultant mutant was more stable than the
wild-type protein and exhibited all characteristics of a native protein four helix
up-down-up-down bundle with the
coiled-coil helix packing of "knobs into
holes" (230). This mutant is less active in
binding RNA than the wild type proteins
but the activity can be restored to the
mutant by reversing the Ala and Leu
residues in the second and seventh
hydrophobic layers (231). Other mutations in the hydrophobic core which place
Leu in all the a positions and Ala in all the
d positions or Ala in the a positions and
Ile in the d positions result in coiled-coil
bundles that are more stable than the
mutant with Ala in a and Leu in d but have
lost all RNA binding activity. Replacing

the Leu in postion d with Met or Val in the
protein with Ala in a and Leu in d destabilizes the coiled-coil to the extent that
only the protein with Met in position d
forms a very weak dimer. Finally, placing
Leu in all the hydrophobic a and d positions of the Rop protein destroys the
coiled-coil tetramer and replaces it with a
very stable oligomer which appears to be a
tetramer but the structure of this product
is still under investigation (231). The
dimeric Rop can be redesigned to create a
monomeric  four-helix-coiled-coil  by
insertion of glycine rich loops with the
creation of a stable, native-like protein
(232). The length of the loop regions
between helices is important to the correct
folding of the designed four-helix-coiled-
coil.

   A  detailed  examination  of  the
antiparallel packing of coiled-coils in the
dimer, trimer, and tetramer arrangements
indicate that the interhelical interaction
patterns differ considerably from the corresponding parallel coiled coils (233).
Most noticeably is that the antiparallel
coiled-coil helices are not in register as seen
in the parallel coiled-coils. As a result, the
dimer and tetramer have a-d pairing where
as the trimer has two helices in register
with pairing identical to the parallel three
stranded coiled-coil and one out of register
with alternating a-d and d-a pairing. These
differences in register of the helices produces differences in the positions of the
interfacial residues which are known to
assist in stabilizing the parallel coiled-
coils. The e-e and g-g residues are in position to interact in the antiparallel dimer,
the e-e, g-g and g-e positions are in position
to interact in the trimer, and the b-e and
g-c positions could interact to help
stabilize the antiparallel coiled-coil
tetramer (Fig. 29) (233). Using a single
peptide with three heptad repeats and
keeping Leu in the d position of the model
heptad repeat the position a was varied
with Val, Leu, Ala, and Thr. Only the
Val/Leu combination resulted in a native

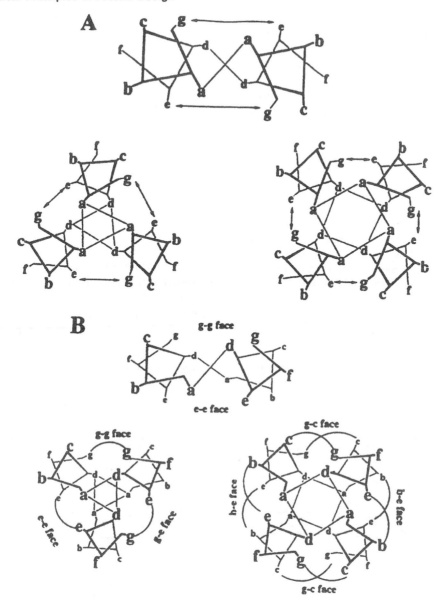

**Fig. 29** (A) Heptad arrangements in parallel two, three, and four α-helical bundles. (B) Heptad arrangements in antiparallel two, three and four α-helical bundles. The heptad is labeled a through g. Note that the interhelical faces differ between parallel and antiparallel packing. (Reprinted with permission from SF Betz, WF DeGrado. Biochemistry 35: 6955-6962. Copyright 1997 American Chemical Society).

like tetramer while the Leu/Leu combination formed a molten globule like tetramer and the Ala/Leu and Thr/Leu combinations failed to form 4-helix coiled coil bundles.

A similar study was carried out on a 51 residue helix-turn-helix peptide with identical helix sequences of three heptad repeats designed to dimerize into a four-helix bundle similar to Rop (234). In this case of formation of Rop like structure it is important to assure that the monomers will dimerize in either the left or right-handedness with the *anti* orientation as opposed to the *syn* orientation of the connecting loops (Fig. 30). This is done through judicious choice of the c-g and b-e residues of the repeat heptad to:

1.  Favor the c-g *anti* left-handed interaction between helices 1/1′ and 2′/2 and the b-e *anti* left-handed interaction between helices 1/2 and 1′/2′.
2.  Destabilize c-g *syn* right-handed tetramer interactions between 1/2 and 1′/2′ and the b-e *syn* right-handed tetramer interactions between 1/2′ and 2/1′.
3.  Destabilize the c-g *anti* right-handed tetramer interactions between 1/2 and 1′/2′ and the

b-e *anti* right-handed tetramer interactions between 1/1′ and 2/2′.
4.  Destabilize the c-g *syn* left-handed tetramer interactions between 1/2′ and 2/1′ as well as the b-e *syn* left-handed interactions between 1/2 and 2′/1′.
5.  Destabilize all possible parallel pairings mainly through unfavorable g-e interactions.

## 2. *β-Structure: Sheets, Sandwiches and Barrels*

The simplest of super-secondary structures involving the β-strand is the two stranded anti-parallel ββ-structure. The isolated motif is seldom found in natural proteins and is probably the most difficult to study due to the obvious aggregation

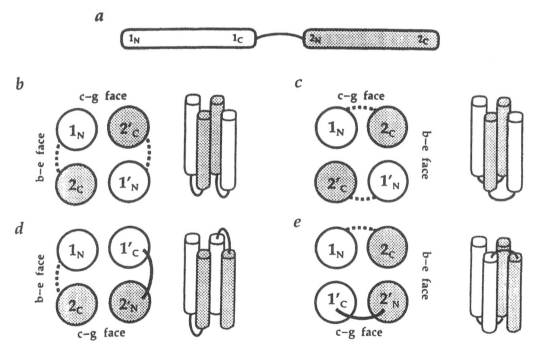

**Fig. 30** Schematic of the loop arrangements and interfacial packing within possible antiparallel (up-down-up-down) four-helix bundle topologies: (a) cartoon of a helix-turn-helix peptide indicating the N and C termini of each helix; (b) syn left-turning; (c) syn right-turning; (d) anti left-turning; (e) anti right-turning. Notice that the interfacial interactions between the syn topologies are identical, while those of the anti topologies are different from each other as well as from the syn topologies. (Reprinted with permission from SF Betz, PA Liebman, WF DeGrado. Biochemistry 36: 2450–2458. Copyright 1997 American Chemical Society).

and solubility problems which arise with the isolated $\beta\beta$-structure. A key substructure in the anti-parallel $\beta\beta$-super-secondary structure is the $\beta$-hairpin (235) which is considered to be a nucleation point for $\beta$-sheet formation (236) and as such, the features of its formation and stability should be considered very important.

It has been demonstrated that an isolated $\beta$-hairpin structure which is not seen in aqueous medium can be induced to form by the addition of organic solvents such as methanol or trifluoroethanol (237, 238). The fact that a short peptide may fold into a $\beta$-hairpin suggests that the structural determinants for this $\beta$-sheet initiator are short-range coded and that the initiation of anti-parallel $\beta$-structure may involve short range interactions in the amino acid sequence similar to the formation of $\alpha$-helix (238). The presence of a structure inducing organic solvent in the medium may prevent aggregation and provide the missing "long-range" interactions that are important in $\beta$-sheet formation. However, the demonstration that a 16 residue peptide fragment of a $\beta$-hairpin region of the B1 domain of Protein G folds into the native-like hairpin structure in water supports the hypothesis that local interactions are important in the initiation of anti-parallel $\beta$-structure (239). The two stabilizing factors in the $\beta$-hairpin formation in this model peptide are the orthogonal packing of the aromatic/hydrophobic residues on one side of the sheet and the large number of threonine residues, five in all, with high $\beta$-sheet propensity. The hydrophobic residues occupy the non-H-bonded sites while the threonine residues are located in the 10-member H-bonded sites which is contrary to the expectation that the $\beta$-branched residues prefer the non-H-bonded site in anti-parallel sheets and that aromatic rings prefer the H-bonded site when pairing with themselves (134). It is also interesting that this $\beta$-hairpin peptide has distinct hydrophilic and hydro-

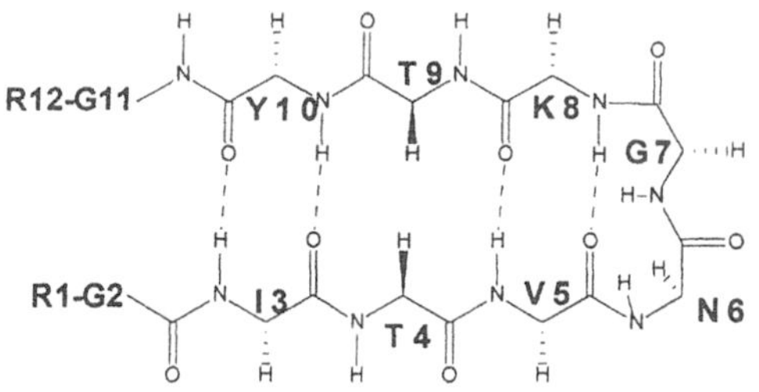

Fig. 31   Schematic of the $\beta$-hairpin structure. Dashed lines correspond to the expected hydrogen bonds. The amino acids are indicated using the one letter code.

phobic sides but aggregation is not a problem.

The $\beta$-turn in the hairpin structure may also play a major role in the nucleation of a $\beta$-sheet during protein folding. A successful preparation of $\beta$-hairpin peptides (240–242) utilized the NPDG turn sequence which is the preferred sequence for a Type I $\beta$-turn (192, 243) however, the region connecting the $\beta$-strands in the examples cited turned out to be a three residue turn which produced a pairing of $\beta$-strands different from the native conformations. Some interesting design features for a $\beta$-hairpin structure that could quite possibly be utilized in the nucleation of an anti-parallel $\beta$-sheet have been delineated (Fig. 31) (244). Type I' and II' $\beta$-turns predominate in the $\beta$-hairpin structure (235) and careful selection of the preferred amino acids in the $i$, $i+1$, $i+2$, and $i+3$ positions (Y, N, G, K, respectively for a Type I' turn) could enhance the probability of a turn being formed and a $\beta$-hairpin being initiated (192). Other features of the *ab initio* designed twelve residue water soluble $\beta$-hairpin include (244):

1.  Threonine in the $-$B2 and $+$B2 strand sites (the two, three residue strands have positions defined as $-$B1, $-$B2, $-$B3 for the N-terminal strand, $+$B1, $+$B2, $+$B3 for the C-terminal strand,

and L1 and L2 for the turn region according to Ref. 235 which is the non-hydrogen bonded site of the anti-parallel strand).

2. The L1 and L2 residues were chosen as Asn-Gly, the most favorable for a Type I′ β-turn.

3. A high tendency for charged residues at +B1 and aromatic residues at +B3, hence Lys and Tyr in these positions, respectively, the Lys is also expected to increase solubility and Tyr can also act as a probe in nmr NOE studies.

4. Ile and Val at positions −B3 and −B1 respectively resulting in a Ile-Tyr interstrand pair which is highly favorable of β-structure when occurring in a hydrogen bonded site (177).

5. Sheet stacking is precluded due to the lack of alternating hydrophobic/hydrophilic residues.

6. Reduction of lateral oligomerization through placing arginine at the N- and C-terminals.

With the design of an efficient β-structure nucleation center such as the β-hairpin described above the next step is to control solvation and precipitation of the β-structure as it forms. One of the key features of a β-strand was utilized in early studies of copolypeptides with alternating hydrophobic and hydrophilic residues (245–249). However, these polymers of (Ala-Gly), (Glu-Ala), (Tyr-Glu), (Lys-Leu-Lys-Leu), and (Leu-Glu-Leu-Glu) were found to be predominately bilayers of β-sheets lacking any β-turns and therefore were stabilized through formation of a hydrophobic core between two β-sheets which exposed the hydrophilic side of the sheets to the aqueous environment. These early studies could be considered as designs of β-sandwiches rather than a study of the isolated ββ super-secondary structure. However, no effort was made to control the formation of β-structure, and no β-turns were formed.

It is obvious that some form of control of packing of the β-structure during or after formation will be required if specific ββ-supersecondary structure is to be prepared. Early attempts at controlling β-sheet formation involved the design of a DDT binding, four strand β-sheet (250) and an opiate receptor mimetic consisting of two, two stranded β-sheets (251). The tools used in the opiate receptor mimetic design consisted of amino acid intrinsic β-strand forming propensities, alternating hydrophobes and hydrophiles, formation of Arg/Glu salt bridges between sheets (a feature chosen due to the stability which arises from the formation of cooperative resonance of the guanidine-carboxylate ionic bonds (252)), and finally, a large number of β-sheet forming threonine residues (13 of the 40 residues) making up the entire hydrophilic sides of the two β-sheets. Extensive use was also made of molecular modeling using CPK models to examine the fit of particular amino acids in the final structure. Other than CD studies, the final structure of the model peptide was not characterized but it was demonstrated to bind endogenous opiate peptides in a specific stereoselective manner.

The design of betabellin (253–255) and betadoublet (256) utilized design criteria which included secondary structure prediction, placement of hydrophobic/hydrophilic residues, preferred residue pairing on adjacent strands, amino acid composition of inside and outside of sheets, preferred strand twist, position specificity for turn residues, preferred side chain conformations, and patterns of internal packing interactions. Betadoublet shares the same β-sandwich motif with betabellin but was designed to improve side chain packing and water solubility of the protein. Molecular modeling was critical to the design of these β-sandwiches and extended polyalanine is the model for an extended conformation with $\phi/\psi$ angles of −139°/139°. The $\phi/\psi$ angles are modified taking the twist, bend, and coiling of $\beta$-sheets into consideration with the resultant $\phi/\psi$ values of $-144°/153°$ found in naturally occurring $\beta$-sheets. Molecular modeling indicates that the minimal number of residues per strand should be six and that the internal strands cannot have an odd number of residues without causing the chain to turn in the wrong direction (254). The Type I′ $\beta$-turn is instituted by Asp-Gly in betadoublet or the D-Ala-D-Lys sequence of betabellin 14D. The choice of turn residues may be made based on the position specificity or on the basis of overall turn preference. The decision is critical and should be made based on the turn position specificity. There may also be difficulties with this selection as seen in the design of the $\beta$-hairpin structure where the NPDG turn sequence (best positional residues for a $\beta$-turn (243)) is not necessarily the ideal sequence when a Type I′ turn is desired (240, 241). The critical nature of the turn is demonstrated by the way in which the Type I turn is detrimental to the formation of right-handed twist in an antiparallel $\beta$-sheet while the more crowded Type I′ turn provides ideal backbone arrangement for right-handed twist to occur (Fig. 32) (254).

The alternating hydrophobic/hydrophilic residues in the sequence form the expected hydrophobic and hydrophilic faces of the $\beta$-sheet. The hydrophobic face is expected to pack in a dimer to form the core of the sandwich, and extensive molecular modeling assists in designing the core formation. It is assumed that the sheets pack in an aligned arrangement with a $-30°$ angle between the sheet axes (6). A key concept introduced into the design of the hydrophobic core of betadoublet is based on Salemme's observation (142) that $\beta$-sheets tend to combine the twist and bend to form a coil with the side chains from the H-bonded site in the antiparallel arrangement tending to be located on the concave side of the coil (see Fig. 16). The

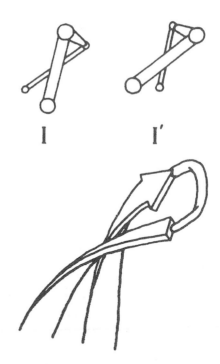

**Fig. 32** Schematic drawing of the $\beta$-hairpin, showing how the Type I′ matches and the Type I fights the $\beta$-sheet twist. Taken from JS Richardson, DC Richardson. In DL Oxender, CF Fox eds. Protein Engineering. New York. 1987, pp 149–163. (Reprinted by permission of Wiley-Liss, Inc., a subsidiary of John Wiley and Sons, Inc.).

betadoublet has the hydrophobic residues located in the H-bonded sites with the intention of encouraging hydrophobic packing on the concave side of the slight coil which then results in a barrel like sandwich when two sheets condense to from the hydrophobic core. Interstrand packing of hydrophobic residues was also facilitated by pairing the $\beta$-branched amino acids (V, T, I) with unbranched hydrophobic residues. Finally, sheet dimerization was facilitated by planning a disulfide bond between sheets in the hydrophobic core of the sandwich. However, this disulfide link did prove problematic in the betabellin design. At the moment there appears to be no better method for design of the hydrophobic core than the use of molecular models and computer software that will provide a dot-surface display of the $\beta$-sheets to permit a rationale for particular hydrophobic residues in particular positions. It is also important to use a number of different algorithms for secondary structure prediction to maximize the design through molecular modeling. As well, it is important to be reasonably flexible in your choices of amino acids since it may be more important to have a particular amino acid in a particular location in the sequence more for its side chain characteristics than its propensity to form a particular structure.

"Negative design" is another concept that can be utilized in protein structure design (132). This approach is making certain that the designed sequence is not able to assume another unwanted structure. For example, the design of $\beta$-sheets should avoid using a string of amino acids with high helix propensity. Another example employed in the betadoublet design was to attempt to prevent the formation of a Greek key arrangement of the $\beta$-strands by choosing very short turns and, for the central hairpin, putting hydrophobic side chains in the non-H-bonded sites rather than in the H-bonded sites which would discourage extensive coiling of the $\beta$-sheet in this region and promote the up-and-down arrangement of strands rather than the coiling of sheets to form the Greek key.

With all these design principles in place there are still problems concerning solubility of the designed protein (257). Mayo has postulated a number of structural design principles with which he has demonstrated limited success in the design of water-soluble $\beta$-sheet forming peptides. These principles are as follows:

1.  Adjusting the overall net charge of the peptide to mostly positively charged amino acids is observed to improve solubility. Therefore

one should maintain the positive (K, R) to negative (E, D) ratio between 4/2 and 6/2.

2.  When the number of polar residues is too high and other stabilizing forces are too low, intramolecular collapse or folding may be opposed by intermolecular peptide-water associations. Therefore, maintain noncharged polar residue (N, Q, T, S) composition less than 20%.

3.  An appropriate hydrophobic residue composition and placement promotes self-association-induced structural collapse and stability. Therefore, maintain aliphatic hydrophobic residue (I, L, V, M, A) composition between 40% to 50%.

4.  To generate a compact fold, side-chain pairing and packing must be optimized. Therefore, careful consideration must be given to placement of the hydrophobic residues in the sequence.

5.  Specific turn character may promote or stabilize a desired fold. This is reminiscent of the $\beta$-hairpin as a nucleation center for $\beta$-sheet formation and the importance of the Type I$'$ turn in stabilizing the $\beta$-hairpin.

### 3.  $\alpha\beta$-Structure: Barrels

A few studies have examined the combination of $\alpha$ and $\beta$ structure using peptide and protein models of $\alpha\beta$ (258, 259), $\beta\beta\alpha$ (260–263), and $\alpha\beta\beta$ (264) structural units producing $\alpha$-helices in combination with an antiparallel $\beta$-sheet. While not providing new design concepts, these studies emphasized the importance of turns and hydrophobic interactions in the stability of interacting $\alpha$ and $\beta$ structure and the use of the disulfide bond in assisting the stabilization of these structures.

The parallel $\beta$-strand arrangement provides a bigger challenge in the field of *ab initio* protein design. The amino acid sequence must loop back upon itself to allow the two β-strands to align in a parallel fashion. This loop region can take on a variety of structures but the most common and easiest to design is the α-helix thereby creating a βαβ super-secondary structural unit common to the αβ-barrel domain found in a number of enzymes. The αβ-barrel is actually composed of repeating α-β structural units. These repeating units align key residues in their sequences that may be important for packing of the helix onto the β-strand of the unit (265). The structural requirements of the αβ-barrel primarily arise from two factors: the right-handed twist of the β-sheet, which dictates the topology of the surface of the β-sheet and its interstrand H-bonded stability, and the requirement to exclude solvent from the barrel interior, which puts constraints on the barrel dimensions (266, 267). Barrels making up the protein core will tolerate very little variability in these two factors. The optimal strand number of eight arises from the above structural restrictions and the fact that adjacent $C_\alpha$ atoms in neighboring strands lying near or at the equatorial plane of the barrel, where the surface curvature is highest, point alternatively inside and outside the barrel. This is a situation of optimal side-chain packing that is a consequence of the orientation of strand axes relative to the barrel, which in turn depends on the strand twist (266).

An eight stranded αβ-barrel called octarellin has been designed based on a 30 amino acid residue repeating βα structural unit which consists of a four residue turn, a six residue β-strand, a second turn of seven residues, and a thirteen residue α-helix (268, 269). The length of the secondary structural elements were averaged from 24 turn-β-turn-α structural units found in three αβ-barrel proteins (triose phosphate isomerase, KDPG aldolase, and xylose isomerase). The original amino acid composition of the structural unit was arrived at through trial and error

using the frequency of residue appearance in each of the secondary structural elements in 16 known βα-unit sequences. The sequence was subsequently modified using the β-strand and α-helix probabilities of Chou and Fasman (172).

Since two different secondary structures are involved, it was important to examine the structural features of packing β-sheets and α-helices. The design of an interface between the two stranded β-sheet and the α-helix took into consideration the recognition that the β-sheet has a right-handed twist when viewed in the direction parallel to the polypeptide chain and the adjacent rows of residues $i$, $i+4$, $i+8$, and $i+1$, $i+5$, $i+9$, form a surface that also has a righthanded twist (Fig. 33) (42). The feature of the helix/sheet packing is that the helix packs onto the sheet with its axis parallel to the sheet strand so that twisted surfaces are complementary (Fig. 33). Four residues of the α-helix, $i+1$, $i+4$, $i+5$, and $i+8$, form a hydrophobic diamond structure that surrounds one β-sheet residue, residue $j$, which is generally a leucine or valine residue (Fig. 33) (270). This appears to be an elaboration of the packing described for the repeating αβ structural unit described above (265). The location of a hydrophobic diamond on the surface of a α-helix can be predicted through maximizing the sum of the weighted non-polar accessible contact areas (φ-area) according to (270):

$$H\phi = 0.5(A_i + A_{i+9}) \\ + 1.0(A_{i+4} + A_{i+5}) \\ + 1.5(A_{i+1} + A_{i+8})$$

where $A_i$ is the φ-area for residue type at position $i$ except if $i$ is not in the helix when $A_i = 0$ (Table 8).

A consequence of this alignment is a steric clash between the side-chain of one helix residue, either $i+1$ or $i+8$, which will point directly at the strand residue $j+2$ or $j-2$. This steric clash may be minimized in a number of ways which

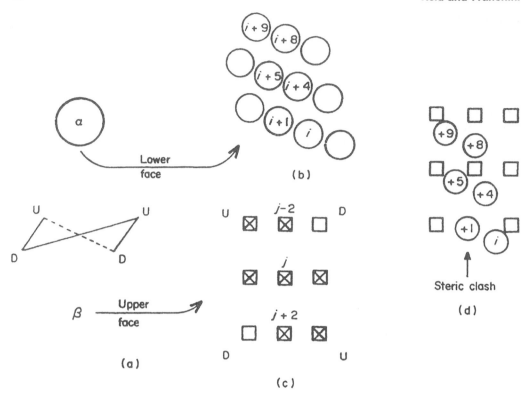

**Fig. 33** The model for $\alpha$-helix/$\beta$-sheet packing. (a), (b) and (c) follow the model proposed by Janin and Chothia (see Ref. 140). The figures show the lower surface of an $\alpha$-helix packing against the upper face of a twisted $\beta$-sheet. U and D denote the raised and lower corners of the twisted sheet; (b) an $\alpha$-helix with the positions of the residues that interact with the $\beta$-sheet being labeled; (c) the up-pointing $\beta$-strand residues (boxes) that enclose a cross to indicate the interacting residue; (d) a detailed packing model with $\alpha$-helical residues $i+1$, $i+4$, $i+5$, $i+8$ surrounding one $\beta$-sheet residue. The steric clash between $\alpha$-helical residue $i+1$ and $\beta$-sheet residue $j+2$ that needs to be prevented is shown. (Reprinted from FE Cohen, MJE Sternberg, WR Taylor. Journal of Molecular Biology 156: 821–862. Copyright 1982 by permission of the publisher Academic Press).

include eliminating the clashing residues from the helix and sheet structures or using small residues such as Gly, Ala, Ser, or Thr in the clashing helix and sheet positions (270).

Attempts were made to introduce conformational "flexibility" into the structural unit because it was known that the backbone conformation of the eight-stranded $\beta$-barrels cannot display an eight-fold symmetry and therefore, some of the repeating units would adopt a distinctly different conformation. This flexibility was created by introducing two

adjacent core residues in the $\beta$-strand and two adjacent hydrophobic diamonds in the $\alpha$-helix which made four distinct conformations possible.

Another series of eight-stranded $\alpha\beta$ barrel designs utilized some different concepts from those described for octarellin (271–273). The proteins were designed on a twofold repeating $\beta_1\alpha_1\beta_2\alpha_2\beta_3\alpha_3\beta_4\alpha_4$ structural unit. This repeating unit eliminates some of the tilt and packing problems associated with other possible repeating units (272). The $\alpha$-helices were designed to be amphipathic. There was the occasional

**Table 8** Distribution and $\phi$-Area Change for $\alpha$ and $\beta$ Residues

| | α-Helical residues | | | | | β-Sheet residues | | |
| | Observed change[a] | | | Model α-helix[b] | Prediction[c] | Observed Change[d] | | |
| | | φ-Area change | | | | | φ-Area change | |
| Residue Type | Fraction out of 174 (%) | Mean (Å²) | Max (Å²) | φ-Area (Å²) | Potential φ-area (Å²) | Fraction out of 180 (%) | Mean (Å²) | Max (Å²) |
|---|---|---|---|---|---|---|---|---|
| Ala | 11.5 | 14 | 21 | 18 | 21 | 3.8 | 11 | 15 |
| Arg | 1.1 | 7 | 8 | 19 | 0 | 1.3 | 11 | 16 |
| Asn | 0.6 | 8 | 8 | 11 | 0 | 0.6 | 5 | 5 |
| Asp | 1.1 | 7 | 7 | 11 | 0 | 1.3 | 6 | 7 |
| Cys | 4.6 | 16 | 26 | 24 | 24 | 3.8 | 10 | 17 |
| Gln | 0.6 | 6 | 7 | 14 | 0 | 0.0 | 0 | 5 |
| Glu | 1.1 | 12 | 12 | 15 | 0 | 1.9 | 17 | 21 |
| Gly | 2.3 | 7 | 9 | 11 | 0 | 0.0 | 0 | 5 |
| His | 1.7 | 17 | 22 | 27 | 11 | 1.9 | 11 | 21 |
| Ile | 12.1 | 19 | 35 | 39 | 35 | 16.9 | 15 | 34 |
| Leu | 18.4 | 17 | 31 | 33 | 34 | 13.8 | 19 | 32 |
| Lys | 5.7 | 11 | 21 | 27 | 23 | 1.3 | 9 | 10 |
| Met | 7.5 | 24 | 35 | 41 | 35 | 3.8 | 15 | 21 |
| Phe | 6.9 | 23 | 39 | 45 | 35 | 6.9 | 22 | 37 |
| Pro | 1.1 | 14 | 15 | 20 | 18 | 1.3 | 20 | 25 |
| Ser | 3.4 | 10 | 13 | 12 | 0 | 5.0 | 10 | 14 |
| Thr | 5.2 | 14 | 25 | 23 | 0 | 3.1 | 9 | 18 |
| Trp | 0.6 | 35 | 35 | 60 | 35 | 1.9 | 34 | 51 |
| Tyr | 2.9 | 19 | 24 | 38 | 35 | 5.0 | 20 | 26 |
| Val | 11.5 | 15 | 25 | 33 | 33 | 26.9 | 15 | 14 |

[a] Only $\alpha$-residues with a $\phi$-area change of greater than 5 Å when 1 $\alpha$-helix docks to a $\beta$-sheet are included, as they mediate the $\alpha\beta$ interaction.
[b] The $\phi$-areas of a residue type X exposed in a model $\alpha$-helix Ala₄XAla₄.
[c] These values are used to predict the site on an $\alpha$-helix that packs against a $\beta$-sheet. The exact values were determined by trial and error.
[d] Only $\beta$-residues with a $\phi$-area change of greater than 5 Å² when all the $\alpha$-helices dock onto the $\beta$-sheet are included.
(Reprinted from FE Cohen, MJE Sternberg, WR Taylor. Journal of Molecular Biology 156: 821–862. Copyright 1982 by permission of the publisher Academic Press).

opportunity for salt bridges in each of the helices and the N-termini carried an N-cap sequence. Special attention was paid to the nature of the loop regions between the $\alpha$ and $\beta$ structures. A common characteristic of the $\alpha\beta$-barrel proteins is that they are enzymes with the active site composed of residues located in the loop regions between the C-termini of the $\beta$-strands and the N-termini of the $\alpha$-helices. These $\beta$ to $\alpha$ loops are unusually long when compared to the usual loop region in $\beta\alpha$ structure. The $\alpha$ to $\beta$ connecting loops, on the other hand, are short and differ in length depending upon whether they connect odd numbered loops to even numbered strands or vice versa (274). Many of the loops from the N-termini of odd numbered $\alpha$-helices to the C-termini of even numbered $\beta$-strands are turns which consist of an initial Gly residue with a positive $\phi$ angle and two following residues with $\beta$ and $\alpha$-conformations, respectively. The loops connecting the N-termini of even numbered $\alpha$-helices to the C-termini of odd numbered $\beta$-strands are primarily characterized by a single Gly residue with a left-handed helical conformation producing a sharp chain reversal. The two fold symmetry of the repeating unit was designed to correct potential "weak points" in the hydrophobic core of the protein (275). The largest intrinsic weak point of the $\alpha\beta$-barrel is located at the center of the barrel where the volume is too large to be filled by the aliphatic amino acid residues. Therefore, the amino acid sequence of the $\beta$-strands were designed such that Phe and Trp would be located in $\beta_1$, Val and Gly in $\beta_2$, Ile and Ala in $\beta_3$, and Ile and Ala in $\beta_4$ which provided complementary packing of the side chains in the center of the barrel. The barrel residues facing outward were also hydrophobic residues to interact with the hydrophobic face of the amphiphilic helices. In spite of these efforts to idealize the packing of the core residues in the barrel, the designed product lacked the unique tight packing found in natural proteins.

## III. ENGINEERING FUNCTION

### A. Peptide Models of Enzyme Active Sites

The design of enzymic activity was, and still is, an appealing concept to those interested in the *ab initio* design of a functional protein. Some key concepts in this functional design were developed through the study of short linear, cyclic, and copolymer peptides as base catalysts, and as models of the active sites of known proteases, and have met with limited success (276–287).

Kopple and Nitecki demonstrated that the histidine catalyzed hydrolysis of p-nitrophenyl, and 2,4-dinitrophenyl acetate could be increased by cyclic hexa, and dipeptides containing histidine, and tyrosine. The side chain functional groups of histidine, and tyrosine were demonstrated to be independent of each other with no hydrogen bonding between residue side chains. However, the O-acetylation of tyrosine was dependent upon the proximity of the histidine imidazole side chain. The acetyl group was demonstrated to be transferred from the intermediate N-acetylimidazole side chain of histidine to the aromatic hydroxyl group of the tyrosine side chain in those cyclic peptides which had the histidyl and tyrosyl side chains in a *cis* stereochemical arrangement (276–278). A copolymer of His-Glu was demonstrated to increase the base catalysis of histidine through the hydrogen bonding of the side chains of histidine and glutamic acid which was a direct result of the formation of an $\alpha$-helical structure by the copolymer (285). Formation of designed $\alpha$-helical structure was also attributed to the glycosidase activity of a decapeptide which placed a glutamic acid residue in a hydrophobic environment preventing normal ionization of the side chain which subsequently protonated a proximally placed glycosidic bond initiating the catalytic activity of the peptide (287).

Sheehan investigated the catalytic activity of a cyclic and a linear peptide containing histidine and serine, two amino acids that had been implicated in the serine protease activity of trypsin and chymotrypsin (279). This study did not find any catalytic activity toward the hydrolysis of p-nitrophenyl ester greater than that expected for a histidine containing peptide. However, when a synthetic peptide with the sequence (TASHD) of the active site of phosphoglucomutase was examined for p-nitrophenyl acetate hydrolysis it was found to have six times the catalytic activity of the histidine and serine containing peptides (280). The addition of aspartic acid, the third member of the serine protease catalytic triad, to the synthetic peptide model was sufficient to elicit catalytic activity as well as a limited stereoselectivity for the L isomer of N-methoxycarbonylphenylalanine p-nitrophenyl ester. Increasing the flexibility between the members of the catalytic triad by insertion of $\gamma$-aminobutyric acid to produce the L-seryl-$\gamma$-aminobutyryl-L-histidyl-$\gamma$-aminobutyryl-L-aspartic acid further increased the catalytic activity. A similar attempt to produce a cysteine protease through the synthesis of a TACHD peptide sequence was unsuccessful (288). Attempts to utilize secondary $\beta$-structure in combination with the catalytic triad of serine proteases to produce a catalytic site with a steric arrangement of the His, Ser, and Asp residues resembling that found at the active site of a number of serine proteases were unsuccessful (289, 290).

A number of attempts at increasing catalytic activity of model peptides through the use hydrophobic pockets built into the peptide were successful. The catalytic activity of histidine toward the hydrolysis of p-nitrophenyl laurate could be increased 20 times by incorporation of histidine into the cyclic dipeptide cyclo(D-Leu-L-His) which was a much better catalyst than the cyclo(L-Leu-L-His) indicating a preference for a particular steric arrangement of the side chains (283). Two cyclic peptides linked by a disulfide bridge to form a hydrophobic pocket was also a successful design as a catalyst for the hydrolysis of long chain fatty acid esters of p-nitrophenol (282, 283).

Graduating from the small peptide models of binding/active sites to the larger models of proteins has developed some interesting approaches to the design of function. The earliest attempts developed from the selective digestion of proteins, isolation of protein fragments, and testing for activity. Although these studies could not be considered an *ab initio* design of functional protein, the studies were preliminary to the *ab initio* studies following on their heels. Gutte utilized the fact that the C-terminal of RNase S-peptide and the N-terminal of RNase S-protein could be shortened without decreasing the enzymic activity of the S protein/peptide complex. The missing region of RNase S was a wide loop at the surface of the protein distant from the active site. He used the concept of deleting several loops and turns distant from the active sited to design and synthesize several shortened analogs of the enzyme (291–294). The deleted regions were replaced with Gly and Ala residues to minimize structural change in the mutants. Although the enzymic activity obtained from the shortened synthetic peptides was considerably less than the RNase S parent protein, the viability of the concept of synthesizing a short polypeptide analog of an enzyme had been established. A similar concept was later developed by Atazzi and termed "surface simulation synthesis" (295, 296). Application of this concept to the design of peptide models of the active sites of trypsin and chymotrypsin resulted in cyclic 29-residue peptides with the activity and selectivity of the natural enzymes, a phenomenal feat that has yet to be duplicated (297–300).

## B.  Scaffolding as the Infrastructure for Functional Design

Molecular size has always been the major problem when dealing with the functional design of proteins in comparison to the smaller functional peptides. However, the approach to this problem has been to reduce size to the lowest common denominator as seen with the simplified studies on enzyme function described above. A very useful concept in this reductionist approach to the design of functional proteins is the domain (14, 19, 22, 143, 301–307). The domain has evolved as the unit of structure in proteins, and is often associated with function. As yet there is no clear definition of the domain and other terms such as module, motif, and fold add to the confusion of discussion on this topic. For the sake of simplicity we will equate domain, module, and fold

defined as a single contiguous portion of the polypeptide chain folded into a compact, local, semi-independent unit. The domain is a unit of function either functional on its own or as part of a functional unit involving one or more domains. The motif is defined as a super-secondary structural arrangement of two or more of the three secondary structural elements. Motifs are rarely associated with a particular biological function and are generally part of a larger structural and functional assembly or domain (20).

Based on this domain concept as the functional unit another concept arises which involves designing a small functionless protein domain that will fold independently into a compact structural unit. This "functionless domain" is called the "scaffolding" upon which one can build any number of functions (Fig. 34). Based on this definition, the four-helix bundle is the most commonly used scaffolding in the design of protein function (Fig. 24) (219).

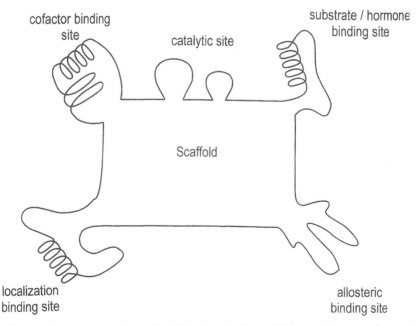

**Fig. 34**  Schematic representation of a "functionless" scaffold carrying a variety of functional domains. Each of the functional domains may be made up of one or more motifs which are characteristic of the domain function.

## 1. Enzyme Design

One approach to enzyme design involves the use of antibodies as enzymes with very high catalytic activity (308–310). Catalytic antibodies are antibodies directed against haptens that are synthetic analogs of the transition states of the chemical reactions to be catalyzed. A second very successful approach to the manipulation of enzyme activity arises from the reengineering of natural enzymes through genetic engineering and mutagenesis technology. Random and site directed mutagenesis of genes for particular enzymes have resulted in enzymes with improved thermostability, activity in artificial environments, altered substrate specificity, and even the catalysis of chemical reactions different from that of the native enzyme (311). This technique, coupled with "directed evolution" has proven to be a very powerful tool for the manipulation of enzyme function (312–314). This chapter will not deal with these methods since neither could be considered *ab initio* techniques.

Engineering of the charge transfer triad of serine proteases into a four helix bundle as part of the design of a serine protease active site is just one example of an *ab initio* protein engineering that uses scaffolding to hold a functional design (315, 316). The design of a 73 residue peptide placed the aspartate, serine, and histidine residues at the amino terminus of separate amphipathic helical peptide chains and the fourth helix terminated in a glutamate residue. The location of the catalytic residues at the ends of the helical chains was supposed to mimic the steric arrangement of the residues in the active site of chymotrypsin. The glutamate residue at the terminus of the fourth helix was designed to serve as the "*oxy-anion hole*" that is designed to stabilize the tetrahedral carbanion intermediate in the catalytic process. The helical chains were linked covalently at their carboxyl ends by a lysine and ornithine template that aligned the helices in a parallel arrangement. The amphiphilic nature of the helices assured that the four-helix bundle would associate to hold the catalytic residues in the proper arrangement as well as provide a hydrophobic binding pocket for the substrates. The leucine zipper concept of interdigitating the hydrophobic residues was used for design of the carboxyl end of the four helix bundle while the amino terminal end of the bundle was designed using smaller hydrophobic residues in order to provide a hydrophobic pocket for the substrate (317). Finally, the helices were also stabilized by the positioning of lysine and glutamic acid residues such that their side chains could form ionic bonds between turns in the helices as well as between the helices themselves. While there is no indication that the designed protein actually formed the predicted structure and the information on the biological activity is sparse (316), the concepts utilized in the design of the enzyme are enlightening and could serve as fundamental design principles for other functional proteins.

While the amphipathic helix (318) is not considered a domain in its own right, it has a tendency to self-associate in solution to form aggregates which frequently involve four helices. This fact was utilized in the *ab initio* design of a 14 residue amphipathic α-helix as the scaffolding to hold a reactive amine in the active site of a metal-independent oxaloacetate decarboxylase (319).

## 2. Metal-binding Protein Design

The four-helix bundle has also been utilized as the scaffolding in the *ab initio* design of metal binding proteins (320, 321). Metals generally play one of two roles in proteins, either catalytic or structural and the binding sites can be either discrete units or composed of two or more side chains from amino acid residues in different areas of the protein. Hence the design of metal binding sites *ab initio* may lead to proteins with novel catalytic activities or greater stabilities. Cysteine or histidine residues have been located in the $i$ and $i+4$ positions of a single $\alpha$-helical peptide (322, 323). This arrangement has been shown to stabilize the structure of the helix in the presence of a metal and was used as a method of peptide purification through metal ion affinity chromatography.

The binding of two moles of zinc to a synthetic four helix bundle formed by the homo-dimerization of a helix-loop-helix forming peptide was demonstrated by Handel and DeGrado who also showed zinc binding to a monomolecular four-helix bundle containing a single tri-coordinate zinc binding site composed of histidine side chains (324). The gene for a tetra-coordinate zinc and cobalt binding four-helix bundle has also been designed to express four identical helices connected by three identical loops (325). The metal chelating site was designed through the use of a computer program which determined the location of the two cysteine and two histidine metal ligands such that at tetrahedral coordination geometry would be obtained. The dissociations constants were estimated at $2.5 \times 10^8$ M for Zn(II) and $1.6 \times 10^5$ M for Co(II). In both examples of metal binding four-helix bundles, the metal binding was demonstrated to stabilize the helices against denaturation, and produced a designed protein product which was less like a molten globule and more like a native protein (326). This scaffolding is currently being utilized in the design of metal binding catalytic sites in enzymes (327).

An iron-sulfur center has also been designed into a four-helix bundle as a first step in using the four-helix bundle as scaffolding in designing a model system for studying the Photosystem I (PSI) reaction center (328). The $\alpha_4$ scaffolding of Regan and DeGrado (210) with four identical amphipathic helices was modified

such that the PRR sequence of loops 1 and 3 were deleted and replaced with the sequence PCDGPGRGGTC that carries the thiolate ligands which are the iron-sulfur cluster coordinating ligands of the $F_x$ domain in the Photosystem I. The loop region is not only predicted to bind the iron-sulfur cluster of PSI but is also the site at which the PsaC subunit interacts with the PsaA and PsaB subunits in the PSI complex possibly through ionic interactions with the arginine residues near the middle of the loops (329). The activity of this model system is just the beginning of the modeling of redox centers relevant to biological energy conversion and demonstrates the value of scaffolding in the *ab initio* design of functional protein.

Metal binding sites have also been designed into the variable regions of antibody light and heavy chains to assist in the design of catalytic antibodies (330–332). In these examples, the $\beta$-structure of the immunoglobulin domain provides the infrastructure for attaching the metal binding residues *via* the variable loop regions of the light and heavy chains.

A designed 61 residue all $\beta$-fragment of the heavy chain variable domain of a mouse immunoglobulin (McPC 603) incorporating three $\beta$-strands from each of the two $\beta$-sheets of the variable heavy domain, and two regions corresponding the the hypervariable loops H1 and H2 has been synthesized as a protein scaffolding called a minibody (333–337). This concept of utilizing the immunoglobulin fold as a natural scaffolding for the presentation of various protein functional topologies is taken to the next step by utilizing a modified fold reduced in size. Three histidine residues were incorporated into the H1 and H2 loop regions to serve as chelating residues. The designed metal binding site was found to be selective for different ions in the order, Cu > Zn $\gg$ Cd > Co similar to that observed for carbonic anhydrase. Initial studies encountered solubility problems that were overcome in later analogs of the minibody through manipulation of the β-sheet sequence and the addition of N-terminal and C-terminal three lysine residue tails (336). The minibody can be prepared either through solid-phase peptide synthesis or via gene construction and expression and has great potential for use as scaffolding for the phage display of variable peptide sequences for affinity-selection or for the production of small proteins with specifically engineered pharmacological activities.

A number of toxins isolated from scorpions (338–340) sea anemones (341–343), insects (344), plants (345), and snails (346–351) have a variety of functions from proteinase inhibition (352) to sodium, potassium, chloride, or calcium channel blockade. The toxins are small (10–70 amino acid residues), compact peptides with a number of cysteine residues forming disulfide bonds that appear to provide a stable scaffolding from which the functional amino acid residue side chains can be presented for biological function. These molecules have been useful tools in probing biochemical pathways as well as the function of membrane proteins. The scorpion toxin, charybdotoxin, has been utilized as a scaffolding for a metal binding site designed to mimic the zinc binding site of carbonic anhydrase B. The 37 amino acid residue peptide forming a short helix and antiparallel triple-stranded β-sheet of βαββ arrangement stabilized by three disulfide bonds had three metal chelating histidines designed into two antiparallel strands of the β-sheet (Fig. 35) (353). The novel peptide was demonstated to bind $Zn^{2+}$, $Cu^{2+}$, $Cd^{2+}$, $Ni^{2+}$, and $Mn^{2+}$ in the order that mirrored metal affinity of carbonic anhydrase B (339). The same scaffolding has also been utilized to design, synthesize and test a chimeric protein containing the functional site of snake curaremimetic toxin linked to the αβ scaffold of the charybdotoxin (354, 355).

**Fig. 35** Scorpion toxin (charybdotoxin) βαββ scaffolding.

The chimera was demonstrated to displace snake curaremimetic toxin α from the acetylcholine receptor at a concentration of $10^{-5}$ M. The success of these two design experiments bodes well for the general application of toxins as scaffolding in the engineering of novel therapeutic peptides and proteins.

Other scaffolding designs have been utilized to hold metal binding sites in *ab initio* designed proteins. The B1 domain of IgG-binding protein G, a 56 residue protein composed of four β strands and a single α-helix in a ββαββ arrangement, has been used as the scaffolding for a high affinity Co(II), Cd(II), and Zn(II) binding site (356, 357). The tetrahedral coordinating site is composed of three histidines and one cysteine residue.

*3. Hemeprotein Design*

The first example of an *ab initio* designed hemeprotein utilized the metallopor-

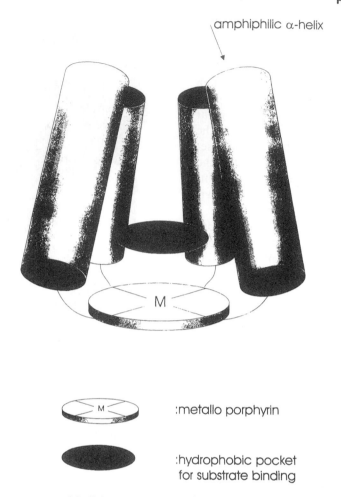

**Fig. 36**  Design concept of helichrome as an artificial hemoprotein. Taken from T Sasaki, ET Kaiser. Biopolymers 29: 79–88, 1990. Synthesis and structural stability of helichrome as an artificial hemeprotein. (Reprinted by permission of Wiley-Liss, Inc., a subsidiary of John Wiley and Sons, Inc.).

phyrin ring system as a template upon which a four-helix bundle was built (358). This so called "helichrome" was designed as a protein mimic of cytochrome P450. The porphyrin ring system therefore served two purposes; (1) as a catalytic center and (2) as a molecular constraint to prevent collapse of the helical bundle. The cavity above the porphyrin ring at the N-terminal end of the parallel α-helices in the bundle was designed to provide the hydrophobic cavity for substrate binding,

however, no function was described for the designed protein (Fig. 36).

Much of the value of the heme groups in proteins is to promote long range electron transfers and charge separation in respiration and photosynthesis. An understanding of the mechanism through which heme promotes these protein functions can be obtained by the use of *ab initio* designed heme binding proteins. A single 22-residue amphipathic helix model of a redox system which incorporated a

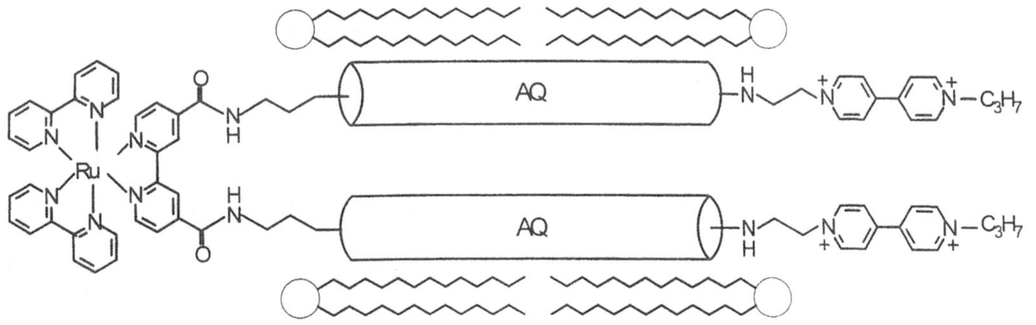

**Fig. 37** Schematic of the designed structure of a hybrid peptide with a redox triad of ruthedium trisbipyridine, anthraquinone and propylviologen. The cylinder represents the helical region of the peptide with the ruthedium complex attached at the C-terminus and the propylviologen attached to the N-terminus and the anthraquinone attached to the middle of the helix via an ornithine side chain amide linkage.

triad of a ruthenium complex of trisbipy-ridine, anthraquinone, and propylviolo-gen was designed to be inserted into a lipid bilayer (Fig. 37) (359). The designed photo-functional peptide was deployed into egg yolk lecithin vesicles but the nature of the structure incorporated into the lipid bilayer was uncertain. A more successful design of a hemeprotein utilized the diheme cytochrome b subunit of cytochrome bc$_1$ as a working model (360). The scaffolding for presentation of the heme binding sites consisted of a four-helix bundle prepared by the synthesis of an amphipathic 31 residue α-helix with a four residue flexible Cys-Gly-Gly-Gly tether at the N-terminus which allowed the forma-tion of a disulfide bond between two parallel aligned helices. Two histidine residues were incorporatedd at positions 10 and 24 of the peptide. The hemes were 13 residues apart so that the helical dis-tance between them would approximate 20 A°. A Phe residue was inserted at pos-ition 17 to separate the heme binding domains and an Arg residue was added at position 27 for the dual purpose of defining the heme-binding site at His 24 and raising the redox potential of the heme through charge interaction. Two other 31 residue models were also prepared with a single

His residue at positions 10 and 24, respectively. The designed peptides associ-ated into four helix bundles after oxidation of the Cys thiol to provide covalently linked dimers. Addition of ferric heme to the four helix bundles resulted in either a two or a four heme protein depending upon whether the four helix bundle was prepared from the two histidine monomer or the one histidine monomer. The planes of the heme groups were nearly parallel and displayed a twofold symmetry. How-ever, it was not known if the four helix bundle was formed by an all-parallel arrangement of the helices with the dis-ulfide links at one end or by an antiparallel arrangement with the disulfide links at opposite ends of the bundle.

The standard four helix bundle (α$_2$ dimer) has been modified through com-puter modeling to incorporate a single iron protoporphyrin IX molecule in the center of the bundle, parallel to the helices (361). A cysteine was introduced at the N-terminus of the α$_2$ peptide and oxidized to form a homodimeric disulfide-linked protein, α$_2$(S-S), thus ensuring an unam-biguous dimeric aggregation state with both loops at one end of the bundle. The standard model α$_2$(S-S) was further modeled for the ideal arrangement of amino acids for incorporation of the protoporphyrin IX molecule giving rise to a peptide with the liganding His residue located at position 25. Leu 22 and 29 had to be replaced by Val and Leu 10 by Ala to prevent the hydrophobic residues overlapping with the Van der Waals surface of the heme (VAVH$_{25}$(S-S)). A model peptide with His inserted into the $\alpha_2$ sequence at positon 25 was also prepared as a control to determine whether a simple substitution of His in the center of the standard sequence would produce a heme binding protein. Finally, a reverse sequence of VAVH$_{25}$(S-S) was synthesized (retro(S-S)) because reversing the sequence of the peptide should convert stabilizing interactions with the helix macrodipole into destabilizing interactions. It was found that only the VAVH$_{25}$(S-S) and retro(S-S) proteins bind heme with a 1:1 stoichiometry and with high affinity in the heme binding pocket. An interesting result was that the retro(S-S) protein was less structured in the absence of the heme molecule and became a more structured compact molecule upon binding the heme group. It appears that this feature of the retro(S-S) protein is advantageous in permitting the molecule to search a large ensemble of conformers which could be compatible with a six-coordinate low-spin heme complex which predominates in the retro(S-S) heme complex, something the less flexible VAVH$_{25}$(S-S) protein is not able to accomplish and hence has difficulty in forming the low-spin complex.

A water soluble hemeprotein in which two short identical 13 residue peptides are covalently linked to the mesoheme propionyl groups via the $\varepsilon$-amino group of lysine has been prepared (362). The complex binds iron via the single histidine residue found in each of the peptides and the complex leads to a large increase in $\alpha$-helicity of the peptide. Removal of iron yields a CD spectrum consistent with a random coil conformation of the peptide.

### 4. Membrane Spanning Protein Design

The most studied membrane spanning proteins with a function involving the membrane spanning region are the ion channels (363, 364). Possible candidates for membrane spanning structure include the $\alpha$-helix, $\pi$-helix, the $\beta$-helix of which the gramicidin channel is probably the most characterized example (365, 366), and the $\beta$-barrel structure of the bacterial porins (367, 368). Early studies on the *ab initio* design of ion channels dealt with the pore forming capacity of Ala-Ala-Gly and Leu-Ser-Leu-Gly oligmers which indicated that the ion transporting capacity of these molecules did not involve the $\beta$-helical structure and likely involved assemblies of several molecules (369). While the membrane spanning aggregated $\beta$-helical structure of gramicidin and the 16 stranded $\beta$-barrel trimer of the bacterial porins provide fascinating examples of pore forming structures, this discussion on the *ab initio* design of membrane spanning proteins will concentrate on the ion channel proteins.

The amphipathic helical structure is a natural candidate for the scaffolding upon which a membrane-spanning protein structure resembling the sodium, potassium, calcium, or chloride ion channel could be designed. It is relatively easy to conceive of the amphipathic helices grouping together in an inside-out arrangement where the hydrophobic side is oriented toward the lipid environment and the hydrophilic side is oriented toward the center of the $\alpha$-helical cluster to produce the aqueous pore environment (Fig. 38). Different sized pores could be produced by different numbers of $\alpha$-helices involved in the pore formation. Model peptide studies have established the 20–23 mer amphipathic $\alpha$-helix as the pore forming motif of the voltage dependent sodium channel (370, 371), the ligand-gated nicotinic acetylcholine

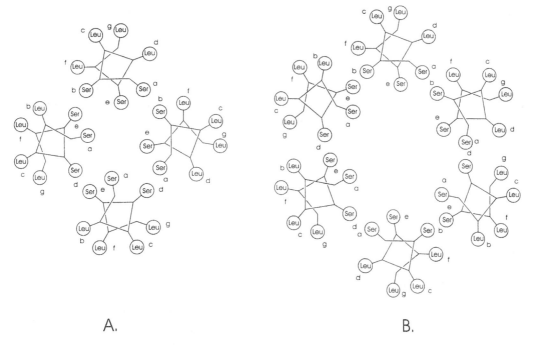

A.                                                                      B.

**Fig. 38** Schematic of an amphipathic (A) four and (B) six "inside-out" helix bundle/coiled-coils forming a transmembrane pore. The heptad repeat shown is LLSSLLS.

receptor sodium channel (371–374), the L-type dihydropyridine sensitive calcium channels (375, 376), and the glycine receptor anion channel (377). The model peptide studies indicated a four helix bundle as a functional aggregate using the template assembled synthetic peptide technology (378, 379). However, larger aggregates are more likely to occur in the natural protein channel (374).

Aggregation of amphipathic α-helical peptides of 20–23 residues in length is likely the mechanism of pore formation. The α-helix is a strong candidate for pore formation mainly due to the ease of partitioning the fully hydrogen bonded structure into the lipid medium from the aqueous environment of the cytoplasm (380). The ideal scaffolding for packing of the helices in the lipid membrane to form the channel pore is suggested to be the coiled-coil (381). The conventional heptad repeat sequence describing the molecular conformation of

the α-fibrous proteins is characterized by a periodic disposition of apolar residues at positions a and d for the heptad repeat (a b c d e f g)$_n$ (222, 206). These two positions are frequently occupied by large hydrophobic residues like leucine such that a "knobs into holes" packing can occur between helices placing the hydrophobic residues in the center of the coiled-coil protected from the aqueous environment. A scaffolding for channel pore functional analysis has been developed utilizing the coiled-coil packing of amphipathic α-helices formed through the synthesis of peptides with the triplet heptads (LSSLLSL)$_3$ and (LSLLLSL)$_3$ (382–386). Leucine was chosen as the hydrophobic residue due to its high helix propensity and serine was chosen as the hydrophilic residue because it is uncharged and will thus also contribute to the requirement that the peptide models have an overall high hydrophobicity. The two different models were voltage dependent and displayed different channel characteristics which was attributed to a difference in the number of coiled-coils packing to form the channel. However, it is not clear on the number of coiled-coil dimers involved in aggregation to form the channels nor is it clear how the helices aggregate to form a dimer which will aggregate to produce the "inside-out" coiled-coil necessary to interact with the lipid membrane and form the channel pore. The amphipatic helices are thought to insert into the membrane parallel to the surface of the phospholipid and dimerization into parallel coiled-coils may occur (385). Voltage dependence of pore formation in the model systems is thought to arise as a result of interaction between the helix dipole moment and the voltage difference across the membrane. A gating mechanism is postulated involving the reorientation of the peptide from horizontal to transmembrane upon application of voltage to align the peptide dipole with the electric field (Fig. 39).

The concept of template-assembled synthetic proteins (TASP) has also been utilized to develop a scaffolding from which to examine the structure/functional relationships in channel pores (Fig. 40) (387, 388). The TASP concept maintained the helices in a parallel arrangement and was an attempt to clarify the number of helices involved in the formation of the pores. Although the model peptides formed pores, the problems of aggregation remained evident. An interesting feature of the TASP pore models was that they were voltage independent, possibly indicating that these models are inserted vertically into the membrane and do not require voltage application to cause a change in orientation to one favoring pore formation.

Finally, an interesting application of the coiled-coil as molecular scaffolding was developed to examine the cytoplasmic domain of the integrin $\alpha_{IIb}\beta_3$ receptor (389). CD spectroscopy indicated the

presence of a coiled-coil in aqueous solution and fluorescence studies of the 126 amino acid model protein indicated interaction between the two cytoplasmic domains. This peptide was suggested to be a good model for the solution study of the interactions between the cytoplasmic domains of a multisubunit cell-surface receptor.

## 5. DNA Binding Proteins

Protein/DNA recognition and binding describes a number of specific protein structural motifs that could be utilized as scaffolding for the design of potential DNA binding therapeutic agents. The very diverse families of DNA binding proteins are classified based on the related structural motifs utilized for DNA recognition. The main classes consist of the helix-turn-helix (HTH) proteins, the homeodomain proteins, zinc finger proteins (Zif), steroid receptors, leucine zipper (bZip) proteins, helix-loop-helix (bHLH) proteins, and the helix-loop-helix-leucine zipper proteins (b/HLH/Zip) (390, 391). The sequence specific recognition of nucleic-acids by proteins arises from structural complementarity due to direct hydrogen bonding and van der Waals interactions between protein side chains usually from $\alpha$-helical regions of the protein structure and the exposed edges of base pairs, primarily in the major groove of B-DNA (390).

These sequence specific recognitions have yet to be delineated for specific DNA/protein interactions. Therefore, a complete understanding of the sequence specificity of DNA binding proteins requires identification of both the factors that enhance interaction with the correct site on DNA and the factors that inhibit interaction with the incorrect site (392). The DNA binding element in most DNA binding proteins is a motif that can function independently of the rest of the protein. In several instances the motif is a stable, folded domain maintained by hydrophobic interactions or metal binding (391). In other cases the motif is fully or partially unfolded in solution and forms a stable conformation when bound to DNA (393). The helix-turn-helix (HTH) motif was the first to be discovered but is not a separate stable domain. It cannot fold or

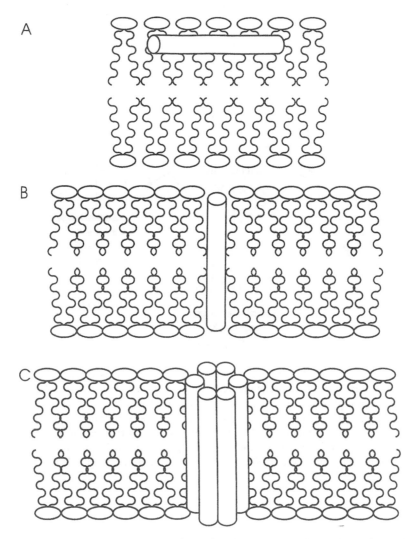

**Fig. 39** Proposed model for the voltage- dependent frequency of channel openings: (A) the major closed state for the peptide incorporated into lipid membranes in the absence of a voltage difference across the bilayer. The peptide is oriented parallel to the plane of the bilayer and located in or close to the lipid head group region of the membrane; (B) change of peptide orientation from horizontal to transmembrane. A reorientation of the peptide occurs upon application of a voltage gradient to the membrane. The peptide "flips" from a surface to a transmembrane orientation due to alignment of the helical dipole moment with the electric field; (C) formation of conducting ion channels by the transmembrane-oriented peptide. The transmembrane-oriented peptide aggregates, sequestering the polar seryl hydroxyls from the lipid acyl chains and creating polar pathways large enough to pass ions. (Reprinted with permission from LA Chung, JD Lear, WF DeGrado. Biochemistry 31: 6608–6616. Copyright 1992 American Chemical Society).

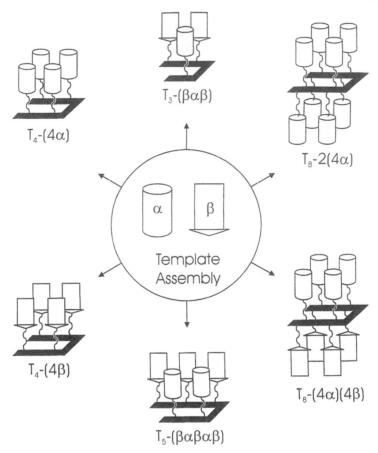

**Fig. 40** Schematic representation of the variety of supersecondary structures available through template assembled proteins. $T_n$ is the template molecule with $n$ as the number of attachment sites on T. $\alpha$ and $\beta$ are amphiphilic peptide blocks with a tendency for helix ($\alpha$) and $\beta$-sheet ($\beta$) secondary structure formation. Taken from M Mutter, R Hersperger. K Gubernator, K Muller. Proteins: Structure, Function, and Genetics 5: 13–21, 1989. The construction of new proteins V: A template-assembled synthetic protein (TASP) containing both a 4-helix bundle and a $\beta$-barrel-like structure. (Reprinted by permission of Wiley-Liss, Inc., a subsidiary of John Wiley and Sons, Inc.).

function on its own and is always found as part of a larger DNA binding domain (391, 394).

The 60 residue homeodomain contains an HTH motif but also consists of a third helix which is roughly perpendicular to the first two helices and packs against helixes 1 and 2 to form part of the protein core (391, 395). This motif forms a stable, folded domain that can bind DNA on its own (396). The 30 residue zinc finger motif contains an antiparallel $\beta$-sheet and an

$\alpha$-helix (391). Two cysteines near the turn in the $\beta$-sheet region and two histidines in the $\alpha$-helix coordinate a central zinc ion which is responsible for holding the secondary structural units in a stable DNA binding domain. The steroid receptor protein contains separate domains for hormone binding, DNA binding, and transcriptional activation (397). The motif, consisting of two $\alpha$-helices, is monomeric in solution but binds DNA as a dimer (398).

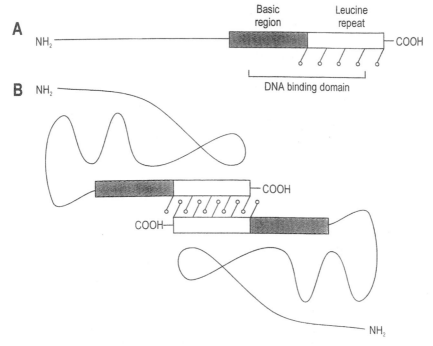

**Fig. 41** Schematic representation of the leucine zipper (A) monomer and (B) dimer. The region necessary for sequence-specific interaction with DNA extends beyond the leucine repeat toward the amino-terminus of the protein, and includes a 30-amino acid region that is highly positively charged (stippled region). (Reprinted with permission from WH Landschultz, PF Johnson, SL McKnight. Science 240: 1759–1764. Copyright 1988 American Association for the Advancement of Science).

The 60–80 residue leucine zipper motif contains two distinct subdomains, the leucine zipper region which is responsible for dimerization and the basic region which makes contact with the DNA binding site (317, 399). The leucine zipper sequence consists of 30–40 residues characterized by a heptad repeat with leucine appearing at every seventh position in the sequence. These α-helical regions form antiparallel coiled-coil dimers with a "knobs into holes" arrangement of the leucine residues which gives the conformation the appearance of a zipper (Fig. 41). The 30 residue basic region is rich in arginines and lysines and is responsible for the DNA sequence preference of the leucine zipper proteins (400). This DNA binding region is a random disordered conformation in solution but α-helical when complexed with DNA.

The helix-loop-helix (bHLH) class of transcription factors contain an N-terminal basic region that contacts the DNA and an helix-loop-helix sequence that mediates dimer formation (401).

A family of eukaryotic transcription factors combines the HLH and Zip domains into the b/HLH/Zip domain. The two potential dimerization interfaces (Zip and HLH) distinguish this class of transcription factors from the bZip and the bHLH classes (402).

The HLH, Zip, and Zif domains present the most stable conformational scaffolding upon which functional DNA binding can be built. The minimalist approach has been used to demonstrate the scaffolding role of the coiled-coil leucine zipper and the DNA seqeunce specificity of the basic region in the b-Zip proteins (403). The leucine zipper serves as a basis for the "scissor grip" or "induced helical fork" models which propose that the zipper is a molecular scaffolding from which the basic region presents the DNA binding arrangement of amino acids (403, 404). The basic region is predicted to form an α-helical structure upon interaction with DNA, consistent with the suggestion that interaction with DNA occurs predominately through α-helices. This induced α-helical structure has also been observed in b/HLH/Zip and truncated b/HLH DNA binding proteins (393, 402, 405).

The versatility of the leucine zipper as a model of the scaffolding concept was demonstrated by Gutte when he utilized the leucine zipper to double the DNA binding activity of a HIV enhancer binding protein that was based on the recognition helix and basic region of bacteriophage 434 repressor protein (406–408).

The zinc finger has served as a scaffolding upon which various DNA binding specificities can be built (409). A consensus sequence for zinc fingers based on 131 zinc finger sequences was utilized as the scaffolding upon which four residues were varied in the DNA recognition region which spans seven consecutive residues in the consensus zinc finger sequence. The study indicated that the zinc fingers can be qualitatively modular in that specificities can be swapped along with individual fingers, but that the zinc fingers are not always quantitatively modular.

The $\beta\beta\alpha$ motif commonly found in zinc fingers was also designed to bind the anticodon triplet of yeast tRNA$^{Phe}$, GAA (410, 411). Since this *ab initio* design of a nucleic acid binding protein was prior to the discovery of the zinc finger, a zinc binding site was not built into the peptide. However, a cysteine residue was placed in one β-strand and another was placed in

the penultimate position of the C-terminal α-helix in order to form a disulfide bond to further stabilize the protein conformation. Chemical synthesis and purification of the oxidized product gave both the disulfide linked monomer and dimer. Both peptides had poor affinity of the substrate for which they were designed. However, both peptides bound cytidylic acid and exhibited ribonuclease activity, the 68 residue dimer being more active that the 34 residue monomer in these cases. It was not demonstrated that the model peptide actually formed the predicted $\beta\beta\alpha$ conformation in solution and one wonders what would have happened to substrate binding activity had a zinc binding site been designed into the sequence.

The consensus sequence of the three helices of the homeodomain with the variable residues filled in as alanine (the "minimAl" approach) was utilized as scaffolding for DNA recognition protein (412). The model recognized DNA but with less than optimal specificity due to the molten globule nature of the model protein. High affinity DNA binding and appropriate tertiary structure were obtained when non consensus residues in the N-terminal arm of the homeodomain were returned to the sequence, indicating that the N-terminal arm not only has a DNA recognition role but also has a function in maintaining homeodomain structure.

## IV.  CONCLUSIONS

Peptides and proteins are well on their way to establishing a prominent role in the therapeutic arsenal. Many peptides like insulin, oxytocin, vasopressin, and growth hormone and proteins like the interleukins, interferon, colony-stimulating factors, and monoclonal antibodies have major roles as important therapeutic agents. With the rapid development of molecular biological technology, the availability of many of these therapeutic agents and analogs is no longer problematic. The fact that proteins and peptides are naturally occurring molecules routinely seen and handled by the mammalian cell is a strength of these molecules as therapeutic agents.

The challenge now is to utilize the techniques of preparing and modifying these agents to develop an understanding of their mechanism of action and pharmacophoric pattern. The development of structure/function relationships in proteins and peptides is progressing rapidly and will permit the understanding and control of the absroption, distribution, metabolism, excretion, and biological function of these molecules which in turn should lead to the design of more potent and more selective agents.

The modular concept of protein structure has reduced protein complexity to super-secondary structural motifs composed of the three elements of protein secondary structure—the helix, sheet, and turn/loop. The domain, as a functional motif, could be the key to an understanding of the relationship between protein structure and function. Structural simplification through the study of functional domains built from structural motifs is providing valuable information in a number of areas, some of which include enzyme function, metal binding, membrane channel function and DNA binding. Ultimately the goal would be to design a protein/peptide to perform a specific therapeutic function and deliver this agent through gene therapy. These studies are paramount to the characterization of natural protein and peptide pharmacophores and the information obtained can be utilized in the subsequent design of non-peptide small molecule therapeutic agents. Finally, model peptides and proteins provide a valuable tool to develop a better understanding of biological systems under normal and pathological conditions.

## ACKNOWLEDGEMENTS

The quality of this manuscript was greatly improved by the criticisms and corrections provided by Grant Mauk and Lawrence McIntosh and for which we are deeply grateful. However, the responsibility for any remaining errors and omissions remains with us.

## REFERENCES

1. JD Watson, FHC Crick. Molecular structure of nucleic acids: A structure for deoxyribose nucleic acid. Nature 171: 737–738, 1953.
2. JD Watson. The human genome project: Past, present, and future. Science 248: 44–49, 1990.
3. DJ Weatherall. Science and the Quiet Art—The Role of Medical Research in Health Care. New York, London: WW Norton and Company, 1995.
4. CB Anfinsen. Principles that govern the folding of protein chains. Science 181: 223–230, 1973.
5. TE Creighton. Protein Folding. Biochem J 270: 1–16, 1990.
6. C Chothia, AV Finkelstein. The classification and origins of protein folding patterns. Ann Rev Biochem 59: 1007–1039, 1990.
7. GD Rose, R Wolfenden. Hydrogen bonding, hydrophobicity, packing, and protein folding. Ann Rev Biomol Struct 22: 381–415, 1993.
8. CR Matthews. Pathways of protein folding. Ann Rev Biochm 62: 653–683, 1993.
9. KA Dill, S Bromberg, K Yue, KM Fiebig, DP Yee, PD Thomas, HS Chan. Principles of protein folding—A perspective from simple exact models. Protein Science 4: 561–602, 1995.
10. KA Dill, HS Chan. From Levinthal to pathways to funnels. Nature Structural Biology 4: 10–19, 1997.
11. CI Bränden. Relation between structure and function of $\alpha/\beta$ proteins. Q Rev Biophys 13: 317–338, 1980.
12. J Janin, SJ Wodak. Structural domains in proteins and their role in the dynamics of protein function. Prog Biophys Molec Biol 42: 21–78, 1983.
13. C Chothia, AM Lesk. Helix movements in proteins. Trends in Biochemical Sciences 10: 116–118, 1985.

14. S Cox, E Radzio-Andzelm, SS Taylor. Domain movements in protein kinases. Current Opinion in Structural Biology 4: 893–901, 1994.

15. WG Hol. The role of the α-helix dipole in protein function and structure. Prog. Biophys Molec Biol 45: 149–195, 1985.

16. IUPAC-IUB Commission on Biochemical Nomenclature. Abbreviations and symbols for the description of the conformation of polypeptide chains. J Mol Biol 52: 1–17, 1970.

17. GN Ramachandran, C Ramakrishnan, V Sasisekharan. Stereochemistry of polypeptide chain configurations. J Mol Biol 7: 95–99, 1963.

18. J Janin, S Wodak, M Levitt, B Maigret. Conformation of amino acid side-chains in proteins. J Mol Biol 125: 357–386, 1978.

19. JS Richardson. Introduction: Protein motifs. FASB J 8: 1237–1239, 1994.

20. CI Brändén, J Tooze. Introduction to Protein Structure. New York and London: Garland Publishing, Inc., 1991.

21. TS Rao, MG Rossman. Comparison of supersecondary structures in proteins. J Mol Biol 76: 241–256, 1973.

22. RF Doolittle. The multiplicity of domains in proteins. Ann Rev Biochem 64: 287–314, 1995.

23. KA Dill. Dominant forces in protein folding. Biochemistry 29: 7133–7155, 1990.

24. PK Ponnuswamy. Hydrophobic characteristics of folded proteins. Prog Biophys Molec Biol 59: 57–103, 1993.

25. EP Baldwin, BW Matthews. Core-packing constraints, hydrophobicity and protein design. Current Opinion in Biotechnology 5: 396–402, 1994.

26. AR Fersht. Mapping the structures of transition states and intermediates in folding: delineation of pathways at high resolution. Phil Trans R Soc Lond B 348: 11–15, 1995.

27. RL Baldwin. How does protein folding get started? Trends in Biochemical Sciences 14: 291–295, 1989.

28. OB Ptitsyn. Sequential mechanism of protein folding. Dokl Acad Nauk SSSR 210: 1213–1215, 1973.

29. OB Ptitsyn. Protein folding: hypothesis and experiments. J Prot Chem 6: 273–293, 1987.

30. OB Ptitsyn. How does protein synthesis give rise to the 3D-structure. FEBS Letts 285: 176–181, 1991.

31. M Karplus, DL Weaver. Protein folding dynamics—the diffusion-collision model and experimental data. Protein Science 3: 650–668, 1994.

32. DB Wetlaufer. Nucleation, rapid folding, and globular intrachain regions in proteins. Proc Natl Acad Sci USA 70: 697-701, 1973.

33. KA Dill. Theory for the folding and stability of globular proteins. Biochemistry 24: 1501–1509, 1985.

34. KA Dill, DOV Alonso, K Hutchinson. Thermal stabilities of globular proteins. Biochemistry 28: 5439–5449, 1989.

35. S Kamtekar, JM Schiffer, H Xiong, JM Babik, MH Hecht. Protein design by binary patterning of polar and nonpolar amino acids. Science 262: 1680–1685, 1993.

36. CI Brändén, TA Jones. Between objectivity and subjectivity. Nature 343: 687–689, 1990.

37. EE Lattman, GD Rose. Protein folding—What's the question? Proc. Natl Acad Sci USA 90: 439–441, 1993.

38. M Groß, KW Plaxco. Reading, writing and redesigning. Nature 388: 420–421, 1997.

39. OB Ptitsyn. The molten globule state. In: TE Creighton, ed. Protein Folding. New York: WH Freeman and Company,1992, pp 243–300.

40. WA Lim, DC Farruggio, RT Sauer. Structural and energetic consequences of disruptive mutations in a protein core. Biochemistry 31: 4324–4333, 1992.

41. WGJ Hol, PT Van Duijnen, HJC Berendsen. The α-helix dipole and the properties of proteins. Nature 273: 443–446, 1978.

42. C Chothia. Principles that determine the structure of proteins. Ann Rev Biochem 53: 537–72, 1984.

43. L Pauling, RB Corey, HR Branson. The structure of proteins. Proc Natl Acad Sci USA. 37: 205–211, 1951.

44. JC Kendrew, RE Dickerson, BE Strandberg, RG Hart, DR Davies, DC Phillips, VC Shore. Structure of myoglobin. Nature 185: 422–427, 1960.

45. DJ Barlow, JM Thornton. Helix geometry in proteins. J Mol Biol 201: 601–619, 1988.

46. C Branden, J Tooze. Introduction to Protein Structure. New York and London: Garland Publishing, Inc., 1991, pp 12.

47. EN Baker, RE Hubbard. Hydrogen bonding in globular proteins. Prog Biophys Mol Biol 44: 97–179, 1984.

48. BK Vainshtein, WR Melik-Adamyan, VV Barynin, AI Grebenko, VV Borisov. Three-dimensional structure of catalase from *Penicillium vitale* at 2.0 Å resolution. J Mol Biol 188: 49–61, 1986.

49. CA Hasemann, RG Kurumbail, SS Boddupalli, JA Peterson J Deisenhofer. Structure and function of cytochromes P450: A comparative analysis of three crystal structures. Structure 3: 41–62, 1995.

50. BH Zimm, JK Bragg. Theory of phase transition between helix and random coil in polypeptide chains. J Chem Phys 31: 526–535, 1959.

51. R Lifson, A Roig. On the theory of helix-coil transitions in polypeptides. J Chem Phys 34: 1963–1974, 1961.

52. H Qian, JA Schellman. Helix-coil theories: A comparative study for finite length polypeptides. J Phys Chem 96: 3987-3994, 1992.

53. PY Chou, G Fasman. Conformational parameters for amino acids in helical, beta-sheet, and random coil regions calculated from proteins. Biochemistry 13: 211–222, 1974.

54. S Padmanabhan, S Marqusee, T Ridgeway, TM Laue, RL Baldwin. Relative helix-forming tendencies of nonpolar amino acids. Nature 344: 268–270, 1990.

55. G Merutka, W Lipton, W Shalongo, SH Park, E Stellwagen. Effect of central-residue replacements on the helical stability of a monomeric peptide. Biochemistry 29: 7511–7515, 1990.

56. PC Lyu, MI Liff, LA Marky, NR Kallenbach. Side chain contributions to the stability of alpha-helical structure in peptides. Science 250: 669–673, 1990.

57. A Chakrabartty, JA Schellman, RL Baldwin. Large differences in the helix propensities of alanine and glycine. Nature 351: 586–588, 1991.

58. A Chakrabartty, T Kortemme, RL Baldwin. Helix propensities of the amino acids measured in alanine-based peptides without helix-stabilizing side-chain interactions. Protein Sci 3: 843–852, 1994.

59. PJ Gans, PC Lyu, MC Manning, RW Woody, NR Kallenbach. The helix-coil transition in heterogeneous peptides with specific side-chain interactions: theory and comparison with CD spectral data. Biopolymers 31: 1605–1614, 1991.

60. DS Kemp, JG Boyd, CC Muendel. The helical s constant for alanine in water derived from template-nucleated helices. Nature 352: 451–454, 1991.

61. E Stellwagen, SH Park, W Shalongo, A Jain. The contribution of residue ion pairs to the helical stability of a model peptide. Biopolymers 32: 1193–1200, 1992.

62. KT O'Neil, WF DeGrado. A thermodynamic scale for the helix-forming tendencies of the commonly occurring amino. Science 250: 646-651, 1990.

63. M Blaber, XJ Zhang, BW Matthews. Structural basis of amino acid alpha-helix propensity. Science 260: 1637–1640, 1993.

64. A Horovitz, JM Matthews, AR Fersht. Alpha-helix stability in proteins. II. Factors that influence stability at an internal position. J Mol Biol 227: 560–568, 1992.

65. M Blaber, XJ Zhang, JD Lindstrom, SD Pepiot, WA Baase, BW Matthews. Determination of alpha-helix propensity within the context of a folded protein. Sites 44 and 131 in bacteriophage T4 lysozyme. J Mol Biol 235: 600–624, 1994.

66. TP Creamer, GD Rose. Side-chain entropy opposes alpha-helix formation but rationalizes experimentally determined helix-forming propensities. Proc. Natl. Acad. Sci. U.S.A 89: 5937–5941, 1992.

67. MJ McGregor, SA Islam, MJE Sternberg. Analysis of the relationship between side chain conformation and secondary structure in globular proteins. J. Mol. Biol. 198: 295–310, 1987.

68. G Nemethy, SJ Leach, HA Scheraga. The influence of amino acid side chains on the free energy of helix-coil transitions. J Phys Chem 70: 998–1004, 1966.

69. A Chakrabartty, RL Baldwin. Stability of α-helicies. Adv Prot Chem 46: 141–176, 1995.

70. JS Richardson, DC Richardson. Amino acid preferences for specific locations at the ends of α-helices. Science 240: 1648–1652, 1988.

71. B Forood, EJ Feliciano, KP Nambiar. Stabilization of alpha-helical structures in short peptides via end capping. Proc Natl Acad Sci USA 90: 838–842, 1993.

72. MD Bruch, MM Dhingra, LM Gierasch. Side chain-backbone hydrogen bonding contributes to helix stability in peptides derived from an alpha-helical region of carboxypeptidase A. Proteins 10: 130–9, 1991.

73. PC Lyu, DE Wemmer, HX Zhou, RJ Pinker, NR Kallenbach. Capping interactions in isolated alpha helices: position-dependent substitution effects and structure of a serine-capped peptide helix. Biochemistry 32: 421–5, 1993.

74. L Serrano, AR Fersht. Capping and alpha-helix stability. Nature 342: 296–299, 1989.

75. L Serrano, J Sancho, M Hirshberg, AR Fersht. Alpha-helix stability in proteins. I. Empirical correlations concerning substitution of side chains at the N and C-caps and the replacement of alanine by glycine or serine at solvent-exposed surfaces. J Mol Biol 227: 544–559, 1992.

76. JA Bell, WJ Becktel, U Sauer, WA Baase, BW Matthews. Dissection of helix capping in T4 lysozyme by structural and thermodynamic analysis of six amino acid substitutions at Thr 59. Biochemistry 31: 3590–3596, 1992.

77. L Serrano, JL Neira, J Sancho, AR Fersht. Effect of alanine versus glycine in alpha-helices on protein stability. Nature 356: 453–455, 1992.

78. AJ Doig, A Chakrabartty, TM Klinger RL Baldwin. Determination of free energies of N-capping in alpha-helices by modification of the Lifson–Roig helix-coil therapy to include N- and C- capping. Biochemistry 33: 3396–3403, 1994.

79. LG Presta, GD Rose. Helix signals in proteins. Science 240: 1632–1641, 1988.

80. E Harper, GD Rose. Helix stop signal in proteins and peptides: The capping box. Biochemistry 32: 7605-7609, 1993.

81. N Yumoto, S Murase, T Hattori, H Yamamoto, Y Tatsu, S Yoshikawa. Stabilization of alpha-helix in C-terminal fragments of neuropeptide Y. Biochem Biophys Res Commun 196: 1490–1495, 1993.

82. A Chakrabartty, AJ Doig, RL Baldwin. Helix capping propensities in peptides parallel those in proteins. Proc Natl Acad Sci USA 90: 11332–11336, 1993.

83. AJ Doig, RL Baldwin. N- and C-capping preferences for all 20 amino acids in alpha-helical peptides. Protein Sci 4: 1325-1336, 1995.

84. R Thapar, EM Nicholson, P Rajagopal, EB Waygood, JM Scholtz, RE Klevit. Influence of N-cap mutations on the structure and stability of *Escherichia coli* HPr. Biochemistry 35: 11268–11277, 1996.

85. EA Zhukovsky, MG Mulkerrin, LG Presta. Contribution to global protein stabilization of the N-capping box in human growth hormone. Biochemistry 33: 9856–9864, 1994.

86. AM Gronenborn, GM Clore. Identification of N-terminal helix capping boxes by means of 13C chemical shifts. J Biomol NMR 4: 455–458, 1994.

87. JW Seale, R Srinivasan, GD Rose. Sequence determinants of the capping box, a stabilizing motif at the N-termini of alpha-helices. Protein Sci 3: 1741–1745, 1994.

88. M Petukhov, N Yumoto, S Murase, R Onmura, S Yoshikawa. Factors that affect the stabilization of alpha-helices in short peptides by a capping box. Biochemistry 35: 387–397, 1996.

89. MA Jimenez, V Munoz, M Rico, L Serrano. Helix stop and start signals in peptides and proteins. The capping box does not necessarily prevent helix elongation. J Mol Biol 242: 487–496, 1994.

90. HX Zhou, P Lyu, DE Wemmer, NR Kallenbach. Alpha-helix capping in synthetic model peptides by reciprocal side chain-main chain interactions: evidence for an N-terminal "capping box". Proteins 18: 1–7, 1994.

91. AJ Doig, MW MacArthur, BJ Stapley, JM Thornton. Structures of N-termini of helices in proteins. Protein Science 6: 147–155, 1997.

92. R Aurora, R Srinivasan, GD Rose. Rules for alpha-helix termination by glycine [published erratum appears in Science 1994 Jun 24; 264(5167):1831] [see comments]. Science 264: 1126–30, 1994.

93. G Esposito, B Dhanapal, P Dumy, V Varma, M Mutter, G Bodenhausen. Lysine as helix C-capping residue in a synthetic peptide. Biopolymers 41: 27–35, 1997.

94. J Prieto, L Serrano. C-capping and helix stability: the Pro C-capping motif. J Mol Biol 274: 276–88, 1997.

95. RP Sheridan, LC Allen. The electrostatic potential of the alpha-helix (electrostatic potential/alpha-helix/secondary structure /helix dipole). Biophys Chem 11: 133–136, 1980.

96. RP Sheridan, RM Levy, FR Salemme. Alpha-helix dipole model and electrostatic stabilization of 4-alpha-helical proteins. Proc Natl Acad Sci USA 79: 4545-4549, 1982.

97. D Sali, M Bycroft. AR Fersht. Stabilization of protein structure by interaction of alpha-helix dipole with a charged side chain. Nature 335: 740–743, 1988.

98. J Sancho, L Serrano, AR Fersht. Histidine residues at the N- and C-termini of alpha-helices: perturbed pKas and protein stability. Biochemistry 31: 2253–2258, 1992.

99. R Fairman, KR Shoemaker, EJ York, JM Stewart, RL Baldwin. Further studies of the helix dipole model: effects of a free alpha-NH3+ or alpha-COO− group on helix stability. Proteins 5: 1–7, 1989.

100. EK Bradley, JF Thomason, FE Cohen, PA Kosen, ID Kuntz. Studies of synthetic helical peptides using circular dichroism and nuclear magnetic resonance. J Mol Biol 215: 607–622, 1990.

101. BM Huyghues-Despointes, JM Scholtz, RL Baldwin. Effect of a single aspartate on helix stability at different positions in a neutral alanine-based peptide. Protein Science 2: 1604–1611, 1993.

102. KR Shoemaker, PS Kim, DN Brems, S Marqusee, EJ York, IM Chaiken, JM Stewart, RL Baldwin. Nature of the charged-group effect on the stability of the C-peptide helix. Proc Natl Acad Sci U S A 82: 2349–2353, 1985.

103. H Nicholson, DE Anderson, S Dao-pin, BW Matthews. Analysis of the interaction

between charged side chains and the alpha-helix dipole using designed thermostable mutants of phage T4 lysozyme. Biochemistry 30: 9816–9828, 1991.

104. VG Eijsink, G Vriend, B van den Burg, JR van der Zee, G Venema. Increasing the thermostability of a neutral protease by replacing positively charged amino acids in the N-terminal turn of alpha-helices. Protein Eng 5: 165–170, 1992.

105. KM Armstrong, RL Baldwin. Charged histidine affects alpha-helix stability at all positions in the helix by interacting with the backbone charges. Proc Natl Acad Sci USA 90: 11337–11340, 1993.

106. S Walter, B Hubner, U Hahn, FX Schmid. Destabilization of a protein helix by electrostatic interactions. J Mol Biol 252: 133–143, 1995.

107. A Bierzynski, PS Kim, RL Baldwin. A salt bridge stabilizes the helix formed by isolated C-peptide of RNase A. Proc Natl Acad Sci USA 79: 2470–2474, 1982.

108. M Rico, E Gallego, J Santoro, FJ Bermejo, JL Nieto, J Herranz. On the fundamental role of the Glu 2-···Arg 10+ salt bridge in the folding of isolated ribonuclease A S-peptide. Biochem Biophys Res Commun 123: 757–763, 1984.

109. A Bierzynski. Deprotonation of Glu9 destabilizes the alpha-helix in C-peptide of RNase A. Int J Pept Protein Res 32: 256–261, 1988.

110. PC Lyu, PJ Gans, NR Kallenbach. Energetic contribution of solvent-exposed ion pairs to alpha-helix structure. J Mol Biol 223: 343–350, 1992.

111. L Serrano, A Horovitz, B Avron, M Bycroft, AR Fersht. Estimating the contribution of engineered surface electrostatic interactions to protein stability by using double-mutant cycles. Biochemistry 29: 9343–9352, 1990.

112. B Odaert, F Baleux, T Huynh-Dinh, JM Neumann, A Sanson. Nonnative capping structure initiates helix folding in an annexin I fragment. A 1H NMR conformational study. Biochemistry 34: 12820–12829, 1995.

113. BM Huyghues-Despointes, JM Scholtz, RL Baldwin. Helical peptides with three pairs of Asp-Arg and Glu-Arg residues in different orientations and spacings. Protein Science 2: 80–85, 1993.

114. S Marqusee, RL Baldwin. Helix stabilization by Glu⁻···Lys⁺ salt bridges in short peptides of *de novo* design. Proc Natl Acad Sci U S A 84: 8898–8902, 1987.

115. BM Huyghues-Despointes, RL Baldwin. Ion-pair and charged hydrogen-bond inter-

116. DF Stickle, LG Presta, KA Dill, GD Rose. Hydrogen bonding in globular proteins. J Mol Biol 226: 1143–1159, 1992.

117. BMP Huyghues-Despoints, TM Klinger, RL Baldwin. Measuring the strength of side-chain hydrogen bonds in peptide helices: The Gln·Asp ($i, i + 4$) Interaction. Biochemistry. 34: 13267–13271, 1995.

118. V Munoz, L Serrano. Analysis of $i, i + 5$ and $i, i + 8$ hydrophobic interactions in a helical model peptide bearing the hydrophobic staple motif. Biochemistry 34: 15301–15306, 1995.

119. JP Segrest, RL Jackson, JD Morrisett, AM Gotto Jr. A molecular theory for protein-lipid interactions in plasma lipoproteins. Febbs Letts 38: 247–253, 1974.

120. ET Kaiser, FJ Kezdy. Secondary structures of proteins and peptides in amphiphilic environments. Proc Natl Acad Sci USA 80: 1137–1140, 1983.

121. AW Bernheimer, B Ruby. Interactions between membranes and cytolytic peptides. Biochim. Biophys Acta 864: 123–141, 1986.

122. M Zasloff, B Martin, HC Chen. Antimicrobial activity of synthetic magainin peptides and several analogues. Proc Natl Acad Sci USA 85: 910–913, 1988.

123. P Kanellis, AY Romans, BJ Johnson, H Kercert, R Chiovetti Jr., TM Allen, JP Segrest. Studies of synthetic peptide analogs of the amphipathic helix: Effect of charged amino acid topography on lipid affinity. J Biol Chem 255: 11464–11472, 1980.

124. GM Anantharamaiah, JL Jones, CG Brouillette, CF Schmidt, BH Chung, TA Hughes, AS Bhown, JP Segrest. Studies of synthetic peptide analogs of the amphipathic helix. J Biol Chem 260: 10248–10255, 1985.

125. I Glover, I Haneef, J Pitts, S Wood, D Moss, I Tickle, T Blundell. Conformational flexibility in a small globular hormone: x-ray analysis of avian pancreatic polypeptide at 0.98 Å resolution. Biopolymers 22: 293–304, 1983.

126. TL Blundell, D Barlow, N Borkakoti, JM Thorton. Solvent-induced distortions and the curvature of α-helices. Nature 306: 281–283, 1983.

127. R Parthasarathy, S Chaturvedi, K Go. Design of alpha-helical peptides: their role in protein folding and molecular biology. Prog Biophys Mol Biol 64: 1–54, 1995.

128. JW Bryson, SF Betz, HS Lu, DJ Suich, HX Zhou, KT O'Neil, WF DeGrado. Protein

design: A hierarchic approach. Science 270: 935–941, 1995.

129.  CL Nesloney, JW Kelly. Progress towards understanding β-sheet structure. Bioorganic and Medicinal Chemistry 4: 739–766, 1995.

130.  L Pauling, RB Corey. Configurations of polypeptide chains with favored orientations around single bonds: Two new peated sheets. Proc Natl Acad Sci USA 37: 729–740, 1951.

131.  MJE Sternberg, JM Thornton. On the conformation of proteins: An analysis of β-pleated sheets. J Mol Biol 110: 285–296, 1977.

132.  JS Richardson, DC Richardson, NB Tweedy, KM Gernert, TP Quinn, MH Hecht, BW Erickson, Y Yan, RD McClain, ME Donlan, MC Surles. Looking at proteins: Representations, folding, packing, and design. Biophys J 63: 1186–2109, 1992.

133.  FR Salemme, DW Weatherford. Conformational and geometrical properties of β-sheets in proteins II, Antiparallel and mixed β-sheets. J Mol Biol 146: 119–41, 1981.

134.  MA Wouters, PMG Curmi. An analysis of side chain interactions and pair correlations within antiparallel β-sheets: The differences between backbone hydrogen-bonded and non-hydrogen-bonded residue pairs. Proteins: Structure, Function, and Genetics 22: 119–131, 1995.

135.  MJE Sternberg, JM Thornton. On the conformation of proteins: Hydrophobic ordering of strands in β-pleated sheets. J Mol Biol 115: 1–17, 1977.

136.  JS Richardson. β-sheet topology and the relatedness of proteins. Nature 268: 495–500, 1977.

137.  JS Richardson, ED Getzoff, DC Richardson. The β-bulge: A common small unit of nonrepetitive protein structure. Proc Natl Acad Sci USA 75: 2574–2578, 1978.

138.  AWE Chan, EG Hutchinson, D Harris, JM Thornton. Identification, classification, and analysis of beta-bulges in proteins. Protein Science 2: 1574–1590, 1993.

139.  EJ Milner-White, R Poet. Loops, bulges, turns, and hairpins in proteins. Trends in Biochemical Sciences 12: 189–192, 1987.

140.  C Chothia, J Janin. Orthogonal packing of β-pleated sheets in proteins. Biochemistry 21: 3955-3965, 1982.

141.  C Daffner, G Chelvanayagam, P Argos. Structural characteristics and stabilizing principles of β-strands in protein tertiary architectures. Protein Science 3: 876–882, 1994.

142.  FR Salemme. Structural properties of protein β-sheets. Prog Biophys Molec Biol 42: 95–133, 1983.

143.  JS Richardson. The anatomy and taxonomy of protein structure. In: CB Anfinson, JT Edsall, FM Richards, ed. Advances in Protein Chemistry. New York: Academic Press, 1981, pp 167–439.

144.  N Vtyurin. The role of local tight packing of hydrophobic groups in β-structure. Proteins: Structure, Function, and Genetics 15: 62–70, 1993.

145.  L Wang, T O'Connell, A Tropsha, J Hermans. Molecular simulations of β-sheet twisting. J Mol Biol 262: 283–293, 1996.

146.  K-C Chou, G Nemethy, HA Sheraga. Role of interchain interactions in the stabilization of the right-handed twist of β-sheets. J Mol Biol 168: 389–407, 1983.

147.  K Nishikawa, HA Scheraga. Geometrical criteria for formation of coiled-coil structures of polypeptide chains. Macromolecules 9: 395–407, 1976.

148.  FR Salemme, DW Weatherford. Conformational and geometrical properties of β-sheets in proteins I. Parallel β-sheets. J Mol Biol 146: 101–117, 1981.

149.  C Chothia. Coliing of β-pleated sheets. J Mol Biol 163: 107–117, 1983.

150.  JS Richardson. Handedness of crossover connections in β-sheets. Proc Natl Acad Sci USA 73: 2619–2623, 1976.

151.  JS Richardson. β-Sheet topology and the relatedness of proteins. Nature 268: 495–500, 1977.

152.  MJE Sternberg, JM Thornton. On the conformation of proteins: The handedness of the β-strand-α-helix-β-strand unit. J Mol Biol 105: 367–382, 1976.

153.  CM Dobson, AR Fersht, eds. Protein Folding. Cambridge University Press, 1995.

154.  U Carlsson, B-H Jonsson. Folding of β-sheet proteins. Current Opinion in Structural Biology 5: 482–487, 1995.

155.  L Lins, R Brassuer. The hydrophobic effect in protein folding. FASEB J 9: 535–540, 1995.

156.  GD Fasman, editor. Prediction of Protein Structure and The Principles of Protein Conformation. New York, Plenum Press, 1989.

157.  CN Pace, BA Shirley, M McNutt, K Gajiwala. Forces contributing to the conformational stability of proteins. FASEB J. 10: 75–83, 1996.

158.  KA Dill, KM Fiebig, HS Chan. Cooperativity in protein-folding kinetics. Proc Natl Acad Sci USA 90: 1942–1946, 1993.

159. GD Rose, RH Winters, DB Wetlaufer. A testable model for protein folding. FEBS Letts. 63: 10–16, 1976.

160. GM Crippen. The tree structural organization of proteins. J Mol Biol 126: 315–332, 1978.

161. GD Rose, S Roy. Hierarchic organization of domains in globular proteins. J Mol Biol 134: 447–470, 1979.

162. GD Rose, S Roy. Hydrophobic basis of packing in globular proteins. Proc Natl Acad Sci USA 77: 4643–4647, 1980.

163. EP Garvey, J Swank, CR Matthews. A hydrophobic cluster forms early in the folding of dihydrofolate reductase. Proteins: Structure, Function, and Genetics 6: 259–266, 1989.

164. PA Evans, KD Topping, DN Woolfson, CM Dobson. Hydrophobic clustering in nonnative states of a protein: Interpretation of chemical shifts in nmr spectra of denatured states of lysozyme. Proteins: Structure, Function, and Genetics 9: 248–266, 1991.

165. D. Neri, M Billeter, G Wider, K. Wüthrich. NMR determination of residual structure in a urea-denatured protein, the 434-repressor. Science 257: 1559–1563, 1992.

166. IJ Ropson, C Frieden. Dynamic nmr spectral analysis and protein folding: Identification of highly populated folding intermediate of rat intestinal fatty acid-binding protein by $^{19}$F nmr. Proc N atl Acad Sci USA 89: 7222–7226, 1992.

167. Z-P Liu, J Rizo, LIVI Gierasch. Equilibrium folding studies of cellular retinoic acid binding protein, a predominatly β-sheet protein. Biochemistry 33: 134–142, 1994.

168. HJ Dyson, G Merutka, JP Waltho, RA Lerner, PE Wright. Folding of peptide fragments comprising the complete sequence of proteins. Models for initiation of protein folding I. Myohemerythrin. J Mol Biol 226: 795–817, 1992.

169. HJ Dyson, JR Sayre, G Merutka, H-C Shin, RA Lerner, PE Wright. Folding of peptide fragments comprising the complete sequence of proteins. Models for initiation of protein folding II. Plastocyanin. J Mol Biol 226: 819–836, 1992.

170. SE Radford, CM Dobson, PA Evans. The folding of hen lysozyme involves partially structured intermediates and multiple pathways. Nature 358: 302–307, 1992.

171. M Kotik, SE Radford, CM Dobson. Comparison of the refolding of hen lysozyme from dimethyl sulfoxide and guanidinium chloride. Biochemistry 34: 1714–1724, 1995.

172. PY Chou, GD Fasman. Prediction of the secondary structure of proteins from their amino acid sequence. Adv Enzymol 47: 45–108, 1978.

173. CA Kim, JM Berg. Thermodynamic β-sheet propensities measured using a zinc-finger host peptide. Nature 362: 267–270, 1993.

174. DL Minor Jr., PS Kim. Measurement of the β-sheet-forming propensities of amino acids. Nature 367: 660–663, 1994.

175. DL Minor Jr., PS Kim. Context is a major determinant of β-sheet propensity. Nature 371: 264–267, 1994

176. DL Minor Jr., PS Kim. Context-dependent secondary structure formation of a designed protein sequence. Nature 380: 730–734, 1996.

177. CK Smith, L Regan. Guidelines for protein design: The energetics of β-sheet side chain interactions. Science 270: 980–982, 1995.

178. CK Smith, JM Withka, L Regan. A thermodynamic scale for the β-sheet forming tendencies of the amino acids. Biochemistry 33: 5510–5517, 1994.

179. H Xiong, BL Buckwalter, H-M Shieh, MH Hecht. Periodicity of polar and nonpolar amino acids is the major determinant of secondary structure in self-assembling oliqomeric peptides. Proc Natl Acad Sci USA 92: 6349–6343, 1995.

180. S Kamtekar, JM Schiffer, H Xiong, JM Babik, MH Hecht. Protein design by binary patterning of polar and nonpolar amino acids. Science 262: 1680–1685, 1993.

181. MW West, MH Hecht. Binary patterning of polar and nonpolar amino acids in the sequences and structures of native proteins. Protein Science 4: 2032–2039, 1995.

182. M Bajaj, T Blundell. Evolution and the tertiary structure of proteins. Ann Rev Biophys Bioeng 13: 453–492, 1984,

183. P Argos. Analysis of sequence-similar pentapeptides in unrelated protein tertiary structures. Strategies for protein folding and a guide for site-directed mutagenesis. J Mol Biol 197: 331–348, 1987.

184. EJ Milner-White. Situations of gamma-turns in proteins. Their relation to alpha-helices, beta-sheets and ligand binding sites. J Mol Biol 216: 385–397, 1990.

185. BL Sibanda, TL Blundell, JM Thornton. Conformation of β-hairpins in protein structures. A systematic classification with applications to modelling by homology, electorn density fitting and protein engineering. J Mol Biol 206: 759–777, 1989.

186. CS Ring, DG Kneller, R Langridge, FE Cohen. Taxonomy and conformational analysis of loops in proteins. J Mol Biol 224: 685–699, 1992.

187. JS Fetrow. Omega loops: nonregular secondary structures significant in protein

function and stability. FASEB J 9: 708–717, 1995.

188. GD Rose, LM Gierasch, JA Smith, Turns in peptides and proteins. Adv Prot Chem 37: 1–109, 1985.

189. CN Venkatachalam. Stereochemical criteria for polypeptides and proteins. V. Conformation of a system of three linked peptide units. Biopolymers 6: 1425–1436, 1968.

190. PN Lewis FA Momany, HA Scheraga, Chain reversals in proteins. Biochim Biophys Acta 303: 211–229, 1973.

191. CM Wilmot, JM Thornton. β-turns and their distortions: a proposed new nomenclature. Protein Engineering 3: 479–493, 1990.

192. EG Hutchinson, JM Thornton. A revised set of potentials for β-turn formation in proteins. Protein Science 3: 2207–2216, 1994.

193. DC Rees, M Lewis, WN Lipscomb. Refined crystal structure of carboxypeptidase A at 1.54 Å resolution. J Mol Biol 168: 367–387, 1983.

194. JA Smith, LG Pease. Reverse turns in peptides and proteins. CRC Crit Rev Biochem 8: 315–399, 1980.

195. AS El Hawrani, KM Moreton, RB Sessions, AR Clarke, JJ Holbrook. Engineering surface loops of proteins—a preferred strategy for obtaining new enzyme function. Trends in Biotechnology 12: 207–211, 1994.

196. JF Leszczynski, GD Rose. Loops in globular proteins: A novel category of secondary structure. Science 234: 849–855, 1986.

197. C Chothia, AM Lesk, A Tramontano, M Levitt, SJ Smith-Gill, G Air, S Sheriff, EA Padlan, D Davies, WR Tulip, PM Colman, S Spinelli, PM Alzari, RJ Poljak. Conformations of immunoglobulin hypervariable regions. Nature 342: 877–883, 1989.

198. M Saraste, PR Sibbald, A Wittinghofer. The P-loop—a common motif in ATP- and GTP-binding proteins. Trends in Biochemical Science 15: 430–434, 1990.

199. PA Baeuerle, F Lottspeich, WB Huftner. Purification of yolk protein 2 of *Drosophila melanogaster* and identification of its site of tyrosine sulfation. J Biol Chem 263: 14925–14929, 1988.

200. E Bek, R Berry. Prohormonal cleavage sites are associated with Ω-loops. Biochemistry 29: 178–183, 1990.

201. OP Kuipers, MMGM Thunnissen, P deGeus, BW Dikjstra, J Drenth, HM Verheij, GH deHaas. Enhanced activity and altered specificity of phospholipase A2 by deletion of a surface loop. Science 244: 82–85, 1989.

202. PT Jones, PH Dear, J Foote, MS Neuberger, G Winter. Replacing the complementarity-determining regions in human antibody with those from a mouse. Nature 321: 522–525, 1986.

203. L Hedstrom, L Szilagyi, W. Rutter. Converting trypsin to chymotrypsin: the role of surface loops. Science 255: 1249–1253, 1992.

204. FHC Crick. The packing of α-helices: Simple coiled-coils. Acta Crystallographia 6: 689–697, 1953.

205. C Chothia, M Levitt, D Richardson. Helix to helix packing in proteins. J Mol Biol 145: 215–250, 1981.

206. C Cohen, DAD Parry. α-Helical coiled coils and bundles: How to design an α-helical protein. Proteins: Structure, Function, and Genetics 7: 1–15, 1990.

207. L Pauling, RB Corey. Compound helical configurations of polypeptide chains: Structure of proteins of the α-keratin type. Nature 171: 59–61, 1953.

208. D Eisenberg, W Wilcox, SM Eshita, PM Pryciak, SP Ho, WF DeGrado. The design, synthesis, and crystallization of an alpha-helical peptide. Proteins: Structure, Function, and Genetics 1: 16–22, 1986.

209. SP Ho, WF DeGrado. Design of a 4-helix bundle protein: Synthesis of peptides which self-associate into a helical protein. J Am Chem Soc 109: 6751–6758, 1987.

210. L Regan, WF DeGrado. Characterization of a helical protein designed from first principles. Science 241: 976–978, 1988.

211. PTP Kaumaya, KD Berndt, DB Heidorn, J Trewhella, FJ Kezdy, E Goldberg. Synthesis and biophysical characterization of engineered tipographic immunogenic determinants with αα topology. Biochemisty 29: 13–23, 1990.

212. MH Hecht, JS Richardson, DC Richardson, RC Ogden. *De novo* design, expression , and characterization of Felix: A four-helix bundle protein of native-like sequence. Science 249: 884–891, 1990.

213. DR Raleigh, WF DeGrado. A *de novo* designed protein shows a thermally induced transition from a native to a molten globule-like state. J Am Chem Soc 114: 10079–10081, 1992.

214. CE Schafmeister, LJW Miercke, RM Stroud. Structure at 2.5 Å of a designed peptide that maintains solubility of membrane proteins. Science 262: 734–738, 1993.

215. WF DeGrado, ZA Wasserman, JD Lear. Protein design, a minimalist approach. Science 243: 622–643, 1989.

216. S Kamtekar, MH Hecht. The four-helix bundle: what determines a fold? FASEB J 9: 1013–1022, 1995.

217. Y Fezoui, DL Weaver, JJ Osterhout. *De novo* design and structural characterization of an α-helical hairpin peptide: A model system for the study of protein folding intermediates. Proc Natl Acad Sci USA 91: 3675–3679, 1994.

218. AP Brunet, ES Huang, ME Huffine, JE Loeb, RJ Weltman, MH Hecht. The role of turns in the structure of an α-helical protein. Nature 364: 355–358,1993.

219. PC Weber, FR Salemme. Structural and functional diversity in 4-α-helical proteins. Nature 287: 82–84, 1980.

220. R Pressnell, FE Cohen. Topological distribution of four α-helix bundles. Proc Natl Acad Sci USA 86: 6592–6596, 1989.

221. AV Efimov. Role of constrictions in formation of protein structures containing four helical regions. J Mol Biol 16: 218–227, 1982.

222. JG Adamson, NE Zhou, RS Hodges. Structure, function and application of the coiled-coil protein folding motif. Current Opinion in Biotechnology 4:428–437, 1993.

223. KJ Lumb, PS Kim. Measurement of interhelical electrostatic interactions in the GCN4 leucine zipper. Science 268: 436–439, 1995.

224. Y Yu, OD Monera, RS Hodges, PL Privalov. Ion pairs significantly stabilize coiled-coils in the absence of electrolyte. J Mol Biol 255: 367–372, 1996.

225. P Lavigne, FD Sönnichsen, CM Kay, RS Hodges. Interhelical salt bridges, coiled-coil stability, and specificity of dimerization. Science 271: 1136–1137, 1997.

226. EK O'Shea, JD Klemm, PS Kim, T Alber. X-ray structure of the GCN4 leucine zipper, a two-stranded, parallel coiled coil. Science 254: 539–544, 1991.

227. PB Harbury, T Zhang, PS Kim, T Alber. A switch between two-, three-, and four-stranded coiled coils in GCN4 leucine zipper mutants. Science 262: 1401–1407, 1993.

228. PB Harbury, PS Kim, T Alber. Crystal structure of an isoleucine-zipper trimer. Nature 371: 80–83, 1994.

229. DW Banner, M Kokkinidis, D Tsernoglou. Structure of the ColE1 Rop protein at 1.7 Å resolution. J Mol Biol 196: 657–675, 1987.

230. M Munson, R O'Brien, JM Sturtevant, L Regan. Redesigning the hydrophobic core of a four-helix-bundle protein. Protein Science 3: 2015–2022, 1994.

231. M Munson, S Balasubramanian, KG Fleming, AD Nagi, R O'Brien, JM Sturtevant, L Regan. What makes a protein a protein? Hydrophobic core designs that specify stability and structural properties. Protein Science 5: 1584–1593, 1996.

232. PF Predki, L Regan. Redesigning the topology of a four-helix-bundle protein: Monomeric Rop. Biochemistry 34: 9834–9839, 1995.

233. SF Betz, WF DeGrado. Controlling topology and native-like behavior of *de novo*-designed peptides: Design and characterization of antiparallel four-stranded coiled coils. Biochemistry 35: 6955–6962, 1997.

234. SF Betz, PA Liebman, WF DeGrado. *De novo* design of native proteins: Characterization of proteins intended to fold into antiparallel, Rop-like, four-helix bundles. Biochemistry 36: 2450–2458, 1997.

235. BL Sibanda, JM Thornton. β-hairpin families in globular proteins. Nature 316: 170–174, 1985.

236. OB Ptitsyn. Protein folding: General physical model. FEBS Letts 131: 197–202, 1981.

237. JPL Cox, PA Evans, LC Packman, DH Williams, DN Woolfson. Dissecting the structure of a partially folded protein—Circular dichroism and nuclear magnetic resonance studies of peptides from ubiquitin. J Mol Biol 234: 483–492, 1993.

238. FJ Blanco, MA Jiménez, A Pineda, M Rico, J Santoro, JL Nieto. NMR solution structure of the isolated N-terminal fragment of protein-G B$_1$ domain. Evidence of trifluoroethanol induced native-like β-hairpin formation. Biochemistry 33: 6004–6014, 1994.

239. FJ Blanco, G, Rivas, L. Serrano. A short linear peptide that folds into a native stable β-hairpin in aqueous solution. Structural Biology 1: 584–590, 1994.

240. FJ Blanco, MA Jiménez, J Herranz, M Rico, J Santoro, JL Nieto. NMR evidence of a short linear peptide that folds into a β-hairpin in aqueous solution. J Am Chem Soc 115: 5887–5888, 1993.

241. MS Searle, DH Williams, LC Packman. A Short linear peptide derived from the N-terminal sequence of ubiquitin folds into a water-stable non-native β-hairpin. Structural Biology 2: 999–1006, 1995.

242. E de Alba, FJ Blanco, MA Jiménez, M Rico, JL Nieto. Interactions responsible for the pH dependence of the β-hairpin conformational population formed by a designed linear peptide. Eur J Biochem 233: 283–292, 1995.

243. CM Wilmot, JM Thornton. Analysis and prediction of the different types of β-turn in proteins. J Mol Biol 203: 221–232, 1988.

244. M Ramirez-Alverado, FJ Blanco, L Serrano. *De novo* design and structural analysis of a model β-hairpin peptide system. Nature Structural Biology 3: 604–612, 1996.

245. A Brack, A Caille. Synthesis and β-conformation of copolypeptides with alternating hydrophilic and hydrophobic residues. Int J Peptide Protein Res 11: 128–139, 1978.

246. A Brack, G. Spach. Synthesis and conformations of periodic copolypeptides of L-alanine and glycine. Biopolymers 11: 563–586, 1972.

247. WB Rippon, HH Chen, AG Walton. Spectroscopic characterization of poly(Glu-Ala). J Mol Biol 75: 369–375, 1973.

248. A Brack, LE Orgel. β structures of alternating polypeptides and their possible prebiotic significance. Nature 256: 383–387, 1975.

249. S Brahms, J Brahms, G Spach, A Brack. Identification of β,β-turns and unordered conformations in polypeptide chains by vacuum ultraviolet circular dichroism. Proc Natl Acad Sci USA 74: 3208–3212, 1977.

250. R Moser, RM Thomas, B Gutte. An artificial crystalline DDT-binding polypeptide. FEBS Letts 157: 247–251, 1983.

251. W Kullman. Design, synthesis, and binding characteristics of an opiate receptor mimetic peptide. J Med Chem 27: 106–115, 1984.

252. S Lewin. Order of strength of ionic linkages in protein interactions. Biochemistry 114: 83–84, 1969.

253. JS Richardson, DC Richardson. The *de novo* design of protein structures. Trends in Biochemical Sciences 14: 304–309, 1989.

254. JS Richardson DC Richardson. Some Design Principles: Betabellin. In: DL Oxender, CF Fox, eds. Protein Engineering. New York: Alan R. Liss,1987, pp 149–163.

255. Y Yan, BW Erickson. Engineering of betabellin 14D: Disulfide-induced folding of a β-sheet protein. Protein Science 3: 1069–1073, 1994.

256. TP Quinn, NB Tweedy, RW Williams, JS Richardson, DC Richardson. Betadoublet: *De novo* design, synthesis, and characterization of a β-sandwich protein. Proc Natl Acad Sci USA 91: 8747–8751, 1994.

257. KH Mayo, E Ilyina, H Park. A recipe for designing water-soluble β-sheet-forming

258. DJ Butcher, MD Bruch, GR Moe. Design and characterization of a model αβ peptide. Biopolymers 35: 109–120, 1995.

259. DJ Butcher, GR Moe. Role of hydrophobic interactions and desolvation in determining the structural properties of a model αβ peptide. Proc Natl Acad Sci USA 93: 1135–1140, 1996.

260. B Gutte, M Däumigen, E. Wittschieber. Design, synthesis and characterization of a 34-residue polypeptide that interacts with nucleic acids. Nature 281: 650–655, 1979.

261. TG Oas, PS Kim. A peptide model of a protein folding intermediate. Nature 336: 42–48, 1988.

262. DY Kwon, PS Kim. The stabilizing effects of hydrophobic cores on peptide folding of bovine-pancreatic-trypsin-inhibitor folding-intermediate model. Eur J Biochem 223: 631–636, 1994.

263. MD Struthers, RP Chong, B Imperiali. Design of a monomeric 23-residue polypeptide with defined tertiary structure. Science 271: 342–345, 1996.

264. AN Fedorov, DA Dolgikh, VV Chemeris, BK Chernov, AV Finkelstein, AA Schulga, YuB Alakhov, MP Kirpichnikov, OB Ptitsyn. *De novo* design, synthesis and study of albebetin, a polypeptide with a predetermined three-dimensional structure. J Mol Biol 225: 927–931, 1992.

265. PA Rice, A Goldman, TA Steitz. A helix-turn-strand structural motif common in α-β proteins. Proteins: Structure, Function, and Genetics 8: 334–340, 1990.

266. I Lasters, SJ Wodak, P Alard, E VanCutsem. Structural principles of parallel β-barrels in proteins. Proc Natl Acad Sci USA 85: 3338–3342, 1988.

267. K-C Chou, L Carlacci, GG Maggiora. Conformational and geometrical properties of idealized β-barrels in proteins. J Mol Biol 213: 315–326, 1990.

268. K Goraj, A Renard, JA Martial. Synthesis, purification and initial characterization of octarellin, a *de novo* polypeptide modelled on the α/β barrel proteins. Protein Engineering 3: 259–266, 1990.

269. M Beauregard, K Goraj, V Goffin, K Heremans, E Goormaghtigh, J-M Ruysschaert, JA Martial. Spectroscopic investigation of structure in octarellin (a *de novo* protein designed to adopt the α/β-barrel packing). Protein Engineering 4: 745–749, 1991.

270. FE Cohen, MJE Sternberg, WR Taylor. Analysis and prediction of the packing of α-helices against a β-sheet in the tertiary

structure of globular proteins. J Mol Biol 156: 821–862, 1982.

271. T Tanaka, H Kimura, M Hayashi, Y Fujiyoshi, K-I Fukuhara, H Nakamura. Characteristics of a *de novo* designed protein. Protein Science 3: 419–427, 1994.

272. T Tanaka, M Hayashi, H Kimura, M Oobatake, H Nakamura. *De novo* design and creation of a stable artificial protein. Biophysical Chemistry 50: 47–61, 1994.

273. T Tanaka, Y Kuroda, H Kimura, S-I Kidokoro, H Nakamura. Cooperative deformation of a *de novo* designed protein. Protein Engineering 7: 969–976, 1994.

274. J-PY Scheerlinck, I Lasters, M Claessens, M DeMaeyer, F Pio, P Delhaise, SJ Wodak. Recurrent $\alpha\beta$ loop structures in TIM barrel motifs show a distinct pattern of conserved structural features. Proteins: Structure, Function, and Genetics 12: 299–313, 1992.

275. AV Finkelstein, H Nakamura. Weak points of antiparallel $\beta$-sheets. How are they filled up in globular proteins? Protein Engineering 6: 367–372, 1993.

276. KD Kopple, DE Nitecki. Reactivity of cyclic peptides I. Transannular histidine-O-acetyltyrosine interaction. J Am Chem Soc 83: 4103–4104, 1961.

277. KD Kopple, DE Nitecki. Reactivity of cyclic peptides II. cyclo-L-tyrosyl-L-histidyl and cyclo-L-tyrosyltriglycyl-L-histidylglycyl. J Am Chem Soc 84: 4457–4464, 1962.

278. KD Kopple, DE Nitecki. Reactivity of cyclic peptides III. Reaction of isomeric histidine, tyrosine peptides with *p*-nitrophenyl acetate. Biochemistry 2: 958–964, 1963.

279. JC Sheehan, DN McGregor. Synthetic peptide models of enzyme active sites I. Cyclo-glycyl-L-histidyl-L-serylglycyl-L-histidyl-L-seryl. J Am Chem Soc 84: 3000-3005, 1962.

280. P Cruickshank, JC Sheehan. Synthetic peptide models of enzyme active sites II. L-threonyl-L-alanyl-L-seryl-L-histidyl-L--aspartic acid, an active esterase model. J Am Chem Soc 86: 2070–2071, 1964.

281. JC Sheehan, GB Bennett, JA Schneider. Synthetic peptide models of enzyme active sites. III. Stereoselective esterase models. J Am Chem Soc 88: 3455–3456, 1966.

282. K Nakajima, K Okawa. Studies of the synthetic model enzyme. The synthesis of cyclo(His-Glu-Cys-D-Phe-Gly)$_2$ as the esterase model. Bull Chem Soc Jpn 46: 1811–1816, 1973.

283. Y Imanishi, M Tanihara, T Sugihara, T Higashimura. Histidine-containing cyclic dipeptides as catalysts in the hydrolysis of carbonic acid *p*-nitrophenyl esters. Biopolymers 16: 2203–2215, 1977.

284. M Tanihara, Y Imanishi, T Higashimura. Solution conformation and hydrolytic activity of cyclo(D-Leu-L-His). Biopolymers 16: 2217–2229, 1977.

285. HJ Goren, T Fletcher, M Fridkin, E Katchalski-Katzir. Poly(L-histidyl-L-alanyl-$\alpha$-L-glutamic acid). II. Catalysis of *p*-nitrophenyl acetate hydrolysis. Biopolymers 17: 1679–1692, 1978.

286. Y Murakami, A Nakano, K Matsumoto, K Iwamoto. Macrocyclic enzyme model systems. Catalytic activity of cyclic peptides involving hydrophobic segments. Bull Chem Soc Jpn 51: 2690–2697, 1978.

287. PK Chakravarty, KB Mathur, MM Dhar. The synthesis of a decapeptide with glycosidase activity. Experientia 29: 786–788, 1973.

288. I Photaki, V Bardakos, AW Lake, G Lowe. Synthesis and catalytic properties of the pentapeptide, Thr-Ala-Cys-His-Asp. J Chem Soc (C): 1860–1864, 1968.

289. Y Trudelle. Synthesis, conformation, and reactivity towards *p*-nitrophenyl acetate of polypeptides incorporating aspartic acid, serine, and histidine. Int J Peptide Protein Res 19: 528–535, 1982.

290. V Sieber, GR Moe. Interactions contributing to the formation of a $\beta$-hairpin-like structure in a small peptide. Biochemistry 35: 181–188, 1994.

291. B Gutte. A synthetic 70-amino acid residue analog of ribonuclease S-protein with enzymic activity. J Biol Chem 250: 889–904, 1975.

292. B Gutte. Study of RNase A mechanism and folding by means of synthetic 63-residue analogs. J Biol Chem 252: 663–670, 1977.

293. B Gutte. Synthetic 63 residue RNase A analogs. J Biol Chem 253: 3837–3842, 1978.

294. B Gutte. Effect of various nucleotides on folding and enzymic properties of a synthetic 63-residue analog of ribonuclease A and natural ribonuclease A. Eur J Biochem 92: 403–410, 1978.

295. MZ Atassi. Precise determination of protein antigenic structures has unravelled the molecular immune recognition of proteins and provided a prototype for synthetic mimicking of other protein binding sites. Mol Cell Biochem 32: 21–43, 1980.

296. MZ Atassi. Surface-simulation synthesis of the substrate-binding site of an enzyme. Biochemical J 226: 477–485, 1985.

297. MZ Atassi, T Manshouri. Design of peptide enzymes (pepzymes): Surface-simulation synthetic peptides that mimic the chymotrypsin and trypsin active sites

exhibit the activity and specificity of the respective enzyme. Proc Natl Acad Sci USA 90: 8282–8286, 1993.

298. BW Matthews, CS Crak, H Neurath. Can small cyclic peptides have the activity and specificity of proteolytic enzymes? Proc Natl Acad Sci USA 91: 4103–4105, 1994.

299. JA Wells, WJ Fairbrother, J Otlewski, M Laskowski Jr, J Burnier. A reinvestigation of a synthetic peptide (TrPepz) designed to mimic trypsin. Proc Natl Acad Sci USA 91: 4110–4114, 1994.

300. DR Corey, MA Phillips. Cyclic peptides as proteases: A reevaluation. Proc Natl Acad Sci USA 91: 4106–4109, 1994.

301. CA Orengo, DT Jones, JM Thornton. Protein superfamilies and domain superfolds. Nature 372: 631–634, 1994.

302. M Baron, DG Norman, ID Campbell. Protein modules. Trends in Biochemical Sciences 16: 13–17, 1991.

303. ID Campbell, AK Downing. Building protein structure and function from modular units. Trends in Biotechnology 12: 168–172, 1994.

304. T Pawson. Protein modules and signalling networks. Nature 373: 573–579, 1995.

305. L Patthy. Introns and exons. Current Opinion in Structural Biology 4: 383–392, 1994.

306. TW Traut. Are proteins made of modules? Mol Cell Biochem 70: 3–10, 1986.

307. TW Traut. Do exons code for structural or functional units in proteins? Proc Natl Acad Sci USA 85: 2944–2948, 1988.

308. A Tramontano, KD Janda, RA Lerner. Catalytic antibodies. Science 234: 1566–1570, 1986.

309. SJ Pollack, JW Jacobs, PG Schultz. Selective chemical catalysis by an antibody. Science 234: 1570–1573, 1986.

310. PG Schultz, RA Lerner. From molecular diversity to catalysis: Lessons from the immune system. Science 269: 1835–1842, 1995.

311. KT Douglas. Alteration of enzyme specificity and catalysis. Current Opinions in Biotechnology 3: 370–377, 1992.

312. Z Shao, FH Arnold. Engineering new functions and altering existing functions. Current Opinion in Structural Biology 6: 513–518, 1996.

313. O Kuchner, FH Arnold. Directed evolution of enzyme catalysts. Trends in Biotechnology 15: 523–530, 1997.

314. FH Arnold. Directed evolution: Creating biocatalysts for the future. Chemical Engineering Science 51: 5091–5102, 1996.

315. KW Hahn, WA Klis, JM Stewart. Design and synthesis of a peptide having chymotrypsin-like esterase activity. Science 248: 1544–1547, 1990.

316. MJ Corey, E Hallakova, K Pugh, JM Stewart. Studies on chymotrypsin-like catalysis by synthetic peptides. Applied Biochemistry and Biotechnology 47: 199–210, 1994.

317. WH Landschultz, PF Johnson and SL McKnight. The leucine zipper: a hypothetical structure common to a new class of DNA binding proteins. Science 240: 1759–1764, 1988.

318. JP Segrest, H De Loof, JG Dohlman, CG Brouillette, GM Anantharamaiah. Amphipathic helix motif: Classes and properties. Proteins: Structure, Function, and Genetics 8: 103–117, 1990.

319. K Johnsson, RK Allemann, H. Widmer, SA Benner. Synthesis, structure and activity of artificial, rationally designed catalytic polypeptides. Nature 365: 530–532, 1993.

320. L Regan. The design of metal-binding sites in proteins. Ann Rev Biophys Biomol Struct 22: 257–281, 1993.

321. L Regan. Protein design: novel metal-binding sites. Trends in Biochemistry 20: 280–285, 1995.

322. MR Ghadiri, C Choi. Secondary structure nucleation in peptides. Transition metal ion stabilized $\alpha$-helices. J. Am. Chem. Soc. 112: 1630–1632, 1990.

323. SS Suh, BL Haymore, FH Arnold. Characterization of His-$X_3$-His sites in $\alpha$-helices of synthetic metal-binding bovine somatotropin. Protein Engineering 4: 301–305, 1991.

324. T Handel, WF DeGrado. *De novo* design of a $Zn^{2+}$-binding site. J Am Chem Soc 112: 6710–6711, 1990.

325. L Regan, ND Clarke. A tetrahedral zinc(II)-binding site introduced into a designed protein. Biochemistry 29: 10878–10883, 1990.

326. TM Handel, SA Williams, WF DeGrado. Metal ion-dependent modulation of the dynamics of a designed protein. Science 261: 879–885, 1993.

327. M Klemba, L Regan. Characterization of metal binding by a designed protein: Single ligand substitutions at a tetrahedral $Cys_2His_2$ site. Biochemistry 34: 10094–10100, 1995.

328. MP Scott, J Biggins. Introduction of a $[4Fe-4S (S-cys)4]^{+1,+2}$ iron-sulfur center into a four $\alpha$-helix protein using design parameters from the domain of the $F_x$ cluster in the photosystem I reaction center. Protein Science 6: 340–346, 1997.

329. SM Rodday, S-S Jun, J Biggins. Interaction of the $F_AF_B$-containing subunit with the

photosystem 1 core heterodimer. Photosyn Res 36: 1–9, 1993.

330. VA Roberts, BL Iverson, SA Iverson, SJ Benkovic, RA Lerner, ED Getzoff, JA Tainer. Antibody remodeling: A general solution to the design of a metal-coordination site in an antibody binding pocket. Proc Natl Acad Sci USA 87: 6654–6658, 1990.

331. BL Iverson, SA Iverson, VA Roberts, ED Getzoff, JA Tainer, SJ Benkovic, RA Lerner. Metalloantibodies. Science 249: 659–662, 1990.

332. WS Wade, JS Koh, N Han, DM Hoekstra, RA Lerner. Engineering metal coordination sites into the antibody light chain. J Am Chem Soc 115: 4449–4456, 1993.

333. DM Segal, EA Padlan, GH Cohen, S Rudikoff, M Potter, DR Davies. The three-dimensional structure of a phosphorylcholine-binding mouse immunoglobulin Fab and the nature of the antigen binding site. Proc Natl Acad Sci USA 71: 4298–4302, 1974.

334. A Pessi, E Banchi, A Crameri, S Venturini, A Tramontano, M Sollazzo. A designed metal-binding protein with a novel fold. Nature 362: 367–369, 1993.

335. E Bianchi, M Sollazzo, A Tramontano, A Pessi. Chemical synthesis of a designed β-protein through the flow-polyamide method. Int J Peptide Protein Res 41: 385–393, 1993.

336. E. Bianchi, S Venturini, A Pessi, A Tramontano, M. Sollazzo. High level expression and rational mutagenesis of a designed protein, the minibody: From an insoluble to a soluble molecule. J Mol Biol 236: 649–659, 1994.

337. Y Satow, GH Cohen, EA Padlan, DR Davies. Phosphocholine binding immunoglobulin Fab McPC603. J Mol Biol 190: 593–604, 1986.

338. PN Strong. Potassium channel toxins. Pharmac Ther 46: 137–162, 1990.

339. C Vita, C Roumestand, F Toma, A Ménez. Scorpion toxins as natural scaffolds for protein engineering. Proc Natl Acad Sci USA 92: 6404–6408, 1995.

340. F Dreyer. Peptide toxins and potassium channels. Rev Physiol Biochem Pharmacol 115: 93–136, 1990.

341. M Dauplais, A Lecoq, J Song, J Cotton, N Jamin, B Gilquin, C Roumestand, C Vita, CLC de Medeiros, EG Rowan, AL Harvey, A Ménez. On the convergent evolution of animal toxins. Conservation of a diad of functional residues in potassium channel-blocking toxins with unrelated structures. J Biol Chem 272: 4302–4309, 1997.

342. JE Tudor, PK Pallaghy, MW Pennigton, RS Norton. Solution structure of ShK toxin, a novel potassium channel inhibitor from a sea anemone. Nature Structural Biology 3: 317–320, 1996.

343. H Schweitz, T Bruhn, E Guillemare, D Moinier, J-M Lancelin, L Béress, M Lazdunski. Kalicludines and kaliseptine. Two different classes of sea anemone toxins for voltage-sensitive K+ channels. J Biol Chem 270: 25121–25126, 1995.

344. G Mer, C Kellenberger, P Koehl, R Stote, O Sorokine, A Van Dorsselaer, B Luu, H Hietter, J-F Lefèvre. Solution structure of PMP-D2, a 35 residue peptide isolated from the insect *Locusta migratoria*. Biochemistry 33: 15397–15407, 1994.

345. M Bruix, MA Jiménez, J Santoro, G González, FJ Colilla, E Méndez, M Rico. Solution structure of $\gamma$1-H and $\gamma$1-P thionins from barley and wheat endosperm determined by $^1$H-nmr: A structural motif common to toxic arthropod proteins. Biochemistry 32: 715–724, 1993.

346. BM Olivera, J Rivier, C Clark, CA Ramilo, GP Corpuz, FC Abogadie, EE Mena, SR Woodward, DR Hillyard, LJ Cruz. Diversity of *Conus* neuropeptides. Science 249: 257–263, 1990.

347. WR Gray, BM Olivera. Peptide toxins from venomous *Conus* snails. Ann Rev Biochem 57: 665–700, 1988.

348. CA Ramilo, GC Zararalla, L Nadasdi, LG Hammerland, D Yoshikami, WR Gray, Ramasharma Kristipati, J Ramachandran, G Miljanich, BM Olivera, LJ Cruz. Novel α- and ω-conotoxins from *Conus striatus* venom. Biochemistry 31: 9919–9926, 1992.

349. JH Davis, ER Bradley, GP Miljanich, L Nadasdi, J Ramachandran, VJ Basus. Solution structure of ω-conotoxin GVIA using 2 D nmr spectroscopy and relaxation matrix analysis. Biochemistry 32: 7396–7405, 1993.

350. RG Simmonds, DE Tupper, JR Harris. Synthesis of disulfide-bridged fragments of ω-conotoxins GVIA and MVIIA. Int J Peptide Protein Res 43: 363–366, 1994.

351. DJ West, EB Andrews, D Bowman, AR McVean, MC Thorndyke. Toxins from some poisonous and venomous marine snails. Comp Biochem Physiol 113C: 1–10, 1996.

352. M Laskowski Jr, I Kato. Protein inhibitors of proteinases. Ann Rev Biochem 49: 593–626, 1980.

353. B Pierret, H Virelizier, C Vita. Synthesis of a metal binding protein designed on the α/β scaffold of charybdotoxin. Int J Peptide Protein Res 46: 471–479, 1995.

354. E Drakopoulou, S Zinn-Justin, M Guenneugues, B Gilquin, A Ménez, C Vita. Changing the structural context of a functional $\beta$-hairpin. Synthesis and characterization of a chimera containing the curaremimetic loop of a snake toxin in the scorpion $\alpha/\beta$ scaffold. J Biol Chem 271: 11979–11987, 1996.

355. S Zinn-Justin, M Guenneugues, E Drakopoulou, B Gilquin, C Vita, A Ménez. Transfer of a $\beta$-hairpin from the functional site of snake curaremimetic toxins to the $\alpha/\beta$ scaffold of scorpion toxins: Three-dimensional solution structure of the chimeric protein. Biochemistry 35: 8535–8543, 1996.

356. AM Gronenborn, DR Filpula, NZ Essig, A Achari, M Whitlow, PT Wingfield, GM Clore. A novel, highly stable fold of the immunoglobulin binding domain of *Streptococcal* protein G. Science 253: 657–661, 1991.

357. M Klemba, KH Gardner, S Marino, ND Clarke, L Regan. Novel metal-binding proteins by design. Nature Structural Biology 2: 368–373, 1995.

358. T Sasaki, ET Kaiser. Synthesis and structural stability of helichrome as an artificial hemeprotein. Biopolymers 29: 79–88, 1990.

359. H Mihara, N Nishino, R Hasegawa, T Fujimoto, S Usui, H Ishida, K Ohkubo. Design of a hybrid two $\alpha$-helix peptides and ruthenium trisbipyridine complex for photo-induced electron transfer system in bilayer membranes. Chem Letts 1992: 1813–1816, 1992.

360. DE Robertson, RS Farid, CC Moser, JL Urbauer, SE Mulholland, R Pidikiti, JD Lear, AJ Wand, WF DeGrado, PL Dutton. Design and synthesis of multi-haem proteins. Nature 368: 425–432, 1994.

361. CT Choma, JD Lear, MJ Nelson, PL Dutton, DE Robertson, WF DeGrado. Design of a heme-binding four-helix bundle. J Am Chem Soc 116: 856–865, 1994.

362. DR Benson, BR Hart, X Zhu, MB Doughty. Design, synthesis and circular dichroism investigation of a peptide-sandwiched mesoheme. J Am Chem Soc 117: 8502–8510, 1995.

363. M Montal. Design of molecular function: Channels of communication. Ann Rev Biophys Biomol Struct 24: 31–57, 1995.

364. M Montal. Molecular anatomy and molecular design of channel proteins. FASEB J 4: 2623–2635, 1990.

365. RE Koeppe II. Engineering the gramicidin channel. Ann Rev Biophys Struct 25: 231–258, 1996.

366. MR Ghadiri, JR Granja, LK Buehler. Artificial transmembrane ion channels from self-assembling peptide nanotubes. Nature 369: 301–304, 1994.

367. MS Weiss, U Abele, J Weckesser, W Welte, E Schiltz, GE Schultz. Molecular architecture and electrostatic properties of a bacterial porin. Science 254: 1627–1630, 1991.

368. SW Cowan, T Schirmer, G Rummel, M Steiert, R Ghosh, RA Pauptit, JN Jansonius, JP Rosenbusch. Crystal structures explain functional properties of two *E. coli* porins. Nature 358: 727–733, 1992.

369. G Spach, Y Trudelle, F Heitz. Peptides as channel-making ionophores: Conformational aspects. Biopolymers 22: 403–407, 1983.

370. S Oiki, W Danho, M Montal. Channel protein engineering: Synthetic 22-mer peptide from the primary structure of the voltage-sensitive sodium channel forms ionic channels in lipid bilayers. Proc Natl Acad Sci USA 85: 2393–2397, 1988.

371. S Oiki, V Madison, M Montal. Bundles of amphipathic transmembrane $\alpha$-helices as a structural motif for ion-conducting channel proteins: Studies on sodium channels and acetylcholine receptors. Proteins: Structure, Function, and Genetics 8: 226–236, 1990.

372. M Montal, MS Montal, JM Tomich. Synporins—synthetic proteins that emulate the pore structure of biological ionic channels. Proc Natl Acad Sci USA 87: 6929–6933, 1990.

373. M Oblatt-Montal, LK Buhler, T Iwamoto, JM Tomich, M Montal. Synthetic peptides and four-helix bundle proteins as model systems for the pore-forming structure of channel proteins. J Biol Chem 268: 14601–14607, 1993.

374. M Oblatt-Montal, T Iwamoto, JM Tomich, M Montal. Design, synthesis and functional characterization of a pentameric channel protein that mimics the presumed pore structure of the nicotinic cholinergic receptor. FEBS Letts 320: 261–266, 1993.

375. A Grove, JM Tomich, M Montal. A molecular blueprint for the pore-forming structure of voltage-gated calcium channels. Proc Natl Acad Sci USA 88: 6418–6422, 1991.

376. A Grove, JM Tomich, T Iwamoto, M Montal. Design of a functional calcium channel protein: Inferences about an ion channel-forming motif derived from the primary structure of voltage-gated calcium channels. Protein Science 2: 1918–1930, 1993.

377. GL Reddy, T Iwamoto, JM Tomich, M Montal. Synthetic peptides and four-helix

bundle proteins as model systems for the pore-forming structure of channel proteins. J Biol Chem 268: 14608–14615, 1993.

378. M Mutter. The construction of new proteins and enzymes—a prospect for the future? Angew Chem Int Ed 24: 639–653, 1985.

379. M. Mutter, R Hersperger, K Gubernator, K Muller. The construction of new proteins: V. A template-assembled synthetic protein (TASP) containing both a 4-helix bundle and a $\beta$-barrel-like structure. Proteins: Structure, Function, and Genetics 5: 13–21, 1989.

380. DM Engelman, TA Steitz. On the folding and insertion of globular membrane proteins. In: DB Wetlaufer, ed. The Protein Folding Problem. AAAS Selected Symposium 89, Boulder CO, Westview Press Inc, 1984, pp 87–113.

381. C Cohen, DAD Parry. $\alpha$-helical coiled coils—a widespread motif in proteins. TIBS 11: 245–248, 1986.

382. JD Lear, ZR Wasserman, WF DeGrado. Synthetic amphiphilic peptide models for protein ion channels. Science 240: 1177–1187, 1988.

383. WF DeGrado, JD Lear. Coformationally constrained $\alpha$-helical peptide models for protein ion channels. Biopolymers 29: 205–213, 1990.

384. KS Åkerfeldt, JD Lear, ZR Wasserman, LA Chung, WF DeGrado. Synthetic peptides as models for ion channel proteins. Acc Chem Res 26: 191–197, 1993.

385. LA Chung, JD Lear, WF DeGrado. Fluorescence studies of the secondary structure and orientation of a model ion channel peptide in phospholipid vesicles. Biochemistry 31: 6608–6616, 1992.

386. PK Kienker, WF DeGrado, JD Lear. A helical-dipole model describes the single-channel current rectification of an uncharged peptide ion channel. Proc Natl Acad Sci USA 91: 4859–4863, 1994.

387. A Grove, M Mutter, JE Rivier, M Montal. Template-assembled synthetic proteins designed to adopt a globular, four-helix bundle conformation form ionic channels in lipid bilayers. J Am Chem Soc 115: 5919–5924, 1993.

388. KS Åkerfeldt, RM Kim, D Camac, JT Groves, JD Lear, WF DeGrado. Tetraphilin: A four-helix proton channel built on a tetraphenylprophyrin framework. J Am Chem Soc 114: 9656–9657, 1994.

389. TW Muir, MJ Williams, MH Ginsberg, SBH Kent. Design and chemical synthesis of a neoprotein structural model for the cytoplasmic domain of a multisubunit cell-surface receptor: Integrin $\alpha_{IIb}\beta_3$ (platelet GPIIb-IIIa). Biochemistry 33: 7701–7708, 1994.

390. TA Steitz. Structural studies of protein-nucleic acid interaction: the sources of sequence-specific binding. Quarterly Reviews of Biophysics 23: 205–280, 1990.

391. CO Pabo, RT Sauer. Transcription factors: Structural families and principles of DNA recognition. Ann Rev Biochem 61: 1053–1095, 1992.

392. B Cuenoud, A Schepartz. Design of a metallo-bZIP protein that discriminates between CRE and AAPI target sites: Selection against API. Proc Natl Acad Sci USA 90: 1154–1159, 1993.

393. TE Ellenberger, CG Brandl, K Struhl, SC Harrison. The GCN4 basic region leucine zipper binds DNA as a dimer of uninterrupted $\alpha$-helices: Crystal structure of the protein DNA complex. Cell 71: 1223–1237, 1992.

394. WF Anderson, DH Ohlendorf, Y Takeda, BW Matthews. Structure of the *cro* repressor from bacteriophage $\lambda$ and its interaction with DNA. Nature 290: 754–758, 1981.

395. A Laughton, MP Scott. Sequence of a Drosophila segmentation gene: protein structure homology with DNA-binding proteins. Nature 310: 25–31, 1984.

396. YQ Qian, M Billeter, G Otting, M Müller, WJ Gehring, K Wüthrich. The structure of the *Antennapedia* homeodomain determined by nmr spectroscopy in solution: Comparison with prokaryotic repressors. Cell 59: 573–580, 1989.

397. AD Frankel, CO Pabo. Fingering too many proteins. Cell 53: 675, 1988.

398. BF Luisi, WX Zu, Z Otwinowski, LP Freedman, KR Yamamoto, PB Sigler. Crystallographic analysis of the interaction of the glucocortoid receptor with DNA. Nature 352: 497–505, 1991.

399. EK O'Shea, R Rutkowski, PS Kim. Evidence that the leucine zipper is a coiled-coil. Science 243: 538–542, 1989.

400. P Agre, PF Johnson, S McKnight. Cognate DNA binding specificity retained after leucine zipper exchange between GCN4 and C/EBP. Science 246: 922–925, 1989.

401. C Murre, PS McCaw, D Baltimore. A new DNA binding and dimerization motif in immunoglobulin enhancer binding daughterless, Myo D, and Myc proteins. Cell 56: 777–783, 1989.

402. AR Ferré-D'Amaré, P Pognonec, RG Roeder, SK Burley. Structure and function of the b/HLH/Z domain of USF. EMBO J 13: 180–189, 1994.

403. KT O'Neil, RH Hoess, WF DeGrado. Design of DNA-binding peptides based on the leucine zipper motif. Science 249: 774–778, 1990.

404. CR Vinson, PB Sigler, SL McKnight. Scissor grip model for DNA recognition by a family of leucine zipper proteins. Science 246: 911–916, 1989.

405. DE Fisher, LA Parent, PA Sharp. High affinity DNA-binding Myc analogs: Recognition by an α-helix. Cell 72: 467–476, 1993.

406. T Hehlgans, M Stolz, S Klauser, T Cui, P Salgam, SB Verca, M Widmann, A Leiser, K Städler, B Gutte. The DNA-binding properties of an artificial 42 residue polypeptide derived from a natural repressor. FEBS Letts 315: 51–55, 1993.

407. K Städler, N Liu, L. Trotman, A Hiltpold, G Caderas, S Klauser, T Hehlgans, B Gutte. Design, synthesis, and characterization of HIV-1 enhancer-binding polypeptides derived from bacteriophage 434 repressor. Int J Peptide Protein Res 46: 333–340, 1995.

408. N. Liu, G Caderas, B Gutte, RM Thomas. An artificial HIV enhancer-binding peptide is dimerized by the addition of a leucine zipper. Eur Biophys J 25: 399–403, 1997.

409. JR Desjarlais, JM Berg. Use of a zinc-finger consensus sequence framework and specificity rules to design specific DNA binding proteins. Proc Natl Acad Sci USA 90: 2256–2260, 1993.

410. B Gutte, M Daümigen, M Wittschieber. Design, synthesis and characterization of a 34-residue polypeptide that interacts with nucleic acids. Nature 281: 650–655, 1979.

411. R Jaenicke, B Gutte, U Glatter, W Strassburger, A Wollmer. Conformation of a synthetic 34-residue polypeptide that interacts with nucleic acids. FEBS Letts 114: 161–164, 1980.

412. Z Shang, VE Issac, H Li, L Patel, KM Catron, T Curran, GT Montelione, C Abate. Design of a "minimAl" homeodomain: The N-terminal arm modulates DNA binding affinity and stabilizes homeodomain structure. Proc Natl Acad Sci USA 91: 8373–8377, 1994.

# 2

# Peptidomimetic and Nonpeptide Drug Discovery: Chemical Nature and Biological Targets

**Tomi K. Sawyer**
*ARIAD Pharmaceuticals, Cambridge, Massachusetts*

## I. INTRODUCTION

Peptide drug discovery challenges synthetic, computational and biophysical chemists, biochemists, pharmacologists and drug delivery scientists to identify lead compounds that exhibit the requisite potency, selectivity, metabolic stability, and *in vivo* pharmacological efficacy to warrant further development as drug candidates. In retrospect, the discovery of a vast number of peptide hormones, neurotransmitters, growth factors and cytokines first provided knowledge of primary structure and related 3-D molecular architecture for naturally-occuring, biologically-active peptides (Table 1). Over the past two decades a highly focused effort in both industry and academia has advanced the rational transformation of "first generation" peptide lead compounds to significantly modified analogs having minimal peptide-like chemical structure (1–18).

Peptide-based drug design first evolved from receptor-targeted research and was not able to take advantage of the 3-D structure of the target protein. Nevertheless, a hierarchial approach of peptide $\longrightarrow$ peptidomimetic/nonpeptide

drug design has evolved over the past two decades as related to systematic transformation of the peptide scaffold or functional group elaboration of nonpeptide templates (Fig. 1). Specifically, the use of various backbone and/or sidechain modifications, non-amino acid building block replacements, and novel templates have led to the discovery of chemically-unique compounds (*e.g.*, peptidomimetics and *de novo* designed nonpeptides). This work has typically employed structural biology (x-ray crystallography and/or NMR spectroscopy), computer-assisted molecular modeling, and biological testing to advance iterative structure-based drug design. Finally, the integration of structure-based drug design with high-throughput screening and synthetic combinatorial libraries is markedly re-shaping peptidomimetic and nonpeptide drug discovery research.

## II. PEPTIDE STRUCTURE AND $\phi$-$\psi$-$\chi$ SPACE

Peptides are chemically diverse by virtue of the number of unique combinations of amino acid building blocks which consti-

**Table 1**   Examples of Naturally-Occuring, Biologically-Active Peptides

| *Peptide* | *Primary Structure* |
| --- | --- |
| Thyrotropin-Releasing Hormone | $<$Glu$^1$-His-Pro$^3$-NH$_2$ |
| Enkephalin(Met) | Tyr$^1$-Gly-Gly-Phe-Met$^5$ |
| Cholecystokinin-8 | Asp$^1$-Tyr[SO$_3$H]-Met-Gly-Trp-Met-Asp-Phe$^8$-NH$_2$ |
| Angiotensin II | Asp$^1$-Arg-Val-Tyr-Ile-His-Pro-Phe$^8$ |
| Oxytocin | Cys$^1$-Tyr-Ile-Gln-Asn-Cys-Pro-Leu-Gly$^9$-NH$_2$ |
| Vasopressin | Cys$^1$-Tyr-Phe-Gln-Asn-Cys-Pro-Arg-Gly$^9$-NH$_2$ |
| Bradykinin | Lys$^1$-Arg-Pro-Gly-Phe-Ser-Pro-Phe-Arg$^9$ |
| Gonadotropin-Releasing Hormone | $<$Glu$^1$-His-Trp-Ser-Tyr-Gly-Leu-Arg-Pro-Gly$^{10}$-NH$_2$ |
| Substance P | Arg$^1$-Pro-Lys-Pro-Gin-Gln-Phe-Phe-Gly-Leu-Met$^{11}$-NH$_2$ |
| α-Melanotropin | Ac-Ser$^1$-Tyr-Ser-Met-Glu-His-Phe-Arg-Trp-Gly-Lys-Pro-Val$^{13}$-NH$_2$ |
| Neurotensin | $<$Glu$^1$-Leu-Tyr-Glu-Asn-Lys-Pro-Arg-Arg-Pro-Tyr-Ile-Leu$^{13}$ |
| Somatostatin | Ala$^1$-Gly-Cys-Lys-Asn-Phe-Phe-Trp-Lys-Thr-Phe-Thr-Ser-Cys$^{14}$ |
| Endothelin | Cys$^1$-Ser-Cys-Ser-Ser-Leu-Met-Asp-Lys-Glu-Cys-Val-Tyr-Phe-Cys-His-Leu-Asp-Ile-Ile-Trp$^{21}$ |
| Vasoactive Intestinal Peptide | His$^1$-Ser-Asp-Ala-Val-Phe-Thr-Asp-Asn-Tyr-Thr-Arg-Leu-Arg-Lys-Gln-Met-Ala-Val-Lys$^{21}$-Tyr-Leu-Asn-Ser-Ile-Leu-Asn$^{28}$-NH$_2$ |
| Glucagon | His$^1$-Ser-Gln-Gly-Thr-Phe-Thr-Ser-Asp-Tyr-Ser-Lys-Tyr-Leu-Asp-Ser-Arg-Arg-Ala-Gln-Asp$^{21}$-Phe-Val-Gln-Trp-Leu-Met-Asp-Thr$^{29}$ |
| Galanin | Gly$^1$-Trp-Thr-Leu-Asn-Ser-Ala-Gly-Tyr-Leu-Leu-Gly-Pro-His-Ala-Ile-Asp-Asn-His-Arg-Ser$^{21}$-Phe-His-Asp-Lys-Tyr-Gly-Leu-Ala$^{29}$-NH$_2$ |
| Corticotropin | Ser$^1$-Tyr-Ser-Met-Glu-His-Phe-Arg-Trp-Gly-Lys-Pro-Val-Gly-Lys-Lys-Arg-Arg-Pro-Val-Lys$^{21}$-Val-Tyr-Pro-Asn-Gly-Ala-Glu-Asp-Glu-Ser-Ala-Glu-Ala-Phe-Pro-Leu-Glu-Phe$^{39}$ |
| Neuropeptide-Y | Tyr$^1$-Pro-Ser-Lys-Pro-Asp-Asn-Pro-Gly-Glu-Asp-Ala-Pro-Ala-Glu-Asp-Leu-Ala-Arg-Tyr-Tyr$^{21}$Ser-Ala-Leu-Arg-His-Tyr-Ile-Asn-Leu-Met-Thr-Arg-Gln-Arg-Tyr$^{36}$-NH$_2$ |
| Corticotropin-Releasing Factor | Ser$^1$-Gin-Glu-Pro-Pro-Ile-Ser-Leu-Asp-Leu-Thr-Phe-His-Leu-Leu-Arg-Glu-Val-Leu-Glu-Met$^{21}$-Thr-Lys-Ala-Asp-Gln-Leu-Ala-Gln-Gln-Ala-His-Ser-Asn-Arg-Lys-Leu-Leu-Asp-Ile$^{40}$-Ala$^{41}$-NH$_2$ |

tute their molecular framework. The precise linear arrangement, amino $\longrightarrow$ carboxy ($N \longrightarrow C$) directionality, of amino acids in a peptide is referred to as its primary structure. To a large extent, intramolecular covalent bonding (*e.g.*, disulfide or amide cyclization linkages) and noncovalent bonding (*e.g.*, hydrogen and/or ionic) interactions determine the 3-D architecture of a peptide. The 3-D structural properties of peptides (Fig. 2) are typically defined in terms of torsion angles ($\psi$, $\phi$, $\omega$, $\chi$) between the backbone amine nitrogen ($N^{\alpha}$), the backbone carbonyl carbon ($C'$), the backbone α-carbon ($C^{\alpha}$), and sidechain hydrocarbon func-

Peptide

| Drug Design Concept | Synthetic Chemistry Strategy |
|---|---|
| *Primary Structure and Function* | *Peptide Sequence Reduction* |
| • Contiguous Binding/Active Site | • N-/C-Terminal Truncation |
| • Discontiguous Binding/Active Site | • Internal Deletion |
| *Sidechain Substructure Requirements* | *Amino Acid Substitutions* |
| • Cβ Functionalization | • Ala Scan |
| • Cα Chirality (L vs. D) | • D-Amino Acid Scan |
| *Backbone Substructure Analysis* | *Amide Replacements* |
| • Cis/Trans Isomerization | • N-Alkylation |
| • 3-D Flexibility; H-Bonding | • Amide Surrogates |
| *Secondary Structure Analysis* | *Peptide Sidechain/Backbone Modification* |
| • Local Conformational Constraint | • Constrained Amino Acid Replacement |
| • Local Topographical Constraint | • Peptide Substructure Constraints (Cyclization) |
| *Pharmacophore Model Development* | *Peptide → Peptidomimetic/Nonpeptide* |
| • Key Backbone Functionalities | • Peptide Scaffold-Based Lead |
| • Key Sidechain Functionalities | • Nonpeptide Template-Based Lead |

Peptidomimetic and/or Designed Nonpeptide

**Fig. 1**  Hierarchial approach in peptide → peptidomimetic/nonpeptide drug design.

*Backbone/Side-Chain Dihedral Angles of Phe in a Peptide*  *Cis and Trans Peptide Bond Isomers of Phe-Pro*

*Backbone/Side-Chain Cα–Cβ Spatial Projections of Phe*  *Intramolecular H-Bonding in Peptide Secondary Structure*

γ–Turn (n = 1)
β–Turn (n = 2)
α–Helix (n = 3)

*Amino Acid*

*Conformationally-Restricted Phe Analogs*

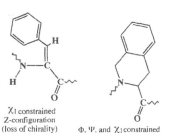

**Fig. 2**  Three-dimensional structural properties of peptides: backbone and sidechain.

tionalization (*e.g.*, $C^\beta$, $C^\gamma$, $C^\delta$, $C^\varepsilon$ of Lys) derived from the constituent amino acids. A Ramachandran plot ($\psi$ versus $\phi$) defines the preferred combinations of torsion angles for ordered conformations (secondary structures), such as $\alpha$-helix, $\beta$-turn, $\gamma$-turn, or $\beta$-sheet. With respect to the backbone amide bond torsion angle ($\omega$) the *trans* geometry is more energetically-favored for most typical dipeptide substructures, however, when the C-terminal partner is Pro or other N-alkylated (including cyclic) amino acids the *cis* geometry is possible and may further stabilize $\beta$-turn or $\gamma$-turn conformations. In summary, even modest chemical modifications by $N^\alpha$-methyl, $C^\alpha$-methyl or $C^\beta$-methyl can have significant consequences on molecular flexibility and peptide 3-D structure (e.g., Phe analogs in Fig. 2).

## III.  PEPTIDE SCAFFOLD CHEMICAL MODIFICATIONS

Sophisticated synthetic chemistry approaches have contributed to contemporary peptide-based molecular design, including well-established applications of unusual amino acids, dipeptide surrogates, amongst other types of chemical modifications (1–8). Such backbone or sidechain modifications may afford stability of the parent peptide to peptidases, and have provided impetus for yet more sophisticated drug design strategies. A few noteworthy dipeptide surrogates of the generic structure $Xxx\Psi[Z]Yyy$ have been advanced (Fig. 3), and exemplify a number of key amide bond isosteres, including: aminomethylene or $\Psi[CH_2 NH]$, **1**; ketomethylene or $\Psi[COCH_2]$, **2**; ethylene or $\Psi[CH_2CH_2]$, **3**; olefin or $\Psi[CH{=}CH]$, **4**; ether or $\Psi[CH_2O]$, **5**; thioether or $\Psi[CH_2S]$, **6**; tetrazole or $\Psi[CN_4]$, **7**; thiazole or $\Psi[thz]$, **8**; retroamide or $\Psi[NHCO]$, **9**; thioamide or $\Psi[CSNH]$, **10**; and ester or $\Psi[CO_2]$, **11**.

These amide bond surrogates provide insight into the conformational and H-bonding properties that may be requisite for peptide molecular recognition and/or biological activity at receptor targets. Furthermore, such backbone replacements can impart metabolic stability towards peptidase cleavage relative to the parent peptide. The discovery of yet other nonhydrolyzable amide bond isosteres has particularly impacted the design of protease inhibitors, and these include: hydroxymethylene or $\Psi[CH (OH)]$, **12**; hydroxyethylene or $\Psi[CH (OH)CH_2]$ and $\Psi[CH_2CH(OH)]$, **13** and **14** respectively; dihydroxyethylene or ($\Psi[CH(OH)CH(OH)]$, **15**, hydroxyethylamine or $\Psi[CH(OH)CH_2N]$, **16** and $C_2$-symmetric hydroxymethylene **17** and dihydroxyethylene **18**.

Both peptide backbone and sidechain modifications may provide prototypic leads for the design of secondary structure mimicry (19–29) as typically suggested by the substitution of *D*-amino acids, $N^\alpha$-Me-amino acids $C^\alpha$-Me-amino acids and/or dehydroamino acids which may induce or stabilize regiospecific $\beta$-turn, $\gamma$-turn, $\beta$-sheet or $\alpha$-helix conformations. To date, a variety of secondary structure mimetics have been designed and incorporated in peptides or peptidomimetics (Fig. 4). The $\beta$-turn has been of particular interest to the area of receptor-targeted peptidomimetic drug discovery. This secondary structural motif exists within a tetrapeptide sequence in which the first and fourth $C\alpha$ atoms are $\leq 7\,\text{Å}$ separated, and they are further characterized as to occur in a non-helical region of the peptide sequence and possessing a 10-membered intramolecular H-bond between the $i$ and $i + 4$ amino acid residues. One of the first noteworthy approaches to the design of $\beta$-turn mimetics was the monocyclic dipeptide-based template **19** (19) which employs sidechain-to-backbone constraint at the $i + 1$ and $i + 2$ sites. Over the past decade, many other monocyclic or bicyclic templates have been developed as β-turn mimetics, including: **20** (20), **21** (21), **22** (22), **23** (23) and **24** (24). Recently, the monocyclic β-turn mimetic **25** has been described (25) as a novel scaffold capable of multiple sidechain-like modifications $i, i+1, i+2$ and $i+3$ positions) as well as

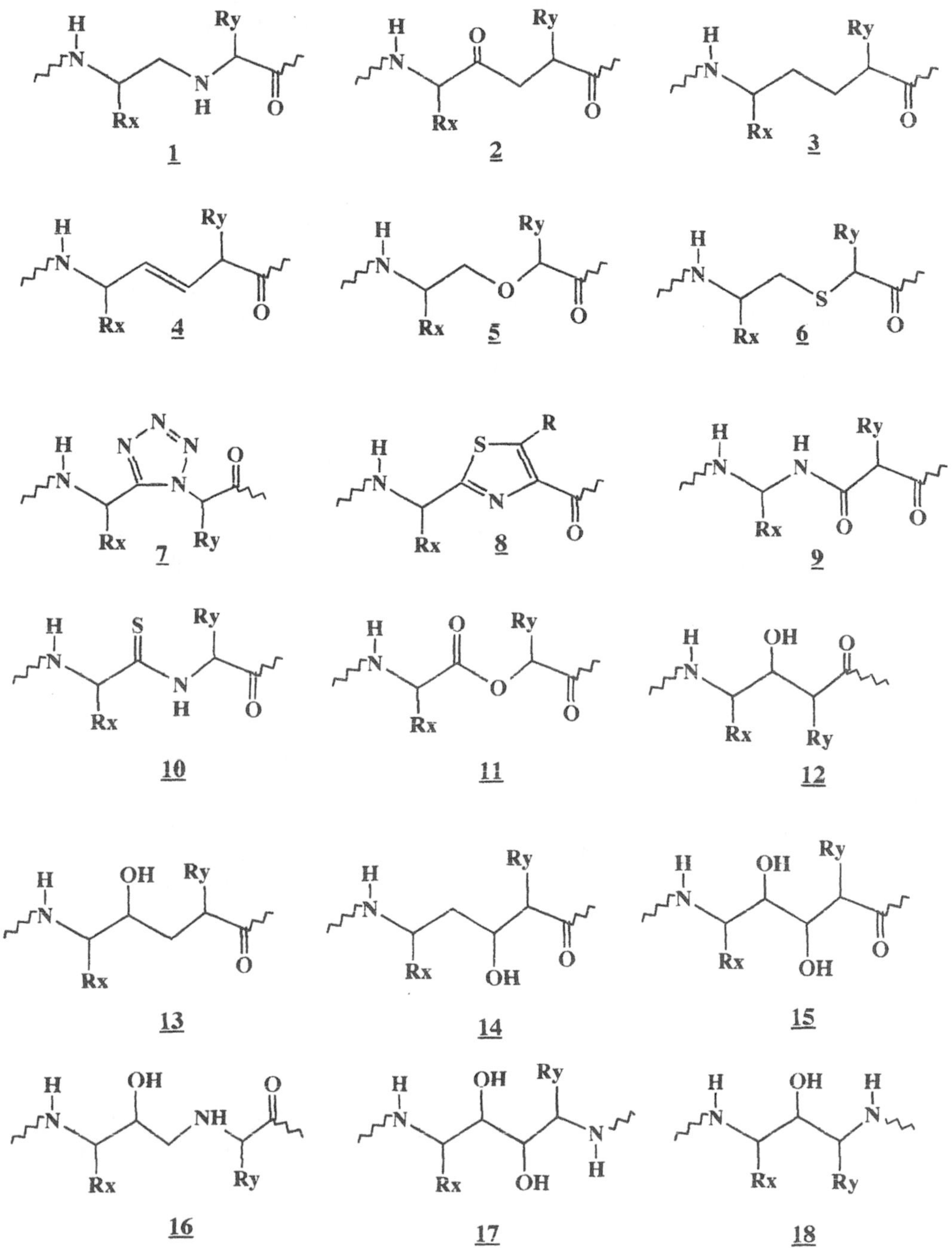

**Fig. 3** Backbone amide bond surrogates: $\psi$[CONH] replacements.

five of the eight NH or C=O func-tionalities relative to a parent tetrapeptide sequence. Similarly, the benzodiazepine template **26** has shown (26, 27) utility as a β-turn mimetic that may be multisubstituted to simulate sidechain functionalization, particularly at the i and

i + 3 positions, of the corresponding tetrapeptide sequence modeled in type I–VI β-turn conformations. A recently reported (28) γ-turn mimetic **27** illustrates an innovative approach to incorporate a retroamide surrogate between the i and i + 1 amino acid residues with an ethylene

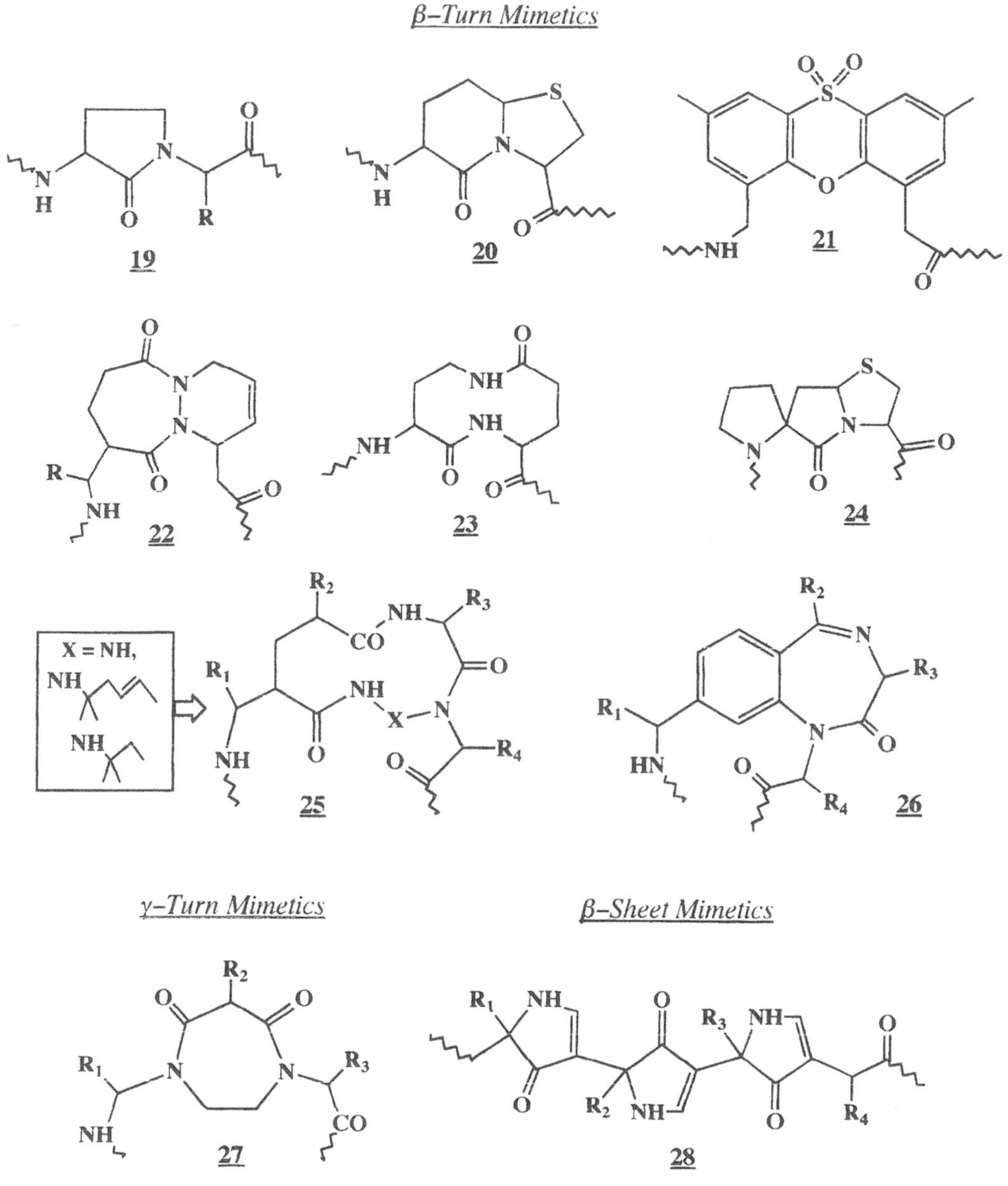

*β–Turn Mimetics*

*γ–Turn Mimetics*

*β–Sheet Mimetics*

**Fig. 4**  Secondary structure modifications: β/γ-turn and β-sheet scaffolds.

bridge between the N′ (*i.e.*, nitrogen replacing the carbonyl C′) and N atoms of the *i* and *i* + 2 positions, and this template allows the possibility for all three sidechains of the parent tripeptide sequence. Finally, the design of a $\beta$-sheet mimetic **28** provides an attractive template to constrain the backbone of a peptide to that simulating an extended conformation (29). The $\beta$-sheet is of particular interest to the area of protease-targeted peptidomimetic drug discovery.

## IV. PEPTIDOMIMETICS AND *DE NOVO* DESIGNED NONPEPTIDES

More extensive, or radical, structural transformation of a peptide to provide analogs having essentially no naturally-occurring amino acids or dipeptide substructure has advanced (30–38) as based on the premise that such compounds might be more well-suited for drug development in terms of chemical and/or biological properties (*e.g.*, low molecular mass, metabolic stability, and bioavailability by oral administration). Such designed peptidomimetics and nonpeptides (Fig. 5) have historical precedence in the discovery of orally-effective and sustained-acting inhibitors of angiotensin-converting enzyme (ACE) such as enalapril, **29** (39). Other noteworthy designed peptidomimetics or nonpeptides include: thyrotropin-releasing hormone (TRH) agonist, **30** (32); fibrinogen (GPIIa/IIIb) antagonists, **31** (40) and **32** (41); endothelin antagonist, **33** (42); CCK$_B$/gastrin antagonist, **34** (43); somatostatin agonist, **35** (44, 45); substance-P (NK$_1$) antagonist, **36** (46); neurokinin-A (NK$_2$) antagonist, **37** (47); Ras farnesyl protein transferase inhibitors, **38** (48) and **39** (49); HIV protease inhibitors, **40** (50), **41** (51) and **42** (52); renin inhibitors, **43** (53), **44** (54), **45** (55); and src homology-2 (SH2) inhibitors, **46** (56), **47** (57), and **48** (58), and **49** (59).

## V. SCREENING-DERIVED NONPEPTIDES AND PHARMACOPHORE MODELING

In synchrony with the above peptide structure-based design strategy efforts has been discovery efforts focused on identifying novel lead compounds by screening, including mass screening technologies, of a variety of natural product, chemical file, or synthetic combinatorial libraries. Precedence for the success of nonpeptide drug discovery can be traced to the identification of morphine (**49** Fig. 6) as nonpeptide natural product agonist at $\mu$-opioid peptide receptors (60). To date, the robust momentum of nonpeptide drug discovery continues to be accelerated by sophisticated high-throughput screening technologies. The scope of molecular diversity as well as therapeutic targets for such screening-based nonpeptide ligand lead compounds includes the following examples (Fig. 6): substance P (NK$_1$) antagonist, **51** (61); angiotensin AT$_1$ antagonist, **52** (62); growth hormone-releasing peptide (GHRP) agonist, **53** (63); cholecystokinin CCK$_A$ antagonist, **54** (64); CCK$_B$/gastrin antagonist, **55** (65); CCK$_A$ agonist, **56** (66); endothelin antagonist, **57** (67); gonadotropin-releasing hormone (GnRH) antagonist, **58** (68); vasopressin V$_1$ antagonist, **59** (69); gastrin-releasing peptide antagonist, **60** (70); glucagon antagonist, **61** (71), neurotensin antagonist, **62** (72); angiotensin AT$_1$ agonist, **63** (73), oxytocin antagonist, **64** (74); and HIV protease inhibitor, **65** (75).

Historically, research on opioid G-protein coupled receptor targets (*e.g.*, $\mu$, $\delta$, $\kappa$) has provided insight to explore the pharmacophores of both agonist and antagonists derived from endogenous peptides (*e.g.*, endorphin, endorphin, dynorphin) or nonpeptides (*e.g.*, the $\mu$-receptor selective agonist morphine and its N-allyl-substituted antagonist derivative naloxone). Relative to the

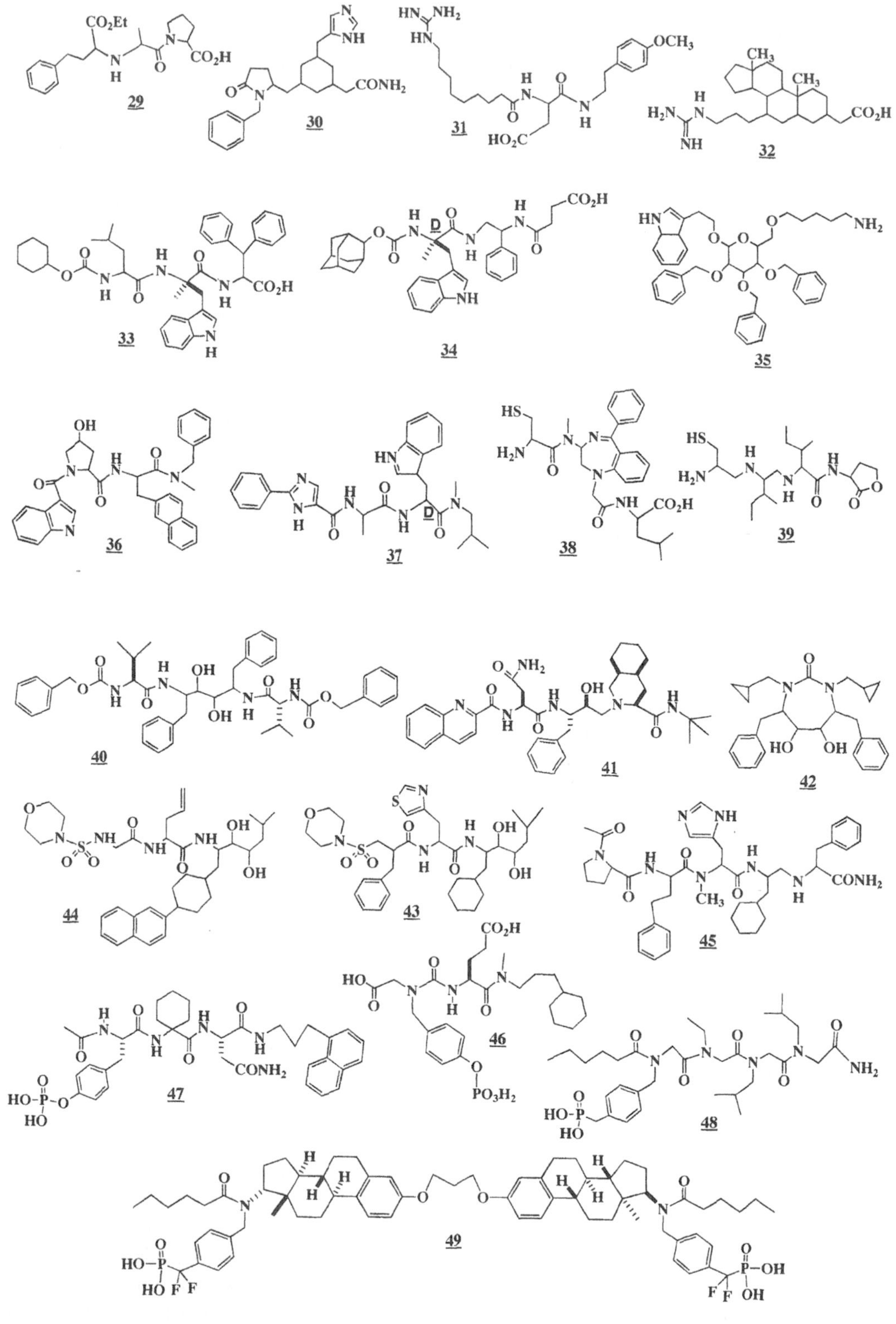

**Fig. 5** Peptidomimetic/nonpeptide leads from structure-based and *de novo* design.

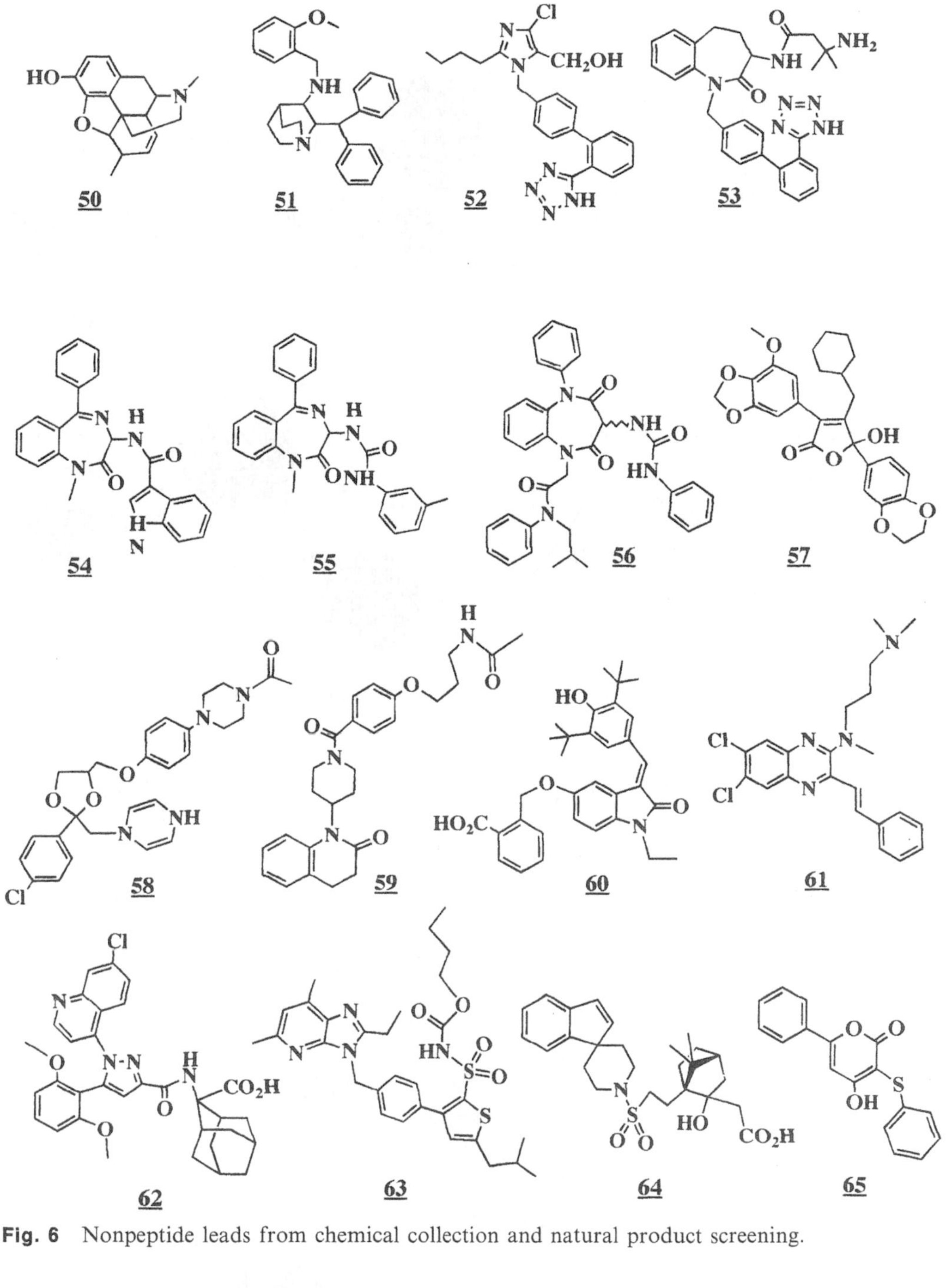

**Fig. 6** Nonpeptide leads from chemical collection and natural product screening.

Met-endorphin it has been described (76) that the N-terminal Tyr sidechain and α-amino functionalities are likely common pharmacophoric features relative to the N-methyl-tyramine, substructure of mor-

phine **49** (Fig. 7) for μ-receptor binding. In the case of angiotensin II receptor antagonist drug discovery, both peptide and nonpeptide (screening-derived) ligand structure-based design studies have

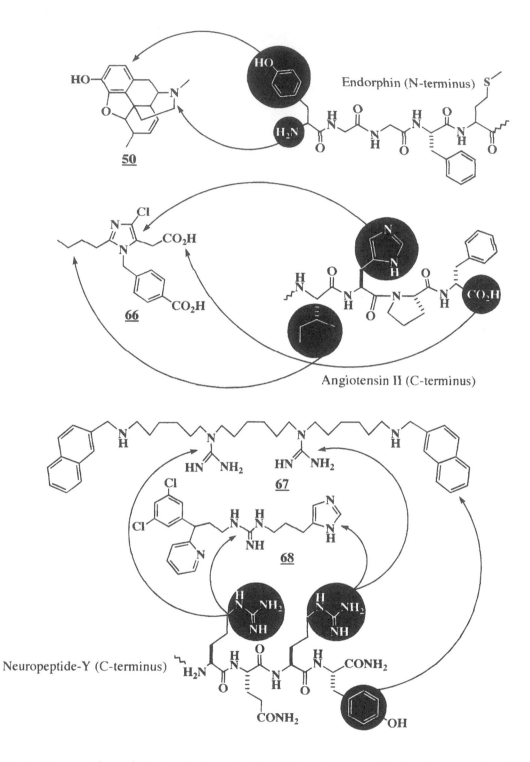

**Fig. 7** Comparative substructural elements of peptide and nonpeptide ligands: examples of μ-opioid receptor agonists, and angiotensin and neuropeptide-Y receptor antagonists.

lead to concepts of possible common pharmacophoric features. A proposal (77) describing the possible substructural relationships between the C-terminal His-Pro-Phe-OH sequence of angiotensin II and nonpeptide **66** is illustrated in Fig. 7, and such studies impacted the discovery of the potent antagonist **51** (Lorsartan). Another case where correlation between peptide and nonpeptide pharmacophore models has become apparent is that of neuropeptide-Y (NPY) versus the benextramine-based derivative **67** (78) or arpromidine-based derivative **68** (79) as illustrated in Fig. 8. In both cases, the C-terminal Arg-Gln-Arg-Tyr-NH$_2$ sequence of NPY was modeled relative to the nonpeptide structures such that the guanido functionalities were superimposed upon the corresponding basic (*i.e.*, guanido or imidazole) substructural elements of either **67** or **68**.

Two final examples of nonpeptide design illustrate the use of x-ray crystallographic information to propose correlations between screening-based leads and known peptide or peptidomimetic ligands. As shown in Fig. 8, pharmacophore models of both a cyclic hexapeptide oxytocin antagonists and conformationally-constrained, tolylpiperazine camphorsulfonamide nonpeptide antagonist (**64**) suggest the likelihood of common substructural elements important for molecular recognition at the oxytocin receptor, and led to the design of a highly potent derivative **69** (80). In yet another example, the nonpeptide HIV protease inhibitor **65** (75), which was originally identified from screening a chemical collection, was superimposed onto the C2-symmetric, peptidomimetic inhibitor **40** (50) of HIV protease to identify common key substructural elements (Fig. 8). This led to iterative structure-based design strategies to optimize the potency of this novel nonpeptide class of HIV protease inhibitors.

## VI. PEPTIDE-BASED COMBINATORIAL LIBRARIES AND MOLECULAR DIVERSITY

In recent years, synthetic combinatorial libraries have significantly impacted peptide-based drug discovery (81). A few standard approaches have been developed to generate or optimize peptide-based leads, including: (a) the multipin (MP) technique (82); (b) the teabag (TB) method (83); and (c) the split-couple-mix (SCM) strategy (84). The MP technique has been used for parallel synthesis of peptide and peptidomimetic libraries as well as to optimize organic synthetic chemistry on tips of polypropylene pins using classical solid phase procedures (82). The TB method the solid support is partitioned into polypropylene "tea bags"; hence, the resin can be immersed in a reaction mixture containing a single or multiple activated amino acids to create libraries of unique compounds or mixtures, respectively. The SCM strategy represents a mixture approach to combinatorial chemistry, and with relatively simple screening the so-called "one peptide (compound)-one bead" (84) variation of the SCM approach has been used to generate large combinatorial libraries of peptides or peptidomimetics ($>10^5$–$10^6$ compounds). Other peptide-based combinatorial library technologies have been developed (85). One non-chemical technology which has significantly impacted the generation of peptide (or protein) libraries for receptor- or enzyme-targeted drug discovery is phage display which utilizes recombinant DNA methods (86).

Using a peptide-based combinatorial library approaches, significant molecular diversity has been exploited to advance the discovery of novel antibacterial peptides (87–89), peptide vaccines (90–96), T cell epitopes (97), leukocyte antigen MHC-11 (98), bradykinin B$_2$ receptor antagonists (99), gpIIb/IIIa receptor antagonists (100,

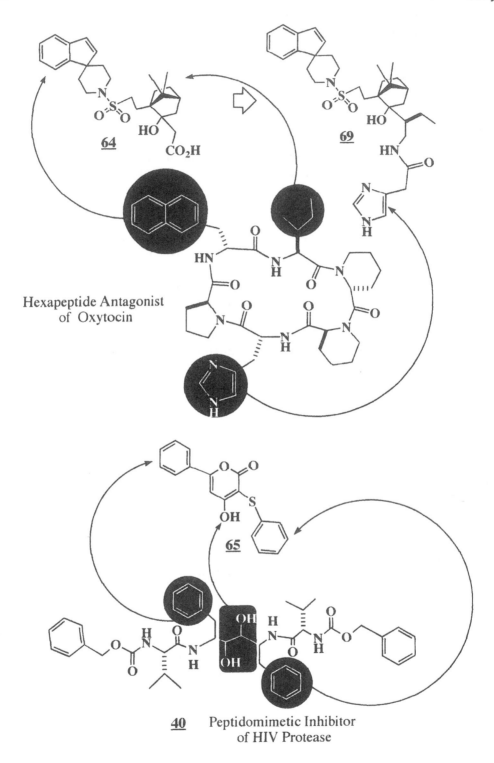

**Fig. 8** Comparative substructural elements of peptide and nonpeptide ligands: examples of oxytocin receptor antagonists and HIV protease inhibitors.

101), $\mu$-opioid receptor antagonists (102), Factor Xa inhibitors (103), MMP-3 and MMP-8 inhibitors (104), Src homology domain (SH3 and SH2) antagonists (105, 106), and Ras farnesyl protein transferase inhibitors (107).

## VII. PEPTIDOMIMETIC AND NONPEPTIDE DRUG DISCOVERY: CASE STUDIES

Peptide, peptidomimetic and nonpeptide drug discovery has emerged as a powerful approach in many areas of pharmaceutical research, including receptor agonists and antagonists, protease inhibitors and, more recently, in a rapidly developing area of signal transduction proteins. A non-comprehensive listing of such receptor, protease and signal transduction protein targets is shown in Table 2.

## A. Receptor-Targeted Drug Discovery

With respect to receptor targeted drug discovery, noteworthy success has been attained for G-protein coupled receptor agonists and antagonists as well as, more recently, cell adhesion integrin receptor antagonists (*vide infra*). Specific examples which illustrate peptide scaffold- and nonpeptide template-directed drug design strategies are shown in Fig. 9, and include: $\mu$-opioid endorphin agonist, **70** (108), thyrotropin-releasing hormone agonist, **30** (13); fibrinogen (GPIIa/IIIb) antagonists, **71** (109) and **72** (110); CCK$_A$ antagonist, **73** (111), CCK$_B$-gastrin antagonist, **34** (43); growth hormone secretagogue (GHRP), **74** (112); somatostatin agonists, **35** (41) and **75**(113), substance-P (NK$_1$) antagonist, **36** (46); neurokinin-A (NK$_2$) antagonists, **37** (47); and neurokinin-B (NK$_3$) antagonists, **76** (114). Such compounds have been advanced as the result of extensive structure-activity studies and 3-D

structural studies (*e.g.*, NMR) on a conformationally-constrained, linear or cyclic peptide lead or series. Such structure-conformation-activity studies have led to the development of a pharmacophore models to guide iterative structure-based design strategies.

A first case study is that involving the use of a glucopyranoside nonpeptide template (41, 45) for systematic functionalization to create novel peptidomimetic agonists at somatostatin (SRIF) and substance-P (NK$_1$) receptors. As illustrated in Fig. 10, a cyclic hexapeptide SRIF agonist provided a macrocyclic lead compound that was transformed to a glucopyranoside template designed to mimic a postulated $\beta$-turn about the Tyr-$D$-Trp-Lys-Thr substructure of the parent peptide ligand. The prototype peptidomimetic **35** was found to be a moderately potent SRIF-like agonist (partial) in cellular assays (41). This discovery extends previous studies on TRH (see peptidomimetic **30**, Fig. 6) which utilized a cyclohexane ring system as a nonpeptide template to functionalize with the <Glu and His sidechains as well as the C-terminal carboxamide group of the parent peptide ligand (13). Interestingly, further structure-activity studies of analogs of the SRIF-mimetic **35** it was also found that N-acetylation of the Lys-like sidechain moiety yielded a potent antagonist **77** of substance-P (NK$_1$ receptor). This indicated that slightly different functionalization of the nonpeptidic glucopyranoside template was quite compatible with NK$_1$ receptor molecular recognition. Very recently, an integrated approach of combinatorial chemistry and pharmacophore modeling has led to the discovery (113) of nonpeptide SRIF-mimetics, including receptor subtype-selective leads such as **75**. Chemical similarities to the Tyr-$D$-Trp-Lys substructure of a parent cyclic hexapeptide is apparent for these novel, nonpeptide SRIF-mimetics.

**Table 2** Examples of Receptor, Protease and Signal Transduction Protein Targets

| *G Protein-Coupled Receptor* | *Aspartic Protease* | *Tyrosine Kinase* |
|---|---|---|
| $AT_1$ (Angiotensin II) | Pepsin | Abl (SH3-SH2-Kinase) |
| $AT_2$ (Angiotensin II) | Renin | Src (SH3-SH2-Kinase) |
| $B_1$ (Bradykinin) | Cathepsin-D | Lck (SH3-SH2-Kinase) |
| $B_2$ (Bradykinin) | Cathepsin-E | Fyn (SH3-SH2-Kinase) |
| $CCK_A$ (Cholecystokinin) | HIV-1 Protease | Syk (SH2-SH2-Kinase) |
| $CCK_B$ (Gastrin; Cholecystokinin) | HIV-2 Protease | Zap-70 (SH2-SH2-Kinase) |
| $ET_A$ (Endothelin) | | |
| $ET_B$ (Endothelin) | *Serinyl Protease* | *Serine/Threonine Kinase* |
| MC1R ($\alpha$-Melanotropin) | Trypsin | cAMP-Dependent Protein Kinase |
| MC2R (Adrenocorticotropin) | Thrombin | Raf |
| NK1 (Substance P) | Chymotrypsin-A | CDK2 |
| NK2 (Neurokinin-A) | Kallikrein | |
| NK3 (Neurokinin-B) | Elastase | *Dual Specificity Kinases* |
| $\delta$-opioid (Enkephalin) | Tissue Plasminogen | Mitogen-Activated Protein |
| $\mu$-opioid (Endorphin) | Activator | Kinase |
| $\kappa$-opioid (Dynorphin) | Cathepsin-A | |
| OT (Oxytocin) | Cathepsin-G | *Tyrosine Phosphatases* |
| $V_{1A}$ (Vasopressin) | Cathepsin-R | PTP1B |
| $V_{1B}$ (Vasopressin) | | Syp (SH2-SH2-phosphatase) |
| $V_2$ (Vasopressin) | *Cysteinyl Protease* | |
| $Y_1$ (Neuropeptide-Y) | Papain | *Serine/Threonine Phosphatase* |
| $Y_2$ (Neuropeptide-Y) | Cathepsin-B | PP-1 |
| | Cathepsin-H | Calcineurin |
| *Growth Factor Receptor* | Cathepsin-L | |
| EGF (Epidermal Growth Factor) | Cathepsin-M | *Dual Specificity Phosphatases* |
| FGF (Fibroblast Growth Factor) | Cathepsia-N | VH1 |
| Insulin | Cathepsin-S | |
| Insulin-Like Growth Factor | Cathepsin-T | *Adapter Proteins* |
| NGF (Nerve Growth Factor) | Proline Endopeptidase | Grb2 (SH3-SH2-SH3) |
| PDGF (Platelet-Derived Growth | Interleukin-Converting | Crk (SH3-SH2-SH3) |
| Factor) | Enzyme | IRS-1 (PTB) |
| TGF (Transforming Growth | Apopain (CPP-32) | Shc (SH2-PTB) |
| Factor) | Picornavirus C3 Protease | |
| | Calpains | *Transferase* |
| *Cytokine Receptor* | | Famesyl Transferase |
| $IL_1$ (Interleukin-1) | *Metallo Protease* | Geranyl-Geranyl Transferase |
| $IL_2$ (Interleukin-2) | *Exopeptidase Group* | |
| $IL_3$ (Interleukin-3) | Peptidyl dipeptidase-A | *Ras Exchange Factors* |
| $IL_4$ (Interleukin-4) | Aminopeptidase-M | GAP |
| $IL_5$ (Interleukin-5) | Carboxypeptidase-A | SOS |
| $IL_6$ (Interleukin-6) | *Endopeptidase Group* | |
| $IL_7$ (Interleukin-7) | Thermolysin | *Proline Cis-Trans Isomerases* |
| $IL_8$ (Interleukin-8) | Endopeptidase 24.11 | Cyclophilin |
| | Endopeptidase 24.15 | FKBP-12 |
| *Cell Adhesion Integrin Receptor* | Stromelysin | |
| $\alpha v\beta 3$ (Fibrinogen) | Gelatinase-A | *Lipase* |
| $\alpha IIb\beta 3$ or gpIIaIIIb (Fibrinogen) | Gelatinase-B | Phospholipase-C |
| $\alpha 5\beta 1$ (Fibronectin) | Collagenase | (SH2-SH2-lipase) |
| $\alpha 4\beta 1$ (VCAM-1) | | |

A second case study is that of integrin receptor gpIIb/IIIa antagonists that are structurally-derived from the tripeptide sequence Arg-Gly-Asp which is common to gpIIb/IIIa protein ligands such as fibrinogen, vitronectin, fibronectin, von Willebrand factor, osteopontin, thrombospondin, and the collagens

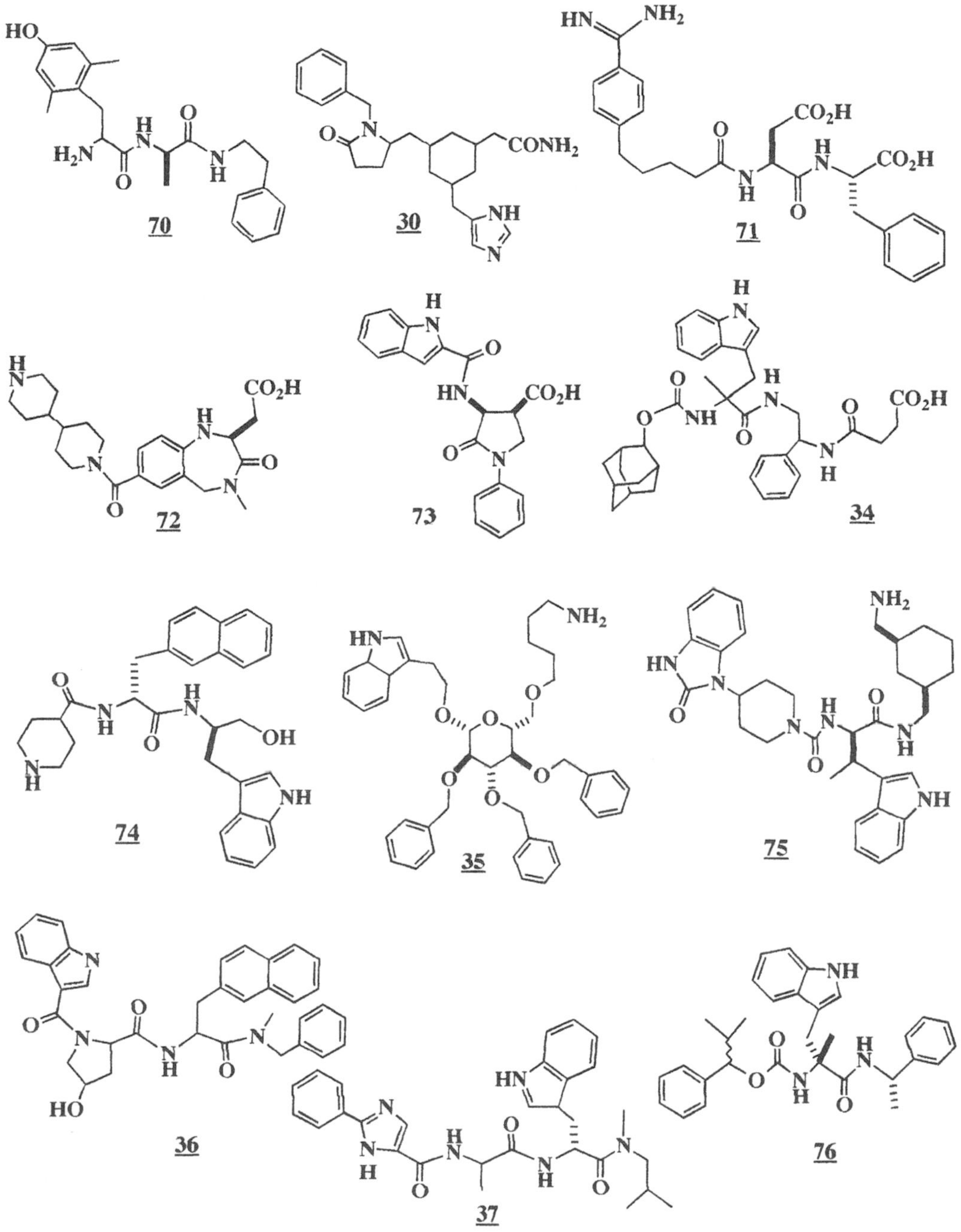

**Fig. 9** Receptor-targeted peptidomimetic/nonpeptide drug discovery.

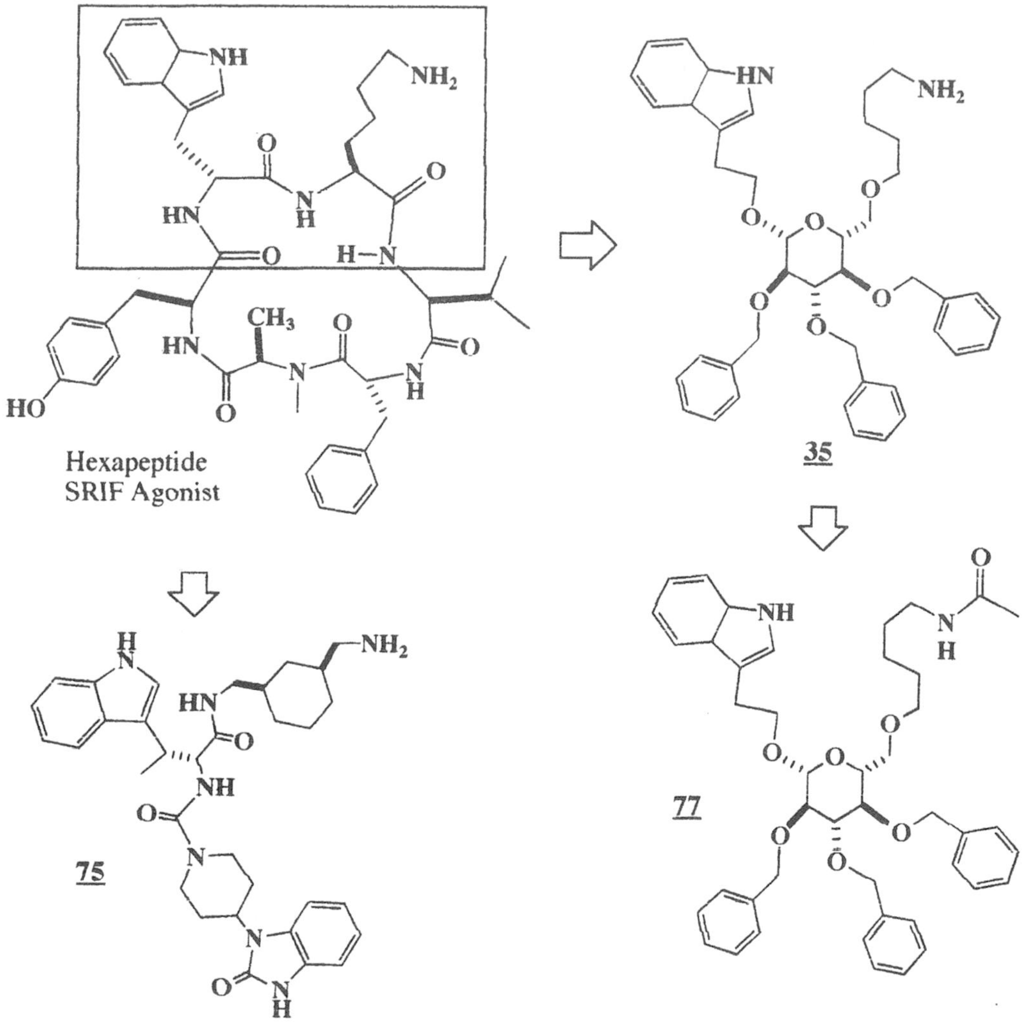

**Fig. 10** Peptidomimetic and nonpeptide somatostatin receptor agonists.

(115). As shown in Fig. 11, transformations of the linear peptide ligand Arg-Gly-Asp-Phe by both peptide scaffold (at the Arg-Gly backbone) modification and substitution of the Arg sidechain by a benzamidine moiety provided the peptidomimetic lead **71** that is active *in vivo* as an antiplatelet agent (109). On the other hand, peptidomimetics such as **72** illustrate nonpeptide template-based design strategies derived from iterative transformations of a cyclic peptide lead Ac-Cys-N-Me-Arg-Gly-Asp-Pen-NH$_2$ in

which a $\gamma$-turn about the Asp residue was implicated in a pharmacophore model for the bioactive conformation (110). Specifically, the benzodiazepinone substructure of **72** may effectively replace this predicted $\gamma$-turn conformation about the Asp, and the N-Me-Arg replacement with piperidine moiety was also compatible to high affinity receptor binding.

The above examples of peptide scaffold- or nonpeptide template-based peptidomimetic agonists or antagonists illustrate various strategies to elaborate

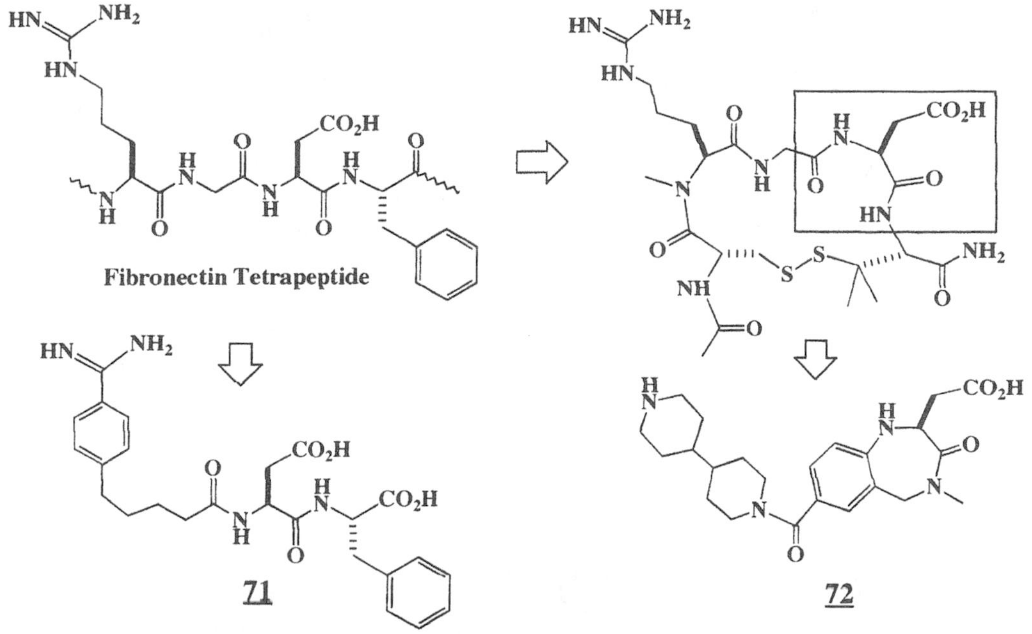

**Fig. 11** Peptidomimetic and nonpeptide gpIIb/IIIa receptor antagonists.

bioactive conformation and/or pharmacophore models of peptide ligands at their receptors.

## B. Peptidase-Targeted Drug Discovery

The development of inhibitors of angiotensin-converting enzyme (ACE) is considered a significant contribution as well as inspiration to contemporary peptidomimetic design. In retrospect, ACE inhibitor research has integrated studies on biochemical mechanisms and pharmacophore modeling approaches to develop a number of orally-bioavailable, metabolically-stable and sustained-acting drugs (116–118). Specific examples of peptidomimetics which illustrate peptide scaffold- and nonpeptide template-directed drug design strategies as applied to protease inhibitors (Fig. 12) include: renin inhibitors, **43** (53) and **45** (55); HIV protease inhibitors **78** (119) and **41** (51); angiotensin-converting enzyme inhibitors

**79** (120) and **29** (39); collagenase inhibitor, **80** (121); gelatinase inhibitor, **81** (122); stromelysin inhibitor, **82** (123); elastase inhibitor, **83** (124); thrombin inhibitors, **84** (125) and **85** (126); and interleukin-converting enzyme inhibitor, **86** (127). In general, the design of protease inhibitors has focused on both the natural substrate structure and the mechanism of substrate cleavage to provide first-generation inhibitors. Also, these initial leads are typically peptide scaffold-based to provide the possibility for β-sheet conformation which may permit "extensive" H-bonding between the backbone amide groups of the inhibitor and complementary H-bond donor or acceptor groups of the enzyme active site. Furthermore, the traditional approach to designing protease inhibitors includes the substitution of nonhydrolyzable amide surrogates (*vide supra*, Fig. 3) at the $P_1$-$P'_1$ cleavage site. Specificity to a particular protease may sometimes be extrapolated directly from the primary structure of the substrate (*e.g.*, human

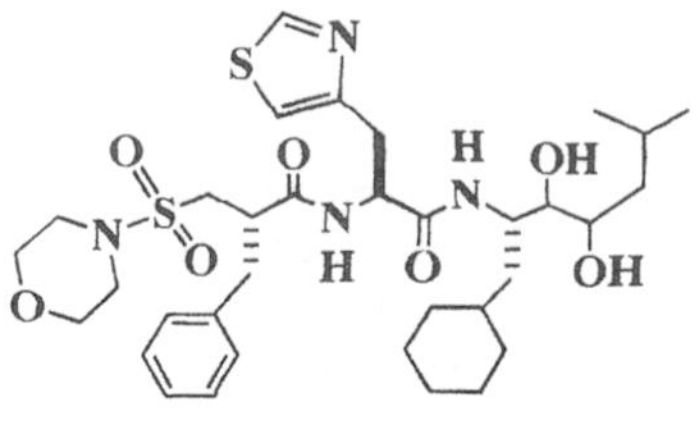
43

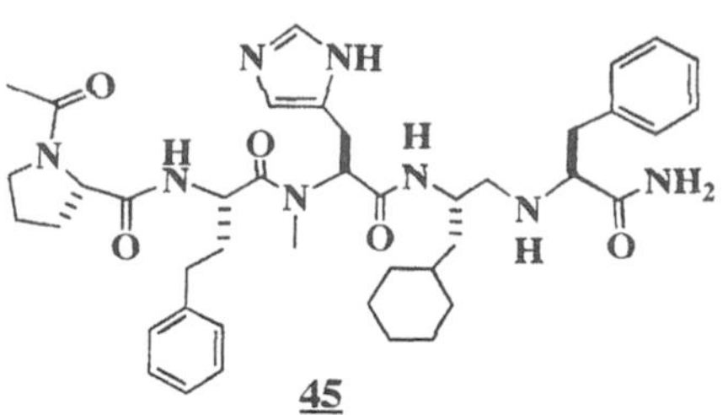
45

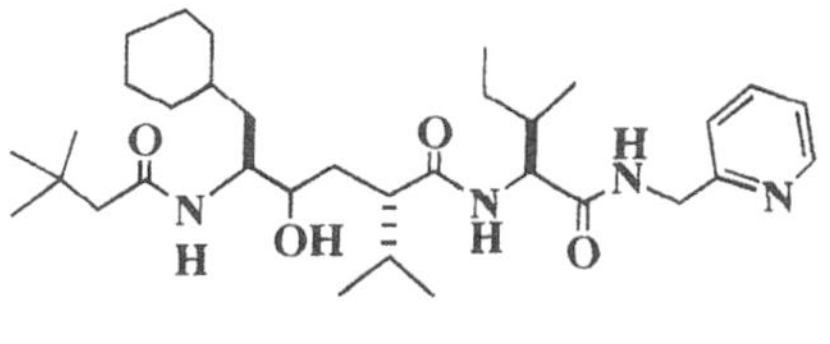
78

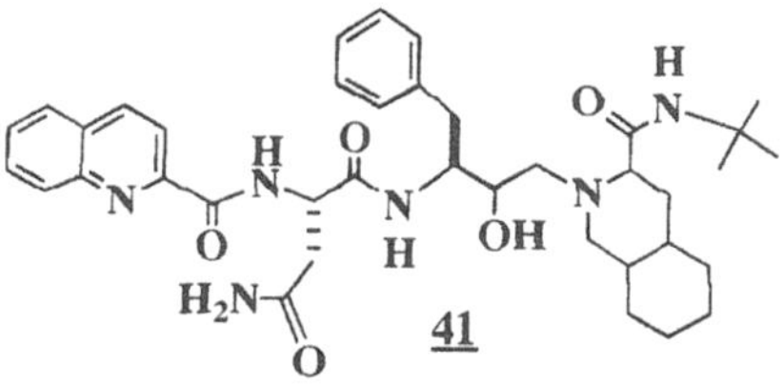
41

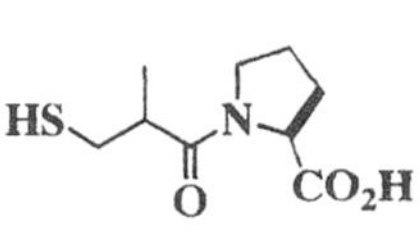
79

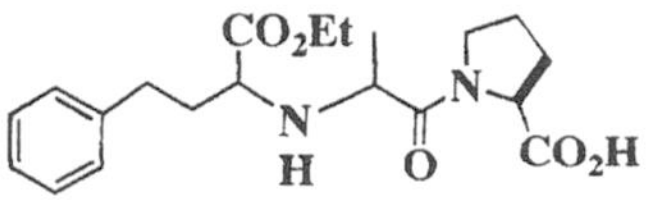
29

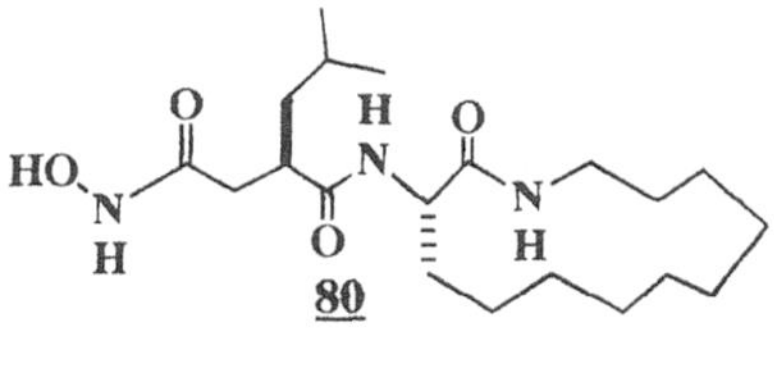
80

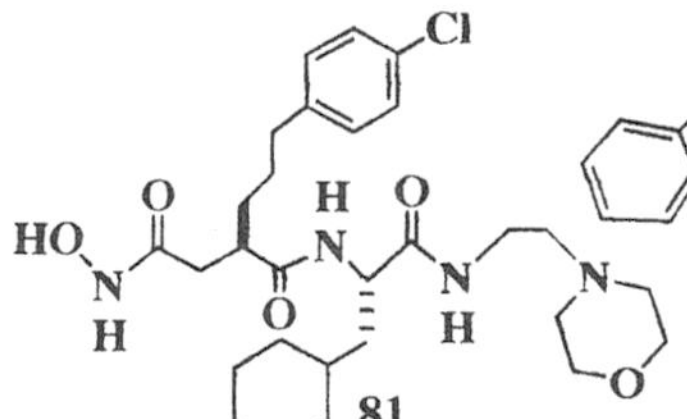
81

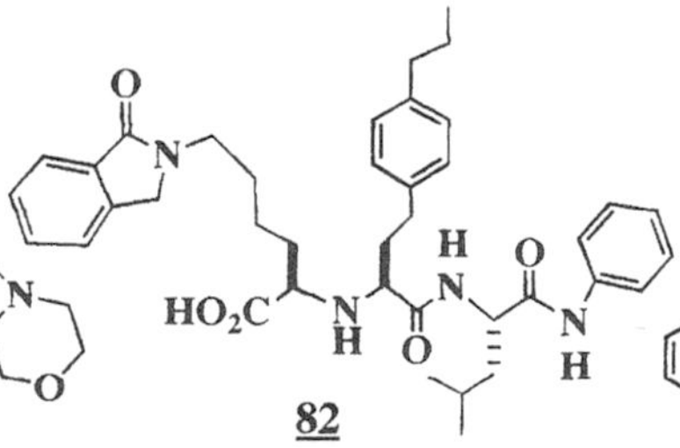
82

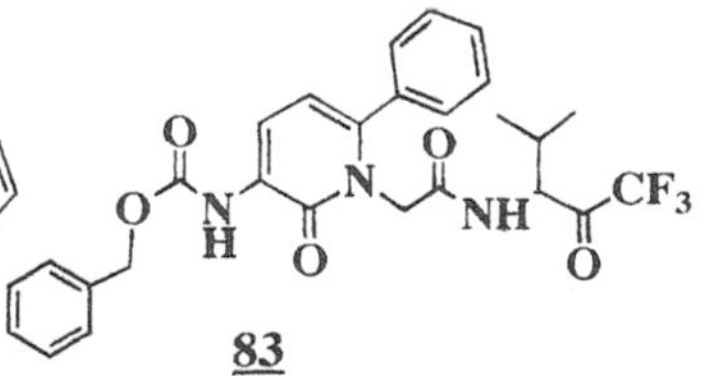
83

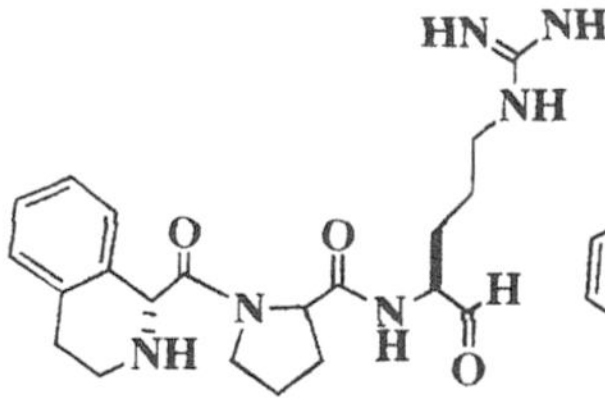
84

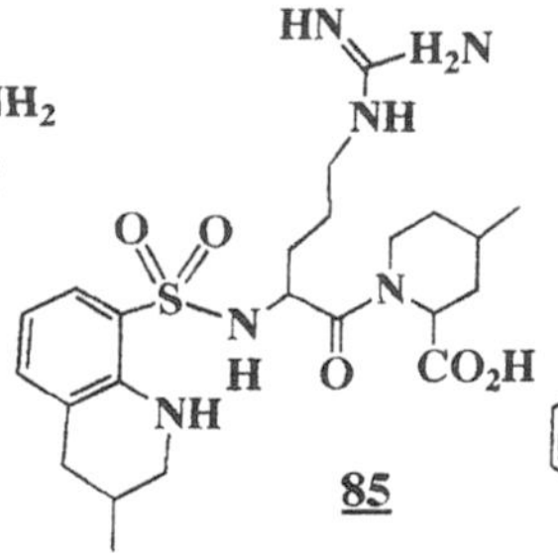
85

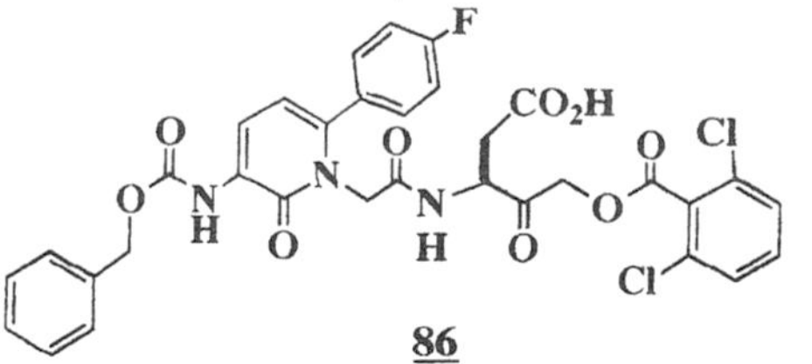
86

**Fig. 12** Protease-targeted peptidomimetic/nonpeptide drug discovery.

renin substrate specificity is conferred from the angiotensinogen N-terminal oetapeptide sequence ~His$^{6}$-Pro-Phe-His-Leu$\downarrow$Val-Ile-Mis$^{13}$~ in which $\downarrow$ refers to the cleavage site). In many cases substitution of the scissile amide (substrate) by a "transition state" bioisostere or an electrophilic ketomethylene moiety have provided tight binding "first generation" pseudopeptide inhibitor leads (128).

A first case study of protease inhibitor design which illustrates the peptide scaffold-based approach is that for HIV protease inhibitors. Albeit over the past several years HIV protease inhibitor research has become a highly advanced example of iterative structure-based drug design (*vide infra*), the first discoveries of pseudopeptide and peptidomimetic inhibitors of this aspartyl protease were not made with knowledge of the 3-D structure of the target enzyme. Specifically, the natural product pseudopeptide pepstatin **87** (Fig. 13), a typical inhibitor of the aspartyl protease family of enzymes, was determined to be weakly potent against HIV-1 protease (129). Relative to pepstatin, the central $P_1$-$P_1'$ statine (*i.e.*, Sta or Leu$\Psi$[CH(OH)]Gly) moiety was further evaluated within the context of an "optimized" N- and C-terminal amino acid sequence using a chemical library strategy (130). As shown in Fig. 13, a resultant pseudo-tetrapeptide **88** (Ac-Trp-Val-Sta-*D*-Leu-NH$_2$) was found to be a relatively potent HIV protease inhibitor. In another approach, a designed renin inhibitor (**89**; Fig. 13) was determined to be a highly potent HIV protease inhibitor (131). Further optimization studies (119) led to the discovery of the first described peptidomimetic inhibitor of HIV protease (**78**; Fig. 13). Compound **78** provided the first evidence of cellular anti-HIV activity and, therefore, supported proof-of-concept studies related to the therapeutic significance of targeting HIV-1 protease. Replacement of the peptide scaffold by a pyrrolidinone-type $\beta$-sheet mimetic in a chemically-related $P_1$-$P_1'$ Phe$\Psi$[CH(OH)CH$_2$] Phe-containing compound has been reported (132) to yield the potent peptidomimetic inhibitor **90** of the HIV-1 protease (Fig. 13). This pyrrolidinone-type lead compound has shown enhanced cellular permeability relative to its peptide backbone-type counterparts. In a third approach guided by HIV substrate-based

design, the cleavage site dipeptide Phe-Pro was substituted by the "transition state" bioisostere to provide the highly potent and selective HIV protease inhibitor **91**, a $P_1$-$P_1'$ Phe$\Psi$[CH(OH)CH$_2$N]Pro-modified heptapeptide (133). As compared to this pseudopeptide, a pioneering effort focused on peptide ligand structure-based design provide a second series of highly potent, selective and cellularly-active HIV protease inhibitors (51) as represented by the FDA-approved anti-HIV drug **41** (Saquinavir). The design of yet other HIV protease inhibitors having novel chemical structures (*e.g.*, C$_2$-symmetric scaffolding, $P_1$-$P_1'$ "transition state" bioisostere cyclization, achiral nonpeptide template replacement) has further progressed at an extraordinary pace [for reviews see (134)], and in a majority of cases such work has been strongly impacted by knowledge of the 3-D structure of the target enzyme and/or inhibitor complexes thereof.

A second case study of protease inhibitor design which well illustrates the peptide scaffold-based approach is that for thrombin inhibitor drug discovery which have led to the identification of highly potent, selective and *in vivo*-effective lead compounds. A member of the serine protease family, thrombin cleaves a number of substrates (*e.g.*, fibrinogen) and activates its platelet receptor (a G protein-coupled receptor) by proteolysis of the extracellular N-terminal domain which results in self-activation [for a review see (135)]. Initial lead inhibitors of thrombin were substrate-based, including the fibrinogen $P_3$-$P_1$, Phe-Pro-Arg sequence (136). And, recent efforts illustrate the efficacy of arylsulfonyl-modified Arg derivatives (137). The natural product cyclothreonide-A, a macrocyclic peptide containing a Pro-Arg ketoamide sequence, provided an inhibitory peptide ligand lead (138). As shown collectively in Fig. 14, compounds **84**, **85**, and **93** illustrate different strategies to advance the design of thrombin

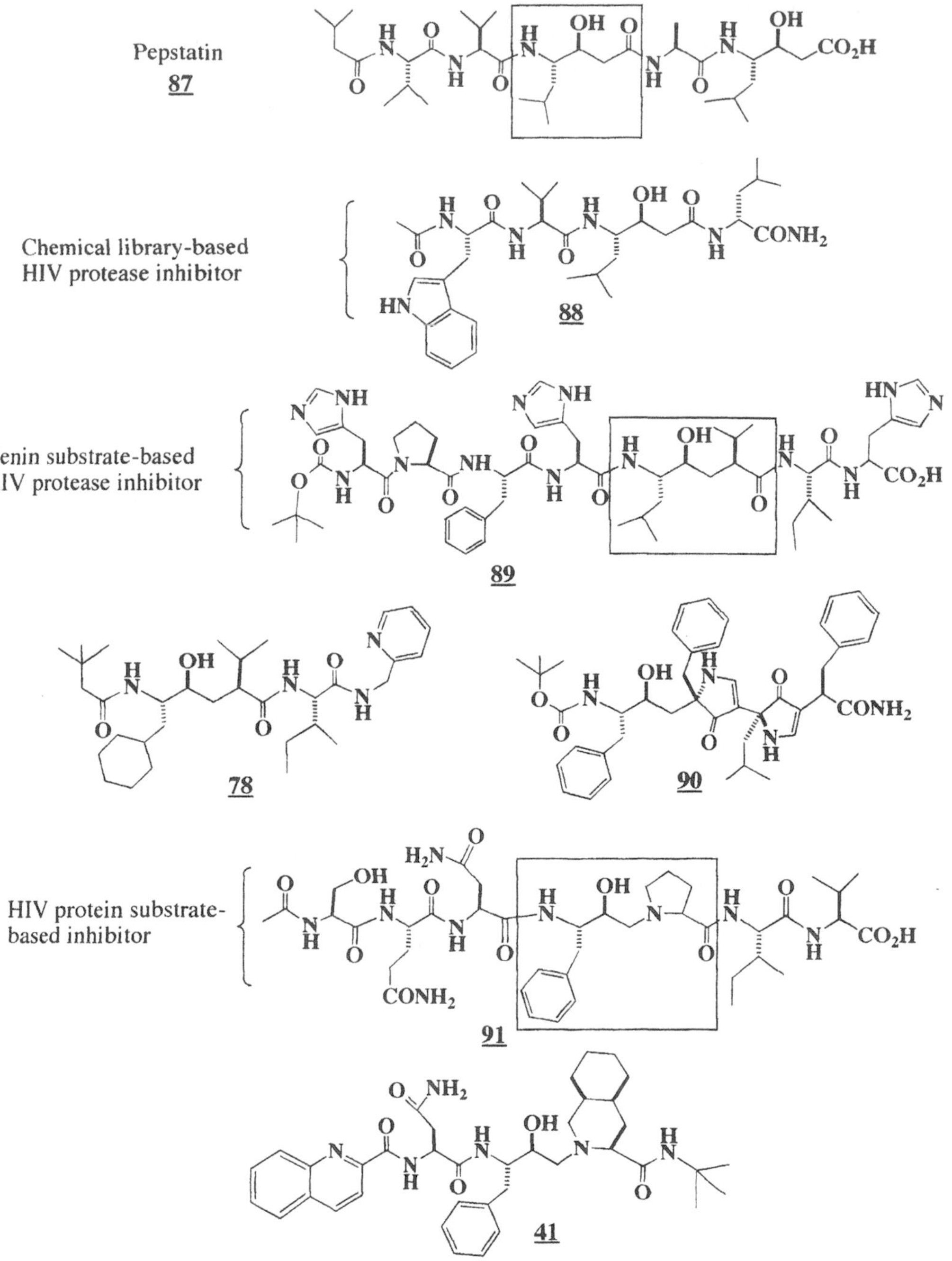

Fig. 13  Peptidomimetic and nonpeptide HIV protease inhibitors.

inhibitors. Particularly noteworthy from these early peptidomimetic lead discovery studies was the design effort (126) which lead to highly potent thrombin inhibitor 85 (Agatroban), a sulfonamide-modified Arg derivative which incorporates an unusual cyclic amino acid substituent C-terminal to the $P_1$ moiety as opposed to

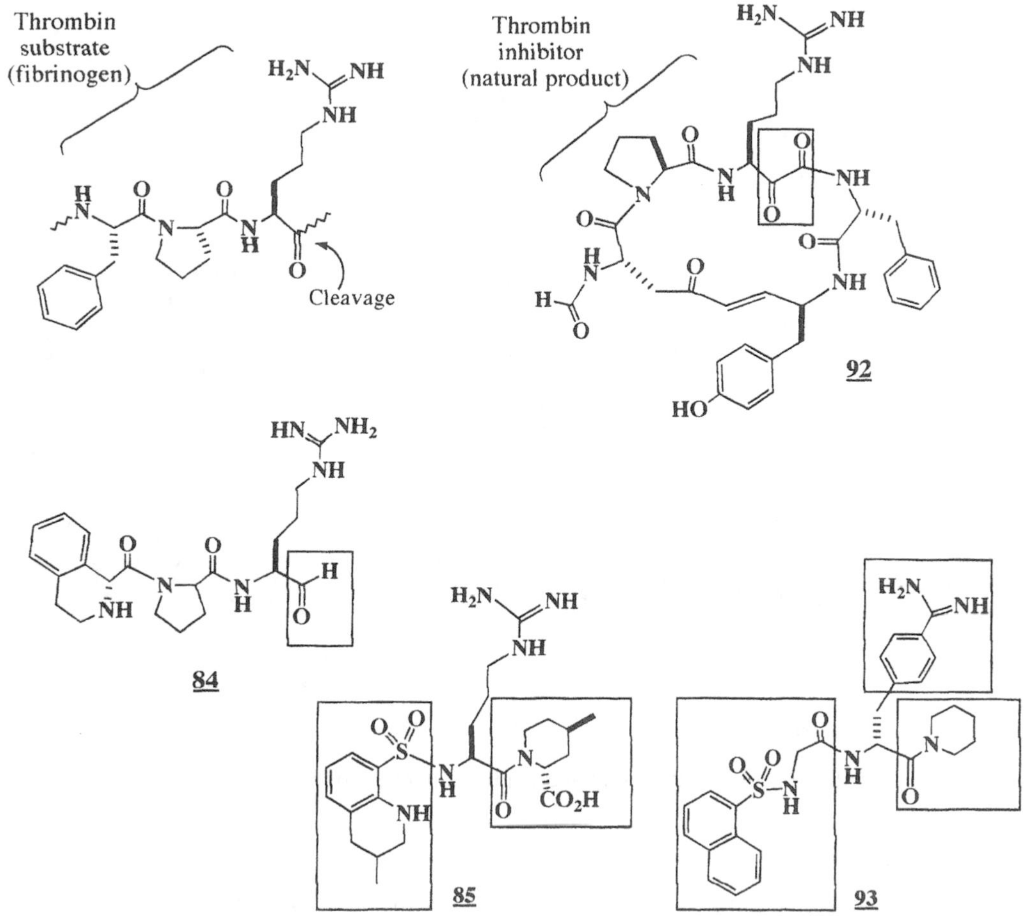

**Fig. 14** Peptidomimetic and nonpeptide thrombin inhibitors.

reactive electrophilic groups (*e.g.*, ketone, aldehyde, or boronic acid). Interestingly, replacement of the Arg sidechain moiety within a structurally similar analog **93** by a amidinobenzyl group was shown (137) to be optimal when the stereochemistry at the $P_1$ $\alpha$-carbon has a *D*-configuration suggesting different modes of binding for **84** and **85**.

## C. Signal Transduction-Targeted Drug Discovery

Beyond receptors and proteases exists the rapidly-emerging area of signal transduction proteintargeted drug discovery

research. To date, a multitude of catalytic and noncatalytic proteins have been identified which are critical components of intracellular signal transduction pathways. These signal transduction proteins provide the molecular basis for communication from extracellular "effectors" (*e.g.*, hormones, neurotransmitters, growth factors, cytokines) to stimulate cells in specific and regulated manner. Signal transduction pathways often involve protein-protein interactions, including examples of enzyme-substrate (*e.g.*, kinases, phosphatases, transferases, isomerases) as well as non-enzymatic complex formation (*e.g.*, "adapter"

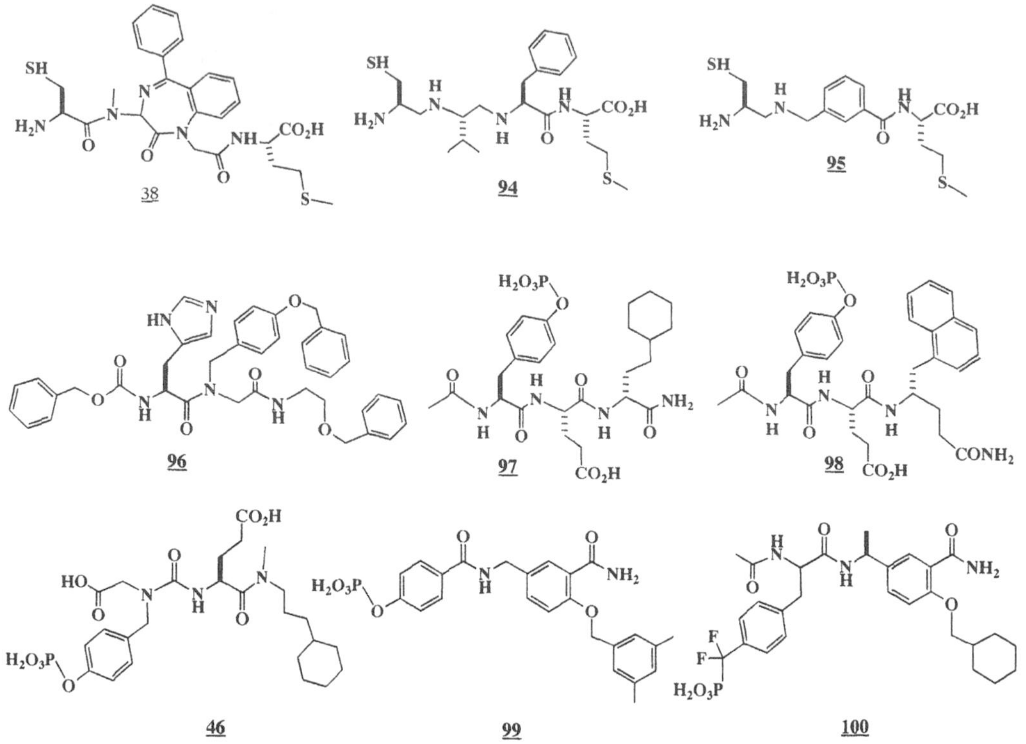

**Fig. 15** Signal transduction protein-targeted peptidomimetic/nonpeptide drug discovery.

proteins, exchange factors, transcription factors). Specific examples which illustrate peptide scaffold- and nonpeptide template-directed drug design strategies are shown in Fig. 15, and include: *Ras* farnesyl transferase inhibitors, **94** (139), **38** (48), **95** (140), and **96** (141); and Src SH2 domain antagonists, **97** (142), **98** (143), **46** (56), **99** (144), and **100** (145).

A first case study of signal transduction protein-targeted inhibitor design which illustrates both peptide scaffold- and nonpeptide template-based approaches is that for *Ras* farnesyl transferase inhibitor discovery. Such compounds show potential promise as new therapeutic agents for *Ras*-related carcinogenesis (146). Substrate sequences for farnesyl transferase have the consensus ~Cys-AA$_1$-AA$_2$-Met motif (AA refers to Val or Ile). Both substrate-based inhibi-

tors (48, 139, 140) and recently reported non-Cys containing peptide inhibitors (141) have led to potent and cellularly-active compounds. As illustrated in Fig. 16, the "collected" substrate-based inhibitor **101** was designed to covalently attach farnesyl to a peptide via a phosphinic acid linker replacement for (147), and this compound has been determined to be both potent against the target enzyme and cellularly-effective. Relative to peptide substrate structure-based design efforts, peptidomimetics incorporating $\Psi[CH_2NH]$-substitutions [*e.g.*, **94**, (139)] or a benzodiazepinone replacement of the central dipeptide moiety [*e.g.*, **38**, (48)] have yielded high affinity inhibitors. Another series of very potent *Ras* farnesyl transferase inhibitors have been designed in which the central dipeptide has been substituted by various isomeric and/or

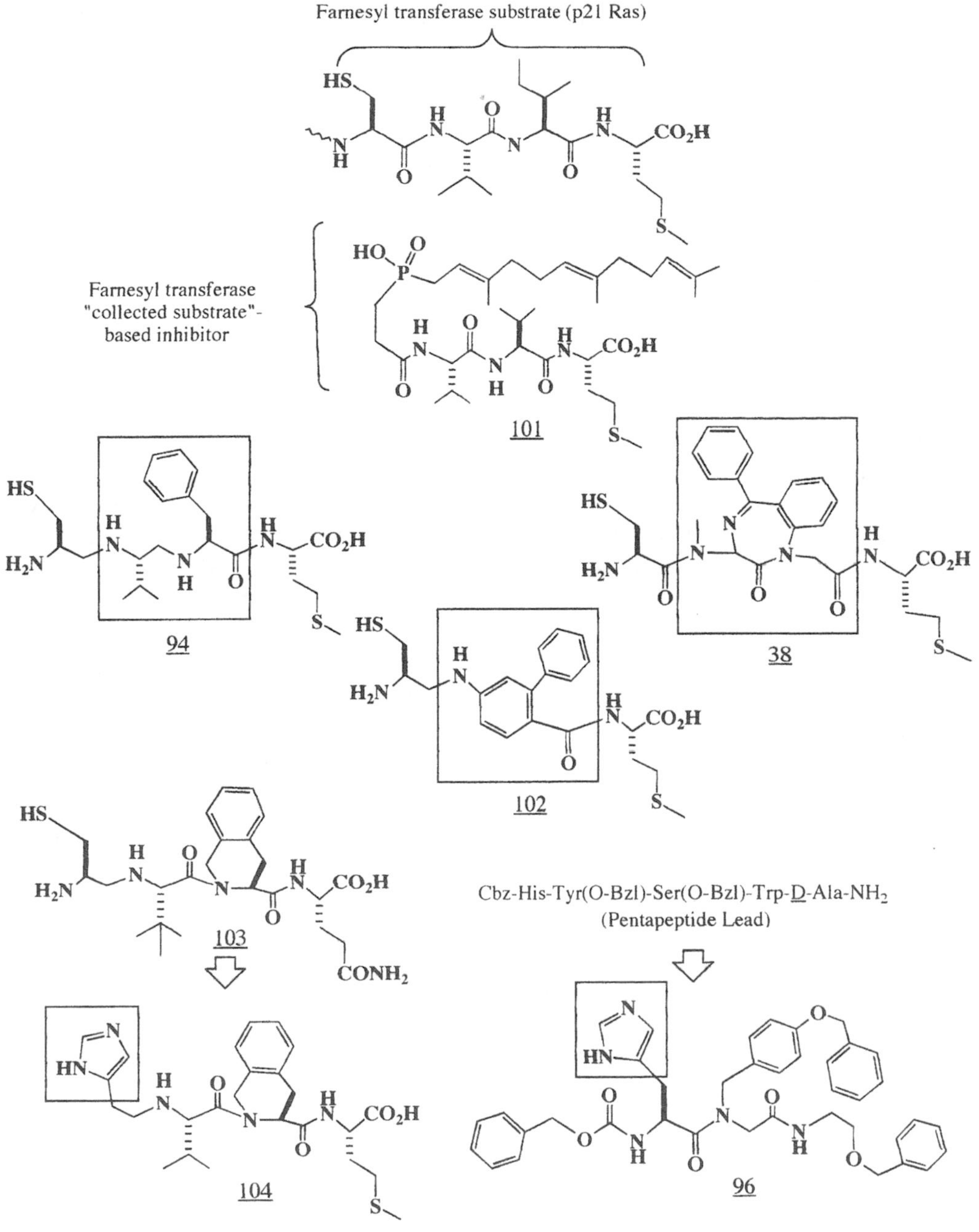

Farnesyl transferase substrate (p21 Ras)

Farnesyl transferase "collected substrate"-based inhibitor

<u>101</u>

<u>94</u>

<u>38</u>

<u>102</u>

<u>103</u>

Cbz-His-Tyr(O-Bzl)-Ser(O-Bzl)-Trp-<u>D</u>-Ala-NH₂
(Pentapeptide Lead)

<u>104</u>

<u>96</u>

**Fig. 16** Peptidomimetic and nonpeptide Ras farnesyl protein transferase inhibitors.

homologated derivatives of amino-benzoic acid [*e.g.*, **95** (140)], including a particularly effective analog biphenyl derivative **102** (148). The above studies indicated that both conformation-ally-flexible or constrained peptide scaffolds as well as nonpeptide template replacements can be used to link the Cys and Met substructures. It is also important to point out that although compounds such as __38__, __94__, and __95__ have free sulfhydryl groups (Cys) there is no evidence that they become farnesylated, and therefore the binding mode and effect on catalytic function of the target enzyme are unique relative to their peptide substrate counterparts. Recently, a novel peptide inhibitor series, as exemplified by Cbz-His-Tyr(O-Bzl)-Ser(O-Bzl)-Trp-*D*-Ala-NH$_2$, have been discovered (149). These inhibitors do not contain a Cys residue and structure-based design efforts have successfully led to a series of peptidomimetics (*e.g.*, __96__) having only one chiral center (141). Interestingly, this novel inhibitor -series has been determined to competitively inhibit farnesyl pyrophosphate binding rather than the binding of peptide substrate to the target enzyme. Another His-substituted peptidomimetic inhibitor of *Ras* farnesyl transferase has been recently reported (150) as exemplified by __104__, which was designed relative to a peptide substrate-based parent analog (__103__, Fig. 16).

A second case study of the signal transduction protein-targeted drug design which illustrates peptide scaffold- and nonpeptide template-based approaches is that for *Src* SH2 domain antagonist discovery. Such compounds show potential promise as new therapeutic agents for *Src* (or *Src* family) related carcinogenesis, osteoporosis, and immune diseases (151). The Src SH2 domain is a prototype example of a super-family of intracellular signal transduction proteins which possess SH2 domains that specifically recognize cognate phosphoproteins in a sequence-dependent manner relative to a critical phosphotyrosine (pTyr) residue (*i.e.*, ~pTyr-AA$_1$-AA$_2$-AA$_3$-AA$_4$~). Recent x-ray crystallographic studies of a several SH2 protein targets (*e.g.*, phosphopeptide complexes) has greatly impacted the opportunity for iterative structure-based drug design in this field of research (151). Relative to *Src* SH2 domain antagonist lead discovery, peptide library studies

(152) have shown that ~pTyr-Glu-Glu-Ile~ as a preferred consensus sequence. Peptide scaffold-based approaches to replace the internal dipeptide, Glu-Glu, by both flexible and rigid linkers have been explored (153) but were unsuccessful in yielding potent analogs. As shown in Fig. 17, prototype peptidomimetic __97__ (142) illustrates an approach in which stereoinversion at the second residue (P$_{+2}$ relative to the pTyr) to the *D*-configuration and sidechain substitution to hydrophobic functionalities (*e.g.*, cyclohexyl, naphthyl) that provides accessibility to the known hydrophobic binding pocket for the P$_{+3}$ Ile sidechain. Such work has led to the structure-based design of novel nonpeptide *Src* SH2 inhibitors __46__ (56), __99__ (144) and __100__ (145). Noteworthy, in this series is that both nonpeptide classes have been determined by x-ray crystallography to bind to the *Src* SH2 domain resulting in the displacement of structural water. Relative to a potent pentapeptide, it has been shown (153) that the phosphate ester of pTyr is particularly critical for molecular recognition, and that significant loss in binding occurs by its replacement with sulfate, carboxylate, nitrosyl, hydroxy, and amino functionalities.

## VIII. FUTURE PERSPECTIVES

The impact of peptidomimetic and nonpeptide research has been significant over recent years and is expected to continues to exist as a major area of drug discovery for the future. Sophisticated computer-assisted molecular modeling methods, structural biology (x-ray crystallography and NMR spectroscopy), combinatorial chemistry and high-throughput screening have become powerful technologies to accelerate the discovery of the next generation of therapeutic agents. Peptidomimetic and nonpeptide

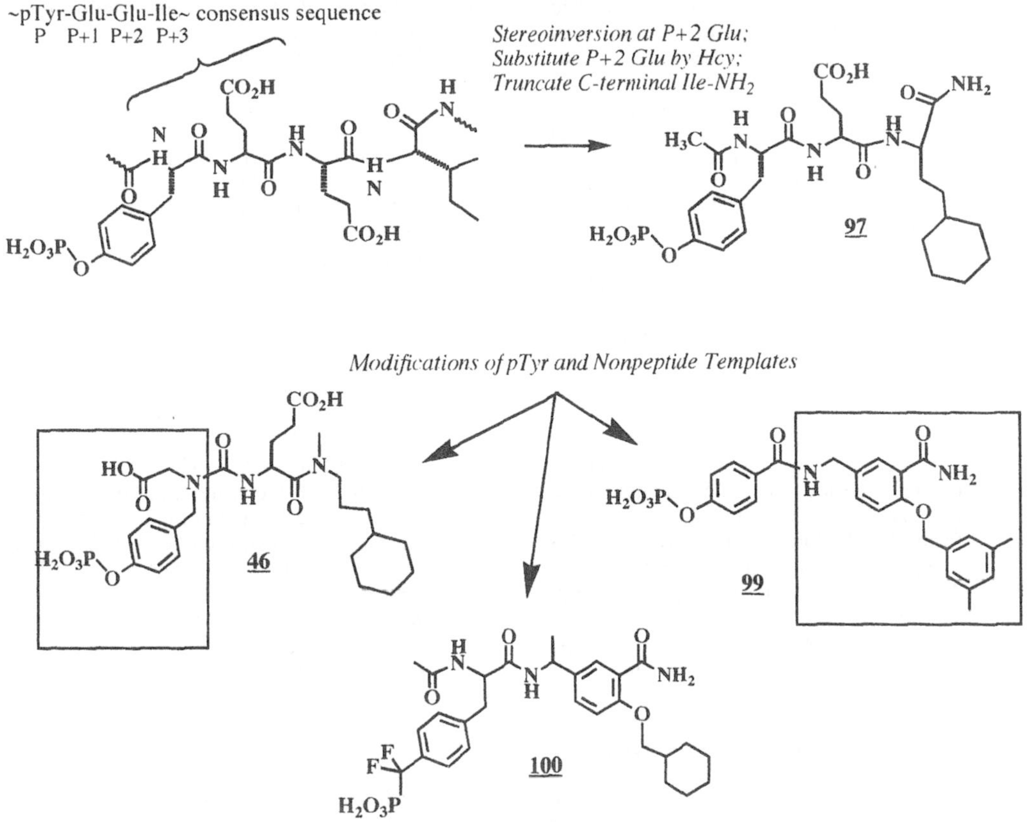

**Fig. 17** Peptidomimetic and nonpeptide Src homology-2 protein antagonists.

drug design is providing tremendous insight to our concepts of molecular recognition and biochemical mechanisms which underlie peptide (or protein) ligand interactions at receptors, proteases, and signal transduction protein targets. With particular regard to molecular diversity, synthetic peptide, peptidomimetic, or nonpeptide-type libraries will provide new lead compounds employing novel scaffolds and templates. Such work is described in this review as related to the peptidomimetics or nonpeptide drug discovery exemplifying the use of D-amino acid, Nα-alkyl or Cα-alkyl amino acid, N-substituted Gly, dipeptide surrogates (e.g., XxxΨ[Z]Yyy replacements incorporating amide bone isosteres), benzodiazepines and amino-benzoic acid modifications.

**REFERENCES**

1. (a) VJ Hruby, FA Al-Obeidi, W Kazmierski, W. Emerging approaches in the molecular design of receptor-selective peptide ligands. Biochem J. 268: 249–262, 1990; (b) VJ Hruby, BM Pettit. Conformation biological activity relationships for receptor-selective, conformationally constrained opioid peptides. In: TJ Perun, CL Propst, eds. Compuuter-Aided Drug Design, Method and Application. New York: Marcel Dekker, 1989, pp. 405–461.

2. J-L Fauchere. Elements for the rational design of peptide drugs. In: B Testa, ed. Advances in Drug Research. vol. 15. London: Academic Press, 1986, pp. 29–69.

3. DJ Ward. Peptide Pharmaceuticals— Approaches to the Design of Novel Drugs. Buckingham, England: Open University Press, 1991.

4. WF DeGrado. Design of peptides and proteins. Adv Protein Chem 39: 51–124, 1988.

5. C Toniolo. Conformationally restricted peptides through short-range cyclizations. Int J Peptide Protein Res 35, 287–300, 1990.

6. M Goodman, S Ro. Peptidomimetics for Drug Design. In: ME Wolff, ed. Medicinal Chemistry and Drug Design. vol. I. Principles of Drug Discovery. 5th ed. New York: John Wiley & Sons, 1994, pp. 803–861.

7. H Kessler, A Haupt, M Will. Design of conformationally restricted cyclopeptides for the inhibition of cholate uptake of hepatocytes. In: TJ Perun, CL Propst, eds. Computer-Aided Drug Design, Method and Application. New York: Marcel Dekker, 1989, pp. 461–484.

8. RS Struthers, AT Hagler, J Rivier. Design of peptide analogs. Theoretical simulation of conformation, energetics and dynamics. In: JA Vida, M Gordon, eds., Conformationally Directed Drug Design. ACS Symposium Series 251. Washington D.C.: American Chemical Society, 1984, pp. 23–261.

9. M Kahn. Peptide secondary mimetics: recent advances and future challenges. Syn Lett 11: 821–826, 1993.

10. RS McDowell, DR Artis. Structure-based design from flexible ligands. Ann Rep Med Chem 30: 265–174, 1995.

11. (a) RM Freidinger. Toward peptide receptor ligand drugs: progress on nonpeptides. Progress Drug Res. 40: 33–98, 1993; and (b) RM Freidinger. Non-peptide ligands for peptide receptors. Trends Pharm Sci 10: 270–274, 1989.

12. BA Morgan, JA Gainor. Approaches to the discovery of non-peptide ligands for peptide receptors and peptidases. Ann Rep Med Chem 24: 243–252, 1989.

13. GL Olson, DR Bolin, NP Bonner, M Bos, CM Cook, DC Fry, BJ Graves, M Hatada, DE Hill, M Kahn, VS Madison, VK Rusiecki, R Sarubu, J Sepinwall, GP Vincent, ME Voss. Concepts and progress in the development of peptide mimetics. J Med Chem 36: 3039–3049, 1993.

14. DP Fairlie, G Abbenante, DR March. Macrocyclic peptidomimetics: forcing peptides into bioactive conformations. Curr Med Chem 2: 65–686, 1995.

15. (a) RA Wiley, DH Rich. Peptidomimetics derived from natural products. Med Res. Rev. 13: 327–284, 1993; and (b) AS Ripka, DH Rich. Peptidomimetic design. Curr Opin Chem Biol 2: 441–452, 1998.

16. A Giannis, T Kolter. Peptide mimetics for receptor ligands: discovery, development and medicinal perspectives. Angew Chem Int Ed Engl 32: 1244–1267, 1993.

17. J Gante. Peptide mimetics: tailor-made enzyme inhibitors. Angew Chem Int Ed Engl 33: 1699–1720, 1994.

18. TK Sawyer. Peptidomimetic design and chemical approaches to peptide metabolism. In: MD Taylor, GL Amidon, eds. Peptide-Based Drug Design: Controlling Transport and Metabolism. Washington D.C.: ACS Professional Books, 1995, pp. 387–422.

19. RM Freidinger, DF Veber, DS Perlow, JR Brooks, J.R., R Saperstein. Bioactive conformation of luteinizing hormone-releasing hormone: evidence from a conformationally constrained analog. Science 210: 656–658, 1980,

20. U Nagai K Sato. Synthesis of a bicyclic dipeptde with the shape of β-turn central part. Tet Lett 26: 647–650, 1985.

21. M Feigl. 2,8-Dimethyl-4-(carboxymethyl)-6-(aminomethyl)phenothalin S-dioxide: an organic substituted for the β-turn in peptides? J Amer Chem Soc 108: 181–182, 1986.

22. M Kahn, S Wilke, B Chen, K Fujita. Nonpeptide mimeticg of β-turns: a facile oxidative intramolecular cycloaddition of an azodicarbonyl system. J Amer Chem Soc 110: 1638–1639, 1988.

23. DS Kemp, WE Stites. A convenient preparation of derivatives of 3(S)-amino-10(R)-carboxy-1,6-diazcyclodeca-2,7-dione, the dilactam of L-α, γ-diaminobutyric acid and D-glutamic acid: a β-turn template. Tet Lett 29: 5057–5060, 1988.

24. MJ Genin, RL Johnson. Design, synthesis, and conformational analysis of a novel spiro-bicyclic systern as a type II β-turn peptidomimetic. J Amer Chem Soc 114: 8778–8783,1992.

25. M Kahn, S Wilke, B Chen, K Fujita, YH Lee, ME Johnson, M.E. The design and synthesis of mimetics of β-turns. J Mol Recognition 1(2): 75–79, 1988.

26. WC Ripka, GV DeLucca, AC Bach II, RS Pottorf, R.S., JM Blaney. Protein β-turn mimetics. design, synthesis and evaluation in model cyclic peptides. Tetrahedron 49: 3593–3608, 1993.

27. WC Ripka, GV DeLucca, AC Bach II, RS Pottorf, JM Blaney. Protein β-turn mimetics: design, synthesis, and evaluation in the cyclic peptide gramacidin S. Tetrahedron 49: 3609–3628, 1993.

28. JR Callahan, KA Newlander, JL Burgess, DS Eggleston, A Nichols, A Wong, WF Huffman. The use of γ-turn mimetics to define peptide secondary structure. Tetrahedron 49: 3479–3488, 1993.

29. (a) AB Smith, TP Keenan, RC Holcomb, PA Sprengeler, MC Guzman, JL Wood,

PJ Caroll, RS Hirschmann. Design, synthesis, and crystal structure of a pyrrolinone-based peptidomimetic posessing the conformation of a $\beta$-strand: potential application to the design of novel inhibitors of proteolytic enzymes. J Amer Chem Soc 114: 10672–10674, 1992; and (b) AB Smith, SD Knight, PA Sprengler, R Hirschmann. The design and synthesis of 2,5-linked pyrrolinones: a potential non-peptide peptidomimetic scaffold. Bioorg Med Chem 4. 1021–1034, 1996.

30. GR Marshall. A hierarchial approach to peptidomimetic design. Tetrahedron 49: 3547–3558, 1993.

31. PS Farmer. Bridging the gap between peptides and nonpeptides: some perspectives in design. In: EJ Ariens, ed., Drug Design, Vol. X. New York: Academic Press, pp. 119–143, 1980.

32. GL Olson, DR Bolin, MP Bonner, M Bos, CM Cook, DC Fry, BJ Graves, M Hatada, DE Hill, M Kahn, VS Madison, VK Rusiecki, R Sarubu, J Sepinwall, GP Vincent, ME Voss. Concepts and progress in the development of peptide mimetics. J Med Chem 36: 3039–3049, 1993.

33. TK Sawyer. Peptidomimetic and non-peptide drug discovery: impact of structure-based design. In: P Veerapandian, ed., Structure-Based Drug Design: Diseases, Targets, Techniques and Developments. New York: Marcel Dekker, 1997, pp. 559–634.

34. DC Horwell (guest editor). Approaches to Nonpeptide Ligands for Peptide Receptor Sites. Bioorg Med Chem Lett (entire issue) 3: No. 5, 1993.

35. BA Morgan, JA Gainor. Approaches to the discovery of non-peptide ligands for peptide receptors and peptidases. Ann Rep Med Chem 24: 243–252, 1989.

36. RM Freidinger. Non-peptide ligands for peptide receptors. Trends Pharm Sci 10: 270–274, 1989.

37. DC Rees. Non-peptide ligands for neuropeptide receptors. Ann Rep Med Chem 28: 59–69, 1993.

38. RA Wiley, DH Rich. Peptidomimetics derived from natural products. Med Res Rev 13: 327–384, 1993.

39. AA Patchett, E Harris, EW Tristam, MJ Wyvratt, MT Wu, D Taub, ER Peterson, TJ Ikeler, J TenBroeke, LG Payne, DL Ondeyka, ED Thorsett, WJ Greenlee, NS Lohr, RD Hoffsommer, H Joshua, WV Ruyle, JW Rothrock, SD Aster, AL Maycock, FM Robinson, R Hirschmann, CS Sweet, EH Ulm, DM Grosse, TC Vassil, CA Stone. A new class of angiotensin-converting enzyme inhibitors. Nature 288: 280–283, 1980.

40. L Alig, A Edenhofer, M Muller, A Trzeciak, T Weller. Preparation of amidino derivatives of peptides and amino acids as drugs. U.S. Patent 5039805, 1991.

41. R Hirschmann, PA Sprengeler, T Kawasaki, JW Leahy, WC Shakespeare, AB Smith III. The first design and synthesis of a steroidal peptidomimetic. The potential value of peptidomimetics in elucidating the bioactive conformation of peptide ligands. J Amer Chem Soc 114: 9699–9701, 1992.

42. JVN Vara Prasad, WL Cody, X-M Cheng, AM Doherty, PL DePue, JB Dunbar Jr., KM Welch, MA Flynn, EE Reynolds, TK Sawyer. Structure-activity and modeling studies of tryptophan-modified peptidyl antagonists of endothelin. In: R Hodges, J Smith, eds., Proceedings of the Thirteenth Amer. Peptide Symp. Ae Leiden: ESCOM Science Publishers, 1995, pp. 637–639.

43. DC Horwell. Development of CCK-B antagonists. Neuropeptides 19 (suppl.): 57–64, 1991.

44. KC Nicolaou, JM Salvino, K Raynor, S Pietranico, T Reisine, R Freidinger, R Hirschmann. Design and synthesis of a peptidomimetic employing $\beta$-D-glucose for scaffolding. In: JE Rivier, GR Marshall, eds., Proceedings of the Eleventh American Peptide Symposium. Ae Leiden: Escom Science Publishers, 1990, pp. 881–884.

45. R Hirschmann, KC Nicolaou, S Pietranico, J Salvino, EM Leahy, PA Sprengeler, G Furst, AB Smith III, CD Strader, MA Cascieri, MR Candelore, C Donaldson, W Vale, L Maechler. Non-peptidal peptidomimetics with $\beta$-D-glucose scaffolding. A partial somatostatin agonist bearing close structural relationship to a potent, selective substance P antagonist. J Amer Chem Soc 114: 9217–9218, 1992.

46. W Schilling, H Bittiger, F Brugger, L Criscione, K Hauser, S Ofner, H-R Olpe, A Vassout, S Veenstra. Approaches towards the design and synthesis of nonpeptidic substance P antagonists. In: B Testa, ed., Perspectives in Medicinal Chemistry, 1993, pp. 207–220.

47. PW Smith. AB McElroy, JM Pritchard, MJ Deal, GB Ewan, RM Hagen, SJ Ireland, D Ball, I Beresford, R Sheldrick, CC Jordan, P Ward, P. Low-molecular weight neurokinin NK2 antagonists. Bioorg Medl Chem Lett 3: 931–936, 1993.

48. NE Kohl, SD Mosser, SJ deSolms, EA Giuliani, DL Pompliano, SL Graham, RL

Smith, EM Scolnick, A Oliff, JB Gibbs. Selective inhibition of ras-dependent transformation by a farnesyltransferase inhibitor. Science 260: 1934–1936, 1993.

49.  GL James, JL Goldstein, MS Brown, TE Rawson, TC Somers, RS McDowell, CW Crowley, BK Lucas, AD Levinson, JC Marsters, Jr. Benzodiazepine peptido-mimetics: potent inhibitors of Ras farnesylation in animal cells. Science 260: 1937–1942, 1993.

50.  DJ Kempf, L Codacovi, XC Wang, WE Kohlbrenner, NE Wideburg, A Saldivear, A Vasavanonda, KC March, P Bryant, HL Sham, B Green, DA Betebenner, J Erickson, DW Norbeck. Symmetry-based inhibitors of HIV protease. Structure-activity studies of acylated 2,4-diamino-1,5-diphenyl-3-hydroxypentane and 2,5-diamino-1,6-diphenylhexane-3, 4-diol. J Med Chem 36: 320–330, 1993.

51.  NA Roberts, JA Martin, D Kinchington, AV Broadhurst, JC Craig, IB Duncan, SA Galpin, BK Handa, J Kay, A Krohn, RW Lambert, JH Merrett, JS Mills, KEB, Parkes, S Redshaw, AJ Ritchie, DL Taylor, GJ Thomas, PS Machin. Rational design of peptide-based HIV proteinase inhibi-tors. Science 248: 358–361, 1990.

52.  CJ Eyermann, PK Jadhav, CN Hodge, CH Chang, JD Rodgers, PYS Lam. The role of computer-aided and structure-based design techniques in the discovery and opti-mization of cyclic urea inhibitors of HIV protease. In: A Abell, Advances in Amino Acid Mimetics and Peptidomimetics. Greenwich: JAI Press, 1997, pp. 1–40.

53.  SH Rosenberg, KP Spina, SL Condon, J Polakowski, Z Yao, P Kovar, HH Stein, J Cohen, JL Barlow, V Klinghofer, DA Egan, KA Tricaro, TJ Perun, WR Baker, HD Kleinert. Studies directed towards the design of orally active renin inhibitors. 2. Developments of the efficacious, bio-available renin inhibitor (2S)-2-benzyl-3-[[(1-methylpiperazin-4-yl)sulfonyl] propionyl]-3-thiazol-4-yl-L-alanine amide of (2S, 3R, 4S)-2-amino-1-cyclohexyl-3, 4-dihydroxy-6-methyheptane (A-72517). J Med Chem 36: 460–467, 1993.

54.  M Plummer, JM Hamby, G Hingorani, BL Batley, ST Rapundalo. Peptidomimetic inhibitors of Renin incorporating topo-graphically modified isosteres spanning the P1(→P3)-P1 sites. Bioorg Med Chem Lett 3: 2119–2124, 1993.

55.  TK Sawyer, LL Maggiora, L Liu, DJ Staples, VS Bradford, B Mao, DT Pals, BM Dunn, R⟩ Poorman, J Hinzmann, AE DeVaux, JA Affholter, CW Smith. Highly potent Ψ[CH₂NH]-modified pseudo-

peptidyl inhibitors of renin: molecular modeling and aspartyl protease selectivity. In: GR Marshall, J Rivier, eds. Proceedings of the Eleventh American Peptide Symposium. Ae Leiden: Escom Science Publishers, 1990, pp. 855–857.

56.  MS Plummer, DR Holland, A Shahripour, EA Lunney, JH Fergus, J.H., Marks, J.S., P McConnell, WT Mueller, TK Sawyer. Design, synthesis and cocrystal structure of a nonpeptide Src SH2 domain ligand. J Med Chem 140: 3719–3725, 1997.

57.  (a) P Furet, B Gay, C Garcia-Echeverria, J Rahuel, H Fretz, J Schoepfer, G Caravatti. Discovery of 3-aminobenytoxycarbonyl as a N-terminal group conferring high affinity to the minimal phosphopeptide sequence recognized by the Grb2-SH2 domain. J Med Chem 40: 3551–3556, 1997; and (b) P Furet, B Gay, G Carabatti, C Garcia-Echeverria, J Rahuel, J Schoepfer, H Fretz. Structure-based design and syn-thesis of high affinity tripeptide ligands of the Grb2-SH2 domain. J Med Chem 41: 3442–3449, 1998.

58.  L Revesz, F Bonne, U Manning, J-F Zuber. Solid phase synthesis of a biased mini-tetrapeptoid library for the discovery of monodentate ITAM mimics as ZAP-70 inhibitors. Bioorg Med Chem Lett 8: 405–408, 1998.

59.  L Revesz, E Blum, U Manning, BJ Demange, A Widmer, J-F Zuber. Non-peptide ITAM mimics as ZAP-70 antagonists. Bioorg Med Chem Lett 7: 2875–2878, 1997.

60.  (a) CB Pert, SH Synder. Opiate receptor. Demonstration in nervous tissue. Science 179: 1011–1014, 1973; and (b) JAH Lord, AA Waterfield, J Hughes, HW Kosterlitz. Endogenous opioid peptides: multiple agonists and receptors. Nature 267: 495–499, 1977.

61.  MR Snider, JW Constantine, JA Lowe, KP Longo, WS Lebel, HA Woody, SE Drozda, MC Desaia, FJ Vinick, RW Spencer, H-J Hess. A nonpeptide antagonist of the subtance P (NK1) receptor. Science 251: 435–437, 1991.

62.  AT Chin, DE McCall, PE Aldrich, PBMWM Timmermans. [³H]DUP753, a highly potent and specific radioligand for the angiotensin II-1 receptor subtype. Biochem Biophys Res Commun 172: 1195-1202, 1990.

63.  RG Smith, K Cheng, WR Schoen, S-S Pong, G Hickey, T Jacks, B Butler, WWS Chan, LYP Chaung, F Judith, J Taylor, MJ Wyvratt, MH Fisher. A nonpeptidyl growth hormone secretagogue. Science 260: 1640–1643, 1993.

64. BE Evans, KE Rittle, MG Bock, RM DiPardo, RM Freidinger, WL Whitter, GF Lundell, DF Veber, PS Anderson, RSL Chang, VJ Lotti, DJ Cerno, TB Chen, PS Kling, KA Kunkel, JP Springer, JJ Hirshfield. Methods for drug discovery: development of potent, selective and orally effective cholecystokinin antagonists. J Med Chem 31: 2235–2246, 1988.

65. MG Bock, RM DiPardo, BE Evans, KE Rittle, WL Whitter, DF Veber, PS Andeson, RM Freidinger. Benzodiazepine gastrin and brain cholecystokinin receptor ligands: L-365,260. J Med Chem 32: 13–16, 1989.

66. CJ Aquino, DR Armour, JM Berman, LS Birkemo, RAE Carr, DK Croom, M Dezube, RW Dougherty, GN Ervin, MK Grizzle, JE Head, GC Hirst, MK James, MF Johnson, LJ Miller, KL Queen, TJ Rimele, DN Smith, EE Sugg. Discovery of 1,5-benzodiazepines with peripheral cholecystokinin (CCK-A) receptor agonist activity. 1. Optimization of the agonist "trigger". J Med Chem 39: 562–569, 1996.

67. (a) AM Doherty, WC Patt, JJ Edmunds, KA Berryman, BR Reisdorph, MS Plummer, A Shahripour, C Lee, X-M Cheng, DM Walker, SJ Haleen, JA Keiser, MA Flynn, KM Welch, H Hallak, DG Taylor, EE Reynolds. Discovery of a novel series of orally active non-peptide endothelin-A (ETA). J. Med. Chem. 38: 1259–1263, 1995; and (b) AM Doherty. Design and discovery of nonpeptide endothelin antagonists. Drug Disc Today 1: 60–70, 1996.

68. B De, JJ Plattner, EN Bush, H-S Jae, G Diaz, ES Johnson, TJ Perun. LH-RH antagonists: design and synthesis of a novel series of peptidomimetics. J Med Chem 32: 2038–2041, 1989.

69. Y Yamamura, H Ogawa, T Chihara, K Kondo, T Onogawa, S Nakamura, T Mori, M Tominaga, Y Yabuuchi. OPC-21268, an orally effective nonpeptide vasopressin V1 receptor antagonist. Science 252: 572–574, 1991.

70. JJ Valentine, S Nakanishi, DL Hageman, RM Snider, RW Spencer, FJ Vinick. CP-0,030 and CP-75,898: the first non-peptide antagonists of bombesin and gastrin-releasing peptide. Bioorg Med Chemi Lett 2: 333–338, 1992.

71. JL Collins, PJ Dambek, SW Goldstein, WS Faraci. CP-99,711: a nonpeptide glucagon receptor antagonist. Bioorg Med Chem Lett 2: 915–918, 1992.A

72. D Gully, M Canton, R Boigegrain, F Jeanjean, JC Molimard, M Poncelete, C Gueudet, M Heaulem R Leris, A Brouard, D Pelaprat, C Labbe-Jullie, J Mazella, P Soubrie, JP Moffrand, W Rostene, P Kitabji, G Le Fur. Biochemical and pharmacogical profile of a potent and selective nonpeptide antagonist of the neurotensin receptor. Proc Natl Acad Sci USA 90: 65–69, 1993.

73. S Perlman, HT Schambye, RA Rivero, WJ Greenlee, SA Hjorth, TW Schwartz. Non-peptide angiotensin agonist. Functional and molecular interaction with the ATI receptor. J Biol Chem 270: 1493–1496, 1995.

74. BE Evans, JJ Leighton, KE Rittle, KF Gilbert, GF Lundell, NP Gould, DW Hobbs, RM DiPardo, DF Veber, DJ Pettitbone, BV Clineschmidt, PS Anderson, RM Freidinger. Orally-active, nonpeptide oxytocin antagonists. J Med Chem 35: 3919–3927, 1992.

75. JVN Vara Prasad, KS Para, EA Lunney, DF Ortwine, JB Dunbar, D Ferguson, PJ Tummino, D Hupe, BD Tait, JM Domagala, C Humblet, TN Bhat, B Liu, DMA Guerin, ET Baldwin, JW Erickson, TK Sawyer. A novel series of achiral, low molecular weight and potent HIV-1 protease inhibitors. J Amer Chem Soc 116: 6989–6990, 1994.

76. PS Portoghese. Selective nonpeptide ligands as probes to explore $\delta$-opioid receptor architecture. Persp Receptor Res 24: 303–312, 1996.

77. RR Wexler, WJ Greenlee, JD Irvin, MR Goldberg, K Prendergast, RD Smith, PBMWM Timmermans. Nonpeptide angiotensin II receptor antagonists: the next generation in antihypertensive therapy. J Med Chem 39: 625–656, 1996.

78. (a) MB Doughty, SS Chu, GA Misse, R Tessel. Neuropeptide Y (NPY) functional group mimetics: design, synthesis, and characterization as NPY receptor antagonists. Bioorg Med Chem Lett 2: 1497–1502, 1992; and (b) C Chaurasia, G Misse, R Tessel, MB Doughty. "Non-peptide peptidomimetic antagonists of the neuropeptide Y receptor: benextramine analogs with selectivity for the peripheral Y2 receptor. J Med Chem 37: 2242–2248, 1994.

79. S Nieps, MC Michel, S Dove, A Buschauer. Non-peptide neuropeptide Y antagonists derived from the histamine H2 agonist arpromidine: role of the guanidine group. Bioorg Med Chem Lett 5: 2065–2070, 1995.

80. (a) PD Williams, RG Ball, BV Clineschmidt, JC Culberson, JM Erb, RM Freidinger, JM Pawluczyk, DS Perlow, DJ Pettibone, DF Veber. Conformationally constrained o-tolylpiperazine camphor-

sulfonamide oxytocin antagonists. Structural modifications that provide high receptor affinity and suggest a bioactive conformation. J Med Chem 2: 971–985, 1994; and (b) PD Williams, PS Anderson, RG Ball, MG Bock, LA Carroll, S-H Lee Chiu, BV Clineschmidt, JC Culberson, JM Erb, BE Evans, SL Fitzpatrick, RM Freidinger, MJ Kaufman, GF Lundell, JS Murphy, JM Pawluczyk, DS Perlow, DJ Pettibone, SM Pitzenberger, KL Thompson, DF Veber. 1-((7,7-dimethyl-2(S)-(2(S)-amino-4-(methylsulfonyl) butyramido)bicyclo[2.2.1]heptan-1(2)-yl-methyl)sulfonyl)-4-(2-methylphenyl)piperzine (L-368, 898): an orally bioavailable, non-peptide oxytocin antagonist with potential utility for managing preterm labor. J Med Chem 37: 565–571, 1994.

81. FA Al-Obeidi, VJ Hruby, TK Sawyer. Peptide and peptidomimetic libraries: molecular diversity and drug design. Molec Biotech 9: 205–223, 1998.

82. (a) HM Geysen, RH Meloen, SJ Barteling. Use of peptide synthesis to probe viral antigens for epitopes to a resolution of a single amino acid. Proc Natl Acad Sci USA 81: 3998–4002, 1994; and (b) AM Bray, DS Chiefari, RM Valerio, NJ Maeji. Rapid optimization of organic reactions on solid phase using the multipin approach: synthesis of 4-aminoproline analogues by reductive amination. Tetrahedron Lett 36: 5081–5084, 1995.

83. (a) RA Houghten, C Pinilla, SE Blondelle, JR Appel, CT Dooley, JH Cuervo. Generation and use of synthetic peptide combinatorial libraries for basic research and drug discovery. Nature 354: 84–86, 1991; and (b) RA Houghten. General method for the rapid solid-phase synthesis of large numbers of peptides: Specificity of antigen-antibody interaction at the level of individual amino acids. Proc Natl Acad Sci USA 82: 5131–5135, 1985.

84. (a) A Furka, F Sebestyen, M Asgedom, G Dibo. General method for rapid synthesis of multicomponent peptide mixtures. Int J Pep Prot Res 37: 487–493, 1991; (b) A Furka, F Sebestyen, M Asgedom, G Dibo. Cornucopia of peptides by synthesis. Abstract of the Fourteenth International Congress of Biochemistry, Prague, Poster-47, 1988; and (c) KS Lam, SE Salmon, EM Hersh, VJ Hruby, WM Kazmierski, RJ Knapp. A new type of synthetic library for identifying ligand binding. Nature 354: 82–84, 1991.

85. JN Ableson, Ed. Combinatorial Chemistry, Vol. 267, Methods in Enzymology. San Diego: Academic Press, 1996.

86. D McGregor. Selection of proteins and peptides from libraries displayed on filamentous bacteriophage. Molec Biotech 6: 155–462, 1996.

87. (a) JM Ostresh, SE Blondelle, B Domer, RA Houghten. Generation and use of nonsupport-bound peptide and peptidomimetic combinatorial libraries. In: JN Ableson, ed., Methods in Enzymology, Vol 267. San Diego: Academic Press, 1996, pp. 220–234; and (b) JM Ostresh, B Domer, SE Blondelle, RA Houghten. Comb Chem, 225–140, 1997.

88. SE Blondelle, RA Houghten. Novel antimicrobial compounds identified usig synthetic combinatorial library technology. Trends in Biotechnology 14: 60–65, 1996.

89. SE Blondelle, E Perea-Paya, RA Houghten. Synthetic combinatorial libraries: novel discovery strategy for identification of antimicrobial agents. Antimicrobial Agents and Chemotherapy 40: 1067–1071, 1996.

90. G Georgiou, C Stathopoulos, PS Daugherty, AR Nayak, BL Iverson, R Curtiss. Display of :heterologous proteins on the surface of microorganisms from the screening of combinatorial libraries to live recombinant vaccines. Nature Biotechnology 15: 29–34, 1997.

91. JW Slootstra, WC Puijk, GJ Ligtvoet, JPM Langeveld, RH Meloen. Structural aspects of antibody-antigen interaction revealed through small random peptide libraries. Molecular Diversity 1: 87–96, 1996.

92. ZJ Yao, MC Chan, MCC Kao, MCM Chang. Linear epitopes of sperm whale myoglobin identified by polyclonal antibody screening of random peptide library. Int J Pept Protein Res 48: 477–485, 1996.

93. B Triantufyllou, G Tribbick, NJ Maeji, HM Geysen. Use of the multipin peptide synthesis technique for the generation of peptide antisera. Cell Biophys 21: 33–52, 1993.

94. CP Homes, CL Adams, LM Kochersperger, RB Mortensen, LA Aldwin. The use of light directed combinatorial peptide synthesis in epitope mapping. Biopolymers (Peptide Sci.) 37: 199–211, 1995.

95. MW Steward, CM Stanley, OE Obeidi. A mimeotope from a solid phase peptide library induces a measles virus neutralizing and protective antibody respone. J Virol 69: 7668–7673, 1995.

96. Y Hirabayashi, H Fukuda, J Kimura, M Miyamoto, K Yasui. Identification of pep-

tides mimicking the antigenecity and immunogenecity of conformational epitopes Japanese encephalitis virus protein using synthetic peptide libraries. J Virol Methods 61: 23–36, 1996.

97. BR Gundlach, K-H Wiesmuller, T Junt, S Kienle, G Jung, P Waldlen. Determination of T cell epitopes with random peptide libraries. J Immunol Methods 192: 149–155, 1996.

98. C Pinilla, JR Appel, P Blane, RA Houghten. Rapid identification of high affinity peptide ligands using positional scanning synthetic peptide combinatorial libraries. Biochemistry 13: 901–905, 1992.

99. S Chakraverty, BJ Mavunkel, R Andy, DJ Kyle. Nonpeptidic bradykinin antagonists from a structurally directed non-peptide combinatorial library. In: PTP Kaumaya, RS Hodges, eds., Proceedings of the Fourteenth American Peptide Symposium. Kingswinford: Mayflower Scientific, 1996, pp. 717–718.

100. WJ Hoekstra, BE Maryanoff, P Andrade-Gordon, JH Cohen, MJ Costanzo, BP Damiano, BJ Haertlein, BD Harris, JA Kauffman, PM Keane, DF McComsey, FJ Villani, Jr., SC Yabut. Solid-phase parallel synthesis applied to lead optimization: discover potent analogues of the GPIIb/IIIa antagonist RWJ-50042. Bioorg Med Chem Lett 6: 2372–2376, 1996.

101. T Harada, J Katada, A Tachiki, T Asari, K Iijima, I Uno, I Ojima, Y Hayashi. Development of the new potent non-peptide GPIIb/IIIa agonist NSL-95301 by utilizing combinatorial technique. Bioorg Med Chem Lett 7: 209–212, 1996.

102. RN Zuckermann, EJ Martin, DC Spellmeyer, GB Stauber, KR Shoemaker, JM Kerr, GM Figliozzi, DA Goff, MA Siani, RJ Simon, SC Banville, EG Brown, L Wang, LS Richter, WH Moos. Discovery of nanomolar ligands for 7-transmembrance G protein-coupled receptors from a diverse N-(substituted)glycine peptoid library. J Med Chem 37: 2678–2685, 1994.

103. JA Ostrem, FA Al-Obeidi, P Safar, A Safarova, S Stringer, M Patek, NT Cross, J Spoonamore, JC LoCascio, P Kasireddy, D Thorpe, N Sepetov, M Lebl, P Wildgoose, P Strop. Discovery of a novel, potent, and specific family of Factor Xa inhibitors via combinatorial chemistry. Biochemistry 37: 1053–1059, 1998.

104. A Rockwell, M Melden, RA Cepeland, K Hardman, CP Decicco, WF DeGrado. Complementarity of combinatorial chemistry and structure-based ligand design: applications to the discovery of novel

inhibitors of matrix metalloproteinases. J Amer Chem Soc 118: 10337–10338, 1996.

105. JT Nguyen, CW Turck, FE Cohen, RN Zuckermann, WA Lim. Exploiting the basis of proline recognition by SH3 and WW domains: design of N-substituted inhibitors. Science 282: 2088–2092, 1999.

106. K Muller, FO Gombert, U Manning, F Grossmuller, P Graff, H Zaege, JF Zuber, F Freuler, C Tschopp, G Baumann. Rapid identification of phoshopeptide ligands for SH2 domains. J Biol Chem 271: 16500–16505, 1996.

107. A Wallace, KS Koblan, K Hamilton, DJ Marquis-Omer, PJ Miller, SD Mosser, CA Omer, MD Schaber, R Cortese, A Oliff, JB Gibbs, A Pessi. Selection of potent inhibitors of farnesyl protein from a synthetic tetrapeptide combinatorial library. J Biol Chem 271: 31306–31311, 1996.

108. BS Pitzele, RW Hamilton, KD Kudla, S Taymbalov, A Stapefield, MA Savage, M Clare, DL Hammond, DW Hansen. Enkephalin analogs as systemically active antinociceptive agents: O- and N-alkylated derivatives of the dipeptide amide L-2,6-dimethyltyrosyl-N-(3-phenylpropyl)-D-alaninamide. J Med Chem 37: 888–896, 1994.

109. JA Zablocki, M Miyano, B Garland, D Pireh, L Schretzman, SN Rao, RJ Lindmark, S Panzer-Knodle, N Nicholson, B Taite, A Salyers, L King, L Feigen. Potent in vitro and in vivo inhibitors of platelet aggregation based upon the Arg-Gly-Asp-Phe sequence of fibrinogen. A proposal on the nature of the binding interaction between the Arg-guanidine of RGDX mimetics and the platelet GPIIb-IIIa receptor. J Med Chem 36: 1811–1819, 1993.

110. J Samanen, FE Ali, L Barton, W Bondinell, J Burgess, J Callahan, R Calvo, W Chen, L Chen, K Erhard, R Heyes, S-M Hwang, D Jakas, R Keenan, T Ku, C Kwon, C-P Lee, W Miller, K Newlander, A Nichols, C Peishoff, G Rhodes, S Ross, A Shu, R Simpson, D Takata, TO Yellin, I Uzsinskas, J Venslavasky, A Wong, C-K Yuan, W Huffman. GPIIb/IIIa antagonist with long oral duration designed from cyclic peptides. In: PTP Kaumaya, RS Hodges, eds. Proceedings of the Fourtenth American Peptide Symposium. Kingswinford: Mayflower Scientific Ltd., 1996, pp. 679–681.

111. DL Flynn, Cl Villamil, DP Becker, GW Gullikson, C Moummi, D-C Yang. 1,3,4-Trisubstituted pyrrolidinones as scaffolds for construction of pepti-

domimetic cholecystokinin antagonists. Bioorg Med Chem Lett 2: 1251–1256, 1992.

112. RS McDowell, KS Elias, MS Stanley, DJ Burdick, CP Burnier, KS Chan, WJ Fairbrother, RG Hammonds, GS Ingle, NE Jacobsen, DL Mortensen, TE Rawson, WB Won, RG Clark, TC Somers. Growth hormone secretagogues: characterization, efficacy, and minimal bioactive conformation. Proc Nad Acad Sci USA 92: 1165–1169, 1995.

113. PW Smith. AB McElroy, JM Pritchard, MJ Deal, GB Ewan, RM Hagen, SJ Ireland, D Ball, I Beresford, R Sheldrick, CC Jordan, P Ward. Low-molecular-weight neurokinin NK2 antagonists. Bioorg Med Chem Lett 3: 931–935, 1993.

114. SP Rohrer, ET Birzin, RT Mosley, SC Berk, SM Hutchins, D-M Shen, Y Xiong, EC Hayes, RM Parmar, F Foor, SW Mitra, SJ Degrado, M Shu, JM Klopp, S-J Cai, A Blake, WWS Chan, A Paternak, L Yang, AA Parchett. RG Smith, KT Chapman, JM Schaeffer. Rapid identification of subtype-selective agonists of the somatostatin receptor through combinatorial chemistry. Science 282: 737–740, 1998.

115. JA Zablocki, NS Nicholson, LP Feigin. Fibrinogen receptor antagonists. Expert Opin Invest Drugs 3: 437–448, 1994.

116. EW Petrillo Jr., MA. Angiotensin-converting enzyme inhibitors: medicinal chemistry and biological actions. Med Res Rev 2: 1–41, 1982.

117. AA Patchett, EH Cordes. The design and properties of N-carboxyalkyldipeptide inhibitors of angiotensin-converting enzyme. Adv Enzymol 57: 1–84, 1985.

118. MRW Ehlers, JF Riordan. Angiotensin-converting enzyme: new concepts concerning its biological role. Biochemistry 28: 5311–95318, 1989.

119. (a) TJ McQuade, AG Tomasselli, L Liu, V Karacostas, B Moss, TK Sawyer, RL Heinrikson, WG Tarpley. A synthetic human immunodeficiency virus protease inhibitor with potent antiviral activity arrests HIV-like particle maturation. Science 247: 454–456, 1990; and (b) TK Sawyer, JF Fisher, JB Hester, CW Smith, AG Tomasselli, WG Tarpley, PS Burton, JO Hui, TJ McQuade, RA Conradi, VS Bradford, L Liu, JH Kinner, J Tustin, DL Alexander, AW Harrison, DE Emmert, DJ Staples, LL Maggiora, Y-Z Zhang, RA Poorman, BM Dunn, C Rao, PE Scarbourogh, TT Lowther, C Craik, D DeCamp, J Moon, WJ Howe, RL Heinrikson. Peptidomimetic inhibitors of human immunodeficiency virus protease: design, enzyme binding ,and selectiva ant-

iviral efficacy and in vitro intestinal transport studies. Bioorg Med Chem Lett 3: 819–1993. 824

120. MA Ondetti, B Rubin, DW Cushman. Design of specific inhibitors of angiotensin-converting enzyme: a new class of orally active antihypertensive agents. Science 196: 441–444, 1977.

121. J Bird, GP Harper, I Hughes, DJ Hunter, EH Karran, RE Markwell, AJ Miles-Williams, SS Rahman, RW Ward. Inhibitors of human collagenase: dipeptide mimetics with lactam and azalactam moicties at the P2'/P3' position. Bioorg Med Chem Lett 5: 2593–2598, 1995.

122. JR Porter, NRA Beeley, BA Boyce, B Mason, A Millican, K Miller, J Leonard, JR Morphy, JP O'Connell. Potent and selective inhibitors of gelatinase-A. 1. Hydroxamic acid derivatives. Bioorg Med Chem Lett 4: 2741–2746, 1994.

123. KT Chapman, PL Durette, CG Caldwell, KM Sperow, LM Niedzwiecki, RK Harrison, C Saphos, AJ Christen, JM Olszewski, VL Moore, M MacCoss, NWK Hagmann. Orally active inhibitors of stromelysin-1 (MMP-3). Bioorg Med Chem Lett 6: 803–806, 1996.

124. PD Edwards, DW Andiskik, AM Strimpler, B Gomes, PA Tuthill. Non-peptidic inhibitors of human neutraphil elastase. 7. Design, synthesis and in vitro activity of a series of pyridopyrimidine ttifluoromethyl ketones. J Med Chem 39: 1112–1124, 1996.

125. RT Shuman, RB Rothenberger, CS Campbell, GF Smith, DS Gifford-Moore, PD Gesellechen. Highly selective tripeptide thrombin inhibitors. J Med Chem 36: 314–319, 1993.

126. S Okamoto, K. Kinjo, A. Hijikata, R Kikumoto, Y Tamao, K. Ohkubo, Tonomura, S. Thrombin inhibitors. 1. Ester derivatives of Nα-(arylsulfonyl)-L-arginine. J Med Chem 23: 827–830, 1980.

127. RE Dolle, CP Prouty, CVC Prasad, E Cook, A Saha, TM Ross, JM Salvino, CT Helaszek, MA Ator. First examples of peptidomimetic inhibitors of interleukin-1β converting enzyme. J Med Chem 39: 2438–2440, 1996.

128. (a) DH Rich. Effect of hydrophobic collapse on enzyme-inhibitor interactions. Implications for the design of peptidomimetics. Perspect. Med Chem pp. 15–25, 1993; (b) G Lawton, PM Paciorek, JF Waterfall. The design and biological profile of ACE inhibitors. Adv Drug Res 23: 12–220, 1992; and (c) MJ Wyvratt, AA Patchett. Recent developments in the

design of angiotensin-converting enzyme inhibitors. Med Res Rev 5: 483–531, 1985.

129. HG Krausslich, H Schneider, G Zybarth, CA Carter, E Wimmer. Processing of in vitro-synthesized gag precursor proteins of human immunodeficiency virus (HIV) type-1 by HIV proteinase generated in Escherichia coli. J Virol 2: 4394–4397, 1988.

130. RA Owens, PD Gesellchen, BJ Houchins, RD DiMarchi. The rapid identification of HIV protease inhibitors through the Znthesis and screening of defined peptide mixtures. Biochem Biophys Res Comm 181: 401–408, 1991.

131. AD Richards, R Roberts, BM Dunn, MC Graves, J Kay. Effective blocking of HIV-1 proteinase activity by characteristic inhibitors of aspartic proteinases. FEBS Lett 247: 113–117, 1989.

132. AB Smith, R Hirschmann, A Pasternak, MC Guzman, A Yokoyama, PA Sprengeler, PL Darke, EA Emini, WA Schleif. Pyrrolinone-based HIV protease inhibitors. Design, synthesis, and antiviral activity: Evidence for improved transport. J Amer Chem Soc 117: 11113–11123, 1995.

133. DH Rich, J Green, MV Toth, GR Marshall, SBH Kent. Hydroxyethylamine analogs of the p17/p24 substrate cleavage site are tight-binding inhibitors of HIV protease. J Med Chem 33: 1288–1295, 1990.

134. D Kempf, HL Sham. HIV protease inhibitors. Curr Pharm Design 2: 225–246, 1996.

135. J Das, SD Kimball. Thrombin active site inhibitors. Bioorg Med Chem 3: 999–1007, 1995.

136. S Bajusz, E Szell, D Bagdy, E Barbas, G Horvath, M Dioszegi, Z Fittler, G Szabo, A Juhasz, E Tomori, G Szilagyi. Highly active and selective anticoagulants: D-Phe-Pro-Arg-H, a free tripeptide aldehyde prone to spontaneous inactivation, and its stable N-methyl derivative, D-MePhe-Pro-Arg-H. J Med Chem 33: 1729–1735, 1990.

137. DR St. Laurent, N Balasubramanian, WT Han, A Trehan, ME Federici, NA Meanwell, JJ Wright, SM Seiler. Active site-directed thrombin inhibitors. II. Studies related to arginine/guanidine bioisosteres. Bioorg Med Chem 3: 1145–1156, 1995.

138. J Sturzebecher, F Markwardt, B Voigt, G Wagner, P Walsmann. Cyclic amides of Nα-arylsulfonylaminoacylated 4-amidinophenylalanine: Tight binding inhibitors of thrombin. Thromb Res 29: 635–642, 1983

139. NE Kohl, SD Mosser, SJ deSolms, EA Giuliani, DL Pompliano, SL Graham, RL Smith, EM Scolnick, A Oliff, JB Gibbs. Selective inhibition of ras-dependent transformation by a farnesyltransferase inhibitor. Science 260: 1934–1936, 1993.

140. JY Qian, MA Blaskovich, C-M Seong, A Vogt, AD Hamilton, SM Sebti. Peptidomimetic inhibitors of p21ras farnesyltransferase: hydrophobic functionalization leads to disruption of p21ras membrane association in whole cells. Bioorg Med Chem Lett 21: 2579–2584, 1994.

141. JD Scholten, K Zimmerman, GM Oxender, J Sebolt-Leopold, R Gowan, D Leonard, D Hupe. Inhibititors of farnesyl:protein transferase: A possible cancer chemotherapeutic. Bioorg Med Chem 4: 1537–1543, 1996.

142. MS Plummer, EA Lunney, KS Para, JVN Vara Prasad, A Shahripour, J Singh, CJ Stankovic, C Humblet, JH Fergus, JS Marks, TK Sawyer. Hydrophobic D-amino acids in th design of phosphopeptide ligands for the pp60$^{src}$ SH2 domain. Drug Design Discovery 13: 75–81, 1996.

143. M Rodriguez, R Crosby, K Alligood, T Gilmer, J Berman. Tripeptides as selective inhibitors of src-SH2 phosphoprotein interactions. Lett Peptide Sci 2: 1–6, 1995.

144. EA Lunney, KS Para, MS Plummer, A Shahripour, D Holland, JR Rubin, C Humblet, J Fergus, J Marks, S Hubbell, R Herrera, AR Saltiel, TK Sawyer. Structure-based design of a novel series of nonpeptide ligands that bind the pp60$^{src}$ SH2 domain. J Amer Chem Soc 119: 12471–12476, 1998.

145. K Para, EA Lunney, M Plummer, CJ Stankovic, A Shahripour, D Holland, JR Rubin, C Humblet, JH Fergus, JS Marks, S Hubbell, R Herrera, AR Saltiel, TK Sawyer. Structure-based de novo design and discovery of nonpeptide antagonists of the pp60$^{src}$ SH2 domain. In: J Tam, P Kamaya, Proceedings of the Fifteenth American Peptide Symposium, 173–175. (1999).

146. (a) SM Sebti, AD Hamilton. New approaches to anticancer drug design based on the inhibition of farnesyltransferase. Drug Disc Today 3: 26–33, 1998; (b) Y Qian, SM Sebti, AD Hamilton. Farnesyltranserase as a target for anticancer drug design. Biopolymers (Pept. Sci.) 43: 25–41, 1997; and (c) JE Buss, JC Marsters, Jr. Farnesyl transferase inhibitors: The successes and surprises of a new class of

potential chemotherapeutics. Chem BioI 2: 787–791, 1995.

147. V Manne, N Yan, JM Carboni, AV Tuomari, CS Ricca, JG Brown, ML Andahazy, RJ Schmidt, D Patel, R Zahler, R Weinmann, CJ Der, AD Cox, JT Hunt, EM Gordon, M Baracid, BR Seizinger. Bisubstrate inhibitors of farnesyltransferase: A novel class of specific inhibitors of ras transformed cells. Oncogene 10: 1763–1779, 1995.

148. EC Lerner, Y Qian, MA Blaskovich, RD Fossum, A Vogt, J Sun, AD Cox, JD Der, AD Hamilton, SM Sebti. Ras CAAX peptidomimetic F171-277 selectively blocks oncogenic Ras signalling by inducing cytoplasmic accumulation of inactive Ras-Raf complexes. J Biol Chem 270: 26802–26806, 1995.

149. DM Leonard, KR Shuler, CJ Poulter, SR Eaton, TK Sawyer, JR Hodges, JD Scholten, RC Gowan, JS Sebolt-Leopold, AM Doherty. Structure-activity relationships of cysteine-lacking pentapeptide derivatives that inhibit Ras farnesyl protein transferase. J Med Chem 40: 192–200, 1997.

150. JT Hunt, VG Lee, K Leftheria, B Seizinger, J Carboni, J Mabus, C Ricaa, N Yan, V Manne. Potent, cell active, non-thiol tetrapeptide inhibitors of farnesyltransferase. J Med Chem 39: 353–358, 1996.

151. (a) TK Sawyer. Src homology-2 domains: structure, mechanisms, and drug discovery. Peptide Science (Biopolymers) 47: 243–261, 1998; (b) MC Botfield, J Green. SH2 and SH3 domains: choreographers of multiple signaling pathways. Ann Rep Med Chem 30: 227–237, 1995; and (c) J Kuriyan, D Cowbum. Structures of SH2 and SH3 domains. Curr Opin Struct Biol 3: 828–837, 1993.

152. Z Songyang, LC Cantley. Recognition and specificity in protein tyrosine kinase-mediated signalling. Trends Biochem Sci 20: 471–475, 1995.

153. T Gilmer, M Rodriguez, S Jordan, R Crosby, K Alligood, M Green, M Kimery, C Wagner, D Kinder, P Charifson, AM Hassell, D Willard, M Luther, D Rusnak, DD Sternbach, M Mehrotra, M Peel, L Shampine, R Davis, J Robbins, IR Patel, D Kassel, W Burkhart, M Moyer, T Bradshaw, J Berman. Peptide inhibitors of Src SH3-SH2-phosphoprotein interactions. J Biol Chem 269: 31711–31719, 1994.

# 3

# Structure Prediction and Molecular Modeling

**Robin E. J. Munro\* and William R. Taylor**
*National Institute for Medical Research, London, United Kingdom*

## I. INTRODUCTION

The ultimate aim of protein structure prediction is to take a protein with unknown structure and from its sequence alone predict the tertiary, or 3D, structure. Despite the simplicity with which the basic problem can be stated, no method, over the 30 or so years that people have been considering it, has ever proved to be generally (some would say, even partly) successful. The intellectual challenge of the problem, despite its apparent intractability, has ensured that many have been (and still are) willing to look at it. Although no general method has resulted, all this effort has not been in vain as there are now many methods that, although they cannot predict a full tertiary structure, can provide insight into the sort of structure that a sequence might adopt. In the current situation, in which sequence data is being elucidated at an amazingly rapid rate, any methods that can extract any structural features from sequence data alone is of great value.

The two major types of biological molecule, protein and nucleic acid (for simplicity, we just refer to DNA), perform radically different functions; that of active-agents and data-archive respectively,

and this contrast is also manifest in their structure. DNA is regular, stable and inert, whereas proteins are asymmetric, plastic and active. This contrast was quite unexpected at the time the first structure was solved. The structure was that of myoglobin (a protein containing only alpha helix structure) solved by John Kendrew and co-workers at the Medical Research Council Unit for Molecular Biology: the same institute where only three years earlier the Watson and Crick model of DNA was proposed. In his paper on the x-ray model of myoglobin, Kendrew said that "Perhaps the most remarkable features of the molecule are its complexity and lack of symmetry. The arrangement seems to be almost totally lacking in the kind of regularities which one instinctively anticipates, and it is more complicated than has been predicted by any theory of protein structure" (1). This complexity and flexibility of proteins is perhaps a fundamental necessity to permit the innumerable roles that they must fulfill in life as too much inherent regularity would probably restrict the structures a protein might adopt. While necessary for life, complexity makes the job of structure prediction difficult.

Since myoglobin more than 6,000 structures have been solved by either crystallography or NMR. There has however been a relative explosion in the number of protein sequences without

*\*Current affiliation: LION Bioscience AG, Heidelberg, Germany*

structure. Structure prediction goes some of the way to redress the balance of sequences over structure and develops faster methods for the solution of structure than those presently available in the "Wet lab". Structures can play a large role in aiding scientists in the direction their work should take: allowing the ability to design drugs, to modify or interfere with proteins and the ways in which they work. These procedures are all facilitated by having a protein structure to work on. If the protein predictioner can only provide a rough model of a structure, then information can be gained which might be vital for the fast development of a new drug. Only when scientists, from all fields of research, work together will the problem at hand be solved.

## A. The Current State of the Field

At the beginning of 1998 there were just under 290,000 non-redundant sequenced proteins publicly available in various databases (All non-redundant GenBank CDS translations + PDB + SwissProt + PIR). Compared with around 6,900 structures in the PDB (http://www.pdb.bnl. gov/) Many of these, however, are very similar; some are theoretical and others are very low resolution or just backbone atoms. Various groups have classified protein structures into a non-homologous database of, in some cases, up to 1000 proteins but more usually around 300 (2, 3). These are proteins that have little sequence and structural similarity and make up a good representation of the possible structurally solved folds. Using many computational methods the aim of protein structure prediction is to redress the imbalance between the number of protein sequences and structure.

An unbiased survey of the power of the various prediction and modeling methods is provided in a prediction contest in which crystallographers (and others) tell in advance if they have, or are about to,

solve a protein structure. Having provided only the sequence of their protein, it is then up to the predictors to determine the structure—by whatever means they can. The first gathering to assess such prediction results, called the Critical Assessment of Techniques for Protein Structure Prediction (CASP) was held at Asilomar state park, California, at the end of 1994 with a repeated and a similar meeting held in 1996 as a culmination to the second experiment. The assessment covered the major areas of protein structure prediction: Comparative Modeling, Fold Recognition and *Ab initio* and also the prediction of associations between ligands and proteins (Docking). More details can be found in special issues of Proteins: Structure, Function and Genetics (4, 5).

Throughout this chapter, we will survey the various fields, as identified for the CASP competition, and draw examples from our own work, that was submitted to the last CASP meeting.

## II. SEQUENCE SIMILARITY

The identification of any clear sequence similarity with a protein of known structure is the most certain way to infer the structure of a protein from just its sequence. This is a strong principle since, through evolution, the amino acid sequence can change much more (through conservative substitutions) than the structure itself. Any sequence similarity therefore implies a similar structure and explains the importance of developing methods that can detect the most elusive of similarities, as even from these, some structural inference can be made.

Even without similarity to a protein of known structure, the alignment of other sequences can still be very helpful as they reveal the evolutionary constraints imposed at each position of the sequence. As these constraints are often directly related to the local structure, then some

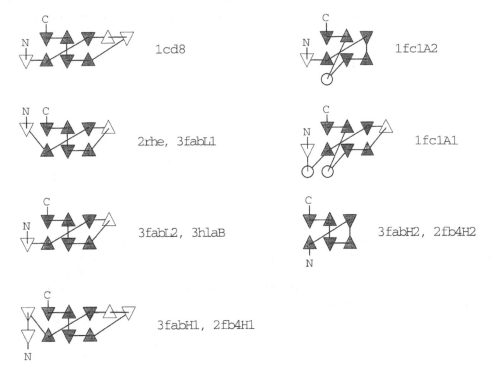

**Fig. 1** Topology Diagrams. Illustration of different domains from the immunoglobulin family of proteins. The folds represented as topology diagrams [TOPS representation (74)]. The conserved core of the domains, characteristic of the Ig family is shaded. Triangles represent a strand, with circles for helices; the lines show connectivity either above or below the secondary structure.

idea of that structure can be inferred. Often the more sequences (across a wide phylogenetic range) that can be aligned, then (assuming a common structure), the greater is the information which can be gleaned about each position and the better chance there is of predicting the structure.

Using methods to identify sequences with similarities to the sequence of interest can also give insight about function. For example the protein may be similar to a well characterized family of proteins about which the function, if not the structure, is known.

## A. Pairwise Sequence Alignment

For the structure of a protein to remain consistent with its functionality, a reasonable assumption which can be made is that, the main core of a protein should remain well conserved, and that its secondary structure elements are similarly arranged across a family of proteins. This arrangement of secondary structure elements is referred to as the protein architecture (i.e. the spatial orientation of the elements). The order of connectivity of these elements is the fold or topology. The connections between secondary structure elements are referred to as loops and turns. These can be highly variable in length, and are areas in which substitutions and deletions of sequence preferentially occur. Whole secondary structure elements can be inserted or deleted within a family of folds, but the main "defining" core still exists. This can be best illustrated by Fig. 1, which shows a series of topological representations of a family of proteins. Loops are sometimes vital for protein function and may bind ligands and nucleotides. In such cases these loops are well conserved; one such example is the 'P' loop which is important in binding adenosine tri-phosphate (6).

### 1. Amino Acid Similarity

When comparing two protein sequences (represented as strings of characters) some measure of amino acid similarity must be known. For nucleic acids, simply counting identical matches is sufficient, but with amino acids, the variety of chemical and physical properties that they exhibit requires a more graded matching scheme. Many have been devised over the years but those most commonly used are based on empirical counts of observed substitutions between related proteins.

When sequences are evolutionarily distant, the problem of identifying the similarity (if indeed it exists) between the two sequences is a challenging one. The simplest model would be to construct an identity matrix where a high matching score would be achieved if, at a particular point in the comparison, the same amino acid type was found in each sequence. The more self–self matches achieved then the better the alignment. This idea is all very well for reasonably similar sequences (more than 50% similarity), however when trying to identify the best pairwise alignment, when the sequences have a lower similarity than this; then something different is called for. A more discriminating matrix system has been devised where the evolutionary likelihood of a particular amino-acid mutating into a different amino-acid has been calculated from sequences.

The most widely known series of substitution matrices are the PAM matrices (7) although a more recently developed series called the BLOSUM matrices are now also widely used (8). An updated PAM matrix called JTT (9) based on more recent sequence data is also sometimes used. In general all matrices contain scores for substitutions which are higher if size and hydrophobicity are conserved.

### 2. Gap Penalty

For more distantly related sequences it is necessary to introduce relative insertions and deletions into both sequences to attain a maximum matching of amino acids. However, the inclusion of such gaps cannot be be allowed to occur without some cost to the score, otherwise, to take an extreme example, a short protein aligned with very long sequence could insert gaps between every residue and eventually find a perfect match for every position.

To avoid this unbiological trend, but still allow the possibility of insertions and deletions (indels) we introduce a gap-penalty. Sometimes this is a fixed penalty for a gap of any length (10), but in most cases the penalty is made up of a gap creation penalty and a gap extension penalty. Usually a gap insertion penalty is applied and then once the gap is opened a lower, but still incremental, penalty is added for each further gap – this is the gap extension penalty. Hence the gap penalty can be written as: $g = a + bn$, (where $g$ is the applied penalty, $a$ and $b$ are the opening and extending parameters and $n$ the number of residues in the gap). Other methods have used a gap penalty for opening a gap and then penalize by the logarithm of the length of the gap.

When a gap penalty is too high the pairwise alignment lacks sensitivity. If too low, the alignment will be very dispersed making it hard to recognize a sensible evolutionary relationship. There has been much work on the significance a gap penalty plays in the comparison of sequences but the only general rule is that the size of the penalty is dependent on the size of the numbers in the amino acid substitution scoring scheme. Further aspects and some alternative gap penalties are discussed in a reference (11).

## B. Alignment Algorithms

When aligning two sequences to gain an idea of the evolutionary relationship between them a large number of gaps might need to be inserted. For one gap in a sequence of four letters there are five possible options. For two gaps there are 12 possible alignments to be considered. So generally there are many possible combinations and for large sequences and arbitrary gap sizes we have a combinatorial explosion. Fortunately, to cope with this problem, a very simple and effective computer algorithm has been developed called Dynamic Programming (DP) and is used (in one form or another) in almost all methods that align sequences (10) [and even structures (12, 13)].

Using DP, the placement of insertions and deletions (sometimes referred to as indels, but more commonly just gaps), as well as the similarity of different residues, can be taken into account. Thus, using DP, we calculate the highest scoring alignment between two sequences in a time proportional to the product of the lengths of the sequences.

To find the optimum pairwise alignment a matrix is constructed where one sequence is placed along each axis. For each element in the matrix a score is calculated based on the match between amino acids. DP will find the highest scoring path through the matrix (taking into account gaps), and in theory the best alignment for the given scoring scheme.

The weakness in sequence alignment is not the DP algorithm but the uncertainty in what constitutes the best gap penalty and substitution matrix. Given these inherent errors, there is a limit to the information which can be gained from a pairwise comparison; so, if possible, a multiple alignment should be considered (see Sec. II.D).

## C. Homologous Sequence Searching

Some fast methods for searching for sequences which are homologous to a sequence of interest have been developed over the last 10 to 15 years. An early program was FASTA (14–16) and has since been followed by BLAST (17). Both these programs contain shortcuts to reduce the computational time they take to run. With the advent of larger computers and parallelization of the algorithms it is now possible to run a full dynamic programming search over the sequence databank. This has been implemented as BLITZ and, in theory, this would be the best method to search for similar sequences. However, there even exists BLAST servers run in parallel making it incredibly fast.

### 1. FASTA

FASTA uses a fast technique to roughly locate regions of similarity, within which dynamic programming is then used to extract an alignment. The program uses hash coding of typically dipeptides in a query and scans the database counting the hits on each diagonal of the alignment. It then re-scores the areas of high score by allowing some amino acid substitution and shorter lengths of identity and then joins the best scoring regions from different diagonals. This is followed by a dynamic programming alignment to extract the alignment.

### 2. BLAST

BLAST (which stands for Basic Local Alignment Search Tool) has a specific protein version (BLASTP) while other variants include BLASTN—for nucleotide searches, and BLASTX—for nucleotide to peptide conversion before a peptide search, and TBLASTN to search for a peptide in a nucleotide database which is converted to peptide. The program uses a very rapid search algorithm, not unlike FASTA, but more flexible. Developed by Altschul *et al.*, the basic algorithm is simple and robust and offers an increase in speed over FASTA. Recent developments in BLAST take gaps into account and searches can be performed iteratively where the results of one search are used as a basis for the second search. These programs are called Gapped BLAST and PSI-BLAST, respectively (18).

*3. BLITZ*

BLITZ uses the Smith and Waterman algorithm (19) and no pre-filtering of the data. A similarity matrix is used to compare the target sequence with those in the database, the algorithm searches for the highest local match and takes into account a gap penalty. The best results are ranked but only the highest match for any one sequence is given. Other searchers are also available which run on massively parallel machines (MasPar) (20) or Biocellerators (BIC's).

Generally, all the methods give similar results when searching for proteins that are moderately related to the probe. It is worth submitting the sequence to several search methods so that the likelihood of missing a relevant match is reduced.

BLAST can be performed interactively on the WWW but the other methods generally need a search form which sends the results back by e-mail once the jobs have been completed on a remote computer.

**D.  Multiple Alignment**

To gain further information from protein sequences and the evolutionary information that they contain, a multiple alignment can provide a wealth of information. Several methods exist for going beyond a simple pairwise alignment: probably the most common method available for building multiple alignments is that of Tree or

Hierarchical methods. The method assumes that a multiple alignment can be built from successive pairwise alignments. The first step in the method performs a comparison of all the sequence pairs to align. So for N sequences there are $[N \times (N-1)]/2$ pairwise comparisons. Cluster analysis of the comparisons is performed to give a *tree* or *hierarchy* of the sequences from which a multiple alignment is constructed by the pairwise comparisons indicated by the tree. The Neighbor-Joining method (21) is most often used for the cluster analysis as it gives a reasonable and fast comparison.

In theory another method would be to extend a simple pairwise comparison, using DP, to a multiple comparison. So instead of constructing a 2D matrix and finding the optimal scoring path through it and hence a pairwise alignment. The method would build a 3D or nD matrix (one dimension per sequence) and find the best route through the matrix, also using DP. In practice this is possible in 3D, but the time is proportional to $N^5$. Higher dimensional comparisons would take far too long to be practical, although with the use of shortcuts it is possible to an extent. This method incorporating the shortcuts is used by the program MSA (22), which can handle alignments of up to 8 sequences, each 200–300 amino acids in length.

The major packages for multiple alignment are CLUSTALW or CLUS-TALX (23, 24) (formerly CLUSTALV — ftp://ftp-igbmc.u-strasbg.fr/pub/Clustal X/), PILEUP (25) (which is part of GCG—http://www.gcg.com/) and CAMELEON (which incorporates MULTAL (26)—http://www.oxmol.com /prods/cameleon/). It is more widely becoming the case that programs for multiple alignment (particularly CUSTALW) are available over the WWW. Sites such as SRS (Sequence Retrieval Service— http://srs.ebi.ac.uk:5000/) are incorporating CLUSTALW into form style interfaces. Although these interfaces do not give the full range of options found on command line based versions, they are nevertheless a good starting point for constructing a multiple alignment. CLUSTAL is also available in an X interface format (CLUSTALX).

## III. SECONDARY STRUCTURE PREDICTION

Using a multiple alignment, identifying well-conserved areas, and in particular areas of conserved hydrophobicity can give a good indication of secondary structure. Secondary structure prediction methods have been much improved since the simple statistical methods of Chou and Fasman, particularly when they include the information gained by multiple sequence alignments. In the next section are some of the most widely used and available methods for predicting secondary structure.

In a globular protein in solution the more hydrophobic amino acids in a sequence will tend to lie towards the center of a protein, away from the surrounding water. The hydrophobic side chains will pack into the interior of the molecule. For proteins of any reasonable size, this creates a problem, because the backbone has polar atoms and is therefore hydrophilic, and not easily buried in a hydrophobic protein core. To overcome this problem the amino (N—H) and carbonyl (C=O) groups in the backbone form hydrogen bonds, and so their partial charges are neutralized. (The NH group is a H-bond donor and the C=O a H-bond acceptor). In the core of the protein hydrogen bonds can form in two distinct ways. (1) a helix where the CO of residue $n$ bonds to the NH of residue $n+4$; (2) a sheet, which is comprised of extended strands forming parallel or anti-parallel interconnections often between distant parts of the sequence. These are referred to as secondary structure.

Depending on how buried the secondary structure elements are in a protein, the pattern of conserved hydrophobic amino acids (indicating structural importance) can indicate what type of structure is present. For example a buried beta sheet will have a run of hydrophobic residues, whereas a partially exposed beta will contain alternating hydrophobic/hydrophilic residues. This is a slight over simplification, but it is possible to successfully predict secondary structure in this way.

### A. Statistical Methods

Analysis of proteins with known structures gives some idea of the propensities for different amino acids to occur in different secondary structures. For example GLY and PRO are often found at or in turns and at the ends of alpha helices (27).

#### 1. Chou-Fasman Method

Statistical methods for predicting secondary structure by Chou and Fasman who conducted a statistical study of protein structure to attempt to map the secondary structure from sequence (28, 29). They looked whether not only a residue prefers to be in an alpha or beta state but also classed the residues into six classes depending on their likelihood to form or disrupt an alpha or beta structure.

#### 2. GOR Method

The method, developed by Garnier, Osgothorpe and Robson in 1978, used four likelihood profiles to represent an alpha, beta, turn or coil (30, 31). Each likelihood profile is a function of a 17 amino acid window around the position of interest. To compute a probability for a position the 17 values, and the corresponding positions of the surrounding residues are added up to give a score for each of the four states.

## B.   Neural Networks

One of the earlier methods to use neural networks was based on a non-linear model (32). Probably the most widely used method for predicting the secondary structure of proteins is the predict protein server based at EMBL, Heidelberg (33). The method, called PHD, is based upon a series of trained neural networks. PHD is particularly useful as it constructs a multiple alignment and uses this additional information to predict the secondary structure. Additionally, several formats of multiple alignments can be submitted to PHD to give a secondary structure prediction of a specific alignment. Another neural net based method is NNPREDICT (34).

## C.   Other Secondary Structure Prediction Programs

A fast and simple approach to secondary structure prediction from multiple alignments is available as SSPRED (http://www.embl-heidelberg.de/sspred/sspred info.html) (35). The method uses aligned homologous sequences and structures to derive residue exchange statistics for each secondary structure type. The prediction for a given multiple alignment is calculated by correlating particular types of mutations known to prefer one of the scondary structure states, based on the derived statistics.

Another approach which uses the basic, but important, aspects of secondary structure prediction and then combines them using statistics is DSC (Discrimination of Secondary structure Class —http://bonsai.lif.icnet.uk/bmm/dsc/dsc_read_align.html) (36). DSC combines GOR potentials, gaps in the multiple alignment, mean moment of conservation, mean moment of hydrophobicity and some other attributes to give a prediction. This approach has the advantage of being easy to understand and simple to run, when compared with the "black-box" workings of neural network methods.

## IV.   STRUCTURE COMPARISON/ CLASSIFICATION

Most proteins have similarities with other proteins, and many structural similarities are conserved better than their amino acid sequences. In general indels in a sequence occur within loop regions and do not alter the overall fold. Therefore fold families have related structure, but not necessarily highly related sequence. Similar folds without any significant sequence similarity are termed analogous, suggesting that the same fold has been arrived at from different evolutionary starting points; whereas if common evolutionary origin is implied by clear sequence similarity, the term homologous is used. It is usually the case that sequences with >30% similarity adopt the same fold.

Structure comparison between proteins can be carried out computationally. For different comparisons, the better methods involve the characterization of a local structural environment for each position in a protein (37, 38). In one of these methods a vector set of all inter-atomic distances for each point in the two structures was compared, and from that the relative similarity of their respective positions. A similarity was derived for all pairs, and from this an optimum alignment of the structures obtained (37). A modification of this algorithm (SAP) is available at http://mathbio.nimr.mrc.ac.uk/.

## A.   Structural Classification Databases

There are four major classes of proteins: all alpha, all beta, alternating alpha/beta, and mixed alpha and beta (39). Further sub groups and classifications can be made and the study of how proteins are related is kept up to date in publicly available databases.

There are about a few hundred classified folds we know of with an expected maximum estimate of 1000 from sequence analysis. Bearing in mind structural comparison a more conservative estimate of between 500 and 700 folds has been proposed. There a several ways in which protein structures have been classified. Two widely available databases on the subject are SCOP and CATH.

SCOP is a highly comprehensive description of the structural and evolutionary relationships between all known protein structures (40, 41). The hierarchical arrangement is constructed by mainly visual inspection in conjunction with a variety of automated methods. The principal levels are Family, Superfamily and Fold. The Family category contains proteins with clear evolutionary relationships. The Superfamily have probable common origin; there may be low sequence identity, but structure and functional details suggest a common origin. The Fold level groups together proteins with the same major secondary structure elements in the same arrangement with topological connections the same.

CATH is a hierarchical domain classification of only NMR and crystal structures with resolution of 3.0 Å or less (42). The database is constructed wherever possible by automatic methods. The four levels in the hierarchy are: class(C), architecture(A), topology(T) and homologous superfamily(H) a further level called sequence(S) families is sometimes included. "C" classifies proteins into mainly alpha, mainly beta and alpha-beta, which includes both $\alpha/\beta$ and $\alpha + \beta$. "A" describes the shape of the structure, or fold. "T" describes the connectivity and shape. "H" indicates groups thought to have a common ancestor, i.e. homologous. "S" describes structures clustered on sequence identity.

SCOP and CATH are both accessed via. the WWW. Many mirrors exist for SCOP around the world, but here are two starting points for SCOP and CATH respectively:

http://scop.mrc-lmb.cam.ac.uk/scop/
http://www.biochem.ucl.ac.uk/bsm/cath/

## B. Topological Prediction

Predicting the way in which the secondary structure elements of a protein fold are connected can be difficult. For any one fold there are a number of ways to connect the elements together. One rule to observe, in general, is that of the chirality (or handedness) of the connections between secondary structure within proteins. In the majority of protein folds a right handed topology is maintained throughout.

## V. FOLD RECOGNITION

Fold recognition, or threading, is a process whereby a sequence with unknown structure is compared to a database of structures with different folds. In making the comparison the structure which the unknown sequence finds a best fit for is taken to be the fold that the sequence will most likely form.

At the moment, where not every possible fold has been identified, one of the hardest areas of threading is to recognize when there is no fold similar for the sequence. In such a case other methods such as *ab initio* prediction could be employed.

Fold recognition falls broadly into two categories, one method which uses pairwise energy/interaction potentials and the other which performs a 1D to 3D comparison.

## A. Pairwise Energy Potentials

Pairwise potentials are any measure which can be used to classify a residue:residue interaction, or atom:atom interaction.

THREADER is a program which takes an empirical potential map of a protein and threads the target sequence onto the structure of the known protein (43, 45, 46). The targets are compared to a database of non homologous proteins, this is performed in 3-Dimensional space. The THREADER output for the target sequence can be ranked according to several scores and the structures which score significantly well may have the correct fold.

Another method derives knowledge-based force fields from known structures. The sequence is then compared to a fold database and the corresponding energies calculated to give an indication of the predicted best fit fold (47, 48).

All the above methods use a single sequence. To use the information in multiple sequences, a Multiple Sequence Threading (MST) has been developed which compares multiple sequence information with a database of structures to determine the correct fold (49, 50).

### B.  1D/3D Comparison

Multiple sequence information is also used in the simpler 1D/3D fold recognition methods which perform a secondary structure prediction on the sequence of interest and then compares that secondary structure with all the known secondary structures in known sequences to find a possible match.

Methods which fall into this category include: TOPITS (51), MAP (52) and H3P2 (53).

### VI.  COMPARATIVE MODELING

Sometimes in a multiple alignment it is possible to align sequences which already have a known structure. If this is the case then the alignment of the sequence with unknown structure and that of the known structure can be directly or indirectly aligned by the multiple sequence alignment. This method is also frequently referred to as homology modeling. The first application of this idea was by Browne and co-workers (54) and later refined by Greer *et al.* (55) and Blundell *et al.* (56).

Fragment based homology modelling is a technique where models of proteins can be constructed from separate fragments of other proteins. Areas where there are inserted residues, and no structure in the homologue, can be built using fragment matching (57).

MODELLER, a comparative modelling program, optimally satisfies structural restraints derived from an alignment with one or more structures. These restraints are expressed as probability density functions (pdfs) for each feature, where a feature may be solvent accessibility, hydrogen bonding, secondary structure, etc. at residue positions, and between residues (58).

Based on a multiple alignment with known homologous structures distance restraints can also be derived for a sequence with unknown structure, and then solved using distance geometry (59–64).

A web-based homology modelling package is also available, and can give some very good models based on a single sequence. Further refinement can also be made, such as adjusting the alignment and specifying which specific structures to use when model building (65).

Due to the large number of already solved folds it can be expected that more and more sequences will in fact have homologous structures. Hence, using these methods will play an ever more important role in model building.

### VII.  *AB INITIO* MODELING

*Ab initio* modeling, or *de novo* folding is not based on any template structures, but rather a secondary structure assignment and various sets of constraints. Several approaches to the problem of *ab initio* modeling have been considered (66, 67). Probably the most common of these is the combinatorial approach where first the secondary structure is predicted, followed by simulations of the possible combinations of folds imposed by the secondary structure prediction. The most successful of these combinatorial approaches modeled α-helix proteins. Several different combinatorial packing algorithms have been used (68, 69).

Many *ab initio* approaches involve simplifying the model of the protein to make it easier to handle. Once a rough model of the protein has been created, a series of further steps can be applied to build the protein into a full atom representation.

One of the main advantages to any *ab initio* method is that the models are not restricted to a known fold, and can be used to model proteins with no known fold in the databank. Computing power and the complexity of the problem still limit the uses and success of *ab initio* in protein structure prediction.

### A. Combinatorial Method

One of the first steps in any prediction is to identify the secondary structure elements. The combinatorial methods try to explore all the combinations of arrangements of the secondary structure elements. These methods try to pack hydrophobic residues in the core of the protein (if it is a globular protein) (68–70).

### B. Distance Geometry

Distance geometry based techniques have occasionally been applied to *ab initio* folding, but more frequently to homology modeling as seen in Sec. VI.3. One such method is incorporated into a program called DRAGON (64, 71). A simplified model chain is folded by projecting it into gradually decreasing dimensional spaces whilst subjecting it to a set of defined restraints, primarily secondary structure. In this way the geometry space is successfully explored to produce a protein backbone. The method generates many folds in a short time due to the embedding algorithm incorporated into the program (72).

We applied DRAGON to several of the CASP2 targets. The examples that follow illustrate the use of this method. Further information on DRAGON, and executables are available at http://mathbio.nimr.mrc.ac.uk.

## VIII. EXAMPLES

### A. Retroviral Protease

With the initial whirlwind of research into AIDS and HIV during the 1980s a great emphasis was put on getting high resolution crystal structures as fast as possible. An obvious protein to study is the HIV protease which cleaves several of the component proteins required in the construction of the HIV virus. If this protein could be disrupted in some way then the proliferation of the virus could be slowed or even halted. To gain some idea of what drugs would make good targets for interfering with the protease a structure was needed.

Prediction and modeling work carried out by Pearl and Taylor (73) was able to produce a structure for HIV protease before any high resolution crystal structure was available, and in this way aid in the development of a widely used AIDS drug.

As this chapter has suggested, the best way to go about this was to find similar sequences and construct an alignment. The multiple alignment was able to show that the HIV protease had an evolutionary relationship with a family of aspartic proteases. In particular an Asp-Thr-Gly motif conserved in the active sites of both families. One of the aspartic proteases had a known structure which enabled the construction of a model using an HIV protease sequence (homology modeling). The sequence was aligned on the backbone structure and several possible secondary structural elements were predicted and aligned with the homologue. This rough fitting method was used to find what they thought was the best alignment considering structure and was an early precursor to the threading method discussed earlier.

The final sequence was modeled onto the carbon backbone of the aspartic protease to give a model for the HIV protease.

Analysis of this model after a high resolution structure was published showed a good RMSD of 2.1 Å for all atoms in the model. Particularly high accuracy was achieved at the important substrate binding site (1.2 Å) (74).

## B. Modeling by Threading and Distance Geometry

Often in structure prediction there is no readily identifiable homologue with your sequence of interest. With the development of better threading methods it is now possible to identify a fold which a sequence may form even when there is no detectable similarity at the sequence level. It follows therefore that a combination of threading to identify a possible fold combined with homology modeling of the sequence on that identified fold can result in a good model.

This hybrid method is best illustrated in an example of a model we generated using multiple sequence threading and distance geometry. The CASP2 target T0004 [polyribonucleotide nucleotidyltransferase S1 motif (PNS1) from *Escherichia coli*, PDB code: 1SRO] had no obvious structural homologues. Using MST we identified several folds and then generated models, using DRAGON, based on the structural equivalences taken from MST.

Three possible folds were identified by threading: 1HRH chain A, 1LTS chain D and 2SNS. The scores obtained from threading indicated that 2SNS was a more

likely candidate than 1LTSD, followed by 1HRHA. Comparison of the known structure (1SRO) with the models based on these folds show that the correct fold was the model based on 1LTS chain D. In fact by structure comparison the core region of 2SNS showed a better fit but the overall "correct" fold is that of 1LTSD. The model (see Fig. 2) based on 1LTSD was found to be more similar to the experimental structure than the template ($C_\alpha$ RMSD of 6.2 Å compared with 6.4 Å).

This example clearly shows the benefits of a combined approach. A fold recognition method can identify a more remote similarity between proteins than by their sequences alone. Couple this with a fast model generating method like DRAGON and the initial structures produced can be good candidates for further protein modeling (44).

## C. NK Lysin

The NMR structure of NK Lysin (a tumor lysing and antimicrobial protein) was one of the structures available to test out the current methods of *ab initio* prediction, at CASP2. Using just a multiple alignment, secondary structure, disulphide bonding data as well as the built in restraints of what is known about proteins: simple low-resolution $C_\alpha$ and $C_\beta$ models were constructed using distance geometry (incorporated in DRAGON).

One of these models was submitted for assessment to the CASP2 experiment. A slightly modified SSAP algorithm (12) was used to compare the model structures with the NMR coordinates.

The correct fold was successfully identified and the 10 best models were found to be quite similar to the experimental structure with an average $C_\alpha$ RMSD of 5.38 Å.

Further modeling using DRAGON was carried out after the structural coordinates of the target sequence was known. Using the correct, rather than

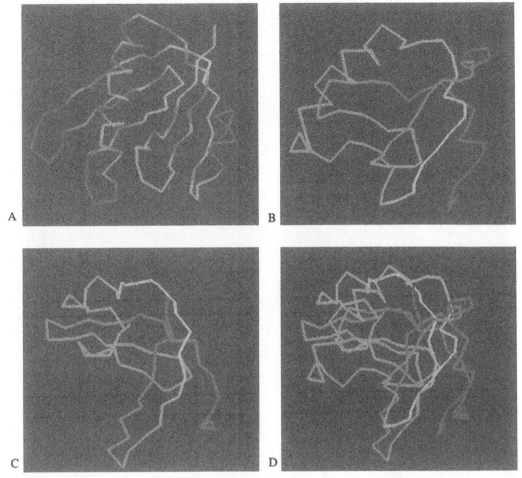

**Fig. 2** 1LTSD model. A) Threading (dark = deletion, light = insertion), B) threaded model, C) NMR model, D) superposition of B on C.

predicted, secondary structure can improve models with RMSD of 4.65 Å, see Fig. 3. From this it is clear that as secondary structure prediction methods improve, so will the ability to fold them into tertiary models.

The results show DRAGON as an efficient and reasonably accurate method for the *ab initio* prediction of tertiary structure. This distance geometry approach can give potential models when there are no homologous models in the PDB and where no putative structure can be found by threading methods.

The use of DRAGON as a method for building *ab initio* models shows con-

siderable promise. The results showed that most models are plausible including many close to the correct fold. Taking all the DRAGON ranking criteria into account a close, purely *ab initio*, model compares well with the NMR data supplied by the experimentalists at CASP2.

## IX.  CONCLUSIONS

Protein structure prediction can and does offer insight into the whole field of molecular biology and protein chemistry. Due to the lengthy x-ray crystallographic process new methods to predict protein structure

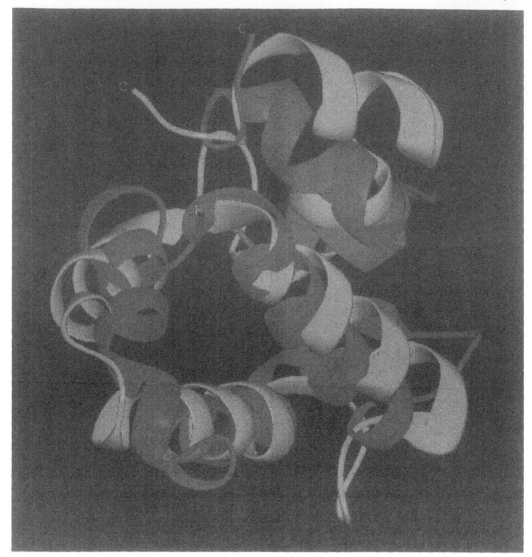

**Fig. 3** NK-lysin. Predicted model superposed on NMR model.

are essential if we are to keep up with the huge influx of sequence data. With faster and better computer prediction we can hope to address the ever increasing need to solve the tertiary structure of proteins.

Combining techniques to predict protein structure enables us to see a wider picture than each individual part of the problem could possibly convey. Homology modeling gives very accurate insight into the 3D structure of sequences with no

solved crystal structure. The models produced can greatly aid in the understanding of detailed protein structure. This works well where there are similar proteins with known fold in the database. Unfortunately this approach falls over when a close homologue is not available. By using a threading method a likely fold, or folds, can be identified and used as a template for the modeling. Failing to identify a potential template structure leads to *ab initio* prediction and modelling. Perhaps when all the possible fold types have been recognized there will be no need for purely *ab initio* prediction. Current and future methods are likely to be concerned with more and more genomic data and facilitating the pharmaceutical industry to determine and design better and better solutions for the myriad of diseases affecting our lives.

## REFERENCES

1. JC Kendrew, G Bodo, HM Dintzis, RG Parrish, H Wyckoff, DC Phillips. A three-dimensional model of the myoglobin molecule obtained by X-ray analysis. Nature 181: 662–666, 1958.
2. CA Orengo. Classification of protein folds. Curr Op Struct Biol 4: 429–440, 1994.
3. L Holm, C Sander. Touring protein fold space with dali/fssp. Nucleic Acids Res 26: 316–319, 1998.
4. CASP special issue. Prot. Struct. Funct. Genet., 23(3), 1995.
5. CASP2 special issue. Prot Struct Funct Genet suppl. 1, 1997.
6. M Saraste, PR Sibbald, A Wittinghofer. The p-loop – a common motif in ATP- and GTP-binding proteins. Trends Biochem Sci 15: 430–434, 1990.
7. MO Dayhoff, RM Schwartz, BC Orcutt. A model of evolutionary change in proteins. In MO Dayhoff, ed., Atlas of Protein Sequence and Structure, pages 345–352. Nat Biomed Res Foundation, Washington DC, USA, 1978. Volume 5, Supplement 3.
8. S Henikoff, JG Henikoff. Amino acid substitution matrices from protein blocks. Proc Natl Acad Sci USA 89: 10915–10919, 1992.
9. DT Jones, WR Taylor, JM Thornton. The rapid generation of mutation data matrices from protein sequences. CABIOS, 8: 275–282, 1992.
10. SB Needleman, CD Wunsch. A general method applicable to the search for similarities in the amino acid sequence of two proteins. J Mol Biol 48: 443–453, 1970.
11. WR Taylor. Multiple protein sequence alignment: algorithms for gap insertion. In RF Doolittle, editor, Computer methods for macromolecular sequence analysis, volume 266 of Meth Enzymol, pages 343–367. Academic Press, Orlando, FA, USA., 1996.
12. WR Taylor, CA Orengo. Protein structure alignment. J Mol Biol 208: 1–22, 1989.
13. CA Orengo, WR Taylor. SSAP: sequential structure alignment program for protein structure comparison. In RF Doolittle, ed. Computer methods for macromolecular sequence analysis, volume 266 of Meth Enzymol, pages 617–635. Academic Press, Orlando, FA, USA., 1996.
14. DJ Lipman, WR Pearson. Rapid and sensitive protein similarity searches. Science, 227: 1435–1441, 1985.
15. WR Pearson. Rapid and sensitive sequence comparison with FASTP and FASTA. In RF Doolittle, editor, Molecular Evolution: computer analysis of protein and nucleic acid sequences, volume 183 of Methods Enzymol., chapter 5, pages 63–98. Academic Press, Inc., 1990.
16. WR Pearson, W Miller. Dynamic programming algorithms for biological sequence comparison. In L Brand and ML Johnson, eds. Numerical Computer 27 Methods, volume 210 of Methods Enzymol, chapter 27, pages 575–601. Academic Press Inc., N.Y., 1992.
17. SF Altschul, W Gish, W Miller, EW Myers, DJ Lipman. Basic local alignment search tool. J Mol Biol 214: 403–410, 1990.
18. SF Altschul, TL Madden, AA Schaffer, J Zhang, Z Zhang, W Miller, DJ Lipman. Gapped BLAST and PSI-BLAST: A new generation of protein database search programs. Nuc Acids Res 25: 3389–3402, 1997.
19. TF Smith, MS Waterman. Comparison of bio-sequences. Adv Appl Math 2: 482–489, 1981.
20. JF Collins, AFW Coulson. Significance of protein sequence similarities. volume 183 of Methods Enzymol, pages 474–487. Academic Press, Inc., 1990.
21. N Saitou, M Nei. The neighbor-joining method: A new method for reconstructing phylogenetic trees. Mol Biol Evol 4: 406–425, 1987.
22. DJ Lipman, SF Altschul, JD Kececioglu. A tool for multiple sequence alignment. Proc Natl Acad Sci USA, 86: 4412–4415, 1989.
23. JD Thompson, DG Higgins, TJ Gibson. Clustal-W: improving the sensitivity of progressive multiple sequence alignment through sequence weighting, position-specific gap penalties and weight matrix choice. NAR, 22: 4673–4680, 1994.
24. JD Thompson, TJ Gibson, F Plewniak, F Jeanmougin, DG Higgins. Clustal-X windows interface: Flexible strategies for multiple sequence alignment aided by quality analysis tools. Nuc Acids Res 25: 4876–4882, 1997.
25. DF Feng, RF Doolittle. Progressive sequence alignment as a prerequisite to cor-

rect phylogenetic trees. J Mol Evol 25(4): 351–360, 1987.

26. WR Taylor. A flexible method to align large numbers of biological sequences. J Mol Evol 28: 161–169, 1988.

27. JS Richardson, DC Richardson. Amino acid preferences for specific locations at the ends of alpha helices. Science, 240: 1648–1652, 1988.

28. PY Chou, GD Fasman. Prediction of protein conformation. Biochemistry, 13: 222–245, 1974.

29. PY Chou, GD Fasman. Prediction of the secondary structure of proteins from their amino acid sequence. Adv Enzymol 47: 45–148, 1978.

30. J Garnier, DJ Osguthorpe, B Robson. Analysis of the accuracy and implications of simple methods for predicting the secondary structure of globular proteins. J Mol Biol 120: 97–120, 1978.

31. J-F Gibrat, J Garnier, B Robson. Further developments of protein secondary structure prediction using information theory— new parameters and consideration of residue pairs. J Mol Biol 198: 425–443, 1987.

32. N Qian, TJ Sejnowski. Predicting the secondary structure of globular proteins using neural network models. J Mol Biol 202: 865–884, 1988.

33. B Rost, C Sander. Prediction of protein secondary structure at better than 70-percent accuracy. J Mol Biol 232: 584–599, 1993.

34. DG Kneller, FE Cohen, R Langridge. Improvements in protein secondary structure prediction by an enhanced neural network. J Mol Biol 214: 171–182, 1990.

35. PK Mehta, J Heringa, P Argos. A simple and fast approach to prediction of protein secondary structure from multiply aligned sequences with accuracy above 70%. Protein Science 4: 2517–2525, 1995.

36. RD King, MJE Sternberg. Identification and application of the concepts important for accurate and reliable protein secondary structure prediction. Protein Science 5: 2298–2310, 1996.

37. WR Taylor, CA Orengo. Protein structure alignment. J Mol Biol 208: 1–22, 1989.

38. A Šali, TL Blundell. Definition of general topological equivalence in protein structures: a procedure involving comparison of properties and relationship through simulated annealing and dynamic programming. J Mol Biol 212: 403–428, 1990.

39. M Levitt, C Chothia. Structural patterns in globular proteins. Nature 261: 552–558, 1976.

40. AG Murzin, SE Brenner, T Hubbard, C Chothia. SCOP: a structural classification of proteins database for the investigation

of sequences and structures. J Mol Biol 247: 536–540, 1995.

41. TJP Hubbard, AG Murzin, SE Brenner, C Chothia. SCOP: a structural classification of proteins database. Nuc Acids Res 25: 236–239, 1997.

42. CA Orengo, AD Michie, S Jones, DT Jones, MB Swindells, JM. Thornton. CATH – A hierarchic classification of protein domain structures. Structure 5: 1093–1108, 1997.

43. DT Jones, WR Taylor, JM Thornton. A new approach to protein fold recognition. Nature 358: 86–89, 1992.

44. TP Flores, DS Moss, JM Thornton. An algorithm for automatically generating protein topology cartoons. Protein Engineering 7: 31–37, 1994.

45. DT Jones, CA Orengo, WR Taylor, JM Thornton. Progress towards recognising protein folds from amino acid sequence. Protein Engineering 6 (supplement): 124, 1993. (abstract).

46. DT Jones, JM Thornton. Protein fold recognition. J Comp. Aided Mol Desig 7: 439–456, 1993.

47. M Hendlich, P Lackner, S Weitckus, H Floeckner, R Froschauer, K Gottsbacher, G Casari, M J Sippl. Identification of native protein folds amongst a large number of incorrect models—the calculation of low-energy conformations from potentials of mean force. J Mol Biol 216: 167–180, 1990.

48. MJ Sippl, S Weitckus. Detection of native-like models for amino-acid-sequences of unknown 3-dimensional structure in a data-base of known protein conformations. Proteins-structure Function Genetics 13: 258–271, 1992.

49. WR Taylor. Multiple sequence threading: an analysis of alignment quality and stability. J Mol Biol 269: 902–943, 1997.

50. WR Taylor, REJ Munro. Multiple sequence threading: Conditional gap placement. Folding and Design 2, suppl.: S33–S39, 1997.

51. B Rost. TOPITS: Threading one-dimensional predictions into three-dimensional structures. In C Rawlings, D Clark, R Altman, T Hunter, L Lengauer, S Wodak, eds. The third international conference on Intelligent Systems for Molecular Biology (ISMB), pages 314–321. AAAI Press, Menlo Park, CA, USA., 1995. Cambridge, U.K., Jul 16–19.

52. RB Russell, RR Copley, GJ Barton. Protein fold recognition by mapping predicted secondary structures. J Mol Biol 259: 349–365, 1996.

53. D Rice, D Eisenberg. A 3D-1D substitution matrix for protein fold recognition that includes predicted secondary structure of the sequence. J Mol Biol 267: 1026–1038, 1997.

54. WJ Browne, ACT North, DC Phillips, K Brew, TC Vanaman, RL Hill. A possible three-dimensional structure of bovine alpha-lactalbumin based on that of hen's egg-white lysozyme. J Mol Biol 42: 65–86, 1969.

55. J Greer. Comparative model-building of the mammalian serine proteases. J Mol Biol 153: 1027–1042, 1981.

56. TL Blundell, BL Sibanda, MJE Sternberg, JM Thornton. Knowledge-based prediction of protein structures and the design of novel molecules. Nature 326: 347–352, 1987.

57. TA Jones, S Thirup. Using known substructures in protein model building and crystallography. EMBO J 5: 819–822, 1986.

58. A Sali, TL Blundell. Comparative protein modelling by satisfaction of spatial restraints. J Mol Biol 234: 779–815, 1993.

59. TF Havel, ME Snow. A new method for building protein conformations from sequence alignments with homologs of known structure. J Mol Biol 217: 1–7, 1991.

60. TF Havel. Predicting the structure of the avodoxin from escherichia-coli by homology modeling, distance geometry and molecular-dynamics. Mol Simulation 10: 175–210, 1993.

61. S Srinivasan, CJ March, S Sudarsanam. An automated method of modeling proteins on known templates using distance geometry. Protein Science, 2: 277–289, 1993.

62. SM Brocklehurst, RN Perham. Prediction of the three-dimensional structures of the biotinylated domain from yeast pyruvate carboxylase and of the lipoylated h-protein from the pea leaf glycine cleavage system: A new automated method for the prediction of protein tertiary structure. Protein Science 2: 626–639, 1993.

63. S Sudarsanam, CJ March, S Srinivasan. Homology modeling of divergent proteins. J Mol Biol 241: 143–149, 1994.

64. A Aszódi, WR Taylor. Homology modelling by distance geometry. Folding and Design, 1: 325–334, 1996.

65. MC Peitsch. ProMod and Swiss-Model: Internet-based tools for automated comparative protein modelling. Biochem Soc Trans 24: 274–279, 1996.

66. PS Kim, RL Baldwin. Specific intermediates in the folding reactions of small proteins and the mechanism of protein folding. Ann Rev Biochem 51: 459–489, 1982.

67. KA Dill. Theory for the folding and stability of globular proteins. Biochemistry 24: 1501–1509, 1985.

68. FE Cohen, MJE Sternberg, WR Taylor. Analysis and prediction of protein $\beta$-sheet structures by a combinatorial approach. Nature 285: 378–382, 1980.

69. MJE Sternberg, FE Cohen, WR Taylor. A combinatorial approach to the prediction of the tertiary fold of globular proteins. Biochem Soc Trans 10: 299–301, 1982.

70. WR Taylor. Protein fold refinement: building models from idealised folds using motif constraints and multiple sequence data. Protein Engineering 6: 593–604, 1993.

71. A Aszódi, MJ Gradwell, WR Taylor. Global fold determination from a small number of distance restraints. J Mol Biol 251: 308–326, 1995.

72. A Aszódi, WR Taylor. Hierarchical inertial projection: A fast distance matrix embedding algorithm. Computers Chem 21: 13–23, 1997.

73. LH Pearl, WR Taylor. A structural model for the retroviral proteases. Nature 329: 351–354, 1987.

74. IT Weber. Evaluation of homology modeling of HIV preotease. Prot Struct Funct Genet 7: 172–184, 1990.

# 4

# Chemical Synthesis of Peptides

**Henriette A. Remmer[1] and Gregg B. Fields[2]**
[1]*University of Illinois, Urbana, Illinois*
[2]*Florida Atlantic University, Boca Raton, Florida*

## I. INTRODUCTION

Peptide synthesis is the most practical approach for creating biomolecules containing unnatural or chemically modified amino acids. Peptide or pseudo-peptide drugs with specific functional groups, tailored conformations and diminished enzymatic susceptibility can be designed and produced. For example, synthesis of novel peptide drugs may include introduction of (i) posttranslational modifications, such as hydroxylation, phosphorylation, sulfation, glycosylation and disulfide bridges; (ii) unnatural structures, such as methylation of the peptide amide bond and cyclic conformations produced by lactam formation; and (iii) specific chemical properties, such as reduced, retro-inverso, or thio-peptide bonds. Also, single amino acids or specific sequences can be substituted by the corresponding D-amino acids or unnatural amino acids in order to minimize enzymatic cleavage or prevent side reactions. Most of the biologically or medicinally important peptides which are the targets for useful structure-function studies by chemical synthesis comprise less than 50 amino acid residues. Synthetic approaches can also lead to important conclusions about small proteins in the 100-residue size range. Methods for synthesizing peptides are divided conveniently into two categories: solution (classical) and solid-phase. In this chapter we present the most contemporary and accessible procedures for solid-phase peptide synthesis and discuss them critically.

## II. THE CONCEPT OF SOLID-PHASE PEPTIDE SYNTHESIS (SPPS)

The concept of SPPS is to retain chemistry proven in solution, but to add a covalent attachment step that links the nascent peptide chain to an insoluble polymeric support (Fig. 1). Subsequently, the anchored peptide is extended by a series of addition cycles. It is the essence of the solid-phase approach that reactions are driven to completion by the use of excess soluble reagents, which can be removed by simple filtration and washing without manipulative losses. Because of the speed and simplicity of the repetitive steps, which are carried out in a single reaction vessel at ambient temperature, the major portion of the solid-phase procedure is readily amenable to automation. Once chain elongation has been accomplished, it is necessary to release the crude peptide from the support under conditions that are minimally destructive towards

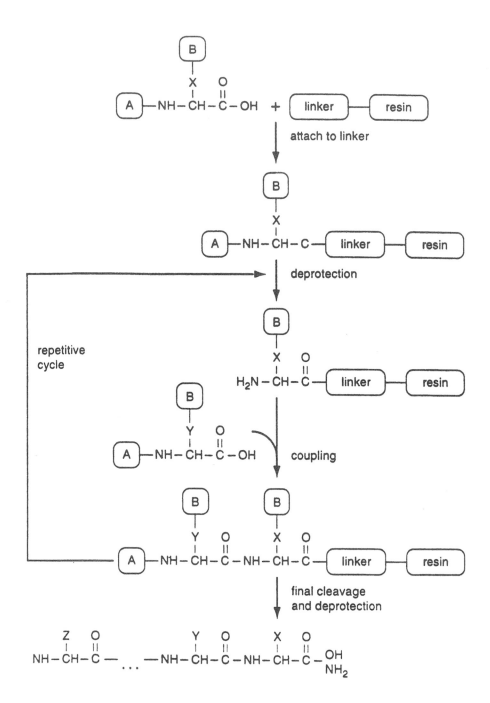

**Fig. 1** General scheme for stepwise solid phase peptide synthesis. X,Y,Z: functional groups of amino acid side-chains; A: α-amino (temporary) protecting group; B: side-chain (permanent) protecting groups.

sensitive residues in the sequence. Finally, there must follow purification and characterization of the synthetic peptide product, in order to verify that the desired structure is indeed the one obtained.

## A.  The Solid-Phase

An appropriate polymeric support (resin) must be chosen that has adequate mechanical stability, as well as desirable physicochemical properties that facilitate solid-phase synthesis. In practice, such supports include those that exhibit significant levels of swelling in useful reaction/wash solvents. Swollen resin beads are reacted and washed batch-wise with agitation, and filtered either with suction or under positive nitrogen pressure. Alternatively, solid-phase synthesis may be carried out in a continuous-flow mode, by pumping reagents and solvents through resins that are packed into columns.

The resin support is quite often a polystyrene suspension polymer cross-linked with 1% of 1,3-divinylbenzene. The chemistry of solid-phase synthesis takes place within a well-solvated gel containing mobile and reagent-accessible chains (1, 2). Polymer supports have also been developed based on the concept that the insoluble support and peptide backbone should be of comparable polarities (3). A resin of copolymerized dimethylacrylamide, $N,N'$-bisacryloyl-ethylenediamine, and acryloyl sarcosine methyl ester, commercially known as polyamide or Pepsyn, has been synthesized to satisfy this criteria (4). Increasing popularity has been seen for polyethylene glycol (PEG)-polystyrene graft supports (0.1–0.4 mmol/g), which swell in a range of solvents and have excellent physical and mechanical properties for both batch-wise and continuous-flow SPPS (5–9). Heterobifunctional polyethylene glycol derivatives can be cleanly attached to polystyrene resins in defined ratios generating PEG-PS supports with specific swelling properties (10). PEG-PS resins were shown to be suitable for SPPS using acetonitrile as the solvent for coupling and washing steps (10). Also applicable for batch and continuous flow methods are the hydrophilic PEGA resins (11), which consist of bis-2-acrylamidoprop-1-yl polyethylene glycol and dimethyl acrylaminde crosslinked with acryloyl sarcosine methyl ester. PEGA has excellent swelling properties in a wide variety of solvents, and is freely permeable to macromolecules. The latter property makes PEGA resins particularly useful for the preparation and assaying of peptide libraries. The recently developed cross-linked ethoxylate acrylate resin (CLEAR) (12) is suitable for both peptide and organic solid-phase synthesis due to its excellent swelling behavior in polar and nonpolar solvents ranging from water to DCM.

Regardless of the structure and nature of the polymeric support chosen, it must contain appropriate functional groups onto which the first amino acid can be anchored. This is achieved by the use of a "linker," which is a bifunctional spacer that on one end incorporates features of a cleavable protecting group. The other end of the linker contains a functional group, often a carboxyl, that can be activated to allow coupling to functionalized supports.

## B.  Assembly of the Peptide Chain

In the vast majority of solid-phase syntheses, suitably $N^\alpha$- and side-chain protected amino acids are added *stepwise* in the $C \rightarrow N$ direction. This strategy proceeds, in general, with only negligible levels of racemization. A "temporary" protecting group is removed quantitatively at each step to liberate the $N^\alpha$-amine of the peptide-resin, following which the next incoming protected amino acid is introduced with its carboxyl group suitably activated (see "Formation of the Peptide Bond"). It is frequently worthwhile to verify that the coupling has gone to completion by some monitoring technique (see "Monitoring").

Once the desired linear sequence has been assembled, the anchoring linkage must be cleaved. Depending on the chemistry of the original handle and on the cleavage reagent selected, the product from this step can be a *C*-terminal peptide acid, amide, or other functionality. Typically, final deprotection is carried out concurrent to cleavage; in this way, the released product is directly the free peptide.

## C. Protection Schemes

At least two levels of protecting group stability are required for SPPS, insofar as the "permanent" groups used to prevent branching or other problems on the side-chains must withstand repeated applications of the conditions for quantitative removal of the "temporary" $N^\alpha$-amino protecting group. On the other hand, structures of "permanent" groups must be such that conditions can be found to remove them with minimal levels of side reactions that affect the integrity of the desired product.

### 1. Tertiary-*Butyloxycarbonyl (Boc)-Based Chemistry*

The so-called "standard Merrifield" system is based on graduated acid lability. The acidolyzable "temporary" $N^\alpha$-*tertiary* butyloxycarbonyl (Boc) group is stable to alkali and nucleophiles, and removed rapidly by inorganic and organic acids (13). Boc removal is usually carried out with trifluoroacetic acid (TFA) (20–50%) in DCM for 20–30 min. Deprotection with neat (100%) TFA, which offers enhanced peptide-resin solvation compared to TFA-DCM mixtures, proceeds in as little as 4 min (14, 15). Following acidolysis, a rapid diffusion-controlled neutralization step

with a tertiary amine, usually 5–10% triethylamine (Et$_3$N) or *N,N*-diisopropylethylamine (DIEA) in DCM for 3–5 min, is interpolated to release the free $N^\alpha$-amine. Alternatively, Boc-amino acids may be coupled without prior neutralization by using *in situ* neutralization, i.e., coupling in the presence of DIEA or *N*-methylmorpholine (NMM) (16, 17). By using HBTU/DIEA as coupling reagent, activation of the protected amino acid and *in situ* neutralization take place concurrently resulting in shorter cycle times for synthesis (18).

"Permanent" side-chain protecting groups are ether, ester, and urethane derivatives based on benzyl alcohol. Alternatively, ether and ester derivatives based on cyclopentyl or cyclohexyl alcohol are sometimes applied, as their use moderates certain side reactions. These "permanent" groups are sufficiently stable to repeated cycles of Boc removal, yet cleanly cleaved in the presence of appropriate scavengers by use of liquid anhydrous hydrogen fluoride (HF) at 0°C or trimethylsilyl trifluoromethanesulfonate (TMSOTf)/TFA at 25°C. The phenylacetamidomethyl (PAM; for producing peptide acids) or 4-methylbenzhydrylamine (MBHA; for producing peptide amides) anchoring linkages have been designed to be cleaved at the same time (Fig. 2).

HF cleavages are always carried out in the presence of a carbonium ion scavenger. TMSOTf/TFA has also been used for strong acid cleavage and deprotection reactions, which are accelerated by the presence of a "soft" nucleophile.

### 2. *9-Fluorenylmethoxycarbonyl (Fmoc)-Based Chemistry*

A mild orthogonal alternative protection scheme is constructed using Carpino's base-labile but acid stable 9-fluorenylmethyloxycarbonyl (Fmoc) group for tem-

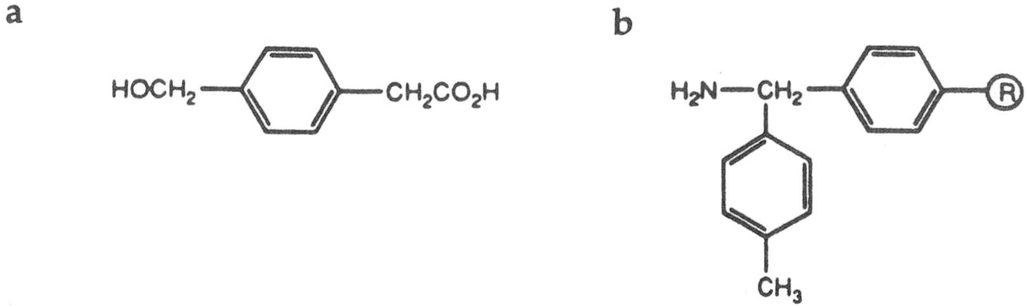

a

b

**Fig. 2** Structures of (a) phenylacetamidomethyl (PAM) linker and (b) 4-methyl benzhydrylamine (MBHA) resin used in Boc-based SPPS.

porary blocking of the $\alpha$-amino group. "Permanent" protection is provided primarily by ether, ester, and urethane derivatives based on *tertiary*-butanol. These derivatives are cleaved at the same time as appropriate anchoring linkages, by use of TFA at 25°C. Scavengers must be added to the TFA to trap the reactive carbocations which form under the acidolytic cleavage conditions. Removal of the Fmoc group is achieved usually with 20–55% piperidine in DMF or NMP for 10–18 min (19). Piperidine scavenges the liberated dibenzofulvene to form a fulvene-piperidine adduct. 2% 1,8-diazabicyclo[5.4.0]undec-7-ene (DBU)-DMF can also be used for Fmoc removal (20). The dibenzofulvene intermediate does not form an adduct with DBU and thus must be washed rapidly from the peptide-resin to avoid reattachment of dibenzofulvene (20). A solution of DBU-piperidine-DMF (1:1:48) is effective for batch syntheses, as the piperidine component scavenges the dibenzofulvene (21). After Fmoc removal, the liberated $N^{\alpha}$-amine of the peptide-resin is free and ready for immediate acylation.

The TFA-labile 4-hydroxymethylphenoxy (HMP; for producing peptide acids), 2-chlorotrityl (for producing peptide acids), 5-(4-aminomethyl-3,5-dimethoxyphenoxy)valeric acid (PAL; for producing peptide amides), or 4-(2′, 4′- dimethoxyphenylaminomethyl) phenoxy (Rink amide) anchoring linkages are used in conjunction with Fmoc chemistry (Fig. 3).

## D. Attachment of the First Amino Acid

Almost all syntheses by the solid-phase method are carried out in the $C \rightarrow N$ direction, and therefore generally start with the intended $C$-terminal residue of the desired peptide being linked to the support either directly or via a suitable handle. Such linkers have been designed so that eventual cleavage provides either a free acid or amide at the $C$-terminus, although, in specialized cases, other useful end groups can be obtained.

All of the anchoring linkages that ultimately provide peptide acids are esters; rates and yields of reactions for ester bond formation are less than those for corresponding methods for amide bond formation. Consequently, compromises are needed to achieve reasonable loading reaction times and substitution levels, while ensuring that the extent of racemization remains acceptably low. The well-established cesium salt method (22) allows loading of $N^{\alpha}$-protected amino acids to chloromethyl linkers and resins with low levels of racemization (23, 24) while effectively preventing alkylation of susceptible residues (Cys, His, Met). The most generally applicable ester bond formation methods for $N^{\alpha}$-protected amino

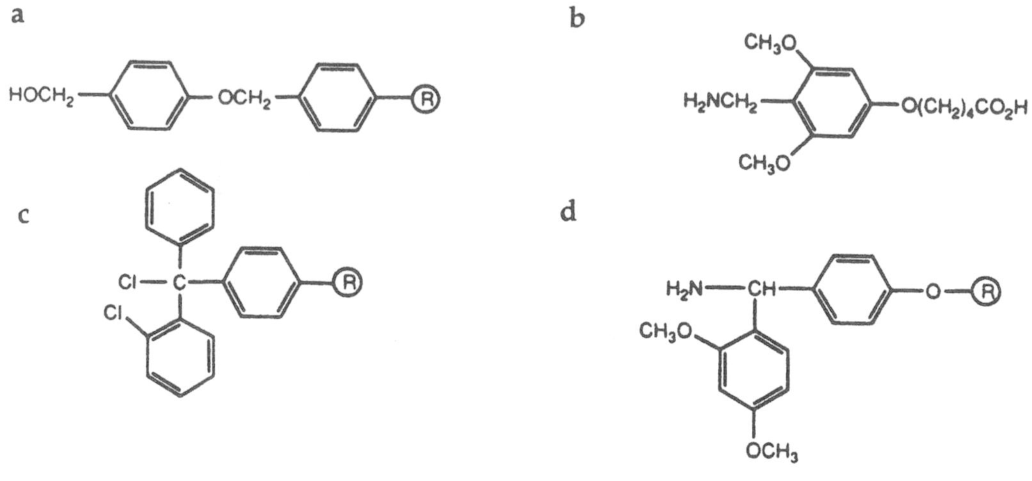

a

b

c

d

**Fig. 3** Structures of (a) 4-hydroxymethyl-phenoxy (HMP) resin, (b) 5-(4-aminomethyl-3, 5-dimethoxyphenoxy) valeric acid (PAL) linker, (c) 2-chlorotrityl (Clt) chloride resin and (d) 4-(2′,4′-dimethoxyphenylaminomethyl) phenoxy (Rink amide) resin used in Fmoc-SPPS.

acids are *in situ* with carbodiimide (DCC or DIPCDI) in the presence of catalytic amounts of DMAP (0.06–0.1 equiv) or with *N,N*-dimethylformamide dineopentyl acetal. For esterifications in the presence of DMAP, time and temperature should be carefully mediated, and HOBt (1-2 equiv) may be included.

Fmoc-His and Fmoc-Cys derivatives are particularly difficult to load efficiently while suppressing racemization. Low racemization loadings have been documented using the cesium salts of Fmoc-His(Trt) (0.4% D-His) and Fmoc-Cys(Acm) (0.5% D-Cys) (24), and for the side-chain anchoring of Fmoc-Cys(2-Xal$_4$)-O$t$Bu via attachment to PEG-PS resin (25). For Fmoc-His (Bum), an efficient loading procedure involves Fmoc-His(Bum) (2 equiv) esterified *in situ* by *N,N′*-diisopropyl-carbodiimide (DIPCDI) (2 equiv) and DMAP (0.16 equiv) in DCM-DMF (1:3) for 1 h (26).

Esterification of $N^\alpha$-protected Asn and Gln can be sluggish (13, 26-28). Fmoc-Asp(OH)-O$t$Bu or Fmoc-Glu

(OH)-O$t$Bu have been coupled (via an unprotected $\beta$- or $\gamma$-carboxyl side-chain) to amide-forming linker-resins, with TFA cleavage yielding peptides containing *C*-terminal Asn or Gln (26, 29, 30).

**E. Side-Chain Protecting Groups**

Choices for the permanent side chain protection are made in the context of potential side reactions, which should be minimized. Problems may be anticipated either during the coupling steps or during final cleavage/deprotection. For certain residues (e.g., Cys, Asp, Glu, Lys), side-chain protection is absolutely essential, whereas for others, an informed decision should be made depending upon the length of the synthetic target and other considerations. Most solid-phase syntheses follow maximal rather than minimal protection strategies. Almost all of the useful $N^\alpha$-Boc and $N^\alpha$-Fmoc protected derivatives are commercially available.

The side-chain carboxyls of Asp and Glu are protected by the benzyl (OBzl) group in Boc chemistry or by *tertiary*-butyl (O*t*Bu) esters in Fmoc chemistry. To minimize the imide/$\alpha \rightarrow \beta$ rearrangement side reaction (see "Side Reactions"), Fmoc-Asp may be protected with the 1-adamantyl (O-1-Ada) group (31) and Boc-Asp with either the 2-adamantyl (O-2-Ada) (31) or cyclohexyl (OcHex) (32) groups. The N[1-(4, 4,-dimethyl-2,6-dioxocyclohexylidene)-3--methylbutyl]aminobenzyl (Dmab) group can be used for quasi-orthogonal protection of Asp, as Dmab is stable to 20% piperidine and can be selectively removed with 2% hydrazine in DMF (33). Fully orthogonal protection is achieved with allyl esters, which are selectively removed by tetrakistriphenylphosphine-palladium (0) (34, 35; see "Lactams").

The side-chain hydroxyls of Ser, Thr, and Tyr are protected as Bzl (Boc strategy) and *t*Bu (Fmoc strategy) ethers. In strong acid, the Bzl protecting group can migrate to the 3-position of the Tyr aromatic ring (36). This side reaction is decreased greatly when Tyr is protected by the 2,6-dichlorobenzyl (2,6-Cl$_2$Bzl) (36) or 2-bromobenzyloxycarbonyl (2-BrZ) (37) group; consequently, the latter two derivatives are much preferred for Boc SPPS.

The ε-amino group of Lys is best protected by the the 2-chlorobenzyloxy-carbonyl (2-ClZ) or Fmoc group for Boc chemistry and reciprocally by the Boc group for Fmoc chemistry. Orthogonal protection for Fmoc-Lys is provided by the palladium-sensitive allyloxycarbonyl (Aloc) group (38, 39). In addition, Fmoc-Lys(Dde) is an excellent choice for selective deprotection and subsequent preparation of cyclic peptides via lactam formation. It has also been successfully employed for the synthesis of branched peptides and peptide templates (40, 41). Dde (1-(4,4-dimethyl-2,6-dioxocyclohex-1-ylidene)ethyl) is stable to 20% piperidine and can be selectively cleaved with 2% hydrazine in DMF (41). Another selectively removable side-chain pro-

tecting group for Lys is Mtt, which is labile to 1% TFA/triisopropylsilane in DCM (42, 43).

The highly basic trifunctional guanidino side-chain group of Arg may be protected by appropriate benzenesulfonyl derivatives, such as the 4-toluenesulfonyl (Tos) or mesitylene-2-sulfonyl (Mts) groups in conjunction with Boc chemistry and 4-methoxy-2,3,6-trimethylbenzene-sulfonyl (Mtr), 2,2,5,7,8-pentamethyl-chroman-6-sulphonyl (Pmc), or 2,2,4,6, 7-pentamethyldihydro-benzofuran-5-sul-fonyl (Pbf) groups for Fmoc chemistry. These groups most likely block the ω-nitrogen of Arg, and their relative acid lability is Pbf > Pmc > Mtr (44–46). The Mtr group may require extended TFA-thioanisole treatment (2–8 h) for removal, while Arg(Pmc) is deprotected readily by 50% TFA (<2 h). A number of other Arg protecting groups have been proposed, particularly $N^{\omega}$-mono or $N^{\delta,\omega}$-bis-urethane derivatives. However, based on current information, the afore-mentioned benzenesulfonyl derivatives seem to offer the best prospects of clean incorporation of Arg without ornithine contaminants (47).

Activated His derivatives are uniquely prone to racemization during stepwise SPPS, due to an intramolecular abstraction of the proton on the optically active α-carbon by the imidazole π-nitrogen (48). Racemization could be suppressed by either blocking the τ- or π-imidazole nitrogen. The Tos group blocks the $N^{\tau}$ of Boc-His, and is removed by strong acids. However, the Tos group is also lost prematurely during SPPS steps involving 1-hydroxybenzotriazole (HOBt); this allows acylation or acety-lation of the imidazole group, followed by chain termination due to $N^{im} \rightarrow N^{\alpha}$-amino transfer of the acyl or acetyl group (49, 50). Therefore, HOBt should never be used during couplings of amino acids once a His(Tos) residue has been incorporated into the peptide-resin. An HF-stable, orthogonally removable, $N^\tau$-protecting group for Boc strategies is the 2,4-dinitrophenyl (Dnp) function. Final Dnp deblocking is best carried out at the peptide-resin level prior to the HF cleavage step, by use of thiophenol in DMF. Since Dnp and Trt $N^\tau$-protection do not allow preformed symmetrical anhydride coupling with low racemization, it is recommended that the appropriate derivatives be coupled as preformed esters or *in situ* with carbodiimide in the presence of HOBt (51,52). Boc-His(Tos) is coupled efficiently using benzotriazol-1-yl-oxy-tris(dimethylamino)phosphonium hexafluorophosphate (BOP) (3 equiv) in the presence of DIEA (3 equiv); these conditions minimize racemization and avoid premature side-chain deprotection by HOBt (53).

The $\tau$-nitrogen of Fmoc-His can be protected by the Boc and triphenylmethyl (Trt) groups. When His is $N^\tau$-protected by the Boc group, the basicity of the imidazole ring is reduced sufficiently so that acylation by the preformed symmetical anhydride (PSA) method proceeds with little racemization (3). The Trt group reduces the basicity of the imidazole ring (the $pK_a$ decreases from 6.2 to 4.7), although racemization by the PSA method is not eliminated completely (52). Since Dnp and Trt $N^\tau$-protection do not allow PSA coupling with low racemization, it is recommended that the appropriate derivatives be coupled as preformed esters or *in situ* with carbodiimide in the presence of HOBt (51, 52).

Blocking of the $\pi$-nitrogen of the imidazole ring has been shown to be effective in reducing His racemization (54). The $N^\pi$ of His can be protected by the *tertiary*-butoxymethyl (Bum) group. However, TFA deprotection of His(Bum) liberates formaldehyde, which can modify susceptible side-chains (see "Cleavage").

The carboxamide side-chains of Asn and Gln are often left unprotected in SPPS, but this approach leaves open the danger of dehydration to form nitriles upon activation with *in situ* reagents. On the other hand, acylations by activated esters result in minimal side-chain dehydration (13, 55, 56) (see "Formation of the Peptide Bond"). Nitrile formation is also inhibited during *in situ* carbodiimide acylations when HOBt is added (55, 56) (see "Formation of the Peptide Bond"). However, the presence of HOBt does not effectively inhibit $N^\alpha$-protected Asn dehydration during BOP *in situ* acylations (56).

At the point where an $N^\alpha$-amino protecting group is removed from Gln, the possibility exists for an acid-catalyzed intramolecular aminolysis which displaces ammonia and leads to pyroglutamate formation (13, 57, 58). Cyclization occurs primarily during couplings, as $N^\alpha$-protected amino acids and HOBt promote this side reaction (57). Side-chain protecting groups such as 2,4,6-trimethoxybenzyl (Tmob), and Trt minimize the occurence of dehydration (59, 60) and pyroglutamate formation, and may also inhibit interchain hydrogen bonding that otherwise leads to secondary structures which substantially reduce coupling rates. Unprotected Fmoc-Asn and -Gln have poor solubility in DCM and DMF; solubility is improved considerably by the Tmob or Trt side-chain protected derivatives.

The thioether side-chain of Met survives cycles of Fmoc chemistry, but protection during Boc chemistry is often advisable. The reducible sulfoxide function is applied under these circumstances. Smooth deblocking of methionine sulfoxide occurs in 20–25% HF in the presence of dimethylsulfide (61).

The highly sensitive side chain of Boc-Trp is best protected by the formyl (For) group (62). Trp(For) is deprotected at the peptide-resin level by treatment with piperidine-DMF (9:91), 0°C, 2 h, *prior* to HF cleavage. In Fmoc chemistry, the indole function of Trp is most efficiently protected with the Boc group (63). Cleavage with TFA generates an $N^{in}$

carboxy-indole which protects the Trp from alkylation (63-65) and sulfonation (63, 66, 67). The $N^{in}$ carboxy group is then removed during normal work-up of the peptide in aqueous solution.

The most challenging residue to manage in peptide synthesis is Cys. Compatible with Boc chemistry are the 4-methylbenzyl (Meb), acetamidomethyl (Acm), *tertiary*-butylsulfenyl (S$t$Bu), and 9-fluorenylmethyl (Fm) $\beta$-thiol protecting groups. The Meb group is optimized for removal by strong acid (36); Cys(Meb) residues may also be directly converted to the oxidized (cystine) form by thallium (III) trifluoroacetate [Tl(Tfa)$_3$], although some cysteic acid forms at the same time. Cys(Fm) is stable to acid, and cleaved by base. Compatible with Fmoc chemistry are the Acm, S$t$Bu, Tmob, and Trt groups. The Trt and Tmob groups are labile in TFA; due to the tendency of the resultant stable carbonium ions to realkylate Cys (68, 69), effective scavengers are needed. The Acm group is acid- and base-stable and removed by mercuric (II) acetate, followed by treatment with H$_2$S or excess mercaptans to free the $\beta$-thiol. Mercuric (II) acetate can modify Trp, and thus should be used in the presence of 50% acetic acid for Trp-containing peptides (70). In multiple Cys(Acm)-containing peptides, mercuric (II) acetate may not be a completely effective removal reagent (71). Alternatively, Cys(Acm) residues are converted directly to disulfides by treatment with I$_2$ or Tl(Tfa)$_3$ on-resin (72, 73). Finally, the acid-stable S$t$Bu group is removed by reduction with thiols or phosphines.

Cys is also susceptible to racemization during SPPS. When applying protocols for Cys incorporation which include phosphonium and aminium salts as coupling reagents (BOP, HBTU, HATU, and PyAOP), as well as preactivation in the presence of suitable additives (HOBt and HOAt) and tertiary amine bases (DIEA and NMM), 5–33%

racemization is observed (74). These high levels of racemization are generally reduced by avoiding preactivation, using a weaker base (such as collidine), and switching to the solvent mixture DMF-DCM (1:1). Couplings with less than 1% racemization were successfully performed using either BOP, HBTU, or HATU and HOAt/collidine without preactivation in DMF-DCM. Alternatively, the Pfp-ester of a suitable Fmoc-Cys derivative can be used. Under these optimized conditions Cys protecting groups Tmob, Trt, Xan (75), and Acm do not enhance the risk of racemization (74).

## F. Formation of the Peptide Bond

There are currently four major categories of coupling techniques that serve well for the *stepwise* introduction of $N^{\alpha}$-protected amino acids for solid-phase synthesis. In the solid-phase mode, coupling reagents are used in excess to ensure that reactions reach completion.

### 3. In Situ *Reagents*

The classical example of an *in situ* coupling reagent is $N,N'$-dicyclohexylcarbodiimide (DCC) (76, 77). The related $N,N$-diisopropylcarbodiimide (DIPCDI) is more convenient to use under some circumstances, as the resultant urea co-product is more soluble in DCM. The generality of carbodiimide-mediated couplings is extended significantly by the use of 1-hydroxybenzotriazole (HOBt) as an additive, which accelerates carbodiimide-mediated couplings, suppresses racemization, and inhibits dehydration of the carboxamide side-chains of Asn and Gln to the corresponding nitriles (55, 78, 79). Recently, protocols involving benzotriazol-1-yl-oxy-tris(dimethylamino)phosphonium hexafluorophosphate (BOP), 2-(1H-benzotriazol-1-yl)-1,1,3,3-tetramethyluronium hexafluorophosphate (HBTU), 2-(1H-benzotriazol-1-yl)-1,1,3,

3-tetramethyl- uronium tetrafluoroborate (TBTU), and 2-(2-oxo-1(2H)-pyridyl)-1,1,3,3-bispentamethylene-uronium tetrafluoroborate (TOPPipU) have deservedly achieved popularity. BOP, HBTU, TBTU, and TOPPipU require a tertiary amine such as NMM or DIEA for optimal efficiency (80–86). HOBt has been reported to accelerate further the rates of BOP- and HBTU-mediated couplings (85, 87). *In situ* activations by excess HBTU or TBTU can cap free amino groups (88, 89); it is not known whether HOBt can suppress this side reaction. Acylations using BOP result in the liberation of the carcinogen hexamethylphosphoramide. The modified BOP reagent benzotriazole-1-yl-oxy-tris-pyrrolidinophosphonium hexafluorophosphate (PyBOP) liberates potentially less toxic by-products (90). Protocols have been reported for the use of BOP to incorporate side-chain unprotected Thr and Tyr (81, 91).

1-Hydroxy-7-azabenzotriazole (HOAt) and its uronium salt derivative O-(7-azabenzotriazol-1-yl)-1,1,3,3-tetramethyluronium hexafluorophosphate (HATU) have been found to enhance coupling yields and reduce racemization more efficiently than HOBt (92). The efficiency of the additive HOAt and HATU has been demonstrated by the Fmoc SPPS of acyl carrier protein fragment 65–74 (92). HATU/HOAt couplings have been successfully applied for incorporation of sterically hindered amino acids such as Aib (92).

## 4. *Active Esters*

A long-known but steadfast coupling method involves the use of active esters. For SPPS, workers have concentrated on pentafluorophenyl (OPfp), HOBt, and 3-hydroxy-2,3-dihydro-4-oxo-benzotriazine (ODhbt) esters. Although OPfp esters alone couple slowly, the addition of HOBt (1-2 equiv) increases the reaction rate (93, 94). Fmoc-Asn-OPfp allows for efficient

incorporation of Asn with little side-chain dehydration (56). HOBt esters of Fmoc-amino acids are rapidly formed (with DIPCDI) and highly reactive (95, 96). $N^\alpha$-protected amino acid ODhbt esters suppress racemization and are highly reactive, in similar fashion to HOBt esters (97).

## 5. *Amino Acid Halides*

Fmoc-amino acid chlorides are basically stable, highly reactive reagents for peptide bond formation (98). Unfortunately, trifunctional Fmoc amino acids with *tertiary*-butyl side-chain protection are not converted to the corresponding chlorides (99). Therefore, the use of Fmoc-amino acid chlorides is rather limited. In contrast, Fmoc-amino acid fluorides with *tertiary*-butyl side-chain protection could be obtained without difficulties. They have proven to be efficient coupling reagents inducing low levels of racemization, and are especially suited for incorporation of sterically hindered amino acids such as Aib (100). Fmoc-amino acid fluorides have been successfully applied in the solid phase synthesis of alamethicin F30 and F50, saturnisporin SIII, and trichtoxin A 50-J (101).

## 6. *Preformed Symmetrical Anhydrides and N-Carboxyanhydrides*

Preformed symmetrical anhydrides (PSAs) are favored by some workers because of their high reactivity. They are generated *in situ* from the corresponding $N^\alpha$-protected amino acid (2 or 4 equiv) plus DCC (1 or 2 equiv) in DCM; following removal of the urea by filtration, the solvent is exchanged to DMF for optimal couplings.

Preformed Fmoc-*N*-carboxyanhydrides (NCAs) are stable derivatives which do not have to be generated *in situ*. Their reactivity is comparable to PSAs. In addition, NCAs are thermally stable, allowing for couplings at elevated temperatures. Such protocols can be advantageous for coupling sterically hindered amino acids. NCAs are sensitive to moisture and should therefore be used in dry solvents (102, 103).

## III. OPTIMIZING A SYNTHESIS

The aim of a synthesis is to obtain the desired peptide in high yields and high purity. Assembling a peptide on a solid support means performing a multistep synthesis without isolating or purifying intermediate products. Therefore, it is desirable to minimize the generation of deletion sequences and other side products by applying optimized conditions for each reaction during synthesis. Avoiding deletion sequences during the synthesis of longer peptides is especially important, as deletion peptides often have physical properties very similar to the desired peptide and are difficult to remove during purification (104). It can be advantageous to monitor each step during synthesis, and a capping step may be performed after each coupling to terminate poorly reactive peptide chains.

## A. Solvation and Aggregation

Effective solvation of the peptide-resin is one of the most crucial conditions for efficient chain assembly (3). Under proper solvent conditions, there is no decrease in synthetic efficiency up to 60 amino acid residues in Boc SPPS (1). The ability of the peptide-resin to swell increases with increasing peptide length due to a net decrease in free energy from solvation of the linear peptide chains (1). Therefore, there is no theoretical upper limit to efficient amino acid couplings, provided that proper solvation conditions exist (105). In practice, obtaining these conditions is not always straightforward.

"Difficult sequences" during SPPS have been attributed to poor solvation of the growing chain within the peptide resin complex resulting in slow or incomplete Fmoc-deprotection and coupling reactions (3, 106). Infrared and NMR spectroscopies have shown that intermolecular $\beta$-sheet aggregates are responsible for lowering coupling efficiencies (107–109).

Aggregation also occurs in regions of apolar side-chain protecting groups, sometimes resulting in a collapsed gel structure (110, 111). In cases where aggregation occurs due to apolar side-chain protecting groups, increased solvent polarity may not be sufficient to disrupt the aggregate. A relatively unstudied problem of Fmoc chemistry is that the lack of polar side-chain protecting groups could, during the course of an extended peptide synthesis, inhibit proper solvation of the peptide-resin (110, 112). To alleviate this problem, the use of solvent mixtures containing both a polar and nonpolar component, such as THF-NMP (7:13) or TFE-DCM (1:4), is recommended (112). The addition of DMSO (112, 113), chaotropic salts (114), or a solvent mixture containing detergents (115) can be effective for disrupting such aggregates.

A different approach to circumvent potential aggregation is the use of reversible protection of the amino acid amide nitrogen. For this purpose, $N^{\alpha}$-Fmoc,$N^{\alpha}$(2-Fmoc-hydroxy-4-methoxybenzyl) [Fmoc(Fmoc-Hmb)] derivatives have been prepared (65). Formation of $\beta$-sheet stucture can be minimized by using amide nitrogen protection every sixth to eighth amino acid (65).

## B. Side Reactions

Side-reactions that occur during SPPS have been reviewed extensively (3, 13, 14, 116–118). The present discussions focus on new appproaches for alleviating well established side-reactions.

## 1. Diketopiperazine Formation

The free $N^{\alpha}$-amino group of an anchored dipeptide is poised for a base-catalyzed intramolecular attack of the C-terminal carbonyl (13, 119, 120). Base deprotection (Fmoc) or neutralization (Boc) can thus release a cyclic diketopiperazine while a hydroxymethyl-handle leaving group remains on the resin. With residues that can form *cis* peptide bonds, e.g., Gly, Pro, N-methylamino acids, or D-amino acids, in either the first or second position of the $C \rightarrow N$ synthesis, diketopiperazine formation can be substantial (120–122). For most other sequences, the problem can be adequately controlled by *in situ* neutralization (see "Protection Schemes: Boc-Based Chemistry"). For susceptible sequences being addressed by Fmoc chemistry, the use of piperidine-DMF (1:1) deprotection for 5 min (120), or deprotection for 2 min with a 0.1 M solution of tetrabutylammonium fluoride in DMF ("quenched" by MeOH) (123) has been recommended to minimize cyclization. Alternatively, the second and third amino acids may be coupled as a preformed $N^{\alpha}$-protected-dipeptide, avoiding the diketopiperazine-inducing deprotection/neutra-lization at the second amino acid. The steric hindrance of the 2-chlorotrityl linker may minimize diketopiperazine formation of susceptible sequences during Fmoc chemistry (124, 125).

## 2. Aspartimide Formation

A sometimes serious side reaction with protected Asp residues involves an intramolecular elimination to form an aspartimide, which can then partition in water to the desired $\alpha$-peptide and the undesired by-product with the chain growing from the $\beta$-carboxyl (13, 32, 126) (Fig. 4). Aspartimide formation is sequence dependent, with Asp(OBzl)-Gly, -Ser, -Thr, -Asn, and -Gln sequences showing the greatest tendency to cyclize under basic conditions (126–128); the same sequences are also quite susceptible in strong acid (127). Sequences containing Asp(OtBu)-Gly are somewhat susceptible to base-catalyzed aspartimide formation (11% after 4 h treatment with 20% piperidine in DMF) (128), but do not rearrange at all in acid (129).

Piperidine catalysis of aspartimide formation from side-chain protected Asp residues can be rapid, and is dependent upon the side-chain protecting group. Aspartimide formation can also be conformation dependent (130, 131). This side-reaction can be minimized by including 0.1 M HOBt in the piperidine solution (132) or by using an amide backbone protecting group (i.e., 2-hydroxy-4-methoxybenzyl) for the residue in the X position of an Asp-X sequence (131).

## 3. Deamidation

Asn (and even Gln) can undergo deamidation in (acidic) solution (133–135) or when stored in the solid state in the presence of residual acid (136). Therefore, Asn- and Gln-containing peptides should never be stored in a solid form with residual acid present, and samples stored in solution should be monitored carefully for deamidation and decomposition. In basic solution, the deamidation is accompanied by aspartimide formation and $\alpha \rightarrow \beta$ rearrangement (137). Succinimide formation is largely sequence dependent, with Asn-Gly showing the greatest tendency to rearrange (133).

## 4. Racemization of Cys Residues

C-terminal esterified (but not amidated) Cys residues are racemized by repeated piperidine deprotection treatments during Fmoc SPPS. Following 4 h exposure to piperidine-DMF (1:4), the extent of racemization found was 36% D-Cys from Cys(StBu), 12% D-Cys from Cys(Trt), and 9% D-Cys from Cys(Acm) (138).

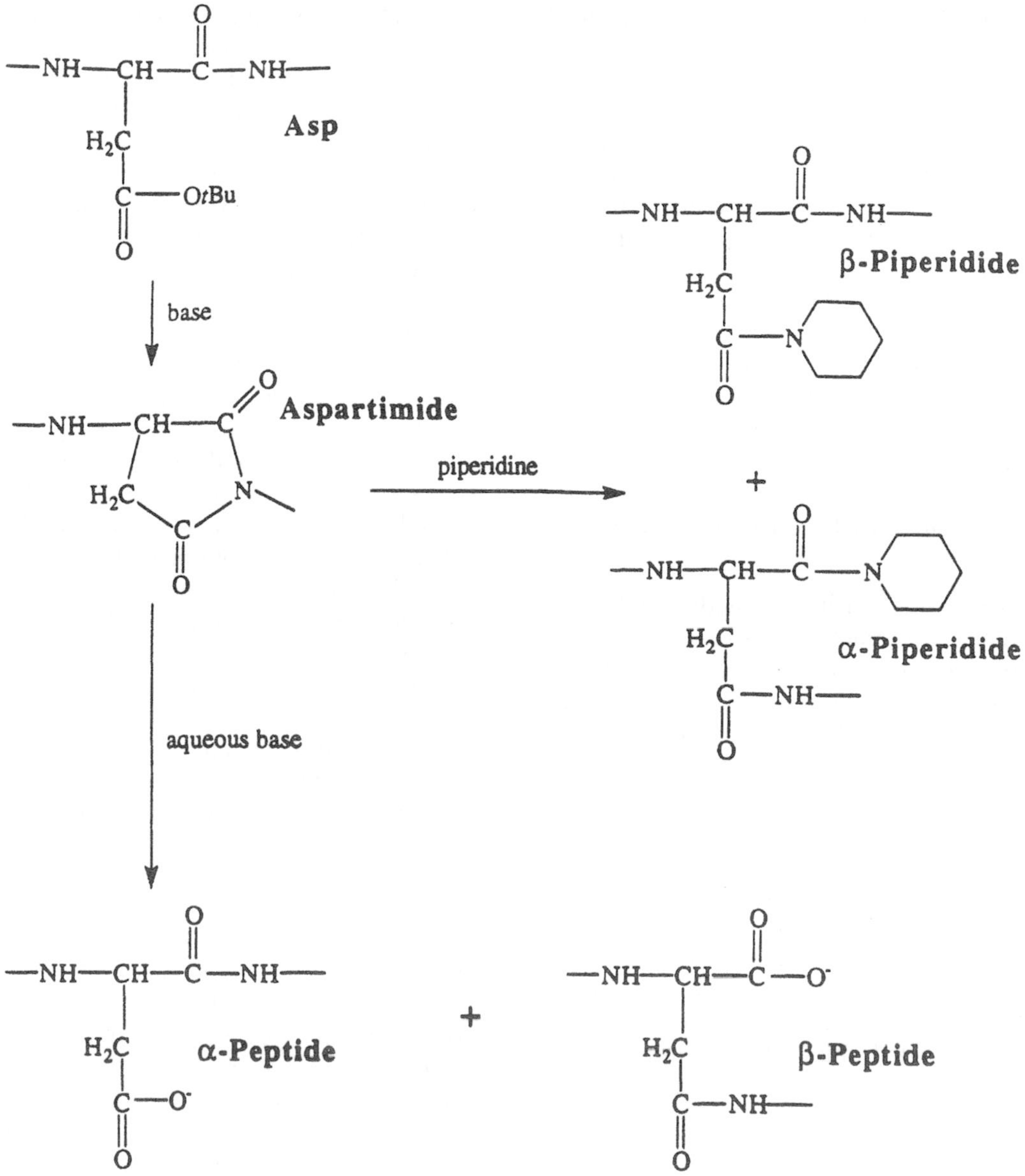

**Fig. 4** Formation and ring opening reactions of aspartimide.

Racemization of esterified Cys(Trt) was reduced from 11.8% with 20% piperidine-DMF to only 2.6% with 1% DBU-DMF after 4 h treatment (20, 138). Additionally, the steric hindrance of the 2-chlorotrityl linker has been shown to minimize racemization of C-terminal Cys residues (139).

Reduced racemization may be achieved by anchoring Cys to the resin via its side-chain. For this purpose, Fmoc-Cys-OtBu is condensed with 5-(9-hydroxyxanthen-2-oxy)valeric acid (Xal) to provide a preformed handle, which is subsequently coupled to an amino functionalized PEG-PS resin (25). Preformed handle coupling was performed using DIPCDI/HOBt (4 equiv. each) or BOP/HOBt/NMM (4, 4, and 8 equiv., respectively) for 2 hours.

At least two examples have been provided where the use of Cys(S$t$Bu) as the $C$-terminal residue esterified to an HMP-type resin was entirely incompatible with formation of the desired peptide. Instead, TFA cleavage gave reduction-resistant by-products (structures not fully determined) retaining the $t$Bu group but evidently missing two molecules of water based on mass spectrometric evidence (140).

## 5. Acid-Sensitive Side-Chains and Bonds

Trp is quite sensitive to acidic conditions, undergoing reactions with carbonium ions and molecular oxygen (see "Peptide-Resin Cleavage"). Synthesis of Trp-containing peptides is thus best approached via Fmoc chemistry, where acidolysis is kept at a minimum. Gramicidins A, B, and C (where either three or four of the 15 residues are Trp) have been synthesized efficiently by Fmoc chemistry; acid was avoided entirely throughout the synthesis and final nucleophilic cleavage was achieved with ethanolamine (96, 141). Fmoc chemistry has also been suggested for incorporation of $^2$H-labeled amino acids, as the repetitive acidolyses of Boc chemistry can exchange out the $^2$H label (117, 142). Finally, Fmoc chemistry may be the better choice for the synthesis of peptides containing the acid-labile Asp-Pro bond.

## C.  Monitoring

A crucial issue for stepwise solid-phase peptide synthesis is the repetitive yield per deprotection/coupling cycle. The efficiency of these steps may be monitored by several methods. The best known qualitative monitoring methods are the ninhydrin (143) and isatin (144) tests for free $N^\alpha$-amino and -imino groups, respectively, where a positive colorimetric response to an aliquot of peptide-resin indicates the presence of unreacted $N^\alpha$-amino/imino groups. These tests are easy, reliable, and require only a few minutes to perform, allowing the chemist to make a quick decision on how to proceed. A highly accurate quantitative modification of the ninhydrin procedure has been developed (118).

Other monitoring techniques exist that are generally non-destructive (non-invasive) and can therefore be carried out on the total batch of resin. Resin-bound $N^\alpha$-amino groups can be titrated with picric acid, 4,4'-dimethoxytrityl chloride, bromophenol blue dye, and quinoline yellow dye. Picric acid is removed from resin-bound amines with 5% DIEA in DCM, and the resulting chromophore quantitated at $\lambda = 362$ nm (145, 146). For trityl monitoring, 4,4'-dimethoxytrityl- chloride and tetra-$n$-butylammonium perchlorate are reacted with the resin, released with 2% dichloroacetic acid in DCM, and quantitated at $\lambda = 498$ nm (147, 148). The effect of the dilute acid on Fmoc-amino acid side-chain protecting groups and linkers has not been reported. For bromophenol blue and quinoline yellow monitoring, the dye is bound to free amino groups following deprotection, then displaced as acylation proceeds. Quantitative monitoring can be carried out at $\lambda = 600$ and 495 nm for bromophenol blue and quinoline yellow, respectively (104, 149, 150).

As an alternative to quantitating resin-bound species, soluble reactants or co-products can be analyzed. Continuous measurement of electrical conductivity can be used to evaluate coupling and Fmoc deprotection efficiencies (151–153). The progress of Fmoc chemistry can be evaluated by observing at $\lambda = 300–312$ nm the decrease of absorbance when Fmoc-amino acids are taken up during coupling, and by the increase in absorbance when the Fmoc group is released with piperidine (3, 111, 154, 155). Monitoring a decrease in Fmoc-amino acid concentration at $\lambda = 300$ nm can be complicated when OPfp esters are utilized (156). More straightforward acylation monitoring is possible when Fmoc-amino acid ODhbt esters are used. During the coupling of Fmoc-amino acid ODhbt esters, the liberated Dhbt-OH component binds to free $N^\alpha$-amino groups, producing a bright yellow color, which diminishes as acylation proceeds (157). Deprotection of the Fmoc group can proceed slowly in certain sequences (111, 158). By monitoring deprotection as the synthesis proceeds, one can extend base deprotection times and/or alter solvation conditions as necessary (159).

Invasive monitoring of both synthetic efficiency and amino acid composition of peptide-resins can be achieved by a powerful quantitative variation of the Edman sequential degradation, called "preview analysis" (160–162). Crude peptide-resins are sequenced directly; each Edman degradation cycle serves to identify a primary amino acid residue and "preview" the next amino acid in the sequence. Because preview is cumulative, quantitation of peaks after a number of cycles indicates the average level of deletion peptides and thus the overall synthetic efficiency. Preview sequence analysis of Fmoc synthesized peptide-resins requires initial TFA cleavage, followed by immobilization (covalent or non-covalent) of the crude peptide on a suitable support (163, 164). Most side-chain protecting groups used in conjunction with Fmoc chemistry are not stable to the conditions of Edman degradation; hence, the usual free side-chain phenylthiohydantoin standards can be used.

Recently, on-resin MS via the MALDI technique has been developed (165). This method can be used as a semi on-line monitoring technique during SPPS (166). An especially elegant approach for on-resin MS is to use the photolabile $\alpha$-methylphenacyl-ester linker, which is cleaved directly upon laser photolysis and MALDI ionization (165, 167). On-resin monitoring may also be achieved by NMR using the magic angle spinning method with a nanoprobe (168). Magic angle spinning allows for the possible recording of high resolution NMR spectra of the resin-bound growing peptide chain and analyis of the synthesis.

The technique of "internal reference amino acids" (IRAA) is often very useful to measure accurately yields of chain assembly and retention of chains on the support during synthesis and after cleavage (3, 169, 170). In addition, amino acid analysis of peptide-resins may be used to monitor synthetic efficiency; the advent of microwave hydrolysis technology may permit rapid analysis (171).

## D. Capping

Some workers choose to "cap" unreacted chains, thereby substituting a family of terminated peptides for a family of deletion peptides. In either case, these by-products must ultimately be separated from the desired product. Intentional termination of chains may be carried out when there is an indication of unreacted sites. In the simplest case, capping is carried out with reactive acetylating agents, such as acetic anhydride ($Ac_2O$) or $N$-acetylimidazole in the presence of tertiary base (116, 172).

## IV. SYNTHESIS OF MODIFIED PEPTIDES

In order to obtain potent peptide analogs with improved receptor selectivity and metabolic stability, the natural peptide sequence can be altered by modification of side-chains or the backbone in various ways. One can duplicate, by chemical synthesis, post-translational modifications achieved in nature including formation of half-cysteine residues in disulfide bonds. We can also consider a set of unnatural structures which are of considerable interest for peptide drug design, namely cyclic peptides acheived via lactams.

## A. Hydroxylated Residues

Hydroxyproline (Hyp) has been incorporated successfully without side-chain protection in both Boc (173, 174) and Fmoc (175, 176) SPPS. Alternatively, the usual hydroxyl protecting groups Bzl (177) for Boc and $t$Bu (178) for Fmoc can be used. Hydroxylysine (Hyl) can be incorporated in SPPS as Fmoc-Hyl(Boc,$O$-Tbdms (179). Recently, Fmoc-Hyl(Aloc)-OH was synthesized starting from pure L-Hyl by simply introducing the Aloc group in $N^\varepsilon$-position to copper complexed Hyl (180).

## B. γ-Carboxyglutamate

Acid-sensitive $\gamma$-carboxyglutamate (Gla) residues have been identified in a number of diverse biomolecules, such as prothrombin and the "sleeper" peptide from the venomous fish-hunting cone snail (*Conus geographus*). Fmoc chemistry has been utilized for the efficient SPPS of the 17-residue "sleeper" peptide, with the 5 Gla residues incorporated as Fmoc-Gla (O$t$Bu)$_2$ (181). Cleavage and side-chain deprotection of the peptide-resin by TFA-DCM (2:3) for 6 h resulted in no apparent conversion of Gla to Glu.

## C. Phosphorylation

Incorporation of side-chain phosphorylated Ser and Thr by SPPS is especially challenging, as the phosphate group is decomposed by strong acid, and lost with base in a $\beta$-elimination process

(182). Boc-Ser(PO$_3$Ph$_2$) and Boc-Thr (PO$_3$Ph$_2$) have been used, where HF or hydrogenolysis cleaves the peptide-resin, and hydrogenolysis removes the phosphate phenyl groups (183, 184). Fmoc-Ser(PO$_3$Bzl,H) can be used in conjuction with Fmoc chemistry with some care. Alternatively, peptide-resins that were built up by Fmoc chemistry to include unprotected Ser or Thr side-chains may be treated with a suitable phosphorylating reagent, e.g., $N,N$-diisopropyl-bis (4-chlorobenzyl)phosphoramidite or dibenzylphosphochloridate. The desired phosphorylated peptide is then obtained in solution following simultaneous deprotection and cleavage with TFA in the presence of scavengers (185, 186).

Side-chain phosphorylated Tyr is less susceptible to strong acid decomposition, and is not at all base-labile. Thus, SPPS has been used to incorporate directly Fmoc-Tyr(PO$_3$Me$_2$) (187), Fmoc-Tyr (PO$_3$Bzl$_2$) (188), Fmoc-Tyr(PO$_3$$t$Bu$_2$) (189), Fmoc-Tyr(PO$_3$H$_2$) (190, 191), and Boc-Tyr(PO$_3$H$_2$) (192). Syntheses incorporating Fmoc-Tyr(PO$_3$Bzl$_2$) use 2% DBU in DMF for $N^\alpha$-amino deprotection, as piperidine was found to remove the benzyl protecting groups from phosphate (188) whereas the *tertiary*-butyl phosphate group is inert to piperidine-mediated dealkylation during Fmoc deprotection steps (193). BOP/PyBOP and HBTU have been sucessfully applied for coupling of Fmoc-Tyr(PO$_3$Me$_2$), Fmoc-Tyr(PO$_3$$t$Bu$_2$) and Fmoc-Tyr(PO$_3$Bzl$_2$). DCC and DIPCDI are not recommended, since interaction with the phosphorodiester caused undesired activation (193). TFMSA or TMSBr can be used for peptide-resin cleavage and removal of the methyl phosphate groups without $O$-dephosphorylation (187, 192), while TFA is used for removal of the benzyl and *tertiary*-butyl phosphate groups (188).

Phosphorylation can be performed after chain assembly by phosphite triester phosphorylation using the dialkyl $N,N$-diethylphosphoamidates $(tBuO)_2$ $PNR'_2$, $(BzlO)_2PNR'_2$, or $(TmseO)_2 PNR'_2$ ($R'$ = ethyl or isopropyl) followed by MCPBA, $tBuOOH$, or aqueous iodine/THF oxidation. If Cys or Met residues are present in the peptide sequence, the use of $tBuOOH$ is preferedly used to minimize oxidation of these residues (193, 194). Phosphorylation may be accomplished *on-line*, directly after incorporation of the Tyr, Ser, or Thr residue but prior to assembly of the whole peptide.

## D. Sulfation

Gastrin, cholecystokinin, and related hormones contain sulfated Tyr. Synthesis of Tyr sulfate-containing peptides is difficult, due to the substantial acid lability of the sulfate ester; also, most sulfating agents are more reactive towards the hydroxyls of Ser or Thr with respect to the phenol of Tyr. Side-chain unprotected Tyr can be incorporated by Boc or Fmoc chemistry, and sulfation is achieved by use of pyridinium acetyl sulfate (91). Base or acid-promoted deprotection/cleavage follows, under conditions carefully optimized to avoid or minimize desulfation. Alternatively, sulfated Tyr can be incorporated directly by use of $Fmoc\text{-}Tyr(SO_3^- Na^+)\text{-}OPfp$, Fmoc-Tyr $(SO_3^- Na^+)$, or $Fmoc\text{-}Tyr(SO_3^- Ba_{1/2}^{2+})$ *in situ* with BOP/HOBt (195–198). A brief and carefully optimized acidolytic cleavage/deprotection is then used to minimize desulfation. PEG-PS resins functionalized with xanthenyl amide handles are the solid supports of choice for this application, as cleavage requires only 1–5% TFA. Cholecystokinin containing a sulfonated Tyr (CCK-8) was synthesized sucessfully using a 15-min treatment with 5% TFA, yielding the desired peptide 95% pure (75).

*O*-sulfonated Ser and Thr residues can be obtained by simply incorporating $Fmoc\text{-}Ser/Thr(SO_3H)$ into the peptide chain. If an Mtr or Pmc protected Arg is present in the peptide, the sulfonated Ser and Thr can alternatively be obtained by cleaving the peptide with 50%TFA in DCM without any scavengers. Under these conditions the sulfate moiety of the Arg-protecting group will be transferred to the hydroxyl groups in a yield of up to 60%. The sulfonated and unsulphonated peptides can then be separated via HPLC or ion -exchange chromatography (199).

## E. Glycosylation

Methodology for site-specific incorporation of carbohydrates during chemical synthesis of peptides has developed rapidly. The mild conditions of Fmoc chemistry are more suited for glycopeptide syntheses than Boc chemistry, as repetitive acid treatments can be detrimental to sugar linkages (200). Fmoc-Ser, -Thr, -Hyp, and -Asn have all been incorporated successfully with glycosylated side-chains. Side-chain glycosylation is performed with glycosyl bromides or glycose-$BF_3Et_2O$ for Ser, Thr, and Hyp, and glycosylamines for Asp (to produce a glycosylated Asn). The side-chain glycosyl is usually hydroxyl protected by either benzyl or acetyl groups (201–208), although some SPPS have been successful with no protection of glycosyl hydroxyl groups (208–210). Glycosylated residues are incorporated as preformed Pfp esters or *in situ* with DCC/HOBt (201, 203, 204, 206, 207, 209–211). These sugars are relatively stable to Fmoc deprotection by piperidine, morpholine, or DBU (201, 203, 204, 206, 207, 209–212), brief treatments (<2 hours) with TFA for side-chain deprotection and peptide-resin cleavage (201, 203, 204, 206, 207, 209–211), and palladium treatment for peptide-resin cleavage from HYCRAM$^{TM}$ (213). Deacetylation and debenzylation are performed with hydrazine-MeOH (4:1) prior to glycopeptide-resin cleavage (200, 206). Some side reactions may occur upon removal of acetyl groups, including

$\beta$-elimination and Cys-induced degradation of the peptide backbone (201, 214-216). Acid-labile protecting groups have been developed for the carbohydrate moiety, such as *tertiary*-butyldimethylsilyl (TBDMS), *tertiary*-butyldiphenylsilyl (TBDPS), and isopropylidene (214). The use of DBU in combination with silyl protection should be avoided due to DBU-induced decomposition of the glycopeptide (217).

## F.   Disulfide Bond Formation

In the majority of cases, intramolecular disulfides or simple intermolecular homodimers have been formed from purified linear precursors by non-specific oxidations in dilute solutions. An even number of Cys residues are brought to the free thiol form by removal of the same $S$-protecting group; disulfide formation is subsequently mediated by molecular oxygen (from air), potassium ferricyanide [$K_3Fe(CN)_6$], DMSO, or others from a lengthy catalogue of reagents (116, 218–220). Accomplishing the same end, but proceeding by a different mechanism, the polythiol can be treated with a mixture of reduced and oxidized glutathione, which catalyze the net oxidation by thiol-disulfide exchange reactions (221–223). These procedures, which require scrupulous attention to experimental details, have often given the desired disulfide-containing peptide products in acceptable yields. However, even under the best conditions, significant levels of dimeric, oligomeric, or polymeric material are observed. The non-monomeric material has usually proven to be difficult to "recycle" by alternating reduction and reoxidation steps.

A more sophisticated approach, which also requires dilute solutions, involves selective pairwise co-oxidations of two designated free or protected sulfhydryl groups. The prototype oxidative deprotections involve $I_2$ treatments on Cys(Trt) or Cys(Acm) residues (224). These reactions are carried out in neat or mixtures of the solvents TFE, MeOH, 1,1,1,3,3,3-hexafluoroisopropanol, HOAc, DCM, and chloroform, and often proceed in modest to high yield; however, side reactions have been observed at Trp residues resulting in Trp-2'-thioethers (225) and $\beta$-3-oxindolylalanine (226). Pairwise oxidative removal of appropriate Acm or Tacm Cys protecting groups with $Tl(Tfa)_3$ or methyltrichlorosilane (in the presence of diphenylsulphoxide) also furnishes the disulfide directly (227–229). However, Trp and Met must be side-chain protected during such treatments. As a final example in this category, Cys(Fm) residues form disulfides directly upon treatment with piperidine (230, 231).

Experiments aimed at controlled formation of multiple disulfide bridges require at least two classes of selectively removable Cys protecting groups. An over-riding concern with all such chemical approaches is to develop conditions that avoid scrambling (disulfide exchange). Oxidation by $I_2$ in TFE allows for selective disulfide bond formation between Cys(Trt) residues in the presence of Cys(Acm) residues; in DMF, $I_2$ oxidation is preferred between Cys(Acm) residues in the presence of Cys(Trt) residues (224). Direct $I_2$ oxidation of Cys(Acm) or Cys(Trt) residues is particularly advantageous in that existing disulfides are not exchanged (13, 111, 218, 224, 231, 232). Since the Acm group is essentially stable to HF, an Acm/Meb combination of protecting groups facilitates selective disulfide formation in Boc chemistry (232, 233).

The alternative of carrying out deprotection/oxidation of the Cys residues while the peptide chain remains anchored to a polymeric support is of obvious interest, and has received some recent attention. Such an approach takes advantage of *pseudo-dilution*, which is a kinetic phenomenon expected to favor facile intramolecular processes and thereby minimize dimeric and oligomeric by-products (13). Disulfide bond formation on peptide-resins has been demonstrated by air, $K_3Fe(CN)_6$, dithiobis (2-nitrobenzoic acid), or diiodoethane oxidation of free sulfhydryls, direct deprotection/oxidation of Cys(Acm) residues by Tl(Tfa)$_3$ or I$_2$, direct conversion of Cys(Fm) residues by piperidine, and nucleophilic attack by a free sulfhydryl on either Cys(Npys) or Cys(Scm) (72, 140, 232, 234–238, and references to earlier work cited in these papers). The most generally applicable and efficient of these methods is direct conversion of Cys(Acm) or Cys(Trt) residues by I$_2$ (10 equiv in DMF), Cys(Acm) residues by Tl(Tfa)$_3$ (1.5 equiv in DMF) (72, 73), and Cys(Fm) residues by (a) piperidine-DMF (1:1) for 3 h at 25 C (72, 231) or (b) piperidine-DMF-2-mercaptoethanol (10:10:0.7) treatment for 1 h at 25°C, followed by air oxidation in pH 8.0 DMF for 1 h at 25°C (72). The best solid-phase yields were at least as good, and in some cases better than, the results from corresponding solution oxidations.

## G. Lactams

Intra-chain lactams are formed between the side-chains of Lys or Orn and Asp or Glu to conformationally constrain synthetic peptides, with the goal of increasing biological potency and/or specificity. Lactams can also be formed via side-chain to head, side-chain to tail, or head to tail cyclization. The residues used to form intra-chain lactams must be selectively side-chain deprotected, while side-chain protecting groups of other residues remain intact. Selective deprotection is best achieved by using orthogonal side-chain protection, such as Fmoc and Fm for Lys/Orn and Asp/Glu, respectively, in combination with a Boc/Bzl strategy (239–241). A more complicated, but equally efficient approach, is to use side-chain protection based on graduated acid-lability (241–243). In the Fmoc strategy, orthogonality is achieved with Aloc or Dde protection for Lys and allyl or Dmab protection for Asp/Glu, respectively (38, 40). Optimized automated conditions for removal of Aloc/allyl groups are Pd(Ph$_3$)$_4$ (1 equiv.) in chloroform-acetic acid-NMM (37:2:1) for 2 h at 25°C, followed by washing steps with a solution of 0.5% DIEA and 0.5% sodium diethyldithiocarbamate in DMF (244). Cyclization is carried out most efficiently with BOP (3-6 equiv, 2 h, 20°C) in the presence of DIEA (6–7.5 equiv) while the peptide is still attached to the resin (239, 245), taking advantage of the *pseudo-dilution* phenomenon discussed in the previous section.

The three dimensional orthogonal protection scheme of Fmoc/*t*Bu/allyl protecting groups is the strategy of choice for head to tail cyclizations (34, 35, 39, 246). In these examples, an amide linker was used for side-chain attachment of a *C*-terminal Asp/Glu and the α-carboxyl group was protected as an allyl ester (34, 35).

For side-chain to head cyclizations, the Lys/Orn and Asp/Glu protecting groups should be either very acid sensitive or base labile. The *N*-terminal amino acid (head) can simply be introduced as an $N^\alpha$-Boc or $N^\alpha$-Fmoc derivative while the peptide-resin linkage and the other side-chain protecting groups are stable to dilute acid or carry a third dimension of orthogonality.

## H. Peptidomimetics with Peptide Bond Modifications

Main chain peptidomimetics are based upon modification of the peptide bond and inversion of chirality to yield retro- and retro-inverso peptides. The specific characteristics of peptidomimetics reveal essential information about the structure/function relationship of the peptide molecule leading to further modification and development of a tailored peptide (or non-peptide) drug.

All-D peptides are expected to be advantageous compared with standard all-L peptides due to their increased proteolytic stability. Peptides with (partially) inversed chirality can easily be synthesized by incorporating the protected D-amino acid derivatives. This strategy has been successfully applied in the cases of a D-HIV-1 protease (247), an active morphine receptor hexapeptide agonist (248), and a tumor cell inhibitory collagen analog (249).

Retro peptides are molecules in which the direction of the linear sequence is reversed. In retro inverso peptides the chirality of each amino acid is inverted in addition to reversion of sequence direction. These peptides are also easy to synthesize via SPPS and are very useful to probe antigenicity and immunogenity (250).

Partially retro-inverso peptides contain select regions where the sequence is reversed and chirality is inverted. The reversed sequence is connected to the original sequence via malonyl residues. The C-2 of these malonyl moieties is extremely prone to epimerisation. This lability of configuration renders synthesis, purification and conformation analysis of partial retro-inverso peptides complicated. To circumvent this problem, 2-methyl-2-alkyl-malonyl residues have been introduced successfully into peptides (251).

## V.  PEPTIDE-RESIN CLEAVAGE

Boc SPPS is designed primarily for simultaneous cleavage of the peptide anchoring linkage and side-chain protecting groups with strong acid (HF or equivalent), while Fmoc SPPS is designed primarily to accomplish the same cleavages with

moderate strength acid (TFA or equivalent). In each case, careful attention to cleavage conditions (reagents, scavengers, temperature, and times) is necessary in order to minimize a variety of side reactions.

Treatment with HF simultaneously cleaves PAM and MBHA linkages and removes the side-chain protecting groups commonly applied in Boc chemistry (61). HF cleavages are always carried out in the presence of a carbonium ion scavenger, usually 10% anisole. For cleavages of Cys-containing peptides, further addition of 1.8% 4-thiocresol is recommended. TMSOTf/TFA has also been used for strong acid cleavage and deprotection reactions, which are accelerated by the presence of thioanisole as a "soft" nucleophile (252, 253). TMSOTf (1 M)-thioanisole (1 M) in TFA (a.k.a. DEPRO$^{TM}$) efficiently cleaves PAM and MBHA linkages (253, 254).

The combination of side-chain protecting groups, e.g., *t*Bu (for Asp, Glu, Ser, Thr, and Tyr), Boc (for His and Lys), Tmob (for Cys), and Trt (for Asn, Cys, Gln, and His), and anchoring linkages, e.g., HMP or PAL, commonly used in Fmoc chemistry are simultaneously deprotected and cleaved by TFA. Such cleavage of *t*Bu and Boc groups results in *tertiary*-butyl cations and *tertiary*-butyl trifluoroacetate formation (255–260). These species are responsible for *tertiary*-butylation of the indole ring of Trp, the thioether group of Met, and, to a very low degree (0.5–1.0%), the 3′-position of Tyr. Modifications can be minimized during TFA cleavage by utilizing effective *tertiary*-butyl scavengers. An early comprehensive study showed the advantages of 1,2-ethanedithiol (EDT) (259); this thiol has the additional virtue of protecting Trp from oxidation that occurs due to acid-catalyzed ozonolysis (261). To avoid acid-catalyzed oxidation of Met to its sulfoxide, a thioether scavenger such as dimethylsulfide, ethylmethylsulfide, or thioanisole should be added (262–264). TFA deprotection of Cys(Trt) is reversible in the absence of a scavenger, and can occur readily following TFA cleavage if solutions of crude peptide in TFA are concentrated by rotary evaporation or lyophilization (68). EDT, triethylsilane, or triisopropylsilane are recommended as efficient scavengers to prevent Trt reattachment to Cys (265); this recommendation extends to prevent reattachment of Tmob as well (69). TFA deprotection of His(Bum) liberates formaldehyde, in similar fashion to HF deprotection of His(Bom) (266). Cyclization of $N$-terminal Cys residues to a thiazolidine is only partially (60%) inhibited by even complex TFA/scavenger mixtures, such as reagent K (see discussion below).

The indole ring of Trp can be alkylated irreversibly by Mtr and Pmc groups from Arg (95, 263, 264, 267, 268), Tmob groups from Asn, Gln, or Cys (56, 60), and even by some TFA-labile ester and amide linkers (93, 164, 170, 269–271). Cleavage of the Pmc group may also result in $O$-sulfation of Ser, Thr, and Tyr (199, 268). Three efficient cleavage "cocktails" (264) for Mtr/Pmc/Tmob quenching, and preservation of Trp, Tyr, Ser, Thr, and Met integrity, are TFA-phenol-thioanisole-EDT-$H_2O$ (82.5:5:5:2.5:5) (reagent K) (263), TFA-thioanisole-EDT-anisole (90:5:3:2) (reagent R) (170), and TFA-phenol-water-triisopropylsilane (88:5:5:2) (reagent B) (272). Studies on Trp preservation during amino acid analysis (273) have led to the development of reagent K', where EDT is replaced by 1-dodecanethiol. Water is an essential component of reagents K and K', but phenol is necessary only with multiple Trp-containing peptides (263). Thioanisole, a soft nucleophile, accelerates TFA deprotection of both Arg(Mtr) and Arg(Pmc). Triethylsilane (4 equiv.) in MeOH-TFA (1:9) has been reported to efficiently cleave and scavenge Pmc groups (274). Given a choice for Arg protection,

Pmc is preferred because it is more labile, and gives less Trp alkylation during unscavenged TFA cleavage; the recommendation for Pmc is particularly appropriate for sequences containing multiple Arg residues (95, 263, 275). The Trt group, instead of Tmob, is suggested for Asn/Gln side-chain protection in Trp-containing peptides, as Trt cations are easier to scavenge (60).

## VI. FROM THE CLEAVED PEPTIDE TO THE PURE PRODUCT

Following peptide-resin cleavage, the crude product usually contains peptides, peptide side products, and residual non-peptide scavenger-adducts. The aim of purification is to isolate the desired peptide in high yield by applying few and effective operations which ideally differ in their separation principles. Combining the results of different (orthogonal) analytical methods yields a more complete picture of the final peptide, its purity, identity, and structure. Knowledge of peptide integrity is essential for reliable and unambigious interpretation of functional bioassays. For synthetic peptides, HPLC and mass spectrometry (MS) have become the best tools for effective minimum characterization. In addition, however, amino acid analysis or sequencing are highly recommended. RP-HPLC is the most commonly used method for both peptide analysis and separation, as it can reveal the complexity of the crude product and separate side-products from the desired peptide. In cases of complex crude products which lack baseline separation of peaks, it is advisable to apply a two-step HPLC purification using different columns, solvent systems, or gradients. Since RP-HPLC separates peptides according to their hydrophobic character, compounds which are similar in this respect but have different net charges might be overlooked due to coelution.

One can confirm the homogeneity of an HPLC purified product by capillary electrophoresis (CE) or run the sample in a different solvent system on HPLC. Peptides which coelute in one solvent system may become separated in a different one (199). A detailed discussion of HPLC and CE is given elsewhere in this volume.

If the peptide contains Cys residues, the crude product often needs to be treated with reducing agents to avoid intra- and intermolecular crosslinking during purification. Suitable reducing agents are 100 mM DTT or 0.1–0.3 M $\beta$-mercaptoethanol in Tris-HCl or EDTA buffers (276). Subsequent purification should be carried out at acidic pH to minimize oxidation and/or disulfide exchange. If one desires to form disulfides in solution after resin cleavage, it is advantageous to first purify the peptide rather than oxidizing the crude product (276).

Following purification, several analytical methods are available to elucidate the structure and integrity of a peptide and at the same time confirm the homogenity of a sample by showing the absence of side products. MS is a versatile analytical method which reveals maximum information about the identity, homogeneity, and, with tandem MS, the sequence of a peptide (277). Undesired modifications such as protecting groups still attached or reattached to amino acid side-chains or side-chain modifications such as oxidation of Met may be identified by MS. If acid or base sensitive structural modifications have been incorporated during synthesis, MS may be the only routinely applied method for detection. Glycosyl, sulfate, or phosphate groups may not be identified by sequencing or amino acid analysis but are easily identified by MS. Aspartimide formation or the generation of deletion peptides result in products of lower molecular masses than desired, but are also readily identified by MS.

The products of deamidation of Asn and Gln, namely Asp and Glu, respectively, differ by only one mass unit from their amidated counterparts, which is within the range of accuracy for most MS methods and instruments. Information about this side reaction can be drawn from analysis of the isotope pattern of the molecular ion peak. A very elegant technique for analyzing crude synthetic peptides is LC-MS, where HPLC or CE is coupled on-line to the mass spectrometer. Thus, HPLC or CE peaks can be directly identified. Although one may apply on-resin MS (see "Monitoring") for evaluating the crude product, it has to be kept in mind that additional side products formed during resin cleavage will have been overlooked.

MS techniques are capable of clearly assessing the presence of disufide bonds versus reduced Cys residues (mass difference of 2 amu × number of disulfides). This is done by comparison of the mass spectra of the disulfide peptide, the reduced peptide, and the alkylated species. With tandem MS, intra- and intermolecular disulfide bonds can be distinguished by the number of molecular ions present after reduction (276).

Edman degradation sequencing evaluates the assembly of a peptide. After completion of a synthesis, the identification of deletion sequences by Edman chemistry helps to design an effective purification strategy for removal of these by-products. Sequencing requires a free $N$-terminal amino group. In some cases, the amino terminus is blocked by an acetyl group or by an intramolecular lactam formation. Specific enzymes can be applied to provide a free amino group. Although peptide modifications can not be detected, Edman sequencing is excellent for assessing the identity, integrity and homogentiy of a peptide (278).

Amino acid analysis (AAA) reveals the amino acid composition of a peptide, and is the most reliable method for quantitation of each amino acid. The peptide is generally hydrolysed in 6 N HCl/0.1% phenol for 24 h at 110°C. The resultant amino acids are derivatized prior to or after column separation (pre- or post-column derivatization) and the derivatives are detected by UV-VIS or fluorescence. A variety of methods for derivatization and detection are available. Most commonly, ion-exchange chromatography in combination with post-column ninhydrin derivatization or precolumn o-phtalaldehyde derivatization followed by HPLC separation is used (278). During acid hydrolysis, Cys, Trp, Asn and Gln are completely destroyed, whereas Ser, Thr, and Met are partially destroyed. Cys and Met can be detected as cysteic acid and methionine sulfone, respectively, or by oxidation with performic acid prior to hydrolysis. To prove the unchanged presence of Asn and Gln in a peptide, enzymatic hydrolysis prior to amino acid analysis using an exopeptidase is the method of choice. Aminopeptidase M

(E.C. 3.4.11.2.) cleaves the N-terminal amino acid but does not cleave β-Asp or aspartimide moieties (279). This analysis reveals indirectly the presence of Asn.

Another condition for establishing integrity of a synthetic peptide is its optical purity. It is advisable to check for racemization if His and/or Cys are present in the peptide. Racemization tests can be performed by gas chromatography, RP-HPLC, or MS. Peptides are first hydrolyzed, then either derivatized and separated on the chiral "chirasil Val" glass capillary and identified (280) or treated with Marfey's reagent and analyzed by RP-HPLC (281). Alternatively, enantioselective enzymatic hydrolysis with subsequent MALDI-MS analysis can be used to evaluate racemization. This MS approach has been successfully demonstrated on short synthetic model peptides containing either L-His and D-His following incubation with carboxypeptidase A (282).

## ABBREVIATIONS

Abbreviations used for amino acids and the designations of peptides follow the rules of the IUPAC-IUB Commission of Biochemical Nomenclature in J Biol Chem 247: 977–983 (1972). The following additional abbreviations are used: AA, amino acid; Acm, acetamidomethyl; Ada, adamantyl; Al, allyl; Alloc, allyloxycarbonyl; (BzlO)$_2$PNR$_2'$, dibenzylalkylphosphoamidate; Boc, tert-butyloxycarbonyl; Boc-ON, 2-tert-butyloxycarbonyloximino-2-phenylacetonitrile; Bom, benzyloxymethyl; BOP, benzotriazolyl N-oxytrisdimethylaminophosphonium hexafluorophosphate; 2-BrZ, 2-bromobenzyloxycarbonyl; Bum, tert-butoxymethyl; Bzl, benzyl; cHex, cyclohexyl; Cs, cesium salt; 2,6-Cl$_2$Bzl, 2,6-dichlorobenzyl; CLEAR, cross-linked ethoxylate acrylate resin; DBU, 1,8-diazabicyclo[5.4.0]undec-7-ene; DCBC, 2,6-dichlorobenzoyl chloride; DCC, N,N'-dicyclohexylcarbodiimide; DCE, 1,2-dichloroethane; DCM, dichloromethane (methylene chloride); Dde, 1-(4,4-dimethyl-2,6-dioxo-cyclohex-1-ylidene) ethyl; DhbtOH, 3,4-dihydro-4-oxobenzotriazine; DIEA, N,N-diisopropylethylamine; DIPCDI, N,N'-diisopropylcarbodiimide; DMA, N,N-dimethylacetamide; Dmab, N[1-(4,4-dimethyl-2,6-dioxocyclohexylidene)-3-methylbutyl]amino benzyl; DMAP, 4-dimethylaminopyridine; DMF, N,N-dimethylformamide; Dnp, 2,4-dinitrophenyl; DMSO, dimethyl sulfoxide; Dod, 4-(4'-methoxybenzhydryl)phenoxyacetic acid; EDT, 1,2-ethanedithiol; ES-MS electrospray-mass spectrometry; Et$_3$N, triethylamine; FAB-MS fast atom bombardment mass spectrometry; Fm, 9-fluorenylmethyl; Fmoc, 9-fluorenylmethyloxycarbonyl; Fmoc(Fmoc-Hmb), fluorenylmethyloxycarbonyl-Nα-(2-fluorenylmethyloxycarbonyl-oxy-4-methoxybenzyl); Fmoc-OSu, fluorenylmethyl succinimidyl carbonate; HAL, 5-(4-hydroxymethyl-3,5-dimethoxyphenoxy)valeric acid; HATU, O-(7-azabenzotriazol-1-yl)-1,1,3,3-tetramethyluronium hexafluorophosphate; HBTU, 2-(1H-benzotriazol-1-yl)-1,1,3,3-tetramethyl uronium hexafluorophosphate; HF, hydrogen fluoride; HMFA, 9-(hydroxymethyl)-2-fluoreneacetic acid; HMP, 4-hydroxy-methylphenoxy; HOAc, acetic acid; HOAt, 1-hydroxy-7-azabenzotriazole; HOBt, 1-hydroxybenzotriazole; HOPip, $N$-hydroxypiperidine; HPLC, high performance liquid chromatography; Hpp, 1-(4-nitrophenyl)-2-pyrazolin-5-one; HYCRAM$^{TM}$, hydroxy-crotonylaminomethyl; $K_3Fe(CN)_6$, potassium ferricyanide; MALDI-TOF-MS, matrix assisted laser desorption time of flight mass spectrometry; MBHA, 4-methyl-benzhydrylamine (resin); MCPBA, $m$-chloroperoxybenzoic acid; Meb, 4-methylbenzyl; MeIm, 1-methylimidazole; MeOH, methanol; MMA, $N$-methylmercaptoacetamide; Mob, 4-methoxybenzyl; Mts, mesitylene-2-sulfonyl; NCA, $N$-carboxyanhydride; NMM, $N$-methylmorpholine; NMP, $N$-methylpyrrolidone; NMR, nuclear magnetic resonance; Nonb, 3-nitro-4-aminomethylbenzoic acid; NPE, 4-(2-hydroxyethyl)-3-nitrobenzoic acid; Npp, 3-methyl-1-(4-nitrophenyl)-2-pyrazolin-5-one; N pys, N-pyridinesulfenyl; N.R., not reported; ODhbt, 1-oxo-2-hydroxydihydrobenzotriazine; ONb, 2-nitrobenzyl ester; ONp, 4-nitrophenyl; OPfp, pentafluorophenyl; Orn, ornithine; OSu, $N$-hydroxysuccinimidyl; OTDO, 2,5-diphenyl-2,3-dihydro-3-oxo-4-hydroxythiophene dioxide; PAB, 4-alkoxy-benzyl; PAL, 5-(4-aminomethyl-3,5-dimethoxyphenoxy)valeric acid; PAM, phenylace-tamidomethyl; Pbf, 2,2,4,6,7-pentamethyldihydrobenzofuran-5-sulfonyl; Pd(Ph$_3$)$_4$, tetrakistriphenylphosphine-palladium; PEG, polyethylene glycol; PEGA, polyethylene glycol dimethyl acrylamide; Pmc, 2,2,5,7,8-pentamethylchroman-6-sulfonyl; Pnp, 3-phenyl-1-(4-nitrophenyl)-2-pyrazolin-5-one; PS, polystyrene; PSA, preformed sym-metrical anhydride; PyAOP, 7-azabenzotriazol-1-yl-oxy-trispyrrolidinophosphonium hexafluorophosphate; PyBOP, benzotriazole-1-yl-oxy-tris-pyrrolidinophosphonium hexafluorophosphate; RP-HPLC, reversed-phase high performance liquid chromatog-raphy; SASRIN$^{TM}$, 2-methoxy-4-alkoxybenzyl alcohol; Scm, $S$-carboxymethylsulfenyl; SPPS, solid-phase peptide synthesis; StBu, $tertiary$-butylsulfenyl; Tacm, tri-methylacetamidomethyl; TBDMS, $tertiary$-butyldimethylsilyl; TBDPS, $tertiary$-butyl-diphenylsilyl; tBu, $tertiary$-butyl; tBuOOH, $tertiary$-butyl peroxide; TFA, trifluoroacetic acid; TFE, 2,2,2-trifluoroethanol; TFMSA, trifluoromethanesulfonic acid; THF, tetrahydrofuran; Tl(Tfa)$_3$, thallium (III) trifluoroacetate; Tmob, 2,4,6-trimethoxybenzyl; TMSBr, trimethylsilyl bromide; Tmse, trimethylsilylethyl; TMSOTf, trimethylsilyl trifluoromethanesulfonate; TOPPipU, 2-(2-oxo-1(2H)-pyridyl)-1,1,3,3-bispentamethyl eneuronium tetrafluoroborate; Tos, 4-toluenesulfonyl; Trt, triphenylmethyl; Xal, 5-(9-aminoxanthen-2-oxy)valeric acid; Xan, 9-xanthenyl; Z, benzyloxycarbonyl. Amino acid symbols denote the L-configuration where applicable, unless indicated otherwise.

## REFERENCES

1. VK Sarin, SBH Kent, RB Merrifield. Properties of swollen polymer networks: Solvation and swelling of peptide-containing resins in solid-phase peptide synthesis. J Am Chem Soc 102: 5463–5470, 1980.

2. DH Live, SBH Kent. Fundamental aspects of the chemical applications of cross-linked polymers. In: JE Mark, ed. Elastomers and Rubber Elasticity. Washington: American Chemical Society, 1982, pp 501–515.

3. E Atherton, RC Sheppard. Solid Phase Peptide Synthesis: A Practical Approach. Oxford, U.K.: IRL Press, 1989.

4. R Arshady, E Atherton, DLJ Clive, RC Sheppard. Peptide synthesis, part 1: Preparation and use of polar supports based on poly(dimethylacrylamide). J Chem Soc Perkin Trans I: 529–537, 1981.

5. H Hellermann, H-W Lucas, J Maul, VNR Pillai, M Mutter. Poly(ethylene glycol)s grafted onto crosslinked polystyrenes, 2: Multidetachably anchored polymer systems for the synthesis of solubilized peptides. Makromol Chem 184: 2603–2617, 1983.

6. S Zalipsky, F Albericio, U Slomczynska, G Barany. A convenient general method for synthesis of $N^\alpha$- or $N^\omega$-dithiasuccinoyl (Dts) amino acids and dipeptides: Application of polyethylene glycol as a carrier for funtional purification. Int J Peptide Protein Res 30: 740–783, 1987.

7. E Bayer, W Rapp. New polymer supports for solid-liquid-phase peptide synthesis. In: W Voelter, E Bayer, YA Ovchinnikov, VT Ivanov, eds. Chemistry of Peptides and Proteins, Vol. 3. Berlin: Walter de Gruyter, 1986, pp 3–8.

8. E Bayer, K Albert, H Willisch, W Rapp, B Hemmasi. $^{13}$C NMR relaxation times of a tripeptide methyl ester and its polymer-bound analogues. Macromolecules 23: 1937–1940, 1990.

9. G Barany, F Albericio. Peptide synthesis for biotechnology in the 1990's. In: K North, ed. Biotechnology International 1990/1991. London: Century Press Ltd, 1991, pp 155–163.

10. S Zalipsky, JL Chang, F Albericio, G Barany. Preparation and applications of polyethylene glycol-polystyrene graft resin supports for solid-phase peptide synthesis. Reactive Polymers 22: 243–258, 1994.

11. M Meldal. PEGA: A flow stable polyethylene glycol dimethyl acrylamide copolymer for solid phase synthesis. Tetrahedron Lett 33: 3077–3080, 1992.

12. M Kempe, G Barany. CLEAR: A novel family of highly cross-linked polymeric supports for solid-phase peptide synthesis. J Am Chem Soc 118: 7083–7093, 1996.

13. G Barany, RB Merrifield. Solid-phase peptide synthesis. In: E Gross, J Meienhofer, eds. The Peptides, Vol. 2. New York: Academic Press, 1979, pp 1–284.

14. SBH Kent. Chemical synthesis of peptides and proteins. Ann Rev Biochem 57: 957–989, 1988.

15. CJA Wallace, P Mascagni, BT Chait, JF Collawn, Y Paterson, AEI Proudfoot, SBH Kent. Substitutions engineered by chemical synthesis at three conserved sites in mitochondrial cytochrome c. J Biol Chem 264: 15199–15209, 1989.

16. K Suzuki, K Nitta, N Endo. Suppression of diketopiperazine formation in solid phase peptide synthesis. Chem Pharm Bull 23: 222–224, 1975.

17. M Schnölzer, P Alewood, A Jones, D Alewood, SBH Kent. In situ neutralization in Boc-chemistry solid phase peptide synthesis: Rapid, high yield assembly of difficult sequences. Int J Peptide Protein Res 40: 180–193, 1992.

18. P Alewood, D Alewood, L Miranda, S Love, W Meutermans, D Wilson. Rapid in situ neutralization protocols for Boc and Fmoc solid phase synthesis. In: GB Fields, ed. Methods in Enzymology, Volume 289. Orlando: Academic Press, 1997, pp 14–29.

19. GB Fields. Methods for removing the Fmoc group. In: MW Pennington, BM Dunn, eds. Methods in Molecular Biology, Vol. 35: Peptide Synthesis Protocols. Totowa, NJ: Humana Press Inc, 1994, pp 17–27.

20. JD Wade, J Bedford, RC Sheppard, GW Tregear. DBU as an $N^\alpha$-deprotecting reagent for the fluorenylmethoxycarbonyl group in continuous flow solid-phase peptide synthesis. Peptide Res 4: 194–199, 1991.

21. CG Fields, DJ Mickelson, SL Drake, JB Mc Carthy, GB Fields. Melanoma cell adhesion and spreading activities of a synthetic 124-residue triple-helical "minicollagen". J Biol Chem 286: 14153–14160, 1993.

22. BF Gisin. The preparation of Merrifield-resins through total esterification with cesium salts. Helv Chim Acta 56: 1476–1482, 1973.

23. R Colombo, E Atherton, RC Sheppard, V Woolley. 4-Chloromethylphenoxyacetyl polystyrene and polyamide supports for solid-phase peptide synthesis. Int J Peptide Protein Res 21: 118–126, 1983.

24. M Mergler, R Nyfeler, R Tanner, J Gosteli, P Grogg. Peptide synthesis by a combination of solid-phase and solution methods II: Synthesis of fully protected peptide fragments on 2-methoxy-4-alkoxy-benzyl alcohol resin. Tetrahedron Lett 29: 4009–4012, 1988.

25. Y Han, J Vagner, G Barany. A new side chain anchoring strategy for solid phase synthesis of peptide acids with C-terminal cysteine. In: R Epton, ed. Solid Phase Synthesis and Combinatorial Libraries. Birmingham, U.K.: Maylower Scientific Ltd, 1996, pp 385–386.

26. CG Fields, GB Fields. New approaches to prevention of side reactions in Fmoc solid phase peptide synthesis. In: JE Rivier GR Marshall, eds. Peptides: Chemistry, Structure, and Biology. Leiden: Escom, 1990, pp 928–930.

27. CH Li, S Lemaire, D Yamashiro, BA Doneen. The synthesis and opiate activity of β-endorphin. Biochem Biophys Res Commun 71: 19–25, 1976.

28. CR Wu, JD Wade, GW Tregear. β-Subunit of baboon chorionic gonadotropin: Continuous flow Fmoc-polyamide synthesis of the C-terminal 37-peptide. Int J Peptide Protein Res 31: 47–57, 1988.

29. F Albericio, R Van Abel, G Barany. Solid-phase synthesis of peptides with C-terminal asparagine or glutamine. Int J Peptide Protein Res 35: 284–286, 1990.

30. G Breipohl, J Knolle, W Stüber. Facile SPS of peptides having C-terminal Asn and Gln. Int. J. Peptide Protein Res. 35: 281–283, 1990.

31. Okada, S Iguchi. Amino acid and peptides, part 19: Synthesis of β-1- and β-2-adamantyl aspartates and their evaluation for peptide synthesis. J Chem Soc Perkin Trans I: 2129–2136, 1988.

32. JP Tam, MW Riemen, RB Merrifield. Mechanisms of aspartimide formation: The effects of protecting groups, acid, base, temperature and time. Peptide Res 1: 6–18, 1988.

33. WC Chan, BW Bycroft, DJ Evans, PD White. A novel 4-aminobenzyl ester-based carboxy-protecting group for synthesis of atypical peptides by Fmoc-Bu$^t$ solid phase chemistry. J Chem Soc Chem Commun 2209-210, 1995.

34. A Trzeciak, W Bannwarth. Synthesis of "head to tail" cyclized peptides on solid support by Fmoc chemistry. Tetrahedron Lett 33: 4557–4560, 1992.

35. SA Kates, NA Sole, F Albericio, G Barany. Solid-phase synthesis of cyclic peptides. In: C Basava, GM Anantharamaiah, eds. Peptides: Design, Synthesis and Biological Activity. Boston: Birkhaeuser, 1994, pp 39–57.

36. BW Erickson, RB Merrifield. Acid stability of several benzylic protecting groups used in solid-phase peptide synthesis: Rearrangement of O-benzyltyrosine to 3-benzyltyrosine. J Am Chem Soc 95: 3750–3756, 1973.

37. D Yamashiro, CH Li. Protection of tyrosine in solid-phase peptide synthesis. J. Org Chem 38: 591–592, 1973

38. MH Lyttle, D Hudson. Allyl based side-chain protection for SPPS. In: JA Smith, JE Rivier, eds. Peptides: Chemistry and Biology. Leiden, The Netherlands: Escom, 1992, pp 583–584.

39. F Albericio, G Barany, GB Fields, D Hudson, SA Kates, MH Lyttle, N Solé. Allyl-based orthogonal solid-phase peptide synthesis. In: CH Schneider, AN Eberle, eds. Peptides 1992. Leiden, The Netherlands: Escom, pp 191–193, 1993.

40. CG Fields, CM Lovdahl, AJ Miles, VL Matthias Hagen, GB Fields. Solid-phase synthesis and stability of triple-helical peptides incorporating native collagen sequences. Biopolymers 33: 1695–1707, 1993.

41. BW Bycroft, WC Chang, SR Chhabra, ND Hone. A novel lysine protecting procedure for continuous flow solid phase synthesis of branched peptides. J Chem Soc Chem Commun: 778–779, 1993.

42. K Barlos, D Gatos, O Chatzi, S Koutsogianni, W Schäfer. Solid phase synthesis using trityl type side chain protecting groups. In: CH Schneider, AN Eberle, eds. Peptides 1992. Leiden, The Netherlands: Escom, pp 283–284, 1993.

43. A Aletras, K Barlos, D Gatos, S Koutsogianni, P Mamos. Preparation of the very acid-sensitive Fmoc-Lys(Mtt)-OH. Int J Peptide Protein Res 45: 488–500, 1995.

44. M Fujino, M Wakimasu, C Kitada. Further studies on the use of multi-substituted benzenesulfonyl groups for protection of the guanidino function of arginine. Chem Pharm Bull 29: 2825–2831, 1981.

45. J Green, OM Ogunjobi, R Ramage, ASJ Stewart, S McCurdy, R Noble. Application of the N$^G$-(2,2,5,7,8-pentamethylchroman-6-sulphonyl) derivative of Fmoc-arginine to peptide synthesis. Tetrahedron Lett 29: 4341–4344, 1988.

46. LA Carpino, H Shroff, SA Triolo, E-SME Mansour, H Wenschuh, F Albericio. The 2,2,4,6,7-pentamethyldihydrobenzofuran--5-sulfonyl group (Pbf) as arginine side

chain protectant. Tetrahedron Lett 34: 7829–7832, 1993.

47. H Rink, P Sieber, F Raschdorf. Conversion of $N^G$-urethane protected arginine to ornithine in peptide solid phase synthesis. Tetrahedron Lett 25: 621–624, 1984.

48. JH Jones, WI Ramage, MJ Witty. Mechanism of racemization of histidine derivatives in peptide synthesis. Int J Peptide Protein Res 15: 301–303, 1980.

49. T Ishiguro, C Eguchi. Unexpected chain-terminating side reaction caused by histidine and acetic anhydride in solid-phase peptide synthesis. Chem Pharm Bull 37: 506–508, 1989.

50. M Kusunoki, S Nakagawa, K Seo, T Hamana, T Fukuda. 1990. A side reaction in solid phase synthesis: Insertion of glycine residues into peptide chains via $N^{im} \rightarrow N^\alpha$ transfer. Int J Peptide Protein Res 36: 381–386, 1988.

51. B Riniker, P Sieber. Problems and progress in the synthesis of histidine-containing peptides. In: B Penke and A Torok, eds. Peptides: Chemistry, Biology, Interactions with Proteins. Berlin: Walter de Gruyter & Co, 1988, pp 65–74.

52. P Sieber, B Riniker. Protection of histidine in peptide synthesis: A reassessment of the trityl group. Tetrahedron Lett 28: 6031–6034, 1987.

53. M Forest, A Fournier. BOP reagent for the coupling of pGlu and Boc-His(Tos) in solid phase peptide synthesis. Int J Peptide Protein Res 35: 89–94, 1990.

54. AR Fletcher, JH Jones, WI Ramage, AV Stachulski. The use of the $N(\pi)$-phenacyl group for the protection of the histidine side chain in peptide synthesis. J Chem Soc Perkin Trans I: 2261–2267, 1979.

55. S Mojsov, AR Mitchell, RB Merrifield. A quantitative evaluation of methods for coupling asparagine. J Org Chem 45: 555–560, 1980.

56. H Gausepohl, M Kraft, RW Frank. Asparagine coupling in Fmoc solid phase peptide synthesis. Int J Peptide Protein Res 34: 287–294, 1989.

57. RD DiMarchi, JP Tam, SBH Kent, RB Merrifield. Weak acid-catalyzed pyrrolidone carboxylic acid formation from glutamine during solid phase peptide synthesis. Int J Peptide Protein Res 19: 88–93, 1982.

58. A Orlowska, E Witkowska, J Izdebski. Sequence dependence in the formation of pyroglutamyl peptides in solid phase peptide synthesis. Int J Peptide Protein Res 30: 141–144, 1987.

59. D Hudson. 2,4,6-Trimethoxybenzyl (Tmob) protection for asparagine and glutamine in Fmoc solid-phase peptide synthesis. Biosearch Technical Bulletin 9000–01, MilliGen/Biosearch Division of Millipore, Bedford, MA., 1988.

60. P Sieber, B Riniker. Side-chain protection of asparagine and glutamine by trityl: Application to solid-phase peptide synthesis. In: R Epton, ed. Innovation and Perspectives in Solid Phase Synthesis. Birmingham, U.K: Solid Phase Conference Coordination, Ltd, 1990, pp 577–583.

61. JP Tam, RB Merrifield. Strong acid deprotection of synthetic peptides: Mechanisms and methods. In: S Udenfried, J Meienhofer, eds. The Peptides, Vol 9. New York: Academic Press, 1987, pp 185–248.

62. M Ohno, S Tsukamoto, S-I Sato, N Izumiya. Improved solid phase synthesis of tryptophan containing peptides; II. Use of $N^\alpha$-$t$-butyloxycarbonyl-$N^{in}$-formyl-tryptophan. Bull Chem Soc Jpn 46: 3280–3285, 1973.

63. P White. Fmoc-Trp(Boc)-OH: A new derivative for the synthesis of peptides containing tryptophan. In: JA Smith, J Rivier, eds. Peptides, Chemistry and Biology. Leiden, The Netherlands: Escom, 1992, pp 537–538.

64. B Riniker A Floersheimer, H Fretz, P Sieber, B Kamber. A general strategy for the synthesis of large peptides. The combined solid-phase and solution approach. Tetrahedron 49: 9307–9320, 1993.

65. T Johnson, M Quibell, D Owen, RC Sheppard. A reversible protecting group for the amide bond in peptides. Use in the synthesis of "difficult sequences". J Chem Soc Chem Commun: 369–372, 1993.

66. H Choi, JV Aldrich. Comparison of methods for Fmoc solid phase synthesis and cleavage of a peptide containing both tryptophan and arginine. Int J Peptide Protein Res 42: 58–63, 1993.

67. CG Fields, GB Fields. Minimization of tryptophan alkylation following 9-fluorenylmethoxy-carbonyl solid-phase peptide synthesis. Tetrahedron Lett 34: 6661–6664, 1993.

68. I Photaki, J Taylor-Papadimitriou, C Sakarellos, P Mazarakis, L Zervas. On cysteine and cystine peptides, part V: S-trityl- and S-diphenylmethyl-cysteine and -cysteine peptides. J Chem Soc (C): 2683–2687, 1970.

69. MC Munson, C García-Echeverría, F Albericio, G Barany. S-2,4,6-trimethoxybenzyl (Tmob): A novel cysteine protecting group for the $N^\alpha$-9-fluorenylmethyloxycarbonyl (Fmoc) strategy of peptide synthesis. J Org Chem 57: 3013–3018, 1992.

70. H Nishio, T Kimura, S Sakakibara. Side reaction in peptide synthesis: Modification of tryptophan during treatment with mercury(II) acetate/2-mercaptoethanol in aqueous acetic acid. Tetrahedron Lett 35: 1239–1242, 1994.

71. GW Kenner, IJ Galpin, R Ramage. Synthetic studies directed towards the synthesis of a lysozyme analog. In: E Gross, J Meienhofer, eds. Peptides: Structure and Biological Function. Rockford, IL: Pierce Chemical Co, 1979, pp 431–438.

72. F Albericio, RP Hammer, C García-Echeverría, MA Molins, JL Chang, MC Munson, M Pons, E Giralt, G Barany. Cyclization of disulfide-containing peptides in solid-phase synthesis. Int J Peptide Protein Res 37: 402–413, 1991.

73. WB Edwards, CG Fields, CJ Anderson, TS Pajeau, MJ Welch, GB Fields. Generally applicable, convenient solid-phase synthesis and receptor affinities of octreotide analogs. J Med Chem 37: 3749–3757, 1994.

74. Y Han, F Albericio, G Barany. Occurrence and minimization of cysteine racemization during stepwise solid-phase peptide synthesis. J Org Chem 62: 4307–4312, 1997.

75. Y Han, SL Botems, P Heygens, MC Munson, CA Minor, SA Kates, F Albericio, G Barany. Preparation and application of xanthenylamide (Xal) handles for solid phase synthesis of C-terminal peptide amides under particularly mild conditions. J Org Chem 61: 6326-6339, 1996.

76. DH Rich, J Singh. The carbodiimide method. In: E Gross, J Meienhofer, eds. The Peptides, Vol 1. New York: Academic Press, 1979, pp 241–314.

77. RB Merrifield, J Singer, BT Chait. Mass spectrometric evaluation of synthetic peptides for deletions and insertions. Anal Biochem 174: 399–414, 1988.

78. W König, R Geiger. Eine neue Methode zur Synthese von Peptiden: Aktivierung der Carboxylgruppe mit Dicyclohexylcarbodiimid unter Zusatz von 1-Hydroxybenzotriazolen. Chem Ber 103: 788–798, 1970.

79. W König, R Geiger. N-hydroxyverbindungen als Katalysatoren für die Aminolyse aktivierter Ester. Chem Ber 106: 3626–3635, 1973.

80. V Dourtoglou, B Gross, V Lambropoulou, C Zioudrou. O-benzotriazolyl-N,N,N′, N′-tetramethyluronium hexafluorophosphate as coupling reagent for the synthesis of peptides of biological interest. Synthesis: 572–574, 1984.

81. A Fournier, CT Wang, AM Felix. Applications of BOP reagent in solid phase peptide synthesis: Advantages of BOP reagent for difficult couplings exemplified by a synthesis of [Ala$^{15}$]-GRF(1-29)-NH$_2$. Int J Peptide Protein Res 31: 86–97, 1988.

82. D Ambrosius, M Casaretto, R Gerardy-Schahn, D Saunders, D Brandenburg, H Zahn. Peptide analogues of the anaphylatoxin C3a; synthesis and properties. Biol Chem Hoppe-Seyler 370: 217–227, 1989.

83. H Gausepohl, M Kraft, R Frank. In situ activation of Fmoc-amino acids by BOP in solid phase peptide synthesis. In: G Jung E Bayer, eds. Peptides 1988, Berlin: Walter de Gruyter & Co, 1989, pp 241–243.

84. R Seyer, A Aumelas, A Caraty, P Rivaille, B Castro. Repetitive BOP coupling (REBOP) in solid phase peptide synthesis: Luliberin synthesis as model. Int J Peptide Protein Res 35: 465–472, 1990.

85. CG Fields, DH Lloyd, RL Macdonald, KM Otteson, RL Noble. HBTU activation for automated Fmoc solid-phase peptide synthesis. Peptide Res 4: 95–101, 1991.

86. R Knorr, A Trzeciak, W Bannwarth, D Gillessen. 1,1,3,3-Tetramethyluronium compounds as coupling reagents in peptide and protein chemistry. In: E Giralt, D Andreu, eds. Peptides 1990 Leiden, The Netherlands: Escom, pp 62–64, 1991.

87. D Hudson. Methodological implications of simultaneous solid-phase peptide synthesis 1: Comparison of different coupling procedures. J Org Chem 53: 617–624, 1988.

88. H Gausepohl, U Pieles, RW Frank. Schiffs base analog formation during in situ activation by HBTU and TBTU. In: JA Smith & JE Rivier, eds. Peptides: Chemistry and Biology, Leiden, The Netherlands: Escom, 1992, pp 523–524.

89. SC Story, JV Aldrich. Side-product formation during cyclization with HBTU on a solid support. Int J Peptide Protein Res 43: 292–296, 1994.

90. J Coste, D Le-Nguyen, B Castro. PyBOP: A new peptide coupling reagent devoid of toxic by-product. Tetrahedron Lett 31: 205–208, 1990.

91. A Fournier, W Danho, AM Felix. Applications of BOP reagent in solid phase peptide synthesis III: Solid phase peptide synthesis with unprotected aliphatic and aromatic hydroxyamino acids using BOP reagent. Int J Peptide Protein Res 33: 133–139, 1989.

92. LA Carpino, A El-Faham, CA Minor, F Albericio. Advantageous application of azabenzotriazole (triazolopyridine)-based coupling reagents to solid-phase peptide

synthesis. J Chem Soc Chem Commun: 201–203, 1994.

93. E Atherton, LR Cameron, RC Sheppard. Peptide synthesis, part 10: Use of pentafluorophenyl esters of fluorenylmethoxycarbonylamino acids in solid phase peptide synthesis. Tetrahedron 44: 843–857, 1988.

94. D Hudson. Methodological implications of simultaneous solid-phase peptide synthesis: A comparison of active esters. Peptide Res 3: 51–55, 1990.

95. JL Harrison, GM Petrie, RL Noble, HS Beilan, SN McCurdy, AR Culwell. Fmoc chemistry: Synthesis, kinetics, cleavage, and deprotection of arginine-containing peptides. In: TE Hugli, ed. Techniques in Protein Chemistry. San Diego: Academic Press, pp 506–516, 1989.

96. CG Fields, GB Fields, RL Noble, TA Cross. Solid phase peptide synthesis of [$^{15}$N]-gramicidins A, B, and C and high performance liquid chromatographic purification. Int J Peptide Protein Res 33: 298–303, 1989.

97. W König, R Geiger. Racemisierung bei Peptidsynthesen. Chem Ber 103: 2024–2033, 1970.

98. LA Carpino, BJ Cohen, KE Stephens, Jr, SY Sadat-Aalaee, J-H Tien, DC Langridge. ((9-Fluorenylmethyl)oxy)carbonyl (Fmoc) amino acid chlorides: Synthesis, characterization, and application to the rapid synthesis of short peptides. J Org Chem 51: 3732–3734, 1986.

99. LA Carpino, D Sadat-Aalaee, HG Chao, RH DeSelms. ((9-Fluorenylmethyl)oxy) carbonyl (Fmoc) amino acid fluorides: Convenient new peptide coupling reagents applicable to the Fmoc/tert-butyl strategy for solution and solid-phase syntheses. J Am Chem Soc 112: 9651–9652, 1990.

100. H Wenschuh, M Beyermann, E Krause, M Brudel, R Winter, M Schuemann, LA Carpino, M Bienert. Fmoc amino acid fluorides: Solid phase assembly of peptides incorporating sterically hindered residues. J Org Chem 59: 3275–3280, 1994.

101. H Wenschuh, M Beyermann, H Haber, JK Seydel, E Krause, M Bienert. Stepwise automated solid phase synthesis of naturally occuring peptaibols using Fmoc amino acid fluorides. J Org Chem 60: 405–410, 1995.

102. WD Fuller, NJ Krotzer, FR Naider, C-B Xue, M Goodman. Urethane-protected amino acid $N$-carboxyanhydrides: Stability, reactivity and solubility. In: CH Schneider, AN Eberle, eds. Peptides 1992. Leiden, The Netherlands: Escom, 1993, pp 229–230.

103. TT Romoff, M Goodman. Urethane-protected N-carboxyanhydrides (UNCAs) as unique reactants for the study of intrinsic racemization tendencies in peptide synthesis. J Peptide Res 49: 281–292, 1997.

104. M Flegel, RC Sheppard. A sensitive, general method for quantitative monitoring of continuous flow solid phase peptide synthesis. J Chem Soc Chem Commun: 536–538, 1990.

105. S Pickup, FD Blum, WT Ford. Self-diffusion coefficients of Boc-amino acid anhydrides under conditions of solid phase peptide synthesis. J Polym Sci Polym Chem Ed 28: 931–934, 1990.

106. J Bedford, C Hyde, T Johnson, W Jun, D Owen, M Quibell, RC Sheppard. Amino acid structure and "difficult sequences" in solid phase peptide synthesis. Int J Peptide Protein Res 40: 300–307, 1992.

107. DH Live, SBH Kent. Correlation of coupling rates with physicochemical properties of resin-bound peptides in solid phase synthesis. In: VJ Hruby, DH Rich, eds. Peptides: Structure and Function. Rockford, IL: Pierce Chemical Co, 1983, pp 65–68.

108. M Mutter, K-H Altmann, D Bellof, A Flörsheimer, J Herbert, M Huber, B Klein, L Strauch, T Vorherr. The impact of secondary structure formation in peptide synthesis. In: CM Deber, VJ Hruby, KD Kopple, eds. Peptides: Structure and Function. Rockford, IL: Pierce Chemical Co, 1985, pp 397–405.

109. AG Ludwick, LW Jelinski, D Live, A Kintanar, JJ Dumais. Association of peptide chains during Merrifield solid-phase peptide synthesis: A deuterium NMR study. J Am Chem Soc 108: 6493–6496, 1986.

110. E Atherton, V Woolley, RC Sheppard. Internal association in solid phase peptide synthesis: Synthesis of cytochrome C residues 66-104 on polyamide supports. J Chem Soc Chem Commun: 970–971, 1980.

111. E Atherton, RC Sheppard, P Ward. Peptide synthesis, Part 7: Solid-phase synthesis of conotoxin G1. J Chem Soc Perkin Trans I: 2065-2073, 1985.

112. CG Fields, GB Fields. Solvation effects in solid phase synthesis. J Am Chem Soc 113: 4202–4207, 1991.

113. C Hyde, T Johnson, RC Sheppard. Internal aggregation during solid phase peptide synthesis. Dimethyl sulfoxide as a powerful dissociating solvent. J Chem Soc Chem Commun: 1573–1575, 1992.

114. WA Kils, JM Stewart. Chaotropic salts improve SPPS coupling reactions. In: JE Rivier, GR Marshall, eds. Peptides:

Chemistry, Structure and Biology. Leiden, The Netherlands: Escom, 1990, pp 904–906.

115. L Zhang, C Goldhammer, B Henkel, G Panhaus, F Zuehl, G Jung, E Bayer. "Magic mixture", a powerful solvent system for solid-phase synthesis of difficult peptides. In: R Epton, ed. Innovation and Perspectives in Solid Phase Synthesis. Birmingham, U.K.: Mayflower Worldwide, Ltd, 1994, pp 711–712.

116. JM Stewart, JD Young. Solid Phase Peptide Synthesis, 2nd Ed. Rockford, IL: Pierce Chemical Co, 1984.

117. GB Fields, RL Noble. Solid phase peptide synthesis utilizing 9-fluorenylmethoxycarbonyl amino acids. Int J Peptide Protein Res 35: 161–214, 1990.

118. GB Fields, Z Tian, G Barany. Principles and practice of solid-phase peptide synthesis In: GA Grant, ed. Synthetic Peptides: A User's Guide. New York: WH Freeman and Co, 1992, pp 77–183.

119. BF Gisin, RB Merrifield. Carboxyl-catalyzed intramolecular aminolysis: A side reaction in solid-phase peptide synthesis. J Am Chem Soc 94: 3102–3106, 1972.

120. E Pedroso, A Grandas, X de las Heras, R Eritja, E Giralt. Diketopiperazine formation in solid phase peptide synthesis using p-alkoxybenzyl ester resins and Fmoc-amino acids. Tetrahedron Lett 27: 743–746, 1986.

121. F Albericio, G Barany. Improved approach for anchoring $N^\alpha$-9-fluorenylmethyloxy carbonylamino acids as p-alkoxybenzyl esters in solid-phase peptide synthesis. Int J Peptide Protein Res 26: 92–97, 1985.

122. M Gairi, P Lloyd-Williams, F Albericio, E Giralt. Use of BOP reagent for the suppression of diketopiperazine formation in Boc/Bzl solid-phase peptide synthesis. Tetrahedron Lett 31: 7363–7366, 1990.

123. M Ueki, M Amemiya. Removal of 9-fluorenylmethyloxycarbonyl (Fmoc) group with tetrabutyl-ammonium fluoride. Tetrahedron Lett 28: 6617–6620, 1987.

124. K Barlos, D Gatos, J Hondrelis, J Matsoukas, GJ Moore, W Schäfer, P Sotiriou. Darstellung neuer säureempfindlicker Harze vom sek.-Alkohol-Typ und ihre Anwendung zur Synthese von Peptiden. Liebigs Ann Chem: 951–955, 1989.

125. K Barlos, D Gatos, J Kallitsis, G Papaphotiu, P Sotiriu, Y Wenqing, W Schäfer. Darstellung geschützter peptidfragmente unter einsatz substituierter Tri-

phenylmethyl-Harze. Tetrahedron Lett 30: 3943–3946, 1989.

126. M Bodanszky, JZ Kwei. Side reactions in peptide synthesis VII: Sequence dependence in the formation of aminosuccinyl derivatives from $\beta$-benzyl-aspartyl peptides. Int J Peptide Protein Res 12: 69–74, 1978.

127. M Bodanszky, JC Tolle, SS Deshmane, A Bodanszky. Side reactions in peptide synthesis VI: A reexamination of the benzyl group in the protection of the side chains of tyrosine and aspartic acid. Int J Peptide Protein Res 12: 57–68, 1978.

128. E Nicolas, E Pedroso, E Giralt. Formation of aspartimide peptides in Asp-Gly sequences. Tetrahedron Lett 30: 497–500, 1989.

129. GW Kenner, JH Seely. Phenyl esters for C-terminal protection in peptide synthesis. J Am Chem Soc 94: 3259–3260, 1972.

130. Y Yang, WV Sweeney, K Scheider, S Thörnqvist, BT Chait, JP Tam. Aspartimide formation in base-driven 9-fluorenylmethoxycarbonyl chemistry. Tetrahedron Lett 35: 9689–9692, 1994.

131. M Quibell, D Owen, LC Packman, T Johnson. Suppression of piperidine-mediated side product formation for Asp(OBu$^t$)-containing peptides by the use of N-(2-hydroxy-4-methoxy benzyl)(Hmb) backbone amide protection. J Chem Soc Chem Commun: 2343–2344, 1994.

132. JL Lauer, CG Fields, GB Fields. Sequence dependence of aspartimide formation during 9-fluorenylmethoxycarbonyl solid-phase peptide synthesis. Lett Peptide Sci 1: 197–205, 1995.

133. RC Stephenson, S Clarke. Succinimide formation from aspartyl and asparaginyl peptides as a model for the spontaneous degradation of proteins. J Biol Chem 264: 6164–6170, 1989.

134. K Patel, RT Borchardt. Chemical pathways of peptide degradation II: Kinetics of deamidation of an asparaginyl residue in a model hexapeptide. Pharm Res 7: 703–711, 1990.

135. K Patel, RT Borchardt. Chemical pathways of peptide degradation III: Effect of primary sequence on the pathways of deamidation of asparaginyl residues in hexapeptides. Pharm Res 7: 787–793, 1990.

136. PBW Ten Kortenaar, BMM Hendrix, JW van Nipsen. Acid-catalyzed hydrolysis of peptide-amides in the solid state. Int J Peptide Protein Res 36: 231–235, 1990.

137. R Tyler-Cross, V Schirch. Effects of amino acid sequence, buffers and ionic strength on the rate and mechanism of deamidation

of asparagine residues in small peptides. J Biol Chem 266: 22549–22556, 1991.

138. E Atherton, PM Hardy, DE Harris, BH Matthews. Racemization of C-terminal cysteine during peptide assembly. In: E Giralt, D Andreu, eds. Peptides 1990. Leiden: Escom, 1991, pp 234–244.

139. Y Fujiwara, K Akaji, Y Kiso. Racemization-free synthesis of C-terminal cysteine-peptide using 2-chlorotrityl resin. Chem Pharm Bull 42: 724–726, 1994.

140. R Eritja, JP Ziehler-Martin, PA Walker, TD Lee, K Legesse, F Albericio, BE Kaplan. On the use of S-t-butylsulphenyl group for protection of cysteine in solid-phase peptide synthesis using Fmoc-amino acids. Tetrahedron 43: 2675–2680, 1987.

141. GB Fields, CG Fields, J Petefish, HE Van Wart, TA Cross. Solid phase synthesis and solid state NMR spectroscopy of [$^{15}$N-Ala$_3$]-Val-gramicidin A. Proc Natl Acad Sci USA 85: 1384-1388, 1988.

142. RS Prosser, JH Davis, FW Dahlquist, MA Lindorfer. $^2$H nuclear magnetic resonance of the gramicidin A backbone in a phospholipid bilayer. Biochemistry 30: 4687–4696, 1991.

143. Kaiser, RL Colescott, CD Bossinger, PI Cook. Color test for detection of free terminal amino groups in the solid-phase synthesis of peptides. Anal Biochem 34: 595–598, 1970.

144. E Kaiser, CD Bossinger, RL Colescott, DB Olsen. Color test for terminal prolyl residues in the solid-phase synthesis of peptides. Anal Chim Acta 118: 149–151, 1980.

145. RS Hodges, RB Merrifield. Monitoring of solid phase peptide synthesis by an automated spectrophotometric picrate method. Anal Biochem 65: 241–272, 1975.

146. O Arad, RA Houghten. An evaluation of the advantages and effectiveness of picric acid monitoring during solid phase peptide synthesis. Peptide Res 3:42–50, 1990.

147. M Horn, C Novak. A monitoring and control chemistry for solid-phase peptide synthesis. Am Biotech Lab 5, September/October: 12–21, 1987.

148. MP Reddy, PJ Volker. Novel method for monitoring the coupling efficiency in solid phase peptide synthesis. Int J Peptide Protein Res 31: 345–348, 1988.

149. V Krchnák, J Vágner, P Safár, M Lebl. Noninvasive continuous monitoring of solid-phase peptide synthesis by acid-base indicator. Coll Czech Chem Commun 53: 2542–2548, 1988.

150. JD Young, AS Huang, N Ariel, JB Bruins, D Ng, RL Stevens. Coupling efficiencies of amino acids in solid phase synthesis of peptides. Peptide Res 3: 194–200, 1990.

151. CS Nielson, PH Hansen, A Lihme, PMH Heegaard. Real time monitoring of acylations during solid phase peptide synthesis: A method based on electrochemical detection. J Biochem Biophys Methods 20: 69–75, 1989.

152. J Fox, R Newton, P Heegard, C Schafer-Nielsen. A novel method of monitoring the coupling reaction in solid phase synthesis. In: R Epton, ed. Innovation and Perspectives in Solid Phase Synthesis. Birmingham, U.K.: Solid Phase Conference Coordination, Ltd., pp 141–153, 1990.

153. NV McFerran, B Walker, CD McGurk, FC Scott. Conductance measurements in solid phase peptide synthesis I: Monitoring coupling and deprotection in Fmoc chemistry. Int J Peptide Protein Res 37: 382–387, 1991.

154. CD Chang, AM Felix, MH Jimenez, J Meienhofer. Solid-phase peptide synthesis of somatostatin using mild base cleavage of N$^\alpha$-fluorenylmethyloxycarbonylamino acids. Int J Peptide Protein Res 15: 485–494, 1980.

155. R Frank, H Gausepohl. Continuous flow peptide synthesis. In: H Tschesche, ed. Modern Methods in Protein Chemistry, Vol 3. Berlin: Walter de Gruyter, 1988, pp 41–60.

156. E Atherton, JL Holder, M Meldal, RC Sheppard, RM Valerio. Peptide synthesis, Part 12: 3,4-Dihydro-4-oxo-1,2,3-benzotriazin-3-yl esters of fluorenylmethoxycarbonyl amino acids as self-indicating reagents for solid phase synthesis. J Chem Soc Perkin Trans I: 2887–2894, 1988.

157. LR Cameron, JL Holder, M Meldal, RC Sheppard. Peptide synthesis, Part 13: Feedback control in solid phase synthesis: Use of fluorenylmethoxycarbonyl amino acid 3,4-dihydro-4-oxo-1,2,3-benzotriazin-3-yl esters in a fully automated system. J Chem Soc Perkin Trans I: 2895–2901, 1988.

158. BD Larsen, C Larsen, A Holm. Incomplete Fmoc-deprotection in solid phase synthesis. In: E Giralt, D Andreu, eds. Peptides 1990. Leiden, The Netherlands: Escom, 1991, pp 183–185.

159. O Ogunjobi, R Ramage. Ubiquitin: Preparative chemical synthesis, purification and characterization. Biochem Soc Trans 18: 1322–1323, 1990.

160. GW Tregear, J van Rietschoten, R Sauer, HD Niall, HT Keutmann, JT Potts Jr. Synthesis, purification, and chemical characterization of the amino-terminal 1-34 fragment of bovine parathyroid

hormone synthesized by the solid-phase procedure. Biochemistry 16: 2817–2823, 1977.

161. GR Matsueda, JM Stewart. A p-methylbenzhydrylamine resin for improved solid-phase synthesis of peptide amides. Peptides 2: 45–50, 1981.

162. SBH Kent, M Riemen, M LeDoux, RB Merrifield. A study of the Edman degradation in the assessment of the purity of synthetic peptides. In: M Elzinga, ed. Methods in Protein Sequence Analysis. Clifton, New Jersey: Humana Press, 1982, pp 205–213.

163. ML Kochersperger, R Blacher, P Kelly, L Pierce, DH Hawke. Sequencing of peptides on solid phase supports. Am Biotech Lab 7(3): 26–37, 1989.

164. CG Fields, VL VanDrisse, GB Fields. Edman degradation sequence analysis of resin-bound peptides synthesized by 9-fluorenylmethoxycarbonyl chemistry. Peptide Res 6: 39–46, 1993.

165. BJ Egner, M Cardno, M Bradley. Linkers for combinatorial chemistry and reaction analysis using solid phase in situ mass spectrometry. J Chem Soc Chem Commun: 2163–2164, 1995.

166. G Talbo, JD Wade, N Dawson, M Manoussios, GW Tregear. Rapid semi on-line monitoring of Fmoc solid-phase peptide synthesis by matrix-assisted laser desorption/ionization mass spectrometry. Lett Peptide Sci 4: 121–127, 1997.

167. MC Fitzgerald, K Harris, CG Shevlin, G Siuzdak. Direct characterization of solid phase resin-bound molecules by mass spectrometry. Bioorg Med Chem Lett 6: 979–982, 1996.

168. WL Fitch, G Detre, CP Holmes. High-resolution 1H-NMR in solid phase organic synthesis. J Org Chem 59: 7955-7956, 1994.

169. GR Matsueda, E Haber. The use of an internal reference amino acid for the evaluation of reactions in solid-phase peptide synthesis. Anal Biochem 104: 215–227, 1980.

170. F Albericio, N Kneib-Cordonier, S Biancalana, L Gera, RI Masada, D Hudson, GBarany. Preparation and application of the 5-(4-(9-fluorenylmethyloxycarbonyl)amino-methyl-3,5-dimethoxyphenoxy)valeric acid (PAL) handle for the solid-phase synthesis of C-terminal peptide amides under mild conditions. J Org Chem 55: 3730–3743, 1990.

171. H-M Yu, S-T Chen, S-H Chiou, K-T Wang. Determination of amino acids on Merrifield resin by microwave hydrolysis. J Chromatogr 456: 357–362, 1988.

172. E Bayer, H Eckstein, K Haegele, WA Koenig, W Bruening, H Hagenmaier, W Parr. Failure sequences in the solid phase synthesis of polypeptides. J Am Chem Soc 92: 1735–1738, 1970.

173. AM Felix, MH Jiminez, R Vergona, MR Cohen. Synthesis and biological studies of novel bradykinin analogues. Int J Peptide Protein Res 5: 201–206, 1973.

174. JM Stewart, JW Ryan, AH Brady. Hydroxyproline analogs of bradykinin. J Med Chem 17: 537–539, 1974.

175. GB Fields, HE Van Wart, H Birkedal-Hansen. Sequence specificity of human skin fibroblast collagenase: Evidence for the role of collagen structure in determining the collagenase cleavage site. J Biol Chem 262: 6221–6226, 1987.

176. S Netzel-Arnett, GB Fields, H Birkedal-Hansen, HE Van Wart. Sequence specificities of human fibroblast and neutrophil collagenases. J Biol Chem 266: 6747–6755, 21326, 1991.

177. LJ Cruz, G Kupryszewski, GW LeCheminant, WR Gray, BM Olivera, J Rivier. $\mu$-Conotoxin GIIIA, a peptide ligand for muscle sodium channels: Chemical synthesis, radiolabeling, and receptor characterization. Biochemistry 28: 3437–3442, 1989.

178. S Becker, E Atherton, RD Gordon. Synthesis and characterization of $\mu$-conotoxin IIIa. Eur J Biochem 185: 79–84, 1989.

179. B Penke, J Zsigo, J Spiess. Synthesis of a protected hydroxylysine derivative for application in peptide synthesis. The Eleventh American Peptide Symposium Abstracts, The Salk Institute and University of California at San Diego, La Jolla, CA, P-335, 1989.

180. HA Remmer, GB Fields. Problems encountered with the synthesis of a glycosylated hydroxylysine derivative suitable for Fmoc-solid phase peptide synthesis. In: JP Tam, PTP Kaumaya, eds. Peptides: Frontiers of Peptide Science (Proceedings of the Fifteenth American Peptide Symposium, June 14–19, 1997, Nashville, TN). Dordrecht, The Netherlands: Kluwer Academic Publishers, 1999, pp 287–288.

181. J Rivier, R Galyean, L Simon, LJ Cruz, BM Olivera, WR Gray. Total synthesis and further characterization of the $\gamma$-carboxyglutamate-containing "sleeper" peptide from Conus geographus venom. Biochemistry 26: 8508–8512, 1987.

182. JW Perich. Modern methods of O-phosphoserine- and O-phosphotyrosine-containing peptide synthesis. In: BE Kemp, ed. Peptides and Protein Phosphorylation.

Boca Raton, FL: CRC Press, 1990, pp 289–314.

183. JW Perich, RM Valerio, RB Johns. Solid-phase synthesis of an O-phospho-seryl-containing peptide using phenyl phosphorotriester protection. Tetrahedron Lett 27: 1377–1380, 1986.

184. A Arendt, K Palczewski, WT Moore, RM Caprioli, JH McDowell, PA Hargrave. Synthesis of phosphopeptides containing O-phosphoserine or O-phosphothreonine. Int J Peptide Protein Res 33: 468–476, 1989.

185. L Otvös Jr, I Elekes, VM-J Lee. Solid-phase synthesis of phosphopeptides. Int J Peptide Protein Res 34: 129–133, 1989.

186. HBA de Bont, JH van Boom, RMJ Liskamp. Automatic synthesis of phophopeptides by phosphorylation on the solid phase. Tetrahedron Lett 31: 2497–2500, 1990.

187. EA Kitas, JW Perich, JD Wade, RB Johns, GW Tregear. Fmoc-polyamide solid phase synthesis of an O-phosphotyrosine-containing tridecapeptide. Tetrahedron Lett 30: 6229–6232, 1989.

188. EA Kitas, JD Wade, RB Johns, JW Perich, GW Tregear. Preparation and use of $N^\alpha$-fluorenylmethoxycarbonyl-O-dibenzylphosphono-L-tyrosine in continuous flow solid phase peptide synthesis. J Chem Soc Chem Commun: 338–339, 1991.

189. JW Perich, EC Reynolds. Fmoc/solid-phase synthesis of Tyr(P)-containing peptides through t-butyl phosphate protection. Int J Peptide Protein Res 37: 572–575, 1991.

190. EA Ottinger, LL Shekels, DA Bernlohr, G Barany. Synthesis of phosphotyrosine-containing peptides and their use as substrates for protein tyrosine phosphatases. Biochemistry 32: 4354–4361, 1993.

191. ASH Chan, JL Mobley, GB Fields, Y Shimizu. CD7-mediated regulation of integrin adhesiveness on human T cells involves tyrosine phosphorylation-dependent activation of phosphatidylinositol 3-kinase. J Immunol 159: 934–942, 1997.

192. G Zardeneta, D Chen, ST Weintraub, RJ Klebe. Synthesis of phosphotyrosyl containing phosphopeptides by solid-phase peptide synthesis. Anal Biochem 190: 340–347, 1990.

193. JW Perich. Synthesis of phosphopeptides using modern chemical approaches. In: GB Fields, ed. Methods in Enzymology, Volume 289. Orlando: Academic Press, 1997, pp 245–266.

194. JW Perich, RB Johns. Di tert-butyl-N,N-diethylphosphoamidite and dibenzyl-N,N-diethylphosphoamidite; highly reactive reagents for the "phosphite triester" phosphorylation of serine containing peptides. Tetrahedron Lett 29: 2369–2372, 1988.

195. B Penke, J Rivier. Solid-phase synthesis of peptide amides on a polystyrene support using fluorenylmethoxycarbonyl protecting groups. J Org Chem 52: 1197–1200, 1987.

196. B Penke, L Nyerges. Preparation and application of a new resin for synthesis of peptide amides via Fmoc-strategy. In: G Jung, E Bayer, eds. Peptides 1988. Berlin: Walter de Gruyter and Co, 1989, pp 142–144.

197. B Penke, L Nyerges. Solid phase synthesis of porcine cholecystokinin-33 in a new resin via Fmoc-strategy. Peptide Res 4: 289–295, 1991.

198. RJ Bontems, P Hegyes, SL Bontems, F Albericio, G Barany. Synthesis and applications of XAL, a new acid-labile handle for solid-phase synthesis of peptide amides. In: JA Smith, JE Rivier, eds. Peptides, Chemistry and Biology. Leiden, The Netherlands: Escom, 1992, pp 601–602.

199. E Jaeger, HA Remmer, G Jung, J Metzger, W Oberthuer, KP Ruecknagel, W Schaefer, J Sonnenbichler, I Zetl. Side reactions in peptide synthesis V: O-sulfonation of serine and threonine during removal of the Pmc and Mtr protecting groups from arginine residues in Fmoc solid phase synthesis. Biol Chem Hoppe-Seyler 347: 349–362, 1993.

200. H Kunz. Synthesis of glycopeptides: Partial structures of biological recognition components. Angew Chem Int Ed Engl 26: 294–308, 1987.

201. H Paulsen, G Merz, U Weichert. Solid-phase synthesis of O-glycopeptide sequences. Angew Chem Int Ed Engl 27: 1365–1367, 1988.

202. JL Torres, I Haro, G Valencia, F Reig, JM Garcia-Anton. Synthesis of $O^{1.5}$-($\beta$-D-galacto pyranosyl)[DMet$^2$,Hyp$^5$] enkephalin amide, a new highly potent analgesic enkephalin-related glycosyl peptide. Experientia 45: 574–576, 1989.

203. H Paulsen, G Merz, S Peters, U Weichert. Festphasensynthese von O-glycopeptiden. Liebigs Ann Chem: 1165–1173, 1990.

204. AM Jansson, M Meldal, K Bock. The active ester N-Fmoc-3-O-[Ac$_4$-$\alpha$-D-Man$p$-(1 → 2)-Ac$_3$-$\alpha$-D-Man$p$-1-]-threonine-O-Pfp as a building block in solid-phase synthesis of an O-linked

dimannosyl glycopeptide. Tetrahedron Lett 31: 6991–6994, 1990.

205. BG de la Torre, JL Torres, E Bardají, P Clapés, N Xaus, X Jorba, S Calvet, F Albericio, G Valencia. Improved method for the synthesis of o-glycosylated Fmoc amino acids to be used in solid-phase glycopeptide synthesis. J Chem. Soc Chem Commun: 965–967, 1990.

206. E Bardaji, JL Torres, P Clapes, F Albericio, G Barany, RE Rodriguez, MP Sacristan, G Valencia. Synthesis and biological activity of O-glycosylated morphiceptin analogs. J Chem Soc Perkin Trans I: 1755–1759, 1991.

207. L Biondi, F Filira, M Gobbo, B Scolaro, R Rocchi, F Cavaggion. Synthesis of glycosylated tuftsins and tuftsin-containing IgG fragment undecapeptide. Int J Peptide Protein Res 37: 112–121, 1991.

208. L Otvos Jr, K Wroblewski, E Kollat, A Perczel, M Hollosi, GD Fasman, HCJ Ertl, J Thurin. Coupling strategies in solid-phase synthesis of glycopeptides. Peptide Res 2: 362–366, 1989.

209. L Otvös Jr, L Urge, M Hollosi, K Wroblewski, G Graczyk, GD Fasman, J Thurin. Automated solid-phase synthesis of glycopeptides: Incorporation of unprotected mono- and disaccharide units of N-glycoprotein antennae into T cell epitopic peptides. Tetrahedron Lett 31: 5889–5892, 1990.

210. F Filira, L Biondi, F Cavaggion, B Scolaro, R Rocchi. Synthesis of O-glycosylated tuftsins by utilizing threonine derivatives containing an unprotected monosaccharide moiety. Int J Peptide Protein Res 36: 86–96, 1990.

211. M Meldal, KJ Jensen. Pentafluorophenyl esters for the temporary protection of the α-carboxy group in solid phase glycopeptide synthesis. J Chem Soc Chem Commun: 483–485, 1990.

212. J Kihlberg, M Elofsson, LA Salvador. Direct synthesis of glycosylated amino acids from carbohydrate preacetates and Fmoc-amino acids: solid-phase synthesis of biomedicinally interesting glycopeptides. In: GB Fields, ed. Methods in Enzymology, Volume 289. Orlando: Academic Press, 1997, pp 245–266.

213. H Kunz, B Dombo. Solid phase synthesis of peptides and glycopeptides on polymeric supports with allylic anchor groups. Angew Chem Int Ed Engl 27: 711–713, 1988.

214. M Elofsson, LA Salvador, J Kihlberg. Preparation of TN and sialyl TN building blocks used in Fmoc-solid phase synthesis of glycopeptide fragments from HIV GP120. Tetrahedron 53: 369–390, 1997.

215. T Vuljanic, KE Bergquist, H Clausen, S Roy, J Kihlberg. Piperidine is preffered to morpholine for Fmoc cleavage in solid phase glycopeptide synthesis as exemplified by preparation of glycopeptides related to HIV GP120 and mucins. Tetrahedron 52: 7983–8000, 1996.

216. S Peters, TL Lowary, O Hindsgaul, M Meldal, K Bock. Solid phase synthesis of a fucosylated glycopeptide of human factor IX with a fucose-α-(1 → O)-serine linkage. J Chem Soc Perkin Trans 1: 3017–3022, 1995.

217. I Christiansen-Brams, M Meldal, K Bock. Silyl protection in the solid phase synthesis of N-linked neoglycopeptides. One step deprotection of fully protected neoglycopeptides. Tetrahedron Lett 34: 3315–3318, 1993.

218. RG Hiskey. Sulfhydryl Group Protection in Peptide Synthesis. In: E Gross, J Meienhofer, eds. The Peptides, Vol. 3. New York: Academic Press, 1981, pp 137–167.

219. SN McCurdy. The investigation of Fmoc-cysteine derivatives in solid phase peptide synthesis. Peptide Res 2: 147–151, 1989.

220. JP Tam, C-R Wu, W Liu, J-W Zhang. Disulfide bond formation in peptides by dimethyl sulfoxide: Scope and applications. J Am Chem Soc 113: 6657–6662, 1991.

221. AK Ahmed, SW Schaffer, DB Wetlaufer. Nonenzymic reactivation of reduced bovine pancreatic ribonuclease by air oxidation and by glutathione oxidoreduction buffers. J Biol Chem 250: 8477–8482, 1975.

222. Y-Z Lin, G Caporaso, P-Y Chang, X-H Ke, JP Tam. Synthesis of a biological active tumor growth factor from the predicted DNA sequence of Shope fibroma virus. Biochemistry 27: 5640–5645, 1988.

223. MW Pennington, SM Festin, ML Maccecchini. Comparison of folding procedures on synthetic ω-conotoxin. In: E Giralt, D Andreu, eds. Peptides 1990. Leiden, The Netherlands: Escom, 1991, pp 164–166.

224. B Kamber, A Hartmann, K Eisler, B Riniker, H Rink, P Sieber, W Rittel. The synthesis of cystine peptides by iodine oxidation of S-trityl-cysteine and S-acetamidomethyl-cysteine peptides. Helv Chim Acta 63: 899–915, 1980.

225. P Sieber, B Kamber, B Riniker, W Rittel. Iodine oxidation of S-trityl and S-acetamidomethyl-cysteine-peptides containing tryptophan: Conditions leading to

the formation of tryptophan-2-thioesters. Helv Chim Acta 63: 2358–2363, 1980.

226. R Casaretto, R Nyfeler. Isolation, structure and activity of a side product from the synthesis of human endothelin. In: E Giralt ‿, eds. Peptides 1990. Leiden, The ‿ands, Escom, 1991, pp 181–182.

‿ rujii, A Otaka, S Funakoshi, K Bessho, T Watanabe, K Akaji, H Yajima. Studies on peptides CLI: Synthesis of cystine-peptides by oxidation of S-protected cysteine-peptides with thallium (III) trifluoroacetate. Chem Pharm Bull 35: 2339–2347, 1987.

228. Y Kiso, M Yoshida, Y Fujiwara, T Kimura, M Shimokura, K Akaji. Trimethylacetamido methyl (Tacm) group, a new protecting group for the thiol function of cysteine. Chem Pharm Bull 38: 673–675, 1990.

229. K Akaji, T Tatsumi, M Yoshida, T Kimura, Y Fujiwara, Y Kiso. Synthesis of cystine-peptide by a new disulphide bond-forming reaction using the silyl chloride-sulphoxide system. J Chem Soc Chem Commun: 167–168, 1991.

230. M Ruiz-Gayo, F Albericio, M Pons, M Royo, E Pedroso, E Giralt. Uteroglobin-like peptide cavities I: Synthesis of antiparallel and parallel dimers of bis-cysteine peptides. Tetrahedron Lett 29: 3845–3848, 1988.

231. B Ponsati, E Giralt, D Andreu. Solid-phase approaches to regiospecific double disulfide formation: Application to a fragment of bovine pituitary peptide. Tetrahedron 46: 8255–8266, 1990.

232. WR Gray, A Luque, R Galyean, E Atherton, RC Sheppard, BL Stone, A Reyes, J Alford, M McIntosh, BM Olivera, LJ Cruz, J Rivier. Contoxin GI: Disulfide bridges, synthesis, and preparation of iodinated derivatives. Biochemistry 23: 2796–2802, 1984.

233. JP Tam, W Liu, J-W Zhang, M Galantino, R de Castiglione. D-Amino acid and alanine scans of endothelin: An approach to study refolding intermediates. In: E Giralt, D Andreu, eds. Peptides 1990. Leiden, The Netherlands: Escom, 1991, pp 160–163.

234. AW Mott, U Slomczynska,G Barany. Formation of sulfur-sulfur bonds during solid-phase peptide synthesis: Application to the synthesis of oxytocin. In: B Castro, J Martinez, eds. Forum Peptides Le Cap d'Agde 1984. Nancy, France: Les Impressions Dohr, 1984, pp 321–324.

235. R Buchta, E Bondi, M Fridkin. Peptides related to the calcium binding domains II and III of calmodulin: Synthesis and calmodulin-like features. Int J Peptide Protein Res 28: 289–297, 1986.

236. O Ploux, G Chassaing, A Marquet. Cyclization of peptides on a solid support: Application to cyclic analogs of substance P. Int J Peptide Protein Res 29: 162–169, 1987.

237. PBW Ten Kortenaar, JW van Nispen. Formation of open-chain asymmetrical cystine peptides on a solid support: Synthesis of pGlu-Asn-Cyt-Pro-Arg-Gly-OH. Coll Czech Chem Commun 53: 2537–2541, 1988.

238. C García-Echeverría, F Albericio, M Pons, G Barany, E Giralt. Convenient synthesis of a cyclic peptide disulfide: A type II $\beta$-turn structural model. Tetrahedron Lett. 30: 2441–2444, 1989.

239. AM Felix, CT Wang, EP Heimer, A Fournier. Applications of BOP reagent in solid phase synthesis II: Solid phase side-chain cyclization using BOP reagent. Int J Peptide Protein Res 31: 231–238, 1988.

240. AM Felix, EP Heimer, CT Wang, TJ Lambros, A Fournier, TF Mowles, S Maines, RM Campbell, BB Wegrzynski, V Toome, D Fry, VS Madison. Synthesis, biological activity and conformational analysis of cyclic GRF analogs. Int J Peptide Protein Res 32: 441–454, 1988.

241. VJ Hruby, F Al-Obeidi, DG Sanderson, DD Smith. Synthesis of cyclic peptides by solid phase methods. In: R Epton, ed. Innovation and Perspectives in Solid Phase Synthesis. Birmingham, U.K.: Solid Phase Conference Coordination, Ltd., 1990, pp 197–203.

242. PW Schiller, TM-D Nguyen, J Miller. Synthesis of side-chain cyclized peptide analogs on solid supports. Int J Peptide Protein Res 25: 171–177, 1985.

243. EE Sugg, AM de L Castrucci, ME Hadley, G van Binst, VJ Hruby. Cyclic lactam analogues of Ac-[Nle$^4$]$\alpha$-MSH$_{4-11}$-NH$_2$. Biochemistry 27: 8181–8188, 1988.

244. SA Kates, SB Daniels, F Albericio. Automated allyl cleavage for continuous flow synthesis of cyclic and branched peptides. Anal Biochem 212: 303–310, 1993.

245. S Plaué. Synthesis of cyclic peptides on solid support: Application to analogs of hemagglutinin of influenza virus. Int J Peptide Protein Res 35: 510–517, 1990.

246. SA Kates, NA Solé, F Albericio, G Barany. Solid-phase synthesis of cyclic peptides. In: C Basava, GM Anantharamaiah, eds. Peptides: Design, Synthesis, and Biological Activity. Boston: Birkhäuser, 1994, pp 39–58.

247. RC deL Milton, SCF Milton, SBH Kent. Total chemical synthesis of a D-enzyme: The enantiomers of HIV-1 protease show demonstration of reciprocal chiral substrate specifity. Science 256: 1445–1448, 1992.

248. CT Dooley, NN Chung, BC Wilkes, PW Schiller, JM Bidlack, GW Pasternak, RA Houghten. An all D-amino acid opioid peptide with central analgesic activity from a combinatorial library. Science 266: 2019–2022, 1994.

249. AJ Miles, APN Skubitz, LT Furcht, GB Fields. Promotion of cell adhesion by single-stranded triple helical peptide models of basement membrane collagen α1(IV)531–543. J Biol. Chem 269: 30939–30945, 1994.

250. M Chorev, M Goodman. Recent developments in retro peptides and proteins – an ongoing topochemical exploration. Tibtech 13: 438–445, 1995.

251. A Dal Pozzo, E Laurita. Configurationally stable retro-inverso peptides containing α 2-methyl-2-alkyl-malonyl residue. In: HCS Maia, ed. Peptides 1994. Leiden, The Netherlands: Escom, pp 714–715, 1995.

252. H Yajima, N Fujii, S Funakoshi, T Watanabe, E Murayama, A Otaka. New strategy for the chemical synthesis of proteins. Tetrahedron 44: 805–819, 1988.

253. M Nomizu, Y Inagaki, T Yamashita, A Ohkubo, A Otaka, N Fujii, PP Roller, H Yajima. Two-step hard acid deprotection/cleavage procedure for solid phase peptide synthesis. Int J Peptide Protein Res 37: 145–152, 1991.

254. K Akaji, N Fujii, F Tokunaga, T Miyata, S Iwanaga, H Yajima. Studies on peptides CLXVIII: Syntheses of three peptides isolated from horseshoe crab hemocytes, tachyplesin I, tachyplesin II, and polyphemusin I. Chem Pharm Bull 37: 2661–2664, 1989.

255. E Jaeger, P Thamm, S Knof, E Wünsch, M Löw, L Kisfaludy. Nebenreaktionen bei Peptidsynthesen III: Synthese und Charakterisierung von N^in-tert-butylierten Tryptophan-Derivaten. Hoppe-Seyler's Z Physiol Chem 359: 1617–1628, 1978.

256. E Jaeger, P Thamm, S Knof, E Wünsch. Nebenreaktionen bei peptidsynthesen IV: Charakterisierung von C- und C,N-tert-butylierten tryptophan-derivaten. Hoppe-Seyler's Z Physiol Chem 359: 1629–1636, 1978.

257. M Löw, L Kisfaludy, E Jaeger, P Thamm, S Knof, E Wünsch. Direkte tert-butylierung des Tryptophans: Herstellung von 2,5,7-tri-tert-Butyl-Tryptophan. Hoppe-Seyler's Z Physiol Chem 359: 1637–1642, 1978.

258. M Löw, L Kisfaludy, P Sohár. tert-Butylierung des Tryptophan-Indolringes während der Abspaltung der tert-butyloxycarbonyl-Gruppe bei Peptidsynthesen. Hoppe-Seyler's Z Physiol Chem 359: 1643–1651, 1978.

259. BF Lundt, NL Johansen, A Volund, J Markussen. Removal of t-butyl and t-butoxycarbonyl protecting groups with trifluoroacetic acid. Int J Peptide Protein Res 12: 258–268, 1978.

260. Y Masui, N Chino, S Sakakibara. The modification of tryptophyl residues during the acidolytic cleavage of Boc-groups I: Studies with Boc-tryptophan. Bull Chem Soc Jpn 53: 464–468, 1980.

261. E Scoffone, A Previero, CA Benassi, P Pajetta. Oxidative modification of tryptophan residues in peptides. In: L Zervas, ed. Peptides 1963. Oxford, U.K.: Pergamon Press, 1966, pp 183–188.

262. St Guttman, RA Boissonnas. Synthése de l'α-mélanotropine (α-MSH) de porc. Helv Chim Acta 42: 1257–1264, 1959.

263. DS King, CG Fields, GB Fields. A cleavage method for minimizing side reactions following Fmoc solid phase peptide synthesis. Int J Peptide Protein Res 36: 255–266, 1990.

264. CA Guy, GB Fields. Trifluoroacetic acid cleavage and deprotection of resin-bound peptides following synthesis by Fmoc chemistry. In GB Fields, ed. Orlando: Academic Press, 1997, pp 67–83.

265. DA Pearson, M Blanchette, ML Baker, CA Guindon. Trialkylsilanes as scavengers for the trifluoroacetic acid deblocking of protecting groups in peptide synthesis. Tetrahedron Lett 30: 2739–2742, 1989.

266. J-C Gesquiére, J Najib, E Diesis, D Barbry, A Tartar. Investigations of side reactions associated with the use of Bom and Bum groups for histidine protection. In: JA Smith, JE Rivier, eds. Peptides: Chemistry and Biology. Leiden, The Netherlands: Escom, 1992, pp 641–642.

267. P Sieber. Modification of tryptophan residues during acidolysis of 4-methoxy-2,3,6-trimethyl benzenesulfonyl groups: Effects of scavengers. Tetrahedron Lett 28: 1637–1640, 1987.

268. B Riniker, A Hartmann. Deprotection of peptides containing Arg(Pmc) and tryptophan or tyrosine: Elucidation of by-products. In: JE Rivier, GR Marshall, eds. Peptides: Chemistry, Structure and Biology. Leiden, The Netherlands: Escom, 1990, pp 950–952.

269. B Riniker, B Kamber. Byproducts of Trp-peptides synthesized on a p-benzyloxybenzyl alcohol polystyrene resin. In: G

Jung, E Bayer, eds. Peptides 1988. Berlin: Walter de Gruyter and Co, 1989, pp 115–117.

270. PD Gesellchen, RB Rothenberger, DE Dorman, JW Paschal, TK Elzey, CS Campbell. A new side reaction in solid-phase peptide synthesis: Solid support-dependent alkylation of tryptophan. In: JE Rivier, GR Marshall, eds. Peptides: Chemistry, Structure and Biology, Leiden, The Netherlands: Escom, 1990, pp 957–959.

271. GB Fields, SA Carr, DR Marshak, AJ Smith, JT Stults, LC Williams, KR Williams, JD Young. Evaluation of peptide synthesis as practiced in 53 different laboratories. In: RH Angletti, ed. Techniques In Protein Chemistry IV. San Diego: Academic Press, 1993, pp 229–238.

272. NA Solé, G Barany. Optimization of solid-phase synthesis of [Ala$^8$]-dynorphin A. J Org Chem 57: 5399–5403, 1992.

273. M Bozzini, R Bello, N Cagle, D Yamane, D Dupont. Tryptophan recovery from auto hydrolyzed samples using dodecanethiol. Applied Biosystems Research News, February 1991, Applied Biosystems, Inc., Foster City, CA.

274. WC Chan, BW Bycroft. Deprotection of Arg(Pmc) containing peptides using TFA-trialkylsilane-methanol-EMS; application to the synthesis of propeptides of nisin. In: JA Smith, JE Rivier, eds. Peptides: Chemistry and Biology. Leiden, The Netherlands: Escom, 1992, pp 613–614.

275. J Green, OM Ogunjobi, R Ramage, ASJ Stewart, S McCurdy, R Noble. Application of the N$^G$-(2,2,5,7,8-pentamethylchroman-6-sulphonyl) derivative of Fmoc-arginine to peptide synthesis. Tetrahedron Lett 29: 4341–4344, 1988.

276. D Andreu, F Albericio, NA Sole, MC Munson, M Ferrer, G Barany. Formation of disulfide bonds in synthetic peptides and proteins. In: MW Pennington, BM Dunn, eds. Methods in Molecular Biology, Vol 35. Totowa, NJ: Humana Press, 1994, pp 91–169.

277. DF Hunt, J Shabanowitz, JR Yates, PR Griffin, NZ Zhu. Protein sequence analysis by tandem mass spectrometry. In: CN McEvan, BS Larsen, eds. Mass Spectrometry of Biological Materials. New York: Marcel Dekker, Inc, 1990, pp 169–195.

278. GA Grant. Evaluation of the finished product. In: GA Grant, ed. Synthetic Peptides: A User's Guide. New York: WH Freeman & Co, 1992, pp 185–258.

279. RJ Delange, EL Smith. Aminopeptidase M. In: PD Boyer, ed. The Enzymes, Vol. III. New York: Academic Press, 1971, pp 104–105.

280. H Frank, W Woiwode, G Nicholson, E Bayer. Determination of the rate of acidic catalyzed racemization of protein amino acids. Liebigs Ann Chem 354–365, 1981.

281. JG Adamson, T Hoang, A Crivici, GA Lajoie. Use of Marfey's reagent to quantitate racemization upon anchoring of amino acids to solid supports for peptide synthesis. Anal Biochem 202: 210–214, 1992.

282. WT Moore. Laser desorption mass spectrometry. In: GB Fields, ed. Methods in Enzymology, Volume 289. Orlando: Academic Press, pp 520–542, 1997.

# 5

# Chemical Synthesis of Proteins

**Graham J. Cotton, Maria C. Pietanza and Tom W. Muir***
*The Rockefeller University, New York, New York*

## I. INTRODUCTION

The ability to precisely alter the covalent structure of a protein through chemical modification has enormous potential to further our understanding of protein structure-activity relationships, generate new tools for biomedical research, and develop new therapeutic agents. Several methodologies have been developed which complement site-directed mutagenesis of recombinantly cloned genes by allowing the introduction of unnatural amino acids or molecules into polypeptide chains. These approaches principally include unnatural amino acid mutagenesis via *in vitro* expression [reviewed in (1)]; site-specific protein modification [reviewed in (2)]; protein semi-synthesis [reviewed in (3)]; and protein total chemical synthesis [reviewed in (4, 5)]. Although this article will focus mainly on the total chemical synthesis of proteins, it is important to stress that each of these methods has its own set of synthetic advantages and disadvantages which make each approach more or less well suited to studying the structure and function of a given protein molecule. At the present time no single technique has been refined to the point where it provides a general solution to the problem of how to routinely chemically engineer a protein molecule regardless of its size or the nature/number of the modifications to be made.

Total chemical synthesis is unmatched in terms of the level of synthetic control it provides over the chemical structure of the protein molecule. In principle, all aspects of the covalent structure of a protein can be systematically varied using total synthesis, including the polypeptide backbone and the stereochemistry of the chiral carbon centers. Interest in the field of protein total synthesis continues to be fueled by the numerous successes of synthetic chemistry in studying the structure-activity relationships of small bioactive peptides (6, 7). Although the ability to perform similar chemical mutagenesis studies within proteins is enormously appealing, the large size of most proteins has historically posed a significant hurdle to chemists interested in this area. Nonetheless, the last several years have seen the emergence of new synthetic techniques which have greatly

*To whom correspondence should be addressed: Tom W. Muir, Assistant Professor/Head of Laboratory, Laboratory of Synthetic Protein Chemistry, The Rockefeller University, Box 223, 1230 York Ave, New York, NY10021. Tel: (212) 327 7368. Fax: (212) 327 7358. email: muirt@rockvax.rockefeller.edu

extended the size of polypeptide chains accessible to total chemical synthesis. This chapter will review these new developments in protein total synthesis, and will attempt to illustrate the utility of some of these techniques through a series of specific examples from our own work.

## A. Classical Strategies for Protein Total Synthesis

Given its potential applications, protein total synthesis has been the subject of intense study for many years (4). Efforts have generally fallen into one of two synthetic strategies, namely; linear protein synthesis or convergent protein synthesis. In the former strategy, stepwise solid phase peptide synthesis (SPPS) (8) is directly used to produce the longer peptide chain lengths required for protein synthesis. Optimized versions of SPPS (9–11) have been developed which now allow the routine assembly of small proteins and protein domains of up to circa 50 amino acids in length (12–19). Classical convergent synthetic strategies involve the fragment condensation of small protected peptide segments, themselves accessible via solution or solid-phase methods. Procedures have been developed which allow suitably activated fragments to be condensed (via there mutually reactive N- and C-termini) either in solution or on the solid phase, in both cases to yield a fully protected polypeptide sequence [reviewed in (20, 21)]. A number of proteins have been successfully synthesized by this approach (22–27), which negates the intrinsic problems associated with the linear synthesis of long peptide sequences by SPPS (see below). Classical convergent synthesis does, however, require the production of fully protected peptide fragments, which are difficult to both purify and characterize due to their solubility properties. Another limitation of the strategy is that strong activation of the carboxy-terminus of a peptide frag-

ment can lead to epimerization of the C-terminal amino acid, although this can be overcome by choosing glycine or proline as the C-terminal residue.

The ability to generate small protein domains directly by optimized SPPS is typified by the total synthesis of the 55 amino acid SH3 (Src Homology type 3) domain of the c-Abl non-receptor protein tyrosine kinase. An analog of this protein domain, in which the single cysteine residue in the sequence was replaced by the isosteric amino acid $\gamma$-aminobutyric acid, was assembled by stepwise SPPS using the Boc $N^\alpha$ protection strategy. Use of *in situ* neutralization/HBTU activation protocols (9) allowed single 10 minute coupling times to be used throughout the synthesis, which as a consequence was extremely rapid (1 day using an automated synthesizer or 3 days by manual synthesis). Following cleavage and global deprotection using liquid HF at 0–4°C, the crude preparation was purified by preparative HPLC and the covalent structure of the protein characterized by electrospray mass spectrometry (Fig. 1).

At this point it is necessary to emphasize the enormous impact that modern biological mass spectrometry (28) has had on synthetic peptide and protein chemistry over the last few years. Both electrospray mass spectrometry (ESMS) and matrix-assisted laser desorption ionization mass spectrometry (MALDI MS) have become indispensable tools to researchers interested in properly characterizing large biopolymers. In addition, these techniques have also played an integral role in the development of the protein chemistries described later in this article, and have without question raised the bar considerably on what is deemed a homogeneous polypeptide sample!

An important difference between peptide synthesis and protein synthesis is that the latter involves an additional step in which the nascent polypeptide chain is folded into a native protein conformation.

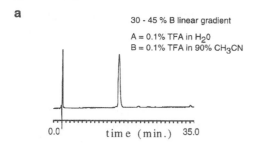

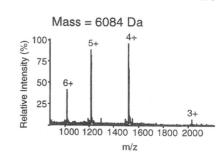

**Fig. 1** Synthesis of Abl-SH3 by stepwise SPPS. (a) Reverse phase HPLC of purified synthetic Abl-SH3, (b) Electrospray mass spectrum of purified synthetic Abl-SH3. Expected Mass = 6085 Da (average isotope composition).

Although the theoretical basis of protein folding is still far from clear, experience suggests that folding of most of the small proteins generated through total synthesis is relatively straightforward in practice. This should not be surprising given Christian Anfinsen's realization that the conformation of a protein is encoded solely within the primary sequence of the molecule (29) (the ability to fold a synthetic protein is perhaps the ultimate vindication of Anfinsen's work). Moreover, it should be pointed out that unlike folding *in vivo*, it is possible to precisely control the folding conditions used in the test tube, particular with respect to the concentration of the protein solution and the presence or absence of chaotropic agents. Another potentially difficulty in this area involves the formation of multiple disulfide bonds, always a problem when dealing with small unstructured peptides (30). In reality, this turns out to be quite easy in a number of small proteins, undoubtedly because correct disulfide formation is being directed by the folding of the polypeptide chain. Good examples of this folding-assisted disulfide formation can be seen in the cytokine and chemokine protein families whose members are found to efficiently form the correct disulfide patterns under mildly denaturing conditions (31, 32).

The Abl-SH3 domain contains no disulfide bonds, and was found to spontaneously fold by simply dissolving the lyophilized sample in phosphate buffer at pH 7.0 to a final concentration of anywhere up to 1 mM. NMR studies showed that the synthetic Abl-SH3 folded in this way adopted the same global structure as the recombinantly-derived domain. Consistent with this structural data, a fluorescence-based assay indicated that the synthetic domain was functionally similar to the recombinant protein with respect to ligand binding (Fig. 2). The overall yield of the folded synthetic protein domain was on the order of 10%.

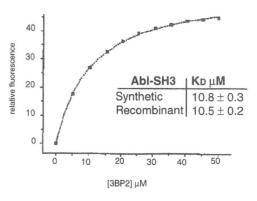

**Fig. 2** Ligand binding isotherm for the interaction of Abl-SH3 with the proline rich sequence from the protein 3BP2, [PPAYPPPPVP]. The Kd values were obtained by monitoring the change in the intrinsic tryptophan fluorescence of both synthetic and recombinant Abl-SH3 as a function of ligand concentration.

Extrapolating stepwise SPPS to the synthesis of molecules of around 100 amino acids, the typical size of protein domains, has proved more problematic. The large number of impurities produced by incomplete couplings are compounded by various side-reactions, especially during the cleavage/global deprotection step. This often leads to an overall heterogeneous synthesis, giving products which are extremely difficult to purify and characterize. Nonetheless, both Boc and Fmoc methodologies have yielded a number of successful stepwise syntheses such as those of ribonuclease A (33), interleukin-3 (31), insulin-like growth factor (34), epidermal growth factor (13), interleukin 8 (32), HIV-1 protease (35, 36), rat Cpn10 (37), ubiquitin (38, 39) and monellin (40). Although synthesizing the wild-type sequence is often a significant accomplishment, the real power of the chemical approach is in its ability to probe structure-activity relationships in a protein molecule. Again stepwise SPPS has made important contributions in this area, including the synthesis and functional analysis of IL-3 disulfide analogs (41); the synthesis of a monomeric version of IL-8 (42); the generation of an analog of HIV-1 protease containing a $\beta$-turn mimic (43); the synthesis of both side-chain and backbone engineered mutants of the N-terminal SH3 domain of Crk (18); and chemical engineering of the ubiquitin hydrophobic core by unnatural amino acid replacement (44). The utility of total synthesis is also illustrated through the generation of the D-enantiomers of the proteins rubredoxin (45), brazzein (46), 4-oxalocrotonate tautomerase (19) and HIV-1 protease (47). Each molecule was assembled from D-amino acids and exhibited a far uv CD spectrum which was the mirror image of the L-form. Significantly, the D-versions of HIV-1 protease and 4-oxalocrotonate tautomerase also demonstrated reciprocal chiral activities.

In spite of the above examples, stepwise SPPS has not offered a generally applicable route to the systematic modification of larger polypeptide chains. The inherent statistical limitations of linear chemical synthesis have proved difficult to overcome, and increasingly researchers have turned their attentions back to the convergent synthetic strategy. In the following sections we will focus on the contemporary fragment condensation approaches which have emerged as a consequence.

## II. MODERN APPROACHES TO FRAGMENT CONDENSATION

With stepwise SPPS having a size restriction of approximately 50 amino acids it became clear that convergent strategies held the key to reproducible chemical protein synthesis. However, the intractable problem with classical fragment condensation techniques has always been the poor solubility associated with maximally protected peptide fragments. Over the last 10–15 years new strategies in protein synthesis have evolved which directly address these solubility issues. Instead of condensing fully protected peptide fragments the trend has been to minimalize the protecting groups, overcoming the solubility demon and allowing reactions to be performed in aqueous solution.

An early sign of this shift was the minimal protection approach pioneered by Blake (48, 49) and then successfully extended to many systems by Yamashiro (50–52). The key to this approach is the use of novel SPPS resins which upon acidolytic cleavage yield an unprotected peptide-$^{\alpha}$COSH. Citraconyl groups are used to reprotect all free amino functions except the desired N-terminal $\alpha$-amino group. The thioacid is then activated as the thioester and reacted with the free $N^{\alpha}$amino function of the second fragment using Ag$^+$ as a catalyst. The citraconyl groups are then removed under mild conditions to yield the target polypeptide. Several polypeptides have been generated by this approach most notably apocytochrome c (49), α-inhibin-92 (50) and the HBs protein (52). The elegant thiol-capture technique developed by Kemp also allows regiospecific condensation of minimally protected peptide fragments (53). In this pioneering approach the α-carboxyl of one fragment is derivatized as the 4-mercaptodibenzofuran ester, instigating disulfide bridge formation with an amino terminal cysteine on a second fragment. The carboxy and amino termini are therefore brought into close proximity, promoting amide bond formation via an O → N acyl shift.

The logical progression of this philosophy is to use completely unprotected peptide fragments as the raw materials for protein synthesis. This generates new synthetic challenges if the fragments are to be assembled regioselectively. The two general solutions to this problem have been enzymatic ligation and more recently chemical ligation.

## A.  Enzymatic Ligation Techniques

In this convergent strategy, proteolytic enzymes are used to direct amide bond formation between unprotected peptide fragments in a regioselective manner. Classically, this process involved altering the reaction conditions such that aminolysis of an acyl-enzyme intermediate is favoured over hydrolysis. This "reverse-proteolysis" is typically achieved by including high concentrations of organic solvents such as glycerol, DMF or acetonitrile in the reaction medium. Under these conditions the acyl-enzyme intermediate will undergo aminolysis with the second peptide fragment leading to the formation of an amide-linked product (54, 55). Useful results have been obtained using a variety of enzymes including sub-

tilisin, trypsin and papain. The power of reverse-proteolysis in protein synthesis is illustrated by the recent generation of novel glycoforms of ribonuclease B (56). Using a series of enzyme-mediated oligosaccharide and polypeptide ligation steps, this group was able to site-specifically introduce the branched oligosaccharide, sialyl Lexis X, at a single site in the protein molecule.

Significant progress in the area of enzyme-mediated peptide ligation has been achieved by genetically engineering the active site of proteolytic enzymes. This approach was pioneered in the mid-1980's by Kaiser and co-workers who chemically incorporated a thiol group into the active site of the serine protease subtilisin. The resulting thiolsubtilisin analog possessed improved acylation activity relative to the wild-type enzyme (57). In an extremely elegant extension of this work, Wells and co-workers further engineered the active site of subtilisin affording an enzyme capable of efficiently catalyzing the ligation of peptide fragments (58). Their double mutant form of the enzyme, termed subtiligase, functions as an effective acyl transferase and importantly has vastly reduced proteolytic activity compared to the wild-type enzyme. Mutation of the active site serine to a cysteine allows the enzyme to be acylated with peptides esterified at their C-terminus with a glycolate phenylalanyl amide ester group. A second Pro ⇒ Ala mutation within the active site relieves steric crowding, allowing efficient aminolysis of this enzyme-peptide thioester intermediate by the amino-terminus of a second peptide fragment. Subtiligase has been used to assemble ribonuclease A analogs from six synthetic peptide fragments (58). These protein analogs contained site-specifically incorporated 4-fluorohistidine residues, allowing new insights into the enzymatic mechanism of ribonuclease to be determined. More recently, subtiligase has been used to ligate biotinylated or

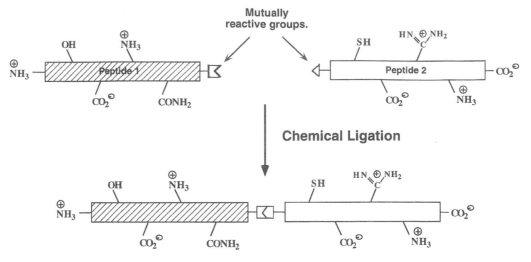

**Fig. 3**  The principle of chemical ligation. Appropriate functionalities are synthetically incorporated at the C- and N-termini of the unprotected peptide fragments, such that under aqueous conditions they chemoselectively react to give the desired ligation product.

heavy atom derivatized peptide sequences onto the N-terminus of human growth hormone (59, 60) and to produce head to tail cyclized peptides from their linear precursors (61).

## B.  Chemical Ligation

The most direct way to regioselectively link two fully unprotected peptide fragments together is to incorporate mutually reactive functional groups at the N- and C-terminus, respectively, of each peptide. Under appropriate conditions,

these peptide fragments would then be expected to chemoselectively react with one another to form the desired protein product (Fig. 3). Despite the simplicity of this "chemical ligation" approach, it is only in the last decade that the chemistries needed to allow the controlled and selective ligation of unprotected peptide fragments have been developed. All of the chemical ligation methodologies listed in Table 1 are, in principle, applicable to any size of peptide fragment and, unlike enzymatic ligation techniques, are compatible with a range of reaction conditions

**Table 1**  Reactive functionalities used in the chemical ligation of peptide fragments.

| Chemistry of ligation | Nucleophile | Electrophile | Ligation conditions |
| --- | --- | --- | --- |
| Thioester [69] | Thioacid | Bromoacetyl | pH 3–7 |
| Thioether [76, 77] | Thiol | Bromoacetyl, maleimide | pH 6–8 |
| Hydrazone [66] | Hydrazide | Glyoxyl | pH 4.6 |
| Oxime [75] | Amino-oxy | Glyoxyl, levulinic acid | pH 4.6 |
| Thiozolidine [79] | Cysteine | Aldehyde | pH 3–5 |
| Amide [81] | Cysteine | Thioester | pH 5–8 |
| Amide [86] | Histidine | Thioester | pH 5.7 |
| Amide [85] | Oxyethanethiol | Thioester | pH 7.5 |

including the presence of chaotropic agents such as urea or guanidinium HCl. This latter point is of particular importance since it allows the peptides to be brought into solution at high concentrations, thereby improving the kinetics of the chemical reaction. Moreover, large unprotected peptides are frequently insoluble in neutral aqueous buffers (due to aggregation), a problem which can usually be overcome by including chaotropes in the reaction buffer.

The earliest ligation chemistries were born from the field of protein semi-synthesis, where proteolytically or chemically cleaved fragments of natural proteins are used as the building blocks for the resynthesis of the target molecule (3). It was noted that CNBr fragments of pancreatic trypsin inhibitor can spontaneously ligate to reform the native peptide bond between them (62), an effect which is also observed with cytochrome c (63). In both cases, the two fragments produced after CNBr cleavage cooperatively refold, bringing the homoserine lactone at the carboxy-terminus of one fragment in close proximity to the free amino terminus of the other. The high local concentration effect of these groups and the mild activation produced by the homoserine lactone result in spontaneous amide bond formation. This autocatalytic fragment religation approach has been used to incorporate natural and unnatural amino acids into cytochrome c (64, 65). In this example, the N-terminal CNBr cleavage fragment of the recombinant protein, is ligated to chemically synthesized derivatives of the C-terminal fragment to provide ready access to a series of semi-synthetic protein mutants for use in structure-function studies.

The elegant protein semi-synthesis strategy developed by Offord and co-workers uses a combination of reverse proteolysis and chemical ligation (66). In this approach, trypsin is first used to cleave at a unique site (e.g. Lys-Ser) within the protein sequence. Reverse proteolysis is then used to introduce a hydrazide group at the C-terminus of the N-terminal fragment, and mild periodate oxidation is used to convert the N-terminal serine residue within the C-terminal fragment to a glyoxyl moiety. These two derivatized fragments can then be chemoselectively ligated at around pH 5 to give a hydrazone bond-containing product. This semi-synthesis strategy has been used to generate backbone engineered analogs of the growth factor G-CSF (67), and an analogous oxime-based approach has recently been used to site specifically incorporate an aminooxy pentane molecule to the N-terminus of the human RANTES chemokine (68). The resulting semi-synthetic protein (AOP-RANTES) was found to be a potent CCR5 receptor antagonist and is thus a promising anti-HIV therapeutic.

The first reported total chemical synthesis of a protein by a chemical ligation approach was the 99 amino acid residue HIV-1 protease. This was assembled from two ~50 residue unprotected synthetic peptides using a thioester bond forming chemical reaction performed in aqueous solution at pH 7.0 (69). This thioester ligation approach has proved highly successful and a number of mechanistic and functional questions have now been probed using protein derivatives synthesized in this manner (70–74). The original thioester ligation chemistry has now been joined by oxime (75), thioether (76–78) and thiozolidine/oxazolidine (79, 80) based approaches for selectively joining unprotected peptide fragments via the introduction of non-natural linkages (Table 1).

The native chemical ligation approach developed by Kent and co-workers (81) has significantly extended the scope of protein total synthesis via chemical ligation (Fig. 4). In this strategy, two fully unprotected synthetic peptide fragments, one containing a C-terminal thioester

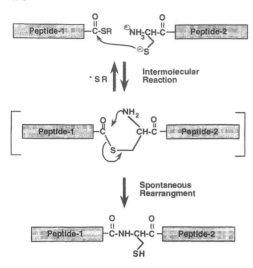

**Fig. 4** The reaction mechanism of native chemical ligation. The initial step is a reversible transthioesterification involving the C-terminal thioester and the sulfhydryl group of the N-terminal cysteine residue. This intermediate spontaneously rearranges to form the natural amide linkage with regeneration of the cysteine residue at the ligation site.

group and the other an N-terminal cysteine residue, are chemically ligated under neutral aqueous conditions with the formation of a native peptide bond at the reaction site. The initial step in native chemical ligation involves formation of a thioester-linked intermediate generated via a transthioesterification reaction involving the carboxy terminal thioester of one fragment and the N-terminal cysteine sulfhydryl group of the other. This intermediate then spontaneously re-arranges to produce a native peptide bond at the ligation site with the restoration of the cysteine sulfhydryl group. This type of thioester-based acyl rearrangement chem-istry was pioneered in the 1950's by Wieland using small peptides (82), and is reminiscent of the thiol capture method of Kemp described earlier (53). However, the thiol capture strategy, as well as a related method (83), requires regiospecific protection of any cysteine sulfhydryls

prior to the ligation step. In contrast, native chemical ligation is compatible with all naturally occurring side-chain groups including the sulfhydryl group of cysteine.

The compatibility of native chemical ligation with multiple cysteine sulfhydryls is highlighted by the syntheses of interleukin 8 (81), and human type II secretory phospholipase A(2) (84) which contain four and 14 Cys residues, respectively. In both systems, the unpro-tected peptide segments each contain sev-eral cysteine sulfhydryl groups all of which can react with the thioester group in the N-terminal peptide fragment to pro-duce thioester linked products. However, the sole reaction intermediate is that involving the N-terminal cysteine residue since only this can rearrange to the final amide bond. The remaining thioesters are unproductive and can be simply converted back to starting materials by including small amounts of a thiol co-factor in the reaction mixture (Fig. 4).

The general approach used in the native chemical ligation strategy has been extended to include peptide fragments which contain $N^\alpha$(oxyethanethiols) (85), or N-terminal histidine residues (86). In each case the mechanism is analogous with that outlined for an N-terminal cysteine (Fig. 4) except that the preformed thioester is initially attacked by the thiol group of the $N^\alpha$-oxyethanethiol or the imidazole nitrogen of histidine. S-alkylation of a pep-tide $^\alpha$thioacid by a second peptide con-taining an N-terminal $\beta$-bromoalanine residue, in a manner similar to the original ligation strategies (69), has provided yet another route to forming the requisite thioester intermediate (87). A cysteine residue is again generated at the ligation site after the S to N acyl transfer.

Native chemical ligation has pro-vided a robust and straightforward route for the synthesis of small proteins and pro-tein domains (74, 88). Routine synthetic access to these systems has enabled researchers to undertake sophisticated

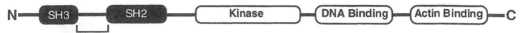

**Linker Region Important in Regulation**

**Fig. 5**    The domain structure of cellular Abl protein tyrosine kinase.

structure-function studies involving modification of the protein molecule impossible via site-directed mutagenesis techniques. This is exemplified in the synthesis of backbone engineered mutants of the serine protease inhibitor protein, OMTKY3 (89). This work involved replacing a specific backbone amide in the protein with an oxyester group and allowed the energy of a key hydrogen bond between the proteinase and the protein inhibitor to be quantified.

As an example of how the total chemical synthesis of proteins via native chemical ligation can be used to address a specific biological problem, we will turn to some recent work from our own laboratory. We are studying the regulatory mechanism of the cellular signalling protein c-Abl, one of the few non-receptor protein tyrosine kinases directly linked to human malignancies (90). Like the Src family of protein kinases it contains both Src Homology 2 (SH2) and Src Homology 3 (SH3) domains, which are involved in regulating the transforming activity, but the mechanisms are thought to be entirely different. (Fig. 5). The SH3 domain suppresses the transforming activity of c-Abl, possibly through interactions with other protein targets (91), whereas the transforming activity of the activated form of this proto-oncogene is dependent on the SH2 domain (92). These studies have shown that changing the relative position of the SH3 domain within the context of the whole protein modifies its transforming activity. This effect was even observed when only a small alteration was made to the length of the linker region between the SH2 and SH3 domains.

We are interested in determining how the activity of the protein is related to the length of this linker region. This can be probed by systematically varying the length of the linker and studying the effects on the structure and function of the enzyme. To this end we must first define which residues in the primary sequence constitute the linker region and which are considered part of the SH3 and SH2 domains. These boundaries can be precisely identified by gradually truncating from the C- and N- termini respectively of each individual domain to locate the minimum primary sequence which still possesses native structure and function. Previous studies in our laboratory have shown that the C-terminus of SH3 domain can not be truncated past Val-119 without losing the structure or function of the protein.

A systematic approach was adopted to identify the minimum N-terminal sequence requirements for a fully functional Abl SH2 domain. This involved using a native chemical ligation strategy to assemble a defined set of protein domains. Note, preliminary studies had shown that the 101 amino acid residue Abl SH2 domain (residues 120–220 in the protein sequence) could be readily synthesized from two approximately 50 residue fragments using native chemical ligation. To determine the minimal sequence of the SH2 domain we required synthetic access to a family of N-terminally truncated mutants of the Abl-SH2 protein sequence. From multidimensional NMR studies (93) as well as sequence alignment analysis, it seemed likely that the minimum N-terminus of the Abl-SH2

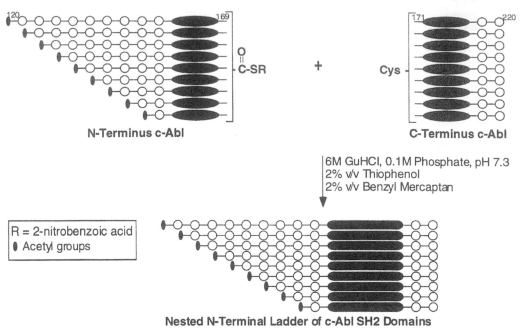

**Fig. 6** Production of an N-terminally truncated ladder of the Abl-SH2 domain. All the N-terminal fragments are produced from one synthesis by removal of small aliquots of resin at the appropriate point. After conversion into the corresponding thioester, each member is purified, and then ligated to the common C-terminal fragment containing an amino-cysteine functionality.

protein lay somewhere in the region $Asn^{120}$–$Trp^{127}$. In principle, native chemical ligation of a ladder of N-terminal peptide fragments (each one residue longer than the other) to a C-terminal fragment of fixed length, allows the generation of the desired truncated array of protein domains with a minimum amount of synthetic effort (Fig. 6).

The Abl-SH2 sequence does not contain any cysteine residues meaning that a single X/Cys substitution needs to be made to facilitate native chemical ligation. Our previous studies had indicated that $Arg^{170}$ could be mutated to a cysteine without altering the structure or function of the protein domain. Consequently two sets of peptide fragments were chemically synthesized, namely: $Cys^{170}$[171-220]Abl-SH2 and Abl-SH2[X-169]$^{\alpha}$COSH, where $X = N^{120}$-$Y^{128}$.

The C-terminal fragment $Cys^{170}$[171-220]Abl-SH2 was synthesized on an MBHA resin using the optimized *in situ* neutralization/HBTU activation protocol for Boc SPPS (9) and purified by reverse phase HPLC after deprotection and cleavage. The truncated ladder of N-terminal fragments Abl-SH2[X-169] $^{\alpha}$COSH was synthesized on a preformed Boc-Gly-S-resin (94) using the same optimized Boc SPPS protocols as above. Note that the entire family of N-terminal peptides was obtained from a single synthesis by removing aliquots of the peptide-resin at appropriate points in the chain assembly. After cleavage, each crude peptide-thioacid was thioesterified with 5,5′-dithio-bis(2-nitrobenzoic acid) as previously described (81), and the desired α-thioester derivative purified by HPLC. For each of the nine chemical ligation reactions the appropriate purified N-terminal fragment (10 mg/mL) was mixed with the C-terminal peptide segment (10 mg/mL) in a 6 M GdmCl, 0.1 M

**Table 2** Characterization of the family of Abl-SH2 N-terminal truncates prepared by native chemical ligation.

| Truncation Position | Observed Mass (Da) | Theoretical Mass Monoisotopic (Da) | Theoretical Mass Av. Isotope Comp. (Da) |
|---|---|---|---|
| Asn120 | $11345.5 \pm 5.0$ | 11342.7 | 11349.6 |
| Ser121 | $11234.4 \pm 1.7$ | 11228.6 | 11235.5 |
| Leu122 | $11148.0 \pm 1.5$ | 11141.6 | 11148.4 |
| Glu123 | $11029.1 \pm 2.5$ | 11028.5 | 11035.3 |
| Lys124 | $10905.4 \pm 1.4$ | 10899.5 | 10906.1 |
| His125 | $10776.5 \pm 2.9$ | 10771.4 | 10778.0 |
| Ser126 | $10640.4 \pm 2.9$ | 10634.3 | 10640.8 |
| Trp127 | $10553.6 \pm 3.6$ | 10547.3 | 10553.7 |
| Tyr 128 | $10367.4 \pm 1.2$ | 10361.2 | 10367.5 |

phosphate buffer at pH 7.5 containing 2% benzylmercaptan and 4% thiophenol. The aromatic thiols were added to keep the mixture under reducing conditions as well as to accelerate the ligation reaction through transthioesterification (95). An added benefit to including these thiol co-factors in the mixture is that they help prevent hydrolysis of the thioester component during the course of the ligation reaction. Using these ligation conditions the reactions were typically complete after 48 h at which point each protein was purified by preparative HPLC and folded by controlled dialysis from denaturant (Table 2).

Having generated the desired set of Abl-SH2 domains, the effect of systematic truncation on the structure and function of the protein was evaluated (Fig. 7). A short phosphotyrosine containing peptide derived from the signaling protein 2BP1 is known to be a natural ligand for the Abl-SH2 domain (96). A fluorescence based ligand binding assay was used to study the affinity of each of the nine synthetic Abl-SH2 domains for a synthetic 2BP1 peptide. This revealed that the dissociation constants ($K_d$) for constructs Asn[120] through Glu[123] were similar to the reported value for the recombinant protein [$K_d = 2.5 \pm 0.3 \, \mu M$ (96)]. However, truncation beyond Glu[123] led to a rapid loss of

function until no specific affinity for the ligand remained by Ser[126] (Fig. 7a). In order to determine the effect of systematic N-terminal truncation on overall secondary structure content, each of the nine constructs was also studied by circular dichroism (CD) spectroscopy. Consistent with the biochemical data, truncation beyond Glu[123] lead to a clear structural transition in which the protein goes from a native folded state to a misfolded state (although not entirely unstructured) with lower $\alpha$-helical content (Fig. 7b).

These results indicate that the N-terminus of the c-Abl SH2 domain can not be truncated beyond Glu[123] without affecting its native structure and function. Thus, total chemical synthesis by native chemical ligation has proved an extremely efficient method for tackling this structure-function problem.

Another example of the power of native chemical ligation is in the construction of proteins possessing circular topology. Although originally developed as a way of joining peptide fragments in an intermolecular fashion, recent studies have indicated that the chemistry works just as well in the intramolecular situation (97, 98). Intermolecular *versus* intramolecular ligation simply depends on whether the reactive moieties are present within the same or different polypeptide

a.
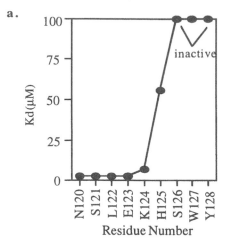

b.
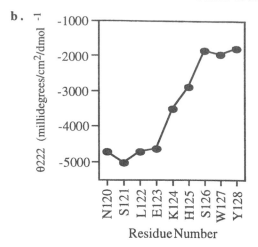

**Fig. 7** Effect on (a) Function and (b) Structure of systematic N-terminal truncation of the Abl-SH2 domain. A fluorescence based assay was used to determine the binding affinities of each of the truncates for the Abl-SH2 ligand 2BP1 (PVpYENV-NH$_2$, pY = phosphotyrosine). The overall secondary structure of each truncate was determined using far UV circular dichroism spectroscopy. The residue elipticity at 222nm is used as a convenient method for comparing the structural differences between each of the compounds.

chains. Intramolecular native chemical ligation has recently been applied to a large polypeptide system resulting in the generation of a head-to-tail cyclized protein, i.e. a protein containing neither an N- nor a C- terminus (99). The WW domain from the human yes kinase associated protein (YAP) was initially synthesized as the unprotected linear precursor polypeptide, containing a cysteine at the N-terminus and a C-terminal thioester. The proximity of the N- and C-termini within the native fold of this protein domain was expected to enhance the rate of the intramolecular ligation. Under the appropriate conditions this molecule rapidly and quantitatively cyclized to yield a circular protein domain which had a similar global fold and ligand binding function as its linear counterpart. The rate of cyclization was found to significantly decrease when initiated under unfolding conditions, indicating that the reaction is conformationally assisted under native

folding conditions. Interestingly, the denatured circular WW domain could refold to adopt the same global conformation as the native domain, suggesting that the folding process is not inhibited by the circular topology.

Native chemical ligation is also playing an important role in the emerging field of peptide self-replication (100–103). Here, the tendency of certain peptide sequences to form supersecondary structures such as coiled-coil dimers, is used as the basis of an autocatalytic native chemical ligation reaction. Significant rate increases are observed for the intermolecular native chemical ligation in the presence of a catalytic amount of a peptide template. This rate enhancement is somewhat analogous to that observed in the folding-assisted protein cyclization reaction described above, although this is not a catalytic process. Recent studies in this area suggest that peptide self-replication may be a useful way of studying molecular symbiosis and origin-of-life theories (102).

## III.  MULTIPLE CHEMICAL LIGATIONS

The ability to perform several chemical ligation reactions in a controlled and directed fashion allows synthetic access to much larger and complex polypeptide systems. This is illustrated by the synthesis of the heterodimeric transcription factor c-Myc-Max from four synthetic peptide fragments and involving two orthogonal ligation chemistries (73). Thioester-forming ligation reactions were first used to assemble the individual c-Myc and Max domains, which were then linked together using the oxime-based ligation chemistry. The resulting covalently linked 20 kDa c-Myc-Max construct was found to specifically bind promoter DNA, allowing the biochemistry of this system to be explored for the first time. Another elegant use of multiple chemical ligation reactions has been in the generation of template assembled synthetic proteins (TASPS) (104). The recent work of Mutter and co-workers involves the use of mutually exclusive ligation chemistries to regioselectively assemble peptide sequences onto a synthetic template (105). This approach, in principle, overcomes the inherent problems associated with the folding of long polypeptide chains to create synthetic proteins of defined topologies, and is exemplified by the construction of artificial proteins which possess a four-helix-bundle topology, using both homologous and heterologous $\alpha$-helical subunits (106). This TASP approach has also been used to generate both glyco- and lipo-peptides (107), and for the *de novo* design of synthetic protein constructs which mimic receptor and antibody binding sites (105, 108).

We have recently developed a novel strategy which allows sequential native chemical ligation reactions to be performed directly from a solid-support (109) (Fig. 8). Using a combination of Boc-SPPS on a 3-mercaptopropionamide-polyethyleneglycol-poly-($N,N$-dimethylacrylamide) copolymer resin (HS-PEGA) the desired fully protected peptide $^{\alpha}$thioester resin is initially generated. Treatment of this resin with HF then yields the corresponding fully unprotected peptide still tethered to the solid support via a reactive $^{\alpha}$thioester linkage. Such peptide-resins can now be directly used to perform both intermolecular and intramolecular (cyclization) native chemical ligation reactions (109).

Crucial to this sequential solid-phase ligation strategy is the production of the immobilized central peptide fragment containing both an N-terminal cysteine residue and a C-terminal thioester. In the absence of any $N^{\alpha}$ or cysteine sulfhydryl protection this construct would simply cyclize to form the circular product (97, 98), and hence prevent the intermolecular ligation reaction. To avoid this problem we use the base labile 2-(Methylsulfonyl)-ethyloxycarbonyl (Msc) group to reversibly protect the $N^{\alpha}$ functionality. After the desired intermolecular ligation the Msc group is simply removed by briefly raising the pH $\approx 13$.

Sequential solid-phase chemical ligation has been used to assemble a number of proteins within our lab, and is illustrated here by the assembly of a novel 79 amino acid receptor-ligand construct from three peptide segments, involving two separate ligation steps. This protein construct contains both the Abl-SH3 domain and its natural proline rich ligand, 3BP2, connected by a 12 residue poly glycine linker region. Sequential native chemical ligation has previously been used to synthesize a related construct which differs only in the sequence of the linker region (109). Due to the modularity of the sequential ligation approach the total synthesis of this structural variant is now relatively straightforward. Only the small peptide fragment containing the linker region is resynthesized by SPPS. This is then selectively coupled to the existing

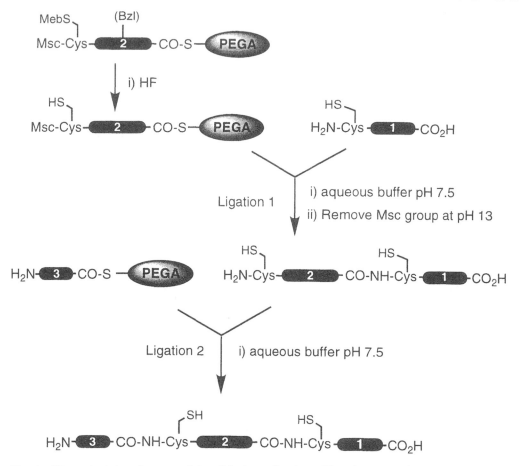

**Fig. 8** The principle of sequential solid-phase ligation. The deprotected central fragment is tethered to the solid support via an $^{\alpha}$thioester linkage and contains a $N^{\alpha}$ protected amino terminal cysteine. In the first native chemical ligation step this peptide is reacted with the purified C-terminal fragment which contains an N-terminal cysteine. Briefly raising the pH >12 removes the Msc group. This newly synthesized fragment is then reacted in a second native ligation step with the N-terminal segment of the construct, which is also tethered to a solid support via an $^{\alpha}$thioester linkage.

fragments to give the desired protein construct.

Initially the N-terminal fragment of the Abl-SH3 domain, containing an Msc protected N-terminal cysteine, was generated on a solid support such that it was immobilized through $^{\alpha}$thioester (Fig. 8, fragment 2). In the first solid-phase ligation step, this immobilized peptide $^{\alpha}$thioester was reacted with the purified C-terminal fragment of Abl-SH3 containing an N-terminal cysteine residue

(Fig. 8, fragment 1); only the desired ligation reaction took place as the $N^{\alpha}$ group of the immobilized N-terminal fragment was protected with the Msc group. On completion of the ligation reaction the pH of the crude ligation mixture was briefly raised to facilitate complete removal of the $N^{\alpha}$–Msc group. The resulting fully deprotected ligation product (Abl-SH3 with an N-terminal Cys) was purified by HPLC in preparation for the second ligation step; the reaction of Cys-Abl SH3

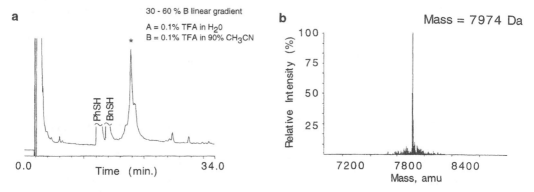

**Fig. 9** (a) Reverse phase HPLC of the ligation between the 3BP2 ligand-linker peptide $^\alpha$thioester and synthetic cys-Abl-SH3 after 6h. The peak labelled * corresponds to the desired product. (b) Electrospray mass spectrum of purified receptor ligand construct. Expected Mass = 7975 Da (average isotope composition).

with the immobilized 3BP2 ligand-linker peptide $^\alpha$thioester (Fig. 8, fragment 3). HPLC was used to monitor the progress of this second ligation which was deemed to be complete after six hours (Fig. 9a). The ligation product was confirmed as the desired contiguous receptor-ligand construct using electrospray mass spectrometry (Fig. 9b). The addition of thiol cofactors to the ligation mixture is known to accelerate the ligation reaction via transthioesterification (95). Hence both PhSH and BnSH were included in the reaction buffers. These reagents also ensure that the product from each ligation is recovered in its fully reduced form.

This sequential ligation strategy provides an effective method for the synthesis of large polypeptide sequences. In principle, it can be extended to more than two ligation steps, allowing the chemoselective assembly of multiple synthetic peptide building blocks. The inherent modularity of the approach also permits the primary structure of a peptide sequence to be altered with a minimal amount of synthetic effort, which is highly desirable for systematic protein structure/function studies.

## IV. CONCLUSIONS

The ability to routinely synthesize proteins has always been an attractive proposition. Complete synthetic control over the assembly of the target sequence allows the protein to be manipulated in an unprecedented manner. There have been many elegant examples over the years which highlight the development of this approach, but recent advances in the related technologies now mean that total synthesis exists as a true complement to recombinant methodologies. This is due to a number of criteria, but the optimization of solid phase peptide synthesis and the advent of chemical ligation strategies have been essential in achieving this goal. The potential applications are enormous, and one can only imagine the possibilities as we enter into the uncharted waters of synthetic protein chemistry.

## ACKNOWLEDGMENTS

The authors wish to acknowledge the support of Gryphon Sciences (G.J.C.), the Pew Scholars Program in the Biomedical Sciences (T.W.M) and the US National Institutes of Health (GM55843-01, T.W.M).

## REFERENCES

1. VW Cornish, D Mendel, PG Schultz. Probing protein structure and function with an expanded genetic code. Angew Chem Int Ed Engl 34: 621–633, 1995.

2. GT Hermanson. Bioconjugate techniques. San Diego: Academic Press, Inc., 1996.

3. CJA Wallace. Peptide ligation and semisynthesis. Curr Opin Biotech 6: 403–410, 1995.

4. TW Muir, SBH Kent. The chemical synthesis of proteins. Curr Opin Biotech 4: 420–427, 1993.

5. TW Muir. A chemical approach to the construction of multimeric protein assemblies. Structure 3: 649–652, 1995.

6. VJ Hruby, F Al-Obeidi, W Kazmierski. Emerging approaches in the molecular design of receptor-selective peptide ligands: conformational, topographical and dynamic considerations. Biochem J 268: 249–262, 1990.

7. VJ Hruby. Conformational restrictions of biologically active peptides via amino acid side chain groups. Life Sciences 31: 189–199, 1982.

8. RB Merrifield. Solid Phase Peptide Synthesis. I. The synthesis of a tetrapeptide. J Am Chem Soc 85: 2149–2154, 1963.

9. M Schnölzer, P Alewood, A Jones, D Alewood, SBH Kent. In situ neutralization in Boc-chemistry solid phase peptide synthesis: rapid, high yield assembly of difficult sequences. Int J Pept Protein Res. 40: 180–193, 1992.

10. P Alewood, D Alewood, L Miranda, S Love, W Meutermans, D Wilson. Rapid in situ neutralization protocols for Boc and Fmoc solid phase chemistries. Methods Enzymol 289: 14–29, 1997.

11. DA Wellings, E Atherton. Standard Fmoc protocols. Methods Enzymol 289: 44–67, 1997.

12. W Lu, MA Qasim, SBH Kent. Comparative total synthesis of turkey ovomucoid third domain by both stepwise solid phase synthesis and native chemical ligation. J Am Chem Soc. 118: 8518–8523, 1996.

13. WF Heath, RB Merrifield. A synthetic approach to structure-function relationships in the murine epidermal growth factor molecule. Proc Natl Acad Sci USA 83: 6367–6371, 1986.

14. M Ferrer, C Woodward, G Barany. Solid-phase synthesis of bovine pancreatic trypsin inhibitor (BPTI) and two analogues. Int J Peptide Protein Res 40: 194–207, 1992.

15. YS Yang, C Garbay, M Duchesne, F Cornille, N Jullian, N Fromage, B Tocque, BP Roques. Solution structure of GAP SH3 domain by $^1$H NMR and spatial arrangement of essential Ras signaling-involved sequence. EMBO J 13: 1270–1279, 1994.

16. N Goudreau, F Cornille, M Duchesne, F Parker, B Tocque, C Garbay, BP Roques. NMR structure of the N-terminal SH3 domain of GRB2 and its complex with a proline rich peptide from Sos. Nature struct Biol 1: 898–907, 1994.

17. MC Pietanza, Z Hou, JA Camarero, TW Muir. Peptides 1996. Proceedings of the Twenty-fourth European Peptide Symposium, Edinburgh, September 1996, 721–722.

18. TW Muir, PE Dawson, MC Fitzgerald, SBH Kent. Probing the chemical basis of binding activity in an SH3 domain by protein signature analysis. Chemistry and Biology 3: 817–825, 1996.

19. MC Fitzgerald, I Chernushevich, KG Standing, SBH Kent, CP Whitman. Total chemical synthesis and catalytic properties of the enzyme enantiomers L- and D-oxalocrotonate tautomerase. J Am Chem Soc 117: 11075–11080, 1995.

20. P Lloyd-Williams, F Albericio, E Giralt. Convergent solid-phase peptide synthesis. Tetrahedron 48: 11065–11133, 1993.

21. F Albericio, P Lloyd-Williams, E Giralt. Convergent solid-phase peptide synthesis. Methods Enzymol 289: 313–336, 1997.

22. K Akaji, N Fujii, H Yajima, K Hayashi, K Mizuka, M Aono, M Moriga. Studies on peptides. 127. Synthesis of a tripentacontapeptide with epidermal growth factor activity. Chem Pharm Bull 33: 184–201, 1985.

23. P Sieber, B Kamber, A Hartmann, A Johl, B Riniker, W Rittel. Total synthesis of human insulin by directed formation of disulphide bridges. Helv Chem Acta 57: 2617–2621, 1974.

24. H Yajima, N Fujii. Studies on Proteins. 103. Chemical Synthesis of a Crystalline Protein with the Full Enzymatic Activity of Ribonuclease A. J Am Chem Soc 103: 5867–5871, 1981.

25. T Kimura, M Takai, T Morikawa, S Sakakibara. Total Synthesis of Human Parathyroid Hormone (1–84). In ST Osaka, ed. Peptide Chemistry 1981, Protein Research Foundation: Osaka, 1981.

26. T Kimura, M Takai, K Yoshizawa, S Sakakibara. Solution synthesis of [Asn76]-Human parathyroid hormone (1–84). Biochem Biophys Res Comm 114: 493–499, 1983.

27. K Barlos, D Gatos, W Schafer. Synthesis of Prothymosin α(ProTα)-a Protein Consisting of 109 Amino Acid Residues. Angew Chem Int Ed Engl 30: 590–593, 1991.

28. BT Chait, SBH Kent. Weighing naked proteins: Practical high accuracy mass measurement of peptides and proteins. Science 257: 1885–1894, 1992.

29. CB Anfinsen. Principles that govern the folding of protein chains (Nobel Lecture). Science 181: 223–230, 1973.

30. D Andreu, F Albericio, NA Sole, MC Munson, M Ferrer, G Barany. Formation of disulphide bonds in synthetic peptides and proteins, In MW Pennington and BM Dunn, eds. Methods in Molecular Biology, Vol 35. Human Press Inc.: Totowa, 1994, pp. 91–169.

31. I Clark-Lewis, R Aebersold, H Ziltener, JW Schrader, LE Hood, SBH Kent. Automated chemical synthesis of a protein growth factor for hemopoietic cells, Interleukin-3. Science (231), 134–139, 1986.

32. I Clark-Lewis, B Moser, A Walz, M Baggiolini, GJ Scott, R Aebersold. Chemical synthesis, purification and characterization of two inflammatory proteins, neutrophil activating peptide 1(Interleukin-8) and neutrophil activating peptide 2. Biochemistry 30: 3128–3134, 1991.

33. B Gutte, RB Merrifield. The synthesis of Ribonuclease A. J Biol Chem 246: 1922–1941, 1971.

34. CH Li, D Yamashiro, D Gospodarowicz, SL Kaplan, G Van Vliet. Total synthesis of insulin-like growth factor 1 (somatomedin c). Proc Natl Acad Sci USA 80: 2216–2220, 1983.

35. J Schneider, SBH Kent. Enzymatic activity of a synthetic 99 residue protein corresponding to the putative HIV-1 protease. Cell 54: 363–368, 1988.

36. RF Nutt, SF Brady, PL Darke, TM Ciccarone, D Colton, EM Nutt, JA Rodkey, CD Bennett, LH Waxman, IS Sigal, PS Anderson, DF Veber. Chemical synthesis and enzymatic activity of a 99-residue peptide with a sequence proposed for the human immunodeficiency virus protease. Proc Natl Acad Sci USA 85: 7129–7133, 1988.

37. HL Ball, P Mascagni. Chemical synthesis and purification of proteins: a methodology. Int J Peptide Protein Res 48: 31–47, 1996.

38. R Ramage, J Green, OM Ojunjobi. Solid phase synthesis of ubiquitin. Tetrahedron Lett 30: 2149–2152, 1989.

39. R Ramage, J Green, TW Muir, OM Ogunjobi, S Love, K Shaw. Synthetic, structural and biological studies on the ubiquitin system: the total chemical synthesis of ubiquitin. Biochem J 299, 151–158, 1994.

40. M Kohmura, N Nio, Y Ariyoshi. Solid phase synthesis of crystalline monellin, a sweet protein. Agricul Biol Chem 55: 539–545, 1991.

41. I Clark-Lewis, LE Hood, SBH Kent. Role of disulphide bridges in determining the biological activity of interleukin-3. Proc Natl Acad Sci USA 85: 7897–7901, 1988.

42. K Rajarathnam, BD Sykes, CM Kay, B Dewald, T Geiser, M Baggiolini, I Clark-Lewis. Neutrophil activation by monomeric interleukin-8. Science 264: 90–92, 1994.

43. M Baca, PF Alewood, SBH Kent. Structural engineering of the HIV-1 protease molecule with a β-turn mimic of fixed geometry. Protein Sci 2: 1085–1091, 1993.

44. SG Love, TW Muir, R Ramage, KT Shaw, D Alexeev, L Sawyer, SM Kelly, NC Price, JE Arnold, MP Mee, RJ Mayer. Synthetic, structural and biological studies of the ubiquitin system: synthesis and crystal structure of an analogue containing unnatural amino acids. Biochem J 323: 727-734, 1997.

45. LF Zawadzke, JM Berg. A racemic protein. J Am Chem Soc 114: 4002–4003, 1992.

46. H Izawa, M Ota, M Kohmura, Y Ariyoshi. Synthesis and characterization of the sweet protein brazzein. Biopolymers 39: 95–101, 1996.

47. RC Milton, SC Milton, SBH Kent. Total chemical synthesis of a D-enzyme: the enantiomers of HIV-1 protease show reciprocal chiral substrate specificity. Science 256: 1445–1448, 1992.

48. J Blake, CH Li. New segment-coupling method for peptide-synthesis in aqueous solution -application to synthesis of human [Gly17]-beta endorphin. Proc Natl Acad Sci 78: 4055–4058, 1981.

49. J Blake. Total synthesis of S-carbamoylmethyl bovine apocytochrome c by segment coupling. Int. J. Pept. Protein Res. 27: 191–200, 1986.

50. D Yamashiro, CH Li. New segment synthesis of α-inhibin-92 by the acyl disulphide method. Int J Pept Protein Res 31: 322-334, 1988.

51. HC Cheng, D Yamashiro. Chemical synthesis of β-endorfin (1-27) analogs by peptide segment coupling. Int J Peptide Protein Res 38: 70–78, 1991.

52. H Hojo, S Aimoto. Polypeptide synthesis using the S-alkyl thioester of a partially protected peptide segment. Synthesis of the DNA-binding domain of c-Myb pro-

tein (142–193)-NH2. Bull Chem Soc Jpn 64: 111–117, 1991.

53. DS Kemp, RI Carey. Synthesis of a 39-peptide and a 25-peptide by thiol-capture ligations: observation of a 40-fold rate acceleration of the intramolecular O,N-acyl transfer reaction between peptide fragments bearing only cysteine protecting groups. J Org Chem 58: 2216–2222, 1993.

54. GA Homandberg, M Laskowski. Enzymatic resynthesis of the hydrolysed peptide bond(s) in ribonuclease S. Biochemistry 18: 586–592, 1979.

55. K Vogel, J Chmielewski. Rapid and efficient resynthesis of proteolyzed triose phosphate isomerase. J Am Chem Soc 116: 11163–11164, 1994.

56. K Witte, P Sears, R Martin, CH Wong. Enzymatic glycoprotein synthesis: Preparation of ribonuclease glycoforms via enzymatic glycopeptide condensation and glycosylation. J Am Chem Soc 119: 2114–2118, 1997.

57. T Nakatsuka, T Sasaki, ET Kaiser. Peptide segment coupling catalysed by the semisynthetic enzyme thiolsubtilisin. J Am Chem Soc 109: 3808–3810, 1987.

58. DY Jackson, J Burnier, C Quan, M Stanley, J Tom, JA Wells. A designed peptide ligase for total synthesis of ribonuclease A with unnatural catalytic residues. Science 266: 243–247, 1994.

59. TK Chang, DY Jackson, JP Burnier, JA Wells. Subtiligase: A tool for semisynthesis of proteins. Proc Natl Acad Sci 91: 12544–12548, 1994.

60. AC Braisted, JK Judice, JA Wells. Synthesis of proteins by subtiligase. Methods Enzymol 289: 298–313, 1997.

61. DY Jackson, JP Burnier, JA Wells. Enzymatic cyclization of linear peptide esters using subtiligase. J Am Chem Soc 117: 819–820, 1995.

62. DF Dyckes, T Creighton, RC Sheppard. Spontaneous re-formation of a broken peptide chain. Nature 247: 202–204, 1974.

63. CJA Wallace. Understanding cytochrome c function: engineering protein structure by semisynthesis. FASEB J 7: 505–515, 1993.

64. CJA Wallace, I Clark-Lewis. Functional role of heme ligation in cytochrome c. J Biol Chem 267: 3852–3861, 1992.

65. CJA Wallace, I Clark-Lewis. A rationale for the absolute conservation of Asn70 and Pro71 in mitochondrial cytochromes c suggested by protein engineering. Biochemistry 36: 14733–14740, 1997.

66. HF Gaertner, K Rose, R Cotton, D Timms, R Camble, RE Offord. Construction of protein analogues by site-specific condensationof unprotected peptides. Bioconjugate Chem 3: 262–268, 1992.

67. HF Gaertner, RE Offord, R Cotton, D Timms, R Camble, K Rose. Chemo-enzymic backbone engineering of proteins. J Biol Chem 269: 7224–7230, 1994.

68. G Simmons, PR Clapham, L Picard, RE Offord, MM Rosenkilde, TW Schwartz, R Buser, TNC Wells, AEI Proudfoot. Potent inhibition of HIV-1 infectivity in macrophages and lymphocytes by a novel CCR5 antagonist. Science 276: 276–279, 1997.

69. M Schnölzer, SBH Kent. Constructing proteins by dovetailing unprotected synthetic peptides: backbone-engineered HIV protease. Science 256: 221–225, 1992.

70. MJ Williams, TW Muir, SBH Kent, MH Ginsberg. Chemical synthesis of a $\beta$-sandwich domain by chemoselective ligation: The tenth type 3 module from fibronectin. J Am Chem Soc 116: 10797–10798, 1994.

71. M Baca, SBH Kent. Catalytic contribution of flap-substrate hydrogen bonds in HIV-1 protease explored by chemical synthesis. Proc Natl Acad Sci USA 90: 11638–11642, 1993.

72. M Baca, TW Muir, M Schnolzer, SBH Kent. Chemical ligation of cysteine-containing peptides: Synthesis of a 22 kDa tethered dimer HIV-1 protease. J Am Chem Soc 117: 1881–1887, 1995.

73. LE Canne, AR Ferre-D'Amare, SK Burley, SBH Kent. Total chemical synthesis of a unique transcription factor-related protein: cMyc-Max. J Am Chem Soc 117: 2998–3007, 1995.

74. TW Muir, PE Dawson, SBH Kent. Protein synthesis by chemical ligation of unprotected peptides in aqueous solution. Methods Enzymol 289: 266–298, 1997.

75. K Rose. Facile synthesis of homogeneous artificial proteins. J Am Chem Soc 116: 30–33, 1994.

76. TW Muir, MJ Williams, MH Ginsberg, SBH Kent. Design and chemical synthesis of a neoprotein structural model for the cytoplasmic domain of a multisubunit cell-surface receptor: Integrin $\alpha 11b\beta 3$ (Platelet GP11b-111a). Biochemistry 33: 7701–7708, 1994.

77. A Nefzi, X Sun, M Mutter. Chemoselective ligation of multifunctional peptides to topological templates via thioether formation for TASP synthesis. Tetrahedron Lett 36: 229–230, 1995.

78. DR Englebresten, BG Garnham, DA Bergmann, PF Alewood. A novel thioether linker: Chemical synthesis of a HIV-1 protease analogue by thioether ligation. Tetrahedron Lett 36: 8871–8874, 1995.

79. C-F Liu, JP Tam. Chemical ligation approach to form a peptide bond between unprotected peptide segments. Concept and model study. J Am Chem Soc 116: 4149–4153, 1994.

80. C-F Liu, JP Tam. Peptide segment ligation strategy without use of protecting groups. Proc Natl Acad Sci USA 91: 6584–6588, 1994.

81. PE Dawson, TW Muir, I Clark-Lewis, SBH Kent. Synthesis of proteins by native chemical ligation. Science 266: 776–779, 1994.

82. T Wieland. Sulfur in Biomimetic Peptide Synthesis. In VD Kleinkauf Jaeniche, ed. The Roots of Modern Biochemistry. Berlin, New York: Walter de Gruyter & Co, 1988, pp 213–221.

83. CF Liu, C Rao, JP Tam. Acyl disulfide-mediated intramolecular acylation for orthogonal coupling between unprotected peptide segments. Mechanism and applications. Tetrahedron Lett 37: 933–936, 1996.

84. TM Hackeng, CM Mounier, C Bon, PE Dawson, JH Griffin, SBH Kent. Total chemical synthesis of enzymatically active human type II secretory phospholipase A(2). Proc Natl Acac Sci USA 94: 7845–7850, 1997.

85. L Canne, S Bark, SBH Kent. Extending the applicability of native chemical ligation. J Am Chem Soc 118: 5891–5896, 1996.

86. L Zhang, JP Tam. Orthogonal coupling of unprotected peptide segments through Histidyl amino terminus. Tetrahedron Lett 38: 3–6, 1997.

87. JP Tam, Y-A Lu, C-F Liu, J Shao. Peptide synthesis using unprotected peptides through orthogonal coupling methods. Proc Natl Acad Sci USA 92: 12485–12489, 1995.

88. SBH Kent, SJ Bark, LE Canne, PE Dawson, MC Fitzgerald, T Hackeng, W Lu, T Muir. Peptides 1996. Proceedings of the Twenty-fourth European Peptide Symposium, Edinburgh, September 1996, 187–192.

89. W Lu, MA Qasim, M Laskowski, SBH Kent. Probing intermolecular main chain hydrogen bonding in serine proteinase-protein inhibitor complexes: Chemical synthesis of backbone engineered turkey ovomucoid third domain. Biochemistry 36: 673–679, 1997.

90. SM Feller, R Ren, H Hanafusa, D Baltimore SH2 and SH3 domains as molecular adhesives: the interactions of Crk and Abl. Trends Biochem Sci 19: 453–459, 1994.

91. AM Pendergast. Nuclear tyrosine kinases: from Abl to WEE 1. Curr Opin Cell Biol 8: 174–181, 1996.

92. BJ Mayer, D Baltimore. Mutagenic analysis of the roles of SH2 and SH3 domains in regulation of the Abl tyrosine kinase. Mol Cell Biol 14: 2883–2894, 1994.

93. M Overduin, CB Rios, BJ Mayer, D Baltimore, D Cowburn. Three-dimensional solution structure of the Src homology 2 domain from c-Abl. Cell 70: 697–704, 1992.

94. LE Canne, SM Walker, SBH Kent. A general method for the synthesis of thioester resin linkers for use in the solid phase synthesis of peptide-α-thioacids. Tetrahedron Lett 36: 1217–1220, 1995.

95. PE Dawson, MJ Churchill, MR Ghadiri, SBH Kent. Modulation in Native Chemical Ligation through the use of thiol additives. J Am Chem Soc 119: 4325–4329, 1997.

96. YQ Gosser, J Zheng, M Overduin, BJ Mayer, D Cowburn. The solution structure of Abl SH3, and its relationship to SH2 in the SH(32) construct. Structure 3: 1075–1086, 1995.

97. JA Camarero, TW Muir. Chemoselective backbone cyclization of unprotected peptides. J Chem Soc Chem Commun 1997: 1369–1370, 1997.

98. JP Tam, YA Lu. Synthesis of large cyclic cystine-knot peptide by orthogonal coupling strategy using unprotected peptide precursor. Tetrahedron Lett 38: 5599–5602, 1997.

99. JP Camarero, J Pavel, TW Muir. Chemical synthesis of a functional circular protein domain: Evidence for folding assisted cyclization. Angew Chem Eng Int Ed 37: 347–348, 1998.

100. DH Lee, JR Granja, JA Martinez, K Severin, R Ghadiri. A self-replicating peptide. Nature 382: 525–528, 1996.

101. K Severin, DH Lee, AJ Kennan, R Ghadiri. A synthetic peptide ligase. Nature 389: 706–709, 1997.

102. DH Lee, K Severin, Y Yokobayashi, R Ghadiri. Emergence of symbiosis in peptide self-replication through a hypercyclic network. Nature 390: 591–594, 1997.

103. S Yao, I Ghosh, R Zutshi, J Chmielewski. A pH-modulated, self replicating peptide. J Am Chem Soc. 119: 10559–10560, 1997.

104. M Mutter, S Vuilleumier. A chemical approach to protein design – Template assembled synthetic proteins (TASP). Angew Chem Int Ed Engl 28: 535–554, 1989.

105. M Mutter, P Dumy, P Garrouste, C Lehmann, M Mathieu, C Peggion, S

Peluso, A Razaname, G Tuchscherer. Template assembled synthetic proteins (TASP) as functional mimetics of proteins. Angew Chem Int Ed Engl 35: 1482–1485, 1996.

106. M Mutter, GG Tuchscherer, C Miller, K-H Altmann, RI Carey, DF Wyss, AM Labhardt, JE Rivier. Template-assembled synthetic proteins with four-helix-bundle topology. Total chemical synthesis and conformational studies. J Am Chem Soc 114: 1463–1470, 1992.

107. S Cervigni, P Dumy, M Mutter. Synthesis of glycopeptides and lipopeptides by chemoselective ligation. Angew Chem Int Ed Engl 35: 1230–1232, 1996.

108. S Peluso, P Dumy, IM Eggleston, P Garrouste, M Mutter. Protein mimetics (TASP) by sequential condensation of peptide loops to an immobilised topological template. Tetrahedron 53: 7231–7236, 1997.

109. JA Camarero, GJ Cotton, A Adeva, TW Muir. Chemical ligation of unprotected peptides directly from a solid support. J Pep Res 51: 303–316, 1998.

# 6

# Introduction of Unnatural Amino Acids into Proteins

**Scott A. Lesley**
*Novartis Institute for Functional Genomics, La Jolla, California*

## I. INTRODUCTION

In recent years, it has become common-place to modify the amino acid sequences of proteins via mutagenesis. New techniques have made site-directed mutagenesis a straightforward and efficient task (1). High quality synthetic oligonucleotides can be obtained rapidly and inexpensively to facilitate these methods. Amplification techniques for wholesale substitutions and gene shuffling allow millions of alterations to be created in a very short time (2). Provided that a means exists for evaluating these alterations, substantial changes in the protein's activity can be obtained from this great diversity of amino acid combinations. The limitation to creating even greater combinatorial diversity, and therefore functionality, derives from the limited set of naturally occurring amino acids forming the building blocks of proteins. If additional unnatural amino acids could be incorporated into proteins, even greater changes to protein activities would be possible.

One typically presumes that the natural world produces proteins comprising the 20 L-amino acids. This is not an absolute truth, however, as a variety of rare substitutions and post-translational modifications expand the palette of protein composition. While the limited set of natural amino acids is sufficient for nature's needs, it does not satisfy our human nature. Our desire to perturb protein structure to evaluate its effect or create new activities leads us to try to expand on this repertoire. The natural amino acids have given us a rather limited set of functional groups and geometries that serves as the basis of enzyme activity. Organic chemistry provides a much broader range of functionalities which hold the potential to perform reactions in proteins that are unprecedented. In addition to the limitations of the sidechain functionalities, the chirality and structure of the amino acid backbone is limited to L-, alpha substituted amino acids. D-amino acids are found in macromolecules, such as peptidoglycan, but their incorporation is outside the normal realm of the protein synthesis machinery. Alternatives to L-, alpha substituted amino acids could lead to proteins with greatly improved temperature stability, solubility, and/or protease resistance. The limitation, of course, is achieving incorporation of these novel building blocks using biological systems that have evolved exquisite selectivity for the natural amino acids or by using synthetic methods that cannot efficiently construct such large polymers. This chapter deals with biological and combined biological semi-synthetic methods which have been used to produce proteins with non-canonical amino acid content.

## II. BIOLOGICAL SYSTEMS

The universal genetic code is based on triplets of ribonucleotide bases to form a codon which specifies a particular amino acid. There are 4 nucleotides which comprise RNA, and therefore, 64 possible codons. Relegating one codon for a stop signal there is still room to encode 63 amino acids without altering the basic paradigm of the genetic code. Subtracting the 20 naturally occurring amino acids, there are still 43 "spare" codons which could be used. These codons are not really extra, however, as the code has evolved to be degenerate and redundant. Although the genetic code is extremely well conserved, there are exceptions. In addition to its role as a stop codon, the triplet UGA can code for tryptophan in mitochondria or cysteine in *Euplotes octocarinatus* (3). Probably the best example of alternative coding is insertion at amber (UAG) stops by suppressor mutants in *Escherichia coli* (4). Of the 20 amino acids, 13 can be inserted at amber stops by simply expressing the protein in the different suppressor hosts (5, 6). Yet, these all are examples of incorporation of natural amino acids which could easily by achieved by standard site-directed mutagenesis techniques. Using a biological system to generate a protein containing unnatural amino acids would necessitate using the existing protein synthesis machinery. It is tempting to try to hijack one of the "spare" codons for our own use, but this is not a simple matter. The examples of biological systems which incorporate non-canonical amino acids

typically do so by exploiting the context surrounding the codon, which in effect provides a new codon.

### A. Unusual Naturally Occurring Amino Acids

The most common example of a non-canonical amino acid is formyl-methionine. While methionine is only encoded by a single triplet (AUG) and most proteins have multiple occurrences of the codon, formyl-methionine is only incorporated at the initiating AUG (or occasionally GUG or UUG). A post-translational deformylation typically converts the formyl-methionine into methionine leaving only the canonical amino acids in the mature protein (7). Although it is incorporated, its absence in the mature protein makes formyl-methionine a poor example of a non-canonical amino acid.

Selenocysteine is a better example of incorporation of a non-canonical amino acid in a biological system (reviewed in 8). Structurally similar to cysteine, this amino acid contains the more reactive selenium rather than sulfur. Incorporation of selenocysteine utilizes the UGA codon which is a standard stop codon. UGA codons are relatively rare in *E. coli* (9), which may explain the choice of this codon for dual interpretation. Further specifying the location of insertion are elements of RNA secondary structure within the message which promote efficient insertion at the UGA codon (10–12).

Free selenocysteine is not incorporated at specific sites. An alternate pathway does allow some mischarging of tRNA$^{cys}$ to give random incorporation in place of cysteine (13). However, specific insertion at UGA follows a different pathway. Selenocysteine is incorporated into proteins using a unique tRNA$^{sec}$ which is the product of the *selC* gene in *E. coli* (14). The first step towards incorporation of selenocysteine is the enzymatic esterification of this tRNA with serine and subsequent reaction of the seryl-tRNA with selenophosphate in an enzymatic reaction to yield selenocysteine (15, 16). Insertion of this amino acid into proteins also requires yet another unique protein. The product of the *sel*B gene in *E. coli* is a selenocysteine-specific equivalent to the translation factor EF-Tu and functions to shuttle the charged tRNA into the translating ribosome (17, 18). Thus, for incorporation of a single non-canonical amino acid, many steps are involved and multiple specific factors needed to evolve.

## B.  Post-translational Modifications

There are many examples of post-translational modifications which result in non-canonical amino acid derivatives. Such modifications are an important part of cell physiology, playing key roles in regulation and signal transduction pathways. Examples of glycosylation, phosphorylation, ribosylation, and methylation abound in the scientific literature. For the most part, these post-translational modifications are enzyme-specific and of little use for those interested in a general approach to incorporate novel chemical moieties into proteins. Phage display technology has given us the potential to incorporate some of the biological pathways for protein modification into a more general system. Examples have been published which utilize phage display to probe the substrate specificity of the modification enzymes and identify alternate peptide sequences which can be recognized and modified. One such example involves *in vivo* biotinylation. Biotin is an essential cofactor required for fatty acid synthesis and is covalently attached to a specific lysine residue of the carboxylase, or carrier protein, via biotin ligase in an ATP-dependent reaction (19, 20). Deletion studies identified a small domain of approximately 120 amino acids which was recognized by the biotin ligase and efficiently biotinylated (21). Subsequent work by Shatz (22) identified a much smaller amino acid sequence from phage display panning of potential sequences. This smaller sequence is efficiently biotinylated and is much more amenable to site-specific incorporation into existing protein structures. Similar studies have identified short sequences that are recognized by protein kinases and phosphorylated as gene fusions (23). While there still will be some constraints on recognition of these sequences by the cognate modification enzymes, use of these short modification tags allows additional chemical functionality to be incorporated into proteins. Further coupling with enzyme evolution techniques on modification enzymes may allow even more chemical diversity to be incorporated into target proteins.

## C.  Direct Incorporation of Unnatural Amino Acids

There are many barriers to the *in vivo* incorporation of an unnatural amino acid into a cell. First the compound must be able to enter the cell. For relatively small compounds in bacterial systems, this is not a huge hurdle. The primary selectivity against incorporation is at the level of the tRNA synthetase. In the cell, these enzymes must discriminate between very similar compounds (leucine vs. isoleucine and tyrosine vs. phenylalanine for example) and in some cases will even deacylate those tRNAs which have been misacylated (24). The ribosome and associated translational machinery is far less discriminating but still can pose a barrier to incorporation. In the *in vitro* systems discussed later, there are examples of amino acids which can be ligated to a tRNA but not incorporated into the translating protein (25). Even if an unnatural amino acid makes it into a protein, study of the protein may not be possible.

Such analogs can be highly toxic to the cell or the target protein may be rapidly degraded.

While the biological systems for protein synthesis have evolved exquisite selectivity for their natural substrates, there are examples of tricking the systems by using related amino acids. Tryptophan derivatives long have been used as probes for following protein folding and protein-protein interactions. These analogs contain fluorine at position four or five of the tryptophan. The analog 4-fluorotryptophan is non-fluorescent and incorporation into proteins allows other chromophores within or interacting with the protein to be studied (26). The analog 5-fluorotryptophan has been used as a probe for following proteins by $^{19}$F NMR (27, 28). Combining standard site directed mutagenesis techniques to specifically place tryptophan resides with incorporation of this analog allows individual locations within a protein to be monitored (29).

In these direct incorporation studies, tryptophan is typically preferred by the cell over the analog, thus steps must be taken to improve its uptake. The cellular pathway for tryptophan synthesis is typically deleted, forcing the resulting auxotroph to utilize the analog supplemented in the growth media. An alternative approach is to block aromatic amino acid synthesis with glyphosate and provide the analog for uptake (30). *Bacillus*, *E. coli*, and *Sacchromyces cerevisiae* have all been shown to incorporate tryptophan analogs, although there may be some species preference for particular analogs (26–28).

Fluorotyrosine is another example of an analog which can readily be incorporated into proteins. Like the tryptophan analog, an auxotrophic strain is used and over 80% of the tyrosine has been shown to be substituted by the 3-fluoro analog (31). Like fluorotryptophan, fluorotyrosine can be used as an NMR probe

but halogenation also can be used to affect the pKa of the hydroxyl group.

While direct incorporation of unnatural amino acids is appealing, the scope of analogs which can be readily incorporated is very limited. Even if efficient incorporation were achieved, there is little control to achieve a unique site of incorporation. The number of cellular proteins also incorporating such an analog would lead to toxicity problems and generally require purification of the protein for evaluation.

## III. *IN VITRO* SYSTEMS

The ideal system for incorporating unnatural amino acids into proteins would be one that is both efficient and generic. The nature of enzymatic reactions puts these two parameters at odds. While *in vivo* methods require the re-engineering of multiple parts of the translation machinery, *in vitro* methods allow some of these parts to be circumvented thereby leaving only the basic translational machinery to be addressed. Most *in vitro* methods use partially purified extracts from biological systems that contain all of the components necessary for translation except for nucleic acid template and small molecules, such as amino acids and nucleotides. The process can be further simplified by providing the unnatural amino acid already charged onto an appropriate tRNA. This greatly reduces the discrimination against the unnatural amino acid as the ribosomes are much more permissive than the synthetases. The tRNA can be acylated with a natural amino acid and further modified chemically (32–34), or acylated directly with the unnatural amino acid (35). Both approaches will be explored later in this chapter. The tRNAs used are either amino acid specific, therefore potentially incorporating at every instance of that amino acid, or suppressor tRNAs which will incorporate only at a designated site within the template.

    *In vitro* systems allow for manipulation of translation conditions for optimization of unnatural amino acid incorporation. Additional components can be added or removed to study protein-protein interactions. Proteolytic or other activities which interfere with the final assay can be minimized. Most importantly, the proteins which are synthesized are determined by the template added. This allows for straight-forward control reactions and the ability

to specifically incorporate a radiolable into the protein. In short, *in vitro* systems are simpler, more defined, and more flexible than *in vivo* systems for unnatural amino acid incorporation. What they lack, however, is the ability to readily produce even milligram amounts of protein. This deficiency is typically far outweighed by the advantages of such systems.

## A. *In Vitro* Translation Extracts

The translation process is complex and shown broadly outlined in Fig. 1. Trans-

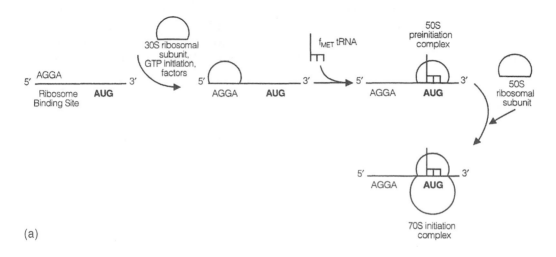

(a)

(b)

**Fig. 1** Translation initiation. (a) Prokaryotic translation initiation requires a functional ribosome binding site appropriately located upstream from the AUG initiation site. (b) Eukaryotic translation initiation begins from the first AUG encountered from the 5'-end of the message. Message capping and upstream sequences can effect translation efficiency.

lation extracts are typically crude lysates from actively growing cellular sources. Because the translation machinery is so complex, it is not effective to fractionate this machinery away from the remaining cellular proteins. These extracts are prepared by lysis of the cells and centrifugation to remove membrane and other cellular debris. *In vitro* translation, then, takes place within the context of the majority of the cellular proteins. Because the *in vitro* translation reaction may only produce several ng to a ug of synthesized protein in roughly 200 ug of total protein, the cellular proteins are an important consideration when designing experiments. Synthesized proteins are rarely visible by 1-dimensional SDS PAGE and contaminating enzymatic activity from the cellular components can mask the target activity. Fortunately, a simple "no template" control is sufficient to determine the source of the activity in question and the newly synthesized protein easily can be visualized by incorporation of radiolabeled amino acids such as $^{35}$S-methionine or $^{14}$C-leucine.

There are many sources of translation extracts. The phylogenic kingdoms are well represented with extracts from *E. coli*, yeast, rabbit reticulocyte, and wheat germ. The most popular lysates for incorporation of unnatural amino acids are derived from *E. coli* and rabbit reticulocytes. *E. coli* translation extracts have been well studied and were used to determine the genetic code (36). The method, described by Zubay (37), is the basis for a coupled transcription-translation system which is both more convenient and more efficient than simple translation systems. The convenience is preparation of plasmid rather than mRNA. The increased efficiency of translation is due to nuclease-protection of the message by the translating ribosomes. Extracts can be prepared such that PCR products (38) or templates containing phage promoters, such as from T7 (39), can be utilized. The

expression requirements for the *in vitro* system are similar to *in vivo* expression. Plasmid DNA containing the gene of interest is the typical template. The gene is located downstream of a functional promoter such as *tac* or $\lambda P_L$. A functional ribosome binding site is located approximately eight bases immediately upstream of the AUG start. Polycistronic messages can be translated by the system, provided that additional and appropriately placed ribosome binding sites are included and the gap between genes is not too large.

Rabbit reticulocyte lysate is also a very popular extract for *in vitro* translation. Standard reticulocyte extracts utilize mRNA, however recent improvements now permit a coupled reaction using phage RNA polymerases to transcribe from plasmid or PCR products (40). Translation in reticulocytes typically is monocystronic from the first AUG encountered and does not require a specific ribosome binding site. Sequences important for initiation are reviewed by Kozak (41).

Most of the work developing systems for unnatural amino acid incorporation has been done in the *E. coli* system, however eukaryotic systems like rabbit reticulocyte lysate also can be utilized. Each system has its own advantages. *E. coli* benefits from the vast history of genetics research which has allowed investigators to customize some of the translational machinery for unnatural amino acid incorporation. The bacteria are easily grown and the translation extract typically produces higher levels protein than the reticulocyte extract. The rabbit reticulocyte system is preferable for translation of some eukaryotic genes. These genes can contain internal sequences that are recognized in the *E. coli* system as ribosome binding sites and cause internal initiation of translation. Both systems, as well as a wheat germ system, are commercially available.

For incorporation of unnatural

amino acids via *in vitro* translation, the amino acid is provided as a charged tRNA. As described below, this tRNA is either a chemically modified charged-tRNA or generated by a semi-synthetic method. In either case, the charged tRNA is added to the translation extract along with a buffer primarily containing salts, standard amino acids, and an ATP/GTP energy source. The reaction is started by the addition of the desired transcription/translation template. Reactions are complete in 30–60 minutes at 37°C. Reaction products can be visualized by incorporation of radio-labeled amino acid, enzymatic activity, immunochemical, or other non-isotopic detection methods. Purification of proteins from the reactions can be performed, but typically requires either an affinity purification method for efficient isolation of the rare protein or a large-scale reaction to create sufficient product for purification.

The primary limitation of *in vitro* translation systems is the small amount of protein produced relative to the large amount of cellular proteins present. Numerous attempts have been made at increasing the level of protein synthesis in these extracts (42, 43). Simply adding more substrates has minimal effect. It appears that accumulation of small molecule inhibitors is largely responsible for the shutdown of protein synthesis after a relatively short period of time. These inhibitors can be removed without loss of the translation machinery by the use of a semi-permeable membrane with appropriate size cutoff (42). This method has not been used widely with success, thus most researchers prefer to simply increase the scale of the reaction to provide sufficient protein for analysis.

## B. Chemically Modified Acylated-tRNAs

As Described above, the specificity of the tRNA synthetases in large part prevents the direct acylation of unnatural amino acids to tRNAs. Chemical modification of tRNAs charged with an amino acid is one method by which unnatural amino acids can be introduced into proteins (32–34). This method takes advantage of unique chemical functionalities on some amino acid sidechains, which are not found in the tRNA, to achieve selective labeling. The primary amine of the lysine sidechain and the sulfhydryl group of cysteine provide unique handles which may be chemically modified to produce tRNAs acylated with unnatural amino acids.

The acyl-bond by which the amino acid is attached to the tRNA is quite labile and will readily deacylate under basic conditions. Reagents used to label sidechains should be selective and efficient to avoid unwanted modifications and deacylation of the tRNA. Modifications by this method are limited to derivatives of lysine or cysteine, but several substitutions have been reported (44). Modification of lysine residues in proteins with NHS-biotin esters is a common method of labeling proteins for detection. The difference in pKa for the alpha-amine (pKa=7.4) and epsilon-amine (pKa=11.1) of lysine permits selective labeling of the sidechain by NHS-esters. In the procedure described by Johnson (32), $tRNA_{lys}$ is chromatographically purified and enzymatically coupled with lysine. The coupled lysine is reacted with NHS-biotin or NHS-linker-biotin to result in a tRNA primarily labeled with biotinyl-lysine at the epsilon position. This acyl-tRNA can be added to *in vitro* translation extracts (both prokaryotic and eukaryotic) where it is incorporated in place of lysine. Proteins labeled in this way can easily be detected by streptavidin-enzyme conjugates and can readily be isolated by incubation with avidin or streptavidin particles (45). As an affinity purification tool, proteins labeled in this way with biotin cannot be efficiently released from the particles. Multiple incorporation events per protein also do not allow efficient release even by lower affinity monomeric avidin resins (46).

Fluorescent groups also readily can be incorporated via this scheme (47). Such groups are useful for detecting proteins and have additional potential for monitoring protein-protein interactions via fluorescent polarization techniques. Fluorescent reporters such as tryptophan are commonly used to measure the folding state of a protein (48). Since there are many lysine residues present in most natural proteins, the multiple incorporation of fluorescent probes by this method makes interpretation difficult. Modification of cysteine residues offers an improvement. There are typically fewer cysteine residues than lysine residues in proteins thereby offering more specificity of incorporation. Combining site-directed mutagenesis techniques to further reduce the number allows even greater specificity.

While these methods are useful for producing proteins with detection labels and cross-linkers, they suffer from some limitations. The biggest restriction is the requirement of a unique functionality on the sidechain which can be chemically modified while attached to the tRNA. Many reagents can be used to modify free amino or sulfhydryl groups in this fashion, but the requirement is a restriction on what classes of unnatural amino acids may be incorporated. A second limitation is the selectivity of incorporation. There is a competition for incorporation at any lysine or cysteine codon between the unnatural and the natural amino acid. Chemical modifications are not complete and endogenous lysine or cysteine residues also can be readily incorporated. For evaluation of particular positions and functionalities for enzyme function or protein folding and stability studies, an alternative approach described below is better suited.

## C.  Semi-Synthetic Synthesis of Charged tRNAs

A method first described by Schultz and co-workers (35) is a general technique for incorporating a wide variety of amino acid analogs. Reviews describing the application of semi-synthetic synthesis of charged tRNAs have been presented elsewhere (49, 50). Briefly, this method utilizes a mischarged tRNA for incorporation in an *in vitro* system. However, rather than chemical modification of an acyl-tRNA, this technique involves chemical synthesis of a dinucleotide acylated to a protected amino acid analog that is subsequently ligated onto a truncated tRNA. After removal of the protecting groups, the resulting acylated-tRNA is added to an *in vitro* translation reaction for incorporation into the translated protein. The process for generating acylated tRNAs is illustrated in Fig. 2.

It is a very significant point that specificity in both the location of the analog incorporation and the homogeneity of the incorporation are controlled in this method. Specificity in location is achieved by the use of an amber suppressor tRNA. The stop codon UAG can be efficiently suppressed in *E. coli* by the use of mutant tRNAs that have altered (CUA) anticodons and are still recognized by their cognate synthetases (51). Such suppressor tRNAs are not useful for directed incorporation of amino acid analogs since any free tRNA could be acylated by synthetases found in the translation extract resulting in heterogeneity of incorporation at the UAG site. The solution to this problem was found by utilizing a suppressor tRNA derived from yeast tRNA$^{phe}$, which is not recognized by the endogenous *E. coli* synthetases (52). Semi-synthetic acylation of this tRNA, then, results in homogeneous incorporation of the acylated amino acid analog at the UAG site. Suppression of the UAG

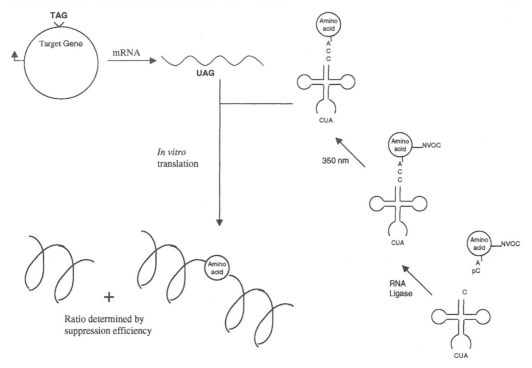

**Fig. 2** Incorporation of unnatural amino acids via semi-synthetic synthesis of charged tRNAs. Protected amino acids are coupled to the dinucleotide pCpA or pdCpA and ligated onto a truncated suppressor tRNA. After deprotection of the amino acid, the tRNA is added to an *in vitro* translation reaction. The unnatural amino acid is incorporated at a specific site through suppression of the UAG stop codon.

stop by this tRNA is not absolute. The absence of suppression results in a termination of translation and therefore a protein truncated at the site of the UAG stop. Such truncated proteins typically are inactive but must be considered when interpreting results.

RNA ligase is used in place of the tRNA synthetase for coupling the amino acid analog to the tRNA. While the synthetase has very high selectivity for the cognate amino acid, the ligase activity is virtually independent of the context of the tRNA and amino acid analog. This is the basis for the tremendous flexibility of the method. Most tRNA acceptor stems, including tRNA$^{phe}$, terminate at the 3' end with the sequence CCA. The tRNA$^{phe}$ suppressor is produced by runoff

transcription of a template encoding the gene lacking the terminal pCpA residues. A dinucleotide of pCpA or pdCpA is reacted with a cyanomethyl ester of the desired, appropriately protected, amino acid analog which couples to the terminal 2' or 3' hydroxyl group of the dinucleotide (53). The aminoacyl-pdCpA is attached to the tRNA$^{phe}$ (-CA) suppressor tRNA by the RNA ligase enzyme.

Protecting groups typically are required for synthesis of the acyl-pdCpA containing the desired unnatural amino acid. These groups are removed immediately prior to use. Deprotection schemes must take into account the labile nature of the acyl-tRNA bond. Photocleavable protecting groups are most convenient. Nitroveratryloxy carbamate (NVOC) is used to protect the alpha-amino groups and reactive sidechains. After ligation to the tRNA, deprotection is performed by irradiation at 350 nm under acidic conditions (49) and the tRNA can be added directly to the reaction. Other protecting groups such as biphenylisopropyloxy-carbonyl (BPOC) can be used for those unnatural amino acids which are light sensitive (53, 54). With these schemes, a tremendous variety of amino acid analogs can be prepared for introduction into proteins.

Some examples of unnatural amino acids incorporated by this method are shown in Fig. 3. Over 70 amino acid analogs have reportedly been incorporated by this method. These can be classified as alternative sidechain, alpha-disubstitued, or amino-substituted. Much of enzyme functionality is due to the functional groups present in the amino acid sidechains. With the ability to select functional groups, one can select the pKa, alter hydrogen bonding, or even change the mechanism of a reaction. While D-amino acids appear to be prohibited in the system (55, 56), alpha-disubstituted amino acids can be incorporated and are useful for studying protein folding and stability. The amino group can be substituted to add a methyl group or replaced with a hydroxyl or sulfhydryl group to yield proteins with alternative backbone structures (56–58).

Amino acid analogs have been used for a variety of structure and function studies of proteins. T4 lysozyme is a well studied and classic example for structural studies. Unnatural amino acids have found many applications for the study of this protein. Extensive mutagenesis has been performed on this protein and virtually every possible natural substitution already has been examined in this protein (59). These results have been extended far beyond the natural repertoire of substitutions by the use of this method. For example, attempts to stabilize T4 lysozyme by using conventional mutagen-

esis (L133F and A129V) to increase the core packing density actually resulted in decreased thermostability (60). Use of the analogs one and two at position 133 shown in Fig. 4 permitted better packing of the core of the enzyme with a corresponding increase in stability of 0.6 kcal mol$^{-1}$ and 1.2 kcal mol$^{-1}$, respectively (61). Isostructural sidechain substitutions, as well as various iterations of addition or removal of the sidechain methylene groups, can give insight into the role various elements of the amino acid in protein structure. In short, the ability to incorporate such a wide range of analogs allows one to ask very refined questions, such as what role an individual hydrogen bond plays in stability.

Biophysical probes as sidechains can be incorporated to evaluate a protein's structure. Cornish et al. (62) have incorporated a spin label and fluorescent probes. Isotopically labeled amino acids also have also been incorporated by this means (63) and serve as a convenient tag for NMR studies of a protein.

Enzyme mechanisms routinely are studied by conventional mutagenesis and the ability to incorporate amino acid analogs adds yet another dimension to these studies. One protein which serves as an example is Staphylococcal nuclease. The protein shows a tremendous ($10^{16}$ fold) enhancement in the hydrolysis of the phosphodiester bonds in nucleic acids. Conventional mutagenesis has shown Glu43 in this enzyme to be essential for efficient activity (64) and led to a proposal that it acts as a general base in the enzyme mechanism. Use of the nitro-equivalent to glutamate (S-4-nitro-2-aminobutyric acid), a poorer base, demonstrated that an alternate mechanism is occurring and that the native glutamate most likely plays a structural role in the protein. The ability to incorporate amino acid analogs into enzymes provides bountiful opportunities to explore catalytic mechanisms and substrate specificity.

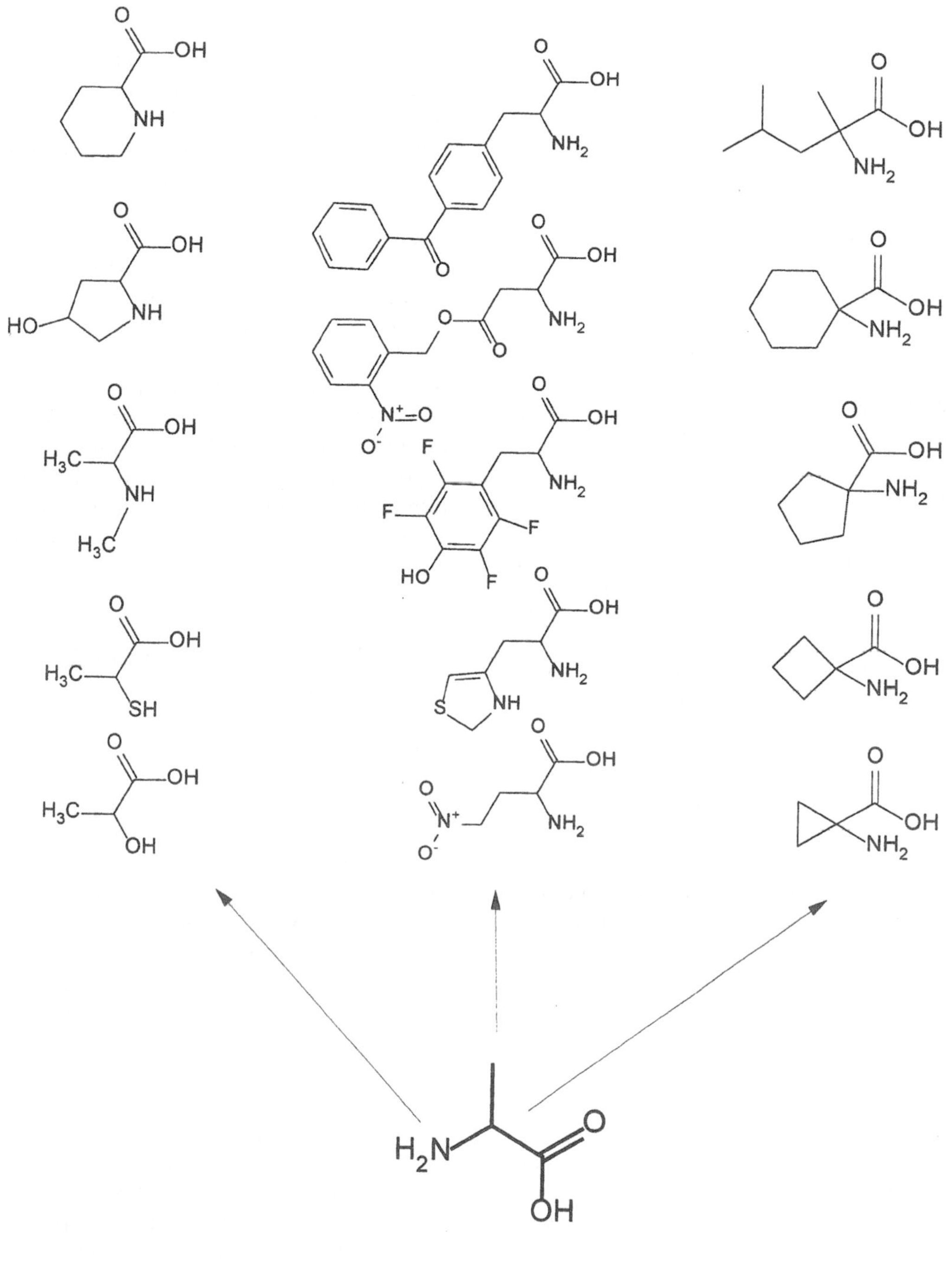

**Fig. 3** Examples of unnatural amino acids incorporated into proteins. Over 70 different unnatural amino acids have been incorporated into proteins. Variations in sidechains, alpha-substitution, and amino-substitution can be introduced.

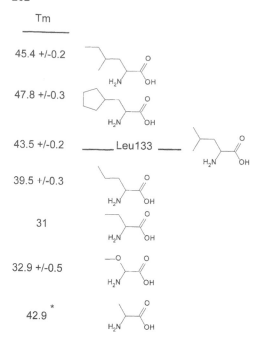

**Fig. 4** Effects of unnatural amino acid incorporation on protein stability. The amino acids indicated were incorporated into T4 lysozyme at position 133 (60, 61). The thermostability relative to the wild-type amino acid (leucine) is indicated for each substitution.

## IV. *IN VIVO* OPPORTUNITIES

The semi-synthetic method of creating acyl-tRNAs also can be applied to *in vivo* experimentation. Such tRNAs have been injected into oocytes (65) to follow the synthesis and the fate of proteins within a cell. The major barrier to performing such experiments is delivery of the acyl-tRNA to the translation machinery of the cell. While direct injection is an option in some cases, the wide range of reagents for transfection offers an attractive, although untested, alternative.

Progress also is being made toward an *in vivo* amber suppression system for unnatural amino acids. The difficulty in creating such a system is as immense as the utility. Transport of the amino acid analog and its potential toxicity are a con-

cern. Specific tRNAs and synthetases must be evolved (66) which are both selective and efficient. Specificity for the site of incorporation is achieved through suppression of the amber stop codon. Therefore, the translation machinery also must utilize this novel acyl-tRNA efficiently. Since, such systems would be specific for the particular unnatural amino acid, the utility also would be limited. A general solution to this problem is being pursued through the use of a keto-containing amino acid analog (67). Once an *in vivo* system for incorporating this analog is developed, the unique chemical reactivity of the sidechain can be used to subsequently attach any number of functionalities, provided that the sidechain is exposed. Such a system would have tremendous utility for producing large amounts of proteins containing unnatural amino acids.

## V. COMBINED SYNTHETIC AND BIOLOGICAL

The most direct method for incorporating an unnatural amino acid into a protein might be to chemically synthesize the polypeptide. Limitations to chemical synthesis make this impossible for almost all proteins of biological interest. An interesting approach which combines peptide synthesis with *in vivo* expression has been described (68). The method uses a mutant form of the protein subtilisin which is defective in one of the catalytic residues responsible for proteolysis. While defective in hydrolysis, the protein is capable of performing the reverse reaction and ligating two peptides under appropriate conditions. This method has been used to ligate a variety of synthetic peptides containing unnatural amino acids onto human growth hormone and atrial natriuretic peptide (69). The method has also been used to construct full length and active RNaseA from a series of synthetic peptides containing amino acid analogs (70). While some technical limitations currently exist for this technology, it offers the potential for larger scale *in vitro* synthesis than *in vitro* translation methods and for incorporation of D-amino acids and other compounds not efficiently incorporated by ribosomes.

## VI. SUMMARY

While the general utility of the various methods for introducing unnatural amino acids into proteins has yet to be fully realized, the concept of one codon for one amino acid appears to be an oversimplified paradigm. Clearly, the natural world values diversity and can violate this basic tenet. Such is the case for incorporation of selenocysteine and formyl-methionine, through the use of alternate pathways and translation context or through post-translational modification of proteins. Why nature has chosen to maintain the redundancy in the genetic code at the expense of additional diversity is unclear. With the numerous methods described, it is now possible to introduce any number of functional groups into proteins. The limitation remaining is in the imagination and resources of the investigator.

## REFERENCES

1. MK Trower. In Vitro Mutagenesis Protocols. Methods in Molecular Biology. Vol. 57. Totowa, NJ: Humana Press, 1996.
2. WP Stemmer. DNA shuffling by random fragmentation and reassembly: in vitro recombination for molecular evolution. Proc Natl Acad Sci USA 91: 10747–10751, 1994.
3. D Hatfield, A Diamond. UGA: split personality in the genetic code. Trends Genet. 9: 69–70, 1993.
4. J Normanly, LG Kleina, J-M Masson, JH Miller. Construction of Escherichia coli amber suppressor tRNA genes III. Determination of tRNA specificity. J Mol Biol 213: 719–726, 1990.
5. LG Kleina, J-M Masson, J Normanly, J Abelson, JH Miller. Construction of Escherichia coli amber suppressor tRNA genes II. Synthesis of additional tRNA genes and improvement of suppressor efficiency. J Mol Biol 213: 705–717, 1990.
6. S Lesley. Analysis of point mutations by use of amber stop codon suppression. in In Vitro Mutagenesis Protocols. Methods in Molecular Biology. Vol. 57. Totowa, NJ: Humana Press, 1996, pp 65–73.
7. UL Rajbhandary. Initiator transfer RNAs. J Bacteriol 176: 547–552, 1994.
8. TC Stadtman. Selenocysteine. Ann Rev Biochem 65: 83–100, 1996.
9. PM Sharp, E Cowe, DG Higgins, DC Shields, KH Wolfe, F Wright. Codon usage patterns in Escherichia coli, Bacillus subtilis, Saccharomyces cerevisiae, Schizosaccharomyces pombe, Drosophila melanogaster and Homo sapiens; a review of the considerable within-species diversity. Nucl Acids Res 16: 8207–8211, 1988.
10. F Zinoni, J Heider, A Bock. Features of the formate dehydrogenase mRNA necessary for decoding of the UGA codon as selenocysteine. Prod Natl Acad Sci USA 87: 4660–4664, 1990.
11. J Heider, C Baron, A Bock. Coding from a distance: dissection of the mRNA determinants required for the incorporation of selenocysteine into protein. EMBO J 11: 3759–3766, 1992.
12. G-FT Chen, L Fang, M Inouye. Effect of the relative position of the UGA codon to the unique secondary structure in the fdhF mRNA on its decoding by selenocysteinyl tRNA in Escherichia coli. J Biol Chem 268: 23128–231, 1993.
13. PA Young, II Kaiser. Aminoacylation of Escherichia coli cysteine tRNA by selenocysteine. Arch Biochem Biophys 171: 483–489, 1975.
14. W Leinfelder, E Zehelin, M-A Mandrand-Berthelot, A Bock. Gene for a novel tRNA species that accepts L-serine and cotranslationally inserts selenocysteine. Nature 331: 723–735, 1988.
15. K Forchhammer, A Bock. Selenocysteine synthase from Escherichia coli. Analysis of the reaction sequence. J Biol Chem 266: 6324–6328, 1991
16. RS Glass, WP Singh, W Jung, Z Veres, TD Scholz, TC Stadtman. Monoselenophosphate: synthesis, characterization, and identity with the prokaryotic biological selenium donor, compound SePX. Biochemistry 32: 12555–12559, 1993.
17. K Forchhammer, W Leinfelder, A Bock. Identification of a novel translation factor necessary for the incorporation of seleno-

cysteine into protein. Nature 342: 453–456, 1989.

18. K Forchhammer, P Rucknagel, A Bock. Purification and biochemical characterization of SELB, a translation factor involved in selenoprotein synthesis. J Biol Chem 265: 9346–9350, 1990.

19. D Samols, CG Thornton, VL Murtif, GK Kumar, FC Hasse, HG Wood. Evolutionary conservation among biotin enzymes. J Biol Chem 263: 6641–0000, 1988.

20. KP Wilson, LM Shewchuk, RG Brennan, AJ Otsuka, BW Matthews. Escherichia coli biotin holoenzyme synthetase/bio repressor crystal structure delineates the biotin and DNA-binding domains. Proc Natl Acad Sci USA 89: 9257–0000, 1994.

21. JE Cronan Jr. Biotinylation of proteins in vivo. A post-translational modification to label, purify, and study proteins. J Biol Chem 265: 10327–0000, 1990.

22. PJ Schatz. Use of peptide libraries to map the substrate specificity of a peptide- modifying enzyme: a 13 residue consensus peptide specifies biotinylation in Escherichia coli. Biotechnology 11: 1138–1143, 1993.

23. R Schmitz, G Baumann, H Gram. Catalytic specificity of phosphotyrosine kinases Blk, Lyn, c-Src and Syk as assessed by phage display. J Mol Biol 260: 664–77, 1996

24. P Schimmel. An operational RNA code for amino acids and variations in critical nucleotide sequences in evolution. J Mol Evol 40: 531–36, 1995.

25. JA Ellman, D Mendel, PG Schultz. Site-specific incoroporation of novel backbone structures into proteins. Science 255: 197–200, 1992.

26. PM Bronskill, JT Won. Suppression of fluorescence of tryptophan residues in proteins by replacement with 4-fluorotryptophan. Biochem J 249: 305–308, 1988.

27. SP Williams, PM Haggie, KM Brindle. 19F NMR measurerments of the rotational mobility of proteins in vivo. Acta Otorhinolaryngol Belg 490–498, 1997.

28. LA Luck, JE Vance, TM O'Connell, RE London. 19F NMR relaxation studies on 5-fluortryptophan and tetradeutero-5-fluorotryptophan-labeled E. coli glucose/galactose receptor. J Biomol NMR 7: 261–272, 1996.

29. HT Truong, EA Pratt, GS Rule, PY Hsue, C Ho. Inactive and termperature-sensitive folding mutants generated by tryptophan substitutions in the membrane-bound d-lactate dehydrogenase of Escherichia coli. Biochem 30: 10722–10729, 1991.

30. HW Kim, JA Perez, SJ Ferguson, ID Campbell. The specific incorporation of labelled aromatic amino acids into proteins through growth of bacteria in the presence of glyphosate. Application to fluorotryptophan labelling to the H(+)-ATPase of Escherichia coli and NMR studies. FEBS Lett 272: 34–36, 1990.

31. M Ring, RE Huber. The properties of beta-galactosidases (Escherichia coli) with halogenated tyrosines. Biochem. Cell Bio. 71: 127–132, 1993.

32. AE Johnson, WR Woodward, E Herbert, JR Menninger. Nepsilon-acetyllysine transfer ribonucleic acid: a biologically active analogue of aminoacyl transfer ribonucleic acids. Biochemistry 15: 569–575, 1975.

33. UC Krieg, P Walter, AE Johnson. Photocrosslinking of the signal sequence of nascent preprolactin to the 54-kilodalton polypeptide of the signal recognition particle. Proc Natl Acad Sci USA 83: 8604–8608, 1986.

34. TV Kurzchalia, M. Wiedmann, H Breter, W Zimmermann, E Bauschke, TA Rapoport. tRNA-mediated labelling of proteins with biotin. Eur J Biochem 172: 663–668, 1988.

35. CJ Noren, SJ Anthony-Cahill, MC Griffity, PG Schultz. A general method for site-specific incorporation of unnatural amino acids into proteins. Science 244: 182–188, 1989.

36. M Nirenberg, JH Matthaei. The dependence of cell-free protein synthesis in E. coli upon naturally occurring or synthetic polyribonucleotides. Proc Natl Acad Sci USA. 47: 1588–1602, 1961.

37. G Zubay. In vitro synthesis of protein in microbial systems. Ann Rev Genet 7: 267–287, 1973.

38. SA Lesley, MAD Brow, RR Burgess. Use of in vitro protein synthesis from polymerase chain reaction-generated templates to study interaction of Eschericia coli transcription factors with core RNA polymerase and for epitope mapping of monoclonal antibodies. J Biol Chem 266: 2632–2638, 1991.

39. DE Nevin, JM Pratt. A coupled in vitro transcription-translation system for the exclusive synthesis of polypeptides expressed from the T7 promoter. FEBS Lett 291: 259–263, 1991.

40. D Craig, MT Howell, CL Gibb, T Hunt, RJ Jackson. Plasmid cDNA-directed protein synthesis in a coupled eukaryotic in vitro transcription-translation system. Nucl Acids Res 20: 4987–4995, 1993.

41. M Kozak. Comparison of initiation of protein synthesis in prokaryotes, eukaryotes and organelles. Microb Rev 47: 1–45, 1983.

42. AS Spirin, VI Baranov, LA Ryabova, SY Ovodov, YB Alakhov. A continuous cell-free translation system capable of

producing polypeptides in high yield. Science 242: 1162–1164, 1988.

43. E Resto, A Iida, MD Van Cleve, SM Hecht. Amplification of protein expression in a cell free system. Nucl Acids Res 20: 5979–5983, 1992.

44. J Brunner. New photolabeling and cross-linking methods. Ann Rev Biochem 62: 483–514, 1993.

45. Reviewed in Meth Enzymol v: 184, 1990.

46. NM Green, EJ Toms. The properties of subunits of avidin coupled to Sepharose. Biochem J 133: 687–698, 1973.

47. BD Hamman, JC Chen, EE Johnson, AE Johnson. The aqueous pore through the translocon has a diameter of 40-60 A during cotranslational protein translocation at the ER membrane. Cell 89: 535–544, 1997.

48. L Brand, B Witholt Meth Enzymol 11: 776-000, 1967.

49. J Ellman, D Mendel, S Anthony-Cahill, CJ Noren, PG Schultz. Biosynthetic method for introducing unnatural amino acids site-specifically into proteins. Meth Enzymol 202: 301–336, 1991.

50. D Mendel, VW Cornish, PG Schultz. Site-directed mutagenesis with an expanded genetic code. Ann Rev Biophys Biomol Struct 24: 435–462, 1995.

51. JH Miller. Genetic studies of the lac repressor XI. On aspects of lac repressor structure suggested by genetic experiments. J Mol Biol 131: 249–258, 1979.

52. Y Kwok, JT Wong. Evolutionary relationship between Halobacterium cutirubrum and eukaryotes determined by use of aminoacyl-tRNA synthetases as phylogenetic probes. Can J Biochem 58: 213–218, 1980.

53. SA Robertson, JA Ellman, PG Schultz. A general and efficient route for chemical aminoacylation of transfer RNAs. J Am Chem Soc 113: 2722–2729, 1991.

54. JD Bain, CG Glabe, TA Dix, AR Chamberlin, ES Diala. Biosynthetic site-specific incorporation of a non-natural amino acid into a polypeptide. J Am Chem Soc 111: 8013–8014, 1989.

55. JD Bain, Da Wacker, EE Kuo, AR Chamberlin. Site-specific incorporation of non-natural residues into peptides: effect of residue structure on suppression and translational efficiencies. Tetrahedron 47: 2389-2400, 1991.

56. D Mendel, J Ellman, PG Schultz. Protein biosynthesis with conformationally restricted amino acids. J Am Chem Soc 115: 4359–4360, 1993.

57. S Fahnestock, H Neumann, V Shashoua, A Rich. Ribosome-catalyzed ester formation. Biochemistry 9: 2477–2483, 1970.

58. VW Cornish, D Mendel, PG Schultz. Probing protein structure and function with an expanded genetic code. Angew Chem Int Ed Engl, ???, 1995

59. D Rennell, SE Bouvier, LW Hardy, AR Poteete. Systematic mutation of bacteriophage T4 lysozyme. J Mol Biol 222: 67–88, 1991.

60. M Karpusas, WA Baase, M Matsumura, BW Matthews. Hydrophobic packing in T4 lysozyme probed by cavity-filling mutants. Proc Natl Acad Sci USA 86: 8237–8241, 1989.

61. D Mendel, JA Ellman, Z Chang, DL Veenstra, PA Kollman, PG Schultz. Probing protein stability with unnatural amino acids. Science 256: 1798–1802, 1992.

62. VW Cornish, DR Benson, CA Altenbach, K Hideg, WL Hubbell, PG Schultz. Site-specific incorporation of biophysical probes into proteins. Proc Natl Acad Sci USA 91: 2910–2914, 1994.

63. JA Ellman, BF Volkman, D Mendel, PG Schultz, DE Wemmer. Site-specific isotopic labeling of proteins for NMR studies. J Am Chem Soc 114: 7959-7961, 1992.

64. DW Hibler, NJ Stolowich, MA Reynolds, JA Gerlt, JA Wilde, PH Bolton. Site-directed mutants of staphylococcal nuclease. Detection and localization by 1H NMR spectroscopy of conformational changes accompanying substitutions for glutamic acid 43. Biochemistry 26: 6278–6286, 1987.

65. MW Nowak, PC Kearny, JR Sampson, ME Saks, GC Lavarca. Side chain contributions at the nicotinic receptor binding site probed with unnatural amino-acid incorporation in intact cells. Science 268: 439–442, 1995.

66. DR Liu, TJ Magliery, M Pastrnak, PG Schultz. Engineerining a tRNA and aminoacyl-tRNA synthetase for the site-specific incorporation of unnatural amino acids into protein in vivo. Proc Natl Acad Sci USA 94: 10092–10097, 1997.

67. VW Cornish, KM Hahn, PG Schultz. Site-specific protein modification using a ketone handle. J Am Chem Soc 118: 8150–8151, 1996

68. RE Offord. Protein engineering by chemical means? Protein Eng 1: 151–157, 1987.

69. TK Chang, DY Jackson, JP Burnier, JA Wells. Subtiligase: a tool for semisynthesis of proteins. Pro Natl Acad Sci USA 91: 12544–12548, 1994.

70. DY Jackson, J Burnier, C Quan. M Stanley, J Tom, JA Wells. A designed peptide ligase for total synthesis of ribonuclease A with unnatural catalytic residues. Science 266: 243–247, 1994.

# 7

# Combinatorial Chemistry and the Structure–Function Analysis of Proteins

**Rachel L. Winston[1] and Michael C. Fitzgerald[2]***
[1] *The Scripps Research Institute, La Jolla, California*
[2] *Duke University, Durham, North Carolina*

## I. INTRODUCTION

Combinatorial chemistry methods provide an attractive means to study the structural and functional properties of proteins. Using combinatorial methods large numbers of compounds including small organic molecules, peptides, other proteins, oligonucleotides, and oligosaccharides can be screened for their interaction with proteins. This process permits a comprehensive evaluation of protein binding and protein recognition properties producing a wealth of information about structure-function relationships in proteins. Moreover, screening such diverse compound libraries for protein function can uncover information not readily acquired by the rational design of protein ligands.

The first combinatorial syntheses involved the construction of peptide and oligonucleotide libraries, exploiting well developed protocols for the chemical synthesis of these compounds. These applications were extended to drug discovery efforts, as medicinal chemists developed synthetic methodologies to create chemically diverse, small molecule libraries. More recently, combinatorial strategies have been extended to other molecules whose chemical synthesis has been more technically challenging. For example, the combinatorial synthesis of proteins and carbohydrates have recently been reported.

Critical to the success of combinatorial approaches has been the ability to unambiguously characterize individual library members. Classical analytical techniques, such as nuclear magnetic resonance (NMR) spectroscopy, infrared (IR) spectroscopy, and mass spectrometry have enjoyed some success in the characterization of multi-component libraries. However, as libraries become larger and more complex their analytical characterization becomes exceedingly difficult. To this end, a variety of encoding schemes and pooling strategies have been developed. In particular, molecular tagging schemes have revolutionized the

*To whom correspondance should be addressed: Michael C. Fitzgerald, Department of Chemistry, Box 90346, Duke University, Durham, North Carolina 27708-0346
Phone: (919)660-1547 Fax: (919)660-1605
Email: mfitz@chem.duke.edu

way in which one-bead, one-compound libraries are currently characterized. These tagging schemes have made it possible to rapidly accumulate information about a wide range of protein ligand interactions.

Here we summarize the most popular strategies for the combinatorial synthesis and characterization of small molecule, peptide, protein, oligonucleotide, and carbohydrate libraries. We also highlight studies in the current literature that have used combinatorial chemistry to define binding motifs, and to characterize unknown biological activities in proteins.

## II.  SMALL MOLECULE LIBRARIES

The ability to chemically synthesize and screen large libraries of small organic molecules for their interaction with proteins has led to the discovery of new drugs that can modulate a wide range of biological processes. These studies have also contributed to our fundamental understanding of structure-activity relationships in proteins. Here, we discuss popular synthetic approaches and encoding strategies used to generate small molecule libraries [for a more comprehensive review see (1)]. We also highlight several examples from the current literature where such libraries have been used to study protein function.

### A.  Solution-Phase Synthesis and Characterization

Solution-phase strategies for the synthesis of small molecule libraries have relied on the parallel construction of multiple sub-libraries to generate complex mixtures of compounds. The identification of active compounds in these mixtures is accomplished through an iterative process of synthesis and screening. For example, the compounds in a combinatorial, solution-phase synthesis are generally

synthesized in multiple pools. After screening these different pools for biological activity, the compounds in an active pool are re-synthesized in a number of smaller pools and re-screened for activity. Ultimately, this process makes possible the identification of active compounds by decreasing the absolute number of compounds per pool. After several rounds of "sub-pooling," discrete compounds are eventually synthesized and screened for their biological activity.

### 1.  Chemical Indexing

In order to facilitate the identification of active compounds in solution-phase libraries, systematic approaches to sub-library construction have been developed. In one approach known as chemical indexing (2), multiple libraries are generated from a series of defined sub-libraries. This approach is outlined in Fig. 1 for a case in which two sets of building blocks, A and B, are used in a combinatorial synthesis. In this example, A and B are each defined by a different set of 20 related compounds. Two libraries containing 20 spatially separate pools of 20 compounds are created. In one library, 20 sub-libraries are generated by reacting each compound from A with a mixture of all 20 compounds from B. In a second library, 20 sub-libraries are prepared by reacting each compound from B with a mixture of all 20 compounds from A. By screening each one of the 20 sub-libraries in the first library, it is possible to obtain the best building block from B. Likewise, by screening each one of the 20 sub-libraries in the second library, it is possible to obtain the best building block from A. Once the optimal building blocks are defined, the corresponding compound can be synthesized. This approach has been used by Pirrung and coworkers to synthesize a 54 member carbamate library from nine alcohols and six isocyanates (2), and to synthesize a 72 tetrahydroacridine

# Library One

## (20 sub-libraries)

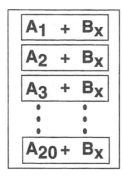

**Bx= all 20 compounds from B**

# Library Two

## (20 sub-libraries)

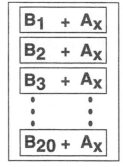

**Ax= all 20 compounds from A**

**Fig. 1** Chemical Indexing. Two libraries are generated by reacting molecules from one set of related compounds, A, with another set of related compounds, B. Library one consists of 20 sub-libraries, where a single compound from A is reacted with all 20 compounds from B. Library two also consists of 20 sub-libraries, where a single compound from B is reacted with all 20 compounds from A.

library from 12 cyclohexanones and six *o*-cyanoaniline derivatives (3). From these libraries, novel acetylcholinesterase inhibitors were identified. Similarly, Smith and coworkers have applied this strategy to create a 1600 member library of esters and amides (4) by reacting 40 acid chlorides with 40 alcohols and 40 amines, respectively. From these libraries, two biologically active compounds were identified.

## 2. NMR-Based Approaches

In an alternative solution-phase approach, high affinity ligands are "built-up" by linking together individually optimized small molecule fragments. Termed "SAR by NMR," this approach relies on NMR techniques to screen small molecule libraries for binding with a target protein (5, 6). As a first step, the target [15]N-labeled protein is screened against a library of low molecular weight compounds, using a heteronuclear NMR experiment to identify molecules that bind the protein. To

optimize binding of the newly discovered ligand, a series of ligand analogues are individually prepared and assayed. The tightest binding analogue becomes the first "link" in this molecular building block approach. Next, a nearby sight on the protein is screened for ligand binding in the same way, by monitoring a different set of [15]N amide chemical shift changes. The optimal compound from this second screen becomes the second "link" in the chain. In order to optimize the connection between building blocks one and two, the three dimensional structure of the protein with both ligands bound is determined. In a final step, a series of tethered compounds containing both building blocks are designed and screened for optimal binding. In this way, only a limited number of tethered molecules are required for building block selection and optimization. The feasibility of this technique was first demonstrated with FK506 binding protein (FKBP). In under two months, two ligands with micromolar affinities were identified and converted to five tethered compounds, all of which exhibited nanomolar affinities for FKBP. The approach has also been used to identify a tight binding inhibitor to a metalloproteinase (7).

Recently, another NMR-based method was developed to identify active ligands from small molecule libraries (8). This approach exploits changes in the relaxation or diffusion rates of a small molecule that occur upon protein binding. By subtracting the NMR spectrum obtained for a set of *ligands* in the presence and in the absence of protein, a bound ligand can be identified from a small mixture of compounds (Fig. 2). The subtraction spectrum (Fig. 2D) depicts signals of only the bound ligand, simplifying the determination of bound ligand structures. This strategy requires only a micromolar quantity of unlabeled protein, in contrast to the millimolar quantity of $^{15}$N-labeled protein needed for conventional protein NMR analyses. Also, protein size is not a factor because the method relies on detection of the ligand NMR signal. Currently, the approach is useful for the analysis of relatively small libraries containing a single active compound.

## B. Solid-Phase Synthesis and Characterization

In contrast to solution-phase methods for the synthesis of small molecule libraries, solid-phase methods are amenable to the construction of large libraries and to the straightforward identification of individual library components. Support-bound compounds can be prepared in spatially addressable formats, such as on the tips of fine pins (9), on polymeric membranes (10), or on resin beads in the wells of microtiter plates (11). Such parallel synthesis strategies are convenient because library compounds are easily identified by their particular location. However, their application to more extensive libraries (>10,000 components) can be exceedingly tedious and impractical. In an effort to

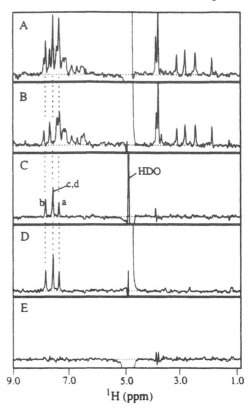

**Fig. 2** Combinatorial analysis of ligand binding to FKBP using NMR. Relaxation-edited proton NMR of a the nine component small molecule library in the absence (A) and in the presence (B) of protein. A difference spectrum obtained by subtracting the spectrum in B from A (C). Reference spectrum of 2-phenylimidazole (D). A difference spectrum obtained in an analogous fashion to the spectrum shown in C, but only on the eight non-binding compounds in the library (E). (Reprinted with permission from PJ Hajduk, ET Olejniczak, SW Fesik. One dimensional relaxation- and diffusion-edited NMR methods for screening compounds that bind to macromolecules. J Am Chem Soc 119: 12259–12261, 1997. Copyright 1997 American Chemical Society.

overcome these limitations, Fodor and co-workers pioneered a parallel synthesis strategy that employs photolithographic methods and that is amenable to the synthesis of over 100,000 spatially separate compounds (12).

## 1. One-Bead, One-Compound Libraries

Solid-phase methods are also amenable to split resin procedures. One of the more popular strategies, pioneered by Furka, relies on a "split-and-mix" protocol (13, 14). In this approach, resin beads are split into equal portions before coupling individual elements of chemical diversity (Fig. 3). After each portion of beads is reacted with a different building block, the beads from all the reactions are recombined. Repeating this procedure produces a combinatorial library in which each resin bead contains a unique compound.

Classical analytical techniques such as mass spectrometry (15–19), NMR (20, 21), and IR (22–25) have been used to directly characterize resin-bound compounds. For example, matrix-assisted laser desorption/ionization (MALDI) mass spectrometry has been used to directly characterize compounds linked to a solid support through a photocleaveable linkage (19). Fig. 4 shows a MALDI mass spectrum of bromoacetamide attached to a polystyrene resin through a multifunctional linker. The combination of a

photocleavable linker and an ionization tag permits the MALDI analysis of individual compounds while still covalently attached to resin beads. In the MALDI experiment, ultraviolet laser light photolytically cleaves the analyte from the resin and, simultaneously, promotes its gas phase ionization for subsequent mass spectral analysis.

As libraries become larger and more complex, their analytical characterization becomes exceedingly difficult. In large libraries the NMR, IR, and mass spectrometry signals of closely related compounds are not easily resolved. Thus, these analytical techniques have only lent themselves to the analysis of relatively small, structurally diverse libraries.

## 2. Encoding Strategies

As the complexity and size of "one-bead, one-compound" libraries increase, a convenient encoding strategy is required to characterize individual library members. The first encoding strategies that were developed involved the use of amino acids (26) or nucleotides (27–29) to encode the resin-bound compounds. As diversity was introduced into the library, sequences of amino acids or nucleotides were used to define a particular chemical building block. Ultimately, the complete peptide or DNA sequence defined the resin-bound compound. Subsequent microsequencing of the peptide, or PCR amplification and sequencing of the oligonucleotide was used to identify library members. One drawback to this methodology is that it is limited to library chemistry that is compatible with peptide or oligonucleotide synthesis.

In 1993, Still and co-workers developed a more chemically robust method for encoding small molecule libraries (30). Their approach involved the use of chemically inert molecular tags in a "binary encoding" scheme, which minimized the number of tags needed to

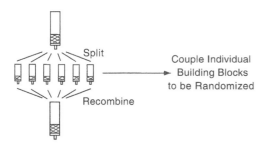

**Fig. 3** Split and mix strategy. Resin beads are divided into equal portions and placed in separate reaction vessels. Each portion of beads is reacted with a unique building block and then recombined into a single reaction vessel.

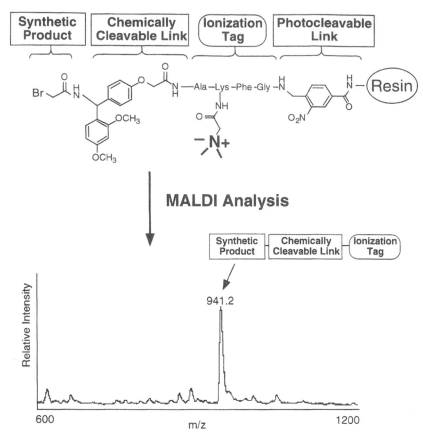

**Fig. 4** Direct analysis of a resin-bound compound using MALDI mass spectrometry. The bromoacetamide molecule in covalently linked to the resin by a multifunctional linker consisting of a chemically cleavable linker, an ionization tag, and a photocleavable linkage. The photocleavable linkage is cleaved during MALDI analysis to generate a molecular ion of the bromoacetamide, the chemical linker, and the ionization tag as an intact species.

encode a particular synthesis. In their seminal work, Still and co-workers constructed a solid-phase combinatorial library encoded by 18 haloaromatic compounds (T1–T18). These compounds were arranged in order of their retention time on a gas chromatography (GC) column, such that T1 had the longest retention time and T18 had the shortest retention time. The molecular tags were used to encode both the step number of the synthesis and the chemical reagent used in that step. This was accomplished by assigning a three bit binary code to each reagent: Reagent 1: 100, Reagent 2: 010, Reagent 3: 001,

Reagent 4: 110, Reagent 5: 011, Reagent 6: 101, and Reagent 7: 111. After completing their six step synthesis, each library member had an 18 bit code. The first three tags (T1–T3) were used to define bits in the first synthetic step, such that T1 was placed at position one, T2 at position 2, and T3 at position 3 of the reagent's 3 bit code, arranged from right to left. T4–T6 encoded the reagent used in the second step, T7–T9 encode the reagent used in the third step, etc., such that 18 tags were used to represent all 18 bits in the library code. When a "1" was present in the reagent code, then a small portion of the resin was functionalized with the appropriate tag. When a "0" was present, the tag was omitted. For example, when Reagent 3 (with a binary code of 001) was used in the first synthetic step, then only the molecular tag T1 was reacted with the resin (T2 and T3 were absent). After the split and mix synthesis was complete, the reaction history of each bead in the library could be obtained by GC analysis of the cleaved tags (Fig. 5). This approach has been widely adopted by a number of researchers for encoding small molecule libraries. We highlight a recent application below.

The utility of molecular tagging to encode combinatorial libraries has inspired the development and application of different tagging schemes. For example,

stable isotopes readily analyzed by mass spectrometry have been used as tags (31). Radiofrequency tags have also been used to encode resin-bound compounds (32–34). Radiofrequency Encoded Combinatorial (REC$^{TM}$) chemistry uses a semiconductor memory device (SMART) to record a unique radiofrequency signal for each molecular building block in a split and mix synthesis. To date, this strategy has been used to encode combinatorial peptide libraries and a taxol-based library. In another approach, optical lasers have been employed to graft a bar code encoding the reaction history of individual polystyrene chips (35). These emerging encoding strategies have potential applications for virtually any library synthesis

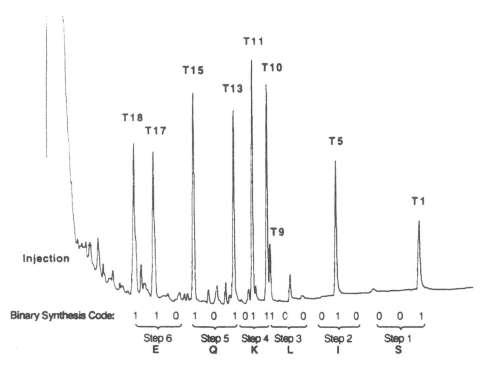

**Fig. 5** Gas chromatogram of the molecular tags (T1–T18) cleaved from a single bead. The presence or absence of tag at its expected retention time represents a "1" or "0" in the binary code used to define each reagent and each step in the synthesis. (Reprinted with permission from MJH Ohlmeyer, RN Swanson, LW Dillard, JC Reader, G Asouline, R Kobayashi, M Wigler, WC Still. Complex Synthetic Chemical Libraries Indexed with Molecular Tags. Proc Natl Acad Sci USA 90: 10922–10926, 1993. Copyright 1993 National Academy of Sciences, USA).

because the encoding signals do not chemically interfere with library synthesis.

## C.  Applications

### 1.  Small Molecule Ligands for the Src SH3 Domain

In order to better understand the properties of SH3 domains, Schreiber and co-workers applied a combinatorial approach to find a family of ligands capable of binding to the SH3 domain of Src [reviewed by (36)]. They synthesized a combinatorial library containing a polyproline sequence as a base sequence, and then incorporated 32 small molecule substituents during three consecutive cycles of split and mix synthesis (37). This library was encoded using the binary encoding scheme described above. Screening a library of over million compounds for binding activity identified fifteen ligands specific for the Src SH3 domain. Interestingly, none of these compounds would have been predicted to bind Src based on analysis of its three-dimensional structure.

In a follow-up study, NMR was used to investigate the molecular basis of how the small molecule part of several of these ligands (NL1, NL2, VSL12) interacted with Src (38). From this work, it was found that the polyproline region was responsible for directing ligand binding to the "specificity pocket" of the protein. However, the non-peptide part of each ligand was found to interact uniquely with the protein. For example, NL1 interacted primarily through hydrophobic contacts and $\pi$-$\pi$ packing. In contrast, VSL12 had no aromatic interactions and was stabilized by a salt bridge worth an estimated 2.1 kcal/mol of free energy. Significantly, NL1 and NL2 are the first contiguous ligands which interact with the RT loop, an important regulatory region of Src. Through their unique interactions, these non-peptide ligands may be useful probes for investigating Src regulation.

### 2.  Future Directions

Combinatorial chemistry is beginning to find applications in cell-based assays. Schreiber and co-workers have described a new approach called "chemical genetics," where protein function is altered directly through the use of a small molecule ligand (39). As part of this work, they pioneered a nanospray technology that allows resin beads to be sprayed with cells into a fine mist (39–41), such that droplets contain a controlled number of cells and a single resin bead. In a model experiment, rapamycin, a small molecule which binds the target protein FKBP12, was linked to resin beads via a photo-cleavable linker. Rapamycin was released from the resin by UV irradiation, and assayed for cell growth, or inhibition of cell growth using genetically engineered yeast cells. Using this approach, they were able to detect rapamycin binding to FKBP12. A particularly exciting application of this technique will be to screen small molecule libraries for protein interaction. In this way, ligands capable of entering the cell and binding a protein of interest can be identified.

## III.  PEPTIDE LIBRARIES

The preparation of peptide libraries by either chemical or biological means has been invaluable in obtaining important information about how proteins recognize specific peptide sequences. Many proteins such as proteolytic enzymes, antibodies, and a host of cellular receptors bind to peptide sequences to carry out a myriad of biological process. Characterizing the binding motifs of these proteins can help elucidate their biological activity.

The successful generation of peptide libraries by chemical methods has relied on solid-phase peptide synthesis (SPPS) approaches, while phage display has been the most popular recombinant

DNA-based approach. Phage display technology has proven immensely powerful in the identification of peptide recognition sequences for a wide variety of protein systems; these methods have been extensively reviewed elsewhere (42, 43). Here, we will focus on novel synthetic methodologies and analytical methods used for the synthesis and characterisation of chemically generated combinatorial peptide libraries. We will also highlight several examples where combinatorial-based approaches have provided unique insight into describing the molecular basis of protein-peptide interactions.

## A. Chemical Synthesis and Characterization

Solid-phase peptide synthesis (SPPS) methods are well suited for the preparation of combinatorial libraries. Two notable synthesis strategies are the split and mix procedure, described above (13, 14), and the positional scanning procedure described below (44–46).

### 1. Positional Scanning

In the positional scanning procedure, a "library of libraries" is generated, where one position in the polypeptide sequence is defined with an amino acid, while all other positions contain an approximately equimolar mixture of amino acids (47). Significantly, there is no encoding strategy. Rather, multiple libraries are synthesized in which the invariant amino acid is systematically scanned throughout the sequence. Fig. 6 illustrates an example in which the positional scanning approach is used to generate a combinatorial tri-peptide library. In this example, three libraries are prepared such that each library contains one position with a defined amino acid. The other two positions are randomized with all twenty amino acids, creating a mixture of 400

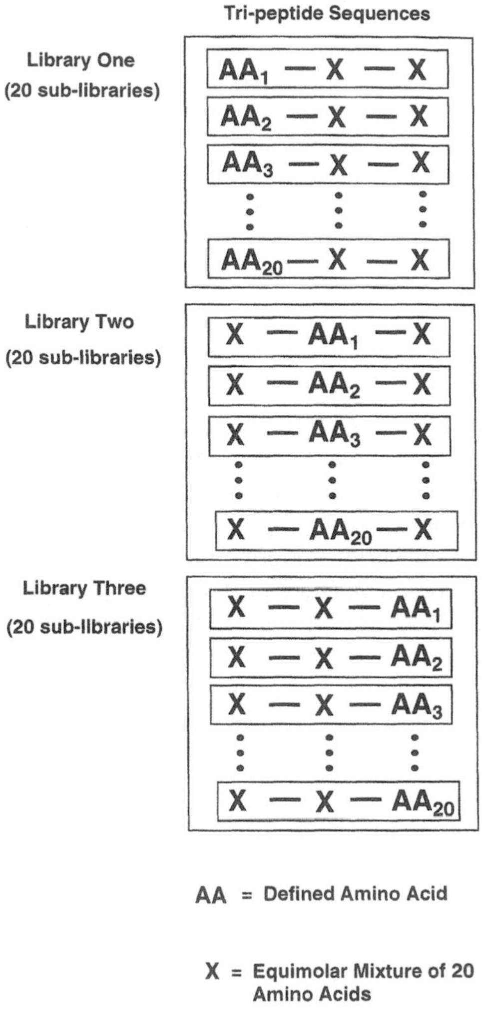

**Fig. 6** Positional scanning for a tri-peptide library. Library one contains a defined amino acid at position one, with a mixture of all 20 amino acids at positions two and three of the tri-peptide. Libraries two and three are generated in a similar fashion by placing the defined amino acid at positions two and three, respectively. A different amino acid is incorporated at the defined position in each of the sub-libraries.

compounds $(1 \times 20 \times 20)$. To create additional diversity, the identity of the defined amino acid in each library is altered, such that all 20 amino acids are represented in a parallel fashion. This process ultimately generates 20 sub-libraries, or a total of 8000 compounds, in each library (20 sub-libraries containing 400 compounds each). By screening each sub--library, amino acids that are tolerated at each position in the sequence are identified. This approach has been used recently to identify substrate-analogue inhibitors (48) and to determine the substrate specificity of a proteolytic enzyme (49).

## 2. *Encoding Strategies*

Peptides from one-bead, one compound libraries can be sequenced by classical Edman degradation chemistry (50). However, this approach requires a relatively large amount of material (ca. 25 picomoles), and it fails when the peptide is capped at the N-terminus. To circumvent these limitations, several different encoding strategies have been developed specifically for use with peptide libraries. Youngquist and co-workers have presented a strategy which uses mass spectrometry to read out sequence information (51). In this approach, a small amount of the growing polypeptide chain is capped at each step, generating a peptide ladder encoding the polypeptide sequence (Fig. 7). The peptide ladder is analyzed by MALDI mass spectrometry, and the sequence is determined from the mass differences between termination products.

More recently, an encoding strategy, termed "encoded amino acid scanning," was reported by Muir and co-workers (52). As part of this new approach, standard SPPS methods are used to prepare a small array of peptides which differ by a single mutation. In order to encode both the position and the chemical nature of the modification, a fluorenylmethoxycarbonyl (Fmoc) protected amino acid is attached to the peptide through a selectively cleavable bond. Each amino acid tag has a unique HPLC retention time and can be conveniently monitored at a wavelength of 300 nm. After functional analysis of the

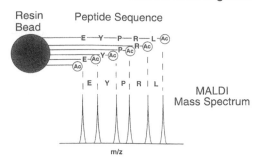

**Fig. 7** Ladder encoding strategy for peptide libraries. A small portion of the peptide is capped with acetylated alanine during amino acid coupling, generating a peptide ladder. These truncated sequences are characterized by MALDI-MS.

peptide library, the tags are selectively cleaved and subjected to HPLC analysis. Because there is a 1:1 ratio of peptide to tag, the results of HPLC analysis correlate directly with the amount of peptide present in the mixture. In this way, it is possible to determine the relative binding affinities of the entire peptide array in a single step.

## C. Applications

### 1. *SH3 Domain Ligand Specificity*

In the early 1990's, there was substantial evidence that SH3 domains had a general affinity for polyproline-rich ligands; however, there was no known high affinity SH3 ligand. To elucidate the general principles of SH3-ligand binding, Schreiber and co-workers prepared a biased combinatorial peptide library, where all peptides contained a core sequence of PPXP (X denotes any amino acid, P denotes proline) (36, 53, 54). The library was screened for binding activity against two structurally unique SH3 domains. Fluorescently labeled PI3K and Src SH3 domains were incubated with resin beads, and the brightest beads were characterized by automated Edman degradation. Out of approximately ten million peptides, two distinct classes of ligands appeared: RXLPP(R/L)PXX and XXXPPLPXR. This discovery led to the hypothesis that there were two different SH3-ligand binding modes.

With two high affinity ligands in hand, they went on to solve the structures of the protein-ligand complexes by NMR, and ultimately piece together the molecular details of SH3-ligand interactions. Interestingly, they found that these two ligand classes do indeed bind in opposing orientations, an unprecedented protein-ligand interaction at that time. These reports illustrate how combinatorial chemistry can compliment structural studies to dissect protein recognition properties.

## 2.  TAP Transporter Peptide Selectivity

In another study, combinatorial peptide chemistry was used to determine the specificity of TAP, a protein that is responsible for transporting peptides as part of the antigen presentation pathway (55). In this work, a combinatorial peptide library was assayed to obtain biologically relevant ligands for TAP. Previous studies investigating TAP-peptide selectivity suggested that TAP was "unselective." However, the results of the combinatorial assay identified peptides with a wide range of binding affinities (137 nM to >1 mM). Screening a second peptide library containing D-amino acids revealed that TAP-peptide binding is stabilized primarily through backbone hydrogen bonds, with significant contributions from side chains at the N- and C-terminal positions. The results of these studies suggest that TAP is indeed selective, and that its mode of binding differs from other peptide-bearing receptors.

## 3.  Antibody Binding Promiscuity

In a recent report, the molecular basis for binding of a monoclonal antibody to HIV p24 (CB4-1) was investigated (56).

Although monoclonal antibodies are thought to have a preferred peptide recognition sequence, the authors identified five unrelated peptides that compete with each other for binding. They screened for CB4-1 binding using a synthetic positional scanning combinatorial peptide library. This search detected 225 peptide mixtures from a total of 68,590 pools. After subsequent iteration and deconvolution, a peptide similar in sequence to a known epitope, three unrelated sequences, and a peptide containing all D-amino acids were isolated. From these analyses, it was shown that each peptide bound to the antibody in a unique way. The authors also designed secondary libraries, based on the sequences of two of these interacting peptides, and assayed them for binding activity. Surprisingly, CB4-1 bound thousands of these peptides. These sequences were subjected to database searching, and matched to known protein sequences. Several of these proteins were assayed for antibody binding, and all those that were tested bound the antibody with micromolar or submicromolar dissociation constants. These results demonstrate the binding promiscuity of the CB4-1 monoclonal antibody, and suggest that antibody recognition is not a strict readout of a unique peptide sequence.

## IV.  PROTEIN LIBRARIES

The combinatorial approaches presented in the previous sections have focused on altering *ligand* structure to collect information about functional properties of a protein. However, a complete understanding of protein function requires studies on structural variants of the protein molecule itself. By altering the covalent structure of a protein in a systematic way, and characterizing the functional effects of such modifications, it is possible to learn more about how individual residues contribute to ligand binding. For example, some amino acids may confer ligand specificity, while others make contributions to ligand affinity. Historically, site directed mutagenesis has been used to pinpoint key residues involved in protein function (57). However, the individual synthesis, purification, and subsequent characterization of multiple protein analogues can be time consuming, and may provide a narrow perspective of how the protein really works.

Recently, two different combinatorial approaches have been developed for the construction of *protein* libraries. One approach relies on genetic methods to randomly introduce mutations throughout the protein sequence. A second approach relies on SPPS methods to introduce mutations into the protein molecule more systematically. The genetic methods have been largely used as part of *in vitro* evolution approaches to improve or alter protein function; they have been discussed elsewhere (58–61). Here, we discuss the chemical methods used for the synthesis and characterization of combinatorial protein libraries.

## A.  Chemical Synthesis and Characterization

The chemical synthesis of an ever growing number of proteins has become possible because of improved protocols (SPPS) (62), and because of the development of various peptide ligation strategies (63–65), (see also this volume, chapter by Cotton and Muir). Currently, stepwise SPPS methods permit the routine preparation of polypeptide chains 50 to 100 amino acids in length. The total chemical synthesis of larger protein constructs (up to 200 amino acids in length) has also been possible using chemoselective ligation strategies (66), such as dovetailing (64) and native chemical ligation (63).

Because protein sequences can be assembled on the solid-phase, split resin strategies can be exploited to generate combinatorial protein libraries. For example, the split and mix approach (outlined above) can be used to generate chemical diversity at a particular position in the polypeptide chain of a protein. However, the application of combinatorial chemistry to proteins has been limited by difficulties associated with characterizing complex protein mixtures. Many chromatographic techniques commonly used for the analytical characterization of proteins (i.e. electrophoresis and HPLC) are not capable of resolving proteins with only slight variation in their covalent structure.

One technique that lends itself to the analysis of complex protein mixtures is MALDI mass spectrometry. In cases where each member of a combinatorial protein library has a unique mass, MALDI mass spectrometry can be used for library characterization. For example, a small library of deletion analogues of a basic-helix-loop-helix (bHLH) transcription factor is easily resolved by MALDI mass spectrometry (Fig. 8). Each peak in the mass spectrum corresponds to a single component of the bHLH protein mixture.

## B.  Protein Signature Analysis

Recently, a new method for the chemical synthesis and readout of "self-encoded" arrays of protein analogues was reported (67, 68). In this approach, termed protein signature analysis (PSA), a modified split resin procedure (69) is used to prepare a mixture of protein analogues. Each analogue in the mixture contains a specific modification at a unique and defined position in the polypeptide chain, encoded by incorporating a selectively cleavable bond into the modification. Following a chemical cleavage step, MALDI analysis of the resulting families of N- and C-terminal

A.

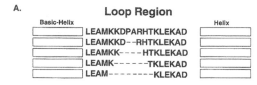

B.

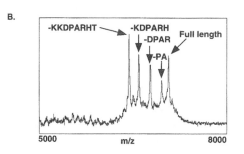

**Fig. 8** MALDI-MS readout of a combinatorial protein library. The wild-type and four deletion analogues of the bHLH domain of Deadpan were synthesized by SPPS as a single product mixture (A). The MALDI mass spectrum obtained for the five component protein mixture (B).

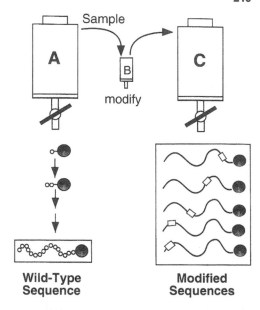

**Fig. 9** Resin shuffling procedure for the synthesis of protein analogue arrays (From Ref. 69).

segments permits the identification of each polypeptide in the library.

The synthetic procedure for generating such defined arrays of protein analogues is shown in Fig. 9. Peptide resin is shuffled between three reaction vessels. Standard SPPS is performed in reaction vessels A and C, and modifications to the sequence are introduced in reaction vessel B. At each point in the synthesis that a modification is desired, a portion of resin is transferred from the primary reaction vessel A, to reaction vessel B. In reaction vessel B, a mutation is made by coupling a "self-encoded" amino acid (i.e. a dipeptide unit with a chemically cleavable bond) to the growing polypeptide chain. This modified peptide resin is transferred to a third reaction vessel C, where the synthesis is completed. Repeated cycles of this procedure result in the synthesis of an array of polypeptides, where each member of the array contains a single mutation in the protein sequence. Following this synthesis, the mixture of synthetic proteins is

subjected to a functional selection (for example, folding or ligand binding) that divides it into two pools—one that is active, and one that is not active. Ultimately, MALDI mass spectrometry is used to unambiguously identify the library components in the active and inactive pools.

## C.   Applications

### 1.   Binding Specificity for the SH3 Domain of c-Crk

Initially, PSA was used to investigate the functional properties of the SH3 domain of c-Crk (68). While several labs have used combinatorial *peptide* libraries to study SH3-ligand interactions, this work involved the synthesis of an SH3 domain *protein* library. Using PSA, a mixture of 19 SH3 analogues were prepared in a single synthetic procedure. During the synthesis, a glycine-S-$\beta$-alanine dipeptide unit was systematically introduced into a twenty amino acid stretch of the SH3 domain.

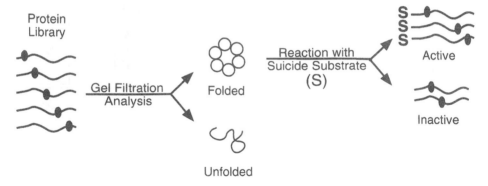

**Fig. 10** Selection assay for 4OT folding and catalysis. An array of 4OT analogues is subjected to gel filtration to separate folded and unfolded material. Folded material is reacted with a suicide substrate to separate catalytically active sequences from inactive sequences.

This modification effectively deleted individual amino acid side chains, two at a time, throughout a region of the protein known to be important for ligand binding. Functional mutants were pulled out of the mixture by affinity chromatography with an immobilized polyproline-rich ligand. The results of the binding assay showed that the dipeptide unit was tolerated at only 8 different positions in the 20 amino acid region that was studied. Moreover, this study pinpointed Asp150 as a key residue involved in ligand binding, a result not easily obtained from the x-ray crystallographic data alone (70).

### 2. Folding and Catalytic Properties of 4-Oxalocrotonate Tautomerase

Currently, we are using PSA to study the molecular determinants of catalysis in a small enzyme, 4-oxalocrotonate tautomerase (4OT). 4OT is made up of a 62 amino acid polypeptide chain that folds into a 40 kDa homo-hexameric complex. The question of how only 62 amino acids can define the fold and function of this isomerase is particularly intriguing. In order to determine which residues are important for 4OT folding and catalysis, we have set up a two-step selection assay

(Fig. 10). In an initial study, we have synthesized an array of four 4OT analogues in which a Gly-O-Gly dipeptide unit was systematically placed over a region thought to be important for substrate binding (Fig. 11). Gel filtration was used to separate folded, hexameric species from unfolded monomers. MALDI analysis of this four analogue array after gel filtration revealed that all four analogues were folded (Fig. 11). This result was surprising because the missing amino acid side chains did not prevent the enzyme from folding into a hexamer. In order to identify which analogues are catalytically active, work is in progress to react the folded pool of four analogues with an active site directed, irreversible inhibitor of 4OT. In this way, we expect to correlate changes in the chemical structure of 4OT with their effect on both folding and catalysis.

### IV.  OLIGONUCLEOTIDE LIBRARIES

Protein-oligonucleotide interactions are essential biochemical recognition processes that regulate a myriad of cellular functions including DNA replication,

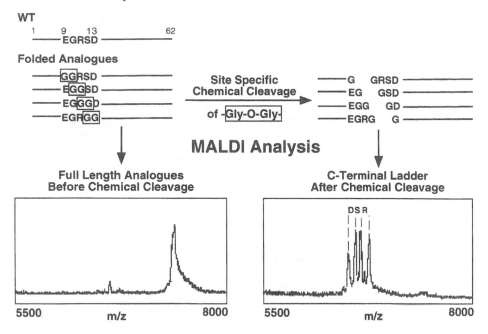

**Fig. 11**  MALDI mass spectrum of 4OT analogues that were selected for folding. Spectrum of analogues before (A) and after (B) chemical cleavage of the analogue unit. All four analogues were able to fold as hexamers.

DNA repair, gene expression, and protein translation. An important characteristic of proteins which interact with oligonucleotides is their ability to recognize specific base pair sequences. For example, most transcription factors recognize a specific DNA sequence of 6–10 base pairs. The first step in the characterization of a DNA-binding protein often involves defining the DNA sequence or sequences it binds. In order to determine the core nucleotide sequence recognized by a protein, a large number of oligonucleotides with varied sequences must be assayed for binding before a consensus sequence can be defined. This process can be especially tedious if oligonucleotides are assayed for protein binding individually. Recently, a new combinatorial approach, commonly referred to as SELEX (Systematic Evolution of Ligands by EXponential enrichment) (71–73) has made it possible to identify protein binding sites by screening

large pools of nucleic acids for binding [reviewed in (74–76)]. This *in vitro* evolution approach has been exploited in a wide range of studies to characterize the RNA and DNA binding sites of hundreds of different proteins.

## A.  SELEX

A basic overview of the strategy employed by the SELEX technique is shown in Fig. 12. First, a randomized oligonucleotide library is prepared by solid-phase synthesis using standard phoshoramidite chemistry. Sequence diversity in this library is generated by coupling an equimolar mixture of all four nucleotides at desired steps in the synthesis. Usually, these random sequences are flanked with a "fixed" sequence for subsequent PCR amplification. A completely randomized library covering 25 bases or less, with approximately $10^{14}$–$10^{15}$ sequences can be

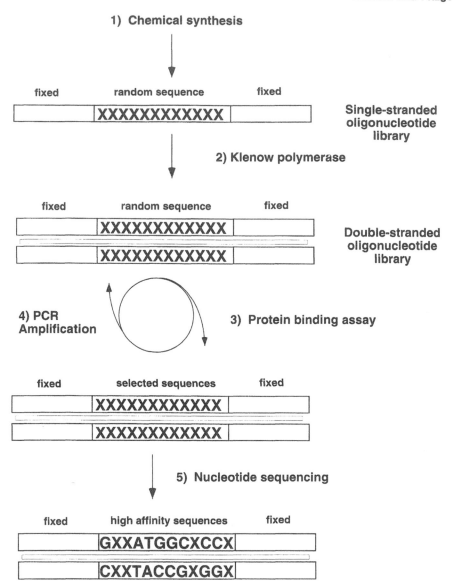

**Fig. 12** SELEX procedure. Step 1: random DNA sequences flanked by a fixed sequence are generated by chemical synthesis. Step 2: single stranded DNA is converted to double-stranded DNA by a fill-in reaction. Step 3: the random double-stranded DNA library is assayed for protein binding. Step 4: bound sequences are amplified by PCR. Steps 3 and 4 are repeated to obtain high affinity sequences that are ultimately identified using conventional nucleotide sequencing protocols.

constructed (77). Next, either a fill-in reaction with the Klenow fragment of DNA polymerase or a PCR reaction is used to convert this library of single-stranded DNA to a library of double-stranded DNA. The selection starts by incubating the double-stranded DNA library with the protein of interest. Bound sequences are isolated by one of several methods, including a filter binding assay, an affinity column, or an electrophoretic mobility shift assay

(EMSA). For example, when EMSA is used in the selection procedure, a radiolabeled oligonucleotide library is equilibrated with the target protein, and the entire mixture is subjected to native polyacrylamide electrophoresis. Oligonucleotide sequences that bind the protein are retarded in the gel matrix, or "shifted" with respect to the unbound sequences. The shifted band is excised from the gel, and the DNA is recovered in an elution procedure. Functional sequences are amplified by PCR, and subjected to further rounds of selection and amplification. In order to reduce the time and effort involved in sequencing, the selected oligonucleotides can be sequenced in batches to assess the convergence of selected sequences. A series of sequencing reactions corresponding to the sequences from different selection rounds is shown in Fig. 13. After round 4, specific base pairs which are important for binding the target protein dominate the library population. Individual clones can then be analyzed using standard nucleotide sequencing techniques.

When the goal is to study protein-RNA interactions, the DNA library can be converted to an RNA library by *in vitro* transcription. The RNA library is then subjected to a functional selection as described for the DNA library. Once the RNA is isolated, it must be converted back to DNA by reverse transcriptase for PCR amplification. After several rounds of selection, RNA sequences that interact with the target protein are enriched in the library population. The final RNA sequences are determined from sequencing the converted DNA population.

## B.  SELEX Theory

As in other combinatorial methods, SELEX is a competition experiment. However, the libraries used in these experiments are orders of magnitude more

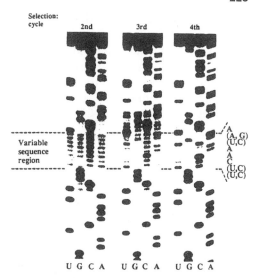

**Fig. 13**  Batch sequencing of RNA library over the course of selection. Before selection, all four bases are degenerate. By round four, a consensus sequence of A(A/G)(U/C)AAC (U/C)(U/C) can be seen. (Reprinted with permission from Ref. 105. Copyright 1990 American Association for the Advancement of Science.)

complex than the combinatorial libraries discussed here so far. Out of a library of $10^{15}$ variants, the goal in SELEX is to identify the top 50 to 100 optimal sequences—truly a needle in a haystack. In order to guarantee that the highest affinity sequences will not get lost in the "haystack" of low affinity sequences, it is necessary to use high levels of protein and DNA during initial rounds of selection. After several selection and amplification rounds, the sequence population will be enriched with high affinity binders. At this point, these high affinity ligands are sufficient in number to survive a competition experiment, so DNA levels are reduced in subsequent rounds of selection. After several rounds of selection with decreasing DNA levels, rapid enrichment of the optimal sequences is achieved.

The enormous impact that the SELEX approach has had on the identification of protein binding sites is evident in the number of labs using the technique. Hundreds of oligonucleotide binding proteins have been characterized using the selection and amplification strategy described above. In the following paragraphs, we highlight a few recent examples.

## C. Applications

### 1. Nucleosome Positioning Sequences

In a recent study, SELEX was used to identify DNA sequences that are preferentially packaged into nucleosome particles (78). In order to produce nucleosome particles, histone proteins were equilibrated with random DNA sequences in high salt, and then dialyzed against low salt buffers. Sequences that bound histone octamers were isolated in a sucrose gradient, amplified, and forwarded to further selection rounds. After fifteen rounds of selection, sequences that were effective in packaging DNA into nucleosomes were identified. The relative affinity of these sequences was almost 3 kcal/mol better than any known "positioning" sequence. From these experiments, rules were determined for the composition of base pairs in a given sequence which attract histone octamers. This study is significant because these high affinity sequences provide useful tools for systematically studying the effects of nucleosome packaging in gene regulation and chromatin structure.

### 2. Transcriptional Activation

The SELEX procedure is specifically designed to identify DNA sequences that have a high affinity for a target protein. However, the DNA sequences that are identified in this way may not always be biologically relevant targets. For example, SELEX identified several high affinity sequences that bound MyoD, a bHLH transcription factor, but that were unable to support transcriptional activation (72).

In an effort to obtain transcriptionally relevant binding sites for MyoD, a modified SELEX procedure was developed where the protein was allowed to bind random DNA sequences in the context of a mammalian cell (79). Only sequences that were able to activate transcription *in vivo* were selected for subsequent rounds of amplification and selection. Interestingly, the sequences identified using this modified procedure yielded sequences distinct from the optimal sequences identified in the initial SELEX experiment. Moreover, the *in vivo* selected sequences resembled functional sequences found in wild-type promoters. These experiments suggest that optimal binding is not equivalent to optimal transcriptional activation *in vivo*. It appears that the DNA binding activity of a transcription factor is only one determinant of its ability to promote transcription.

### 3. Nucleolin-pre-rRNA Binding

In a final example, we discuss how SELEX was used to probe protein-RNA interactions. Unlike protein DNA interactions, where proteins bind to a predictable B-form DNA helix, RNA can form complex tertiary structures that determine protein recognition. Bouvet and co-workers were interested in how nucleolin, an essential protein in ribosome biogenesis, interacted with nascent pre-rRNA molecules (80). Through the use of standard biochemical techniques, a 175 base pair (nt 294–469) fragment was identified as a high affinity sequence. In a parallel set of experiments, SELEX was used to isolate a high affinity RNA sequence from a pool of RNAs randomized at 25 positions. After nine rounds of selection for nucleolin binding, functional RNA sequences were fished out of the library. Analysis of the selected RNA sequences revealed a consensus sequence of 18 nucleotides predicted to form a stem-loop structure. Within this loop forming sequence was the sequence UCCCGA, a motif which was shown to be necessary for protein-RNA interaction in the wild type 175 base pair fragment. Interestingly, this UCCCGA binding site had to be displayed in the context of the hairpin structure for protein binding to occur. This study illustrates that SELEX can be used to obtain RNA molecules which possess structural features important for binding, as well as high affinity sequences.

## VI. CARBOHYDRATE LIBRARIES

The carbohydrate moieties of glycoproteins modulate protein stability and folding, cell adhesion, and other intercellular recognition processes (81, 82). Protein-carbohydrate interactions have also been implicated in a variety of disease states including chronic inflammation, viral and bacterial infection, and cancer (83–85). Understanding how proteins recognize carbohydrate signals has been especially challenging because of the difficulties associated with carbohydrate and glycoprotein synthesis. For example, it has been difficult to apply recombinant DNA technology to the production of glycoproteins because the enzymes required for glycoconjugate biosynthesis have a high error rate, and naturally produce proteins with multiple glycoforms. In addition, the staggering diversity of linkages within oligosaccharides makes the chemical synthesis of complex carbohydrates and glycopeptides technically demanding. However, with the advent of new chemical approaches for the preparation of oligosaccharides and glycoproteins, combinatorial chemistry methods are being used to examine protein-carbohydrate interactions in detail. Here we summarize these advances and highlight a few examples where carbohydrate libraries have been used to study protein function.

## A. Solution-Phase Synthesis and Characterization

Initial attempts to prepare oligosaccharide libraries involved the solution- phase synthesis of small libraries from a limited repertoire of saccharide monomers. For example, Hindsgaul and co-workers presented a random glycosylation method to generate small pools of di- or tri-saccharides (86). In this approach, a glycosyl donor with one type of protecting group is reacted with a glycosyl acceptor in which all hydroxyl groups are unprotected. This reaction generates a mixture of oligosaccharides in which the donor molecule randomly attaches to the hydroxyl groups of the acceptor. However, the resulting oligosaccharide library is not well-defined. Moreover, isolation of active compounds from this random mixture relies on multiple chromatographic separations.

In another solution-phase approach, a "latent-active" glycosylation method was used to prepare a small tri-saccharide library (87). In this approach, glycosyl donor and acceptor molecules are derived from a common monosaccharide (Fig. 14). For example, removal of an acetyl protecting group of the starting material yields the glycosyl acceptor, while isomerization of the allyl ether moiety to a vinyl group yields the glycosyl donor. Reaction of this first series of donor and acceptor molecules produces a di-saccharide which contains acetyl and allyl protecting groups. These protecting groups can be removed or altered as before to create a new pair of donor and acceptor molecules. A tri-saccharide library is generated by reacting donor and acceptor monomers with each set of di-saccharide donor and acceptors.

A limitation to these solution-phase approaches is that tri-saccharides need to be purified from excess monomers present in the solution. In addition, the utility of these approaches are limited to a narrow range of glycosyl donor and acceptor

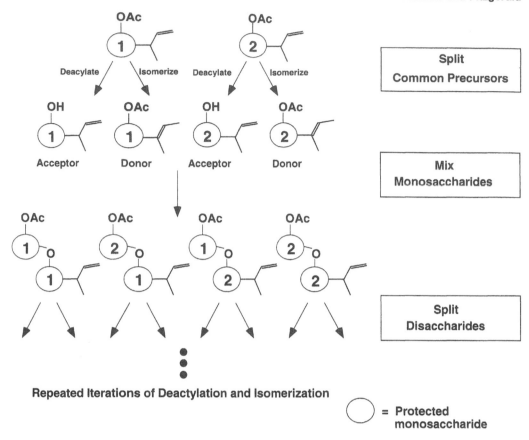

**Fig. 14** Latent glycosylation approach for combinatorial carbohydrate synthesis. Two precursors are converted to glycosyl donors and acceptors by an isomerization and a deacetylation reaction, respectively. Each donor and acceptor pair are reacted in a combinatorial fashion to produce a library of disaccharides (From Ref. 87).

pairs, because the glycosylation rate is highly dependent on the steric and electronic environment of the alcohol acceptor.

## B. Solid-Phase Synthesis and Characterization

Recently, Kahne and co-workers overcame many of the synthetic obstacles associated with the synthesis of oligosaccharides and presented a novel method for generating a solid-phase oligosaccharide library (88). In their approach, a sulfoxide glycosylation strategy was used to uniformly activate

glycosyl donors (Fig. 15). By using anomeric sulfoxides as glycosyl donors, the donors were activated instantaneously at low temperatures, regardless of the hydroxyl protecting groups present. In addition, the carbohydrates were synthesized from the reducing to the nonreducing end, allowing an excess of glycosyl donors to drive the reaction to completion. This approach was very successful, providing almost quantitative yields for a wide range of glycosyl donor-acceptor pairs. Kahne and co-workers exploited this chemistry in a split and mix procedure to produce a 1300 member library composed of di- and

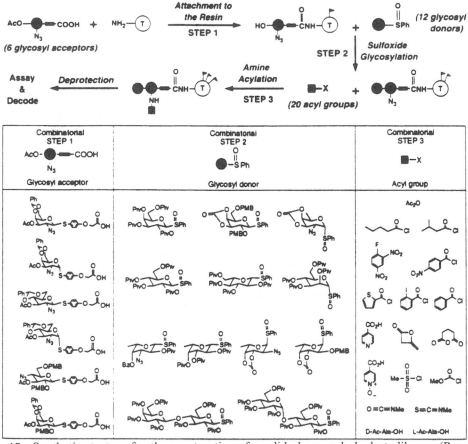

**Fig. 15** Synthetic strategy for the construction of a solid phase carbohydrate library. (Reprinted with permission from R Liang, L Yan, J Loebach, M Ge, Y Uozumi, C Thompson, A Smith, K Biswas, WC Still, D Kahne. Parallel synthesis and screening of a solid phase carbohydrate library. Science 274: 1520–1522, 1996. Copyright 1996 American Association for the Advancement of Science).

tri-saccharides. This one-bead, one compound library was encoded with chemical tags (30, 89) incorporated during the synthesis to record the reaction history of each bead.

## C. Applications

### 1. Lectin-Carbohydrate Recognition

Lectin was used as a model protein to screen a 1300 member carbohydrate library (88). It was thought that *Bauhinia purpurea* lectin would provide a good model system for studying protein-carbohydrate recognition because it binds carbohydrates on the surfaces of erythroctyes. The carbohydrate library was screened for lectin binding. Surprisingly, two ligands which bound more tightly than the natural ligand were identified. Moreover, the relative affinities of the two ligands differed compared to solution-based studies (90). The authors reasoned that the solid-phase surface presentation of the carbohydrate library provided cell surface-like conditions, allowing a more biologically relevant assay for studying proteins which interact with cell surface carbohydrates. To support this hypothesis, Liang and co-workers showed that carbohydrate-linked beads mimic cell agglutination behavior in the presence of low levels of lectin. This work demonstrated that, despite the low affinity of proteins for individual carbohydrates in solution, there is considerable specificity directed via polyvalent carbohydrates present on cell surfaces.

### 2. Hemaglutinin-Cell Surface Interactions

As an alternative approach to using "native" oligosaccharide libraries to study glycoprotein function, Whitesides and co-workers have recently developed a method for generating and analyzing polyacrylic-based carbohydrate libraries (91). In a model experiment, they designed libraries to search for inhibitors against influenza infection. During influenza infection, the viral coat protein hemaglutinin (HA) mediates cell-viral interactions by binding to clusters of N-acetylneuraminic acid (NeuAc) on the cell surface (Fig. 16). In order to disrupt this interaction, polymers displaying multiple NeuAc side chains were generated and assayed for viral inhibition. These derivatized polymers were prepared by sonicating a mixture of poly(acrylic anhydride) and NeuAc, spiked with various natural and unnatural amino acids. Diversity was created by changing the molecular equivalents of the reactive amines to anhydride groups. The reactions were performed directly in microtiter plates, producing a "quasi-solid-phase" library of NeuAc and amino acid polymers in each well. These libraries were assayed directly with no further manipulation.

Although the polyacrylic backbone is an unnatural modification not found in cellular NeuAc clusters, a $K_i$ of 0.5 nM was obtained for the most potent inhibitor, which was prepared by mixing poly(acrylic anhydride) with 0.10 mol equivalents of NeuAc-L with $R = (CH_2)_3S(CH_2)_2NH_2$

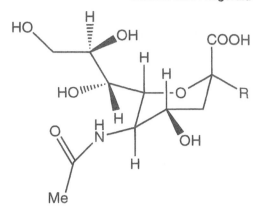

N-Acetylneuraminic acid (NeuAc)

**Fig. 16** Structure of N-acetylneuramic acid (NeuAc).

and 0.12 equivalents of 1-3-(2′-naphthyl) alanine. One trend that was observed in these studies was that polymer chains containing aromatic side chains showed improved inhibition over polymer chains containing only NeuAc derivatives. The authors postulate that these side chains are involved in nonspecific binding to hydrophobic sites on the surface of the virus, while the multivalent NeuAc groups bind to HA sites. These interactions are thought to be stabilized by the gel-like nature of the water-swollen polymer. This straightforward synthetic method may prove useful in future studies of viral protein-oligosaccharide recognition.

## VII. FUTURE DIRECTIONS

There has been increasing interest to generate homogeneous carbohydrate-linked peptides and proteins. Important progress in chemical synthesis methodologies have made possible the preparation of defined, homogeneous N-linked glycopeptides both in solution (92) and on the solid-phase (93). More recently, Bertozzi and co-workers have reported a chemoselective strategy for the convergent, solid-phase synthesis of O-linked glycopeptides (94). Several groups have also used an enzymatic approach, where glycosidic enzymes have been tricked into performing oligosaccharide synthesis (95) or covalently linking oligosaccharides to peptides (96) and proteins (97). These approaches provide important stepping stones for the *de novo* synthesis of complex homogeneous glycoproteins.

Several groups have also been active in "remodeling" cell surface carbohydrates (98–104). Perhaps most promising, Bertozzi and co-workers have achieved the expression of a cell surface oligosaccharide containing an unnatural ketone functional group, allowing chemoselective ligation to produce a defined, carbohydrate-decorated cell surface (102). In the near future, application of combinatorial chemistry to the above mentioned techniques may enable the creation of defined glycoprotein libraries expressed on cell surfaces.

## REFERENCES

1.  LA Thompson, JA Ellman. Synthesis and applications of small molecule libraries. Chem Rev 96: 555–600, 1996.
2.  MC Pirrung, J Chen. Preparation and screening against acetylcholinesterase of a nonpeptide indexed combinatorial library. J Am Chem Soc 117: 1240–1245, 1995.
3.  MC Pirrung, JH-L Chau, J Chen. Discovery of a novel tetrahydroacridine acetylcholinesterase inhibitor through an indexed combinatorial library. Chem Biol 2: 621–626, 1995.
4.  PW Smith, JYQ Lai, AR Whittington, B Cox, JG Houston. Synthesis and biological evaluation of a library containing potentially 1600 amide esters—A strategy for rapid compound generation and screening. Bioorg Med Chem Lett 4: 2821–2824, 1994.
5.  SB Shuker, PJ Hajduk, RP Meadows, SW Fesik. Discovering high-affinity ligands for proteins: SAR by NMR. Science 274: 1531–1534, 1996.
6.  PJ Hajduk, RP Meadows, SW Fesik. Discovering high-affinity ligands for proteins. Science 278: 497–498, 1997.
7.  PJ Hajduk, G Sheppard, DG Nettesheim, ET Olejniczak, SB Shuker, RP Meadows, DH Steinmen, GM Carrera, Jr., PA Marcotte, J Severin, K Walter, H Smith, E Gubbins, R Simmer, TF Holzman, DW Morgan, SK Davidsen, JB Summers, SW Fesik. Discovery of potent nonpeptide inhibitors of stromelysin using SAR by NMR. J Am Chem Soc 119: 5818–5827, 1997.
8.  PJ Hajduk, ET Olejniczak, SW Fesik. One-dimensional relaxation- and diffusion-edited NMR methods for screening compounds that bind to macromolecules. J Am Chem Soc 119: 12257–12261, 1997.
9.  HM Geysen, RH Meloen, SJ Barteling. Use of peptide synthesis to probe viral antigens for epitopes to a resolution of a single amino acid. Proc Natl Acad Sci 81: 3998–4002, 1984.
10. R Frank. Spot synthesis: an easy technique for the positionally addressable, parallel chemical synthesis on a membrane support. Tetrahedron 48: 9217–9232, 1992.
11. HV Meyers, GJ Dilley, TL Durgin, TS Powers, NA Winssinger, H Zhu, MR Pavia. Multiple simultaneous synthesis of phenolic libraries. Mol Divers 1: 13–20, 1995.
12. SPA Fodor, JL Read, MC Pirrung, L Stryer, AT Lu, D Solas. Light-directed, spatially addressable parallel chemical synthesis. Science 251: 767–773, 1991.
13. A Furka, F Sebestyen, M Asgedom, G Dibo. General method for rapid synthesis of multicomponent peptide mixtures. Int J Pept Protein Res 37: 487–493, 1991.
14. F Sebestyen, G Dibo, A Kovacs, A Furka. Chemical synthesis of peptide libraries. Bioorg Med Chem Lett 3: 413–418, 1993.
15. CL Brummel, INW Lee, Y Zhou, SJ Benkovic, N Winograd. A mass-spectrometric solution to the address problem of combinatorial libraries. Science 264: 399–402, 1994.
16. CL Brummel, JC Vickerman, SA Carr, ME Hemling, GD Roberts, W Johnson, J Weinstock, D Gaitanopoulos, SJ Benkovic, N Winograd. Evaluation of mass spectrometric methods applicable to the direct analysis of non-peptide bead-compound combinatorial libraries. Anal Chem 68: 237–242, 1996.
17. BJ Egner, GJ Langley, M Bradley. Solid-phase chemistry—direct monitoring by matrix-assisted laser-desorption ionization time-of-flight mass-spectrometry—a tool for combinatorial chemistry. J Org Chem 60: 2652–2653, 1995.
18. MC Fitzgerald, K Harris, CG Shevlin, G Siuzdak. Direct characterization of solid

phase resin-bound molecules by mass spectrometry. Bioorg Med Chem Lett 6: 979–982, 1996.

19. MR Carrasco, MC Fitzgerald, Y Oda, SBH Kent. Direct monitoring of organic reactions on polymeric supports. Tet Lett 38: 6331–6334, 1997.

20. J Chin, B Fell, S Pochapsky, MJ Shapiro, JR Wareing. 2D SECSY NMR for combinatorial chemistry. High-resolution MAS spectra for resin-bound molecules. J Org Chem 63: 1309–1311, 1998.

21. SK Sarkar, RS Garigipati, JL Adams, PA Keifer. An NMR method to identify nondestructively chemical compounds bound to a single solid-phase-synthesis bead for combinatorial chemistry applications. J Am Chem Soc 118: 2305–2306, 1996.

22. K Russell, DC Cole, FM McLaren, DE Pivonka. Analytical techniques for combinatorial chemistry: quantitative infrared spectroscopic measurements of deuterium-labeled protecting groups. J Am Chem Soc 118: 7941–7945, 1996.

23. B Yan, G Kumaravel, H Anjaria, A Wu, RC Petter, CF Jewell, Jr., JR Wareing. Infrared spectrum of a single resin bead for real-time monitoring of solid-phase reactions. J Org Chem 60: 5736–5738, 1995.

24. B Yan, G Kumaravel. Probing solid-phase reactions by monitoring the IR bands of compounds on a single "flattened" resin bead. Tetrahedron 52: 843–848, 1996.

25. DE Pivonka, TR Simpson. Tools for combinatorial chemistry: real-time single-bead infrared analysis of a resin-bound photocleavage reaction. Anal Chem 69: 3851–3853, 1997.

26. V Nikolaiev, A Stierandova, V Krchnak, B Seligmann, KS Lam, SE Salmon, M Lebl. Peptide-encoding for structure determination of nonsequenceable polymers within libraries synthesized and tested on solid-phase supports. Peptide Res 6: 161–170, 1993.

27. S Brenner, RA Lerner. Encoded combinatorial chemistry. Proc Natl Acad Sci USA 89: 5381–5383, 1992.

28. J Nielsen, S Brenner, KD Janda. Synthetic methods for the implementation of encoded combinatorial chemistry. J Am Chem Soc 115: 9812–9813, 1993.

29. MC Needles, DG Jones, EH Tate, GL Heinkel, LM Kochersperger, WJ Dower, RW Barrett, MA Gallop. Generation and screening of an oligonucleotide-encoded synthetic peptide library. Proc Natl Acad Sci, USA 90: 10700–10704, 1993.

30. MHJ Ohlmeyer, RN Swanson, LW Dillard, JC Reader, G Asouline, R Kobayashi, M Wigler, WC Still. Complex synthetic chemical libraries indexed with molecular tags. Proc Natl Acad Sci USA 90: 10922–10926, 1993.

31. HM Geysen, CD Wagner, WM Bodnar, CJ Markworth, GJ Parke, FJ Schoenen, DS Wagner, DS Kinder. Isotope or mass encoding of combinatorial libraries. Chem Biol 3: 679–688, 1996.

32. EJ Moran, S Sarshar, JF Cargill, MM Shahbaz, A Lio, AMM Mjalli, RW Armstrong. Radiofrequency tag encoded combinatorial library method for the discovery of tripeptide-substituted cinnamic acid inhibitors of protein tyrosine phosphatase PTP1B. J Am Chem Soc 117: 10787–10788, 1995.

33. X-Y Xiao, Z Parandoosh, MP Nova. Design and synthesis of a taxoid library using radiofrequency encoded combinatorial chemistry. J Org Chem 62: 6029–6033, 1997.

34. KC Nicolaou, X-Y Xiao, Z Parandoosh, A Senyei, MP Nova. Radiofrequency encoded combinatorial chemistry. Angew Chem Int Ed Engl 34: 2289–2291, 1995.

35. X-Y Xiao, C Zhao, H Potash, MP Nova. Combinatorial chemistry with laser optical encoding. Angew Chem Int Ed Engl 36: 780–782, 1997.

36. JK Chen, SL Schreiber. Combinatorial synthesis and multidimensional NMR spectroscopy: an approach to understanding protein-ligand interactions. Angew Chem Int Ed Engl 34: 953–969, 1995.

37. AP Combs, TM Kapoor, S Feng, JK Chen, LF Daudé-Snow, SL Schreiber. Protein structure-based combinatorial chemistry: discovery of non-peptide binding elements to Src SH3 domain. J Am Chem Soc 118: 287–288, 1996.

38. S Feng, TM Kapoor, F Shirai, AP Combs, SL Schreiber. Molecular basis for the binding of SH3 ligands with non-peptide elements identified by combinatorial synthesis. Chem Biol 3: 661–670, 1996.

39. AJ You, RJ Jackman, GM Whitesides, SL Schreiber. A miniaturized arrayed assay format for detecting small molecule-protein interactions in cells. Chem Biol 4: 969–975, 1997.

40. A Borchardt, SD Liberles, SR Biggar, GR Crabtree, SL Schreiber. Small molecule-dependent genetic selection in stochastic nanodroplets as a means of detecting protein-ligand interactions on a large scale. Chem Biol 4: 961–968, 1997.

41. J Huang, SL Schreiber. A yeast genetic system for selecting small molecule inhibitors

of protein-protein interactions in nanodroplets. Proc Natl Acad Sci USA 94: 13396–13401, 1997.

42. R Cortese, P Monaci, A Luzzago, C Santini, F Bartoli, I Cortese, P Fortugno, G Galfre, A Nicosia, F Felici. Selection of biologically active peptides by phage display of random peptide libraries. Curr Opin Biotech 7: 616–621, 1996.

43. GP Smith, VA Petrenko. Phage display. Chem Rev 97: 391–410, 1997.

44. JM Ostresh, SE Blondelle, B Dörner, RA Houghten. Generation and use of nonsupport-bound peptide and peptidomimetic combinatorial libraries. Methods Enzymol 267: 220–234, 1996.

45. C Pinilla, JR Appel, P Blanc, RA Houghten. Rapid identification of high affinity peptide ligands using positional scanning synthetic peptide combinatorial libraries. Biotechniques 13: 901–905, 1992.

46. CT Dooley, NN Chung, BC Wilkes, PW Schiller, JM Bidlack, GW Pasternak, RA Houghten. An all D-amino acid opioid peptide with central analgesic activity from a combinatorial library. Science 266: 2019–2022, 1994.

47. JM Ostresh, JH Winkle, VT Hamashin, RA Houghten. Peptide libraries: determination of relative reaction-rates of protected amino acids in competitive couplings. Biopolymers 34: 1681–1689, 1994.

48. J Eichler, RA Houghten. Indentification of substrate-analog trypsin inhibitors through the screening of synthetic peptide combinatorial libraries. Biochemistry 32: 11035–11041, 1993.

49. TA Rano, T Timkey, EP Peterson, J Rotonda, DW Nicholson, JW Becker, KT Chapman, NA Thornberry. A combinatorial approach for determining protease specificities: application to interleukin-1$\beta$ converting enzyme (ICE). Chem Biol 4: 149–155, 1997.

50. KS Lam, SE Salmon, EM Hersh, VJ Hruby, WM Kazmierski, RJ Knapp. A new type of synthetic peptide library for identifying ligand-binding activity. Nature 354: 82–84, 1991.

51. RS Youngquist, GR Fuentes, MP Lacey, T Keough. Generation and screening of combinatorial peptide libraries designed for rapid sequencing by mass spectrometry. J Am Chem Soc 117: 3900–3906, 1995.

52. JA Camarero, B Ayers, TW Muir. Studying receptor-ligand interactions using encoded amino acid scanning. Biochemistry 37: 7487–7495, 1998.

53. H Yu, JK Chen, S Feng, DC Dalgarno, AW Brauer, SL Schreiber. Structural basis for the binding of proline-rich peptides to SH3 domains. Cell 76: 933–945, 1994.

54. JK Chen, WS Lane, AW Brauer, A Tanaka, SL Schreiber. Biased combinatorial libraries- novel ligands for the SH3 domain of phosphatidylinositol 3-kinase. J Am Chem Soc 115: 12591–12592, 1993.

55. S Uebel, W Kraas, S Kienle, K-H Wiesmüller, G Jung, R Tampé. Recognition principle of the TAP transporter disclosed by combinatorial peptide libraries. Proc Natl Acad Sci USA 94: 8976–8981, 1997.

56. A Kramer, T Keitel, K Winkler, W Stöcklein, W Höhne, J Schneider-Mergener. Molecular basis for the binding promiscuity of an anti-p24 (HIV-1) monoclonal antibody. Cell 91: 799–809, 1997.

57. M Smith. Synthetic DNA and biology. Angew Chem Int Ed Engl 33: 1214–1221, 1994.

58. L Giver, FH Arnold. Combinatorial protein design by in vitro recombination. Curr Opin Chem Biol 2: 335–338, 1998.

59. PA Patten, RJ Howard, WPC Stemmer. Applications of DNA shuffling to pharmaceuticals and vaccines. Curr Opin Biotechnol 8: 724–733, 1997.

60. O Kuchner, FH Arnold. Directed evolution of enzyme catalysts. Trends Biotechnol 15: 523–530, 1997.

61. A Crameri, SA Raillard, E Bermudez, WPC Stemmer. DNA shuffling of a family of genes from diverse species accelerates directed evolution. Nature 391: 288–291, 1998.

62. M Schnölzer, P Alewood, A Jones, D Alewood, SBH Kent. In situ neutralization in Boc-chemistry solid phase peptide synthesis. Int J Peptide Protein Res 40: 180–193, 1992.

63. PE Dawson, TW Muir, I Clark-Lewis, SBH Kent. Synthesis of proteins by native chemical ligation. Science 266: 776–779, 1994.

64. M Schnölzer, SBH Kent. Constructing proteins by dovetailing unprotected synthetic peptides: backbone-engineered HIV protease. Science 256: 221–225, 1992.

65. C-F Liu, JP Tam. Chemical ligation approach to form a peptide bond between unprotected peptide segments. Concept and model study. J Am Chem Soc 116: 4149–4153, 1994.

66. TW Muir. A chemical approach to the construction of multimeric protein assemblies. Structure 3: 649–652, 1995.

67. TW Muir, PE Dawson, MC Fitzgerald, SBH Kent. Protein signature analysis: a

practical new approach for studying structure-activity relationships in peptides and proteins. Methods Enzymol 289: 545–564, 1997.

68. TW Muir, PE Dawson, MC Fitzgerald, SBH Kent. Probing the chemical basis of binding activity in an SH3 domain by protein signature analysis. Chem Biol 3: 817–825, 1996.

69. PE Dawson, MC Fitzgerald, TW Muir, SBH Kent. Methods for the chemical synthesis and readout of self-encoded arrays of polypeptide analogues. J Am Chem Soc 119: 7917–7927, 1997.

70. X Wu, B Knudsen, SM Feller, J Zheng, A Sali, D Cowburn, H Hanafusa, J Kuriyan. Structural basis for the specific interaction of lysine-containing proline-rich peptides with the N-terminal SH3 domain of c-Crk. Structure 3: 215–226, 1995.

71. C Tuerk, L Gold. Systematic evolution of ligands by exponential enrichment: RNA ligands to bacteriophage T4 DNA polymerase. Science 249: 505–510, 1990.

72. TK Blackwell, H Weintraub. Differences and similarities in DNA-binding preferences of MyoD and E2A protein complexes revealed by binding site selection. Science 250: 1104–1110, 1990.

73. R Green, AD Ellington, JW Szostak. In vitro genetic analysis of the tetrahymena self-splicing intron. Nature 347: 406–408, 1990.

74. L Gold, B Polisky, O Uhlenbeck, M Yarus. Diversity of oligonucleotide functions. Ann Rev Biochem 64: 763–797, 1995.

75. TK Blackwell. Selection of protein binding sites from random nucleic acid sequences. Methods Enzymol 254: 604–618, 1995.

76. RC Conrad, L Giver, Y Tian, AD Ellington. In vitro selection of nucleic acid aptamers that bind proteins. Methods Enzymol 267: 336–367, 1996.

77. M Famulok, JW Szostak. In vitro selection of specific ligand-binding nucleic acids. Angew Chem Int Ed Engl 31: 979–988, 1992.

78. PT Lowary, J Widom. New DNA sequence rules for high affinity binding to histone octamer and sequence-directed nucleosome positioning. J Mol Biol 276: 19–42, 1998.

79. J Huang, TK Blackwell, L Kedes, H Weintraub. Differences between MyoD DNA binding and activation site requirements revealed by functional random sequence selection. Mol and Cell Biol 16: 3893–3900, 1996.

80. L Ghisolfi-Nieto, G Joseph, F Puvion-Dutilleul, F Amalric, P Bouvet. Nucleolin is a sequence-specific RNA-binding protein: characterization of targets on pre-ribosomal RNA. J Mol Biol 260: 34–53, 1996.

81. A Varki. Biological roles of oligosaccharides: all of the theories are correct. Glycobiology 3: 97–130, 1993.

82. LA Lasky. Selectin-carbohydrate interactions and the initiation of the inflammatory response. Ann Rev Biochem 64: 113–139, 1995.

83. SD Rosen, CR Bertozzi. The selectins and their ligands. Curr Opin Cell Biol 6: 663–673, 1994.

84. N Sharon, H Lis. Carbohydrates in cell recognition. Sci Am 268: 82–89, 1993.

85. P Sears, C-H Wong. Intervention of carbohydrate recognition by proteins and nucleic acids. Proc Natl Acad Sci USA 93: 12086–12093, 1996.

86. O Kanie, F Barresi, Y Ding, J Labbe, A Otter, LS Forsberg, B Ernst, O Hindsgaul. A strategy of "random glycosylation" for the production of oligosaccharide libraries. Angew Chem Int Ed Engl 34: 2720–2722, 1995.

87. G-J Boons, B Heskamp, F Hout. Vinyl glycosides in oligosaccharide synthesis: a strategy for the preparation of trisaccharide libraries based on latent-active glycosylation. Angew Chem Int Ed Engl 35: 2845–2847, 1996.

88. R Liang, L Yan, J Loebach, M Ge, Y Uozumi, K Sekanina, N Horan, J Gildersleeve, C Thompson, A Smith, K Biswas, WC Still, D Kahne. Parallel synthesis and screening of a solid phase carbohydrate library. Science 274: 1520–1522, 1996.

89. HP Nestler, PA Bartlett, WC Still. A general method for molecular tagging of encoded combinatorial chemistry libraries. J Org Chem 59: 4723–4724, 1994.

90. R Liang, J Loebach, N Horan, M Ge, C Thompson, L Yan, D Kahne. Polyvalent binding to carbohydrates immobilized on an insoluble resin. Proc Natl Acad Sci USA 94: 10554–10559, 1997.

91. S Choi, M Mammen, GM Whitesides. Generation and in situ evaluation of libraries of poly(acrylic acid) presenting sialosides as side chains as polyvalent inhibitors of influenza-mediated hemagglutination. J Am Chem Soc 119: 4103–4111, 1997.

92. ST Cohen-Anisfeld, PT Lansbury. A practical, convergent method for glycopeptide synthesis. J Am Chem Soc 115: 10531–10537, 1993.

93. JY Roberge, X Beebe, SJ Danishefsky. A strategy for a convergent synthesis of

N-linked glycopeptides on a solid support. Science 269: 202–204, 1995.

94. EC Rodriguez, KA Winans, DS King, CR Bertozzi. A strategy for the chemoselective synthesis of O-linked glycopeptides with native sugar-peptide linkages. J Am Chem Soc 119: 9905–9906, 1997.

95. LF Mackenzie, Q Wang, RAJ Warren, SG Withers. Glycosynthases: mutant glycosidases for oligosaccharide synthesis. J Am Chem Soc 120: 5583–5584, 1998.

96. M Schuster, P Wang, JC Paulson, C-H Wong. Solid-phase chemical-enzymatic synthesis of glycopeptides and oligosaccharides. J Am Chem Soc 116: 1135–1136, 1994.

97. K Witte, P Sears, R Martin, C-H Wong. Enzymatic glycoprotein synthesis: preparation of ribonuclease glycoforms via enzymatic glycopeptide condensation and glycosylation. J Am Chem Soc 119: 2114–2118, 1997.

98. C Hällgren, O Hindsgaul. An amidated GDP-fucose analog useful in the fucosyltransferase catalyzed addition of biological probes onto oligosaccharide chains. J Carbohydr Chem 14: 453–464, 1995.

99. G Srivastava, KJ Kaur, O Hindsgaul, MM Palcic. Enzymatic transfer of a pre-assembled trisaccharide antigen to cell surfaces using a fucosyltransferase. J Biol Chem 267: 22356–22361, 1992.

100. RE Kosa, R Brossmer, H-J Grob. Modification of cell sufaces by enzymatic introduction of special sialic acid analogues. Biochem Biophys Res Commun 190: 914–920, 1993.

101. LK Mahal, CR Bertozzi. Engineered cell surfaces: fertile ground for molecular landscaping. Chem Biol 4: 415-422, 1997.

102. LK Mahal, KJ Yarema, CR Bertozzi. Engineering chemical reactivity on cell surfaces through oligosaccharide biosynthesis. Science 276: 1125–1128, 1997.

103. S Tsuboi, Y Isogai, N Hada, JK King, O Hindsgaul, M Fukuda. 6′-Sulfo Sialyl Le$^x$ but not 6-sulfo Sialyl Le$^x$ expressed on the cell surface supports L-selectin-mediated adhesion. J Biol Chem 271: 27213–27216, 1996.

104. KJ Yarema, CR Bertozzi. Chemical approaches to glycobiology and emerging carbohydrate-based therapeutic agents. Curr Opin Chem Biol 2: 49–61, 1998.

105. C Teurk, L Gold. Systematic Evolution of Ligands by Exponential Enrichment: RNA Ligands to Bacteriophage T4 DNA polymerase. Science 249: 505–510, 1990.

# 8

# Folding Polypeptides for Drug Production and Discovery*

**Elizabeth Strickland,[1] Philip J. Thomas[1] and Ming Li[2]**
[1]*The University of Texas Southwestern Medical Center at Dallas, Dallas, Texas*
[2]*Prince of Wales Hospital, The Chinese University of Hong Kong, Hong Kong*

## I. INTRODUCTION

Protein folding is no longer a purely academic research topic as it was twenty years ago. Instead, understanding how polypeptides fold is becoming crucial in both industrial research and biotechnology for two reasons. First, the biotechnology industry must produce large quantities of correctly folded polypeptides in recombinant expressions systems and often must fold these polypeptides *in vitro* after their expression and accumulation into inclusion bodies. Second, an increasing number of human diseases are being found to be due to protein folding defects (see Table 1) (1, 2). This chapter will address both of the themes introduced above: the role of polypeptide folding in biotechnology and the development of novel pharmaceutical agents based on protein folding. First, a summary of our current understanding of how polypeptides fold will be presented to provide some practical guidance for those attempting to fold polypeptides *in vitro*. Second, several stratagems will be introduced in an attempt to catalyze creative thinking about how polypeptide folding might be capitalized upon to intervene therapeutically in human disease.

Advances in molecular biology which enable almost any foreign or heterologous gene to be expressed in another host cell line have been the basis for the success of modern biotechnology. In the past fifteen years the biotechnology industry has utilized both *in vivo* and *in vitro* folding to successfully produce several recombinant proteins which are currently approved for therapeutic use; insulin and tissue plasminogen activator (TPA) are two of these (3, 4). These successes, however, obscure the reality that translation of a recombinant gene does not always lead to the generation of a folded, fully active polypeptide. Frequently the barrier to successful production of polypeptides is not expression, but rather, their ability to fold properly. In place of an active polypeptide, an insoluble and inactive one is often produced. In order to make a useful product, this insoluble polypeptide must be solubilized and artificially refolded. Effective *in vitro* refolding strategies prevent the translation of polypeptides that misfold from becoming a fruitless exercise

*Supported by NIH-NIDDK and the Welch Foundation

**Table 1**  Protein Folding Pathologies

| Protein folding disease | Protein involved |
| --- | --- |
| Cystic fibrosis | CFTR |
| Alzheimer's disease | APP/Aβ-amyloid |
| Heinz-body Anemia | globin |
| Wilson's Disease | Copper transporting P-type ATPase |
| Scrapie/Creutzfeldt-Jakob disease/Familial amyloidosis | Prion |
| von Willebrand Disease type 3 | von Willebrand factor |
| Marfan syndrome | Fibrillin |
| Huntington's Disease | Huntingtin |
| Familial hypercholesterolemia | LDL-receptor |
| Familial amyloidosis | Transthyretin / Lysozyme |
| Congenital Hypothyriod Goiter | Thyroglobin |
| Amyotropic lateral sclerosis | Superoxide dismutase |
| Scurvy | Collagen |
| Hypercysteinemia | many proteins |
| Hypothyriodism/Dwarfism | many proteins |
| Carbohydrate-deficient Glycoprotein Syndrome | many proteins |
| Maple syrup urine disease | α-Ketoacid dehydrogenase complex |
| Cancer | p53 |
| $\alpha_1$-Antitrypsin deficiency | $\alpha_1$-Antitrypsin |
| Tay-Sachs disease | β-Hexosaminindase |
| Cataracts | Crystallins |
| Ovarian dysgenesis | Folicle Stimulating Hormone Receptor |
| Hyperphenylalanemia/Phenylketonuria | Phenylalanine Hydroxylase |
| CNS dysmyelination/Pelizaen's Disease | Proteolipid protein |

and transform recombinant polypeptide expression into a billion-dollar business for the biotechnology and pharmaceutical industries. The continuing success of polypeptides or proteins as pharmaceutical agents will depend, in part, on the ability to produce these new products inexpensively on an industrial scale. Strategies for large-scale production of recombinant polypeptides will require either optimized expression of correctly folded polypeptides *in vivo* or, as will be discussed in this chapter, the development of advanced techniques for polypeptide folding *in vitro*.

A clear understanding of polypeptide folding is important not only for the utilitarian aspects of recombinant protein production, but also for therapeutic intervention in the increasing number of human diseases that are linked to protein misfolding. These methods may include specific small molecules that inhibit misfolding, drugs that would provide a scaffold to promote proper folding or small molecules that hinder the folding of a polypeptide with an unwanted activity.

## A. Primary Sequence Determinants of Polypeptide Folding

To understand the protein folding problem is to understand how a polypeptide adopts a specific conformation. The information required for the wide structural and functional diversity of proteins is contained in a linear primary sequence of twenty different amino acids that genetically codes for specific secondary and tertiary structures. Although a variety of factors, including prosequences, signal sequences, cofactors, molecular chaperones and environmental conditions, all impact polypeptide chain folding, the primary amino acid sequence remains the critical determinant of folding and native protein structure. The fundamental significance of the primary sequence in the unfolding and folding of proteins became clear through the pioneering studies of Anfinsen in the 1950's (5). By showing that an unfolded polypeptide chain could spontaneously refold *in vitro* to form a native protein with full biological activity in the absence of other factors, it was concluded that the primary sequence alone contains all the information necessary to define the three dimensional structure of a protein and, therefore, its biological function (5). Thus, although accessory factors such as molecular chaperones may assist a polypeptide in reaching its native state, they apparently do not impart any structural information on the polypeptide in addition to that already encoded in its own primary sequence. Thus, protein folding is a second interpretation of the genetic code (Fig. 1).

As if to emphasize the importance of primary sequence in determining protein structure, the mutation or deletion of even one amino acid in the primary sequence can have a significant impact on a polypeptide's folding and stability and, therefore, on its biological function. Such is the case for the cystic fibrosis transmembrane conductance regulator (CFTR), where deletion of one residue out of the 1480 amino acids in the sequence, prevents the protein from efficiently folding (6). This seemingly subtle change in the primary sequence results in a fatal disease (7–10).

## B. Primary Intramolecular Forces Driving Polypeptide Folding

The folding of a disordered polypeptide chain to a highly ordered, compact structure occurs as the system reaches its kinetically accessible free energy minimum. Many intramolecular forces must be balanced against each other to achieve a stable polypeptide structure (11). These forces include, but are not limited to, steric hindrance of some conformations, favorable *van der Waals* interactions,

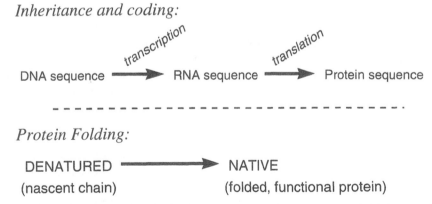

**Fig. 1** Genetic coding for a functional protein structure. The initial steps in the transmission of information from DNA involve the transcription of the sequence into RNA and the subsequent translation into a protein sequence. This linear protein sequence contains subtle information that directs the chain to fold into a specific native conformation that allows it to perform its molecular function. The way in which the information is deciphered in the initial steps is straightforward and well understood. In contrast, the manner in which the protein sequence encodes the information to fold is complex, and the code has not been deciphered.

cooperative hydrogen bonding, ionic interactions and the disordering of the solvent. Entropic forces resulting from interactions between the polypeptide and the surrounding solvent are the dominant forces in polypeptide folding.

## II.  ANALYSIS OF POLYPEPTIDE FOLDING

Although much is known about the forces that contribute to the stability of the native fold, less is known about the specifics of the complicated process by which the native fold is achieved. In fact, solving the conundrum of how primary sequence leads to one specific native conformation out of the myriad of possible stable structures is one of the great intellectual challenges of the past fifty years. A large number of experimental and theoretical studies have added to our knowledge, but the problem remains unsolved.

To date, theoretical work has focused on simple polymer systems or on extracting rules by examining sequences with known three-dimensional structure (12–15). This computational work is currently restricted to fairly simple force fields due to the available computational capacity. Extracting rules about how proteins fold from observations of known three-dimensional structures assumes that the ability to fold is dictated by the same forces that provide native state stability and largely neglects the contribution of the primary sequence in defining the kinetics of folding by affecting unfolded and partially folded structures, as well as the native state. Experimental work has provided a wealth of information on the energetics and kinetics of folding of specific proteins, but such studies are, by necessity, limited to a small fraction of sequence space.

Two main views of the protein folding problem that have arisen from this work are presented in Fig. 2. One view stresses the commonalties in the process, while the other stresses the diversities. Traditionally, experimentalists have represented folding as a stepwise process from the denatured state through intermediate states to the native state. This representation considers folding to be much like a sequential chemical reaction and highlights the conformational relatedness of intermediate states, making them accessible to experimental characterization. It has long been recognized that the polypeptide chain has a huge entropic penalty to pay during folding as the multitude of conformations open to the denatured protein collapses to the unique conformation of the ordered, native structure. Thus, even though only a limited number of paths to the native state can exist for folding to occur on a biologically relevant time scale, it is naive to consider the early steps part of a discrete pathway. Differences between the multiple denatured conformations and the initial heterogeneity in the pathway are highlighted in a theoretical representation of folding as an energy landscape. The depth, width and surface characteristics of this energy landscape are determined by the primary sequence (16, 17). This view of the problem is a useful thought tool because it emphasizes the diversity of initial steps in folding, which occur as the compact molten globule intermediate is formed (Fig. 2). As the later, more discrete, conformational steps occur, the two views merge.

The stability of the native state can be measured by titration of denaturant into the system and the determination by a spectroscopic or biochemical assay of the equilibrium constant between the folded and denatured state under each condition (Fig. 3A). The Gibbs free energy for folding is determined from this equilibrium constant ($\Delta G^{0,\text{fold}} = -RT\ln K_{eq}$) and then plotted against the denaturant concentration. Extrapolation of the line to zero denaturant provides a measure of $\Delta G^{0,\text{fold}}$ in the absence of the perturbation.

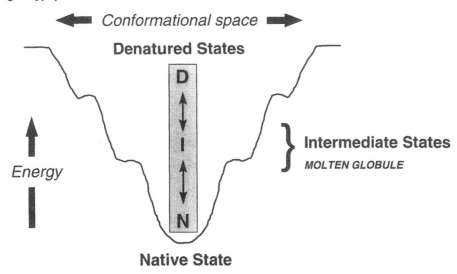

**Fig. 2** Two views of the protein folding problem. The process of folding from the multitude of conformations that comprise the denatured state (D) through related intermediates (I) to the native state (N) typically has been represented as a reaction pathway (shaded area) by experimentalists. This view stresses the commonalties of the intermediates, which are amenable to measurement, and the limited number of pathways by which the chain can assume the native fold. The thermodynamics and kinetics of the interconversions of these conformers provide insight into the structural basis of folding. An alternate way to look at the process is as an energy landscape, or folding funnel, whose height represents an energy function and whose width represents a configurational function. This view stresses the large number of conformations that are available to the denatured chain and the diversity of paths to the native state.

The slope of this line, the *m* value, reflects the surface area which is buried upon formation of the native state (18). The point at which the line crosses the x axis, where $\Delta G^{0,fold} = 0$, is the melting concentration, $C_m$. The details of these thermodynamic methods have been covered extensively elsewhere (19, 20).

To measure the kinetics of the folding process, the protein is exposed to a denaturant concentration sufficient to stabilize the denatured state (*i*, Fig. 3A). To initiate refolding, the denaturant concentration is then reduced by dilution to a concentration where the native state is stabilized (*ii*, Fig. 3A). Typically folding occurs on a time scale of milliseconds to hundreds of seconds. In the case of rapid folding, sophisticated mixing techniques must be employed (21, 22).

In kinetic studies, the fraction of the protein that exists in various conformations is typically monitored spectroscopically (Fig. 3B). The presence of multiple exponentials during the kinetic refolding process is indicative of multiple steps. These steps may be the result of either sequential or parallel events, and these models must be distinguished kinetically. The denaturant dependence of the refolding and unfolding rates can be determined for individual steps. Extrapolation to zero denaturant concentration, when *ln*k is plotted *versus* denaturant concentration provides an estimate of the folding and unfolding rates for each of the steps in the absence of denaturant. The equilibrium constant and, thus, $\Delta G^{0,fold}$ for each of the intermediates can then be calculated from these derived rate con-

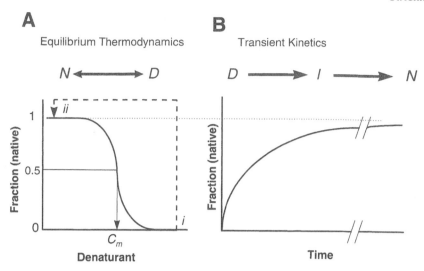

**Fig. 3** Thermodynamics and kinetics of folding. The stability of a protein can be measured by monitoring the fractional population folded at equilibrium as a function of a denaturant such as urea or guanidinium hydrochloride. The kinetics of the folding process can be determined experimentally by monitoring the fractional population folded as a function of time after a perturbation of the denaturant concentration from denaturing to native conditions.

stants (21, 22). In addition, the slope of the denaturant dependence of the unfolding rate constant provides an indication of the height of the transition barrier for the conformational conversion (23, 24). Alternatively, the temperature dependence of the rate constant can be used to determine the height of the barrier (25). However, the non-linearity of the temperature dependence at high temperatures, which is due to the destabilization of necessary intermediates, may complicate this interpretation. Kinetic studies, such as those described above, performed on a small number of proteins have begun to define the shape of some representative folding energy landscapes.

Understanding the relationship between the kinetics and energetics of folding and the effects of denaturants is important for the design of successful *in vitro* refolding protocols. As shown in Fig. 4, not only do the relative populations of the various conformational states change

as a function of denaturant concentration, but so does the protein solubility. Agents that tend to be excluded from the protein surface lead to preferential hydration and, thus, increased stability, but decreased solubility (discussed further in the following section). In contrast, denaturants preferentially interact with (bind to) the surface of the protein. Therefore, both polar and apolar side chains are more soluble in denaturants, and the native state is destabilized due to its reduced accessible surface relative to the unfolded chain. As such, at high denaturant concentrations the protein is most soluble—*albeit* in denatured form (*segment d*, Fig. 4). At denaturant concentrations corresponding to the transition region between folded and unfolded conformers the protein is preferentially hydrated as the denaturant is removed, leading to folding. However, the polypeptide's solubility is low under these conditions as the folded conformation is only

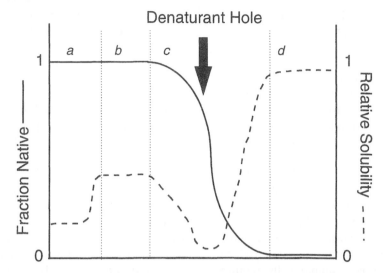

**Fig. 4** Protein solubility and conformation at various denaturant conditions. Protein solubility and conformation vary with denaturant concentration in a complex way. Proteins are most soluble at high denaturant concentrations (*d*) where both polar and apolar side chains are soluble. However, under these conditions, and for the same reasons, the protein is unfolded. At the lowest denaturant concentrations (*a*) the native state is stable, but so might be partially folded intermediates which tend to self-associate and precipitate from the solution. Therefore, many times it is most effective to fold a protein at a denaturant concentration (*b*) adequate to stabilize the native state but not partially folded intermediates. Moreover, these conditions also may serve to increase the solubility of the native conformer by interacting with any exposed apolar surfaces that would otherwise cause increased solvent order. It is also wise to avoid conditions near the transition region (*c*) during refolding, where both native and denatured conformers are of limited solubility.

partially populated, and the amount of denaturant available to interact with the unfolded surfaces is limiting (*segment c*, Fig. 4). Thus, exposure to these conditions should be minimized in optimal refolding protocols. Not only is solubility low, but folding is slow, thus maximizing the transient accumulation of partially-folded states. The most successful folding will occur at denaturant concentrations inadequate to destabilize the native state (*segment b*, Fig. 4). Under these conditions partially-folded states are not populated, and the denaturant present can further aid in the solubility of the native state, especially if the polypeptide in question is a domain with an exposed non-polar surface. In contrast, at the lowest denaturant concentrations, partially folded

intermediate states can become significantly populated relative to the native state (*segment a*, Fig. 4). These intermediates may be prone to self-association, thereby reducing solubility.

## III. METHODS OF FOLDING POLYPEPTIDES *IN VITRO*

Several steps are necessary for the *in vitro* folding of polypeptides that do not fold in heterologous expression systems (26–28). These steps include synthesis of the polypeptide, its solubilization if it has accumulated in inclusion bodies, its purification and its folding *in vitro* by reducing the concentration of denaturant. The section that follows will address some practical issues for *in vitro* protein folding.

## A. Recombinant Polypeptide Expression Systems

There are four major expression systems for recombinant polypeptide production: bacteria, yeast, mammalian cells and insect cells. Although mammalian cells might, at first glance, appear to be the natural choice for the production of most polypeptides of therapeutic interest because they can correctly glycosylate polypeptides and more easily fold large, multi-domain polypeptides, the majority of recombinant polypeptides are produced in *Escherichia coli*. Eukaryotic expression systems have several potential limitations that reduce their utility. First, animal cells require longer growth times to achieve maximum cell density and achieve a much lower cell density ($1-10 \times 10^6$ cells/ml) than do *E. coli* ($1 \times 10^8$ cells/ml). In addition, animal cells are often grown in expensive media. The high costs, technical difficulties and often low polypeptide expression levels in animal cells have made *E. coli*-based products the preference of manufacturers.

In order to avoid the problems that accompany animal cell expression, bacterial systems are chosen frequently because they usually provide a high level of expression of the desired polypeptide at low cost. Bacterial systems such as *E. coli*, however, lack the ability to secrete large polypeptides into the culture medium and the cellular machinery to facilitate proper folding in cases where glycosylation and multi-domain folding must occur. Therefore, the expression of polypeptides with these characteristics, and overexpression in general, usually results in the formation of insoluble inclusion bodies containing misfolded conformations of the recombinant protein (29). The compromise to be made in using low cost and high expression bacterial systems compared to eukaryotic systems is that strategies for solubilizing inclusion bodies and correctly folding polypeptides *in vitro* are often required.

## B. Solubilization of Unfolded or Misfolded Polypeptides

Successful solubilization of inactive aggregates, which frequently accumulate during overexpression of a recombinant polypeptide, is clearly the first step in *in vitro* refolding (26, 28, 30). Which chemical agents are used to solubilize the aggregates influences the choice of refolding strategy. If the polypeptide retains substantial native conformation when embedded within an aggregate, then it might seem logical to choose a weak denaturant for solubilization, a denaturant that disrupts intermolecular interactions without completely disrupting the protein structure. However, weak denaturants are frequently ineffective, and most insoluble polypeptides must be solubilized with solvents that completely denature the polypeptide. Thus, choosing a strongly denaturing solubilizing agent such as 6 M guanidinium hydrochloride (GdnHCl) or 8 M urea, which completely denatures the protein, requires the development of a refolding protocol (31). The addition of reducing agents such as dithiothreitol (DTT) or glutathione may also be necessary for solubilization of polypeptide aggregates containing intermolecular or non-native disulfide bonds (30).

The ideal denaturant to choose will vary for any given polypeptide; however, certain criteria that allow downstream processing should be met, including:

> Ability to solubilize the polypeptide aggregates rapidly.
> Lack of chemical reactivity towards labile amino acid residues in the polypeptide.
> Prevention of oxidation of thiol groups in the polypeptide or their protection.
> Compatibility with subsequent separation methods such as column chromatography.
> Inhibition of proteolytic degradation.

## C.  Strategies for Polypeptide Folding In Vitro

Successful folding of polypeptides into a stable, functional native state is a critical step in modern biotechnology. The determinants of native state stability in aqueous solution are the amino acid sequence of the polypeptide, as well as the variable conditions of pH, temperature, protein concentration, ionic strength and the concentration of other additives (32, 33). For successful *in vitro* folding of a polypeptide with a fixed primary sequence, the goal is to adjust the variable folding conditions to shift the conformational equilibrium toward the folded state. As shown in Fig. 4, this can be accomplished by removing the denaturant so that the polypeptide is at low denaturant concentration and favors the native state. Although, unfortunately, there are no simple, general recipes for successful *in vitro* folding of a given polypeptide, some strategies that may be successful and some examples of polypeptides with which these strategies have worked will be discussed below.

### 1.  Single Stage Reduction of the Denaturant Concentration

Once soluble polypeptide has been recovered from aggregate particles under denaturing conditions, such as 8 M urea or 6 M GdnHCl plus 10 mM DTT, any insoluble material remaining is removed by centrifugation prior to initiating re-folding. The completely unfolded and linearized polypeptide is then refolded by decreasing the concentration of the denaturant sharply by rapid dilution with a refolding buffer. The refolding buffer should contain dithiothreitol (DTT), $\beta$-mercapto-ethanol or reduced and oxidized glutathione if the polypeptide contains multiple cysteine residues, and other additives may be included in the refolding buffer to increase the folding

yield (see below) (26). Folding is then allowed to proceed at low temperature and low polypeptide concentration in order to minimize higher order off-path-way processes. An advantage of the rapid dilution method is that it avoids conditions near the denaturant hole (Fig. 4). The protein can then be concentrated after refolding is complete.

The final polypeptide concentration at which folding is carried out is an important factor for successful *in vitro* folding. Obviously, the ideal scenario would be to fold at a high concentration of polypeptide in a small volume; this would be particularly useful for industrial production of polypeptides. However, when denaturant is diluted out at high polypeptide concentration, the frequent result is aggregation of unfolded and partially folded conformers. Unlike folding, which is an intramolecular first order process, aggregation has higher order reaction kinetics and is highly concentration dependent. Therefore, when folding *in vitro* one must always balance the desire to achieve high polypeptide concentrations in the folding reaction against the threat of increased aggregation. A typical starting concentration would be 20 $\mu$M or less for proteins that fold slowly or have particularly association-prone folding intermediates.

### 2.  Staged Reduction of the Denaturant Concentration

Proteins also have been renatured by a process that involves denaturant removal in two or more stages. An example of the successful use of staged dilution in renaturation is the folding of reduced chymotrypsinogen from GdnHCl, a protein refractory to refolding by other methods (34). For chymotrypsinogen, initial dilution to below 0.5 M GdnHCl or above 3 M GdnHCl leads to sharply increased precipitation. The midpoint for the equilibrium unfolding of the oxidized protein is 2.5 M GdnHCl, and at 1 M GdnHCl the oxidized protein is fully folded. Orsini and Goldberg, thus, successfully refolded chymotrypsinogen by dilution from 6 M GdnHCl into 2 M urea followed by removal of the residual denaturant by dialysis (35).

A variation on the theme of staged dilution of a protein from its initial denaturant is to dilute the protein into a different denaturant prior to refolding. Two examples of this are the successful refolding of chymotrypsinogen by dilution from 6 M GdnHCl into 2 M urea followed by dialysis and the successful refolding of GdnHCl-denatured aldolase by dilution into 2.3 M urea, a solvent that prevented aggregation of folding intermediates (35, 36). These examples demonstrate that the buffer component initially used to denature the aggregated protein need not necessarily be the same solubilizing agent used in the intermediate refolding step.

In general, the success of the staged dilution method likely is due to the inclusion of low denaturant concentrations, which destabilize partially folded intermediates, although they are not sufficient to destabilize the final native conformation (Fig. 4). Moreover, low denaturant concentrations may increase the solubility of the partially folded forms that are populated during refolding. A dilution step to a low denaturant concentration where the stability of both transient and equilibrium folding intermediates is low should be the first step in removing the denaturant. When designing a staged refolding strategy, care should be taken to avoid denaturant concentrations near the low solubility denaturant hole (Fig. 4). After equilibrium is achieved the remaining denaturant is then removed by dialysis. As is also the case in the rapid dilution method, the efficiency of folding will be maximized at low protein concentrations, which favor the first order folding reaction over higher order off-pathway reactions.

## 3. The Use of Refolding Buffer Additives to Reduce Aggregation and Promote Folding

The refolding yield of many polypeptides is increased by the presence of specific compounds in the refolding buffer, which either stabilize the native state or increase the solubility of the transient, partially folded forms. L-arginine, salts, organic solvents, the non-detergent sulphobetaines (NDSBs), some detergents, cellular osmolytes and other miscellaneous compounds have all been used with success to improve folding yields (37, 38). A partial list of these is shown in Table 2. Some of these additives may be removed once folding has occurred.

Improved refolding of recombinant pro-urokinase was first achieved by including 0.2 M arginine in the refolding buffer (39). Subsequently, arginine at concentrations of 0.2 M to 0.8 M dramatically improved recoveries of tissue plasminogen activator (TPA) (4), soluble human IL-6 receptor (40), immunoglobulin fragments (41) and recombinant CFTR NBD1 (10). L-arginine is an amphiphile that may serve to protect exposed hydrophobic surfaces during folding in addition to promoting preferential hydration of the polypeptide.

Because ionic interactions are involved in protein folding, varying the ionic strength of the refolding buffer may improve the folding of some proteins. For example, at pH 12.7 and low ionic strength, creatine kinase denatures gradually; however, with the addition of salt the poorly-structured unfolded enzyme undergoes a conformational change at the same pH to assume a compact conformation typical of hydrophobic collapse (42).

Organic solvents have improved folding in some systems. Ethanol (20%), for example, improved the folding of recombinant human insulin-like growth factor I (IGF-I) by varying the polarity of the solvent (43). Trifluoroethanol (TFE)

**Table 2** Some Small Molecules that Enhance Protein Folding

| Additive | Putative mechanism of action |
| --- | --- |
| Glycerol | Preferential hydration of protein/Increases solvent viscosity |
| L-Arginine | Amphiphile/Osmolyte |
| Glycyl betaine/sorbitol | Preferential hydration of protein |
| Arabanose | Increases solvent viscosity |
| Xylitol | Increases solvent viscosity |
| Ethanol | Modulates solvent polarity |
| DMSO | Modulates solvent polarity |
| Zwitterionic detergents | Protects exposed non-polar surfaces |
| Triton X-100 | Protects exposed non-polar surfaces |
| Non-detergent sulfobetaine | Protects exposed non-polar surfaces |
| Sucrose | Preferential hydration of protein/Increases solvent viscosity |
| Trimethylamine N-oxide (TMAO) | Preferential hydration of protein |
| Trifluoroethanol (TFE) | Promotes formation of secondary structure |
| Low guanidinium hydrochloride | Destabilizes intermediates/Increases solubility of the native state |
| Low urea | Destabilizes intermediates/Increases solubility of the native state |
| Ligands | Stabilize native state |
| Polyethylene glycol (PEG) | Protection of molten globule/Increases solvent viscosity |

induces increased alpha-helical structure in many proteins in aqueous solution (44). For proteins where folding is limited by the formation of helices or by misfolding due to excessive beta structure, TFE may improve the folding yield.

To minimize aggregation on the folding pathway, it may be desirable to include solubilizing agents in the refolding buffer. The non-detergent sulphobetaines (NDSBs) are a new family of amphiphilic solubilizing molecules that dramatically improve the folding yield of some proteins (45). In the cases of hen egg white lysozyme and bacterial $\beta$-D-galactosidase, NDSB improved the yield of native lysozyme up to 12-fold and the yield of $\beta$-D-galactosidase up to 80-fold (45). NDSBs strongly affect the balance between aggregation and folding and are effective because of their interactions with folding intermediates. NDSBs with multiple structural attributes have been developed, and some NDSBs may be more useful than others for the refolding of specific proteins.

Because hydrophobic surfaces are important in the structure and function of proteins, detergent micelles and liposomes may be useful in some cases, particularly with membrane-associated proteins (46). For example, extensive hydrophobic interactions in the interdomain region of rape seed oleosins and rhodanese stabilize the intact protein structure and are important for function (46–49). Recovery of active proteins such as these after unfolding or after heterologous expression is often kinetically limited by aggregation of transient folding intermediates with exposed hydrophobic surfaces. However, both rape seed oleosins (47, 48) and the enzyme rhodanese (46, 49) were efficiently refolded with the addition of detergent micelles, liposomes or artificial oil-bodies, presumably by preventing aggregation due to the interaction of transiently exposed hydrophobic surfaces.

Some reagents known to stabilize proteins in their native conformation, such as the cellular osmolytes glycerol and trimethylamine N-oxide (TMAO), have been referred to as chemical chaperones because of their influence on protein folding (37, 50, 51). These osmolytes are thought to stabilize the folded state of polypeptides through surface tension effects, which lead to preferential solvation of the protein (38). The addition of xylose to refolding *staphylococcal* nuclease significantly increased the rate of refolding, presumably by stabilizing the native conformer (52). Another osmolyte, glycerol, at a concentration of 10%, inhibits the aggregation of a nucleotide binding domain (NBD1) of CFTR *in vitro* (53). Glycerol (10%–40%) has improved the folding yield of other polypeptides as well (54–56). The role of glycerol may be to increase the viscosity of the folding solution and retard higher order associations in addition to preferentially hydrating the protein and, thereby, stabilizing its native structure (50, 57, 58). The effectiveness of osmolytes in promoting folding is illustrated in an *in vitro* model of prion disease where the fundamental event may involve a conformational change in the alpha-helices of the cellular prion protein (PrP$^C$) into beta-sheets during the formation of the pathogenic isoform (PrP$^{Sc}$). The chemical chaperones (osmolytes) mentioned above interfere with the formation of PrP$^{Sc}$ from newly synthesized PrP$^C$ (59).

Other miscellaneous compounds also have been shown to increase folding. The folding yield of bovine carbonic anhydrase B could be increased by the addition of polyethylene glycol (PEG) to refolding buffers (60, 61). Glycine (0.1 M) aided the recovery of relaxin (62), while the addition of heme and calcium improved the refolding of horseradish peroxidase (63). Not surprisingly, inclusion of substrates and cofactors in a refolding mix stabilizes the native state and often improves the refolding yield.

### 4. Chemically Modifying or Mutating the Protein to Make the Refolding Species More Soluble During Renaturation

In cases where protein self-association is responsible for the lack of functional monomer, genetic engineering of the polypeptide at the association interface may minimize self-association. By mutating two residues known to be at the interface of the self-association in insulin, for example, it is possible to minimize the oligomerization process and increase the yield of soluble monomer (64). In another example, engineered mutations in the death-effector domain of the apoptotic protein FADD dramatically increased its solubility to levels acceptable for NMR studies (0.7 mM) (65).

### 5. Reducing Aggregation with Physical Barriers to Intermolecular Interaction

Because intermolecular interactions leading to aggregation of misfolded polypeptides are a significant hindrance to proper folding, strategies that minimize these interactions may increase the yield of correctly folded protein. One way to accomplish this is to physically limit the diffusion of refolding monomers by binding them to a resin. Monomeric α-glucosidase fused with a hexa-arginine tag, which is prone to aggregation under normal folding conditions, was bound, for example, to a resin while solubilized in denaturant. Then the denaturant was removed, leading to folding at high yield (66). A second example is the matrix-assisted folding of the chloroplast membrane proteins Toc75 and LHCP recombinantly expressed in *E. coli*. Both were bound by their hexa-histidine tags to a nickel column and folded by the exchange of urea and a strong detergent for a mild detergent (67). The success of matrix-supported folding of these two proteins is particularly encouraging since integral membrane proteins are typically difficult to fold *in vitro*.

Care must be taken in matrix-assisted folding to optimize the system. For example, because folding occurs on the solid-state matrix, the matrix itself must not interfere with folding. Thus, the ionic strength of the column buffer must inhibit interactions between the fusion peptide and the matrix itself, particularly if folding is occurring on an ion exchange resin (66). If the polypeptide is to be bound to the matrix through a fused tag, then fusion to either the N- or the C-terminus may be preferable. The length of the linker between the tag and the polypeptide to be folded may also be varied to improve the results.

### 6. Molecular Chaperone Assisted Folding

In cells, many polypeptides fold with assistance. Partly because of the complications of folding in a solution of high protein concentration and partly because during synthesis the polypeptide is incrementally exposed as it emerges from the ribosome, cellular folding of polypeptide chains often occurs with the assistance of molecular chaperones and folding catalysts, such as prolyl isomerases and disulfide isomerases (68). Prokaryotic and eukaryotic cells fold polypeptides differently and, therefore, require different chaperone machinery to facilitate folding. In prokaryotes folding of most polypeptides is post-translational (69). Completed proteins are released from the ribosome before folding begins. For small domains this is frequently a sufficient mechanism for folding; however, many larger polypeptides synthesized in prokaryotes are assisted in their folding by binding to molecular chaperones. Chaperones of the Hsp60 class (GroEL/ES), which shelter partially folded polypeptides in their central cage until folding can be completed, are required for the folding of larger polypeptides by prokaryotes (70). The size of the cage restricts GroEL/ES to substrates 50 kDa or less in size. In eukaryotes the protein translation and folding process is different (69). At least

some polypeptides begin to fold co-translationally, while still bound to the ribosome. Hydrophobic portions of the polypeptide sequence are bound by Hsp70 class chaperones, which prevent inappropriate interactions of partially folded chains (68, 71). Co-translational folding is helpful for multi-domain protein folding, which would likely be difficult if the partially folded polypeptide is free in solution.

Molecular chaperones and folding catalysts may provide a biological alternative to the use of osmolytes as chemical chaperones in the folding process. These proteins might be employed to aid in the folding of heterologous substrates in the same way that they normally assist the folding of polypeptides into native structures. Chaperone proteins may be used to promote folding in two ways. First, in several examples, overexpression of GroEL in E. coli increased the proportion of soluble folded protein by disfavoring the formation of inclusion bodies (72–74). A notable example of this is the use of GroEL to increase the folding of Rubisco in E. coli (75). In these cases, the need for in vitro folding is avoided from the outset. Alternatively, chaperone molecules may be added in vitro to increase folding. Hsc70, for example, improves the in vitro folding of CFTR NBD1 (76).

A carefully developed refolding protocol is characterized by optimal recombinant polypeptide expression, solubilization of inclusion bodies and removal of denaturant to promote folding. Each of the steps above, as well as possible protein engineering of the polypeptide to be folded or the addition of either chemical agents or proteins that promote folding, must be empirically optimized for each protein of interest. The result of this effort, however time-consuming it may be, is the production of potentially large quantities of correctly folded polypeptides for biochemical experimentation or therapeutic use.

## IV. PERSPECTIVES AND CONCLUSION

### A. Protein Folding Diseases

The molecular phenotype of a growing number of human diseases is a defect in protein folding. In protein folding diseases, mutant polypeptides do not fold correctly, but instead, are destined for either accumulation into insoluble aggregates or proteolysis. A list of prominent protein folding diseases is given in Table 1. Notably, in several cases the mutant protein is functional when forced into the native conformation. In other cases the misfolded protein is toxic to the cell. The implication for drug design is clear: new strategies for the treatment of disease must be developed which can facilitate proper protein folding.

As illustrated in Fig. 5, disease-causing mutations may destabilize the mutant protein compared to the wild type protein. This is the case for the Z type mutation in α-1-antitrypsin, which is associated with emphysema (77); fibrillin, which leads to Marfan syndrome (78); and misfolding of the LDL-receptor, which leads to familial hypercholesterolemia (79). These diseases demonstrate that correct functioning of a protein is dependent upon its ability to form a stable native state as dictated by the primary amino acid sequence. One folding disease characterized from a mechanistic perspective is cystic fibrosis where the prevalent disease-causing ΔF508 mutation in CFTR is temperature sensitive for folding (*tsf*) (6). An isolated CFTR domain, NBD1, which contains the ΔF508 mutation, was shown to have the same native state stability, but

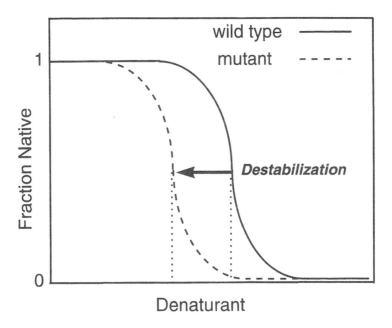

**Fig. 5** Mutational effects on protein stability. A genetic disease may be caused by a mutation that leads to aberrant protein folding in several ways. As shown in the figure, a mutation may lead to a loss of thermodynamic stability. Thus, under physiological conditions the mutant does not fold into a stable structure. Other mutants may prevent efficient folding of a protein that would be stable under physiological conditions by altering the kinetics of the process. These mutants may or may not alter the thermodynamic stability of the protein.

different folding yields for the wild type and mutant, reflecting the kinetic nature of the defect (10). This demonstration that the temperature-sensitive defect is on the protein folding pathway explained the original observations that the mutation destabilized a peptide model of a folding intermediate of CFTR NBD1 (80) and that the ΔF508 CFTR chloride channel only folds in cells grown at low temperature but retains function after a shift to 37°C (8).

Protein folding diseases may be classified according to the manner by which misfolding leads to the disease state. For diseases such as cystic fibrosis the disease is caused, in most cases, by a lack of functional protein (6, 7). However, in other diseases such as Alzheimer's disease, the pathology is related to the accumulation of toxic misfolded proteins (1, 2, 81). In either case, however, the goal of drug design is the same—to promote folding to the functional native state rather than formation of potentially toxic off-pathway conformers.

## B. Implications for Drug Design

One can envision that protein folding reactions can be beneficially manipulated to treat human disease. Several possibilities for this are outlined below and fall into two classes. In the first category, it may be possible to ameliorate the pathology of a disease by enhancing the folding of proteins that misfold in the disease state. One way to accomplish this may be through the use of low molecular weight compounds that increase folding. Glycerol and TMAO, which, for example, are known to stabilize proteins in their native conformation (see section above on additives), are effective in correcting the temperature-sensitive protein folding defect associated with ΔF508 CFTR in cells (51, 82, 83). These observations indicate that protein stabilizing agents are effective for correcting protein folding

abnormalities associated with human diseases *in vivo*. However, it is important to note that these agents lack specificity and may have many effects on other cellular systems; their beneficial effects may be indirect. These compounds might cause cells to upregulate a stress response that increases chaperone levels and protein folding, for example.

The folding yield may also be increased by shifting the conformational equilibrium toward the native state. Native state stabilization may occur in the presence of a specific ligand, as illustrated in Fig. 6. For example, by adding the ligand thyroxin to transthyretin, the formation of off-pathway fibrils was inhibited (84). When the native state tetramer is stabilized, transthyretin is less likely to populate the monomer intermediate state prone to fibril formation (84). In another example, the maturation of human P-glycoprotein mutants is increased upon the addition of its substrates (85). Ligand stabilization may not, however, always be successful, and depends on the nature of the mutation's effect on folding. In the case of CFTR, for example, the addition of ATP to CFTR NBD1 did not increase the folding yield *in vitro*, presumably due to the kinetic nature of the defect (53). Therefore, in the case of CFTR NBD1, the conformation that binds ligand must be kinetically isolated from the step in the folding process that is affected by the mutation.

Peptide scaffolds, which functionally replace mutated sequences during the folding process, might be used to facilitate folding of mutant proteins. Several examples have been reported of peptides that interact with a protein either to enhance or inhibit the folding of the protein from which they arose. Taniuchi and Anfinsen demonstrated that two proteolytic fragments (A and B in Fig. 7) of the single domain protein *staphylococcal* nuclease are unable to fold in isolation (86). However, when fragment A and

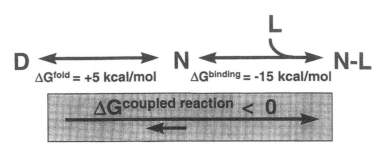

*Protein Folding - Ligand Binding*

**Fig. 6** Energetic coupling to promote protein folding. When a mutation causes a disease by destabilizing the thermodynamic stability of a protein, as shown in Fig. 5, coupling of an energetically favorable reaction to the unfavorable folding reaction may prove effective in promoting formation of the native state. Ligands that bind with high affinity to the native state but do not interact with the denatured chain would provide one avenue to the development of such a drug. However, in cases in which the folding defect is kinetic in nature such an approach would only prove effective when the defective step is not kinetically isolated from the binding conformation. Similar order-disorder transitions also provide a means of generating protein-ligand interactions that are specific but of low affinity, as is required for many regulatory functions. These apparently opposed characteristics can be achieved when a large interaction interface is utilized for specificity and an unfavorable folding (ordering) reaction is coupled to binding, thereby reducing affinity. Conceivably, a second ligand which stabilizes the ordered binding conformation by interacting at a remote site could be used to increase the affinity and prolong the regulatory signal.

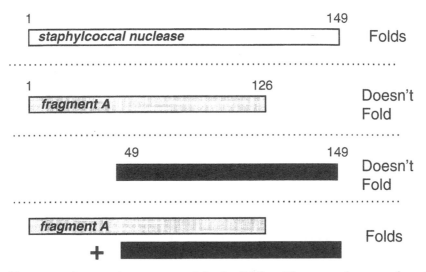

**Fig. 7** Trans-complementation to correct defective folding. There are a large number of examples of truncated proteins which cannot fold being repaired by providing the missing sequence on a separate polypeptide chain. The figure, drawn from the data of Taniuchi and Anfinsen (95), offers a particularly dramatic example in which two overlapping fragments of *staphylococcal* nuclease interact for proper folding. Such trans-complementation has long been used in genetic screens and could, conceivably, find utility in the treatment of folding diseases.

fragment B are combined and folding is initiated, active nuclease is formed. Further evidence indicates that the fragments compliment each other by forming a heterodimer in which the extraneous sequences from each fragment are excluded from the newly folded catalytic domain. This is probably the first example of subdomain swapping, an emerging mechanism for oligomer assembly (87). The complementation of S-peptide and S-protein to form ribonuclease A (88) and the complementation of fragments of $\beta$-galactosidase (89, 90) are additional examples of trans-complementation leading to folding.

The use of either a ligand or a peptide scaffold to promote native structure formation also may be utilized to modulate other order-disorder transitions in a polypeptide of interest (Fig. 6). For example, several polypeptides, which become ordered upon binding to a partner molecule, as in the case of p21 binding to Cdk2 (91) or retinoid × receptor (RXR) binding to DNA (92), apparently utilize the favorable conformational energy to achieve specific binding and a rapid off rate as required for effectiveness as regulatory molecules. Promoting this conformational change with a second interaction at a distinct site could be used to prolong binding to the original partner molecule.

The second category of beneficial ways to manipulate protein folding reactions in treating human disease would be to prevent the formation of a functional conformation when the native protein has a detrimental activity. Peptide scaffolds that detour polypeptides off the productive pathway might accomplish this. Several well-studied examples of peptides inhibiting the folding of the protein from which they were derived have been published. For example, a peptide corresponding to residues 102 to 154 of dihydrofolate reductase (DHFR) was shown to inhibit refolding of the enzyme by greater than 80% (93). Two other DHFR peptides had no effect, indicating that there is specificity in the process. In addition, Brems and coworkers reported that a 38 amino acid fragment of bovine growth hormone decreased the kinetics and yield of growth hormone refolding by interacting with an intermediate on the folding pathway (94).

The pitfall of many of these strategies is, of course, that specificity is typically required for effective drug design. Specificity must be carefully considered in the strategies discussed here because the polypeptide conformation to be acted upon might have little or no higher order structure. From this perspective, ligand stabilization of a native structure seems the most promising approach.

Basic research into the mechanism of folding polypeptides *in vitro* may provide guiding principles that are necessary for the design of agents either to promote or prevent the folding of polypeptides *in vivo*. Alternatively, in the absence of a structure-based approach, the search for possible therapeutic agents could be based on a high throughput assay for protein folding/misfolding. For example, the kinetic partitioning of $\Delta$F508-CFTR NBD1 between folding and aggregation can be assayed easily by monitoring turbidity as the misfolded mutant protein aggregates (53, 76). Such assays are particularly amenable to adaptation to a 96 well plate format for screening a library for various agents that improve mutant protein folding.

A clear understanding of how polypeptides follow a kinetic pathway and thermodynamic rules to reach their native structures is increasingly important, not only for successful *in vitro* folding of polypeptides, but also for the development of novel strategies to manipulate protein folding processes to treat human disease. The success of both of these ventures is directly related to our level of understanding of the protein folding problem. As we learn in greater detail the rules polypeptides use to fold correctly, we will be better equipped to shift the conformational equilibrium toward (or away from, as the case may be) the native state, and we will be capable of treating new classes of human disease.

## REFERENCES

1. PJ Thomas, B-H Qu, PL Pedersen. Defective protein folding as a basis of human disease. TIBS 20: 456–459, 1995.
2. JW Kelly. Altenative conformations of amyloidogenic proteins govern their behavior. Cur Opin Struc Biol 6: 11–17, 1996.
3. DV Goeddel, DG Kleid, F Bolivar, HL Heyneker, DG Yansura, R Crea, T Hirose, A Kraszewski, K Itakura, AD Riggs. Expression in Escherichia coli of chemically synthesized genes for human insulin. Proc Natl Acad Sci USA. 76: 106–110, 1979.
4. S Fischer, R Rudolph, R Mattes. European Patent Application (0 393 725A1). 1986.
5. CB Anfinsen. Principles that govern the folding of protein chains. Science 181: 223–230, 1973.
6. B-H Qu, E Strickland, PJ Thomas. Cystic Fibrosis: A Disease of Altered Protein Folding. J Bioenerg Biomem 29: 483–490, 1997.
7. SH Cheng, RJ Gregory, J Marshall, S Paul, DW Souzǎ, GA White, CR O'Riordan, AE Smith. Defective intracellular transport and processing of CFTR is the molecular basis of most cystic fibrosis. Cell 63: 827–834, 1990.
8. GM Denning, MP Anderson, JF Amara, J Marshall, AE Smith, MJ Welsh. Processing of mutant cystic fibrosis transmembrane conductance regulator is temperature-sensitive. Nature 358: 761–764, 1992.
9. PJ Thomas, P Shenbagamurthi, J Sondek, JM Hullihen, PL Pedersen. The cystic fibrosis transmembrane conductance regulator. J Biol Chem 267: 5727–5730, 1992.
10. B-H Qu, PJ Thomas. Alteration of the cystic fibrosis transmembrane conductance regulator folding pathway: effects of the ΔF508 mutation on the thermodynamic stability and folding yield of NBD1. J Biol Chem 271: 7261–7264, 1996.
11. KA Dill. Dominant forces in protein folding. Biochemistry 29: 7133–7155, 1990.
12. KA Dill, S Bromberg, K Yue, KM Fiebig, DP Yee, PD Thomas, HS Chan. Principles of protein folding–a perspective from simple exact models. Protein Science 4: 561–602, 1995.
13. D Shortle. Structure prediction: folding proteins by pattern recognition. Current Biol 7: R151–R154, 1997.
14. M Karplus, A Sali. Theoretical studies of protein folding and unfolding. Cur Opin Struc Biol 5: 58–73, 1995.
15. A Sali, E Shakhovish, M Karplus. How does a protein fold? Nature 369: 248–251, 1994.
16. PG Wolynes. Folding funnels and energy landscapes of larger proteins within the capillarity approximation. Proc Natl Acad Sci USA 94: 6170–6175, 1997.
17. HS Chan, KA Dill. Protein folding in the landscape perspective: Chevron plots and non-Arrhenius kinetics. Proteins 30: 2–33, 1998.
18. D Shortle. Staphlococcal nulcease: a showcase of m-value effects. Adv Protein Chem 46: 217–247, 1995.
19. BA Shirley. Urea and guanidinium hydrochloride denaturation curves. In BA Shirley, ed. Protein Stability and Folding: Theory and Practice. Totowa, NJ: Humana Press, 1995, pp 177–190.
20. CN Pace, JM Scholtz. Measuring the comformational stability of a protein. In TE Creighton, ed. Protein Structure: A Practical Approach. Oxford: Oxford University Press, 1997, pp 299–321.
21. T Keifhaber. Protein Folding Kinetics. In BA Shirley, ed. Protein Stability and Folding: Theory and Practice. Totowa, NJ: Humana Press, 1995, pp 313–343.
22. MR Eftink, MC Shastry. Fluorescence methods for studying kinetics of protein-folding reactions. Methods Enzymol. 278: 258–286, 1997.
23. A Matouschek, AR Fersht. Application of physical organic chemistry to engineered mutants of proteins: Hammond postulate behaviour in the transition state of protein folding. Proc Natl Acad Sci USA 90: 7814–7818, 1993.
24. A Matouschek, DE Otzen, LS Itzhaki, SE Jackson, AR Fersht. Movement of the position of the transition state in protein folding. Biochemistry 34: 13656–13662, 1995.
25. S-I Segawa, M Sugihara. Characterization of the transition state of lysozyme unfolding. I. Effect of protein-solvent interactions on the transition state. Biopolymers 23: 2473–2488, 1984.

26. R Rudolph, H Lilie. In vitro folding of inclusion body proteins. FASEB J 10: 49–56, 1996.

27. R Seckler, R Jaenicke. Protein folding and protein refolding. FASEB J 6: 2545–2552, 1992.

28. Buchner, J Rudolph, R. Routes to active proteins from transformed micro-organisms. Cur Opin Biotech 2: 532–538, 1991.

29. T Kiefhaber, R Rudolph, H-H Kohler, J Buchner, Protein aggregation in vitro and in vivo: a quantitative model of the kinetic competition between folding and aggregation. Biotechnology 9: 825–829, 1991.

30. B Fischer, I Summer, P Goodenough. Isolation, renaturation, and formation of disulfide bonds of eukaryotic proteins expressed in Eschericia coli as inclusion bodies. Biotechnol Bioeng 1: 3–13, 1992.

31. NK Puri, M Cardamone. A relationship between the starting secondary structure of porcine growth hormone solubilized from inclusion bodies and the yield of native monomeric protein after in vitro refolding. FEBS Lett 305: 177, 1992.

32. T Alber. Mutational effects on protein stability. Ann Rev Biochem 58: 765–798, 1989.

33. CN Pace. Conformational stability of globular proteins. TIBS 15: 14–17, 1990.

34. G Orsini, C Skrzynia, ME Goldberg The renaturation of reduced polyalanyl chymotrypsinogen and chymotryposinogen. Eur.J.Biochem. 59: 433–440, 1975.

35. G Orsini, ME Goldberg. The renaturation of reduced chymotrypsinogen A in guanidine HCl. Refolding versus aggregation. J Biol Chem 253: 3453–3458, 1978.

36. WW Chan, JS Mort, DK Chong, PD MacDonald. Studies on protein subunits. 3. Kinetic evidence for the presence of active subunits during the renaturation of muscle aldolase. J Biol Chem 248: 2778–2784, 1973.

37. WJ Welch, CR Brown. Influence of molecular and chemical chaperones on protein folding. Cell Stress & Chaperones 1: 109–115, 1996.

38. CS Schein. Solubility as a function of protein structure and solvent components. Bio/Technology 8: 308–317, 1990.

39. ME Winkler, M Blaber. Purification and characterization of recombinant single chain urokinase produced in Escherichia coli. Biochemistry 25: 4041–4045, 1998.

40. T Stoyan, U Michaelis, H Schooltink, M Van Dam, R Rudolph, PC Heinrich, S Rose-John. Recombinant soluble human interleukin-6 receptor. Expression in Escherichia coli, renaturation and purification. Eur J Biochem 216: 239–245, 1993.

41. R Rudolph, J Buchner, H Lenz. Process for activation of recombinant protein produced by prokaryotes. (5077392). US Patent 1991.

42. HP Yang, HN Zhong, S Li, HM Zhou. Salt-induced folding of alkaline denatured creatine-kinase under high pH conditions. Biochemistry and Molecular Biology International 41: 257–267, 1997.

43. RA Hart, DM Giltinan, PM Lester, H Reifsnyder, JR Ogez, SE Builder. Effect of environment on insulin-like growth factor I refolding selectivity. Biotechnol Appl Biochem 20 (Pt 2): 217–232, 1994.

44. AI Arunkumar, TK Kumar, G Jayaraman, D Samuel, C Yu. Induction of helical conformation in all beta-sheet proteins by trifluorethanol. J Biomol Struc Dyn 14: 381–385, 1996.

45. ME Goldberg, N Expert-bezancon, L Vuillard, T Abilloud. Non-detergent sulfobetaines—a new class of molecules that facilitate in vitro protein renaturation. Folding and Design 1: 21–27, 1996.

46. G Zardeneta, P Horowitz. Protein refolding at high concentrations using detergent/phospholipid mixtures. Anal Biochem 218: 392–398, 1994.

47. M Li, LJ Smith, DC Clark, R Wilson, DJ Murphy. Secondary structures of a new class of lipid body proteins from oilseeds. J Biol Chem 267: 8245–8253, 1992.

48. M Li, JS Keddie, LJ Smith, DC Clark, DC Murphy. Expression and characterization of the N-terminal domain of an oleosin protein from sunflower. J Biol Chem 268: 17504–17512, 1993.

49. JH Ploeghman, G Drent, KH Kalk, WG Hol, RL Heinrikson, P Keim, L Weng, J Russell. The covalent and tertiary structure of bovine liver rhodanese. Nature 273: 124–129, 1978.

50. T Arakawa, SN Timasheff. The stabilization of proteins by osmolytes. Biophys J 47: 411–414, 1985.

51. CR Brown, LQ Hong-Brown, MJ Welch. Correcting temperature-sensitive protein folding defects. J Clin Invest 99: 1432–1444, 1997.

52. KJ Frye, CA Royer The kinetic basis for the stabilization of staphylococcal nuclease by xylose. Protein Science 6: 789–793, 1997.

53. B-H Qu, E Strickland, PJ Thomas. Localization and suppression of a kinetic defect in cystic fibrosis transmembrane conductance regulator folding. J Biol Chem 272: 15739–15744, 1997.

54. H Sawano, Y Koumoto, K Ohta, Y Sasaki, S Segawa, H Tachibana. Efficient in vitro folding of the three-disulfide derivatives of hen lysozyme in the presence of glycerol. FEBS Lett 303: 11–14, 1992.

55. S Damodaran. Influence of solvent conditions on refolding of bovine serum albumin. Biochim Biophy Acta 914: 114–121, 1987.

56. GH Snyder. Free energy relationships for thiol-disulfide interchange reactions between charged molecules in 50% methanol. J Biol Chem 259: 7468–7472, 1984.

57. K Gekko, SN Timasheff. Mechanism of protein stabilization by glycerol: preferential hydration in glycerol-water mixtures. Biochemistry 20: 4667–4676, 1981.

58. K Gekko, SN Timasheff. Thermodynamic and kinetic examination of protein stabilization by glycerol. Biochemistry 20: 4677–4686, 1981.

59. J Tatzelt, SB Prusiner, WJ Welch. Chemical chaperones interfere with the formation of scrapie prion protein. EMBO J 15: 6363–6373, 1996.

60. JL Cleland, DI Wang. Co solvent assisted protein refolding. Biotechnology 8: 1274–1278, 1990.

61. JL Cleland, C Hedgepeth, DIC Wang. Polyethylene glycol enhanced refolding of bovine carbonic anhydrase B. J Biol Chem 267: 13327–13334, 1992.

62. JP Burnier PD Johnston. Method of Chain Combination. (0251615). European Patent.

63. AT Smith, N Santama, S Daceys, M Edwards, RC Bray, RN Thornley, JF Burke. Expression of a synthetic gene for horseradish peroxidase C in Escherichia coli with folding and activation of the recombinant enzyme with Ca2 + and heme. J Biol Chem 265: 1335–1343, 1990.

64. DN Brems, LA Alter, MJ Beckage, RE Chance, RD DiMarchi, LK Green, HB Long, AH Pekar, JE Shields, BH Frank. Altering the association properties of insulin by amino acid replacement. Protein Engineering 5: 527–533, 1992.

65. M Eberstadt, B Huang, Z Chen, RP Meadows, S-C Ng, L Zheng, MJ Lenardo, SW Fesik. NMR structure and mutagenesis of the FADD (Mort1) death-effector domain. Nature 392: 941–945, 1998.

66. G Stempfer, B Holl-Neugebauer, R Rudolph. Improved refolding of an immobilized fusion protein. Nature Biotechnology 14: 329–334, 1996.

67. H Rogl, K Kosemund, W Kuhlbrandt, I Collinson. Refolding of Escherichia coli produced membrane protein inclusion bodies immobilised by nickel chelating chromotagraphy. FEBS Lett 432: 21–26, 1998.

68. FU Hartl. Molecular Chaperones in Cellular Protein Folding. Nature 381: 571–580, 1997.

69. WJ Netzer, F-U Hartl. Recombination of protein domains facilitated by co-translational folding in eukaryotes. Nature 388: 343–349, 1997.

70. WA Fenton, AL Horwich. GroEL-mediated protein folding. Protein Science 6: 743–760, 1997.

71. M-J Gething, J Sambrook. Protein folding in the cell. Nature 355: 33–45, 1992.

72. JG Wall, A Pluckthun. Effects of overexpressing folding modulators on the in vivo folding of heterologous proteins in E. coli. Cur Opin Biotech 6: 507–516, 1995.

73. P Caspers, M Steiger, P Burn. Overproduction of bacterial chaperones improves the solubility of recombinant protein tyrosine kinases in Escherichia coli. Cell Molec Biol 40: 635–644, 1994.

74. KE Amrein, B Takacs, M Steiger, J Molnos, NA Flint, P Burn. Purification and characterization of recombinant human p50csk protein-tyrosine kinase from an Escherichia coli expression system overproducing the bacterial chaperones GroES and GroEL. Proc Natl Acad Sci USA 92: 1048–1052, 1995.

75. P Goloubinoff, AA Gatenby, GH Lorimer. GroE heat-shock proteins promote assembly of foreign prokaryotic ribulose biphosphate carboxylase oligomers in Escherichia coli. Nature 337: 44–47, 1989.

76. E Strickland, B-H Qu, L Millen, P Thomas, The molecular chaperone Hsc70 assists the in vitro folding of the N-terminal nucleotide-binding domain of the cystic fibrosis transmembrane conductance regulator. J Biol Chem 272: 25421–25424, 1997.

77. MH Yu, KN Lee, J Kim. The Z type variation of human alpha 1-antitrypsin causes a protein folding defect. Nat Struct Biol 2, 363–367, 1995.

78. Y-S Wu, VLH Bevilacqua, JM Berg. Fibrillin domain folding and calcium binding: significance to Marfan syndrome. Chem Biol 2: 91–97, 1995.

79. SC Blacklow, PS Kim. Protein folding and calcium binding defects arising from familial hypercholesterolemia mutations of the LDL receptor. Nat Struct Biol 3: 758–762, 1996.

80. PJ Thomas, PL Pedersen. Effects of the $\Delta$F508 mutation on the structure, function, and folding of the first nucleotide-binding

domain of CFTR. J Bioenerg Biomem 25: 11–19, 1993.

81. DJ Selkoe. Normal and abnormal biology of the beta-amyloid precursor protein. Ann Rev Neurosci 17: 489–517, 1994.

82. CR Brown, LQ Hong-Brown, J Biwersi, AS Verkman, WJ Welch. Chemical chaperones correct the mutant phenotype of the deta-F508 cystic fibrosis transmembrane conductance regulator protein. Cell Stress Chap 1: 117–125, 1996.

83. S Sato, CL Ward, ME Krouse, JJ Wine, RR Kopito. Glycerol reverses the misfolding phenotype of the most common cystic fibrosis mutation. J Biol Chem 271: 635–638, 1996.

84. GJ Miroy, Z Lai HA Lashuel, SA Peterson, C Strang, JW Kelly. Inhibiting transthyretin amyloid fibril formation via protein stabilization. Proc Natl Acad Sci USA 93: 15051–15056, 1996.

85. TW Loo, DM Clarke. Correction of defective protein kinesis of human P-glycoprotein mutants by substrates and modulators. J Biol Chem 272: 709–712, 1997.

86. H Taniuchi, CB Anfinsen. Simultaneous formation of two alternative enzymology active structures by complementation of two overlapping fragments of staphylococcal nuclease. J Biol Chem 246: 2291–2301, 1971.

87. MJ Bennett, MP Schlunegger, D Eisenberg. 3D domain swapping: a mechanism for oligomer assembly. Protein Science 4: 2455–2468, 1995.

88. FM Richards, PJ Vathayathil. The preparation of subtilisin-modified ribonuclease and the separation of the peptide and protein components. J Biol Chem 234: 1459–1465, 1959.

89. JK Welply, JK Fowler, I Zabin. Beta-galactosidase alpha-complementation. J Biol Chem 256: 6804–6810, 1981.

90. A Ullmann, F Jacob, J Monod. Characterization by in vitro complementation of a peptide corresponding to an operator-proximal segment of the beta-galactosidase structural gene of *Escherichia coli.* J Mol Biol 24: 339–343, 1967.

91. RW Kriwacki, L Hengst, L Tennant, SI Reed, PE Wright. Structural studies of p21Waf1/Cip1/Sdi1 in the free and Cdk2-bound state: conformational disorder mediates binding diversity. Proc Natl Acad Sci USA 93: 11504–11509, 1996.

92. SM Holmbeck, MP Foster, DR Casmiro, DS Sem, HJ Dyson, PE Wright. High-resolution solution structure of the retinoid x receptor DNA-binding domain. J Mol Biol 281: 271–284, 1998.

93. JG Hall, C Frieden. Protein fragments as probes in the study of protein folding mechanisms: differential effects of dihydrofolate reductase fragments on the refolding of the intact protein. Proc Natl Acad Sci USA 86: 3060–3064, 1989.

94. MR DeFelippis, LA Alter, AH Pekar, HA Havel, DN Brems. Evidence for a self-associating equilibrium intermediate during folding of human growth hormone. Biochemistry 32: 1555–1562, 1993.

95. H Taniuchi, CB Anfinsen. Simultaneous formation of two alternative enzymology active-structures by complementation of two overlapping fragments of staphylococcal nuclease. J Biol Chem 246: 2291–2301, 1971.

# 9

# Protein and Peptide Chemical and Physical Stability

**Bernard N. Violand and Ned R. Siegel**
*Monsanto/Searle Company, St. Louis, Missouri*

## I. INTRODUCTION

An understanding of peptide and protein degradation pathways at the molecular level is important for the development of stable molecules, both as reagents and as pharmaceutical drugs. The instability of proteins and peptides can be divided into two separate pathways involving chemical and physical processes. Chemical instability refers to any reaction which involves the breaking or formation of covalent bonds and results in the generation of a new chemical entity. Because peptides and proteins are composed of a large variety of chemical groups, they are susceptible to a diverse number of chemical degradation processes. These include deamidation, isoaspartate formation, racemization, oxidation, covalent oligomer formation, diketopiperazine formation and β-elimination. The relative importance each of these chemical modifications will have in determining stability will depend both on the structure of the molecule and its environmental conditions.

Physical instability refers to an alteration of the three-dimensional (conformation) structure. Generally, conformational stability is more important for proteins than peptides since higher order structures generally are not as important for peptides. Conformational modifications may lead to diminished activity as well as other deleterious physical processes which may impact the peptide or protein. These processes include aggregation, precipitation and surface adsorption.

Even though chemical and physical routes of degradation occur through different mechanisms they are closely related to each other. A chemical modification may make a protein more susceptible to physical instability. This is possible since chemical modifications may affect the conformation of the protein and hence its physical properties. Similarly, a physical alteration may decrease a protein's chemical stability because this conformational alteration may expose to the environment a previously buried residue into a new environment in which it is unstable . The interplay of these chemical and physical processes will be the overall factor in determining the stability of these molecules.

It is difficult to predict *a priori* the stability of a protein or peptide with labile amino acids since many external factors are critical. Some of these factors are pH, temperature, physical state of the molecule, concentration, and the nature of the buffer or other additives. Sections II and III of this chapter will describe the chemical and physical instability of proteins and peptides. Section IV will describe strategies for improving the stability of these molecules.

## II.  CHEMICAL DEGRADATION OF PROTEINS AND PEPTIDES

### A.  Non-Enzymatic Degradation of Asparaginyl, Aspartyl and Glutaminyl Residues

The non-enzymatic deamidation of asparaginyl and glutaminyl residues, the formation of isoaspartyl residues from both asparaginyl and aspartyl residues, racemization of asparaginyl and aspartyl residues, and hydrolysis of aspartyl-prolyl bonds are major routes of degradation for peptides and proteins. The goal of this section is to summarize current knowledge on the instability of these residues in peptides and proteins. Detailed descriptions of these modifications are also summarized in several excellent reviews (1, 2).

### 1.  Succinimide Formation and Deamidation of Asparaginyl Residues in Peptides

The major route of deamidation of asparaginyl residues at neutral to alkaline pH is through the formation of a succinimide intermediate (Fig. 1). This succinimide is formed when the main-chain peptide nitrogen from the adjacent $n+1$ residue attacks the side chain amide carbonyl, resulting in the formation of a succinimidyl residue with concurrent release of ammonia. Kinetic studies in peptides have shown that succinimide formation involves deprotonation of the main chain nitrogen followed by its nucleophillic attack on the carbonyl amide (3). This succinimide can subsequently be hydrolyzed at either carbonyl group to

yield both the isoaspartyl and aspartyl peptides in approximately a 3:1 ratio, respectively. The rate limiting step for this deamidation reaction is the formation of the succinimide (3). For the isoaspartyl product the peptide backbone now passes through the previous side chain carbonyl group which results in the inclusion of an extra methylene group in the peptide backbone. Formation of succinimides has been shown to occur during the hydrolysis of aspartyl-Beta-ester derivatives (4) and in the alkaline hydroxylamine cleavage of asparaginyl-glycyl residues (5).

The reactions involved in succinimide formation and hydrolysis can be accompanied by racemization of the alpha-carbon which results in a mixture of both L- and D-amino acids in each of the products (Fig. 1). Extensive studies on deamidation of the model peptide, Val-Tyr-Pro-Asn-Gly-Ala, yielded the equilibrium concentration at pH 7.4 and 37°C for each of the seven possible components (6). The concentrations determined were the original L-Asn peptide (0%), L-succinimidyl peptide (0.14%), D-succinimidyl peptide (0.14%), L-Asp peptide (15.3%), L-isoAsp peptide (55.8%), D-Asp peptide (7.0%) and the D-isoAsp peptide (21.7%). The majority of the racemization appears to be derived from interconversion of the L- and D-succinimidyl residues (6). These data demonstrate the complexity of products that can be generated during the deamidation of just one asparaginyl residue. This level of microheterogeneity demonstrates that the analysis and characterization of deamidated peptides and proteins represents a major challenge.

Small peptides have been utilized to determine the key factors affecting deamidation of asparaginyl and glutaminyl residues. Early studies demonstrated that the sequence of the adjacent amino acids to asparaginyl and glutaminyl residues had a profound effect on the rate of deamidation (7). These same studies

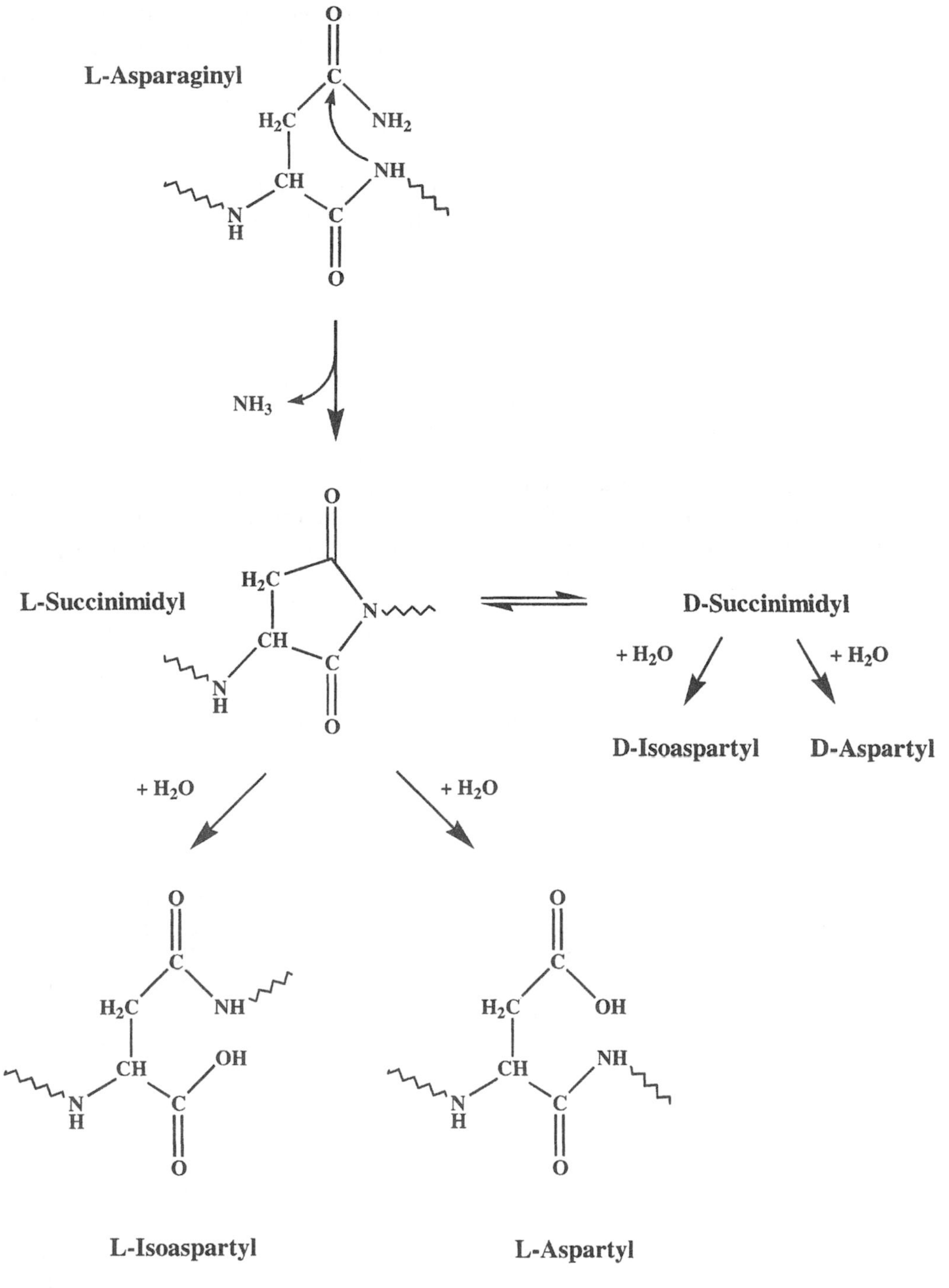

**Fig. 1** Mechanism of deamidation of L-asparaginyl residue through a succinimidyl intermediate. An L-succinimidyl residue is formed by elimination of ammonia from the side chain of the L-asparaginyl residue. Hydrolysis of the L-succinimidyl residue can lead to either the L-isoaspartyl or L-aspartyl residue. Racemization of the L-succinimidyl residue also lead to the formation of D-succinimidyl, D-isoaspartyl and D-aspartyl residues.

showed that asparaginyl residues generally deamidate significantly faster than glutaminyl residues. The median half-time for deamidation of 34 asparaginyl residues was 70 days and for 30 glutaminyl residues was 410 days (7).

Numerous studies analyzing the deamidation of asparaginyl residues in peptides have shown that the $n+1$ residue has a critical role in determining the rate of this reaction (6–12). Analyses of peptides of various sequences demonstrated that peptides with glycinyl residues in the $n+1$ position had deamidation rates 6–65 times faster compared to the same peptide containing a larger amino acid in this position (6, 10–12). These and other studies have demonstrated that the size of the $n+1$ residue is critically important for the rate of this reaction in peptides. However, it is now generally recognized that size alone cannot always explain the observed rates of deamidation for all peptides. For example, it has been found that Ser, His and Cys in the $n+1$ position result in faster deamidation rates compared to hydrophobic residues of similar or even smaller size (6, 10-12). Electron induction and general acid catalysis appear to be other important factors which can affect the rate of deamidation (1, 2).

There have been fewer studies on the effects of residues in other positions on the rate of asparagine deamidation. Two studies using model peptides have demonstrated that the $n-1$ residue has little if any effect on this reaction (11, 12). The $n-1$ residues in these studies included acidic, basic, and both small and large amino acids. It thus appears that the $n+1$ but not the $n-1$ residue is critical for determining the rate of deamidation of asparaginyl residues in peptides.

The deamidation of asparaginyl residues is accelerated by increasing pH since the deprotonation of the main chain peptide bond nitrogen is required for formation of the succinimidyl residue. This has been verified in studies on deamidation

rate vs pH with the rate of deamidation between pH 5–12 found to be dependent on the hydroxide ion concentration (11–14). These same studies showed that increasing temperature and ionic strength also correlated with increased deamidation rates. Buffer catalysis of deamidation was demonstrated for phosphate (pH 7.0 and 7.5), Tris (pH 8.0 and 8.5), and bicarbonate (pH 10.0) (14).

## 2.  Isoaspartate Formation from Aspartyl Residues in Peptides

Aspartyl residues can also form succinimides through the same mechanism as that for asparaginyl residues with release of water (Fig. 2) (1, 2, 15). The same breakdown products, such as isoaspartyl and racemized residues will be generated from an aspartyl derived succinimide as from an asparaginyl succinimide (Fig. 1). Rate constants for racemization of three model L-aspartyl peptides derived from alpha-A-crystallin have been determined (16). These studies demonstrated that the racemization rate was affected by the amino acid on the C-terminal side of the aspartyl residue.

Because isoaspartyl formation from aspartyl residues does not result in a change in overall charge, the formation of these residues may not be detected as frequently compared to their generation from asparaginyl residues. Peptide mapping has been the most frequently used method for detecting these modifications but it is not a highly sensitive technique. A specific method for detecting two of the products from succinimide formation and hydrolysis involves use of the enzyme D-aspartyl/L-isoaspartyl methyltransferase to measure L-isoaspartyl and D-aspartyl residues (17–19). This enzyme catalyzes the methyl esterification of these modified amino acids and can be used to detect small amounts of these residues which previously could not be detected by other less sensitive procedures. Utilization of this

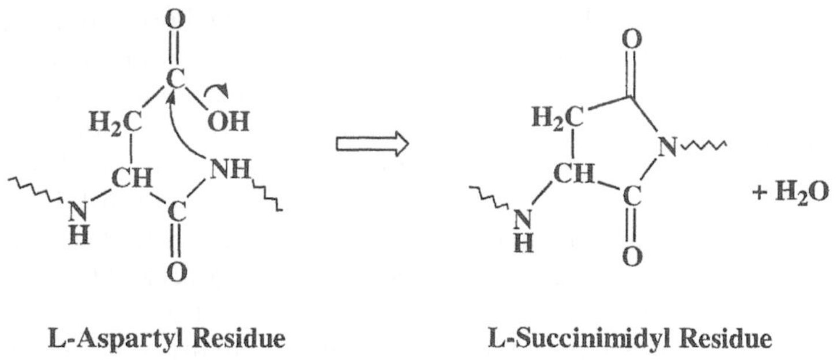

**L-Aspartyl Residue**          **L-Succinimidyl Residue**

**Fig. 2** Formation of a succinimidyl residue through dehydration of an aspartyl residue. Succinimide formation from an aspartyl residue occurs through dehydration. The subsequent hydrolysis and racemized products from this succinimidyl residue are the same as that shown for a succinimidyl residue derived from an asparaginyl residue as illustrated in Fig. 1.

enzyme has shown that a large proportion of isoaspartyl residues in proteins and peptides are derived not only from asparaginyl but also from aspartyl residues (1, 2).

The rate of succinimide formation at pH 7.4 has been compared for a series of identical peptides except for replacement of asparagine by aspartate (10). These data showed that the rate of formation of succinimidyl residues in the aspartyl containing peptides was 13-36 times slower than for the same peptides containing asparaginyl residues. This study was performed at pH 7.4 which can be used to explain the slower succinimidyl formation rate for the aspartyl peptides. The carboxyl group must be protonated to be an efficient leaving group for succinimide formation. Because its pKa is 3.9, only 1 out of 3000 carboxyl groups will be protonated at pH 7.4, so the rate of succinimide formation will be very slow since most of the carboxyls are unprotonated at this pH. Studies on small peptides demonstrated that esterification of the aspartyl side chain carboxyl group resulted in a 10,000 to 20,000 times faster rate of succinimide formation at pH 7.4 compared to the same unesterified aspartyl peptides, thereby demonstrating the importance of the leaving group on the rate of this reaction (10). These results

illustrate that the rate of isoaspartyl formation from aspartyl residues is dependent not only on the primary sequence and conformation of this portion of the peptide or proteins but also on the pKa of the aspartyl group. Since the microenvironment can alter the pKa of aspartyl residues, this factor plays an important role in the rate of succinimidyl formation. This may be an especially important factor in proteins where the pKa of aspartyl residues can vary considerably.

The effect of the $n-1$ and $n+1$ residues on succinimide formation from aspartyl residues has not been studied as exhaustively as it has for asparaginyl residues. Generally, similar conclusions have been made with regard to the effect these groups have on succinimide formation from aspartyl peptides. However, a recent study using a model peptide from ACTH has provided evidence that an $n-1$ histidinyl residue has an important effect on succinimide formation while a $n+1$ histidinyl residue has no effect. This study demonstrated that histidine in the $n-1$ position increased the rate of succinimide formation from an aspartyl residue approximately 10-fold, yet when it was in the $n+1$ position it had no effect on the rate of this reaction compared to other amino acids (20). These authors proposed that the histidine was able to protonate the side chain of the aspartyl residue resulting in acceleration of this reaction.

## 3. Deamidation and Isoaspartate Formation in Proteins

The basic "rules" of succinimide formation established for peptides are not always predictive for proteins. This is because the conformation of asparaginyl and aspartyl residues in a protein can have a major effect and in fact may be more important in controlling succinimide formation than the primary sequence. Analyses of X-ray crystallography data have shown that asparaginyl and aspartyl residues in native proteins usually exist in conformations where the $n + 1$ peptide bond nitrogen cannot approach the side chain carbonyl to form the succinimide ring (21). Furthermore, this non-reactive conformation generally represents the lowest energy state for the orientations of these amino acids in proteins. These data may help to explain why the rates of deamidation and isoaspartate formation in proteins are usually lower than comparable peptides of the same sequence. In proteins, there are many examples where Asn-Ser and Asn-Gly sites are stable against deamidation, yet in peptides these sequences are generally labile (21, 22).

A compilation of the major sites of deamidation in numerous proteins illustrates that it is very difficult to predict from sequence alone whether a particular site in a protein will be readily susceptible to succinimide formation (22). Since deamidation has been proposed to occur only in regions where there is considerable flexibility in the polypeptide chain, sequences which are labile in peptides are often stable in protein environments most likely because they exist in rigid regions. Two examples which demonstrate the importance of protein conformation on deamidation are RNAse and calbindin-9K. The rate of deamidation in native vs unfolded RNAse demonstrated

that the unfolded protein deamidated at a 30-fold higher rate compared to the native protein, apparently because of the greater flexibility in the unfolded protein (23). Similarly, the rate of deamidation of calbindin-9K at an Asn-Gly site has been found to be accelerated significantly after denaturation by urea (24).

Whereas succinimides have been identified in many studies with peptides, the number of protein succinimide species isolated are minimal. The first demonstration of succinimide formation in a protein was for human somatotropin (hST). Heating lyophilized hST at 45°C for four months generated a significant amount of a succinimide species (25). Ion-exchange chromatography was used to isolate this modified species. RP-HPLC was used to isolate porcine somatotropin containing a succinimide at residue 129 generated from a solution incubated at pH 5.0 (26). This succinimide formed from an Asp129-Gly 130 sequence. At pH > 7.0 no formation of this modified species was observed. In order to isolate the succinimide containing peptide it was necessary to perform trypsin digestion at pH 6.0 with elevated amounts of trypsin, since the cyclic imide in this peptide readily hydrolyzed under normal incubation conditions between pH 7–8. Subsequent hydrolysis of the isolated succinimidyl peptide to its aspartyl and isoaspartyl species was used to verify the presence of the succinimide. Lysozyme also formed a succinimide from an Asp101-Gly102 sequence when it was incubated under acidic conditions (pH 4.0) (27). The small number of isolated succinimide derivatives in proteins may be due to the fact that, during either isolation or analysis of the modified protein, exposure to pH > 6.0 may subsequently have hydrolyzed the succinimide moiety.

Methods of analyzing for succinimide residues in proteins without first isolating the modified protein have been developed. One procedure utilizes hydroxylamine treatment at pH 9.0 at

45°C for two hours. This treatment results in approximately 50% cleavage on the C-terminal side of succinimidyl residues. Amino acid sequencing can then be used to determine the positions of the succinimidyl residues in the protein (28). A second method for detecting succinimide formation involves reducing the succinimide with sodium borohydride and subsequent identification of the generated homoserine and iso-homoserine using amino acid analysis (29). These procedures should facilitate identification of succinimide residues in proteins.

### 4. Succinimide Mediated Peptide Bond Cleavage

Another reaction which occurs at asparaginyl residues is cleavage of the peptide bond between the asparaginyl residue and the residue on its C-terminal side (Fig. 3). In this reaction, the side-chain amide nitrogen of the asparaginyl residue attacks the peptide bond carbonyl group resulting in release of the carboxyl-side peptide and formation of a C-terminal succinimide on the original N-terminal polypeptide. This reaction has been verified in several peptides (6, 11, 12). The rate of this reaction has been shown to increase with increasing pH between 7.4–13.8 (30). It appears that both succinimide-mediated cleavage and succinimide-mediated deamidation of asparaginyl residues can occur in most peptides, with deamidation usually being the predominant reaction. However, for peptides which have a proline on the C-terminal side of asparagine, succinimide mediated deamidation does not occur since there is no deprotonatable peptide

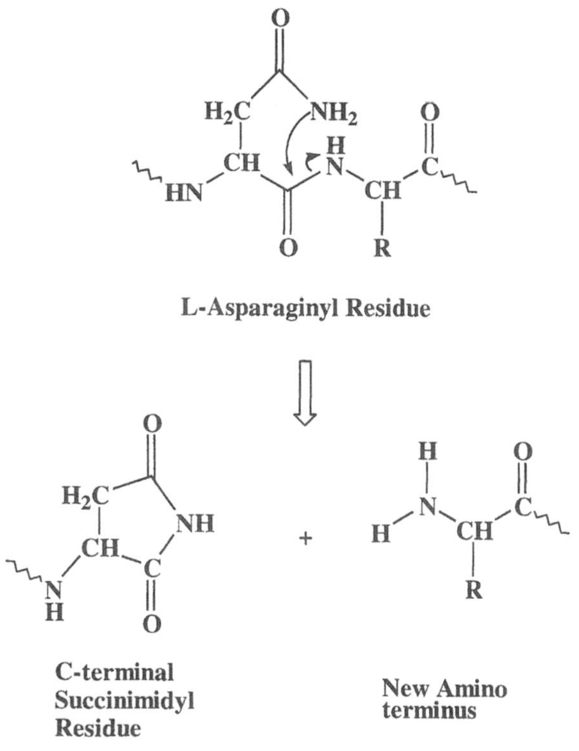

**L-Asparaginyl Residue**

**C-terminal Succinimidyl Residue**

**New Amino terminus**

**Fig. 3** Succinimide mediated cleavage of the peptide bond between an asparaginyl residue and its carboxyl-side residue. Cleavage of the peptide bond C-terminal to an asparaginyl residue occurs through the attack of the side chain nitrogen on its main chain carbonyl group resulting in generation of a C-terminal succinimidyl residue and release of a new peptide.

bond nitrogen on the proline residue which is a prerequisite for this reaction (1). Consequently, the cleavage reaction will be the only reaction observed for peptides or proteins at an Asn-Pro sequence. This cleavage reaction was also shown to be a major degradation pathway for a peptide containing an Asn-Leu sequence (6).

There are several instances where succinimide mediated cleavage after an asparaginyl residue has been demonstrated in proteins. These examples include alpha-crystallin (31) and bovine (bST) and porcine (pST) somatotropin (32). Both succinimide-mediated deamidation and cleavage occur at Asn99-Ser100 in pST in the pH range 7–11. These two modified species can be readily separated and quantified by RP-HPLC (32). The rate of both of these reactions increased with increasing pH over the examined range of 7–11 with the deamidation reaction being the major degradation pathway. The ratio of Isoaspartate99 formation to cleavage product at Asn99 was approximately 10:1 at each of the pHs analyzed (32).

### 5.  Deamidation of Asparaginyl Residues by Other Mechanisms

There are several examples of asparaginyl residues deamidating through different mechanisms than a succinimide. It has been established that the major route of asparagine deamidation in peptides at pH 1–2 is through direct hydrolysis of the side chain amide leading to formation of only aspartyl residues (Fig. 4) (11, 12). The deamidation of the peptide Val-Asn-Gly-Ala at pH 7.0 was ascertained to occur through the attack of the N-terminal Val alpha-amino group on the asparaginyl side-chain carbonyl to form a seven membered ring (Fig. 5) (8). Interestingly, at pH > 9.0 this same peptide deamidated through the previously described succinimide reaction. These authors demonstrated that the conformation of this tetrapeptide changed between pH 7.5 and 9.0 and proposed that the alternative deamidation mechanisms could in turn explain the observed alterations.

Another unique mechanism for deamidation can occur for C-terminal asparaginyl residues. This mechanism involves the formation of a five-membered cyclic anhydride (Fig. 6). This reaction has been shown to occur in the insulin A-chain at Asn-21 through the nucleophillic attack of the C-terminal carboxyl group on the asparaginyl carbonyl side-chain (33–35). Subsequent hydrolysis of the anhydride yields an aspartyl residue.

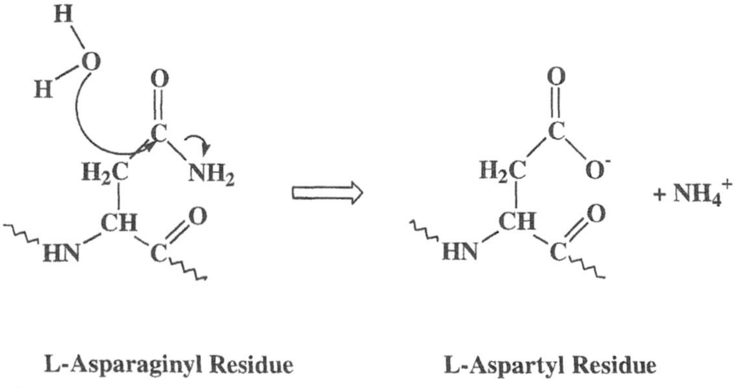

L-Asparaginyl Residue                    L-Aspartyl Residue

Fig. 4  Deamidation of asparaginyl residue by direct hydrolysis of the side chain amide group. Strongly acidic conditions lead to direct hydrolysis of the side chain amide bond generating an aspartyl residue.

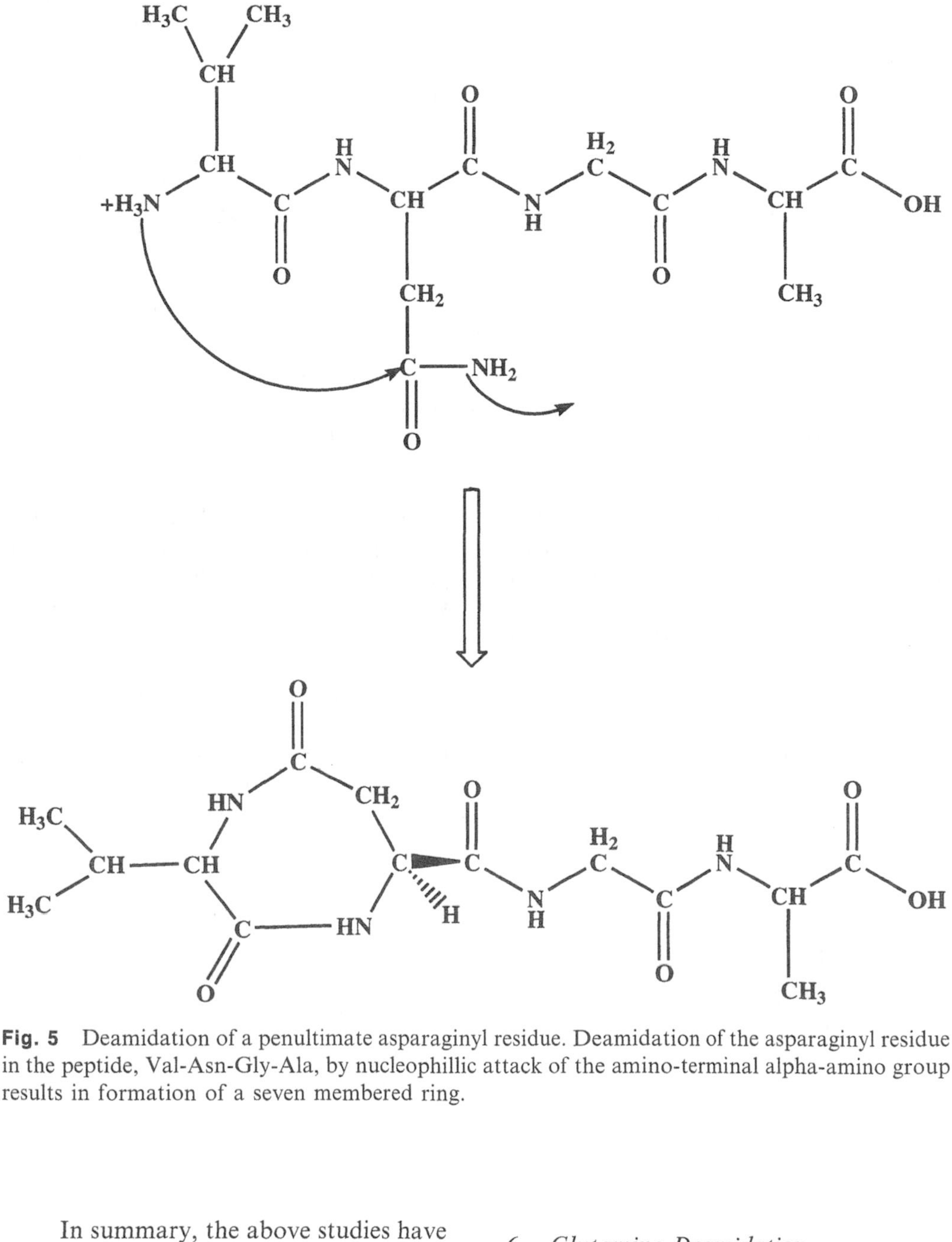

**Fig. 5** Deamidation of a penultimate asparaginyl residue. Deamidation of the asparaginyl residue in the peptide, Val-Asn-Gly-Ala, by nucleophillic attack of the amino-terminal alpha-amino group results in formation of a seven membered ring.

In summary, the above studies have shown at least four different mechanisms for non-enzymatic deamidation of asparaginyl residues. The mechanism which predominates will depend on many factors which include pH and the position of the asparaginyl residue in the polypeptide chain.

### 6. Glutamine Deamidation

Glutamine residues at the N-terminus of peptides or proteins can be deamidated with subsequent formation of pyroglutamic acid. This reaction occurs through nucleophillic attack of the N-terminal amino group on the side-chain

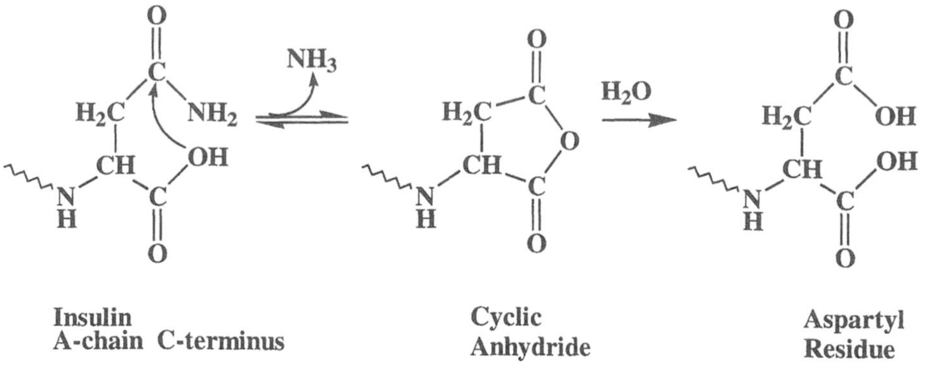

**Fig. 6** Deamidation of the C-terminal asparaginyl residue in the Insulin A-chain. This deamidation mechanism involves formation of an intermediate cyclic anhydride through cleavage of the side chain amide by the C-terminal carboxyl group of this asparagine.

carbonyl group resulting in formation of a five-membered ring (Fig. 7) (36). This reaction readily occurs for N-terminal glutaminyl residues and can be accelerated by acid and base. N-terminal asparaginyl residues do not deamidate through this mechanism since the smaller side chain would require formation of an unfavorable four-membered ring.

Most instances of non-enzymatic deamidation in proteins and peptides under weakly acidic or alkaline con-ditions, pHs 4–10, have shown that asparaginyl residues are more labile than glutaminyl residues. In fact, there are very few instances demonstrating glutaminyl deamidation is favored over asparaginyl deamidation. In granulocyte colony stimulating factor (G-CSF), the only sites of deamidation are at glutaminyl residues but this protein contains seventeen glutaminyl residues and no asparaginyl residues (37).

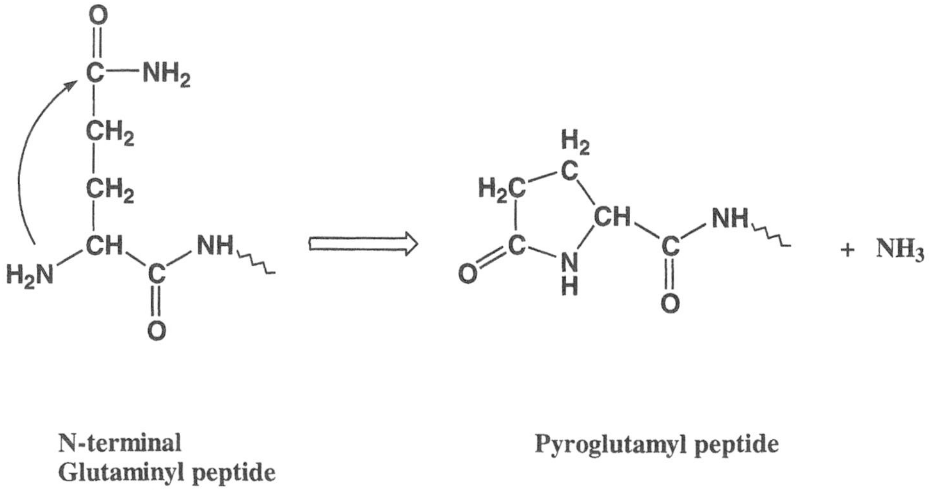

**Fig. 7** Deamidation of N-terminal glutaminyl residue. Deamidation of N-terminal glutaminyl residues can rapidly occur with concurrent formation of a pyroglutamyl residue.

### 7. Hydrolysis of Aspartyl-Prolyl Bonds

The peptide bond between aspartyl-prolyl residues can be cleaved under mildly acidic conditions. Aspartyl-prolyl bonds have been shown to be 8–20 times more labile in 0.015M HCl at 100°C than other aspartyl-X or X-prolinyl peptide bonds (38). Several studies have also shown the lability of this sequence in proteins. Incubation of basic fibroblast growth factor (b-FGF) at pH 5 for 14 days at 25°C resulted in cleavage at an Asp28-Pro29 site, demonstrating that even under these mild conditions cleavage had occurred (39). Stability studies on recombinant interleukin-11 (IL-11) at elevated temperatures demonstrated that under acidic conditions the major degradation pathway was cleavage at Asp133-Pro134 (40). Moreover, the rate of cleavage at this Asp-Pro site was significantly faster than at the other Asp-Pro site in this protein. Similar selectivity toward cleavage at only two of the three Asp-Pro sites was also observed in studies on the stability of recombinant macrophage colony-stimulating factor (M-CSF) (41). These data have shown that each Asp-Pro site will be cleaved at a different rate depending on its conformation, similar to the effect that conformation may have on the rates of deamidation.

### 8. The Effect of Deamidation and Isoaspartate Formation on Bioactivity

It is not possible to predict *a priori* the effect deamidation and isoaspartate formation will have on peptide and protein activities. The overall effects of these reactions may be complex since, for asparaginyl residues, deamidation results in both a change in the overall charge of the protein while the generation of an isoaspartate residue also results in the insertion of an extra methylene group into the main chain backbone. Furthermore, these reactions can also result in the formation of the D-isomers of isoaspartate and aspartate which may in turn affect the overall conformation of the peptide or protein. A recent review has described the effect of protein deamidation on the activity of several proteins (42). Some examples where deamidation has had a negative effect on protein activity are human adrenocorticotropin hormone (43), cytochrome C (44), calmodulin (45–47), soluble CD4 (48), OKT-3 monoclonal antibody (49), hGH releasing factor (50), triosephosphate isomerase (51, 52), porcine adrenocorticotropin (53), interleukin-1β (54), lysozyme (55), and human epidermal growth factor (56). Examination of these examples illustrates many different ways that deamidation can affect protein activity.

Instances where no change in activity has been observed after deamidation include bovine seminal ribonuclease (57), interleukin-1 alpha (58, 59), insulin (60) aspartate aminotransferase (61–65), hGH (66, 67), hirudin (68), and interleukin-2 (69).

Parathyroid hormone is an interesting example where its binding affinity as measured by renal and skeletal cytochemical activities was reduced. However, in both renal and skeletal adenylate assays, the deamidated molecule had equipotent activity (70). This example demonstrates that separate activities for a protein may be differentially affected by deamidation and stresses the importance of evaluating each activity.

### B. Oxidation Reactions

Peptides and proteins readily undergo oxidation both in solution and in the solid state. These oxidations are generally catalyzed by transition metal ions, light, peroxides and free radicals (71, 72). Amino acid residues which can be oxidized non-enzymatically include Met, Cys, Trp,

His and Tyr, however, under typical isolation and storage conditions Met and
Cys appear to be the predominant residues
which are oxidized.

There are several reactive oxygen
species which may lead to oxidation. These
include hydrogen peroxide ($H_2O_2$),
hydroxyl radicals (HO$\cdot$ and HOO$\cdot$), halide
oxygen compounds (ClO$^-$), singlet oxygen
($^1O_2$) and superoxide radical ($O_2^-$) (72–74).
The exact role each of these compounds
has in peptide and protein oxidation is not
always clear because of limited detailed
studies on this topic.

The rate of oxidation of amino acids
in proteins can vary extensively depending
on relative position and exposure to the
environment. Chemically-induced oxidation has been used to probe the environment of residues, and these studies have
established that the rate of oxidation
depends on whether or not vulnerable
amino acids are fully exposed, partially
exposed or buried in the interior of the protein (73, 74). There are many examples
illustrating the effect of the methionyl residue environment on its ability to be
oxidized.

The effect of primary sequence on
oxidation, however, is not completely
understood because no comprehensive
studies have been performed on this
subject. One recent study evaluated the
effect of primary sequence on methionyl
oxidation in small peptides containing
from 2–6 residues (75). This study demonstrated that enhancement of methionine
oxidation occurred if histidine was present
in the peptide, however, the number of
other amino acids analyzed was small.

Factors which have been determined
to be important for oxidation include pH,
light exposure, temperature, peroxides,
free-radical initiators and the presence of
metals such as Cu(II) and Fe(III) (73,74).
Precautions to minimize oxidation include
temperature reduction, removal of trace
metals and potential peroxide containing
reagents, protection from light, removal

or limitation of exposure to oxygen and
the addition of antioxidants. This section
summarizes common oxidation reactions
in peptides and proteins and the effects of
these modifications on biological activity.

## 1. Oxidation of Methionyl Residues

Mild oxidation of methionyl residues generates methionine sulfoxide, Met(O),
while only under highly oxidative conditions will methionyl residues oxidize to
the sulfone derivative (Fig. 8). The
oxidation of methionyl residues to Met(O)
is a reversible reaction, whereas methionine sulfone formation is irreversible. Generation of the more hydrophilic and less
flexible sulfoxide results in an asymmetric
sulfur atom with two diastereoisomers of
Met(O) being formed.

The reduction of Met(O) to Met can
be accomplished both chemically and
enzymatically. Chemical reduction of
Met(O) to Met can be accomplished by
both sulfhydryl and non-sulfhydryl compounds but harsh reaction conditions are
generally required (76). Enzymatic reduction can be accomplished by a Met(O)
reductase which is present in a wide
variety of sources, including prokaryotes,
eukaryotes and plants (73, 74, 77). This
enzyme appears to have an important
physiological role of keeping methionyl
residues in a reduced state to prevent
inactivation. For proteins which utilize
methionine as a methyl donor, prevention
of oxidation is critical since the methyl
group cannot be donated after Met(O)
formation.

There are numerous studies on
autooxidation and chemically-induced
oxidation of methionyl residues in peptides and proteins (73, 74). Several recent
studies have been performed on various
factors affecting methionine oxidation in
formulations of human insulin-like growth
factor I (hIGF-I) (78–80). Conclusions
from these studies indicate that light
exposure, storage temperature and oxygen

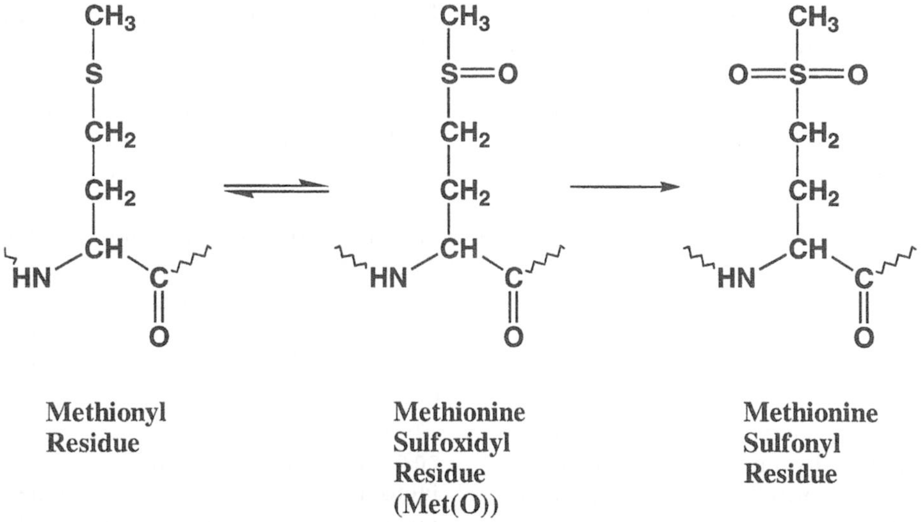

**Fig. 8** Oxidation of a methionyl residue. Oxidation of methionyl residues first leads to addition of an oxygen to the side chain sulfur, resulting in formation of methionine sulfoxide, Met(O). Under strongly oxidizing conditions the sulfoxide can be oxidized to the methionyl sulfone.

content all had a significant impact on the rate of methionine oxidation in this protein. Interestingly, no differences in reaction rate were observed between solution and lyophilized preparations of this protein.

Relaxin A was established to be sensitive to fluorescent light exposure in a formulation consisting of 3% methylcellulose in 10 mM sodium citrate, pH 5.0 (81). Exposure to light at 5°C in this formulation resulted in a significant loss of bioactivity with Met(O) being the major degradation product formed. Addition of methionine at a 500:1 molar ratio to this formulation resulted in resistance to light-induced loss of bioactivity. Because the added Met was probably a scavenger, it would have been interesting to determine the amount of free Met converted to Met(O).

Excipients which have the potential for containing peroxides, such as polyethylene glycols and surfactants should be used with caution during isolation and formulation of peptides and proteins since they may contain trace quantities of peroxides. The presence of peroxides in polysorbate-20 and polysorbate-80 were shown to be responsible for the oxidation of methionyl residues in human granulocyte-colony stimulating factor (hG-CSF) (82).

Methionine oxidation may result in decreased bioactivity in the modified peptide or protein, however, there are many cases where no effect is observed. (73, 74, 83). Oxidation of methionyl residues in human (hST) and bovine somatotropin (bST) appears to have no effect on their bioactivity. The oxidation of methionyl residues in lyophilized hST was shown to occur both in solution and in the lyophilized state (84, 85). Other studies which induced oxidation of hST by $H_2O_2$ resulted in oxidation of Met14 and Met125 but not Met170, showing that all methionyl residues are not oxidized at the same rate (86). This partially-oxidized protein showed no loss of activity in the rat tibia growth bioassay. A similar study with bovine somatotropin in which three of the four methionyl residues were oxidized with $H_2O_2$ also showed no loss of biological activity (87). An enkephalin analog was determined to be more active after its Met was converted to Met(O) (88). These results demonstrate that it is not possible *a priori* to predict if oxidation of methionyl residues will affect the bioactivity of a protein or peptide.

## 2. Oxidation of Cysteinyl Residues and Thiol-Disulfide Exchange

Cysteine residues can be oxidized to form either intra- or intermolecular disulfide bonds which may impact the activity, immunogenicity and other properties of peptides and proteins. This reaction can occur through air oxidation by dissolved oxygen, and it is greatly accelerated by metal ions such as Cu(II) and Fe(III). This reaction may generate oligomers which usually are less active and more immunogenic than the monomeric forms of the native molecule.

Thiol-disulfide exchange can lead to mispaired disulfide residues and covalent oligomeric species. This reaction involves nucleophilic attack of a thiolate ion on a disulfide bond resulting in generation of a new disulfide bond and a different thiolate ion:

$$R_1S^- + R_2SSR_3 \Leftrightarrow R_1SSR_2 + R_3S^-$$

The new thiolate can now react with another disulfide bond leading to an accumulation of many different mispaired disulfide bonds. This can eventually lead to the formation of large oligomers or potentially monomeric species with many mispaired disulfides.

The rate of this reaction depends on numerous factors which include pH, conformation and electrostatic environment. Since the attacking sulfhydryl must be ionized, increasing pH will accelerate this reaction assuming all other factors are the same. The conformation of the protein may also affect the propensity for this reaction since the disulfide bond must be accessible to the attacking thiolate ion.

Finally, the electrostatic environment around the reacting species can influence this reaction since nearby positive and negative charges either accelerate or inhibit the reaction, respectively (89).

There are numerous examples illustrating loss of bioactivity after cysteine oxidation or thiol-disulfide exchange. These reactions were determined to be responsible for loss of interferon activity even when the protein was stored at −70°C (90). In T4 lysozyme, oxidation of cysteinyl residues was shown to be a major cause of thermal inactivation of this enzyme (91). Basic fibroblast growth factor (bFGF) contains four cysteines and no cystines, so prevention of the oxidation of these cysteines to form oligomeric complexes represents a major challenge. There are two surface accessible cysteines and two buried cysteines. Oxidization of the two exposed cysteines caused undesirable oligomerization and loss of activity (92). Inclusion of EDTA and storage at low pH 5.0 helped to minimize this oxidation reaction in bFGF (93).

Acidic fibroblast growth factor contains three cysteines which are easily oxidized. The formation of either intra- or intermolecular disulfide bonds from these cysteines causes a dramatic loss in activity of this protein (94). Subsequent reduction of these disulfides back to cysteinyl residues resulted in complete restoration of normal activity. These examples demonstrate that cysteine oxidation and thiol-disulfide interchange can be important degradative reactions which affect the structures and activities of peptides and proteins.

## 3. Oxidation of Histidinyl Residues

Oxidation of histidinyl residues in proteins has been shown to occur through light- and metal-catalyzed reactions (71, 72). Under typical isolation and storage conditions the oxidation of histidinyl residues is usually not a common degradative pathway. Most observed histidinyl oxidations instead have occurred under stressed conditions using either high light intensity or chemical induction.

Several studies have been performed on histidyl oxidation in human somatotropin (hST). Photooxidation of hST resulted in oxidation of only one histidinyl residue which contained either one, two or three atoms of oxygen (95). Oxidation of hST with ascorbate/Cu(II)/O2 resulted in oxidation of both His18 and His21 to the mono 2-oxo derivatives (96). These two amino acids are part of the metal binding site in this protein, and the integrity of this site was shown to be essential for the oxidation of these two histidinyl residues. Disruption of this metal binding site in hST resulted in prevention of this oxidation, thereby demonstrating the relative catalytic importance of the metal for this particular reaction.

## C. Beta-Elimination

Incubation of peptides and proteins at alkaline pH can result in $\beta$-elimination in cystinyl, cysteinyl, seryl and threonyl residues. The initial step is extraction of the $\alpha$-hydrogen by an hydroxide ion resulting in the formation of a carbanion. This carbanion can either add back a proton to give racemization of the amino acid, or alternatively, the leaving group can be expulsed with formation of the unsaturated derivative, dehydroalanine. The nature of the leaving group has a large effect on the rate of this reaction.

Beta-elimination of cystine is one of the most common reactions occurring in strongly alkaline conditions with resultant formation of dehydroalanine and a persulfide (Fig. 9) (97–99). This persulfide can then react with OH⁻ to generate hydrosulfide which can participate in disulfide scrambling through thiol-disulfide

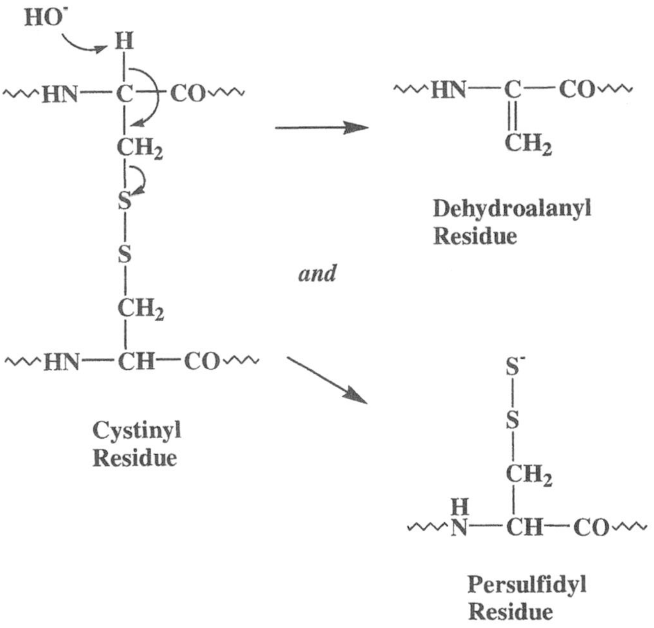

**Fig. 9** Beta-elimination of a cystinyl residue. Beta-elimination of cystinyl residues occurs through extraction of the hydrogen from the $\alpha$-carbon resulting in formation of dehydroalaninyl and persulfidyl residues.

interchange. This can lead to mispairing of disulfides and covalent aggregation which may have negative consequences on the activity and stability of the protein or peptide. Beta-elimination also readily occurs from glycosylated and phosphorylated serinyl residues because both of these respective modifications represent good leaving groups.

Dehydroalanine will react with nucleophilic groups such as the ε-amino group of lysine or the thiol of cysteine, resulting in the formation of lysinoalaninyl and lanthionyl residues, respectively. These reactions may be intermolecular which will result in formation of oligomeric species. The formation of dehydroalanine has generally been observed only under strongly alkaline or elevated temperature conditions (100).

## D. Diketopiperazine Formation

Peptides which contain glycine as the third amino acid and are not blocked at their N-terminus can undergo diketopiperazine (DKP) formation (101, 102). The rate of this reaction is enhanced if Pro or Gly is the first or second residue of the peptide. The mechanism of DKP formation involves the nucleophilic attack of the N-terminal alpha-amino nitrogen on the peptide bond carbonyl group between the second and third amino acids (Fig. 10). This reaction is accelerated by increasing pH since the nucleophilicity of the attacking amino group increases with pH up to the point where it is no longer protonated. This reaction has been observed in hST, resulting in formation of a truncated form of this protein (103). The amino-terminus of hST is Phe-Pro-Thr so it does not have

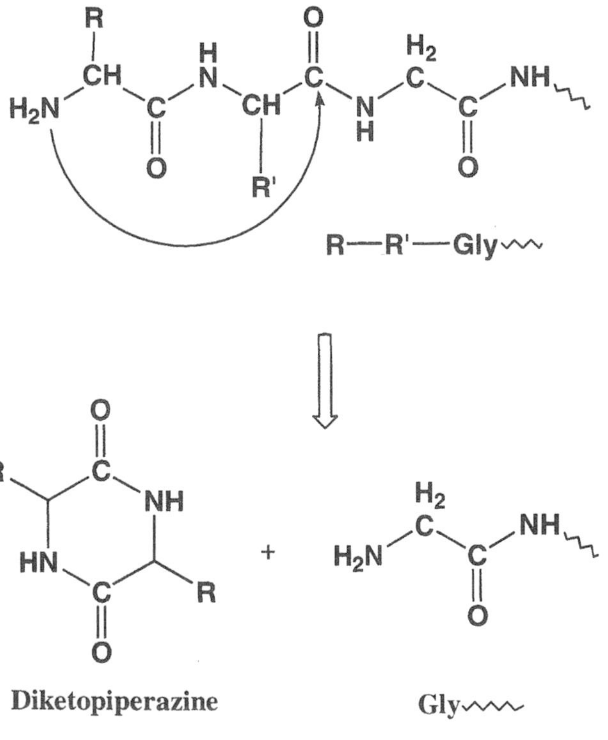

**Fig. 10** Diketopiperazine formation. Diketopiperazine can form through attack of the α-amino group on the main chain carbonyl of the second amino acid. This reaction results in formation of a diketopiperazine and release of the remainder of the polypeptide chain.

a glycine at the third amino acid position but the presence of a proline at the penultimate position may help to promote the observed cleavage.

## E. Cross-Linking Through C-Terminal Anyhdride Formation

An unusual reaction which has been shown to result in dimerization in insulin involves the nucleophilic attack of either the A or B chain amino-terminal amino group on a C-terminus cyclic anhydride (Fig. 11). This reaction is one of the major degradation pathways for insulin stored at pH 3–5. The cyclic anhydride formation was previously discussed in Section II.A.5. since this is a deamidation reaction. The reactive anhydride can either be opened through hydrolysis or reaction with the nucleophilic N-terminal amino group.

## III. PHYSICAL STABILITY

The previous section focused on the chemical instability associated with peptides and proteins. These processes result in specific modifications to the primary structure (amino acid sequence) associated with a given polypeptide. The alterations in turn may or may not induce local changes in the protein's native globular structure (conformation) which may affect its physical stability. Physical instability may be introduced when the aforementioned chemical modifcations have impacted either the folding patterns of these secondary structural units or, more typically, the spatial dispositions of the respective side chains on the normal versus altered component amino acids.

Physical instability denotes the disruption of the native conformation of a peptide or protein and represents another major degradation pathway. However, since peptides normally assume only limited structural elements, physical instability is more often associated only with proteins and this section will concentrate on proteins and their physical properties. Physical instability is generally manifested as a disruption in the conformation displayed by the protein, and the conformation often is critical to its function. The folding patterns of the secondary structural units in concert with the spatial arrangements of the side chain moieties constitute the tertiary structure of the given protein. A fourth structural level is present when the given protein is comprised of multiple subunits, with the spatial arrangement of these subunits termed the protein's quaternary structure (104–108). The loss of globular structure, typically in the form of distortions to the tertiary structure of the protein, is referred to as denaturation. The transformations induced during denaturation processes are often followed by aggregation, precipitation, and/or surface adsorption as the three primary benchmarks representing physical instability.

## A. Denaturation

The native conformation (N) defines the given specificity and biological function associated with a protein. The denatured (D) state, on the other hand, refers to any non-native conformation assumed by the protein (6). This denaturation process may be induced by a number of physical or chemical stressors which alter the conformational stability of the native protein. These stressors typically include temperature, pH, and organic solvent content (108, 110–111); as well as salts (112) including those of guanidinium, i.e. Gd-HCl, as a commonly used denaturant (113), destabilizing (e.g. chaotropic) agents such as urea (114), and surfactants of primarily ionic nature, i.e. anionic sodium dodecyl sulfate (SDS) (104, 115). Denaturation can also be introduced by cold (116) and/or freezing (113), by dehydration during the lyophilization process (117), as well as by pressure, shear

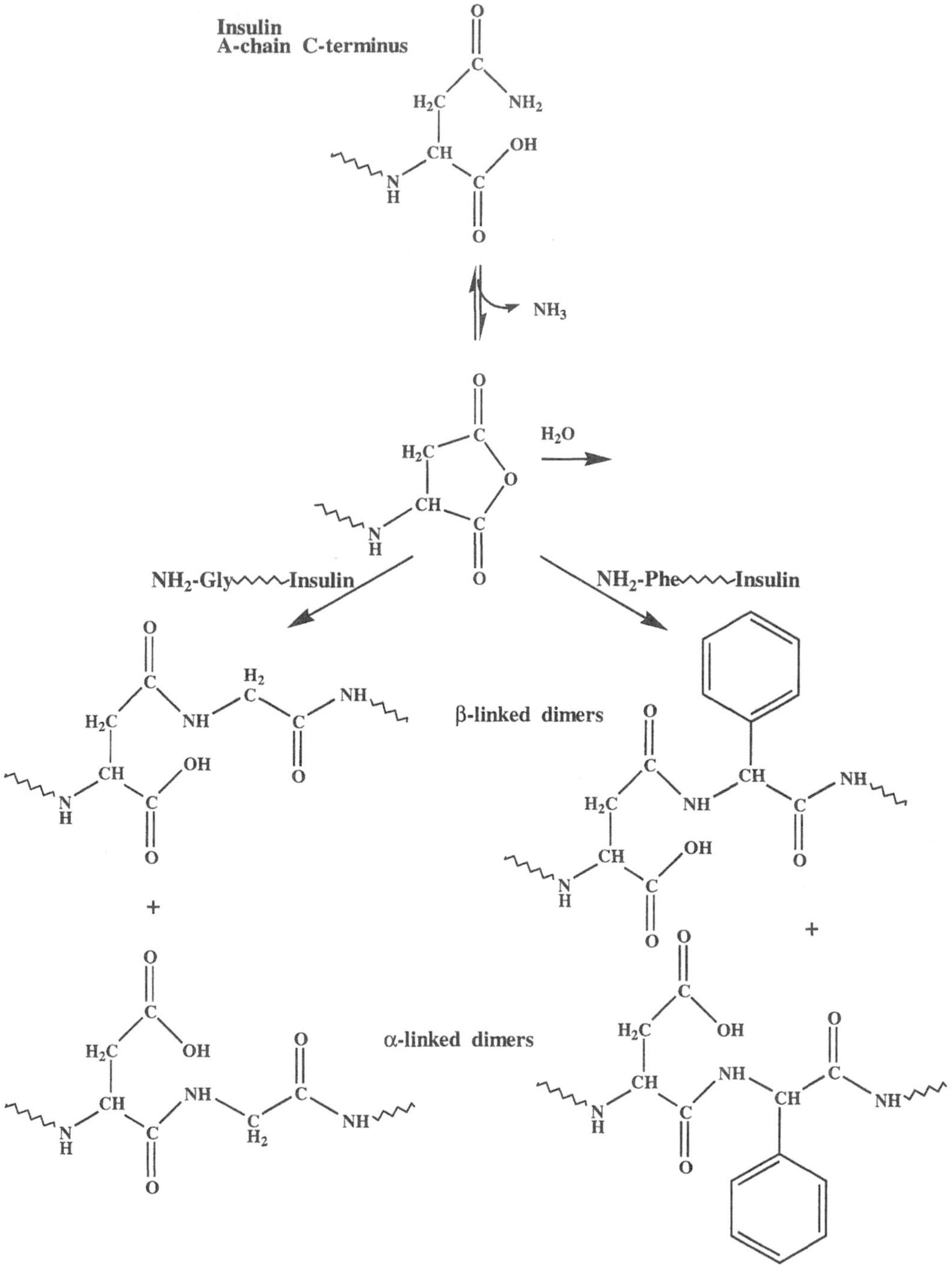

**Fig. 11** Formation of dimeric insulin species. Dimeric forms of insulin are formed through nucleophilic attack of the N-terminal amino-groups from each of the two chains against a cyclic anhydride which can form by deamidation of the C-terminal asparaginyl residue in the A-chain of insulin. Since the amino-group of the N-terminal residues can react with either carbonyl group in the cyclic anhydride, both beta and alpha-linked dimers can be formed.

forces, and by surface forces associated with interfacial sorption (113).

The simplest models (104, 109, 113) regarding the conformational stability of proteins in solution assume that denaturation processes are reversible and that the an unfolded protein can return to its native state (118–119). Furthermore, the assumption continues that the two conformational states are in dynamic equilibrium as described by the following equation:

Native conformation (N) $\Leftrightarrow$
Denatured state (D)

In its simplest form, then, this two-state model assumes that denaturation to the D state and renaturation to the N state are cooperative, all-or-none reversible processes, and moreover that they are not accompanied by the presence of any intermediate states reaching appreciable concentrations. Simple equilibrium thermodynamics can then be applied and the process quantified using an equilibrium constant (K) defined by the following equation:

$$K = [D]/[N]$$

based upon the relative concentrations of the respective states. The Gibbs free energy associated with the process can in turn be calculated as:

$$\Delta G = (G_D - G_N) = -RT\ln K$$

$\Delta G$ can ultimately by determined either directly using calorimetric techniques or indirectly by experimentally determining the equilibrium constant for any physical or biological property independently for the two states as the intensity of a given denaturant is increased (113). Most typically, indirect measurements are made by assessing spectroscopic properties, bioassay activities, chromatographic properties, etc., as a function of the intensity of the given perturbant employed. If temperature is used as the chosen stressor, the midpoint of this process is typically defined as the protein's $T_m$, or melting temperature (118). The $T_m$ represents the temperature at which 50% of the molecules are unfolded, i.e. where $\Delta G = 0$. Fig. 12. illustrates the unfolding, or denaturation, process for a recombinant protein when exposed either to increasing temperature or concentration of urea, as followed by a

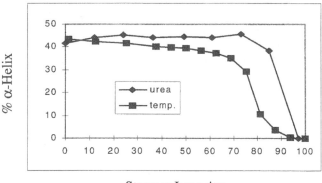

Stressor Intensity

**Fig. 12** The Denaturation of a Protein as Assessed using Secondary Structure Determinations. Optical spectroscopy (fUV circular dichroism) was used to determine the secondary structure of a recombinantly-expressed protein. The denaturation process, i.e. physical instability, is observed as either a function of the concentration of urea ($0 \rightarrow 8$ M) or as the temperature ($1 \rightarrow 80°$C) is increased in order to disrupt the global folding patterns, ultimately resulting in a loss of secondary structure in the form of $\alpha$-helical content.

loss in secondary structure as assessed using fUV circular dichroism.

Finally, it should be noted that chemical denaturation, in practice, is actually often promoted in order to enhance desirable attributes such as solubility. This practice is commonly used during the purification and re-folding of recombinantly-expressed proteins after their sequestration into refractile bodies. In this case, the foreign proteins are present in an insoluble form largely as denatured proteins in a reducing environment. In this case, strong denaturants, e.g. urea, alkaline pH, or guanidine-HCl, are used to re-solubilize the proteins prior to renaturation processes (120–122).

## B.  Aggregation, Precipitation, and Surface Adsorption

Although proteins may be readily soluble and exist as single molecules in solution, changes in the environment (i.e. pH, temperature, solvent composition, agitation, etc.) or variations in the protein concentration may increase the likelihood that protein-protein interactions occur. While these interactions may result in the formation of dimers, trimers, or other higher-order polymers, it is this association which leads to the aggregation phenomenon. The nature of the attractive versus repulsive forces as well as the balance of interactions among the solvent and solute molecules present ultimately determine whether or not the joined states are energetically favorable (123–133).

In many such instances, the simple two-state model for protein physical stability illustrated above represents only a crude approximation for the actual course of denaturation. This is especially evident in the instances where intermediate states have been isolated and identified. In the case of $\gamma$-Interferon (130) and bGH (134), it has been postulated that these intermediates ultimately lead to aggregation and biological inactivation. Multi-

state denaturation is then better described (113) as follows:

$$N \rightarrow I \rightarrow D_1 \quad OR \quad N \rightarrow D_1 \rightarrow D_2$$
$$\downarrow$$
$$D_2$$

One or the other of these steps might be fast or slow and/or reversible or irreversible. When irreversible, aggregation is likely to follow largely because the changes introduced as the result of global unfolding often result in the exposure of hydrophobic surfaces (135). The extent to which these hydrophobic surfaces tend to interact either with exposed surfaces, interfaces, or with other protein molecules ultimately determines the physical stability of the protein (136–139). Although many physicochemical interactions can lead to protein aggregation, it is primarily these noncovalent hydrophobic interactions which represent the major cause of aggregation from the denatured protein state (109, 129, 140–144). If these proteins also contain cysteine or cystine, intermolecular thiol-disulfide interchange can occur (104) as described previously.

While aggregates, once formed, can retain their solubility in a given solvent, when aggregation proceeds to a macroscopic scale it is often manifested as a precipitation. The size and number of particles that form is dependent upon the amount of protein that comes out of solution and the competition between the growth of existing particles and the nucleation of new particles. Both the particle growth and the nucleation processes are concentration dependent (144, 145). Light scattering-turbidity, chromatographic (size exclusion), and analytical ultracentrifugation procedures are often employed to determine the extent of protein aggregation as it relates to the formation of aggregate species and ultimately particles.

As mentioned previously, another factor which can promote protein denaturation is surface interaction. Indeed, adsorption studies have indicated that proteins will bind tightly to hydrophobic surfaces, including slightly charged hydrophobic silica surfaces associated with glass containers (146–150). After adsorption, a gradual unfolding of the protein may occur, with the denatured protein remaining adhered to the surface. As buried hydrophobic residues become more and more exposed and in turn adsorbed, gradual dissociation from the surface can result in aggregation followed by precipitation. Unlike many native proteins, insulin has exposed hydrophobic surfaces (109) and the concomitant adsorption, aggregation and precipitation of insulin has been well-documented due to difficulties encountered with its physical instability in delivery systems and its tendency to form a finely-divided precipitate film on storage containers (118, 151–153).

Disruption of a protein's native structure through denaturation, aggregation and surface adsorption generally leads to decreased activity, increased immunogenicity, and greater susceptibility to chemical modifications. Physical instability caused by extremes of pH, temperature and other factors can usually be minimized by controlling the external conditions to which the protein is exposed.

## IV. STRATEGIES FOR IMPROVING STABILITY

The chemical and physical stability of peptides and proteins is determined by many factors described in the prior two sections of this chapter. Strategies for improving the stability of these molecules involves both chemical changes to their structures as well as the inclusion of stabilizing additives. Chemical alterations to the primary structure can be made either through site-directed mutagenesis or by direct chemical modification (e.g., PEGylation). This section will summarize some of these strategies (118, 154, 155).

One alternative for improving physical and chemical stability is to alter the primary sequence through mutagenesis. Rational design can be used to replace amino acids which have been found to be particularly susceptible to chemical modifications, such as methionyl, asparaginyl and aspartyl residues. Likewise, physical stability can sometimes be improved by the introduction of new disulfide bonds, ion-binding sites, or by changing amino acids to improve hydrophobic interactions (118, 154).

Replacement of an important methionyl residue which is very susceptible to oxidation stabilized subtilisin against oxidative inactivation resulting from formation of Met(O) (156). Serine, alanine and leucine were all determined to be effective replacements for methionine in this protein. Another example of altering the primary sequence to improve stability is the replacement in epidermal growth factor of an aspartyl residue prone to degradation to an isoaspartyl (56). Substitution of this aspartyl by a glutamyl residue prevented formation of this modified species which only possessed 20% of the native activity.

Another process for stabilizing peptides and proteins is through the use of stabilizing additives. Salts, polyalcohol compounds, such as sugars and glycerol, detergents, amino acids and cofactors have all been used to stabilize peptides and proteins (109, 118, 154, 155). These compounds have been effective for improving stability both in solution and in the solid state. Salts exhibit their stabilization through specific ion binding sites on proteins and have been demonstrated to improve thermal stability and prevent aggregation and precipitation (157). Polyalcohol compounds may stabilize proteins against denaturation and are frequently used as excipients during lyophilization to improve stability. Detergents appear to increase stability through several mechanisms including the inhibition of aggregation, precipitation, and adsorption to surfaces (118, 154). Combinations of the above methodologies may be used to determine the most efficient manner for stabilizing peptides and proteins against the many chemical and physical processes which can inactivate these molecules.

## V.  CONCLUSIONS

This chapter has illustrated that the instability of peptides and proteins involves many diverse processes which are often interrelated. While it may not be possible to completely prevent all of the chemical and physical degradation processes described, many can be minimized through the appropriate choice of purification and storage conditions. The ever-increasing utilization of peptides and proteins as therapeutic agents makes the minimization of degradation and an improvement in the understanding of stability even more important than had previously been the case. Determining the correct balance between the options described for minimizing degradative processes and improving stability will be a major challenge for successful production of new peptide and protein reagents and drugs.

## REFERENCES

1.  DW Aswad. Deamidation and isoaspartate formation in peptides and proteins. Boca Raton, FL: CRC Press, 1995.
2.  S Clarke, RC Stephenson, JD Lowenson. Lability of asparagine and aspartic residues. In: TJ Ahern, MC Manning, eds. Proteins and peptides: Stability of Protein Pharmaceuticals, Part A. New York: Plenum Press, 1992, pp 1–29.
3.  S Capasso, L Mazzarella, F Sica, A Zagari, S Salvadori. Kinetics and mechanism of succinimide ring formation in the deamidation process of asparagine residues. J Chem Soc, Perkin Trans 2: 679–682, 1993.
4.  SA Bernhard, A Berger, JH Carter, E Katchalski, M Sela, Y Shalitin. Cooperative effects of functional groups in

peptides. I. Aspartyl-serine derivatives. J Am Chem Soc 84: 2421–2434, 1962.
5.  P Bornstein, G Balian. Cleavage at Asn-Gly bonds with hydroxylamine. In: CH Hirs, SN Timasheff, eds. Methods Enzymol Vol 47. New York: Academic Press, 1977, pp 132–145.
6.  T Geiger, S Clarke. Deamidation, isomerization, and racemization at asparaginyl and aspartyl residues in peptides: Succinimide-linked reactions that contribute to protein degradation. J Biol Chem 262: 785-794, 1987.
7.  AB Robinson, CJ Rudd. Deamidation of glutaminyl and asparaginyl residues in peptides and proteins. Curr Top Cell Regul 8: 247–254, 1974.
8.  R Lura, V Schirch. Role of peptide conformation in the rate and mechanism of deamidation of asparaginyl residues. Biochemistry 27: 7671–7677, 1988.
9.  M Bodanszky, JZ Kwei. Side reactions in peptide synthesis. VII. Sequence dependence in the formation of aminosuccinyl derivatives from beta-benzyl-aspartyl peptides. Int J Peptide Protein Res 12: 69–74, 1978.
10.  RC Stephenson, S Clarke. Succinimide formation from aspartyl and asparaginyl peptides as a model for the spontaneous degradation of proteins. J Biol Chem 264: 6164–6170, 1989.
11.  K. Patel, RT Borchardt. Chemical pathways of peptide degradation. III. Effect of primary sequence on the pathways of deamidation of asparaginyl residues in hexapeptides. Pharm Res 7: 787–793, 1990.
12.  R Tyler-Cross, V Schirch. Effects of amino acid sequence, buffers, and ionic strength on the rate and mechanism of deamidation of asparagine residues in small peptides. J Biol Chem 266: 22549–22556, 1991.
13.  NP Bhatt, K Patel, RT Borchardt. Chemical pathways of peptide degradation. I. Deamidation of adrenocorticotropic hormone. Pharm Res 7: 593–599, 1990.
14.  K Patel, RT Borchardt. Chemical pathways of peptide degradation. II. Kinetics of deamidation of an asparaginyl residue in a model hexapeptide, Pharm Res 7: 703–711, 1990.
15.  S Capasso, AJ Kirby, S Salvadori, F Sica, A Zagari. Kinetics and mechanism of the reversible isomerization of aspartic acid residue in tetrapeptides. J Chem Soc Perkin Trans 2: 437–442, 1995.
16.  N Fujii, Y Momose, K Harada. Kinetic study of racemization of aspartyl residues in model peptides of A-crystallin. Int J Peptide Protein Res 48: 118–122, 1996.

17. S Clarke. Protein carboxyl methyltransferases: Two distinct classes of enzymes. In: CC Richardson, PD Boyer, IB Dawid, A Meister, eds. Ann Rev Biochem Vol 54. Palo Alto, CA, 1985, pp 479–506.

18. JD Lowenson, S Clarke. Structural elements affecting the recognition of L-isoaspartyl residues by the L-isoaspartyl/ D-aspartyl protein methyltransferase. J Biol Chem 266: 19396–19406, 1991.

19. JD Lowenson, S Clarke. Recognition of D-aspartyl residues in polypeptides by the erythrocyte L-isoaspartyl/D-aspartyl protein methyltransferase. J Biol Chem 267: 5985–5995, 1992.

20. TV Brennan, S Clarke. Effect of adjacent histidine and cysteine residues on the spontaneous degradation of asparaginyl- and aspartyl-containing peptides. Int J Pept Protein Res 45: 547–553, 1995.

21. S Clarke. Propensity for spontaneous succinimide formation from aspartyl and asparaginyl residues in cellular proteins. Int J Peptide Protein Res 30: 808–821, 1987.

22. AA Kossiakoff. Tertiary structure is a principal determinant to protein deamidation. Science 240: 191–194, 1988.

23. SJ Wearne, TE Creighton. Effects of protein conformation on rate of deamidation: Ribonuclease A. Proteins:Struct Funct Genet 5: 8–12, 1989.

24 WJ Chazin, J Kordel, E Thulin, T Hofmann, T Drakenberg, S Forsen. Identification of an isoaspartyl linkage formed upon deamidation of bovine calbindin-D9k and structural characterization by 2D 1H NMR. Biochemistry 28: 8646–8653, 1989.

25 G Teshima, JT Stults, V Ling, V. E Canova-Davis. Isolation and characterization of succinimide variant of methionyl human growth hormone. J Biol Chem 266: 13544–13550, 1991.

26. BN Violand, MR Schlittler, EW Kolodziej, PC Toren, MA Cabonce, NR Siegel, KL Duffin, JF Zobel, CE Smith, JS Tou. Isolation and characterization of porcine somatotropin containing a succinimide residue in place of aspartate129. Prot Science 1: 1634–1641, 1992.

27 H Tomizawa, H Yamada, T Ueda, T Imoto. Isolation and characterization of 101-succinimide lysozyme that possesses the cyclic imide at Asp101-Gly102. Biochemistry 33: 8770–8774, 1994.

28. MY Kwong, RJ Harris. Identification of succinimide sites in proteins by N-terminal sequence analysis after alkaline hydroxylamine cleavage. Protein Sci 3: 147–149, 1994.

29. DA Carter, PN McFadden. Trapping succinimides in aged polypeptides by chemical reduction. J Prot Chem 13: 89–96, 1994..

30. S Capasso, J Mazzarella, AJ Kirby, A Salvadori. Succinimide-mediated pathway for peptide bond cleavage: Kinetic study on an Asn-Sar containing peptide. Peptides 17: 1075–1077, 1996.

31. CE Voorter, WA de Haard-Hoekman, PJ van den Oetelaar, H Bloemendal, WW de Jong. Spontaneous peptide bond cleavage in aging alpha-crystallin through a succinimide intermediate. J Biol Chem 263: 19020–19023, 1988.

32. BN Violand, MR Schlittler, PC Toren, NR Siegel. Formation of isoaspartate-99 in bovine and porcine somatotropins. J Protein Chem 9: 109–117, 1990.

33. FH Carpenter. Relationship of structure to biological activity of insulin as revealed by degradative studies. Am J Med 40: 750–758, 1966.

34. RT Darrington, BD Anderson. The role of intramolecular nucleophilic catalysis and the effects of self-association on the deamidation of human insulin at low pH. Pharm Res 11: 784–793, 1994.

35. RT Darrington, BD Anderson. Evidence for a common intermediate in insulin deamidation and covalent dimer formation: effects of pH and aniline trapping in dilute acidic solutions. J Pharm Sci 84: 275–282, 1995.

36. J Melville. Labile glutamine peptides and their bearing on the origin of the ammonia set free during the enzymatic digestion of proteins. Biochem J 29: 7671–7675, 1935.

37. AC Herman, TC Boone, HS Lu. Characterization, formulation and stability of Neupogen, a recombinant human granulocyte-colony stimulating factor. In: R. Pearlman, Y.J. Wang, eds. Formulation, characterization, and stability of protein drugs. New York: Plenum Press, 1996, pp. 324–325.

38. T Marcus. Preferential cleavage at aspartyl-prolyl peptide bonds in dilute acid. Int J Pept Protein Res 25: 542–546, 1985.

39. JY Wang, Z Shahrokh, S Vemuri, G Eberlein, I Beylin, M Busch. Characterization, stability, and formulations of basic fibroblast growth factor. In: R. Pearlman, Y.J. Wang, eds. Formulation, characterization, and stability of protein drugs, Case Histories. New York: Plenum Press. 1996 pp 159–163.

40. RA Kenley, NW Warne. Acid-catalyzed peptide bond hydrolysis of recombinant

human interleukin 11. Pharm Res 11: 72–76, 1994.

41. JA Schrier, RA Kenley, R Williams, RJ Corcoratn, Y Kim, RP Northey, D D'Augusta, B Huberty. Degradation pathways for recombinant human macrophage colony-stimulating factor in aqueous solutions. Pharm Res 10: 933–944, 1993.

42. YJ Wang, R Pearlman. Stability and Characterization of Protein and Peptide Drugs, Case Histories. New York: Plenum Press, 1993.

43. L Graf, G Hajos, A Patthy, G Cseh. The influence of deamidation on the biological activity of porcine adrenocorticotropic hormone (ACTH), Horm Metab Res 5: 142–143, 1973.

44. T Flatmark. Multiple forms of bovine heart cytochrome C: a comparative study of their physicochemical properties and their reactions in biological systems. J Biol Chem 242: 2454–2459, 1967.

45. BA Johnson, JM Shirokawa, DW Aswad. Deamidation of calmodulin at neutral and alkaline pH: quantitative relationships between ammonia loss the susceptibility of calmodulin to modification by protein carboxyl methyltransferase. Arch Biochem Biophys 268: 276–286, 1989

46. BA Johnson, EL Langmack, DW Aswad. Partial repair of deamidation-damaged calmodulin by protein carboxyl methyltransferase. J Biol Chem 262: 12283–12287, 1987

47. SM Potter, WJ Henzel, DW Aswad. In vitro aging of calmodulin generates isoaspartate at multiple Asn-Gly and Asp-Gly sites in calcium-binding domains II, III and IV. Protein Sci 2: 1648–1663, 1993.

48. G Teshima, J Porter, K Yim, K, V Ling, A Guzzetta. Deamidation of soluble CD4 at asparagine-52 results in reduced binding capacity for the HIV-1 envelope glycoprotein gp120. Biochemistry, 30: 3916–3922, 1991.

49. DJ Kroon, A Baldwin-Ferro, P Lalan. Identification of sites of degradation in a therapeutic monoclonal antibody by peptide mapping. Pharm Res 9: 1386–1393, 1992.

50. J Bongers, EP Heimer, T Lambros, Y-CE Pan, RM Campbell, AM Felix. Degradation of aspartic acid and asparagine residues in human growth hormone releasing factor. Int J Pept Protein Res 39: 364–374, 1992

51. JL Castal, TJ Ahern, RC Davenport, GA Petsko, AM Klibanov. Subunit interface of triosephosphate isomerase: site directed mutagenesis and characterization of the altered enzyme. Biochemistry 26: 1258–1264, 1987.

52. TJ Ahern, JI Casal, GA Petsko, AM Klibanov. Control of oligomeric enzyme thermostability by protein engineering. Proc Natl Acad Sci USA 84: 675–679, 1987.

53. L Graf, G Hajos, A Patthy, G Cseh. The influence of deamidation on the biological activity of porcine adrenocorticotropic hormone (ACTH). Horm Metab Res 5: 142–147, 1973.

54. GO Daumy, CL Wilder, JM Merenda, AS McColl, KF Geoghegan, IG Otterness. Reduction of biological activity of murine recombinant interleukin-1-beta by selective deamidation at asparagine-149. FEBS Lett 278: 98–101, 1991.

55. TJ Ahern, AM Klibanov. The mechanism of irreversible enzyme inactivation at 100 degrees C. Science 228: 1280–1286, 1985.

56. C George-Nascimento, J Lowenson, M Borissenko, M Calderon, A Medina-Selby, J Kuo, S Clarke, S Randolph. Replacement of a labile aspartyl residue increases the stability of human epidermal growth factor. Biochemistry 29: 9584–9591, 1990.

57. A Di Donato, G D'Alessio. Heterogeneity of bovine seminal ribonuclease. Biochemistry 20: 7232–7236, 1981.

58. PT Wingfield, M Payton, P Graber, K Rose, JM Dayer, AR Shaw, U Schmeissner. Purification and characterization of human interleukin-1 alpha produced in Escherichia coli. Eur J Biochem 165: 537–544, 1987.

59. PT Wingfield, RJ Mattaliano, HR McDonald, S Craig, GM Clore, AM Gronenborn, U Schmeissner. Recombinant-derived interleukin-1 alpha stabilized against specific deamidation. Protein Eng 1: 413–418, 1987.

60. J Brange, B Skelbaek-Pederson, L Langkjaer, U Damgaard, H Ege, S Havelun, LG Heding, KH Orgensen, J Lykkeberg, J Markussen, M Pingel, E Rasmussen, P Galenics. Insulin: The Physicochemical and Pharmaceutical Aspects of Insulin and Insulin Preparations. Berlin: Springer-Verlag, 1987.

61. M Martinez-Carrion, F Riva, C Turano, P Fasella. Multiple forms of supernatant glutamate-aspartate transaminase from pig heart, Biochem Biophys Res Comm 20: 206–212, 1965.

62. M Martinez-Carrion, C Turano, E Chiancone, R Bossa, A Giartosio, F Riva, P Fasella. Isolation and characterization of multiple forms of glutamate-aspartate transaminase from pig heart. J Biol Chem 242: 2397–2403, 1967.

63. LH Bertland, NO Kaplan. Chicken heart soluble aspartate amino transferase: purification and properties. Biochemistry 7: 134–141, 1968.

64. CM Michuda, M Martinez-Carrion. Mitochondrial aspartate transaminase. II. Isolation and characterization of multiple forms. Biochemistry 8: 1095–1102, 1969.

65. RA John, RE Jones. The nature of multiple forms of cytoplasmic aspartate aminotransferase from pig and sheep heart. Biochem J 141: 401–406 1974.

66. A Skottner, A Forsman, B Skoog, JL Kostyo, CM Cameron, NA Damfio, KG Thorngren, M Hagerman. Biological characterization of charge isomers of human growth hormone. Acta Endocrinol. 118: 14–19, 1988.

67. GW Becker, PM Tackitt, WW Bromer, DS Lefeber, and RM Riggin. Isolation and characterization of a sulfoxide and a desamido derivative of biosynthetic human growth hormone. Biotech Appl Biochem 10: 326–337, 1988.

68. R Bischoff, P Lepage, M Jauinod, G Cauet, M Acker-Klein, D Clesse, M Laporte, A Bayol, A Van Dorsselaer, C Roitsch. Sequence-specific deamidation: isolation and biochemical characterization of succinimide intermediates of recombinant hirudin. Biochemistry 32: 725–731, 1993.

69. K Sasaoki, T Hiroshima, S Kusumoton, K Nishi. Deamidation at asparagine-88 in recombinant interleukin-2. Pharm Bull 40: 976–980, 1992.

70. G Zaman, PW Saphier, N Loveridge, T Kimura, S Sakakibara, SM Bernier, GN Hendy. Biological properties of synthetic human parathyroid hormone: effect of deamidation at position 76 on agonist and antagonist activity. Endocrinology 128: 2583–2590, 1991.

71. J Swarbrick, JC Boylan. Autooxidation and antioxidants, in Encyclopedia of Pharmaceutical Technology, Vol. 1, Absorption of Drugs to Bioavailability and Bioequivalence. New York: Marcel Dekker, 1988.

72. L Packer, AN Glazer. Methods in Enzymology, Vol. 186. New York: Academic Press, 1990.

73. N Brot, H Weissbach. Biochemistry of methionine sulfoxide residues in proteins. BioFactors 3: 91–96, 1991.

74. W Vogt. Oxidation of methionyl residues in proteins: tools, targets and reversal. Free Rad Biol Med 18: 93–105, 1995.

75. C Schoncich, F Zhao, GS Wilson, RT Borchardt. Iron-thiolate induced oxidation of methionine to methionine sulfoxide in small model peptides. Intramolecular catalysis by histidine. Biochem Biophys Acta 1158: 307–322, 1993.

76. RT Houghten, CH Li. Reduction of sulfoxides in peptides and proteins. In: CH Hirs, SN Timasheff eds. Methods Enzym, Vol 91. New York: Academic Press, 1983, pp 549–559.

77. N Brott, L Weissbach, J Werth, H Weissbach. Enzymatic reduction of protein-bound methionine sulfoxide. Proc Natl Acad Sci 78: 2155–2158, 1981.

78. J Fransson, E Florin-Robertsson, B Axelsson, C Nyhlen. Oxidation of human insulin-like growth factor I in formulation studies: kinetics of methionine oxidation in aqueous solution and solid state. Pharm Res 13: 1252–1257, 1996.

79. J Fransson, A Hagman. Oxidation of human insulin-like growth factor I in formulation studies, II. Effects of oxygen, visible light, and phosphate on methionine oxidation in aqueous solution and evaluation of possible mechanisms. Pharm Res 13: 1476–1481, 1996.

80. JR Fransson. Oxidation of human insulin-like growth factor I in formulation studies. 3. Factorial experiments of the effects of ferric ions, EDTA, and visible light on methionine oxidation and covalent aggregation in aqueous solution. J Pharm Sci 86: 1046–1050, 1997.

81. TH Nguyen, SJ Shire. Stability and characterization of recombinant human relaxin. In: R Pearlman, YJ Wang, eds. Formulation, characterization, and stability of protein drugs. Formulation, characterization, and stability of protein drugs, New York: Plenum Press, 1996, pp 234–238.

82. AC Herman, TC Boone, HS Lu. Characterization, formulation and stability of Neupogen, a recombinant human granulocyte-colony stimulating factor. In: R Pearlman, YJ Wang, eds. Formulation, characterization, and stability of protein drugs, New York: Plenun Press, 1996, pp 324–325.

83. JL Cleland, MF Powell, SJ Shire. The development of stable protein formulations: A close look at protein aggregation, deamidation, and oxidation, in Critical Reviews in therapeutic drug carrier systems, 10: 307–377, 1993.

84. GW Becker, PM Tackitt, WW Bromer, DS Lefeber, RM Riggin. Isolation and characterization of a sulfoxide and desamido derivative of biosynthetic human growth hormone. Biotechnol Appl Biochem 10: 326–333, 1988.

85. MJ Pikal, KM Dellerman, ML Roy, RM Riggin. The effects of formulation

variables on the stability of freeze-dried human growth hormone. Pharm Res 8: 427–434, 1991.

86. RA Houghten, CB Glaser, CH Li. Human somatotropin reaction with hydrogen peroxide. Arch Biochem Biophys 178: 350–355, 1977.

87. CB Glaser, CH Li. Reaction of bovine growth hormone with hydrogen peroxide. Biochemistry 13: 1044–1047, 1975.

88. JA Kirsty-Roy, SK Chan, ET Iwanmoto. Methionine oxidation enhances opioid activity of an enkephalin analog. Life Sci 32: 889–893, 1983.

89. Z Shaked, RP Szajewski, GM Whitesides. Rates of thiol-disulfide interchange reactions involving proteins and kinetic measurements of thiol pKa values. Biochemistry 19: 4156–4166, 1980.

90. DF Mark, SD Lu, AA Creasey, R Yamamoto, LS Lin. Site specific mutagenesis of the human fibroblast interferon gene. Proc Natl Acad Sci USA 81: 5662–5666, 1984.

91. LJ Perry, R Wetzel. The role of cysteine oxidation in the thermal inactivation of T4 lysozyme. Protein Eng 1: 101–105, 1984.

92. SA Thompson, JC Fiddes. Chemical characterization of the cysteines of basic fibroblast growth factor. Ann NY Acad Sci 638: 78–88, 1992.

93. LC Foster, SA Thompson, SJ Tarnowski. Methods and formulations for stabilizing fibroblast growth factor. International Patent Appl WO/91. 15509–15521, 1991.

94. DL Lindemeyer, JG Menke, LJ Kelly, J DiSalvo, D Soderman, MT Schaeffer, S Ortega, G Gimenez-Gallego, KA Thomas. Disulfide bonds are neither required, present, nor compatible with full activity of human recombinant acidic fibroblast growth factor. Growth Factors 3: 287–298, 1990.

95. SH Chang, GM Techima, T Milby, B Gillece-Castro, E Canova-Davis. Metal-catalyzed photoxidation of histidine in human growth hormone. Anal Biochem 244: 221–227, 1997.

96. R Zhao, E Ghezzo-Schoneich, GI Aced, J Hong, T Milby, CJ Schoneich. Metal catalyzed oxidation of histidine in human growth hormone. Mechanism, isotope effects and inhibition by a mild denaturing alcohol. J Biol Chem 272: 9019–9029, 1997.

97. TM Florence. Degradation of protein disulphide bonds in dilute alkali. Biochem J 189: 507–520, 1980.

98. TJ Ahern, AM Klibanov. The mechanism of irreversible enzyme inactivation at 100 degrees C. Science 228: 1280–1284, 1985.

99. DB Volkin, AM Klibanov. Thermal destruction processes in proteins involving cystine residues. J Biol Chem 262: 2945–2950, 1987.

100. JJ Correia, LD Lipscomb, S Lobert. Nondisulfide crosslinking and chemical cleavage of tubulin subunits: pH and temperature dependence. Arch Biochem Biophys 300: 105–114, 1993.

101. JP Greenstein, M Winitz. Chemistry of the amino acids, Vol 2. Florida: RE Publishing Co. 1991.

102. NF Sepetov, MA Krymsky, MV Ovchinnikov, ZD Bespalava, OL Isakova, M Soucek, M Lebl. Rearrangement, racemization and decomposition of peptides in aqueous solution. Pept Res 4: 308–313, 1991.

103. JE Battersby, WS Hancock, E Canova-Davis, J Oeswein, B O'Connor. Diketopiperazine formation and N-terminal degradation in recombinant human growth hormone. Int J Pept Protein Res 44: 215–222, 1994.

104. D Wong, J Parasrampuria. Pharmaceutical Excipients for the Stabilization of Proteins. BioPharm. 1997, pp 52–61.

105. L Stryer. Protein Structure and Function. In: L Stryer, ed. Biochemistry. New York: W H Freeman and Company, 1988, pp 15–42.

106. D Voet, JG Voet. Amino Acid. In: D Voet, JG Voet, eds. Biochemistry. New York: John Wiley and Sons, 1990, pp 59–74.

107. D Voet, JG Voet. Covalent Structures of Proteins. In: D Voet, JG Voet, eds. Biochemistry. New York: John Wiley and Sons, 1990, pp 108–143.

108. D Voet, JG Voet. Three-Dimensional Structures of Proteins. In: D Voet, JG Voet, eds. Biochemistry. New York: John Wiley and Sons, 1990, pp 144–192.

109. JL Cleland, MF Powell, SJ Shire. The Development of Stable Protein Formulations: A Close Look at Protein Aggregation, Deamidation, and Oxidation. Critical Reviews in Therapeutic Drug Carrier Systems 10(4): 307–377, 1993.

110. PL Privalov, NN Khechinashvili. A Thermodynamic Approach to the Problem of Stabilization of Globular Protein Structure: A Calorimetric Study. J Mol Biol 86: 665–684, 1974.

111. VV Filimonov, W Pfeil, TN Tsakova, PL Privalov. Thermodynamic Investigations of Proteins IV. Calcium Binding Protein Parvalbumin. Biophy Chem 8: 117–122, 1978.

112. PH von Hippel, KY Wong. On the conformational stability of globular proteins. The effects of various electrolytes and

non-electrolytes on the thermal ribonuclease transition. J Biol Chem 240(10): 3909–3923, 1965.

113. F Franks. Conformational Stability of Proteins. In: F Franks, ed. Protein Biotechnology. New Jersey: The Humana Press, 1993, pp 395–436.

114. F Franks, D Eagland. The role of solvent interactions in protein conformation. Crit Rev Biochem 3(2): 165–219, 1975.

115. YJ Wang, MA Hanson. Parenteral Formulations of Proteins and Peptides: Stability and Stabilizers. J Parenteral Sci Tech 42: S4–S24, 1988.

116. F. Franks. Biophysics and Biochemistry at Low Temperatures. Cambridge: Cambridge University Press, 1985.

117. SJ Prestrelski, N Tedeschi, JF Carpenter, T Arakawa. Dehydration-induced conformational transitions in proteins and their inhibition by stabilizers. Biophys J 65(2): 661–671, 1993.

118. MC Manning, K Patel, RT Borchardt. Stability of Protein Pharmaceuticals. Pharm Res 6(11): 903–916, 1989.

119. D Rozema, SH Gellman. Artificial Chaperone-Assisted Refolding of Carbonic Anhydrase B. J Biol Chem 271(7): 3478–3487, 1996.

120. FAO Marston, DL Hartley. Solubolization of Protein Aggregates. In: MP Deutscher, ed. Methods in Enzymology, Vol 182. Guide to Protein Purification. San Diego: Academic Press, 1990, pp 264–276.

121. FAO Marston. In: D Glover, ed. DNA Cloning, Vol 3. Oxford: IRL Press, 1987, p 59.

122. FAO Marston. The purification of eukaryotic polypeptides synthesized in Escherichia coli. Biochem J 240(1): 1–12, 1986.

123. WR Liu, R Langer, AM Klibanov. Moisture-Induced Aggregation of Lyophilized Proteins in the Solid State. Biotech Bioeng 37: 177–184, 1991.

124. CE Glatz. Modeling of Aggregation-Precipitation Phenomena. In: TJ Ahern, MC Manning, eds. Stability of Protein Pharmaceuticals, Part A. Chemical and Physical Pathways of Protein Degradation. New York and London: Plenum Press, 1992, pp 135–166.

125. S Yoshioka, Y Aso, K Izutsu, S Kojima. Is Stability Prediction Possible for Protein Drugs? Denaturation Kinetics of $\beta$-Galactosidase in Solution. Pharm Res 11(12): 1721–1725, 1994.

126. LC Gu, EA Erdos, HS Chiang, T Calderwood, K Tsai, GC Vizor, J Duffy, WC Hsu, LC Foster. Stability of Interleukin 1$\beta$ (IL-1$\beta$) in Aqueous Solution: Analytical Methods, Kinetics, Products, and Solution Formulation Implications. Pharm Res 8(4): 485–490, 1991.

127. T Arakawa, R Bhat, SM Timasheff. Preferential Interactions Determine Protein Solubility in Three-Component Solutions: The MgCl$_2$ System. Biochemistry 29: 1914–1923, 1990.

128. TF Holtzman, JJ Dougherty Jr., DN Brems, NE MacKenzie. pH-Induced Conformational States of Bovine Growth Hormone. Biochemistry 29(5): 1255–1261, 1990.

129. S Tandon, PM Horowitz. Detergent-associated refolding of guanidinium chloride-denatured rhodanese: the effect of lauryl maltoside. J Biol Chem 262: 4486–4491, 1987.

130. T Arakawa, Y-R Hsu, DA Yphantis. Acid Unfolding and Self-Association of Recombinant Escherichia coli Derived Human Interferon. Biochemistry 26: 5428–5432, 1987.

131. H Krebs, FX Schmid, R Jaenicke. Native-like Folding Intermediates of homologous Ribonucleases. Biochemistry 24: 3846–3852, 1985.

132. T Arakawa, SN Timasheff. Mechanism of Protein Salting In and Salting Out by Divalent Cation Salts: Balance Between Hydration and Salt Binding. Biochemistry 23: 5912–5923, 1984.

133. T Arakawa, SN Timasheff. Preferential Interactions of Proteins with Salts in Concentrated Solutions. Biochemistry 21: 6545–6552, 1982.

134. DN Brems, SM Plaisted, HA Havel, C-SC Tomich. Stability of an associated folding intermediate of bovine growth hormone by site-directed mutagenesis. Proc Nat Acad Sci USA 85(10): 3367–3371, 1988.

135. D Shortle, HS Chan, KA Dill. Modeling the effects of mutations on the denatured states of proteins. Protein Science 1(2): 201–215, 1992.

136. RT Sauer, WA Lim. Mutational analysis of protein stability. Curr Opin Struct Biolo 2: 46–54, 1992.

137. WA Lim, RT Sauer. The role of internal packing interactions in determining the structure and stability of a protein. J Mol Biol 219(2): 359–376, 1991.

138. AE Eriksson, WA Baase, X-J. Zhang, DW Heinz, M Blaber, EP Baldwin, and BW Matthews. Response of a protein structure to cavity-creating mutations and its relation to the hydrophobic effect. Science 255: 178–183, 1992.

139. KA Dill. Dominant forces in protein folding. Biochemistry 29(31): 7133–7155, 1990.

140. JL Cleland, DIC Wang. In Vitro Protein Folding. In: GN Stephanopuolos, ed. Bioprocessing, Vol 3. Weinheim, Germany:VCH Publishers, 1993, chapter 23.

141. A Mitraki, J King. Protein folding intermediates and inclusion body formation. Biotechnology 7:690-699, 1989.

142. WW Fish, A Danielsson, K Nordling, SH Miller, CF Lam, I Bjork. Denaturation behavior of antithrombin in guanidinium chloride: irreversibility of unfolding caused by aggregation. Biochemistry 24(6): 1510–1517, 1985.

143. Y Fuke, M Sekiguchi, H Matuoka. Nature of stem bromelian treatment on aggregation and gelation of soybean proteins. J Food Sci 50: 1283–1290, 1985.

144. A Ikai, S Tanaka, H Noda. Reactivation kinetics of guanidine denatured bovine carbonic anhydrase B. Arch Biochem Biophys 190(1): 39–45, 1978.

145. FJ Reithel. The dissociation and association of protein structures. Adv Protein Chem 18: 121–140, 1963.

146. A Sadana. Protein adsorption and inactivation on surfaces. Influence of heterogeneities. Chem Rev 92: 1799, 1992.

147. JD Andrade, V Hlady, A-P Wei, C-H Ho, AS Lea, SI Jeon, YS Lin, E Stroup. Proteins at interfaces: principles, multivariate aspects, protein resistant surfaces, and direct imaging and manipulation of adsorbed proteins. Clin Mater 11: 67, 1992.

148. BL Steadman, KC Thompson, CR Middaugh, K Matsuno, S Vrona, EQ Lawson, RV Lewis. The effects of surface adsorption on the thermal stability of proteins. Biotech Bioeng 40: 8, 1992.

149. W Norde, J Lyklema. Why proteins prefer interfaces. J Biomater Sci Polym Ed 2(3): 183–202, 1991.

150. W Norde, ACI Anusiem. Adsorption, Desorption, and Readsorption of Proteins on Solid Surfaces. Colloids and Surfaces 66: 73–80, 1992.

151. V Sluzky, J Tamada, AM Klibanov, R Langer. Kinetics of insulin aggregation in aqueous solution on agitation in the presence of hydrophobic surfaces. Proc Nat Acad Sci USA 88(21): 9377–9381, 1991.

152. H Thurow, K Geisen. Stabilization of dissolved proteins against denaturation at hydrophobic interfaces. Diabetologia 27: 212, 1984.

153. WD Lougheed, AM Albisser, HM Martindale, JC Chow, JR Clement. Physical stability of insulin formulations. Diabetes 32: 424, 1983.

154. TJ Ahern, MC Manning. Stability of protein pharmaceuticals, Part B, In vivo pathways of degradation and strategies for protein stabilization. New York: Plenum Press, 1992.

155. CO Fagain. Understanding and increasing protein stability. Biochim Biophys Acta. 1252: 1–14, 1995.

156. DA Estell, TP Graycar, JA Wells. Engineering an enzyme by site-directed mutagenesis to be resistant to chemical oxidation. J Biol Chem 260: 8518–8521, 1985.

157. R Palmieri, R W-K Lee, MF Dunn. 1H fourier transform NMR studies of insulin: coordination of $Ca^{+2}$ to the Glu(B13) site drives hexamer assembly and induces a conformation change. Biochemistry 27: 3387-3397, 1988.

# 10

# Bioinformatics and Computational Chemistry in Molecular Design: Recent Advances and Their Applications

**Jin Li[1] and Barry Robson[2]**
[1]*Proteus Molecular Design Ltd., Macclesfield, Cheshire, United Kingdom*
[2]*IBM Thomas J. Watson Research Center, Hawthorne, New York*

## I. INTRODUCTION

Over the past decade, the use of computer-assisted molecular design (CAMD) techniques in drug discovery research has increased considerably (1–3). Many technologies have been developed to assist the design of molecules as putative therapeutics agents (4–6).

In this chapter, some recent developments in the key areas of CAMD, i.e., aspects of bioinformatics and computational chemistry, will be discussed. It is impossible to be exhaustive in this field because it covers a wide range of research and applications. Nonetheless, there is increasing interest in characterizing the most profitable work-flows in drug discovery; an attempt has been made to identify those areas of CAMD development which have shown promising applications, and will be important for drug discovery research.

## II. CLINICAL AND COMMERCIAL SIGNIFICANCE OF BIOINFORMATICS IN DRUG DISCOVERY

The current chapter aims to be primarily scientific. That is, it is important to distinguish such a purely technical analysis from the assessments of the market and venture capitalist sector. However, necessity is the mother of invention, and some introductory comments are needed to help place the significance of the current review, and certain specific issues mentioned, in clinical and commercial perspective.

Despite the steady progression of technical development which we would argue has taken place in the computational area, there have been significant swings in the popularity of CAMD methods and these are only now reaching a point of proper equilibrium. In the early 1990s, computational methods were just begin-

[1] Proteus Molecular Design Limited, Beechfield House, Lyme Green Business Park, Macclesfield, Cheshire SK11 0JL, UK. Tel: (UK) 1625 500555. Fax: (UK) 1625 500666. email: J.Li@proteus.co.uk
[2] Computational Biology Center, IBM Thomas J. Watson Research Center, 30 Saw Mill River Road, Hawthorne, NY 10532. Tel: (US) 914 784 7816. Fax: (US) 914 784 6307. email: robsonb@us.ibm.com

ning to bear fruit when attention of the industry was directed more and more to the benefits of automated chemistry, combinatorial chemistry and high throughput screening methods. In our view this direction also included the erroneous perception of this as an independent and competitive approach. This aspect is considered in Sec. VII.

To appreciate the significance of these developments, one needs to recall that pharmaceutical development processes are complex, full of pitfalls: they proceed in spiralling cycles of identifying therapeutic indications, discovering molecular targets, validating targets, developing assays, generating diversity of compounds to screen against these targets, identifying lead compounds, and refining these compounds for entering clinical trials. Previously, in the 1970s to late 1980s, the Centre for Medicines Research reported some 45–70 new medicines introduced annually, but at an annual expenditures of 5–15 billion dollars. Moreover, it might take some 15 years to realise a given project as a drug in the market place. Primarily, the problem was that intermediate yield of compounds for study was just too slow: the traditional organic chemists might synthesise just 2000 compounds in a lifetime and the pharmacological team might screen some 100 natural and synthetic compounds a week. Hence, the lure of any potential new automated and high throughput technology for enhancing drug discovery was obviously likely to be considerable. When applied, this technology seemed to deliver. Robotic syntheses around 1996 could generate 10,000–100,000 compounds a week and high throughput analyses could screen some 500,000 corporate compounds in a year.

Combinatorial chemistry is described in Sec. VII., and has been particularly influential. A typical report (7) at this time described the specific approach of combinatorial chemistry delivering more than 100-fold increase of compounds, and high throughput screening of generating 50 fold increase in detection of possible leads.

One would thus conclude that we should be experiencing several fold increase in productivity. Sadly, however, the promise of all this has *not* been born out in commercial terms. As described by the Centre for Medicines Research in reports from 1995 to 1997, The number of new medicines has fallen to around 45 new entries per annum or less, with more than a doubling of annual R and D expenditure. To inject some 3–5 new drugs into the marketplace, a pharmaceutical major now needs to generate some 200 new targets a year and have some 20 reaching the clinic at any one time. Moreover, despite high expectations in the early 1990s in many other organisations, it is remarkable that, despite all the technological development, it *still* takes about 15 years to bring a pharmaceutical conception to market.

Where then are the bottlenecks now? Actually there are several, and it is noteworthy for the purposes of this review that all are subject to being potentially alleviated by the application of computers to molecular information management.

Many are identified as the human element, associated with data management and steps which currently involve complex decisions. The scientist in particular has not the time to stop and analyze results, nor to design new experiments. Indeed, he is faced with a veritable Angst as results from combinatorial chemistry and high throughput screening rain incessantly, and with gathering force, down upon him. In this, one already sees the merits of turning to computer-aided molecular information management.

Importantly, however, he is faced with the obligation to consider the additional (and notoriously overwhelming) torrent of data from the genome projects, an area which ironically was earlier of little interest to the medicinal chemist save perhaps as an intellectual pursuit.

This need arises as follows. Both "virtual screening" of drugs by computer and experimental screening methods represent powerful co-operative technologies. Each seeks to discover new drugs as "molecular keys", to turn on or turn off physiological and cytological functions in order to help maintain or enhance the well being of the organism. A key feature of the initiative of George Poste at SmithKline Beecham was to appreciate that in proper application of the new technologies (experimental and computational) to the pharmaceutical industry, the choice of molecular keys benefits hugely from the identification of specific molecular locks (especially receptors, enzymes, ion channels), and their essential components, to which to find a fit, and from interpretation of the data which transforms activation and inhibition of such targets into physiological effect. These targets are primarily proteins which for the most part will be (a) examined by structure determination methods such as crystallisation and x-ray diffraction to provide a rational informed basis for computational drug design and to aid experimental design and characterisation, (b) isolated or manufactured for in vitro screening methods, or (c) identified and monitored as CDNA or *in vivo* in large scale systems.

Important, then, is the progressive unveiling of the human three billion base pairs of the human genome with its potential of some 100,000 or more molecular targets, primarily protein molecules, signifying a potential increase in the molecular locks to fit by more than a hundred-fold. Also important is the similar elucidation of the countless billions of base pairs in the genomes of animals and plants and not least of viruses, bacteria and other parasites as disease pathogens, against the protein products of which will be directed the molecular keys as "magic

bullets". These human, animal, plant and pathogen studies are an international effort undertaken by both academic and industrial organisations (8). Importantly for this programme, the above mentioned commercial considerations, and particularly the SmithKline initiative, established the worth of the capture and management and use of biological information, especially genomics and computational proteomic information, as an overall strategy within the industry. It is estimated that before long, over many hundreds of thousands protein sequences will be available for analysis in biological research and drug discovery. There are still important questions to ask about how exactly one may select potential therapeutics targets from this flood of biological sequence information. In particular, better ways to gain knowledge of the tertiary structure of a protein would be of invaluable importance for drug discovery research both in terms of understanding a protein's functions and designing specific ligands to interact with the target protein.

The information management and analysis component of all this genomic and proteomic research, including structural aspects such as protein structure modelling and computational study of protein-ligand interactions, constitutes *bioinformatics*. This term seems to have been first formally used in 1984 in COM 84 Final of the European Commission. It is believed that this document arose in response to a memo to the EC from the White House, to the effect that Europe needed to integrate its biotechnology effort in order to compete with America and Japan. Within bioinformatics, the specific question of the management of the information in order to select and interpret molecular targets is of mounting importance (9).

In fact, the information management challenges of *bioinformatics* may yet well outweigh the problem of management of chemical information data, ie. *chemo-*

*informatics,* by orders of magnitude. Combined with data from other organisms, from polymorphisms, and from forced artificial molecular evolution of peptides and proteins as in combinatorial peptide chemistry and phage display, we might within the next five years expect hundreds to thousands of gigabytes of data, potentially conferring a competitive commercial advantage, to be pouring in nightly. We are already routinely addressing hundreds of gigabytes of gene data, which represented on a flat-file would require many hours to insert scientific annotation. Such aspects have promoted the recent development of relational and object-relational database methods. Yet, even if we fully automate the operations of analysis, certain important bioinformatics operations such as BLAST (used to discover relations of new sequences to those already on the data base) is still in most hands operating on flat-files, and the preparation of such *already* takes many hours! Soon, there will be little opportunity for human or machine error: one stumble and it may be impossible to catch up.

With the above concerns in mind, there are clearly many crucial and pressing issues of automated pipeline analysis, high-speed algorithm and hardware development, security and system crash recovery strategy, that need to be addressed in bioinformatics. For the most part, these "engineering" aspects lie outside our present scope. Here we concentrate on some key areas of science which represent to various extents potential bottlenecks in full application of bioinformatics: that is, there has been some controversy as to the extent to which these have or can become practical, and routine technology. These address the question of what is plausible, but the plausibility, nonetheless, of potentially vital tools in the repertoire of analysis. Indeed, if at least some of the aspects considered were of limited scientific worth, it may be argued that the above important "engineering"

considerations may become purely academic.

## III.  PROTEIN STRUCTURE PREDICTION

After the initial classical folding studies of Anfinsen (showing that proteins could be unfolded and spontaneously refold) (10) and the initial determination of some protein structures using x-ray crystallography in 1960s (11), great effort has been put into developing theories and algorithms for protein folding and protein structure prediction.

Anfinsen, Scheraga and many others have long maintained that the problem of predicting protein structure can be to a large extent independent of any biological considerations. That is, proteins can fold in a manner directly determined by their sequence alone, and that structure could in principle be calculated *ab initio,* from first principles, considering the protein sequence alone. This kind of approach is considered below in regard to *de novo* methods, where we follow the tendency of authors to use "*de novo*" rather than "*ab initio*" when some kind of extra information are used (as discussed in that section: in effect, *de novo* methods are for practical adaptations of, and compromises to, the *ab initio* philosophy).

The validity of *ab intio* approaches and the worthwhile character of *de novo* approaches obviously depends on the validity of the above idea that proteins do indeed generally, or at least in most cases, dictate their own folding. Several complications arise in discussing this: these include the distinction between kinetic and thermodynamic dictation of folding, and others relate to the implications and consequences of post-translational modifications. There are also nuances on the finer details of the folding process (see Sec. IV: The "New View"). Relatively poorly discussed in the literature is the simple aspect that many authors seem to have little difficulty in imagining that the small entities (say of less than 50 residues) can rapidly adopt low energy conformers; however, they balk at the idea that larger structures may do so without folding up in the presence of the ribosome or chaperonins. This may be coupled with the fact that it is naturally easier to model successfully such smaller peptides, and that despite the still considerable complexity of the potential energy surface, one can in the mind's eye accept how the molecule could "feel its way" fairly readily into a deep stable state.

Nonetheless, while there is still some debate, there seems little doubt that even proteins with well over 200 residues (though with the probable exception of very large, multi-domain proteins) can be folded without the presence of ribosomes or chaperonins or any other cellular machinery. This is with the caveat that great care is taken to avoid aggregation (12). This signifies that they contain their own information for nucleating and directing their own folding, at least in many cases. Following on the synthetic chemistry developed by Steve Kent at The Scripps Research Institute, this has been further emphasised by synthesising and folding a large enzyme (superoxide dismutase, 153 residues) in mirror image which are made purely of D-amino acids, yet which emerges as fully functional in the case when the ligand (e.g. substrate) is achiral in the time frame of the study (13–15). Related studies show that such mirror image proteins will also act on chiral substrates provided they have similarly reflected chirality and indeed experimental procedures exist by which one can restore interaction of a D-protein with a selected biological target (13, 15). In all such studies it is unlikely that any significant contaminating trace of biological machinery for enhancing folding can be present, or that the unfolded protein "cheats" by retaining some con-

formational features. Moreover, this proven ability to make functional D-proteins with mirror folds, plus the feasibility of an experimental route to generate folds to bind to biological targets (13) underlines the idea that, *novel, stable and functional folds can be produced if the appropriate required amino acid sequence could but be elucidated.*

The problem of protein structure prediction from amino acid sequence (and hence ultimately from genomic data) was once the *bête noire* of academic researchers; it might be said that it now dogs the full and effective application of bioinformatics to drug discovery. *However,* it has also recently often been said that this problem is hugely diminished in practice by the vigorous determination of protein structures, which, even if not relating directly to the protein of interest, provide important related information with which to commence the modelling (see Sec. A, *Protein Homology Modeling* below). This is plausible because of the conservative character of biological evolution, which reuses many protein folds, both for proteins in entirety and as components of other folded proteins. Thus one need not derive the three dimensional structure *ab initio,* i.e. from sequence alone, but by reference to the other structures.

The number of protein structures determined, despite outstanding progress, is still two orders of magnitude smaller than the number of protein sequences determined. This is unlikely to contain all folds. Indeed, it is unlikely that some small, specified number of protein fold exists. Rather, there will probably be an exponential-like or Zippf's Law distribution of types, with many common ones but, showing up quite commonly over the next ten years at least, a seemingly endless variety of rare types. In addition, use of anything but a very close relative to make an accurate determination of the structure of a protein of interest to a resolution and accuracy suitable for studying ligand binding is still a formidable task. As it happens, it is one which benefits greatly from exactly the kind of technology being developed for *ab initio* types of protein folding study, since these would not depend on available data. Finally, comments like the above in regard to the diminishing significance of the *ab initio* folding problem actually refer to the biological realm of proteins: it is precisely the ability to design a variety of unnatural proteins and their analogues, not confined by the conservative character of evolution, which is of interest to some arms of the pharmaceutical and biotechnology industry. From these considerations, it seems reasonably clear that we should purse the technology of both approaches, pushing the ab initio approach at least as far as it will go.

There is a wide range of methods available to analyse and predict protein structure. These methods have been made possible largely because of the increased number of experimentally determined structures [there are over ten thousand entries in the 3DB or structural PDB (16), previously known as the Brookhaven Protein DataBank], combined with increased computing power. The methods for predicting tertiary structures can be classified into the following three main categories:

> Homology modeling.
> Sequence threading (inverse folding).
> *De novo* folding (e.g., lattice folding).

## A.   Protein Homology Modeling

The goal of homology modeling (also known as comparative protein-structure modeling) is to construct detailed 3D structures of proteins on the basis of sequence homology to proteins of known structure. After alignment of the amino acid sequence of a test protein with the sequences of a family of homologues, the closest family member with a high resolution structure is used as a template for placing the most highly conserved amino acid segments. This typically involves secondary structure elements such as α-helices and β-strands. On to these fixed segments are added the intervening loops (which are structurally variable regions), whose conformations can be modeled from other proteins (loop libraries) or by generating conformations *de novo*. Typically, the core residues in the test protein are repacked around substitutions (i.e., differences from the template sequence) by searching for more favourable side-chain rotamers. Finally the full atomic model structure is refined by molecular mechanics methods. In most CAMD software packages, these steps are generally carried out by sophisticated and in some cases automated protocols. Indeed, they represented the first area in which automated pipelines, urgently required for bioinformatics analysis, were developed.

Homology modelling algorithms have been developed to address mainly the following aspects:

> How to construct highly accurate structures for test proteins which have strong sequence homology with template proteins (e.g., $> 60\%$).
> How to construct reasonable models for test proteins which have weak sequence homology with template proteins (e.g., $< 40\%$).

It was mentioned above that modeling a protein from a related one is still far from trivial unless homology is quite high. Correspondingly, it has been a generally held view that homology modeling with high sequence homology is an essentially solved problem. However even this may be optimistic: although homology modeling has improved in terms of both accuracy and speed over the years (17), there are major research and development efforts needed, even in close-homology modelling, in order to make the method reliable for drug design. Apparent disagreement between authors largely relates to the level or accuracy at the required level of resolution which is being discussed. Unfortunately, for purposes of drug design, and especially for smaller less information-rich ligands, quite fine levels of resolution, less than 1 Å, may ideally be required. Conversely, courser levels of resolution are often adequate for function-related studies, studying interactions with larger ligands with a higher information content, and most notably for immunological applications such as development of a diagnostic or vaccine.

In the area of modeling using weaker homology or as a first step in more detailed modeling, there has been interesting progress. Highlighted below are three important examples of progress made in recent years which may have broad applications:

## 1. Hidden Markov Models (HMM)

HMMs are a well studied approach to the modeling of sequence data. They can be viewed as a stochastic generalization of finite-state automata, where both the transitions between states and the generation of output symbols are governed by probability distribution. Briefly, an HMM consists of states and transitions like a Markov chain. It generates strings by performing random walks between an initial and a final state, outputting symbols at every state in between. HMMs have been important in speech recognition, cryptography, and recently in protein classification and sequence alignment (18). HMMs are emerging as a powerful, adaptive and modular tool for computational biology. They have been shown to produce multiple sequence alignments which agree closely with those produced by using structural alignment methods [(19), discussed in relation to threading methods below].

## 2. Distance Geometry Methods

Distance geometry methods have long been used in NMR-based structure determination, molecular conformational searching and structure comparison. Instead of directly using atomic coordinates from template protein to build the 3D structure of a test protein, distance geometry uses the interatomic distance matrix to generate the structure. This approach could benefit particularly well from certain optimisation algorithms ("GLOBEX") and related conformational search features employed in our laboratory in the 1980s and was an intrinsic feature of the LUCIFER suite ["Logical Use of Conformational information and Fast Energy Routines" (20)]. In this period we carried out a large number of homology modeling and related studies using these same principles [see for example (21–23)]. Subsequently these early efforts were found to be tolerably promising on comparison with x-ray results (24, 25). These and particularly the good results for many smaller peptides inspired more widespread use in both academic and industrial laboratories at least of the rigid-geometry interatomic parameters used, especially when solvent is refined (26), though the use of the distance methods employed have not been widely exploited perhaps because they functioned best with the unusual optimisation methods employed rather than other currently available methods. Recently however this kind of distance-based approach has been substantially refined for more general application (27, 28). As with our earlier methods, the distance matrix is constructed by computing interatomic distances for topologically equivalent atoms which are identified from a given sequence alignment between a test protein and a template protein. For atoms which are not covered in the alignment, standard valence and experimental information, such as disulphide bridges, are used to construct relevant elements in the distance matrix. Particularly encouraging results have been published using this method to model divergent proteins (29).

### 3.  Sequence Threading

In recent years, the question of protein structure prediction has been approached from a different angle compared with traditional methods. Instead of asking how a given sequence folds into a specific 3D structure, methods have been developed to assess how compatible a given protein sequence is with known protein structures. These methods are referred to as sequence threading, inverse protein folding or fold recognition methods (30–33). They are based on the idea of threading a sequence through a known 3D structure and calculating a compatibility score between a sequence and a structure. The driving forces behind this development have been the increased number of protein structures resolved, the observation that some proteins with weak sequence homology have structurally very similar folds, and the need to understand the functions of a large number of protein sequences produced by modern molecular biology. Two assumptions behind these methods are the reasonable one that evolution may have conserved protein folds to a greater extent than the sequences, and the more presumptive one that natural proteins may have a relatively small number of discrete folds.

Most threading algorithms employ a score function which contains a pairwise interaction energy term operating between residues. This term models the network of pairwise interactions which hold a protein in a particular fold. Various potential functions have been proposed and which is the best for fold recognition is still an open question. The derivation of the potential usually follows what is called a knowledge-based method, i.e., the inter-residue interaction parameters are derived from protein structure databases by the Boltzman principle (34):

$$E = -kT \ln [f]$$

where $r$ is the distance (or some other parameter describing coordinates) between two atoms, $E$ is the energy at $r$, $f$ is the probability density at $r$, $k$ is the Boltzmann constant, and $T$ is the absolute temperature.

Although these methods are relatively new, encouraging results have been obtained. For example, the structural similarity of actin and heat-shock protein 70 can be recognised, even though their sequence homology is low. Accurate threading results have also been reported in cases of low sequence similarity such as globins and phycocyanin or immunoglobulin domains. Sequence threading approach also performed well for fold recognition in a blind structure prediction competition (35).

More recently attempts have been made to strip these algorithms down to basic essentials to see what are the important factors. The FORREST method (17) is wholly based on secondary structure and makes use of Hidden Markov models to assign the structure to aligned sequences. There is also an increased tendency to develop and adapt these methods to make less and less dependence on using conformational data, but by virtue of this they become, in effect, methods for the detection of weak homology.

### B.  *de novo* Folding

In parallel to homology modelling and sequence threading, another line of interesting and exciting research has been in the area of *de novo* folding. Methods have been developed to construct three dimensional protein structures using the *ab initio* philosophy made more approachable by powerful optimisation techniques, such as Monte Carlo simulated annealing and genetic algorithms, and rendered more practical by a slight compromise to pure *ab initio* philosophy by using a reduced representation of protein structures and knowledge-based potentials.

It seems apparent that, if the folding problem is to be addressed in the context of a dynamical simulation, the enormity of conformational space to be explored must be reduced considerably. This has generally been approached by reduced representation of protein structures (36, 37). One particular approach has been to represent protein structure on a discrete lattice (36). By way of simplifying protein structure at the amino acid residue level, much longer effective timescales become accessible, compared with more detailed full atom schemes, permitting the observation of folding during the course of a simulation.

Successes of the *de novo* folding algorithms include reasonably accurate predictions of the tertiary structure of a Greek key $\beta$-barrel protein, apoplastocyanin and triose phosphate isomerase-type $\alpha/\beta$ barrel proteins. More recent applications have given rise to correct fold predictions for the B domain of staphylococcal protein A (2.25 Å $C\alpha$ r.m.s.) and crambin (3.18 Å r.m.s.) (38).

These folding studies feature a hierarchy of lattice Monte Carlo models with differing granularity, i.e., coarse lattices are employed initially for fast assembly of protein topology, followed by simulation on a finer lattice for improved geometric accuracy. This approach has some advantages over other folding algorithms. It uses a flexible and accurate lattice representation. The lattice Monte Carlo scheme is two orders of magnitude faster than off-lattice Monte Carlo or Brownian dynamics algorithms. The knowledge-based potential is augmented by the inclusion of cooperative protein-like interactions which seemingly improve the quality of results.

It would seem most appropriate to apply a *de novo* folding method when use of sequence alignment and threading techniques fails. In the first instance, standard sequence alignment techniques may detect homology of a given protein sequence to other members of the same family whose structures may already be solved. Then, generation of a suitable model is best car-

ried out through a homology modelling procedure. Where there is little sequence homology to any of the known structures available, one may approach the problem by threading the sequence onto protein structures in a fold library. In this way, very distant relationships or purely structural homology may be identified. At present, *de novo* folding is perhaps best considered the last resort to structure determination by computational methods.

## IV. NEW PARADIGMS AND PERSPECTIVES ON PROTEIN FOLDING

Although our discussion regarding the experimental aspects of protein has above been limited, they did indicate that the experimental view is important in setting the paradigms and philosophy of modeling methods.

An important area of protein folding research in the past few years is the emergence of a so-called "New View" of protein folding kinetics (39, 40).

To appreciate this, we need to return to the classical view. For a number of years, the basic view of protein folding contained the famous "Levinthal paradox". According to the experimental observations of Anfinsen and co-workers (41, 42), proteins can fold reversibly to stable states, which was taken as evidence that folding is "path independent". However, Levinthal argued that proteins would fold too slowly if the vast number of conformational states has to be searched without any direction, and concluded that protein folding must follow a "pathway", which is "a well-defined sequence of events which follow one another so as to carry the protein from the unfolded random coil to a uniquely folded metastable state" (43).

In contrast, the "New View" of protein folding provides a more consistent view of protein folding by replacing the concept of "folding pathway" with "energy landscapes". (44–49). According to the energy landscape perspective, protein folding consists of two aspects:

> Reaching a global minimum in free energy (satisfying Anfinsen's and Scheraga's explanation— thermodynamic requirement).
>
> Folding by multiple routes on funnel-like energy landscapes (satisfying Levinthal's concerns —kinetic argument).

Instead of viewing folding as a process in which all protein chains perform essentially the same sequence of events to reach the native state, the new view regards folding as representing the ensemble average of a process in which each protein molecule may follow its own individual folding trajectory on an energy landscape, and it is this multitude of folding routes which give rise to the kinetic realisation of observed protein folding.

Of course, there is "nothing new under the sun" and with the exception of one caveat below, this "New View" falls happily within the frameworks of statistical mechanics to which the Scheraga school of protein folding in particular closely adhered. In fact, more recent simulations demonstrated the preference of certain pathways on the folding energy landscape (50). The emphasis on model is however somewhat different. Pathway routes are much broader, and entropy considerations are of enormous importance. The approach may be described as "Levinthal with more entropy": a multiple path between two states implies a lower entropic free energy of activation (higher entropy of activation). Increasing the width n-fold, or opening n equivalent trajectories, decreases the entropic free energy by $Tkln(N)$. In addition, attention shifts to a manifold-like surface rather than a narrow, trajectory constrained path, and one has in mind's eye evolutions of the folding pathway resembling the quantum mechanical superposition of

states, or in the Newtonian limit, a broadly spread bundle (Liouville volume) of states evolving classically in phase space. The latter view also reminds us that this shift is in essence a shift to the population, rather than considering a specific molecule. However, the former is not necessarily merely allegorical: indeed, from some viewpoints, the quantum mechanical superposition of states is the correct and relevant model and there is no sharp distinction for the one canonical molecule time averaged or population averaged ensemble. There may be a material distinction based on the energies of the processes, however: adapting the view of Penrose, the wave function may remain pure and uncollapsed for a duration time $\Delta t = \nabla E / \hbar$ in crossing a barrier of energy E. We would trust however that the macroscopic and microscopic views will "come out in the wash" as the same, and that these new views would not seem alien to the Scheraga school who took great pains to emphasise in fact that the starting state, the random coil, was already a broad ensemble.

The caveat to the above is that, oddly enough, a rigorous simple treatment in both the classical and quantum mechanical approaches is hampered. It is hampered by the fact that at the crest of a transition, the dimensionality of the highest free energy state is one less than in the rest of the configuration space, and in first efforts to put pen to paper and treat the problem mathematically, this may allow some lee-way in abandoning classical approaches and thinking of the new view as distinct. However, there is no reason to believe that this issue is fundamental, merely that the correct treatment is laborious.

The "New View" also has its ancestry in a protein state which is reminiscent of the random coil ensemble but clearly a transition intermediate. This is the molten globule as normally attributed to Ptitsyn, but clearly described earlier by Robson and Pain (51) as the "Monomolecular

Liquid Micelle" (MLM model). Indeed Robson and Pain (12) sought this state experimentally in the folding of penicillinase and, failing to find it except for a weak association of helices at their hydrophophobic faces, developed a thermodynamic description consistent with the thermodynamic and kinetic data. This showed that the two states, the very broad micelle state and the tighter helix assembly pathway, could be envisaged as always co-existing but with different relative free energies. Later, this MLM view was reflected in the computer-simulated folding models by Robson and colleagues, the tendency to form such globules being introduced as a bias (22), and giving the behaviour seen in simulation studies by the advocates of the "New View".

It is nonetheless held that the "New View" as a broadly accepted view will affect the protein structure prediction research (52). From the above, it would be correctly concluded that we see the view as a convenient working model for thinking about the problem, rather than what is implemented in the calculations. Nonetheless, this mind-set has inspired some interesting approximate models like those of Dill and colleagues. Although the ultimate goal to build a reliable and accurate 3D structure for a given protein sequence is still some distance away, these recently developed methods and new insights into protein folding mechanisms have shown some encouraging results and with the increased number of experimentally determined protein structures and the increased computing power, they will no doubt continue to improve.

## V. *DE NOVO* DRUG DESIGN METHODS

In recent years there has been a sharp rise in the number of protein structures solved at high-resolution using x-ray crystallography and Nuclear Magnetic Resonance (NMR) spectroscopy. This increase has enabled structure-based techniques to be applied on a wider scale in drug discovery research (53). Among these structure-based drug design (SBDD) methods, a novel class of computational programs has emerged to carry out what has been referred to as *de novo* drug design, i.e., the design of novel molecules to satisfy a set of steric and/or chemical constraints. Much research has been carried out in the field of *de novo* design and many programs have been developed by various CAMD research groups in academia and industry (54–60). In addition, the Ludi program has been commercialised by Biosym and MCSS/HOOK is similarly marketed by MSI. Two of the other software houses, Tripos and Chemical Design Ltd., have developed their own *de novo* design methodologies, LeapFrog and Chem-Novel, respectively.

Readers interested in a detailed review of available techniques are referred to the publications quoted above and some recent reviews (61, 62). However, a brief overview of the general ideas underlying the majority of *de novo* design programs is presented here. In general, such programs operate in three stages.

### A. Constraint Definition

Hereby the steric and/or chemical constraints of the design problem are delineated and supplied to the program in an appropriate form. In many cases, these constraints will be derived from the active site of the macromolecule of interest but constraints derived from pharmacophore models, CoMFA models, or just a single molecule are equally valid.

### B. Structure Generation

Here, molecular structures are assembled which attempt to meet as many of the imposed constraints as possible. Two main classes of structure generation algorithm have been described: atom-by-atom construction and fragment/template joining. In the latter class, distinction may be made between 'build-up' strategies in which fragments are joined sequentially to each other and 'outside-in' strategies in which fragments are placed at 'hotspots' and then bridges sought between them in a subsequent phase. One of the apparent features of these *de novo* programs is the speed at which new chemical structures are generated to fit the specified constraints. It is not uncommon that a *de novo* design program can generate many hundreds structures overnight on an entry level workstation. The structure generation can usually be controlled to allow the generation of specific types of molecular structures, e.g., peptides, peptoids or small organic molecules.

## C.  Structure Evaluation

Here the generated structures are assessed and prioritised for further study either by computation or chemical synthesis.

Many of the published papers in the field of *de novo* molecular design have sought to demonstrate the validity of these approaches with various test cases (63–67). In general, such examples have demonstrated the ability of *de novo* design programs to reproduce known ligands or to suggest reasonable alternatives to them and such results are usually accepted as sufficient indication of the programs' capability. It is believed that real-life applications of *de novo* design programs will emerge in the near future, and their effectiveness in drug discovery research will be assessed.

## VI.  PREDICTING MOLECULAR ACTIVITIES

Once a structural model is built, novel molecular structures can be designed with programs such as the above mentioned *de novo* drug design programs or indeed many other methods (e.g., database searching) at a rapid speed. However, how to select molecular structures for synthesis from a potentially large number of designs has been a major subject of research in CAMD. Ideally, the molecules chosen for synthesis should be relatively easy to synthesise (especially at the lead generation stage), have strong binding affinity towards the intended target, be specific to the target, have no toxicity, be metabolically stable and have good oral bioavailability. Computational methods have been developed to address some of these issues, particularly binding affinity prediction. The general features of these methods and some of their applications are briefly described below. The summary should not be considered exhaustive because of the enormity of this field.

### A.  Statistical Methods

Essentially, these methods utilise statistical methods to construct a model which correlates a set of chemical descriptors and the observed activities or properties of the individual molecules. The best example of this class of methods is the well known CoMFA approach from Tripos, although this is just one of a number of methods collectively referred to as 3D QSAR (68, 69). Because of their general nature, these methods are not only used to correlate chemical structures with the observed activities, such as $IC_{50}$s, but they are also used to construct statistical models which describe the correlation between the chemical characteristics of a set of molecules with their observed properties, such as toxicity and solubility, provided that the data sets satisfy certain criteria. In practice, it is important to ensure that the model has true predictability and to avoid chance correlation. A number of interesting developments have taken place in the QSAR area over the last few years. One of the methods uses the idea of genetic algorithms to optimise the functional which describes the correlation between chemical descriptors and the observed activities, and good performance has been obtained for difficult QSAR data sets (70). Another development has been in the area of pseudo-receptor modelling using essentially the same approach. According to this approach, a pseudo-receptor model consisting of a finite set of atoms is generated to interact with individual ligands, and the receptor models are evolved by a genetic algorithm such that the final model produced is able to correlate the observed activities with the interaction energies between the receptor model and the ligands (71). Another recent development in this field is what is called the Predicted Activity Contribution (PAC) method. Instead of describing observed activities of the ligands in terms of chemical groups or regions qualitatively, PAC describes the activity of a ligand in terms of quantitative contributions from individual atoms in the ligand. It is believed that PAC gives more insight into the chemical basis of a ligand's activity and provides a more explicit link into molecular design programs (65).

## B. Energy Function and Molecular Simulation Approach

A fundamental requirement for progress in many drug design areas is an improved understanding into the underlying physical forces which determine molecular structure and function. It is generally believed that quantum mechanics provides a valid description of molecular systems. However, as biological macromolecules are too large to allow rigorous quantum mechanical studies, most computational methods use classical approximations such as potential functions or forcefields to describe the energetics. In classical approaches, a mathematical function is usually used to describe the relationship between the energetics of a molecular system and its conformation in terms of

atomic radii, charge distribution, and valence descriptors. Examples of these include AMBER, MM3, CHARMM, CFF, MMF94 (72). These potential functions have been tested extensively through their ability to reproduce experimental data, but despite many successes they do have deficiencies. Recent development in this area has been to parameterise force fields based on the use of higher level quantum mechanical calculations, so that the potential functions become more accurate and can be applied to a wider range of molecular systems. Once a good force field is available which can describe molecular conformation and molecular interactions well, molecular docking and free energy simulation (73–76) can be applied to study the binding behaviours of different ligands towards a target protein. It has been demonstrated that free energy simulation methods can produce results which are within 1 kcal/mol of experimentally determined values. Since biological molecules are surrounded by solvents, it is important to describe such effects correctly in order to predict the activities of novel molecular designs. Major progress has been made in this area in the form of the Poisson-Boltzmann equation (77), which describes electrostatic interactions in a continuum. The method has been applied to problems such as molecular conformation, salt effects on ligands and protein-DNA interactions.

Recently, the use of empirical binding scoring functions to estimate the binding affinity of ligands in receptor complexes has become very popular. Empirical scoring functions for predicting ligand binding affinity have emerged as a useful way of filtering out a large number of molecular designs generated by structure-based molecular design methodology (78–80). These empirical scoring functions usually consist of several terms which are believed to be important for drug binding to receptors. Each term can be calculated rapidly for a given protein ligand complex and its conformation, and the scoring is usually calibrated against a set of protein ligand complexes with high resolution to reproduce their experimentally determined binding affinities. The accuracy of such functions is moderate, e.g., within two orders of magnitude. A significant advantage of such scoring functions over the force field approach is its speed of calculation, and its robustness across a range of ligand chemical structures.

Another important part of structure-based molecular design is the use of molecular docking to assess the binding mode of proposed ligands to a protein structure. The background of molecular docking has been thoroughly reviewed (81–86). An important feature of any docking method is an energy function which is capable of predicting binding modes. The minimum values of the function should correspond to the preferred binding mode(s) of the ligand. Another important issue in molecular docking study is how to treat molecular flexibility. A combination of better conformational sampling tools, such as Simulated Annealing (87), Evolutionary Programming (88), and Genetic Algorithms (89, 90) with better calibrated energy functions such as the empirical binding scoring function mentioned above (91, 92) has led to improved molecular docking approaches. The most impressive results published include a docking study of 100 diverse protein-ligand complexes. The docking program achieved 71% success rate. It is clear that improved speed and accuracy will make the molecular docking approach more widely used in practical structure-based drug and combinatorial library design applications (85).

## C.  Knowledge-based Methods

The methods described so far share a common basis, i.e., attempting to compute or predict molecular properties by physical principles which describe specific molecu-lar structure and mechanisms. But some areas of drug discovery research, such as drug metabolism, toxicity, solubility and bioavailability (i.e., A.D.M.E.), are still largely beyond the capability of precise computational approach. These properties of molecules are essential for them to progress into the clinic. Empirical data sets have been collected by pharmaceutical industry and academia for those important drug properties, particularly for toxicity and metabolism. Computational (or computer-based) methods to treat this type of application are known as knowledge-based methods. These methods have been developed to treat situations which are essentially a black box, i.e., certain inputs generate certain outputs via a complex process whose precise mechanism and stages are poorly understood. These methods fall into the following two broad categories.

### a.   Rule-based Systems

These systems are generally referred to as Expert Systems or Knowledge-based Systems. They attempt to solve problems in a way similar to how a human expert would solve them by incorporating *rules of thumb* that experts in the field have developed through many years of data collection and observation. The problems attacked are not necessarily procedural, they are often vague, complex and can contain incomplete or inexact information.

Expert systems contain three basic components: **a knowledge base, a problem solving and interference engine**, and **a user interface**. The knowledge base contains the information which the program uses to reach decisions. A key difference between expert systems and classical computational programs is the fact that the knowledge is separated from the program. The inference engine is the program that processes the knowledge base to reach these decisions by logical programming or intelligent database searching. The user interface such as a natural language processor allows the program and user to communicate with each other in an effective way.

There are many application examples of expert systems in chemistry (93). The expert systems which have shown practical importance in drug discovery research include programs such as DEREK for toxicity assessment of molecular structures (94) and META for evaluating metabolic transformation of chemical compounds (95, 96).

*b. Rule Learning Systems*

For problems where rules for reaching a solution are not known even on the basis of *rules of thumb* and hence cannot be approached by expert system methods, other methods have been developed in recent years. These methods are usually called adaptive learning systems, and include methods such as rule induction programs and artificial neural networks (ANN). Rule induction programs (97) have been developed in the field of AI to extract rules from data sets for use in expert systems (98). They have recently been applied to QSAR problems (99). ANNs have gained popularity in recent years for their wide range of applicability and the ease of dealing with complex datasets. ANNs have arisen out of simulating biological neural systems, and have demonstrated the ability to learn complex associations between data sets and to generalise, i.e. to predict. There are now many forms of ANNs, but essentially they have three basic components: processing elements, connect topology between the processing elements and the learning method. The applications of ANNs in chemistry have been reviewed recently (100).

## VII. CAMD AND COMBINATORIAL CHEMISTRY

As mentioned in Sec. II., the most visible development in drug discovery in the past few years is undoubtedly combinatorial chemistry (101–107) and high throughput screening (108, 109). Combinatorial chemistry offers a way to synthesize very large numbers of chemical compounds in short periods of time by using straightforward synthetic routes between a diverse set of chemical compounds. The pioneering work has been in peptide synthesis. For example, a combinatorial library consisting of hexapeptides made from the 20 naturally occurring amino acid residues may contain up to 64,000,000 peptides. This type of library can be made in a relatively short period of time by today's synthesis technique. Coupled with high throughput screening technologies, combinatorial chemistry has the potential to accelerate drug discovery considerably. Because of the A.D.M.E deficiencies associated with the peptide leads, the more important development in combinatorial chemistry will be to synthesize drug-like small organic molecules combinatorially. For CAMD developers, it is important to recognize the performance and potential of combinatorial chemistry methods. There are two aspects which CAMD technologies can make considerable contribution to the development and application of combinatorial chemistry technologies.

### A. Library Design

Although some people view that the main attraction of combinatorial chemistry is its ability to make vast number of molecules quickly, it is important to recognize that this is not the purpose of drug discovery. The purpose of drug discovery is to make the right molecule(s) for the right biological target by the most efficient and effective methods. A term which has been developed in this context is called molecular diversity. It is a measure how much chemical space (more specifically drug-like properties) a combinatorial library covers. There is no generally accepted way of either the definition of molecular diversity or how it should be calculated (122). However, according to one diversity measure, a set of 200 marketed drugs has far greater molecular diversity than some combinatorial libraries consisting of vast numbers of compounds. Recently, there have also been a number of studies attempting to characterize what features drug molecules possess in order to design better chemical libraries (123–125). In general terms, this is where CAMD technologies should be developed and applied to the quantification of the combinatorial libraries and the design of novel libraries which would satisfy a particular set of pharmacophoric constraints or structural constraints with the most diversity (110, 111). Considerable opportunities exist for CAMD and combinatorial chemistry methods to be used synergistically (112, 113).

A very interesting development in this area is what is called virtual screening of combinatorial libraries (114–116, 126). A combinatorial library of small organic molecules typically consists of a common central template or scaffold onto which different sets of substituents are attached at several attachment sites. The substituents can be derived from available chemical reagents and custom synthesized reagents. The scaffold is either an available reagent, is easily synthesised, or is formed in the synthesis of library members. In many cases, the synthesis of full libraries may be problematic because of the enormous numbers of library members, and the difficulties associated with establishing robust synthesis routes suitable to a wide range of substituents. Powerful new approaches have emerged which utilise the 3D structure of target enzymes (or enzymes structurally related to the target) in order to customise the design of combinatorial libraries. Virtual combinatorial libraries are constructed in silico, i.e. in the computer, and are screened computationally for 3D complementarity to the target enzyme. The advantage is that diversity can be focused into relevant areas of molecular property space. This allows much smaller libraries to be constructed with the potential to deliver improved hit rates, and hence accelerate the process of drug discovery. The advantage over existing structure-based drug design approaches is that designed molecules are chosen to be members of a virtual combinatorial library, and are therefore amenable to synthesis. As it is possible to synthesize quickly a considerable number of designs to explore different aspects of molecular recognition, the inherent inaccuracies of many computational methods become less of a liability. Also, the process can efficiently yield QSAR sets to stimulate further designs.

## B. Data Management and Processing

Of current concern is the tidal wave of data generated by the new data capture technologies. A practical aspect of combinatorial library synthesis and screening is the fact that vast amount of data can be generated quickly. How to manage them and especially how to derive important information to assist the identification of active leads and the design of better libraries has been the focus of several CAMD software vendors. Although no single solution has been universally accepted, considerable practical experience has been accumulated which is pointing the way forward for the design and development of better automated tools for such applications (117, 118). Generally, it is important to consider the following aspects:

Data storage and management: the system needs to be robust, and can accommodate large volumes of, and ever-increasing amounts of data. Valuable experience can be borrowed from the data warehousing techniques and systems used in the financial, tele-communication and transport industry.

Data processing and information analysis: this is probably the most important aspect of such a data management system in drug discovery because it helps to maximize the value available in the data. The most obvious technique to use is to make SAR methods robust and high throughput so that researchers can rapidly obtain useful insight into the data generated, and have new directions for designing novel and better compound libraries for synthesis and screening.

User-interface or information delivery: this is an important element of an information management system in drug discovery since a drug discovery environment usually consists of a diverse range of scientific experts. How to make the access to large volume of data easy and versatile for all or most of the people involved in discovery research is a major challenge facing computational and information technology personnel. It is clear that internet/intranet and World Wide Web technologies have opened the way forward for this aspect.

## VIII. INTEGRATION: THE ELECTRONIC PHARMACEUTICAL COMPANY

The views of the authors about the topic of integrated environments is well reported in the literature. In the early days of the Proteus group we set out with Control Data Corporation to visit many major pharmaceutical companies and to discuss with them their vision for an integrated approach to drug design. To many, this would be part of a larger "Electronic Pharmaceutical Company" analogous to the current "electronic office". This led to the development of the GLOBAL high level language and the Prometheus system based primarily upon it. We have described the experimental and developmental forms of this system in a variety of reviews starting with the perception of the system as a "polymorphic programming environment" (119) and as a "Big Hammer Approach" integrating many approaches in an automated, Expert System style environment (120). This system and its even earlier forms have played substantial roles in the development of real clinical, veterinary and diagnostic products.

A number of the issues discussed in these papers, but taken outside the specific Prometheus context to discussion of more widely available systems, have been described elsewhere (15). Emphasis reflected the author's opinion that the current nature of bioinformatics is such, and that there is such widespread software development in the field, that in practice computer languages reflect the primary interface by which knowledge is embedded in bioinformatics systems. Notably, this generalisation to other systems outside Prometheus obliged one author (BR) to attempt to assess a rich variety of programming languages and bioinformatics software command languages. To get a fair "feel", the author made an attempt to write at least one useful little routine in each of the languages studied, albeit with not always the same functionality. As it happened, variety of form and role of languages are such that a really meaningful comparative analysis is neither possible nor useful, nor fair. Further, as the article took some time to write, new approaches were appearing throughout the preparation of the manuscript, making the task a moving target. This promoted a plea for the unification of the "Babel" of , startlingly, some 14 language systems (UNIX, PERL, C, C++, SQL, HTML, JAVA, JAVASCRIPT...) of which the bioinformaticist must have at least nodding acquaintance, combined with a discussion of what features such a *lingua franca* might have. Note that the above is the "reasonably widely used list", not including valuable specialised environments for molecular work such as ISIS-PL, forms of Python, MOE (Molecular Operating Enviroment), BioWidget interfaces, and so on. At the same time, there was an interesting example of the disadvantages of non- uniformity when the final printing of the paper was meaninglessly scrambled in part of the text, apparently by a type setter faced with a new type setting environment and control system (as a consequence of the acquisition of the publishing company during the course of the work)! The argument that there should be a single polymorphic programming environment is certainly consistent with the same view of the authors which gave rise to GLOBAL and Prometheus (119), and in fact the above attempt to make a more democratic assessment of the field has only strengthened the earlier conviction of one of the authors.

Though the two present authors have from time to time differed in their views of the relative importance of good interfaces versus automated approaches, there is no doubt that both views have emerged in bioinformatics as important. On the one hand there is a widespread interest in user-friendly, interactive, web based tools and in Java applets, while at the same time there is enhanced recognition of the pressing need for scripting tools and automated pipelines for bioinformatics analysis of incoming genomic data, and indeed chemoinformatics tools for rapidly deriving SAR models from large volumes of data generated by high throughput screening systems.

## IX. SUMMARY

In the past few years, several important developments have taken place in the basic CAMD technologies, notably protein sequence threading, *de novo* drug design technology, advanced molecular simulation methods such as free energy perturbation and Poisson-Boltzman equation, improved protein-ligand binding affinity prediction and molecular docking approaches. These developments, to a large extent, have been made possible by the rapidly increased amount of bioinformatics data, particularly 3D structural information on proteins and their complexes with ligands. The increased information resources can be used to gain more insights into the physical forces governing molecular structures and molecular recognition, and to improve the performance of computational algorithms by better calibrating against more experimental data (121). Those methodological developments will no doubt be applied to practical drug discovery problems and be continuously enhanced. New challenges have also arisen for CAMD technology research and applications because of the development in combinatorial chemistry and high throughput screening. These areas have enjoyed alternating swings of popularity, but they are not distinct. The integrated application of CAMD and combinatorial chemistry will potentially offer a very effective drug discovery technology.

## REFERENCES

1. RS Bochek, C McMartin, WC Guida. The art and practice of structure-based drug design. Med Res Rev 16: 3–50, 1996.
2. ID Kuntz, CE Meng, BK Shoichet. Structure-based molecular design. Acc Chem Res 27: 117–123, 1994.
3. CLMJ Verline, WGJ Hol. Structure-based drug design: progress, results and challenges. Structure 2: 577–587, 1994.
4. KB Lipkowitz, DB Boyd. Reviews in Computational Chemistry, Vol 1–11. Wiley-VCH, New York, pp 1990–1997.
5. CH Reynolds, MK Holloway, HK Cox. Computer-Aided Molecular Design. ACS Symposium Series 589, 1995.

6. H van de Waterbeemd. Advanced Computer-Assisted Techniques in Drug Discovery. VCH, Weinhaim, 1994.

7. McKinsey and Co., *In Vivo*, May 1996.

8. ED Green, RH Waterston. The human genome project. Prospects and implications for clinical medicine. JAMA, 14: 1966–1975, 1991.

9. RJ Robbins. Bioinformatics: essential infrastructure for global biology. J Comp Biol 3: 465–478, 1996.

10. CB Affinsen, E Haber, M Sela, Fh White Jr. The kinetics of formation of native ribonuclease during oxidation of the reduced polypeptide chain. Proc Natl Acad Sci USA, 47: 1309–1314, 1961.

11. M Perutz. Protein structures: new approaches to disease and therapy. WH Freeman and Co., New York, 1992.

12. B Robson, RH Pain. The mechanism of folding of globular proteins. Equilibria and kinetics of conformational transitions of penicillinase from *Staphylococcus aureus* involving a state of intermediate conformation. Biochem J 155: 325–330 (1976).

13. B Robson. Doppelganger proteins as drug leads. Nature Biotechnology 14: 892–893, 1996.

14. GM Figliozzi, MA Siani, LE Canne, B Robson, RJ Simon. Chemical synthesis and activity of D-superoxide dismutase. Protein Science 5, suppl. 1, 72, 1996.

15. B Robson. Computer languages in pharmaceutical designs. In: WF van Gunsteren, PK Weiner, AJ Wilkinson, ed. Computer Simulation of Biomolecular Systems. Theoretical and Experimental Applications., Vol III. Kluwer-ESCOM, 1997, pp. 494–562,

16. EE Abola, NO Manning, J Prilusky, DR Stampf, JL Sussman. The Protein Data Bank: current status and future challenges. J Res Natl Inst Stand Technol 101: 231–241, 1996.

17. S Mosimann, R Meleshko, MNG James. A critical assessment of comparative molecular modeling of tertiary structures of proteins. Proteins: Struct Funct Genet 23: 301–317, 1996.

18. A Krogh, M Brown, IS Mian, K Sjolander, D Hussler. Hidden Markov Models in computational biology. Applications to protein modeling. J Mol Biol 235: 1501–1531, 1994.

19. V Di Fransco, JA Garnier, PJ Munsen. Protein topology recognition from secondary structure sequences: applications of the Hidden Markov Models to the alpha class proteins. J Mol Biol 267: 446–463, 1997.

20. B Robson, E Platt. Refined models for computer calculations in protein engineering. Calibration and testing of atomic potential functions compatible with more efficient calculations. J Mol Biol 188: 259–281, 1986.

21. RV Fishleigh, DJ Ward, EC Griffiths, B Robson. Conformational study of neurotensin and some of its analogs. Biochem Soc Trans 14: 1259–1260, 1986.

22. B Robson, E Platt, RV Fishleigh, A Marsden, P Millard. Expert system for protein engineering: its application in the study of chloramphenicol acetyltransferase and avian pancreatic polypeptide. J Mol Graphics 5: 8–17, 1987.

23. B Robson, E Platt. Modeling of alpha-lactalbumin from the known structure of hen egg white lysozyme using molecular dynamics. J Comp Aided Molecular Design 1: 17–22, 1987.

24. B Robson, E Platt. Comparison of the x-ray structure of baboon α-lactalbumin and the tertiary predicted computer models of human α-lactalbumin. J Comp Aided Mol Design 4: 369–379, 1990.

25. E Platt, B Robson. Proc. R. Soc. Edinburgh 99B, 1992, 1/2: pp. 123–136.

26. VP Collura, PJ Greaney, B Robson. A method for rapidly assessing and refining simple solvent treatments in molecular modelling. Example studies on the antigen-combining loop H2 from FAB fragment McPC603. Protein Engineering 7: 221–233, 1994.

27. TF Havel, ME Snow. A new method for building protein conformations from sequence alignments with homologues of known structures. J Mol Biol 217: 1–7, 1991.

28. S Srinivasan, CJ March, S Sudarsarnam. An automated method for modeling proteins on known templates using distance geometry. Protein Science. 2: 277–289, 1993.

29. S Sudarsanam, S., CJ March, S Srinivasan. Homology modeling of divergent proteins. J Mol Biol 241: 143–149, 1994.

30. DT Jones, WR Taylor, JM Thorton. A new approach to protein fold recognition. Nature 358: 86–89, 1992.

31. JU Bowie, D Eisenberg. Inverted protein structure prediction. Curr Opin Struct Biol 3: 437–444, 1993.

32. 30. MJ Sipple. Knowledge-based potentials for proteins. Curr Opin Struct Biol 5: 229–235, 1995.

33. SH Bryant, SF Altschul. Statistics of sequence-structure threading. Curr Opin Struct Biol 5: 236–244, 1995.

34. MJ Sipple. Calculation of conformational ensembles from potentials of mean force. An apptoach to the knowledge-based prediction of local structures in globular proteins. J Mol Biol 213: 859–853, 1990

35. DT Jones, RT Miller, JM Thornton. Successful protein fold recognition by optimal sequence threading validated by rigorous blind test. Proteins: Struct Funct Genet 23: 387–397, 1995.

36. J Skolnick, A Kolinski. Dynamic Monte Carlo simulations of a new lattice model of globular protein folding, structure and dynamics. J Mol Biol 221: 499–531, 1991.

37. R Srinivasan, GD Rose. LINUS: a hierachic procedure to predict the fold of a protein, Proteins: Struct Func Genet 22: 81–99, 1995.

38. A Kolinski, J Skolnick. Monte Carlo simulations of protein folding. II. Application to protein A,ROP, and crambin. Proteins: Struc Func Gen 18: 353–366, 1994.

39. RL Baldwin. Matching speed and stability. Nature 369: 183–184, 1994.

40. RL Baldwin. The nature of protein folding pathways: the classical versus the new view. J Biomol NMR 5: 103–109, 1995.

41. CB Affinsen, E Haber, M Sela, F H White Jr. The kinetics of formation of native ribonuclease during oxidation of the reduced polypeptide chain. Proc Natl Acad Sci USA, 47: 1309–1314, 1961.

42. CB Affinsen. Principles that govern the folding of protein chains. Science 181: 223–230, 1973.

43. C Levinthall, Are there pathways for protein folding? J Chim Phys 65: 44–45, 1968.

44. JD Bryngelson, JN Onuchic, ND Socci, PG Wolynes. Funnels, pathways and the energy landscapes of protein folding: A synthesis. Proteins Struct Funct Genet 21: 167–195, 1995.

45. KA Dill, S Bromberg, K Yue, KM Fiebig, DP Yee, PD Thomas, HS Chan. Principles of protein folding—a perspective from simple exact models. Protein Science 4: 561–602, 1995.

46. M Karplus. M Sali. Theoretical studies of protein folding and unfolding. Curr Opin Struct Biol 5: 58–73, 1995.

47. PG Wolynes, JN Onuchic, D Thirumalai. Navigating the folding routes. Science 267: 1619–1620, 1995.

48. D Thirumalai, SA Woodson. Kinetics of folding of proteins and RNA. Acc Chem Res 29: 43–439, 1996.

49. KA Dill, H S Chan. From Levinthal to pathways to funnels. Nat Struct Biol 4: 10–19, 1997.

50. T Lazaridis, M Karplus. "New View" of protein folding reconuciled with the old through multiple unfolding simulations. Science 278: 1928–1931, 1997.

51. B Robson, RH Pain. Analysis of the code relating sequence to conformation in proteins. Possible implications for the mechanism of formation of helical regions. J Mol Biol 58: 237–256, 1971.

52. E I Shaknovich. Theoretical studies of protein-folding thermodynamics and kinetics. Curr Opin Struct Biol 7: 29–40, 1997.

53. PM Colman. Structure-based drug design. Curr Opin Struct Biol 4: 868, 1994.

54. RA Lewis. Rational methods for site-directed drug design: novel approaches for the discovery of potential ligands. Biochem Soc Trans 19: 883, 1991.

55. PL Chau, PM Dean. Automated site-directed drug design: the generation of a basic set of fragments to be used for automated structure assembly. J Comp Aïded Mol Design 6: 385, 1992.

56. JB Moon, WJ Howe. Computer design of bioactive molecules: a method for receptor-based de novo ligand design. Proteins: Struct Funct Genet 11: 314, 1991.

57. H-J Böhm. The computer program LUDI: a new method for the de novo design of enzyme inhibitors. J Comp Aided Mol Design 6: 61, 1992.

58. SH Rotstein, MA Murcko. GroupBuild: a fragment-based de novo drug design. J Med Chem 36: 1700, 1993.

59. V Gillet, AP Johnson, P Mata, S Sike, P. Williams. SPROUT: a program for structure generation. J Comp Aided Mol Design 7: 127, 1993.

60. DE Clark, D Frenkel, SA Levy, J Li, CW Murray, B Robson, B Waszkowycz, DR Westhead. PRO_LIGAND: an approach to de novo molecular design. 1. application to the design of organic molecules. J Comp Aided Mol Design 9: 13, 1995.

61. M A Murcko. Recent Advances in Ligand Design Methods. In: KB Lipkowitz, DB Boyd, ed. Reviews in Computational Chemistry, Vol. 11, Wiley-VCH 1997, pp 1–66.

62. DE Clerk, CW Murray, J Li. Current Issues in De Novo Molecular Design, In: KB Lipkowitz, DB Boyd, ed. Reviews in Computational Chemistry. Vol. 11, Wiley-VCH, 1997, pp 67–126.

63. H-J Böhm. LUDI: rule-based automatic design of new substituents for enzyme inhibitor leads. J Comp Aided Mol Design 6: 593, 1992.

64. DE Clark, CW Murray. PRO_LIGAND: an approach to de novo molecular design. 5. tools for the analysis of generated

structures. J Chem Inf Comput Sci 35: 914, 1995.

65. B Waszkowycz, DE Clark, D Frenkel, J. Li, CW Murray, B Robson, DR Westhead. PRO_LIGAND: an approach to de novo molecular design. 2. design of novel molecules from molecular field analysis. J Med Chem 37: 3994, 1994.

66. DR Westhead, DE Clark, D Frenkel, J. Li, CW Murray, B Robson, B. Waszkowycz. PRO_LIGAND: an approach to de novo molecular design. 3. a genetic algorithm for structure refinement. J Comp Aided Mol Design 9: 139, 1995.

67. D Frenkel, DE Clark, J Li, CW Murray, B Robson, B Waszkowycz, DR Westhead. PRO_LIGAND: an approach to de novo molecular design. 4. application to the design of peptides. 9: 213, 1995.

68. H Kubinyi. 3D-QSAR in Drug Design. In: H Kubinyi, ed. Theory, Methods and Applications, ESCOM, 1993.

69. H Kubinyi. QSAR. In: H Kubinyi, ed. Hansch Analysis and Related Appraoches, VCH, Weinhaim, 1994.

70. D Rogers, AJ Hopfinger. Application of genetic function approximation to quantitative structure-activity relationships quantitative structure-property relationships. J Chem Inf Comput Sci 34: 854, 1994.

71. DE Walter, RM Hinds. Genetically evolved receptor models: a computational approach to construction of receptor models. J Med Chem 37: 2527, 1994.

72. T Halgren. Potential energy functions. Curr Opin Struc Biol 5: 205–210, 1995.

73. P Kollman. Free energy calculations: applications to chemical and biological phenomena. Chem Rev 93: 2395–2417, 1993.

74. WF van Gunsteren, PK Weiner. Computer Simulation of Biomolecular Systems. In: WF van Gunsteren, PK Weiner eds. Theoretical and Experimental Applications, Vol. 1. ESCOM, 1989.

75. WF van Gunsteren, PK Weiner, AJ Wilkinson. Computer Simulation of Biomolecular Systems. In: WF van Gunsteren, PK Weiner, AJ Wilkinson eds. Theoretical and Experimental Applications, Vol. 2. ESCOM, 1993.

76. WF van Gunsteren, PK Weiner, AJ Wilkinson. Computer Simulation of Biomolecular Systems. In: WF van Gunsteren, PK Weiner, AJ Wilkinson eds. Theoretical and Experimental Applications, Vol. 3. ESCOM, 1997.

77. KA Sharp, B Honig. Electrostatic interactions in macromolecules—theory and applications. Ann Rev Biophys Chem 19: 301–332, 1990.

78. H-J Bohm. The development of a simple empirical scoring function to estimate the binding constant for a protein-ligand complex of known three-dimensional structure. J Comp Aided Mol Design 8: 243–256, 1994.

79. RD Head, ML Smythe, TI Oprea, CL Waller, SM Green, GR Marshall. VALIDATE—a new method for the receptor-based prediction of binding affinities of new ligands. J Am Chem Soc 118: 3959–3969, 1996.

80. MD Eldridge, CW Murray, TR Auton, GV Paolini, RP Mee. Empirical scoring function: I. The development of a fast empirical scoring function to estimate the binding affinity of ligands in receptor complexes. J Comp Aided Mol Design 11: 425–445, 1997.

81. JM Blaney, JS Dixon. A good ligand is hard to find: automatic docking methods. Perspect Drug Discov Res 1: 301–319, 1993.

82. G Jones, P Willett. Docking small-molecule ligands into active sites. Curr Opin Biotech 6: 652–656, 1995.

83. TP Lybrand. Ligand-protein docking and rational drug design. Curr Opin Struct Biol 5: 224–228, 1995.

84. R Rosenfield, S Vajda, C DeLisi. Flexible docking and design. Ann Rev Biophys Biomol Struct 24: 677–700, 1995.

85. DA Gschwend, AC Good, ID Kuntz. Molecular docking towards drug discovery. J Mol Recognition 8: 175–186, 1996.

86. T Lengauer, M Rarey. Computational methods for biomolecular docking. Curr Opin Struct Biol 6: 402–406, 1996.

87. GM Morris, DS Goodsell, R Huey, AJ Olson. Distributed automated docking of flexible ligands to proteins: parallel applications of AutoDock2.4. J Comp Aided Mol Design 10: 293–403, 1996.

88. DK Gehlhar, GM Verkhivker, PA Rejto, CJ Sherman, DB Fogel, LJ Fogel, ST Freer. Molecular recognition of the inhibitor AG-1343 by HIV-1 protease: conformationally flexible docking by evolutionary programming. Chem Biol 2: 317–324, 1995.

89. RS Judson, EP Jaeger, AM Treasurywala. A genetic algorithm method for molecular docking flexible molecules. J Mol Struct 308: 191–206, 1994.

90. G Jones, P Willet, RC Glen, AR Leach, R Taylor. Development and validation of a genetic algorithm for flexible docking. J Mol Biol 267: 727–748, 1997.

91. M Rarey, B Kramer, T Lengauer, G Klebe. Predicting receptor-ligand interactions by an incremental construction algorithm. J Mol Biol 261: 470–489, 1996.

92. W Welch, J Ruppert, AN Jain. Hammerhead: fast, fully automated docking of flexible ligands to protein binding sites. Chem Biol 3: 449–462, 1996.

93. TH Pierce, BA Hohne. ed., Artificial Intelligence Applications in Chemistry. ACS, Washington DC, 1986.

94. JE Ridings, MD Barratt, R Carey, CG Earnshaw, CE Egginton, MK Ellis, PN Judson, JJ Langowski, CA Marchant, MP Payne, WP Watson, TD Yih. Computer prediction of possible toxic action from chemical structure: an update on the DEREK system. Toxicology 106: 267–279 1996.

95. G Klopman, M Dimayuga, J Talafous. META. 1. A program for the evaluation of metabolic transformation of chemicals. J Chem Inf Comput Sci 34: 1320–1325, 1994.

96. G Klopman, M Tu, J Talafous. META. 3. A genetic algorihm for metabolic transform priorities optimization. J Chem Inf Comput Sci 37: 329–334, 1997.

97. R King, S Muggleton, A Srinivasan, MJE Sternberg. Structure activity relationships derived by machine learning. Proc Natl Acad Sci 93: 438–442, 1996.

98. PH Winston. Artificial Intelligence, 3rd Ed. Addison-Wesley, 1992.

99. M A-Razzak, RC Glen, Applications of rule-induction in the derivation of quantitative structure-activity relationships. J. Comp Aided Mol Design 6: 349–383, 1992.

100. J Zupan, J Gasteiger. Neural Networks for Chemists. VCH, Weinheim, 1993.

101. MA Gallop, RW Barnett, WJ Dower, SPA Fodor, EM Gordon. Applications of combinatorial technologies to drug discovery. 1. Background and peptide combinatorial libraries. J Med Chem 37: 1233–1251, 1994.

102. KD Janda. Tagged versus untagged libraries: methods for the generation and screening of combinatorial chemical libraries. Proc Natl Acad Sci USA, 91: 10779–10785, 1994.

103. EM Gordon, MA Gallop, DV Patel. Strategy and tactics in combinatorial organic synthesis, Applications to drug discovery. Acc Chem Res 29: 144–154, 1996.

104. E Martin, JM Blaney, MA Siani, DC Spellmeyer, AK Wong, WH Moos. Measuring diversity: experimental design of combinatorial libraries for drug discovery. J Med Chem 38: 1431–1436, 1995.

105. LA Thompson, JA Ellman. Synthesis and application of small molecule libraries. Chem Rev 96: 555–600, 1996.

106. NK Terrett, M Gardner, DW Gordon, RJ Kobylecki, J Steele, Combinatorial synthesis–the design of compound libraries and their application to drug discovery. Tetrahedron 51: 8135, 1995.

107. RA Houghten. Combinatorial libraries. In: RA Houghten, ed. The theory and practice of combinatorial chemistry. ESCOM, 1995.

108. JJ Burbaum, NH Signal. New technolgies for high throughput screening. Curr Opin Chem Biol 1: 72–78, 1997.

109. GS Sittampalam, SD Kahl, WP Janzen. High throughput screening—advances in assay technologies. Curr Opin Chem Biol 1: 384–391, 1997.

110. RP Sheridan, SK Kearsley. Using a genetic algorithm to suggest combinatorial libraries. J Chem Inf Comput Sci 35: 310, 1995.

111. A Gobbi, D Poppinger, B Rohde. Developing an in-house system to support combinatorial chemistry. In: P Willett, ed. Computational methods for analysis of molecular diversity. Kluwer/ESCOM, 1997, pp 131–158.

112. EJ Martin, DC Spellmeyer, RC Critchlow Jr., JM Blaney. Does combinatorial chemistry obviate computer-aided drug design. In KB Lipkowitz, DB Boyd, ed. Rev Comp Chem Vol. 10, 1997, pp 75–100.

113. J Li. CAMD in modern drug discovery. Drug Discovery Today 1: 311–312, 1996.

114. J Li, CW Murray, B Woaszkowycz, SC Young. Targeted molecular diversity in drug discovery—integration of structure-based design and combinatorial chemistry. Drug Discovery Today 3: 105–112, 1998.

115. CW Murray, DE Clark, TR Auton, MA Firth, J Li, RA Sykes, B Waszkowycz, DR Westhead, SC Young. PRO_SELECT: combining structure-based drug design and combinatorial chemistry for rapid lead discovery. 1. Technology. J Comp Aided Mol Design 11: 193–207, 1997.

116. EK Kick, DC Roe, AG Skillman, GC Liu, TJA Ewing, YX Sun, ID Kuntz, JA Ellman. Structure-based design and combinatorial chemistry yield low nanomolar inhibitors of cathepsin D. Chem Biol 4: 297–307, 1997.

117. BA Leland, BD Christie, JG Nourse, DL Grier, RE Carhart, T Maffett, SM Welford, DH Smith. Managing the combinatorial explosion. J Chem Inf Comput Sci 37: 62–70, 1997.

118. AC Good, RA Lewis. New methodology for profiling combinatorial libraries and screening sets: cleaning up the design process with HARPick, J Med Chem 40: 3926, 1997.

119. J Ball, RV Fishleigh, PJ Greaney, A Marsden, E Platt, JL Pool J.L, B Robson. In: D Bawden, EM Mitchell, eds. 'Chemical Information Systems—Beyond the Structure Diagrams.' Ellis Horwood Press, 1990, pp 107–123.

120. B Robson, E Platt, J Li. In: David L Beveridge, Richard Lavery, eds. 'Theoretical Biochemistry and Molecular Biophysics' 2 Proteins. Adenine Press, 1992, pp 207–222.

121. H-J Böhm, G Klebe. What can we learn from molecular recognition in protein-ligand complexes for the design of new drugs? Angew Chem Int Ed Engl 35: 2588–2614, 1996.

122. DK Agrafiotis, JC Myslik, FR Salemme. Advances in diversity profiling and combinatorial series design. Mol Diversity 4: 1–22, 1999.

123. DJ Cummins, CW Andrews, JA Bentley, M Corey. Molecular diversity in chemical databases: comparison of medicinal chemistry knowledge bases and databases of commercially available compounds. J Chem Inf Comput Sci 36: 750–763, 1996.

124. Ajay, WP Walters, MA Murcko. Can we learn to distinguish between "drug-like" and "nondrug-like" molecules? J Med Chem 41: 3314–3324, 1998.

125. J Sadowski, H Kubinyi. A scoring scheme for discriminating between drugs and nondrugs. J Med Chem 41: 3325–3329, 1998.

126. WP Walters, MT Stahl, MA Murcko. Virtual screening–an overview. Drug Dis Today 3: 160–179, 1998.

# 11

# High Performance Liquid Chromatography of Peptides and Proteins

**Marie-Isabel Aguilar**
*Monash University, Clayton, Victoria, Australia*

## I. INTRODUCTION

The introduction of high performance liquid chromatography (HPLC) to the analysis of peptides and proteins some 20 years ago revolutionised the biological sciences by enabling the rapid and sensitive analysis of peptide and protein structure in a way that was inconceivable 30 years ago.

Today, HPLC in its various modes has become the pivotal technique in the characterization of peptides and proteins and has therefore played a critical role in the development of peptide and protein-based pharmaceuticals. The extraordinary success of HPLC can be attributed to a number of factors. These include 1. the excellent resolution that can be achieved under a wide range of chromatographic conditions for very closely related molecules as well as structurally quite distinct molecules; 2. the experimental ease with which chromatographic selectivity can be manipulated through changes in mobile phase characteristics; 3. the generally high recoveries and hence high productivity and 4. the excellent reproducibility of repetitive separations carried out over a long period of time, which is due partly to the stability of the sorbent materials under a wide range of mobile phase conditions.

Reversed phase chromatography (RPC) is by far the most commonly used mode of separation for peptides, although ion-exchange (IEC) and size exclusion (SEC) chromatography also find application. The three dimensional structure of proteins can be sensitive to the often harsh conditions employed in RPC, and as a consequence, RPC is employed less for the isolation of proteins where it is important to recover the protein in a biologically active form. IEC, SEC, and affinity chromatography are therefore the most commonly used modes for proteins, but RPC and hydrophobic interaction (HIC) chromatography are also employed.

HPLC is extremely versatile for the isolation of peptides and proteins from a wide variety of synthetic or biological sources. The complexity of the mixture to be chromatographed will depend on the nature of the source, and the degree of preliminary clean-up that can be performed. In the case of synthetic peptides, RPC is generally employed both for the initial analysis and the final large scale purification. The isolation of proteins from a biological cocktail however, often requires a combination of techniques to produce a homogenous sample. HPLC techniques are then introduced at the later stages following initial precipitation, clarification and preliminary separations using soft gels. Purification protocols therefore need to be tailored to the specific target molecule.

This chapter deals with the different HPLC techniques which are commonly employed for the analysis, and purification of peptides and proteins. A brief overview of the theory of each mode of chromatography will be presented and then discussed in terms of the parameters that control resolution and illustrated with relevant examples. The interested reader is also referred to a number of recent publications which provide a comprehensive theoretical and practical overview of this topic (1-5).

## II. THEORETICAL CONSIDERATIONS

An appreciation of the factors that control the resolution of peptides and proteins in interactive modes of chromatography can assist in the development and manipulation of separation protocols to obtain the desired separation. The capacity factor $k'$ of a solute can be expressed in terms of the retention time $t_r$, through the relationship

$$k' = (t_r - t_0)/t_0 \qquad (1)$$

where $t_0$ is the retention time of a non-retained solute. The practical significance of $k'$ can be related to the selectivity parameter $\alpha$, defined as the ratio of the capacity factors of two adjacent peaks as follows

$$\alpha = k'_i/k'_j \qquad (2)$$

which allows the definition of a chromatographic elution window in which retention times can be manipulated to maximise the separation of components within a mixture.

The optimisation of high resolution separations of peptides and proteins involves the separation of sample components through manipulation of both retention times and solute peak shape. The second factor which is involved in defining the quality of a separation is therefore the peak width $\sigma_t$. The degree of peak broadening is directly related to the efficiency of the column, and can be expressed in terms of the number of theoretical plates, N, as follows

$$N = (t_r)^2/\sigma_r^2 \qquad (3)$$

N can also be expressed in terms of the reduced plate height equivalent h, the column length L, and the particle diameter of the stationary phase material $d_p$, as

$$N = hL/d_p \qquad (4)$$

The resolution, $R_s$, between two components of a mixture therefore depends on both selectivity, and bandwidth according to

$$R_s = 1/4\sqrt{N}(\alpha - 1)[1/(1+k')] \qquad (5)$$

This equation describes the relationship between the quality of a separation and the relative retention, selectivity, and the bandwidth, and also provides the formal basis upon which resolution can be manipulated to achieve a particular level of separation. Thus, when faced with an unsatisfactory separation, the aim is to improve resolution by one of three possible strategies: the first is to increase $\alpha$, the second is to vary $k'$ within a defined range normally $1 < k' < 10$, or thirdly to increase N, usually by using very small particles in microbore columns.

The challenge facing the scientist who wishes to analyse and/or purify their peptide or protein sample is the selection of the initial separation conditions and subsequent optimisation of the appropriate experimental parameters. This chapter provides an overview of the different techniques used for the analysis and purification of peptides, and proteins, and the experimental options available to achieve a high resolution separation of a peptide or protein mixture.

## III. REVERSED PHASE CHROMATOGRAPHY

### A. Introduction

Reversed phase chromatography (RPC) is now an indispensable tool for the high performance separation of complex mixtures of peptides, and proteins, and is used for both analytical, and preparative applications (6, 7). Analytical applications range from the assessment of purity of peptides following solid-phase peptide synthesis (8), to the analysis of tryptic maps of proteins (9). Preparative RPC is also used for the micropurification of protein fragments for sequencing (10) to large scale purification of synthetic peptides (8). An example of the high resolution analysis of a tryptic digest of a protein is shown in Fig. 1 (11). This figure, in which 150 peaks were

identified, demonstrates the highly selective separation that can be achieved with enzymatic digests of proteins using RPC as part of the quality control or structure determination of a recombinant or natural protein. The chromatographic separation was obtained with a C2/C18 stationary phase packed in a column of dimensions 10 cm × 4.6 mm internal diameter. Separated components can then be directly subjected to further analysis such as automated Edman sequencing or electrospray mass spectrometry.

The purification of synthetic peptides usually involves an initial separation on an analytical scale to assess the complexity of the mixture followed by large-scale purification, and collection of the target product. A sample of the purified material can then be subjected to RPC analysis under different elution conditions to check for purity.

The extensive use of RPC for the purification of peptides, small polypeptides with molecular weights up to

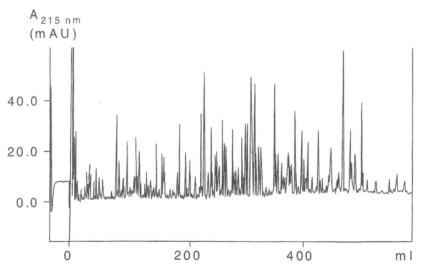

**Fig. 1** High resolution reversed phase chromatographic separation of a tryptic digest of a 165 kDa protein on a μRPC C2/C18 ST 4.6/100 column, (dimensions 10 cm × 4.6 mm ID, 3 μm particle size, 12 nm pore size). Eluent A: 0.065% trifluoroacetic acid (TFA) in water, eluent B: 0.050% TFA in 84% acetonitrile. Gradient elution was carried out with 0% B for 2 column volumes (CV), 0–50% for 392 CV (650 min); 50–100% B for 55 CV (91 min); 100% B for 10 CV (17 min), flow rate of 1 ml/min and detection was at 215 nm (from Ref. 11, courtesy of Amersham Pharmacia Biotech).

10,000 Da, and related compounds of pharmaceutical interest has not been replicated to the same extent for larger polypeptides (MW > 10,000 Da), and globular proteins. The combination of the traditionally used acidic buffering systems and the hydrophobicity of the $n$-alkylsilica supports which can result in low mass yields or the loss of biological activity of larger polypeptides, and proteins have often discouraged practitioners from using RPC methods for large scale protein separations. The loss of enzymatic activity, the formation of multiple peaks for compositionally pure samples and poor yields of protein can all be attributed to the denaturation of protein solutes during the separation process using RPC (12–15). Whilst these features detract from the use of RPC as a technique of choice in preparative purification protocols with proteins, these same characteristics can provide a unique opportunity to study protein folding and stability. Thus, the widespread practical application of RPC has been recently accompanied by a significant improvement in our understanding of the molecular basis of the retention process and its impact on conformational stability of both peptides and proteins. As a result, RPC can now also be used as a physicochemical tool for the analysis of the dynamic behaviour of peptides and proteins at hydrophobic surfaces (16–22).

## B. Theoretical Considerations

The mechanism by which peptides and proteins are retained in RPC involves the hydrophobic expulsion of the solute from the polar mobile phase followed by adsorption onto a non-polar sorbent (23). Peptides and proteins are thus retained to different extents depending on their surface hydrophobicity, the eluotropicity of the mobile phase, and the nature of the hydrophobic ligands.

The physicochemical basis of RPC lies in the hydrophobic interaction between a peptide or protein and the hydrophobic matrix, and can be described in terms of the solvophobic theory where the isocratic retention factor can be expressed as (23)

$$\ln k' = \log k_0 - (N\Delta A_h + 4.836N^{1/3}$$
$$[\kappa^e - 1]V^{2/3})\gamma/RT, \tag{6}$$

where N is Avogadro's number, $\Delta A_h$ is the hydrophobic contact area of the interacting solute and $\gamma$ is the surface tension of the mobile phase. The parameter $\kappa^e$ is the ratio of the energy required to create a cavity for a solvent molecule and the energy required to extend the planar surface of the liquid by the surface area of the solute molecule. Thus, simply stated, elution in RPC is achieved through a decrease in the microscopic surface tension associated with the solute-sorbent interface. Experimentally this is achieved through changes in the water content by variation in the mole fraction of organic solvent in RPC. However, peptides and proteins are generally separated by gradient elution conditions. Under these conditions, the experimentally observed retention data can be analyzed according to the linear solvent strength model (LSS) as follows (24–27)

$$\log \bar{k} = \log k_0 - S\bar{\varphi} \tag{7}$$

where $\bar{k}$ is the solute median capacity factor, and $\bar{\varphi}$ is the corresponding organic mole fraction. The $\log k_0$ value is the affinity of the solute for the sorbent in the absence of organic solvent. By analogy with the solvophobic equation (6) above, the S-value is related to the hydrophobic contact region established between the solute and the sorbent surface. Both S and $\log k_0$ can be derived from plots of $\log \bar{k}$ versus $\bar{\varphi}$ and the LSS model provides the computational basis for the rational optimisation of peptide and protein separations. An example of these plots is shown in Fig. 2 for a series of peptide analogues related to neuropeptide Y-

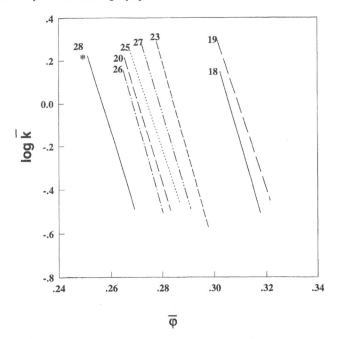

**Fig. 2** Plots of log $\bar{k}$ versus $\bar{\varphi}$ for a series of D-analogues of neuropeptide Y-[18-36] separated on C18 silica with acetonitrile as the organic modifier at 4°C. Column: Bakerbond (JT Baker, Phillipsburg, NJ) RP-C18, 25 cm × 4.6 mm ID, 5 μm particle size, 30 nm pore size. The retention plot for each analogue is designated by the residue position of the D-amino acid substitution. The plots were derived from best fit analysis to the data points (from Ref. 16).

[18–36] which differ in sequence only by the substitution of a single D-amino acid residue (16). These plots illustrate the ability of RPC to resolve very small differences in peptide structure and also illustrate how RPC can act as a molecular probe of peptide surface topography. The data presented in these plots also provide a clear example of how to optimize separations, whereby maximal optimization will be achieved by selecting elution conditions where there is the greatest separation between the retention plots (24–27). The elution profiles of five of the NPY-[18–36] analogues are shown in Fig. 3 and demonstrate that the order of elution of the peaks corresponds to the relative position of the retention plots shown in Fig. 2 and also illustrates the corresponding degree of resolution that is obtained (16).

Peptides and proteins are retained in RPC through hydrophobic interactions. As depicted in Fig. 4A, the contact region for small peptides involves the contribution from all or a large proportion of the peptide structure. As a consequence, the retention time for small peptides can be predicted on the basis of the amino acid composition through summation of the hydrophobicity coefficients for the constituent amino acid residues (28, 29). In contrast, it has been well-established that proteins interact with the chromatographic surface in an orientation-specific manner, in which their retention time is determined by the molecular composition of specific contact regions (30–32). For larger polypeptides and proteins which adopt a significant degree of secondary and tertiary structure, there is no correlation between retention times, and the

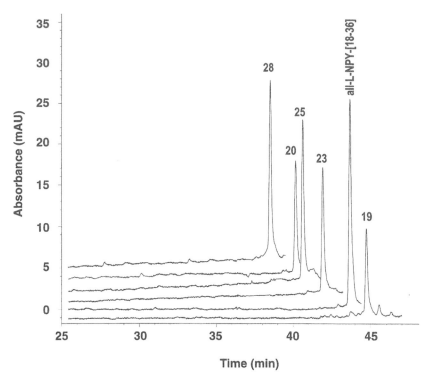

**Fig. 3** Chromatograms showing the separation of the all-L-NPY-[18-36] peptide and NPY-[18-36]- analogues with D-subsitutions at positions 19, 20, 23, 25 and 28 separated on a C18 silica [Bakerbond, (JT Baker, Phillipsburg, NJ) RP-C18, 25 cm × 4.6 mm ID, 5 μm particle size, 30 nm pore size) with a 60 minute gradient from 0–50% acetonitrile at 1 ml/min. The separation was used to derive the retention plots shown in Fig. 2 (from Ref. 16).

summated hydrophobicity coefficients of all constituent amino acid residues. In these cases, the chromatographic contact region comprises a small proportion of the total molecular surface as shown in Fig. 4B and 4C. It is also generally observed that the retention time increases with solute molecular weight as a result of increases in hydrophobic surface area. However, it is the molecular composition of the contact region that determines the retention and bandwidth properties of the peptide or protein solute. Thus, in the presence of any degree of preferred secondary structure, no simple relationship will exist between the retention time and the summated retention coefficients unless the identity of the contact region can be established.

## C.  Factors Influencing Retention in RPC

### 1.  Stationary Phases

The most commonly employed experimental procedure for the RPC analysis of peptides and proteins generally involves the use of an octadecylsilica based sorbent and a mobile phase. The chromatographic packing materials that are generally used are based on microparticulate porous silica which allows the use of high linear flow velocities resulting in favourable mass transfer properties and rapid analysis times (33–35). The silica is chemically modified by a derivatised silane bearing an *n*-alkyl hydrophobic ligand. The most commonly used ligand is *n*-octadecyl (C18), while *n*-butyl (C4) and *n*-octyl (C8)

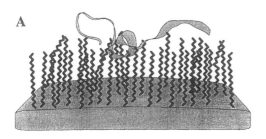

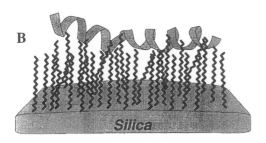

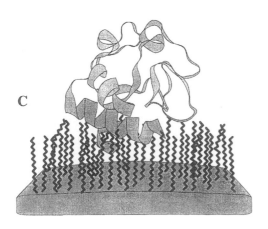

**Fig. 4** Schematic representation of the binding of a polypeptide to an interactive chromatographic sorbent. The shaded areas on the polypeptide indicate the chromatographic contact region which will be defined by the nature of the sorbent material. In reversed phase, and hydrophobic interaction chromatography, the areas will be hydrophobic while in ion-exchange chromatography, the contact regions will represent the surface regions of highest charge density.

also find important application and phenyl and cyanopropyl ligands can provide different selectivity (36). The process of chemical immobilisation of the silica surface results in approximately half of the surface silanol group being modified. The sorbents are therefore generally subjected to further silanisation with a small reactive silane to produce an end-capped packing material. The type of $n$-alkyl ligand significantly influences the retention of peptides, and proteins, and can therefore be used to manipulate the selectivity of peptide and protein separations. While the detailed molecular basis of the effect of ligand structure is not fully understood, a number of factors including the relative hydrophobicity, and ligand chain length, flexibility and the degree of exposure of surface silanols all play a role in the retention process (37–39). An example of the effect of chain length on peptide separations can be seen in Fig. 5 (6). It can be seen that the peaks labeled $T_3$ and $T_{13}$ are fully resolved on the C4 packing but cannot be separated on the C18 material. In contrast, the peptides $T_5$ and $T_{18}$ are unresolved on the C4 column but fully resolved on the C18 material. In addition to effects on peptide selectivity, the choice of ligand type can also influence protein recovery and conformational integrity of protein samples. Generally higher protein recoveries are obtained with the shorter and less hydrophobic $n$-butyl ligands. However, proteins have also been obtained in high yield with $n$-octadecyl silica (40–42).

The denaturation of proteins by RPC sorbents can also be controlled by using silica which has been coated with polymethacrylate-based polymers which provide a series of sorbents with varying surface hydrophobicity, and in which the surface silanol groups have been masked (43, 44). Silica-based packings are also susceptible to cleavage at pH values greater than seven. This limitation can severely restrict the utility of these materials for

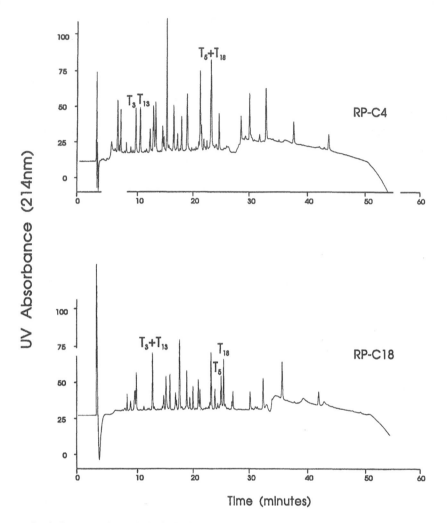

**Fig. 5** The influence of *n*-alkyl chain length on the separation of tryptic peptides derived from porcine growth hormone. Top: Bakerbond (JT Baker, Phillipsburg, NJ) RP-C4, 25 cm × 4.6 mm ID, 5 μm particle size, 30 nm pore size. Bottom: Bakerbond (JT Baker, Phillipsburg, NJ) RP-C18, 25 cm × 4.6 mm ID, 5 μm particle size, 30 nm pore size. Conditions, linear gradient from 0–90% acetonitrile with 0.1% TFA over 60 minutes, flow rate of 1 ml/min, 25°C (from Ref. 6).

separation which require basic pH conditions to effect resolution. In these cases, alternative stationary phases have been developed including cross-linked polystyrene divinylbenzene (45, 46) and porous zirconia (47, 48) which are all stable to hydrolysis at alkaline pHs.

The geometry of the particle in terms of the particle diameter, and pore size, is also an important feature of the packing material. As predicted by Eq. (4), improved resolution can be achieved by decreasing the particle diameter $d_p$. The most commonly used range of particle diameters for analytical scale RPC is 3–5 μm. There are also examples of the use of nonporous particles of smaller diameter (49). For preparative scale separations, 10–20 μm particles are utilised. The pore size of RPC sorbents is also an important factor that must be considered. For peptides, the pore size generally ranges between 100 – 300 Å depending on the size of the peptides. Porous materials of ≥300 Å pore size are necessary for the separation of proteins, as the solute molecular diameter must be at least one-tenth the size of the pore diameter to avoid restricted diffusion of the solute and to allow the total surface area of the sorbent material to be accessible. The recent development of particles with 6000–8000 Å pores with a network of smaller pores of 500–1000 Å has also allowed very rapid protein separations to be achieved (50, 51).

## 2. Mobile Phases

One of the most powerful characteristics of RPC is the ability to manipulate solute retention, and resolution through changes in the composition of the mobile phase. In RPC, peptide, and protein retention is due to muiti-site interactions with the ligands. The practical consequence of this is that high resolution isocratic elution of peptides and proteins can rarely be achieved as the experimental window of solvent concentration required for their elution is very narrow. Mixtures of peptides, and proteins are therefore routinely eluted by the application of a gradient of increasing organic solvent concentration. RPC is generally carried out with an acidic mobile phase, with trifluoroacetic acid (TFA) the most commonly used additive due to its volatility. Phosphoric acid, perchloric acid, formic acid, hydrochloric acid, acetic acid and heptafluorobutyric acid have also been used (52–55). The effect of ion-pairing reagents on peptide separation is illustrated in Fig.6 for a series of peptide standards separated on a C18 column. Alternative additives such as non-ionic detergents can be used for the isolation of more hydrophobic proteins such as membrane proteins (56).

The three most commonly employed organic solvents are acetonitrile, methanol, and 2-propanol which all exhibit high optical transparency in the detection

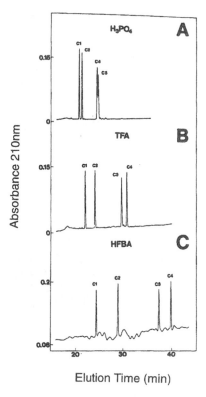

**Fig. 6** The influence of ion-paring reagent on the separation of a mixture of synthetic peptides in reversed phase chromatography using a SynChropak C18, 25 cm × 4.6 mm ID, 6.5 μm particle size, 30 nm pore size (SynChrom, Linden, IN). Conditions: linear gradient from 0–100% acetonitrile containing A. 0.1% $H_3PO_4$; B, 0.1% TFA; and C; 0.1% HFBA; flow rate of 1 ml/min, 26°C. Peptide sequences:
C1=Ac-GGGLGGAGGLK-amide,
C2 = Ac-KYGLGGAGGLK-amide.
C3 = Ac-GGALKALKGLK-amide,
C4 = Ac-KYALKALKGLK-amide
(Reprinted with permission from Ref. 55. Copyright CRC Press, Boca Raton, Florida).

wavelengths used for peptide and protein analysis. Acetonitrile provides the lowest viscosity solvent mixtures and 2-propanol is the strongest eluent. An example of the influence of organic solvent is shown in Fig. 7 where changes in selectivity can be observed for a number of peptide peaks in the tryptic map. In addition to the

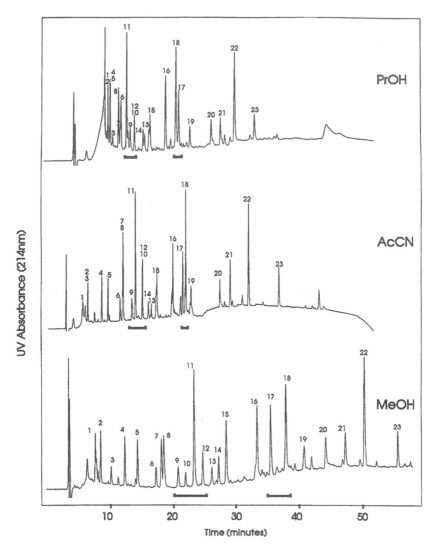

**Fig. 7** The influence of organic solvent on the reversed phase chromatography of tryptic peptides derived from porcine growth hormone. Column: Bakerbond (JT Baker, Phillipsburg, NJ) RP-C4, 25 cm × 4.6 mm ID, 5 μm particle size, 30 nm pore size. Conditions, linear gradient from 0–90% 2-propanol (top), acetonitrile (middle) or methanol (bottom) with 0.1% TFA over 60 minutes, flow rate of 1 ml/min, 25°C (from Ref. 6).

eluotropic effects, the nature of the organic solvent can also influence the conformation of both peptides and proteins, and will therefore have an additional effect on selectivity through changes in the structure of the hydrophobic contact region (14, 57). In the case of proteins, this may also impact on the level of recovery of biologically active material.

### 3. Column Geometry

The desired level of efficiency, and sample loading size determines the dimension of the column to be used. For small peptides and proteins, increased resolution will be obtained with increases in column length. Thus, for applications such as tryptic mapping, column lengths between

15–25 cm and internal diameter (ID) of 4.6 mm are generally employed. However, for larger proteins, low mass recovery and loss of biological activity may result with these columns due to irreversible binding and/or denaturation. In these cases, shorter columns of between 2 and 20 cm in length can be used. For preparative applications in the 1–500 mg scale, such as the purification of synthetic peptides, so-called semi-preparative columns of dimensions 30 cm × 1 cm ID and preparative columns of 30 cm × 2 cm ID can be used.

The selection of the internal diameter of the column is based on the sample capacity and detection sensitivity. While most analytical applications are carried out with columns of internal diameter of 4.6 mm ID (as shown in Figs 1, 3 and 5), for samples derived from previously unknown proteins where there is a limited supply of material, the task is to maximise the detection sensitivity. In these cases, the use of narrow bore columns of 1 or 2 mm ID can be used which allow the elution and recovery of samples in much smaller volumes of solvent (10). Capillary chromatography is also finding increasing application where capillary columns of internal diameter between 0.2–0.4 mm, and column length of 15 cm result in the analysis of fmole of sample as shown in Fig. 8 (58). The effect of decreasing column internal diameter on detection sensitivity is shown in Fig. 9 for the analysis of lysozyme on a C18 material packed into columns of 4.6 mm, 2.1 mm and 0.3 mm ID (59).

### 4.  Operating Parameters

There are several operating parameters that can be changed in order to manipulate the resolution of peptide and protein mixtures in RPC. These parameters include the gradient time, the gradient shape, the mobile phase flow rate and the operating temperature. The typical exper-

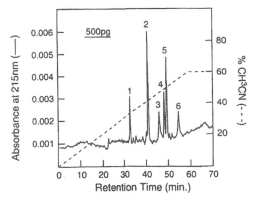

**Fig. 8**  The separation of proteins by reversed phase capillary chromatography. Column; Brownlee RP-300, 15 cm × 0.2 mm ID, 5 $\mu$m particle size, 30 nm pore size. Conditions; linear gradient from 0–60% acetonitrile over 60 minutes, flow rate of 1.4 $\mu$l/min. Protein; 1 = ribonuclease B, 2 = lysozyme, 3 = bovine serum albumin, 4 = carbonic anhydrase, 5 = myoglobin, 6 = ovalbumin (from Ref. 59).

iment with an analytical scale column would utilise a linear gradient from 5% organic solvent up to between 50–100% solvent over a time range of 20–120 minutes while flow rates are between 0.5–2.0 ml/min. With microbore columns [1–2 mm ID] flow rates of 50–250 $\mu$l/min are used, while for capillary columns of 0.2–0.4 mm ID, flow rates of 1–4 $\mu$l/min are applied. At the preparative end of the scale with columns of 10–20 mm ID, flow rates between 5–20 ml/min are required.

The choice of gradient conditions will depend on the nature of the molecules of interest. The influence of gradient time on the separation of a series of ribosomal proteins is shown in Fig. 10 (60). Generally the use of longer gradient times provides improved separation. However, these conditions also increase the residence time of the peptide or protein solute at the sorbent surface, which may then result in an increase in the degree of denaturation.

The operating temperature can also be used to manipulate resolution. While the separation of peptides and proteins is

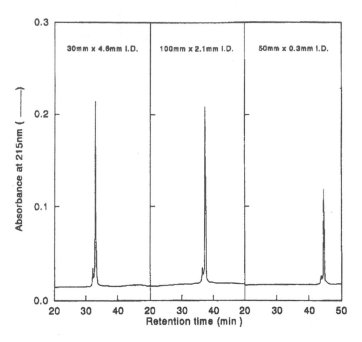

**Fig. 9** Effect of column internal diameter on detector sensitvity. Column: Brownlee RP-300 C8 (7 $\mu$m particle size, 30 nm pore size), 3 cm × 4.6 mm ID and 10 cm × 2.1 mm ID (Applied Biosystems) and 5 cm × 0.32 mm ID. Conditions: linear gradient from 0–60% acetonitrile with 0.1% TFA over 60 minutes, 45°C. Flow rates, 1 ml/min, 200 $\mu$l/min and 4 $\mu$l/min for the 4.6, 2.1 and 0.32 mmID columns respectively. Sample loadings, lysozyme, 10 $\mu$g, 4 $\mu$g and 0.04 $\mu$g for the 4.6, 2.1 and 0.32 mmID columns respectively (Reprinted from Ref. 58, with kind permission from Elsevier Science—NL, Sara Burgerhartstraat 25, 1055 KV Amsterdam, The Netherlands).

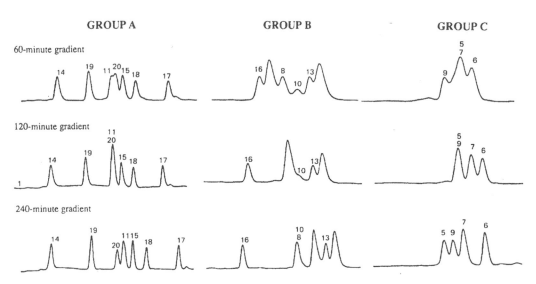

**Fig. 10** The effect of gradient time on the resolution of ribosomal proteins by reversed phase chromatography. Column: Du Pont Bioseries Protein PLUS. Conditions: 25–46% acetonitrile with 0.1% triethylamine and 0.042% TFA, over 60 minutes (top), 120 minutes (middle) or 240 minutes (bottom), at a flow rate of 0.7 ml/min (from Ref. 60).

normally carried out at ambient temperature, solute retention in RPC is influenced by temperature through changes in solvent viscosity. In addition to this, peptide and protein conformation can also be manipulated by temperature. In the case of proteins, where biological recovery is not important, increasing temperature can be used to modulate retention via denaturation of the protein structure (12–15). For peptides, it has been shown that secondary structure can actually be enhanced through binding to the hydrophobic sorbent (16–22). Changes in temperature can therefore also be used to manipulate the structure, and retention of peptide mixtures.

Chromatographic recovery of proteins from polyacrylamide gels is another important application of RPC. Inverse gradient elution chromatography has been successfully utilized for the micropreparative isolation of proteins from SDS-PAGE electroeluates (61). This approach is based on observations that certain RPC packings display strong interactions with proteins at high organic solvent concentrations. This allows the loading of the electroeluate under conditions where the protein is retained while the SDS and other gel-related contaminants are washed through the column. The protein is then recovered in high yield by a gradient of decreasing organic solvent.

Detection of peptides and proteins in RPC, and in all modes of chromatography, generally involves detection at between 210–220 nm which is specific for the peptide bond, or at 280 nm which corresponds to the aromatic amino acids tryptophan and tyrosine. The use of photodiode array detectors can enhance the detection capabilities by the on-line accumulation of complete solute spectra. The spectra can then be used to identify peaks specifically on the basis of spectral characteristics, and for the assessment of peak purity (62–64). In addition, second derivative spectroscopy can provide information on the conformational integrity of proteins following elution (65, 66)

One of the most significant recent advances in bioanalytical technology is the advent of mass spectrometry for the analysis and measurement of peptide and protein molecular mass (67). In particular, the development of on-line electrospray mass spectrometry following RPC (LC-ES-MS) has provided a powerful detection system for the rapid analysis of peptide, and proteins (68). Figure 11 shows the LC-MS analysis of an Arg-C digest of plasminogen activator separated on a C18 column, with the total ion current in the upper trace and the elution profile detected at 214 nm in the lower trace. The availability of on-line mass spectrometry thus significantly facilitates the identification of the peptide fragments. Other important applications involve the identification of posttranslational modifications of peptides and proteins (69, 70), assignment of disulfide bonds (71, 72), and the identification of peptides bound to major histocompatibility complex molecules (73).

In summary, RPC is now firmly entrenched as the central tool for the analysis of peptides and proteins and thus plays a pivotal role in the pharmaceutical industry providing the core analytical technique at all stages of the development of peptide, and protein-based therapeutics.

## IV. HYDROPHOBIC INTERACTION CHROMATOGRAPHY

Hydrophobic interaction chromatography (HIC) is a valuable technique for the separation of proteins under non-denaturing conditions (74–76). HIC involves the use of high salt concentrations to promote hydrophobic interactions between the protein and the hydrophobic stationary phase. Solutes are then eluted in order of increasing hydrophobicity

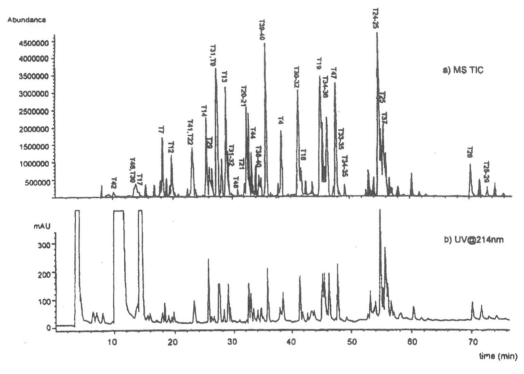

**Fig. 11** LC-MS of a tryptic digest of single-chain plasminogen activator. Column: Vydac C18, 5 μm particle size, 30 nm pore size. Conditions: linear gradient from 0–60% acetonitrile with 0.1% TFA over 90 minutes, 45°C, flow rate 0.2 ml/min. (a) electrospray mass spectrometry total ion current, (b) detection at 214 nm. (Reprinted from Ref. 68, with kind permission from Elsevier Science—NL, Sara Burgerhartstraat 25, 1055 KV Amsterdam, The Netherlands).

though the application of a descending salt gradient which weakens the hydrophobic interactions between the protein and the sorbent material. The ligands used for HIC materials are less hydrophobic than those in RPC. Thus, in contrast to the denaturing effects of low pH and organic solvent present in RPC systems, the mobile phases used in HIC generally stabilise protein structure.

## A. Theoretical Considerations

The primary factors which control protein retention in HIC are similar to those factors which contribute to retention in RPC, i.e. the surface tension of the mobile phase and the exposed hydrophobic sur-face area of the protein solute. The use of the solvophobic theory has been extended to describe the effects of neutral salts on protein solubility, and retention in HIC (77, 78). In particular, the dependence of the logarithmic capacity factor on salt concentration $m$, can be expressed as

$$\log k' = \log k_0 \atop + [(-D\mu + v + N\Delta A_h\sigma)m]/RT \tag{8}$$

where $D\mu$ and $v$ are terms related to the dielectric constant of the medium and the protein dipole moment. Thus retention in HIC is dependent on the contact surface area $\Delta A_h$, established between the solute and the sorbent and the molal surface tension increment $\sigma$ of the eluting salt solution. Selectivity optimization in HIC

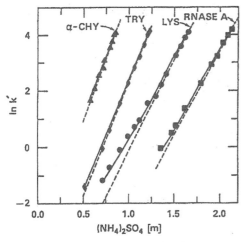

**Fig. 12** The dependence of log $k'$ on ammonium sulphate concentration for a series of proteins separated by HIC during isocratic elution. Column; TSK-phenyl (Toyo Soda). Conditions: 50 mM phosphate buffer, pH 7, flow rate 3 ml/min, 45°C. Proteins; α-CHY = α-chymotrypsinogen, TRY = trypsinogen, LYS = lysozyme, RNASE A = ribonuclease A (from Ref. 79).

therefore involves changes in protein solubility characteristics through either changes in the structure of the stationary phase ligand or changes in the mobile phase characteristics through the use of salts of different molal surface tension increments. A linear relationship between the capacity factor of a protein and the corresponding salt concentration is generally observed, and Eq. (8) can be simplified to

$$\log k' = \lambda m + C \qquad (9)$$

where C incorporates all the salt-independent terms and $\lambda$ is the slope of plots of $\log k'$ versus salt concentration, $m$. The linear dependence of protein retention on salt concentration is illustrated in Fig. 12 for a series of globular proteins (79). The term $\lambda$ has also been shown to be linearly related to the protein surface area and the molal surface tension increment which is consistent with the predictions of Eq. (8) (77, 79).

## B. Factors Influencing Retention in HIC

### 1. Stationary Phase

Both polymeric, and silica based HIC supports have been produced, and a range of mildly hydrophobic ligands are available to perform HIC (74). In particular both alkyl and aryl ligands have proven to be successful in obtaining high levels of selectivity. Figure 13 shows the influence of a range of ligands on the retention behaviour of a series of globular proteins and demonstrates that protein retention increases in the order hydroxypropyl < methyl < benzyl = propyl < isopropyl < phenyl < pentyl (80). This figure clearly illustrates the influence of the ligand structure on the retention of proteins in HIC.

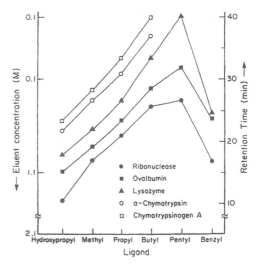

**Fig. 13** The effect of ligand type on the elution of proteins in hydrophobic interaction chromatography. Column: SynChropak hydroxypropyl, propyl, methyl, butyl, pentyl and benzyl columns, 25 cm × 4.6 mm ID, 6.5 μm particle size, 30 nm pore size (SynChrom, Linden, IN). Conditions; linear gradient from 2-0M ammonium sulphate in 0.1M potassium phosphate, pH 6.8, over 30 minutes at a flow rate of 1 ml/min (Reprinted from Ref. 80, with kind permission from Elsevier Science—NL, Sara Burgerhartstraat 25, 1055 KV Amsterdam, The Netherlands).

Other ligand types that can be used include silica-based ether-bonded alkyl phases, neopentylagarose and phenylagarose.

## 2. Mobile Phase

Selectivity in HIC can be manipulated by changes in the nature of the eluting salt, salt concentration, pH, temperature and the addition of mobile phase modifiers. Protein retention is strongly dependent on the type of salt employed. Salts defined as being kosmotropic or structure-making are used as they enhance hydrophobic interactions through a salting-out mechanism. This effect is formally described in Eq. (8) in terms of the molal surface tension increment $\sigma$, the values of which are listed in Table 1 for a selection of commonly used salts in HIC. The salts with a higher $\sigma$-value will result in longer protein retention times. There have been a number of studies which have documented the influence of different salts on protein retention in HIC (77, 81, 82). An example of the effect of salt on the retention of lysozyme in HIC is shown in Fig. 14. $(NH_4)_2SO_4$ is the most commonly used salt in HIC, while $Na_2SO_4$ and NaCl also find application. $(NH_4)_2SO_4$ has a high solubility and low UV absorbance, and is readily available in high purity required for HPLC. Initial salt concentrations usually range from 1–3 M, and the starting concentration can also influence selectivity of a separation.

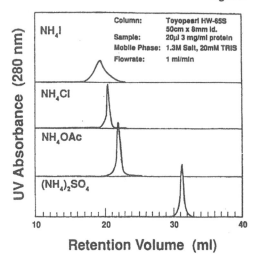

**Fig. 14** The effect of salt on the elution of lysozyme in HIC. Column; Toyopearl HW-65S (Toyo Soda), 50 cm × 8 mm ID, 30 $\mu$m particle size, 100 nm pore size. Conditions; isocratic elution with 20 mM TRIS, pH 7, containing 1.3M ammonium iodide, chloride, acetate or sulphate, at a flow rate of 1 ml/min (from Ref. 82).

The pH of mobile phases used in HIC is typically between 5–7, and buffered with sodium or potassium phosphate. The influence of pH on protein retention is dependent on the particular protein as the manipulation of charges located in or near the hydrophobic binding domain will have a profound effect on the affinity of the protein for the sorbent material. Thus changes in pH represents a useful parameter to modulate selectivity (81).

The gradient shape is an additional parameter which can be used to manipulate selectivity in HIC. An example of the influence of gradient time on the HIC separation of proteins is illustrated in Fig. 15 (83). The LSS model can also be applied to the retention behaviour of proteins in HIC and generally reveals a linear dependence of retention on salt concentration (79). The selection of elution conditions to maximise resolution between components therefore follows the same

**Table 1** Molal Surface Tension Increment for Salts Used in HIC*

| Salt | $\sigma$ ($\times 10^3$ dyn-g/cm-mol) |
| --- | --- |
| $Na_3PO_4$ | 2.88 |
| $Na_2SO_4$ | 2.74 |
| $(NH_4)_2SO_4$ | 2.17 |
| $Mg_2SO_4$ | 2.06 |
| $Na_2HPO_4$ | 2.03 |
| NaCl | 1.64 |
| KCl | 1.50 |

* *Source*: Ref. 78.

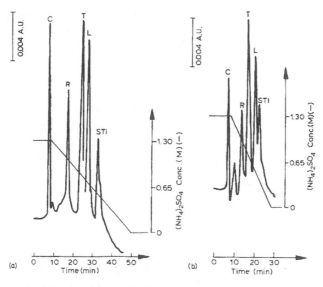

**Fig. 15** The effect of gradient time on the resolution of proteins in HIC. Column: Octyl agarose, $60 \times 6$ mm ID, $5$–$7\mu$m particle size. Conditions: linear gradient from $1.33$–$0$ M ammonium sulphate in $0.05$ M sodium phosphate, pH 7, over (a) 40 minutes or (b) 20 minutes at a flow rate of $0.2$ ml/min, detection at 280 nm. Proteins: C = cytochrome c, R = ribonuclease A, T = transferrin, L = lysozyme and STI = soybean trypsin inhibitor (Reprinted from Ref. 83, with kind permission from Elsevier Science—NL, Sara Burgerhartstraat 25, 1055 KV Amsterdam, The Netherlands)

rationale as described for RPC in the previous section.

The addition of other solvent modifiers has also been shown to affect retention in HIC through changes in the surface tension of the mobile phase. These include detergents such as Triton X-100 and CHAPS, organic solvents such as 5–20% methanol, acetonitrile or even ethylene glycol and urea or guanidine hydrochloride at concentrations of 1–2 M (84–87). In all cases, it is possible that these additives may cause denaturation of the target proteins, so care is needed to minimize protein conformational changes when introducing these additives to the mobile phase.

Temperature can also be used to manipulate selectivity in HIC through changes in protein conformation (88–90). Depending on the protein solute, chromatography in the range 15–50°C can be used to sharpen individual peaks shapes and hence improve resolution as can be

seen for myoglobin in Fig 16B (88). However, significant band broadening can also be observed as a result of slow conformational interconversions which results in a decline in resolution as is evident in Fig 16A for cytochrome c.

Overall, HIC is a powerful tool for the purification of proteins in a biologically active form. Moreover, protein structure and conformation play a crucial role in the chromatographic behaviour of proteins, and subtle changes in selectivity can be achieved through changes in the relative solubility and three dimensional structure of the protein solute.

## V. ION-EXCHANGE CHROMATOGRAPHY

High performance ion-exchange chromatography (IEC) is now extensively used in the analysis of proteins, and also

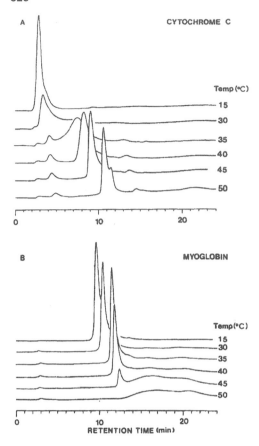

**Fig. 16** The influence of temperature on the retention of (A) cytochrome c and (B) myoglobin separated by HIC. Column; Bio-Gel TSK-Phenyl-5-PW, 7.5 cm × 7.5 mm ID, 10 μm particle size, 100 nm pore size (Bio-Rad, Richmond, CA). Conditions; linear gradient from 1.7-0M ammonium sulphate in 0.1M sodium phosphate, pH 7, over 15 minutes and a flow rate of 1 ml/min (Reprinted from Ref. 88, with kind permission from Elsevier Science—NL, Sara Burgerhartstraat 25, 1055 KV Amsterdam, The Netherlands).

to a lesser extent for the analysis of peptides (91, 92). The early stages of protein purification generally utilise solubility-based techniques to carry out the initial fractionation. Differences in size and shape of the proteins are then exploited through application of size exclusion or preparative electrophoretic techniques. Adsorptive techniques, including IEC and

RPC are then introduced to allow rapid increases in the level of resolution, recovery, and product purity.

A significant advantage of IEC over the other adsorptive modes of chromatography is the non-denaturing effects of the solutions used to elute proteins from the ion-exchange sorbents. Thus, while gross conformational changes can be observed in RPC, these are not commonly found in IEC of proteins.

## A. Theoretical Considerations

Protein retention in ion-exchange chromatography arises from electrostatic interactions between the peptide or protein and the charged sorbent material and solutes are eluted by increases in the concentration of a displacer salt. As a consequence, the "net charge" concept is widely used to predict the retention characteristics of proteins with both anion and cation-exchange materials (93, 94). According to this model, and as illustrated in Fig. 17, a protein will be retained on a cation-exchange column if the solvent pH is lower than the pI of the protein. Conversely, a protein will be retained on an anion-exchange column if the pH is above the pI of the protein. With mobile phases operating at a pH equal to the protein's pI, the surface of the protein is considered to be overall electrostatically neutral, and under these conditions, the protein should not be retained on either cation- or anion-exchangers.

While this model can be used to predict the retention behaviour of peptides in IEC, this classical model is recognized as a simplistic approach to describing protein retention (94–96). The amphoteric nature of proteins results in the existence of localized areas of electrostatic charge at different pHs, which can allow the protein to be retained even under conditions where the protein may be at its isoelectric point (95, 96).

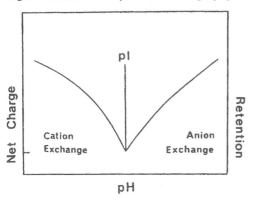

**Fig. 17** The theoretical relationship between protein net charge, and chromatographic retention in ion-exchange chromatography.

A number of more detailed models have been developed to describe protein retention in ion-exchange chromatography. These include the stoichiometric displacement model (94, 95), and different approaches to solving the Poisson-Boltzmann equation for the interaction of charged surfaces (97–99). Each of these approaches attempts to provide a mechanistic description of the retention of proteins in IEC, and provide a theoretical basis for the experimentally observed linear relationship between the solute retention and either (a) the reciprocal of the displacing salt concentration $c$, as follows

$$\log k' = \log K_0 + Z_c \log(1/c) \qquad (10)$$

or (b) the corresponding ionic strength, I, according to

$$\log k' = \log K_0 + B(1/\sqrt{I}) \qquad (11)$$

where $K_0$ is a constant which incorporates a number of terms including a binding constant, the phase ratio of the packing material and various charge valency terms, and can be determined by linear regression of plots of $\log k'$ versus $\log(1/c)$ or plots of $\log k'$ versus $1/\sqrt{I}$. An example of the linear dependency of protein retention on $\log(1/c)$ is shown in Fig. 18 for the isocratic and gradient elution of ovalbumin, and carbonic anhydrase by

strong anion exchange chromatography (100). $Z_C$ is the slope of these plots, and can be related to the size of the electrostatic contact area established between the solute and the stationary phase (100–102). In addition, the LSS model can be used as the basis to optimise separations as described previously for RPC, through establishing the experimental conditions which maximize the separation of these plots for different components in a mixture (103). The variation in protein retention with changes in displacer salt concentration can be seen in Fig.19 for the isocratic separation of myoglobin, carbonic anhydrase, and ovalbumin using sodium chloride as the displacer salt (92).

Overall, it has now been established that the magnitude of electrostatic interactions between proteins and the charged sorbent material in IEC depends on the number and distribution of charged sites on the solute molecule that define the electrostatic contact area of the protein, the charge density of the sorbent and the mobile phase composition (91, 92, 94). In summary, the magnitude of the electrostatic interactions, and hence retention, in IEC are dependent on the following structural and chromatographic parameters.

1. The number and distribution of charged sites on the solute molecule that constitute the electrostatic contact area.
2. The charge density of the immobilized charged ligand.
3. The mobile phase composition.

It is these parameters which represent the factors that can be used to manipulate peptide and protein surface charge to allow optimisation of selectivity in IEC.

## B. Factors Influencing Retention in IEC

### 1. Stationary Phases

The support materials available for high performance ion-exchange chroma-

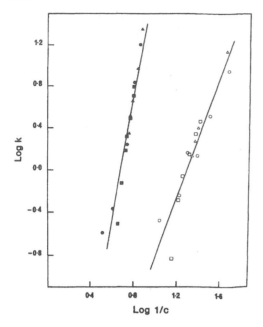

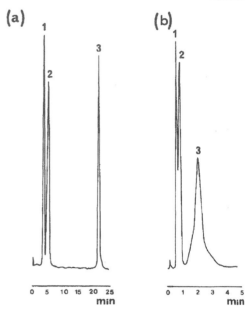

**Fig. 18** Retention plots for the isocratic, and gradient separation of ovalbumin (filled symbols), and carbonic anhydrase (open symbols) by strong anion-exchange chromatography. Column; Mono-Q HR 5/5, (Pharmacia, Uppsala, Sweden) Conditions; 0.02 M piperazine at pH 9.6 with sodium bromide as the displacer (Reprinted from Ref. 100, with kind permission from Elsevier Science—NL, Sara Burgerhartstraat 25, 1055 KV Amsterdam, The Netherlands).

**Fig. 19** Isocratic elution profile of proteins separated by anion-exchange chromatography. Column; Mono-Q HR 5/5, (Pharmacia, Uppsala, Sweden) Conditions; 0.02 M piperazine at pH 9.6 with sodium chloride as the displacer at a concentration of (a) 60 mM and (b) 180 mM. Proteins; (1) = myoglobin, (2) = carbonic anhydrase, (3) = ovalbumin (Reproduced with permission from Ref. 92. Copyright John Wiley and Sons Ltd).

tography are generally silica-based or polymer-based materials (33, 91). A novel tentacle-type material is also available where the charged ligands are attached to long chains which are attached to either silica or polymer-based material (105–108). An ion-exchange material is selected on the basis of the desired particle and pore size, swelling characteristics at the operational pH range. However, the major factor is the capacity of the ion-exchanger which depends on the nature of the charged functional groups and the charge density as well as the pore size of the material, and the charge distribution on the protein solute.

The two classes of ion-exchangers are cation exchangers which contain negatively charged functional groups and anion exchangers which contain positively charged functional groups. The most commonly encountered ligands are listed in Table 2. Strong cation exchangers normally contain sulfonic acid groups while strong anion exchangers contain quaternary ammonium functional groups. The charged ligands in weak cation exchangers generally contain carboxylic groups while weak anion-exchangers are primary, secondary or tertiary amines. The terms strong and weak refer to the degree of ionisation with pH as strong ion-exchangers are completely ionised over a much wider pH range than weak ion-exchangers. The physical properties

**Table 2** Commonly Encountered Ligands for Ion-Exchange Chromatography

| Ion Exchanger | Functional Group | |
|---|---|---|
| Strong cation exchanger (SCX) | Methylsulphonate | $-CH_2SO_3^-$ |
| Weak cation exchanger (WCX) | Carboxymethyl | $-OCH_2COO^-$ |
| Strong anion exchanger (SAX) | Methyl trimethyl ammonium | $-CH_2N^+(CH_3)_3$ |
| Weak anion exchanger (WAX) | Diethylaminoethyl | $-OCH_2H_2N^+H(CH_2CH_3)_2$ |

of some typical anion and cation exchangers used in protein chromatography are listed in Table 3 and a more comprehensive list of columns is given in Ref. 91.

## 2. Mobile Phases

The selectivity of proteins in IEC can be manipulated by variation in solution pH and ionic strength which alters the electrostatic surface potential of the protein solutes, and the charged ligand thereby influencing the strength of the electrostatic interactions. Changes in the nature of the displacer ion, and the buffer species represent additional methods by which protein retention can be modified.

While NaCl is the most commonly used ionic displacer, a number of other monovalent and multivalent salts can be used (94, 100, 110, 111). The ions may influence retention through specific interactions with the ion-exchange ligand, thereby changing their ionic properties. In addition, specific salts may alter the conformation of proteins which in turn will influence their retention behaviour. At fixed ionic strengths, anions can be ranked in terms of solute retention as follows

$$F^- < CH_3COO^- < Cl^- < HPO_4^{2-} < SO_4^{2-}$$

Similarly, cations are ranked according to the series

$$K^+ < Na^+ < NH_4^+ < Ca^{2+} < Mg^{2+}$$

KCl, NaOAc, MgCl$_2$, Mg$_3$(SO$_4$)$_2$ have all been used for the analysis and purification of a wide range of proteins. The effect of different displacer salts on protein retention in weak anion-exchange chromatography is shown in Fig. 20, which illustrates the profound influence that the nature of the salt can exert on the electrostatic interactions between proteins, and ion-exchange materials (112).

The selection of buffer depends on the pH range required to adsorb the protein to the stationary phase. While selection of a pH can be a straightforward task based on the known pI of the protein, for proteins of unknown pI or closely related proteins such as isoforms or recombinant muteins, a map of retention versus pH can assist in the selection of mobile phase pH. An example of the changes in selectivity which can occur over a given pH range is shown in Fig. 21 for the separation of a series of lysozymes by cation-exchange chromatography (104). Once the pH range is established, additional changes in selectivity can be obtained through changes in the nature of the buffer species. The most commonly employed buffer species include phosphate and tris buffers. A range of buffers which are commonly used in IEC of proteins is listed in Table 4.

## 3. Operating Parameters

While isocratic elution can be used to separate proteins in IEC, gradient elution is generally employed to obtain high resolution separations of proteins in IEC. Linear elution over 16–120 minutes is generally applied between ionic strengths

**Table 3** Properties of Typical Anion and Cation Exchangers Used in Protein Chromatography*

| Property | Anion Exchangers | | | Cation Exchangers | | |
|---|---|---|---|---|---|---|
| | TSKgel DEAE-5PW (Toyo Soda) | Mono-Q (Pharmacia) | SynChropak AX-300 (Synchrom) | TSKgel SP-5PW (Toyo Soda) | Mono-Q (Pharmacia) | SynChropak CM-300 (Synchrom) |
| Support | Macroreticular polymer with hydrophilic surface | Macroreticular polymer with hydrophilic surface | Silica | Macroreticular polymer with hydrophilic surface | Macroreticular polymer with hydrophilic surface | Silica |
| Particle diameter ($\mu m$) | 10 | 10 | 10 | 10 | 10 | 10 |
| Charged ligand | $-CH_2CH_2N(C_2H_5)_2$ | $-CH_2N^+(CH_3)_3$ | $-NHCH_2CH_2$ | $-CH_2CH_2CH_2SO_3^-$ | $-CH_2SO_3^-$ | $-CH_2COO^-$ |
| Operational pH range | 2–12 | 2–12 | 2–8 | 2–12 | 2–12 | 2–8 |
| pKa | 11.3 | 1.4 | NA | 2.5 | 2.6 | NA |
| Binding Capacity (mg/ml) | 30 | 65 | 30 | 40 | 75 | NA |
| Exclusion limit for polyethylene glycol (Da) | 1,000,000 | 500,000 | 150,000 | 1,000,000 | 500,000 | 150,000 |

*Source: Ref 109.

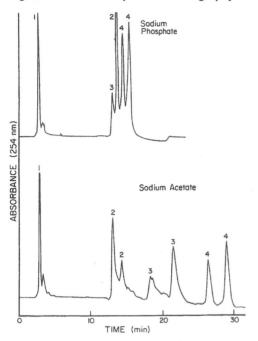

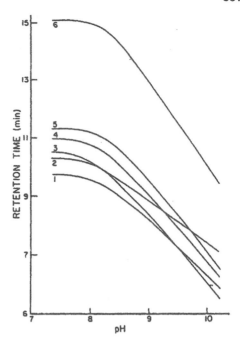

**Fig. 20** The effect of different displacer salts on the the retention of proteins separated by weak anion-exchange chromatography. Column, SynChropak AX-300, 25 cm × 4.1 mm ID, 6.5 μm particle size, 30 nm pore size (SynChrom, Linden, IN). Conditions; Linear gradient from 0–1 M salt over 30 minutes at a flow rate of 1 ml/min, detected at 254 nm. Top panel = sodium phosphate, lower panel = sodium acetate. Proteins; 1 = myoglobin, 2 = conalbumin, 3 = ovalbumin, 4 = b-lactoglobulins B and A (Reprinted with permission from Ref. 112. Copyright CRC Press, Boca Raton, Florida).

**Fig. 21** The effect of pH on the retention of lysozymes in cation-exchange chromatography. Column; SynChropak CM-300, 25 cm × 4.1 mm ID, 6.5 μm particle size, 30 nm pore size (SynChrom, Linden, IN). Conditions; linear gradient from 0–1 M sodium chloride in 10 mM phosphate buffer (pH < 8) or 10 mM borate buffer (pH > 8), at a flow rate of 1 ml/min over 20 minutes. Proteins; 1 = hen egg white lysozyme, 2 = japanese quail, 3 = ring necked pheasant. 4 = duck A, 5 = duck B, 6 = duck C (Reprinted from Ref. 104, with kind permission from Elsevier Science—NL, Sara Burgerhartstraat 25, 1055 KV Amsterdam, The Netherlands).

ranging from 0–0.5 M salt. Buffer concentrations usually range between 20–50 mM.

A number of additional materials can be added to the mobile phase to further enhance selectivity. For example, hydrophobic interactions may contribute to peptide, and protein retention in IEC due to the nature of the stationary phase material (113–114). It has been reported that solutes cannot be eluted with some ion-exchangers without the addition of acetonitrile or methanol to the mobile phase (113–115). The percentage organic

modifier is usually in the range of 10–40%. Higher levels of solvent can cause salt precipitation, and may also affect protein conformation.

## C. Peptides

For peptide applications where incomplete separation is observed in RPC, ion-exchange chromatography represents a very useful alternative separation mode (116). At neutral pH, basic peptides can

**Table 4**  Buffers Commonly Used in IEC

| pH range | Buffer |
|----------|--------|
| **Cation Exchange** | |
| 1.5–2.5 | Maleic acid |
| 2.6–3.6 | Citric acid |
| 3.6–4.3 | Lactic acid |
| 4.8–5.2 | Acetic acid |
| 5.0–6.0 | Malonic acid |
| 6.7–7.6 | Phosphate |
| 7.6–8.2 | HEPES |
| | |
| **Anion Exchange** | |
| 4.5–5.0 | N-methyl piperazine |
| 5.0–6.0 | Piperazine |
| 5.8–6.4 | bis-Tris |
| 7.3–7.7 | Triethanolamine |
| 7.6–8.0 | Tris |
| 8.5–9.0 | 1,3-diaminopropane |
| 9.5–9.8 | Piperazine |

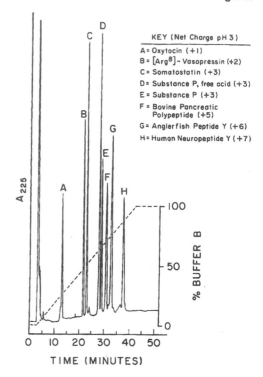

**Fig. 22**  Separation of a series of peptides by cation-exchange chromatography. Column; PolySULFOETHYL Aspartamide, 20 cm × 4.6 mm ID, 5 μm particle size, 30 nm pore size (PolyLC, Columbia, MD). Conditions; linear gradient from 0–0.25 M potassium chloride in 5 mM potassium phosphate pH 3 and 25% acetonitrile at a flow rate of 0.7 ml/min. The peptides vary in net charge from $+1 \rightarrow +7$ as indicated in the figure (Reprinted from Ref. 116, with kind permission from Elsevier Science—NL, Sara Burgerhartstraat 25, 1055 KV Amsterdam, The Netherlands).

be separated by cation-exchange chromatography, and acidic peptides can be analysed by anion-exchange chromatography. However, ion exchange of peptides is more commonly carried out at acidic pH, in the range 2.5–3.0, where most peptides are positively charged and hence cation-exchange is applicable. At this pH range, the negative charges associated with aspartate, and glutamate residues and the C-terminus are neutralized, while arginine, lysine, histidine residues, and the N-terminus are positively charged.

Commonly used solvents for peptide IEC are usually based on phosphate buffers with NaCl or KCl as the displacer ion. For peptides up to approximately 50 residues in length with no significant secondary structure, retention is governed by the number of positive charges as illustrated in Fig. 22 which shows the separation of eight peptides ranging in overall net charge of $+1 \rightarrow +7$ (116). Thus peptides differing by a single charge are generally well-resolved. However peptides with the same charge but different amino acid composition can also be separated due to differences in overall charge density. In addition, hydrophobic interactions may also contribute to the retention of peptides with ion-exchange resins (117). In these cases organic solvent can be added to the mobile phases to further modulate selectivity. 10–40% v/v of either methanol or acetonitrile can be used, and an example of the selectivity changes with the addition of acetonitrile is shown in Fig. 23 for a series of peptides separated by strong cation-exchange chromatography (118).

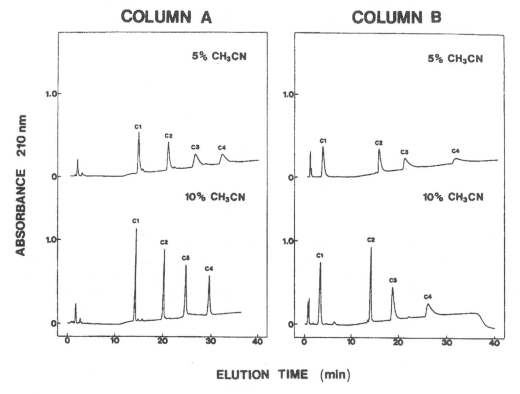

**Fig. 23** The effect of organic solvent on the retention of peptides separated by cation-exchange chromatography. Column A: PolySULFOETHYL Aspartamide, 20 cm × 4.6 mm ID, 5 mm particle size, 30 nm pore size (PolyLC, Columbia, MD). Column B; Mono S HR 5/5 (Pharmacia, Uppsala, Sweden). Conditions, linear gradient from 0-1M sodium chloride 5 mM potassium phosphate, pH 6.5, at a flow rate of 1 ml/min, 26C. Peptide sequences: C1 = Ac-GGGLGGAGGLK-amide, C2 = Ac-KYGLGGAGGLK-amide. C3 = Ac-GGALKALKGLK-amide, C4 = Ac-KYALKALK-GLK-amide (Reprinted with permission from Ref. 118. Copyright CRC Press, Boca Raton, Florida).

In summary, high performance ion-exchange chromatography continues to be an important technique for the analysis and purification of proteins under mild non-denaturing conditions, and also provides a very useful selectivity alternative for the analysis of peptide samples.

## VI. SIZE EXCLUSION CHROMATOGRAPHY

Size exclusion chromatography (SEC) is frequently used as the first step in the isolation of proteins from complex mixtures where separation is carried out according to molecular size and shape (119, 120). SEC can also be used for desalting samples through buffer exchange, and has also found application in the analysis of peptides (121). SEC has also been established as a physicochemical tool for estimating molecular size, and shape of proteins (122, 123) and has provided insight into protein folding mechanisms by monitoring changes in protein size as a function of changes in the concentration of chemical denaturants (66, 124, 125).

## A. Theoretical Considerations

Separation in SEC is based on differences in molecular size in solution. Porous stationary phases are used with defined pore diameters, and elution conditions are used which minimise interaction between the solute molecules, and the stationary phase material. The larger the molecule the smaller the amount of accessible pore volume. Molecules which are larger than the largest pore diameter cannot penetrate into the stationary phase pores, and pass through the column with the fastest retention time. These molecules are eluted with $V_0$, while the smallest molecule is eluted in the $V_t$, the total volume of the column. $V_t$ is the sum of the void volume, and the interstitial volume $V_i$, i.e.,

$$V_t = V_0 + V_i \qquad (12)$$

Solute elution volume in SEC is denoted by $V_e$ which should be between $V_0$ and $V_t$ and can be expressed as follows

$$V_e = V_0 + K_d V_i \qquad (13)$$

where $K_d$ is the distribution coefficient which defines the fraction of internal volume which is accessible to the protein solute.

## B. Experimental Conditions

The packing materials available for high performance SEC are generally silica-based or polymeric (119, 126). The pore diameter of SEC supports determines the exclusion limits of the material. Columns are characterized in terms of the molecular weight range which can be adequately separated which is dependent on the pore diameter. Generally pore diameter ranges between $100 - 500\,\text{Å}$. The pore volume is also an important property of an SEC material which must be sufficiently large to provide a high peak capacity, i.e., the ability to separate seven peaks with a resolution of one. Column efficiency in high performance SEC supports is particularly important as solutes are eluted iso-cratically and therefore do not exhibit band sharpening which occurs with gradi-

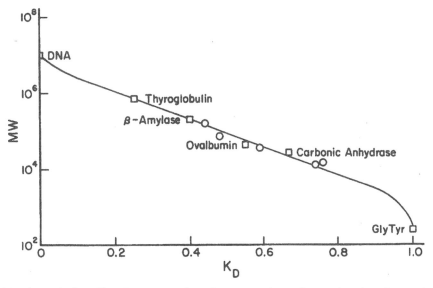

**Fig. 24** A typical calibration curve for the separation of proteins by size exclusion chromatography. Column; SynChropak GPC300, 30 cm × 7.8 mm ID, 5 mm particle size, 30 nm pore size (SynChrom, Linden, IN). Conditions; 0.1 M potassium phosphate, pH 7, flow rate 1 ml/min (Reprinted with permission from Ref. 126. Copyright CRC Press, Boca Raton, Florida)

ent elution. SEC supports generally have particle diameters between 5–10 $\mu$m.

The hydrodynamic shape and volume rather than molecular weight *per se* is the physical property of proteins which causes separation in SEC. In order to achieve accurate estimations of molecular weight, a column must be calibrated with molecules which have the same overall shape. Under ideal conditions, $K_d$ will be proportional to the logarithm of the molecular weight of the protein as illustrated by a typical calibration curve shown in Fig. 24 for the protein separation shown in Fig. 25 (119). While the majority of applications of SEC have involved the analysis, and purification of proteins,

SEC has also recently been used in the analysis of peptides. An example of the high level of resolution which can be achieved for peptides is shown in Fig. 26, which demonstrates excellent separation in the MW range of 75–12,500Da (127).

Mobile phase selection is important in SEC to minimize non-specific interactions between the support and the solutes and also to avoid mobile phase induced changes in solute molecular shape. Both ionic and hydrophobic interactions can also contribute to the elution volume of proteins in SEC as most packings are weakly acidic (due to residual negative charges), and/or mildly hydrophobic (128). Ionic interactions can be minimized by increasing the ionic strength of the mobile phase through the addition of up to 0.5 M NaCl. However, increasing ionic strength also results in the enhancement of hydrophobic interactions. Thus a balance is necessary to minimize both undesired interactions. A common mobile phase composition which is employed for SEC of proteins is phosphate buffer at pH7 with ionic strength of 0.05–0.1 Thus by optimizing pH, and ionic strength, secondary interactions can be almost excluded. However, ionic or hydrophobic interactions can also be used to manipulate selectivity of protein separations. The example shown in Fig. 27 clearly demonstrates the effect of salt on the separation of peptides at two different pHs (129).

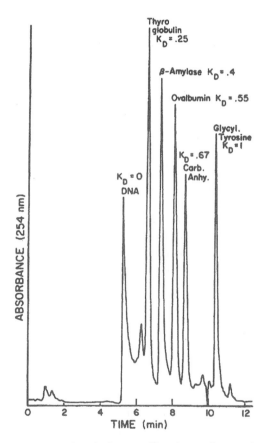

**Fig. 25** The elution profile of proteins used to create the calibration curve shown in Fig. 22 (Reprinted with permission from Ref. 126. Copyright CRC Press, Boca Raton, Florida).

## VII. FUTURE PROSPECTS

The number of applications of HPLC in peptide and protein purification continue to expand at an extremely rapid rate. Solid phase peptide synthesis and recombinant DNA techniques have allowed the production of large quantities of peptides and proteins which need to be highly purified. The design of multidimensional purification schemes to achieve high levels of product purity highlight the power of HPLC

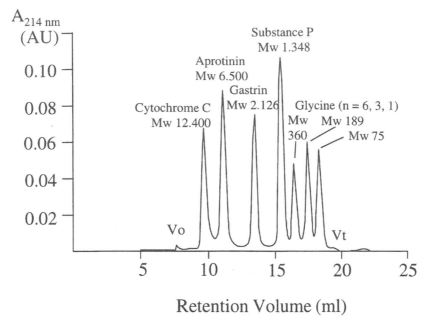

**Fig. 26** The separation of peptides by size exclusion chromatography. Column; Superdex Peptide HR 10/30 (Pharmacia, Uppsala, Sweden). Conditions; 0.25 M NaCl in 0.02 M phosphate buffer pH 7.2, flow rate 0.25 ml/min. (from Ref 127, courtesy of Amersham Pharmacia Biotech).

techniques in the production of peptide, and proteins-based therapeutics. One such example is shown in Fig. 28 where the judicious use of SEC, IEC and RPC resulted in the efficient purification of murine epidermal growth factor (130) whereby manipulation of the sample between stages has been minimized and selectivity has been maximized.

Following purification, mass spectrometry can be used to confirm the structural identity of synthetic peptides, while recombinant proteins require further structural analysis by high resolution analytical fingerprinting to confirm the amino acid sequence. RPC will therefore continue to be the central method for the characterization, and quality control analysis of synthetic peptides and recombinant proteins. Moreover, the coupling of mass spectrometry to allow on-line identification of samples will become routine, and continue to expand

the analytical power of HPLC. Other coupled techniques such as LC-CE (capillary electrophoresis) and LC-biosensor will also allow more rapid on-line analysis of sample purity and bioactivity.

Other areas of separation technology in which significant advances are anticipated to emerge are in the areas of miniaturisation and high speed analysis to allow the efficient purification of femtomolar to attomolar levels of material. These techniques will have important impact in the discovery of new bioactive peptides and novel proteins as potential candidates for new therapeutics.

The development of fully mechanistic models which describe the underlying thermodynamic and kinetic processes involved in the interaction of peptides and proteins with hydrophobic, and charged stationary phase surfaces has lagged behind the frenetic pace in the growth of HPLC applications. This is pre-

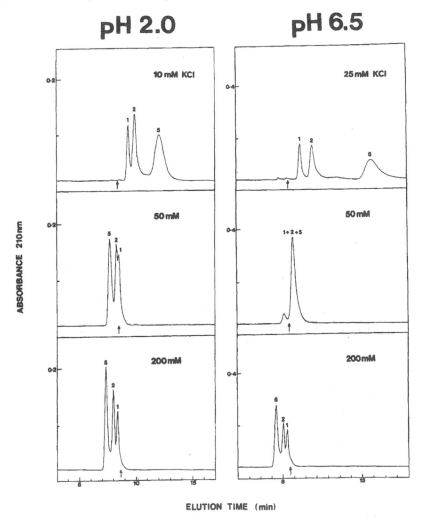

**Fig. 27** Effect of salt on non-specific interactions in size exclusion chromatography at pH 2 (left), and pH 6.5 (right). SynChropak GPC60, 30 cm × 7.8 mm ID, 5 mm particle size, 30 nm pore size (SynChrom, Linden, IN). Conditions; left—0.1% TFA containing 10, 50 or 200 mM potassium chloride; right—5 mM $KH_2PO_4$ containing 25, 50 or 200 mM potassium chloride, flow rate 1 ml/min Peptide sequences: 1 = AcGLGAKGAGVG-Amide, 2 = Ac(GLGAKGAGVG)$_2$-amide, 3 = Ac(GLGAKGAGVG)$_5$- amide (Reprinted with permission from Ref. 128. Copyright CRC Press, Boca Raton, Florida).

dominantly due to the complex nature of peptide and protein structure. However, both spectroscopic and molecular modelling techniques are starting to provide detailed molecular information on peptide and protein surface interactions which will not only allow further advances in the rational design of new sorbent materials but will further establish HPLC as a physicochemical tool for the analysis of peptide, and protein-surface interactions.

Finally, there has been a strong synergy between the biotechnology industry and the field of biomacromolecular HPLC as several significant recent advances in separation technology have been driven

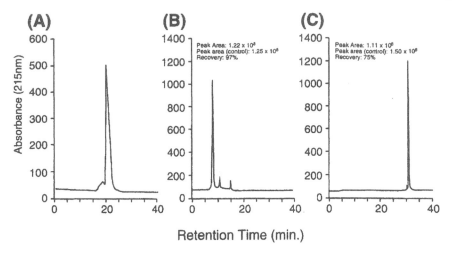

**Fig. 28** Multidimensional micropreparative HPLC of murine epidermal growth factor (mEGF), using size exclusion, anion-exchange, and reversed phase chromatography. (A) mEGF (4.5 $\mu$g, 750 pmol) was chromatographed on a Superose 12 PC 3.2/30 SEC column (Pharmacia, Uppsala, Sweden), 30 cm × 3.2 mm ID, using a mobile phase of 1% ammonium bicarbonate/0.02% Tween 20 at a flow rate of 0.1 ml/min. (B) The major peak in (A) was collected and applied to a Mono-Q PC 1.6/5 anion-exchange column (Pharmacia), 5 cm × 1.6 mm ID, using a linear gradient from 0-1M sodium chloride in 20 mM Tris, pH 7.5 over 50 minutes at a flow rate of 0.1 ml/min. (C) The major peak from (B) was collected and further purified by RPC using a Brownlee RP-300 column, 30 cm × 2.1 mm ID, and a linear gradient from 0-60% acetonitrile in 0.15% TFA over 60 minutes at a flow rate of 0.1 ml/min. The integrated peak areas and the calculated recovery between stages is indicated in (B) and (C) (Reproduced with permission from Ref. 130. Copyright John Wiley and Sons Limited).

by the stringent and continually-evolving regulatory requirements of the biotechnology industry. HPLC will therefore continue to be at the heart of the analytical techniques with which scientists in both academia and the pharmaceutical industry must arm themselves to be able to fully characterize the identity, purity and potency of peptides and proteins.

## REFERENCES

1.  Methods in Enzymology, vol 270, 1996, entire volume.
2.  Methods in Enzymology, vol 271, 1996, entire volume.
3.  CT Mant, RS Hodges. High Performance Liquid Chromatography of Peptides and Proteins. In: CT Mant, RS Hodges, eds. Separation, Analysis and Conformation. Boca Raton, Fl.: CRC Press, 1991.
4.  MTW Hearn. HPLC of Proteins, Peptides and Polynucleotides. In: MTW Hearn, ed. Contemporary Topics and Applications. New York: VCH, 1991.
5.  KM Gooding, FE Regnier. HPLC of Biological Macromolecules. In: KM Gooding, FE Regnier, eds. Methods and Applications. New York: Dekker, 1990.
6.  MI Aguilar, MTW Hearn. High resolution reversed phase high performance liquid chromatography of peptides and proteins. Methods Enzymol 270: 3–26, 1996.
7.  CT Mant, RS Hodges. Analysis of peptides by high performance liquid chromatography. Methods Enzymol 271: 3–50, 1996.
8.  C Miller, J Rivier. Peptide Chemistry: Development of High Performance Liquid Chromatography and Capillary Zone Electrophoresis. Bioploymers 40: 265–318, 1996.
9.  ER Hoff, RC Chloupek. Analytical peptide mapping of recombinant DNA-derived proteins by reversed phase high perform-

ance liquid chromatography. Methods Enzymol 271: 51–67, 1996.

10. E Nice. Micropreparative High Performance Liquid Chromatography of Proteins and Peptides: Principles and Applications. Bioploymers 40: 319–341, 1996.

11. Pharmacia Biotech. Efficient isolation of protein fragments for structure analysis. ÄKTA™purifier Application Note No 18-1119-53, 1996.

12. AW Purcell, MI Aguilar, MTW Hearn. Conformational effects in the RP-HPLC of polypeptides. II: The role of insulin A and B chains in the chromatographic behaviour of insulin, J Chromatogr 711: 71–79, 1995.

13. KL Richards, MI Aguilar, MTW Hearn. The effect of protein conformation on experimental bandwidths in RP-HPLC, J Chromatogr 676: 33–41, 1994.

14. P Oroszlan, S Wicar, G Teshima, S-L Wu, WS Hancock, BL Karger. Conformational effects in the reversed phase chromatographic behaviour of recombinant human growth hormone (rhGH) and N-methionyl recombinant human growth hormone (met-hGH). Anal Chem 64: 1623–1631, 1992.

15. S Lin, BL Karger. Reversed phase chromatographic behaviour of proteins in different unfolded states. J Chromatogr 499: 89–102, 1990.

16. E Lazoura, J Maidonis, E Bayer, MTW Hearn, MI Aguilar. The conformational analysis of NPY-[18–36] analogues at hydrophobic surfaces, Biophys J 72: 238–246, 1997.

17. T-Z Lee, PT Thompson, MTW Hearn, MI Aguilar. Conformational stability of a type II′ β-turn motif in human growth hormone [6–13] peptide analogues at hydrophobic surfaces, J Pep Res 49: 394–403, 1997.

18. AW Purcell, MI Aguilar, REH Wettenhall, MTW Hearn. The induction of amphipathic helical structures in peptides during RP-HPLC, Peptide Res 8: 160–170, 1995.

19. SE Blondelle , RA Houghten. Design of model peptides having potent antimicrobial activities. Biochemistry 31: 12688–12694, 1992.

20. SE Blondelle, K Buttner, RA Houghten. Evaluation of peptide-peptide interactions using reversed phase high performance liquid chromatography. J Chromatogr 625: 199–206, 1992.

21. NE Zhou, CT Mant, RS Hodges. Effect of preferred binding domains on retention behaviour in reversed-phase chromatography. Pept Res 3: 8–20, 1990.

22. RS Hodges, B-Y Zhu, NE Zhou, CT Mant. Reversed -phase liquid chromatography as a useful probe of hydrophobic interactions involved in protein folding and protein stability. J Chromatogr 676: 3–15, 1994.

23. C Horvath, W Melander, I Molnar. Solvophobic interactions in liquid chromatography with non-polar stationary phases. J Chromatogr 125: 129–156, 1976.

24. L Snyder. Gradient Elution. In: C Horvath, ed. HPLC-Advances and Perspectives, New York: Academic Press, 1980, vol 1, pp 207–316.

25. MA Stadalius, HS Gold, LR Snyder. Optimisation model for the gradient elution separation of peptide mixtures by reversed phase high performance liquid chromatography. J Chromatogr 296: 31–59, 1984.

26 LR Snyder, Automated method development in high performance liquid chromatography. Meth Enzymol 270: 151–175, 1996.

27. MI Aguilar, AN Hodder, MTW Hearn. Studies on the optimisation of the reversed phase gradient elution of polypeptides: Evaluation of retention relationships with β-endorphin-related polypeptides. J Chromatogr., 327: 115-138, 1985.

28. D Guo, CT Mant, AK Taneja, RS Hodges. Prediction of peptide retention times in reversed phase high performance liquid chromatography. II. Correlation of observed and predicted peptide retention times and factors influencing the retention times of peptides. J Chromatogr 359: 519–532, 1986.

29. MCJ Wilce, MI Aguilar, MTW Hearn. The physicochemical basis of hydrophobicity scales: Evaluation of four new sets of amino acid coefficients derived from reversed phase high performance liquid chromatography, Anal Chem 67: 1210–1219, 1995.

30. FE Regnier. The role of protein structure in chromatographic behaviour. Science 238: 319–323, 1987.

31. KL Richards, MI Aguilar, MTW Hearn. A comparative study of the retention behaviour and stability of cytochrome c in RP-HPLC, J Chromatogr 676: 17–31, 1994.

32. AW Purcell, MI Aguilar, MTW Hearn. Conformational effects in the RP-HPLC of polypeptides. I: The resolution of insulin variants. J Chromatogr 711: 61–70, 1995.

33. KK Unger, B Anspach, R Janzen, G Jilge, KD Lork. Bonded Silica Phases for the Separation of Biopolymers by Means of Column Liquid Chromatography. In C Horvath, ed. HPLC-Advances and perspectives, New York: Academic Press, 1988, vol 5, pp 2–93.

34  KK Unger. Silica as a Support. In: KM Gooding, FE Regnier, eds. HPLC of Biological Macromolecules: Methods and Applications. New York: Marcel Dekker, 1990, pp 3–24.

35.  M Henry. Design requirements of silica-based matrices for biopolymer chromatography. J Chromatogr 544: 413–443, 1991.

36.  NE Zhou, CT Mant, JJ Kirkland, RS Hodges. Comparison of silica-based cyanopropyl and octyl reversed phase packings for the separation of peptides and proteins. J Chromatogr 548: 179–193, 1991.

37.  I Yarovsky , MI Aguilar, MTW Hearn. Influence of the chain length and surface density on the conformation and mobility of n-alkyl ligands chemically immobilised to a silica surface, Anal Chem 67: 2145–2153, 1995.

38.  I Yarovsky, MTW Hearn, MI Aguilar. Molecular simulation of peptide interactions with an RP-HPLC sorbent, J Phys Chem B, 101: 10962–10970, 1997.

39.  K Albert, E Bayer. Characterisation of bonded phases by solid-state NMR spectroscopy. J Chromatogr 544: 345–370, 1991.

40.  FJ Moy, Y-C Li, P Rauenbeuhler, ME Winkler, HA Scheraga, GT Montelione. Solution structure of human type-α transforming growth factor determined by heteronuclear NMR spectroscopy and refined by energy minimisation with restraints. Biochemistry 32: 7334–7353, 1993.

41.  MA Chlenov, EI Kandyba, LV Nagornaya, IL Orlova, YV Volgin. High performance liquid chromatography of human glycoprotein hormones. J Chromatogr. 631: 261–267, 1993.

42.  BS Welinder, HH Sorenson, B Hansen. Reversed-phase high performance liquid chromatography of insulin. Resolution and recovery in relation to column geometry and buffer components. J Chromatogr 361: 357–363, 1986.

43.  M Hanson, KK Unger, CT Mant, RS Hodges. Polymer-coated reversed-phase packings with controlled hydrophobic properties. I. Effect on the selectivity of protein separations. J Chromatogr 599: 65–76, 1992.

44.  M Hanson, KK Unger, CT Mant, RS Hodges. Polymer-coated reversed-phase packings with controlled hydrophobic properties. II. Effect on the selectivity of peptide separations J Chromatogr 599: 77–86, 1992.

45.  N Tanaka, K Kimata, Y Mikawa, K Hosoya, T Araki, Y Ohtsu, Y Shiojima, R Tsuboi, H Tsuchiya. Performance of wide-pore silica- and polymer-based packing materialsin polypeptide separation: effect of pore size and alkyl chain length. J Chromatogr 535: 13–32, 1990.

46.  BS Welinder. Use of polymeric reversed-phase columns for the characterisation of polypeptides extracted from human pancreata. II. Effect of the stationary phase. J Chromatogr 542: 83–99, 1991.

47.  H-J Wirth, K-O Eriksson, P Holt, MI Aguilar, MTW Hearn. Ceramic based particles as chemically stable chromatographic supports. J Chromatogr 646: 129–141, 1993.

48.  L Sun, PW Carr. Chromatography of proteins using polybutadiene-coated zirconia. Anal Chem 67: 3717–3721, 1995.

49.  G Jilge, R Janzen, KK Unger, KN Kinkel, MTW Hearn. Evaluation of advanced silica packings for the separation of biopolymers by high performamce liquid chromatography. III. Retention and selectivity of proteins and peptides in gradient elution on non-porous monodisperse 1.5 $\mu$m reversed phase silicas. J Chromatogr 397: 71–80, 1987.

50.  NB Afeyan, NF Gordon, I Mazsaroff, L Varady, SP Fulton, YB Yang, FE Regnier. Flow-through particles for the high-performance liquid chromatographic separations of biomolecules: perfusion chromatography. J Chromatogr 519: 1–29, 1990.

51.  SK Paliwal, M deFrutos, FE Regnier. Rapid separations of Proteins by liquid chromatography. Meth Enzymol 270: 133–151, 1996.

52.  PM Young, TE Wheat. Optimisation of high performance liquid chromatographic peptide separations with alternative mobile and stationary phases. J Chromatogr 512: 273–281, 1990.

53.  DJ Poll, DRK Harding. Formic acid as a milder alternative to trifluoroacetic acid and phosphoric acid in two dimensional peptide mapping. J Chromatogr 469: 231–239, 1989.

54.  G Thevenon, FE Regnier. Reversed-phase liquid chromatography of proteins with strong acids. J Chromatogr 476: 499–511, 1989.

55.  CT Mant, RS Hodges. The Effects of Anionic Ion-pairing Reagents on Peptide Retention in Reversed Phase Chromatography. In: CT Mant, RS Hodge, eds. High Performance Liquid Chromatography of Peptides and Proteins: Separation, Analysis and Conformation.

Boca Rato, Fl: CRC Press, 1991, pp 327–341.

56. GW Welling, R Van der Zee, S Welling-Wester. Column liquid chromatography of integral membrane proteins. J Chromatogr 418: 223–243, 1987.

57. MI Aguilar, S Mougos, J Boublik, J Rivier, MTW Hearn. The effect of D-amino acid substitutions on the RP-HPLC behaviour of neuropeptide-Y [18–36] analogues, J Chromatogr 646: 53–65, 1993.

58. RL Moritz, RJ Simpson. Application of capillary reversed phase high performance liquid chromatography to high senstivity protein sequence analysis. J Chromatogr 599: 119–130, 1992.

59. RL Moritz, RJ Simpson. Capillary liquid chromatography. In: K Imahori, F Sakiyama, eds. A tool for protein structural analysis, in Methods in Protein Sequence Analysis. New York: Plenum Press, pp 3–10, 1993.

60. BFD Ghrist, LR Snyder, BS Cooperman. In: KM Gooding, FE Regnier, eds. HPLC of Biological Macromolecules: Methods and Applications. New York: Marcel Dekker, 1990, pp 403–427.

61. RJ Simpson, RL Moritz, EC Nice, B Grego. A high performance liquid chromatography procedure for recovering subnanomole amounts of protein from SDS-gel electroeluates for gas-phase sequence analysis. Eur J Biochem 165: 21–29, 1987.

62. J Frank, A Braat, JA Duine. Assessment of protein purity by chromatography and multiwavelength detection. Anal Biochem 162: 65–73, 1987.

63. F Nyberg, C Pernow, U Moberg, RB Eriksson. High performance liquid chromatography and diode array detection for the identification of peptides containing aromatic amino acids in studies of endorphin-degrading activity in human cerebrospinal fluid. J Chromatogr 359: 541–551, 1986.

64. GP Rozing. Diode array detection. Meth Enzymol 270: 201–234, 1996.

65. XM Lu, K Benedek, BL Karger. Conformational effects in the high-performance liquid chromatography of proteins. Further studies of the reversed phase chromatographic behaviour of ribonuclease A. J Chromatogr 359: 19–29, 1986.

66. M Fridman, MI Aguilar, MTW Hearn. A comparative study of the equilibrium refolding of bovine, porcine and human growth hormone using size exclusion chromatography. J Chromatogr 512: 57–75, 1990.

67. SA Carr, ME Hemling, MF Bean, GD Roberts. Integration of mass spectrometry in analytical biotechnology. Anal Chem 63: 2802–2824, 1991.

68. A Apffel, J Chakel, S Udiavar, WS Hancock, C Souders, E Pungor Jr. Application of capillary electrophoresis, high performance liquid chromatography, on-line electrospray mass spectrometry and matrix-assisted laser desorption ionization-time of flight mass spectrometry to the characterization of single-chain plasminogen activator. J Chromatogr A, 717: 41–60, 1995.

69. J Ding, W Burkhart, DB Kassel. Identification of phosphorylated peptides from complex mixtures using negative-ion orifice-potential stepping and capillary liquid chromatography/electrospray ionisation mass spectrometry. Rapid Commun Mass Spectrom 8: 94–98, 1994.

70. Z Randhawa, HE Witkowska, J Cone, JA Wilkins, P Hughes, K Yamanishi, S Yashuda, Y Masui, P Arthur, C Kletke, F Bitsch, CHL Shackleton. Incorporation of norleucine at methionine positions in recombinant human macrophage colony stimulating factor (m-CSF, 4-153). Biochemistry 33: 4352–4362, 1994.

71. K Mock, M Hail, I Mylchreest, J Zhou, K Johnson, I Jardine. Rapid high sensitivity peptide mapping by liquid chromatography-mass spectrometry. J Chromatogr 646: 169–174, 1993.

72. DL Smith, ZR Zhou. Strategies for locating disulphide bonds in proteins. Meth Enzymol 193: 374–389, 1990.

73. DF Hunt, RA Henderson, J Shabanowitz, K Sakaguchi, H Michel, N Sevilir, AL Cox, E Appella, V H Engelhard. Characterisation of peptides bound to the class I MHC molecule HLA-A2.1 by mass spectrometry. Science. 255: 1261–1263, 1992.

74. SL Wu, B Karger. Hydrophobic Interaction Chromatography of Proteins. Meth Enzymol 270: 27–47, 1996.

75. MI Aguilar, MTW Hearn, Reversed phase and hydrophobic interaction chromatography of proteins', in M T W Hearn (ed), HPLC of Proteins, Peptides and Polynucleotides—Contemporary Topics and Applications. Deerfield, FL: VCH 1991, pp 246–290.

76. RA Ingraham. Hydrophobic interaction chromatography of proteins in CT Mant, RS Hodges, eds. High Performance Liquid Chromatography of Peptides and Proteins: Separation, Analysis and Conformation. Boca Rato, Fl: CRC Press, 1991, pp 425–435.

77. WR Melander, D Corradini, C Horvath. Salt-mediated retention of proteins in hydrophobic interaction chromatography. J Chromatogr 317: 67–85, 1984.

78. W Melander, C Horvath. Salt effects on hydrophobic interactions in precipitation and chromatography of proteins: an interpretation of the lyotropic series. Arch Biochem Biophys 183: 200–215, 1977.

79. A Katti, Y-F Maa, Cs Horvath. Protein surface area and retention in hydrophobic interaction chromatography. Chromatographia, 24:646–650, 1987.

80. DL Gooding, MN Schmuck, MP Nowlan, KM Gooding. Optimisation of preparative hydrophobic interaction chromatographic purification methods. J Chromatogr 359: 331–337, 1986.

81. JL Fausnaugh, FE Regnier. Solute and mobile phase contributions to retention in hydrophobic interaction chromatography of proteins. J Chromatogr 359: 131–146, 1986.

82. BF Roettger, JA Myers, MR Ladisch, FE Regnier. Mechanisms of Protein Retention in Hydrophobic Interaction Chromatography. In: MR Ladisch, RC Willson, CC Painton, SE Builder, eds. Protein Purification: From Molecular Mechanisms to Large Scale Processes. Washington DC: American Chemical Society, 1990, pp 80–92.

83. S Hjertén, K Yao, K-O Ericksson, B Johansson. Gradient and isocratic high-performance-hydrophobic interaction chromatography of proteins on agarose columns. J Chromatogr 359: 99–109, 1996

84. DB Wetlaufer, MR Koenigbauer. Surfactant-mediated protein hydrophobic interaction chromatography. J Chromatogr 359: 55–60, 1986.

85. JJ Buckley, DB Wetlaufer. Use of the surfactant 3-(3-cholamidopropyl)-dimethylammoniopropane sulfonate in hydrophobic interaction chromatography of proteins. J Chromatogr 464: 61–71, 1989.

86. Y Kato, T Kitamura, T Hashimoto. Operational variables in high performance hydrophobic interaction chromatography of proteins on TSK Gel Phenyl-5PW. J Chromatogr 298: 407–414, 1984.

87. ML Heinitz, L Kennedy, W Kopaciewicz, FE Regnier. Chromatography of proteins on hydrophobic interaction and ion-exchange chromatographic matrices: mobile phase contributions to selectivity. J Chromatogr 443: 173–182, 1988.

88. RH Ingraham, SYM Lau, AK Taneja, RS Hodges. Denaturation and the effects of temperature on hydrophobic interaction and reversed phase high performance liquid chromatography of proteins: Biogel-TSK-Phenyl-5PW column. J Chromatogr 327: 77–92, 1985.

89. S-L Wu, K Benedek, B L Karger. Thermal behaviour of proteins in high performance hydrophobic-interaction chromatography. On-line spectroscopic and chromatographic characterisation. J Chromatogr 359: 3–17, 1986.

90. S-L Wu, A Figueroa, BL Karger. Protein conformational effects in hydrophobic interaction chromatography. Retention characterisation of and the role of mobile phase additives and stationary phase hydrophobicity. J Chromatogr 371, 3–27, 1986.

91. G Choudhary, C Horvath. Ion-Exchange Chromatography, Meth Enzymol 270: 47–82, 1996.

92. MI Aguilar, AN Hodder, MTW Hearn. High performance ion-exchange chromatography of proteins. In: MTW Hearn, ed. HPLC of Proteins, Peptides and Polynucleotides—Contemporary Topics and Applications, Deerfield, FL: VCH 1991, pp 199–245.

93. NK Boardman, SM Partridge, Separation of neutral proteins on ion-exchange resins. Biochem J 59: 543–552, 1955.

94. W Kopaciewicz, MA Rounds, J Fausnaugh, FE Regnier, Retention model for hig-performance ion-exchange chromatography. J Chromatogr 266: 3–21, 1983.

95. A Velayudhan, C Horvath. On the stoichiometric model of electrostatic interaction chromatography for biopolymers. J Chromatogr 367: 160–162, 1986.

96. MTW Hearn, AN Hodder, PG Stanton, MI Aguilar, Evaluation of retention and bandwidth relationships for proteins separated by isocratic anion-exchange chromatography. Chromatographia 24: 769–776, 1987.

97. J Stahlberg, B Jonsson. Influence of charge regulation in electrostatic interaction chromatography of proteins. Anal Chem 68: 1536–1544, 1996

98. CM Roth, KK Unger, AM Lenhoff. Mechanistic model of retention in protein ion-exchange chromatography. Reversed phase chromatographic behaviour of proteins in different unfolded states. J Chromatogr A, 726: 45–56, 1996.

99. V Noinville, C Vidal-Madjar, B Sebille. Modelling of protein adsorption on polymer surfaces—computation of adsorption potential. J Phys Chem 99: 1516–1522, 1995.

100. AN Hodder, MI Aguilar, MTW Hearn, The influence of different displacer salts on the retention properties of proteins separated by gradient anion exchange chromatography. J Chromatogr 476: 391–411, 1989.

101. AN Hodder, KJ Machin, MI Aguilar, MTW Hearn. Identification and characterisation of coulombic interactive regions on hen lysozyme by high performance liquid anion-exchange chromatography and computer graphic analysis. J Chromatogr 517: 317–331, 1990.

102. AN Hodder, KJ Machin, MI Aguilar, MTW Hearn. Identification and characterisation of coulombic interactive regions on sperm whale myoglobin by high performance anion-exchange chromatography and computer graphic analysis. J Chromatogr 507: 33–44, 1990.

103. RW Stout, SI Sivakoff, RD Ricker, LR Snyder. Separation of proteins by gradient elution from ion-exchange columns. Optimizing experimental conditions. J Chromatogr 353: 439–463, 1986.

104. JF Pollit, G Thevenon, L Janis, FE Regnier. Chromatographic resolution of lysozyme variants. J Chromatogr 443: 221–228, 1988.

105. W Muller. New ion-exchangers for the chromatography of biopolymers. J Chromatogr 510: 133–140, 1990.

106. MTW Hearn, AN Hodder, FW Fang, MI Aguilar. Retention behaviour of proteins with macroporous tentacle-type anion-exchangers. J Chromatogr 548: 117–126, 1991.

107. J Frenz, J Cachia, CP Quan, MB Sliwkowski, M Vasser. Protein sorting by high performance liquid chromatography of recombinant deoxyribonuclease I on polyionic stationary phases. J Chromatogr 634: 229–239, 1993.

108. FW Fang, MI Aguilar, MTW Hearn, The influence of temperature on the retention behaviour of proteins in cation-exchange chromatography, J Chromatogr 729: 49–66, 1996.

109. Y Kato. Study: Comparison of Ion-Exchange Columns. Tosoh Corporation, Montgomeryville, PA, 1993.

110. W Kopaciewicz, FE Regnier. Mobile phase selection for the high performance ion-exchange chromatography of proteins. Anal Biochem 133: 251–259, 1983.

111. MA Rounds, FE Regnier. Evaluation of a retention model for high performance ion-exchange chromatography using two different displacer salts. J Chromatogr 283: 37–45, 1984.

112. MP Nowlan, KM Gooding. High performance ion-exchange chromatography of proteins. In: CT Mant, RS Hodges, eds. High Performance Liquid Chromatography of Peptides and Proteins: Separation, Analysis and Conformation. CRC Press, Boca Raton, Fl: CRC Press, 1991, pp 203–213.

113. ML Heinitz, L Kennedy, W Kopaciewicz, FE Regnier. Chromatography of proeins on hydrophobic interaction and ion-exchange chromatographic matrices: mobile phase contributions to selectivity. J Chromatogr 443: 173–182, 1988.

114. WR Melander, Z El Rassi, C Horvath. Interplay of hydrophobic and electrostatic interactions in biopolymer chromatography. Effect of salts on the retention of proteins. J Chromatogr 469: 3–27, 1989.

115. F Fang, MI Aguilar, MTW Hearn. Influence of temperature on the retention behaviour of proteins in cation-exchange chromatography. J Chromatogr A, 729: 49–66, 1996.

116. AL Alpert, PC Andrews. Cation-exchange chromatography of peptides on poly(2-sulfoethyl aspartamide)-silica. J Chromatogr 443: 85–96, 1988.

117. TWL Burke, CT Mant, JA Black, RS Hodges. Strong cation-exchange high performance liquid chromatography of peptides. Effect of non-specific hydrophobic interactions and linearization of peptide retention behaviour. J Chromatogr 476: 377–389, 1989.

118. CT Mant, R S Hodges. The Use of Peptide Standards for Monitoring Ideal and Non-ideal Behaviour in Cation-Exchange Chromatography. In: CT Mant, RS Hodges, eds. High Performance Liquid Chromatography of Peptides and Proteins: Separation, Analysis and Conformation. Boca Raton, Fl: CRC Press, 1991, pp 171–185.

119. KM Gooding, FE Regnier. Size Exclusion Chromatography. In: KM Gooding, FE Regnier, eds. HPLC of Biological Macromolecules: Methods and Applications. New York: Marcel Dekker, 1990, pp 47–75.

120. H Engelhardt, UM Schom. Optimal conditions for size exclusion chromatography of proteins. Chromatographia 22: 388–394, 1986.

121. CT Mant, JMR Parker, RS Hodges. Size-exclusion high performance chromatography of peptides. Requirement for peptide standards to monitor column performance and non-ideal behaviour. J Chromatogr 397: 99–112, 1987.

122. M Potschka. Universal calibration of gel permeation chromatography and determination of molecular shape in solution. Anal Biochem 162: 47–64, 1987.

123. M leMaire, A Veil, JV Moller. Size exclusion chromatography and universal calibration of gel columns. Anal Biochem 177: 50–56, 1989.

124. VN Uversky. Use of fast protein size exclusion liquid chromatography to study the unfolding of proteins which denature through the molten globule. Biochemistry 32: 13288–13298, 1993.

125. M Herold, B Leistler. Unfolding of truncated and wild type aspartate aminotransferase studied by size exclusion chromatography. J Chromatogr 539: 383–391, 1991.

126. KM Gooding, HH Freiser. High Performance Size Exclusion Chromatography of Proteins. In: CT Mant, RS Hodges, eds. High Performance Liquid Chromatography of Peptides and Proteins: Separation, Analysis and Conformation. Boca Raton, Fl: CRC Press, 1991, pp 135–144.

127. Pharmacia Biotech. Superdex® Peptide HR 10/30 Superdex Peptide PC 3.2/30. High performance size exclusion columns Data File No 18-1106-06, 1994.

128. B Anspach, HU Gierlich, KK Unger. Comparative study of Zorbax Bio Series GF 250 and GF 450 and TSK-Gel 3000 SW and SWXL columns in size exclusion chromatography of proteins. J Chromatogr 443: 45–54, 1988.

129. CT Mant, RS Hodges. Requirement for Peptide Standards to Monitor Ideal and Non-Ideal Behaviour in Size-Exclusion Chromatography. In: CT Mant, RS Hodges, eds. High Performance Liquid Chromatography of Peptides and Proteins: Separation, Analysis and Conformation. Boca Raton, Fl: CRC Press, 1991, pp 125–134.

130. EC Nice, L Fabri, A Hammacher, K Andersson, U Hellman. Micropreparative ion-exchange HPLC. Applications to microsequence analysis. Biomed Chromatogr 7: 104–111, 1993.

# 12

# Practical Aspects of Peptide and Protein Analysis by Capillary Electrophoresis

**Nicole J. Munro, Andreas Hühmer\* and James P. Landers**
*University of Pittsburgh, Pittsburgh, Pennsylvania*

There is a growing interest in the use of capillary electrophoresis (CE) for the analysis of biopolymers, including peptides and proteins. The ability to obtain rapid, automated, high efficiency separations under native or denaturing conditions has made CE an attractive alternative to conventional gel electrophoresis and complementary to HPLC methods. This chapter will provide an overview of peptide and protein analysis via a number of CE-based modes. Within the framework of discussing several exemplary peptide separations, we will emphasize the importance of buffer pH, ionic strength, organic additives, surfactants and capillary coatings in altering selectivity and optimizing resolution. With regard to protein separations, we will discuss the importance of buffer choice, complexing reagents, capillary coatings and dicationic alkyl additives in effecting high resolution separations.

## I. INTRODUCTION

Protein separations provide a unique challenge to CE. While similar in prin-

ciple to slab gel electrophoresis, electrophoresis in a capillary is, from a practical point of view, a great departure from the traditional method. Unlike most of the separations in the slab gel format, researchers are drawn to CE for its ability to electrophorese protein in free solution under nondenaturing conditions. This presents the ability to electrophoretically resolve proteins varying in size and charge in complex mixtures or to resolve closely-related glycoforms of a microheterogeneous glycoprotein. While powerful, CE is fraught with challenges not encountered with its slab gel counterpart. These include the predilection of proteins for interacting with the capillary wall or with other proteins under nondenaturing conditions, and having to optimize for separation based on charge-to-mass ratio and not just mass. However, there are a number of well-defined steps that can be employed to deal with these aspects of CE. These will be covered here with specific reference to reports in the literature. These sections will illustrate that, once solved, the power of CE for protein analysis is unparalleled for application to diverse areas such as vaccine production, epitope mapping and glycoform analysis.

\* *Current affiliation*: Texas Instruments Incorporated, Dallas, Texas

## A. Why Use Capillary Electrophoresis (CE) for Drug Discovery?

Proteins are the expression of genetic information and the agents of biological function. The important role that proteins and peptides play in almost every aspect of cell structure and function is substantiated by the fact that more than 50% of the cellular dry weight is protein. Clearly, understanding protein structure and function relies on the ability to accurately and quantitatively access the composition of peptide and protein mixtures. There are a variety of techniques available for this purpose, many of them involving some form of separation. Among the numerous analytical separation techniques amenable to peptide and protein analysis, capillary electrophoresis (CE) is the most recent arrival. This technique, a simplistic approach to separation involving free solution electrophoresis in hollow glass capillaries, has provided multiple new modes of separation selectivity within a single electrophoretic platform. Consequently, quantitative compositional information can be obtained about protein/peptide mixtures, and structural information from purified preparations, in a manner that augments the information retrievable with existing techniques like high performance liquid chromatography (HPLC).

The interest in this relatively new separation technique is multi-faceted. It is not restricted to the ability to attain fast separation times, the fact that only nanoliters of sample and microliters of reagent are consumed, or the capability for on-line detection of an electrophoretic separation. Many researchers, both industrial and academic, involved in peptide and protein analyses are excited about the capability for carrying out analytical separations under non-denaturing conditions without the mandatory involvement of organic solvents. Native electrophoretic separations of peptides at low or high pH and of proteins at slightly acidic or basic pH provide a new tool for carrying out a diverse array of analyses including the ability to evaluate peptide/ protein structure (such as post-translational modification) (1, 2) and correlate it with biological function (3). Such information not only enables us to understand normal cellular structure and function, but also provides a foundation for understanding abnormal cellular function in disease and the key for novel design and development of therapeutics for those pathological states.

## B. Main CE Techniques

CE encompasses a diverse family of techniques all born of the basic principle of electrophoretic separation (Table 1). These offspring include capillary zone electrophoresis (CZE), capillary gel electrophoresis (CGE), capillary isoelectric focusing (CIEF), micellar electrokinetic chromatography (MEKC) and capillary isotachophoresis (CITP). A basic understanding of the subtleties within, and differences between, each is important to optimizing, or even achieving, a separation. A parallel jump from one technique to another can be as simple as adding a surfactant to the run buffer, but can be the key to achieving the desired separation. In this section, CE will be discussed primarily under its most commonly employed variant, capillary zone electrophoresis (CZE). The principle differences of the other CE variants that are useful for peptide and protein separations will also be discussed. [For a more comprehensive overview of CE and its variations the reader is referred to the references in Table 2 and a number of books on the topics (24–26)].

**Table 1** Modes CE and Their Corresponding Characteristics

| Technique | Abbreviation | Separation Mechanism | Conventional Comparison | Applications | Proteins | Peptides |
|---|---|---|---|---|---|---|
| Capillary Zone Electrophoresis | CZE | charge-to-mass ratio | non-denaturing slab-gel electrophoresis | mapping, quantification, purity checks, heterogeneity | — | — |
| Capillary Isoelectric Focusing* | CIEF | isoelectric point | slab or tube gel Isoelectric Focusing | quantification, purity checks, determining pI values, heterogeneity identity of monoclonal antibodies | — | |
| Capillary Gel Electrophoresis | CGE | relative molecular mass and/or charge | SDS-slab-gel electrophoresis | quantification, purity checks, heterogeneity, MW estimation | — | |
| Capillary Isotachophoresis[†] | CITP | mainly used as a sample loading tool | | large volume sample loading | — | |
| Micellar Electrokinetic Capillary Chromatography | MEKC | relative hydrophobicity | Liquid Chromatography | quantification, purity checks and heterogeneity | | — |

* Decreased sensitivity over other CE techniques—UV absorbance at 280 nm due to absoptivity of ampholytes.
[†] Can also be used as a "stand alone" technique.

**Table 2**  Select Reviews and the Subjects Covered

| Technique | Subject | Reference |
|-----------|---------|-----------|
| General CE | Peptides | (4) |
|  | Proteins | (5) (6) |
|  | Protein and peptide | (7) (8) |
|  | Glycoproteins and glycopeptides | (9) (10) |
|  | Recombinant proteins | (11) |
|  | Biopolymers | (12) |
|  | Proteins and Drugs in body fluids | (13) (14) |
|  | Drug-related impurities analysis | (15) |
|  | Drug protein binding study techniques | (16) |
|  | Comparison with slab gel electrophoresis | (17) |
| CIEF | Proteins | (18) (19) |
|  | Antibodies, peptides and proteins of pharmaceutical interest | (20) |
| MECK |  | (7) |
| CGE | Protein | (20) (21) (22) |
| ACE |  | (23) |

## 1. Capillary Zone Electrophoresis (CZE) and the Basic CE Platform

Capillary zone electrophoresis (CZE) is the most commonly applied CE technique, and separates species based on their charge-to-mass ratio (Fig. 1). Separations occur in a narrow diameter (10–200 $\mu$m inner diameter) fused silica capillary. As with more traditional slab gel electrophoresis approaches, a potential field (voltage) is applied along the capillary. In the simplest embodiment, charged species within this field travel towards the electrode of opposite charge. Their relative progress along the capillary towards the respective electrodes is dependent on their charge and mass, which ultimately influence their electrophoretic mobility. There are obvious fundamental differences between CE and the traditional slab gel technique: CZE is performed in a free-solution i.e. there is no absolute requirement for the presence of a gel matrix support system, although one can be used, and of course, it is carried out in a capillary. These basic differences favorably impact the separation. The capillary permits more rapid heat diffusion which allows higher currents to be used without substantial Joule heating. The capillary also allows easy detection via spectroscopic methods which provides distinct advantages over the stain-destain detection approach inherent with slab gels. Other advantages of CZE include easy quantification, low sample and buffer use, and fast analysis times. A CZE analysis can take 10 to 20 minutes while the same slab-gel analysis can take over one day. Unfortunately, CZE is currently limited to serial sampling. However, efforts are underway to overcome this limitation (see Sec. IV.B).

One of the distinct advantages of CZE is that positive, negative, and uncharged analytes can be separated and detected under nondenaturing conditions in a single run. Positive, negative, and uncharged analytes will clearly separate based on their inherent electrophoretic mobilities but this does not guarantee that they will reach the detector. Under normal polarity, using an uncoated fused silica

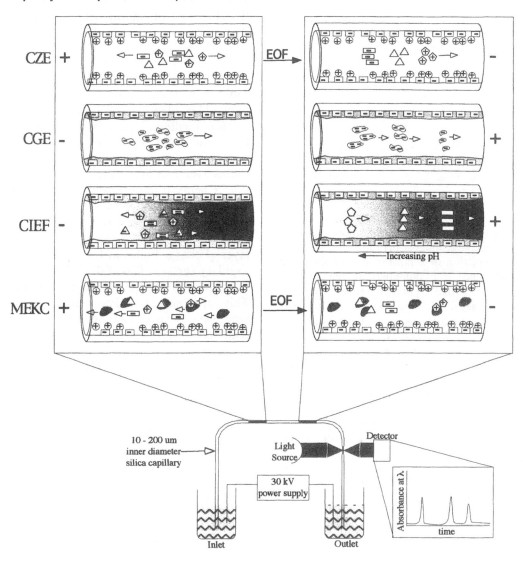

**Fig. 1** General schematic of a CE instrument illustrating the different modes of separation.

capillary, the capillary inlet will be positive (anode) and the outlet will be negative (cathode), electroosmotic flow (EOF) is generated as a result of cations that accumulate at the negatively charged silica surface (Fig. 2A). The cations comprising this positive mobile layer move towards the capillary outlet, i.e. the cathode. These cations are surrounded by hydration spheres which are also driven towards the negative electrode, and thus, eventually, the bulk solution is "dragged" to the outlet by the electromotive force of the cations. This EOF, essentially an electric field-driven bulk flow of solution from the anode to the cathode, allows that all molecules, regardless of charge, will move towards the outlet and pass the detector. Through this mechanism, even neutral and negatively charged species will

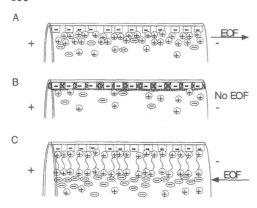

**Fig. 2** Capillary surface coverage effect on EOF. (A) a fused silica capillary has a negative surface charge in most pH ranges, resulting in the formation of a positive double ion layer (and hence a $\zeta$ potential) at the surface. Under normal polarity (inlet—positive, outlet—negative), the positive ions will migrate towards the outlet inducing an endoosmotic flow (EOF) towards the detector. (B) If the surface is coated with a nonionic layer no surface charge ($\zeta$ potential) will be established, therefore, there is no EOF. Because there is reduced or no EOF, electrode polarity must be chosen so that the molecules of interest have the opposite charge of the outlet electrode. (C) Capillary coatings imparting a positive charge to the inner capillary surface will attract a layer of negative ions. Here the EOF will be in the direction of the positive electrode.

migrate towards the detector since the EOF is typically larger than the most electrophoretically mobile negatively-charged molecule. Fig. 1 illustrates this concept: the negative ions migrate initially towards the positive electrode, while, after the EOF is established, they are pulled towards the negative electrode.

The role of capillary surface charge in determining EOF provides one with the ability to optimize separations by altering the properties of the capillary surface. If the surface is rendered neutral, the EOF will be greatly reduced or even eliminated (Fig. 2B). This can be achieved via either covalent modification of the silica func-

tional groups, or coating with a substance that binds to the surface electrostatically. The advantages to having the ability to easily modify the capillary surface treatments include reduction of analyte-capillary wall interactions which can lead to band broadening, and a stationary separation medium which is desirable for CGE and CIEF applications. Reducing the EOF will also slow the flow increasing the necessary separation time.

Chemical modification or coating with an electrostatically binding substance can also render a positive charge on the capillary wall. Fig. 2C shows a negative layer of charge builds at the now positive capillary surface. The EOF will now be in the direction of the positive electrode. This approach is advantageous when separating positive analytes that interact regularly with the bare negatively charged capillary surface.

Several modified techniques exist based on the CZE platform. Each technique brings with it a different separation mechanism and arms the separator with a magnitude of options for perfecting each particular separation.

## 2.  Capillary Gel Electrophoresis (CGE, SDS-CGE)

Capillary gel electrophoresis in the presence of sodium dodecyl sulfate (SDS) can provide separations similar to slab gel electrophoresis using SDS. With traditional slab gel electrophoresis, SDS, an anionic surfactant, binds hydrophobically to proteins at a ratio of 1.4 g per 1 g protein to produce species with similar charge per mass ratios. The proteins are then separated through a polymerized sieving gel matrix, typically crosslinked polyacrylamide, based on mass differences, where smaller proteins migrate through the gel matrix faster then the larger proteins. All proteins will migrate towards the cathode due to the negative charges contributed by the bound SDS (Fig. 1).

With CGE, similar conditions can be created within the capillary by using either a "chemical gel," a permanently crosslinked polymer (e.g., polyacrylamide), or a "physical gel" which is a pumpable polymer matrix that can be used for separation and removed (e.g., methyl cellulose). Coated capillaries that provide reduced or negligible EOF are usually required to prevent the sieving matrix from moving during the electrophoresis. This technique can generally be performed reproducibly on proteins in the 10–200 kDa range although, with highly charged proteins, (particularly small proteins), the native charge may effect the analysis.

The capillary approach has several advantages over the traditional slab gel approach. Detection is on-line and versatile (see Sec. I.B.6.). UV absorbance detection at 200 nm is routinely used and is effective for the analysis of protein preparations having a total concentration of >100 $\mu$g/ml and allows for easy quantitation (compared with semi-quantitation on gels). Fluorescence detection, which typically requires that the proteins be fluorescently-tagged, can boost detection limits by 2–3 orders of magnitude. Perhaps the most significant advantage is time and labor saved with reduced sample preparation, increased speed of analysis and automation. With certain applications, these may outweigh the advantages of the traditional approach which allows for simultaneous multiple sample analysis with preparative capabilities, including sample recovery.

An example of an area where CE is advantageous over the slab gel approach is for determining the molecular mass of post-translationally modified proteins, e.g., glycosylated, phosphorylated, sulfonated. Post-translational modification can adversely affect the charge-to-mass ratio of proteins under denaturing conditions. This is particularly true with glycosylated proteins where the carbohydrate moieties do not bind SDS, causing lower charge-to-mass ratios, resulting in decreased migration and an over-estimation of molecular mass on polyacrylamide gels. Under these conditions, Ferguson plots can often be used for molecular weight estimation (27). With the Ferguson plot, the mobility of a protein during electrophoresis in different sieving matrices (e.g., varying acrylamide concentrations) allows for a more accurate determination of molecular weight. Since the composition of the sieving matrix can be easily changed in CE by simply refilling the capillary with a physical gel at differing concentrations, the data required for generating a Ferguson plot can be obtained in an automated fashion (28).

In general, there are a variety of capillary coating methodologies and sieving matrices that may be applicable to CGE separations—these are discussed in more detail in section III.

### 3. Capillary Isoelectric Focusing (CIEF)

Like conventional isoelectric focusing on slab-gels, capillary isoelectric focusing (CIEF) separates proteins based on their isoelectric point, pI. The difference is that it is performed in a "physical gel" as opposed to a solid gel matrix with the traditional method. In CIEF, a capillary is either filled with an ampholyte solution containing a polymer (e.g., methyl cellulose) or the ampholytes are added to the sample matrix. As the voltage is applied over the capillary, the ampholytes will migrate to form a pH gradient (Fig. 1). Simultaneously, positively-charged proteins will migrate towards the anode experiencing an environment that incrementally increases in pH until the pH equal to their pI is reached, at which point they become zwitterionic (net neutral) and stop migrating. Similarly, negatively-charged proteins will migrate towards the positive electrode until they reach a pH region corresponding to their isoelectric point. CIEF is a self-focusing technique, meaning that if a protein migrates out of its zone it will obtain a charge and, consequently, start migrating back to its proper zone.

Once the focusing has occurred, the zones must migrate past the detector in order for the analysis to be fruitful. Two methods are commonly used to induce "mobilization"; a one-step method and a two-step method. In the one step method, the mobilization occurs simultaneously with the focusing. The EOF drives the focused zones past the detector. Unfortunately, zone broadening can occur, and the experiment must be carefully planned to prevent the zones from passing the detector until they are adequately separated (29). EOF and zone broadening can be reduced in a two-step process using a coated capillary. Focusing is carried out followed by mobilization, which is induced chemically by adding, acids, bases or salts as a second step. Pressure can also be applied to "push" the zones past the detector. With both these methods, an electric field is applied to retain tightly focused zones [mobilization techniques are discussed with respect to their advantages and disadvantages in a review by Schwer (19)].

As with any technique, CIEFs has advantages and limitations. The resolution attainable with this CE modality provides a clear advantage over other methods of protein separation. Resolution of proteins with pI values differing by as little as 0.02 pH units has been described (30). A disadvantage associated with CIEF stems from the requirement for detection to be limited to 280 nm or higher as a result of the presence of ampholytes in the separation buffer. This decreases detection sensitivity by roughly 10-fold over that at 200 nm and restricts the use of this modality for the analysis of proteins present in mixtures at high concentrations (mg/ml or higher).

### 4. Micellar Electrokinetic Chromatography (MEKC)

Micellar electrokinetic chromatography is a hybrid of electrophoresis and chromatography in that it employs a pseudo-stationary phase to facilitate electrophoretic separations. The incorporation of a surfactant into the run buffer allows for the formation of micelles when the concentration is above the critical micelle concentration (CMC). The analytes partition into the micelles as they would a stationary phase for HPLC (Fig. 1). Separation is, therefore, based on the partition coefficients of the protein with respect to the micelles, in addition to the normal charge-to-mass effect. Organic solvents (acetonitrile, methanol, etc.) can be added to change the partitioning, as well as, increase the solubility of the analytes in the run buffer. Larger molecules, such as proteins, can not partition into the micelles so this technique remains limited to smaller molecule (e.g., peptide) applications. Since MEKC involves hydrophobic partitioning, it is particularly useful for separating peptides that vary in degree of hydrophobicity but only subtly in charge. Under these conditions, separation will be based predominantly on the partitioning of the analytes into the micelles, since charge-to-mass differences may be minimal.

Many surfactants have been used successfully and are discussed in section II.A.3.

### 5. Capillary Isotachophoresis (CITP) and other Procedures Used for Stacking

Several advantages of the CE techniques, like small sampling size and efficient heat dissipation via the use of narrow-bore capillaries, become limitations when addressing limits of detection. On-column concentration techniques, such as field amplified stacking (31), can increase detection limits up to 5-fold. Other approaches include the use of pH gradients or exploitation of capillary isotachophoresis (CITP) as an injection mode.

CITP is another offspring of CE that utilizes the inverse proportionality of field strength to the conductivity of a solute. While CITP has been shown to be useful for several "stand alone" protein separations [for example see (32) or (33)], it is used mainly as a focusing technique for other CE approaches, and will be discussed here in this context. Several reports describe the full theory, implementation, and potential applications of CITP including Wanders and Everaerts (34) and Weinberger (25).

With CITP as a sample injection mode, conditions must be chosen so that cations or anions, specifically, can be stacked. This requires a leading buffer containing cations (anions if focusing anions) with higher mobilities than the cationic analyte (Fig. 3). The run buffer can be used as the trailing buffer if its cations have a lower mobility than the analytes. The components stack according to their mobilities, such that the analytes are sandwiched between the leading buffer and the run buffer. As the solutions mix, the CZE separation takes over and the analyte bands separate. This approach has been successful for peptides (35) and proteins (36, 37). This exemplifies one approach, but several other CITP methodologies have been demonstrated successfully (36, 38).

Another stacking procedure, involves modification of the sample matrix ionic strength and is termed field amplified stacking (39). Stacking can occur when the ionic strength of the sample matrix is lower than that of the run buffer (Fig. 4A). Under these conditions, the charged analytes will have a high mobility and "race" towards the sample/run buffer interface. As molecules reach the interface, they are slowed due to the higher ionic strength of the separation buffer. This

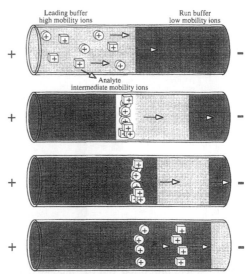

**Fig. 3** Isotachophoretic stacking. Analytes of the same charge are injected in a buffer of higher mobility ions. The run buffer contains ions of the lowest mobility. As electrophoresis is initiated the ions migrate towards the outlet, after the higher mobility ions overtake the analyte and run buffer ions the three groups of ions will migrate in zones. As the sample zone diffuses into the separation buffer the electrophoretic separation mechanism will take over.

method can be used for negatively or positively charged analytes, the only difference being the side of the sample plug on which the analytes will stack. An example of the effectiveness of this method for peptide analysis is shown in Fig. 5.

Differences in the run buffer and sample pH can also be used for stacking (Fig. 4B). The pH conditions are chosen such that the analyte(s) of interest will have different charges in the sample matrix and the run buffer (41). Fig. 4B shows two species both of which are positively charged in the sample matrix. These species migrate towards the anode, eventually reaching the sample/run buffer interface. In the higher pH buffer, the analytes will adopt a negative charge and start migrating in the opposite direction.

A                                                              B

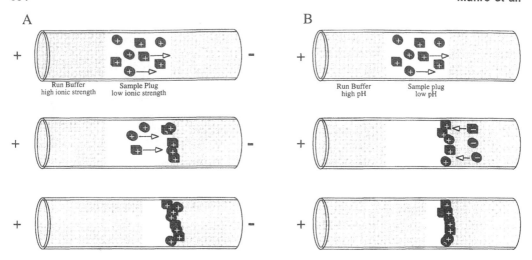

**Fig. 4** Field amplified and pH-mediated stacking mechanisms. (A) With field amplified stacking, the analytes are dissolved in a matrix of lower ionic strength than the run buffer. The analytes will have a higher mobility in the sample plug and will move quickly towards the sample plug-run buffer interface. At the interface, the analytes are slowed in the higher ionic strength run buffer. As the sample zone diffuses into the separation buffer, separation will proceed. (B) In pH mediated stacking, analytes are dissolved in a matrix at a pH where they posses a different charge than in the pH of the run buffer. As the analytes enter the run buffer they adopt the opposite charge and migrate in the opposite direction, resulting in a focusing effect. As the sample zone diffuses into the separation buffer the electrophoretic separation proceeds.

Thus, the analytes will be "focused" at the sample/run buffer interface until, through diffusion, the pH gradient diffuses and electrophoresis ensues.

## 6. Detection Methods

Two main areas of detection exist for CE techniques: optical (25, 42) and electrochemical (25, 43). The optical techniques are, by far, more common due to commercial availability and ease. Common optical techniques include absorbance, fluorescence and chemiluminescence, many more specialized and equipment intensive detection techniques have also shown utility (25, 42). Recently, detection of analytes via mass spectrometry has shown the possibility of yielding a new dimension of information for CE while being compatible under certain separation conditions (44).

Absorption detection is the most universal of CE detection techniques. The peptide bond absorbs below 220 nm, and many common buffers allow detection in this region (Table 3). The main limitation of this method is the sensitivity; detection limits are typically around $10^{-8}$-$10^{-6}$ M. Fluorescence detection, another common and commercially available detection method, has better detection limits, around $10^{-12}$ M. Unfortunately, only the amino acids tryptophan, tyrosine and phenylalanine fluoresce naturally, so most proteins need to be fluorescently-tagged for detection via fluorescence [for example see (45)]. Mass spectrometry (MS) is developing into an information rich means of CE detection, although currently limited practically by stability issues and expense. Many modalities of MS have shown usefulness for CE applications. CE-electrospray ionization (46, 47), ion

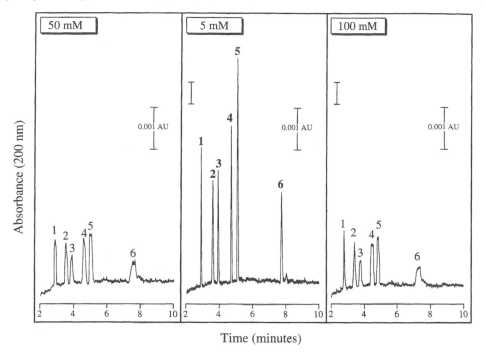

**Fig. 5** An example of sample "stacking." Separation of a standard peptide mixture (peptide calibration kit, Bio-Rad, Richmond, CA) in phosphate buffer (pH 2.5) at varying concentrations: left panel—50 mM, middle panel—5 mM and right panel—100 mM. Separation conditions: 5 s hydrostatic injection; 10 kV running voltage; 50 μm × 27 cm (20 cm to detector) capillary. Labeled peaks: 1 = bradykinin, 2 = angiotensin, 3 = a-melanin stimulating hormone, 4 = thyrotropin-releasing hormone, 5 = leutinizing hormone releasing hormone, 6 = leucine enkephalin, 50 μg/mL each. From Ref. 40.

**Table 3** Commonly Used CE Buffers and Their Associated Properties

| Buffer | Useful pH range | Minimum useful (nm) |
|---|---|---|
| Phosphate | 1.14–3.14 | 195 |
| Formate | 2.75–4.75 | 200 |
| Acetate | 3.76–5.76 | 200 |
| Citrate | 3.77–4.77 | 200 |
| MES* | 5.15–7.15 | 230 |
| Citrate | 5.40–7.40 | 200 |
| PIPES* | 5.80–7.80 | 215 |
| Phospate | 6.20–8.20 | 195 |
| HEPES* | 6.55–8.55 | 230 |
| Tricine* | 7.15–9.15 | 230 |
| Tris | 7.30–9.30 | 220 |
| Borate | 8.14–10.14 | 180 |
| CAPS* | 9.70–11.10 | 220 |

* zwitterionic buffers.
  From Ref. 40.

trap storage/reflectron time-of-flight MS (48), and matrix assisted laser desorption time-of-flight MS (49) have been used recently for protein analysis. Glycosylated proteins separated via CE have been analyzed on-line using matrix-assisted laser desorption ionization (50), MALDI/time-to-flight MS (51), and electrospray ionization-MS (51–53). Biomolecules have even been characterized at the attomole level with this technique (54).

## C. Separation Strategies for Peptides vs. Proteins

Peptides and proteins require very different separation strategies even though the latter are composed of the former. The separation of small chains of amino acids

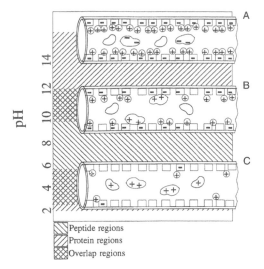

Peptide regions
Protein regions
Overlap regions

**Fig. 6** Diagram of capillary surface charge character in different pH ranges. (A) At high pH, the capillary is highly negatively charged and attracts a large positive ion layer inducing a strong EOF. Proteins and peptides will not be positively charged and therefore not attracted to the capillary wall. (B) At neutral pH the capillary wall is negatively charged and an EOF will exist due to the positive ion layer, but proteins and peptides may also be positively charged and attracted to the capillary wall. (C) At low pH the capillary will be mostly nonionic (silanol groups protonated) and the EOF will be diminished. Proteins and peptides will, therefore, not be electrostatically attracted to the capillary wall.

(below 5–7 kDa) are relatively predictable based on properties of their primary sequence. However, larger chains, proteins, are less predictable due to a dramatic increase in structural complexity. As discussed earlier, electrophoretic separations are predicted using analyte charge-to-mass ratio. An estimate of each of these values can usually be obtained for peptides, but not as easily for most proteins.

For peptides, an estimate of the pKa value can be obtained through either the isolated pKa values of the amino acids or adjusted pKa values. The latter are, apparently, more accurate for peptides

(55). The accuracy of this measurement will depend on the primary sequence and any tertiary structure the peptide may adopt. Other methods such as titration or computer programs can also estimate pKa values. The pKa is an important value for determining a starting pH for the separation buffer. Since separation occurs mainly from electrophoretic mobility differences (a parameter affected by both charge and size), pH conditions can be chosen so that the analytes will adopt the greatest difference in charge. Later, if necessary, peptide ionization can also be altered through buffer additives. Figure 6 illustrates the pH regions most commonly used for peptide separations. For acidic residue peptides the common pH is usually pH 2–7 and for basic residue peptides, pH 8–10. Peptides, unlike proteins, can use a diverse range of pH's and buffers, since they are not as prone to capillary wall adsorption.

To successfully separate proteins, a different strategy must be utilized. The main factors affecting the electrophoretic mobility of a protein include mass, charge, protein-capillary wall attraction, and the protein's hydrodynamic radius. Attempting to predict how any of these impact the protein's electrophoretic mobility under the chosen separation starting conditions is difficult. The fundamental problem lies in the present inability to measure a protein's effective net charge and hydrodynamic radius. As knowledge in these areas increases through fundamental protein studies, accurately predicting protein migration may be possible. Currently, initial separation conditions can be chosen through consulting the accumulating literature describing successful protein separations.

The one aspect of a CE based protein separation that must be addressed is the minimization of protein-capillary wall interactions. This is often challenging since separation conditions that reduce protein adsorption onto the capillary wall, might not be optimal for separation. Fig. 6B illustrates why protein adsorption is a problem at mid-range pHs: The capillary wall is negatively charged and the positive ions in the buffer along with the positive sites on the proteins are attracted. Under these conditions adsorption of proteins bearing large positive charges may render the separation ineffective. Due to this phenomenon, the optimization of protein separations typically revolves around adjusting the separation conditions to reduce protein-wall interactions.

Two CE-based modes have proven most useful for peptide separations, CZE and MEKC (Table 1). Protein separations have been more successful using CZE, CIEF, and CGE. While working under the same general idea, each of these techniques is associated with advantages and disadvantages with any potential application.

## II. PEPTIDE AND GLYCOPEPTIDE SEPARATIONS

Peptides play a very important role in biological systems functioning in a diverse array of roles ranging from effecting neural transmission to controlling systemic blood pressure. Analysis of peptides either as complex physiological mixtures or as pure synthetic products is of obvious importance. CE has been found to be useful in this context as well as in heterogeneity and peptide structural analysis. As a means of providing the reader with the tools required for carrying out standard and esoteric peptide analysis, the following sections detail the importance of buffer choice, additives and capillary surface characteristics in the CE-based separation of peptides.

### A. Optimizing Peptide Separations

#### 1. Buffer and Buffer pH

Separation buffer optimization focuses on three areas: buffer pH, buffer components

and buffer concentration. Initially, the separation buffer pH should be chosen using the guidance of the charge-to-mass ratio. A good starting pH range, as mentioned earlier, is usually pH 2–7 for peptides with more acidic residues and pH 8 to 10 for peptides with more basic residues (Fig. 6). It is in these pH ranges that the largest differences in charge-to-mass will exist between the analyte peptides. Figure 7 illustrates the pH dependence of a separation of five peptides composed of the same twelve amino acids

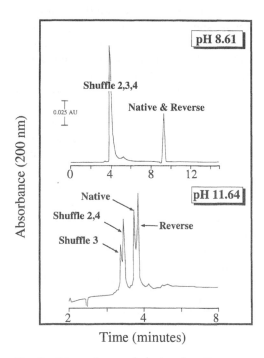

**Fig. 7** Dependence of electrophoretic separation and resolution on buffer pH. Five peptides containing the same twelve amino acids but in varied sequence were analyzed at pH 8.61 and 11.64. Sequence of native peptide = KTNYCTKPQKSY, reverse peptide = YSKQPKTCYNTK, shuffle 2 = QPSKK-TYCKTYN, shuffle 3 = KPTQYNKSTYKC, shuffle 4 = KKNKPYCTQTSY. Experimental conditions: 100 mM borate buffer with 10 mM diaminopentane, adjusted to pH with NaOH; 25 kV running voltage; 50 $\mu$m $\times$ 27 cm (20 cm to detector). From Ref. 40.

but varying in primary sequence (amino acid order).

Considering the buffering components, several areas should be taken into account. First, the species must buffer well at the desired pH. For most common peptide pH ranges, several buffer choices exist and these are detailed in Table 3. It is important to be aware of the fact that the mobility of the buffering components also affect the separation. If the mobility of the buffering component is too high, the current will increase and Joule heating becomes a problem. The mechanism for detection must also be considered. If the most common detection method, low UV absorbance ($< 210$ nm), is employed, a buffer species that does not absorb significantly in this region should be used (Table 3). If peptides are to be derivatized for laser-induced fluorescence (LIF) or absorbance detection at longer wavelengths, many more buffer choices exist. This is due to the fact that most derivatizing agents react with primary or secondary amines, thus negating cationic residues of the peptide, enabling separation at neutral pH without concern about peptide wall interactions. Finally, the concentration of the buffer should be considered. Normal buffer concentrations for CE are typically low, ranging from 10 mM to 100 mM for most buffers. With the exception of borate which can be as high as 500 mM, high ionic strength buffers generate excessively high currents that produce Joule heating that is destructive to the separation. In contrast, buffers at extremely low concentrations can lead to problems with separation-to-separation reproducibility.

## 2. Organic Solvents

After buffer pH and salts have been trialed and the desired separation not attained, organic additives may be evaluated for optimizing the separation. Organic additives can affect the solvation of the peptide, as well as the pKa of ionizable groups. These two factors can alter the mobility of the peptide, thus, affecting the overall separation. Organic additives also slow the EOF through alteration of the effective charge at the wall as well as through changing the buffer viscosity. While this produces longer analysis times, the on-capillary time may be required for the separation. If more separation time is not necessary, the capillary can always be shortened. Some common organic additives include acetonitrile, methanol, and tetrahydrofuran (THF), all of which are added to the separation buffer in concentrations ranging from 1–40% (v/v) (4, 56).

## 3. Surfactants (MEKC)

MEKC involves the use of surfactants at concentrations above their critical micelle concentration (CMC) to augment the separation buffer. As discussed earlier, a different separation mechanism exists in MEKC as opposed to other CE techniques, which gives one another method to chose from to optimize a peptide separation (see Sec. I.B.4).

Many surfactants have been used successfully for peptide separations. The cationic, anionic, zwitterionic, and non-ionic surfactants commonly used are given in Table 4 along with the corresponding CMC values. Surfactants are usually added at concentrations roughly 2-times their CMC values. Fig. 8 provides an example wherein a zwitterionic surfactant is exploited to obtain separation of two closely related peptides. Assuming a CMC value of 3.7 mM in the buffer used, the authors tested several concentrations to optimize the separation. It is noteworthy that the organic solvents described in the previous section can be used in conjunction with surfactants in a "multiple buffer additive" strategy as a means of altering solute partitioning between the pseudo-stationary phase and the buffer.

**Table 4** Commonly Used Surfactants, Their CMC and Aggregation (N) Values, and Literature Examples

| | | CMC (mM) | N | References |
|---|---|---|---|---|
| *Anionic* | | | | |
| | Sodium dodecyl sulfate (SDS) | 8.27 | 62 | (57–59) |
| | Sodium taurocholate (STC)* | | | (59) |
| | Sodium cholate (SC)* | 13 | | (59) |
| | Sodium taurodeoxycholate (STDC)* | 9 | | (60) |
| | Pentanesulfonic acid | | | (59) |
| *Cationic* | | | | |
| | Cetyltrimethylammonium chloride (CTAC) or bromide (CTAB) | 13 | 78 | (61) |
| *Non-Ionic* | | | | |
| | Polyoxyethylene-23 lauryl ether (Brij-35) | | | (59) |
| | Polyoxyethylene-t-octylphenol (Triton X-100) | 0.24 | 143 | (59) |
| | Tween 20 | | | (59, 62) |
| *Zwitterionic* | | | | |
| | N-alkyl-N,N-dimethylammonio-1-propane sulfonate (SB3-8) | | | (63) |
| | N-dodecyl-N,N-dimethyl-3-ammonio-1-propanesulfonate (DAPS) | 3.7 | | (64) |
| | 3-(N,N-Dimethylhexadexylammonium) propanesulphonate (PAPS) | | | (65) |

* bile salt
Modified from Ref. 66.

## 4. Capillary Coatings

For most applications, peptide separations can be performed at low pH in uncoated capillaries, reducing unnecessary monetary and time investments. However, problems can be encountered with very basic peptides which cannot be resolved at low pH or hydrophobic peptides both of which may present problems with capillary-wall interactions. In these instances, capillary coatings must be pursued and can either be covalent or dynamic in nature. Covalent coatings are permanent derivations of the capillary wall functional groups that result in changing the charge and/or the hydrophobic character of the wall. A wide variety of established and novel coatings have been reported in the literature as being effective for peptide CE (Table 5).

Dynamic coatings, on the other hand, are typically buffer additives that dynamically passivate the capillary surface. As given in Table 5, a number of compounds have been exploited effectively. Two such compounds are noteworthy in this regard: polybrene (hexadimethrine bromide) (91) and PDMAC (polydiallyldimethyl-ammonium chloride) (92). While both polybrene and PDMAC provide good resolution of peptides that are problematic using standard conditions (93), our experience with these dynamic coatings has shown that PDMAC is the more reliable. Studies with insulin-like growth factors-I and -II, 7 kDa peptides prone to interacting with the capillary wall, were shown to be reproducibly separated over 30–40 runs before recoating was required (94). Under the same conditions polybrene required recoating every 12 runs.

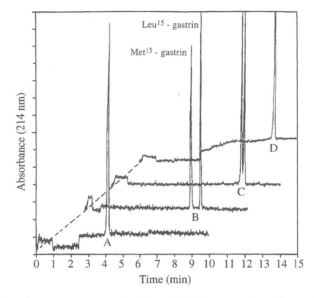

**Fig. 8** Effects of surfactant concentration in a MEKC separation. Separation of two gastrins, Met[15]- and Leu[15]-gastrin, in various DAPS concentrations: (A) 3.5 mM, (B) 7.5 mM, (C) 20 mM, (D) 70 mM. Experimental conditions: 20 mM Tris-Taps buffer at pH 8.0 with 10% acetonitrile; 30 kV running voltage, 8 μA. Abbreviations: DAPS - - N-dodecyl-N, N-dimethyl-3-ammonio-1-propanesulfonate; Tris - - 2-amino-(hydroxymethyl)-1,3-propanediol and Taps - - [N-tris(hydroxymethyl) methyl-3-amino-propanesulfonic acid. From Ref. 64.

## B.  Glycopeptide Separations

As the biological importance of glycoproteins is realized, the need to resolve and characterize the glycopeptides from such molecules will escalate. In an elegant study, Rush et. al. (95) demonstrated the power of CE for the analysis of glycopeptides. A trypic map of recombinant human erythropoietin (r-hu-EPO) was separated into over 30 peaks. The peptides were resolved into two main groups: nonglycosylated and glycosylated peptides (Fig. 9). Heptane sulfonic acid was used as an ion-pairing agent in this particular separation and, through the combined use of HPCE and HPLC, they were able to identify several glycosylation sites on the r-hu-EPO.

For a more comprehensive review on CE separations of glycopeptides and glycoproteins and also released oligosaccharide separations see Kakehi and Honda (10) or Olechno and Nolan (96).

## III.  PROTEIN AND GLYCOPROTEIN SEPARATIONS

Protein separations provide a unique challenge to CE. While similar in principle to slab gel electrophoresis, electrophoresis in a capillary is, from a practical point of view, a great departure from the traditional method. Unlike most of the separations in the slab gel format, researchers are drawn to CE for its ability to electrophorese protein in free solution under nondenaturing conditions. This presents the ability to electrophoretically resolve proteins varying in size and charge in complex mixtures or to resolve closely-related glycoforms of a microheterogeneous glycoprotein. While powerful, CE is fraught with challenges not encountered with its slab gel counterpart. These include the predilection of proteins for interacting with the capillary wall or with other proteins under nondenaturing conditions, and having to optimize for sep-

**Table 5**  Covalent and Dynamic Capillary Coatings Commonly Used and Literature Examples

| Coating | Mechanism* | Proteins | Peptide | References |
|---|---|---|---|---|
| *COVALENT* | | | | |
| polyacrylamide | hydrophilic, suppresses EOF | | | (67) |
| | | — | — | (68) |
| polyacrylamide (crosslinked with formaldehyde) | hydrophilic, suppresses EOF | — | — | (69) |
| "fuzzy" polyether | hydrophilic, eliminates EOF | — | | (70) |
| interlocked polyether | hydrophilic, eliminates EOF | — | | (70) |
| methylcellulose | hydrophilic, eliminates EOF | — | | (67) |
| | | — | | (71) |
| | | — | — | (72) |
| dextran | hydrophilic, eliminates EOF | — | | (71) |
| dextran (crosslinked with PEG) | hydrophilic, suppresses EOF | — | | (73) |
| C18 w/nonionic surfactants | hydrophilic, suppresses EOF | — | | (74) |
| | | — | | (75) |
| multilayer | reverse EOF | — | — | (76) |
| epoxy resin (crosslinked) | moderate EOF | — | | (77) |
| poly(acryloylamino-ethoxyethanol) | hydrophilic, suppresses EOF | — | | (78) |
| poly(ethylene-propylene glycol) | hydrophilic, suppresses EOF | — | | (79) |
| *DYNAMIC* | | | | |
| Polybrene | ionic, reverses EOF | — | | (80) |
| | | — | | (81) |
| | | — | — | (82) |
| | | — | | (67) |
| | | — | — | (83) |
| poly(methoxyethoxyethyl) ethylenimine | ionic, reverses EOF | — | | (80) |
| poly(diallyldimethyl ammonium chloride) | reverses EOF | — | | (84) |
| poly(ethlene oxide) | hydrophilic, suppresses EOF | — | | (85) |
| polyethylenimine | ionic, reverses EOF | — | | (80) |
| | | — | | (86) |
| CTAB | ionic, reverses EOF | — | | (86) |
| putrescine | reduces EOF | | — | (87) |
| Fluorinated | ionic, reverses EOF | — | | (88) |
| N,N,N',N',-tetromethyl-1,3-butanediamine | pH dependent effect | — | | (89) |
| PDMAC | reverses EOF | — | | [68, 70, 94] |
| cationic polymer | reverses EOF | — | | (90) |

* The distinction between "eliminates" and "suppresses" EOF was not necessarily demonstrated experimentally in these studies.

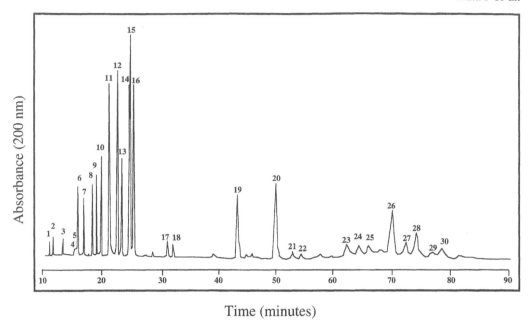

**Fig. 9** Capillary electrophoretic separation of peptides and glycopeptides from recombinant human erythropoietin. The faster migrating peaks (1–18) represent nonglycosylated peptides, while the remainder represent those that are glycosylated. Experimental conditions: 40 mM sodium phosphate buffer with 100 mM heptanesulfonic acid at pH 2.5; 16 kV running voltage, 110 $\mu$A; 50 $\mu$m × 75 cm (50 cm to detector). From Ref. 95.

aration based on charge-to-mass ratio and not just mass. However, there are a number of well-defined steps that can be employed to deal with these aspects of CE. These will be covered here with reference to specific reports in the literature. These sections will illustrate that, once solved, the power of CE for protein analysis is unparalleled for application to diverse areas such as vaccine production, epitope mapping and glycoform analysis.

## A.  Optimizing Protein Separations

### 1.  Buffer

As with any electrophoretic process, the buffer is critical for effective separation. This is particularly true with CE since many separations are carried out in free solution under nondenaturing conditions (not just separation by size). Con-

sequently, the buffer type and pH has a significant impact on protein structure including protein shape, size and charge, all of which impact electrophoretic mobility. There are many buffers that are suitable for capillary electrophoresis and some of these are given in Table 3.

### a.  pH

As stated earlier, one of the largest obstacles in protein separation is the tendency for protein to seek interaction with the capillary wall. As shown in Fig. 6, unfavorable protein-wall interactions can be minimized by electrophoresing at pH extremes, below pH ~ 4 or above the proteins pI. Under the first condition, the wall silanol groups will be protonated and the proteins cationic, therefore, minimizing the electrostatic attraction between the two (Fig. 6C). While this approach is effective for peptide separations, the pH extreme used for peptide separations (as low as pH 2.0) can cause problems with proteins in the form of denaturation and/or precipitation. Consequently, protein separations may be carried out in slightly acidic buffers to minimize protein-wall interactions but without inducing denaturation. For example, low pH phosphate buffers have been shown in several reports to be effective for certain protein separations [for example, (97)].

Protein separations can also be performed at higher pHs, typically above the pI of the protein of interest. Under these conditions, the protein will be anionic (Fig. 6A) and will be repelled from the anionic capillary wall, therefore, reducing protein adsorption. This approach may also suffer from the same hydrophobic binding and protein denaturation issues as does the low pH method. Furthermore, some proteins have very basic pI values, greater than pH 13, in pH regions detrimental to the protein and the capillary surface.

Both high and low pHs buffers will reduce protein-wall interactions and have been found successful for several protein separations. Although extreme pH buffers will prohibit studies requiring the proteins to be in their native conformation, such as the study of protein-antigen complexation, antibody-receptor interactions or noncovalent multimeric protein separations. These alkaline or acidic pH methodologies may also provide unacceptable separations. In these cases, the buffer modification strategy should avoid large changes in pH and focus on the use of additives or coated capillaries. Interestingly, a recent review of the glycoprotein literature (9) indicates that the majority (> 70%) of CE-based protein separations are carried out in the slightly acidic or basic pH range (pH 4–6 and pH 8–10) with the minority in the neutral (pH 6–8) range.

b. *Additives*

Buffer additives can enhance the electrophoretic separation of proteins through a variety of mechanisms depending on the nature of the specific additive (Table 6). For example, sulfonic acid additives can ion-pair with peptides or proteins, changing their net charge (98). In addition to reducing the cationic character and, subsequently, the possibility of analyte-wall interactions, this additive can alter selectivity as a result of differential affinity for specific components. Butanesulfonic acid has been shown to be effective for peptide and protein applications (99) while pentanesulfonic acid up through decanesulfonic acid appear to be the more common choices. In the latter cases, the chain length provides an additional degree of optimization. Phytic acid has also been noted to ion-pair with very basic proteins allowing effective separations. Okafo et. al. (100) illustrated this showing the dramatic effect this additive can have on the separation of basic proteins (Fig. 10).

2. *Ionic Strength*

High ionic strength buffers often facilitate the separation of protein mixtures not attainable with the use of low ionic strength buffers. This is likely due to a reduction in protein-capillary wall interactions, protein-protein interactions, or both. Inherent with the use of higher ionic strength buffers is an increase in Joule heating [see (101) for an extensive discussion on temperature generation and control in CE]. Consequently, the applied field often must be limited so as not to exceed a power of 1–2 W/meter beyond which the generated heat cannot be dissipated. The Joule heating problem can also be circumvented by narrowing the capillary internal diameter, but this strategy will decrease allowable sample injection size and compromise the detection limit. An alternative approach is to bolster

**Table 6**  Common Buffer Additives in CE and Their Effects

| Additive | Function |
| --- | --- |
| Inorganic Salts | Protein conformational changes |
| | Protein hydration |
| | Reduced wall interactions |
| | Modify EOF |
| Zwitterions | Reduce wall interactions |
| Metal ions | Modify protein mobility |
| Organic solvents | Solubilize proteins |
| | Modify EOF |
| | Reduce wall interactions |
| Urea | Solubilize proteins |
| | Modify EOF |
| Sulfonic acids | Ion pairing agents |
| | Surface charge modifier |
| | Hydrophobic interaction |
| Cellulose polymers | Modify EOF |
| | Provide sieving medium |
| Amine modifiers | Charge reversal on capillary wall |
| | Ion pairing agents |
| Cationic surfactants | Charge reversal on capillary wall |

Modified from Ref. 40.

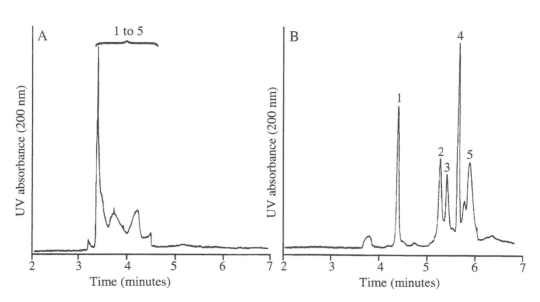

**Fig. 10**  Electrophoretic separation dependence on an ion pairing agent. Separation of five basic proteins in the absence (A) and presence (B) of the sodium salt of phytic acid (40 mM). Experimental conditions: 500 mM borate buffer at pH 8.4, 5.0 kV, 75 μm × 45 cm capillary. Labeled peaks: 1 = lysozyme; 2 = trypsinogen; 3 = α-chymotrypsin; 4 = ribonuclease A; 5 = cytochrome c. From Ref. 100.

ionic strength with zwitterionic species [examples: glycine, betaine; (102)], which increase the ionic strength but are low mobility species and, hence, low conductivity.

One practical consideration when increasing ionic strength, and choosing buffer species in general, is the detection technique. Small increases in ionic strength have been shown to increase separation efficiency and should, therefore, be considered first (103). However, if higher ionic strength buffers are to be explored, the user should be aware that some buffer species may have some absorbance at the detection wavelength (see section I.B.6), decreasing the signal-to-noise ratio with increasing concentration (Table 3).

### 3. Capillary Coatings

Coating capillary walls is another strategy that can be used to decrease protein-wall interactions. The coatings function by either reversing the charge on the capillary wall (making it cationic) or rendering it neutral reducing protein attraction for certain applications. This approach allows the buffer pH to be manipulated for optimization of the separation. As with peptides, two types of capillary coating approaches have been shown to be effective: dynamic and covalent.

### a. Dynamic

Dynamic coatings involve coating the normal fused silica capillary with a temporary, but stable, material. The coating masks or reverses the anionic charge at the wall so protein adsorption is reduced. Dynamic coatings are either applied prior to the separation or can be added to the separation buffer depending on the specific coating species. A clear advantage associated with this approach is that the coating can easily be removed and generating the original bare silica capillary.

Several approaches using cationic surfactants as capillary coatings have shown promise (Table 5). The cationic head of these materials align with the anionic groups of the capillary so their hydrophobic tails are extended into the capillary lumen. Another layer of the cationic surfactant then aligns in the opposite orientation with their hydrophobic tails bound hydrophobically. This produces a pseudo double layer, yields the capillary wall cationic and reverses the EOF. Two cationic materials exhibiting this effect are cetyltrimethyl-ammonium bromide (CTAB) (82) and hexadimethrine bromide (polybrene) (80, 81, 90, 91). Under these conditions, separations are performed at a pH below the protein's pI so the cationic proteins will be repelled from the cationic wall surface. These materials can be coated onto the capillary surface before the separation, and remain stable for many runs after which the capillary wall can be recoated (83).

Córdova et. al. (80) compared the effects of several cationic polymers on the separation of a lysozyme charge ladder. Fig. 11 shows separations performed in capillaries coated with three different polymers. Polybrene and polyethyleni-mine proved to be stable and provided efficient separations of the charge ladder. Separation under the same conditions in an uncoated capillary did not provide resolution of the charge ladder.

Fluorinated surfactants have also been shown to perform in a similar manner. Emmer et. al. (88) used Fluorad FC 134 (3M Company) in the run buffer to separate basic proteins. The authors observed improved separations and good reproducibility they attributed to an increase in the coating stability due to the more stable bilayer formed by the extremely nonpolar tail of the fluorsurfactant.

Dynamic coatings have garnered considerable interest due to their relative

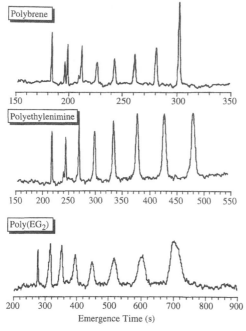

**Fig. 11** Comparison of dynamic coatings for an electrophoretic separation. Separation of a lysozyme charge ladder using polybrene, PEI, and poly(EG$_2$) dynamic polymer coatings. Differences in intensities are not significant. New capillaries were rinsed for 15 min with 0.1 mM NaOH then 15 min with water. The capillary was coated by flushing it with 7.5% (w/v) polymer solution prepared in 25 mM Tris-192 mM Gly buffer at pH 8.3 for 15 min. The capillary was then rinsed with electrophoresis buffer (absence of polymer) for 5 min. Experimental conditions: 25 mM Tris-192 mM Gly at pH 8.3; vacuum injection 8 nL; 30 kV, 5-6 μA, negative polarity; 50 μm × 72 cm (38 cm to detector) capillary. Abbreviation: poly(EG$_2$) - Poly(methoxyethoxyethyl)ethylenimine. From Ref. 80.

ease of implementation, cost-effectiveness and the potential for obtaining effective separations without having to work at pH extremes or high ionic strengths. However, we still have much to learn regarding generation of a pH stable capillary surface for high efficiency separation of proteins.

*b.  Covalent*

Covalently coated capillaries, also called functionalized capillaries, attempt to produce permanently-modified capillary surfaces which provide reproducible and efficient separations. Unlike HPLC, the coatings should not provide any solute interaction and ideally the surface should be hydrophilic to reduce the potential of hydrophobic binding. Several commercially-available coated capillaries as well as many "homemade" capillaries have been shown to be effective (Table 5). Potential disadvantages of this approach for reducing protein-wall interactions include cost, production time, stability, efficiency, and the requirement for the use of additional buffer additives.

The first materials used as capillary coatings were polyacrylamide and, separately, methylcellulose by Hjertén (67). These materials seemed the obvious choice due to their well established history with protein separation but proved to have stability, efficiency and lifetime limitations. Improvements in coating procedures have decreased these problems and polyacrylamide-coated capillaries are now available commercially. A consideration when using these capillaries is the potential for hydrophobic interactions between the proteins and the capillary surface. This can be circumvented through the use of surfactants added to the buffer which function to mask the hydrophobic wall sites (74, 75). The EOF will also be reduced, as with the use of all nonionic coated capillaries, and reversal of electrode polarity may also be necessary depending on the separation conditions.

An interesting approach to permanent capillary coating has been developed by Van Tassel et. al. (104). These authors irreversibly bound the protein fibrinogen to the inside of a fused silica capillary using a thermal treatment. After thermal treatment, desorption of the coating when exposed to an electric field was

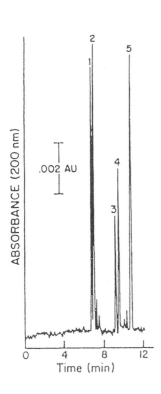

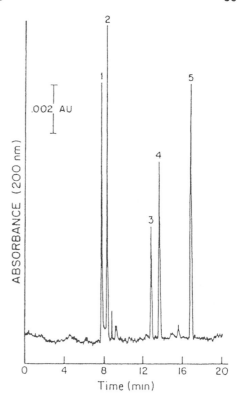

**Fig. 12** Capillary electrophoretic separation of model basic proteins in surfactant coated capillaries. Covalently coated capillaries (alkylsilane) were dynamically coated with (A) Tween-20 and (B) Brij-35 nonionic surfactants. Experimental conditions: 10 mM phosphate buffer at pH 7.0; 300 V/cm; 75 μm × 50 cm. Labeled peaks: 1 = lysozyme, 2 = cytochrome, 3 = ribonuclease A, 4 = α-chymotrypsinogen, 5 = myoglobin. From Ref. 74.

eliminated. The authors are investigating the effectiveness of this coating for CE separations.

An interlocked and "fuzzy" polyether coating produced by Nashabeh and El Rassi was used successfully for the separation of a variety of proteins (70). The reduction in EOF could be adjusted using higher molecular weight polyethylene glycol for a larger reduction without any change in analyte-wall interactions. Capillary lifetime in this case was several weeks when using a separation buffer pH in the 6–7 range.

Columns hydrophobically-coated can be used in conjunction with surfactant buffer additives to facilitate the separation of acidic or basic proteins. A hydrophilic capillary surface is created which prevents both ionic and hydrophobic interactions. Figure 12 illustrates the effect this type of capillary coating can have on a separation. Common surfactants are Brij-35 and tweens which have been used at concentrations of 0.01% in the run buffer (74).

### 4. Sieving Matrices (CGE)

Sieving matrixes are the fundamental separation medium when performing capillary gel electrophoresis (CGE). This technique parallels slab gel electrophoresis which has been a fundamental tool for protein analysis for decades (as discussed in section I.B.2.) which exploits differences in mass for separation.

**Table 7**  Commonly Used Gels in CGE and Select Examples from the Literature

|  | Notes | References |
|---|---|---|
| Chemical Gel |  |  |
| Polyacrylamide | 3-5% T, proteins ≤ 97.4 kD | (105) |
|  | 2-9% T, 5% C, proteins ≤ 116kD | (106) |
| Physical gels |  |  |
| Polyacrylamide | 10% | (107) |
|  | 4%, uncoated capillaries | (108) |
|  | ProSort™ SDS-Protein Analysis Kit (Applied Biosystems), Proteins ≤ 200 kD | (109) |
|  | 1.5%, polyacrylamide coated capillary | (110) |
| Dextran (branched) | 10% dextran (MW 2,000,000) | (111) |
| Dextran (oligomeric) | 10% dextran (MW ~ 1000) | (112) |
| Polyethylene glycol | 3% (MW 100,000) | (111) |
|  | eCAP SDS-200 kit (Beckman Instruments) | (113) |
|  | eCAP SDS 14–200 kit (Beckman Instruments) | (114, 115) |
| Polyethylene oxide | MW 100,000 | (116) |
| Pullulan | 7%, polyacrylamide coated capillary | (117) |
|  |  | (118) |
|  | 3–10%, polyacrylamide coated capillary | (119) |

As discussed earlier, sieving matrices for CE-based protein separations fall into two categories, chemical and physical gels. Chemical gels are crosslinked polymers that are created in the capillary, while physical gels involve entangled, low viscosity polymers that can easily be pumped in and out of the capillary. The latter usually require a coated capillary to prevent EOF effects.

Table 7 provides a list of some of the gels used in CGE and reference examples. Although chemical gels consistently provide high resolution separations even on smaller proteins (< 10 kD), they are associated with many practical problems. The chemical gels are prone to bubble formation under high applied voltages and injection must be performed electrokinetically with samples that are reasonably clean. Since the gels are polymerized in-capillary, degradation of the gel, which occurs over the course of 20–30 runs, requires replacing the entire capillary, not just the gel. Producing a satisfactory chemical gel-filled capillary is sometimes difficult but

fortunately, various gel prefilled capillaries are available from several manufacturers (e.g., Beckman, Perkin-Elmer/Applied Biosystems, BioRad, and SRI).

Physical gels can be produced from branched or unbranched polymers and are pressure-filled into a capillary which, typically, is coated. These gels allow pressure and electrokinetic injections and have the advantage of being easily replaceable. Physical gels that have provided successful separations include acrylamide, dextrans, polyethylene glycol (PEG), polyethylene oxide (PEO), and a variety of cellulose derivatives (see Table 7, several SDS-CGE kits employing physical gels are available from commercial sources such as Beckman, BioRad, and Perkin-Elmer/ABI). As mentioned earlier, a disadvantage of physical gels is that a coated capillary is typically necessary to minimize the EOF preventing the sieving matrix from flowing towards the outlet changing the matrix's structure. Wu and Regnier (108) compared different acrylamide concentrations in coated and uncoated

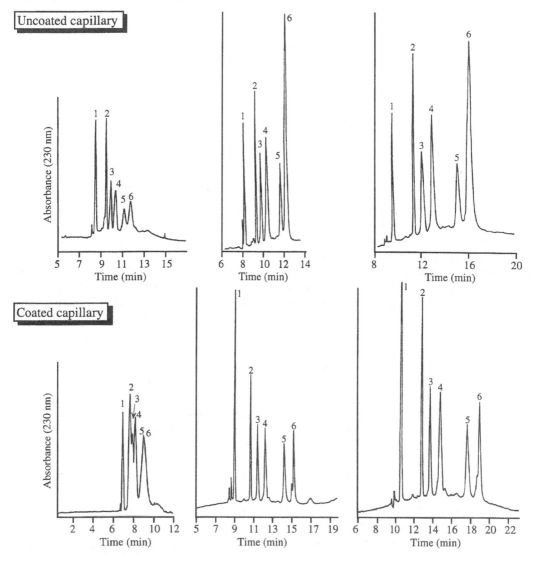

**Fig. 13** Capillary gel electrophoretic separation dependence on capillary coating and acrylamide concentration. Separation of protein molecular mass standards in polyacrylamide coated and uncoated capillaries with 3% (left), 4% (middle) and 6% (right) linear acrylamide physical gels. Experimental conditions: 100 mM Tris-250 mM borate buffer at pH 8.1; 0.1% SDS; 10 s electrophoretic injection; separation potential, 12 kV; 75 μm × 45 cm (25 cm to detector). Labeled peaks: 1 = α-lactalbumin (MW = 14,400); 2 = carbonic anhydrase (MW = 29,000); 3 = glyceraldehyde-3-phosphatedehydrogenase (MW = 36,000); 4 = albumin (chicken egg) (MW = 45,000); 5 = albumin (bovine) (MW = 66,000); 6 = conalbumin (MW = 78,000). From Ref. 108.

capillaries (Fig. 13). They found above a viscosity of 100 centipoise (cP), the potential for EOF was overcome by the viscosity of the polymer, and no longer affected the efficiency of the separations.

## B. Glycoprotein Separations

Glycoproteins add another degree of complexity to the CE process due to the presence of oligosaccharide chains. These carbohydrates typically do not possess any charge to aid in CZE or CIEF separations. Nonetheless, using CE as a separation tool, structural studies and even fraction collection have been successful. With an increased database of separation protocols, CE is becoming an efficient alternative to many traditional techniques, particularly for glycoproteins that are difficult to resolve by traditional slab gel approaches [for an extensive listing of applications see ref. (9)].

The same CE techniques for protein analysis, CZE, CIEF, and CGE have been used for glycoprotein analysis. CZE has been used predominantly as a separation technique for glycoproteins but methods for studying antibody-antigen binding (120–122) and glycoprotein structural studies [via CE-MS, (53)] have also been reported. A common traditional technique for the analysis of glycoproteins is isoelectric focusing (IEF) as a result of its high resolution capabilities. The CE counterpart of IEF, capillary IEF (CIEF), has been successfully used for the separation of the glycoforms of transferrin from human serum (123), recombinant human tissue plasminogen activator [r-tPA, (124, 125)] and recombinant tissue plasminogen activator [r-TPA, (126)]. CGE of SDS-denatured glycoproteins has also been performed (28, 109) but required a Ferguson plot (27) for accurate molecular weight determination.

A method has been developed exploiting the known affinity of borate for compounds containing cis-diol groups. Borate is routinely used as a CE buffer but it was noted by Landers et. al. (127) and Traverna et. al. (128) that used in conjunction with diaminoalkanes (DAA) resolution of glycoproteins was significantly enhanced. Landers et. al. (127) showed only incomplete resolution was achieved when only either the borate or DAA was used for separation. The authors concluded the borate may be complexing with the diol groups on the glycoproteins, thus,

imparting an additional negative charge to the borate-protein complex while the EOF was slowed via the diaminobutane (DAB) adsorbed on the capillary wall. Oda et. al. (129) also determined that using hexamethonium bromide and decamethonium bromide, $\alpha,\omega$-bis-quaternary amines, at lower concentrations then DAB (100–300 iM vs. 1 mM) produced separations of comparable resolution in shorter time periods. The authors indicate these results suggest the borate of the borate-glycoprotein complex also interacts with the capillary wall absorbed DAA to facilitate glycoprotein separations.

There have been a number of interesting studies on this topic. Highly sialylated glycoproteins have been separated using a borate-DAA buffer and a borate buffer where it was found the glycoforms eluted based on increasing number of silaic acid residues (128, 130). Using a borate buffer Hoffstetter-Kuhn et. al. (131) were able to separate oligosaccharide-mediated microheterogeneous monoclonal antibodies where a phosphate buffer failed. Watson and Yao (132) used a tricine-borate buffer with 7M urea to optimize a denaturing separation of glycoforms of a recombinant human erythropoietin, rhEPO.

## IV. THE FUTURE DIRECTION OF CE

### A. Drug Binding Studies

CE was first used for protein binding studies in 1989 by Grossman et. al. (133) when the full potential of CE in this area was just being recognized. The relatively recent adoption of traditional high performance liquid chromatography binding study techniques and the development of CE specific techniques have increased the utility of CE for binding studies.

Techniques showing utility of CE for binding studies include the Hummel-Dreyer, vacancy peak, frontal analysis, affinity capillary electrophoresis (ACE), and vacancy affinity capillary electrophoresis methods [see (23, 134, 135) and the references therein for an inclusive review]. Busch et. al. (134) detailed and contrasted these five methods with respect to the bovine serum albumin-warfarin and human serum albumin-warfarin systems. Frontal analysis was found to be the most favorable due to ease and low sample consumption while agreeing well with accepted binding information. The authors do point out, however, that due to limitations in each technique the five methods are more complimentary than competitive.

Gao et. al. (136) have developed another approach for binding studies, frontal analysis continuous capillary electrophoresis (FACCE), which utilizes continuous sampling. Using the $\beta$-lactoglobulin/sodium poly(styrenesulfonate) system, they show that FACCE can provide lower detection limits and has utility in studying slow binding processes.

Many other areas of research are aiding in advancing binding study analysis via CE. Combinatorial library screening (137) and analysis of glycopeptide antibiotics (138) by ACE have been demonstrated. Approaches for improving detection sensitivity have been explored using capillary electrophoresis with laser-induced fluorescence detection for binding studies with proteins tagged using traditional methods (139, 140) and naturally fluorescent proteins generated recombinantly (141). Ligand concentration effects on binding constant determination have also been studied (142) while others have combined ACE with MEKC to analyze the bioaffinity of actin-avidin and biotin (143). As new methods continue to be developed, such as the multiple-plug binding assay developed by Gomez et. al. (144), it is clear that we have only begun to realize the potential of this technology for binding assay applications.

## B. Electrophoretic Microchip CE

In parallel to the developments in conventional capillary electrophoresis, the exploration of a miniaturized CE technique has emerged. The use of microfabricated techniques to construct chemical measurement systems was first reported about ten years ago (145). Miniaturization in the separation sciences is not a foreign concept and has been realized in almost all areas of the separation techniques e.g., chromatography and electrophoresis (146). Since then, developments, especially in the area of liquid measurement systems, i.e., planar CE, has progressed dramatically. The idea to miniaturize is not new, but a rather logical development of basic concepts and theories that predict that an overall reduction in the dimensions of the flow geometry should result in the enhancement of analytical performance. Microchip CE, originally termed planar CE, comprises the production and use of small structures fabricated by micromachining techniques (147). Micromachining is a combination of film deposition, photolithography and precise etching and bonding techniques to fabricate diverse three-dimensional structures on a suitable substrate. Fabrication of the microchip devices begins with the exposure of a photoresist mask which depicts the design of the channel geometry and is followed by etching and bonding techniques to achieve the devised three-dimensional structure. A variety of materials as substrates have been used in the past, ranging from glass and quartz to plastic and silicon. Recent alternative approaches have utilized sol-gel techniques to fabricate microchip devices as well as other methods which have explored the direct production of chips through laser abrasion from polymer substrates (148–151). The resultant structures have channel architecture in the micrometer scale and, subsequently, are not dramatically different than the channels used in capillary electrophoresis. In contrast to conventional capillary electrophoresis, the higher surface-to-volume ratio in the microchip devices allows for better heat dissipation and, therefore, separations at higher field strengths.

In addition to the predicted enhancement in separation efficiency and performance of such analytical devices, the microchip approach shows a number of other important advantages over current separation technology. The compact design of the devices brings standard liquid separation methods, e.g., liquid chromatography and capillary electrophoresis, into a format that is about 10 times smaller than conventional systems. The reduction in size enables the use of several such miniaturized capillary-based separation methods on one chip device. This puts to rest the stigma associated with the conventional single capillary apparatus that operates in a batch-wise mode, which is considered to be a major disadvantage of this technique in the current laboratory environment. Clearly the ability for fabricating many separation channels with identical geometry into one chip for massive parallel processing of samples, harnesses tremendous analytical power. The high throughput capabilities of these devices are currently very attractive for the needs in the pharmaceutical industry where vast amounts of unknown compounds from combinatorial chemistry techniques need to be explored for their biological and physiochemical properties. Additionally, the mass fabrication of such devices would be relatively inexpensive and, consequently, disposable and easy to use. However, the concept as initially described by Manz and coworkers is much broader, and envisions a miniaturized total analysis system ($\mu$TAS) (145). The concept foresees that laboratory functions other than separation techniques could be integrated into the chip platform. Sample treatment steps, such as concentration steps, cleanup procedures and derivatization protocols preceding or following the actual separation step could be performed on the chip. The realization of the concept would involve advantages in time and cost savings that are at least similar or better than savings experienced with the use of robotic workstations today.

The manipulation and transport of analytes in the microchip devices is realized using electrokinetic phenomena, e.g. electrophoretic and electroosmotic effects. Buffer and sample flows within the channel network can be precisely controlled through the applied potentials at the reservoirs. The compact design of the microchip device, a system with virtually zero dead volume, allows manipulation of picoliter volumes with high precision, which ultimately leads to separation features equivalent or exceeding current separation techniques. Elaborate injection schemes have been worked out allowing either injection in a continuously-proportional mode or injection with a fixed plug length (152–156). Consequently, all CE-based methods described in the preceding chapters can be transferred to the chip platform taking advantage of the efficiency of the separation on the chip and its other virtues. However, the reduction in channel dimensions on the chip has serious consequences in the detection sensitivity of analytes that are monitored by absorbance detection. Detection of proteins in CE and liquid chromatography is usually achieved by absorbance in the UV-visible wavelength range with the detection limit given by the pathlength of the absorbance cell. It is obvious that further reduction in channel diameter in microchip devices are complicating the detection of the analytes by absorbance techniques. Microchip electrophoresis of underivatized proteins using UV absorbance detection was, therefore, not introduced until recently (157). The integration of fiberoptics into the chip device together with the construction of an U-shaped micromachined absorbance cell has proven to be effective to increase the detection limits in absorbance detection methods (158). Although, absorbance detection is more accepted in the detection of analytes in conventional instrumentation due to its wider applicability, the detection of analytes on microchips rely mainly on LIF. Fluorescence detection, routinely employed in CE analysis for detection of DNA is obviously less practical in the detection of proteins and peptides, unless the analytes have natural fluorescence or are derivatized with a fluorescent label.

One particular strength of the microchip is that other laboratory functionalities can also be miniaturized and coupled to the separation without compromising the separation efficiency. The ability to make closely matched channel geometry and junctions of various types has given designers the flexibility to construct chemical reaction chambers and/or channels in which chemical reactions can be performed. It was demonstrated recently that derivatization of amino acids with the fluorescence label *o*-phthaldialdehyde (OPA) was readily achieved in a section of a channel. The reaction, as well as separation conditions on the microchip, were sufficient to obtain fluorescencent-labeled amino acids in a continuous mode with high reproducibility (159). Labeling of proteins and peptides with the appropriate fluorescence tag often introduces heterogeneity that is difficult to resolve and to interpret. In this case, the postcolumn addition of the derivatizing reagent becomes the method of choice. In one example, a postcolumn reactor directly connected to the separation channel was employed together with a sophisticated injection scheme that allowed the labeling of the amino acids residues without extreme band distortion of the preceding separation (160). As an alternative to the chemical derivatization of analytes, the separation of the compounds by microchip electrophoresis combined with the detection and identification by mass spectrometry may be favorable. The minute sample requirements and small flow rates employed in microchip electrophoresis are the ideal prerequisites for the coupling of the microchip with the ion-source of an electrospray ionization mass spectrometer (ESI-MS). This type of detection technique using microfabricated multiple-channel glass chips interfaced to an ESI-MS was successfully demonstrated for the identification of peptides and proteins (161). The combination of an ESI-MS and a high throughput chip device is a very powerful solution for the identification and characterization of new peptides and proteins from combinatorial libraries or natural sources. Other methods suitable for the detection of analytes that, thus far, have not been demonstrated explicitly, await their debut in this field. The use of optical waveguides as detectors in microchip devices using the surface plasmon resonance effect can be named as one of the promising candidates, especially for the detection of protein-protein interactions.

A number of recent reports have demonstrated the application of microchip devices for chemical/biochemical analysis. Enzyme assays, usually very labor intensive at the "benchtop scale," can easily be performed on a microchip device using current microfluidic systems. The precise control of fluid streams was applied to kinetic studies of the reagents $\beta$-galactosidase and resorufin $\beta$-D-galactopyranoside and its inhibition by three different inhibitors. The reagent concentrations were dynamically changed within a single experiment and the hydrolysis product resorufin was monitored by fluorescence detection (162). A clinically relevant immunological assay for the determination of immunoglobulin G and the drug theophylline was also demonstrated on-chip. Protein separations, particularly separations of antibodies and their complexes, are made difficult by adsorption on the capillary wall. The combination of using zwitterionic buffers and surface-active compounds was sufficient to suppress undesired antibody channel-surface interactions and provide reproducible migration times for up to 4 days with this system (163).

The work accomplished to date on microchip electrophoresis suggests that planar capillary electrophoresis is a promising technique that could result in the development of electrophoretic systems to carry out complex sample processing and analysis. The studies carried out so far indicate that further improvements in separation efficiency at the level of planar CE will unlock the true potential of capillary electrophoretic separations. However, in order to apply the methods currently practiced in protein analysis in conventional CE, solutions for problems described above, such as low-level protein detection and undesired protein/peptide capillary wall interaction, must be found. The advantages of micromachined CE systems established up to this time in terms of total analysis times will then be complimented by more sophisticated analysis methods: the total analysis of proteins and peptides, the probing of protein-protein interactions and possibly the characterization of protein drug interactions. These types of protein analyses in low-cost microchip systems will open up the possibility to carry out high throughput screening of drug candidates, making drug discovery an "industrial process."

## C.  Pre-concentration for Proteins and Peptides

Many of the advantages associated with CE arise from the microscalar dimensions of the capillary. However, the small dimensions also provide limitations. The most important of these is detection sensitivity, which is affected by two factors. With UV-VIS absorbance detection, the most universally used mode of detection with CE, sensitivity is dependent on path length at the detector. Unlike HPLC, the path length in CE is the internal diameter of the capillary itself. Having a path length restricted to the maximum capillary i.d. suitable for electrophoretic separation, which is typically $100\,\mu m$, severely handicaps sensitivity. Compounding this problem is the fact that, as a result of the capillaries having total volumes in the microliter range, injected sample volumes are limited to the low nanoliter range. Combined, these factors induce the prerequisite of relatively high sample concentrations (minimum $10^{-6}$–$10^{-8}$ M) for analysis by CE.

It is for this reason that methodology-based manipulations, such as sample stacking (39) and isotachophoresis (ITP) (164), have had a profound effect on the applicability of CE in research and clinical laboratories. Without physically-modifying the capillary, these techniques allow for as much a 50% of the capillary to be filled with sample, with some form of sample zone compression leading to increased sensitivity without significant loss of resolution. While these are elegant approaches for enhancing CE detection sensitivity, they are restricted to sample volumes less than the total capillary volume (i.e., as much as 500 nL depending on the capillary length and diameter). There is only one approach that will allow for sample volumes substantially larger than the total capillary volume and this involves physically-modifying the capillary with an on-line solid phase extraction (SPE) device.

As a result of a variety of terms in the literature which fail to adequately define this approach for on-line sample concentration, we have termed this technique solid phase extraction-capillary electrophoresis (SPE-CE) (165–167). This technique is simple in design and involves incorporation of a concentrator device containing a particulate solid phase, on-line, with a CZE capillary. This concept is, in fact, not new. Significant advancements in this area had previously been reported by Kasicka and Prusik (168) who attached an "adsorption element" to an isotachophoretic apparatus and Guzman et. al. (169) who constructed an "Analyte Concentrator" which was coupled to a CZE capillary. Both of these devices utilized a packed bed of an immunoaffinity resin for specific concentration of analytes. Debets et. al. (170) described a "micro precolumn" (reversed phase, C8) retained within a switching valve that could be switched on-and off-line to a CZE capillary while Schwartz and Merion (171) showed the utility of the Accusep$^{TM}$ C/PRP capillary (Waters) which was similar in construction to the "analyte concentrator" except containing a C18 reversed phase packing instead of an immunoabsorbant.

The SPE-CE technique conforms to the concentration strategy of these earlier studies but is easier to implement, less costly and requires no additional or exotic instrumentation for fabrication of the concentration device. The SPE-CE capillary is constructed from common commercially-available materials and can enhance UV detection sensitivity up to 500-fold. This technology has been shown useful for a variety of analytes including peptides. Strausbauch et al. (166, 167) have shown that 100-500 fold enhancement in sensitivity is possible with this technique and have elucidated the idiosyncrasies associated with SPE-CE of peptides (167). They (166) have also shown the utility for SPE-CE to allow for entire HPLC fractions to be injected, concentrated on-line for electrophoretic analysis. Using an in-gel tryptic digest of BSA, RP-HPLC analysis yielded a chromatogram containing greater than 20 peaks. A series of peaks early in the chromatogram were collected as five individual $100 \, \mu L$ fractions and the acetonitrile evaporated reducing the volume to $55 \, \mu L$. The entire $55 \, \mu L$ fraction was loaded onto the SPE device and then desorbed (using acetonitrile/TFA) and electrophoresed. The second dimensional analysis by CE showed that one of the major peaks in one fraction was not a peptide but, in fact, an acrylamide contaminant resulting from extraction of the peptides after the in-gel digestion. This demonstrates the importance of CE for not only allowing enhancement in detection sensitivity, but also for allowing CE to be used as a second dimensional analysis using a selectivity different than the first dimension technique (in this case HPLC).

## V. CONCLUDING COMMENTS

The arena defining the application of CE to peptide and protein analysis is vast. This chapter was written with the aim of providing the reader with a flavor for the potential of CE as a new analytical tool. It is, by no means, a comprehensive review of the literature but has attempted to utilize, where appropriate, reports in the literature that illustrate the basic concepts associated with this technique. Other pertinent literature that may be of interest to the reader can be found in Table 8.

With the examples referenced, the power of CE for peptide and protein analysis should be clear. Although not discussed, the fact that most CE separations are based on mass-to-charge ratio positions this technology perfectly as complimentary to standard gel electrophoresis and HPLC. While CE has begun to mature as a useful analytical technique, there is still much room for growth. This will come in the form of improvements in method development and instrumentation. It is clear that no universal conditions exist for peptide or protein separations and that individual optimization is required for different analytes based on their structure, charge,

**Table 8**  Pertinent Literature on Separations of Glycosylated or Nonglycosylated Peptides or Proteins

| Subject | Notes* | Technique | Reference | Proteins | Peptides |
|---|---|---|---|---|---|
| Human growth hormone digest | extreme pHs | CZE | (79) | | — |
| Metallothionein isoform separation | | MECK | (57) | | — |
| Insulin receptor peptide phosphorylated isomers | polyacrylamide coated capillary | | (58) | | — |
| Separation of enkephalin-related peptides | aminopropyl coated capillary | MECK | (60) | | — |
| Immunoglobulin A immunoassay | fluorescein-labeled antibody | CZE | (121) | — | |
| Immunoglobulin G isoelectrotype separation | | CZE | (122) | — | |
| Monoclonal antibodies against phosphotyrosine | | CZE | (120) | — | |
| Angiotensin II analog separation | | CZE | (62) | | — |
| Tranferrin isoform analysis | pH 2.0, Tween 20 | CIEF | (123) | — | |
| Prolidase deficiency analysis | pentanesulfonate micelles | MECK | (59) | | — |
| Cerebrospinal fluid low MW proteins | | CGE | (114) | — | |
| Direct urine and serum analysis | | CZE | (172) | — | |
| Human plasma/serum protein separations | non-denaturing conditions | CIEF | (173) | — | |
| B-trace protein | | CZE, CGE, CIEF | (115) | — | |
| PEG modified proteins | dynamic coating | CZE | (82) | — | — |
| CTLA4Ig fusion protein conformational transition kinetic studies | | MECK | (174) | — | |
| Metalloproteins (conalbumin, transferrin and metallothionein) | | CIEF | (175) | — | |
| Various recombinant proteins | Ferguson method | CGE | (116) | — | |

| | | | | |
|---|---|---|---|---|
| Recombinant insulin-like growth factor I variants | zwitterionic detergents | MEKC | (176) | — |
| Human recombinant erythropoietin glycoform separation | tricine-DAB/urea buffer | | (132) | — |
| Recombinant tissue plasminogen activator glyco-isoform separation | denatured (4M urea) dynamic coating (HPMC or PEG) | CIEF | (126) | — |
| Recombinant human erythropoietin drug preparation separations | 1 mM nickel chloride in buffer | CZE | (177) | — |
| Hepatitis C viral recombinant protein mapping | | CZE | (49) | — |
| Recombinant human tissue plasminogen activator (r-tPA₁) glycoprotein isoforms | | CZE CIEF | (124) | — |
| Recombinant tissue plasminogen activator glycoprotein analysis | | CZE CIEF CGE | (178) | — |
| *Lipoproteins* | | | | |
| Separation of apolipoprotein B 48 (and others) | | Dynamic sieving CE | (179) | — |
| Analysis of apolipoprotein A-I | | CZE | (180) | — |

*Where information was available.

hydrophobicity and mass characteristics. In light of this, successful analysis of any particular protein or peptide will require access to an arsenal of proven strategies, such as those discussed in this review. It is also likely that instrumentation will be improved. As more sensitive detection systems are devised and the instrumentation simplified, it is not unreasonable to expect that the separation protocols for proteins and peptides will eventually become a robust, turn-key operation applied to a diverse array of applications. The possibility exists that electrophoresis on glass or plastic chips will provide that "simple" platform for capillary electrophoresis in the next millenium.

## REFERENCES

1. TE Creighton. Proteins: Structure and Molecular Properties. San Francisco: WH Freeman and Co, 1983, p 515.
2. DS Goodsell, AJ Olson. Soluble proteins: size, shape and function. Trends in Biochemical Sciences 18: 65–68, 1993.
3. TW Rademacher, RB Parekh, RA Dwek. Glycobiology. Ann Rev Biochem 57: 785–838, 1988.
4. C Miller, J Rivier. Peptide chemistry: development of high performance liquid chromatography and capillary zone electrophoresis. Biopolymers 40(3): 265–317, 1996.
5. T Wehr, R Rodriguez-Diaz, CM Liu. Capillary electrophoresis of proteins. Advances Chromatogr 37: 237–361, 1997.
6. TA van de Goor. Capillary electrophoresis of proteins. Pharma Biotech 7: 301–327, 1995.
7. E Szökö. Protein and peptide analysis by capillary zone electrophoresis and micellar electrokinetic chromatography. Electrophoresis 18: 74–81, 1997.
8. R Lehmann, W Voelter, HM Liebich. Capillary electrophoresis in clinical chemistry. J Chromatogr B 697: 3–35, 1997.
9. RP Oda, BJ Madden, JP Landers. Capillary electrophoretic analysis of glycoproteins and glycoprotein-derived oligosaccharides. In: PR Brown, E Grushka, eds. Advances in Chromatography v. 36. New York: Mercel Dekker, 1996, pp 163–199.
10. K Kakehi, S Honda. Analysis of glycoproteins, glycopeptides and glyco-protein-derived oligosaccharides by high-performance capillary electrophoresis. J Chromatogr A 720: 377–393, 1996.
11. KA Denton, SA Tate. Capillary electrophoresis of recombinant proteins. J Chromatogr B 697: 111–121, 1997.
12. MV Novotny. Capillary biomolecular separations. J Chromatogr B 689: 55–70, 1997.
13. DK Lloyd. Capillary electrophoretic analyses of drugs in body fluids: sample pretreatment and methods for direct injection of biofluids. J Chromatogr A 735: 29–42, 1996.
14. RP Oda, R Clark, JA Katzmann, JP Landers. Capillary electrophoresis as a clinical tool for the analysis of protein in serum and other body fluids. Electrophoresis 18: 1715–1723, 1997.
15. KD Altria. Determination of drug-related impurities by capillary electrophoresis. J Chromatogr A 735: 43–56, 1996.
16. J Oravcova, B Bohs, W Lindner. Drug protein binding studies: New trends in analytical and experimental methodology. J Chromatogr B 677: 1–28, 1996.
17. PAHM Wijnen, MP von Dieijen-Visser. Capillary electrophoresis of serum proteins: Reproducibility, comparison with agarose gel electrophoresis and a review of the literature. European J Clin Chem Clin Biochem 34: 535–545, 1996.
18. TJ Pritchett. Capillary isoelectric focusing of proteins. Electrophoresis 17: 1195–1201, 1996.
19. C Schwer. Capillary isoelectric focusing: A routine method for protein analysis? Electrophoresis 16: 2121–2126, 1995.
20. X Liu, Z Sosic, IS Krull. Capillary isoelectric focusing as a tool in the examination of antibodies, peptides and proteins of pharmaceutical interest. J Chromatogr A 735: 165–190, 1996.
21. BL Karger, F Foret, J Berka. Capillary electrophoresis with polymer matrices: DNA and protein separation and analysis. Methods Enzymol 271: 293–319, 1996.
22. A Guttman. Capillary sodium dodecyl sulfate-gel electrophoresis of proteins. Electrophoresis 17: 1333–1341, 1996.
23. K Shimura, K Kasai. Affinity capillary electrophoresis: A sensitive tool for the study of molecular interactions and its use in microscale analyses. Anal Biochem 251: 1–16, 1997.
24. JP Landers, ed. Handbook of Capillary Electrophoresis, 2nd ed. Boca Raton: CRC Press, 1997.
25. R Weinberger. Practical Capillary Electrophoresis. San Diego: Academic Press, Inc., 1993.

26. PD Grossman, JC Colburn. Capillary Electrophoresis. In PD Grossman, JC Colburn, eds. Theory and Practice. San Diego: Academic Press, Inc. 1992, pp 162.

27. KA Ferguson. Starch-gel electrophoresis-Application to the classification of pituitary proteins and polypeptides. Metabolism 13: 985–1002, 1964.

28. WE Werner, DM Demorest, JE Wiktorowicz. Automated Ferguson analysis of glycoproteins by capillary electrophoresis using a replaceable sieving matrix. Electrophoresis 14: 759–763, 1993.

29. JR Mazzeo, IS Krull. Capillary isoelectric focusing of proteins in uncoated fused-silica capillaries using polymeric additives. Anal Chem 63: 2852–2857, 1991.

30. J Wu, J Pawliszyn. Fast analysis of proteins by isoelectric focusing performed in capillary array detected with concentration gradient imaging system. Electrophoresis 14: 469–474, 1993.

31. RL Chien, DS Burgi. Sample stacking of an extremely large injection volume in high-performance capillary electrophoresis. Anal Chem 64: 1046–1050, 1992.

32. AW Stowers, KJ Spring, A Saul. Preparative scale purification of recombinant proteins to clinical grade by isotchophoresis. Bio Technology 13: 1498–1503, 1995.

33. J Caslavska, P Gebauer, W Thormann. Fractionation of human serum proteins by capillary and recycling isotachophoresis. Electrophoresis 15: 1167–1175, 1994.

34. BJ Wanders, FM Everaerts. Isotachophoresis in Capillary Electrophoresis. In: JP Landers, ed. Handbook of Capillary Electrophoresis. Boca Raton: CRC Press, 1994, pp 111–128.

35. DT Witte, S Nagard, M Larsson. Improved sensitivity by on line isotachophoretic preconcentration in the capillary zone electrophoretic determination of peptide like solutes. J Chromatogr A 687: 155–166, 1994.

36. F Foret, E Szoko, BL Karger. On-column transient and coupled column isotachophoretic preconcentration in capillary zone electrophoresis. J Chromatogr 608: 3–12, 1992.

37. F Foret, E Szoko, BL Karger. Trace analysis of proteins by capillary zone electrophoresis with on-column transient isotachophoretic preconcentration. Electrophoresis 14: 417–428, 1993.

38. D Kaniansky, J Marak. On-line coupling of capillary isotachophoresis with capillary zone electrophoresis. J Chromatogr 498: 191–204, 1990.

39. R Chien, DS Burgi. On-column sample concentration using field amplification in CZE. Anal Chem 64(8): 489A–496A, 1992.

40. RP Oda, JP Landers. Introduction to capillary electrophoresis. In: JP Landers, ed. Handbook of Capillary Electrophoresis, 2nd ed. New York: CRC Press, 1997, pp 1–47.

41. R Aebersold, H Morrison. Analysis of dilute peptide samples by capillary zone electrophoresis. J Chromatogr 516: 79–88, 1990.

42. SL Pentoney Jr., JV Sweedler. Optical detection techniques for capillary electrophoresis. In: JP Landers, ed. Handbook of Capillary Electrophoresis. 2nd ed. New York: CRC Press, 1997, pp 379–423.

43. C Haber. Electrochemical detection in capillary electrophoresis. In: JP Landers, ed. Handbook of Capillary Electrophoresis, 2nd ed. New York: CRC Press, 1997, pp 425–447.

44. JH Wahl, DC Gale, RD Smith. Sheathless capillary electrophoresis-electrospray ionization mass spectrometry using 10 $\mu$m I.D. capillaries: analyses of tryptic digests of cytochrome c. J Chromatogr A 659: 217–222, 1994.

45. OW Reif, R Lausch, T Scheper, R Freitag. Fluorescein isothiocyanate-labeled protein G as an affinity ligand in affinity/immunocapillary electrophoresis with fluorescence detection. Anal Chem 66: 4027–4033, 1994.

46. M Girard, HP Bietlot, TD Cyr. Characterization of human serum albumin heterogeneity by capillary zone electrophoresis and electrospray ionization mass spectrometry. J Chromatogr A 772: 235–242, 1997.

47. DP Kirby, JM Thorne, WK Gotzinger, BL Karger. A CE/ESI-MS interface for stable, low-flow operation. Anal Chem 68: 4451–4457, 1996.

48. J Wu, MG Qian, MX Li, L Liu, DM Lubman. Use of an ion trap storage/reflectron time-of-flight mass spectrometer as a rapid and sensitive detector for capillary electrophoresis in protein digest analysis. Anal Chem 68: 3388–3396, 1996.

49. MA Winkler, S Kundu, TE Robey, WG Robey. Comparative peptide mapping of a hepatitis C viral recombinant protein by capillary electrophoresis and matrix-assisted laser desorption time-of-mass spectrometry. J Chromatogr A 744: 177–185, 1996.

50. N Bihoreau, C Ramon, M Lazard, JM Schmitter. Combination of capillary electrophoresis and matrix-assisted laser desorption ionization mass spectrometry

for glycosylation analysis of a human monoclonal anti-Rhesus(D) antibody. J Chromatogr B 697: 123–133, 1997.

51. R Bonfichi, C Sottani, L Colombo, JE Coutant, E Riva, D Zanette. Preliminary investigation of glycosylated proteins by capillary electrophoresis and capillary electrophoresis/mass spectrometry using electrospray ionization and by matrix-assisted laser desorption ionization/time-to-flight mass spectrometry. Rapid Comm Mass Spectrometry Spec No: S95–S106, 1995.

52. B Yeung, TJ Porter, JE Vath. Direct isoform analysis of high-mannose-containing glycoproteins by on-line capillary electrophoresis electrospray mass spectrometry. Anal Chem 69: 2510–2516, 1997.

53. JF Kelly, SJ Locke, L Ramaley, P Thibault. Development of electrophoretic conditions for the characterization of protein glycoforms by capillary electrophoresis-electrospray mass spectrometry. J Chromatogr A 720: 409–427, 1996.

54. GA Valaskovic, NL Kelleher, FW McLafferty. Attomole protein characterization by capillary electrophoresis-mass spectrometry. Science 273: 1199–1201, 1996.

55. VJJ Hilser, GD Worosila, SE Rudnick. Protein and peptide mobility in capillary zone electrophoresis. A comparison of existing models and further analysis. J Chromatogr 630: 329–336, 1993.

56. N Onyewuenyi, P Hawkins. Separation of toxic peptides (microcystins) in capillary electrophoresis, with the aid of organic mobile phase modifiers. J Chromatogr 749: 271–277, 1996.

57. JH Beattie, MP Richards. Analysis of metallothionein isoforms by capillary electrophoresis: optimization of protein separation conditions using micellar electrokinetic capillary chromatography. J Chromatogr A 700: 95–103, 1995.

58. T Tadey, WC Purdy. Capillary electrophoretic resolution of phosphorylated peptide isomers using micellar solutions and coated capillaries. Electrophoresis 16: 574–579, 1995.

59. R Grimm, G Zanaboni, S Viglio, KM Dyne, G Cetta, P Iadarola. Effect of different surfactants on the separation by micellar electrokinetic chromatography of a complex mixture of dipeptides in urine of prolidase-deficient patients. J Chromatogr B 698: 47–57, 1997.

60. M Thorsteinsdóttir, R Isaksson, D Westerlund. Performance of amino-silylated fused-silica capillaries for the separation of enkephalin-related peptides by capillary zone electrophoresis and micellar electrokinetic chromatography. Electrophoresis 16: 557–563, 1995.

61. J Varghese, RB Cole. Cetyltrimethylammonium chloride as a surfactant buffer additive for reversed polarity capillary electrophoresis electrospray mass spectrometry. J Chromatogr A 652: 369–376, 1993.

62. N Matsubara, K Koezuka, S Terabe. Separation of eleven angiotensin II analogs by capillary electrophoresis with a nonionic surfactant in acidic media. Electrophoresis 16: 580–583, 1995.

63. BY Gong, JW Ho. Effect of zwitterionic surfactants on the separation of proteins by capillary electrophoresis. Electrophoresis 18: 732–735, 1997.

64. KF Greve, W Nashabeh, BL Karger. Use of zwitterionic detergents for the separation of closely related peptides by capillary electrophoresis. J Chromatogr A 680: 15–24, 1994.

65. HK Kristensen, SH Hansen. Separation of polymyxins by micellar electrokinetic capillary chromatography. J Chromatogr 628: 309–315, 1993.

66. MJ Sepaniak, AC Powell, DF Swaile, RO Cole. Fundamentals of micellar electrokinetic capillary electrophoresis. In: PD Grossman, JC Colburn, eds. Capillary Electrophoresis: Theory and Practice. San Diego: Academic Press, Inc. 1992, p 162.

67. S Hjertén. High performance electrophoresis: Elimination of electroendoosmosis and solute adsorption. J Chromatogr 347: 191–198, 1985.

68. KA Cobb, V Dolnik, M Novotny. Electrophoretic separations of proteins in capillaries with hydrolytically stable surface structures. Anal Chem 62: 2478-2483, 1990.

69. D Schmalzing, CA Piggee, F Foret, E Carrilho, BL Karger. Characterization and performance of a neutral hydrophilic coating for the capillary electrophoretic separation of biopolymers. J Chromatogr A 652: 149–159, 1993.

70. W Nashabeh, ZE Rassi. Capillary zone electrophoresis of proteins with hydrophilic fused-silica capillaries. J Chromatogr 559: 367–383, 1991.

71. S Hjertén, K Kubo. A new type of pH- and detergent-stable coating for elimination of electroendoosmosis and adsorption in (capillary) electrophoresis. Electrophoresis 14: 390–395, 1993.

72. M Huang, J Plocek, M Novotny. Hydrolytically stable cellulose-derivative coatings for capillary electrophoresis of

peptides, proteins and glyconjugates. Electrophoresis 16: 396–401, 1995.

73. Y Mechref, ZE Rassi. Fused-silica capillaries with surface-bound dextran layer crosslinked with diepoxypolyethylene glycol for capillary electrophoresis of biological substances at reduced electroosmotic flow. Electrophoresis 16: 617–624, 1995.

74. JK Towns, FE Regnier. Capillary electrophoretic separations of proteins using nonionic surfactant coatings. Anal Chem 63: 1126-1132, 1991.

75. MX Huang, D Mitchell, M Bigelow. Highly efficient protein separations in high performance capillary electrophoresis, using hydrophilic coatings on polysiloxane bonded columns. J Chromatogr B 677: 77–84, 1996.

76. JT Smith, ZE Rassi. Capillary zone electrophoresis of biological substances with fused silica capillaries having zero or constant electroosmotic flow. Electrophoresis 14: 396–406, 1993.

77. Y Liu, F Ruonong, J Gu. Epoxy resin coatings for capillary zone electrophoretic separation of basic proteins. J Chromatogr A 723: 157–167, 1996.

78. M Chiari, M Nesi, JE Sandoval, JJ Pesek. Capillary electrophoretic separation of proteins using stable, hydrophilic poly-(acryloylaminoethoxyethanol)-coated columns. J Chromatogr A 717: 1–13, 1995.

79. X Ren, Y Shen, ML Lee. Poly(ethylene-propylene glycol)-modified fused-silica columns for capillary electrophoresis using epoxy resin as intermediate coating. J Chromatogr A 741: 115–122, 1996.

80. E Cordova, J Gao, GM Whitesides. Noncovalent polycationic coating for capillaries in capillary electrophoresis of proteins. Anal Chem 69: 1370–1379, 1997.

81. MX Li, L Liu, J Wu, DM Lubman. Use of polybrene capillary coating in capillary electrophoresis for rapid analysis of hemoglobin variants with on-line detection via an ion trap storage/reflectron time-of-flight mass spectrometer. Anal Chem 69: 2451–2456, 1997.

82. RL Cunico, V Gruhn, L Kresin, DE Nitecki, JE Wiktorowicz. Characterization of polyethylene glycol modified proteins using charge-reversed capillary electrophoresis. J Chromatogr 559: 467–477, 1991.

83. P Thibault, C Paris, S Pleasance. Analysis of peptides and proteins by capillary electrophoresis/mass spectrometry using acidic buffers and coated capillaries. Rapid Commun Mass Spectrom 5: 484–490, 1991.

84. M Morand, D Blass, E Kenndler. Reduction in capillary zone electrophoresis of a basic single-chain antibody fragment by a cationic polymeric buffer additive. J Chromatogr B 691: 192–196, 1997.

85. N Iki, ES Yeung. Non-bonded poly(ethylene oxide) polymer-coated column for protein separation by capillary electrophoresis. J Chromatogr 731: 273–282, 1996.

86. A Cifuentes, MA Rodríguiz, FJ García-Montelongo. Separation of basic proteins in free solution capillary electrophoresis: effect of additive, temperature and voltage. J Chromatogr A 742: 257–266, 1996.

87. FS Stover, BL Haymore, RJ McBeath. Capillary zone electrophoresis of histidine-containing compounds. J Chromatogr 470: 241–250, 1989.

88. A Emmer, M Jansson, J Roeraade. Improved capillary zone electrophoretic separation of basic proteins, using a fluorosurfactant buffer additive. J Chromatogr 547: 544–550, 1991.

89. D Corradini, G Cannarsa. N,N,N',N'-Tetramethyl-1,3-butanediamine as effective running electrolyte additive for efficient electrophoretic separation of basic proteins in bare fused-silica capillaries. Electrophoresis 16: 630–635, 1995.

90. JE Wiktorowicz, JC Colburn. Separation of cationic proteins via charge reversal in capillary electrophoresis. Electrophoresis 11: 769–773, 1990.

91. K Tsuji, RJ Little. Charge-reversed, polymer-coated capillary column for the analysis of a recombinant chimeric glycoprotein. J Chromatogr 594: 317–324, 1992.

92. N Cohen, E Grushka. Controlling Electroosmotic Flow In Capillary Zone Electrophoresis. J Cap Ele 1: 167–175, 1994.

93. M Dong, RP Oda, MA Strausbauch, PJ Wettstein, JP Landers, LJ Miller. Hydrophobic peptide mapping of clinically relavent heptathelical membrane proteins by capillary electrophoresis. Electophoresis 19: 1767–1774, 1997.

94. ME Roche, MA Anderson, RP Oda, LB Riggs, MA Strausbauch, R Okazaki, PJ Wettstein, JP Landers. Capillary electrophoresis of insulin-like growth factors: enhanced uv detection using dynamically-coated capillaries and on-line solid phase extraction (SPE-CE). Anal Biochem, 258: 87–95, 1998.

95. RS Rush, PL Derby, TW Strickland, MF Rohde. Peptide mapping and evaluation of glycopeptide microheterogeneity derived

form endoproteinase digestion of erythropoietin by affinity high-performance capillary electrophoresis. Anal Chem 65: 1834–1842, 1993.

96. JD Olechno, JA Nolan. Carbohydrate analysis by capillary electrophoresis. In JP Landers, ed. Handbook of Capillary Electrophoresis, 2nd ed. Boca Raton: CRC Press, 1997, pp 297–345.

97. RM McCormick. Capillary zone electrophoretic separation of peptides and proteins using low pH buffers in modified silica capillaries. Anal Chem 60: 2322–2328, 1988.

98. GM McLaughlin, JA Nolan, JL Kindahl, RH Palmieri, KW Anderdson, SC Morris, JA Morrison, TJ Bronzert. Pharmaceutical drug separations by HPCE: practical guidelines. J Liq Chromatogr 15: 961–1021, 1992.

99. MK Weldon, CM Arrington, PL Runnels, JF Wheeler. Selectivity enhancement for free zone capillary electrophoresis using conventional ion-pairing agents as complexing additives. J Chromatogr A 758: 293–302, 1997.

100. GN Okafo, A Vinther, T Kornfelt, P Camilleri. Effective ion-pairing for the separation of basic proteins in capillary electrophoresis. Electrophoresis 16: 1917–1921, 1995.

101. RJ Nelson, DS Burgi. Temperature control in capillary electrophoresis. In: JP Landers, ed. Handbook of Capillary Electrophoresis. New York: CRC Press, 1994, pp 549–562.

102. MM Bushey, JW Jorgenson. Capillary electrophoresis of proteins in buffers containing high concentrations of zwitterionic salts. J Chromatogr 480: 301–310, 1989.

103. RG Nielsen, EC Rickard. Method optimization in capillary zone electrophoretic analysis of hGH tryptic digest fragments. J Chromatogr 516: 99–114, 1990.

104. PA van Tassel, D Miras, A Hagege, M Leroy, J Voegel, P Schaaf. Control of protein adsorption in capillary electrophoresis via an irreversibly bound protein coating. J Col Inter Sci 183: 269–273, 1996.

105. K Tsuji. High-performance capillary electrophoresis of proteins: Sodium dodecyl sulphate-polyacrylamide gel-filled capillary column for the determination of recombinant biotechnology-derived proteins. J Chromatogr 550: 823–830, 1991.

106. T Manabe. Sodium dodecyl sulfate-gel electrophoresis of proteins employing short capillaries. Electrophoresis 16: 1468–1473, 1995.

107. A Widhalm, C Schwer, D Blaas, E Kenndler. Capillary zone electrophoresis with a linear, non-cross-linked polyacrylamide gel: separation of proteins according to molecular mass. J Chromatogr 549: 446–451, 1991.

108. D Wu, FE Regnier. Sodium dodecyl sulfated-capillary gel electrophoresis of proteins using non-cross-linked polyacrylamide. J Chromatogr 608: 349–356, 1992.

109. WE Werner, DM Demorest, J Stevens, JE Wiktorowicz. Size-dependent separation of proteins denatured in SDS by capillary electrophoresis using a replaceable sieving matrix. Anal Biochem 212: 253–258, 1993.

110. K Hebenbrock, K Schügerl, R Freitag. Analysis of plasmid-DNA and cell protein of recombinant Escherichia coli using capillary gel electrophoresis. Electrophoresis 14: 753–758, 1993.

111. K Ganzler, KS Greve, AS Cohen, BL Karger, A Guttman, N Cooke. High-performance capillary electrophoresis of SDS-protein complexes using UV-transparent polymer networks. Anal Chem 64: 2665–2671, 1992.

112. MR Karim, JC Janson, T Takagi. Size-dependent separation of proteins in the presence of sodium dodecyl sulfate and dextran in capillary electrophoresis: effect of molecular weight of dextran. Electrophoresis 15: 1531–1534, 1994.

113. A Guttman, J Nolan, N Cooke. Capillary sodium dodecyl sulfate gel electrophoresis of proteins. J Chromatogr 632: 171–175, 1993.

114. A Hiraoka, R Arato, I Tominaga, N Eguchi, H Oda, Y Urade. Analysis of low-molecular-mass proteins in cerebrospinal fluid by sodium dodecyl sulfate capillary gel electrophoresis. J Chromatogr B 697: 141–147, 1997.

115. A Hiraoka, T Arato, I Tominaga, A Anjyo. Capillary electrophoresis analysis of $\beta$-trace protein and other low molecular weight proteins in cerebrospinal fluid from patients with central nervous system diseases. J Pharmaceutical Biomed Analysis 15: 1257–1263, 1997.

116. K Benedek, S Thiede. High-performance capillary electrophoresis of proteins using sodium dodecyl sulfate-poly(ethylene oxide). J Chromatogr A 676: 209–217, 1994.

117. M Nakatani, A Shibukawa, T Nakagawa. High-performance capillary electrophoresis of SDS-proteins using pullulan solution as separation matrix. J Chromatogr A 672: 213–218, 1994.

118. M Nakatani, A Shibukawa, T Nakagawa. Separation mechanism of pullulan solution-filled capillary electrophoresis of sodium dodecyl sulfate-proteins. Electrophoresis 17: 1584–1586, 1996.

119. M Nakatani, A Shibukawa, T Nakagawa. Effect of temperature and viscosity of sieving medium on electrophoretic behavior of sodium dodecyl sulfate-proteins on capillary electrophoresis in presence of pullulan. Electrophoresis 17: 1210–1213, 1996b.

120. NHH Heegaard. Determination of antigen-antibody affinity by immuno-capillary electrophoresis. J Chromatogr 680: 405–412, 1994.

121. FTA Chen. Characterization of charge-modified and fluorescein-labeled antibody by capillary electrophoresis using laser-induced fluorescence: Application to immunoassay of low level immunoglobulin A. J Chromatogr A 680: 419–423, 1994.

122. BJ Compton. Electrophoretic mobility modeling of proteins in free zone capillary electrophoresis and its application to monoclonal antibody microheterogeneity analysis. J Chromatogr 559: 357–366, 1991.

123. F Kilár, S Hjertén. Fast and high resolution analysis of human serum transferrin by high performance isoelectric focusing in capillaries. Electrophoresis 10: 23–29, 1989.

124. KW Yim. Fractionation of the human recombinant tissue plasminogen activator (rtPA) glycoforms by high-performance capillary zone electrophoresis and capillary isoelectric focusing. J Chromatogr 559: 401–410, 1991.

125. KG Moorhouse CA Rickel AB Chen. Electrophoretic separation of recombinant tissue-type plasminogen activator glycoforms: Validation issues for capillary isoelectric focusing methods. Electrophoresis 17: 423–430, 1996.

126. J Kubach, R Grimm. Non-native capillary isoelectric focusing for the analysis of the microheterogeneity of glycoproteins. J Chromatogr A 737: 281–289, 1996.

127. JP Landers, RP Oda, BJ Madden, TC Spelsberg. High-performance capillary electrophoresis of glycoproteins: the use of modifiers of electroosmotic flow for analysis of microheterogeneity. Anal Biochem 205: 115–124, 1992.

128. M Traverna, A Balliet, D Biou, M Schulter, R Werner, D Ferrier. Analysis of carbohydrate-mediated heterogeneity and characterization of N-linked oligosaccharides of glycoproteins by high performance capillary electrophoresis. Electrophoresis 13: 359–366, 1992.

129. RP Oda, BJ Madden, TC Spelsberg, JP Landers. α,ù-Bis-quaternary ammonium alkanes as effective buffer additives for enhanced capillary electrophoretic separation of glycoproteins. J Chromatogr A 680: 85–92, 1994.

130. E Watson, F Yao. Capillary electrophoretic separation of recombinant granulocyte-colony-stimulating factor glycoforms. J Chromatogr 630: 442–446, 1993.

131. S Hoffstetter-Kuhn, G Alt, R Kuhn. Profiling of oligosaccharide-mediated microheterogeneity of a monoclonal antibody by capillary electrophoresis. Electrophoresis 17: 418–422, 1996.

132. E Watson, F Yao. Capillary electrophoretic separation of human recombinant erythropoietin (r-HuEPO) glycoforms. Anal Biochem 210: 389–393, 1993.

133. PD Grossman, JC Colburn, HH Lauer, RG Nielsen, RM Riggin, GS Sittampalam, EC Rickard. Application of free-solution capillary electrophoresis to the analytical scale separation of proteins and peptides. Anal Chem 61: 1186-1194, 1989.

134. MHA Busch, LB Carels, HFM Boelens, JC Kraak, H Poppe. Comparison of five methods for the study of drug-protein binding in affinity capillary electrophoresis. J Chromatogr A 777: 311–328, 1997.

135. FA Robey. Use of capillary electrophoresis for binding studies. In: JP Landers, ed. Handbook of Capillary Electrophoresis. New York: CRC Press, 1997, pp 591-609.

136. JY Gao, PL Dubin, BB Muhoberac. Measurement of the binding of proteins to polyelectrolytes by frontal analysis continuous capillary electrophoresis. Anal Chem 69: 2945–2951, 1997.

137. YM Dunayevskiy, J Lai, C Quinn, F. Talley. Mass spectrometric identification of ligands selected from combinatorial libraries using gel filtration. Rapid Commun Mass Spectrom 11: 1178-1184, 1997.

138. D LeTourneau, NE Allen. Use of capillary electrophoresis to measure dimerization of glycopeptide antibiotics. Anal. Biochem 146: 62–66, 1997.

139. L Tao, RT Kennedy. Measurement of antibody-antigen dissociation constants using fast capillary electrophoresis with laser-induced fluorescence detection. Electrophoresis 18: 112–117, 1997.

140. JK Abler, KR Reddy, CS Lee. Post-capillary affinity detection of protein microheterogeneity in capillary zone electrophoresis. J Chromatogr A 759: 139–147, 1997.

141. GM Korf, JP Landers, DJ O'Kane. Capillary electrophoresis with laser-induced fluorescence detection for the analysis of free and immune-complexed green fluorescent protein. Anal Biochem 251: 210–218, 1997.

142. S Bose, J Yang, DS Hage. Guidelines in selecting ligand concentrations for the determination of binding constants by affinity capillary electrophoresis. J Chromatogr B 697: 77–88, 1997.

143. VM Okun, U Bilitewski. Analysis of biotin-binding protein actinavidin using affinity capillary electrophoresis. Electrophoresis 17: 1627–1632, 1996.

144. FA Gomez, JN Mirkovich, VM Dominguez, KW Liu, DM Macias. Multiple-plug binding assays using affinity capillary electrophoresis. J Chromatogr A 727: 291–299, 1996.

145. A Manz, N Garber, HD Widmer. Miniaturized total chemical analysis systems: A novel concept for chemical sensing. Sensor and actuators B1: 244–248, 1990.

146. S Seiler, DJ Harrison, A Manz. Planar glass chips for capillary electrophoresis: Repetitive sample injection, quantification, and separation efficiency. Anal Chem 65: 1481–1488, 1993.

147. A Manz, DJ Harrison, EMJ Verpoorte, JC Fettinger, A Paulus, H Lüdi, HM Widmer. Planar chips technology for miniaturization and integration of separation techniques into monitoring systems: Capillary electrophoresis on a chip. J Chromatogr A 593: 253–258, 1992.

148. CS Effenhauser, A Manz, HM Widmer. Glass chips for high-speed capillary electrophoresis separations with submicrometer plate heights. Anal Chem 65: 2637–2642, 1993.

149. RM McCormick, RJ Nelson, MG Alonso-Amigo, DJ Benvegnu, HH Hooper. Microchannel electrophoretic separations of DNA in injection-molded plastic substrates. Anal Chem 69: 2626–2630, 1997.

150. CS Effenhauser, GJM Bruin, A Paulus, M Ehrat. Integrated capillary electrophoresis on flexible silicone microdevices: Analysis of DNA restriction fragments and detection of single DNA molecules on microchips. Anal Chem 69 3451-3457, 1997

151. MA Roberts, JS Rossier, H Girault. UV laser machined polymer substrates for the development of microdiagnostic systems. Anal Chem 69: 2035–2042, 1997.

152. ZH Fan, DJ Harrison. Micromachining of capillary electrophoresis injectors and separators on glass chips and evaluation of flow at capillary intersections. Anal Chem 66: 177–184, 1994.

153. SC Jacobson, R Hergenröder, LB Koutny, RJ Warmack, JM Ramsey. Effects of injection schemes and column geometry on the performance of microchip electrophoresis devices. Anal Chem 66: 1107–1113, 1994.

154. K Seller, ZH Fan, K Fluri, J Harrison. Electroosmotic pumping and valveless control of fluid flow within a manifold of capillaries on a glass chip. Anal Chem 66: 3485–3491, 1994.

155. DJ Harrison, A Manz, ZH Fan, H Lüdi, HM Widmer. Capillary electrophoresis and sample injection systems integrated on a planar glass chip. Anal Chem 64: 1926–1932, 1992.

156. K Seiler, DJ Harrsion, A Manz. Planar glass chips for capillary electrophoresis: Repetitive sample injection, quantitation, and separation efficiency. Anal Chem 65: 1431–1488, 1993.

157. LB Koutny, D Schmalzig, TA Taylor, M Fuchs. Microchip electrophoretic immunoasay for serum cortisol. Anal Chem 68: 18–22, 1996.

158. Z Liang, N Chiem, G Ocvirk, T Tang, K Fluri, DJ Harrison. Microfabrication of a planar absorbance and fluoresence cell for integrated capillary electrophoresis devices. Anal Chem 68: 1040–1046, 1996.

159. SC Jacobson, R Hergenröder, AW Moore, Jr., JM Ramsey. Precolumn reactions with electrophoretic analysis integrated on a microchip. Anal Chem 66: 4127–4132, 1994.

160. SC Jacobson, LB Koutny, R Hergenröder, AW Moore, Jr., JM Ramsey. Microchip capillary electrophoresis with an integrated postcolumn reactor. Anal Chem 66: 3472–3476, 1994.

161. Q Xue, F Foret, YM Dunayevsky, PM Zavracky, NE McGruer, BL Karger. Multichannel microchip electrospray mass spectrometry. Anal Chem 69: 426–430, 1997.

162. AG Hadd, DE Raymond, JW Halliwell, SC Jacobson, JM Ramsey. Microchip device for performing enzyme assays. Anal Chem 69: 3407–3412, 1997.

163. N Chiem, DJ Harrison. Microchip-based capillary electrophoresis for immunoassays: analysis of monoclonal antibodies and theophylline. Anal Chem 69: 373–378, 1997.

164. F Foret, V Stustacek, P Bocek. On-line isotachophoretic sample preconcentration for enhancement of zone detectability in capillary zone electrophoresis. J Microcol Sep 2: 229–233,1990.

165. M Strausbauch, JP Landers, PJ Wettstein. Mechanism of peptide separations by solid phase extraction-capillary electrophoresis (SPE-CE) at low pH. Anal Chem 68(2): 306–314, 1996.

166. MA Strausbauch, BJ Madden, PJ Wettstein, JP Landers. Sensitivity enhancement and second dimensional information from the SPE-CE analysis of entire HPLC fractions. Electrophoresis 16: 541–548, 1995.

167. MA Strausbauch, SZ Xu, JE Ferguson, M Nunez, D Machacek, GM Lawson, PJ Wettstein, JP Landers. Concentration and analysis of hypoglycemic drugs using solid phase extraction-capillary electrophoresis. J Chromatogr 717: 279–291, 1995.

168. V Kasicka, Z Prusik. Isotachophoretic electrodesorption of proteins from an affinity adsorbent on a microscale. J Chromatogr 273: 117–128, 1983.

169. NA Guzman, MA Trebilcok, JP Advis. The use of a concentration step to collect urinary components separated by capillary electrophoresis and further characterization of collected analytes by mass spectrometry. J Liquid Chromatogr 14(5): 997–1015, 1991.

170. AJJ Debets, M Mazereeuw, WH Voogt, DJ Van Iperen, H Lingeman, KP Hupe, UAT Brinkman. Switching valve with internal micro precolumn for on-line sample enrichment in capillary electrophoresis. J Chromatogr 608: 151–158, 1992.

171. ME Schwartz, M Merion. On-line sample preconcentration on a packed-inlet capillary for improving the sensitivity of capillary electrophoretic analysis of pharmaceuticals. J Chromatogr 632: 209–213, 1993.

172. E Jellum, H Dollekamp, C Blessum. Capillary electrophoresis for clinical problem solving: analysis of urinary diagnostic metabolites ad serum proteins. J Chromatogr B 683: 55–65, 1996.

173. T Manabe, A Iwasaki, H Miyamoto. Separation of human plasma/serum proteins by capillary isoelectric focusing in the absence of denaturing agents. Electrophoresis 18: 1159–1165, 1997.

174. KF Greve, DE Hughes, P Richberg, M Kats, BL Karger. Liquid chromatographic and capillary electrophoretic examination of intact and degraded fusion protein CTLA4Ig and kinetics of conformational transition. J Chromatogr A 723: 273–284, 1996.

175. MP Richards, TL Huang. Metalloproteins analysis by capillary isoelectric focusing. J Chromatogr B 690: 43–54, 1997.

176. W Nashabeh, KF Greve, D Kirby, F Foret, BL Karger, DH Reifsnyder, S Builder. Incorporation of hydrophobic selectivity in capillary electrophoresis: Analysis of recombinant insulin-like growth factor I variants. Anal Chem 66: 2148–2154, 1994.

177. HP Bietlot, M Girard. Analysis of recombinant human erythropoietin in drug formulations by high-performance capillary electrophoresis. J Chromatogr A 759: 177–184, 1997.

178. JM Thorne, WK Goetzinger, AB Chen, KG Moorhouse, BL Karger. Examination of capillary zone electrophoresis, capillary isoelectric focusing and sodium dodecyl sulfate capillary electrophoresis for the analysis of recombinant tissue plasminogen activator. J Chromatogr A 744: 155–165, 1996.

179. SD Proctor, JCL Mamo. Separation and quantitation of apolipoprotein B 48 and other apolipoproteins by dynamic sieving capillary electrophoresis. J Lipid Research 38: 410–414, 1997.

180. HM Liebich, R Lehmann, AE Weiler, G Grübler, W Voelter. Capillary electrophoresis, a rapid and sensitive method for routine analysis of apolipoprotein A-I in clinical samples. J Chromatogr A 717: 25–31, 1995.

# 13

# Gel Electrophoresis of Peptides and Proteins

**Michael J. Dunn**
*Imperial College School of Medicine, London, United Kingdom*

## I.  INTRODUCTION

### A.  Peptides and Proteins Are the Primary Functional Molecules of a Cell

The peptides and proteins encoded by the genome of an organism are expressed through the process of gene transcription of DNA into mRNA, followed by translation by the ribosomal machinery of mRNA into the corresponding polypeptide chains. The old maxim of molecular biology that one gene encodes only one polypeptide has, of course, been shown to be an oversimplification. In many organisms, especially in the eukaryotes, processes of alternative codon usage during transcription and alternative splicing of primary RNA transcripts into more than one mature mRNA species result in the generation of polypeptide diversity. The situation is further complicated by processes of co- and post-translational modification which result in polypeptides being modified by the addition of other groups. These modifications include phosphorylation, sulfation, glycosylation, hydroxylation, N-methylation, carboxymethylation, acylation, prenylation, and N-myristoylation.

The resulting complex diversity of peptides and proteins are the primary functional molecules within a cell and the cell's properties (or phenotype) are determined largely by the proteins which it expresses. The functional properties of peptides and proteins are remarkably diverse, including (a) enzymic activity responsible for catalyzing the majority of cellular chemical reactions, (b) mechanical support, (c) the control of membrane permeability, transport, and storage processes, (d) hormones, receptors and signal transducers, (e) immune protection, (f) motility, and (g) regulation of gene expression. This natural diversity of polypeptide function, together with its usually exquisite specificity, offers the pharmaceutical scientist an opportunity to develop highly potent and specific therapeutic agents. However, the complexity of these molecules can present severe problems for their accurate chemical, structural and functional analysis, essential pre-requisites for the characterization and validation of potential therapeutic agents.

### B.  Physical Properties of Peptides and Proteins Which Can Be Exploited for Their Separation

Peptides and proteins are composed of a linear array of the 20 different naturally occurring amino acids, each being connected to its neighbor by a peptide bond formed between the amino group of one amino acid and the carboxyl group of another. It is this linear sequence of amino acids which is specifically encoded by the genome of an organism and determines the primary structure of a protein. Various regions of the polypeptide chain, determined by common structural elements such as the α-helix, the β-pleated sheet and the β-turn, are then folded to form the secondary structure. The folding and condensation of these features into compact shapes (domains), involving both covalent and non-covalent bonds, then determines the tertiary structure of the protein. Finally, individual polypeptides (either the same or different) often associate to form the mature, functional protein, this level of organization being termed quaternary structure.

These varied structural features of polypeptides confer a variety of physical properties on proteins which can be exploited by the analyst for their effective separation. Certain of the constituent amino acids contain side chains which are either negatively (glutamic acid and aspartic acids) or positively (lysine, arginine and histidine) charged at near neutral pH values. Peptides and proteins therefore exist as zwitterions, with their charge at a particular pH being determined both by their relative content of the charged amino acids and by their degree of ionization at that pH. While the majority of proteins are negatively charged at physiological pH (i.e. near neutrality), there are important groups of proteins which are positively charged under these conditions (e.g. histones, ribosomal proteins). Finally, for each protein there is a pH value at which the negative and positive charges of the amino acid residues are in balance, with the result that the protein carries no net charge. This is known as the isoelectric point (pI). The electrical properties of peptides and proteins represent a major feature which can be exploited for their effective separation for analysis and characterization.

The number of amino acids comprising each polypeptide, the degree of its post-translational modification, and its association with other polypeptides to form mutlimeric protein complexes are all factors which can contribute to the size of a particular peptide or protein. This property of protein size can also be exploited for protein separation.

The relative composition of the different amino acids (charged, uncharged, hydrophobic) of a polypeptide and the nature of its structural features (secondary, tertiary and quaternary) result in mature proteins having different properties of hydrophobicity and again these can be exploited to effect protein separation. Finally, mature proteins have differential affinities for a variety of ligands (e.g. lectins, specific receptor molecules, inorganic ions, antibodies) and this property can also be used as the basis of protein separation.

## C. Analytical Methods for the Separation of Peptides and Proteins

There are three principal groups of methods which are currently used for the analytical separation of peptides and proteins. Each of these approaches have a range of specific techniques which have been developed to exploit the physicochemical properties of proteins (i.e. charge, mass, hydrophobicity and affinity) to achieve their effective separation. These three groups of methods comprise high pressure liquid chromatography, known as HPLC (this volume, chapter by Aguilar), capillary electrophoresis, often abbreviated to CE (this volume, chapter by Landers), and gel electrophoresis which forms the subject of this chapter.

## II. THEORY OF ELECTROPHORESIS

### A. Basic Principles

Electrophoresis is defined as the migration of charged particle in a D.C. electric field under aqueous conditions. The phenomenon of electrophoretic transport has been known for around 180 years, but electrophoresis did not become a practical reality until the work of Arne Tiselius in the 1930's. Using moving boundary electrophoresis, Tiselius was able to separate human serum into four components, serum albumin and the α, β, and γ globulins, (1) and this pioneering work was recognised in his Nobel Prize of 1948. The following will be a brief introduction to the theory of electrophoresis, but those wanting a more rigorous mathematical treatment of the subject should consult (2).

Under conditions of constant velocity, the driving force on a charged particle is the product of the charge on the particle and the applied potential gradient. This force is counteracted by the frictional resistance of the separation medium, and in free solution this relationship obeys Stokes' law. However, if a gel medium is used the situation becomes much more complicated, such that additional factors such as gel density, and particle size also have an effect.

Electrophoretic mobility is defined as the velocity of a charged particle per unit field strength. Thus it is clear that the choice of field strength, which is dependent on both the applied voltage and the length of the separation path, are important factors in determining the time required for a particular electrophoretic separation. As we have seen in Sec. I.B, proteins act as zwitterions, so that the net charge on a protein will depend on the pH chosen for the separation medium. The operative pH will, therefore, exert a profound influence on protein mobility during electrophoresis. If the pI of the protein is the same as the pH of the medium, then that protein will not migrate during electrophoresis. At pH values below its pI, the protein will migrate towards the cathode, while at pH values above its pI, it will migrate towards the anode.

The ionic strength of the separation medium also has a marked influence on the efficiency of electrophoretic separations. Low ionic strengths permit high rates of migration, while high ionic strengths result in slower rates of migration but generally result in more sharply separated zones. The choice of the ionic strength of the buffer system to be used is, therefore, an important factor in determining the speed and efficiency of an electrophoretic separation.

During electrophoresis, electrical energy is transformed into heat. This effect, known as Joule heating, impairs protein separation due to increased rates of diffusion and can even result in damage to the equipment itself. The limitation and dissipation of Joule heating is a major consideration in the design of equipment, and the choice of the electrophoretic conditions. The choice of the buffer system to be used is very important, as the higher the ionic strength of the buffer, the greater the conductivity and the greater the amount of heat generated. Electrical resistance during electrophoresis depends on the type of buffer system used. In the case of continuous buffers (see Sec. IV.A), resistance will remain constant during electrophoresis so that Joule heating can be controlled by selection of the appropriate power input. In contrast, during electrophoresis using a discontinuous buffer system (see Sec. IV.B), the electrical resistance of the gel increase as the moving boundary migrates through the gel, due to a decrease in conductance. Thus, if constant current conditions are used, voltage and Joule heating increases as the run progresses. On the other hand, if constant voltage conditions are employed, the current and Joule heating will decrease with time, but at the cost of an increased separation time and a concomitant loss of resolution due to increased protein diffusion.

The best power supplies for electrophoresis are those that continuously monitor both voltage and current and are then able to provide a constant power output during electrophoresis. This ensures constant heat generation during the separation, but additional factors such as the buffer system, applied voltage, apparatus design, and cooling system are critical for the control of Joule heating. It must be remembered that the reproducibility of electrophoretic separations is dependent on the use of conditions of constant temperature.

## B.   Zone Electrophoresis

As proteins are charged at any particular pH value (other than their pI), when placed in an electric field they will migrate at a rate dependent on their charge density (defined as the ratio of mass to charge). Thus, when a mixture of proteins is placed in an electric field, the component molecules will migrate at different rates to either the cathode (positively charged molecules) or anode (negatively charged molecules) dependent on their net charge at the pH of the separation medium. Separation of the proteins is not possible if they are present throughout the separation medium. However, if the sample is present as an initial narrow zone, then the different proteins will migrate differentially dependent on their mobility, and thus separate during electrophoresis. This method is termed zone electrophoresis (Fig. 1).

## C.   Support Media

Zone electrophoresis can be performed in free solution and this method was used extensively in the early days of electrophoresis. However, this approach has severe disadvantages including the effects of Joule heating and diffusion

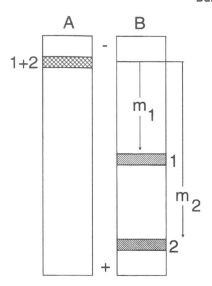

**Fig. 1** Schematic diagram of zone electrophoresis. (A) The protein sample, containing two components (1 and 2) is applied as a narrow zone. (B) In the presence of an electric field, the proteins migrate towards one of the electrodes depending on their charge (here the proteins are in the form of anions). Their rate of migration is dependent on their mobility (here the mobility of component 2 ($m_2$) is greater than that of component 1 ($m_1$).

which can easily result in broadening or disruption of the protein zones, leading to poor separations. It should be noted, however, that zone electrophoresis can be a very satisfactory method of protein separation when carried out in a capillary format (this volume, chapter by Landers).

The efficiency of zone electrophoresis can be significantly improved by performing the electrophoretic separation in a support medium which acts to minimize convection, and diffusion. Porous paper was the first support to be used for electrophoresis, but this was soon supplanted by alternative supports such as cellulose acetate membranes, which have very large pores which exert little sieving effect of proteins (3). This technique is still widely used for routine clinical analysis of serum proteins and for separating native isoenzymes (4). However, for most electrophoretic separations of peptides and proteins, these methods have largely been replaced by gel-based supports such as starch, agarose and polyacrylamide. Starch gels, which were introduced by Smithies (5), are no longer used because they have poor reproducibility, and they are difficult to handle. Agarose gels are used for the separation of very large molecules or complexes, for example nucleic acids (6) and nucleoproteins, and agarose is the medium of choice in the majority of immunoelectrophoretic techniques (7). Agarose and starch gels are also used extensively for the analysis of enzymes (4). However polyacrylamide, which was introduced as an electrophoretic support nearly 40 years ago (8), has become the medium of choice for electrophoresis of proteins (9).

Although the use of a support medium is a significant advantage in protein separations, it is important to realise that there can be associated problems. All gel-based media contain pores which can influence protein migration, so that the resulting separation depends on both the charge density, and the size of the proteins being analyzed. The pore size of starch and agarose gels is relatively large (e.g. 150 nm for 1% agarose; 500 nm for 0.16% agarose) making them suitable for the separation of large molecules and even protein complexes. However, the pore size of polyacrylamide gels is relatively small and is a variable which can be controlled during gel preparation (see Sec. III.B). The molecular sieving effect which then occurs during electrophoresis can be exploited to improve protein separation.

Another factor which must be considered is the possible effect of the charged groups which can be present on the support. The most important of these effects is known as electroendosmosis. Negatively charged groups which are usually present on the matrix are attracted towards the anode during electrophoresis.

However, as these groups cannot migrate due to their physical attachment to the support, hydrated protons ($H_3O^+$) migrate towards the cathode. This results in a flow of water towards the cathode, which is often deleterious to the separation. However, this effect can be exploited to improve separation, for example in some immunoelectrophoretic methods (7) and in particular in capillary electrophoresis (this volume, chapter by Landers). Electroendosmosis is very pronounced in the case of paper supports. In the case of agarose, preparations with different, defined electroendosmotic properties are available to suit the particular separation method. Fortunately, polyacrylamide is a medium with rather low inherent electroendosmotic properties, but these can still be deleterious to protein separations, particularly in isoelectric focusing (IEF) (see Sec. IX).

## III. PROPERTIES OF POLYACRYLAMIDE GELS

### A. Properties of Acrylamide

Perhaps the first thing to note concerning acrylamide is that it is an accumulative neurotoxin, so that care should be exercised when handling reagents in powder form. This problem has been largely overcome with the general availability of ready-made reagents, either in the form of pre-weighed powders or solutions, and the use of these is recommended whenever possible. The added advantage of these products is that they are in a suitable purified form for electrophoresis, so that recrystallization of reagents prior to their use for electrophoresis is not necessary. However, it is important to remember that all stocks of acrylamide will break down on prolonged storage. In particular, the production of charged acrylic acid is deleterious to most electrophoretic separations and to IEF separations in particular. While some manufacturers claim that their products have a long shelf life due to the addition of inhibitors, it is relatively easy to remove charged breakdown products from an acrylamide solution. Our routine method is to mix each 100 mL of acrylamide solution with 1 g of Amberlite IRN-150L mono-bed ion exchange resin (BDH) for at least 1 h. The resin is easily removed by filtration and the solution then used for gel polymerisation.

## B. Gel Polymerisation

A polyacrylamide gel is formed by the polymerisation of acrylamide monomers (Fig. 2a) with monomers of a suitable bifunctional cross-linking agent. The agent most commonly used is N,N'-methylene-bis-acrylamide (Fig. 2b), usually referred to as Bis. A variety of alternative cross-linking agents have been described [reviewed in (10)]. Some of these have advantages of generating gels with a large effective pore size for the separation of large molecules, for example (1,2-dihydroxyethylene)bis-acryalmide (DHEBA) which has the added advantage that the 1,2-diol structure renders gels susceptible to cleavage (i.e. dissolution) with periodic acid. The recently described cross-linking agent, piperazine diacrylyl (PDA), results in gels of superior physical strength, improved resolution of proteins, reduced background with silver staining (see Sec. XI.D) (11), and improved microsequencing results with possibly less N-terminal blockage.

Gel polymerisation is usually initiated with ammonium persulphate and the reaction is accelerated by the addition of a catalyst, typically N, N, N', N'-tetramethylethyene-diamine (TEMED). Alternative initiator and catalyst systems are available. Of these, 1,4-dimethylpiperazine (DMPIP) is reported to give less background with ammoniacal silver staining (see Sec. XI.C)

than TEMED, but it is not as potent a catalyst (12). Ultaviolet (UV) light activated polymerisation with riboflavin is used when proteins which are sensitive to persulphate ions are to be separated. Dissolved oxygen inhibits gel polymerisation, so that degassing of solutions is recommended. Polymerisation is a temperature dependent process, the optimal temperature being in the range 25–30°C. Under these conditions the reaction occurs within minutes, but it is customary to leave the gels for at least 2 h (or overnight) before electrophoresis. The use of elevated temperatures (>50°C) is not recommended as it can result in the formation of rigid gels composed of short polymer chains. Precooling of acrylamide solutions is a good idea if polymerization time is to be prolonged, such as during the casting of gradient gels (see Sec. III.D).

The result of gel polymerization is the formation of a three-dimensional lattice [but this is an over-simplification (10, 13, 14)] by the cross-linking of growing linear polyacrylamide chains by a mechanism of vinyl polymerisation (Fig. 2c). The concentration of the acrylamide monomer determines the average length of the linear chains of the polymer, while the concentration of Bis determines the extent of cross-linking. Both of these parameters are, therefore, important in determining the physical properties of the gel (pore size, elasticity, density, mechanical strength). A distinct advantage of polyacrylamide gels is that they are transparent, a property which is particularly important for the visualisation of the protein zones after electrophoresis.

The most important parameter of a polyacrylamide gel for protein separation is its effective pore size. This depends both on the total concentration of acrylamide monomers plus cross-linking agent, and on the concentration of the cross-linking agent itself. Gel composition is usually defined in terms of %T (total amount of acrylamide monomer + cross-linking agent expressed as a percentage) and %C (percentage by weight of the total monomer which is cross-linking agent). Pore size can be increased by reducing %T at a fixed %C, but very dilute gels ($<2.5\%$ T) are mechanically unstable. The largest effective pore size that can be produced by this approach is about 80 nm, suitable for separating molecules with a molecular weight of up to $10^6$ kDa. At the

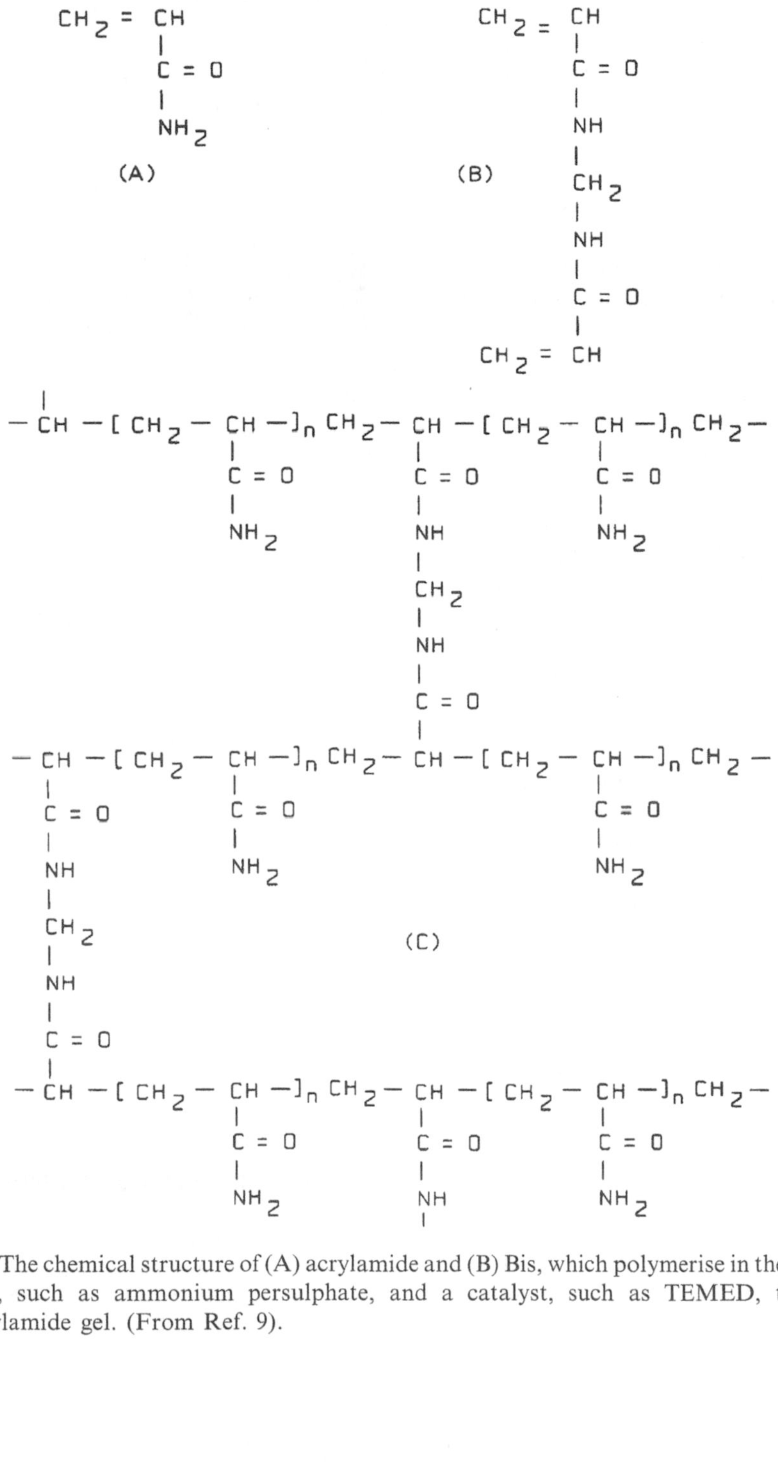

Fig. 2 The chemical structure of (A) acrylamide and (B) Bis, which polymerise in the presence of an initiator, such as ammonium persulphate, and a catalyst, such as TEMED, to form (C) a polyacrylamide gel. (From Ref. 9).

other extreme, gels of very high %T (> 30% T) have very small pore sizes which can impede the migration of proteins and peptides with molecular weights as low a $2 \times 10^3$ kDa. An alternative approach to decreasing the effective pore size of polyacrylamide gels is to increase %C at a fixed %T. However, the pore size attains a minimum at a particular %C, which is dependent on the value of %T. A further increase in %C then results in an increase in pore size, probably due to the formation of bead-like structure within the gel (14). It is possible to make gels of exceptionally high pore size (ca. 250 nm), but at concentration of Bis in excess of 30% C the gels become opaque, hydrophobic and prone to collapse.

## C. Gel Format

The original format for zone electrophoresis in polyacrylamide gels utilised cylindrical gels polymerized in glass tubes. Such "rod" gels are seldom used today, except for the first, IEF dimension of two-dimensional electrophoresis, although even here they are now largely being superseded (see Sec. X.E). Slab gels are currently the preferred format for gel electrophoresis as many samples can be electrophoresed simultaneously on the same gel under identical conditions, resulting in good reproducibility of separations (Fig. 3). Such slab gels are normally cast in a gel cassette consisting of two glass plates separated by spacers which determine the thickness of the gel (Fig. 4). This should normally be between 0.5 and 1.5 mm for most analytical applications. While making gels in-house gives the greatest flexibility with regard to gel size, composition (%T and %C), additives and buffer systems, an increasing range of ready-made slab gels can be purchased from commercial suppliers.

The most popular type of apparatus holds the slab gels in a vertical position between the two electrode reservoirs. In

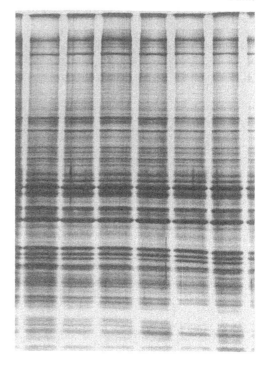

**Fig. 3** Separation of total proteins of human skin fibroblasts by slab gel electrophoresis.

the typical standard format the gels are around 14 cm wide by 16 cm long (Fig. 5), while miniaturised formats (gels around 7 cm or less in length) (Fig. 6) have the advantage of rapid separations, of great

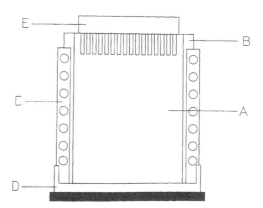

**Fig. 4** A cassette for casting polyacrylamide slab gels. (A) Glass plates, (B) spacer, (C) clamps, (D) stand, (E) plastic comb to form sample wells. (From Ref. 9).

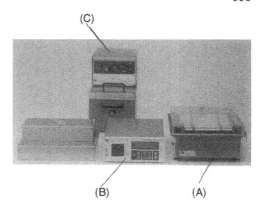

**Fig. 5** Typical vertical slab gel apparatus for use with standard format (14 × 16 cm) polyacrylamide gels. (1) Tank with lid, (2) central cooling core, (3) casting stand, (4) sandwich clamps, (5) alignment card, (6) sample well combs. (Courtesy of Bio-Rad Laboratories Inc, Hercules, California, USA).

**Fig. 7** Typical horizontal slab gel electrophoresis apparatus. (A) Electrophoresis chamber, (B) power pack, (C) themostatic circulating cooler. (Courtesy of Amersham Pharmacia Biotech, Uppsala, Sweden).

benefit in high throughput screening applications. The alternative approach is to use slab gels on a horizontal, flat-bed apparatus (Fig. 7) as described by Görg and her colleagues (15, 16) and commercialized by Pharmacia (now Amersham Pharmacia Biotech). Gels for use with horizontal systems can be easily made using appropriate cassettes (Fig. 8) and these are usually cast on a plastic

support such as GelBond PAG™ film (Amersham Pharmacia Biotech) which facilitates subsequent handling. A range of ready-made gels for use with horizontal electrophoresis systems are available from Amersham Pharmacia Biotech. Horizontal slab gel systems are the method of choice for IEF separations (see Sec. IX) and are also often used for agarose electrophoresis (6) and immunoelectro-

**Fig. 6** Typical mini-format vertical slab gel appratus. (Courtesy of Amersham Pharmacia Biotech, Uppsala, Sweden).

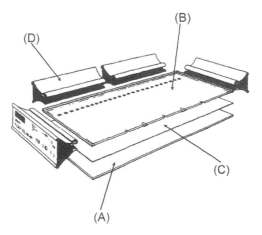

**Fig. 8** Cassette for casting a horizontal slab gel. (A) Lower glass plate, (B) upper glass plate with well formers attached, (C) plastic backing for gel, (D) clips. (From Ref. 9).

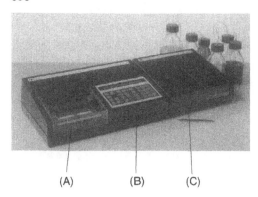

(A)        (B)        (C)

**Fig. 9** PhastSystem automated electrophoresis apparatus. (A) Separation compartment, (B) control unit, (C) staining compartment. (Courtesy of Amersham Pharmacia Biotech, Uppsala, Sweden).

phoresis (7). In addition, a fully automated computer-controlled electrophoresis system, the PhastSystem (Amersham Pharmacia Biotech) (Fig. 9), based on horizontal gel technology is available (17, 18). All of the major electrophoretic techniques are implemented on this system and it also incorporates a module for automated staining of the gels.

### D.   Homogeneous Versus Gradient Gels

In polyacrylamide gel electrophoresis (PAGE), proteins are separated both on the basis of their charge properties and by molecular sieving in the gel matrix. Thus, consideration must be given to the choice of the appropriate acrylamide concentration to attain the required separation of the proteins being analyzed. At too high a concentration the proteins may not enter the separating gel, while at too low a concentration the proteins will not separate and will run together with the buffer front. For complex mixtures containing proteins with widely varying molecular weights a compromise must usually be sought to achieve the best separation. As a guideline, a 5% T gel can separate proteins of molecu-

lar weight in excess of 100 kDa, while a 15% T gel can separate small proteins of less than 15 kDa. For many protein samples, gels of between 10 and 15% are suitable as they can separate proteins in the range 10–150 kDa.

The effective separation range of polyacrylamide gels can be extended using gels containing a linear or non-linear gradient of acrylamide concentration. For example, a linear 5–20% T gradient polyacrylamide gel can separate a mixture of protein in the range 8 to >200 kDa. Gradient gels also can result in sharper protein zones due to the decreasing pore size encountered by the protein during their passage through the gel. However, the major disadvantage of this approach is that gel preparation is more complex, due to the requirement for forming a gradient, and is more difficult to perform reproducibly. These problems can be overcome using the range of ready-made gradient gels of different polyacrylamide concentrations and buffer systems which are available commercially.

### IV.   ELECTROPHORESIS UNDER NATIVE CONDITIONS

Electrophoresis under native conditions is normally employed if it is required to preserve the native structure and conformation of the proteins to be separated. Additionally, the subunit structure of polymeric proteins will be maintained, thereby maximising the likelihood that the separated proteins will retain their biological activity. For this reason it is often possible to investigate the properties of the separated proteins (e.g., enzymic activity, antibody and ligand binding) following electrophoresis. The principal disadvantage of native electrophoresis is that the proteins to be separated must be soluble, and that they must not precipitate or aggregate during the electrophoretic separation. It has already been pointed out in Sec. III.D that in PAGE, proteins are separated both on the basis of their charge properties and by molecular sieving in the gel matrix. It is thus essential to choose the appropriate conditions to optimise the separation of a given protein mixture. The use of homogenous or gradient polyacrylamide gels to optimise protein separations on the basis of size has already been discussed in Sec. III.D. Optimization of separations on the basis of protein charge is achieved by the appropriate choice of the pH and components of the buffer system used for electrophoresis.

## A. Continuous Buffer Systems

The simplest form of PAGE is to use a so-called continuous buffer system in which the same buffer components are used throughout (i.e. in the sample, electrode chambers, and gel). The pH of this buffer system can then be selected to optimise the separation. Theoretically, any pH between pH 2 and 11 can be used, but the extremes (below pH 3 and above pH 10) should be avoided to minimise the potential for protein deamidation and hydrolysis. It is usually best to select a pH which is not close to the pI of the proteins to be separated, as many proteins

precipitate under these conditions. The additional advantage of this strategy is that the increased net charge ensures that the proteins have a high mobility, resulting in reduced separation times. In practice, the majority of proteins have pI values in the range pH 4–7, so that the most popular buffer systems are in the range pH 8–9. As most proteins are negatively charged under these conditions, they are applied at the cathode, with the consequence that any basic (i.e. positively charged) proteins will not enter the gel. If the sample is known (or suspected) to contain basic proteins, then it is best to carry out the separation under acidic conditions where the proteins (applied at the anode) will migrate as cations.

The other important parameter in the choice of a buffer for electrophoresis is its ionic strength. Low ionic strength favours higher rates of protein migration, while higher ionic strength can result in sharper protein zones. It is also necessary to use a sufficiently high ionic strength to maintain the effective pH of the buffer system and to maintain protein solubility, but it is important that the ionic strength selected does not result in excessive Joule heating. A selection of commonly used continuous buffer systems is given in Table 1.

**Table 1** Continuous Buffer Systems for Gel Electrophoresis Covering the Range pH 2–12. (From Ref. 213)

| Effective pH range | Primary buffer constituent | Adjust pH with |
|---|---|---|
| 2.4–6.0 | 0.1 M citric acid | 1 M NaOH |
| 2.8–3.8 | 0.05 M formic acid | 1 M NaOH |
| 4.0–5.5 | 0.05 M formic acid | 1 M NaOH or Tris |
| 5.2–7.0 | 0.05 M maleic acid | 1 M NaOH or Tris |
| 6.0–8.0 | 0.05 M $KH_2PO_4$ or $NaH_2PO_4$ | 1 M NaOH |
| 7.0–8.5 | 0.05 M diethyl barbiturate | 1 M HCl |
| 7.2–9.0 | 0.05 M Tris | 1 M HCl or glycine |
| 8.5–10.0 | 0.015 M $Na_2B_4O_7$ | 1 M HCl or NaOH |
| 9.0–10.5 | 0.05 M glycine | 1 M NaOH |
| 9.0–11.0 | 0.025M $NaHCO_3$ | 1 M NaOH |

## B.  Discontinuous Buffer Systems

While discontinuous buffer systems have
the advantage of simplicity, they suffer
considerable disadvantages, including
relatively low resolution capacity and their
inability to handle dilute samples. For
these reasons, continuous buffer systems
have generally been replaced with methods
using discontinuous (multiphasic) buffer
systems. This approach, often referred to
as disc electrophoresis, has the ability to
produce high resolution separations of
dilute protein samples due to the concentration of the sample components into a
narrow starting zone (stack).

The four important components of a
discontinuous PAGE system, based on
the original buffer system of Ornstein (19)
and Davis (20), are illustrated in Fig. 10.
There is a large pore stacking gel (C),
typically around 3% T, on which the
sample (B) is applied, while the separating
gel (D) has a higher acrylamide concentration (homogeneous or gradient) which
should be appropriate for the size range of
the proteins to be separated. The electrode
buffer (A) contains glycinate ions, while
the sample, stacking and separating gels
contain chloride ions. Finally, the pH of
the separating gel (pH 8.9) is higher than
that of the stacking gel (pH 6.7). At the
latter pH, glycinate is poorly ionised and
has a low mobility, while chloride is highly
ionised and has a high mobility. Sample
proteins will generally have mobilities
which are intermediate between these two
extremes.

During the initial phase of electrophoresis, the rapidly migrating chloride
(leading) ions migrate away from the
slowly moving glycinate (trailing) ions,
resulting in a zone of low conductivity.
The steep voltage gradient in this region
(conductivity is inversely proportional to
field strength) then accelerates the trailing
ions so that they migrate immediately
behind the leading ions. As the resulting
moving boundary sweeps through the

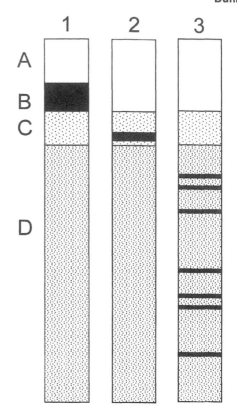

**Fig. 10**  Mechanism of stacking using the
Ornstein-Davis discontinuous buffer system.
(A) Cathodic buffer containing glycine, (B)
sample containing Cl⁻ at pH 6.7, (C) stacking
gel containing Cl⁻ at pH 6.7, (D) separating
gel containing Cl⁻ at pH 8.9. (1) Sample proteins loaded on the stacking gel, (2) proteins
stacked in the low conductivity zone in the
stacking gel, (3) protein zones unstacked and
separated according to their mobility in the
separating gel.

sample, the sample proteins are concentrated (stacked) into sharp zones according to their relative mobilities. When the
moving boundary reaches the separating
gel it encounters both an elevated pH and
decreased pore size. The increase in pH
acts to further ionize the glycinate ions so
that they have a higher mobility than the
proteins. Both types of buffer ion then
migrate ahead of the proteins which are

further unstacked due to the decrease in gel pore size. The subsequent separation then depends on protein size and charge as already discussed (see Sec. III.D).

While the majority of separations are still carried out using the original Ornstein-Davis (19, 20) buffer system, it is worth pointing out that a very large number of alternative buffer systems have been described which have the potential to result in improved separations (13).

## V. ELECTROPHORESIS IN THE PRESENCE OF ADDITIVES

### A. Reagents Which Cleave Disulphide Bonds

One of the main limitations of gel electrophoresis under native conditions is its inability to deal with proteins with inherently low solubility or which aggregate or precipitate during the separation. This problem is solved by the use of suitable additives to the electrophoresis system. Disulphide bonds, which often participate in protein aggregation, can be readily reduced with an excess of reagents such as 100 mM 2-mercaptoethanol or 20 mM dithiothreitol (DTT). Reoxidation and reformation of disulphide bonds can occur during electrophoresis. This can be prevented by adding low concentrations (10 mM) of DTT (21) or thioglycollate to the upper electrolyte reservoir (they cannot be added to the gel as they inhibit polymerisation). An alternative approach is to convert the free thiols generated by protein reduction into non-reactive thioethers by alkylation with an agents such as iodoacetic acid or iodoacetamide (22). Other reagents which can be used for alkylation include substituted maleimides, vinyl pyridine and acrylamide (23). Alkylation is generally compatible with post-separation chemical characterization methods (see Sec. XIV), such as microsequencing, amino acid compositional analysis and mass spectrometry.

### B. Urea

A common additive to gel systems which improves sample solubility is urea, which acts by unfolding and denaturing the sample proteins in a concentration dependent manner. The native state can often be maintained at very low urea concentrations, while 8 M (or higher) urea is used to achieve denaturation. Urea must be added to both the sample and the gel, but is not required in the electrolyte solutions. A major problem with the use of urea is that it breaks down at alkaline pH to form cyanate ions which react with the amino groups of proteins. This process, known as carbamylation results in modification of the proteins, with the potential to significantly modify the resulting separation profile. The reaction is temperature dependent, so that heating of samples containing urea must be avoided. Cyanate ions can be readily removed from stock solutions of urea, along with other charged breakdown products, and impurities, using a suitable cationic or mixed-bed ion exchange resin (see Sec. III.A).

### C. Detergents

The inclusion of a detergent in both the sample and gel is the most effective approach to the electrophoretic separation of proteins with limited solubility properties. Non-ionic detergents such Triton X-100 and Nonidet P-40 (NP-40) are generally considered to be mild in their action and are used where it is desired to retain the biological or biochemical activity of the sample proteins. A further advantage of this group of detergents is that the native charge properties of the sample proteins are preserved, of particular importance in IEF (see Sec. IX). Non-ionic detergents are used in the range 0.5 to 5% and they can be used in combination with gels containing high concentrations of urea. Zwitterionic (amphoteric) detergents, such as CHAPS (3-[(cholamidopropyl)di-methylammonio]-1-propane sulphonate) (24, 25), are also considered be mild in their action and have no effect on the net charge of the sample proteins. CHAPS in combination with a high urea concentration has become very popular for IEF (see Sec. IX). A variety of other non-ionic, zwitterionic and non-detergent additives have been used to improve protein solubilization for electrophoretic analysis, and are reviewed in (22). Sample solubilisation is a critical parameter in two-dimensional polyacrylamide gel electrophoresis (2-DE) and this topic is discussed in Sec. X.C.

The charged detergents are strongly denaturing in their action. The most important reagent in this group is the anionic detergent, sodium dodecyl sulphate (SDS), and SDS-PAGE is discussed below (Sec. VI). However, other anionic (e.g. sodium deoxycholate) and cationic [e.g. cetyltrimethylammonium bromide (CTAB)] detergents have been used for gel electrophoresis.

## D.  Blue Native Electrophoresis of Membrane Proteins

Blue native electrophoresis is a system that has been developed for the separation of membrane proteins under native conditions following their isolation using non-ionic or zwitterionic detergents (26). In this method Coomassie brilliant blue (CBB) G-250 is added to the cathode buffer, and to the protein solution before it is applied to an acrylamide gradient gel using a Tris-tricine buffer system buffered at pH 7. Negatively charged CBB dye binds to hydrophobic protein surfaces, and maintains the proteins soluble in the detergent-free gel during electrophoresis. An additional advantage of the method is that the blue protein zones can be observed during the separation. Protein migration stops when a limiting pore size of the gradient gel is reached, so that the proteins are

separated according to their molecular masses. This technique has been combined with a second Tricine-SDS-PAGE step to provide a two-dimensional separation of the subunits of protein complexes (27, 28).

## VI.  SDS-PAGE

## A.  Basis of SDS-PAGE

SDS-PAGE has become the single most popular technique for the analytical separation of proteins. This is due to the ability of SDS to solubilise, denature and dissociate most multimeric proteins into their constituent polypeptides. It is usually used in conjunction with a disulphide bond cleaving reagent (see Sec. V.A) to reduce any intra- and intra-molecular disulphide bonds. The property of SDS on which its success as a reagent for gel electrophoresis depends is that it masks the intrinsic charge of polypeptide chains by binding to proteins at a constant mass ratio (1.4 g SDS / g protein) (29). The binding of SDS results in the net charge per unit mass of the polypeptides being constant, so that the electrophoretic separation is dependent only on relative molecular mass ($M_r$) and occurs through molecular sieving by the gel matrix. Although there are some exceptions (see Sec. VI.E), this relationship between $M_r$ and electrophoretic migration holds true for a large number of proteins (30) and is largely independent of the amino acid composition and sequence. Other, related detergents, such as decyl- or tetradecyl-sulphate, and decane, dodecane or tetradecane sulphonate do not share this property (22) and their use results in poor electrophoretic separations (31).

As protein separation during SDS-PAGE depends solely on molecular sieving by the gel matrix, the choice of the gel concentration is very important for optimal separation of the components of a particular protein mixture. As a rough guide, a 5% T gel will separate proteins in the range 20–350 kDa, while a 15% T gel is effective in the range 10–150 kDa. As in the case of native PAGE (see Sec. III.D), a gradient gel can be useful for the separation of a mixture containing proteins with a wide range of masses.

## B. Buffer Systems

As in the case of native PAGE either continuous or discontinuous buffer systems can be used for SDS-PAGE, with the same attendant advantages and disadvantages (see Sec. IV.B). The original SDS-PAGE procedure described by Weber (30) used a sodium phosphate, pH 7.2, continuous buffer system. The main disadvantage of this approach is that, unless the sample can be applied as a narrow concentrated zone, low resolution separations with broad protein zones will result. Nowadays, the discontinuous buffer system first described by Laemmli (32), which is essentially the Ornstein-Davis (19, 20) system with the addition of 0.1% w/v SDS, is almost universally used today for SDS-PAGE.

## C. Separation of Small Proteins and Peptides

It is important to realise that small proteins, and peptides with a mass less than 14 kDa cannot be separated by the conventional SDS-PAGE technique using the Laemmli buffer system. This is a result of the SDS-protein complexes formed by small polypeptides all having similar size and charge properties so that they migrate together and do not separate during electrophoresis. One approach to this problem is to include high concentrations of urea which act to decrease the size of the detergent micelles (33, 34). The use of 7 M urea in combination with 10–18% T linear gradient gels with a high crosslinkage ratio (acrylamide:bis of 20:1)

has been reported to separate proteins down to 1.5 kDa (35). An alternative method which avoids the use of high concentrations of urea has been described (36). In this method an additional spacer gel is used, the molarity of the buffer is increased and tricine is used instead of glycine as the trailing ion. This system is claimed to be capable of separating proteins in the range 1–100 kDa.

## D. Sample Preparation

Solid samples, such as tissue, can be dissolved directly in sample buffer, while liquid samples should be diluted 1:1 with double strength sample solution. The solution usually used is that originally described by Laemmli (32) and contains 2% w/v SDS, 10% w/v glycerol in 0.125 M Tris, pH 6.8. A reagent such as 20 mM DTT is usually added to achieve cleavage of any disulphide bonds, but this can be omitted for the analysis of proteins under non-reducing conditions. It is also customary to include a tracking dye (usually 0.001% w/v Bromophenol blue) in the sample buffer, as this migrates with the buffer front during electrophoresis. Some samples may require concentration prior to analysis, and any appropriate method, such as lyophilisation or precipitation with ammonium sulphate, trichloroacteic acid (TCA) or cold acetone, can be used. All types of sample should be treated to ensure optimal reaction with SDS, as the ratio of binding of SDS to protein forms the basis of the technique (see Sec. VI.A). This is best achieved by heating the sample in sample buffer for at least 3 min at 100°C. The amount of sample which must be applied for any particular protein mixture is usually determined empirically, but as a rule of thumb about 50–100 $\mu$g of a complex mixture is generally adequate for detection with CBB R-250. A typical protein separation by SDS-PAGE is shown in Fig. 11.

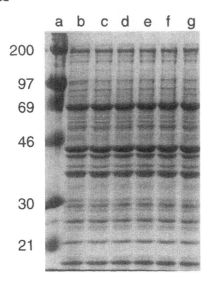

**Fig. 11** SDS-PAGE separation of human heart proteins (lanes b–g). Lane (a) contains the $M_r$ marker proteins and the scale at the left indictes $M_r \times 10^{-3}$.

### E. Limitations of SDS-PAGE

As discussed in Sec. V.A, the usefulness of SDS-PAGE depends on its ability to bind to proteins at a constant ratio (i.e. 1.4 g SDS / 1 g protein). While this holds true for the majority of proteins, there are notable exceptions resulting in proteins which appear to migrate anomalously during SDS-PAGE. The best known exceptions are proteins such as glycoproteins and lipoproteins. In these cases the non-protein part of the molecule cannot be saturated with SDS resulting in a reduced charge to mass ratio. The SDS-protein complex then migrates more slowly during electrophoresis resulting in anomalously high apparent molecular mass. In the case of glycoproteins this effect can be minimised by substituting the Laemmli buffer system with a Tris-borate-EDTA system (37).

Other types of proteins which can behave anomalously in SDS-PAGE are proteins with a high proportion of negatively or positively charged amino acids (e.g. histones) or those with an unusual amino acid composition (e.g. collagen peptides with a high proline content).

### VII. ESTIMATION OF MOLECULAR MASS

### A. Ferguson Plots

Native proteins are separated during electrophoresis on the basis of both their charge and size properties, so that the effects of charge must be negated if their molecular mass is to be determined. One approach to this is the use of the Ferguson plot which depends on the linear relationship between gel concentration (%T), $K_R$ and the logarithm of relative mobility ($R_f$):

$$\log R_f = \log Y_0 - K_R T,$$

where $R_f$ is defined as the mobility of the protein of interest measured with reference to the buffer front (detected by a tracking dye), $Y_0$ is the mobility of the protein at 0% T (i.e. the mobility of the protein in free solution), and $K_R$ is the retardation coefficient. Examples of different types of Ferguson plot which can be obtained are shown in Fig. 12.

In order to use this method to estimate the molecular mass of a protein, gels of at least seven different %T values should be used to give accurate values of $K_R$ and $Y_0$. The $K_R$ value is then used to estimate the molecular mass of the unknown using a standard curve constructed by plotting $K_R$ versus molecular mass for a series of standards whose molecular masses have been determined by other methods. Despite several inherent problems, such as the dependence of $K_R$ on several factors, the dependence of the method on the structure of the proteins used to construct the standard curve, and interference by non-protein groups such as carbohydrate or lipid, the method has been widely applied. The method is dis-

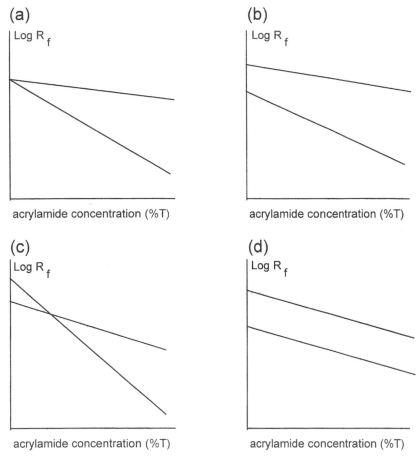

**Fig. 12**  Schematic examples of Ferguson plots. (a) The two proteins differ in size but not in net charge, (b) the smaller protein has the higher net charge and free-solution mobility, (c) the larger protein has the higher net charge, and (d) the two proteins differ in charge but are of the same molecular size. (From Ref. 9).

cussed in detail in (13) where a computer programme, PAGE-PACK, to automate this procedure is described.

## B.  Pore Gradient Electrophoresis

The use of linear or non-linear gradient gels to extend the effective separation range in PAGE was discussed in Sec. III.D. The decreasing pore size encountered by the protein zones as they migrate through the gel also results in a band sharpening effect. For any particular protein this effect becomes increasingly pronounced until it reaches its pore limit, after which its further migration occurs at a slow rate proportional to time. This concept has been termed pore limit electrophoresis and results in well resolved and stable separations. Once the pore limit for a particular protein has been reached, there is then a mathematical relationship between the distance travelled and its molecular mass. The various mathematical procedures which have been developed to exploit this approach for the measurement of protein mass are discussed in (10).

## C. SDS-PAGE

As we have already discussed (Sec. VI.A), the majority of proteins when subjected to SDS-PAGE separate according to their $M_r$. In gels of a single polyacrylamide concentration, it has been found that there is a linear relationship between $\log_{10} M_r$ and $R_f$. In order to use this method to estimate the molecular mass of unknown proteins, suitable marker proteins (kits of these are available from commercial suppliers) are subjected to SDS-PAGE on the same gel as the samples containing the unknown proteins. After electrophoresis, the marker and sample proteins are visualised (Fig. 11) and the position of the tracking dye front is marked. The $R_f$ values calculated for the marker proteins are then used to construct a plot of $\log_{10} M_r$ versus $R_f$ (Fig. 13). The $R_f$ values of the protein bands of interest are then calculated and used to determine their respective $M_r$ values from the standard curve. However, it is important to remember that for any particular gel concentration the linear relationship of $\log_{10} M_r$ and $R_f$ is valid only for a limited range of molecular mass and accurate measurement must be made in the linear region of the curve. For gradient gels the situation is more complex with several alternative mathematical relationships having been proposed to allow measurement of $M_r$ values of unknown proteins (10).

## VIII. PEPTIDE MAPPING BY PAGE

Peptide mapping is a very powerful method for examining relationships between the primary sequences of proteins. This approach involves the cleavage of the intact polypeptide into peptides by hydrolysis of specific peptide bonds using either enzymic or chemical methods (Table 2). The resulting peptides are then separated by an appropriate gel electrophoresis method, usually either the standard SDS-PAGE system of Laemmli

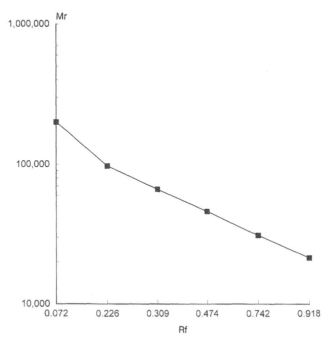

**Fig. 13** SDS-PAGE calibration curve of $M_r$ (log scale) *versus* $R_f$. Note the deviation from linearity at high $M_r$. (From Ref. 9).

**Table 2** Enzyme and Chemical Cleavage Reagents for Polypeptides

| Cleavage reagent | Preferred cleavage site |
|---|---|
| *Enzymic* | |
| Staphylococcus aureus V8 protease | Asp-X, Glu-X |
| Trypsin | Lys-X, Arg-X |
| Chymotrypsin | Trp-X, Tyr-X, Phe-X. Leu-X |
| Endoproteinase Arg-C | Arg-X |
| Endoproteinase Asp-N | Asp-X |
| Endoproteinase Lys-C | Lys-X |
| Endoproteinase Glu-C | Glu-X |
| Papain | Lys-X, Arg-X, Leu-X, Gly-X |
| Subtilisin | Broad specificity |
| Pronase | Broad specificity |
| *Chemical* | |
| Cyanogen bromide | Met-X |
| BNPS-skatole | Trp-X |
| N-chlorosuccinimide | Trp-X |
| Hydroxylamine | Asn-Gly |
| Formic acid | Asp-Pro |

(32) (Sec. VI.B) or by one of the modified SDS-PAGE procedures for the separation of small protein and peptides (Sec. VI.C), or by HPLC (see this volume, chapter by Aguilar) to generate a characteristic separation profile known as a peptide map. Comparison of peptide maps generated from different samples can then be used to indicate the degree of similarity between the proteins. More recently mass spectrometry (MS) (see this volume, chapter by Garner) has been applied to the analysis of peptides and a technique known as peptide mass profiling has been developed. This method makes it possible to identify unknown proteins by comparison of their peptide mass maps with the expected peptide mass maps generated *in silico* from nucleotide, and protein sequence databases. This method is described further in Sec. XIV.D.

The above approach to peptide mapping is only possible if the unknown proteins are available in a sufficiently pure form. This problem can be readily overcome if peptide mass profiling by MS is to be used, as it is now relatively simple to isolate the relatively small amounts of protein required for this technique by micro-preparative 2-DE (see Sec. X.I.). Prior to the availability of this methodology, a technique known as *in situ* peptide mapping was developed to overcome the problem of protein purity (38, 39). Here the protein sample is separated by any suitable PAGE method. Following staining, the protein bands of interest are cut out of the gel, equilibrated in a buffer containing 0.1% w/v SDS and the gel pieces applied in the sample wells of a standard 15% T SDS-PAGE gel. A suitable protease (e.g. V8 protease) is then applied to the wells on top of the gel pieces. Electrophoresis is then started for a short time to allow the proteins to elute from the gel pieces, and together with the protease, enter the stacking gel. Electrophoresis is then halted for 30 min to allow digestion to occur. Upon resumption of electrophoresis, the resulting peptides are separated and subsequently visualised. This method is also compatible with chemical cleavage, but in this case the cleavage reaction must be carried out before the gel pieces are applied to the SDS-PAGE gel.

## IX.  ISOELECTRIC FOCUSING

Isoelectric focusing (IEF) is a high resolution method in which proteins are separated in the presence of a continuous pH gradient. Under these conditions, proteins will migrate according to their charge properties until they reach the pH value at which they have no net charge (i.e. their pI). The proteins will then cease to migrate and be focused into narrow zones. Any proteins which subsequently move away from this zone will become charged and be forced to migrate back into the zone. In this way a steady state is attained so that IEF, at least in theory, results in stable, high resolution protein separations.

As IEF separates proteins on the basis of their charge properties, it is essential to use gels of a low acrylamide concentration, typically between 3 and 4% T, to minimise the effects of molecular sieving. IEF can be performed under native conditions. Alternatively any suitable non-ionic (e.g. Triton X-100, NP-40) or zwitterionic detergent (e.g. CHAPS) detergent, usually in the range 0.5–5%, can be added to the gel to increase sample solubility. If the use of fully denaturing conditions is required then high concentrations of urea (typically 8 M) can be used, but precautions should be taken against possible protein carbamylation (see Sec. V.B). It is possible to combine urea with the use of non-ionic or zwitterionic detergents but it should be noted that highly charged detergents, such as SDS, are not compatible with IEF.

### A.  Generation of pH Gradients Using Synthetic Carrier Ampholytes

Kolin in 1954 (40) was the first to describe the separation of proteins by electrophoresis in a pH gradient, but it was Svensson (who later changed his name to Rilbe) (41) who introduced the concept of natural pH gradients which established the theoretical basis for IEF in its presently used form. However, theory only became a reality with the work of Vesterberg (42) describing the synthesis of carrier ampholytes (SCA) by the reaction of oligoamines with acrylic acid. Alternative procedures for the synthesis of SCA have been subsequently described [reviewed in (43)]. These synthetic procedures result in the synthesis of very diverse mixtures containing as many as several thousand different amphoteric compounds with pI values covering the range pH 2–11.

Preparations of SCA are available from several commercial suppliers, and can be used to generate pH gradients for IEF covering either a wide range (up to pH 3.5–10) or restricted to a narrow range (spanning 1 pH unit) to resolve proteins whose pI values differ by only 0.01 pH units. The appropriate SCA preparation is included, together with the other required reagents (e.g. detergent, urea), into the IEF gel when it is polymerised. Alternatively, ready made gels covering a range of pH gradients are available commercially from Amersham Pharmacia Biotech. IEF gels are usually run using an horizontal, flat-bed electrophoresis apparatus provided with efficient cooling. The latter is essential as IEF is run under high voltage conditions, so that it is also beneficial to use thin (0.5 mm) gels to improve heat dissipation.

At the start of electrophoresis, the SCA are randomly distributed throughout the gel (Fig. 14a). As the run proceeds the SCA molecules migrate according to their individual charge properties until they are concentrated into narrow zones where their net charge is zero, forming a continuous pH gradient (Fig. 14b). Thus, when a protein mixture is applied, the individual components will migrate (Fig. 14c) until they reach the zone in the gel where the local pH is the same as the pI of that particular protein, where they become focused into narrow zones (Fig. 14d).

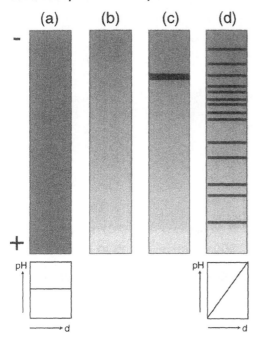

**Fig. 14** Schematic diagram of isoelectric focusing. (a) At the start of electrophoresis the synthetic carrier ampholytes (SCA) are randomly distributed throughout the gel. (b) As the run proceeds, the SCA migrate according to their individual charge properties, becoming concentrated into narrow zones where their net charge is zero and establishing a linear pH gradient across the gel. (c) The protein mixture is applied as narrow zone. (d) The individual proteins migrate until they reach a zone in the gel where the local pH is the same as the pI of that particular protein, where they become focused into narrow zones.

Further details of IEF using SCA can be found in (9, 43). However, this method of IEF has now been largely replaced by the use of immobilised pH gradients (IPG) (see Sec. IX.B) because of several inherent problems with the method. The main Achilles heel of the method results from the nature of the SCA molecules themselves. Their diversity which is fundamental to the generation of pH gradients results from a complex synthesis [described in (43)]. This process is difficult to control reproducibly and results in con-

siderable batch to batch variability, limiting the reproducibility and consistency of separations. Perhaps more importantly SCA are relatively small molecules, which are not fixed within the IEF gel. Thus, the inevitable electroendosmotic flow of water occurring during the IEF run (see Sec. II.C) results in migration of the SCA molecules, usually towards the cathode. This process, which is known as cathodic drift, leads to degradation of the pH gradient, with progressive loss of the separated proteins toward the cathode, thus undermining IEF as a stable, equilibrium technique.

## B. Generation of pH Gradients Using Immobilised pH Gradients

The limitations of IEF with SCA described above have been largely overcome with the development of reagents, known as Immobilines (Amersham Pharmacia Biotech), allowing polyacrylamide gels containing immobilised pH gradients (IPG) to be prepared (44). A detailed review of the basic theory, and methodology of IPG IEF can be found in (45), and an overview of more recent developments in (46).

The Immobiline reagents are a series of 8 acrylamide derivatives with the structure $CH_2 = CH\text{-}CO\text{-}NH\text{-}R$, where R contains either a carboxyl or tertiary amino group, and giving a series of buffers with different pK values distributed throughout the pH 3–10 range. The appropriate IPG reagents, calculated according to published recipes, are added to the mixture used for gel polymerization. Thus, during polymerisation, the buffering groups which will form the pH gradient are covalently attached via vinyl bonds to the polyacrylamide backbone. IPG generated in this way are, therefore, immune to the effects of electroendosmosis, so that they provide the opportunity to carry out IEF separations which are extremely stable,

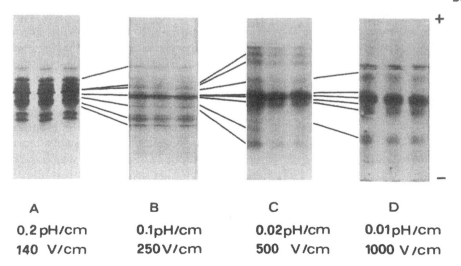

+

| A | B | C | D |
|---|---|---|---|
| 0.2 pH/cm | 0.1pH/cm | 0.02pH/cm | 0.01pH/cm |
| 140 V/cm | 250V/cm | 500 V/cm | 1000 V/cm |

**Fig. 15** Resolving power of IPG IEF. (A) Ovalbumin focused using SCA IEF (Ampholine, pH 4–6), and using IPG IEF with varying pH slopes (B–D). (From Ref. 211).

allowing the true equilibrium state to be attained.

An extensive range of recipes have been published for linear (47, 48) and non-linear (49) pH gradients covering the range pH 4–10, and these recipes have been collected together in (45). The use of a strongly acidic ($pK = 3.1$) and a strongly basic ($pK = 10.3$) acrylamide derivatives (50) in addition to the 8 Immobiline reagents allows more extended pH gradients spanning the range pH 2.5–11 to be constructed (51, 52). A non-linear, sigmoidal IPG covering the range pH 3–10 has been designed specifically for use in two-dimensional polyacrylamide gel electrophoresis (2-DE) (see Sec. X.E) (53). Very alkaline IPG's have also been described which allow the separation of basic proteins and again these have been applied particularly for 2-DE separations (see Sec. X.E) (54, 55).

Using IPG technology it is possible to generate narrow and ultra-narrow pH gradients spanning from as little as 0.1–1 pH unit. Such narrow and ultra-narrow pH gradients have an extremely high resolving power and are claimed to be able to separate proteins with a difference in pI of as little as 0.001 pH units (45, 46). The

resolving power that can be achieved using progressively narrow pH gradients is illustrated in Fig. 15.

IPG IEF gels are prepared as thin (0.3–0.5 mm) slab gels on plastic supports in a similar way to SCA IEF gels. The added complication is that it is necessary to form a gradient of the Immobiline reagents required to form the desired pH gradient. The gradient recipe can either be selected from published sources (45) or generated using special computer programmes (56, 57). This is not difficult to perform using a simple two-chamber gradient mixing device, but is more reproducible using a more sophisticated computer-controlled device (58, 59). Fortunately, a range of ready made IPG IEF gels is available from Amersham Pharmacia Biotech (Table 3).

IPG IEF gels are run using a horizontal, flat-bed apparatus (Fig. 7) in the same way as SCA IEF gels. The best method of sample application is to use a silicon rubber applicator strip containing wells to hold the samples. A useful property of IPG IEF gels is that they are relatively tolerant of samples containing salts (this should be kept below 40 mM, and the salts of strong acids and bases should be

**Table 3** Ready Made IPG IEF gels (Immobiline DryPlates) Available from Amersham Pharmacia Biotech

| Immobiline DryPlate, pH range | Primary application |
|---|---|
| 4–7 | General |
| 4.2–4.9 | α-1-antitrypsin |
| 4.5–5.4 | Gc-globulin |
| 5.0–6.0 | Transferrin |
| 5.6–6.6 | Phosphoglucomutase |

**Table 4** Suggested Running Conditions for 0.5 mm Thick IPG IEF Gels

| Phase of run | Voltage | Current | Power | Time |
|---|---|---|---|---|
| Sample entry | 300 V | 5 mA | 5 W | 30–60 min |
| Separation | 3,000 V | 5 mA | 5 W | 4–16 h |

avoided) and much higher sample loadings can be used (up to several mg of protein has been used in micro-preparative 2-DE applications, see Sec. X.E). It is not recommended to prerun IPG gels prior to sample application as this results in less efficient sample entry and can cause lateral spreading of protein zones.

The initial phase of the IPG IEF run should be performed at a relatively low voltage (around 300 V) to allow the sample proteins to enter the gel. Then, much higher voltages (typically around 3000 V, while as high as 10,000 V has been used) can be applied than in the case of SCA IEF gels, thereby reducing run times and generating sharply focused zones. It is important to realize that the narrower the pH gradient, the longer the time required for the proteins to reach equilibrium. Some suggested running conditions for various types of IPG IEF gel are given in Table 4.

## X. TWO-DIMENSIONAL POLYACRYLAMIDE GEL ELECTROPHORESIS (2-DE)

### A. The Requirement for 2-DE

The one-dimensional gel electrophoretic methods which we have discussed so far are, if run under optimal conditions, capable of generating high resolution separations of complex mixtures of proteins. In practice, however, these techniques are limited to the resolution of 100 or so zones, and are therefore not suitable for the analysis of complex mixtures containing several thousands of proteins, such as total protein homogenates of whole cells and tissues. A further major limitation of 1-DE methods is that they are in general only able to separate proteins on the basis of a single physico-chemical property. For example, the observation of a particular zone following SDS-PAGE does not imply protein heterogeneity, but simply indicates that any proteins present in that zone have nearly identical size properties, while their charge (and other) properties could be very different. This represents a particular problem for the analyst who is asked to determine whether a particular peptide or protein sample is homogeneous or whether it contains other, contaminating species. Very often this problem is overcome by subjecting the same sample to a sequential series of 1-DE separation methods, each examining a different property of the sample proteins. This approach, while effective, is exceedingly laborious and requires substantial amounts of the sample.

The best approach to this problem is to combine two different 1-DE methods into a 2-DE procedure able to separate the sample proteins according to different properties in each dimension. In fact 2-D protein separations date back to the beginning of electrophoresis using a support medium. In 1956, Smithies and Poulik described a combination of paper and starch gel electrophoresis for the separation of serum proteins (60). Since that time, subsequent advances in electrophoresis, such as the use of polyacrylamide, discontinuous buffer systems, gradient gels, SDS-PAGE, and IEF have all resulted in the parallel development of improved 2-D electrophoresis methods [reviewed in (9, 61)]. However, the most useful and commonly used approach for 2-DE is the combination of a first dimension separation by IEF under denaturing conditions with a second dimension separation by SDS-PAGE first described independently by several groups in the mid-1970's (61–64).

### B. The O'Farrell Method of 2-DE

The method described by O'Farrell in 1975 (64) has formed the basis of almost all subsequent developments in 2-DE, and several thousand papers have been published using this technique in the subsequent 25 years. This method was developed for separating more than 500 proteins from whole *Escherichia coli* (*E. coli*) and was based on a combination of IEF in cylindrical gels (cast in glass capillary tubes) containing 8 M urea and 2% w/v of the non-ionic detergent, Nonidet P-40 (NP-40), with the SDS-PAGE system of Laemmli (32). A typical 2-DE separation using this method is shown in Fig. 16.

### C. Sample Preparation and Solubilisation

Different types of sample present different challenges for their preparation and effec-

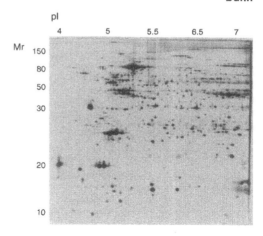

**Fig. 16** A 2-DE separation of dog heart proteins using the O'Farrell method.

tive solubilisation on which the success of the subsequent 2-DE separation depends. Liquid samples containing a relatively high protein concentration (e.g. serum, plasma) require little or no pre-treatment. An example of a 2-DE separation of serum proteins is shown in Fig. 17. However, less concentrated solutions [e.g. urine, cerebrospinal fluid (CSF), amniotic fluid] often require concentration by methods such as lyophilisation, or precipitation with trichloroacteic acid (TCA) or acetone

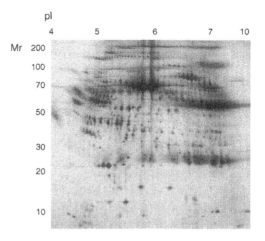

**Fig. 17** A 2-DE separation of human serum proteins. (From SWISS-2DPAGE, http://www.expasy.ch/ch2d/ch2d-top.html).

(61). Solid tissue samples usually require disruption in the presence of solubilization solution. For small samples this is readily achieved by crushing the sample in liquid nitrogen using a pestle, and mortar, while larger tissue samples must be homogenised using a suitable device. A separation of dog heart proteins prepared in this way is shown in Fig. 16. Cell suspensions can be readily harvested by centrifugation, while cells adherent to a substrate (e.g. tissue culture flask or dish) should be collected by scraping (the use of proteases should be avoided to prevent possible sample degradation). Alternatively, the cells can be detached by lysis directly in a small volume of sample solubilization solution. An example of a 2-DE separation of the proteins of a cultured cell line prepared in this way is shown in Fig. 18.

The most popular method for protein solubilization for 2-DE is that described in the original O'Farrell method (64), using a mixture of 9.5 M urea, 4% w/v NP-40, 1% w/v DTT and 2% w/v SCA. While this method works well for the majority of samples, it is not universally applicable, with membrane proteins representing a particular

challenge. A wide range of methods have been described for the solubilization of particular types of sample. Notable among these is the use of the zwitterionic detergent, CHAPS, which is claimed to be particularly effective for the solubilization of membrane proteins (65). More recently, Rabilloud has tested a variety of reagents for their efficacy in solubilising membrane proteins for 2-DE (22, 66). In particular, the use of 2 M thiourea in combination with 8 M urea and 4% CHAPS was found to give good results for membrane proteins (66, 67). In tests of alternative non-ionic and zwitterionic detergents, the most effective agents were found to be linear sulphobetaines, such as SB 3–10 or SB 3–12 (66). However, these are not compatible with high concentrations of urea (68). This can be overcome by using these reagents (e.g. 2% SB 3–10) in combination with 5 M urea, 2 M thiourea and 2% CHAPS (66).

For the preparation of samples for 2-DE, disulphide bonds are normally reduced with free-thiol containing reagents such as DTT or $\beta$-mercaptoethanol. However, reagents such as DTT are charged so that they migrate out of the gel during IEF. This leads to reoxidation of the sample proteins, and can result in a loss of sample solubility, particularly those proteins which are rich in disulphide bonds, such as keratins from hair, and wool (69). Replacing the thiol containing reducing agents with a non-charged reducing agent such as tributyl phosphine (TBP) was found to greatly increase protein solubility during the IEF dimension and result in increased transfer to the second dimension gel (69).

The presence of nucleic acids can be problematical during IEF. This is due to an increase in the viscosity of the sample and in some cases formation of complexes with the sample proteins, leading to artefactual migration and streaking (66). If problems of this type are suspected, it is best to degrade the nucleic acid by the

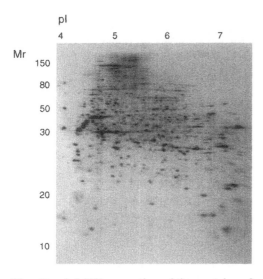

**Fig. 18** A 2-DE separation of the proteins of cultured human skin fibroblast cells.

addition of a suitable pure (i.e protease free) endonuclease to the sample solubilisation solution.

### D.  Limitations of the O'Farrell Method

As we have already discussed in Sec. IX.A, pH gradients generated using SCA are subject to cathodic drift, resulting in pH gradient instability and loss of proteins from the gel. This effect is very pronounced using tube gels due to the negatively charged groups present on the walls of the glass capillaries. As a consequence, pH gradients using the O'Farrell method rarely extend far beyond pH 7 (Fig. 16), with the resultant loss of the basic proteins. This problem was recognized by O'Farrell, who developed an alternative procedure, known as non-equilibrium pH gradient electrophoresis (NEPHGE), for the 2-DE separation of basic proteins (70). In this method, separation occurs on the basis of protein mobility in the presence of a rapidly forming pH gradient. However, the reproducibility of this method is extremely difficult to control. Other approaches to this problem have been developed [reviewed in (61)], but this problem has been solved with the advent of IPG's.

### E.  2-DE using IPG IEF

#### 1.  Method

As discussed in Sec. IX.BS, problems inherent in the use of SCA for the generation of pH gradients during IEF can be overcome using the technique of IPG IEF. This made it a potentially attractive option for use in 2-DE. However, early attempts to implement the IPG technology to 2-D separations encountered several problems. Fortunately, largely due to the work of Görg and her colleagues, these problems have been solved and IPG IEF has become the method of choice for the first dimension separation of 2-DE (71, 72).

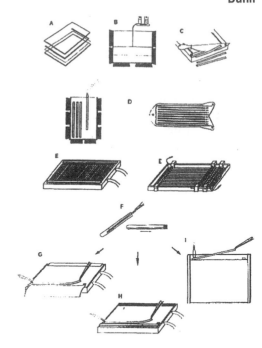

**Fig. 19**  Schematic diagram of the procedure of 2-DE using IPG IEF. (A) Assembly of the polymerisation cassette for the preparation of IPG and SDS gels cast on plastic backings, (B) casting of IPG and gradient SDS gels, (C) cutting of washed and dried IPG gels into individual IPG strips, (D) rehydration of IPG strips, (E) IEF in individual IPG strips, (F) equilibration of IPG strips prior to SDS-PAGE, (G) transfer of IPG strip onto surface of laboratory-made horizontal SDS gel along cathodic wick, (H) transfer of IPG strip onto surface of commercial horizontal SDS gel along cathodic buffer strip, (I) loading of IPG strip onto the surface of a vertical SDS gel. (Courtesy of A. Görg, Technical University, Munich, Germany).

The method is outlined in Fig. 19 and described in detail in (71, 72). Briefly, IPG slab gels of the desired pH range are cast in the normal way (see Sec. IX.B). After polymerisation, the gels are washed, dried and stored at $-20°C$. The required number of gel strips (3–5 mm wide) for 2-DE are cut off of the slab using a paper cutter. Alternatively, a range of ready-made

**Table 5**  Ready Made Immobiline DryStrips for the First, IPG IEF Dimension of 2-DE Available from Amersham Pharmacia Biotech. L, Linear pH Gradient; NL, Non-linear pH Gradient

| DryStrip length (mm) | pH 4–7 L | pH 3–10 L | pH 3–10 NL |
|---|---|---|---|
| 70  | * | * | * |
| 110 | * | * |   |
| 130 | * | * | * |
| 180 | * | * | * |

strips are available commercially from Amersham Pharmacia Biotech (Table 5).

IPG strips of any desired length can be used, but it should be remembered that, in general, the larger the separation area of a 2-D gel, the more proteins can be resolved (see Sec. X.H). Strips of 18 cm are usually employed for high-resolution separations, while shorter strips (e.g. 7 cm) are used for rapid screening applications. The choice of the pH gradient used must also be considered. A choice of a linear pH 3.5–10 gradient is often useful for the initial analysis of a new type of sample. However, for many samples this can result in loss of resolution in the region pH 4–7, in which the pI values of many proteins occur. This problem can be to some extent overcome with the use of a non-linear pH 3.5–10 IPG IEF gel, in which the pH 4–7 region contains a much flatter gradient than in the pH 7–10.5 region (53). This allows good separation in the pH 4–7 region while still resolving the majority of the more basic species (Fig. 20). However, use of a pH 4–7 IPG IEF gel will result in even better protein separation (Fig. 21). IPG strips are available commercially for these pH ranges (Table 5). Laboratory-made IPG strips with either very narrow pH gradients (spanning 1 pH or less) can be useful for separating components with very similar pI values, while very basic pH gradients can be used to advantage for certain types of sample, such as ribosomal proteins, nuclear proteins (54) and other basic proteins (55).

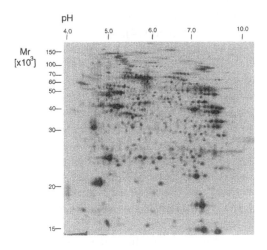

**Fig. 20**  A 2-DE separation of heart proteins using a non-linear pH 3.5 to 10 IPG IEF gel in the first dimension.

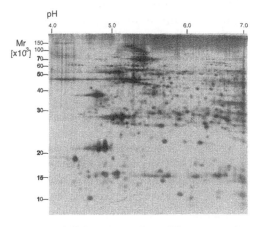

**Fig. 21**  A 2-DE separation of heart proteins using a linear pH 4 to 7 IPG IEF gel in the first dimension.

**Table 6**  Suggested Running Conditions for 18 cm IPG Strips for the First, IEF Dimension of 2-DE. The Strips Should be Run at 0.05 mA per Strip (2 mA maximum total), 5.0 W Maximum, 20°C

| Voltage (maximum) | IPG strip (pH range) | Time |
|---|---|---|
| 150 | All | 30 min |
| 300 | All | 60 min |
| 1,500 | All | 60 min |
| 3,500 | 4–7 | 42,000 Vh |
|  | 4–8 | 35,000 Vh |
|  | 4–9 | 30,000 Vh |
|  | 6–10 | 35,000 Vh |
|  | 3–10.5 | 25,000 Vh |

For use in 2-DE, the strips are rehydrated in a special reswelling cassette (Fig. 19) in a solution containing 8 M urea, 0.5% non-ionic (e.g. NP-40, Triton X-100) or zwitterionic (CHAPS) detergent, 15 mM DTT and 0.2% SCA of the appropriate pH range. The strips can then be placed directly on the surface of the cooling plate of a horizontal flat-bed electrophoresis apparatus (Fig. 19). A convenient alternative is to use the special strip tray available from Pharmacia (Fig. 19). This tray is fitted with a corrugated plastic plate which contains grooves allowing easy alignment of the IPG strips. In addition, the tray is fitted with bars carrying the electrodes and a bar fitted with sample cups allowing application of samples at any desired point on the gel surface. This tray is filled with silicon oil which protects the gel from the effects of the atmosphere during IEF. Horizontal streaking can often be observed at the basic end of 2-D protein profiles, particularly when IPG 6–10 is used for the first dimension. This problem can be resolved by applying an extra electrode strip soaked in 15 mM DTT on the surface of the IPG strip alongside the cathodic electrode strip (72). This has the advantage that the DTT within the gel, which migrates towards the anode during IEF, is replenished by the DTT released from the strip at the anode. An alternative approach is to use the non-charged reducing agent, TBP (see Sec. X.C), which does not migrate during IEF and has been found to greatly improve protein solubility during IEF (69).

## 2. Sample Application and Running Conditions

Samples are usually applied either at the anodic or cathodic end of the IPG strips, the optimal position being determined empirically for each type of sample. As in the case of 1-D IEF, it is important to limit the initial applied voltage to achieve maximal sample entry into the gels. The voltage is initially limited to 150 V (30 min) and then progressively increased until 3500 V is attained (74). The time required for the run depends on several factors, including the type of sample, the amount of protein applied, the length of the IPG strips used, and the pH gradient being used. The IEF run should be performed at 20°C, as at lower temperatures there is a risk of urea crystallization and higher temperatures have been found to result in alterations in the relative positions of some proteins on the final 2-D patterns (75). Some typical running conditions are given in Table 6.

While the method of sample application described above generally works well for most samples, there can be a tendency for certain types of sample to

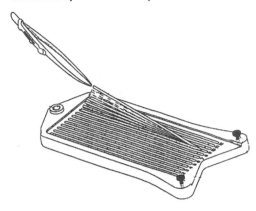

**Fig. 22** Reswelling tray for in-gel sample rehydration of IPG strips. (Courtesy of A. Görg, Technical University, Munich, Germany).

precipitate during sample entry, and this effect is more pronounced when increased amounts of protein (1 or more mg) are applied for micro-preparative purposes. To overcome this problem, a method has been described in which the IPG strips are reswollen directly in a solution containing the protein sample to be analysed (76, 77). A special reswelling tray is available from Amersham Pharmacia Biotech for this purpose (Fig 22). This method has been found to work well for many types of sample, and very high protein loads (>10 mg) have been successfully separated. However, there can be a selective loss of high molecular weight, very basic and membrane proteins.

Recently a new integrated instrument, named the IPGphor (Amersham Pharmacia Biotech), has been developed to simplify the IPG IEF dimension 2-DE (78). This instrument features a strip holder that provides for rehydration of individual IPG strips with or without sample, optional separate sample loading, and subsequent IEF, all without handling the strip after it is placed in the ceramic strip holder. The instrument can accommodate up to 12 individual strip holders and incorporates Peltier solid-state

cooling and a programmable 8000 V, 1.5 mA power supply.

After the IPG IEF dimension, strips can be used immediately for the second dimension. Alternatively, strips can be stored between two sheets of plastic film at −80°C for periods of several months (72).

### 3. Low Abundance Proteins

While the method of sample application by in-gel rehydration can be used for high protein loadings (see Sec. X.E.2), there can still be problems with the detection of low abundance proteins following 2-D separations of whole cell lysates. It has recently been calculated that if 10 mg total cellular proteins are loaded, then only those proteins present at more than $10^3$ copies per cell (0.17 pmol) can be detected by silver staining (see Sec. XI.C) (79). Immunoblotting using high affinity monoclonal antibodies followed by detection with enhanced chemiluminescence (ECL) (see Sec. XIII.C.3) may be able to detect as few as 10 copies per cell (1.7 fmol) (79). There is, therefore, a need to be able to enrich samples for such low abundance proteins to enable their separation, identification and characterization by 2-DE.

One approach to this problem is to use classical subcellular fractionation techniques based on density gradient separation, followed by 2-DE analysis of the proteins isolated from the resulting organelle fractions. An alternative method for organelle preparation is to use free-flow electrophoresis. Several examples of the use of these approaches can be found in (80).

Preparative IEF can also be used to pre-fractionate protein samples prior to 2-DE, thereby achieving enrichment of proteins whose pI values fall within a certain range. One such instrument, known as the Rotofor (Bio-Rad), fractionates proteins into defined pH ranges by liquid-phase IEF with pH gradient generation using SCA. It is claimed to be able to achieve 500-fold purifications of protein from complex mixtures. More than 1 gm of plasma proteins have been efficiently separated with this instrument and the fractions used for subsequent 2-DE analysis (81). A preparative IEF instrument based on IPG technology has also been described (82), and is available commercially as the IsoPrime (Amersham Pharmacia Biotech). Another preparative electrophoresis instrument, called the Gradiflow, has recently been described and this can separate proteins by charge and/or size under mild electrophoretic conditions without denaturing additives such as SDS or reducing conditions (83).

## F.  Equilibration Between Dimensions

In the past many people new to 2-DE found the art of extruding IEF gels from glass capillary tubes to be a major factor limiting their success at the technique! However, there is no such problem using IPG IEF gels as these are firmly bound on the plastic support film. It is essential that after IEF, the gels are equilibrated to allow the separated proteins to interact fully with SDS, thereby guaranteeing that they will migrate properly during the second, SDS-PAGE dimension (Fig. 19). IPG IEF gels are incubated for 15 min in 50 mM Tris buffer, pH 8.8 containing 2% w/v SDS, 1% w/v DTT, 6 M urea and 30% w/v glycerol. The urea and glycerol are used to reduce electroendosmotic effects which otherwise result in reduced protein transfer from the first to the second dimension (71). This is followed by a further 15 min equilibration in the same solution containing 5% w/v iodoacetamide in place of DTT. The latter step is used to alkylate any free DTT, as otherwise this migrates through the second-dimension SDS-PAGE gel, resulting in an artefact known as point-streaking which can be observed after silver staining

(84). An alternative procedure, allowing equilibration to be achieved in a single step, is to replace the DTT in the equilibration buffer with 5 mM TBP, which is uncharged and so does not migrate during SDS-PAGE (69).

## G.  The Second Dimension

After equilibration, the first dimension IEF gels are applied directly to the surface of the second-dimension SDS-PAGE gels. Again, this could represent a considerable handling problem using cylindrical tube gels, but is a relatively simple procedure using IPG gels on plastic supports. The SDS-PAGE gels can be of any appropriate single or gradient polyacrylamide composition as described in Sec. VI.A, and can be used either in a vertical or horizontal format (Fig. 19). The use of vertical formats enables multiple gels to be run simultaneously, which improves reproducibility (see Sec. X.H). The use of horizontal, 0.5 mm thin SDS gels cast on plastic supports improves the ease of handling the gels and gives rapid separations (71, 72). A range of precast homogeneous and gradient SDS gels for use on horizontal systems are available (ExcelGel, Amersham Pharmacia Biotech), while such ready-made gels are becoming available for an increasing variety of types of vertical electrophoresis apparatus.

## H.  Resolution and Reproducibility of 2-DE

As already mentioned above (Sec. X.E), the resolving capacity of 2-D gels is usually considered to be proportional to the total gel area available for the separation. Using 18 cm long IPG IEF gels in combination with 20 cm long second-dimension SDS-PAGE gels, around 2000 proteins can be readily resolved (Fig. 20). Only a few hundred proteins can be separated using mini-gel formats, but these are much quicker to run and can be useful for rapid screening purposes. For maximal resolution of very complex mixtures, very large format gels (>30 cm in each dimension) can be used. These are reported to be able to separate as many as 5000 to 10,000 proteins from whole cell lysates (85, 86), but this is achieved at the expense of the ease of gel handling and processing.

The reproducibility of 2-DE separations is a very important parameter, and until recently was a major problem limiting the more widespread application of the method. Using the tube gel technique of O'Farrell, it was often difficult to obtain reproducible separations of a particular type of sample even within a single laboratory, while comparison of 2-DE separation patterns generated in different laboratories was often considered to be impossible. However, this problem has now been largely overcome. This is due firstly to the development of dedicated apparatus, such as the ISO-DALT system (87, 88), available from Amersham Pharmacia Biotech (Fig. 23), and the Investigator (89) system, available from ESA Inc (Fig. 24) systems. Such equipment allows the simultaneous electrophoresis of large numbers (between five and 20) of 2-D gels under reproducibly controlled conditions.

The second major advance has been the adoption of IPG IEF technology for

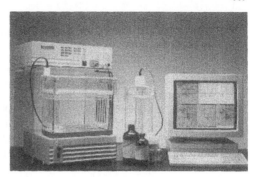

**Fig. 24** The Investigator 2-DE system. (Courtesy of ESA, New York, USA).

the first dimension of 2-DE. This has been shown to result in highly reproducible 2-D protein separations by inter-laboratory studies of heart (90), barley (90), and yeast (91) proteins. In both of these studies a larger number of protein spots (340 heart, 200 barley, 470 yeast) were analysed. The mean standard deviation of spot position in the IPG IEF dimension was 1 mm in the study of heart and barley proteins (90) and 1.8 mm in the study of yeast proteins (91). The figures for the SDS-PAGE dimension were 1.5 mm (90) and 2.8 mm (91) respectively. Quantitative reproducibility was also investigated in the study of yeast proteins and this was found to be relatively high (91).

## I. Proteomics

2-DE has now matured into a technique which is capable of separating reproducibly thousands of proteins present in samples such as cells, tissues and even whole organisms. This has made 2-DE an ideal tool to use in studies designed to determine the nature and function of the large number of structural genes being identified in various genome initiatives. The genomes of several micro-organisms are now complete (see Table 7), while significant progress is being made for several eukaryotic species (see Table 7). It is becoming clear from these

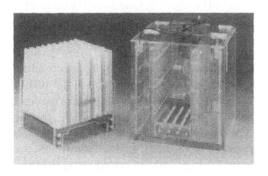

**Fig. 23** 10-place DALT SDS-PAGE gel unit with Western blotting accessory. (Courtesy of Amersham Pharmacia Biotech, Uppsala, Sweden).

**Table 7**   Some Organisms Whose Genomes Have Completely Sequenced and Others Which Are the Subject of Active Genome Sequencing Programmes. ORF's, Open Reading Frames

| Organism | Size (Mb) | ORF's | Year completed |
|---|---|---|---|
| *Micro-organisms* | | | |
| Mycoplasma genitalium | 0.58 | 470 | 1995 |
| Ureaplasma urealyticum | 0.76 | 640 | 1997 |
| Mycoplasma pneumoniae | 0.82 | 679 | 1996 |
| Borrelia burgdorferi | 0.91 | 843 | 1996 |
| Treponema pallidum | 1.05 | 1000 | 1997 |
| Aquifex aeolicus | 1.55 | 1512 | 1998 |
| Helicobacter pylori | 1.67 | 1590 | 1997 |
| Methanococcus jannaschii | 1.67 | 1738 | 1996 |
| M. thermoautotrophicum | 1.75 | 1855 | 1997 |
| Haemophilus influenzae | 1.83 | 1743 | 1995 |
| Pyrobaculum aerophilum | 1.8 | 1900 | 1997 |
| Streptococcus pyogenes | 1.8 | 1900 | 1997 |
| Archaeoglobus fulgidis | 2.18 | 2436 | 1997 |
| Nisseria gonorrhoreae | 2.2 | 2100 | 1997 |
| Synechocystis PCC6803 | 3.57 | 3168 | 1996 |
| Bacillus subtilis | 4.21 | 4100 | 1997 |
| Escherichia coli | 4.67 | 4288 | 1997 |
| *Eukaryotes* | | | |
| Saccharomyces cerevisiae | 12.1 | 5885 | 1996 |
| Dictyostelium doscoideum | 70 | 12500 | |
| Arabidopsis thalania | 70 | 14000 | |
| Caenorhabditis elegans | 80 | 17800 | |
| Drosophila melanogaster | 170 | 30000 | |
| Homo sapiens | 2900 | 50000 | |

genome programmes that it is usually impossible to attribute even putative functions to as many as 40% of the structural genes within a particular organism. In addition, as many as 30% of the open reading frames are assigned putative functions on the basis of homology to genes encoding proteins of known function. There is, therefore, a necessity to resort to the use of techniques of protein biochemistry to resolve this problem. 2-DE currently represents the best interface between protein biochemistry and molecular biology, and this area of endeavour has become known as "proteomics." The term "proteome" was first coined in 1995 (92) and is defined as the protein complement of the genome of a cell, tissue or organism.

The proteomic approach involves the use of 2-DE as the primary method for the separation of the protein components of the cell, tissue or organism being investigated. Quantitative computer analysis (see Sec. XII.C) is then used to construct a database of protein expression for the particular type of sample. Such databases can, therefore, contain information on protein expression under different physiological conditions, in response to pharmacological agents, during development, and as a result of disease processes. Proteins in these databases are identified and characterized using the various highly sensitive techniques now available which are capable of analyzing the small amounts (a few picomoles or less) of protein present in spots on 2-D gels. These

**Table 8** 2-DE Protein Databases Accessible via the WWW on the Internet

| 2-DE protein database | WWW address |
| --- | --- |
| *Federated 2-DE databases* | |
| WORLD-2DPAGE | http://www.expasy.ch/ch2d/2d-index.html |
| SWISS-2DPAGE | http://www.expasy.ch/ch2d/ch2d-top.html |
| HSC-2DPAGE | http://www.harefield.nthames.nhs.uk/nhli/protein/ |
| HEART-2DPAGE | http://www.chemie.fu-berlin.de/~pleiss/dhzb.html |
| HP-2DPAGE | http://www.mdc-berlin.de/~emu/heart/ |
| | |
| *Other 2-DE databases* | |
| Argonne PMG | http://www.anl.gov/CMB/PMG/ |
| YEAST 2D-PAGE | http://yearst-2dpage.gmm.gu.se/ |
| YPD | http://www.proteome.com/YPDhome.html |
| YPM | http://www.ibgc.u-bprdeaux2.fr/YPM/ |
| LSB | http://www.lsbc.com/2dmaps/patterns.htm |
| Maize 2D | http://moulon.moulon.inra.fr/imgd/ |
| Drosophila 2D-PAGE | http://tyr.cmb.ki.se/ |
| ECO2DBASE | http://pcsf.brcf.med.umich.edu/eco2dbase/ |
| Sub2D | http://pc13mi.biologie.uni-greifswald.de:80/sub2D/sub2d.htm |
| PPDB | http://www-lecb.ncifcrf.gov/phosphoDB/ |
| LECB | http://www-lecb.ncifcrf.gov/ips-databases.html |
| TMIG-2DPAGE | http://www.tmig.or.jp/2D/2D$_n$Home.html |
| Cyano2Dbase | http://www.kazusa.or.jp/tech/sazuka/cyano/proteome.html |
| Aberdeen 2D-PAGE | http://www.abdn.ac.uk/~mmb023/2dhome.htm |
| Danish 2D-PAGE | http://biobase.dk/cgi-bin/celis/ |
| UCSF 2D-PAGE | http://rafael.ucsf.edu/2DPAGEhome.html |
| JPSL 2D-PAGE | http://www.ludwig.edu.au/www/jpsl/jpslhome.html |

methods include Western immunoblotting (see Sec. XIII), protein sequence analysis (see Sec. XIV.A and this volume, chapter by Rohde and Lu), mass spectrometry (see Sec.s XIV.D, XIV.E and this volume, chapter by Garner), amino acid compositional analysis (see Sec. XIV.B), and carbohydrate analysis (this volume, chapter by Buckberry).

Many of these 2-D gel protein databases can now be accessed via the World Wide Web (WWW) on the Internet (Table 8). In order to provide optimal interconnectivity between these 2-D gel protein databases and other databases of related information available via the WWW (Fig. 25), it has been suggested that 2-D gel databases are constructed according to five fundamental rules (Table 9) (93). 2-D gel protein databases conforming to these rules are said to be "Federated 2-D Databases," while many of the other databases conform to at least some of the rules (Table 8).

In its widest sense, proteomics is being used to analyse the total protein complement, i.e. the proteome, of whole organisms. This task is currently being undertaken most actively for simpler organisms such as the mycoplasmas, prokaryotes and yeast (the only eukaryote whose genome has been fully sequenced). Recent progress in this area is described in a series of papers in (94). In the case of most eukaryotic organisms, complete characterization of the proteome will require an enormous effort due both to the size of the genome and to co- and post-translation modifications which result in protein diversity far exceeding the complexity of the genome (see Sec. I.A). Nevertheless, a start has been made on the proteomes of organisms such as *Dictyostelium discoideum* (95), *Caenorhabditis elegans* (96), and *Drosophila melanogaster* (97).

The complexity of eukaryotic proteomes has resulted in proteomics being used in a narrower context in which this approach is used to define patterns of

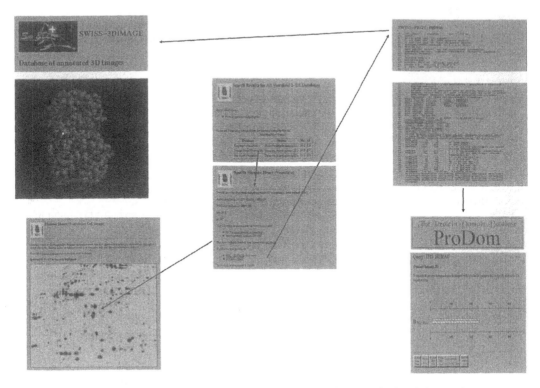

**Fig. 25** Illustration of the interconnectivity of a 2-D gel database and other information resources using the WWW on the Internet. Data for a protein in the HSC-2DPAGE database has been retrieved, the corresponding protein spot has been displayed on the reference 2-DE database image, and links have been made to entries for that protein in the SWIS-PROT, ProDom and SWISS-3DIMAGE databases.

**Table 9** Rules for Federated 2-DE Protein Databases (93)

| | |
|---|---|
| Rule 1: | Individual entries in the database must be accessible by remote keyword search. Other query methods are possible but are not essential, such as full text search. |
| Rule 2: | The database must be linked to other databases through active hypertext cross-references. |
| Rule 3: | There must be a main index which allows all databases to be searched through one unique entry point. |
| Rule 4: | Individual protein entries must be accessible through clickable images. |
| Rule 5: | 2-DE analysis software that has been designed for use with federated databases must be able to directly access individual entries in any federated 2-DE database. |

quantitative protein (gene) expression in particular cells and tissues. This information can then be exploited to characterize biological processes, such as those involved in development, during the cell cycle, during apoptosis, in disease, in response to pharmaceutical intervention, extracellular stimuli and the application of toxic agents. Ultimately, the goal is to decipher the mechanisms controlling gene expression. Space limitation precludes a comprehensive review of these areas and interested readers are referred to the series of papers in (98–101). Readers wishing further general discussion of the rapidly expanding area of proteomics are referred to (102–104).

## XI.   DETECTION METHODS

### A.   Fixation

For the majority of protein detection methods it is necessary to fix (i.e. immobilise) the separated proteins in the gel by a process of precipitation. The best fixative is 20% w/v TCA as it gives effective precipitation of most proteins. Acid methanol (or ethanol), typically a solution containing 10% v/v acetic acid, 45% v/v methanol, and 45% deionized water, is often used for gel fixation, but it should be noted that this can be ineffective for small proteins, basic proteins and glycoproteins.

### B.   Coomassie Brilliant Blue

Coomassie brilliant blue (CBB) R-250, a dye originally developed for use in the textile industry, has become the most popular general stain for visualising protein separated by electrophoresis. Indeed, at one time (before the wearing of laboratory gloves was common) practitioners of the art of electrophoresis were said to belong to the "Society of Blue Fingers!" Staining is usually carried out using 0.1 w/v CBB R-250 in the same acid methanol solution used for fixation (10% acetic

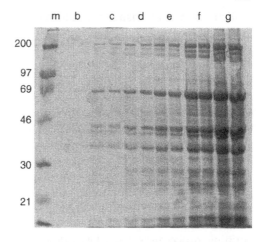

**Fig. 26** SDS-PAGE separation of human heart proteins (lanes b–g). Lane (a) contains the $M_r$ marker proteins and the scale at the left indictes $M_r \times 10^{-3}$. The gel has been stained with Coomassie brilliant blue R-250. The sample protein loadings were (b) 1 $\mu$g, (c) 5 $\mu$g, (d) 10 $\mu$g, (e) 25 $\mu$g, (f) 50 $\mu$g, (g) 100 $\mu$g.

acid, 45% methanol). Depending on gel thickness and polyacrylamide concentration, staining can take from 30 min to several hours. In practice it is often convenient to stain the gel overnight and then destain it by several changes in the same acid methanol solution until intense blue protein zones can be seen against a clear background. This method is able to detect a minimum of around 0.5 ìg protein per band (see Fig. 26), so that for complex mixtures containing several hundred components it is necessary to load relatively high amounts of total protein (>50 $\mu$g).

More sensitive staining (down to 1 ng protein per band) can be achieved using 0.1% w/v CBB G-250 as a colloidal dispersion in 2% w/v phosphoric acid, 10% w/v ammonium sulphate, and 20% v/v methanol (105). An additional advantage of this method is that the colloidal dye only binds to the separated proteins as it is unable to penetrate the gel matrix. This means that no destaining step is required and the intensity of staining can be controlled by visual inspection during the staining process.

## C.  Silver Staining

Silver has been know to be able to develop images for over 200 years, first being usefully exploited in photography and then rapidly adopted for use in histological staining procedures. The ability of silver to detect proteins following their separation by gel electrophoresis was first recognized by Merril and his colleagues in 1979 (106). Subsequently, several silver staining procedures have been described and this group of methods have become the standard approach for the sensitive detection of gel separated proteins, reviewed in [107, 108].

Depending on the method, silver staining is between 20 and 200 times more sensitive than staining with CBB R-250, and is able to detect around 0.1 ng protein per band or spot. There can be problems in using silver staining as a quantitative procedure as it is known to be non-stoichiometric. However, staining intensity is linear over a 50-fold range from 0.04–2 ng protein/mm, comparing well with the 20-fold linear range of CBB R-250 from 10–200 ng protein/mm$^2$ (61). Above this limit the stain becomes non-linear, resulting in saturation and even negative staining of bands and spots at very high protein concentrations, making quantitation of such protein zones impossible. In a 2-DE study of human leukocyte proteins, over 200 spots were observed to have coefficients of variation less than or equal to 15% when data from replicate patters were analyzed (109). In dilution experiments, the majority (>80%) of the proteins were found to have a linear relationship between the amount of protein loaded and the spot volume (109). An additional problem with the quantitation of silver staining is that the relationship between staining intensity and protein concentration may be different for each

protein (107, 109). However, it is often forgotten that this is also the case for staining with CBB R-250 (110).

All silver staining procedure depend on the reduction of ionic silver to its metallic form, but the precise mechanism involved in the staining of proteins has not been fully established. It has been proposed that silver cations complex with protein amino groups, particularly the $\varepsilon$-amino group of lysine (111), and with sulphur residues of cysteine and methionine (112). However, Gersten and his colleagues demonstrated that staining could not be attributed exclusively to specific amino groups and suggested that some component of protein structure is responsible for differential protein staining (113).

Procedures for silver staining can be grouped into two main types depending on the chemical state of the silver when used for impregnating the gel. The first group are alkaline methods based on the use of ammoniacal silver or diamine solution, prepared by adding silver nitrate to sodium-ammonium hydroxide mixture. Copper can be included in these diamine methods to give increased sensitivity, probably by a mechanism similar to the Biuret reaction. The silver ions complexed to proteins in the gel are then developed by reduction to metallic silver with formaldehyde in an acidified environment, usually using citric acid. In the alternative group of methods, silver nitrate in a weakly acidic (around pH 6.0) solution is used for gel impregnation. Development is subsequently achieved by the selective reduction of ionic silver to metallic silver by formaldehyde made alkaline with either sodium carbonate or sodium hydroxide. Any free silver is washed out of the gel prior to development to prevent precipitation of silver oxide which would result in high background staining.

The majority of silver staining procedures are monochromatic, resulting in dark brown to black protein zones. However, if the development time is extended with saturation of the zones of highest protein concentration, then colour effects can be produced (114). In a comparative study of several methods based on both the silver diamine and silver nitrate approaches, the most rapid procedures were found to be generally less sensitive than those which were more time consuming (115). The use of glutaraldehyde pre-treatment of the gel and silver diamine as the silvering agent were found to be the most sensitive, and example of a gel stained with a method of this type is shown in Fig. 27. It is a common experience that silver staining procedures can be problematical to use based on the use of laboratory-prepared reagents. If care is not taken with the use of high purity water, reagents and glassware, then problems of high background staining, surface "mirror" effects and poor reproducibility can be experienced. Many of these problems can be alleviated using one of the commercially available silver staining kits.

## D. Fluorescent Detection Methods

Many of the problems inherent in the quantification of gel separated proteins visualised by silver staining can be overcome using detection methods based on the use of fluorescent compounds. This group of methods are highly sensitive, and generally exhibit excellent linearity and a high dynamic range, making it possible to achieve good quantitative analysis particularly if a suitable imaging device is used (see Sec. XII.A.4).

Two approaches can be used, the first being to couple the proteins with a fluorescently labeled compound prior to electrophoresis. Examples of such compounds are dansyl chloride (116), fluorescamine (4-phenyl-[furan-2(3H),1-phthalan]-3,3'-dione) (117), o-pththaldialdehyde (OPA) (118), and MDPF (2-methoxy-2,4-diphenyl-3(2H)-furanone) (119). The latter reagent has a reported

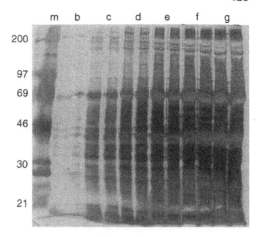

**Fig. 27** SDS-PAGE separation of human heart proteins (lanes b–g). Lane (a) contains the $M_r$ marker proteins and the scale at the left indictes $M_r \times 10^{-3}$. The gel has been silver stained. The sample protein loadings were (b) 1 μg, (c) 5 μg, (d) 10 μg, (e) 25 μg, (f) 50 μg, (g) 100 μg.

sensitivity of 1 ng protein/band and is linear over the range 1–500 ng protein/band.

The main disadvantage of pre-electrophoretic staining procedures is that they can cause protein charge modifications, for example fluorescamine converts an amino group to a carboxyl group when it reacts with proteins. Such modifications usually do not compromise SDS-PAGE unless a large number of additional charged groups are introduced into the protein. However, they result in altered mobility during other forms of electrophoresis, resulting in altered separations by native PAGE, IEF and 2-DE. Recently, compounds which react with cysteine or lysine residues have been described and used successfully for 2-DE separations. The cysteine-reactive reagent monobromobimane (120) has been used to label proteins prior to analysis by 2-DE (121). Using a cooled CCD camera to measure fluorescence (121), the limit of detection was found to be 1 pg protein/spot (122).

In an alternative approach, two amine-reactive dyes (propyl Cy3 and methyl Cy5) have been synthesised and used to label *E. coli* proteins prior to electrophoresis (123). These cyanine dyes have an inherent positive charge, which preserved the overall charge of the proteins after dye coupling. The two dyes have sufficiently different fluorescent spectra that they can be distinguished when they are present together. This allowed two different protein samples, each labeled with one of the dyes, to be mixed together and subjected to 2-DE on the same gel (123). This method, which has been termed difference gel electrophoresis (DIGE) (123), has great potential for improving the efficiency of detection of differences in 2-DE protein profiles between different samples.

For 2-DE, one approach to overcoming the problems associated with charge modification during the IEF dimension is to label the proteins while present in the first dimension gel after IEF, prior to the second dimension separation by SDS-PAGE. Two fluorescent labels which have been used in this way are MDPF (124), and a fluorescent maleimide derivative (69).

The alternative approach, which also overcomes the problem of protein charge modifications, is to label the proteins with fluorescent molecules such as 1-aniline-8-naphthalene sulphonate (ANS) (125), and OPA after the electrophoretic separation has been completed. However, these two methods suffer the disadvantage of relative insensitivity. Recently, two post-electrophoretic fluorescent staining reagents, SYPRO orange and red, have been described [126, 127]. These stains have a very high sensitivity (1-2 ng protein/band) and excellent linearity with a high dynamic range. Using a fluorescent imaging device, the SYPRO dyes have been shown to be linear over three orders of magnitude in protein quantity (127). The other advantage of this

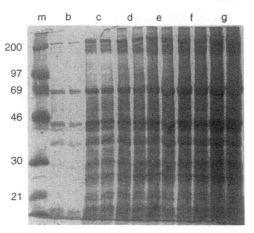

**Fig. 28** SDS-PAGE separation of human heart proteins (lanes b–g). Lane (a) contains the $M_r$ marker proteins and the scale at the left indictes $M_r \times 10^{-3}$. The gel has been stained with SYPRO red. The sample protein loadings were (b) $1\,\mu g$, (c) $5\,\mu g$, (d) $10\,\mu g$, (e) $25\,\mu g$, (f) $50\,\mu g$, (g) $100\,\mu g$.

method is that staining can be achieved in only 30 min, compared with staining with silver, and CBB R-250 which can take from 2 hr to overnight. Gels can be stained without fixation so that they can be subjected to subsequent Western blotting procedures (see Sec. XIII). However, staining with these reagents requires that the proteins are complexed with SDS, so that if the gels are fixed prior to staining or electrophoresis is carried out in the absence of this detergent, then the gels must be incubated in a solution of SDS prior to staining. An SDS-PAGE gel separation visualised using SYPRO red is shown in Fig. 28.

### E. Radioactive Detection Methods

#### 1. Radiolabelling Methods

Metabolic labeling of proteins with a radiolabeled amino acid prior to their separation by gel electrophoresis represents a very sensitive method for the detection of proteins and is ideal for the analysis of protein synthetic events occurring in response to an experimental intervention. This approach is most commonly used in combination with *in vitro* cell culture systems, but it is also possible to radiolabel synthetically the proteins of small pieces of fresh tissue in this way. In this method, the cells or tissue are incubated in the presence of the radiolabeled amino acid for a period of time, normally between 3 and 24 hr. It is important to use tissue culture medium which has been depleted of the amino acid used for radiolabeling. The most commonly used amino acids for radiolabeling are [$^{35}$S]-methionine and [$^{14}$C]-leucine (128). [$^{3}$H]-amino acids can be used, but these are more difficult to detect due to the weak energy of their $\beta$-emissions. Methods are also available for the synthetic radiolabeling to detect specific post-translational modifications of proteins (see Sec. XI.F).

Proteins can also be radiolabeled post-synthetically, prior to their separation by gel electrophoresis, using a variety of methods such as radioiodination with [$^{125}$I] or reductive methylation with [$^{3}$H]-sodium borohydride [reviewed in (129)]. However, most of these methods result in significant charge modification of the target proteins, generally precluding their use for electrophoretic techniques other than SDS-PAGE (see Sec. XI.D).

## 2. Gel Drying

Following electrophoresis of radiolabeled proteins, the gel must normally be dried prior to detection of the radioactive zones. Thin gels cast on plastic supports can be dried, after equilibration in 3% w/v glycerol, in air or in an oven at 40–50°C. It is also possible to air-dry gels which have not been cast on supports. These should be equilibrated in 3% w/v glycerol and placed between two cellophane sheets supported in a plastic frame. The gels are then dried in hot air at 40–50°C; the process usually taking 2 or 3 hr. Apparatus of this type is available com-

**Fig. 29** Apparatus for air-drying polyacrylamide gels. (Courtesy of Amersham Pharmacia Biotech, Uppsala, Sweden).

mercially from Bio-Rad Laboratories and Amersham-Pharmacia Biotech (Fig. 29).

The best method for drying gels which are not on supports is by heating them under vacuum; apparatus for this being available from a number of suppliers (Fig. 30). Gels should be soaked in 3% w/v glycerol prior to drying. Gradient polyacrylamide gels are particularly prone to cracking and these can be protected by soaking in a solution containing 1% w/v glycerol and 2% v/v dimethylsulphoxide (DMSO). Gels can be dried down onto

**Fig. 30** Apparatus for drying polyacrylamide gels under vacuum. (Courtesy of Amersham Pharmacia Biotech, Uppsala, Sweden).

filter paper or onto cellophane. A temperature of 80°C is normally used, but it is better to use a lower temperature (40–60°C) for gels at risk of cracking (i.e. thick, high % T or gradient gels).

### 3. Detection of Radiolabeled Proteins

Radiolabeled proteins are most easily detected by direct autoradiogarphy, in which the dried gel is placed in contact with x-ray film and exposed for the appropriate time. This method works satisfactorily for isotopes such as $^{14}C$, $^{35}S$, $^{32}P$ and $^{125}I$, but us not suitable for $^{3}H$ due to its low energy $\beta$-emissions which are not able to penetrate the gel matrix.

Much more sensitive detection can be achieved using fluorography in which the gel is impregnated with a scintillant, such that low-energy $\beta$-particles excite the fluor molecules to emit photons which can be detected on a suitable (usually blue-sensitive) x-ray film. In the original procedure, 2,5-diphenyloxazole (PPO) which must be dissolved in DMSO, was used (130). However, fluorography with commercially available enhancers is simpler and less tedious than the original PPO-DMSO method, and produces equivalent results (131). Pre-exposure of the x-ray film to a brief flash of light (approximately 1 msec) increases the sensitivity of fluorography by two- or three-fold (132). The use of an intensifying screen and exposure at low temperature (−70°C) also result in a significant increase in sensitivity (132).

Techniques of autoradiograpy and fluorography are simple and require little specialised equipment, apart from the access to darkroom facilities. However, prolonged exposure times are often required to achieve the desired level of sensitivity of protein detection. Moreover, the non-linear response of x-ray film, and its limited dynamic range present severe problems to accurate quantitation. To overcome these problems several devices

for detecting radiolabeled proteins directly in gels have been described (see Sec. XII.A.2).

### F. Detection of Post-translational Modifications of Proteins

### 1. Glycoproteins

Proteins with limited glycosylation can be detected following gel electrophoresis with the general proteins stains such as CBB R-250 and silver. However, such staining gives no direct indication that these proteins are glycosylated and the methods are much less sensitive if the proteins are more highly glycosylated. Proteoglycans are usually stained with cationic dyes, such as Alcian blue or Toluidine blue (133), which bind to the negatively charged glyosaminoglycan side chains. Glycoproteins have generally been detected using variations of the Schiff base reaction, involving oxidation with periodic acid followed by staining with Schiff reagent (134), Alcian blue (135) or a hydrazine derivative (136). A two-fold increase in sensitivity can be achieved with methods in which Alcian blue is used as the primary staining agent followed by subsequent enhancement using a neutral silver-staining protocol (137).

An alternative approach to the analysis of glycosylated proteins is to radiolabel them *in vitro*, followed by gel electrophoretic separation of the radiolabeled proteins, and their detection as described in Sec. XI.E.3. N-linked sugar labelling can be achieved using [$^{3}H$]-mannose and terminal O-linked N-acetylglucosamine can be labelled by galactosyltransferase and UDP-[$^{3}H$]-galactose (128).

Probably the most versatile reagents for the characterization of glycosylated proteins following their separation by electrophoresis are radiolabeled, fluorescent or enzyme-conjugated lectins. Although it is possible to use these directly in the gel matrix, much better results are achieved using Western blotting techniques (see Sec. XIII.C.3).

## 2. *Phosphoproteins*

The most commonly used approach to the analysis of protein phosphorylation is to radiolabel cells in culture with either $[^{32}P]$-orthophosphate (128) or $[\gamma\text{-}^{32}P]$-ATP (138). An alternative approach, which avoids the use of radioactive materials, is to use antibodies which are specific to phosphotyrosine, phosphothreonine and phosphoserine in combination with Western immunoblotting (see Sec. XIII.C.3).

## 3. *Lipoproteins*

Lipoproteins can be stained following electrophoresis with Sudan black B. Prenylated proteins can be radiolabeled prior to electrophoresis with $[^3H]$-mevalono-lactone, while fatty acylated proteins can be radiolabeled with $[^3H]$-palmitic or $[^3H]$-myristic acid (128).

## G. Detection of Enzymes

It is generally considered that specific enzyme activities can only be visualized following gel electrophoresis if native conditions have been used. However, there are several reports demonstrating that SDS-denatured proteins can also be visualized provided that it is possible to achieve at least partial renaturation of their spatial configuration. Such renaturation is most effective if disulphide bonds are not essential for enzymic activity and if the native protein is not composed of subunits of different molecular weights (139). Pre-electrophoresis of gels is usually recommended to remove unreacted acrylamide monomers and catalysts. The subject of renaturation following SDS-PAGE is discussed in detail in (4).

Enzyme staining can be achieved by incubating the unfixed gel in a solution of the appropriate reagents using either fluorogenic or chromogenic substrates. This method works well if the final reaction product is insoluble. However, a soluble reaction product will rapidly diffuse resulting in loss of resolution. It is generally preferable to use a print or gel overlay technique. In this approach, the substrates and other reagents are either impregnated into a filter or included in a thin layer of agarose or polyacrylamide gel cast on a glass or plastic support. The overlay is then placed in direct contact with the surface of the separation gel and following a suitable period of incubation, the enzymic activity is visualized on the overlay. Methods are available for the visualization of a large number of enzyme activities following gel electrophoresis and these are reviewed in detail in (4).

## XII. QUANTITATION

Following electrophoretic separation of a particular protein mixture, it is often a requirement for the analyst to provide an estimate of the relative quantities of the different components. While simple visual inspection of the separation profile can give an impression of the relative abundances of the different components, more rigorous quantitation is usually required.

## A. Gel Imaging

### 1. *Stained Gels*

The first essential step for quantitation is to obtain a digitised image of the stained gel. Until recently, flat-bed scanning densitometers fitted with a laser light source have given the best spatial resolution (down to $50\,\mu m$) combined with a high dynamic range (up to 4 OD). However, such instruments are expensive, and the current range of desktop document scanners can also achieve high resolution (600 dpi is equivalent to $42\,\mu m$) with a high dynamic range (12 bits). CCD cameras are also effective and relatively inexpensive devices for imaging electrophoretic separation patterns, and several gel documentation systems of this type are available commercially.

## 2.  Radioactively Labelled Proteins

Images of radiolabeled proteins separated by gel electrophoresis can be readily captured by autoradiography or fluorography using x-ray film (see Sec. XI.E.3). The resulting film images can then be digitised using any of the devices described above (Sec. XII.A.1). However, accurate quantitation of such film images is complicated both by the limited dynamic range of film, which means that saturated (and therefore non-quantifiable) protein bands or spots are frequently encountered, and by the non-linearity of the film response. These problems can be overcome using densitometry of multiple film exposures together with appropriate calibration samples of dried polyacrylamide containing known amounts of the same radio-isotope with which the sample has been labeled (140).

To overcome these problems, several devices for detecting radiolabeled proteins directly in gels have been described [reviewed in (141)]. The best and most practical of these approaches is that based on the use of photostimulable storage phosphor-imaging screens. The surface of these screens contains a thin layer of $BaFbr:Eu^{2+}$ in a plastic support. The dried gel is placed in contact with the screen in the same way that an x-ray film is exposure is made. During this exposure, the $\beta$-particles emitted by the radiolabeled proteins pass through the layer, converting $Eu^{2+}$ to $Eu^{3+}$. After a suitable exposure time, the screen is transferred to a suitable scanner where light from a high intensity HeNe laser (633 nm) is absorbed, resulting in the emission of blue (390 nm) luminescence proportional to the amount of radiation incident on the screen.

Further details of the physics of this process can be found in (141), and several commercial instruments based on this process are available. The major advantages of this approach compared to conventional autoradiography is that relatively short exposure times are required, it has a high dynamic range and good linearity of response is achieved. The only disadvantage is the high capital cost of the phophorimaging screens and the dedicated imaging device which is required.

## 3.  Fluorescently Stained Gels

Scanning laser densitometers suitable for the imaging gel profiles of fluorescently labeled proteins (Fig. 28) are available from several manufacturers. As an alternative, camera systems fitted with cooled CCD arrays provide good spatial resolution and the low noise inherent in these systems allows images with low light levels to be collected over extended periods, making them suitable for the detection of the separation patterns of fluorescently labeled proteins (121, 122, 142).

## B.  Analysis of 1-D Gel Separation Profiles

Most commercial devices for imaging protein separation profiles generated by 1-D gel electrophoresis are controlled through computer workstations which are also equipped with software that allows automatic acquisition and quantitative analysis of data [reviewed in (143)]. These systems can display the images either as a representation of the protein separation itself (Fig. 31) or in the form of a densitometric trace (Fig. 32). Software running on microcomputer workstations provided with these systems can then detect the individual protein zones and provide quantitative data based either on peak height or integrated area. The width of the individual sample lane used for

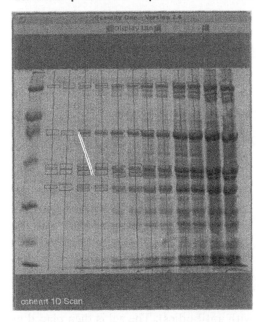

**Fig. 31** Computer analysis of 1-D protein separation profile, displaying the image as a representation of the gel itself. The individual lanes are marked and some of the protein zones selected for quantitation are indicated.

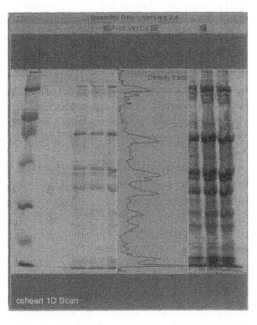

**Fig. 32** Computer analysis of 1-D protein separation profile, with data for one of the lanes being displayed as a densitometric trace.

analysis can usually be adjusted according to the actual width of the separated zones. In most cases, the operator is required to interact with the system to ensure that the protein zones are correctly detected, particularly in the case where "shoulders" on peaks occur.

## C. Analysis of 2-D Gel Separation Profiles

The complexity of the separation patterns resulting from 2-DE analysis of complex protein mixtures has resulted in the development of dedicated automated computer analysis systems. A detailed discussion of this topic is beyond the scope of this chapter, so that only a brief outline will be given. More detailed reviews of this area can be found in (144, 145).

In the first step in 2-DE gel analysis, the digitized gel image is usually subjected to various filtering algorithms and background subtraction procedures to produce a "clean" 2-D gel image. The individual protein components, that is the "spots," are then detected and various procedures, such as thresholding, edge detection or Gaussian fitting, can be applied. Once the individual spots have been detected, it is then a relatively straightforward task to complete the quantitative analysis of each individual gel.

However, in most experiments involving 2-DE a further requirement is that a comparison is made between all of the 2-DE gels involved, which can in some cases involve comparison of large numbers of 2-DE patterns. This process is these days less complex due to the increased spatial reproducibility achievable using the current methodology of 2-DE (see Sec. X.H). Nevertheless, sophisticated algorithms are still used to bring the individual 2-DE gel images into coincidence. The final level of complexity then involves the cross-matching of multiple sets of 2-DE gel patterns to construct large-scale databases of quantitative protein expression, as required in proteomics projects (see Sec. X.I).

In the early days of 2-DE, the computer systems required for such analysis required large, dedicated and expensive hardware. Such pioneering systems were GELLAB (146–148), LIPS (149, 150), Elsie (151) and TYCHO (152). Over the years, this situation has changed along with the rapid progress made in microcomputer technology. Thus, the current generation of 2-DE analysis software is capable of running on small desktop workstations (Unix, PC and Mac), and these systems include Elsie-4 (151), Melanie (153), QUEST (154), GELLAB-II (155). Commercially available versions of four of these software systems are currently available; Melanie II (Bio-Rad Laboratories) (156, 157), BioImage (BioImage Systems Corp), Phoretix (Phoretix International Ltd) and Kepler (LSB Corp).

## XIII.   WESTERN BLOTTING

### A.   Principles of Method

The various methods of gel electrophoresis which have been developed allow direct characterisation of the separated protein zones in terms of their mobility, charge, size, abundance and even hydrophobicity. In addition, specific staining methods can be used to identify specific classes of proteins and characterization of their enzymic properties. The development of the technique of Western blotting (158, 159) has resulted in a variety of possibilities for the identification and characterization of gel-separated proteins (160, 161). In this technique, the separated components are transferred from the restrictive gel matrix onto the surface of a suitable membrane support, such that the proteins are then readily accessible to reaction with any suitable ligand. Specific antibodies (both polyclonal and monoclonal) and lectins have been extensively used in this way. Western blotting techniques have also

been instrumental in the development of several of the methods now available for direct chemical characterisation of proteins separated by gel electrophoresis (see Sec XIV).

### B.   Transfer Methods

The steps involved in Western blotting are shown in Fig. 33. After electrophoresis it is often necessary to equilibrate the gel in the buffer which will be used for the transfer step. This acts to remove reagents such as SDS (which can interfere with the transfer process), minimises changes in gel dimensions during transfer (which can result in distorted zones), and can allow some degree of protein renaturation to occur (which can be beneficial for reaction with antibodies). The gel is then placed in contact with a suitable membrane. Nitrocellulose (NC), which has a high protein binding capacity [249 $\mu g/cm^2$ (162)], is still the most commonly used membrane for most Western blotting protocols. A variety of other membranes are available (161), and in particular polyvinylidene difluoride (PVDF) membranes have become popular due to their high mechanical strength compared with NC and their high protein binding capacity [172 $\mu g/cm^2$ (162)]. These alternative membranes are used in preference to NC in most chemical characterization methods due to their resistance to organic solvents (see Sec. XIV).

Protein transfer can be carried out in various ways such as contact diffusion (163), capillary (164) or vacuum (166) methods. However, electroblotting, developed by Towbin (158), is the most efficient and rapid method, in which the separated proteins are transferred from the gel onto the membrane surface by the application of an electric field perpendicular to the plane of the gel. Electroblotting can be performed in two types of apparatus. The procedure as originally developed uses vertical tanks containing large volumes of

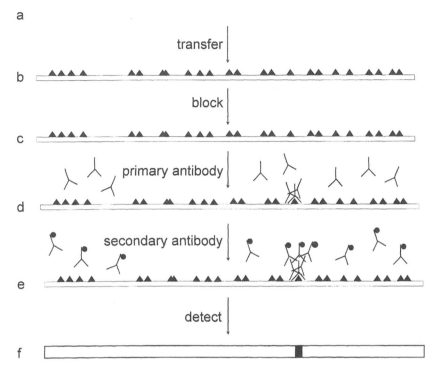

**Fig. 33** Schematic diagram of the steps involved in Western blotting. The protein zones in the polyacylamide gel (a) are transferred to the surface of a blotting membrane (b). The total protein pattern can be detected using a general protein stain. For probing with a specific ligand, unoccupied binding sites on the membrane are blocked (c). The blot is probed with a specific primary antibody (d), then washed and probed with a secondary antibody reactive with the immunoglobulin class and species of the primary antibody (e). After washing, specifically bound antibodies are detected using an appropriate method (see text).

transfer buffer, in which the gel and membrane are held in a sandwich of filter paper and sponge pads placed between two sets of platinum electrodes (166) (Fig. 34). High current (typically 0.5 to 1 mA/gel), low voltage conditions are used and transfer is carried out for several hours. In the second approach, known as semi-dry electroblotting, the gel/membrane/filter paper sandwich is placed in direct contact between flat-plate electrodes, usually of graphite, mounted in a horizontal apparatus (167) (Fig. 35). The advantage of this method is that the electrodes are much closer together than in the buffer tank systems, giving a much higher field strength and resulting in more

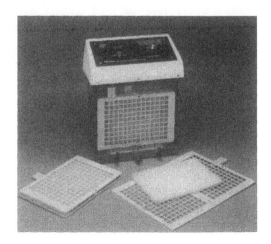

**Fig. 34** Vertical tank apparatus for Western blotting. (Courtesy of Amersham Pharmacia Biotech, Uppsala, Sweden).

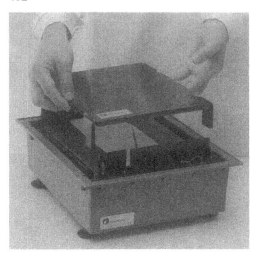

**Fig. 35** Horizontal semi-dry apparatus for Western blotting. (Courtesy of Amersham Pharmacia Biotech, Uppsala, Sweden).

rapid transfer. A maximum of $0.8 \, mA/cm^2$ of membrane surface is recommended and transfer times are restricted to less than 2 hr in order to prevent evaporation of the limited amount of buffer.

The choice of transfer buffer is important to achieve efficient protein transfer combined with its optimum retention on the membrane. That said, the original transfer buffer described by Towbin (158) (25 mM Tris, 192 mM glycine, pH 8.3) is most commonly used. Methanol (10 to 20% v/v) is quite commonly added to the transfer buffer as it promotes dissociation of SDS-protein complexes and promotes the binding of proteins to the membrane. However, the fixative action of methanol can reduce the efficiency of elution of proteins from the gel matrix, resulting in extended blotting times. The addition of small amounts of SDS (up to 0.01%) to the transfer buffer improves the efficiency of protein elution from the gel, but this is at the expense of reduced binding to the membrane (161), and consequently is not generally recommended. A discontinuous buffer system was originally recommended for semi-dry blotting (167),

but its use is not essential and the Towbin buffer generally gives excellent results. It should be realised that use of the Towbin buffer will result in loss of basic proteins. These proteins can be either electroblotted using a very basic transfer buffer [e.g. 25 mM CAPS, pH 10 (168)] or transferred as cations using 0.7% acetic acid. The use of glycine must be avoided if proteins are to be subjected to chemical characterization, and special transfer buffers have been developed for this purpose [see Sec. XIV and (161)].

### C.   Detection Methods

#### 1.   General Protein Stains

Various methods with differing sensitivities have been described which permit visualization of the total protein pattern following Western blotting. Some of these methods together with their relative sensitivities and compatibility with different types of blotting membranes are given in Table 10. Fast green FC and Ponceau S (Fig. 36) are relatively insensitive stains, but can be easily removed from proteins after detection to allow for subsequent immunoprobing or chemical characterisation. Amido black 10B is the most commonly used stain for the rapid and relatively sensitive visualisation of proteins on nitrocellulose membranes (Fig. 37). Coomassie Brilliant Blue R-250 gives a very high background on nitrocellulose, but gives excellent results with PVDF membranes (Fig. 38). Blots stained with Coomassie Brilliant Blue R-250 or Amido black 10B can be used for the chemical characterization of proteins but they are not compatible with immunodetection methods. Instaview Nitrocellulose (BDH Merck, Poole, Dorset, UK) is a rapid and sensitive method of general protein staining (Fig. 39) with the advantage that the membrane can be rapidly destained without loss of immunoreactivity of the proteins which

**Table 10** Methods for the Detection of Total Proteins on Western Blots in Order of Increasing Sensitivity

| Detection reagent | Approximate sensitivity (protein per spot) | Matrix compatibility |
|---|---|---|
| Ponceau S | 5 µg | NC, PVDF |
| Fast green FC | 5 µg | NC, PVDF |
| Amido black 10B | 1 µg | NC, PVDF |
| Coomassie brilliant blue R-250 | 500 ng | PVDF |
| Pyrogallol-red molybdate | 500 ng | NC, PVDF |
| Immobilon-CD stain | 500 ng | Immobilon-CD |
| Instaview Nitrocellulose | 250 ng | NC, PVDF |
| India Ink | 100 ng | NC, PVDF |
| Double metal chelate stain | 50 ng | NC, PVDF |
| Colloidal iron | 30 ng | NC, N, PVDF |
| In situ biotinylation + HRP + avidin | 30 ng | NC, N, PVDF |
| Colloidal gold | 4 ng | NC, PVDF |

can then be used in specific immuno-detection protocols. More sensitive staining can be achieved by staining with India ink (169) or colloidal gold particles (170), but these procedures are more protracted (several hours to overnight). Recently,

several reversible metal chelate stains of varying sensitivity have been developed (171, 172) which are compatible with both subsequent immunodetection protocols and methods for chemical characterization.

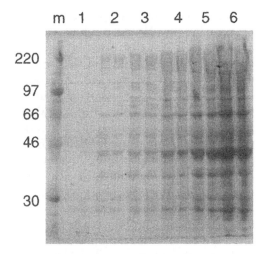

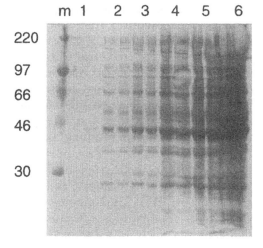

**Fig. 36** SDS-PAGE separation of human heart proteins, electroblotted onto nitrocellulose and visualized by staining with Ponceau S. Lane (m) contains the $M_r$ marker proteins and the scale at the left indicates $M_r \times 10^{-3}$. The sample protein loadings were (1) 1 µg, (2) 5 µg, (3) 10 µg, (4) 25 µg, (5) 50 µg, (6) 100 µg. (From Ref. 212).

**Fig. 37** SDS-PAGE separation of human heart proteins, electroblotted onto nitrocellulose and visualized by staining with Amido black 10B. Lane (m) contains the $M_r$ marker proteins and the scale at the left indictes $M_r \times 10^{-3}$. The sample protein loadings were (1) 1 µg, (2) 5 µg, (3) 10 µg, (4) 25 µg, (5) 50 µg, (6) 100 µg. (From Ref. 212).

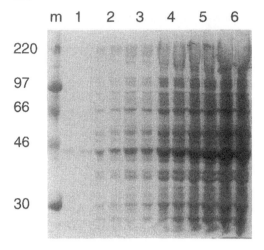

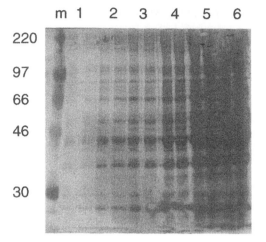

**Fig. 38** SDS-PAGE separation of human heart proteins, electroblotted onto PVDF and visualized by staining with Coomassie brilliant blue R-250. Lane (m) contains the $M_r$ marker proteins and the scale at the left indictes $M_r \times 10^{-3}$. The sample protein loadings were (1) $1\,\mu g$, (2) $5\,\mu g$, (3) $10\,\mu g$, (4) $25\,\mu g$, (5) $50\,\mu g$, (6) $100\,\mu g$. (From Ref. 212).

**Fig. 39** SDS-PAGE separation of human heart proteins, electroblotted onto nitrocellulose and visualised by staining with Instaview Nitrocellulose. Lane (m) contains the $M_r$ marker proteins and the scale at the left indictes $M_r$, $\times 10^{-3}$. The sample protein loadings were (1) $1\,\mu g$, (2) $5\,\mu g$, (3) $10\,\mu g$, (4) $25\,\mu g$, (5) $50\,\mu g$, (6) $100\,\mu g$. (From Ref. 212).

Charged membranes present a particular problem as the membrane surface can bind dyes as avidly as the separated proteins. For example, Immobilon-CD (Millipore, Bedford, MA, USA) is a PVDF matrix that has been modified to carry a cationic charge on the surface and has been optimized for use in internal sequencing protocols. Proteins bind strongly to this membrane during blotting and, after proteolytic cleavage, the resulting peptide fragments can be released from the surface by mild conditions that disrupt the electrostatic interactions (e.g. changes in salt concentration or pH) (173). Immobilon-CD Quick Stain (Millipore) is a negative dye that interacts solely with the membrane and therefore does not interfere with subsequent sequence analysis of the proteins. Nylon membranes are more difficult to stain, but this is possible using a colloidal iron procedure (174) or a method based on *in situ* biotinylation of the blotted proteins followed by visualisation with peroxidase-conjugated avidin (163).

## 2. Blocking

Before methods for the specific detection of proteins on Western blots can be applied, it is usually essential that all unoccupied binding sites on the transfer membrane must be blocked. A protein solution, such as 5% w/v bovine serum in phosphate buffered saline (PBS), in which the membrane is incubated for 1 hr at room temperature is often used. Various other proteins, including gelatin, ovalbumin, casein and haemoglobin have also been used for this purpose (163). A popular alternative blocking procedure, which avoids the use of a protein, is a 3% w/v solution of non-fat dried milk powder in PBS (175). Another alternative blocking agent is the detergent, polyethylene sorbitan monolaurate (Tween 20) (176), which is used at 0.1% w/v in PBS (PBS-T). Some investigators recommend the use of a combination of Tween 20 and non-fat dried milk for blocking, and carry out all the subsequent steps in PBS-T.

**Table 11** Specific Detection Methods for Western Blots Arranged in Order of Increasing Sensitivity

| Method | Approximate sensitivity $(ng/mm^2)$ |
|---|---|
| Peroxidase-protein A | 2.0 |
| Peroxidase-secondary antibody | 1.5 |
| Gold—secondary antibody | 1.5 |
| $^{125}I$—secondary antibody | 1.0 |
| Peroxidase anti-peroxidase | 0.8 |
| Avidin-biotin-peroxidase complex | 0.5 |
| Gold—secondary antibody with silver enhancement | 0.1 |
| Enhanced chemiluminescence | 0.001 |

## 3. Specific Probing

Protein blots can be probed with a wide variety of monoclonal and polyclonal antibodies, lectins, proteins and other ligands. Probing with antibodies is the most common procedure and this has become known as immunoblotting. It is possible to derivatize antibodies (and most other ligands) with a suitable reporter molecule to achieve direct visualization of reactive protein zones on blots. However, this method is not generally recommended as it consumes the primary antibody, which is often expensive or available in restricted quantity, and can result in loss of its specificity or reactivity. This has resulted in the almost universal adoption of an indirect approach employing a labeled secondary (or tertiary) antibody.

Following an appropriate blocking step, the membrane is incubated with a solution containing the appropriately diluted primary antibody. The membrane is then washed and reacted with a solution of the secondary antibody specific for the species and immunoglobulin class of the primary antibody. The secondary antibody can be conjugated with a variety of reporter groups (Table 11), including fluorescent labels [e.g. fluorescein isothiocyanate (FITC) or dichlorotriazynylyamino-fluorescein (DTAF)], radiolabel [usually ($^{125}I$)], or an enzyme [e.g horseradish peroxidase (HRP), alkaline phosphatase (AP), $\beta$-galactosidase or glucose oxidase]. The secondary antibody can be substituted with labeled staphylococcal protein A or streptococcal protein G, which bind to the $F_c$ region of immunoglobulins.

The binding of radiolabeled antibodies is detected by autoradiography (see Sec. XI.E.3) or using a phosphorimaging device (see Sec. XII.A.2). Detection of enzyme-conjugates is traditionally performed using substrates which form insoluble, stable colored reaction products at the sites of antibody binding. The two most popular substrates for use with peroxidase-conjugated antibodies are diaminobenzidine (DAB), resulting in brown zones, and 4-chloro-1-naphthol, giving purple bands or spots. Alkaline phosphatase conjugated antibodies are usually visualised with a mixture of 5-bromo-4-chloro indoxyl phosphate (BCIP) and nitroblue tetrazolium (163).

Several other systems have been developed to increase the sensitivity of detection following immunoprobing of blots. One approach is the use of antibodies or protein A labeled with colloidal gold (177). The stain is visible due to its reddish-pink colur without the necessity for a development step, but sensitivity is considerably increased by a silver enhancement step (178). Other methods which have been used to increase sensitivity of immunodetection on blots include triple antibody probing methods (e.g. peroxidase-antiperoxidase and alkaline phosphatase-antialkaline phosphatase) and the digoxigenin-labeled antibodies detected using an enzyme-conjugated anti-digoxigenin antibody (179). Another method which has been used quite extensively makes use of the specific reaction between the protein, avidin, and its ligand, biotin. In this approach, secondary antibodies are conjugated with biotin and are then detected on the blot in a third step using avidin conjugated with a reporter molecule such as HRP or AP. Even greater sensitivity can be achieved in this third step using preformed complexes of a biotinylated enzyme with avidin, due to the increased signal arising from the presence of large numbers of enzyme molecules.

Recently methods based on the use of enhanced chemiluminescence have become very popular for the detection of immunoreactions on Western blots. Systems are available for use with both HRP- and AP-conjugated antibodies. The system compatible with HRP-conjuagted antibodies is known as Enhanced Chemiluminescenc (ECL, Amersham Pharmacia Biotech) (180). The method is based on the oxidation of luminol in the presence of hydrogen peroxide, resulting in the emission of light. The inclusion of a phenolic enhancer results in up to 1000-fold enhancement of the light emission which is detected using an x-ray film with the appropriate sensitivity (Fig. 40). The recently developed ECL Plus system (Amersham Pharmacia Biotech), which uses acridan-based chemistry rather than luminol, offers up to a 25-fold increase in sensitivity and a more sustained light output compared with the standard ECF system (181). A further advantage is that

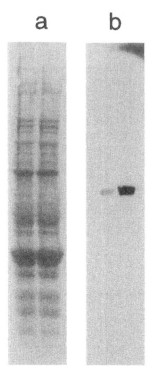

**Fig. 40** SDS-PAGE separation of human heart proteins electroblotted onto nitrocellulose. (a) Total protein staining with Instaview Nitrocellulose. (b) Protein bands reactive with a monoclonal antibody reactive with the inducible form of heat shock protein 70 kDa (hsp70i), visualized using enhanced chemiluminescence. The band of higher intensity (b, right) is in a sample from a heart which had been subjected to heat shock *in vivo*, while the band of low intensity (b, left) is from a normothermic heart.

the emitted light can be detected using a suitable fluorescent imaging device (see Sec. XII.A.4).

For cemiluminescent detection of AP-conjugated antibodies either a dioxetane substrate, disodium 3-(4-methoxyspirol [1,2-dioxetane-3,2'-tricyclo $[3.3.1.1^{3,7}]$decan-4-yl)phenyl phosphate (AMPPD), or its 5-chloro derivative, CSPD is used (182). An alternative approach for use with AP-conjugated secondary antibodies is enhanced chemifluorescence (ECF, Amersham Pharmacia

Biotech), where AP dephosphroylates the ECF substrate to generate a fluorescent product which can be detected in a suitable fluorescent imaging device (see Sec. XII.A.4).

## XIV. CHEMICAL CHARACTERIZATION OF PROTEINS

The methods for gel electrophoresis combined with the use of the various systems for the detection of the separation profiles which we have discussed so far in this chapter allow the separated proteins to be characterized in terms of their pI, $M_r$, and relative abundance. Until relatively recently only time-consuming and laborious methods such as specific staining methods (see Sec. XI), cellular subfractionation and co-electrophoresis with purified proteins were available. However, in recent years a panel of increasingly sensitive methods have been developed for

the identification and characterization of proteins separated by gel electrophoresis (Fig. 41). The first major advance was the development of Western blotting (see Sec. XIII) combined with the availability of specific poly- and monoclonal antibodies which allowed a start to be made on the systematic identification of protein separated by gel electrophoresis. Some groups have used this approach to identify large numbers of proteins separated by 2-DE (183, 184). However, this approach is laborious and depends on the availability of antibodies reactive with the proteins following separation by gel electrophoresis (for example they are denatured following SDS-PAGE and 2-DE).

### A. Protein Sequence Analysis by Edman Degradation

The most direct method of obtaining amino acid sequence information from a

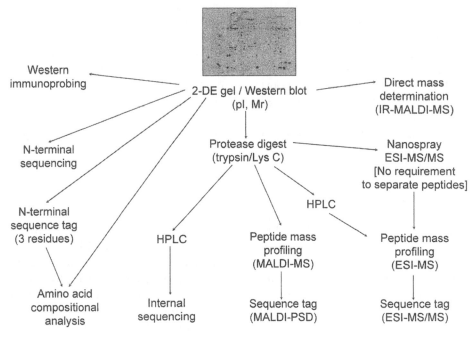

**Fig. 41** Diagram showing some of the methods currently available for the identification and characterisation of proteins separated by gel electrophoresis.

protein is to subject it to N-terminal sequence analysis by the chemical method of Edman degradation. Progress in instrument technology combined with optimization of the chemistry has resulted in the current generation of commercial protein sequenators being able to routinely perform high sensitivity analysis below the 1 pmol level. This level of sensitivity is compatible with the amount of protein typically found in a band or spot following gel electrophoresis, resulting in these methods (particularly 2-DE) being routinely used for the preparation of proteins for sequence analysis (185, 186). The topic of protein sequence analysis by Edman degradation is discussed in detail elsewhere (this volume, chapter by Lu et al).

An alternative approach to protein sequencing takes advantage of the fact that the chemistry of Edman degradation is not quantitative. After several sequencing cycles, a series of truncated peptides remain which differ in mass by the mass of the amino acid released in each step. The peptide sequence can then be determined by accurate mass measurement of the truncated peptides. This approach has been developed into a practical procedure, termed "protein ladder sequencing" (187), in which the partially degraded protein is extracted from the sequencing support after about 20–30 cycles and the truncated peptides analyzed by matrix-assisted laser-desorption ionisation time-of-flight mass spectrometry (MALDI-MS) (this volume, chapter by Garner).

A major problem with protein sequencing by Edman degradation is that many proteins lack a free α-amino group, due to co- or post-translational modification (see Sec. I.A). Such N-terminal blockage occurs typically in up to 50% of eukaryotic proteins and results in no sequence being obtained. This problem can be overcome by subjecting the separated protein, either *in situ* within the gel matrix or after Western blot transfer to a suitable membrane (nitrocellulose,

PVDF), to chemical or enzymatic cleavage to generate shorter peptides which can be isolated and sequenced. The cleavage products are then usually separated by RP-HPLC, and selected peptide fractions directly applied to the protein sequenator. This procedure is highly efficient, but the determination of multiple stretches of sequence usually requires two to three times more protein than does N-terminal protein sequence analysis.

## B. Amino Acid Compositional Analysis

Current methods for the HPLC analysis of pre-column derivatised amino acids are capable of very high (sub-pmol) sensitivity (this volume, chapter by Lu et al). Thus, this method can be applied directly to proteins separated by gel electrophoresis and it has been found to be an excellent method for their rapid identification (188–190). This approach depends on individual proteins having more or less unique amino acid compositions. Data from such an analysis (Fig. 42) is used to interrogate databases of amino acid compositions derived from sequences of known proteins or predicted from nucleotide sequences. A number of search algorithms are available and search data may be filtered by the inclusion of Mr and pI search windows, or species specificity of the target protein (190). A major drawback to this approach is that the end-point is a list of protein identities ranked in order of probability (Fig. 42), but the "correct" protein does not necessarily occur as the first ranked entry. While the use of score patterns to increase confidence of identity has been shown to be useful in protein identification (191), it is usually better to adopt an orthogonal approach and combine amino acid compositional analysis with another rapid method of protein identification, such as peptide mass profiling (190–192). Recently, a method has been developed in

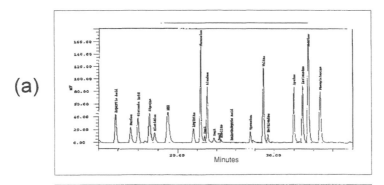

**(b)**

```
AMINO ACID COMPOSITION

              Unknown      Calibration
  Protein:    Protein:     Protein:     Bias:    Weight:
  *---------------------------------------------------------
    Asx        5.70          4.67        1.00      1.00
    Glx        5.25          8.01        1.00      1.00
    Ser        3.26          3.38        1.00      1.00
    His        2.38          3.45        1.00      1.00
    Thr       14.08          6.61        1.00      1.00
    Pro        0.50          2.48        1.00      1.00
    Tyr        1.43          4.85        1.00      1.00
    Arg        2.96          4.38        1.00      1.00
    Val       12.44         11.12        1.00      1.00
    Met        1.45          1.21        1.00      1.00
    Ile       10.17          5.48        1.00      1.00
    Leu       20.84         22.42        1.00      1.00
    Phe       10.92         13.56        1.00      1.00
    Lys        8.63          8.40        1.00      1.00
```

**(c)**

```
Rank Score   Protein    (pI      Mw)   Description
=================================================================
  1    11  FABH_HUMAN    6.34   14727  FATTY ACID-BINDING PROTEIN, HEART (H
  2    17  FABH_MOUSE    5.69   14689  FATTY ACID-BINDING PROTEIN, HEART (H
  3    28  FABH_RAT      5.92   14644  FATTY ACID-BINDING PROTEIN, HEART (H
  4    37  FABL_GINCI    6.79   15079  FATTY ACID-BINDING PROTEIN, LIVER (L
  5    48  FABH_BOVIN    6.91   14648  FATTY ACID-BINDING PROTEIN, HEART (H
  6    56  FABE_RAT      6.73   15059  FATTY ACID-BINDING PROTEIN, EPIDERMAL
  7    69  HBB_CHEKU     7.11   16458  HEMOGLOBIN BETA CHAIN.
  8    70  BTF3_HUMAN    6.85   17699  BTF3B.
  9    73  FABE_HUMAN    6.60   15164  FATTY ACID-BINDING PROTEIN, EPIDERMAL
 10    77  FABE_MOUSE    6.14   15137  FATTY ACID-BINDING PROTEIN, EPIDERMAL
 11    84  RET2_MOUSE    6.12   15478  RETINOL-BINDING PROTEIN II, CELLULAR
 12    89  RET3_HUMAN    5.31   15434  RETINOIC ACID-BINDING PROTEIN I.
 13    89  FABI_HUMAN    6.88   15076  FATTY ACID-BINDING PROTEIN, INTESTINAL
 14    92  RET3_BOVIN    5.31   15460  RETINOIC ACID-BINDING PROTEIN I,
```

**Fig. 42**  The identification of a protein spot from a 2-DE separation by amino acid compositional analysis. (a) HPLC analysis of pre-column derivatized amino acids resulting from hydrolysis of the protein spot. (b) Amino acid composition determined from the HPLC analysis. (c) Result of the amino acid composition database search indicating that the protein is cardiac fatty acid binding protein.

which Edman degradation is used to create a 3 or 4 amino acid N-terminal "sequence tag," following which the proteins are subjected to amino acid compositional analysis (193). The combined amino acid composition and "sequence tag" data is then used for protein identification. A method has also recently been described for the analysis of phosphoamino acids (194).

## C.  Protein Mass

An accurate estimate of the mass of a protein separated by 2-DE can be very useful when used in conjunction with protein identification methods based on sequence database homologies and for the elucidation of post-translational modifications. Estimation of molecular mass by SDS-PAGE is generally considered to have an accuracy of around ±10% and, in addition, many proteins migrate anomalously during SDS-PAGE (see Sec. VI.E). In contrast, pI values as estimated from IPG IEF gels are generally in good agreement with those calculated from their amino acid compositions (195), but this situation is complicated by the presence of any post-translational modifications. A method has been developed, in which proteins separated by 2-DE are electroblotted onto PVDF membranes and analysed by direct insertion of the spots of interest into a MALDI-MS instrument, using either an infra-red (196) or ultraviolet (197) laser source. We have used this approach in a study of human heart protein separated by 2-DE (198). This approach can also be extended to two-dimensional scanning of the sample target (199), thereby generating mass contour images.

## D.  Peptide Mass Fingerprinting

A major breakthrough in rapid protein identification came with the realisation that the set of peptide masses obtained by MS analysis of a protein digest provides a characteristic fingerprint of that protein (Fig. 43). This information is then used to interrogate databases of peptide masses derived from sequences of known proteins or predicted from nucleotide sequences and a number of algorithms have been implemented to facilitate this process [reviewed in (186, 200)]. As in the case of amino acid compositional analysis, this technique of peptide mass profiling is a statistical method with putative protein identities being ranked in order of probability (Fig. 43). The reliability of this approach can be improved by combining peptide mass profiling data from two separate digests (e.g. trypsin, Lys-C) or by adopting an orthogonal approach in combination with amino acid compositional analysis (190–193).

Methods have been developed for peptide mass profiling based on digests generated by enzymatic cleavage either while the protein spot is *in situ* within the gel matrix or after transfer by electroblotting to a suitable membrane. After recovery, the unfractionated peptides can be readily analyzed by MALDI-MS (201). Alternatively, the peptide mixture can be fractionated by RP-HPLC. Systematic screening of HPLC fractions can then performed either by MALDI-MS or by electrospray ionisation (ESI)-MS. Details of the different forms of mass spectrometry can be found elsewhere (this volume, chapter by Garner). ESI-MS can also be coupled on-line with the HPLC separation by splitting the column effluent. This allows simultaneous mass measurement and fraction collection; the peptide present in the fraction then being available for other identification strategies (e.g. protein sequencing, carbohydrate analysis) (202).

## E.  Peptide Fragmentation and Partial Sequence Data by Mass Spectrometry

Partial peptide sequence data used in conjunction with peptide mass database searches is probably the most reliable approach to protein identification. This sequence data can be generated independently either using an automated sequencing instrument based on Edman degradation chemistry or by mass spectrometry (see Sec. XIV.A). Alternatively, two MS-based approaches to the

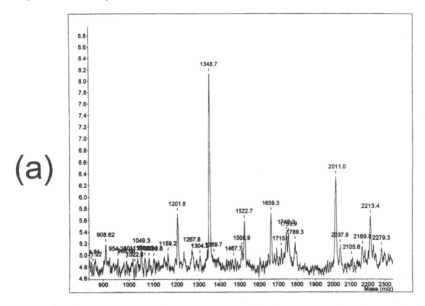

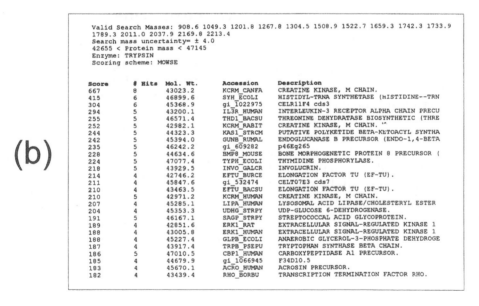

**Fig. 43** The identification of a protein spot from a 2-DE separation by peptide mass fingerprinting. (a) MALDI-MS spectrum of the tryptic digest of the protein spot. (b) Result of the peptide mass database search indicating that the protein is the M-chain of creatine kinase.

identification of proteins have been described which take advantage of the ability of two-stage mass spectrometers [ESI-MS/MS and post-source decay (PSD) MALDI-MS] to induce fragmentation of peptide bonds (202, 203).

The first of these methods, termed peptide sequence tagging, is based upon interpretation of a portion of the ES-MS/MS or PSD-MALDI-MS fragmentation data (204). In this method only a portion of the fragmentation spectrum is interpreted to generate a short partial sequence or "tag," which is used in combination with the mass of the intact parent peptide ion, and provides a significant amount of additional information for the homology search. An elegant extension of this approach has been the development of a nano-electrospray ion source that allows spraying times of more than 30 minutes from about 1 $\mu$l of sample (205). Using this method, multiple peptides can be sequenced from a digest mixture without the need for prior separation by HPLC. This method has been found to be sensitive in the low femtomole range and silver stained 2-DE protein spots containing as little as 5 ng protein have been successfully analyzed (206).

The second method is based on the automated interpretation of ESI-MS/MS fragmentation data which is used to directly search sequence databases (207). In this method, the programme first identifies all those peptides that can be generated from proteins in the sequence database and whose masses match those of the measured peptide ion. In the second step, the programme predicts the fragment ions expected for each of the candidate peptides if they were fragmented under the experimental conditions used. The experimentally determined MS/MS spectrum is then compared with the predicted spectra using cluster analysis algorithms. Each of the comparisons is allocated a score and the highest scoring peptide sequences are reported. The method has been adapted for automated identification of peptide digests analyzed by ESI-MS/MS where the ions subjected to fragmentation are automatically selected during the run and the data are automatically analyzed (208). Proteins present in mixtures can be readily identified with a 30-fold difference in molar quantity and

sensitivity is at the low femtomole level (209).

### F.  Analysis of Glycosylation

Strategies for the analysis of the glycosylation state of proteins separated by gel electrophoresis have been developed (210), but the methods involved have not yet been developed to a high level of sensitivity. As discussed above, analysis of the protein component can be achieved at the low picomole level, while current methods of sugar analysis can only be carried out with tens or hundreds of pmoles of protein. Methods for the carbohydrate analysis of glycoproteins and peptides are reviewed in (210) and in this volume (chapter by Jenkins and Buckberry).

### XV.  CONCLUSION

A diverse range of methods are currently available for the separation of proteins and peptides by gel electrophoresis. These provide the analyst with a set of tools able to characterize the individual components of mixtures in terms of their physicochemical properties (size, charge, hydrophobicity). Sophisticated imaging hardware and dedicated software tools can provide accurate quantitative estimates of the individual components of protein mixtures. Techniques of Western blotting provide the ability to investigate the interaction of the separated proteins to interact with various types of ligand. Finally, the panel of highly sensitive methods for chemical characterization now make it possible to identify and characterize directly proteins separated by gel electrophoresis.

### ACKNOWLEDGEMENT

The author is grateful to the British Heart Foundation for their support of his research.

## REFERENCES

1. A Tiselius. A new apparatus for electrophoretic analysis of colloidal mixtures. Trans Faraday Soc 33: 524–531, 1937.
2. RA Mosher, W Thormann. The Dynamics of Electrophoresis. Weinheim: VCH, 1992.
3. J Kohn. An immuno-electrophoretic technique. Nature 180: 986–987, 1957.
4. GM Rothe. Electrophoresis of Enzymes: Laboratory Methods. Berlin: Springer, 1994.
5. O Smithies. Zone electrophoresis in starch gels: Group variations in serum proteins of normal human adults. Biochem J 61: 629–641, 1955.
6. R Martin. Gel Electrophoresis: Nucleic Acids. Oxford: BIOS Scientific, 1996.
7. TC Bøg-Hansen. Immunoelectrophoresis. In: BD Hames, D Rickwood, eds. Gel Electrophoresis of Proteins: A Practical Approach. Oxford: IRL, 1990, pp 273–300.
8. S Raymond, LS Weintraub. Acrylamide gel as a supporting medium for zone electrophoresis. Science 130: 711–721, 1959.
9. MJ Dunn. Gel Electrophoresis: Proteins. Oxford: BIOS Scientific, 1993.
10. GM Rothe, WD Maurer. One-dimensional PAA-gel electrophoretic techniques to separate functional and denatured proteins. In: MJ Dunn, ed. Gel Electrophoresis of Proteins. Bristol: IOP, 1986, pp 37–140.
11. DF Hochstrasser, A Patchornik, CR Merril. Development of polyacrylamide gels that improve the separation of proteins and their detection by silver staining. Anal Biochem 173: 412–423, 1988.
12. DF Hochstrasser, CR Merril. Catalysts for polyacrylamide gel polymerization and detection of proteins by silver staining. Applied Theoretical Electrophoresis 1: 35–40, 1988.
13. A Chrambach. The Practice of Quantitative Gel Electrophoresis. Weinheim: VCH, 1985.
14. PG Righetti. Isoelectric Focusing: Theory, Methodology and Applications. Amsterdam, Elsevier Biomedical, 1984.
15. A Görg, W Postel, R Westermeier, E Gianazza, PG Righetti. Gel gradient lectrophoresis, isolectric focusing and two-dimensional techniques in horizontal, ultrathin polyacrylamide layers. J Biochem Biophy Methods 3: 273–284, 1980.
16. A Görg, W Postel, P Johann. pH, urea and substrate gradients for the optimization of ultrathin polyacrylamide gel zymograms. J Biochem Biophy Methods 10: 341–350, 1985.
17. I Olsson, UB Axiö-Fredriksson, M Degerman, B Olsson. Fast horizontal electrophoresis. I. Isoelectric focusing and polyacrylamide gel electrophoresis using PhastSystem. Electrophoresis 9: 16–22, 1988.
18. I Olsson, R Wheeler, C Johansson, B Ekström, N Stafström, R Bikhabhai, G Jacobson. Fast horizontal electrophoresis. II. Development of fast autometed staining procedures using Phast Sytsem. Electrophoresis 9: 22–27, 1988.
19. L Ornstein. Disc electrophoresis. I. Background and theory. Ann N Y Acad Sci 121: 321–349, 1964.
20. BJ Davis. Disc electrophoresis. II. Method and application to human serum proteins. Ann N Y Acad Sci 121: 404–427, 1964.
21. JD Fritz, DR Swartz, ML Greaser. Factors affecting polyacrylamide gele electrophoresis and electroblotting of high-molecular-weight myofibrillar proteins. Anal Biochem 180: 205–210, 1989.
22. T Rabilloud. Solubilization of proteins for electrophoretic analyses. Electrophoresis 17: 813–829, 1996.
23. DC Brune. Alkylation of cyteine with acrylamide for protein sequence analysis. Anal Biochem 207: 285–290, 1992.
24. LM Hjelmeland, DW Nebert, JC Osborne. Sulfobetaine derivatives of bile acids: nondenaturing surfactants for membrane biochemistry. Anal Biochem 130: 72–82, 1983.
25. GH Perdew, HW Schaup, DP Selivonchick. The use of a zwitterionic detergent in two-dimensional gel electrophoresis of trout liver microsomes. Anal Biochem 135: 453–455, 1983.
26. H Schägger, G von Jagow. Blue native electrophoresis for isolation of membrane protein complexes in enzymatically active form. Anal Biochem 199: 223–231, 1991.
27. H Schägger. Quantification of oxidative phosphorylation enzymes after blue native electrophoresis and two-diemnsional resolution: Normal complex I protein amounts in Parkinson's disease conflict with reduced catalytic activities. Electrophoresis 16: 763–770, 1995.
28. H Schägger, H Bentlage, W Ruitenbeek, K Pfeiffer, S Rotter, C Rother, A Böttcher-Purkl, E Lodemann. Electrophoretic separation of multoprotein complexes from blood platelets and cell lines: Technique for the analysis of diseases with defects in oxidative phosphorylation. Electrophoresis 17: 709–714, 1996.
29. JA Reynolds, C Tanford. The gross conformation of protein-sodium dodecyl

sulfate complexes. J Biol Chem 245: 5161–5165, 1970.

30. K Weber, M Osborn. The reliability of molecular weight determinations by dodecyl sulfate-polyacrylamide gel electrophoresis. J Biol Chem 244: 4406–4412, 1969.

31. MF Lopez, WF Patton, BL Utterback, N Chung-Welch, P Barry, WM Skea, RP Cambria. Effect of various detergents on protein migration in the second dimension of tow-dimensional gels. Anal Biochem 199: 35–44, 1991.

32. UK Laemmli. Cleavage of structural proteins during the assembly of the head of bacteriophage T4. Nature 227: 680–685, 1970.

33. RT Swank, KD Munkres. Molecular weight analysis of oligopeptides by electrophoresis in polyacrylamide gel with soudium dodecyl sulfate. Anal Biochem 39: 462–477, 1971.

34. BL Anderson, RW Berry, A Telser. A sodium dodecyl sulfate-polyacrylamide gel electrophoresis system that separates peptides and protein in the molecular weight range of 2,500–90,000. Anal Biochem 132: 365–375, 1983.

35. F Hashimoto, T Horigome, M Kanbayashi, K Yoshida, H Sugano. An improved method for separation of low-molecular-weight polypeptides by electrophoresis in sodium dodecyl sulfate-polyacrylamide gel. Anal Biochem 129: 192–199, 1983.

36. H Schägger, G von Jagow. Tricine-sodium dodecyl sulfate-polyacrylamide gel electrophoresis for the separation of proteins in the range from 1–100 kDa. Anal Biochem 166: 368–379, 1987.

37. JF Poduslo. Glycoprotein molecular-weight estimation using sodium dodecyl sulfate-pore gradient electrophoresis: comparison of Tris-glycine and Tris-borate-EDTA buffer systems. Anal Biochem 114: 131–139, 1981.

38. DW Cleveland, SG Fischer, MW Kirschner, UK Laemmli. Peptide mapping by limited proteolysis in sodium dodecyl sulfate and analysis by gel electrophoresis. J Biol Chem 252: 1102–1106, 1977.

39. AT Andrews. Peptide mapping. In: BD Hames, D Rickwood, eds. Gel Electrophoresis of Proteins: A Practical Approach. Oxford: IRL, 1990, pp 301–319.

40. A Kolin. Separation and concentration of proteins in a pH field combined with an electric field. J Chem Phys 22: 1628–1629, 1954

41. H Svensson. Isoelectric fractionation, analysis and characterization of ampholytes in natural pH gradients. I. The differential equation of solute concentrations at a steady state and its solution for simple cases. Acat Chem Scand 15: 325–341, 1961

42. O Vesterberg. Synthesis and isoelectric fractionation of carrier ampholyes. Acta Chem Scand 23: 2653–2666, 1969.

43. PG Righetti. Isoelectric Focusing: Theory, methodology and applications. Amsterdam: Elsevier, 1983.

44. B Bjellqvist, K Ek, PG Righetti, A Görg, R Westermeier, W Postel. Isoelectric focusing in immobilized pH gradients: Principle, methodology and some applications. J Biochem Biophys Methods 6: 317–339, 1982.

45. PG Righetti. Immobilized pH gradients: Theory and methodology. Amsterdam: Elsevier, 1990.

46. PG Righetti, A Bossi. Isoelectric focusing in immobilized pH gradients: Recent analytical and preparative developments. Anal Biochem 247: 1–10, 1997.

47. E Gianazza, F Celentano, G Dossi, B Bjellqvist, PG Righetti. Preparation of immobilized pH gradienst spanning 2–6 pH units with two chamber mixers: Evaluation of two experimental approaches. Electrophoresis 5: 88–97, 1984.

48. E Gianazza, S Astrua-Testori, PG Righetti. Some more formulations for immobilized pH gradients. Electrophoresis 6: 113–117, 1985.

49. E Gianazza, P Giacon, B Sahlin, PG Righetti. Non-linear pH courses with immobilized pH gradients. Electrophoresis 6: 53–56, 1985.

50. M Chiari, PG Righetti. The Immobiline family: From "vacuum" to "plenum" chemistry. Electrophoresis 13: 187–191, 1992.

51. E Gianazza, F Celentano, S Magenes, A Ettori, PG Righetti. Formulations for immobilized pH gradients including pH extremes. Electrophoresis 10: 806–808, 1989.

52. P Sinha, E Köttgen, R Westermeier, PG Righetti. Immobilized pH 2.5–11 gradients for two-dimensional electrophoresis. Electrophoresis 13: 210–214, 1992.

53. B Bjellqvist, C Pasquali, F Ravier, J-C Sanchez, D Hochstrasser. A nonlinear wide-range immobilized pH gradient for two-dimensional electrophoresis and its definition in a relevant pH scale. Electrophoresis 14: 1357–1365, 1993.

54. A Görg, C Obermaier, G Boguth, A Csordas, J-J Diaz, J-J Madjar. Very alkaline immobilized pH gradients for two-dimensional electrophoresis of ribosomal

and nuclear proteins. Electrophoresis 18: 328–337, 1997.

55. S Cordwell, DJ Basseal, B Bjellqvist, DC Shaw, I Humphery-Smith. Characterisation of basic proteins from Spiroplasma melliferum using novel immobilised pH gradients. Electrophoresis 18: 1393–1398.

56. K Altland. IPGMAKER: A program for IBM-compatible personal computers to create and test recipes for immobilized pH gradients. Electrophoresis 11: 140–147, 1990.

57. E Giaffreda, C Tonani, PG Righetti. pH gradient simulator for electrophoretic techniques in a Windows environment. J Chromatogr 630: 313–327, 1993.

58. K Altland, A Altland. Forming reproducible density and solute gradients by computer-controlled cooperation of step-motor-driven burettes. Electrophoresis 5: 143–147, 1984.

59. K Altland, A Altland. Pouring reproducible gradients in gels under computer control: new devices for simultaneous delivery of two independent gradients, for more flexible slope and pH range of immobilized pH gradienst. Clin Chem 30: 2098–2103, 1984.

60. O Smithies, MD Poulik. Two-dimensional electrophoresis of serum proteins. Nature 177: 1033–1035, 1956.

61. MJ Dunn. Two-dimensional polyacrylamide gel electrophoresis. In: A Chrambach, MJ Dunn, BJ radola, eds. Advances in Electrophoresis, Volume 1. Weinheim: VCH, 1987, pp 1–139.

61. J Klose. Protein mapping by combined isoelectric focusing and electrophoresis of mouse tissues. A novel approach to testing for induced point mutations in mammals. Humagenetik 26: 231–243, 1975.

62. GA Scheele. Two-dimensional gel analysis of soluble proteins. Characterization of guinea pig exocrine pancreatic proteins. J Biol Chem 250: 5375–5385, 1975.

63. F Iborra, JM Buhler. Protein subunit mapping. A sensitive high resolution method. Anal Biochem 74: 503–511, 1976.

64. PH O'Farrell. High resolution two-dimensional electrophoresis of proteins. J Biol Chem 250: 4007–4021.

65. GH Perdew. The use of zwitterionic detergent in two-dimensional gel electrophoresis of trout liver microsomes. Anal Biochem 135: 453–455, 1983.

66. T Rabilloud, C Adessi, A Giraudel, J Lunnardi. Improvement of the solubilization of proteins in two-diemensional electrophoresis with immobilized pH gradients. Electrophoresis 18: 307–316, 1997.

67. C Pasquali, I Fialka, LA Huber. Preparative two-dimensional gel electrophoresis of membrane proteins. Electrophoresis 18: 2573–2581, 1997.

68. T Rabilloud, E Gianazza, N Catto, PG Righetti. Amidosulfobetaines, a family of detergents wih improved solubilization properties: Application for isoelectric focusing under denaturing conditions. Anal Biochem 185: 94–102, 1990.

69. BR Herbert, MP Molloy, AA Gooley, BJ Walsh, WG Bryson, KL Williams. Improved protein solubility in 2-D electrophoresis using tributyl phosphine as the reducing agent. Electrophoresis 19: 845–851, 1998.

70. PZ O'Farrell. MM Goodman, PH O'Farrell. High resolution two-dimensional electrophoresis of basic as well as acidic proteins. Cell 12: 1133–1142, 1977.

71. A Görg, W Postel, S Günther. The current state of two-dimensional electrophoresis with immobilized pH gradients. Electrophoresis 9: 531–546, 1988.

72. A Görg, G Boguth, C Obermaier, A Psoch, W Weiss. Two-dimensional polyacrylamide gel electrophoresis with immobilized pH gradienst in the first dimension (IPG-Dalt): The state of the art and the controversy of vertical versus horizontal systems. Electrophoresis 16: 1079–1086, 1995.

73. PG Righetti, A Bossi, A Görg, C Obermaier, G Boguth. Steady-state two-dimensional maps of very alkaline proteins in an immobilized pH 10–12 gradient, as exemplified by histone types. J Biochem Biophys Methods 31: 81–91, 1996.

74. A Görg. Two-dimensional electrophoresis with immobilized pH gradients: current stae. Biochem Soc Trans 21: 130–132, 1993.

75. A Görg, W Postel, C Friedrich, R Kuick, JR Strahler, SM Hanash. Temperature-dependent spot positional vatiability in two-dimensional polypeptide patterns. Electrophoresis 12: 653–658, 1991.

76. T Rabilloud, C Valette, JJ Lawrence. Sample application by in-gel rahydration improves the resolution of two-dimensional electrophoresis with immobilized pH gradients in the first dimension. Electrophoresis 15: 1552–1558, 1994.

77. JC Sanchez, V Rouge, M Pisteur, F Ravier, L Tonella, M Moosmayer, MR Wilkins, DF Hochstrasser. Improved and simplified in-gel sample application using rewelling of dry immobilized pH gradients. Electrophoresis 18: 324–327, 1997.

78. R Islam, C Ko, T Landers. A new approach to rapid immobilised pH gradient IEF for

2-D electrophoresis. Science Tools 3: 14–15, 1998.

79. BR Herbert, JC Sanchez, L Bini. Two-dimensional electrophoresis: The state of the art and future directions. In: MR Wilkins, KL Williams, RD Appel, DF Hochstrasser, eds. Proteome Research: New Frontiers in Functional Genomics. Berlin: Springer, 1997, pp 13–33.

80. LA Huber, ed. Paper Symposium on Cellular Traffic. Electrophoresis 18: pp 2509–2698, 1997.

81. DF Hochstrasser, RW James, D Pometta, D Hochstrasser. Preparative isoelectric focusing and high resolution two-dimensional gel electrophoresis for concentration and purification of proteins. Appl Theor Electrophor 1: 333–337, 1991.

82. PG Righetti, E Wenisch, M Faupel. Preparative protein purification in a multi-compartment electrolyser with Immobiline membranes. J Chromatogr 475: 293–309, 1989.

83. GL Corthals, MP Molloy, BR Herbert, KL Williams, AA Gooley. Prefractionation of protein samples prior to two-dimensional electrophoresis. Electrophoresis 18: 317–323, 1997.

84. A Görg, W Postel, J Weser, S Günther, SR Strahler, SM Hanash, L Somerlot. Elimination of point streaking on silver stained two-dimensional gels by addition of iodoacetamide to the equilibration buffer. Electrophoresis 8: 122–124, 1987.

85. RM Levenson, GM Anderson, JA Cohn, PJ Backshear. Giant two-dimensional gel electrophoresis: Methodological update and comparison with intermediate-format gel systems. Electrophoresis 11: 269–279, 1990.

86. J Klose, U Kobalz. Two-dimensional electrophoresis of proteins: An updated protocol and implications for a functional analysis of the genome. Electrophoresis 16: 1034–1059, 1995.

87. NG Anderson, NL Anderson. Analytical techniques for cell fractions. XXI. Two-dimensional analysis of serum and tissue proteins: multiple isoelectric focusing. Anal Biochem 85: 331–340, 1978.

88. NG Anderson, NL Anderson. Analytical techniques for cell fractions. XXII. Two-dimensional analysis of serum and tissue proteins: multiple gradient-slab gel electrophoresis. Anal Biochem 85: 341–354, 1978.

89. WF Patton, MG Pluskal, WM Skea, JL Buecker, MF Lopez, R Zimmermann, LM Belanger, PD Hatch. Development of a dedicated two-dimensional gel electrophoresis system that provides optimal pattern reproducibility and polypeptide resolution. Biotechnqiues 8: 518–527, 1990.

90. JM Corbett, MJ Dunn, A Posch, A Görg. Positional reproducibility of protein spots in two-dimensional polyacrylamide gel electrophoresis using immobilised pH gradient isoelectric focusing in the first dimension: An interlaboratory study. Electrophoresis 15:, 1205–1211, 1994.

91. A Blomberg, L Blomberg, J Norbeck, SJ Fey, PM Larsen, M Larsen, P Roepstorff, H Degand, M Boutry, A Posch, A Görg. Interlaboratory reproducibility of yeast protein patterns analyzed by immobilized pH gradient two-dimensional gel electrophoresis. Electrophoresis 16: 1935–1945, 1995.

92. VC Wasinger, SJ Cordwell, A Cerpa-Poljak, JX Yan, AA Gooley, MR Wilkins, MW Duncan, R Harris, KL Williams, I Humphery-Smith. Progress with gen-product mapping of the Molicutes: Mycoplasma genitalium. Electrophoresis 16: 1090–1094, 1995.

93. RD Appel, A Bairoch, JC Sanchez, JR Vargas, O Golaz, C Pasquali, DF Hochstrasser. Federated 2-DE database: A simple means of publishing 2-DE data. Electrophoresis 17: 540–546, 1996.

94. I Humphery-Smith. Paper Symposium: Microbial Proteomes. Electrophoresis 18: 1205–1498, 1997.

95. JX Yan JX, L Tonella, JC Sanchez, MR Wilkins, NH Packer, AA Gooley, DF Hochstrasser, KL Williams. The Dictyostelium discoideum proteome: the SWISS-2DPAGE database of the mutlicellular aggregate (slug). Electrophoresis 18: 491–497, 1997.

96. L Bini, H Heid, S Liberatori, G Geier, V Pallini, R Zwilling. Two-dimensional electrophoresis of Caenorhabditis elegans homogenates and identification of protein spots by microsequencing. Electrophoresis 18: 557–562, 1997.

97. C Ericsson, Z Pethö, H Mehlin. An on-line two-dimensional polyacrylamide gel electrophoresis protein database of adult Drosophila melanogaster. Electrophoresis 18: 484–490, 1997.

98. MJ Dunn, ed. From Genome to Proteome: Proceedings of the Second Siena 2-D Electrophoresis Meeting, Siena, Italy, September 16–18, 1996. Electrophoresis 18: 305–662, 1997.

99. DF Hochstrasser. Clinical and biomedical applications of proteomics. In: MR Wilkins, KL Williams, RD Appel, DF Hochstrasser, eds. Proteome Research: New Frontiers in Functional Genomics. Berlin: Springer, 1997, pp 187–219.

100. KL Williams, V Pallini. Biological applications of proteomics. In: MR Wilkins, KL Williams, RD Appel, DF Hochstrasser, eds. Proteome Research: New Frontiers in Functional Genomics. Berlin: Springer, 1997, pp 221–237.

101. RD Appel, MJ Dunn, DF Hochstrasser, eds. Paper Symposium: Biomedicine and Bioinformatics. Electrophoresis 18: 2701–2984, 1997.

102. I Humphery-Smith, SJ Cordwell, WP Blackstock. Proteome research: Complementarity and limitations with respect to the RNA and DNA worlds. Electrophoresis 18: 1217–1242.

103. SR Pennington, MR Wilkins, DF Hochstrasser, MJ Dunn. Proteome analysis: From protein characterization to biological function. Trends Cell Biol 7: 168–173, 1997.

104. MR Wilkins, KL Williams, RD Appel, DF Hochstrasser, eds. Proteome Research: New Frontiers in Functional Genomics. Berlin: Springer, 1997.

105. V Neuhoff, N Arold, D Taube, W Ehrhardt. Improved staining of proteins in polyacrylamide gels including isoelectric focusing gels with clear background at nanogram sensitivity using Coomassie Brilliant Blue G-250 and R-250. Electrophoresis 9: 255–262, 1988.

106. RC Switzer, CR Merril, S Shifrin. A highly sensitive silver stain for detecting proteins and peptides in polyacrylamide gels. Anal Biochem 98: 231–237, 1979.

107. CR Merril. Detection of proteins separated by electrophoresis. In: A Chrambach, MJ Dunn, B Radola, eds. Advances in Electrophoresis, Volume 1. Weinheim: VCH, 1987, pp 111–139.

108. T Rabilloud. Mechanisms of protein silver staining in polyacrylamide gels: a 10-year synthesis. Electrophoresis 11: 785–794, 1990.

109. CS Giometti, MA Gemmell, SL Tollaksen, J Taylor. Quantitation of human leukocyte proteins after silver staining: A study with two-dimensional electrophoresis. Electrophoresis 12: 536–543, 1991.

110. M Tal, A Silberstein, E Nusser. Why does Coomassie Brilliant Blue R interact differently with different proteins? A partial answer. J Biol Chem 260: 9976–9980, 1985.

111. AS Dion, AA Pomenti. Ammoniacal silver staining of proteins: Mechanism of glutaraldehyde enhancement. Anal Biochem 129: 490–496, 1983.

112. J Heukeshoven, R Dernick. Simplified method for silver staining of proteins in polyacrylamide gels and the mechanism of silver staining. Electrophoresis 6: 103–112, 1985.

113. DM Gersten, LV Rodriguez, DG George, DA Johnston, EJ Zapolski. On the relationship of amino acid composition to silver staining of proteins in electrophoresis gels: II. Peptide sequence analysis. Electrophoresis 12: 409–414.

114. DW Sammons, LD Adams, EE Nishizawa. Ultrasensitive silver-based color staining of polypeptides in polyacrylamide gels. Electrophoresis 2: 135–141, 1982.

115. T Rabilloud. A comparison between low background silver diamine and silver nitrate protein stains. Electrophoresis 13: 429–439, 1992.

116. RE Stephens. High-resolution preparative SDS-polyacrylamide gel electrophoresis: fluorescent visualization and electrophoretic elution-concentration of protein bands. Anal Biochem 65: 369–379, 1975.

117. PR Eng, CO Parkes. SDS electrophoresis of fluorescamine-labeled proteins. Anal Biochem 59: 323–325, 1974.

118. E Weidekamm, DF Wallach, R Fluckiger. A new sensitive, rapid fluorescence technique for the determination of proteins in gele electrophoresis and in solution. Anal Biochem 54: 102–114, 1973.

119. BO Barger, RC White, JL Pace, DL Kemper, WL Ragland. Estimation of molecular weight by polyacrylamide gel electrophoresis using heat stable fluors. Anal Biochem 70: 327–335, 1976.

120. DO O'Keefe. Quantitative electrophoretic analysis of proteins labeled with monobromobimane. Anal Biochem 222: 86–94, 1994.

121. VE Urwin, P Jackson. Two-dimensional polyacrylamide gel electrophoresis of proteins labeled with the fluorophore monobromobimane prior to first-dimensional isoelectric focusing: Imaging of the fluorescent protein spot patterns using a charge-coupled device. Anal Biochem 209: 57–62, 1993.

122. SJ Fey, A Nawrocki, MR Larsen, A Görg, P Roepstorff, GN Skews, R Williams, PM Larsen. Proteome analysis of Saccharomyces cerevisia: A methodological outline. Electrophoresis 18: 1361–1372, 1997.

123. M Ünlü, ME Morgan, JS Minden. Difference gel electrophoresis: A single gel method for detecting changes in protein extracts. Electrophoresis 18: 2071–2077, 1997.

124. P Jackson, VE Urwin, CD Mackay. Rapid imaging, using a cooled charge-coupled-device, of fluorescent two-dimensional polyacrylamide gels produced by labelling proteins in the first-dimensional isoelectric

focusing gel with the fluorophore 2-methoxy-2,4-diphenyl-3(2H)furanone. Electrophoresis 9: 330–339, 1988.

125. BK Hartman, S Udenfriend. A method for immediate visualization of proteins in acrylamide gels and its use for preparation of antibodies to enzymes. Anal Biochem 30: 391–394, 1969.

126. TH Steinberg, LJ Jones, RP Haugland, VL Singer. SYPRO orange and SYPRO red protein gel stains: One-step fluorescent staining of denaturing gels for detection of nanogram levels of protein. Anal Biochem 239: 223–237, 1996.

127. TH Steinberg, RP Haugland, VL Singer. Applications of SYPRO orange and SYPRO red protein gel stains. Anal Biochem 239: 238–245, 1996.

128. Patterson SD, Garrels JI. Two-dimensional gel analysis of posttranslational modifications. In: JE Celis, ed. Cell Biology, Volume 3. San Diego: Academic Press, 1994, pp 249–257.

129. D Rickwood. Reagents for the isotopic labelling of proteins. In: BD Hames, D Rickwood, eds. Gel Electrophoresis of Proteins: A Practical Approach. 2nd ed. Oxford: IRL, 1990, pp 346–358.

130. WM Bonner, RA Laskey. A film detection method for tritium-labelled proteins and nucleic acids in polyacrylamide gels. Eur J Biochem 46: 83–88, 1974.

131. A Circolo, A Gulati. Autoradiography and fluorography of acrylamide gels. In: JM Walker, ed. The Protein Protocols Handbook. Totowa: Humana, 1996, pp 235–242.

132. RA Laskey. The use of intensifying screens or organic scintillators for visualizing radioactive molecules resolved by gel electrophoresis. Methods Enzymol 65: 363–371, 1980.

133. D Heinegård, Y Sommarin. Isolation and characterization of proteoglycans. Methods Enzymol 144: 319–372, 1987.

134. RM Zacharius, TE Zell, JH Morrison, JJ Woodlock. Glycoprotein staining following electrophoresis on acrylamide gels. Anal Biochem 30: 148-152, 1969.

135. AH Wardi, GA Michos. Alcian blue staining of glycoproteins in acrylamide disc electrophoresis. Anal Biochem 49: 607-609, 1972.

136. AE Eckhardt, CE Hayes, IJ Goldstein. A sensitive fluorescent method for the detection of glycoproteins in polyacrylamide gels. Anal Biochem 73: 192–197, 1976.

137. H Møller, J Poulsen. Staining of glycoproteins/proteoglycans on SDS-gels. In: JM Walker, ed. The Protein Protocols Handbook. Totowa: Humana, 1996, pp 627–631.

138. J Colyer. Analyzing protein phosphorylation. In: JM Walker, ed. The Protein Protocols Handbook. Totowa: Humana, 1996, pp 501–506.

139. SA Lacks, SS Springhorn. Renaturation of enzymes after polyacrylamide gel electrophoresis in the presence of sodium dodecyl sulfate. J Biol Chem 255: 7467–7473, 1980.

140. JI Garrels. Quantitative two-dimensional gel electrophoresis of proteins. Methods Enzymol 100: 411–423, 1983.

141. JC Sutherland. Electronic imaging of electrohoretic gels and blots. In: A Chrambach., MJ Dunn, BJ Radola, eds. Advances in Electrophoresis, Volume 6. Weinheim: VCH, 1993, pp 3–42.

142. P Jackson, VE Urwin, CD Mackay. Rapid imaging, using a cooled charge-coupled device, of fluorescent two-dimensional polyacrylamide gels produced by labelling proteins in the first-diemensional isoelectric focusing gel with the fluorophore 2-methoxy-2, 4-diphenyl-3 (2H) furanone). Electrophoresis 9:, 330–339, 1988.

143. M Costas. Microcomputers in the comparative analysis of one-dimensional electrophoretic patterns. In: CFA Bryce, ed. Microcomputers in Biochemistry: A Practical Approach. Oxford: IRL Press, 1992, pp 189–213.

144. MJ Miller. Computer-assisted analysis of two-dimensional gel electrophoretograms. In: A Chrambach, MJ Dunn, BJ Radola, eds. Advances in Electrophoresis, Volume 3. Weinheim: VCH, 1989, pp 181–217.

145. MJ Dunn. The analysis of two-dimensional polyacrylamide gels for the construction of protein databases. In: CFA Bryce, ed. Microcomputers in Biochemistry: A Practical Approach. Oxford: IRL Press, 1992, pp 215–242.

146. PF Lemkin, LE Lipkin. GELLAB: A computer system for 2D gel electrophoresis analysis. I. Segmentation of spots and system preliminaries. Comput Biomed Res 14: 272–297, 1981.

147. PF Lemkin, LE Lipkin. GELLAB: A computer system for 2D gel electrophoresis analysis. II. Pairing spots. Comput Biomed Res 14: 355–380, 1981.

148. PF Lemkin, LE Lipkin. GELLAB: A computer system for 2D gel electrophoresis analysis. III. Multiple two-diemensional gels. Comput Biomed Res 14: 407–446, 1981.

149. MM Skolnick, SR Sternberg, JV Neel. Computer programs for adapting two-

dimensional gels to the study of mutation. Clin Chem 28: 969–978, 1982.

150. MM Skolnick. An aproach to completely automatic comparison of two-dimensional electrophoresis gels. Clin Chem 28: 979–986, 1982.

151. AD Olson, MJ Miller. Elsie 4: Quantitative computer analysis of sets of two-dimensional gel electrophoretograms. Anal Biochem 169: 49–70, 1988.

152. NL Anderson, J Taylor, AE Scandora, BP Coulter, NG Anderson. The TYCHO system for computer analysis of two-dimensional gel electrophoresis patterns. Clin Chem 27: 1807–1820, 1981.

153. RD Appel, DF Hochstrasser, M Funk, JR Vargas, C pellegrini, AF Muller, JR Scherrer. The MELANIE project: From biopsy to automatic protein map interpretation by computer. Electrophoresis 12: 722–735, 1991.

154. JI Garrels. The QUEST system for quantitative analysis of two-dimensional gels. J Biol Chem 264: 5269–5282, 1989.

155. PF Lemkin, LE Lipkin, EP Lester. Some extensions to the GELLAB two-dimensional electrophoretic gel analysis system. Clin Chem 28: 840–849, 1982.

156. RD Appel, PM Palagi, D Walther, JR Vargas, JC Sanchez, F Ravier, C Pasquali, DF Hochstrasser. Melanie II—a third generation software package for analysis of two-dimensional electrophoresis images: I. Features and user interface. Electrophoresis 18: 2724–2734.

157. RD Appel, JR Vargas, PM Palagi, D Walther, DF Hochstrasser. Melanie II—a third generation software package for analysis of two-dimensional electrophoresis images: I. Algorithms. Electrophoresis 18: 2735–2748.

158. H Towbin, T Staehelin, J Gordon. Electrophoretic transfer of proteins from polyacrylamide gels to nitrocellulose sheets: Procedure and some applications. Proc Natl Acad Sci USA 76: 4350–4354, 1979.

159. J Renart, J Reiser, GR Stark. Transfer of proteins from gels to diazobenzyloxymethyl-paper and detection with antisera: A method for studying antibody specificity and antigen structure. Proc Natl Acad Sci USA 76: 3116–3120, 1979.

160. BA Baldo, ER Tovey, eds. Protein Blotting: Methodology, Research and Diagnostic Applications. Basel: Karger, 1989.

161. C Eckerskorn. Blotting membranes as the interface between electrophoresis and protein chemistry. In: R Kellner, F Lottspeich,

162. MF Pluskal, MB Przekop, MR Kavonian, C Vecoli, DA Hicks. A new membrane substrate for Western blotting of proteins. Biotechniques 4: 272–282, 1986.

163. DE Garfin, G Bers. Basic aspects of protein blotting. In: BA Baldo, ER Tovey, eds. Protein Blotting: Methodology, Research and Diagnostic Applications. Basel: Karger, 1989, pp 5–42.

164. JM Walker. Protein blotting by the capillary method. In: JM Walker, ed. The Protein Protocols Handbook. Totowa: Humana, 1996, pp 261–262.

165. M Peferoen, R Huybrechts, A DeLoof. Vacuum blotting: A new simple and efficient transfer of proteins from sodium dodecyl sulfate-polyacrylamide gels to nitrocellulose. FEBS Lett 145: 369–372, 1982.

166. M Bittner, P Kupferer, CF Morris. Electrophoretic transfer of proteins and nucleic acids from slab gels to diazobennzyoxymethyl cellulose or nitrocellulose sheets. Anal Biochem 102: 459–471, 1980.

167. J Khyse-Andersen. Electroblotting of multiple gels: A simple apparatus without buffer tank for rapid transfer of proteins from polyacrylamide to nitrocellulose. J Biochem Biophys Methods 10: 203–209, 1984.

168. B Szewczyk, LM Kozloff. A method for the efficient blotting of strongly basic proteins from sodium dodecyl sulfate-polyacrylamide gels to nitrocellulose. Anal Biochem 150: 403–407, 1985.

169. K Hancock, VCW Tsang. India ink staining of proteins on nitrocellulose paper. Anal Biochem 133: 157–162, 1983

170. M Moeremans, G Daneels, J De Mey. Sensitive colloidal metal (gold or silver) staining of protein blots on nitrocellulose membranes. Anal Biochem 145: 315–321, 1985.

171. W Patton, L Lam, Q Su, M Lui, H Erdjument-Bromage, P Tempst. Metal chelates as reversible stains for detection of electroblotted proteins: application to protein microsequencing and immunoblotting. Anal Biochem 220: 324–335, 1994

172. N Shojaee, WF Patton, MJ Lim, D Shepro. Pyrogallol red-molybdate; A reversible, metal chelate stain for detection of proteins immobilized on membrane supports. Electrophoresis 17: 687–693, 1996.

173. SD Patterson, D Hess, T Yungwirth, R Aebersold. High-yield recovery of electroblotted proteins and cleavage fragments from a cationic polyvinylidene fluoride-

based membrane. Anal Biochem 202: 193–203, 1992.

174. M Moeremans, M De Raeymaeker, G Daneels, J De Mey. FerriDye: colloidal iron binding followed by Pearl's reaction for the staining of proteins transferred from sodium dodecyl sulfate gels to nitrocellulose and positively charged nylon membranes. Anal Biochem 153: 18–22, 1986.

175. HP Hauri, K Bucher. Immunoblotting with monoclonal antibodies: Importance of the blocking solution. Anal Biochem 159: 386–389, 1986.

176. B Batteiger, WJV Newhall, RB Jones. The use of Tween 20 as a blocking agent in the immunological detection of proteins transferred to nitrocellulose membranes. Anal Biochem 55: 297–307, 1982.

177. M Moeremans, G Daneels, A Van Dijck, G Langanger, J De Mey. Sensitive visualisation of antigen-antibody reactions in dot blot and blot immune overlay assays with immunogold and immunogold/silver staining. J Immunol Methods 74: 353–360, 1984.

178. SJ Fowler. Protein staining and immunodetection using colloidal gold. In: JM Walker, ed. The Protein Protocols Handbook. Totowa: Humana, 1996, pp 275–287.

179. C Kessler. The digoxigenin: anti-digoxigenin (DIG) technology: A survey on the concept and realization of a novel bioanalytical indicator system. Mol Cell Probes 5: 161–205, 1991.

180. Durrant I. Light-based detection of biomolecules. Nature 346: 297–298.

181. N Latif, MJ Dunn. Enhanced detection of apoptotic bcl-2 signals using ECL Plus. Amerham Life Science News, Issue 22, p 14, 1997.

182. I Bronstein, JC Voyta, OJ Murphy, L Bresnick, LJ Kricka. Improved chemiluminescnet westerm blotting procedure. Biotechniques 12: 748–753, 1992.

183. JE Celis, HH Rasmussen, P Gromov, E Olsen, P Madsen, H Leffers, B Honoré, K Dejgaard, H Vorum, DB Christensen, et al. The human keratinocyte two-dimensional gel protein database (update 1995): Mapping components of signal transduction pathways. Electrophoresis 16: 2177–2240, 1995.

184. JM Corbett, CH Wheeler, CS Baker, MH Yacoub, MJ Dunn. The human mycordial two-dimensional gel protein database: Update 1994. Electrophoresis 15: 1459–1465.

185. R Aebersold. High sensitivity sequence analysis of proteins separated by polyacrylamide gel electrophoresis. In: A Chrambach, MJ Dunn, BJ Radola, eds. Advances in Electrophoresis, Volume 4. Weinheim: VCH, 1987, pp 81–168.

186. SD Patterson. From electrophoretically separated protein to identification: Strategies for sequence and mass analysis. Anal Biochem 221: 1–15, 1994.

187. BT Chait, R Wang, RC Beavis, SBH Kent. Protein ladder sequencing. Science 262: 89–92, 1993.

188. C Eckerskorn, P Jungblut, W Mewes, J Klose, F Lottspeich. Identification of mouse brain proteins after two-dimensional electrophoresis and electroblotting by microsequence analysis and amino acid composition analysis. Electrophoresis 9: 830–838, 1988.

189. U Hobohm, T Houthaeve, C Sander. Amino acid analysis and protein database compositional search as a rapid and inexpensive method to identify proteins. Anal Biochem 222: 202–209, 1994.

190. SJ Cordwell, MR Wilkins, A Cerpa-Poljak, AA Gooley, M Duncan, KL Williams, I Humphery-Smith. Cross-species identification of proteins separated by two-dimensional gel electrophoresis using matrix-assisted laser desorption ionisation/time-of-flight mass spectrometry and amino acid analysis. Electrophoresis 16: 438–443, 1995.

191. MR Wilkins, C Pasquali, RD Appel, K Ou, O Golaz, JC Sanchez, JX Yan, AA Gooley, G Hughes, I Humphery-Smith, KL Williams, DF Hochstrasser. From proteins to proteomes: Large scale protein identification by two-dimensional electrophoresis and amino acid analysis. Bio/Technology 14: 61–65, 1996.

192. CH Wheeler, SL Berry, MR Wilkins, JM Corbett, K Ou, AA Gooley, I Humphery-Smith, KL Williams, MJ Dunn. Characterisation of proteins from 2-D gels by matrix-assisted laser desorption mass spectrometry and amino acid compositional analysis. Electrophoresis 17: 580–587, 1996.

193. MR Wilkins, K Ou, RD Appel, JC Sanchez, JX Yan, O Golaz, V Farnsworth, P Cartier, DF Hochstrasser, KL Williams, AA Gooley. Rapid protein identification using N-terminal "sequence tag" and amino acid analysis. Biochem Biophys Res Commun 221: 609–613, 1996.

194. JX Yan, NH Packer, L Tonella, K Ou, MR Wilkins, JC Sanchez, AA Gooley, DF Hochstrasser, KL Williams. High sample throughput phosphoamino acid analysis of proteins separated by one- and

two-dimensional gel electrophoresis. J Chromatogr A 764: 201–210.

195. B Bjellqvist, GJ Hughes, C Pasquali, N Paquet, F Ravier, JC Sanchez, S Frutiger, DF Hochstrasser. The focusing positions of polypeptides in immobilized pH gradients can be predicted from their amino acid sequences. Electrophoresis 14: 1023–1031, 1993

196. C Eckerskorn, K Strupat, M Karas, F Hillenkamp, F Lottspeich, F. Mass spectrometric analysis of blotted proteins after gel electrophoretic separation by matrix-assisted laser desorption/ionization. Electrophoresis 13: 664–665, 1992.

197. M Schreiner, K Strupat, F Lottspeich, C Eckerskorn. Ultraviolet matrix assisted laser desorption ionization-mass spectrometry of electroblotted proteins. Electrophoresis 17: 954–961, 1996.

198. CW Sutton, CH Wheeler, S U, JM Corbett, JS Cottrell, MJ Dunn. The analysis of myocardila proteins by infrared and ultraviolet laser desorption mass spectrometry. Electrophoresis 18: 424–431.

199. K Strupat, M Karas, F Hillenkamp, C Eckerskorn, F Lottspeich. Matrix-assisted laser desorption mass spectrometry of protein electroblotted after polyacrylamide gel electrophoresis. Anal Chem 66: 464–470, 1994.

200. JS Cottrell. Protein identification by peptide mass fingerprinting. Pept Res 7: 115–124, 1994.

201. CW Sutton, KS Pemberton, JS Cottrell, JM Corbett, CH Wheeler, MJ Dunn, DJ Pappin. Identification of myocardial proteins from two-dimensional gels by peptide mass fingerprinting. Electrophoresis 16: 308–316, 1995.

202. SD Patterson, R Aebersold. Mass spectrometric approaches for the identification of gel-separated proteins. Electrophoresis 16: 1791–1814, 1995.

203. JR Yates. Mass spectrometry and the age of the proteome. J Mass Spectrom 33: 1–19, 1998.

204. M Mann, MS Wilm. Error tolerant identification of peptides in sequence databases by peptide sequence tags. Anal. Chem. 66: 4390–4399, 1994.

205. M Wilm, A Shevchenko, T Houthaeve, S Breit, L Schweigerer, T Fotsis, M Mann. Femtomole sequencing of proteins from polyacrylamide gels by nano-electrospray mass-spectrometry. Nature 379: 466–469, 1996.

206. A Shevchenko, ON Jensen, AV Podtelejnikov, F Sagliocco, M Wilm, O Vorm, P Mortensen, H Boucherie, M Mann. Linking genome and proteome by mass spectrometry: Large-scale identification of yeast proteins from two-dimensional gels. Proc Natl Acad Sci USA 93: 14440–14445, 1996.

207. JK Eng, AL McCormack, JR Yates. An approach to correlate tandem mass spectral data of peptides with amino acid sequences in a protein database. J Am Soc Mass Spectrom 5: 976–989, 1994.

208. JR Yates, JK Eng, AL McCormack, DM Schieltz. Methods to correlate tandem mass spectra of modified peptides to amino acid sequences in the protein database. Anal Chem 67: 1426–1436, 1995.

209. McCormack, DM Schieltz, B Goode, S Yang, G Barnes, D Drubin, JR Yates. Direct analysis and identification of proteins in mixtures by LC/MS/MS and database searching at the low-femtomle level. Anal Chem 15: 767–776, 1997.

210. NH Packer, A Pawlak, WC Kett, AA Gooley, JW Redmond, KL Williams. Proteome analysis of glycoforms: A review of strategies for the microcharacterisation of glycoproteins separated by two-dimensional polyacrylamide gel electrophorsis. Electrophoresis 18: 452–460, 1997.

211. PG Righetti, E Gianazza, B Bjellqvist. Modern aspects of isoelectric focusing: Two-dimensional maps and immobilized pH gradients. J Biochem Biophys Methods 8: 89–108, 1983.

212. MJ Dunn. Detection of total proteins on Western blots of two-dimensional polyacrylamide gels. In: AJ Link, ed. 2-D Protocols for Proteome Analysis. Totowa: Humana, 1999, pp 319–329.

213. AT Andrews. Electrophoresis: Theory, Techniques, and Biochemical and Clinical Applications. Oxford: Oxford University Press, 1986.

# 14

# Carbohydrate Analysis of Glycoproteins and Glycopeptides

**Nigel Jenkins,\* Pir M. Shah[†] and Lorraine D. Buckberry[†]**
\* *Eli Lilly & Co., Indianapolis, Indiana*
[†] *De Montfort University, Leicester, United Kingdom*

## I.  INTRODUCTION

Protein glycosylation is the most complex of all post-translational modifications made to proteins in the eukaryotic cell. It entails the attachment and remodeling of oligosaccharides in the endoplasmic reticulum (ER) and Golgi apparatus prior to secretion of the protein or its appearance at the cell surface (1, 2). The majority of plasma membrane receptors and secreted proteins are glycosylated (with notable exceptions such as insulin, growth hormone, and some interleukins). Major advances in glycobiology in recent years have focused on:

1. Developing methods for the accurate structural analysis of carbohydrates attached to glycoproteins (glycans).
2. Cloning of key glycosyltransferase enzymes in the glycan biosynthetic pathways.
3. Understanding the cellular influences on protein glycosylation.
4. Understanding the varied biological functions of carbohydrates in glycoproteins.

In this chapter the key methodology and findings that underpin the science of glycobiology will be reviewed, with particular focus on their application in the biotechnology industry. Differences in carbohydrate structures that may arise from choosing alternative gene expression systems and cell culture conditions will be discussed (together with their physiological significance), and different methods of carbohydrate analysis will be compared. For information on other aspects of glycobiology the reader should consult reviews on biosynthetic pathways (1, 3–5), the biological properties of carbohydrates (6), and cellular influences on the glycosylation process (7–10).

Three distinct types of glycosylation are known to occur in eukaryotes:

1. N-linked glycosylation of proteins, where oligosaccharides are attached to the amine group of Asn residues.
2. O-linked glycosylation, where oligosaccharides are linked to Ser or Thr residues.
3. A glycosyl-phosphatidylinositol (GPI) anchor, a mechanism by which glucosamine and mannose residues are used to incorporate some proteins into cell membranes.

## A. N-linked Glycosylation

### 1. Common Structures

In N-glycosylation the oligosaccharide moiety is attached to Asn within a tri-peptide core (sequon) of Asn-X-Ser/Thr (where X is any amino acid apart from proline). Residues such as Trp Asp Glu and Leu in position X also have a negative influence on N-glycosylation (11), whereas Thr rather than Ser at position three leads to an increased probability of glycosylation (12). Other recognition sequences are known to exist, but this sequon is the most common. The existence of this motif does not always mean that the protein will be N-glycosylated, since other factors such as protein secondary structure and its folding status within the ER are important influences (13, 14). For example, the immunomodulator interferon-$\gamma$ (IFN-$\gamma$) has two potential N-glycosylation sites (Asn$_{25}$ and Asn$_{97}$) but exists in three glycoforms: doubly-glycosylated (at both Asn$_{25}$ and Asn$_{97}$), singly-glycosylated (at Asn$_{25}$ only), and non-glycosylated (15). Similarly, in the blood protease tissue plasminogen activator (tPA) only two N-glycosylation sites (Asn$_{117}$ and Asn$_{448}$) are fully occupied, whereas Asn$_{184}$ is only 50% glycosylated due to stearic hindrance (16).

N-linked oligosaccharides fall into three structural categories: complex, high/oligo mannose, and hybrid, dependent on the extent of glycan remodelling in the later stages of the N-glycosylation process (Fig. 1). All three categories contain the same core structure of two N-acetylglucosamine (GlcNAc) units attached to the Asn-containing sequon, which is extended by three mannose (Man) units, with a single fucose (Fuc) residue sometimes attached to the innermost GlcNAc. The principal differences lie in the structures on the arms (antennae) of the glycans:

1. Complex oligosaccharides contain two or more antennae attached to the core, and the terms

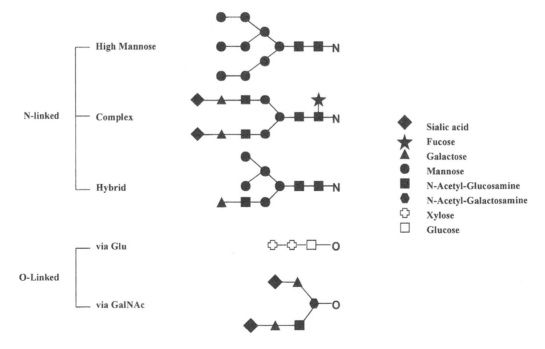

**Fig. 1**  Common N-linked and O-linked glycan structures.

bi-antennary, tri-antennary and tetra-antennary reflect the number of branches. Residues such as GlcNAc, galactose (Gal), N-acetylgalactosamine (GalNAc), Fuc and sialic acids [N-acetyl (NeuAc), and N-glycolyl (NeuGc) neuraminic acids] normally constitute the branches.

2. Oligo (5–9) and high-mannose (>9 Man) structures only contain mannose residues attached to the core structure.

3. Hybrid structures contain at least one complex sugar-containing antenna, with other $\alpha 1$–3 branch containing only mannose residues.

## 2. Biosynthetic Pathways

N-glycosylation occurs in both the endoplasmic reticulum (ER), and Golgi apparatus, and is controlled by a large number of glycosyltransferases and glycosidases (5). Glycosyltransferases are grouped into families based on the type of sugar they transfer to the protein, e.g. galactosyltransferases and sialyltransferases. Theoretically, if all enzymes are fully active in remodeling the nascent glycan core and no structural hindrances are present, all glycans will be fully processed into complex structures terminating in sialic acids. But in practice this is rarely observed, and a mixture of partially and fully processed glycans is more typical of N-linked glycoproteins.

Figure 2 shows a simplified pathway in the biosynthesis of N-glycans. The common core glycan: $Glc_3Man_9GlcNAc_2$ is built on a dolichol pyrophosphate lipid carrier, first in the cytoplasm and then in the ER (1). This oligosaccharide core structure is attached to the target Asn of the nascent polypeptide chain as it enters the ER. The enzyme responsible for catalysing this reaction, oligosaccharyl transferase, is closely associated with the ribosome (step 1). Within the ER lumen the three terminal glucose units are

removed by $\alpha$-glucosidases to give $Man_9GlcNAc_2$ (steps 2 and 3) that is further modified to $Man_8GlcNAc_2$ by $\alpha 2$-mannosidase (step 4). This trimmed structure is no longer recognised by ER chaperones such as calnexin (17), and is transported to the cis-Golgi. The N-glycosylation and protein folding processes are intimately related, and ER carbohydrate-binding proteins such as calnexin and calreticulin are thought to chaperone the nascent glycoprotein, holding it in the ER until folding is completed (18–20), although this is not a factor in O-glycosylation or GPI biosynthesis (21).

In the Golgi the $Man_8GlcNAc_2$ is further trimmed by additional mannosidases until a $Man_5GlcNAc_2$ structure remains (step 5), which is a substrate for GlcNAc transferase I to add GlcNAc to the 1–3 antenna (step 6). The presence of the GlcNAc on this arm of the glycan acts as a signal for further modifications. These are the removal of further Man residues (mannosidase II, step 7), addition of antennary GlcNAc (via GlcNAc transferase II, step 8): and bisecting GlcNAc residues (via GlcNAc-transferase III), creation of further branches (via GlcNAc-transferase IV and V), and core $\alpha 6$-fucosylation (step 9).

The absence of GlcNAc transferase I activity, or a structurally buried core glycan will result in simple oligomannose or hybrid structures being present on the mature glycoprotein. There appears to be no requirement for more complex glycan processing to precede secretion of glycoproteins from the cell or their appearance on the plasma membrane. Certain pathways are mutually exclusive, for example the addition of a bisecting GlcNAc residue via GlcNAc-transferase III prevents further branching of the antennae by GlcNAc-transferase V (22). The final stages of N-glycan processing take place in the trans-Golgi, and involve the addition of galactose, sialic acids (steps

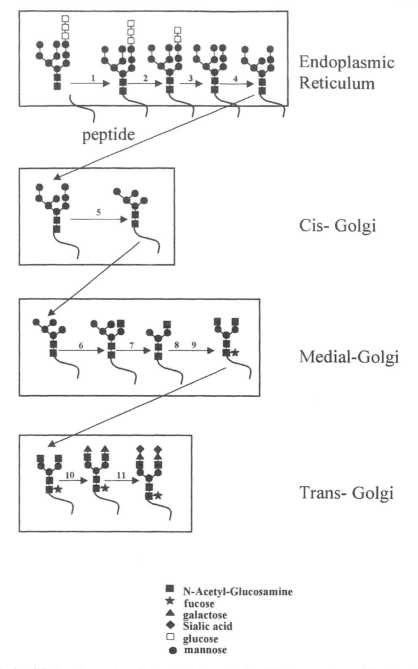

peptide

Endoplasmic
Reticulum

Cis- Golgi

Medial-Golgi

Trans- Golgi

■ N-Acetyl-Glucosamine
★ fucose
▲ galactose
◆ Sialic acid
□ glucose
● mannose

**Fig. 2**  A simplified pathway describing the biosynthesis of N-glycans in vertebrate cells.

10 and 11), and occasionally fucose residues. For further details of this pathway the reader is referred to the reviews of Kornfeld (23) and Schachter (24).

### 3.  Cell-Specific Glycosylation Pathways

In addition to this common pathway, certain cell types possess unusual glycosyltransferases that facilitate the addition of cell-specific glycan structures (see Table 1 for further details). For example, up to 20% of IgG molecules produced by human B-lymphocytes possess a bisecting GlcNAc residue $\beta$1–4 linked to the central $\beta$-linked mannose of the core glycan, which can play a role in antibody-dependent cell-mediated cytotoxicity (ADCC). The GlcNAc transferase III enzyme required for this modification is not active in all rodent cells, indeed only certain cell lines such as the rat Y0 myeloma are able to produce recombinant antibodies containing this bisecting residue (25). Another example is sulfation of antennary GalNAc residues in certain pituitary glycoprotein hormones such as luteinizing hormone (LH), thyrotropin, and pro-opiomelanocortin (26, 27). The specific GalNAc transferase and a terminal GalNAc sulfotransferase that recognise protein motifs in the nascent peptide (e.g. Pro-Leu-Arg) and are mainly restricted to the anterior pituitary gland. Therefore, only cell lines derived from the pituitary gland or endothelium (such as At20 and 293 cells) are able to perform this sulfation (28, 29). A hepatic receptor binds oligosaccharides terminating with $SO_4$-GalNAc residues, and could account for the rapid removal of sulphated LH from the circulation in comparison to follicle-stimulating hormone (FSH) and chorionic gonadotrophin that bear terminal NeuAc residues (30). Mouse C127 cells produce a different pattern of sulphation (NeuAc-$\alpha$3$SO_4$-6Gal) which occurs after the addition of NeuAc (31).

## B.   O-linked Glycosylation

This type of glycosylation is a simpler process involving a smaller number of sugar residues (typically 1–6), and occurs exclusively in the Golgi apparatus (Fig. 3). There is no obvious and invariant peptide sequon surrounding the Ser or Thr attachment site, although a computer algorithm has been devised to predict the likely occupation of potential sites based on sequences in other O-linked glycoproteins (32). Large structural proteins such as mucins have multiple O-linked

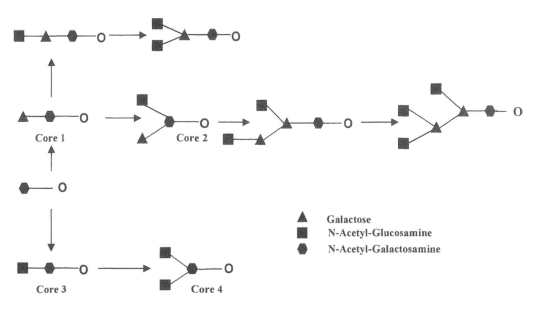

**Fig. 3**  A simplified pathway describing the biosynthesis of O-glycans in vertebrate cells.

oligosaccharides, which may dominate the mass of the mature glycoprotein. These residues protect the extracellular region of the polypeptide from proteolysis and maintain it in a rigid conformation above the cell membrane. Also, regions that contain high proportions of Ser, Thr and Pro have been shown to be extensively glycosylated (33). For example, immunoglobulin class $A_1$ contains a proline-rich hinge region between the Fab and Fc regions. Of the 21 amino acids within this hinge, there are 10 Pro, 5 Ser, and 4 Thr; and the 5 Ser are all O-glycosylated (34).

Several oligosaccharides have been described where different core structures are O-linked to the Ser/Thr (Fig. 3). These cores can be elongated by addition of further Gal, GlcNAc, Fuc and sialic acid residues (35). In contrast, very simple structures such as single fucose residues or Glu-Xyl-Xyl (Xyl stands for xylose) can also be O-linked to Ser and Thr (Fig. 1), and these are common in proteins of the blood clotting and fibrinolytic cascade (36).

## C. Glycosylphosphatidylinositol (GPI) Membrane Anchors

GPI membrane anchors form stable interactions between membranes and proteins. All have the common core structure:

Ethanolamine-$PO_4$-Man-Man-Man-
$GlcNH_2$-myo-inositol-$PO_4$-lipid

Ethanolamine forms a bridge to the membrane protein, and the inositol phospholipid is inserted onto the lipid bilayer of the cell's plasma membrane. The ability of phospholipase C to cleave GPI-anchored proteins facilitates their purification when expressed as recombinant proteins in cell culture (37). There are certain protein-specific and species-specific modifications to this basic structure (38). Several proteins such as Thy-1, the folate receptor and the scrapie prion protein are anchored to the plasma membrane via covalently attached GPI glycolipids containing inositol (39, 40).

## II. INFLUENCES ON THE GLYCOSYLATION PROCESS

Ultimately, the process of glycosylation is controlled by the activity of glycosyltransferases and glycosidases, the accessibility of their glycan substrates on the glycoprotein, and the availability of co-factors, and nucleotide sugar donors (e.g. CMP-NeuAc, GDP-Man, UDP-GlcNAc). At the cellular level, these control points are influenced by the secondary, and tertiary structure of the protein in the ER and Golgi, the host cell type chosen for the recombinant protein expression, and its physiological status. As more recombinant glycoproteins are used therapeutically and for vaccines, the importance of correct glycosylation has become evident. Furthermore, the availability of alternative expression systems for producing recombinant glycoproteins, together with advances in analytical methods for glycans, has enabled comparisons to be made between glycoproteins from different sources. Crucially, the authenticity of recombinant glycoproteins compared to their natural human counterparts can be assessed, and choices made concerning the most appropriate expression system.

### A. Host Cell Line

#### 1. Microbial Cells

Prokaryotic cells are often the first choice for recombinant protein expression, because of their high yields and simple media components. However, they are incapable of expressing large proteins with authentic mammalian-type glycosylation, and common host strains such as *Escherichia coli* have no capacity to glycosylate proteins in a mammalian fashion (41). Hyper-glycosylation (the addition of a large number of mannose residues to the core oligosaccharide), is a common property of most yeast strains (42), and can compromise the efficacy of recombinant proteins such as the hepatitis B vaccine (43). Hyper-glycosylation can be prevented by expressing the polypeptide in mutant yeast strains (e.g. mnn-9 or the temperature-sensitive ngd-29) in which N-glycosylation is confined to core oligosaccharide residues with limited mannose content (e.g. $Man_8GlcNAc_2$), resulting in more effective vaccines (43–45). There is also some evidence to suggest that different O-glycosylation sites are used by yeast and mammalian cells (46).

## 2. Plants

The few studies reporting the production of therapeutic proteins in plants have suggested smaller N-glycan structures are present compared to mammals, lacking outer residues such as sialic acids that may compromise activity (47, 48). For example, erythropoietin (EPO) produced in tobacco cells has no biological activity *in vivo*, presumably because of its high clearance rate (49). Another obstacle may be the presence of potentially allergenic residues such as core $\alpha 1$–3 linked fucose (50) and xylose (51, 52).

## 3. Insects

The baculovirus-infected insect cell expression system has been used to express complex proteins, and has the advantage over stable mammalian expression systems of a short production time. Some studies have shown that this system can attach complex oligosaccharides (53) using Sf9 cells, although this result is controversial as other groups have shown that only simple oligomannose structures can be observed with Sf9 (54). However insect cell lines differ in their glycosylation

capacity, for example the Ea4 and TN-368 cell lines have been shown to add some complex antennary (GlcNAc and Gal) residues to glycoproteins (55, 56).

## 4. Mammals

Although mammalian cell types studied possess the intracellular machinery to perform complex N- and O-linked glycosylation proteins, it has been shown to vary between cells. Mouse and pig cells have a tendency to add NeuGc instead of NeuAc as the terminal sialic acid of glycoproteins. NeuGc levels have been shown to be more prevalent than NeuAc in antibodies derived from mouse or human-mouse hybridomas (57), and account for 1–7% of sialic acids in CHO-derived glycoproteins (58). Low levels of NeuGc (1% of total sialic acids) are tolerated in recombinant proteins such as EPO, but higher levels (e.g. in bovine fetuin containing 7% NeuGc) have the capacity to elicit an anti-NeuGc immune response (59). Furthermore, high levels of terminal NeuGc on a chimeric CT4-IgG fusion protein are correlated with a rapid removal of the molecule from circulation, compared to the same protein bearing terminal NeuAc residues (60).

Mouse hybridomas and C127 cell lines express the enzyme $\alpha 1$,3-galactosyltransferase which generates $Gal\alpha 1$,3-$Gal\beta 1$,4-GlcNAc residues (61). But other lines such as hamster CHO cells, mouse NS0 or rat Y0 myelomas making humanised antibodies do not (25, 62), and humans do not express this enzyme. Since over 1% of human serum IgG is directed against the Gal ($\alpha 1$,3)-Gal-$\beta 1$,4GlcNAc epitope (63) (a consequence of its presence on enteric bacteria), the appearance of this residue in recombinant glycoproteins may be significant. Indeed, strenuous efforts are underway to breed transgenic pigs that lack expression of $\alpha 1$-3-galactosyltransferase, since it is perceived as an antigenic barrier to xenotransplantation (64).

Despite these potential hazards, it must be stated that comparisons of mouse C127, and Chinese hamster ovary (CHO)-derived tPA reveal only minor differences in the pharmacokinetics induced by interaction with these antibodies (61). Furthermore, no major adverse clinical responses have been reported to treatments using proteins with these types of oligosaccharide residues (65). In antibodies, these terminal Galα1,3-Galβ1, 4-GlcNAc or NeuGc residues are more likely to occur where glycosylation occurs in the Fab region, rather than the partially-buried Fc glycosylation site which tends to bear truncated structures (66).

Baby Hamster Kidney (BHK) and CHO cell lines are currently the most favoured mammalian vehicles for the expression of glycoproteins used in human therapy and immunisation. Like human cells, they do not express α1,3-galactosyltransferase, and only manufacture small amounts of the NeuGc residue. However, they do not express a functional gene for α2,6 sialyltransferase (67), therefore only α2,3 linkages are present on sialic acids in recombinant glycoproteins made in these lines, in contrast to the α2,3 and α2,6 linked sialic acids in natural human glycoproteins. The gene for α2,6 sialyltransferase has recently been cloned into CHO, and BHK cells (68, 69), and has been shown to restore human-type sialylation to recombinant glycoproteins expressed in these modified cell lines (58, 70, 71). In addition, the overall sialic acid content is increased, which may be beneficial for the clearance properties of glycoproteins (60). Human cell lines are generally not used for the production of recombinant glycoproteins because of concerns over viral contamination. But for experimental purposes recombinant tPA has been expressed in the human lymphoblastoma Namalwa cell line (72), and displays all the glycosylation characteristics of the natural human glycoprotein.

## 5. Transgenic Animals

Expression of proteins in the milk of transgenic animals is another popular production route for recombinant therapeutics. Evidence from analysing α1-antitrypsin (AAT) expressed in the mouse mammary gland suggests an increased fucosylation tendency compared to natural AAT secreted in the human liver (73). O-glycosylation defects have also been reported in proteins made by this route (74). Other transgenic glycoproteins have been analyzed (75–77) but too few studies have been reported to deduce generic common mammary gland glycosylation characteristics.

In conclusion, the glycosylation properties of the host cell line tend to be species and tissue-specific, but cell lines can be genetically manipulated to more closely resemble the natural human glycosylation processes.

## B. Cell Environment

### 1. Culture Media

The environment in which a cell line is grown can have major effects upon the glycosylation of proteins, especially the type of cell culture conditions and media supplements employed. Mouse monoclonal antibody production in cell culture produces a higher degree of glycosylation consistency than achieved in ascites fluid and IgG$_1$ production in serum-free medium produces higher levels of terminal sialic acid and galactose residues compared to that produced using serum (78–80). The process of adaptation from serum-containing to serum-free medium also leads to more complex glycan production in recombinant IL-2 by recombinant BHK-21 cells (81).

The ambient glucose concentration affects the amount of glycosylation of monoclonal antibody produced by human hybridomas in batch culture (82), and of IFN-$\gamma$ produced by CHO cells in continuous culture (83, 84). However, both lipids and nucleotides are also required for glycan biosynthesis (23). Lipid supplements alone or in combination with lipoprotein carriers can improve the N-glycosylation site occupancy of IFN-$\gamma$ (85,86), and provision of cytidine and uridine can also alter protein glycosylation capacity of hepatocytes by increasing the availability of nucleotide sugars (87). Sodium butyrate is sometimes used to improve protein synthesis, but can change glycosylation by inducing a GlcNAc-transferase involved in O-glycosylation (88) and increasing sialyltransferase activity in recombinant CHO cells (89, 90).

Even the time chosen for the harvesting of the recombinant product can be crucial. For example a monoclonal antibody produced by a mouse NS0 cell line in fed-batch culture showed greater percentages of truncated and oligo-mannose structures at the end of culture compared to mid-growth phase (91). Similar changes have been reported in interferon-$\gamma$ during batch culture of recombinant CHO cells, and even the percentage glycan occupation of N-glycosylation sites was found to decrease with time (84, 92).

## 2. Bioreactor Configuration

In general, the glycosylation profile of recombinant proteins produced from cells in continuous culture (with nutrients supplied constantly and removal of potentially toxic metabolites) is less variable compared to batch or fed-batch culture (93). However growth of CHO cells on microcarriers has been shown to result in less sialylation of human tissue kallikrein compared to suspension culture (94).

High levels of ammonium ions (derived from the degradation of glutamine) result in elevated proportions of tri- and tetra antennary glycans and increased fucosylation of interleukin-2 expressed by BHK cells (95). Increases in

the concentration of ammonium >2 mM may also compromise some Golgi enzymes, resulting in reduced $\alpha$2,6-linked sialic acids in granulocyte-colony-stimulating factor (G-CSF) produced by recombinant CHO cells (96). Increased ammonium ions also reduced the extent of recombinant placental lactogen N-glycosylation by CHO cells, but this was dependent on the pH (97).

Mild hypoxia has minimal effects on the glycosylation of tPA produced by recombinant CHO cells (98), but influences the level of sialylation of recombinant follicle-stimulating hormone, FSH (89). Similarly, pH changes within the range 6.9–8.2 in the cell culture medium do not have a dramatic effect on the glycosylation profile of recombinant placental lactogen expressed in CHO cells, however there was some evidence for under-glycosylation outside this range (99).

## 3. Glycan Degradation

The level of sialylation in glycoproteins can also be compromised by the release of sialidases by dead or dying cells. A number of strategies have been developed to overcome this problem, such as selecting sialidase-deficient CHO lines, inserting antisense to sialidase, or including chemical inhibitors of sialidase in the cell culture medium (100–104). Although CHO cells also produce an $\alpha$-L-fucosidase, this enzyme is incapable of releasing core $\alpha$1,6-fucose from intact recombinant glycoproteins (gp120 or CD4), or the peripheral Fuc-$\alpha$1,3-GlcNAc linkage from serum $\alpha_1$-acid glycoprotein (105). Sialidase, $\beta$-galactosidase, $\beta$-hexosaminidase, and fucosidase can be detected at low levels in supernatants from mouse 293, NS0 and hybridoma cells, but the sialidase activity is much lower than that found in CHO cells (106). In a contrasting study, purified monoclonal antibody incubated with supernatants from various animal cell lines (CHO K1, BHK-21, mouse C127, P3-X63 Ag8.653, human-mouse hetero-hybridoma and the insect cell line SF-21AE) only showed evidence of glycosidase-mediated degradation with the insect supernatant (107). Recent studies have also indicated the presence of natural endoglycosidases which can cleave entire glycan structure from glycoproteins (108) and these may influence the amount of macroheterogeneity found on glycoproteins.

## C.  Regulatory Considerations

Because of these efficacy and safety considerations, both the Food and Drug Administration in the US and the European Medicines Agency now demand more comprehensive carbohydrate analysis before licensing recombinant glycoproteins. It is important to understand that some degree of heterogeneity exists in most natural as well as recombinant glycoproteins. Therefore most current guidelines accept glycan heterogeneity, but it must be shown to be consistent within prescribed boundaries between lots of the finished product. The inherent glycan heterogeneity and its manipulation by genetic or enzymatic engineering poses interesting challenges to the patenting of glycoprotein drugs, and several patents have already been filed with claims to alter the glycosylation process for biotechnological purposes.

## III.  EFFECTS OF GLYCOSYLATION ON GLYCOPROTEIN FUNCTIONS

Although most glycoproteins are influenced by the presence of and the extent of their glycosylation, the particular functions affected tend to be protein and glycan specific. Therefore, data acquired on the structure-functional relationships of glycans present on one recombinant

glycoprotein cannot always be applied to another. However, the main processes where protein glycosylation has been shown to be influential extensively are: protein solubility; protease resistance; immunogenicity (which is carbohydrate residue-specific); pharmacokinetics and clearance *in vivo* (which are particularly influenced by the extent of sialylation).

## A.  Folding and Secretion

Lowering the protein synthesis rate by cycloheximide improves the glycosylation site occupancy of recombinant prolactin produced by C127 cells (109). But studies on tPA synthesis in CHO cells suggest that the rate of protein synthesis by itself has little influence on protein glycosylation (110), with the possible exception of the sialylation steps. Certainly, folding and disulphide bond formation can influence the efficiency of N-linked glycosylation in some proteins. For example, low concentrations of the reducing agent dithiothreitol prevent co-translational disulphide bond formation in the ER and lead to complete glycosylation of a tPA sequon that normally undergoes variable glycosylation (13). The protein backbone itself also influences the glycosylation process, as shown by the many examples of different glycan structures attached to the same protein. More direct evidence for this peptide influence has come from site-directed mutagenesis of $Asp_{243}$ to $Phe_{243}$ in the Fc region of $IgG_3$, where the sialic acid content of the molecule was increased by >60% (111).

If all the glycosylation sites are removed from recombinant IFN-$\gamma$ its secretion rate is compromised by up to 90%, which reflects the importance of glycosylation for the secretory process (112). In transferrin receptors, prevention of glycosylation compromises the ability to form dimers and consequently the molecules are degraded proteolytically in the ER (113).

## B.   Clearance and Stability

The clearance of a given glycoprotein from the bloodstream is highly dependent on its oligosaccharides, particularly those which are situated on the outer arms of the glycan structures (60). There are several receptors for specific oligosaccharide structures that contribute to the clearance of glyco-proteins from the bloodstream, the most significant being the asialoglyco-protein (ASG) receptor (114), which recognizes exposed terminal Gal and GlcNAc residues. This is very important when considering therapeutic drug manufacture, since under-sialylation or de-sialylation at the production stage can result in a rapidly cleared product *in vivo* (115). Other receptors exist in the body such as a hepatic fucose receptor, a mannose-6-phosphate receptor and a Man/GlcNAc receptor of the reticulo-endothelial system (116), but unlike the ASG receptor, their significance in clearing glycoproteins from the body has not been established. Studies in rats using EPO fractions enriched for particular glycoforms indicate that more highly branched glycans (e.g. tri- or tetra-antennary) are less susceptible to renal clearance than biantennary structures (117). There have been many examples showing that de-glycosylated proteins are more susceptible to protease attack (118–120).

## C.   Ligand Binding

In some cases, such as EPO binding to its cellular receptor (121) and IgG binding to Fc$\gamma$ receptors (111), glycosylation directly influences ligand-receptor binding and hence biological activity. The later stage of rheumatoid arthritis are known to be associated with a deficiency of the terminal Gal content on the oligosaccharides of IgG (122). In other molecules (e.g. IFN-$\gamma$) the glycan does not form part of the binding site, nevertheless, glycosylation can affect the protein's efficacy *in vivo* by influencing its clearance.

## IV.   GLYCOSYLATION ANALYSIS

Much of the increased knowledge concerning the influences on the glycosylation pathways and the biological significance of glycan variations has derived from recent advances in complex carbohydrate analysis. It is now possible to unequivocally recognize complex glycan structures and deduce their relative proportions in glycoproteins, by combining chromatographic or mass spectrometric analysis with specific exoglycosidase enzyme arrays. For a complete structural analysis of the oligosaccharides, most glycoproteins must be degraded proteolytically or be made to release their glycans chemically or enzymatically, in order to generate smaller mass units that are more amenable to analysis. There are many different approaches to the determination of the glycosylation profile of a protein, but in recent years mass spectrometry (MS) and high-performance anion exchange chromatography with pulsed amperometric detection (HPAE-PAD) have emerged as the most popular methods.

## A.   Generation and Resolution of Protein Glycoforms

A crude idea of the functions of whole glycan structures can be determined by chemical (hydrazinolysis) or enzymatic (PNGase and/or O-glycanase) de-glycosylation of the protein (123, 124). Alternatively, chemical inhibitors of the glycosylation pathway such as tunicamycin can be introduced into the cell culture medium. A more sophisticated method involves mutating the gene sequence that codes for the amino acid where the glycan joins the protein (i.e.

Asn for N-linked and Ser or Thr for O-linked glycosylation). This can be done by site-directed mutagenesis and can be used to probe the function of individual glycosylation sites (125).

Serial lectin affinity chromatography can be used to prepare different glycoforms of the same intact protein, provided sufficient amounts are available (126). Separation may also be performed using chromatographic and electrophoretic methods, however these may not achieve complete resolution of each glycoform. Polyacrylamide gel electrophoresis (PAGE) with either Coomassie or silver staining can reveal the mass heterogeneity associated with glycosylation, but since some sugars are negatively charged it can lead to inaccurate mass determinations. A rough idea of the mass heterogeneity generated by different glycoforms of the same protein can be deduced from running intact and de-glycosylated proteins in parallel (15), but this approach offers no information on the types of sugars present. The technique can be enhanced by Western blotting using lectins that are able to discriminate between different oligosaccharides (127). Typically, intact proteins are run on SDS-PAGE gels and transferred to nitro-cellulose sheets by Western blotting. The blots are washed, blocked, and then incubated with the desired lectin which is covalently or indirectly bound (e.g. via biotin) to a marker enzyme such as alkaline phosphatase and observed using a chromogenic substrate. Vector Laboratories specialize in the production of conjugated lectins, but they are also available from other suppliers.

The different negative charges on glycans due to differing amounts of sialic acids means that isoelectric focusing (IEF) can be used to separate glycoforms of some proteins, e.g. human granulocyte colony-stimulating factor (128). However, like PAGE, it is impossible to assign precise glycan structures using IEF alone.

Lectins have a use in the preliminary detection of specific sugar residues.

## B. HPLC Methods

Analysis of glycopeptides or carbohydrate chains released from glycoproteins requires high-resolution methods such as HPLC. The most common mode of HPLC for glycopeptide purification is reverse-phase high performance liquid chromatography (rpHPLC) this separates on the basis of peptide hydrophobicity. Other modes employed by HPLC are normal phase partition and ligand exchange. rpHPLC requires the presence of hydrophobic groups such as N-Ac-, deoxy-, -OMe, or OAc groups for longer retention time and better separation. Oligomannose type oligosaccharides (that have only two N-Ac groups) are retained on most rpHPLC columns, and thus are not well separated as compared to complex-type oligosaccharides. However, glycopeptides, released sugars (chemically or enzymically) or derivatized glycans may be separated by this method and can be analyzed subsequently by mass spectrometry. A number of specialized "GlycoSep" columns have recently been developed by Oxford GlycoSciences for the resolution of specific neutral and charged glycans, and for the resolution of sialic acids. Labelling the oligosaccharides with fluorescent tags also allows their separation and quantification by CE or HPLC. Common fluorescent tags are 2-aminobenzamide (129), o-phenylenediamine.2HCl (130), 2-aminoacridone (131), and 1,2-diamino-4,5-methylenedioxybenzene.

Since sialic acids are the most important sugar residues influencing protein clearance, it is extremely important to determine their structure and amounts in any glycoprotein destined for therapeutic use. Sialic acid is a collective term used to describe several derivatives of neuraminic acid that are substituted with an N-acetyl

(NeuAc) or N-glycolyl group (NeuGc) at carbon 5. Trace amounts of sialic acids can also be found O-substituted with further glycolyl, acetyl, lactyl, and methyl groups (132). Sialic acids occur as terminal non-reducing residues in mammalian glycoproteins and have labile glycosidic linkages that can be selectively cleaved with minimal damage to the glycoprotein using linkage-specific sialidases (58). A variety of HPLC and high pressure anion exchange (HPAE) methods have been developed to resolve and quantify the released sialic acids (132).

## C. HPAEC-PAD Methods

Pellicular resin anion exchange columns with pulsed ampometric detection (HPAE-PAD), has been developed by Dionex for the sensitive compositional analysis of released oligosaccharides and glycoconjugates. Neutral oligosaccharides are made to form oxyanion derivatives at high pH (13) and are resolved, together with naturally charged sugars, by anion exchange chromatography and detected using a very sensitive pulsed amperometric detector (detection limit 5 pmol). The system can be used for resolving oligosaccharides or monosaccharides released using exoglycosidase arrays (133), and the fractions eluted can be analyzed by MALDI or electrospray mass spectrometry (25, 134). Early problems encountered using this system from amino acid contaminants and with monosaccharide peak tailing have now been largely resolved (135).

## D. Capillary Electrophoresis

CE can be used for direct analysis of oligosaccharide derivatives intact glycoproteins or glycopeptides, despite the inherent problems of carbohydrate molecules not absorbing or fluorescing and not being readily ionized (136–139). A number of approaches have been employed to render carbohydrates more amenable to analysis including *in situ* complex formation with borate ions and metal cations (140) and the addition of UV-absorbing or fluorescent tags to functional groups. Furthermore, various modes of CE operation such as free zone capillary electrophoresis (CZE), micellar electrokinetic capillary chromatography (MECC), affinity electrophoresis, gel electrophoresis, isoelectric focusing (IEF) and isotacophoresis (141) are available. CE can be used to rapidly "fingerprint" different glycoforms of recombinant proteins that have a relatively high carbohydrate content such as IFN (139) and EPO (142). The glycosylation site occupancy (macroheterogeneity) can be quantified without the need for oligosaccharide release or labelling. Recent reports have indicated that CE can also be coupled with MALDI-MS for full structural determination of glycopeptides (138).

## E. Mass Spectrometry

### 1. Matrix-Assisted Laser Desorption-Ionisation Mass Spectrometry (MALDI-MS)

MALDI-MS is increasingly being used to determine the mass of proteins and polypeptides, confirm protein primary structure, and to characterize post-translational modifications. The technique employs simple time-of-flight analysis of biopolymers which are co-crystallized with a molar excess of a low-molecular weight, strongly UV absorbing matrix (e.g. 2,5-dihydroxybenzoic acid) on a metal sample disk. Both the biopolymer and matrix ions are desorbed by pulses from a nitrogen laser. The molecular ions are detected at the end of a linear flight path, and the time between the initial laser pulse and ion detection is directly proportional to the square root of the molecular ion mass/charge ($m/z$) ratio. For maximum mass accuracy, internal and external protein or peptide calibrants of known molecular mass are required. In addition to this "linear" mode, some instruments offer a "reflectron" mode which effectively lengthens the flight path by redirecting the ions towards an additional ion detector which may enhance resolution, at the expense of decreased sensitivity. Alternatively, a delayed extraction technology is employed by the latest generation of MALDI-MS machines to enhance mass accuracy and resolution (143).

Unlike many analytical techniques, MALDI-MS is tolerant to low (mM) salt concentrations, and can determine the molecular weight of biomolecules in excess of 200 kDa with a mass accuracy of 0.1% with pmol-fmol sensitivity. These properties, combined with its rapid analysis time and ease of use for the non-specialist have made it an attractive technique for the analysis of glycopeptides and released oligosaccharides. Although intact glycoproteins can be analyzed in this system, in practice small glycopeptides (generated by protease digests and resolved on HPLC) of released glycans give superior results, due to the improved accuracy of measuring lower mass values.

MALDI-MS analysis is particularly useful for the identification and characterization of protease-generated glycopeptides following their separation and purification by rp-HPLC (144) enabling site-specific glycosylation data to be obtained (54, 92, 145). The choice of protease used to generate the glycopeptides is crucial: incomplete and non-specific digestion complicates their resolution from non-glycosylated peptide by CE or rpHPLC, and alternative proteases should be tested. Chemical cleavage such as cyanogen bromide may also be employed, provided that the carbohydrate groups and the protein-carbohydrate bonds are stable to the chemical treatment. Before enzymatic or chemical cleavage it is desirable to modify

any disulphide bonds present, by reduction and carboxymethylation in the presence of a denaturant (8 M urea) (75).

MALDI-MS of free N-linked oligosaccharides, after release chemically by hydrazinolysis or enzymatically with an endoglycosidase such as PNGaseF, is quite simple as it requires no prior structural knowledge of the glycoprotein of interest and is ideal for the analysis of underivatized populations of oligosaccharides. However, no site-specific glycosylation data can be generated. Since a pool of released glycans may contain up to 40 different oligosaccharides it is sometimes advantageous to fractionate this oligosaccharide mixture first, e.g. by HPAE-PAD or HPLC. A drawback to the enzymatic release of oligosaccharides is that the presence of SDS is often required to denature the glycoprotein and has to therefore be removed prior to MALDI-MS analysis. Until recently, only desialylated oligosaccharides could be analyzed successfully by MALDI-MS using 2,5 dihydroxybenzoic acid as matrix, or alpha-cyano-4-hydroxy cinnaminic acid (54, 92, 145). Negatively-charged sialic acids interfere with the efficiency of ionisation using this procedure (146, 147). However, acidic oligosaccharides (those containing sialic acids) will ionize effectively as deprotonated molecular ions using 2,4,6 trihydroxyacetophenone as matrix, in the presence of ammonium citrate (148). Further improvements in sensitivity and resolution may be obtained by derivatization of oligosaccharides with fluorophores such as 2-aminobenzamidine (129), and high energy collision-induced fragmentation (149).

## 2. Electrospray Mass Spectrometry (ES-MS)

ES-MS is capable of determining the molecular weight of biopolymers up to 100 kDa with a greater mass accuracy (0.01%) and resolution (2000) than conventional

MALDI-MS. Multiply charged molecular ions are generated by the ionization of biopolymers in volatile solvents, the resulting spectrum being convoluted to produce non-charged peaks. It is better suited to the analysis of whole glycoprotein populations than MALDI-MS, because its superior resolution permits the identification of individual glycoforms (75, 150). ES-MS can also be interfaced with liquid chromatography which permits the separation and on-line identification of glycopeptides from protein digests (150, 151), and the oligosaccharides can be sequenced following digestion with combinations of exoglycosidases (152). However, ES-MS is not generally suited for the analysis of neutral and anionic oligosaccharides which have been chemically or enzymatically released from a glycoprotein of interest as they do not readily form multiply-charged ions.

### 3. *Exoglycosidase Enzyme Arrays*

All the released glycan and glycopeptide analytical methods have benefited from incorporating exoglycosidase enzyme arrays (153, 154). The released glycans or glycopeptides are subjected to a parallel series of controlled digests by enzymes which attack specific carbohydrate linkages (such as sialidases, galactosidases and hexosaminidases), thus trimming the glycan in a predictable fashion. By comparing the resultant mass shift (MS) or elution profile (HPAE-PAD) with defined standards, the precise glycan structures within a pool of oligosaccharides or glycopeptides can be assigned.

1. For glycopeptides, an initial mass determination is made on the intact molecule, which is then digested using PNGaseF (for N-linked glycans) or O-glycanase (for O- linked glycans) and re-analyzed to deduce the mass of the core peptide.

2. Next, simultaneous digestion of purified glycopeptides with linkage-specific exoglycosidase arrays is performed for the sequential removal of oligosaccharides, to permit the sequencing of N-glycans at individual glycosylation sites .

3. Released glycans can be analyzed directly (step 2).

Exoglycosidases recognize predominantly the terminal non-reducing monosaccharide residue enabling a variety of molecules (glycopeptides, released glycans or even intact glycoproteins) to be sequenced. It gives the order of monosaccharides and the anomeric configurations of glycosidic linkages, because the specificity of purified glycosidases is known.

## V. CONCLUSIONS

Major advances in complex carbohydrate analysis have formed the basis of the new science of glycobiology. There are still many challenges and opportunities ahead for scientists working in this field. The initial choice of gene expression system will continue to be of crucial importance. As more recombinant proteins are expressed in different cell lines a pattern of glycoform predictions can be assembled (Table I), although these conclusions remain speculative at present. The influences of the cell culture process (including the effects of scale-up) are less well defined, but as more studies are published generic protocols may emerge to produce consistent (albeit heterogeneous) glycosylation patterns. A more difficult objective for cell technologists will be to produce a single invariant protein glycoform, rather than the current mixture of glycoforms each bearing different characteristics. In future, manufacturers may not be content with achieving the authentic human glycosylation profile, since the product's bioactivity or pharmacokinetics *in vivo*

**Table 1** Glycan Structure Found in Recombinant Proteins Expressed in Different Cell Types

| | | Type of Glycosylation | | | | Saccharide Residues | | | | | | | | | |
| | | | | | | Fucose | | Galactose | | Sialic acids | | | Bisecting | Glyco- |
| Organism | Cell type | O-linked | Oligo-mannose | Hyper-mannose | Complex | α1-6 linked | α1-3 Fucose | Gal α1-3 Gal | SO4-GalNAc | α2,6 linked | α2,3 linked | NeuGc | GlcNAc | sidases |
|---|---|---|---|---|---|---|---|---|---|---|---|---|---|---|
| Bacterium | *E. coli* | 0 | 0 | 0 | 0 | 0 | 0 | 0 | 0 | 0 | 0 | 0 | 0 | ? |
| Yeast | *Sacromyces* | ++ | 0 | ++++ | 0 | 0 | 0 | 0 | 0 | 0 | 0 | 0 | 0 | ? |
| Plants | Tobacco BY2 | ? | ++ | 0 | ? | ? | ++ | ? | 0 | 0 | 0 | 0 | 0 | ? |
| Insect | Sf9 | ++ | ++++ | 0 | D | ++ | + | 0 | 0 | 0 | 0 | 0 | 0 | ++ |
| | Sf21 | ++ | ++++ | 0 | D | ++ | ? | ? | ? | D | D | ? | 0 | ? |
| | *M. brassicae* | ++ | ++++ | 0 | 0 | ++ | ? | ? | ? | ? | ? | ? | 0 | ? |
| Hamster | BHK | ++ | ++ | 0 | ++ | ++ | 0 | + | 0 | 0 | ++ | ? | 0 | ? |
| | CHO | ++ | ++ | 0 | ++ | ++ | D | 0 | 0 | 0 | ++ | + | 0 | ++ |
| Mouse | Hybridoma | ++ | ++ | 0 | ++ | ++ | 0 | ++ | 0 | 0 | ++ | +++ | 0 | + |
| | Myeloma | ++ | ++ | 0 | ++ | ++ | 0 | ++ | 0 | + | + | +++ | 0 | + |
| | C127 | ++ | ++ | 0 | ++ | ++ | 0 | ++ | ++ | ++ | ++ | +++ | 0 | ? |
| | J558L | ++ | ? | 0 | ++ | ++ | 0 | ++ | ? | ++ | ++ | +++ | 0 | ? |
| | Transgenic | ++ | ++ | 0 | ++ | ++ | 0 | ? | ? | + | + | ? | ? | ? |
| Goat and Sheep | Transgenic | ? | ++ | 0 | ++ | ++ | 0 | ? | ? | + | + | ? | ? | ? |
| Human | Liver | ++ | + | 0 | ++ | ++ | 0 | 0 | 0 | ++ | ++ | 0 | 0 | ? |
| | Brain | ++ | ++ | 0 | ++ | ++ | ++ | 0 | 0 | ++ | ++ | 0 | + | ? |
| | Pituitary | ++ | ++ | 0 | + | ++ | 0 | 0 | +++ | + | + | 0 | ++ | ? |
| | B-lymphocyte | ++ | 0 | 0 | + | ++ | 0 | 0 | 0 | + | + | 0 | ++ | ? |
| | Namalwa | ++ | ++ | 0 | ++ | ++ | 0 | 0 | 0 | ++ | ++ | ? | ? | ? |
| Human-Mouse | Hetero-hybridoma | ? | ++ | 0 | ++ | ++ | 0 | 0 | 0 | + | + | + | 0 | ? |

Key: 0 = not detected; ? = not tested; D = disputed (conflicting results reported in different publications). + to ++++: an approximation of the levels of oligosaccharides detected.

can be manipulated by altering specific glycan structures such as the level of terminal sialylation or galactosylation.

Advances are also being made in the chemical production of oligosaccharides, e.g. synthesis of the core N-glycan structure $Man_3GlcNAc_2$ from monosaccharides has been accomplished (155), opening the possibility of adding defined glycan structures after recombinant protein synthesis and secretion. Artificial N-glycosylation sequences (Asn-X-Ser/Thr) have also been introduced into the genes coding for small peptides or synthetic peptides in order to improve their pharmacokinetic properties and to render them resistant to blood proteases (156, 157).

## ACKNOWLEDGEMENTS

The authors are grateful to the Biotechnology and Biological Sciences Research Council of the U.K. and Directorate General XII of the European Commission for their financial support.

## REFERENCES

1. C Abeijon, CB Hirschberg. Topography of glycosylation reactions in the endoplasmic reticulum. Trends Biochem Science 17: 32–36, 1992.

2. N Jenkins, RB Parekh, DC James. Getting the glycosylation right—implications for the biotechnology industry. Nature Biotechnology 14: 975–981, 1996.

3. BK Hayes, GW Hart. Novel forms of protein glycosylation. Curr Opin Struct Biol 4: 692–696, 1994.

4. DH Joziasse. Mammalian glycosyltransferases: genomic organization and protein structure. Glycobiology 2: 271–277, 1992.

5. MC Field, LJ Wainwright. Molecular cloning of eukaryotic glycoprotein and glycolipid glycosyltransferases—a survey. Glycobiology 5: 463–472, 1995.

6. A Varki. Biological roles of oligosaccharides: all of the theories are correct. Glycobiology 3: 97–130, 1993.

7. RB Parekh, TP Patel. Comparing the glycosylation patterns of recombinant glycoproteins. Trends Biotechnol 10: 276–280, 1992.

8. N Jenkins, EM Curling. Glycosylation of recombinant proteins: problems and prospects. Enzyme Microb Technol 16: 354–364, 1994.

9. CF Goochee, TJ Monica. Environmental effects on protein glycosylation. Bio/technology 8: 421–427, 1990.

10. CF Goochee, MJ Gramer, DC Andersen, JB Bahr, JR Rasmussen. The oligosaccharides of glycoproteins—bioprocess factors affecting oligosaccharide structure and their effect on glycoprotein properties. Biotechnology 9: 1347–1355, 1991.

11. SH Shakineshleman, SL Spitalnik, L Kasturi. The amino-acid at the x-position of an Asn-x-Ser sequon is an important determinant of N-linked core-glycosylation efficiency. J Biol Chem 271: 6363–6366, 1996.

12. E Bause. Model studies on N-glycosylation of proteins. Biochem Soc Trans 12: 514–517, 1984.

13. S Allen, HY Naim, NJ Bulleid. Intracellular folding of tissue-type plasminogen-activator—effects of disulfide bond formation on N-linked glycosylation and secretion. J Biol Chem 270: 4797–4804, 1995.

14. Y Gavel, G von Heijne. Sequence differences between glycosylated and non-glycosylated Asn-X-Thr/Ser acceptor sites: implication for protein engineering. Protein Engineering 3: 433–442, 1990.

15. EM Curling, PM Hayter, AJ Baines, AT Bull, K Gull, PG Strange, N Jenkins. Recombinant human interferon-gamma. Differences in glycosylation and proteolytic processing lead to heterogeneity in batch culture. Biochem J 272: 333–337, 1990.

16. RB Parekh, RA Dwek, PM Rudd, JR Thomas, TW Rademacher, T Warren, TC Wun, B Hebert, B Reitz, M Palmier, T Ramabhadran, DC Tiemeier. N-glycosylation and in vitro enzymatic activity of human recombinant tissue plasminogen activator expressed in chinese hamster ovary cells and a murine cell line. Biochemistry 28: 7670–7679, 1989.

17. DN Hebert, B Foellmer, A Helenius. Calnexin and calreticulin promote folding, delay oligomerization and suppress degradation of influenza hemagglutinin in microsomes. EMBO J 15: 2961–2968, 1996.

18. JD Oliver, RC Hresko, M Mueckler, S High. The glut-1 glucose transporter interacts with calnexin and calreticulin. J Biol Chem 271: 13691–13696, 1996.

19. RG Spiro, Q Zhu, V Bhoyroo, HD Soling. Definition of the lectin-like properties of the molecular chaperone, calreticulin, and

demonstration of its co-purification with endomannosidase from rat liver Golgi. J Biol Chem 271: 11588–11594, 1996.

20. DN Hebert, B Foellmer, A Helenius. Glucose trimming and reglucosylation determine glycoprotein association with calnexin in the endoplasmic-reticulum. Cell 81: 425–433, 1995.

21. K Oda, I Wada, N Takami, T Fujiwara, Y Misumi, Y Ikehara. Bip/grp78 but not calnexin associates with a precursor of glycosylphosphatidylinositol-anchored protein. Biochem J 316: 623–630, 1996.

22. E Miyoshi, Y Ihara, N Taniguchi. Trans-fection of n-acetylglucosaminyltransfer-ase-III gene suppresses expression of hepatitis-B virus in a human hepatoma- cell line, hb611. FASEB J 10: 630–630, 1996.

23. R Kornfeld, S Kornfeld. Assembly of asparagine-linked oligosaccharides. Ann Rev Biochem 54: 631–664, 1985.

24. H Schachter. The "yellow brick road" to branched complex N-glycans. Gly-cobiology 1: 453–461, 1991.

25. MR Lifely, C Hale, S Boyce, MJ Keen, J Phillips. Glycosylation and biological activity of CAMPATH-1H expressed in different cell lines and grown under differ-ent culture conditions. Glycobiology 5: 813–822, 1995.

26. RA Siciliano, HR Morris, HPJ Bennett, A Dell. O-glycosylation mimics N-glyco-sylation in the 16-kDa fragment of bovine pro-opiomelanocortin—the major O-glycan attached to $Thr_{45}$ carries $SO_4$-4GalNAc-beta1,4GlcNAc-beta-1, which is the archetypal non-reducing epitope in the N-glycans of pituitary glycohormones. J Biol Chem 269: 910–920, 1994.

27. AA Bergwerff, J Vanoostrum, JP Kamerling, JFG Vliegenthart. The major N-linked carbohydrate chains from human urokinase—the occurrence of 4-O-sulfated, (alpha2,6)-sialylated or (alpha-1,3)-fucosylated N-acetylgalactosa mine (beta-1-4)-N-acetylglucosamine ele-ments. Eur J Biochem 228: 1009–1019, 1995.

28. TP Skelton, S Kumar, PL Smith, MC Beranek, JU Baenziger. Pro-opiomel-anocortin synthesized by corticotrophs bears asparagine-linked oligosaccharides terminating with $SO_4$-4GalNAc-beta1, 4GlcNAc-beta1,2Man alpha. J Biol Chem 267: 12998–13006, 1992.

29. PL Smith, JU Baenziger. Molecular basis of recognition by the glycoprotein hor-mone-specific N-acetylgalactosamine-transferase. Proc Natl Acad Sci USA 89: 329–333, 1992.

30. D Fiete, V Srivastava, O Hindsgaul, JU Baenziger. A hepatic reticuloendothelial cell receptor specific for $SO_4$-4GalNAc-beta1,4GlcNAc-beta1,2Man-alpha that mediates rapid clearance of lutropin. Cell 67: 1103–1110, 1991.

31. G Pfeiffer, KH Strube, R Geyer. Biosynthesis of sulfated glycoprotein-N-glycans present in recombinant human tissue plasminogen activator. Biochem Biophys Res Commun 189: 1681–1685, 1992.

32. JE Hansen, O Lund, JO Nielsen, JES Hansen, S Brunak. O-glycbase—a revised database of o-glycosylated proteins O-glycbase—a revised database of o-glycosylated proteins. Nucl Acid Res 24: 248–252, 1996.

33. N Jentoft. Why are proteins O-glycosylated? Trends Biochem Sci 15: 291–294, 1990.

34. MC Field, S Amatayakulchantler, TW Rademacher, PM Rudd, RA Dwek. Structural-analysis of the N-glycans from human-immunoglobulin $A_1$—comparison of normal human serum immunoglobulin a1 with that isolated from patients with rheumatoid-arthritis. Biochem J 299: 261–275, 1994.

35. EF Hounsell, MJ Davies, DV Renouf. O-linked protein glycosylation structure and function. Glycoconjugate J 13: 19–26, 1996.

36. S Bjoern, DC Foster, L Thim, FC Wiberg, M Christensen, Y Komiyama, AH Pedersen, W Kisiel. Human plasma and recombinant factor VII. Characterization of O- glycosylations at serine residues 52 and 60 and effects of site-directed mutagen-esis of serine 52 to alanine. J Biol Chem 266: 11051–11057, 1991.

37. ML Kennard, JM Piret. Membrane-anchored protein-production from spheroid, porous, and solid microcarrier chinese-hamster ovary cell-cultures. Biotechnol Bioeng 47: 550–556, 1995.

38. J Takeda, T Kinoshita. GPI-anchor biosynthesis. Trends Biochem Sci 20: 367–371, 1995.

39. CA Redman, BN Green, JE Thomasoates, VN Reinhold, MAJ Ferguson. Analysis of glycosylphosphatidylinositol membrane anchors by electrospray-ionization mass-spectrometry and collision-induced dissociation. Glycoconjugate J 11: 187–193, 1994.

40 PT Englund. The structure and bio-synthesis of glycosyl phosphatidylinositol protein anchors. Ann Rev Biochem 62: 121–138, 1993.

41. O Letourneur, S Sechi, J Willettebrown, MW Robertson, JP Kinet. Glycosylation of human truncated Fc-epsilon-RI alpha-chain is necessary for efficient folding in the endoplasmic-reticulum. J Biol Chem 270: 8249–8256, 1995.

42. AO Herscovics, P Orlean. Glycoprotein biosynthesis in yeast. FASEB J 7: 540–550, 1993.

43. PJ Kniskern, A Hagopian, P Burke, LD Schulz, DL Montgomery, WM Hurni, CY Ip, CA Schulman, RZ Maigetter, DE Wampler, D Kubek, RD Sitrin, DJ West, RW Ellis, WJ Miller. Characterization and evaluation of a recombinant hepatitis-B vaccine expressed in yeast defective for N-linked hyperglycosylation. Vaccine 12: 1021–1025, 1994.

44. L Lehle, A Eiden, K Lehnert, A Haselbeck, E Kopetzki. Glycoprotein-biosynthesis in saccharomyces-cerevisiae—ngd29, an N-glycosylation mutant allelic to och1 having a defect in the initiation of outer chain formation. FEBS Lett 370: 41–45, 1995.

45. L Lehle, A Eiden, K Lehnert, A Haselbeck, E Kopetzki. Glycoprotein-biosynthesis in saccharomyces cerevisiae—ngd29, an N-glycosylation mutant allelic to och1 having a defect in the initiation of outer chain formation. FEBS Lett 370: 41–45, 1995.

46. I Kalsner, FJ Schneider, R Geyer, H Ahorn, I Maurerfogy. Comparison of the carbohydrate moieties of re-combinant soluble Fc-epsilon receptor (sFc-ε-RII/scd23) expressed in saccharomyces cerevisiae and chinese hamster ovary cells—different O-glycosylation sites are used by yeast and mammalian cells. Glycoconjugate J 9: 209–216, 1992.

47. P Lerouge, L Faye. Recent developments in structural analysis of N-glycans from plant glycoproteins. Plant Physiol Biochem 34: 263–271, 1996.

48. PR Ganz, AK Dudani, ES Tackaberry, R Sardana, C Sauder, X Cheng, and I Altosaar. Expression of human blood pro-teins in transgenic plants: the cytokine GM-CSF as a model protein. In: MRL Owen, ed., Transgenic plants: a production system for industrial and pharmaceutical proteins. Chichester: J Wiley and Sons Ltd., 1996, pp 281–297.

49. S Matsumoto, K Ikura, M Ueda, R Sasaki. Characterization of a human glycoprotein (erythropoietin) produced in cultured tobacco cells. Plant Mol Biol 27: 1163–1172, 1995.

50. F Altmann, V Tretter, V Kubelka, E Staudacher, L Marz, WM Becker. Fucose in α1-3 linkage to the N-glycan core forms an allergenic epitope that occurs in plant

and in insect glycoproteins. Glyco-conjugate J. 10: 301–301, 1993.

51. MG Yet, F Wold. The distribution of glycan structures in individual N-glycosylation sites in animal and plant glycoproteins. Arch Biochem Biophys 278: 356–364, 1990.

52. G Garcia-Casado, R Sanchezmonge, MJ Chrispeels, A Armentia, G Salcedo, L Gomez. Role of complex asparagine-linked glycans in the allergenicity of plant glycoproteins. Glycobiology 6: 471–477, 1996.

53. DJC Davidson, MJ Fraser, FJ Castellino. Oligosaccharide processing in the expression of human plasminogen cDNA by lepidopteran insect (Spodoptera frugiperda) cells. Biochemistry 29: 5584–5590, 1990.

54. DC James, RB Freedman, M Hoare, OW Ogonah, BC Rooney, OA Larionov, VN Dobrovolsky, OV Lagutin, N Jenkins. N-glycosylation of recombinant human interferon-gamma produced in different animal expression systems. Biotechnology 13: 592–596, 1995.

55. OW Ogonah, RB Freedman, N Jenkins, K Patel, BC Rooney. Isolation and characterization of an insect-cell line able to perform complex N-linked glycosylation on recombinant proteins. Biotechnology 14: 197–202, 1996.

56. TR Davis, HA Wood. Intrinsic glycosylation potentials of insect-cell cul-tures and insect larvae. In Vitro Cell Devel Biol 31: 659–663, 1995.

57. TJ Monica, SB Williams, CF Goochee, BL Maiorella. Characterization of the glycosylation of a human-IgM produced by a human-mouse hybridoma. Glycobiology 5: 175–185, 1995.

58. L Monaco, A Marc, A Eon-Duval, G Acerbis, G Distefano, D Lamotte, JM Engasser, M Soria, N Jenkins. Genetic engineering of alpha 2,6-sialyltransferase in recombinant CHO cells and its effects on the sialylation of recombinant interferon- gamma. Cytotechnology 22: 197–203, 1996.

59. A Noguchi, CJ Mukuria, E Suzuki, M Naiki. Immunogenicity of N-glycolylneur-aminic acid-containing carbohydrate chains of recombinant human erythro-poietin expressed in chinese hamster ovary cells. J Biochem 117: 59–62, 1995.

60. AR Flesher, J Marzowski, WC Wang, HV Raff. Fluorophore labeled glycan analysis of immunoglobulin fusion proteins: corre-lation of oligosaccharide content with in vivo clearance profile. Biotechnol Bioeng 46: 399–407, 1995.

61. J Tsuji, S Noma, J Suzuki, K Okumura, N Shimizu. Specificity of human natural antibody to recombinant tissue-type plasminogen activator (t-PA) expressed in mouse C127 cells. Chem Pharm Bull 38: 765–768, 1990.

62. CC Yu-Ip, WJ Miller, M Silberklang, GE Mark, RW Ellis, LH Huang, J Glushka, H van Halbeek, J Zhu, JA Alhadeff. Structural characterization of the N-glycans of a humanized anti-CD8 murine immunoglobulin G. Arch Biochem Biophys 308: 387–399, 1994.

63. RM Hamadeh, GA Jarvis, U Galili, RE Mandrell, P Zhou, JM Griffiss. Human natural anti-Gal IgG regulates alternative complement pathway activation on bacterial surfaces. J Clin Invest 89: 1223–1235, 1992.

64. JA Lavecchio, AD Dunne, ASB Edge. Enzymatic removal of alpha-galactosyl epitopes from porcine endothelial cells diminishes the cytotoxic effect of natural antibodies. Transplantation 60: 841–847, 1995.

65. CAK Borrebaeck, AC Malmborg, M Ohlin. Does endogenous glycosylation prevent the use of mouse monoclonal antibodies as cancer therapeutics? Immunol Today 14: 477–479, 1993.

66. T Endo, A Wright, SL Morrison, A Kobata. Glycosylation of the variable region of immunoglobulin G: site-specific maturation of the sugar chains. Mol Immunol 32: 931–940, 1995.

67. EU Lee, J Roth, JC Paulson. Alteration of terminal glycosylation sequences on N-linked oligosaccharides of Chinese hamster ovary cells by expression of beta-galactoside alpha2,6-sialyltransferase. J Biol Chem 264: 13848–13855, 1989.

68. E Grabenhorst, A Hoffmann, M Nimtz, G Zettlmeissl, HS Conradt. Construction of stable BHK-21-cells coexpressing human secretory glycoproteins and human Gal (beta-1-4)GlcNAc-R alpha-2,6-sialyltransferase alpha-2,6-linked NeuAc is preferentially attached to the Gal (beta-1-4) GlcNAc (beta-1-2)Man (alpha-1-3)-branch of diantennary oligosaccharides from secreted recombinant beta-trace protein. Eur J Biochem 232: 718–725, 1995.

69. L Monaco, A Marc, A Eon-Duval, G Acerbis, G Distefano, D Lamotte, JM Engasser, M Soria, N Jenkins. Genetic engineering of alpha 2,6-sialyltransferase in recombinant CHO cells and its effects on the sialylation of recombinant interferon-gamma. Cytotechnology 22: 197–203, 1996.

70. E Grabenhorst, A Hoffmann, M Nimtz, G Zettlmeissl, HS Conradt. Construction of stable BHK-21-cells coexpressing human secretory glycoproteins and human Gal (beta-1-4)GlcNAc-r alpha-2,6-sialyltransferase alpha-2,6-linked NeuAc is preferentially attached to the Gal (beta-1-4) GlcNAc (beta-1-2)Man (alpha-1-3)-branch of diantennary oligosaccharides from secreted recombinant beta-trace protein. Eur J Biochem 232: 718–725, 1995.

71. SL Minch, PT Kallio, JE Bailey. Tissue-plasminogen activator coexpressed in chinese-hamster ovary cells with alpha2,6-sialyltransferase contains NeuAc-alpha2,6gal-beta1,4GlcNAc linkages. Biotechnol Prog 11: 348–351, 1995.

72. MW Khan, SC Musgrave, N Jenkins. N-linked glycosylation of tissue-plasminogen activator in Namalwa cells. Biochem Soc Trans 23: S99, 1995.

73. PA Kemp, N Jenkins, AJ Clark, RB Freedman. The glycosylation of human recombinant alpha-1-antitrypsin expressed in transgenic mice. Biochem Soc Trans 24: S339, 1996.

74. M Stromqvist, J Tornell, M Edlund, A Edlund, T Johansson, K Lindgren, L Lundberg, L Hansson. Recombinant human bile salt-stimulated lipase—an example of defective O-glycosylation of a protein produced in milk of transgenic mice. Transgenic Res 5: 475–485, 1996.

75. DC James, MH Goldman, M Hoare, N Jenkins, RWA Oliver, BN Green, RB Freedman. Post-translational processing of recombinant human interferon-gamma in animal expression systems. Protein Science 5: 331–340, 1996.

76. E Higgins, J Pollack, P Ditullio, H Meade. Characterization of the glycosylation on a monoclonal antibody produced in the milk of a transgenic goat. Glycobiology 6: 1211–1211, 1996.

77. A Carver, G Wright, D Cottom, J Cooper, M Dalrymple, S Temperley, M Udell, D Reeves, J Percy, A Scott, D Barrass, Y Gibson, Y Jeffrey, C Samuel, A Colman, I Garner. Expression of human alpha-1 antitrypsin in transgenic sheep. Cytotechnology 9: 77–84, 1992.

78. TP Patel, RB Parekh, BJ Moellering, CP Prior. Different culture methods lead to differences in glycosylation of a murine IgG monoclonal antibody. Biochem J 285: 839–845, 1992.

79. JT Lund, N Takahashi, H Nakagawa, M Goodall, T Bentley, SA Hindley, R Tyler, R Jefferis. Control of IgG/Fc glycosylation: a comparison of oligosaccharides from chimeric human/mouse and mouse

subclass immunoglobulin Gs. Mol Immunol 30: 741–748, 1993.

80. BL Maiorella, J Winkelhake, J Young, B Moyer, R Bauer, M Hora, J Andya, J Thomson, T Patel, RB Parekh. Effect of culture conditions on IgM antibody structure, pharmacokinetics and activity. Biotechnology 11: 387–392, 1993.

81. M Gawlitzek, U Valley, M Nimtz, R Wagner, HS Conradt. Characterization of changes in the glycosylation pattern of recombinant proteins from BHK-21-cells due to different culture conditions. J Biotechnol 42: 117–131, 1995.

82. H Tachibana, K Taniguchi, Y Ushio, K Teruya, H Osada, H Murakami. Changes of monosaccharide availability of human hybridoma lead to alteration of biological properties of human monoclonal antibody. Cytotechnology 16: 151–157, 1994.

83. PM Hayter, EM Curling, ML Gould, AJ Baines, N Jenkins, I Salmon, PG Strange, AT Bull. The effect of dilution rate on CHO cell physiology and recombinant interferon-gamma production in glucose-limited chemostat cultures. Biotechnol Bioeng 42: 1077–1085, 1993.

84. PM Hayter, EM Curling, AJ Baines, N Jenkins, I Salmon, PG Strange, JM Tong, AT Bull. Glucose-limited chemostat culture of chinese hamster ovary cells producing recombinant human interferon-gamma. Biotechnol Bioeng 39: 327–335, 1992.

85. PML Castro, AP Ison, PM Hayter, AT Bull. The macroheterogeneity of recombinant human interferon-gamma produced by Chinese hamster ovary cells is affected by the protein and lipid content of the culture medium. Biotechnol Appl Biochem 87: 87–100, 1995.

86. N Jenkins, PML Castro, S Menon, AP Ison, AT Bull. Effect of lipid supplements on the production and glycosylation of recombinant interferon-gamma expressed in CHO cells. Cytotechnology 15: 209–215, 1994.

87. WRP Rijcken, B Overdijk, DH van den Eijnden, W Ferwerda. The effect of increasing nucleotide sugar concentrations on the incorporation of sugars into glycoconjugates in rat hepatocytes. Biochem J 305: 865–870, 1995.

88. A Datti JW Dennis. Regulation of UDP-GlcNAc:Gal beta 1-3GalNAc-R beta 1-6-N- acetylglucosaminyltransferase (GlcNAc to GalNAc) in Chinese hamster ovary cells. J Biol Chem 268: 5409–5416, 1993.

89. W Chotigeat, Y Watanapokasin, S Mahler, PP Gray. Role of environmental-con- ditions on the expression levels, glycoform pattern and levels of sialyltransferase for hFSH produced by recombinant CHO cells. Cytotechnology 15: 217–221, 1994.

90. CA Gebert, PP Gray. Expression of FSH in CHO cells .2. stimulation of hFSH expression levels by defined medium supplements. Cytotechnology 17: 13–19, 1995.

91. DK Robinson, CP Chan, CY Ip, PK Tsai, J Tung, TC Seamans, AB Lenny, DK Lee, J Irwin, M Silberklang. Characterization of a recombinant antibody produced in the course of a high-yield fed-batch process. Biotechnol Bioeng 44: 727–735, 1994.

92. AD Hooker, MH Goldman, NH Markham, DC James, AP Ison, AT Bull, PG Strange, I Salmon, AJ Baines, N Jenkins. N-glycans of recombinant human interferon-gamma change during batch culture of chinese-hamster ovary cells. Biotechnol Bioeng 48: 639–648, 1995.

93. M Gawlitzek, HS Conradt, R Wagner. Effect of different cell-culture conditions on the polypeptide integrity and N-glycosylation of a recombinant model glycoprotein. Biotechnol Bioeng 46: 536–544, 1995.

94. E Watson, B Shah, L Leiderman, YR Hsu, S Karkare, HS Lu, FK Lin. Comparison of N-linked oligosaccharides of recom- binant human tissue kallikrein produced by chinese-hamster ovary cells on microcarrier beads and in serum-free suspension-culture. Biotechnol Prog 10: 39–44, 1994.

95. M Gawlitzek, U Valley, M Nimtz, R Wagner, HS Conradt. Characterization of changes in the glycosylation pattern of recombinant proteins from BHK-21-cells due to different culture conditions. J Biotechnol 42: 117–131, 1995.

96. DC Andersen, CF Goochee, G Cooper, M Weitzhandler. Monosaccharide and oligosaccharide analysis of isoelectric focusing-separated and blotted granulo- cyte-colony-stimulating factor glycoforms using high-pH anion-exchange chromato- graphy with pulsed amperometric detection. Glycobiology 4: 459–467, 1994.

97. MC Borys, DIH Linzer, ET Papoutsakis. Ammonia affects the glycosylation patterns of recombinant mouse placental lactogen-1 by chinese-hamster ovary cells in a pH- dependent manner. Biotechnol Bioeng 43: 505–514, 1994.

98. AA Lin, R Kimura, WM Miller. Pro- duction of tPA in recombinant CHO cells under oxygen-limited conditions. Biotechnol Bioeng 42: 339–350, 1993.

99. MC Borys, DJH Linzer, ET Papoutsakis. Culture pH affects expression rates and glycosylation of recombinant mouse placental lactogen proteins by chinese hamster ovary (CHO) cells. Biotechnology 11: 720–724, 1993.

100. CF Goochee, MJ Gramer, DV Schaffer, MB Sliwkowski. Potential for extracellular hydrolysis of glycoprotein oligosaccharides by chinese-hamster ovary cell sialidase and fucosidase. J Cell Biochem 263–263, 1994.

101. MJ Gramer, CF Goochee. Glycosidase activities of the 293 and NS0 cell-lines, and of an antibody-producing hybridoma cell-line. Biotechnol Bioeng 43: 423–428, 1994.

102. MJ Gramer, CF Goochee, V Chock, DT Brousseau, MB Sliwkowski. Removal of sialic acid from a glycoprotein in CHO cell supernatant by action of an extracellular CHO cell sialidase. Biotechnology 13: 692–698, 1995.

103. J Ferrari, R Harris, TG Warner. Cloning and expression of a soluble sialidase from chinese- hamster ovary cells – sequence alignment similarities to bacterial sialidases. Glycobiology 4: 367–373, 1994.

104. TG Warner, J Chang, J Ferrari, R Harris, T Mcnerney, G Bennett, J Burnier, MB Sliwkowski. Isolation and properties of a soluble sialidase from the culture fluid of chinese-hamster ovary cells. Glycobiology 3: 455–463, 1993.

105. MJ Gramer, DV Schaffer, MB Sliwkowski, CF Goochee. Purification and characterization of alpha-l-fucosidase from chinese-hamster ovary cell-culture supernatant. Glycobiology 4: 611–616, 1994.

106. MJ Gramer, CF Goochee. Glycosidase activities of the 293 and NS0 cell-lines, and of an antibody-producing hybridoma cell-line. Biotechnol Bioeng 43: 423–428, 1994.

107. M Ackermann, U Marx, V Jager. Influence of cell-derived and media-derived factors on the integrity of a human monoclonal-antibody after secretion into serum-free cell-culture supernatants. Biotechnol Bioeng 45: 97–106, 1995.

108. S Berger, A Menudier, R Julien, Y Karamanos. Do de-N-glycosylation enzymes have an important role in plant cells? Biochimie 77: 751–760, 1995.

109. M Shelikoff, AJ Sinskey, G Stephanopoulos. The effect of protein-synthesis inhibitors on the glycosylation site occupancy of recombinant human prolactin. Cytotechnology 15: 195–208, 1994.

110. NJ Bulleid, RS Bassel-Duby, RB Freedman, JF Sambrook, MJ Gething. Cell-free synthesis of enzymically active tissue-type plasminogen activator. Protein folding determines the extent of N- linked glycosylation. Biochem J 286: 275–280, 1992.

111. J Lund, N Takahashi, JD Pound, M Goodall, R Jefferis. Multiple interactions of IgG with its core oligosaccharide can modulate recognition by complement and human Fc-gamma receptor-I and influence the synthesis of its oligosaccharide chains. J Immunol 157: 4963–4969, 1996.

112. T Sareneva, J Pirhonen, K Cantell, N Kalkkinen, I Julkunen. Role of N-glycosylation in the synthesis, dimerization and secretion of human interferon-gamma. Biochem J 303: 831–840, 1994.

113. BH Yang, MH Hoe, P Black, RC Hunt. Role of oligosaccharides in the processing and function of human transferrin receptors. Effect of the loss of the three N-glycosyl oligosaccharides individually or together. J Biol Chem 268: 7435–7441, 1993.

114. RS Monroe, BE Huber. The major form of the murine asialoglycoprotein receptor— cDNA sequence and expression in liver, testis and epididymis. Gene 148: 237–244, 1994.

115. MN Fukuda, H Sasaki, L Lopez. Survival of recombinant erythropoietin in the circulation: the role of carbohydrates. Blood 73: 84–89, 1989.

116. G Ashwell, J Hartford. Carbohydrate—specific receptors of the liver. Ann Rev Biochem 51: 531–554, 1982.

117. T Misaizu, S Matsuki, TW Strickland, M Takeuchi, A Kobata, S Takasaki. Role of antennary structure of N-linked sugar chains in renal handling of recombinant human erythropoietin. Blood 86: 4097–4104, 1995.

118. MC Manning, K Patel, RT Borchardt. Stability of protein pharmaceuticals. Pharmaceutical Research 6: 903–918, 1995.

119. T Sareneva, J Pirhonen, K Cantell, I Julkunen. N-glycosylation of human interferon-gamma—glycans at Asn-25 are critical for protease resistance. Biochem J 308: 9–14, 1995.

120. PHC Vanberkel, MEJ Geerts, HA Vanveen, PM Kooiman, FR Pieper, HA Deboer, JH Nuijens. Glycosylated and unglycosylated human lactoferrins both bind iron and show identical affinities towards human lysozyme and bacterial lipopolysaccharide, but differ in their

susceptibilities towards tryptic proteolysis. Biochem J 312: 107–114, 1995.

121. E Delorme, T Lorenzini, J Giffin, F Martin, F Jacobsen, TC Boone, S Elliott. Role of glycosylation on the secretion and biological activity of erythropoietin. Biochemistry 31: 9871–9876, 1992.

122. K Furukawa, A Kobata. IgG galactosylation—its biological significance and pathology. Mol Immunol 28: 1333–1340, 1991.

123. GJ Rademaker, J Haverkamp, J Thomasoates. Determination of glycosylation sites in O-linked glycopeptides—a sensitive mass-spectrometric protocol. Organic Mass Spectrometry 28: 1536–1541, 1993.

124. TB Patel, J Bruce, A Merry, C Bigge, M Wormald, A Jaques, RB Parekh. Use of hydrazine to release in intact and unreduced form both N- and O-linked oligosaccharides from glycoproteins. Biochemistry 32: 679–693, 1993.

125. X Liu, D Davis, DL Segaloff. Disruption of potential sites for N-linked glycosylation does not impair hormone binding to the lutropin/choriogonadotropin receptor if Asn173 is left intact. J Biol Chem 268: 1513–1516, 1993.

126. A Kobata T Endo. Immobilized lectin columns: useful tools for the fractionation and structural analysis of oligosaccharides. J.Chromatogr. 597:111-122, 1992.

127. N Sumar, KB Bodman, TW Rademacher, RA Dwek, P Williams, RB Parekh, J Edge, GA Rook, DA Isenberg, FC Hay, et al. Analysis of glycosylation changes in IgG using lectins. J Immunol Methods 131: 127–136, 1990.

128. CLH Clogston, S Hu, TC Boone, HS Lu. Glycosidase digestion, electrophoresis and chromatographic analysis of recombinant human granulocyte colony-stimulating factor glycoforms produced in Chinese hamster ovary cells. J Chromatogr 637: 55–62, 1993.

129. JC Bigge, TP Patel, JA Bruce, PN Goulding, SM Charles, RB Parekh. Non-selective and efficient fluorescent labeling of glycans using 2-amino benzamide and anthranilic acid. Anal Biochem 230: 229–238, 1995.

130. KR Anumula. Rapid quantitative-determination of sialic acids in glycoproteins by high-performance liquid-chromatography with a sensitive fluorescence detection. Anal Biochem 230: 24–30, 1995.

131. P Camilleri, GB Harland, G Okafo. High-resolution and rapid analysis of branched oligosaccharides by capillary electrophoresis. Anal Biochem 230: 115–122, 1995.

132. G Reuter, R Schaur. Determination of sialic acids. Methods Enzymol 230: 168–199, 1994.

133. CJ Edge, TW Rademacher, MR Wormald, RB Parekh, TD Butters, DR Wing, RA Dwek. Fast sequencing of oligosaccharides: the reagent-array analysis method. Proc Natl Acad Sci USA 89: 6338–6342, 1992.

134. D Tetaert, B Soudan, JM Loguidice, C Richet, P Degand, G Boussard, C Mariller, G Spik. Combination of high-performance anion-exchange chromatography and electrospray mass-spectrometry for analysis of the in vitro O-glycosylated mucin motif peptide. J Chromatog B-Biomedical Applications 658: 31–38, 1994.

135. M Weitzhandler. Eliminating monosaccharide peak tailing in HPAE-PAD. Anal Biochem 241: 135–136, 1996.

136. J Frenz. Chromatographic separations in biotechnology. Amer Chem Soc Symp Ser 529: 1–12, 1993.

137. MV Novotny, J Sudor. High-performance capillary electrophoresis of glycoconjugates. Electrophoresis 14: 373–389, 1993.

138. HJ Boss, MF Rohde, RS Rush. Multiple sequential fraction collection of peptides and glycopeptides by high-performance capillary electrophoresis. Anal Biochem 230: 123–129, 1995.

139. DC James, RB Freedman, M Hoare, N Jenkins. High resolution separation of recombinant human interferon-gamma by micellar electrokinetic capillary chromatography. Anal Biochem 222: 315–322, 1994.

140. PM Rudd, IG Scragg, E Coghill, RA Dwek. Separation and analysis of the glycoform populations of ribonuceases B using capillary electrophoresis. Glycoconjugate J 9: 86–91, 1992.

141. MV Novotny. Capillary electrophoresis. Curr Opinion Biotechnol 7: 29–34, 1996.

142. E Watson, F Yao. Capillary electrophoretic separation of human recombinant erythropoietin (r-HuEPO) glycoforms. Anal Biochem 210: 389–393, 1993.

143. E Mortz, T Sareneva, I Julkunen, P Roepstorff. Does matrix-assisted laser-desorption ionization mass-spectrometry allow analysis of carbohydrate heterogeneity in glycoproteins?—a study of natural human interferon-gamma. J Mass Spectrom 31: 1109–1118, 1996.

144. GR Hayes, A Williams, CE Costello, CA Enns, JJ Lucas. The critical glycosylation site of human transferrin receptor contains

a high-mannose oligosaccharide. Glycobiology 5: 227–232, 1995.

145. CW Sutton, JA O'Neill, JS Cottrell. Site specific characterization of glycoprotein carbohydrates by exoglycosidase digestion and laser desorption mass spectrometry. Anal Biochem 218: 34–46, 1994.

146. F Hillenkamp, M Karas, RC Beavis, BT Chait. Matrix-assisted laser desorption-ionization mass spectrometry of biopolymers. Anal Chem 63: 1193A–1203A, 1991.

147. A Tsarbopoulos, M Karas, K Strupat, B Pramanik, TL Nagabhushan, F Hillenkamp. Comparative mapping of recombinant proteins and glycoproteins by plasma desorption and matrix-assisted laser desorption/ionization mass spectrometry. Anal Chem 66: 2062–2070, 1994.

148. DI Papac, A Wong, AJS Jones. Analysis of acidic oligosaccharides and glycopeptides by matrix-assisted laser-desorption ionization time-of-flight mass- spectrometry. Anal Chem 68: 3215–3223, 1996.

149. DJ Harvey, RH Bateman, MR Green. High-energy collision-induced fragmentation of complex oligosaccharides ionized by matrix-assisted laser desorption/ionization mass spectrometry. J Mass Spectrom 32: 167–187, 1997.

150. A Tsarbopoulos, BN Pramanik, TL Nagabhushan, TR Covey. Structural-analysis of the CHO-derived interleukin-4 by liquid- chromatography electrospray-ionization mass-spectrometry. J Mass Spectrom 30: 1752–1763, 1995.

151. D Muller, B Domon, M Karas, J Vanoostrum, WJ Richter. Characterization and direct glycoform profiling of a hybrid plasminogen-activator by matrix-assisted laser-desorption and electrospray mass spectrometry—correlation with high-performance liquid-chromatographic and nuclear-magnetic-resonance analyses of the released glycans. Biol Mass Spectrom 23: 330–338, 1994.

152. PA Schindler, CA Settineri, X Collet, CJ Fielding, AL Burlingame. Site-specific detection and structural characterization of the glycosylation of human plasma-proteins lecithin-cholesterol acyltransferase and apolipoprotein-D using HPLC/electrospray mass- spectrometry and sequential glycosidase digestion. Protein Science 4: 791–803, 1995.

153. CJ Edge, TW Rademacher, MR Wormald, RB Parekh, TD Butters, DR Wing, RA Dwek. Fast sequencing of oligosaccharides: the reagent-array analysis method. Proc Natl Acad Sci USA 89: 6338–6342, 1992.

154. CJ Edge, RB Parekh, TW Rademacher, M Wormald, RA Dwek. Fast sequencing of oligosaccharides using arrays of enzymes. Nature 358: 693–694, 1992.

155. I Matsuo, Y Nakahara, Y Ito, T Nukada, T Ogawa. Synthesis of a glycopeptide carrying a N-linked core pentasaccharide. Bio-Organic Med Chem 3: 1455–1463, 1995.

156. Y Takei, T Chiba, K Wada, H Hayashi, M Yamada, J Kuwashima, K Onozaki. Glycosylated human recombinant interleukin-1-alpha, neo interleukin-1-alpha, with d-mannose dimer exhibits selective activities in-vivo. J Interferon Cytokine Res 15: 713–719, 1995.

157. M Baudys, T Uchio, L Hovgaard, EF Zhu, T Avramoglou, M Jozefowicz, B Rihova, JY Park, HK Lee, SW Kim. Glycosylated insulins. J Controlled Release 36: 151–157, 1995.

# 15

# Protein Sequence Analysis

**Hsieng S. Lu,\* Lee Anne Merewether, and Michael F. Rohde**
*Amgen Inc., Thousand Oaks, California*

## I. INTRODUCTION

Proteins are macromolecules (usually >10,000 Da in molecular weight) which are significantly larger than conventional pharmaceuticals (usually hundreds of mass units in molecular weight). Proteins are complex linear polymers made up of different individual building blocks (i.e., 20 amino acids) linked via peptide bonds and folded into a three-dimensional structure (1). The amino acid sequence, appropriately called primary structure, identifies a protein unambiguously, determines all its chemical properties and biological functions, and specifies the higher orders of protein structure (i.e., secondary, tertiary, and quaternary structures) which are held together by a variety of interactions driven by hydrogen bonding, disulfide bonding, charge-charge interaction, and hydrophobic effect. Table 1 lists the 20 common amino acids and some of their properties. They contain diversified functional side chains including acidic/basic groups, hydrophilic/hydrophobic side chains, polar/neutral species and aromatic rings. These functional groups contribute to the complex nature of the protein molecules. Therefore, almost every protein owns its unique physical and chemical properties, and protein analysis requires multiple methodologies to "fingerprint" each of the protein species. The most basic step in characterizing a protein is therefore to determine its amino acid sequence.

Protein sequence determination can be accomplished by sequencing the protein on an automatic sequencer using chemical methods for successive degradation most often from the amino terminus. The sequence of a protein can also be deduced from its gene sequence at a much faster speed using the established DNA sequencing methodology. In the latter case, the DNA encoding a specific protein has to be cloned by oligonucleotide probes usually designed from partial amino acid sequence information deduced from automatic protein sequencing or by expression cloning in which biological assay can be selected for the detection. Although protein sequencing is considered to be more difficult and slower than DNA sequencing, it offers many features not obtainable by the latter method (see Sec. I.B).

In this chapter we describe methods and approaches for sequence analysis

---

\* Correspondence should be addressed to: Hsieng S. Lu, Ph.D., Department of Protein Structure, Amgen Inc., 1 Amgen Center, Thousand Oaks CA 91320. Tel: 805-447-3092. Fax: 805-499-7464. e-mail: hlu@amgen.com

**Table 1**  The 20 Common Amino Acids with their One- and Three-Letter Symbols, Molecular Weights, Properties of Side Chains, and $pK_a$ Values of Ionizable Side Chains

| Amino acid | Three-letter symbol | One-letter symbol | Residue mass value (daltons) | Properties of side chains | $pK_a$ |
|---|---|---|---|---|---|
| Glycine | Gly | G | 57.06 | aliphatic | — |
| Alanine | Ala | A | 71.08 | aliphatic | — |
| Serine | Ser | S | 87.08 | polar | — |
| Proline | Pro | P | 97.12 | aliphatic | — |
| Valine | Val | V | 99.14 | aliphatic | — |
| Threonine | Thr | T | 101.11 | polar | — |
| Cysteine | Cys | C | 103.14 | sulfur-containing | 9.1–9.5 |
| Leucine | Leu | L | 113.17 | aliphatic | — |
| Isoleucine | Ile | I | 113.17 | aliphatic | — |
| Asparagine | Asn | N | 114.11 | polar | — |
| Aspartate | Asp | D | 115.09 | acidic | 4.5 |
| Glutamine | Gln | Q | 128.14 | polar | – |
| Lysine | Lys | K | 128.18 | basic | 10.4 |
| Glutamate | Glu | E | 129.12 | acidic | 4.6 |
| Methionine | Met | M | 131.21 | sulfur-containing | — |
| Histidine | His | H | 137.15 | basic | 6.2 |
| Phenylalanine | Phe | F | 147.18 | chromophore | — |
| Arginine | Arg | R | 156.12 | basic | ~12 |
| Tyrosine | Tyr | Y | 163.18 | chromophore | 9.7 |
| Tryptophan | Trp | W | 186.12 | chromophore | — |

of proteins with emphasis on their applications in the analysis of protein-based drugs. While excellent books and specific method reviews (2–10) on protein sequence analysis are available, they are not presented from the pharmaceutical perspective. It is our intention to describe methodology of protein sequence analysis that is most current and routinely used in the analysis of protein pharmaceuticals. Those who are interested in more in-depth technical details and knowledge should refer to references selected at the end of this chapter. High sensitivity sequencing improvement using modern instrumentation can be seen in a recent review (11), while technical details in performing contemporary sequence analysis are collected in Current Protocols in Protein Science (12). Development of a variety of contemporary techniques in sequence analysis and description of recipes for sample preparation can be seen in the volumes of

"Techniques in Protein Chemistry" (Academic Press, N.Y.), a collection of presentations in protein analytical techniques selected each year from the Protein Society Symposium since 1988.

## A.  Historical Background

The modern approach to protein sequence analysis originates from careful work by many researchers over several decades. By 1940 it was agreed that proteins were composed of amino acids held together by peptide bonds to form polypeptide chains. In the 1950's, the demonstration that the protein, insulin, has a unique molecular structure by the studies of Sanger and his collaborators (13, 14) has made a significant contribution to the understanding of protein structure. During their studies, several methods for sequence analysis were developed, including amino-terminal labeling with 1-fluoro-2,4-dinitrobenzene, acid-hydrolysis, and partial enzymatic cleavages of peptide bonds (15). The labeled dinitrophenyl amino acids at the amino-terminus are yellow compounds which could be identified by partition or paper chromatography.

A more efficient way to determine peptide sequences was the use of sequential degradation methods (Edman degradation), from the N-terminus using phenylisothiocyanate, PITC (16–18), and from the carboxy-terminus using carboxypeptidases. The manual Edman degradation procedure was originally used to perform protein/peptide sequence determination. Modifications of the Edman sequence method has been adapted as an effort to improve the sensitivity. In 1963, dansyl chloride, a highly fluorescent compound, was introduced to replace fluoro-nitrobenzene as an amino terminal labeling agent due to high fluorescence sensitivity of the dansyl amino acid derivative (19). The manual dansyl-Edman sequencing procedure was subsequently used to sequence a number of proteins (20). Other examples are modification of the Edman reagents such as the use of the colored PITC derivative, 4-N,N-dimethyl-aminoazobenzene-4'-isothiocyanate (21), or others and the use of solid phase sequencing method in which the protein/peptide sample to be sequenced is chemically linked to an inert solid support (22).

Techniques for separation and identification of amino acids and peptides and their derivatives have been improved over the last four decades. Methods of identifying amino acid derivatives released during sequence analysis have a great advancement over the use of such techniques as paper chromatography, electrophoresis, thin layer/micro-thin layer chromatography and gas chromatography, to reverse-phase high sensitivity HPLC. Amino acid analysis by automatic ion-exchange chromatography (23) has also been greatly improved since its introduction, and recently, the classical procedure has faced competition from HPLC analysis of the amino acid derivatives. Reverse-phase HPLC separation of peptides followed by on-line UV detection and/or mass measurement (referred to as on-line LC-MS) has been widely used for isolation and identification of peptides. The HPLC-based peptide mapping procedures represent a greater improvement in both speed and sensitivity when compared to the methods used in earlier years, such as high-voltage paper electrophoresis, gel filtration, ion-exchange chromatography, and thin layer peptide mapping methods (24).

By far the most important technique used for the determination of amino acid sequences in proteins/peptides is the PITC (or Edman) degradation procedure (16–18). Automation of the procedure with the use of the spinning cup sequencer became available in 1967 (25). The gas-phase sequencer, introduced in 1981, represents a major advance in instrumentation (26). In combination with on-line HPLC analysis of the released amino acids the commercially available sequencers become crucial equipments in performing sequence analysis of protein biopharmaceuticals in almost all biotechnology companies. Recent development of various mass spectrometric techniques in the structural analysis of proteins has advanced the accurate mass measurement of peptides and proteins as well as sequencing of small peptides by fragmentation of parent ion (referred to as tandem mass spectrometry or MS/MS) to become very important tools for assessing molecular integrity and modification of proteins (for selected reviews and technical details, see 27–30). In many cases, mass spectrometric analysis of proteins and peptides (see Chapter 21) is regarded as an indispensable tool for peptide/protein primary structural analysis and is highly complementary to the conventional protein structural analysis methods.

## B. Applications for Protein Sequence Analysis

Although protein sequence analysis can be considered to be more difficult and slower than DNA sequence analysis of a cloned gene, it often provides information not obtainable by the latter method. Therefore, the contemporary protein sequencing method still gains wide acceptance as an important technique for characterization and analysis of proteins. Currently, the most common applications for protein sequence analysis include the following.

1. Determination of partial protein sequence information can be used for the design of oligonucleotide probes complementary to the predicted gene sequences. These nucleotide probes have proved to be very useful in cloning novel genes encoding low abundance proteins. In many cases, it has been the only route to the cloning of particular genes, for example, those coding for a number of important pharmaceutical proteins being used in the clinic.

2. The obtained peptide sequences can be used to identify a protein of interest by searching them against protein and translated DNA databases. The peptide sequences may also be used to confirm putative cDNA clones or DNA fragments in the expressed sequence tag (EST) database.

3. Protein sequencing provides direct identification of post-translational modifications which are not predictable from the gene sequence. This is especially important in view of ever increasing numbers of recognized post-translational modifications implicated in the biological functions (31).

4. The sequences may be used to characterize recombinant proteins and fragments of natural and recombinant proteins used in structure and function studies.

5. Protein sequencing may be used to characterize recombinant protein pharmaceuticals in order to confirm that the expressed product conforms to the predicted structure and to evaluate that quality and purity of the product are highly consistent.

## II. INITIAL CHARACTERIZATION OF THE PROTEIN

Recombinant protein biopharmaceuticals have to be produced in highly pure state through a series of stringent large scale purification process steps. To meet regulatory criteria, the products must be void of any contaminants that may be derived from different origins. The finished products can be stored at high concentrations as bulk material or formulated into desired formulation buffers. In the product development stage, they are usually subjected to extensive physicochemical characterization required for investigational new drug (IND) and for biological license application (BLA) filing. These analyses can be seen in various chapters in this book. In this section, only a brief description related to protein sequence analysis is cited.

## A. Polypeptide Chain Separation of Heterooligomeric Proteins

Many proteins contain oligomers of non-identical polypeptide chains, and separation of these is the first step in the determination of protein sequence. The chain may be held together by covalent bonds, such as disulfide bonds (e.g., immunoglobulins/monoclonal antibodies) or by noncovalent interactions (e.g., hemoglobin, cytochrome b). The separation may be achieved by various classical chromatographic procedures including gel filtration, ion-exchange chromatography, affinity chromatography, electrophoresis or reverse-phase HPLC under conditions where the interaction between the chains is abolished (2). Some of the conditions may include the use of denaturants or detergents, such as urea, guanidinium chloride, extreme pH or sodium dodecyl sulfate (SDS). Both $\alpha$- and $\beta$-chains of human hemoglobin can be separated by cationic exchange chromatography in 8 M urea (32). Reduction with dithiothreitol or mercaptoethanol followed by alkylation with iodoacetic acid or iodoacetamide is commonly used to break intermolecular disulfide bonds (Sec. II.D). The separation of light and heavy chains of immunoglobulin after reduction and alkylation can be achieved by gel filtration (33) and by high-performance gel permeation chromatography for more rapid and high resolution separation.

## B. Molecular Weight and Protein Concentration Determination

Determination of the molecular weight of a polypeptide is a routine experiment and is required before sequencing work can be planned. Several methods are available to obtain reliable molecular weight measurement, including ultracentrifugation (see Chapter 27), conventional or HPLC gel filtration (see Chapter 11), SDS-polyacrylamide gel electrophoresis (see Chapter 13), and mass spectrometric analysis (see Chapter 16). Advances in mass spectrometric techniques have made possible the direct determination of polypeptide molecular weight. Much modern equipment, including electrospray ionization mass spectrometer, electrospray quadrupole ion trap mass spectrometer, matrix-assisted laser time-of-flight mass spectrometer with delayed extraction, or electrospray quadrupole time-of-flight mass spectrometer, can provide very precise mass measurement for poly-

peptides. Recombinant human stem cell factor is a non-covalently linked dimer in its native state and has a molecular weight of approximately 37 kDa as determined by ultracentrifugation or by HPLC gel filtration (34). The molecule dissociates into a monomeric form with a molecular weight of approximately 18 kDa upon SDS-PAGE analysis. In electrospray ionization mass spectrometric analysis shown in Fig. 1, the native human stem cell factor gives an average $MH^+$ mass of $18,658.5 \pm 2.3$ which is very close to the theoretical mass of 18,657.6.

Any given protein pharmaceutical requires a precise protein concentration measurement as a basis for quantification in recovery process and dosage calculation. Protein sequence analysis work also requires some estimate of protein concentration. A convenient way to determine precise protein concentration is the measurement of protein extinction coefficient. The extinction coefficient may be determined directly by dry weight analysis if the sample is sufficiently pure and does not contain any low molecular weight tight binding materials such as lipids, detergents, and coenzymes (2). A protein solution is dialyzed exhaustively against distilled, deionized water; and the absorbance at 280 nm of an approximately 1 mg/ml dialyzed solution is determined. An aliquot (10 ml or above) is placed in a vial and dried to constant weight at 105C. Precise weight can be obtained without using a microanalytical balance at this sample quantity. A more sensitive method is the use of the amino acid analyzer (for details, see Sec. II.C).

In amino acid composition and sequence analysis, protein concentration is required to estimate the amount needed for the analysis and to quantify the recovery yield after sequencing. Table 2 illustrates a calculation of protein concentration based on two proteins with theoretical molecular weights of 10,000 and 50,000 daltons. The commonly used concen-

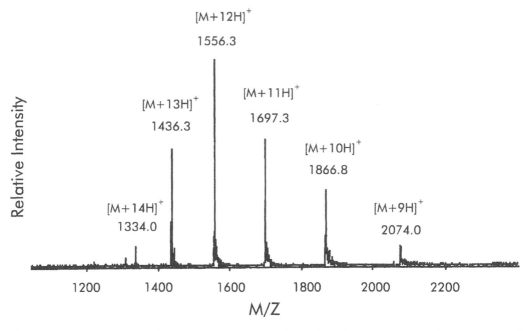

**Fig. 1** Electrospray-ionization mass spectrometry of *E. coli*-derived recombinant human stem cell factor. Molecular ions in various charge states due to multiple protonation are detected and the deconvoluted mass spectrum gives an averaged MH$^+$ mass of $18658.5 \pm 2.3$ (our unpublished data).

tration unit is picomole or nanomole which is related to the weight of a given protein.

## C. Amino Acid Compositional Analysis

Amino acid compositional analysis of a protein or peptide involves two stages: quantitative hydrolysis into amino acids

and the subsequent determination of the released amino acids. An accurate amino acid composition data of a pharmaceutical protein is usually consistent from sample to sample and can be provided as one of the release tests for a GMP (good manufacturing practice) sample. Since the field of amino acid analysis is too large to be covered in detail here, only routinely used methods and the essential points related

**Table 2** Calculation of Protein/Peptide Concentration for Sequence Analysis

|  | Concentration (picomoles) | |
| --- | --- | --- |
| Amount (mg) | Protein mol. weight | |
|  | 10,000 daltons | 50,000 daltons |
| 0.01 | 1 | 0.2 |
| 0.1 | 10 | 2 |
| 1 | 100 | 20 |
| 10 | 1000 | 200 |
| 100 | 10,000 | 2000 |

to primary sequence determination are described here.

## 1. Hydrolysis Methods

The standard hydrolysis procedure employed in routine analysis uses 6 N HCl (freshly distilled-constant boiling HCl) at 110°C for 24 hours in vacuo (35–37). Hydrolysis is usually performed in glass tubes, thoroughly cleaned and free from trace metals and oxidants. When working with acids sealed in glass tubes at high temperatures, precautions should be taken against possible explosions, and eye protection is especially needed. Depending upon the sensitivity of amino acid analyzer, a protein sample amount to be analyzed varies from nmol to pmol ranges. But it is easier to obtain accurate results if larger amounts of protein and preferably more samples are taken for hydrolysis and repetitive analyses performed on the hydrolysates.

When a number of samples are to be hydrolyzed, alternative methods may be preferred. For example, unsealed, individual tubes can be placed together inside a vessel which is itself sealed in vacuo (38, 39). Vapor-phase hydrolysis can be accomplished in a specially designed, compact chamber that is commercially available (40) or may be constructed from thick-walled hard glass with a Teflon tap. In this apparatus, acid is not placed in glass tubes containing protein samples, but rather in the space surrounding the base of the sample tubes. The hydrolysis of the sample is performed with HCl vapor distilled *in situ*. Acid hydrolysis of proteins can also be performed at a much shorter period of time at higher temperature (e.g., 150°C) in pressurized, sealed glass tubes (41). But the method is somewhat dangerous due to the pressures generated in the glass vessel used. Alternatively, an improvement in speed of hydrolysis (in minutes) is achieved by microwave heating in Teflon-sealed Pyrex reusable hydrolysis tubes which contain protein samples and HCl acid (42). Vapor-phase hydrolysis using microwave heating can also be performed in commercially available equipment (43).

The major problem in determining the composition of proteins is the concomitant degradation of certain labile amino acids during complete hydrolysis of peptide bonds. For example, when using the above-mentioned standard hydrolysis method (e.g., the use of 6 N HCl) to hydrolyze peptide bonds, destruction or chemical modification occurs to the amino acids Asn, Gln, Ser, Thr, Tyr, Trp, Met and Cys. Trp is completely destroyed unless a scavenger agent is present or alternative hydrolysis method is used. The amides, asparagine and glutamine, convert into aspartic acid and glutamic acid, respectively. Loss of Thr and Ser is usually around 5 and 10%, respectively, and increases linearly with time of hydrolysis. Low yield of Met may be due to excessive oxidation of the residue. To improve recovery of these labile residues antioxidants can be introduced in the acid during hydrolysis. The introduction of diluted reducing agent such as 0.05% mercaptoethanol into the acid also improves the recovery of these labile amino acids. The yield of Tyr is variable but can be improved in the presence of 0.1% phenol in the acid. Thioglycolic acid can be introduced to improve recovery of Trp as well as other labile amino acids, Ser, Thr, and Met (44). Cystine and cysteine analyses are problematic and inaccurate; they are analyzed after derivatized into oxidized form, cysteic acid, by performic acid oxidation or alkylated form by reduction and iodoacetate alkylation (Sec. II.D) and by 4-vinylpyridine alkylation (45). The use of alternative hydrolysis agents improves dramatically on the quantitative recovery of most amino acids. For example, proteins can be hydrolyzed with 3 M p-toluenesulfonic acid (46) or 4 M methanesulfonic acid (47) containing 0.2%

(w/v) tryptamine. The recovery of all amino acids is satisfactory including ∼90% recovery of Trp. The sulfonic acid derivatives are non-volatile and require neutralization with high concentration of NaOH after hydrolysis.

As a common practice, two methods, standard HCl hydrolysis procedure in the presence of 0.05% mercaptoethanol and 0.1% phenol and 4 M methanesulfonic acid containing 0.2% tryptamine, are used to quantify all amino acids except cysteine or cystine. The latter amino acids are quantified by the analysis of derivatized samples after performic oxidation or alkylation. Performic acid also oxidizes Met into methionine sulfone.

## 2. Automatic Amino Acid Analysis with Post-column Derivatization

Ion-exchange chromatography as developed in the 1950s and subsequently automated (23, 48, 49) has long been the method of choice and standard of comparison for other methods of amino acid analysis. Improvements in cationic exchange resins, the use of narrow-bore columns, more sensitive flow cells, and electronic amplification have made it possible to use the ion-exchange chromatography for amino acid analysis in the subnanomole range (50, 51). The amino acid mixture is separated on columns packed with sulfonated, cross-linked polystyrene cationic exchange resin, using a series of aqueous buffers (generally citrate based) of increasing pH and ionic strength at optimal temperatures. The separated amino acid components in the effluent are detected by mixing with a reagent (usually ninhydrin), allowing reaction to occur in a delay coil, and monitoring by using a flow cell in a colorimeter. The obtained chromatogram is monitored by dual wavelength detection, at 570 nm for amino acids containing primary amine (for most amino acids except proline) and 440 nm for amino

acids containing secondary amine such as proline and hydroxyproline. However, there are drawbacks using ninhydrin as a reagent for detection (2, 49). An alternative reagent such as fluorescamine has been selected (52), but has not been widely used. O-phtalaldehyde (OPA) and 2-mercaptoethanol detection reagent (53, 54) is a standard alternative to ninhydrin, allowing an order of magnitude increase in sensitivity due to the use of fluorescence detection. The major disadvantage of OPA is that secondary amines in proline, hydroxyproline or rarely N-alkylated amino acids are not detected. Detection of proline therefore requires an oxidation step (55).

Modern amino acid analyzers using ninhydrin detection can give very reliable determination at levels of 100 pmol of each amino acid and is still routinely used in many laboratories to obtain accurate compositions of protein drugs. Figure 2 illustrates chromatographic tracing of amino acid analysis of standard amino acids versus an *Escherichia coli*-derived human granulocyte colony-stimulating factor (G-CSF; trade name, Neupogen®) hydrolysate obtained by vapor-phase hydrolysis with 6 N HCl containing 0.05% mercaptoethanol and 0.1% phenol at 110°C for 24 hours. The analysis was performed on a Beckman amino acid analyzer (Model 6300) using ion-exchange chromatography. All amino acids are recovered except Cys and Trp. The standard hydrolysis of Neupogen gives low recovery for Met and no recovery for Cys and Trp. Cys and Met can be quantitatively recovered if they are oxidized to form cysteic acid and methionine sulfone. Carboxymethyl cysteine derivative is also recovered quantitatively, while Trp is recovered in high yield using methanesulfonic acid hydrolysis method. The quantitation for different Neupogen hydrolysates obtained from various sample derivatives and hydrolysis methods is listed in Table 3 (56).

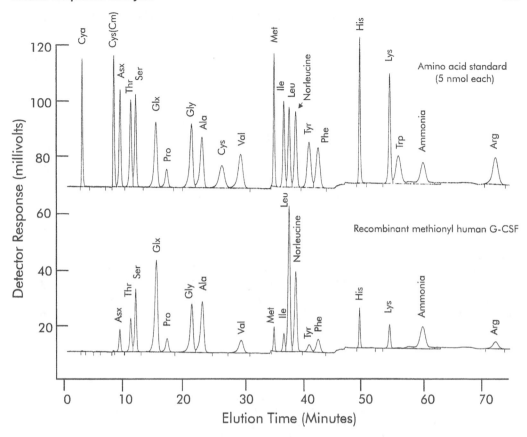

**Fig. 2** Chromatograms of standard amino acid mixture (5 nmol each) and Neupogen (approximately 250 pmol) hydrolysate performed on a Beckman amino acid analyzer (Model 6300) with a Beckman Na-hydrolysate ion-exchange column (4 mm × 12 cm). Total run time is 90 min. Ninhydrin was used as the detection reagent for single column run. The flow rate was 14 ml/hour while the flow rate for post-column ninhydrin reaction was 7 ml/hour. Three Beckman pre-made Na citrate buffers E, F and D, with different pH and ionic strengths, were used for elution. Equal concentration of norleucine used as internal standard was added into the standard and the sample. For conditions of gradient elution see reference manual of the equipment. Our unpublished data was supplied by S Lauren at Amgen Inc.

As described in a previous section (Sec. II.B in this chapter), amino acid analysis is a more sensitive method for extinction coefficient determination of protein. In such analysis, precise amounts of internal standard, norleucine or other stable amino acids not coeluted with any amino acid standard, are added to the protein solution (about 1 mg/ml). The absorbance at 280 nm is determined. Aliquots (0.01-0.1 ml, depending on the sensitivity of the amino acid analyzer) are subjected to acid hydrolysis followed by quantitative amino acid analysis. The concentrations of amino acid residues in the original solution, corrected for the loss during transfer using the value for norleucine, are estimated. The sum of the weights of each amino acid residue is the weight of the protein less any other material, such as carbohydrate or lipid, attached to the molecule, and the absorbance coefficient can be derived using this value.

**Table 3** Amino Acid Composition Analysis of Human Granulocyte Colony Stimulating Factor (Neupogen®)

| Amino acids | Method 1[a] | Method 2[b] | Method 3[c] | Method 4[d] | Theoretical |
|---|---|---|---|---|---|
| Cysteic acid | —[e] | 5.0 | — | — | |
| Carboxymethyl cysteine | — | — | 5.3 | — | |
| Methionine sulfone | — | 3.7 | — | — | |
| Aspartic acid | 4.4 | 4.2 | 4.1 | 4.2 | 4 |
| Threonine | 6.8 | 6.5 | 6.5 | 6.9 | 7 |
| Serine | 12.0 | 12.2 | 11.8 | 12.2 | 14 |
| Glutamic acid | 25.8 | 26.7 | 25.3 | 26.3 | 26 |
| Proline | 14.4 | 13.2 | 12.0 | 13.9 | 13 |
| Glycine | 14.2 | 14.5 | 14.0 | 14.8 | 14 |
| Alanine | 19.4 | 19.7 | 19.4 | 18.9 | 19 |
| Half cystine | ND[f] | — | — | ND | 5 |
| Valine | 7.4 | 7.1 | 7.1 | 6.9 | 7 |
| Methionine | 2.9 | — | 3.0 | 3.6 | 4 |
| Isoleucine | 4.2 | 3.9 | 3.8 | 3.9 | 4 |
| Leucine | 32.7 | 33.8 | 33.1 | 33.4 | 33 |
| Tyrosine | 2.8 | ND | 2.7 | 2.9 | 3 |
| Phenylalanine | 6.4 | 5.6 | 5.9 | 6.0 | 6 |
| Histidine | 5.2 | 4.9 | 5.1 | 5.0 | 5 |
| Lysine | 4.0 | 4.3 | 3.7 | 4.3 | 4 |
| Tryptophan | ND | ND | ND | 1.7 | 2 |
| Arginine | 5.4 | 5.2 | 5.0 | 4.6 | 5 |
| Total | 162.8 | 170.5 | 167.8 | 169.5 | 175 |

[a] Method 1: hydrolysis of Neupogen in 6 M NCl, 0.05% mercaptoethanol and 0.1% phenol.
[b] Method 2: hydrolysis of performic acid-oxidized Neupogen® in conditions identical to Method 1.
[c] Method 3: hydrolysis of carboxymethylated Neupogen® in conditions identical to Method 1.
[d] Method 4: hydrolysis of Neupogen in 4 M methanesulfonic acid containing 0.2% tryptamine.
[e] Not present.
[f] ND: Not determined.

## 3. Pre-column Derivatization Methods for Amino Acid Analysis

As opposed to chromatographic separation of amino acids followed by post-column calorimetric detection, pre-column derivatization of samples containing amino acids or protein hydrolysate following by standard HPLC analysis can be adopted as alternative amino acid analysis methods.

Many reagents have been proposed for the conversion of amino acids to fluorescent, colored (or strong UV absorbing) or volatile derivatives, suitable for subsequent chromatographic analysis. Several reagents have been found to react quantitatively with both primary and secondary amino acids, and generate amino acid derivatives that are stable under the conditions of chromatography. Reverse-phase HPLC has been widely selected for the separation of these derivatives and became the major chromatographic procedure in recent years, while gas chromatographic systems for separation of volatile amino acid derivatives became disfavored. Some reagents were widely used for pre-column derivatization and the obtained derivatives could be separated by reverse-phase HPLC. For example, dansyl chloride yields highly fluorescent derivatives with amino acids that can be detected by a fluorescence detector (57, 58). Dabsyl chloride which is analogous to dansyl chloride yields highly

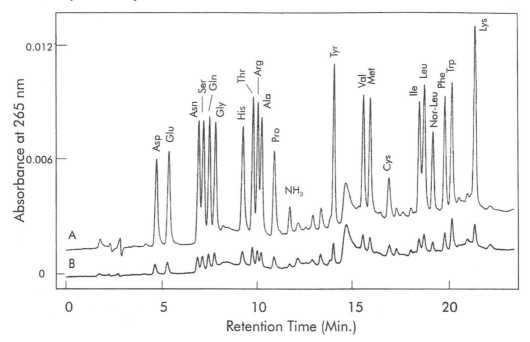

**Fig. 3** HPLC separation of 21 PTC-amino acids using an Altex C18 column (0.2 mm × 25 cm; 5 mm). A: 40 pmol each of standard amino acids including Asn, Gln, Trp, and norleucine internal standard. B: 4 pmol of standard. HPLC was performed in an HP Model 1090 HPLC system with diode-array detection. Data are adapted from Ref. 110 with permission.

intense colored derivatives that can be measured at 436 nm (59). OPA/mercaptoethanol forms fluorescent derivatives with primary amino acids only (60, 61). Phenylisothiocyanate reacts with amino acids to generate UV-absorbing phenylthiocarbamyl (PTC-) amino acid derivatives (40, 62–64). The method has become an important alternative procedure for compositional analysis of proteins. In order to overcome some of the problems resulting from contamination of samples, a fully automated HPLC system incorporating acid hydrolysis, PITC derivatization and microbore reverse-phase HPLC has been available (Applied Biosystems). The other system for the PTC method under the trade name PicoTAG amino acid analysis is commercialized by Waters Associates (62). However, PTC amino acid analysis

can be performed in most of the modern HPLC systems. Technical details of PTC-amino acid analysis is referred to Crabbs et al. (63). Figure 3 shows a typical chromatographic separation of PTC-amino acid standard on a narrow-bore C-18 reverse-phase column (64). Fluorenylmethyl chloroformate (FMOC) which is used extensively in peptide synthesis as a blocking agent yields highly stable and fluorescent fluorenylmethoxycarbonyl amino acids and can be used for amino acid analysis as well (65, 66).

HPLC analysis of pre-column derivatized amino acids usually provides significantly higher sensitivity in detection (low pmol to subpmol levels). Therefore, these methods can be very useful when the quantity of sample is highly limited. These techniques, mainly the PTC method, together with conventional ion-exchange chromatography/post-column detection remain as indispensable tools in protein sequence analysis.

## D. Chemical Derivatization of Proteins for Sequence Analysis

A number of proteins can be completely digested with specific proteases to generate peptide mixtures that are quite soluble in aqueous buffer and can be readily used for peptide separation by HPLC. In these cases, modification of proteins is unnecessary for protein/peptide sequence analysis. Therefore, chemical modification is only performed to prepare specific derivatives of some proteins, especially cysteine and disulfide-containing proteins. Detailed chemical modification methods of proteins was discussed specifically in various book series (67, 68), and many specific modifications of proteins used in protein sequence analysis have been discussed (2).

### 1. Alkylation of Free Cysteines

Cysteine is potentially the most powerful nucleophile in a protein; as a result, it is highly reactive and frequently the easiest to modify with a variety of reagents. Cysteine in many proteins, especially enzymes, is essential and may be involved in substrate binding and catalysis. Review of the literature suggests that dithiobis(2-nitrobenzoic acid) (DTNB, Ellman reagent) has been the most frequently used reagent as the extent of the modification can be easily determined by spectrophotometric measurement (69–71). However, DTNB forms a mixed disulfide with cysteine and is not stable and suitable for protein sequence analysis. Since cysteine residues are easily subject to autooxidation including random formation of disulfide bridges during procedures used for isolation of peptides, stable derivatives have to be made to eliminate formation or exchange of disulfide bonds. Two commonly used reagents include

iodoacetate/iodoacetamide (or related α-haloketo compounds) and 4-vinylpyridine. In the presence of denaturants (6 M guanidinium chloride or 8 M urea) that can completely unfold the protein, iodoacetate (or iodoacetamide) at 2–5 fold molar excess to cysteine reacts with the proteins to form carboxymethyl (or carboxamidomethyl) cysteine derivatives at alkaline pH buffer (pH 8.0–8.5). The pH of the buffer should be kept optimal as His and Lys may be carboxymethylated at lower pH ($<$7.0) and at higher pH ($>$9.5), respectively (72). The excessive reagent can be removed by dialysis or by a simple desalting step using a desalting gel filtration column.

Iodoacetate modification of cysteine introduces an extra negative charge to the protein and may increase solubility of the derivative. The derivatized S-carboxymethyl cysteine can easily be detected during sequence analysis (see subsequent section for sequence analysis). However, iodoacetamide or other compounds may be used instead of iodoacetic acid under the same conditions if introduction of a negative charge at cysteine residues is not desired.

### 2. Reduction and Alkylation of Disulfide Bonds

Cysteines in the polypeptide chain can be oxidized in vivo and in vitro to form disulfide bonds in the protein molecule. As cysteines linked by disulfide bonds are usually distant in the primary structure, the intermolecular and/or intramolecular disulfide formation between them is associated with three-dimensional folding of the polypeptide chain. Unless the determination of disulfide pairing is needed, a number of proteins containing multiple disulfides requires breakage of the disulfide bonds prior to amino acid analysis, proteolytic digestion and peptide separation. Reduction and alkylation is the most frequently used method to break the disulfide bonds. Two techniques, reduction and carboxymethylation (72) as well as reduction and pyridylethylation (73), can be followed. The modification is performed in alkaline pH (250–500 mM Tris-HCl, pH 8.6) under denaturing conditions (6 M guanidinium chloride) (74). In reduction and carboxymethylation, the reducing agent, dithiothreitol, is added to the sample solution (10–500 $\mu$g protein in 50–200 $\mu$l buffer) at 20- to 100-fold molar excess over protein disulfides, which is incubated at 37°C for 2 h under argon. Alkylation of the reduced protein is similar to that described above (see Sec. II.D.1). To prevent carboxymethylation of methionine, an excessive amount of iodoacetate is not recommended (75). For S-pyridylethylation, protein sample (1–10 $\mu$g in 50–200 $\mu$l buffer) is reacted with 2 $\mu$l of 10% mercaptoethanol at 37°C for 2 h under argon. The reduced sample is then reacted with 2 $\mu$l of 4-vinylpyridine at room temperature for 2 h. The prepared protein derivatives are also desalted as described above. Disulfide bonds can also be split by oxidation of proteins with performic acid which converts cysteine and cystine into cysteic acid (76). However, it oxidizes methionine into a sulfone derivative and causes a complete destruction of Trp residue and partial oxidation of several labile amino acids.

## III. STRATEGY AND SAMPLE PREPARATION FOR PROTEIN SEQUENCE ANALYSIS

### A. Strategy

Depending upon the need to analyze pharmaceutical proteins, partial amino acid sequence analysis (providing that the N-terminus is not blocked) can be directly performed on purified protein product to evaluate the N-terminal sequence fidelity as a control lot release test. In the analysis of a particular production GMP (good manufacturing practice) lot to establish a reference lot for future lot-to-lot consistency test, a complete sequence analysis of the lot is required (77). Complete sequence analysis of protein drugs usually follows the classical approach used for many decades, which includes the generation of various sets of internal peptide fragments following by sequence analysis and a complete alignment of the overlapped peptides. Within this exercise, the complete amino acid sequence of a protein drug is established to match the predicted sequence encoded by the genetically engineered gene. Both the N- and C-terminal sequences have to be defined, and correct disulfide bonds and other post-translational modifications, if present, to be determined. This exercise is especially important to the characterization of protein drug reference standard (see Sec. V.A.2). Major or minor heterogeneities are commonly present in the in-process or purified products, and their identification is mandatory. Usually the identification requires exhaustive analysis including the generation and sequence determination of internal peptide fragments associated with the heterogeneity. Moreover, recombinant protein drugs may contain a blocked N-terminus (a number of intracellular proteins in eukaryotic cells have blocked N-terminus) and are refractory to Edman degradation. The sequence data can only be obtained by fragmenting these proteins by chemical or enzymatic digestion and perhaps by mass spectrometry.

### B. Sample Preparation

The key to obtaining protein sequence information is to prepare a sufficient amount of pure material. A variety of approaches toward preparation and handling of a small quantity of protein samples and the subsequent sequence analysis by Edman chemistry, mass spectrometry and computer analysis have been proposed and successfully practiced in many biological samples (78). Such approaches which apply high sensitivity methodologies are very powerful in verifying the identity and novelty of the protein components of interest detected in specific biological systems. Because high sensitivity analysis is feasible, sample preparation or purification protocols must be planned carefully to minimize loss and to eliminate contamination. In contrast to the biological samples isolated in minute quantity with less purity described above, recombinant protein pharmaceuticals are usually prepared in large scale and produced in their highest purity. Sequence analysis can be performed directly from such pharmaceutical grade materials without further sample handling. However, the analysis of pharmaceutical proteins is not limited to the final purified products. In many cases, for sequence analysis of the in-process crude or partially purified products, of products in formulation buffers, and of degraded products after long term storage require similar sample handling and purification steps as those for rare protein components obtained from natural sources. The following describes commonly used procedures for preparation of pure proteins and peptide fragments for sequence analysis of pharmaceutical proteins.

Chromatography and SDS-PAGE are two general approaches for purifying a protein from crude, partially purified, or degraded protein products. The use of HPLC for reverse-phase, gel filtration or ion-exchange separation provides very good resolving power, therefore, little quantity of material ($<1.0$ mg) is needed for the separation. When characterizing a minor form related to a product, more material may be required to isolate enough sample for subsequent analysis. The isolated sample in sequencer-compatible buffer can be directly applied for sequence analysis, otherwise a buffer exchange procedure is used prior to analysis. In many cases, isolated protein components are also subjected to chemical or enzymatic

cleavage to generate internal fragments for sequence analysis. HPLC purification and analysis of peptides and proteins is described in Chapter 11.

Another approach to purify a protein is by SDS gel electrophoresis (79). In many cases, SDS gel electrophoresis can be used to obtain the N-terminal sequence of a single protein component from partially purified samples. The most useful technique is sequencing the protein band excised from a polyvinylidenedifluoride (PVDF) membrane that contains protein bands electrophoretically transferred from the SDS-gel (80). Gel electrophoresis can also be performed in a two-dimensional system yielding a very high resolution of protein separation based on protein sizes and pIs (81). Gel-separated proteins can be subjected to *in-situ* enzymatic digestion in the gels or in the PVDF blot containing electrophoretically transferred protein bands. Advances in these techniques allow the sequence analysis at relatively high sensitivity levels (82–85). Gel electrophoresis of proteins is described in Chapter 13.

## C. Specific Chemical and Enzymatic Cleavages

Internal peptide fragments are generally derived from limited or complete digestion of proteins with specific chemical agents and proteases. Table 4 lists the cleavage agents and their specificity. For applications in proteolytic digestion, proteases with more reliable specificity such as trypsin, endoproteinase Glu-C, endoproteinase Lys-C, and endoproteinase Asp-N are used more frequently than others to generate reproducible digests for HPLC separation. Proteases with less specificity, such as chymotrypsin, pepsin, subtilisin, thermolysin, and papain are occasionally used for isolation of disulfide-containing peptides when proteases with better specificity fail to generate the

**Table 4** Cleavage Sites of Selective Proteolytic Enzymes and Chemical Agents Used in Generation of Protein Internal Peptide Fragments[a]

| Agents | Cleavage site(s)[b] | Source |
|---|---|---|
| *1. Endoproteinases* | | |
| Trypsin (EC 3.4.21.4) | Lys-X | Bovine pancrea |
| | Arg-X | |
| Endoproteinase Lys-C | Lys-X | *Lysobacter enzymogenes* |
| *Achromobacter* protease I | Lys-X | *Achromobacter lyticus* |
| Endoproteinase Glu-C (EC 3.4.21.19) | Glu-X | *Staphylococcus aureus* V8 |
| Endoproteinase Asp-N (EC 3.4.99.30) | X-Asp | *Pseudomonas fragi* |
| Endoproteinase Arg-C | Arg-X | Mouse submaxillaris gland |
| Chymotrypsin | Trp-X; Tyr-X; Phe-X Leu-X; Met-X; etc. | Bovine pancrea |
| Thermolysin | Broad specificty[c] | *Bacillus thermoproteolyticus* |
| Pepsin | Broad specificity[c] | Porcine stomach mucosa |
| Subtilisin | Broad specificity[c] | *Bacillus licheniformis* |
| *2. Chemical agents* | | |
| Cyanogen bromide (CNBr) | Met-X | |
| BNPS-skatole Iodosobenzoic acid | Trp-X[d] | |
| Hydroxylamine | Asn-Gly[d] | |
| Acid | Asp-Pro[d] | |

[a] For detailed references see Ref. 3.
[b] The peptide bond cleaved lies between the amino acid and the "X".
[c] These proteases exhibit broader specificity. For discussion of preferential cleavage sites see Ref. 3.
[d] Only partial cleavage occurs.

desired peptide fragments. CNBr is used most to chemically cleave the Met-X bonds in proteins with satisfactory cleavage efficiency. BNPS-Skatole or iodosobenzoic acid specifically splits Trp-X bonds, however, the cleavage rate is slower than CNBr cleavage. As Met and Trp are present in very low frequency in proteins, these cleavages usually generate larger and sometimes insoluble fragments. Separation of such peptide fragments can be performed using SDS-PAGE, and the peptide band may be transferred onto PVDF membrane for direct sequencing as described in Sec. IV.A.4.

In most cases, peptide fragments in the digests are separated by reverse-phase HPLC methods (referred to as "peptide mapping") with the most frequently used trifluoroacetic acid-acetonitrile elution conditions. The peptides obtained can be directly analyzed for sequence determination. Peptide mapping can be performed in various types of reverse-phase columns (C4, C8, and C18) and provides high sensitivity and high resolution separation with the use of small diameter columns (e.g., 2, 1, and 0.5 mm i.d.). Peptide mapping of proteins can be used to determine the entire primary structure of a protein, as well as providing information such as position of glycosylation sites within the native protein molecule and the position of disulfide linkages or the position of proteolytic processing. Therefore, the method is very important in the quality control of biotechnology products. It is routinely used to compare the peptide profile of a production lot with that of a reference sample (or reference standard, see Sec. V.A.2) to confirm the correct primary protein sequence. The utility of peptide mapping in conjunction with peptide sequence analysis in the verification of protein sequence fidelity is discussed in Sec. V.A.1, and detailed peptide analysis by HPLC is described in Chapter 11.

## IV. SEQUENCE DETERMINATION OF PROTEINS AND PEPTIDES

### A. Amino-terminal Sequence Analysis by Automated Edman Sequencing

#### 1. Chemistry of the Edman Degradation

Proteins/peptides are sequenced by chemical degradation from their N-terminus using Edman reagent, phenylisothiocyanate (PITC). Detailed chemistry and side reactions occurring in the reaction process are referred to in many references (16–18, 86, 87, among others). The reaction or Edman degradation (Fig. 4) is divided into three key steps: coupling, cleavage and conversion. During one cycle of the reaction, the N-terminal residue is removed from the remaining polypeptide chain and identified by reverse-phase HPLC. The shortened polypeptide is left with a free amino terminus that can undergo another cycle of the reaction. Currently, most laboratories perform the Edman degradation using automated sequenators with reasonably good sensitivity (see below for details). The reaction occurs in a reaction device containing a solid support where protein/peptide samples are loaded or bound.

In the coupling reaction, PITC reacts with the free α-amino group of the N-terminal residue to form phenylthiocarbamyl (PTC-) peptide in the presence of base. At pH 8.0–8.5, coupling favors the α-amino group, and the side chain of Lys is less reactive to PITC since the ε-amino group in Lys is protonated in this pH range. Coupling can be partially or completely inhibited when N-terminus modifications like acetylation, formylation and cyclization occur. In the cleavage step, anhydrous acid (usually anhydrous trifluoroacetic acid) is used to cleave the PTC-peptide derivative into two products, an anilinothiazolinone (ATZ-) amino acid and the n-1 polypeptide. The n-1 polypeptide has a reactive N-terminus and can undergo a second cycle of coupling and cleavage, after the ATZ-amino acid is extracted by an organic solvent and transferred to a small flask for the conversion reaction. ATZ-derivatives are not stable, therefore, conversion of the derivatives into a stable phenylthiohydantoin (PTH-) amino acid is necessary. Conversion is a two-step reaction and occurs in aqueous, acidic solutions. The ATZ-amino acid is rapidly hydrolyzed to phenylthiocarbamyl (PTC-) amino acid which then cyclizes to a stable PTH-amino acid.

The degradation reaction shown in Fig. 4 represents a sequencing cycle that results in identification of the N-terminal amino acid present on the protein or peptide at the beginning of that cycle. If each step were 100% efficient, it would be possible to sequence an entire protein in a single sequencing run. However, multiple factors limit the amount of sequence information that can be obtained. In practice, it is feasible to obtain >50–60 residues from a single sequencer run when the amount of the available protein is not limiting. With current technology and sensitivity, it becomes very routine to obtain 20–40 residues of sequence from the N-terminus of proteins and large peptides in the low picomole range (5–50 pmol).

#### 2. Instrumentation

The chemistry as originally described by Edman in 1950 (see Fig. 4) is still in practice today with minor variations in all of

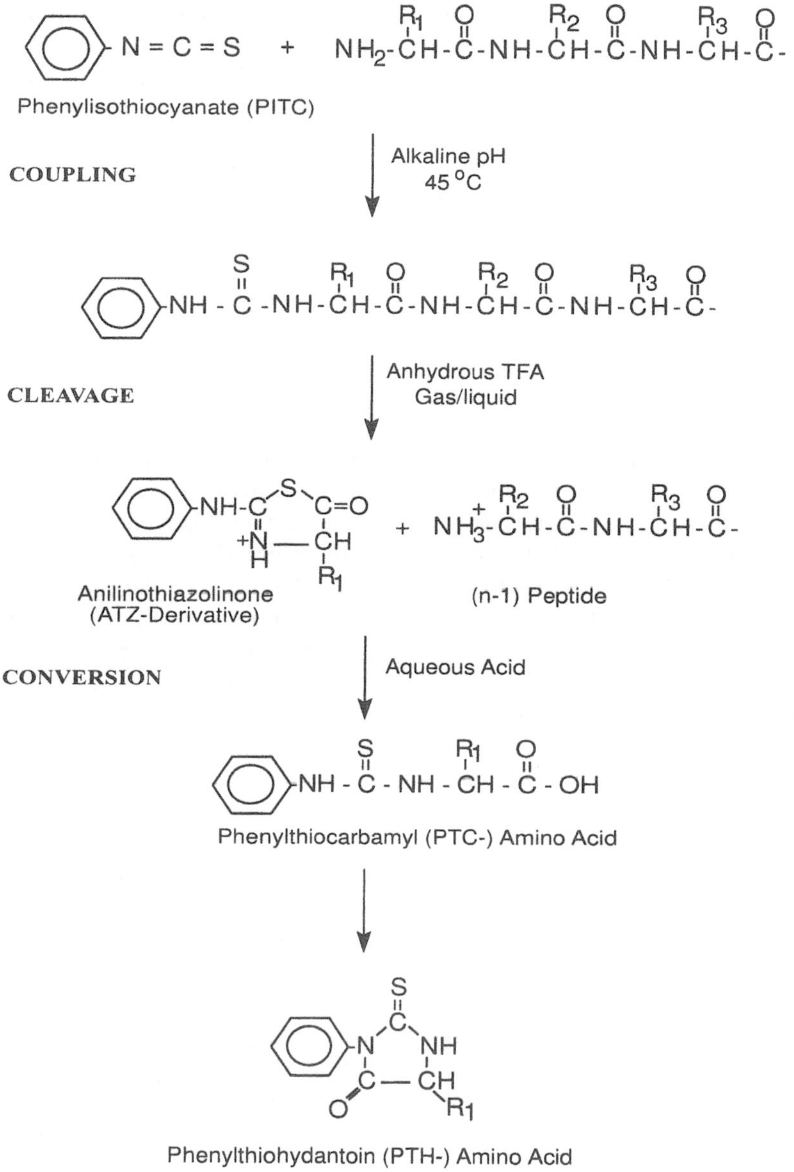

**Fig. 4** Chemistry of the Edman degradation with three key reaction steps, coupling, cleavage, and conversion. In the coupling step PITC reacts with the N-terminal residue of a peptide or protein. Acid cleavage removes the N-terminal residue as an unstable ATZ-derivative and leaves the shortened peptide (n-1) with a reactive N-terminus for next degradation cycle. The ATZ-derivative is converted in the last step to a stable PTH-amino acid for identification.

the commercially available sequenators. The performance routinely expected using today's technology has been primarily achieved through advances in instrumentation. The spinning cup sequenator as originally designed by Edman (25) was capable of sequencing 15 residues in 24 h using a submicromolar amount of protein. Improvement in HPLC detection of PTH amino acids, the use of polybrene carrier, and modification of the spinning cup sequenator allows the sensitivity to be reduced to approximately the 1 nmol level (88–92). The introduction of the gas-phase sequenator (26, 93) together with on-line HPLC detection of the PTH-amino acids using narrow bore reverse-phase columns (2 mm i.d. columns) allows routine sequence analysis at the 10–100 picomole range (94).

Modern, commercially available sequencing instruments require specific sequencing supports for sample loading (see Sec. IV.A.3). Most instruments have been manufactured by Applied Biosystems (Division of Perkin Elmer Inc.), including the first generation equipment Model 470A gas-phase sequencer developed in the early 1980s (26, 93). This instrument has a fraction collector to collect all released PTH amino acids which are manually transferred, dried, and reconstituted into a suitable solvent system for subsequent "off-line" HPLC identification. The latter step actually becomes time-consuming and labor intensive, and sample loss is inevitable due to manual transfer of the released PTH-amino acids for HPLC separation. In the mid 1980's, advances were made possible to adapt an on-line narrow-bore HPLC system for the detection of PTH amino acids (94), which was used in the Perkin Elmer Model 474/477 gas/liquid-phase sequencers. The latest model in this group is the Procise sequencer Model 494, which has four sample cartridges. The availability of multiple sample cartridges allows automated tandem analysis of multiple samples, which substantially increases instrument throughput and flexibility. The second type of sequencer is the Hewlett Packard (HP) sequencer, Model 1000A or 1005A, utilizing a biphasic silica-based sample support with the capability of loading four samples as well. Both types of sequencers contain a modified design of sample cartridges to accommodate sequencing

of proteins bound to a polyvinylidene-difluoride (PVDF) membrane (95). Although there are substantial differences between the two instruments in hardware design and the controlling software, differences in the overall sensitivity and throughput are relatively minor. The adaptation of microbore HPLC for PTH-amino acid detection provides approximately 3-fold enhancement of the limit of detectability (96). Using the Procise Model 494 platform with miniaturized sample cartridges (see below) and a slight modification of hardware and chemistry, a high sensitivity analysis can be routinely performed below 1 pmol detection level in a commercially available sequencer (97).

The two sequencer systems (Perkin Elmer and Hewlett Packard) include integration and coordination of multiple components required for sequence analysis: a computer, a sequencer module, and a dedicated in-line HPLC PTH-amino acid analyzer. The computer controls the overall operation of all components and handling data storage and data analysis from the PTH amino acid analysis. Figure 5 illustrates a schematic of the reagents, solvents and flow paths for a Perkin Elmer Procise model 494 sequencer with four sample cartridges, which highlights the complexity of the instrument and the three major stages of analysis: PITC coupling and cleavage conducted on the sample support (sequencer reactions); conversion of the ATZ amino acid to a stable PTH derivative (conversion reactions); and PTH analysis (HPLC). Each of these three steps takes about the same period of time (30–40 min; about 50 min for high sensitivity sequencer with microbore HPLC). Each of these three stages is performed in parallel: i.e., while the first residue is being separated on HPLC for PTH amino acid identification, the second residue is being converted in the conversion flask, PITC coupling and TFA cleavage of the third residue is occurring

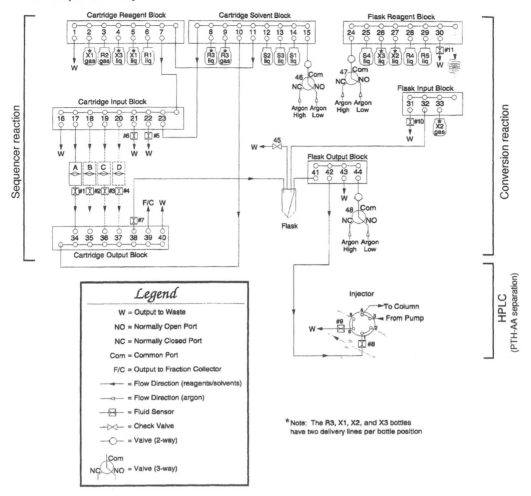

**Fig. 5** Reagent/solvent schematic for Perkin-Elmer Procise 494 sequencer, illustrating the complexity of instrumentation used for automated sequence analysis. Bottles for Edman chemistry involved in the three key reactions (R1, R2, R3, R4, R5, S1, S2, S3, and S4) are detailed in Table 5. Note that trifluoroacetic acid in R3 bottle can be delivered in either the gas phase or as small pulse of liquid reagent (pulse-liquid mode). Additional reagent positions can be used for optional chemistry or solvents in sequence reaction (X1 and X3) and conversion reaction (X2 and X3). Related reagent and solvent valve blocks are in place to regulate the delivery of appropriate amounts of reagents or solvents to the sample cartridges or conversion flask. PTH-amino acid was transferred from the conversion flask to the HPLC column by an HPLC injector. The PTH analyzer is automatically controlled through the sequencer computer during the sequencing process. Adapted with permission from Perkin-Elmer Procise Model 494 User's Manual.

on the sample cartridge (Perkin Elmer) or column (Hewlett Packard). Table 5 lists the reagents and solvents used in these two automated sequencers as well as HPLC solvent systems used for on-line PTH-amino acid analysis. There are other commercial instruments available for N-terminal sequencing (manufactured by Beckman Inc. or Milligen, a division of Waters Associates), but are less frequently used by most protein chemistry laboratories.

**Table 5**  Solvents/Reagents Kit Used for Perkin Elmer Procise (Model 494) and Hewlett Parkard (Model G1000A/G1005A) Automated Sequencer

| Perkin Elmer Procise sequencer | | Hewlett Parkard sequencer | | Usage in Edman chemistry |
|---|---|---|---|---|
| | Sequencer solvents/reagent | | | |
| R1 | 5% PITC in n-heptane | R1 | 5% PITC | Coupling reagent |
| R2B[a] | N-methylpiperidine | R2 | diisopropylethylamine | Coupling buffer |
| R3 | trifluoroacetic acid | R3 | trifluoroacetic acid | Cleavage reagent |
| R4A | 25% trifluoroacetic acid | R4 | 25% trifluoroacetci acid | Conversion reagent |
| R5 | PTH amino acid standard | std | PTH amino acid standard | PTH amino acid identification and quantitation |
| S2B | ethylacetate | S2A | ethylacetate | Washing solvent to remove byproducts |
| S3 | butyl chloride | S2/3 | 23% acetonitrile/77% toluene | Extraction for ATZ amino acid |
| | | S3 | 15% acetonitrile/85% toluene | |
| S4B | 20% acetonitrile | S4 | 10% acetonitrile | Redissolving PTH amino acids for HPLC |
| PTH | analyzer mobile phases | | | |
| | Premix analyzer kit | | | |
| | HPLC solvent A3 (3.5% tetrahydrofuran) | | HPLC solvent A (aqueous triethylamine/acetonitrile) | |
| | HPLC solvent A2 (isopropanol/acetonitrile) | | HPLC solvent B (aqueous triethylamine/propanol) | |
| | Premix buffer concentrate | | HPLC solvent C (acetonitrile) | |

[a] R2 (trimethylamine in methanol) was used previously in early model Perkin Elmer equipments and delivered as a gas-phase basic buffer for coupling reaction. The use of R2 is discontinued due to its discomfort odor.

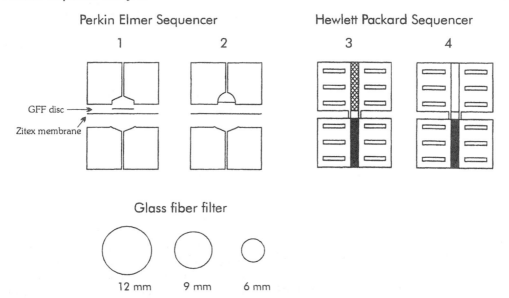

Perkin Elmer Sequencer          Hewlett Packard Sequencer

1          2          3          4

GFF disc →

Zitex membrane

Glass fiber filter

12 mm     9 mm     6 mm

**Fig. 6** Sequencer sample cartridges for Perkin Elmer and Hewlett Packard sequencers. (1). Glass fiber filter sample cartridge that can fits 12 mm, 9 mm, or 6 mm GFF disc depending on size of cartridge; (2). Blott cartridge for Perkin Elmer sequencer; (3). Biphasic reaction cartridge for HP sequencer; and (4). An HP sample cartridge with an empty upper column for samples bound to PVDF.

## 3. Solid Support for Sample Loading

The most commonly used sequencer support is glass fiber filter (GFF). This is used primarily on all Perkin Elmer sequencers (Fig. 6). The sample is loaded onto a polybrene-treated GFF disc, inserted into the top of the reaction cartridge block, and the two halves of the cartridge are sealed with a Zitex (Teflon) membrane. The original design of the GFF disc used in Model 470A or 477 sequencer is 12 mm in width. In Procise equipment, a standard 9 mm GFF is used, while the 6 mm GFF fits into the miniaturized cartridge for the high sensitivity Procise sequencer (Fig. 6). The reduction in total surface area from a twelve mm to a six mm disc also reduces the background level during PTH amino acid analysis. Alternative blot cartridge (top reaction cartridge block) is also used in the Perkin Elmer sequencer to analyze proteins or peptides bound to PVDF membranes, which are inserted into the semicircular slot in the top of the Blott cartridge. A biphasic sequencer reaction cartridge in the Hewlett Packard sequencer is used to analyze samples in solution. The cartridge consists of an upper column unit with a hydrophobic, bonded-phase silica beads, and a lower hydrophilic column unit (98). The sample is loaded on the top half of the column (cross hatched area in Fig. 6) which contains a hydrophobic C18 type of silica beads. After sample loading, the top unit is connected to the bottom half of the unit (solid area shown in Fig. 6). This unique sample support allows the loading of large-volume samples and/or samples contaminated with buffers and reagents that are incompatible with direct sample loading into glass fiber filters. An HP sample cartridge can also be used to analyze samples bound to PVDF membranes. The PVDF strips are inserted into the empty upper column as seen in Fig. 6.

### 4. Sample Loading and Sequence Analysis

The following protocols describe application of proteins or peptides in solution or bound in PVDF membrane to a Perkin Elmer or a Hewlett Packard sequencer. Sample preparation and loading procedures are different depending on the types of sample supports and sequencer used. Great detail in dealing with automatic sequencing and sample handling can be found in Reim and Speicher (99).

A basic Perkin Elmer sequencing system (ABI Technology) includes a sequencer module with glass sample cartridge blocks (a Blott cartridge is used for protein bound to PVDF membrane), an on-line PTH amino acid analyzer (dedicated HPLC system and UV detector) and a computer controller (Macintosh computer). Similar hardware is also required for the Hewlett Packard sequencer, which includes a sequencer module with biphasic sample cartridges, an on-line PTH analyzer (dedicated HPLC and detector) and a computer controller (PC-based computer). A simple sample loading station is used to load liquid sample into the biphasic sample cartridge, which allows an extensive wash of the sample bound on the reverse-phase silica beads for the removal of undesired buffers, salts, or other materials. As listed in Table 5, several solvents and reagents used for the Edman chemistry are similar between the two pieces of equipment. Through dedicated computer control of the on/off valves, reagents and solvents are delivered from an individual solvent/reagent bottle to a sample cartridge or conversion flask for sequencer reactions and conversion reaction (see Fig. 5). The following section describes several basic steps for sample loading with various types of sample cartridges. Detailed precautions and handling are referred to in the user's manual of the instruments for instruction at all steps while many technical tips can be found elsewhere (99).

### a. Analysis of Liquid Samples Loaded on a Glass Fiber Filter (GFF) Disc

Sample preparations. Most recombinant protein therapeutics in liquid formulation buffer can be directly sequenced on various models of Perkin Elmer sequencers (models 477, 491, 492 and 494). If concentration of protein sample is relatively high, it can be aliquoted for dilution with HPLC-grade water before loading. Low concentration of buffers such as NaOAc and ammonium bicarbonate buffers or non-amine reactive salts such as NaCl salt can be tolerated. However, a high concentration of non-volatile buffers can interfere with sequencer performance by altering the actual pHs achieved during the coupling and cleavage steps. Tris buffer and any formulation excipients that react with amine-specific reagents can seriously generate high sequence background. Therefore, prior to sequence analysis certain samples formulated in these buffer conditions require removal of undesired components either by dialysis or by desalting using gel filtration or reverse-phase HPLC. Low concentration samples in larger volumes can be reduced in a Speedvac evaporator. Peptides are usually prepared by 1.0 mm or 2.0 mm i.d. reverse-phase HPLC, and sample aliquots can be directly loaded to the sample cartridge without any pre-treatment.

Precycling of the GFF disc. A TFA-treated GFF disc placed on the top half of the cartridge block is saturated with 9 µL (for 6 mm GFF), 15 µL (for 9 mm GFF) or 30 µL (for 12 mm GFF) of polybrene solution (1 mg/ml in water, containing NaCl). After drying, a Zitex seal is inserted between the two halves of the cartridge which is then assembled and placed onto the sequencer.

Pressure test the cartridge for leaks. Perform 2-3 sequencing cycles using a filter pre-conditioning program.

Sample application and sequence analysis. After precycling, open the block and add 6 to $30\,\mu L$ (depending on filter diameter) of the sample to the GFF. The filter on the block is then dried under argon. Repeat sample application and drying if necessary until the entire sample amount has been applied. Reassemble the cartridge unit back in the sequencer and perform leak tests. Replace all needed reagents, solvents, PTH standard and HPLC mobile phases if the levels of those bottles are low. Program the computer to control an optimized sequencing program using manufacturer's recommendation to perform the sequence run. It is recommended that the quantity of sample loaded be from 100 pmol to as low as several pmol for the Procise model 494 sequencer and 10 pmol or less for the Procise high sensitivity sequencer. Overloading the sample is unnecessary and may cause severe breakdown of sequencing performance or equipment failure.

### b. Analysis of PVDF-bound Samples Using a Blott Cartridge

Preparation of PVDF-bound samples. In many cases, isolation of protein and larger peptide samples for sequence analysis can be better achieved by one-dimensional or two-dimensional gel electrophoresis followed by electroblotting to a high retention PVDF membrane (80, 100–103). This is especially true in analyzing minute quantities of partially purified proteins from a natural source or stability-related degradation products of recombinant protein therapeutics as gel electrophoresis can be a preferred final purification and concentration method. Electrophoresis/electroblotting of samples onto PVDF membrane is a preferred method for sample loading. However, proteins can also be directly loaded onto a PVDF disc in the bottom of a ProSorb device (Applied Biosystems, division of Perkin Elmer).

Sample loading and sequence analysis. The blotted membrane after staining/destaining and cleaning with methanol-water is dried completely in the air and can be stored in a freezer for a long period of time. For sequence analysis, the desired stained protein band is excised from the PVDF membrane using a sharp, clean razor blade on a clean glass plate. Handle the small membrane strip with clean stainless forceps. Insert the membrane(s) in the slot in the top half of the Blott cartridge for the Model 477 or Procise 494 sequencer or on the top half of a regular 6 mm sample cartridge for the high sensitivity Procise sequencer (Fig. 6). After assembling the cartridge unit onto the sequencer, use the procedures of sequence analysis described above. The only difference is the selection of a specific sequencing program for the blotted samples. As the PVDF membrane is highly hydrophobic as compared to the hydrophilic GFF disc described above, sequencing blotted samples requires a specific sequencing program different from that for samples coated on the glass fiber.

### c. Analysis of Liquid Samples on a Biphasic Sequencer

Use of biphasic sample cartridge in the HP sequencer requires quite different sample loading. Since samples are applied onto the hydrophobic part of the cartridge unit, the loading is similar to that involved in the loading of sample to any reverse-phase column. Therefore, a wide range of sample volume can be readily loaded. Most buffers and a large amount of nonvolatile salts can be accommodated as any undesired buffers and salts can be washed out during the loading.

Prior to sample loading, the sample cartridge has to be preconditioned in the sequencer using a column preparation program for one cycle (about 20 min). The next step is to set up the sample loading system by attaching the top half of a pre-conditioned sample cartridge to a washed sample loading funnel which has been washed by a large volume of HPLC-grade water and then inserted into the sample loading station. Wash the sample column by adding 1 ml methanol into the funnel and then applying nitrogen pressure to pass methanol through the sample column until 5–10 $\mu l$ remains in the funnel. Repeat the above step with 1 ml of sample loading solution (2% TFA) to equilibrate the sample column. Aliquot the protein sample and sample solution to make up a 1 ml volume which is then loaded onto the sample column with similar steps as described. After sample loading, the column is washed with 1 ml of sample loading solution. If a high concentration of salt or buffer is present, multiple washes are feasible. At the end, allow the nitrogen to pass through the column for several minutes to ensure complete drying of the column.

Reassemble the hydrophobic and hydrophilic parts of the sample cartridge and insert onto one of the four sample cartridge holders. Perform a sequence run using sequencing programs recommended by the manufacturer.

### d. Analysis of PVDF-bound Samples on a Biphasic Cartridge Sequencer

Preparation of PVDF-bound samples is similar to that described for the Perkin Elmer sequencer (104). PVDF-bound samples sequencing in an HP sequencer are loaded to a reaction cartridge that has an empty top half (i.e., the absence of C18 reverse-phase silica beads) (see Fig. 6).

Precondition and washing of the sample cartridge is identical to that described for the liquid samples. After completion of these steps, the excised PVDF membrane strip is inserted in the top half of the sample cartridge. The reassembled sample cartridge is then inserted into the sequencer for a sequencing run using an appropriate sequencing program.

### 5. Separation of PTH Amino Acids

Successful N-terminal sequencing is dependent on both the sequencer and the on-line HPLC system that identifies all PTH amino acids released from each sequencing cycle. Both systems should operate optimally to obtain maximal sequencing information especially in high sensitivity sequencing of samples in low quantities (low picomole to sub-picomole detection range). The HPLC system requires complete separation of all commonly occurring PTH derivatives and peaks related to sequencer-reagent by-products. Retention times of all peaks should be consistent over the course of a sequencing run.

The HPLC system in the Perkin Elmer sequencer uses dual syringe pumps to deliver the two mobile phases that separate PTH amino acids in a C18 narrow-bore column (2.0 mm i.d. × 25 cm) at a flow rate of 325 $\mu l/min$. Mobile phase A can be easily prepared by mixing a pre-made, diluted aqueous tetrahydrofuran (THF) solution with an aliquot of concentrated buffer Pre-mix (Table 5). A typical separation of standard PTH amino acids at the 4 pmol level is shown in Fig. 7A. This HPLC system is adapted in all Perkin Elmer sequencers except the Procise model 494 high sensitivity sequencer. The latter is equipped with a micro HPLC system with dual syringe micropumps that consistently deliver 40 $\mu l/min$ of mobile phases to separate PTH amino acids in a microbore C18 column (0.8 mm i.d. × 25 cm). Micro HPLC has gained above three-fold of sensitivity as indicated in Fig. 7B. The above HPLC system frequently requires optimization of gradient and solvent components to maximize the resolution of all amino acids (see manufacturer's protocols).

The HPLC system connected to an HP sequencer is a three solvent delivery system (usually an HP liquid

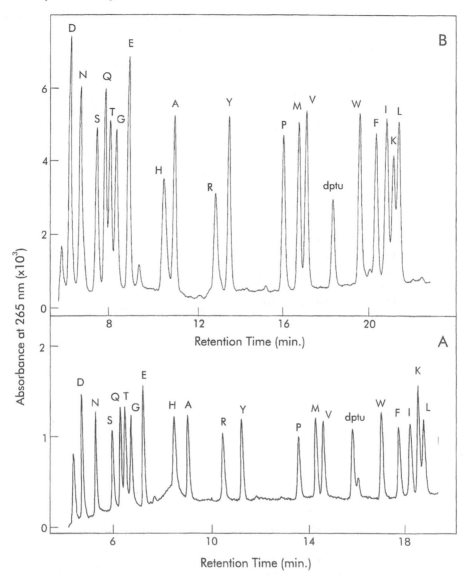

**Fig. 7** Typical separation of standard PTH-amino acids and Edman degradation byproducts for the Perkin-Elmer Procise Model 494 sequencer at 4 pmol level (panel A) and for the Procise high sensitivity sequencer at 1 pmol level (panel B). The high sensitivity sequencer uses a microbore C18 column for the separation which gives an approximately three-fold increase in sensitivity (Ref. 97). The common amino acids are well separated, as are the normal byproducts: dmptu (dimethylphenylthiourea) and dptu (diphenylthiourea). DTT typically present in R4 reagent elutes before aspartic acid. Data are taken from our routine sequence run at Amgen Inc.

chromatograph Model 1090 platform) using preformulated solvents provided by the manufacturer and usually requires little or no adjustment in both gradient and solvent composition.

### 6. Sequence Data Interpretation

After completion of a sequencing run, PTH amino acid separation data from all cycles stored in the computer can be retrieved for sequence assignment. Although assigning some sequences can be straightforward (e.g., from short pure peptide or recombinant protein with enough sample quantity above the 100 pmol level), accurate sequence assignment is complex when interpreting the sequence of larger proteins, proteins/peptides with contaminating or secondary sequences, long protein sequence runs, and analysis of samples with low signals. Perkin Elmer instruments provide a software feature for automatic sequence assignment which is of limited use. Therefore, manual assignment by experienced operators is necessary in most cases. In general, careful manual examination of all tabulated data together with a chromatographic comparison between neighboring cycles will produce the most accurate sequence assignment.

When assigning sequence, several factors that must be considered include the expected recovery of amino acids, the repetitive yield, the initial yield, signal carryover (lag) in subsequent cycles, and increases in background due to nonspecific, acid cleavage of peptide bonds. The expected recovery of PTH amino acids is variable. Most PTH amino acids can be quantitatively recovered, where as others are partially destroyed during sequence analysis or incompletely extracted for subsequent conversion (99). Cysteine is completely destroyed during sequencing unless chemically modified to form a stable derivative such as carboxymethyl cysteine or pyridylethylcysteine (see Sec. II.D). Asn and Gln are partially deamidated and approximately >10% is detected as their respective acid forms. Ser, Thr, Trp, His and Arg are all partially recovered. Ile on a HP sequencer is resolved as two isomers; one isomer migrates with phenylalanine.

The signal observed at each cycle of a sequence relative to that observed in the previous one (usually background corrected) is the repetitive yield. It is easily

calculated from the slope of the best-fit line when the logarithm of the net signal (in pmol) is plotted versus cycle number. A theoretical plot is illustrated in Fig. 8. The sequence signal always steadily decreases as a sequence run progresses due to the inefficiency of the Edman chemistry (slightly less than 100%) and the inevitable loss of sample from sample support (a washout problem). The amount of sequence that can be obtained is closely related to repetitive yield. For example, when an initial yield of 10 pmol and a

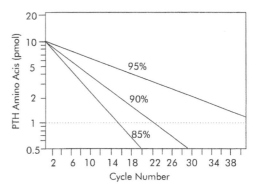

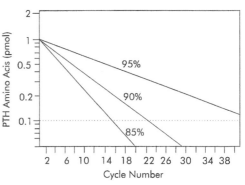

**Fig. 8** Effect of repetitive yield on the number of sequence cycles assigned. Two examples (with an initial yield of 10 and 1 pmol, respectively) of the theoretical PTH-amino acid yield for several different repetitive yields illustrate that the importance of repetitive yield on the amount of sequence that can be determined. Limit of detectability for PTH-amino acids is another factor for obtaining longer sequence assignment in high sensitivity sequencing.

detection threshold of 1 pmol is used, a 95% repetitive yield would allow a sequence assignment of >40 residues, while a 85% repetitive yield only allows assignment of up to 16 residues (Fig. 8). Likewise, the possibility of getting a sequence assignment in a high sensitivity sequence run is also related to initial yield, repetitive yield, and the sensitivity of PTH identification (Fig. 8). Figures 9 and 10 show two typical sequence runs for *E. coli*-derived recombinant human granulocyte-colony stimulating factor (G-CSF) and Chinese hamster ovary cell-derived recombinant human erythropoietin (EPO), respectively. In a Procise 494 sequence run for human G-CSF (50 pmol loaded), an initial yield of 85% and an averaged repetitive yield of 92% were obtained (Fig. 9 and Table 6). In sequencing human EPO (1 pmol loaded) using the high sensitivity Procise sequencer, a 75% initial yield and a 91% repetitive yield were observed. The sequence can be identified at 50–100 femtomole of PTH-amino acid detection at later cycles (15–20 residues). In both sequence runs, the sequence can be readily assigned as the released PTH-amino acids are clearly identified in each cycle with low background.

Carryover (sequence lag), a phenomenon with partial transfer of PTH-amino acid signal into the subsequent cycles can complicate sequence assignment in later cycles. Carryover is derived from incomplete coupling and incomplete cleavage of the N-terminal residue during each sequence cycle. The most obvious example is the significant lag for Pro residues as cleavage at prolines is especially difficult in a sequence run. This is clearly illustrated in Figs 9 and 10. In the sequence analysis of G-CSF, significant lag is detected at cycles four and seven due to the presence of proline at residues three and six, respectively (Fig. 9), and a similar lag is also observed at cycle four due to the presence of two prolines at residues two and three in the analysis of EPO (Fig. 10). Quantitation of sequencing lag can be obtained from Table 6, which lists a computer-calculated data for cycle yields of all amino acids in pmol.

Background, especially glycine and serine, in early sequence cycles can be problematic. Usually background can be derived from sample contamination with free amino acids and small peptides from such sources as airborne contaminants, an ungloved hand, and buffers used in the purification process. In calculating sequence yield, detection of background amino acids may be corrected as illustrated in Table 6.

## B.  C-terminal sequence analysis

Carboxy-terminal (C-terminal) sequence analysis is usually used for direct confirmation of the C-terminal sequence of native and expressed proteins and for detection of protein processing (or post-translational proteolytic cleavages) at the C-terminus. Two approaches, carboxypeptidase digestion and automated chemical sequencing, are desirable to obtain C-terminal sequence information. Alternatively, identification of a C-terminal peptide may be used to obtain C-terminal sequence information if other methods fail. Unlike N-terminal sequence analysis with which significant amount of sequence data can be obtained, C-terminal sequence analysis data provide only several residues of the C-terminal sequence. Since these approaches contain their own inherent difficulty, the C-terminal sequence may not be obtainable for some proteins.

### 1.  Carboxypeptidase Digestion and Amino Acid Analysis

Carboxypeptidases are exopeptidases that remove amino acids, one at a time, from the C-terminus of a protein. After digestion, released amino acids are

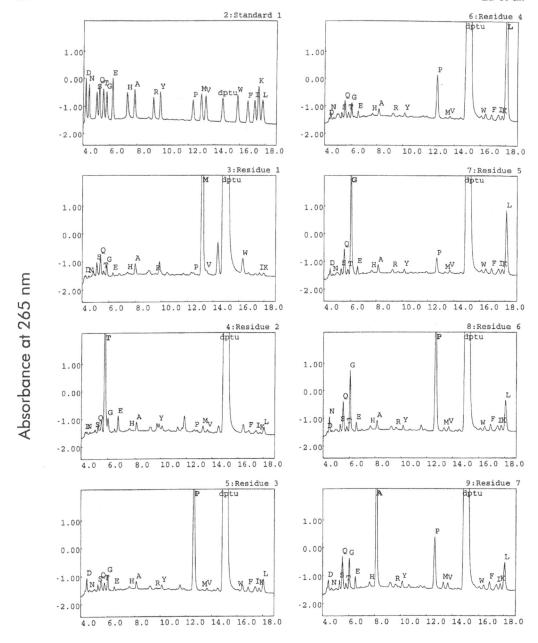

**Fig. 9** A routine N-terminal sequence run of a protein drug, recombinant human methionyl G-CSF, using Procise Model 494 sequencer. A G-CSF sample stored as a purified bulk at 6 mg/ml is serially diluted to 10 mg/ml with HPLC-grade water and a 50 pmol aliquot was spotted onto a precycled polybrene-coated glass fiber disc in a sample cartridge. Fifteen sequence cycles (only seven cycles shown here) are performed to evaluate the N-terminal sequence of the sample which has been stored under −20°C for over five years. The sequence can be easily assigned despite that sequence lag is significantly after Pro residue at residues 3 and 6 (our unpublished data).

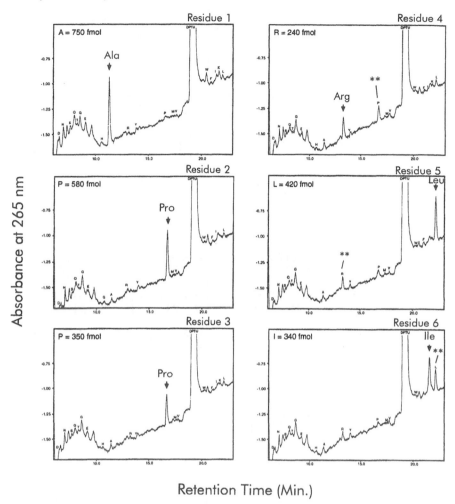

**Fig. 10** Sequence analysis of recombinant human erythropoietin using high sensitivity Procise sequencer. Sample in 1 pmol quantity is loaded to the precycled polybrene-coated glass fiber disc in a miniaturized sample cartridge. Reduction in background signals and the increased detectability of PTH-amino acids using microbore HPLC have made the sequence assignment possible at 0.1 pmol level. Significant lag is also observed after Pro residues. Adapted with permission from Ref. 97.

subjected to amino acid analysis by ion-exchange chromatography coupled with post-column ninhydrin detection or by HPLC after precolumn derivatization (see Sec. II.C). Ideally amino acids are released in order of the position from the C-terminus; thus the C-terminal sequence is deduced from the relative release rate of the observed amino acids over time.

Digestions of proteins with carboxypeptidases A and B (from bovine pancreas) and carboxypeptidase Y (from bakers' yeast) alone or with two enzymes in combination have been mostly cited for polypeptide C-terminal analysis (61, 105–107). Their application in analysis of small peptide sequences was frequently successful. However, it becomes more difficult to cleave some of the higher molecular weight protein substrates by these enzymes, partly due to narrower specificity and lower activity of these enzymes. Car-

**Table 6** PTH Amino Acid Yields from 15-Cycle Sequence Analysis of 50 pmol Recombinant Human Methionyl G-CSF (Neupogen®)[a]

Uncorrected

| Cycle | A | D | E | F | G | H | I | K | L | M | N | P | Q | R | S | T | V | W | Y |
|---|---|---|---|---|---|---|---|---|---|---|---|---|---|---|---|---|---|---|---|
| 1 | 1.45 | 0.63 | 0.30 | 0.00 | 1.70 | 0.27 | 0.45 | 0.29 | 0.00 | **42.21** | 0.60 | 0.44 | 2.54 | 0.13 | 2.27[c] | 0.87 | 0.65 | 0.30 | 0.00 |
| 2 | 1.18 | 0.51 | 1.50 | 0.55 | 1.98 | 0.29 | 0.51 | 0.17 | 1.25 | 0.95 | 0.40 | 0.45 | 1.03 | 0.00 | 0.39 | **24.46** | 0.48 | 0.00 | 1.04 |
| 3 | 1.02 | 1.38 | 0.40 | 0.78 | 2.17 | 0.32 | 0.97 | 0.48 | 2.30 | 0.17 | 0.23 | **29.96** | 1.20 | 0.16 | 0.94 | 1.01 | 0.38 | 0.11 | 0.64 |
| 4 | 1.02 | 0.46 | 0.59 | 0.94 | 2.20 | 0.34 | 0.93 | 0.49 | **32.01** | 0.08 | 1.03 | 8.52 | 2.15 | 0.59 | 1.07 | 0.83 | 0.35 | 0.55 | 0.55 |
| 5 | 1.09 | 0.96 | 0.76 | 0.94 | **25.41** | 0.39 | 0.92 | 0.53 | 11.05 | 0.22 | 0.48 | 2.98 | 2.99 | 0.60 | 1.14 | 0.77 | 0.51 | 0.50 | 0.57 |
| 6 | 1.28 | 0.44 | 0.86 | 1.20 | 8.51 | 0.59 | 0.90 | 0.61 | 5.34 | 0.53 | 1.96 | **27.54** | 3.56 | 0.47 | 1.30 | 0.77 | 0.73 | 0.40 | 0.72 |
| 7 | **30.84** | 1.12 | 1.15 | 1.45 | 4.10 | 0.67 | 0.93 | 0.67 | 4.75 | 1.00 | 0.29 | 10.09 | 3.57 | 0.38 | 1.21 | 0.65 | 0.96 | 0.33 | 0.91 |
| 8 | **10.91** | 1.25 | 1.79 | 1.65 | 3.15 | 0.85 | 0.95 | 0.66 | 4.81 | 0.94 | 0.29 | 3.61 | 3.46 | 3.31 | **10.79** | 0.69 | 0.88 | 0.39 | 1.01 |
| 9 | 4.76 | 1.45 | 2.26 | 1.94 | 3.34 | 0.86 | 0.94 | 0.84 | 5.46 | 0.88 | 0.48 | 2.10 | 3.86 | 4.13 | **12.83** | 0.80 | 1.30 | 0.48 | 1.02 |
| 10 | 3.53 | 1.45 | 2.27 | 2.03 | 3.63 | 0.06 | 0.96 | 0.94 | **25.07** | 0.90 | 0.52 | 1.68 | 3.91 | 1.95 | 5.34 | 0.81 | 1.70 | 0.51 | 1.11 |
| 11 | 3.30 | 1.89 | 2.18 | 1.91 | 3.91 | 0.71 | 0.92 | 0.93 | **13.00** | 1.08 | 0.52 | **16.91** | 3.93 | 0.88 | 2.57 | 0.90 | 1.84 | 0.55 | 1.20 |
| 12 | 3.70 | 1.83 | 2.3 | 2.36 | 4.45 | 0.08 | 1.02 | 1.17 | 8.82 | 1.57 | 0.62 | **9.31** | **18.66** | 1.17 | 2.00 | 1.28 | 2.30 | 0.50 | 1.39 |
| 13 | 4.40 | 1.71 | 4.30[b] | 2.43 | 4.33 | 0.07 | 1.31 | 1.17 | 8.10 | 1.54 | 0.58 | 4.40 | **11.44** | 2.35 | **7.24** | 1.15 | 2.39 | 0.56 | 1.42 |
| 14 | 4.66 | 0.56 | 4.33 | **16.00** | 4.49 | 0.84 | 1.41 | 1.17 | 7.90 | 1.45 | 1.82 | 2.83 | 6.54 | 1.49 | 4.29 | 1.19 | 2.44 | 0.61 | 1.46 |
| 15 | 4.82 | 0.55 | 2.65 | **9.31** | 4.85 | 0.97 | 1.35 | 1.26 | **19.60** | 1.36 | 1.85 | 2.36 | 5.30 | 1.44 | 2.61 | 1.24 | 2.55 | 0.61 | 1.32 |

Background corrected

| Cycle | A | D | E | F | G | H | I | K | L | M | N | P | Q | R | S | T | V | W | Y |
|---|---|---|---|---|---|---|---|---|---|---|---|---|---|---|---|---|---|---|---|
| 1 | 1.12 | 0.00 | 0.00 | 0.00 | 0.04 | 0.00 | 0.00 | 0.05 | 0.00 | **42.30** | 0.22 | 0.00 | 0.27 | 0.08 | 1.62 | 0.26 | 0.58 | 0.00 | 0.00 |
| 2 | 0.52 | 0.00 | 1.00 | 0.01 | 0.10 | 0.00 | 0.00 | 0.00 | 0.00 | 0.92 | 0.00 | 0.00 | 0.00 | 0.00 | 0.00 | **23.82** | 0.23 | 0.00 | 0.65 |
| 3 | 0.04 | 0.57 | 0.00 | 0.06 | 0.06 | 0.00 | 0.27 | 0.09 | 0.23 | 0.01 | 0.00 | **29.10** | 0.00 | 0.25 | 0.02 | 0.32 | 0.00 | 0.00 | 0.16 |
| 4 | 0.00 | 0.00 | 0.26 | 0.04 | 0.00 | 0.00 | 0.18 | 0.03 | **29.31** | 0.00 | 0.59 | 7.45 | 0.00 | 0.16 | 0.00 | 0.10 | 0.00 | 0.20 | 0.00 |
| 5 | 0.00 | 0.00 | 0.10 | 0.00 | **22.85** | 0.00 | 0.12 | 0.00 | 7.73 | 0.00 | 0.02 | 1.71 | 0.38 | 0.00 | 0.00 | 0.00 | 0.00 | 0.13 | 0.00 |
| 6 | 0.00 | 0.00 | 0.60 | 0.00 | 5.73 | 0.11 | 0.04 | 0.00 | 1.40 | 0.35 | 1.48 | **26.06** | 0.21 | 0.00 | 0.00 | 0.00 | 0.00 | 0.00 | 0.07 |
| 7 | **28.56** | 0.00 | 0.52 | 0.02 | 1.09 | 0.14 | 0.02 | 0.00 | 0.19 | 0.16 | 0.00 | 8.40 | 0.00 | 2.58 | 0.00 | 0.00 | 0.00 | 0.07 | 0.08 |
| 8 | **8.31** | 0.00 | 0.00 | 0.13 | 0.00 | 0.28 | 0.00 | 0.00 | 0.00 | 0.00 | 0.00 | 1.71 | 0.14 | 3.30 | **9.21** | 0.00 | 0.00 | 0.08 | 0.08 |
| 9 | 1.84 | 0.11 | 0.00 | 0.03 | 0.00 | 0.25 | 0.00 | 0.02 | 0.00 | 0.00 | 0.00 | 0.00 | 0.01 | 1.02 | **11.12** | 0.00 | 0.00 | 0.01 | 0.00 |
| 10 | 0.29 | 0.02 | 0.10 | 0.00 | 0.00 | 0.01 | 0.00 | 0.04 | **18.64** | 0.00 | 0.00 | 0.00 | 0.00 | 0.00 | 3.50 | 0.00 | 0.06 | 0.01 | 0.00 |
| 11 | 0.00 | 0.38 | 0.10 | 0.00 | 0.00 | 0.00 | 0.00 | 0.00 | 5.95 | 0.00 | 0.00 | **14.39** | 0.00 | 0.04 | 0.60 | 0.00 | 0.01 | 0.03 | 0.00 |
| 12 | 0.00 | 0.22 | 0.20 | 0.00 | 0.31 | 0.00 | 0.00 | 0.13 | 1.15 | 0.29 | 0.03 | 6.59 | **14.40** | 0.04 | 0.00 | 0.21 | 0.30 | 0.00 | 0.10 |
| 13 | 0.18 | 0.02 | 2.20 | 0.00 | 0.00 | 0.00 | 0.07 | 0.06 | 0.00 | 0.14 | 0.00 | 1.47 | 7.00 | 1.13 | **5.01** | 0.04 | 0.21 | 0.04 | 0.04 |
| 14 | 0.12 | 0.00 | 2.10 | **13.28** | 0.00 | 0.01 | 0.12 | 0.00 | 0.00 | 0.00 | 1.20 | 0.00 | 1.92 | 0.17 | 1.92 | 0.04 | 0.09 | 0.01 | 0.00 |
| 15 | 0.00 | 0.00 | 0.00 | **6.41** | 0.04 | 0.09 | 0.00 | 0.00 | **10.06** | 0.00 | 1.20 | 0.00 | 0.49 | 0.03 | 0.11 | 0.05 | 0.02 | 0.00 | 0.00 |

[a] The assigned sequence is indicated by underlines in boldface. The N-terminal sequence of Neupogen is Met-Thr-Pro-Leu-Gly-Pro-Ala-Ser-Ser-Leu-Pro-Gln-Ser-Phe-Leu-...

[b] Glu value has increased due to partial deamidation of Gln in cycle.

[c] Ser contamination is common in cycle 1.

boxypeptidase P from *Penicillium janthinellum* can be purified with no contamination of other proteases, and the observed higher activity and broader specificity for various C-terminal amino acids, including those resistant to carboxypeptidase A, B or Y, has gained more usage in determination of the C-terminal sequence of proteins (108–110). The method in conjunction with amino acid analysis (see Sec. II.C and below) provides a fast analysis of the C-terminal sequence of proteins. Application of digestion using other carboxypeptidases is detailed elsewhere (111). Mass spectrometric analysis of carboxypeptidase-digested peptides ("ladder" generating technique) can also provide C-terminal sequence information (112).

### a.   C-terminal Sequence Analysis by Carboxypeptidase P Digestion (110)

Protein substrates (1–4 nmol) were concentrated by centrifugation using a Centricon-10 (Amicon) microconcentrator with a 10-kilodalton (kD) molecular weight cut-off membrane, and buffer salts were removed by gel filtration over a Pharmacia PD-10 column equilibrated in 10 mM sodium acetate, pH 4.0 or distilled water. In the latter condition, the pH of the desalted sample solution was adjusted to 4.0 by diluted hydrochloric acid. Before adding enzyme, a concentrated solution of Brij-35 was added to the protein sample to a final concentration of 0.05% (w/v). The carboxypeptidase P digestion conditions varied according to the rate of amino acid release obtained from digestion of different substrates. The normal digestion was performed using an enzyme-to-substrate ratio of 1:400 (w/w) at 25°C in 1 h. When such digestion conditions generated a rapid release rate, the reaction could be slowed down by incubating the sample at 4°C. For samples which were difficult to digest, an enzyme-to-substrate ratio of

1:100 for digestion at 37°C in several hours was used. During digestion, samples at various time points including zero time were taken and immediately acidified with trifluoroacetic acid at a final concentration of 10% to stop the digestion. Acidified samples were dried *in vacuo* by a Speedvac centrifuge and subjected to PITC derivatization and subsequent HPLC for PTC amino acid analysis as described above. Norleucine (1–2 nmol) was usually added to sample solutions prior to digestion as an internal standard to calibrate recovery of amino acids.

Separation of the derivatized PTC-amino acids was then carried out by C18 reverse-phase HPLC (2.0 mm i.d. × 25 cm) using a gradient of solvent A (sodium acetate buffer) into solvent B (sodium acetate buffer-acetonitrile-methanol) (see Fig. 3). The pmol amounts of amino acids released at various times are then quantified by calibration with a norleucine internal standard and can be tabulated or plotted versus digestion time. Figure 11 shows a plot obtained from carboxypeptidase P digestion of Chinese hamster ovary cell-derived recombinant human erythropoietin (Epogen®). The digestion was performed in 10 mM sodium acetate, pH 4.0 with an enzyme-to-substrate ratio of 1:100 at 37°C over 60 min. Based on the kinetic release of four amino acids in the order of Asp, Gly, Thr and Arg, the C-terminal sequence of erythropoietin (expected C-terminal sequence:   —Glu-Ala-Cys-Arg-Thr-Gly —Asp-Arg-COOH) was determined to be—Arg-Thr-Gly-Asp-COOH. The determined C-terminal sequence indicates that the arginyl residue predicted to be at the C-terminus according to the gene sequence (113) is missing from the recombinant protein. The digestion is much slower after the fourth residue, because there is a resistant Cys-X bond in the N-side of the C-terminus. C-terminal sequence analysis using this approach can also be successfully performed with other proteins includ-

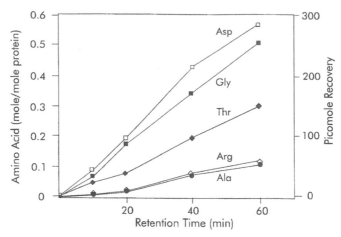

**Fig. 11** Kinetic release of free amino acids after carboxypeptidase P digestion of recombinant human erythropoietin. Sample (2 nmol aliquot in 500 $\mu$l 10 mM sodium acetate buffer, pH 4.0) is digested with carboxypeptidase P at 37°C with an enzyme-to-substrate ratio of 1:100 for 1 h. Aliquot of 100 $\mu$l is taken at 10 min interval for PTC amino acid analysis. Data are adapted with permission from Ref. 110.

ing many recombinant therapeutic proteins such as interleukins, granulocyte colony-stimulating factor, granulocyte-macrophage colony stimulating factor, and interferons, etc. (110).

### b. Limitation of Carboxypeptidase Digestion

A number of peptide bonds are either partially or completely resistant to carboxypeptidase digestion, and the rates of release are strongly influenced by the nature of side chains of both terminal and penultimate amino acids as well as by the general protein structure (111). Depending on the location of these resistant bonds or core structure, in most unfavorable situations there would be no kinetic release of amino acids for sequence determination. For example, Gly-X and Pro-X bonds are usually resistant to carboxypeptidase P digestion (110). Presence of a sequence such as a Cys-X bond, in which Cys is involved in disulfide pairing, frequently slows down the digestion rate; therefore, reduction-and-alkylation of the disulfide bonds is necessary for extended C-terminal analysis.

### 2. Automatic C-terminal Sequencing

Several automated methods for C-terminal sequence analysis of peptides and proteins have been described during the last few years (10, 114). However, automated sequence analysis has not been routinely performed by the protein sequencing laboratories as all the developments involved the use of specialized instrumentation and/or reagents that were not commercially available until recently (114–118).

Hewlett Packard has commercialized a C-terminal sequencing instrument (119), based on the chemistry developed by Bailey et al. (117). The overall chemistry involves two principal chemical events, the coupling reaction and cleavage. The coupling reaction of a sequencing cycle begins with activation of the C-terminal carboxylic acid of the protein with trifluoroacetic acid; this acid treatment promotes the coupling reaction between the C-terminus and the coupling reagent, diphenyl phosphorisothio-cyanatidate (DPP-ITC). Following a pyridine treatment that accelerates the reaction, a stable product of peptidyl thiohydantoin, which bears the chemically modified C-terminal amino acid residue, is derived. The peptidyl thiohydantoin is cleaved with a strong nucleophile, potassium trimethylsilanolate, to release C-terminal thiohydantoin amino acid derivative for HPLC analysis and the shortened peptide for the next cycle of chemical sequencing. Thiohydantoin amino acid has a UV absorption property similar to PTH-amino acids formed during Edman degradation and all 20 amino acid derivatives can be identified by reverse-phase HPLC separation using UV detection (269 nm). Perkin Elmer (ABI Biotechnology) also developed a different C-terminal sequencing chemistry (118) with an automatic prototype sequencer based on the platform of Model 477A N-terminal sequencer (120), now being adapted to the Procise platform. The method generates alkylated thiohydantoin derivatives for HPLC analysis; and the sequence reaction stops when there is a proline present at the C-terminus. Figure 12 illustrates a C-terminal sequencing run of recombinant human interleukin 6 (C-terminal sequence of—Leu-Tyr-Ala-Leu-Tyr-Asn-Met-COOH) directly spotted on the glass fiber disc. Five amino acid residues can be sequenced from the C-terminus of the protein at 1 nmol level. Currently, both automatic procedures can typically sequence 50 pmol-2 nmol of protein samples using various types of sample supports. As the repetitive yield is very low relative to N-terminal sequence analysis, the C-terminal sequence analysis only provides a sequence of 3–5 amino acid residues from the C-terminus. Therefore, the automatic equipment may require further improvement before it gains wider acceptance as a routine method.

### 3. Identification of C-terminal Peptide

Another indirect approach for the analysis of protein C-terminal sequence is the identification of C-terminal peptide. The method takes advantage of high sensitivity HPLC mapping procedure which separates peptides from a protein digest derived from a specific proteolysis. Going through with an extensive sequence and structural characterization of the isolated peptides, C-terminal peptide can be identified and sequenced (by Edman sequencing or mass spectrometry) to verify the location of the actual C-terminal amino acid residue (see Chapter 11 for HPLC analysis of peptides and proteins).

### C. Sequence Analysis by Mass Spectrometry

Mass spectrometry is a powerful analytical technique for forming gas-phase ions from intact, neutral molecules and subsequently determining their molecular masses with great mass accuracy. A variety of high sensitivity mass spectrometers are now available for molecular mass determination of peptides and proteins, which include quadrupole electrospray ionization mass spectrometer (ESI-MS), quadrupole ion-trap mass spectrometer (QIT-MS), matrix-assisted laser desorption mass spectrometer with delayed extraction (MALDI-MS/DE), and quadrupole time-of-flight mass spectrometer (Q-TOF), etc. (see Chapter 16 for details). Accurate molecular mass measurement provides a useful measure of the integrity, purity, and overall state of modification of a peptide or protein. For example, mass spectrometry provides molecular mass information independent of any modifications to structure (e.g., post-translational modifications such as glycosylation, phosphorylation, and sulfation). Mass spectrometry also has the ability to provide amino acid sequence in-

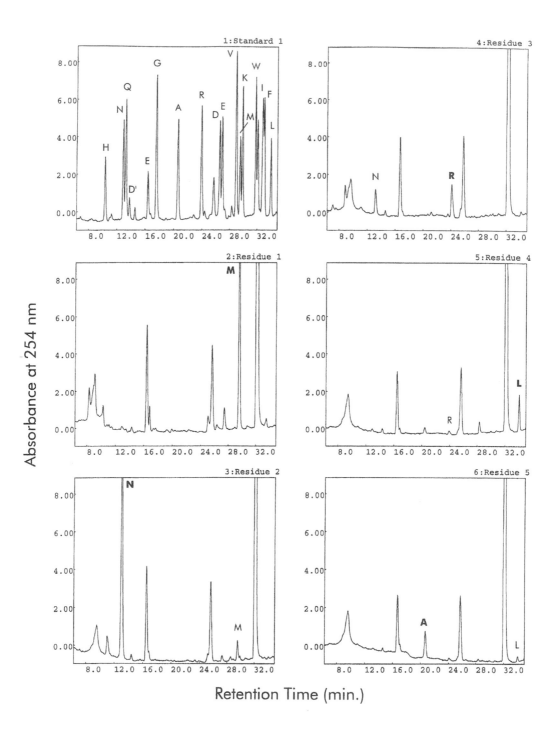

**Fig. 12** Automatic C-terminal sequence analysis of recombinant human interleukin 6. An aliquot of sample in 1 nmol quantity was loaded onto the glass fiber disc and sequenced with a C-terminal sequencer using the platform of Perkin-Elmer protein sequencer Model 477 (our unpublished data).

formation of peptides (usually <25 amino acids) using methodology generically referred to as tandem mass spectrometry or MS/MS. Using MS/MS, complete or partial sequence of peptides can be obtained at the femtomole to picomole levels. Many specific applications of mass spectrometric analysis are valuable in defining sites of post-translational modifications, assignment of disulfide bonds, microheterogeneity of proteins, and carbohydrate structure. More comprehensive discussion of mass spectrometry is described in Chapter 21 together with the references cited.

## V.  UTILITY OF PROTEIN SEQUENCE ANALYSIS IN THE CHARACTERIZATION OF PROTEIN DRUGS

Strict guidelines for validating the identity and purity of proteins for therapeutic use are in place at the U.S. Food and Drug Administration (77, 121, 122). Therefore, the production of pharmaceutical-grade proteins using recombinant DNA technology necessitates a commitment on the part of a manufacturer to analyze and critically characterize these molecules (123). As discussed throughout the chapters in this book, direct physicochemical analysis of the recombinant protein is an absolute necessity in the evaluation of purity, chemical identity, molecular conformation, high order structure and biological potency. Protein sequence analysis described in this chapter offers chemical characterization of recombinant protein for verification of molecular identity and purity. This section describes the utility of protein sequence analysis with illustration of examples to provide critical information relating to the translational fidelity of recombinant protein, the characterization of protein drug reference standard, the determi-

nation of disulfide bonds, and the analysis of post-translational modification.

## A.  Verification of Structural Identity for Recombinant Protein Drugs

### 1.  Sequence Fidelity

#### a.  N-terminal Sequence

After establishing the purity of the recombinant protein drug, it is necessary to confirm that it has the same primary sequence as predicted from the coding cDNA. The most frequent evaluation method is the amino terminal sequence analysis, which is performed to identify the start of the mature protein and to evaluate any potential N-terminal heterogeneity derived from incomplete processing. Usually 15 cycles of automatic Edman degradation of the intact protein is routinely performed for each production lot (preferentially purified bulk material) to establish the identity test. The analytical results usually include a called sequence, initial yield, repetitive yield, and the detection of minor sequences or modified amino acid, if present. Fig. 9 and Table 6 illustrate such an analysis on a purified bulk of recombinant human methionyl G-CSF (trade name: Neupogen®) (124). A single sequence starting with Met followed by the predicted G-CSF sequence can be called, together with an initial yield of 85% and an averaged, repetitive yield of 92%.

Routine N-terminal sequence analysis of the intact protein is a very useful method to evaluate N-terminal heterogeneities of recombinant protein associated with the efficiency of N-terminal processing during or after protein synthesis in an expression system. For example, in bacterial host systems many recombinant genes are constructed with an initiation codon (AUG) one residue upstream from the mature cDNA for the protein to be expressed (referred to as direct expression). However, the degree to which the formyl methionine (f-Met) is removed and the amount of f-Met or Met plus species remaining in the final product appear to vary with the identity of the recombinant protein, the plasmid/host system, and the method of purification. In mammalian or yeast expression, the expressed product is usually secreted into the surrounding medium; evaluation of the N-terminal sequence is therefore important to ensure correct processing of any leader peptide which must be removed to yield the mature protein. Table 7 lists selected examples of recombinant proteins that contain a single or heterogeneous N-terminal sequence(s) in different expression conditions (103, 125–130).

Typically, N-terminal sequence analysis together with other chromatographic and electrophoretic separations would gain sufficient insight for the nature of N-terminal heterogeneity relating to the processing efficiency. However, in the sample that contains N-terminally blocked species, more extended analyses such as electrophoresis, ion-exchange separation, peptide mapping, N-terminal sequence analysis, and mass spectrometry are required. These techniques in combination can be used to identify a minor formyl methionine form in process-related G-CSF preparations (126) and an acetylated des-Met form in the final purified product of consensus interferon-α (130). With the characterization of N-terminally blocked f-Met G-CSF species, a process modification allows this minor species to be completely removed chromatographically from the major methionyl G-CSF form in the final product.

b.  *C-terminal Sequence*

Confirmation of the C-terminal sequence of a recombinant protein is equally important. Individual method or multiple techniques in combination, as discussed in Sec. IV.B, can be used. Carboxypeptidase digestion followed by amino acid analysis is frequently used in a manual analysis. When the C-terminal sequence information is unobtainable from carboxypeptidase analysis, the isolation of C-terminal peptide through peptide map analysis and the identification of the peptide by Edman sequence analysis and mass spectrometry can be performed. Automatic C-terminal sequencing method developed recently can also be applied for the C-terminal sequence verification while the instrument is available.

Most recombinant proteins are produced with properly processed C-terminus (e.g., the observed C-terminus can be predicted from the location of the termination codon in the cDNA sequence). A routine check of the C-terminal sequence of a recombinant protein is to verify that this is indeed the case. However, careful examination should be considered on the occurrence of potential C-terminal processing. Examples discussed below demonstrate that processing occasionally occurs in some recombinant and naturally occurring proteins. For example, recombinant human erythropoietin (trade name: Epogen®) is expressed in Chinese hamster ovary cells as a highly glycosylated protein. C-terminal sequence analysis proved that the mature protein has a loss of a C-terminal Arg predicted from its cDNA sequence (see Fig. 11 for C-terminal sequence analysis and Ref. 110), probably due to cellular processing by carboxypeptidase B-like protease. The polypeptide chain length is 165 instead of 166 amino acid residues. The identification of C-terminal peptide further verifies that the predicted C-terminal Arg is completely removed from the erythropoietin polypeptide chain (our unpublished data).

Important C-terminal processing has been found in several important

**Table 7**  N-terminal Processing of Several Recombinant Human Proteins in Three Expression Conditions

| Recombinant products | Expected sequence | Expression systems used[a] | | | References |
|---|---|---|---|---|---|
| | | Mammalian cells | S. cerevisiae | E. coli | |
| Growth hormone | F-P-T-I — | — | — | Met uncleaved | 125 |
| GM-CSF[b] | A-P-A-R — | — | — | Incomplete processing (Three sequences[b]) | 103 |
| G-CSF | T-P-L-G — | Correctly processed | Correctly processed | Met uncleaved; f-Met present[f] | 126 |
| IL-2 | A-P-T-S — | — | — | Incomplete processing (Two sequences[c]) | 103 |
| Erythropoietin | A-P-P-R — | Correctly processed | — | Incomplete processing (Two sequences[d]) | Footnote d |
| Kallikrein | I-V-G-G — | Propeptide[e] uncleaved | Correctly processed | Met uncleaved | 127, 128 |
| Interferon-α (consensus) | C-D-L-P — | — | Correctly processed | Incomplete processing (Three sequences[g]) | 129, 130 |

[a] Three expression systems are used: expression in mammalian (Chinese hamster ovary) cells, yeast (S. cerevisiae) cells, and bacterial direct expression in E. coli.
[b] GM-CSF, granulocyte macrophage-colony stimulating factor. The three sequences are Met-GM-CSF, des-Met GM-CSF, and des-Met-Ala GM-CSF.
[c] IL-2, interleukin-2; The two sequences identified are Met-IL-2 and des-Met-IL-2.
[d] The two sequences identified are Met-erythropoietin and des-Met erythropoietin; our unpublished data.
[e] A heptapeptide upstream of the mature protein is not processed.
[f] f-Met species has been purified away from the purified bulk. However, the N-terminus with a f-Met residue is not amenable to Edman degradation. Other methods such as ion-exchange HPLC, IEF analysis, peptide mapping with mass spectrometry may be used for the analysis (see Ref. 126).
[g] Three sequences are Met, des-Met and acetylated des-Met interferons. Only two sequences are identified by direct N-terminal sequence analysis. The acetylated des-Met interferon is N-terminally blocked and has to be identified by separation of the isoform by ion-exchange chromatography followed by peptide mapping and sequence analysis (see Ref. 130).

growth factors such as macrophage colony-stimulating factor (131), stem cell factor (132) and *neu* differentiation factor (133). In each case, the full-length unprocessed protein encoded by the cloned gene is a membrane-associated form which contains a short stretch of hydrophobic, membrane-spanning domain. A majority of the precursor form is then proteolytically processed, and removal of the C-terminal hydrophobic region leads to generation of a soluble, secreted factor that can be identified in the circulation or in the culture medium. Protein and peptide C-terminal sequence analyses can be used to determine the precise processing sites. In the case of rat stem cell factor (SCF) produced naturally (132) or recombinant human SCF produced in Chinese hamster ovary cells (134), the membrane-associated precursor which has 248 amino acids is initially expressed and then secreted as a soluble factor after processing. Only a C-terminal amino acid, Ala, is released from the soluble factor as determined by carboxypeptidase P digestion and high sensitivity amino acid analysis (132). Therefore, further studies were performed by HPLC peptide map analysis and isolation of C-terminal peptides from both origins. N-terminal sequence and mass spectrometric analyses of the isolated C-terminal peptides confirm that the soluble factor is consistently processed at two specific sites near the C-terminus of the precursor and a soluble factor is generated, with a ragged C-terminus (-Pro-Pro-Val-Ala and -Pro-Pro-Val-Ala-Ala) and a polypeptide of 164–165 amino acids in length (134). Based on this observation, recombinant human stem cell factor produced in *E. coli* (trade name: Stemgen®) is engineered to be 165 amino acids in length and the purified and biologically functional product has been clinically evaluated.

Another consideration for the importance of performing C-terminal sequence analysis is the identification of the translation of termination codons (often referred to as "readthrough"), especially in the bacterial expression system. The extremely high production yields of recombinant protein in bacterial expression renders a high occurrence rate for readthrough in some proteins, despite that it is often taken for granted that normal translation would start at the initiation codon and stop at the termination codon during gene expression. Termination of peptide-chain synthesis is universally signaled by any of the three termination codons, TAG (amber), TGA (opal) and TAA (ochre), where the readthrough may occur. The C-terminal extension of the readthrough products usually varies and depends upon location of the second stop codon for termination of synthesis. Many recent reports have shown that readthrough products can be isolated and identified by C-terminal sequence analysis of the extended protein and/or by isolation as well as N-terminal sequence and mass spectrometric analyses of the readthrough peptides. For example, Gln readthrough of a UAG stop codon was found during production of bacterially-derived bovine somatotrophin (135), and Trp readthrough of a TGA stop codon was observed in both recombinant human platelet-derived growth factor (136) and recombinant human neurotrophin (137).

In addition to the "readthrough" problem, other forms of C-terminal extension of recombinant protein can occur as well. When murine interleukin 6 is overexpressed in *E. coli*, 5–10% of the recombinant molecules contained a novel C-terminal modification (138). Peptide mapping, mass spectrometry, automated N- and C-terminal sequencing identified a "tag" peptide, Ala-Ala-Asn-Asp-Glu-Asn-Tyr-Ala-Leu-Ala-Ala-COOH, encoded by a small metabolically stable RNA of *E. coli* (10Sa RNA) attached to truncated C-terminii of the recombinant protein.

*c.  Internal Sequence Verification*

Verification of internal sequence for recombinant protein drugs is to establish the fidelity of translation, to characterize the product that has been chemically changed, and to identify degradation sites that are related to the stability of the product. Peptide mapping, a routinely performed quality control test, in conjunction with protein sequence analysis is an essential approach for obtaining internal sequence information. With the capability of on-line MS or MS/MS analysis, peptide mapping can now allow individual peptide peaks to be characterized by precise mass measurement or sequence analysis (see Chapter 16 for details).

Sequence analysis of internal peptide fragments obtained from peptide mapping analysis is essential to identify process-related internal cleavages as illustrated in the study of recombinant human growth hormone produced in an *E. coli* secretion system (139). A two-chain, clipped form was isolated by ion-exchange chromatography and the clipped site was identified by isolation and sequence analysis of peptides near the site of cleavage. Recombinant tissue plasminogen activator which contains a serine protease domain was found to be autolytically clipped upon long term storage. Following sequence determination of the isolated peptides, three internal cleavage sites identified were all lysyl and arginyl bonds (140).

Upon accelerated stability study, proteins usually undergo a degradation process in which chemical cleavage of labile peptide bonds occurs as a function of time, pH and temperature (141). Internal sequence analysis of isolated peptides also allows identification of isomerization of Asp residues (formation of isoaspartyl bond) from preparations of human growth hormone (141), porcine somatotrophin (142) and antibodies (143), and deamidation of Asn residues (formation of cyclic imide, isoaspartyl and/or aspartyl bonds) from human growth hormone (144), CD4 molecule (145) and stem cell factor (SCF) (146). Edman sequence analysis of the isolated peptide containing isoaspartyl residues or cyclic imides will stop at the isomerization site where no further sequence signals can ever be detected. In this case, sequence analysis of the peptide may be performed by tandem mass spectrometry if size of the peptide is small. Oxidation of methionine residues (formation of methionine sulfoxide) is another common degradation pathway in the aging of proteins. Usually Edman sequence analysis fails to analyze this degradation product; however, peptides bearing the oxidation derivatives can be easily separated from the unmodified peptides by HPLC and analyzed by mass spectrometry. For bacterially-derived recombinant protein that contains disulfide bonds, such as recombinant human granulocyte colony-stimulating factor (147) and stem cell factor (148), refolding and oxidation is usually required to recover a functional protein. The folding process usually generates intermediates that contain unpaired disulfide bonds. Peptide sequence analysis can be used to identify the sites of unpaired disulfides and define the folding pathway for process improvement.

Verification of internal sequence is also essential in the identification of microheterogeneities as microheterogeneities may be derived from amino acid substitution in the sequence of final purified products or in-process materials. The identification of amino acid substitution therefore allows further design of the process to remove such undesired heterogeneity from the final product. A common amino acid substitution, as the result of an error in the translation step (mistranslation or misincorporation), is the incorporation of norleucine in place of methionine in the recombinant DNA-derived products including IL-2

(149), bovine somatotrophin (150), and leptin (151). Incorporation of norleucine appears to correlate with the fermentation nutrient conditions in which *E. coli* cells grown under minimal medium would result in the depletion of the methionine supply for protein synthesis and to sustain high protein synthesis rate a de novo synthesis of norleucine is physiologically forced to take place for the misincorporation. Due to a high rate of protein biosynthesis, the translation process in *E. coli* proved to be error-prone (152). In addition to the identification of norleucine incorporation and termination readthrough as described above, verification of internal sequence also revealed various types of mistranslation in recombinant DNA-derived protein drugs present in the in-process materials, including Gln for His substitution in G-CSF (153), Arg for Lys substitution in insulin growth factor (154), and Gln for Tyr substitution in antibody against Her-2 receptor (155). The collection of the above information is important to down-stream process development as an alternative process design can be adapted to remove these minor, undesirable heterogeneities from the major product.

## 2. Characterization of Protein Drug Reference Standard

There are a number of routine analytical tests required for verification of purity of a protein drug (see related chapters in this book). Once the purity has been ascertained, it is imperative to prove that the isolated protein indeed has the correct amino acid sequence. In analyses for the verification of chemical identity, usually a sample protein from a production lot is compared to a standard which may be an authentic protein isolated from natural source. However, an authentic standard used for the comparison is usually scarce and difficult to prepare; therefore, a well-characterized recombinant protein

sample from a production lot can be established as a reference standard.

In addition to the evaluation by all the regular analytical tests, more extensive protein sequence analyses must be performed to characterize the reference standard, which include extended N-terminal sequence analysis, C-terminal sequence analysis, generation of various sets of peptide fragments by high sensitivity peptide mapping, sequence analysis of all of the isolated peptide fragments for overall sequence alignment, and determination of disulfide bonds as well as sites of post-translational modifications or proteolytic processing, if any, etc. The complete amino acid sequence has to be aligned and conformed to the predicted cDNA sequence that is used to express the recombinant protein. Current techniques and instrumentation, especially high sensitivity automatic protein sequencing in conjunction with high sensitivity peptide mapping /MS or MS/MS analysis have dramatically sped up reference standard characterization in a protein chemistry laboratory.

Figure 13 illustrates one set of peptide maps derived from endoproteinase Asp-N digestion of a reference standard for *E. coli*-derived recombinant human stem cell factor. In combination with a second set of the Glu-C peptide fragments, all peptides isolated can be sequenced to delineate the complete sequence of the recombinant protein which contains 165 amino acids and an uncleaved initiator Met upstream of the SCF N-terminus. The two intramolecular disulfide bonds identified in the Asp-N peptide map can be easily assigned by comparing the two maps generated under a native protein state and under reduction with dithiothreitol (see Fig. 13). Figure 14 summarizes overall alignment and complete sequence of the reference standard. The structure has been sequenced at least twice using overlapping fragments. All minor peptide peaks were also isolated

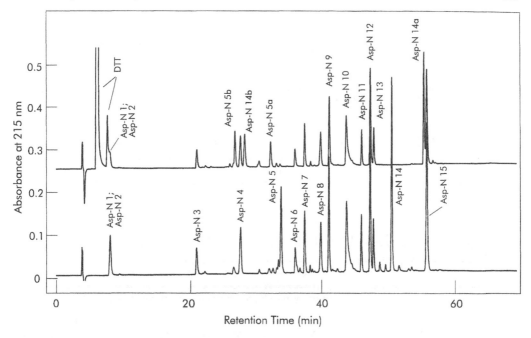

**Fig. 13** HPLC peptide map analysis of recombinant human stem cell factor after endoproteinase Asp-N digestion. Sample digest (25 μg) was directly injected onto a C4 reverse-phase column (2.1 mm × 25 cm) and separated by 0.1% trifluoroacetic acid-acetonitrile gradient elution in an HPLC system (panel A) or the digest is reduced with an excess amount of dithiothreitol to reduce the disulfide-containing peptides prior to peptide mapping (panel B). By this approach two peaks AspN-5 and AspN-14 are identified to be the peptide fragments bearing disulfide linkages. All peptides isolated are sequenced to confirm the structure identity of stem cell factor (our unpublished data).

and sequenced to identify any modified peptides or, in some cases, the unpaired or mis-paired disulfide peptides which may be related to fermentation or purification processes and to detect minor peptides derived from proteins in the cultured cells. An established reference standard should be void of any of these undesired components.

## B.  Assignment of Disulfide Bonds

Correct formation of the disulfide linkages of a recombinant protein is at least as important for function as correct amino acid sequence. The key to success in the assignment of disulfide bonds generally involves digestion of the protein by chemi-

cal or enzymatic means in such a way to cleave the polypeptide chain at least once between successive Cys residues (except where two adjacent Cys residues share a disulfide bond). Generally, the disulfide-containing peptides are detected as peptides whose separation character or other molecular property is changed by cleavage of disulfide bonds. This task is apparently simple when there are only a limited number of disulfides present in the protein. For example, recombinant human SCF contains two disulfide bonds as shown in Figs 13 and 14 ; a single Asp-N peptide mapping procedure generates two disulfide-containing fragments that can be directly used for disulfide bond assignment following DTT reduction. Usually, detec-

**Fig. 14** Amino acid sequence of recombinant human stem cell factor (SCF). A complete sequence analysis of the protein is accomplished by extended N-terminal sequence analysis (up to 50 residues) and by sequence analysis of all peptides generated by two separate peptide maps derived from digestion of SCF with endoproteinases Asp-N and Glu-C. Precise molecular mass for each isolated peptide is measured by electrospray ionization mass spectrometric analysis and used to determine the length of the peptide. Two disulfide-containing fragments, Asp-N 5 and Asp-N 14, or SV8-1 and SV8-12, are used to assign the disulfide bonds.

tion of PTH-cystine can directly assign some disulfide bonds during sequence analysis of a peptide fragment consisting of two peptides linked by a disulfide bond (156, 157). MALDI analysis of disulfide fragments with prompt fragmentation also allows the verification of a single disulfide bond (158, 159).

Difficulties are encountered for disulfide assigment when two Cys residues are very close or juxtaposed (160). It can become even more challenging to obtain suitable, small disulfide peptide subfragments by various cleavage methods, when dealing with molecules that contain multiple sets of disulfide bonds. For example, recombinant human tumor necrosis factor binding protein (TNFbp) contains thirteen disulfide bonds in a 162 amino acids polypeptide chain [(161); see Fig. 15A for TNFbp disulfide arrangement]. A number of the peptide fragments obtained from TNFbp may contain more than two pairs of disulfides for assignment. The development of partial reduction in conditions that resist disulfide exchange followed by isolation and characterization of the partially reduced peptides can allow the elucidation of disulfide linkages of some highly bridged peptides (162). MALDI with postsource decay also allows fragmentation of the peptide backbone between two Cys residues preceding cleavage of disulfide bonds; and the identification of unique fragment ions obtained provide unequivocal evidence for making disulfide assignment for peptides containing two disulfide bonds (163). To make all 13 TNFbp disulfides assignable, it requires two successive proteolytic (trypsin and thermolysin) digestions, identification of PTH-cystine in simple disulfide peptides, MALDI-prompt fragmentation, partial reduction of complex disulfide-containing fragments, and MALDI-post-source decay of disulfide-linked peptides (161, 163). Figure 15B shows a TNFbp peptide fragment containing three peptides linked by two

disulfide bonds. The fragment that cannot be easily assigned by other methods was subjected to analysis by MALDI with postsource decay. The identification of three fragment ions of 825.6, 940, and 1027 daltons (Fig. 15C, indicated by arrows) confirms the disulfide bonding for the peptide and verifies that the disulfide structure in the domain IV of TNFbp is unique from the other three domains, despite that these four domains were commonly recognized to be TNFR/NGFR family cysteine-region signature structures (Fig. 15A).

## C. Detection of Post-translational Modification

It is important to determine the nature and location of all post-translational modifications of recombinant proteins. Unwanted, nonspecific modifications such as Met oxidation, Asp isomerization and Asn/Gln deamidation (their identification is seen in Sec. V.A.3) are degradation products usually related to the stability problems of the protein. Other correct post-translational modifications can be made in recombinant protein, especially if the modification is important for the activity of the proteins. Among the most frequent post-translational modifications are phosphorylation, N-terminal acetylation, glycosylation, farnesylation, and sialic acid capping of oligosaccharides. Selection of expression systems, usually mammalian cell and eukaryotic cell expression systems, is critical to obtain desired modification. Protein sequence analysis in conjunction with mass spectrometric analysis allows the identification of modification sites in most cases, while new approaches by mass spectrometric analysis can also obtain some insights about the structure of the attached moieties (see Chapter 16 for mass spectrometry). Detailed analyses of glycoproteins are described in Chapter 14.

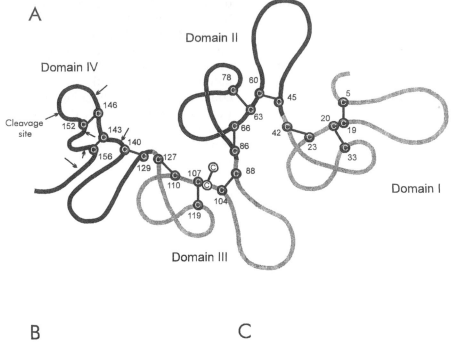

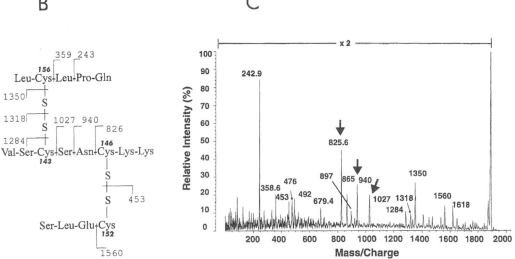

**Fig. 15** (A). Complex disulfide structure of recombinant human tumor necrosis factor binding protein (TNFbp or soluble TNF receptor). The protein is subdivided into four domains based on the pattern of Cys residues, and each domain contains three intramolecular disulfide bonds. In domain I, the artificially introduced Cys-106 forms an extra disulfide bond with a free Cys. (B). A TNFbp disulfide fragment containing three peptides linked together by two disulfide bonds. This fragment is generated by digesting TNFbp with trypsin followed by thermolysin. Cleavage sites are in domain IV (shown by arrows in Fig. 15A). (C). Mass spectrometric analysis of the fragment shown in Fig. 15B by MALDI with postsource decay. Fragmentation of the parent ion (1889.6 daltons) generates a series of sequence ions. Several ions are assigned to the sequence of the fragment shown in Fig. 15B. Three key fragment ions indicated by arrows (825.6, 940, and 1027 daltons) are used to confirm the disulfide assignment for Cys-143- to Cys-156 and Cys-146 to Cys-152. Mass spectrum is adapted with permission from Ref. 162.

## ACKNOWLEDGEMENTS

We are indebted to our colleagues at the Department of Protein Structure, Amgen Inc. for their continual support in the routine maintenance and operation of high performance instrumentation in sequencators and mass spectrometers, to Mrs. Joan Bennett in typing part of the manuscript, and to Dr. Robert Rush for critical reading and comments.

## REFERENCES

1. CR Cantor, PR Schimmel. Biophysical Chemistry-Part I: The conformation of biological molecules. San Francisco: WH Freeman and Company, 1980.
2. G Allen. Sequencing of Proteins and Peptides. Amsterdam: Elsevier, 1989.
3. CHW Hirs, SN Timasheff, eds. Methods in Enzymology: Enzyme Structure, Part E. Vol. 47. New York: Academic Press, 1977.
4. CHW Hirs, SN Timasheff, eds. Methods in Enzymology: Enzyme Structure, Part 1. Vol. 91. New York: Academic Press, 1983.
5. I Kerese, ed. Methods of Protein Analysis. New York: Halsted Press, 1984.
6. SB Needleman, ed. Protein Sequence Determination. Berlin: Springer-Verlag, 1975.
7. KA Walsh, ed. Methods in Protein Sequence Analysis. Clifton, New Jersey: Humana Press, 1987.
8. AS Bhown, ed. Protein/Peptide Sequence Analysis: Current Methodologies. Boca Raton, Florida: CRC Press, 1989.
9. JE Shively, ed. Micromethods for Protein Structural Analysis: A Companion to Methods in Enzymology. New York: Academic Press, 1994.
10. JM Baily. Chemical methods of protein sequence analysis. J Chromatogr A 705: 47–56, 1995.
11. P Tempest, S Geromanos, C Elicone, H Erdjument-Bromage. Improvements in microsequencer performance for low picomole sequence analysis. Methods 6: 248–261, 1994.
12. DF Reim, DW Speicher. N-terminal sequence analysis of proteins and peptides. In: JE Coligan, BM Dunn, HL Ploegh, DW Speicher, PT Wingfield, eds. Current Protocols in Protein Science. New York: John Wiley and Sons, 1995, pp 11.10.1–11.10.38.
13. AP Ryle, F Sanger, LF Smith, R Kitai. Biochem J 60: 541–556, 1955.
14. F Sanger. Science 129: 1340–1344, 1959.
15. F Sanger. Biochem J 39: 507–515, 1945.
16. P Edman. A method for the determination of the amino acid sequence in peptides. Arch Biochem Biophys 22: 475–480, 1949.
17. P Edman. Acta Chem Scand 7: 700–701, 1953.
18. P Edman. Acta Chem Scand 10: 761–768, 1956.
19. WR Gray, BS Hartley. Biochem J 89: 379–380, 1963.
20. BS Hartley. Strategy and tactics in protein chemistry (the 1st BDH lecture). Biochem J 119: 805, 1970.
21. JY Chang, EH Creaser, KW Bently. Biochem J 153: 607–611, 1976.
22. RA Laursen. J Am Chem Soc, 88: 5344–5346, 1966.
23. DH Spackman, WH Stein, S Moore. Automatic recording apparatus for use in the chromatography of amino acids. Anal Chem 30: 1190–1206, 1958.
24. JL Bailey. Techniques in Protein Chemistry (2nd edition). New York: American Elsevier, 1967.
25. P Edman, G Begg. A protein sequenator. Eur J Biochem 1: 80–91, 1967.
26. RM Hewick, MW Hunkapiller, LE Hood, WJ Dreyer. A gas-liquid solid phase peptide and protein sequencer. J Biol Chem 256: 7990–7997, 1981.
27. AL Burlingame, RK Boyd, SJ Gaskell. Mass spectrometry. Anal Chem 66: 634R–683R, 1994.
28. AL Burlingame, RK Boyd, SJ Gaskell. Mass spectrometry. Anal Chem 68: 599R–649R, 1996.
29. SA Carr, RS Annana. Mass spectrometry. In: JE.Coligan, BM Dunn, HL Ploegh, DW Speicher, PT Wingfield, eds. Current Protocols in Protein Sciences. New York: John Wiley and Sons, 1995, pp 16.1.1–16.1.27.
30. K Biemann. Mass spectrometric analysis of peptides and proteins. Ann Rev Biochem 61: 977–1010, 1992.
31. TE Creichton. Postranslational modifications of polypeptide chains. In Protein: Structure and Molecular Properties. New York: WH Freeman, 1993, pp 78–100.
32. JB Clegg, MA Naughton, DJ Weatherall. J Mol Biol 19: 91–108, 1996.
33. BL Chen, RJ Poljak. Biochemistry 13: 1295–1302, 1974.
34. T Arakawa, DA Yphantis, JM Lary, LO Narhi, HS Lu, SJ Prestrelski, CL Clogston, KM Zsebo, EA Mendiaz, J Wypych, KE Langley. Glycosylated and unglycosylated recombinant-derived human stem cell

factors are dimeric and have extensive regular secondary structure. J Biol Chem 266: 18942–18948, 1991.

35. CHW Hirs, WH Stein, S Moore. J Biol Chem 211: 941, 1954.

36. S Moore, DH Spackman, WH Stein. Chromatography of amino acids on sulfonated polystyrene resins. Anal Chem 30: 1185–1190, 1958.

37. S Moore, WH Stein. Chromatographic determination of amino acids by the use of automatic recording equipment. Methods Enzymol. 6: 819–831, 1963.

38. R Knecht, JY Chang. Anal Chem 58: 2375, 1986.

39. NM Meltzer, GI Tous, S Gruber, S Stein. Gas-phase hydrolysis of proteins and peptides. Anal Biochem, 160: 356–361, 1987.

40. BA Bidlingmeyer, SA Cohen, TL Tarvin. Rapid analysis of amino acids using pre-column derivatization. J Chromatogr, 336: 93–104, 1984.

41. FC Westhall, H Hesser. Anal Biochem 61: 610, 1974.

42. SH Chiou, KT Wang. A rapid and novel means of protein hydrolysis by microwave irradiation using Teflon-pyrex tubes. In: JJ Villafranca, ed. Current Research in Protein Chemistry. New York: Academic Press, 1990, pp 1–10.

43. LB Gilman, C Woodward. An evaluation of microwave heating for the vapor phase hydrolysis of proteins. In: JJ Villafranca, ed. Current Research in Protein Chemistry. New York: Academic Press, 1990, pp 23–36.

44 H Matsubara, RM Sasaki. High recovery of tryptophan from acid hydrolysis of proteins. Biochem Biophys Res Commun 35: 175, 1969.

45. AS Inglis. Single hydrolysis method for all amino acids, including tryptophan and cysteine. Methods Enzymol 91: 26–36, 1983.

46. TY Liu, YH Chang. Hydrolysis of proteins with p-toluenesulfonic acid. J Biol Chem 246: 2842, 1971.

47. RJ Simpson, MR Neuberger, TY Liu. J Biol Chem 251: 1936–1940, 1976.

48. S Moore, WH Stein. J Biol Chem 192: 663, 1951.

49. S Moore. The precision and sensitivity of amino acid analysis. In: J Meienhofer, ed. Chemistry and Biology of Peptides. New York: Academic Press, 1972, pp 629–653.

50. PE Hare. Subnanomole-range amino acid analysis. Methods Enzymol 47: 3–18, 1977.

51. JR Benson. Improved ion-exchange resins. Methods Enzymol 47: 18–31, 1977.

52. S Stein, L Brink. The fluorescamine amino acid analyzer. Methods Enzymol, 79: 20–25, 1981.

53. M Roth, A Hampai. J Chromatogr 83: 353–356, 1973.

54. JR Benson, PE Hare. Proc Natl Acad Sci USA 72: 619, 1975.

55. P Bohlen. Analysis of amino acids with o-pthalaldehyde. Methods Enzymol 91: 17–26, 1983.

56. AC Herman, TC Boone, HS Lu. Characterization, formulation, and stability of Neupogen (Filgrastim), a recombinant human granulocyte colony stimulating factor. In R Pearlman and YJ Wang, eds. Formulation, Characterization, and Stability of Protein Drugs. New York: Plenum Press, 1996, pp 303–328.

57. J Speiss, J Rivier, C Rivier, W Vale. Primary structure of corticotropin releasing factor from ovine hypothalamus. Proc Natl Acad Sci USA 78: 6517–6521, 1981.

58. B Oray, HS Lu, RW Gracy. HPLC separation of dansyl amino acid derivatives and applications to peptide and protein structural analyses. J Chromatogr 27: 256–266, 1983.

59. R Knecht, JY Chang. High sensitivity amino acid analysis using dabsyl chloride precolumn derivatization method. In: B Wittmann-Liebold, J Salnikow, VA Erdmann, eds. Advanced Methods in Protein Microsequence Analysis. Berlin: Spring-Verlag, 1986, pp 56–61.

60. DW Hill, FH Walters, TD Wilson, JD Stuart. Anal Chem 51: 1338–1341, 1979.

61. BN Jones. Microsequence analysis by enzymatic methods. In: JE Shively, ed. Methods of Protein Microcharacterization. Clifton, New Jersey: Humana Press, 1986, pp 121–151.

62. SA Cohen, DJ Strydom. Amino acid analysis utilizing phenylisothiocyanate derivatives. Anal Biochem 174: 1–16, 1988.

63 HS Lu, PH Lai. Use of narrow-bore high-performance liquid chromatography for microanalysis of protein structure. J Chromatogr 368: 215–231, 1986.

64. JW Crabb, KA West, WS Dodson, JD Hulmes. Amino acid analysis. In: JE Coligan, BM Dunn, HL Ploegh, DW Speicher, PT Wingfield, eds. Current Protocols in Protein Sciences, New York: J Wiley and Sons, 1995, pp 11.9.1–11.9.43.

65. S Einarsson, B Josefsson, S Lagerkvist. J Chromatogr 282: 609–618, 1983.

66. RA Bank, EJ Jansen, B Beekman, JM te Kopele. Amino acid analysis by reverse-phase high performance liquid chromatography: Improved derivatization and detection conditions with

9-fluorenylmethyl chloroformate. Anal Biochem 240: 167–176, 1996.

67. AN Glazer, RJ DeLanger, DS Sigman. Chemical Modification of Proteins. In: RS Work, E Work, eds. Laboratory Techniques in Biochemistry and Molecular Biology, Amsterdam: North Holland/ American Elsevier, 1975.

68. RL Lundblad. Techniques in Protein Modifications. Boca Raton: CRC Press, 1994.

69. GL Ellman. Tissue sulphydryl groups. Arch Biochem Biophys 82: 70–77, 1959.

70. PW Riddles, RL Blakeley, B Zerner. Reassessment of Ellman's reagent. Methods Enzymol 91: 49-60, 1983.

71. Y Tsukamoto, SJ Wakil. Isolation and mapping of the $\beta$-hydroxyacyl dehydratase activity of chicken liver fatty acyl synthetase. J Biol Chem 263: 16225-16229, 1988.

72. AM Crestfield, S Moore, WH Stein. The preparation and enzymatic hydrolysis of reduced and S-carboxymethylated proteins. J Biol Chem 238: 622, 1963.

73. CS Fullmer. Identification of cysteine-containing peptides in protein digest by HPLC. Anal Biochem 142: 336, 1984.

74. H Charbonneau. Strategies for obtaining partial amino acid sequence data from small quantities of pure and partially purified protein. In: PT Matsudaira, ed. A Practical Guide to Protein and Peptide Purification for Microsequencing. New York: Academic Press, 1989, pp 15–30.

75. MD Jones, LA Merewether, CL Clogston, HS Lu. Peptide mapping of recombinant human granulocyte colony stimulating factor: Side reactions and method optimization. Anal Biochem 216: 135–146, 1994.

76. CHW Hirs. Performic acid oxidation. Methods Enzymol 11: 197–199, 1967.

77. FDA (Food and Drug Administration). Guidance for industry for the submission of chemistry, manufacturing, and controls information for a therapeutic recombinant DNA-derived product or a monoclonal antibody product for in vivo use. Center for Biologics Evaluaiton and Research (CBER) and Center for Drugs Evaluatyion and Research (CDER), FDA, Bethesda, MD, 1996.

78. R Aebersold, SD Patterson. Current problems and technical solutions in protein biochemistry. In: RH Angeletti, ed. Proteins: Analysis and Design. 1998, pp 3–120.

79. UK Laemmli. Cleavage of structural proteins during the assembly of bacteriophage T4. Nature 227: 680–685, 1970.

80. P Matsudaira. Sequencing from picomole quantities of proteins electroblotted onto polyvinylidene difluoride membranes. J Biol Chem 262: 10035–10038, 1987.

81. PH O'Farrell. High resolution two-dimensional gel electrophoresis of proteins. J Biol Chem 250: 4007–4021, 1975.

82. J Fernandez, M DeMott, D Atherson, SM Mische. An improved procedure for enzymatic digestion of polyvinylidene difluoride-bound proteins for internal sequence analysis. Anal Biochem 201: 255-264, 1992.

83. J Rosenfeld, J Capdevielle, JC Guillemot, P Ferrara. In-gel digestion of proteins for internal sequence analysis after one- or two-dimensional gel electrophoresis. Anal Biochem 203: 173–179, 1992.

84. LA Merewether, CL Clogston, SD Patterson, HS Lu. Peptide mapping at the 1 $\mu g$ level: In-gel versus PVDF digestion techniques. In: JW Crabb, ed. Techniques in Protein Chemistry VI. New York: Academic Press, 1995, pp 153–160.

85. U Hellman, C Werstedt, J Gonez, C-H Heldin. Improvement of an in-gel digestion procedure for the micropreparation of internal protein fragments for amino acid sequencing. Anal Biochem 224: 451–455, 1995.

86. P Edman, A Henschen. Edman degradation. In: SB Needleman, ed. Protein Sequence Determination. Berlin: Springer-Verlag, 1975, pp 232–279.

87. GE Tarr. Improved manual sequence analysis methods. Methods Enzymol 47: 335–357, 1977.

88. CL Zimmermann, A Appella, JJ Pisano. Rapid analysis of amino acid phenylthiohy- dantoins by high performance chromatography. Anal Biochem 77: 569–573, 1977.

89. GE Tarr, JF Beecher, M Bell, DJ McKean. Anal Biochem 84: 622–627, 1978.

90. DG Klapper, CE Wilde, JD Capra. Biochem 85: 126–131, 1978.

91. MW Hunkapiller, LE Hood. Biochem 17: 2124–2133, 1978.

92. JE Shively. Sequence determinations of proteins and peptides at the nanomole and subnanomole levels with a modified spinning cup sequenator. Methods Enzymol 79: 31–48, 1981.

93. MW Hunkapiller, JE Strickler, KJ Wilson. Contemporary methodology for protein structure determination. Science 226: 304, 1984.

94. MW Hunkapiller. PTH-amino acid analysis. In JJ L'Italien ed. Proteins: Structure and Function. New York: Plenum Press, 1987, pp 363–381.

95. DG Sheer, S Yuen, J Wong, J Wasson, PM Yuan. A modified reaction cartridge for direct sequencing on polymeric membranes. Biotechniques 11: 526–534, 1991.

96. RW Blacher, JH Wieser. On-line microbore HPLC detection of femtomole quantities of PTH-amino acids. In: RH Angeletti, ed. Techniques in Protein Chemistry IV. New York: Academic Press, 1993, pp 427–433.

97. AE Lavin, LA Merewether, CL Clogston, MF Rohde. Comparison of the high sensitivity and standard version of Applied Biosystems Procise 494 N-terminal protein sequencers using various sequencing supports. In: D Marshak, ed. Techniques in Protein Chemistry VIII. New York: Academic Press, 1997, pp 57–67.

98. CG Miller, HB Bente, S Fisher, J Myerson, G Wagner, R Widmayer, MJ Horm. Fourth Symposium of the Protein Society. Abstract T143. San Diego, 1990.

99. DF Reim, DW Speicher. In: JE Coligan, BM Dunn, HL Ploegh, DW Speicher, PT Wingfield, eds. Current Protocols in Protein Sciences. New York: John Wiley and Sons, 1995, pp 11.10.1–11.10.38.

100. N LeGendre, P Matsudaira. Direct protein microsequencing from Immobilon-P transfer membranes. Biotechniques 6: 154, 1988.

101. M Moos, Jr, NY Nguyen, TY Liu. Reproducible high yield sequencing of proteins electrophoretically separated and transferred to an inert support. J Biol Chem 263: 6005–6008, 1988.

102. DW Speicher. Methods and strategies for the sequence analysis of proteins on PVDF membranes. Methods 6: 262–273, 1994.

103. PR Fausset, HS Lu. Structural analysis of recombinant proteins prepared by one-step purification using semi dry electroblotting after SDS-PAGE. Electrophoresis 12: 22–27, 1991.

104. DF Reim, DW Speicher. A method for high-performance sequence analysis using polyvinylidene difluoride membranes with biphasic reaction column sequencer. Anal Biochem 216: 213–222, 1994.

105. RP Ambler. Enzymatic hydrolysis with carboxypeptidases. Methods Enzymol 11: 155–166, 1967.

106. RP Ambler. Enzymatic hydrolysis with carboxypeptidases. Methods Enzymol 25: 262–272, 1972.

107. R Hayashi. Carboxypeptidase Y in sequence determination of peptides. Methods Enzymol 47: 84–93, 1977.

108. S Yokoyama, A Oobayashi, O Tanabe, E Ichishima. Appl Microbiol 28: 742–747, 1974.

109. S Yokoyama, A Oobayashi, O Tanabe, E Ichishima. Biochem Biophys Acta 397: 443–448, 1975.

110. HS Lu, MK Klein, PH Lai. Narrow-bore high-performance liquid chromatography of phenylthiocarbamyl amino acids and carboxypeptidase P digestion for protein C-terminal sequence analysis. J Chromatogr 447: 351–364, 1988.

111. JM Bailey, CG Miller, B Holmquist, E Fowler. C-terminal sequence analysis. In JE Coligan, BM Dunn, H. Ploegh, DW Speicher, PT Wingfield, eds. Current Protocols in Protein Sciences. New York: John Wiley and Sons, 1995, pp 11.8.1–11.8.14.

112. DH Patterson, G Tarr, FE Regnier, SA Martin. C-terminal ladder sequencing via matrix-assisted laser desorption mass spectrometry coupled with carboxypeptidase Y time-dependent and concentration-dependent digestions. Anal Chem 67: 3971–3978.

113. FK Lin, S Suggs, CH Lin JF Brown, R Smalling, JC Egrie, KK Chen, GM Fox, F Martin, Z Stabinsky, SM Badrawi, PH Lai, E Goldwasser. Cloning and expression of the human erythropoietin gene. Proc Natl Acad Sci USA 82: 7580–7585, 1985.

114. AS Inglis. Chemical procedures for the C-terminal sequencing of peptides and proteins. Anal Biochem 195: 183–196, 1991.

115. JM Bailey, NS Shenoy, M Ronk, JE Shively. Automated carboxy-terminal sequence of peptides. Protein Sci 1: 68–80, 1992.

116. JM Bailey, M Rusnak, JE Shively. Compact protein sequencer for the C-terminal sequencer analysis of peptides and proteins. Anal Biochem 212: 366–374, 1993.

117. JM Bailey, O Tu, G Issai, A Ha, JE Shively. Automated carboxy-terminal sequence sequence analysis of polypeptides containing proline. Anal Biochem 224: 588–596, 1995.

118. VL Boyd, M Bozzini, G Zon, RL Noble, RL Mattaliano. Sequencing of peptides and proteins from the carboxy terminus. Anal Biochem 206: 344–352, 1992.

119. CG Miller, JM Bailey. Biotechnology applications of C-terminal protein sequence analysis. Genet Eng News 15 (12): 12, 1995.

120. M Bozzini, J Zhao, PM Yuan, D Ciolek, YC Pan, J Horton, DR Marshak, VL Boyd. Applications using an alkylation method for carboxy-terminal sequencing. In: JW Crabb, ed. Techniques in Protein

Chemistry VI. New York: Academic Press, 1995, pp 229-237.

121. MJ Geisow. Characterizing recombinant proteins. Bio/Technology 9: 921–924, 1991.

122. RL Garnick, MJ Rose, RA Baffi. Characterization of proteins from recombinant DNA manufacture. In: YYH Chiu and JL Gueriguian, eds. Drug Biotechnology Regulation: Scientific Basis and Practices. New York: Marcel Dekker, 1991, pp 263–313.

123. VR Anicetti, BA Keyt, WS Hancock. Purity analysis of protein pharmaceuticals produced by recombinant DNA technology. TIBTECH 7: 342–349, 1989.

124. LM Souza, TC Boone, PH Lai, KM Zsebo, DC Murdock, VR Chazin, J Bruszewski, HS Lu, KK Chen, J Barendt, E Platzer, J Gabrilove, MAS Moore, R Mertelsman, K Welte. Recombinant pluripotent human granulocyte colony-stimulating factor induces proliferation and differentiation of normal and leukemic myeloid cells. Science 232: 61–68, 1986.

125. DV Goeddel, HL Heyneker, T Hozumi, R Arentzen, K Itakura, DG Yansura, MJ Ross, G Miozzari, R Crea, PH Seeburg. Direct expression in Escherichia coli of a DNA sequence coding for human growth hormone. Nature 281: 544, 1979.

126. CL Clogston, YR Hsu, TC Boone, HS Lu. Detection and quantitation of recombinant methionyl granulocyte colony stimulating factor charge isoforms: Comparative analyses by cationic exchange HPLC, IEF, and peptide mapping, Anal Biochem 202: 375–383, 1992.

127. HS Lu, YR Hsu, LO Narhi, S Karkare, FK Lin. Isolation and characterization of human recombinant tissue prokallikrein and kallikrein expressed in Chinese hamster ovary cells. Protein Expression and Isolation 8: 215–226, 1996.

128. HS Lu, YR Hsu, IL Lu, D Ruff, D Lyons, FK Lin. Isolation and characterization of human tissue kallikrein produced in E. coli: Biochemical comparison to the enzymatically inactive prokallikrein and methionyl kallikrein. Protein Expression and Isolation 8: 227–237, 1996.

129. KM Zsebo, HS Lu, JC Fieschko, L Goldstein, J. Davis, K. Duker, SV Suggs, PH Lai, GA Bitter. Protein secretion from S. cerevisiae directed by prepro-a-factor leader region. J Biol Chem 261: 5858–5865, 1986.

130. ML Klein, TD Bartley, J Davis, D Whiteley, HS Lu. Isolation and characterization of three isoforms of recombinant consensus alpha interferon.

Arch. Biochem. Biophys 276: 531–537, 1990.

131. CW Retenmier. Biosynthesis of macrophage colony-stimulating factor (CSF-1): Differential processing of CSF-1 precursors suggests alternative mechanism of stimulating CSF-1 receptor. Current Topics Microbiol Immunol 149: 129–141.

132. HS Lu, CL Clogston, J Wypych, KM Zsebo, KE Langley. Amino acid sequence and post-translational modification of stem cell factor isolated from buffalo rat liver cell conditioned medium. J Biol Chem 266: 8102–8107, 1991.

133. HS Lu, S Hara, LWI Wong, MJ Jones, V Katta, G Trail, A Zou, D Brankow, S Cole, S Hu, D Wen. Post-translational processing of membrane associated neu differentiation factor proisoforms expressed in mammalian cells. J Biol Chem 270: 4775–4783, 1995.

134. HS Lu, CL Clogston, TD Lee, J Wypych, VP Parker, K Swiderek, RF Baltera, AC Patel, DW Brankow, XD Liu, SG Ogden, S Karkare, S Hu, KM Zsebo, KE Langley. Post-translational processing of membrane-associated recombinant human stem cell factor expressed in Chinese hamster ovary cells. Arch Biochem Biophys 298: 150–158, 1992.

135. BN Violand. Identification of unique modified species recombinant proteins. ABRF Workshop, 8th Protein Society Symposium, San Diego, 1994.

136. KV Lu, MF Rohde, AR Thomason, WC Kenney, HS Lu. Mistranslation of a TGA termination codon as tryptophan in recombinant platelet-derived growth factor expressed in E. coli. Biochem J 309: 411–417, 1995.

137. JO Hui, SY Meng, V Katta, L Tsai, MF Rohde, M Haniu. E. coli-expressed human neurotrophin 3 characterization of a C-terminal extended product. In: JW Crabb, ed. Techniques in Protein Chemistry VI. New York: Academic Press, 1995, pp 341–348.

138. GF Tu, GE Reid, JG Zhang, RL Moritz, RJ Simpson. C-terminal extension of truncated recombinant proteins in Escherishia coli with a 10Sa RNA decapeptide. J Biol Chem 270: 9322–9326, 1995.

139. E Canova-Davis, IP Baldonado, JA Moore, CG Rudman, WF Bennett, WS Hancock. Properties of a cleaved two-chain form of recombinant human growth hormone. Int J Peptide and Protein Res 35: 17, 1990.

140. V Ling, ML Eng, PJ Lee, RG Keck, BA Keyt, E Canova-Davis. Subtleties of pep-

tide mapping in the analysis of protein pharmaceuticals. In: WS Hancock, ed. New Methods in Peptide Mapping for the Characterization of Proteins. Boca Raton: CRC Press, 1996, pp 1–30.

141. G Teshma, JT Stultz, V Ling, E Canova-Davis. Isolation and characterization of of a succinimide variant of methionyl human growth hormone. J Biol Chem 266: 13544, 1991.

142. BN Violand, MR Schlittler, EW Kolodziej, PC Toren, MA Carbonce, NR Siegel, KL Duffin, JF Zobel, CE Smith, JS Tou. Isolation and characterization of porcine somatotrophin containing a succinimide residue in place of aspartate. Protein Science 1: 1634, 1992.

143. J Cacia, R Keck, LG Presta, J Frenz. Isomerization of an aspartic acid residue in the complementarity-determining regions of a recombinant antibody to human IgE: Identification and effect on binding affinity. Biochem 35: 1897–1903, 1996.

144. GW Becker, PM Tackitt, WW Bromer, DS Lefeber, RM Rigin. Isolation and characterization of a sulfoxide and a desamido derivative of biosynthetic human growth hormone. Biotech Appl Biochem 10: 326, 1988.

145. G Teshmer, J Porter, K Yim, V Ling, A Guzzetta. Deamidation of soluble CD4 at asparagine-52 results in reduced binding capacity for the HIV-1 envelope glycoprotein gp120. Biochem 20: 3916, 1991.

146. YR Hsu, WC Chang, EA Mendiaz, S Hara, DT Chow, MB Mann, KE Langley, HS Lu. Selective deamidation of recombinant human stem cell factor during in vitro aging: Isolation and characterization of the aspartyl and isoaspartyl homodimers and heterodimers. Biochem, 37: 2251–2262, 1998.

147. HS Lu, CL Clogston, LA Merewether, LO Narhi, TC Boone. Refolding and oxidation of recombinant human G-CSF: Isolation and characterization of disulfide intermediates and cysteine analogs. J Biol Chem 267: 8770–8777, 1992.

148. MD Jones, LO Narhi, WC Chang, HS Lu. Refolding and oxidation of recombinant human stem cell factor produced in Escherichia coli. J Biol Chem 271: 11301–11308, 1996.

149. HS Lu, LB Tsai, WC Kenney, PH Lai. Identification of unusual replacement of methionine by norleucine in recombinant interleukin-2 produced by E. coli. Biochem Biophys Res Commun 156: 807, 1988.

150. G Bogosian, BN Violand, EJ Dorward-King, WE Workman, PE Jung, JF Kane. Biosynthesis and incorporation into protein of norleucine by Escherichia coli. J Biol Chem 264: 531, 1989.

151. JL Liu, T Eris, SL Lauren, GW Stearns, KR Westcott, HS Lu. Use of LC/MS peptide mapping for characterization of isoforms in $N^{15}$-labeled recombinant human leptin. In: DR Marshak, ed. Techniques in Protein Chemistry VIII. New York: Academic Press, 1997, pp 155–163.

152. MAS Santos, MF Tuite. New insights into mRNA decoding-Implications for heterologous protein synthesis. TIBTECH 11: 500–505, 1993.

153. HS Lu, PR Fausset, L Sotos, CL Clogston, K Stoney, MF Rohde, A Herman. Isolation and characterization of recombinant human granulocyte colony stimulating factor His -> Gln isoforms produced in E. coli. Protein Expression and Isolation 4: 465–472, 1993.

154. CA Scorer, MJ Carrier, RF Rosenberger. Amino acid misincorporation during high-level expression of mouse epidermal growth factor in Escherichia coli. Nucleic Acid Res 19: 3511–3516, 1991.

155. R Harris, AA Murnane, SL Utter, KL Wagner, ET Cox, GD Polastri, JC Helder, MB Sliwkowski. Peptide mapping for assessment of genetic heterogeneity of cell lines, detection of a low level Tyr to Gln sequence variant in a recombinant antibody. Bio/Technology 11: 1293, 1993.

156. HS Lu, ML Klein, RR Everett, PH Lai. Rapid and sensitive determination of protein disulfide bonds. In: J L'Italien, ed. Modern Methods in Protein Chemistry. New York: Plenum Publishing Co., 1986, pp 489–501.

157. M Haniu, C Acklin, WC Kenney, MF Rohde. Direct assignment of disulfide bonds by Edman degradation of selected peptide fragments. Int J Protein Peptide Res 43: 81–86, 1994.

158. SD Patterson, V Katta. Prompt fragmentation of disulfide linked peptides during matrix-assisted laser desorption ionization mass spectrometry. Anal Chem 66: 3727–32, 1994.

159. DL Crimmins, M Saylor, J Rush, RS Thoma. Facile in situ matrix-assisted laser desorption ionization-mass specrometry analysis and assignment of disulfide pairings in heteropeptide molecules. Anal Biochem 226: 355–361, 1995.

160. HR Morris, P Pucci. A new method for rapid assignment of S-S bridges in proteins. Biochem Biophys Res Commun 126: 1122–1128, 1985.

161. WR Gray. Disulfide structure of highly bridged peptides: A new strategy for analysis. Protein Science 2: 1732–1748, 1993.

162. MD Jones, J Hunt, JL Liu, SD Patterson, T Kohno, HS Lu. Determination of tumor necrosis factor binding protein disulfide structure: Deviation of the fourth domain structure from the TNFR/NGFR family cysteine-rich region signature. Biochem, 36: 14914–14923, 1998.

163. MD Jones, SD Patterson, HS Lu. Determination of disulfide bonds in highly bridged disulfide-linked peptides by matrix-assisted laser desorption ionization mass spectrometry with post-source decay. Anal Chem, 70: 136–143, 1998.

# 16
## Mass Spectrometry

**Vic Garner**
*University of Manchester, Manchester, United Kingdom*

## I. INTRODUCTION

In 1966, the late Michael Barber wrote, "One of the outstanding advantages of mass spectrometry is the precision with which it gives the molecular weight of an unknown compound. This is particularly important at high molecular weights since other methods then become increasingly inaccurate." (1). At that time there were few reported examples of mass spectra of pure compounds above mass 600, the two main reasons for this restriction were:

1. The lack of inlet systems that could be used to admit involatile or thermally labile samples;
2. The difficulty of attaining adequate spectrometric resolution at these high masses.

Ten years later Barber and his team invented FABMS, Fast Atom Bombardment Mass Spectrometry, for which he was elected to a Fellowship of the Royal Society. This technique afforded a means whereby mass spectra of involatile and labile samples could be generated routinely and initiated a new era in biological mass spectrometry. It is manifest today by a range of mass spectrometric-based techniques that may be utilised in the analysis and investigation of peptides and proteins. The plethora of techniques now available can be daunting; some techniques are more appropriate than others for particular applications. This chapter aims to provide scientists with interests in the analysis of peptides and proteins with the knowledge to facilitate selection of the most appropriate method or methods (see Fig. 1).

### A. Organic Mass Spectrometry

Mass spectrometry, along with infra-red, nuclear magnetic resonance, and to a lesser extent ultraviolet/visible spectroscopies comprise the main instrumental methods used to elucidate structures and identify organic compounds. Unlike the "spectroscopies" that have developed from the well-defined and predictable interaction between matter and electromagnetic radiation, mass spectrometry concerns the generation and separation of gas phase ions using their relative masses as the distinguishing characteristic. Mass spectrometers are in effect sophisticated weighing machines but they do not use the everyday technology of weight determination. Because of these unique characteristics we refer to mass spectro*metry* rather than mass spectro*scopy*.

How do mass spectrometers work? How can we best interpret the measure-

| 1886 | GOLDSTEIN discovers positive ions |
| 1898 | WEIN analysis by magnetic deflection |
| 1912 | THOMPSON ms of oxygen, nitrogen, carbon oxides; observes negative, multiple charged, metastable ions and isotopes of neon.  Nobel Prize |
| 1918 | DEMPSTER  directional focusing spectrometer |
| 1919 | ASTON velocity focusing spectrometer allows measurement of mass defects.  Nobel Prize |
| 1930 | CONRAD mass spectrometry of organic compounds |
| 1942 | Consolidated Engineering Co produce first commercial instrument for organic analysis |
| 1948 | CAMERON invents TOF |
| 1953 | PAUL *et al* invent QUISTORS |
| 1957 | KRATOS commercial double focussing instrument |
| 1958 | GC coupled to MS |
| 1975 | Capillary GC-MS |
| 1980 | VESTAL Thermospray |
| 1981 | BARBER Fast Atom Bombardment |
| 1985 | HILLENKAMP Matrix Assisted Laser Desorption |
| 1988 | FENN Electrospray |

**Fig. 1**   Some historical developments in mass spectrometry.

ments and data they generate? Can we relate their operation to other aspects of our experience in order to help us with the interpretation? How can we make best use of these instruments to solve the problems that face peptide chemists and biochemists? These questions, and others, we hope to address in the following sections.

## B.   The Nuts and Bolts

A mass spectrometer is used to determine the populations of ions having different mass-to-charge ratios (m/z); these data are presented as a histogram known as the "mass spectrum". In order to obtain such data the mass spectrometer has the following essential functions to perform:

1.   Generate vapour phase ions from compounds that may or may not be volatile;
2.   separate the ions;
3.   detect and record the separated ions.

There are several ways of achieving ion generation, separation and detection (Fig. 2): some of the more common techniques with their abbreviations or acronyms are listed below.

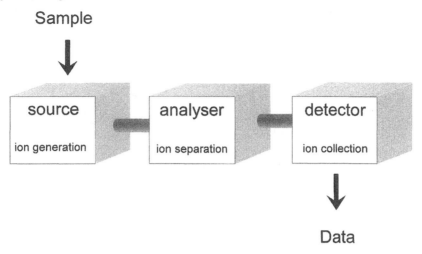

**Fig. 2** Generalized mass spectrometer.

Ion generation

| method | acronym |
|---|---|
| electron impact | EI |
| chemical ionisation | CI |
| fast atom bombardment | FAB |
| electrospray ionisation | ESI |
| matrix assisted laser-desorption ionisation | MALDI |

Ion separation is achieved using the following types of analysers:

| | |
|---|---|
| magnetic field alone | B |
| both magnetic and electric fields | B-E or E-B |
| quistors (quadrupole storage devices utilising radiofrequency and electric fields) | Q |
| time of flight | TOF |

There are also many ways of detecting the ions and generating a signal. The earliest methods depended upon photosensitive paper that was also sensitive to ions. Modern mass spectrometers use either Faraday cups or cascade and array detectors similar to the well-known electron-multipliers. Once an electrical signal has been generated it can be amplified and manipulated before being presented to a galvanometer, a cathode ray oscilloscope or to a computer interface.

### C. Ion Generation

#### 1. Electron Impact Ionization (Fig. 3)

This is perhaps the commonest ion source available and is the historically important method of ion generation that was relied upon more or less exclusively prior to the 1980's. In a typical electron impact source, sample molecules enter in the gas/vapour phase and interact with a beam of electrons. Electron loss is the favored process although electron capture also occurs but at a much lower rate (*ca.* 1%). The energy of the electron beam is usually $70\,eV$ ($\equiv 6.75 \times 10^{6}\,Jmol^{-1}$); less than a fifth of this energy is required to ionise the molecules and form molecular ions, the rest is dissipated through the fragmentation of various bonds to form fragment ions.

The products of electron impact ionisation are thus molecular species that either lack one electron (radical cations $M+\bullet$) or have gained one electron (radical anions, $M-\bullet$); these are odd-electron species. Fragment ions generally arise by loss of a radical and are thus even-electron species, either cations or anions.

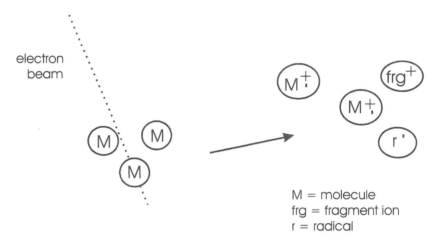

**Fig. 3**   Radical cation formation and fragmentation in gas phase electron impact ionization.

All these ionic products are induced to leave the source (i.e. the ionisation zone) by the application of a small electrical potential. They pass out through an exit slit, and are then subjected to a relatively high potential (four or eight thousand volts in most sector instruments) which accelerates the ions into the analyzer region where separation according to the mass to charge ratio takes place. The whole process takes place in a very high vacuum, of the order of $10^{-7} - 10^{-8}$ torr. The time scale for ionisation is in nanoseconds.

## 2.   *Chemical Ionization* (Fig. 4)

This ionisation method uses the same hardware as electron impact but involves the intermediacy of a reagent gas. Several

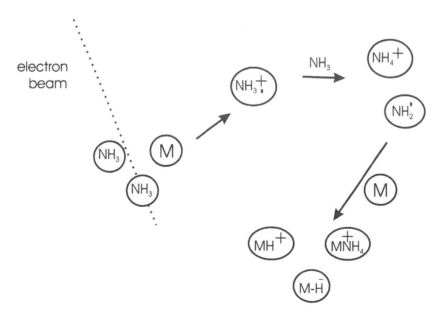

**Fig. 4**   Protonation and cationation by gas phase reagent ions.

reagent gases have been used, ammonia is often preferred because it does not contaminate the ion source as much as methane or isobutane. The reagent gas is introduced into the source to give a pressure of about $10^{-5}$ torr. When the electron beam is turned on, radical cations are formed from the reagent gas molecules; these undergo ion-molecule reactions with the excess reagent gas under the relatively high pressure in the source. The ammonium (or carbonium) ions thus formed are in the gas phase and as such are powerful acids, they protonate the analyte molecules to form cationated species sometimes called "pseudomolecular" ions $[M + H]^+$. This ionisation of the sample molecules requires less energy than that involving electron impact hence once formed, these cationated species tend to remain intact and not undergo fragmentation. Chemical ionisation is thus said to be a "softer" ionisation method than electron impact.

These classical methods of ion generation from organic analytes (EI and CI) require prior volatilization of the sample:

this restricted mass spectrometry to relatively low molecular mass, thermally stable analytes. Later techniques allow gaseous ions to be generated from the condensed phase hence larger molecules that are much less volatile can be analyzed.

### 3. Fast Atom Bombardment Ionization (Fig. 5)

This technique was invented in Manchester in the late 1970s and revolutionized biological mass spectrometry: it is sometimes known as LSIMS (liquid secondary ion mass spectrometry). A beam of "fast, primary" atoms or ions ($\approx 8$–$10\,\mathrm{KeV}$) impinges on the analyte presented as a mixture with a relatively involatile liquid matrix. Positive and negative ions are desorbed together with neutral species. Matrix effects—hydrophobicity especially, are important and cationation is common. Continuous flow FAB involves the continuous replenishment of the sample surface by use of a hollow probe. Although a soft ionization technique in general, fragmentation is

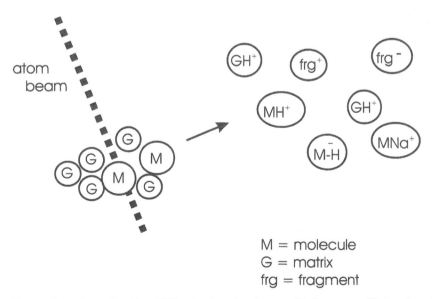

M = molecule
G = matrix
frg = fragment

**Fig. 5** Formation of gas phase ions following bombardment with fast atoms (Xe) or ions (Cs$^+$) in a liquid phase.

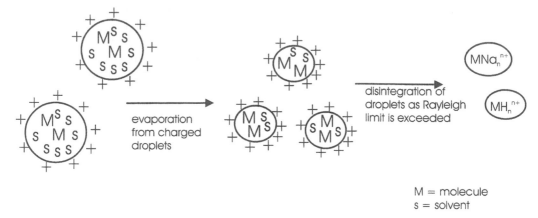

M = molecule
s = solvent

**Fig. 6**   Generation of gas phase cluster ions from charged droplets.

usually observed; the main products are thus "pseudomolecular species" e.g. $[M+H]^+$ and $[M-H]^-$ together with cationic and anionic fragments.

tions cause the droplets to disintegrate or explode. The ions thus formed are multiply charged i.e. $(M+nH)^{n+}$ where n can be quite large e.g. 30 or greater.

### 4.  *Electrospray Ionization* (Fig. 6)

Ionization takes place from electrostatically charged droplets in a nebula produced from a solution of the analyte (normally aqueous methanol or similar). Evaporation of the solvent molecules from the droplets continues until the Rayleigh limit is reached when coulombic interac-

### 5.  *Laser Desorption Ionization* (Fig. 7)

Although desorption ionization methods have been known for some time it is only recently that matrix assisted laser desorption (MALDI) has been applied widely. The matrix is an UV absorbing material such as an aromatic acid that is

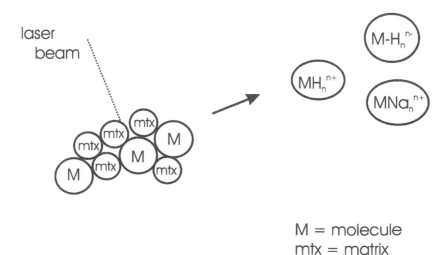

M = molecule
mtx = matrix

**Fig. 7**   Desorption of cluster ions following irradiation of crystalline matrix and sample mixture.

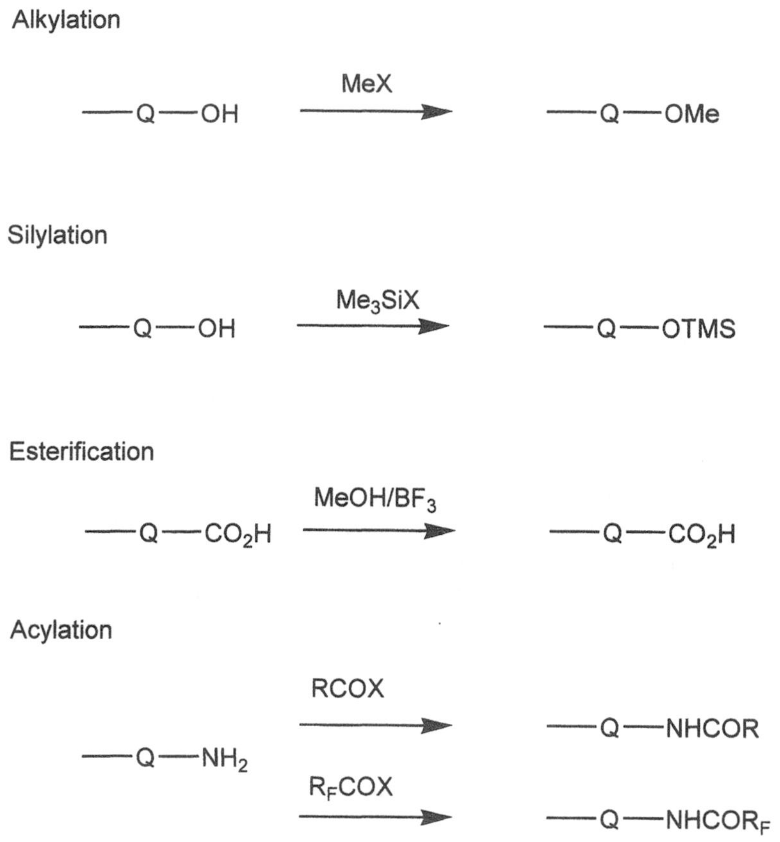

Alkylation

$$—Q—OH \xrightarrow{MeX} —Q—OMe$$

Silylation

$$—Q—OH \xrightarrow{Me_3SiX} —Q—OTMS$$

Esterification

$$—Q—CO_2H \xrightarrow{MeOH/BF_3} —Q—CO_2H$$

Acylation

$$—Q—NH_2 \xrightarrow{RCOX} —Q—NHCOR$$

$$\xrightarrow{R_FCOX} —Q—NHCOR_F$$

**Fig. 8** Derivatization reactions.

intimately mixed with the analyte. Interaction with the laser beam results in improved energy transfer to the analyte either via the matrix itself or a photodegradation product of it. High molar mass compounds can be analyzed by this technique in which cluster ions $[M + nH]^{n+}$ (where n is a low integer) are usually formed.

## D. Ion Separation in Mass Analyzers

High-resolution mass spectrometry requires the use of a double focusing instrument that utilizes magnetic and electric fields to effect a very precise separation of the ions. Such instrumentation is large, expensive and can be relatively slow in operation. The quadrupole analyzers have been developed for their simplicity of manufacture and speed of operation but they lack the precision of the sector instruments. The time of flight analyzers are enjoying resurgence in popularity due to their applicability to the analysis of short events.

### 1. Magnetic Deflection Analyzers

Ions emanating from the source have been accelerated by a large potential to form a coherent beam that now enters a magnetic field. For the ions to reach the collector slit where they will be detected, they have to traverse a circular path of a particular radius (related to the mass to charge ratio) through the magnetic field. By changing the magnetic field strength (B), ions of different m/z ratio separated by the magnetic field can be allowed to reach the detector.

The magnetic analyzer is, in fact, a momentum rather than a mass analyzer thus ions with the same m/z value but having differing translational energies are not brought to a point focus using only a magnetic field. This can be overcome by use of an electrostatic analyzer in series with the magnetic analyzer whereby velocity focusing of the ion beam is achieved. A double focussing instrument thus comprises an electric analyzer then a magnetic analyzer in the "normal geometry"; in "reverse geometry" the ions enter the magnetic field first. The geometry affects the transmission of metastable ions (ions that have been formed by decomposition processes taking place outside the source).

## 2. Quadrupole Ion Storage Devices or Mass Filters

The commonest construction of this type of analyzer involves four parallel rods arranged symmetrically. A composite voltage made up of an electrical component (DC) and a radiofrequency component (RF) is applied between adjacent rods, opposite rods are electrically connected. When ions are accelerated by a small potential (typically 10–20v) into the analyzer, ions within a small range of m/z values undergo stable oscillations and continue along this oscillating path between the rods to the detector. Ions with m/z values outside this small range develop unstable oscillations, and are lost to the rod assembly. Mass separation is thus achieved and a mass spectrum obtained by varying DC and RF whilst maintaining the ratio DC/RF constant. Unlike the magnetic analyzer, small changes in translational energy or deflections due to collisions with gas molecules in the analyzer have little effect upon performance. The "ion trap" mass analyzer is a derivative of this arrangement.

## 3. Time-of-Flight Analyzer

An ion that is accelerated through an electrical potential has a velocity that is characteristic of its m/z ratio. In the time-of-flight analyzer, ions are separated according to their differing velocities and their m/z values are computed from the time taken for the ions to traverse a specified flight path to the detector.

## E. Towards 2000, Some Recent Developments in Mass Spectrometry

The last 20 years have seen an enormous expansion in the analytical applications of mass spectrometry especially in the biological and medical sciences. Before 1980, and the introduction of FABMS, the mass spectrometric analysis of analytes of molar mass greater than 2000 Da was a unique achievement requiring substantial manipulative skill and a fair amount of good luck! FAB led the way to a range of alternative methods for the ionisation of large and labile molecules such that analysis of 300 kDa peptides and proteins has become routine. The increasing availability of tandem mass spectrometers, i.e. two mass spectrometers linked via a collision cell, over the last decade has led to further expansion in the amount of quality data that are becoming available from the application of this technique in the study of biological macromolecules including peptides and proteins.

Developments at the other end of the scale have also taken place: routine methods are now available for the precise determination of isotopic ratios so that metabolic dysfunction can be measured from samples of breath, and adulterated foodstuffs can be identified readily. The extraordinary progress that has taken place has happened so rapidly that few teaching courses have been able to either accommodate or keep up with the changes. Even here we will only be able to identify the major trends and inspect a few illustrative examples.

## II. ELECTRON IMPACT AND CHEMICAL IONIZATION

The electron impact ionization source is widely used because it affords not only molecular mass data but also, through decomposition of the molecular ions to fragment ions, is diagnostic of structure as well. The fragmentations and rearrangements that occur due to the high intrinsic energy of the radical cations or anions first formed follow recognisable pathways that can be rationalized using thermodynamic principles. Frequently the energy transfer is so efficient, and the stability of the molecular ions relatively low, that only fragment ions are observed in the spectrum. This can be overcome in some cases by using a "softer" technique such as chemical ionization that involves transfer of much less energy.

The major disadvantage of both these methods of ion generation is the need for the analyte to be volatile. Thus highly polar molecules and macromolecules are unlikely to possess sufficient volatility for them to enter the electron impact source. Heating the sample can increase its vapour pressure to a certain extent but the likelihood of thermal decomposition or rearrangement taking place is also increased. Another way of addressing this difficulty is to reduce the polarity of the analyte and at the same time increase its volatility and thermal stability by conversion to a chemical derivative. Many derivatisation procedures have been devised (2) that operate on a very small scale, are selective, practically quantitative and provide a stabilized product that retains the key structural features of the original analyte. However, artefacts arising from the derivatization procedure may be formed as a result of either over--reaction or incomplete reaction as well as

reaction at other sites. Some examples of derivatization methods are given in Fig. 8; generally, as small a group as possible is introduced so that the molecular mass of the derivatized product is not increased so much as to exceed the mass range of the instrument.

Electron impact and chemical ionization of derivatized, and non-derivatized compounds is widely used especially in combination with prior separation by gas chromatography. Provided a thermally stable and volatile sample is available, perhaps after derivatization, then coupled gas chromatography mass spectrometry (GC-MS) affords an extremely efficient technique for the separation and characterization of complex mixtures. Increased sensitivity is often observed when the mass spectrometer is operated in CI mode. However, the preparation of volatile derivatives from peptides and proteins is tedious so this approach tends to be restricted to fairly low mass oligopeptides and alternative ionization methods used for the larger and more polar molecules.

Another strategy that made use of the separatory power of GC-MS involves the partial hydrolysis of the peptide or protein using either acid or one or more enzymes. The mixture of di- and tri- peptides thus formed is subjected to reduction (forming polyaminoalcohols) and trimethylsilylation. This mixture of polyaminosilylethers is analyzed using GC-MS, the primary structure of the original peptide can be deduced by identification of the overlapping components. There are many examples of the application of this protocol in the literature, see for example, the determination of the primary structure of a carboxypeptidase (39 residues) described by Nau and Biemann (3).

### A. Examples of EI and CI (Fig. 9)

An essential point to remember when interpreting data from EI and CI spectra

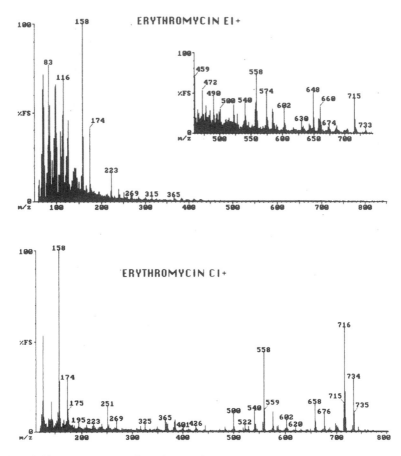

**Fig. 9**  El and Cl mass spectra of erythromycin ($M_R$ 733).

is the different form of the molecular species produced, i.e. in EI it is usually a radical cation whereas in CI a protonated species is formed. Even when derivatized however, there is no guarantee that a molecular ion will be observed under electron impact ionization. Fig. 10 shows the EI spectrum of N-acetylvaline methyl ester, the molecular ion peak is just discernible at 173.

## III.  FAST ATOM BOMBARDMENT AND SECONDARY IONIZATION

The general applicability of mass spectrometry to the determination of primary structures of peptides and proteins was restricted by the necessity to chemically modify these polar compounds in order to impart sufficient volatility for them to enter the EI/CI source. Even then there were problems with particular residues such as arginine which possess polar side chains. Although alternative techniques such as plasma and field desorption mass spectrometry that did not require a volatile sample had been introduced in the 1970s they required special equipment and/or considerable experimental skill on behalf of the operator. The introduction of FAB in the early 1980s overcame these difficulties and eliminated the tedious, time- and material-consuming chemical conversions necessary to prepare derivatives of adequate volatility.

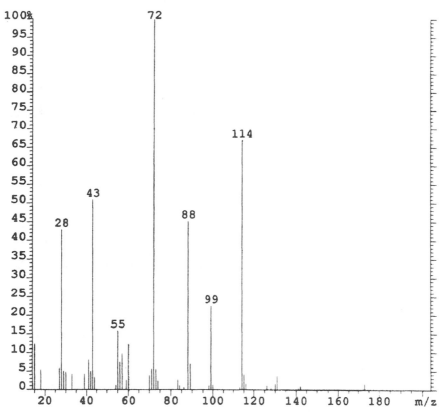

**Fig. 10** EI spectrum of N-acetylvaline methyl ester ($M_R$ 173).

Peptides and proteins were particularly amenable to analysis by the newly introduced technique such that there was a proliferation of applications described in the literature that clearly demonstrated the ionization of progressively larger molecules without the need for prior derivatization. A very early experiment that clearly demonstrated the applicability of FABMS to the analysis of peptides compared two isomeric tripeptides, Ala Leu Gly and Gly Leu Ala. Barber's group showed that the two compounds afforded molecular mass data in both positive and negative ion modes (4). Not only that, but the patterns of fragment ions in the positive ion spectra were clearly different. Inspection of the molecular structures of these two molecules shows that they effectively differ only in the location of a single methyl group (highlighted in Fig. 11) yet this is sufficient for primary sequence information to be deduced. (Figs 12 and 13). N-blocked and cyclic peptides or the inclusion of certain modified amino acids such as gamma carboxy glutamic acid interfere with the operation of automated Edman degradation and sequence determination. FABMS can be applied in all these situations. Figure 14 shows the positive ion FAB mass spectrum of Gramicidin S.

Biological samples frequently contain traces of alkali metal ions; the presence of such contaminants is not detrimental, in fact it is often advantageous by picking out the molecular ion region through identification of peaks at 22 and 38 Daltons above the [M + H] peak (due to [M + Na] and [M + K] species).

Ala Leu Gly

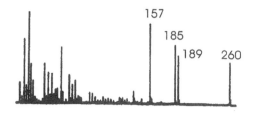

Gly Leu Ala

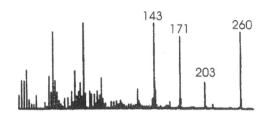

**Fig. 11**   Historical FAB spectra of ALG and GLA.

Since such cationization rarely occurs with fragment ions, this device can be used to advantage by the deliberate addition of alkali metal salts to enhance such signals, see for example Fig. 15.

   An essential feature of FAB is the use of a liquid matrix to support the sample. A wide range of matrices have been

described but for peptides and proteins, glycerol and thioglycerol either singly or mixed together have been widely used. Hydrophobicity effects can modify the behaviour of the analyte profoundly, so much so that in some extreme cases the analyte appears to refuse to sputter and no spectrum is observed. In such cases it

**Fig. 12**   Sequence ions for Ala Leu Gly.

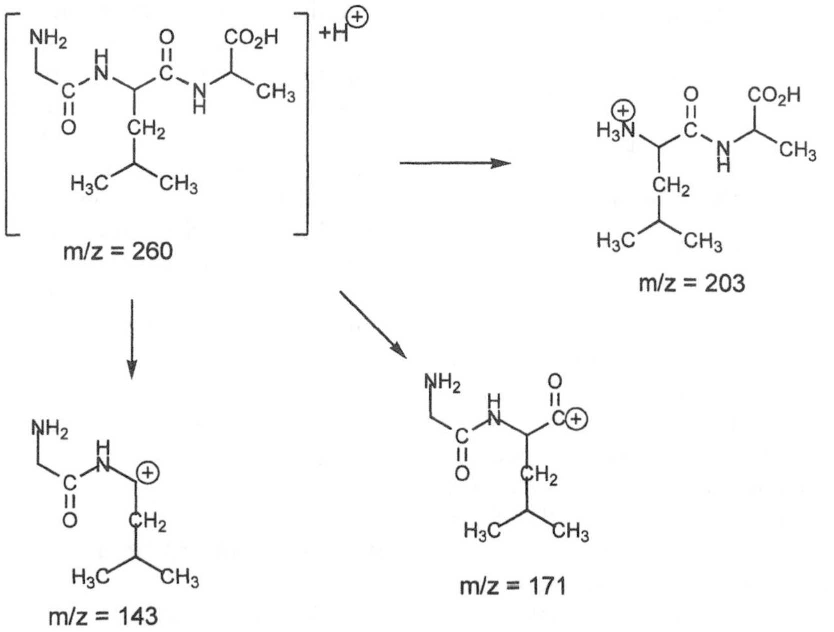

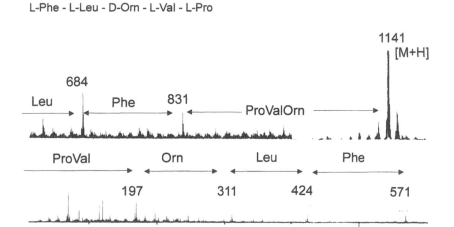

**Fig. 13** Sequence ions for Gly Leu Ala.

becomes necessary to modify the matrix slightly by the addition of reagents such as trifluoroacetic acid or even perform on-probe derivatization with an alkylating or acylating agent. The matrix itself often appears in the spectrum as a series of cluster and cationated ions and these are usually enhanced when the analyte "is unco-operative". These ions are readily observed by use of a mixture of glycerol [G] to which a small amount of sodium iodide has been added whereby peaks cor-

L-Pro - L-Val - L-Orn - L-Leu - D-Phe
    |                          |
L-Phe - L-Leu - D-Orn - L-Val - L-Pro

**Fig. 14** FAB (+ve) mass spectrum of Gramicidin S.

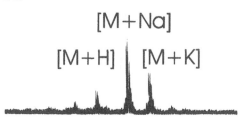

**Fig. 15** Molecular ion region of zervamycin (M 1838).

responding to [nG + H] and [nG + Na] are seen i.e. at (93 + 92*n) and (105 + 92*n), where n = 0 or an integer. A similar system can be used to calibrate the instrument; caesium iodide affords an extensive spectrum of cluster ions of the form $[Cs_{(n+1)}I_n]^+$ shown in Fig. 16.

Continuous flow FAB requires a probe that has a capillary through which the sample in solution can be continuously replenished on the probe tip as it is consumed or evaporates; otherwise the operation is no different. Such an arrangement allows the prior separation of a mixture of peptides, for example arising from an enzyme digest, by liquid chromatography and the column effluent to be directly introduced into the mass spectrometer. Suppression effects, whereby one component interferes with the ionization of other components of the mixture are then minimized or eliminated.

Much of the early development work on FAB was carried out using argon as the bombarding species. It was soon found however that the use of xenon allowed much greater sensitivity despite the higher initial costs. Caesium ion guns also provide enhanced sensitivity and have been used to good advantage in the ionization of macromolecules.

## IV. FRAGMENTATION

The formation of fragment ions is an essential feature of mass spectrometry. The extent of fragmentation is related to the energy transfers involved in the initial ionization and can be quite extensive in some methods. The fragmentation processes themselves are not random but follow defined pathways; with peptides and

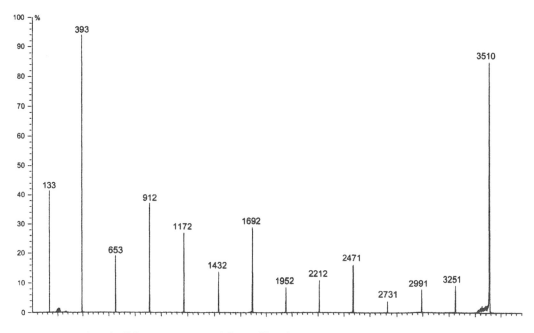

**Fig. 16** Caesium iodide spectrum used for calibration.

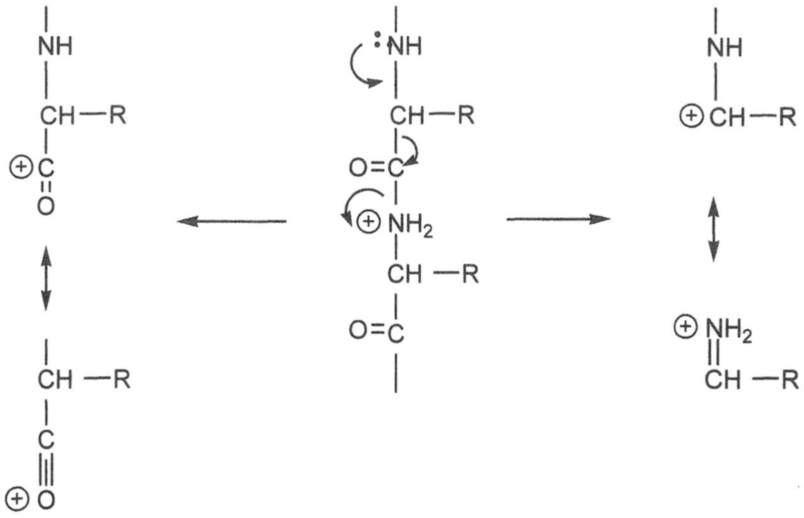

**Fig. 17** Generation of resonance stabilized acylium and aldimine ions by peptide bond fission.

proteins these pathways may involve cleavage of bonds in the side chains or bonds constituting the peptide backbone. When the latter occurs, fragment ions are formed that are diagnostic of the peptide primary structure and are referred to as sequence ions.

Fragmentation involves fission or cleavage of a bond whereby the bonding electrons become disrupted. In the case of a radical cation, homolytic fission of the bond generates a cation and a radical (which is not detected directly in positive ion mode) whereas even-electron species tend to undergo heterolytic fission generating cations and neutral molecules. Thus both acylium and aldimine (or immonium) cations can arise from the peptide chain as exemplified in Fig. 17. Roepstorf, Fohlman (5) developed a nomenclature for the bond fissions that can take place, this is summarized in Fig. 18. The particu-

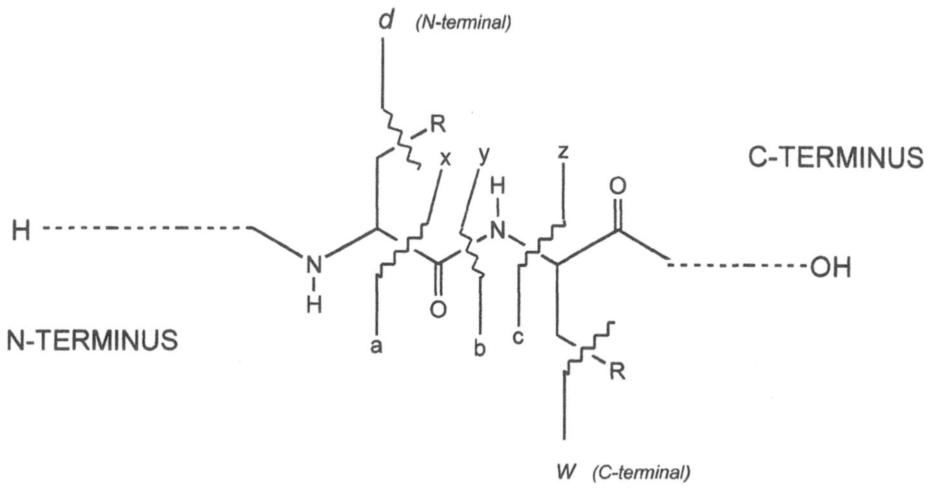

**Fig. 18** Peptide fragmentation nomenclature.

lar pathway or mechanism that is followed depends upon a variety of factors including the chain length, polarity and secondary structural characteristics as well as the ionization mode.

## V.  ELECTROSPRAY IONIZATION

The spray technologies, thermospray, plasmaspray, ionspray, and electrospray have developed as very soft ionization sources that usually generate cluster ions carrying several charges. Ionization is achieved by evaporation of solvent molecules from droplets of analyte-containing solution in a nebula that has been produced using a variety of techniques. Hence the range of sources that developed as interfaces for liquid chromatographs each one using a different or a combination of ways of effecting evaporation from the charged droplets. Again, derivatization is not required and because of the phenomenon of multiple charges in the cluster ions, quite large molecules can be investigated with existing mass analyzers. If an analyzer has a mass limit of say 15 kDa, then singly charged ions of mass to charge ratio (m/z) 15 kDa is the maximum that could be transmitted. However, if the molecule ionizes to form a cluster carrying 10 charges, i.e. protonation occurs at ten sites on the molecule to give $[M + 10H]^{10+}$, the m/z value of the largest ion that can be transmitted is still 15 kDa but because the denominator, z, now equals ten, the molecule can have a molecular mass of 150 kDa, (m/z = 150000/10).

In order to take advantage of this arithmetic factor, it is an essential requirement that the solution being sprayed into the source contains analyte-derived ions. In the case of peptides and proteins, this can be easily achieved by using aqueous methanol (or other low alcohol or acetonitrile) as the solvent to which has been added a small amount of formic, acetic or trifluoroacetic acid so that basic amino

acid residues are protonated. It is essential to avoid the use of involatile buffers that have a habit of crystallizing out just where they are not wanted!

Figure 19 shows the electrospray ionization mass spectrum of normal haemoglobin diluted in acetonitile + aqueous formic acid; two overlapping series of peaks can be discerned and deconvoluted to give values of 15126.6 and 15867.4 Da for the singly charged species.

An important application of electrospray ionization is to determine molecular masses of peptides and proteins to an accuracy that far exceeds that of any other method. This is facilitated by the softness of the technique whereby mainly molecular species are formed with very little fragmentation taking place. However, it is possible to adjust the cone voltage and evaporation rate in the ionization source, and enhance the amount of fragmentation that takes place so that useful structural information can be generated. The fragmentation usually occurs at the peptide link as described earlier.

## VI.  MATRIX ASSISTED LASER DESORPTION IONISATION

Matrix assisted laser desorption ionization has taken over from plasma desorption ionization (PD) for the generation of mass spectra of macromolecules. PD generated ions by means of a plasma formed during the passage of a radioactive decay particle from californium-252 through the sample. Developments in laser technology have provided an alternative energy source to effect similar desorption and ionization.

The sample is mixed in solution with an UV-absorbing matrix and the mixture loaded onto the sample holder where it is evaporated to dryness. A range of matrices is available, synapinic acid and α-cyanocinnamic acid (Fig. 20) have both been found to be good for peptides, proteins

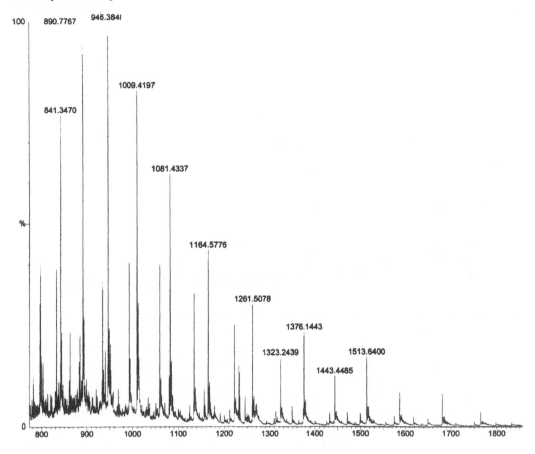

**Fig. 19** Electrospray spectrum (positive mode) of normal hemoglobin.

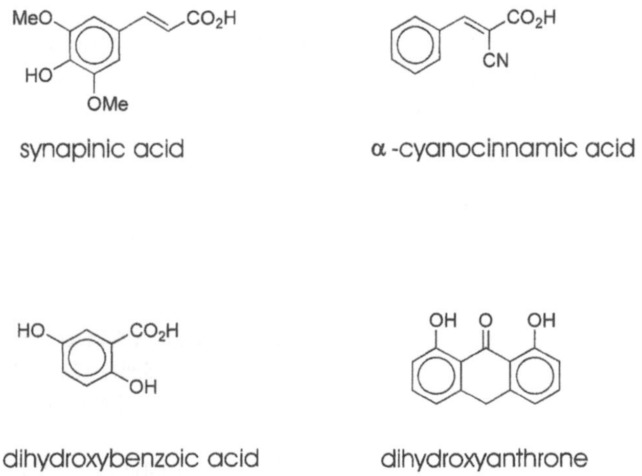

synapinic acid

α -cyanocinnamic acid

dihydroxybenzoic acid

dihydroxyanthrone

**Fig. 20** Matrices used in MALDI-TOF.

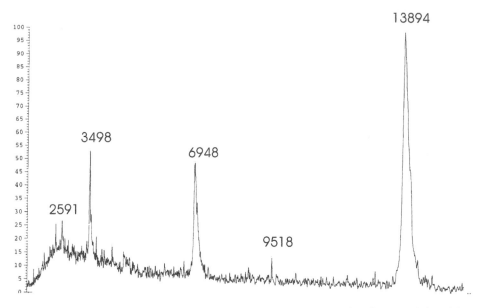

**Fig. 21**   MALDI-TOF spectrum of Malayan Pit Viper venom fraction in synapinic acid.

and digests. The aim is to form a solid solution of the matrix and analyte so that energy transfer is maximized. The nitrogen laser produces a 4 ns pulse at 337 nm, the matrix absorbs the energy that drives decomposition and ionization reactions in both the matrix and sample. The desorbed ions and neutral particles thus formed are accelerated from the source by a high electrical potential into the TOF analyzer. A slight delay in the extraction of the ions gives rise to better quality spectra by effectively energy-focussing the ions as in the electrostatic analyzer of a sector instrument. Delayed extraction of the ions from the source and other electronic procedures, such as incorporation of a Reflectron which effectively doubles the ion path, in the TOF analyzer allow highly refined mass spectra to be recorded for minute samples of macromolecules with masses approaching $10^6$ Da. Figure 21 shows the spectrum generated from a fraction of Malayan Pit Viper venom recorded using synapinic acid as the matrix.

The technique can be applied to single samples of pure peptides and pro-

teins or to highly complex mixtures arising from enzyme digests. Figure 22 shows the spectra obtained for a tryptic digest of myoglobin, use of the Reflectron technique affords excellent resolution. Small amounts of contaminating buffer salts and detergents can be accommodated but too much can interfere with the energy transfer mechanisms. Ladder sequences generated by controlled Edman degradation of single peptides and proteins are another good example of the application of this technique to the analysis of complex mixtures. The ladder corresponds to a series of peptides each differing by one amino acid residue, thus the mass difference between the peaks seen in Fig. 23 corresponds to each particular amino acid residue and thus the sequence can be determined (Fig. 24).

## VII.   HYPHENATED TECHNIQUES AND TANDEM MASS SPECTROMETRY

When two instrumental techniques such as chromatography and mass spectrometry

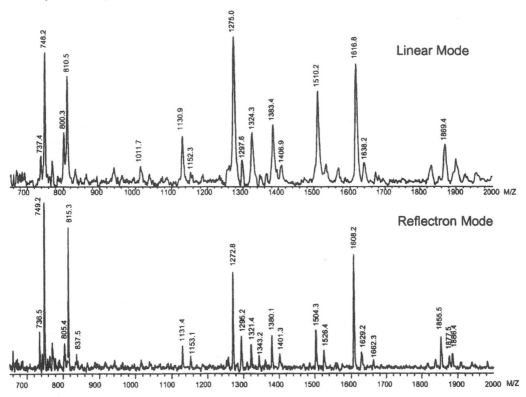

**Fig. 22** Myoglobin (tryptic digest) recorded in normal and reflectron modes.

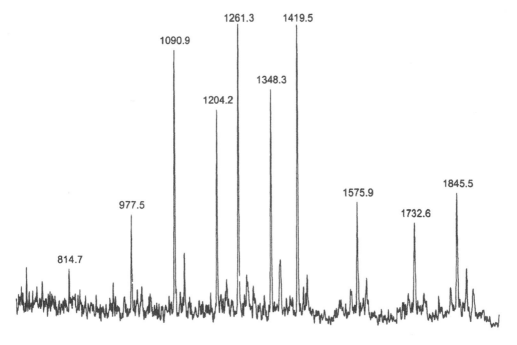

**Fig. 23** Ladder sequence for LRRASGLIYNNPLMAR-NH$_2$.

| Observed ion | Difference | Interpretation |
|---|---|---|
| 1845.5 | | |
| 1732.6 | 112.9 | I/L |
| 1575.9 | 156.7 | R |
| 1419.5 | 156.4 | R |
| 1348.3 | 71.2 | A |
| 1261.3 | 87.0 | S |
| 1204.2 | 57.1 | G |
| 1090.9 | 113.3 | L |
| 977.5 | 113.4 | L |
| 814.7 | 162.8 | Y |
| 700.2 | 114.5 | N |

**Fig. 24** Interpretation of ladder sequence data.

are coupled together, it is referred to as hyphenation; GC-MS has already been mentioned. Hyphenation allows mass spectrometers that are excellent for detection and identification to be linked to systems that afford efficient separation of complex mixtures. The actual coupling is not without problems because the separatory system often involves high pressures and the mass spectrometer requires very high vacuum to operate. However, these problems have been overcome and gas or liquid chromatographic and capillary electrophoresis or electrochromatography systems coupled to mass spectrometers are routinely used for analysis. Additional benefits accrue when two mass spectrometers are used: this arrangement can also be considered to be a hyphenated technique and is called either MS-MS or "Tandem" mass spectrometry.

Tandem instruments consist of two mass spectrometers coupled together in series separated by a collision gas cell. The general arrangement is depicted in Fig. 25. The gas cell can be either a quadrupole (q) or hexapole (h) arrangement similar to an analyzer but without the applied electric field: thus ions are transmitted but not mass analyzed. The mass spectrometers themselves can be of any type: several variations are commercially available e.g. Q-q-Q; EB-q-BE; EBE-q-TOF, Q-q-TOF.

A typical application for structural studies such as the determination of the primary sequence of a peptide requires both molecular mass information and fragment ion data. This is readily achieved

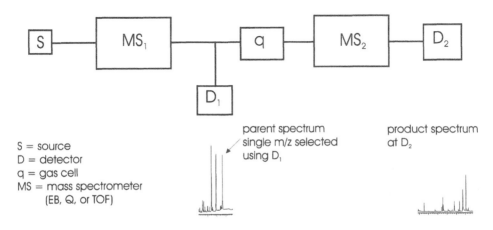

S = source
D = detector
q = gas cell
MS = mass spectrometer
(EB, Q, or TOF)

parent spectrum
single m/z selected
using D₁

product spectrum
at D₂

**Fig. 25** General arrangement of a tandem mass spectrometer.

using MS-MS. The sample is ionized using an appropriate system such as FAB, MALDI or Electrospray and the molecular ion region observed using the first off-axis detector. One peak is then selected by tuning MS1 to allow transmission only of ions having this particular m/z value. These ions pass into the collision cell where they collide with gas molecules and undergo decomposition to form fragment ions. Varying the nature and pressure of the gas in the cell can control the rate and extent of decomposition. This collisionally induced decomposition (CID) generates a mixture of ions that is separated in the second mass spectrometer; i.e. the gas cell effectively acts as a source for MS2. The mass spectrum recorded at the second detector, often referred to as the "product ion spectrum" to distinguish it from the first or "parent spectrum", by scanning MS2 allows the sequence and other ions to be identified.

MS-MS spectra from Peptide Substance P are reproduced in Fig. 26; the upper spectrum (a) shows the sequence ions formed under low energy collision conditions using helium as the collision gas. When the collision gas is changed to xenon, the higher energy conditions lead to much more side chain fragmentation and the formation of "d" type ions (spectrum b). The ability to generate such complementary information is a valuable feature of MS-MS.

The precision of the data generated in MS-MS experiments is startling. Fig. 27 shows the spectrum of Renin Substrate obtained from an MS-MS experiment using xenon as collision gas with a MicroMass AutoSpec-TOF instrument, both a and b ions were observed as well as d ions, Fig. 28 lists the expected and observed masses. Figure 29 shows the product ion spectrum generated from a synthetic fibrinopeptide analogue using an electrospray source fitted to a Micro-Mass Q-Tof instrument. In this MS-MS experiment, the doubly charged ion at m/z = 785.86 was selected for decomposition to give the spectrum shown from which the primary sequence can easily be confirmed.

A wealth of structural information becomes accessible to the peptide scientist through the use of modern mass spectrometers. The increasing availability of sequence databases and developments in computerized file comparison techniques can only lead to further advances in our knowledge and understanding of peptide and protein drugs.

## VIII. ISOTOPE RATIO MASS SPECTROMETRY

Isotopically labeled compounds have found wide use in conjunction with mass spectrometry to trace biochemical (and other) pathways. In the isotope ratioing technique, instead of attempting to ionize, accelerate and detect intact molecular species or isotopomers, the analyte is converted into small simple molecules such as carbon dioxide, nitrogen or water, which can be analyzed with much greater reproducibility and precision. Using this approach the ratio of carbon or nitrogen or other isotopes can be determined and used to investigate the kinetics of physiological processes involving peptides and proteins.

Conversion of the individual samples into the same chemical species (carbon dioxide for example) removes some of the variables arising from different isotopomeric distribution and sample behaviour inherent in normal mass spectrometric analysis. Slight differences in analyzer operation minimize background signal fluctuation by simultaneous recording of the ion beams. Taken together these changes result in a dramatic improvement in the precision of the method and allow determination of isotopic ratios down to about $10^{-5}$ atom %.

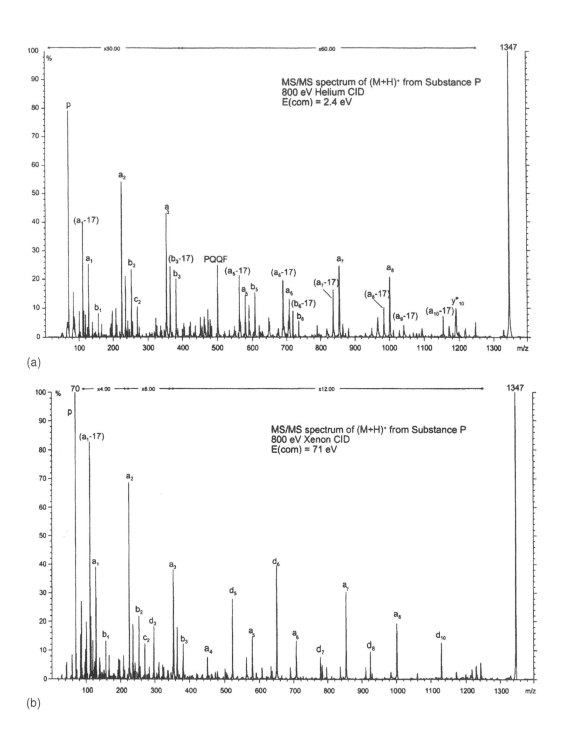

**Fig. 26**   MS-MS spectra of Substance P. (a) Helium CID, (b) xenon CID.

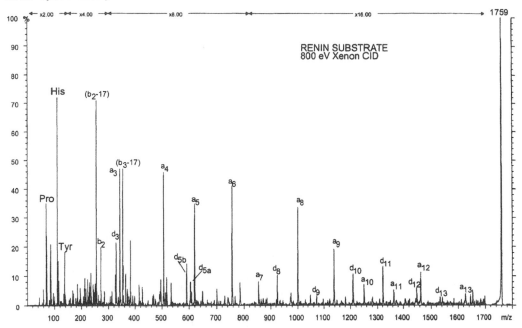

**Fig. 27** Renin substrate FAB MS-MS AutoSpec-TOF.

| | Expected mass | Observed mass | | Expected mass | Observed mass |
|---|---|---|---|---|---|
| His | 110.1 | 110.02 | b5 | 647.35 | 647.28 |
| Tyr | 136.1 | 136.02 | a6 | 756.42 | 756.40 |
| a2 | 244.14 | 244.08 | b6 | 784.41 | 784.46 |
| b2-17 | 255.11 | 255.03 | a7 | 853.47 | 853.41 |
| b2 | 272.14 | 272.07 | d8 | 924.51 | 924.62 |
| d3 | 329.19 | 329.14 | a8 | 1000.54 | 1000.59 |
| a3 | 343.21 | 343.14 | d9 | 1071.57 | 1071.77 |
| b3-17 | 354.17 | 354.13 | a9 | 1137.60 | 1137.74 |
| d4 | 414.25 | 414.11 | d10 | 1208.63 | 1208.55 |
| a4 | 506.27 | 506.20 | a10 | 1250.68 | 1250.70 |
| b4-17 | 517.24 | 517.04 | d11 | 1321.72 | 1321.79 |
| b4 | 534.27 | 534.16 | a11 | 1363.76 | 1364.18 |
| d5a | 591.33 | 591.34 | d12 | 1448.82 | 1448.91 |
| d5b | 605.30 | 605.19 | a12 | 1462.83 | 1462.94 |
| a5 | 619.36 | 619.37 | M+H | 1758.93 | 1758.93 |

**Fig. 28** Product ion spectrum from Renin Substrate (AutoSpec-TOF, 800 eV Xenon CID).

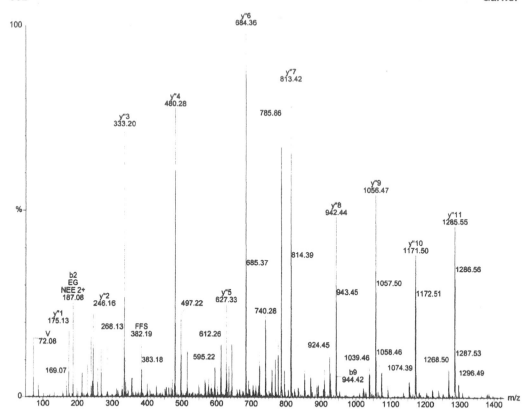

**Fig. 29** MS-MS of a synthetic Glu-fibrinopeptide (EGVNDNEEGFFSAR) recorded on a MicroMass Q-Tof instrument.

The greater availability of amino acids and peptides labeled with stable isotopes of carbon, nitrogen, sulphur, oxygen and hydrogen has led to an increase in the applications of this approach for the study of metabolic processes and dysfunction. Protein turnover rates can be determined by administering a sample of $^{13}C$- or $^{15}N$-labeled amino acid such as leucine or glycine and the metabolic rate assessed by measuring the formation of labeled metabolite from serum, urine, breath gas or biopsy samples. Use of two labeled amino acids e.g. $^{13}C$-leucine and $^{15}N$-phenylalanine allows different metabolic pathways to be distinguished and compared. (Fig. 30).

Breath gas analysis using stable isotopes has become a major application of mass spectrometry in the field of clinical

diagnosis. The improved precision of modern instruments has led to many applications using natural isotope sig-

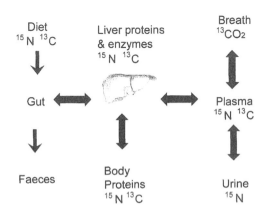

**Fig. 30** General principle for protein turnover determination by SIRMS.

natures in forensic analysis and the detection of adulteration of foodstuffs as well as the classical use of isotopomers in tracer and metabolic studies.

## ACKNOWLEDGEMENTS

The support of Keith Hall of Medimass and Brian Green of Micromass, both based in Manchester, UK, is gratefully acknowledged. The contributions of all the mass spectrometrists who have used the technology to such advantage is also acknowledged and it is not disrespect that has prevented citation of their work.

## REFERENCES

1. M Barber, RM Elliott, TO Merren. Adv Mass Spectrom Proc Conf 3: 717–730, 1966.
2. HR Morris, RJ Dickinson, DH Williams. Biochem Biophys Res Commun 51: 247–255, 1973.
3. H Nau, K Biemann. Anal Biochem 73: 175–186, 1976.
4. M Barber, RS Bordoli, RD Sedgwick, LW Tetler. Org Mass Spectrom 16: 256–260, 1981.
5. P Roepstorff, J Fohlman. Biomed Mass Spectrom 11: 601, 1984, and 12: 631, 1985.

## BIBLIOGRAPHY

There is an extensive range of texts at all levels dealing with the practicalities of mass spectrometers from applications and interpretation to instrumentation and ionization; the reader is recommended to the following selection:

JR Chapman. Practical Organic Mass Spectrometry, A Guide for Chemical and Biochemical Analysis. 2nd ed. Chichester: John Wiley & Sons, 1993.

A Ashcroft, Ionisation Methods in Organic Mass Spectrometry. London, RSC Analytical Spectroscopy Monographs; 1997.

AW Johnstone, ME Rose, Mass Spectrometry for Chemists and Biochemists. 2nd ed. Cambridge University Press, 1996.

# 17

# Protein and Peptide Immunoassays

**Anthony B. Chen**
*Genentech Inc., South San Francisco, California*

## I. INTRODUCTION

Although there are many books and reviews written on the development of methods for quantitating peptides and proteins in clinical chemistry, these are written for analytes that are determined over the normal range observed in humans (1–4). The development of immunoassays to analyze recombinant proteins and peptides in nonclinical and clinical studies has been an extraordinary learning curve for us with many parameters of assay development being discovered by trial and error, sound experimental design and by serendipity (5). In drug development of protein pharmaceuticals, it is necessary to determine the concentration of the administered product in physiological fluids which may contain endogenous identical or cross-reacting proteins or binding proteins which can interfere with quantitation of the administered product. In addition, it may be necessary to quantitate very low concentrations requiring assays of high sensitivity. The scope of this chapter is to provide some of the insights that we have gathered over several years in supporting these studies.

Much of the early work to determine the pharmacokinetics (PK) of proteins and peptides had been based on data obtained from assays which monitored the biological activity of the administered product or monitored the clearance of radiolabeled proteins (6). Later on, immunoassays began to be used in pharmacokinetic studies because of their exquisite specificities and sensitivities (7, 8). Immunoassays have been especially useful in the analyses of therapeutic compounds such as recombinant tissue-type plasminogen activator (t-PA) in physiological fluids of nonclinical models containing similar molecules where bioactivity assays could not distinguish between endogenous and administered activities (9–11). There are some drawbacks to the use of immunoassays. Gloff and Benet (6) point out that a protein may lose a disulfide bond and biological activity thus becoming a metabolite. However the protein may still react in an immunoassay, resulting in a measured concentration that is higher than the actual bioactive concentration. Proteins can also undergo several metabolic pathways *in vivo*, such as proteolytic clips, amino- and carboxy-terminal processing, oxidation, deamidation (12–15), or may become bound to other proteins. Some of these events may result in loss of bioactivity but not necessarily immunoreactivity. Depending on the epitope(s) to which the antibodies are directed the assay may detect processed molecules, which may not be biologically active. Route dependent processing can also occur and result in further complications in the quantitation of PK parameters. Gloff and Benet emphasized that care must be taken to ensure that the measurement obtained is relevant (6).

From the discussion on validation of immunoassay methods, it will become obvious that appropriate assays will be developed in a timely fashion when the groups developing, validating and performing immunoassays are in close collaboration with colleagues who are performing the PK studies because assay development in an *in-vitro* situation does not mimic the nonclinical or clinical situation. In our experience an assay may be initially defined, reagents selected and preliminary work completed, but recovery studies in relevant samples may indicate that additional development work should be performed. For example, the physiological state of patients in a clinical trial may differ from those of healthy subjects and body fluids such as serum may differ in binding protein content and/or the presence of antibodies. Therefore assays developed using serum from healthy subjects may not reflect the assay in these patients.

Because of the enormous impact of the diagnostics industry, scientists developing assays to support PK studies have been able to take advantage of the easy access to a vast body of literature on assay design and applications, enzyme labeling techniques, a multitude of inexpensive equipment, and continuing development of new, faster and more sensitive methodologies. Immunoassays have been used in many nonclinical and clinical PK studies on recombinant human proteins such as growth hormone (GH), (16), t-PA (9–11), gamma interferon (IFN-$\gamma$), (17), erythropoietin (EPO), (18–20), granulocyte colony-stimulating factor (G-CSF), (21, 22), granulocyte-macrophage colony-stimulating factor (GM-CSF), (23), and

deoxyribonuclease (DNase), (24). This chapter will address the development of immunoassays for protein and peptide pharmaceuticals and the issues surrounding validation of these methods. The emphasis will be heavily directed towards enzyme immunoassays because of their high throughput and the experience we have had in this area.

## II.  PRODUCTION OF ANTIBODY REAGENTS

### A.  Immunization

The production of polyclonal and monoclonal antibodies to proteins have become fairly routine methods in the laboratory (4, 25–27) and high affinity antibodies suitable as assay reagents, easy to generate. There are advantages to using monoclonal antibodies, the major one being that the specificity will not change from lot to lot. Whenever possible we have used monoclonal antibodies for assay development. For polyclonal antibody production we have used approximately 0.1 to 1.0 mg of protein mixed with complete Freunds adjuvant for the initial immunization of goats and rabbits followed by subsequent immunizations in incomplete Freunds adjuvant. A simple regimen of boosting every two weeks until the desired titer of antibody has been reached and then boosting and bleeding the animals on alternate weeks has been successful in generating adequate assay antibodies. We have continued immunization for over a year, pooling antisera with high titers and affinity for performing assay development.

With smaller proteins such as transforming growth factor-beta (TGF-$\beta$) 24 kD, and activin A, 28 kD, both disulfide linked dimers of identical subunits that are highly conserved with a high degree of interspecies homology, it was sometimes difficult to produce polyclonal antibodies in a variety of animal species. For activin A, only one of four animals immunized produced antibody that was useful for assay development (26). However, mouse monoclonal antibodies were successfully produced to both these proteins (27, 28). Immunization of chickens has also been used to elicit IgY antibodies to highly conserved proteins such as proliferating cell nuclear antigen (PCNA) and RNA polymerase II (29). Gassman et al (29) showed that 20–30 $\mu$g of PCNA was sufficient to induce an immune response in chickens. Antibodies appeared 20 days after immunization, reached a plateau at 30 days and remained high until at least day 81. These authors obtained 130 mg of specific antibody from 62 eggs from one immunized hen. We have also used chicken antibodies in an ELISA for inhibin A (30). Therefore, it is an advantage to attempt several methods when feasible.

During research stages of product development, when only small quantities of the protein may be available, immunization with 3–10 $\mu$g of protein delivered to the popliteal node of the rabbit every two weeks has been successful in producing antibodies (31–32). The immunization regimen can be closely monitored to produce, high affinity antibodies for increased assay sensitivity. Antibodies can be developed against particular epitopes eg amino or carboxy terminal ends of protein molecules, or larger domains such as the protease or kringle regions of the rt-PA molecule (33). Assays using these antibodies can then dictate the region of the protein molecule which will be followed in the PK study.

## 1. Peptide Conjugation

Depending on the size of the peptide, production of antibodies may require preparation of conjugates usually with a carrier protein such as bovine serum albumin, immunoglobulin or keyhole limpet hemocyanin (KLH). For relaxin, which is 57 amino acids and a very small protein, we have been able to elicit anti-

bodies without conjugation (34). Antibodies were produced to its constituent A and B chains by conjugating these peptides (24 and 33 amino acids in length) to KLH. We consider a polypeptde of fifty amino acids to be the arbitrary minimum number of amino acids for a protein. More sophisticated methods are required to elicit good antibodies to smaller peptides. A method which has produced antibodies to a peptide mimetic is the multiple antigen peptide (MAP) system described by Tam (35). This approach uses a small peptidyl core matrix bearing radially branching synthetic peptides as dendritic arms. One design of a MAP consisted of a core matrix with a heptalysine containing eight dendritic arms of peptides, 9–16 residues in length. The MAP which had a molecular weight >10,000 was used directly as the immunizing antigen. Tam (35) compared the antigenic response of six different MAPs with the same peptides anchored covalently to KLH and found that rabbits immunized with the MAPs produced higher titer of antibodies. Several of these methods should be tried to ensure good quality and titer of antibodies.

## 2. Phage Display

One of the more recent developments in producing antibodies has been the method of using phage display (Fig. 1). This method was described by Smith (36) who made fusions of peptides and the pIII coat protein of a filamentous bacteriophage and showed that the resultant phage could display the peptide on their surface. This approach was extended by Winter and his group who fused an antibody fragment, sFv, onto the pIII protein to generate phage displaying the antibody fragment (37). The next step in the evolution of this method was to generate a library of antibody specificities to obtain phage expressing a large repertoire of antibodies. The library can be obtained by using PCR techniques to amplify the $V_H$ and $V_L$ encoding

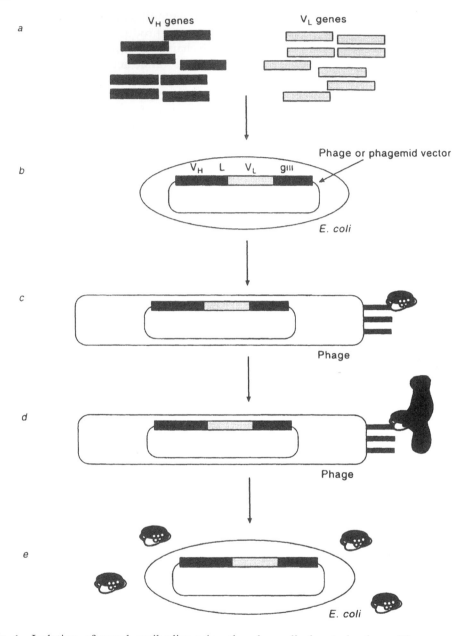

**Fig. 1** Isolation of novel antibodies using the phage display technology. The gene segments encoding the $V_H$ and $V_L$ domains are first isolated (a) using either PCR or cDNA previously-cloned V region segments or artificial sequences. These are cloned as a library in a phage or phagemid vector (b). The diagram illustrates an sFv consisting of a $V_H$ and $V_L$ domain connected by the linker peptide (L) and the gene III encoding pIII. The library is expressed as phage particles with the sFv fused to the pIII minor coat protein (c). The number of sFv expressed on each phage will be dependent on whether a phage or phagemid vectror is used, the nature of the helper phage and the efficiency with which the sFv is expressed. A single sFv is illustrated. Affinity techniques can by used to isolate those phage that express the appropriate antibody specificity (d), thus isolating the genes encoding the sFv. Transfer of the gene constructs into bacteria allows expression of the isolated sFv (e). (Reprinted from Ref. 4).

regions from cDNA derived from spleens of immunized mice. Phage bearing the antibodies can be obtained by using immunoselective techniques such as panning or affinity chromatography. A summary of these methods have been published by Winter (38) and should be useful for obtaining antibodies to highly conserved proteins.

## B. Purification of Antibodies

Assay development requires purification of antibodies from serum components for polyclonal antibodies and from ascites or tissue culture for monoclonal antibodies. There are a variety of methods to accomplish this task including fractional precipitation with ammonium sulfate or other neutral salts, ion exchange chromatography, and protein A purification (25, 39). Purification generally helps to improve assay sensitivity, precision, robustness and in the standardization of the system.

### 1.  Affinity Purification

Since purified recombinant proteins were available to us, we antigen-affinity purified the antibodies for assays developed at Genentech for pharmacokinetic studies (30, 40–42). Early methods for preparation of the affinity column used cyanogen bromide-activated Sepharose (Pharmacia Biotech, Piscataway, NJ) or Affi-gel 10 (Bio-Rad, Richmond, CA) as the column material for coupling the recombinant protein. Coupling and purification were accomplished according to the manufacturers' instructions. For proteins such as TGF-$\beta$1 with limited solubility at neutral or basic pH, coupling was performed at pH 3.5 using aldehyde-activated polyethylene glycol-coated silica, or at pH 4.5 using Affi-gel 102 (27). Recently PerSeptive Biosystems (Framingham, MA) introduced a new support for high-speed liquid chroma-

tography, the POROS bead, which because of its "perfusion" characteristics enables more efficient coupling of antigen and rapid antigen-antibody interactions (43). The POROS® bead can withstand higher pressures than agarose gels and flow rates of 10 mL/min can be used for rapid equilibration of the chromatography system. Affinity purification can be automated with these beads using HPLC systems or the automated immunochemistry systems, BioCad™ or Integral™ (PerSeptive Biosystems) which provide the hardware and software for automating the steps required for affinity purification—equilibration of the affinity column, application of the antiserum, wash steps to remove unbound material, elution of bound antibody and requilibration (44). When affinity purified antibody is used for enzyme or biotin labeling the performance of the assays is improved and our more robust assays are those which use either monoclonal antibodies or affinity purified antibodies. If a sufficient supply of protein is not available to prepare an affinity column, simple Protein A purification of polyclonal antibodies is a rapid method for producing antibodies for an initial assay. Two groups of investigators used the IgG fraction of rabbit anti G-CSF to develop ELISAs with good sensitivity, but this was probably due to the use of Fab fragments for the enzyme-labeled second antibody, see below (21, 22).

### 2.  Fab Fragments

The basic structure of the antibody molecule (Fig. 2) consists of two heavy ($\sim$50 kDa) and two light ($\sim$25 kDa) chains joined by inter and intra-chain disulfide bridges. The enzymes papain and pepsin cleave the IgG molecule into fragments bearing the antigen binding domains, Fab and F(ab')$_2$ and the constant domain of the antibody (Fc). The Fc portion of the antibody molecule is more immunogenic

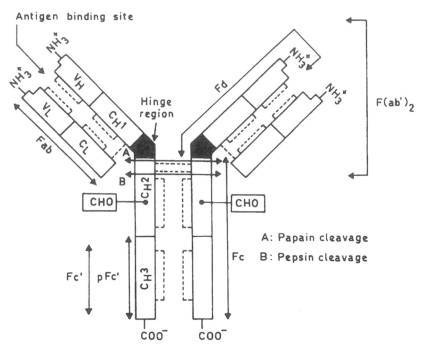

**Fig. 2** The basic four-chain structure of an immunoglobulin showing two identical light (L) poly-peptide chains and two heavy (H) polypeptide chains linked together by disulfide bonds (dotted lines). The antigen binding site is at the N-terminal end of the molecule. The constant (C) and vari-able (V) domains are indicated on both the H and L chains. CHO represents site of attachment for the carbohydrate moiety on the CH2 domain of the H chain. The hinge region is a vaguely defined segment of the H chain between CH1 and CH2 domains. A and B represent the sites at which the immunoglobulin molecule is cleaved by papain and pepsin respectively resulting in the generation of various fragments. (Reprinted from Ref. 3).

than the Fab portion and may be reactive to heterologous antibodies which interfere with assay quantitation. This portion of the antibody molecule is also glycosylated and has been thought to contribute more to non-specific binding in plate assays than the Fab portion. As a consequence, assays have been developed with the Fab or F(ab')$_2$ part of the molecule only. Detailed methods for the preparation of Fab and Fc fragments have been described and simplified procedures are available from commercial sources using immobilized papain and pepsin columns to perform the fragmentation of the IgG molecules. Pro-tein A affinity columns are used to remove the Fc fragments (45). Two groups of investigators (21, 22) developed ELISAs using the IgG fraction of rabbit anti

rG-CSF and the Fab' fragment of these antibodies. The assays had a sensitivity in the pM range (30 and 40 pg/mL). Celniker (46) developed an assay for GH using the Fab fragments of antigen-affinity purified antibody. This assay demonstrated a limit of quantitation of 4 pg/mL (0.2 pM or 0.2 fmoles/mL). Iyer et al (47) affinity purified their assay antibodies and also used Fab fragments in the development of an assay for recombinant hirudin with a sensitivity of 0.2 pmoles/ml. The use of F(ab')$_2$ fragments of a monoclonal anti-body together with a sample diluent con-taining 5% normal mouse serum enabled Jones et al (48) to develop an assay for bov-ine interferon gamma in bovine plasma samples which was virtually free from false positive reactions.

## III.   IMMUNOASSAY METHODS

### A.   Radioimmunoassay (RIA)

We have used RIAs early in product development when the protein and antibodies are in short supply, and switched to an ELISA further on in the product development phase. The RIA methodology is very simple and requires that the protein to be labeled be relatively pure and immunochemically active. Performance of the RIA requires labeled tracer, unlabeled protein for a standard, antisera, and non-specific serum or IgG to assist in the precipitation of specific antibody by a precipitating second antibody. Beads coated with second antibody have been used for precipitation, eliminating the need for carrier protein. Antisera can be used without further purification. Optimization requires block titrations of the tracer and antisera with a dilution series of the reference material or standard to provide an adequate dose response curve. The antigenicity and specific activity of the tracer and the affinity of the antibody govern the sensitivity of the final assay. Assay development time for an RIA is generally shorter than for an ELISA. Polyclonal antibodies have been used in RIAs as it is easier to obtain these antibodies with high binding affinity. Usually a small quantity of antiserum can be used to analyze a considerable number of samples and some high titered antisera can be diluted 10,000 to 100,000 fold.

Development of RIAs have been described in detail (4, 49–51). To prepare iodinated protein for tracer requires 10 $\mu$g of protein or less and methods for iodination of proteins are readily available (25, 45, 51): chloramine T (N-chloro-4-methyl-benzenesulfonamide sodium salt) oxidation and subsequent reduction of the protein with sodium metabisulfite (52); the Iodogen method, 1,3,4,6-tetrachloro-3$\alpha$,6$\alpha$-diphenylglycoluril, which does not require the reduction step (53); and

lactoperoxidase catalysed iodination (54). These methods mainly label the tyrosine residues of proteins. The Bolton-Hunter reagent, 3-(4-hydroxyphenyl) propionic acid N-hydroxysuccinimide ester attaches the phenolic group at primary amines (55). The selection of iodination reagent is based on retention of immunochemical activity and ease of performance of the method. Chloramine T usually provides a tracer with high specific activity, but also tends to cause a reduction in antigenicity compared to the gentler methods of iodination, such as using the Bolton-Hunter reagent or lactoperoxidase catalysis. RIAs have been used to quantitate rhEPO and G-CSF in human serum and have been used for quantitation of IGF-I in PK studies (56). Considerable use is still made of this methodology.

### B.   Two-site Methods

The concept of coating an antigen onto a plastic surface to detect specific antibodies was developed by Engvall and Perlman (57) and termed enzyme-linked immunosorbent assay (ELISA). This method was eventually modified to adsorb antibodies to plastic surfaces and quantitate proteins in solution. When the second antibody was labeled with an enzyme, thereby increasing sensitivity, these assays were also termed—enzyme immunoassays (EIAs). Two-site immunoassay systems employ a sandwich technique with antibodies directed to two-sites on the protein molecule. One antibody is used to capture the protein and this antibody can be coated on a plastic tube, glass or magnetic bead, or microtiter plate. The other antibody is the reporter molecule and can be conjugated to a radiolabel (usually [125]I), enzyme, one of the lanthanide series (terbium or europium) for time resolved fluorescence (58), or more recently ruthenium for electrochemiluminescence (59, 60). Use has also been made of the avidin-biotin system by labeling the reporter antibody with biotin and employing enzyme-labeled streptavidin to generate the signal (61). Many studies described in the literature have been devoted to developing reproducible methods for conjugation and retention of the binding properties of the antibody.

## 1. IRMA

The immunoradiometric assay (IRMA) is a two-site assay with the signal generated by a radiolabeled antibody (62, 63). The IRMA requires antibodies to two determinants unless the protein is polymeric in solution. Some investigators claim that the IRMA can provide higher sensitivity and specificity than the RIA. Ekins (64) introduced the terms excess-reagent and limited-reagent methods to describe immunoassays which require optimal antibody concentrations tending to infinity and zero concentrations, respectively. These terms correspond loosely to noncompetitive and competitive assay formats respectively. Ekins discusses the theoretical concepts when excess-reagent or noncompetitive methods should result in increased sensitivity over limited-reagent or competitive methods such as the RIA. In comparisons of competitive with non-competitive formats, the noncompetitive format appears to provide the more sensitive method (47, 64). With regard to specificity, intuitively two epitopes on the protein must be present and detected in a two-site assay, whereas the RIA requires only one site on the protein to be available and therefore would be less specific than the IRMA. If there is degraded or bound material in the samples to be tested, the RIA may report a higher concentration than the IRMA. Several IRMAs, including an assay for GH, were made commercially available by Hybritech (San Diego, CA). Their assay used coated beads to capture the antigen and $^{125}$I-labeled second antibody to generate the signal. The assay had a sensitivity

of 1–5 ng/mL and was used in PK studies of GH treated subjects (16).

## 2. Enzyme Immunoassay (EIA)

EIAs or ELISAs performed in microtiter plates are the mainstay of the immunoassay labs with methods being developed as sequential, onestep, competitive (65) or multi-layered (66) formats. Porstmann and Kiessig (67) presented a detailed overview of the issues concerning the development of two-site EIAs. When antibodies are adsorbed to the solid phase there is a loss of binding affinity. Therefore the antibody with the higher affinity should be used for coat antibody. The binding constant for the antibody decreases with increasing loading density due to steric hindrance and IgG concentrations of more than $10 \, \mu g/mL$ should not be used for coating the solid phase. Porstmann and Kiessig (67) compared the enzymes that are generally used for EIAs: horseradish peroxidase (HRP), alkaline phosphatase and $\beta$-galactosidase. They also described conjugation procedures and compared the substrates that are available for these enzymes (Table 1). Even though fluorogenic products can be measured with a sensitivity that is 10–1000 times higher than the chromogenic substrates, sensitivity of an assay may be increased only 2–10 fold. Alternative homogenous and heterogenous formats for performing immunoassays were presented and parameters and internal controls which should be used to assess the quality of routine assays in the laboratory were recommended (67).

At Genentech, we have used HRP as the enzyme for labeling antibodies following the method described by Nakane and Kawaoi (68). This enzyme and labeling method have provided us with robust assays with good sensitivity (27, 28, 30, 40–42). Coupling HRP to antibodies was usually accomplished with 10 mg of antibody, using a molar ratio

**Table 1** Enzyme Activities and Detection Limits of Native and IgG Coupled Horseradish Peroxidase (HRP), Alkaline Phosphatase (AP) and β-Galactosidase using Chromogenic and Fluorogenic Substrates

| Parameters | Chromogenic substrates | | | Fluorogenic substrates | | |
|---|---|---|---|---|---|---|
| | HRP | AP | β-Gal | HRP | AP | β-Gal |
| Molar activity (mol/s × 1 × mol) | 2600 (ABTS) | 850 (pNP) | 354 (oNPG) | 196 (HPAA) | 290 (MUP) | 125 (MUG) |
| Specific activity (mmol/s × 1 × g) | 65 (ABTS) | 8.5 (pNP) | 0.68 (oNPG) | 4.9 (HPAA) | 2.9 (MUP) | 0.24 (MUG) |
| Detection limit of enzyme (mol/l) | $10^{-13}$ (ABTS) $10^{-14}$ (oPD) $2 \times 10^{-15}$ (TMB) | $2 \times 10^{-13}$ (pNP) | $2 \times 10^{-13}$ (oNPG) $10^{-13}$ (RG) $3 \times 10^{-14}$ (CPRG) | $5 \times 10^{-14}$ (HPAA) $10^{-15}$ (HPPA) | $10^{-15}$ (MUP) | $5 \times 10^{16}$ (MUP) |
| Specific activity of conjugates (mmol/s × 1 × g) | 9.4 (ABTS) | 2.9 (pNP) | 0.41 (oNPG) | 0.9 (HPAA) | 0.9 (MUP) | 0.15 (MUG) |
| Detection limit of labelled IgG (ng/ml) | 16 (ABTS) 2 (oPD) 0.3 (TMB) | 43 (pNP) | 350 (oNPG) | 10 (HPAA) 0.3 (HPPA) | 0.5 (MUP) | 1.0 (MUG) |

*Key:* ABTS = 2,2′azino-di(3-ethylbenzthiazoline sulphonic acid-6); oPD = o-phenylenediamine; TMB = 3,3′,5,5′-tetramethylbenzidine; pNP = p-nitrophenyl phosphate; oNPG = o-nitrophenyl-β-D-galactopyranoside; RG = resorufin-β-D-galactopyranoside; CPRG = chlorophenolic red-β-D-galactopyranoside; HPAA = p-hydroxyphenylacetic acid; HPPA = 3-(p-hydroxyphenyl) propionic acid; MUP = 4-methylumbelliferyl phosphate; MUG = 4-methylumbelliferyl-β-D-galactopyranoside. (Reprinted from J of Immunological Methods, 150, T Porstmann and ST Kiessig, Enzyme Immunoassay Techniques, 5–21, 1992, with kind permission of Elsevier Science-NL, Sara Burgerhartstraat 25, 1055 KV Amsterdam, The Netherlands. Ref. 67.)

of HRP:IgG of 2–8:1. We have also biotin-labeled the second antibody for the ELISA and made use of commercially available streptavidin-HRP for signal generation. The biotinylation procedure is much simpler than enzyme-labeling methods and may even retain more of the antibody activity since the procedure for biotinylation is potentially less denaturing than the enzyme labeling methods (61).

We have used o-phenylenediamine (OPD) as the substrate for HRP as it is available in easy-to-use tablet form and has provided reproducible and adequate sensitivity. A more recently introduced substrate, 3,3′, 5,5′ tetramethylbenzidine (TMB), (69), is being used with increasing frequency since it can provide greater sensitivity and is also available in ready-to-use form from commercial companies. TMB and OPD have sensitivities of 0.3 and 2.0 ng/mL labeled IgG, respectively (67). This sensitivity seen with labeled IgG is also realized in immunoassays and TMB has provided additional sensitivity (70). Although it has been suggested that assays using fluorogenic substrates would be more sensitive than those using chromogenic substrates, the increase in assay sensitivity has been only 2–10 fold (67). Also, assays using fluorescence tend to be not as robust as assays using chromogenic substrates.

## C.  Miscellaneous Methods

Wild, in his handbook (2) explained the technologies, chemistries and features behind thirty-two representative commercial immunoassay systems used in laboratories, physicians' offices and homes. Depending on the proposed implementation of the assays that are developed, the scientist beginning assay development can choose from a wide array of available assay formats. Wild included a comprehensive troubleshooting guide for manual and automated immunoassay

methods which should be of general application. Wild also introduces several clinical, research and veterinary applications of immunoassays which can prepare the reader for situations that may be encountered during assay development in physiological systems. Another introduction to immunoasay methods can be obtained in the 'Principles and practices of immunoassay' (4) which contains chapters on fluoroimmunoassays, chemiluminescence, light scattering and electrochemical immunoassays. There are also interesting chapters in this book on ingenious commercial devices that are used to perform immunoassays for the diagnostics field.

The Threshold System (Molecular Devices, Sunnyvale, CA) forms a multi-layered complex using biotinylated and fluorescein-labeled antibodies to bind the antigen, and urease-conjugated anti-fluorescein antibodies for signal generation. Streptavidin, is bound to the complex of antibodies and antigen via the biotinylated antibody and this complex is then captured by filtration through a biotinylated nitrocellulose membrane. Urease-conjugated anti-fluorescein antibody is then added to the system and binds to the fluorescein-labeled antibodies on the membrane filter. Urease then reacts with its substrate, urea. The membrane is held tightly against a silicon sensor, which detects a pH change as the urea is hydrolyzed by the urease-conjugated antibody. The system has good sensitivity and has been used in assays to detect low levels of host cell proteins in recombinant products (71). The system appears to be limited by throughput.

### 1.  Electrochemiluminescence

A recently introduced technology uses electrochemiluminescence (ECL) as the detection system for immunoassays (Origen Analyzer, IGEN Inc., Gaithersburg, MD). Light is generated by the

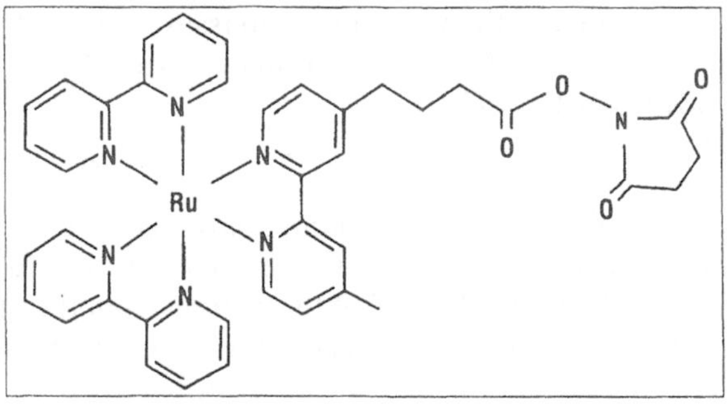

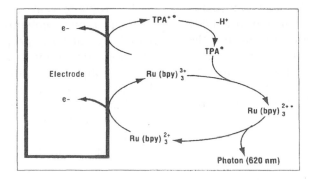

**Fig. 3A** Ruthenium (II) tris-(bipyridine) NHS ester.

**Fig. 3B** Electrochemiluminescent process.

cyclical oxidation-reduction of a ruthenium metal ion [Ru(bpy)$_3$] in the presence of tripropylamine (TPA). This luminescence reaction is triggered when a low voltage is applied to an electrode. In a two-site format, one antibody is conjugated to biotin, the other antibody is conjugated to ruthenium via the tris-(bipyridine) NHS ester (Fig. 3A). In the presence of antigen, a complex is formed which is captured onto streptavidin-coated paramagnetic beads. The mixture containing complexes and unbound proteins is aspirated past a magnet in a flow cell which immobilizes the paramagnetic beads. TPA is introduced into the flow cell and a voltage is applied. The voltage oxidizes both the ruthenium and the TPA simultaneously

(Fig. 3B). TPA loses a proton, becomes a reducing agent and transfers an electron to the ruthenium ion. The electron joins the ruthenium in an excited state and then decays to the ground state releasing a photon in the process. TPA becomes consumed in the process but the reduced ruthenium ion is recyled and continues to produce light. The resultant luminescence is detected and quantitated and is directly proportional to the ruthenium and hence the quantity of analyte (59, 60).

This methodology was compared to a scintillation proximity assay (SPA) and an RIA for detection of insulin in rat serum (72). Data from the ECL assay was comparable to the RIA and SPA, and had a wider dynamic range. It was less susceptible to serum protein interferences than the SPA. Another comparison to an RIA was made which concluded that the ECL assay eliminated a centrifugation step, utilized a smaller sample volume and markedly reduced the assay turnaround time. The ECL assay was successfully used to quantitate RWJ47428-021 in a PK study (73). Compared to ELISA methods, the Origen system in a two-site format can provide similar sensitivity, has fewer wash steps and reagents, and can be more of a stat method since streptavidin coated beads are stable and readily available. ELISA methods generally require coating of microtiter plates with antibody for 12–16 hr, then a step to block nonspecific binding sites and a wash step to remove unbound proteins prior to starting the assay. However, the Origen system does not have the capability for high throughput as it uses a carousel with places for fifty tubes, and does not have the ease of sample handling as ELISA systems.

## 2.   Capillary Electrophoresis

One of the newer methods to be described is capillary electrophoresis (CE) based immunoassays (74, 75). This method is dependent on the relative mobilities of bound and free antigen in a high voltage electric field (kilovolts). Separation takes place within a glass capillary with protein components being detected as they migrate past a window in the capillary. The capillary thus serves as its own detector. Methods have been developed with a competition format using fluorescence labeled antibody or antigen and laser-induced fluorescence for detection. Very small volumes (nanoliters) are required for injection and buffer volumes at the electrodes are typically <4 mL. Separation of the proteins can be achieved rapidly in minutes. The current lack of sensitivity and lack of throughput may limit the utility of CE. However, the introduction of microchips for performing CE immunoassays may address the thoughput issue in the future (76).

## 3.   Immunoaffinity Chromatography

Immunoaffinity chromatography is another method for selection and capture of the protein from the physiological fluid. The affinity column is prepared using specific antibody (or as described below, using a specific binding protein) and methods described for affinity purification of antibodies. The samples to be tested are applied to the affinity column, the column washed to remove non-specific proteins and the protein eluted using a pH change or chaotropic reagent. The protein eluted from the affinity column can be quantitated by intrinsic properties such as ultraviolet absorption or intrinsic fluorescence. Phillips (77) captured IFN-$\alpha$ from the serum of treated patients onto an affinity column composed of glass beads coated with a monoclonal antibody to IFN-$\alpha$. Bound IFN-$\alpha$ was eluted using a gradient of 0–2.5 M sodium thiocyanate. He determined the recovery of IFN-$\alpha$ from various physiological fluids by this method. Plasma and serum samples gave reproducible results with recoveries ranging from 82.2% at the 10 pg level to 97.8% at the 100 pg/ml level. He reported that samples analyzed by immunoaffinity chromatography correlated well with results obtained by a conventional RIA method. Battersby et al (78) made their affinity column with a soluble form of the GH receptor and used this column to capture GH from rat serum. Subsequently reverse phase HPLC was used to concentrate microgram amounts of the GH eluted from the affinity column. The eluted GH was subjected to trypsin digestion and peptide mapping. These authors demonstrated that deamidation and oxidation occurred in rhGH following intravenous (iv) administration. With the minaturization of methods and the increase in sensitivity in analytical methodologies (liquid chromatography and mass spectrometric methods) this type of analysis could be used to characterize the protein molecule that is being detected by immunoassays in PK studies.

### 4. Receptor-Based Methods

The soluble form of a receptor can substitute for one of the antibodies in two-site assays and provide additional selectivity in immunoassays. For example in determining the PK of a humanized antibody, rhuMab HER2, which is directed to a cell surface receptor product of the cERB-b2 oncogene (79), the extracellular domain of the receptor can be used to select and capture the antibody (79). It would be difficult to discriminate the administered antibody from endogenous, circulating IgG molecules in human and primate species. Alternatively to quantitate humanized anti-Tac antibody in human serum, use was made of the bivalent nature of the antibody using the Tac (T activated cell) receptor immobilized on the microtitre plate and the biotinylated receptor to produce a sandwich ELISA (80).

Radio-receptor based methods have been extensively used in the literature for the quantitation of ligands which bind to cell surface receptors. Receptor based methods using non-isotopic methods are also being developed. Such an assay was developed to quantitate a small synthetic cyclic peptide TP-9201 containing the arginine-glycine-aspartic acid (RGD) sequence which inhibits binding of fibrinogen to the GPIIb-IIIa receptor on platelets (81). The method is a competitive enzyme-linked receptor based assay (ELRA) modified from Bachas and associates (81) where the cyclic peptide, TP9201, competes with biotinylated RGD for binding to the solubilized GPIIb-IIIa receptor coated on microtiter plates. Streptavidin-HRP and enzyme substrate were added in sequence and TP9201 in plasma samples quantitated by interpolation on the standard curve (82).

## IV. VALIDATION OF IMMUNOASSAYS FOR PHARMACOKINETIC STUDIES

Development of methods to support PK studies requires a validation of the assay for its intended purpose. Shah et al (83) reported on a meeting sponsored by the American Association of Pharmaceutical Scientists, the US Food and Drug Administration, Federation International Pharmaceutique, and Health Protection Branch (Canada) which provided guiding principles for validation of analytical methods to be used in the bioavailability, bioequivalence and pharmacokinetic studies in humans and animals. These authors emphasized that because these methods play a significant role in evaluation and interpretation of data, that it was essential to use well-characterized and fully validated methods. The International Conference on Harmonization (ICH) also drafted guidelines 'Text on Validation of Analytical Procedures: Definitions and Terminology (84)' and 'Validation of Analytical Procedures: Methodology (85)', which are intended for assay procedures to be used in registrations for the approval of pharmaceutical products. The assays described are used for lot release certificate of analysis testing and stability determination. However these texts contain useful information on terminology, definitions and validation issues which are meant to bridge the differences that often exist between various compendia and regulators of the European Union, Japan and the United States. The use of these guidelines should ensure some uniformity in the usage of terminology. The ICH terminolgy for intra and inter-assay precision (repeatability and intermediate precision), will be used in this chapter.

Method validation includes all of the procedures required to demonstrate that a particular method for the quantitative determination of the concentration of an analyte in a particular biological matrix is reliable for the intended application. The validation of any analytical method can be envisaged to consist of two phases 1) the development phase in which the assay is defined and 2) the validation phase. Since validation is such a lengthy procedure, it is essential to evaluate the assay in the intended physiological fluid before beginning validation. The parameters that are absolutely essential to ensure acceptability of the assay are the accuracy of the method and stability of the protein in the matrix under the storage conditions of the study.

Thus in performing spike recoveries with TNFα, consistent recovery was not obtained using an assay with either two monoclonal antibodies or polyclonal antibodies as coat and conjugate in the sandwich ELISA. When the combination of antibodies was switched to a polyclonal coat and a monoclonal antibody as conjugated antibody, or vice versa the assay then recovered spiked TNFα accurately (41). Similarly in performing an immunoassay for relaxin in serum, goat IgG had to be added to the serum samples to negate the effects of heterologous antibody (5). With t-PA, studies had shown that proteolytic digestion of t-PA might take place in frozen samples during the freeze thaw process prior to analysis (86). In subsequent nonclinical and clinical studies a protease inhibitor, either D-Phe-Pro-Arg-CH$_2$Cl (PPACK, Calbiochem La Jolla CA) or aprotinin, was added to the blood collection tube resulting in elevated concentrations for t-PA and more consistent values for other parameters being quantitated (87, 88).

## A. Definition of the Assay

The following discussion is based on development and validation of sandwich ELISA assays, but with some modification could be applied to any immunoassay in any format used for quantitation of an analyte in a physiological fluid.

## 1. Antibody Optimization

The antibodies used in the assay are optimized by block titration. Coat antibodies are tested at a range of 1–5 μg/mL in 0.05 M bicarbonate buffer, pH 9.6. The standard curve and second antibody are diluted in the assay buffer, which is usually phosphate buffered saline (PBS), pH 7.4, containing 0.5% (w/v) bovine serum albumin (BSA) as a carrier protein and a surfactant such as 0.05% (v/v) polysorbate 20. The standard curve is prepared as a dilution series in assay buffer and a suggested starting range for a protein of molecular weight ~20 kD is 1-32 ng/mL. The second antibody (HRP-labeled) is tested as a series of dilutions to determine a dilution which will provide an adequate dose response curve: the nonspecific binding should be <0.05 optical density (OD) units at A$_{492\,nm}$ for a blank sample (a well containing assay buffer only) and the high end of the standard curve should be ~2.0 OD units at A$_{492\,nm}$ when using OPD as the substrate. Further optimization of the coat and second antibody is then performed to produce the most sensitive assay with a good signal at the high end of the standard curve. At this stage the assay is termed a "buffer" assay to differentiate it from a serum or plasma assay. Buffer assays are used during the development of the manufacturing process of recombinant proteins and are generally available before PK studies are begun. The buffer assay is a starting point for the development of the assay for the protein in physiological fluids.

A draft test procedure should be written and used to continue definition of the assay.

## 2. Accuracy and Determination of the Assay Diluent

Since serum/plasma have been known to present matrix effects (4, 5), it is crucial

**Table 2** Formats for Preparation of Standard Curves and Diluents in Testing Physiological Samples

| Diluent for Standard Curve | Diluent for Serum Sample | Diluent for Over-range Serum Samples |
|---|---|---|
| PBS/BSA (Assay Buffer) | None (Serum tested undiluted) | PBS/BSA |
| PBS/BSA | Serum diluted a minimum dilution eg 1:50 (e.g. for t-PA) | PBS/BSA |
| Serum Diluent | None (Serum tested undiluted) | Serum Diluent |
| Serum Diluent diluted 1:10 in PBS/BSA | Serum diluted 1:10 in PBS/BSA | Serum Diluent diluted 1:10 in PBS/BSA |

to determine how the protein standard should be diluted and samples analyzed to provide good recoveries. The most straightforward approach is if recovery in serum were equivalent to recovery in assay buffer, because the serum can then be tested undiluted. This would provide the most sensitivity with the simplest method for preparing the standard curve and for diluting over-range serum samples. Accuracy is determined by the addition of known quantities of the protein into representative samples of the physiological fluid, eg serum and into assay buffer within the range of the standard curve. Usually several individual samples (3–10) and a pool are used for the accuracy study. The spiked samples are assayed and quantitated against the buffer standard curve and the recovery obtained by comparison of the results. Recoveries of 80–120% are acceptable for this type of assay.

The general experience is that there is an inhibition of signal in serum resulting in under-recovery against the buffer standard curve. What can be done at this stage is to reoptimize the reagents, usually by increasing the concentration of the signal generating antibody and determining if adequate signal can be obtained with a standard curve prepared in undiluted serum. We developed an assay for IFN-$\gamma$ using pooled normal human plasma as the diluent for the standard curve. Patient plasmas were measured undiluted or diluted into the pooled plasma. For many assays, however, the inhibition of signal is severe and the serum is diluted to enhance the signal. Generally, a 1:5 or 1:10 dilution of serum has been satisfactory. In this case the standard curve should also be prepared in the same dilution of serum and over-range samples diluted with the diluted serum. Several formats for preparation of the standard curve and diluents for testing physiological samples are summarized in Table 2.

Motojima et al performed spike recovery experiments by adding three different amounts of rG-CSF to human serum within the standard curve range, obtaining 70–96.4% recovery in a sandwich ELISA (21). They also showed linearity of dilution within the assay range thereby demonstrating lack of serum effects in their assay. To determine activin A levels in serum, samples were diluted 1:5 in assay buffer (28, 30). Further dilutions were made in 20% negative human serum. The standard curve was also diluted in 20% negative human serum. Accuracy of the activin A assay was demonstrated as described by Wong et al (28). Three samples with measured activin A concentrations were mixed with three serum and two control samples of varying activin A concentration. The recovery was determined based on the expected value for the mixtures and ranged from 88.5 to 112% (Table 3). Therefore the activin A

**Table 3**  Accuracy[a]: Activin A ELISA

| Spiked sample[b] type (ng/mL) | Based sample[b] (ng/mL) | Expected[c] (ng/mL) | Observed (ng/mL) | % Recovery |
|---|---|---|---|---|
| Sample A | 0.80 | 0.57 | 0.52 | 91.2 |
| (0.34) | 1.48 | 0.91 | 0.97 | 106.6 |
|  | 2.90 | 1.62 | 1.70 | 104.9 |
|  | 5.66* | 3.00 | 3.36 | 112.0 |
|  | 11.02* | 5.68 | 6.16 | 108.5 |
|  |  |  |  | Ave 104.6 |
| Sample B | 0.80 | 1.08 | 0.99 | 91.7 |
| (1.36) | 1.48 | 1.42 | 1.41 | 99.3 |
|  | 2.90 | 2.13 | 2.09 | 98.1 |
|  | 5.66* | 3.51 | 3.79 | 108.0 |
|  | 11.02* | 6.19 | 6.65 | 107.4 |
|  |  |  |  | Ave 100.9 |
| Sample C | 0.80 | 2.53 | 2.32 | 91.7 |
| (4.26) | 1.48 | 2.87 | 2.54 | 88.5 |
|  | 2.90 | 3.58 | 3.36 | 93.9 |
|  | 5.66* | 4.96 | 4.71 | 95.0 |
|  | 11.02* | 7.64 | 7.89 | 103.0 |
|  |  |  |  | Ave  94.4 |

Average  99.97%

Range  88.5–112.0%

[a] Equal volume of three different (matrix/lot) samples (column #1) added to three serum samples and two serum controls (*) having different concentrations (column #2) and assayed by ELISA.

[b] Activin concentration determined previously by the present assay.

[c] Activin expected concentration = (ng/mL of spiked sample + ng/mL of based sample) ÷ 2 (Reprinted from J of Immunological Methods, 165, WL Wong et al., Monoclonal antibody based ELISAs for measurement of activins in biological fluids, 1–10, 1993, with kind permission of Elsevier Science-NL, Sara Burgerhartstraat 25, 1055 KV Amsterdam, The Netherlands. Ref. 28.)

assay was not influenced by the presence of the serum matrix. If there are factors which interfere with the assay and are present in different concentrations in the serum samples, recovery would have been inadequate.

For t-PA, we were able to use a plasma dilution of 1:50 in a PBS/BSA assay buffer to overcome plasma effects and used a buffer standard curve for PK studies during the development of this protein therapeutic (5). Other investigators have also been able to use BSA as a diluent. Ateshkadi et al (19) used an enzyme immunoassay for Epo and diluted all samples greater than 200 mU/mL with a PBS/BSA buffer. However no data were

presented to demonstrate accuracy or linearity of dilution. Khan et al (66) prepared a standard curve of interferon beta--1b in PBS/1% BSA (1 to 1000 IU/mL) and tested recovery of spiked serum samples diluted 1:2 and 1:5 in PBS/1% BSA. From data presented, it appeared that the assay over-recovered at the lower spiked concentrations of 40 and 80 IU/mL recovering 63 and 92 IU/mL but the recovery was more accurate at 1000 IU/mL, with 1027 IU/mL being the determined concentration. These results were acceptable to Khan et al (66). Kuo el al (70) used a validated ELISA for quantitating human epidermal growth factor and its truncated fragments in rats. Plama samples were diluted in PBS-gelatin when necessary. Good linearity was observed and the limit of quantitation was 7.8 pg-eq/mL. EGF concentrations in rat plasma did not interfere because they were below basal level (70).

For fluids of limited availability such as synovial fluids, appropriate matrices to mimic the fluid may have to be sufficient for initial development work. In some instances, abbreviated validation studies have to be performed on aliquots of the PK samples to determine the characteristics of the assay. Lack of recovery usually demonstrates the presence of endogenous substances which interfere with the quantitation of the analyte in the biological matrix and must be addressed before validation can continue.

### 3. Analyte Stability

The stability of the protein in the proposed biological matrix is the next most important parameter to be evaluated. If the planned matrix is serum, the protein should be spiked into serum samples and stability evaluated under the proposed storage conditions prior to analysis. For example, if samples will be stored frozen, stability should be tested after freezing and thawing. Stability should also be tested at 37°C to mimic what might occur when the protein is in the circulation. The stability is tested by recovery in the assay, with acceptable recovery being 80 to 120% of the expected concentration.

### 4. Sample Collection

Each step in the sample collection method should be investigated to determine the extent to which environmental, matrix, material or procedural variables, from the time of collection of the material up to and including the time of analysis, may affect the estimation of the protein in the matrix (80). Plasma or serum should be evaluated to determine which matrix is more appropriate. Evaluation is accom-

plished by spiking the protein into blood samples and comparing recoveries of the protein from plasma and serum derived from aliquots of the same blood sample (41). If it is known that the protein is susceptible to proteases that may be present when serum is prepared then plasma may be the fluid of choice. The recovery experiment from blood should be performed to demonstrate that the protein will partition completely into serum or plasma and not adhere to the red blood cells or the blood clot (when serum is to be collected). Plasma is simpler to collect and then one should decide between EDTA, citrated or heparin plasma. The anti-coagulant effects of EDTA are more long-lasting than those of the other two anticoagulants. Variability of the matrix due to its physiological state may need to be considered, for example, hemolysed samples may have an effect on the analyte. It was determined that t-PA was unstable in human plasma and samples were collected in the presence of a protease inhibitor (86–88).

### 5. Assay Range

The range of the ELISA is usually determined by the antibody reagents. This range and quantitation limit must be adequate for the needs of the pharmacokineticist, otherwise the assay will not be useful. In some instances, further optimization of the assay can provide additional sensitivity. Parameters which influence sensitivity in a two-site assay are the affinity of the antibodies, incubation times, and non-specific interactions. We have found that in a two-site immunoassay, decreasing the coat antibody concentration and increasing the conjugated or second antibody concentration may provide a two to four fold increase in sensitivity. Incubation of the samples in the microtiter plate for 12–16 hr at 2–8°C may provide another twofold increase in sensitivity.

When the above issues have been adequately addressed, the validation of the assay can begin for its intended purpose.

## B.   Validation of the Method

Pharmacokinetic data have been presented in many manuscripts. Some reports provide adequate validation data on the assays used for quantitation of the administered proteins or refer to companion papers describing validation of the assays (28, 47, 89). Many manuscripts do not provide validation information and this makes it difficult to evaluate differences especially with proteins which may not follow a classical outcome. Commercial companies on the other hand when providing immunoassays in kit form generally perform excellent studies to ensure that their assays will be appropriate for performing analyses in human serum or plasma. They may also provide data for additional species and physiological fluids. Before using these kits, the investigator should perform a qualification of the assay for the intended species and fluid to be tested. Accuracy and linearity of dilution are the most important parameters to test.

### 1.   Accuracy

#### a.   Recovery Intra- and Inter-Species

The same biological matrix as that of the intended samples should be used for validation purposes. For nonclinical studies it is necessary to evaluate every new species that will be used to ensure that the recovery will be the same among the species. It is helpful for the groups developing and performing the assays if a single standard curve can be used to test the various species samples. Some sera, such as mouse serum are not available in large quantities whereas human serum is more readily available. In many instances it is not feasible and the species serum

replaces human serum as the diluent for standard curve preparation. After the initial parameters have been defined some studies should be repeated. Recovery is repeated to evaluate intra-species variability. Generally more variability is expected in primates than in other animal sera. Thus, lack of intra-species differences was confirmed by good recoveries using ten each of individual human and monkey sera in an assay for TNF$\alpha$ (41), and the accuracy of an ELISA for rhDNAse was demonstrated by adding rhDNase to six individual rat sera (24). The percent recovery was 92% $\pm$ 10% (mean $\pm$ sd).

#### b.   Cross-Reacting Endogenous Species

Endogenous proteins which are identical or cross-react with the assay antibodies, can be a problem in determining the PK of the administered product. If the level of the endogenous protein is constant during the PK study, baseline values can be subtracted from the values determined for samples and the PK analysis performed. However, it is necessary to demonstrate linearity of dilution for the endogenous protein to confirm identical or parallel immunoreactivity to the administered product. Woodruff et al. quantitated rh inhibin A in the presence of endogenous rat inhibin using two assays which cross-reacted with the endogenous inhibin. Correction for endogenous levels resulted in identical concentrations to those determined using an assay which was specific for the administered rh inhibin A (30). In an assay for human insulin in rats, Lowe and Temple (90) determined there was 30% cross reactivity. They expressed their results as change in plasma insulin, instead of absolute concentration. When there is pulsatile secretion of the endogenous protein as for human GH, it will not be a simple task to correct for endogenous levels and quantitation of PK parameters for administered GH will be more difficult, especially when low doses are administered.

## c.  Effect of Binding Proteins

Binding protein interactions may affect the protein's PK and elimination, and several examples of binding proteins of protein drugs in development have been described: for IGF-I - IGFBP-1, IGFBP-2 and IGFBP-3 (91); for rt-PA - fast-acting plasminogen activator inhibitors (PAI-1 and PAI-2), alpha 2 macroglobulin and others (91); for human growth hormone—growth hormone binding protein (GHBP), (91); for rhDNase with its binding protein (24); and TNFα with circulating TNFα receptors (92). These binding proteins can interfere with the detection and quantitation of the protein drug if the antibody is directed to the same site as the binding protein. This is more apparent when monoclonal antibodies are used in the assay where reactivity may be completely blocked than when polyclonal antibodies are used as assay antibodies. In some cases, effects due to low affinity binding proteins may be eliminated as in the case of rt-PA where plasma samples were diluted 1:50 (5). This assay used goat and rabbit polyclonal antibodies. It is possible that t-PA which is a protein of molecular weight 64 kD has many epitopes available maximizing the possibility for immunoreactivity when polyclonal antibodies were used. IGF-I is almost tenfold smaller in molecular weight and its binding proteins are bound quite tightly. The affinity of IGF binding proteins to IGF-I is comparatively high ($K_d$ $10^{-10}$ to $10^{-11}$ mol/L) approaching the affinity of antibodies (93). PK studies on IGF-I required information both on the free and total concentrations and assays were developed for total IGF-I using acid ethanol extraction, followed by precipitation to remove the binding proteins, neutralization and then RIA to quantitate IGF-I (56). An alternative procedure was to perform size exclusion chromatography of plasma samples in an acidic mobile phase to ensure complete separation of IGF-I from binding proteins and to measure the IGF-I fractions by RIA (94). In a novel approach, Blum et al (93) circumvented the interference of IGF binding proteins on the quantitation of IGF-I or IGF-II, by blocking the binding proteins with an excess of the non-measured IGF and quantitating the measured IGF with a very specific and sensitive assay. For free IGF-I, plasma samples were subjected to size exclusion chromatography at neutral pH and column fractions which would contain free IGF-I were assayed by RIA (95).

The growth hormone binding protein (GHBP) binds human GH (22 kDa form) with high affinity. The binding protein corresponds to the extracellular domain of the GH receptor (96). Mohler et al reviewed the effects of binding proteins on the PK of GH, and pointed out that the complexed fraction of GH exhibited different PK than the free hormone (91). These authors summarized some of the influences of the GHBP on immunoassays. The GHBP usually inhibits binding of GH in RIAs, whereas in two-site IRMAs, the GHBP non-specifically promotes association of the two antibodies. They added that at physiological concentrations the effects are minor and that for most immunoassays the presence of the GHBP does not substantially alter GH measurements (97). However it is interesting to note that GH assays in various formats have resulted in different quantitation of the same samples (98).

Although there are DNase binding proteins in serum, the ELISA using polyclonal antibodies demonstrated accurate recovery in rat sera (24). Dose-dependent pharmacokinetics for rhDNase was demonstrated and the authors suggested that this could be attributable to saturation of binding proteins in serum at high doses of rhDNAse.

Corti et al showed that different assays for TNFα: a cytolytic bioassay; an ELISA which uses monoclonal and polyclonal anti-TNFα antibodies and detects both oligomeric and monomeric TNFα (OM-ELISA); another ELISA which uses only the monoclonal anti-TNFα antibody and detects only oligomeric TNFα (O-ELISA); and a commercial ELISA are influenced differently by the presence of the soluble form of a specific receptor in in-vitro experiments (92). They found that bioassay activity could be almost completely inhibited at a concentration of sTNF-R1 (soluble form of the TNF receptor 1) where the immunoreactivity was only partially inhibited. The analytical recoveries obtained by the different immunoassays were markedly different (Fig. 4). It is likely that specific binding proteins for TNFα could vary intra-species and the varying values reported for TNFα in biological samples (99) could be a consequence of the different influences on the various assays.

These data suggest that results of assays when binding proteins are present should be carefully evaluated. The format of the assay may influence the interpretation of the disposition of the administered protein.

### d. Interfering Substances

During assay validation, inadequate recoveries or lack of linearity of dilution usually suggests interfering substance(s). Wong et al (28) evaluated specific assays for activin A (a disulfide linked homodimer of $\beta_A$ subunits) and activin B (a homodimer of $\beta_B$ subunits) for interferences by the heterologous activin, rh-inhibin A (a closely related protein which is an $\alpha\beta_A$ heterodimer), rh-follistatin and $\alpha_2$-macroglobulin (two proteins which bind activin and inhibin). Only follistatin interfered in the assays. They found that the addition of a 32-fold molar excess of rh-follistatin caused none

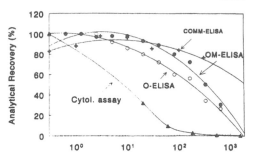

**Fig. 4** Analytical recovery of TNFα from TNFα-sTNF-R1 mixtures by different assays. TNFα, 1 ng/mL in MEM-FCS was incubated for 1 h at 37°C in the presence of sTNF-R1 at various concentrations and tested by cytolytic assay on L-M cells (triangle). Other TNFα-sTNF-R1 mixtures containing 250 pg/mL (+) or 5 ng/mL (open, filled circles) TNFα were prepared in PBS-BSA 1% and tested by O-ELISA (open circle), OM-ELISA (filled circle) and a commercial ELISA (+) from Boehringer Mannheim. Reprinted from J of Immunological Methods, 177, Corti et al., Tumor necrosis factor (TNF)α quantification by ELISA and bioassay: effects of TNFα-soluble TNF receptor (p 55) complex dissociation during assay incubations, 191–198, 1994, with kind permission of Elsevier Science-NL, Sara Burgerhartstraat 25, 1055 KV Amsterdam, The Netherlands, Ref. 92.)

of the activin A spiked into a serum sample to be detected in the activin A ELISA and only 40% of the spiked activin B was detected in the activin B ELISA (28). Lowe and Temple (90) used a 4-fold dilution of rat plasma samples in 4.5% human serum albumin to minimize the interference between rat plasma and the anti-insulin antibody used in their assay. These latter interferences were related to the species of antibodies used in the assay. This phenomenon has been observed previously during assay development of two-site methods (41, 100).

In the analysis of plasma/serum samples for organic molecules, there is generally an extraction step to remove interfering substances and concentrate the analyte of interest. Similar methods can be used in the analysis of small proteins and peptides in physiological fluids. To quantitate rIL-1$\beta$ (17 kD) in plasma, samples were extracted with 40% chloroform (v/v) and after removal of the organic phase each sample was diluted 1:10 with DMEM/5% FBS/Gentamicin and measured in an ELISA (101). The precision was acceptable, 2.30–10.44% RSD for interday precision, and the accuracy was also good with differences between expected and determined values for spiked samples ranging from 7.7 to 2.93%. Isopropanol extraction of plasma samples was used as an initial step to isolate the cyclic peptide, TP9201 from plasma. Apparently biotinylated RGD, used in the competitive assay for TP9201, was causing the plasma samples to clot (82).

### e. Effect of Circulating Antibodies to Administered Protein

Antibodies are, in effect, a specific class of binding proteins (102) and are especially important in nonclinical studies where the administered human proteins are foreign and more immunogenic than the low molecular weight organic entities of classical pharmaceuticals. Antibodies may be neutralizing or non-neutralizing and could have varying effects on the disposition of the drug product. Moreover their presence could cause a low quantitation in plasma due to interference with the assay antibodies. In a small study performed on a dozen monkeys treated with IFN-$\gamma$, the presence of antibodies was associated with both large and small area under the curves (AUCs) and bioavailabilities. There was no consistent pattern to the effect of the presence of antibodies (103). Since administered product in circulating complexes could still be biologically active, it is important to determine when antibodies are present. It is also important for protein disposition studies to determine concentrations of administered protein in the

presence of circulating antibodies. This is an area which needs attention in methods development.

### 2. Linearity of Dilution

Linearity is examined in two ways for PK studies. Assuming that the standard curve is made in serum diluted 1:10 in assay buffer, individual serum samples and a sample of pooled sera are spiked with the protein to a concentration so that a 1:10 dilution will result in a value at the high end of the standard curve. Two-fold dilutions are then made of these spiked serum samples and the spiked pool, each with its own undiluted serum; the spiked pool will be diluted with the pool. Each spiked serum sample along with its dilutions are then further diluted 1:10 in assay buffer and assayed. Linearity is determined for each sample and the pool by plotting the nominal concentrations expected against the observed concentrations (34) and calculating the Pearson correlation coefficient. A coefficient >0.90 indicates adequate linearity in the physiological fluid. There should be linearity of dilution for each sample and the pool. These data would demonstrate that the quantitation of the protein is linear over the range of the standard curve when the serum samples are diluted 1:10 in assay buffer and the standard curve is prepared in a reference serum diluted in the same manner.

Secondly, it is also necessary to demonstrate the assay is linear over the expected range of the PK study. Thus, in the example above, the serum samples and the pooled serum should be spiked with the protein to the highest concentration expected in the study. The spiked samples should first be diluted 1:10 in assay buffer and then diluted with the sample diluent (in this case the normal or negative reference serum diluted 1:10 in assay buffer) to determine recovery over the assay range. We have generally accepted values of

80–120% recovery (although some investigators have used 70% recovery as acceptable). If linearity of dilution is not demonstrated, this suggests that there are reactions taking place in the serum at higher protein concentrations which do not occur at the standard curve concentrations: for example, binding proteins may be saturated at the higher protein concentrations but not at standard curve range.

## 3. Specificity

For PK studies it is necessary to determine the specificity of the assay to ensure that crossreacting species which may be present in the test sample do not react in the assay. For example in preparing to perform PK studies for activin A (homodimer of $\beta_A$ subunits) and inhibin A ($\alpha\beta_A$ heterodimer), assays were developed which were specific and did not cross-react with other related proteins. The antibodies used in the assay were screened to ensure only specific antibodies would be used. With recombinant proteins there is another issue with respect to specificity. Administered proteins are susceptible to proteolytic and other chemical modifications such as deamidation and oxidation in the circulation (12–15, 78) and these protein variants may differ in reactivity in the assay for the parent protein. If these variants are detectable, it is necessary to demonstrate linearity of dilution in the assay for proper quantitation. For example, human IFN-$\gamma$ has a carboxy-terminal sequence consisting of basic residues, that is readily clipped in serum (104). If this form is being detected in the assay, then linearity of dilution should be demonstrated on authentic PK samples. In this instance, *in-vitro* spiked serum samples do not mimic the PK sample as the protein moiety in the PK sample is different from the administered protein (see Sect. XIII below, assays for variants/metabolites). Similarly, terminal

residues on the oligosaccharide chains of glycoproteins can influence clearance (105) and if this is a mechanism of in-vivo processing of the administered protein, it may be necessary to demonstrate identical or parallel activity of such carbohydrate variants.

## 4. Precision

Precision is the most important practical aspect of the immunoassay performance and is typically tested early on in the optimization of the assay to determine that the assay will be acceptable for routine use. Precision can be influenced by parameters that are assay related such as reagent concentrations and timing of binding reactions or operator related such as pipeting and transfer of reagents. Advances in ancillary equipment have aided in the reduction of systematic errors that could occur in enzyme immunoassays due to the number of steps in the method, washing procedures, and reagents that are necessary. Data analyses due to curve fitting techniques can also influence precision at extremes of the standard curve range. In addition to these issues, there is the added imprecision that is possible because of inter- and intra-species differences of samples that are tested.

The ICH terminology for intra and inter-assay precision is repeatability and intermediate precision, respectively (84). Repeatability expresses the precision under the same operating conditions over a short interval of time. Intermediate precision expresses within-laboratory variations: different days, different analysts and different equipment etc. Reproducibility expresses the precision between laboratories as in collaborative studes usually applied to standardization of methodology. Precision is determined as for other assays.

In performing ELISAs, duplicates of each sample, control or standard dilution are usually performed. To determine repeatability, use is made of duplicate values for controls tested over a number of assays and an analysis of variance is performed. This provides a better estimate of repeatability than the calculation of repeatability from a large number of replicates performed on a single day (106). Intermediate precision is determined from the values determined for a sample over several days. In assays where the standard curve is prepared in pooled normal serum and serum samples are assayed, the precision could vary from serum to serum because of intra-species variability. This should be evaluated in a small number of sera to ensure that this contribution to variability will not be significant.

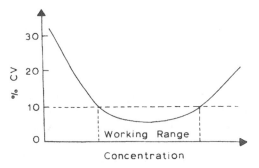

**Fig. 5** The precision profile of an immunoassay illustrates nonuniform error expressed as % coefficient of variation (CV) in the dose estimate as a function of analyte concentration. The working range of an assay can be established as the range where imprecision is below a preset level such as 10% CV in this example. (Reprinted from Ref. 3).

### a. Precision Profile

The precision profile is a means of evaluating the performance of the immunoassay over the standard curve range (107). This is useful for immunoassays as data for concentrations at the extremes of the standard curve are more imprecise than those at the midrange. The precision profile is obtained by performing replicates of each standard concentration and plotting the % CV vs concentration (Fig. 5). The profile can be prepared either by performing a larger number of replicates of each standard or as indicated above, the precision for each standard could be determined over a number of assays (a minimum of six) as this should provide a better estimate of the intermediate precision at each concentration (106). Thus depending on the imprecision the PK analyst can accept (10% CV in the example shown in Fig. 5) reported values are restricted to that part of the standard curve.

### 5. Detection Limit and Quantitation Limit

Detection limit is defined as the lowest amount of analyte in a sample which can be detected but not necessarily quantitated as an exact value (84). Detection limit has been calculated in a variety of ways. It has been calculated as the amount of the analyte which results in a signal that is twice that of the blank signal or results in a signal equivalent to the mean signal + 2 SD of a blank sample. These are most often extrapolated values. The ICH guideline on methodology recommends two to three times the signal to noise for the determination of detection limit (85).

Quantitation limit is defined as the lowest amount of analyte in a sample which can be quantitatively determined with suitable accuracy and precision. This determination is somewhat more arbitrary. The ICH guideline (85) suggests several possibilities: using signal to noise ratio of 10:1; standard deviation of the response and slope of the calibration curve; standard deviation of the blank; and based on the calibration curve. Signal to noise ratio may be appropriate for ELISAs when the background due to a blank sample is < .02 optical units. Multiples of the standard deviation of the blank can also be used. Usually the concentration of the lowest standard is close to the quantitation limit. One approach to determine a quantitation limit could be to prepare control samples close to or at the same concentration as the lowest standard and assay these samples during the course of the validation. The intermediate precision and accuracy for these controls are determined. The accuracy for the expected value should be 80–120% of the nominal value and the precision should be <15% (or <10% if a more precise assay is required). If these criteria are met then the concentration of the lowest control sample is the quantitation limit. It is a good quality control initiative to continue using a control sample at this concentration to ensure that the quantitation limit does not change thoughout the use of the assay.

The detection and quantitation limits should be determined using representative samples and not in assay buffer. When performing PK calculations the quantitation limit is the lowest concentration that is used for analysis.

## 6.  Data Reduction Methods

ELISA data tend to be non-linear and we have fit our standard curve data using a four parameter curve fit program developed at Genentech based on an algorithm developed by Marquardt (108) for least squares estimation of non-linear parameters. Manufacturers of plate readers have software available to fit standard curve data and the calculation of unknown samples are quite easy to perform. An evaluation of five commercial immunoassay software systems was reported by Gerlach et al (109). Comparable results were obtained between the software systems and an independent method. However, numerous deficiencies were observed, including poor documentation and the possibility for incorrect results. Their study should provide guidance to the proper use of these programs (109).

## V.  SPECIAL CONSIDERATIONS FOR BIOLOGICAL FLUIDS

Immunoassays can be affected by the biological matrix and should be examined for each matrix and species (5). Biological fluids other than serum or plasma that have been examined include urine, cerebrospinal fluid (CSF), milk, saliva, synovial fluid, and lymph. Urine has been used as a possible alternative to blood as a fluid to test for levels of GH and IGF-I. Since collection is variable, it is often useful to measure the creatinine content as its excretion is related to body mass and is relatively constant in individuals from day to day. Its concentration is used as a reference point and sample values are based per gm creatinine excreted. Gross changes in creatinine content may throw doubt on the integrity of the urine sample. Urine samples typically require buffering to pH 7.0 to 7.5. In some instances, urine has to be concentrated, however Tonshoff et al (110) quantitated insulin-like growth factors in unconcentrated urine.

CSF is mainly produced by passive filtration from the plasma across the blood-CSF barrier. Proteins are filtered with some selectivity according to size so that the low molecular mass proteins pass more readily than those of high molecular mass, but all plasma proteins commonly measured have been found in the CSF. Milk should be thoroughly mixed before aliquots are taken for storage or for immediate analysis. If the protein to be measured is not lipid soluble, centrifugation could be done to minimize lipid interference with the assay. Phillips (77) using immunoaffinity chromatography determined rIFN-$\alpha$ in urine, CSF and saliva. Good recoveries were obtained from CSF. Saliva proved to be difficult being somewhat more imprecise at the lower concentration levels of 20 and 10 pg/mL (14.48 and 18.77 % CV, respectively) as compared to a %CV of

3.13 at 100 pg/mL. Recoveries were acceptable and ranged from 79.9 to 95.9%. Acid extracts of hypothalamus produced a dose reponse curve which was parallel to that obtained with the synthetic gonadotropin releasing hormone standard (111).

From the work of Supersaxo (112) and Bocci (113), it appears that lymph may play a role in the disposition of proteins following subcutaneous (sc) administration. However, these authors did not comment on any issues on the assay of proteins in lymph samples. Lymph was collected in heparin, centrifuged and stored frozen (113). According to the suggestions in this chapter, when testing lymph samples, the investigator should address issues outlined for validation of an assay in a physiological fluid.

## VI. SPECIAL CHALLENGES TO VALIDATION

### A. Assays for Variants/Metabolites

Partially degraded forms of protein therapeutics may not be immunoreactive in the parent assay. Therefore these metabolites may not be detectable. Alternatively, they may be detectable and present anomalous data. Recombinant IFN-$\gamma$ provides an example of the problems that can arise due to reliance on immunoassay data alone for protein disposition studies. After iv and sc administration of $^{125}$I-labeled rIFN-$\gamma$ to rhesus monkeys, the sc bioavailability was 200% using an ELISA. SDS-PAGE autoradiography of plasma samples showed protein bands with apparent molecular weights of approximately 41–42 kD (presumably nondissociated dimer) for both routes of administration, and 15–16 kD and 14 kD fragments, but in route dependent proportions. After sc adminstration, more of the detectable radioactivity was of lower molecular weight than after iv administration. The positive reactivity of these metabolic forms in the ELISA is a possible reason for the anomalous bioavailability determined (17, 102, 114). This suggests that the PK investigator should attempt to determine what is being detected by the assay in the biological sample during the analysis, when feasible.

Variants are produced as a consequence of production or as part of an active development program to derive variants of therapeutic compounds to enhance inherent biological properties (activity, half life, stability). One example is rIFN-$\gamma$, which is produced as a shortened molecule with four amino acids deleted from the carboxy terminal end of the natural sequence (104). Similarly, the active form of IL-1$\beta$ is a protein consisting of 153 amino acids; the recombinant molecule is the same except that the first two amino acids at the N terminal end are different in rIL-1$\beta$ (101). Assays have to be developed for these variants and use is made of antibodies that are common to the different variants. Kuo et al (70) measured human epidermal growth factor (hEGF1-53) and its five amino acid-truncated fragment (hEGF1-48, C-terminal removal) in plasma and stated that both antibodies reacted equally well with the EGF variants. They showed plasma profiles after intravenous infusion in rats. The limit of quantitation was 7.8 pg-equivalents/mL presumably for both variants.

Polyethylene glycol (PEG) treatment has been suggested as a mechanism for prolonging the half life of administered proteins (42). It was determined that the immunoreactivity of PEG-reacted hGH (PEG-hGH) was not parallel to GH in the immunoassay. Therefore PEG-hGH levels were quantitated using the same assay antibodies but the standard curve was prepared using PEG-hGH (42).

### B. Glycosylated vs. Non-glycosylated and Variably Glycosylated

PK studies have been performed on carbohydrate modified t-PA demonstrating the effects of the carbohydrate moiety on clearance (115). These authors stated that calibration curves of the native and carbohydrate modified variant were superimposable in the ELISA used to determine PK parameters. Alternatively, the assay antibodies can be used with the carbohydrate variant being used to prepare the standard curve (116).

### VII. COMPARISON OF RESULTS FROM DIFFERENT ASSAY FORMATS

In many instances, investigators from various laboratories develop assays for the same protein using different formats. When the results differ for the same sample between investigators or even for the same investigator with two different assays, then the cause of the differences observed should be determined as this often leads to discoveries that are pertinent to the disposition of the protein. Woodruff et al (30) analysed the serum pharmacokinetics of rh inhibin A after iv and sc injection into rats using three different assay formats. The assay formats differ in that one format was specific for rh inhibin A, the other two formats detected endogenous rat inhibin. One of the latter two formats was directed to the $\alpha$ chain and could detect free $\alpha$ chains. The results are shown in Fig. 6. The data with the assay that is specific for rh inhibin A was the most representative of exogenous hormone. However, these authors stated that when corrected for endogenous inhibin all three formats quantitated identical concentrations of inhibin A (30). Interestingly, although both follistatin and $\alpha_2$ macroglobulin are known to associate with rh inhibin A, none

of the assay formats were interfered with by these two binding proteins.

### A. Two-site vs. One-site Assays

Tanaka and Kaneko (117) compared a RIA and ELISA for rhG-CSF in rat serum using rabbit antibody. The ELISA assay was more sensitive than the RIA. Good correlation between these assays and the bioassay were observed with rat serum samples. The authors concluded that there were few fragments produced which contain immunoreactivity and lose bioactivity after rhG-CSF administration.

### B. Immunoassay vs. Bioassay

Whenever possible, a comparison should be made of bioactivity and immunoassay data for the same samples to provide some assurance that what is being quantitated is biologically active. Le Cotonnec et al (63) measured serum concentration-time profiles by immunoassay and *in-vitro* bioassay for FSH and concluded that both assays gave the same trends within the data, but the assays showed non-linearity at the later sampling times. These authors suggested that bioactive FSH concentrations decline more slowly than the immunoreactive FSH concentrations.

Results comparing concentration-time curves by bioassay and immunoassay methods for TNF$\alpha$ appear to be controversial. Prince et al (41) showed parallel PK data for TNF$\alpha$ administered to human subjects. However other investigators showed discrepant results when biological samples were analyzed by both immuno and biological assays (118). These discrepancies might be related to the presence of soluble TNF$\alpha$: TNF receptor complexes that may be detectable by immunoassay but not by bioassay, or the presence of monomeric TNF$\alpha$ or other degradation

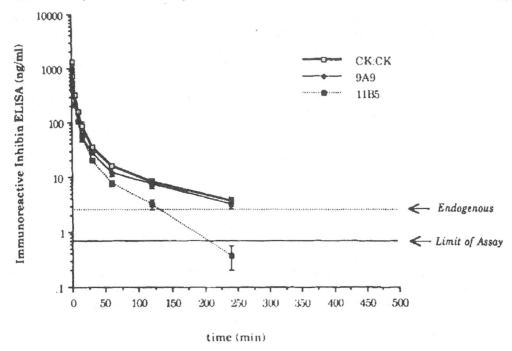

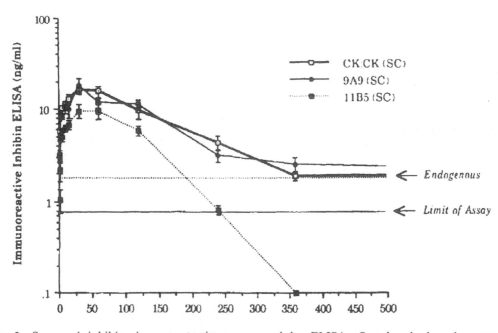

**Fig. 6** Serum rh-inhibin A concentrations measured by ELISA. One hundred and twenty micrograms per kg rh-inhibin A were injected iv or sc into immature female rats. Three assay formats were used to detect the rh-inhibin in the serum: CK: CK ELISA (which detects endogenous inhibin, rh-inhibin A, and free α chain); 9A9 ELISA (inhibin A and rh-inhibin A); and 11B5 ELISA (rh-inhibin A). (Reprinted from Ref. 30).

products (92). Corti and co-workers observed discrepancies between antigenic and bioactive TNFα even when they used a sandwich-ELISA which was unable to detect TNFα monomers based on antibodies that bind epitopes overlapping with the soluble receptor binding site of TNFα. These authors concluded that complex dissociation and differential changes in the TNFα:sTNFR1 bound:free ratio, in different analytical systems, markedly affects TNFα quantitation.

Studies on rIFN-γ showed that after iv adminstration to monkeys, the concentration time curves were similar between the immunoassay and the bioassay. However at later times, samples were negative for biological activity but still exhibited ELISA activity. On the other hand, even though ELISA activity was detected in the circulation after SC administration, these samples did not react in the bioassay (17). In this latter instance the circulating rIFN-γ was an inactive metabolite.

Having completed the assay validation, a report should be generated which summarizes the validation data. Suggested acceptable limits for some validation parameters are summarized in Table 4.

## VIII.  ROUTINE PERFORMANCE OF IMMUNOASSASYS

### A.  Standard ELISA Protocol

The practical aspects of performing immunoassays can be obtained in many of the excellent books written on assay development (1–4). However, we encountered interesting experiences on occasion when using monoclonal antibodies as capture antibodies. Polyclonal antibodies (rabbit and goat) are generally coated at alkaline pH, pH 9.6. Some monoclonal antibodies however, adsorb better to the microtiter plates in a buffer near neutral pH. Therefore, it is recommended that the pH of the coating buffer be evaluated to deter-

mine the optimum pH. The assay should also be carefully checked for plate effects since with some monoclonal antibodies we have observed differential reactivity in the outer wells compared to the inner wells. Plates are coated with 100 $\mu$l per well of affinity purified polyclonal or monoclonal antibody for 12–72 hr at 2–8°C, usually in a bicarbonate butter at pH 9.6. Plates are then washed with three or more cycles of wash buffer (usually PBS containing 0.05% Tween 20). PBS buffer containing 0.5% (w/v) BSA are then added to block unbound sites on the microtiter wells. After 1–2 hr at ambient temperature, the plates are washed to remove excess PBS-BSA. Standards, samples and controls (100 $\mu$l per well) are then added and incubation continued for 1–2 hr at ambient temperature on an orbital shaker. The orbital shaker speeds up the reaction and allows for more uniform reaction in all the wells of the microtiter plate, including the outer rows. In some assays we have found that incubation for 12–16 hr at 2–8°C will provide a doubling of sensitivity. The plates are then washed, second antibody conjugated to HRP (100 $\mu$l per well ), added and incubation continued at ambient temperature for 1–2 hr on the orbital shaker. In performing drift studies for the addition of sample/standard or conjugated antibody to microtiter wells, to determine differences in reactivity due to time of addition to the microtiter plate, we have found that a minimum of 1 hr will eliminate differences due to drift. This is accomplished using multi-channel pipettors and accomplishing the addition of these liquids rapidly to the microtiter wells. The variation in time of 1 to 2 hours provides flexibility to the laboratory performing the assays. Some assays have also been developed where the two antibodies are added simultaneously, thus removing an extra step in the process (28). The plates are then washed and enzyme substrate added (100 $\mu$l per well, OPD or TMB for 15-20

**Table 4** Acceptable Limits for Validation Parameters

| | |
|---|---|
| Accuracy: | |
|     In physiological fluid | 80–120% Recovery in species tested |
| Linearity of Dilution: | |
|     a) Within standard curve | Pearson corr coeff r > 0.90 |
|     b) When spiked at highest expected level | 80–120% Recovery |
| Precision: | |
|     a) Repeatability | <10% at three concentrations within standard curve |
|     b) Intermediate precision | <15% at three concentrations within standard curve |
| Specificity: | |
|     a) Endogenous Compound | Non Reactive or Reactive but dilutes linearly |
|     b) Potential Metabolite | Dilutes linearly |

min). The enzyme reaction is stopped by the addition of 4 N sulfuric or 1 M phosphoric acid, respectively. The plates are read in a microtiter plate reader and data analysed. The volumes have been kept constant for simplicity of performance of the assay. The initial volume of sample can be changed to accomodate rare or insufficient sample or increased for greater sensitivity.

## B. Standards and Assay Controls

In development of immunoassays to support product development in the biopharmaceutical industry, it is not unusual to use different lots of the product to prepare standard curves. Early assays use the protein product that may not be as well purified or characterized as the product in later stages of development. The stability of the standard should always be evaluated and comparative studies should be performed when switching standard preparations. If international standards are available from organizations such as the World Health Organization (WHO) and the National Institute for Biological Standards and Controls (NIBSC), these

standards should be used to calibrate internal standards.

Generally five to six standards, a buffer blank and controls that cover the range of the standard curve are used. Duplicates are performed for each standard and control and mean of duplicates are used for analysis. Data are fit using the four parameter curve fit program developed at Genentech. In most cases this type of curve fitting has resulted in good curve fits. In some instances when using double monoclonal antibodies this fit may not be appropriate. Guidelines are set on the imprecision allowed between duplicates and for when samples should be retested. The values for controls are monitored and acceptability of the assay is based on several parameters and generally follow the Westgard rules (4, 119).

For example, the assay is rejected when:

1. One control measurement exceeds ±3 SD control limits;
2. any two of three consecutive control measurements exceed the same ±2 SD control limits;
3. one control measurement exceeds the +2 SD control limit and another exceeds the −2 SD control limit;

4. three consecutive control observations exceed the same ±1 SD control limit
5. six consecutive control measurements fall on the same side of the mean. Limits are also set on the optical density observed for the highest standard and the buffer blank.

In the routine performance of assays, controls in a buffer matrix may suffice, however inclusion of a control(s) in the matrix when the assay is performed may have added value. Controls in the matrix are useful when reagents have to be replaced and could determine if the specificity of the antibodies have changed, especially when there are binding proteins which may affect antibodies differently.

## C. Dilution into Assay Range

From dose ranging studies, the pharmacokineticist can estimate expected concentrations of product in a PK study and prepare dilutions of each sample to overlap the range of the standard curve. Generally all the values which fall within the range of the standard curve are used in the computation of the concentration for that sample. If the standard curve and sample diluent contains serum, whenever a new lot of serum has to be used it must be qualified for the assay to ensure that the quantitation of unknown samples will be the same. For these purposes, control samples prepared in the physiological fluid should be kept and analysed to ensure equivalence between diluents. Lot to lot consistency in assay performance for some reagents, such as BSA should be tested. For example, some cytokines may interact with BSA and depending on the composition of impurities in the BSA may result in different reactivity with the assay antibodies. One practical observation we have made after years of performing assays, is that errors made in dilution can be a major source of variability. There should be careful training for all individuals performing dilutions and pipetting devices should be routinely calibrated.

The validation process should provide information on the parameters that can affect the long term robustness of the method. If there will be lot to lot variability of reagents, then steps should be taken to characterize these reagents thoroughly.

## IX. CONCLUSION

Information presented in this chapter demonstrate the complexity of assays that are used for determining the disposition of administered protein drugs. It is apparent that there must be continous interactions among groups that develop, perform and interpret the results from these assays. It is not unusual for some aspect of the assay to be reworked following initial PK studies or when evaluating the PK of the protein in a different species. It is important to have control over the assay and determine the molecular species that is being detected to provide assistance in understanding *in-vivo* phenomena. With confidence in the assays, investigators can then draw important conclusions, especially from unexpected data.

Future work in methods development to support PK studies will fall into this area of determining the molecular species being detected in the assay and this will be accomplished coupling affinity techniques with advanced analytical methodologies. Another area that requires further refinement of methods is the quantitation of administered proteins in the presence of circulating antibodies.

## ACKNOWLEDGEMENTS

I would like to acknowledge my colleagues, R Baffi, D Bloedow, S Chen, V Hale and L Myers for reviewing this chapter and providing valuable feed back.

## REFERENCES

1. P Tijssen. Practice and Theory of Enzyme Immunoassays. Amsterdam: Elsevier, 1985.
2. D Wild, ed. The Immunoassay Handbook. Stockton Press, 1994.
3. SS Deshpande. Enzyme Immunoassays: from Concept to Product Development. San Francisco: Chapman Hall, 1996.
4. CP Price, DJ Newman, eds. Principles and Practice of Immunoassay. London: Macmillan Reference Ltd, 2nd Edition, 1997.
5. AB Chen, DL Baker, BL Ferraiolo. Points to Consider in Correlating Bioassays and Immunoassays in the Quantitation of Peptides and Proteins. In: PD Garzone, WA Colburn, M Mokotoff, eds. Pharmacokinetics and Pharmacodynamics Vol 3. Cincinatti: Harvey Whitney Books, 1991, pp 53–71.
6. CA Gloff, LZ Benet, Pharmacokinetics and Protein Therapeutics Adv Drug Delivery Rev 4: 359–386, 1990.
7. RJ Wills, BL Ferraiolo. The role of pharmacokinetics in the development of biotechnologically derived agents. Clin Pharmacokinet 23: 406–414, 1992.
8. NB Modi. Pharmacokinetics and pharmacodynamics of recombinant proteins and peptides. J Controlled Release 29: 269–281, 1994.
9. M Verstraete, H Bounameaux, F de Cock, F Van der Werf, D Collen. Pharmacokinetics and systemic fibrinolytic effects of recombinant human tissue-type plasminogen activator (rt-PA) in humans. J Pharm Exp Ther 235: 506–512, 1985.
10. RA Baughman. Pharmacokinetics of Tissue Plasminogen Activator. In: BE Sobel, D Collen, EB Grossbard, eds. Tissue Plasminogen Activator, New York: Marcel Dekker Inc, 1987, pp 41–53.
11. P Tanswell. Tissue-type Plasminogen Activator. In: AHC Kung, RA Baughman, JW Larrick, eds. Therapeutic Proteins. New York: WH Freeman and Co, 1992, pp 255–281.
12. MC Manning, K Patel, RT Borchardt. Stability of protein pharmaceuticals. Pharm Res 6: 903–917, 1997.
13. HT Wright. Nonenzymatic deamidation of asparaginyl and glutaminyl residues in proteins. Crit Rev Biochem Mol Biol 26: 1–52, 1991.
14. TS Olson, SR Terlecky, JF Dice. Pathways of Intracellular Protein Degradation in Eukaryotic Cells. In: TJ Ahern, MC Manning, eds. Stability of Protein Pharmaceuticals. Part B: In Vivo Pathways of Degradation and Strategies for Protein Stabilization Vol 3. Pharmaceutical Biotechnology, RT Borchardt Series Editor. New York: Plenum Press, 1992.
15. S Li, C Schoneich, GS Wilson, RT Borchardt. Chemical pathways of peptide degradation. V. Ascorbic acid promotes rather than inhibits the oxidation of methionine to methionine sulfoxide in small model peptides. Pharm Res 10: 1572–1578. 1993.
16. SA Chen, AE Izu, J Mordenti, A Rescigno. Bioequivalence of two recombinant human growth hormones in healthy male volunteers after subcutaneous administration. Amer J Therapeutics. 2: 190–195, 1995.
17. J Mordenti, SA Chen, BL Ferraiolo. Pharmacokinetics of interferon-gamma. In: AHC Kung, RA Baughman, JW Larrick, eds. Therapeutic Proteins. New York: WH Freeman and Co, 1992, pp 187–199.
18. AM Cohen. Erythropoietin and G-CSF. In: AHC Kung, RA Baughman, JW Larrick, eds. Therapeutic Proteins. New York: WH Freeman and Co, 1992, pp 165–186.
19. A Ateshkadi, CA Johnson, LL Oxton, TG Hammond, WS Bohenek, SW Zimmerman. Pharmacokinetics of intraperitoneal, intravenous and subcutaneous recombinant human erythropoietin in patients on continuous ambulatory peritoneal dialysis. Am J Kidney Dis 21: 635–642, 1993.
20. JD Jensen, LW Jensen, JK Madsen. The pharmacokinetics of recombinant human erythropoietin after subcutaneous injection at different sites. Eur J Clin Pharmacol 46: 333–337, 1994.
21. H Motojima, T Kobayashi, M Shimane, S Kamachi, M Fukushima. Quantitative enzyme immunoassay for human granulocyte colony stimulating factor (G-CSF). J Immunol Methods 118: 187–192, 1989.
22. H Tanaka, T Tokiwa. Pharmacokinetics of recombinant human granulocyte colony-stimulating factor studied in the rat by a sandwich enzyme-linked immunosorbent assay. J Pharmacol Exp Ther 255: 724–729, 1990.
23. N Stute, WL Furman, M Schell, WE Evans. Pharmacokinetics of recombinant human granulocyte-macrophage colony-stimulating factor in children after intravenous and subcutaneous administration. J Pharm Sci 84: 824–828, 1995.
24. M Mohler, J Cook, D Lewis, J Moore, D Sinicropi, A Championsmith, B Ferraiolo, J Mordenti. Altered pharmacokinetics of recombinant human deoxyribonuclease in

rats due to the presence of a binding protein. Drug Metab Disp 21: 71–75, 1993.

25. E Harlow, D Lane. Antibodies, a Laboratory Manual. New York: Cold Spring Harbor, 1988.

26. JA Lofgren, R Schwall, C Schmelzer, WLT Wong. Generation of polyclonal antibodies against recombinant human activin A. J Immunoassay 12: 565–578, 1991.

27. C Lucas, BM Fendly, VR Mukku, WLT Wong, MA Palladino. Generation of antibodies and assays for transforming growth factor beta. Methods Enzymol 198: 303–316, 1991.

28. WLT Wong, SJ Garg, T Woodruff, L Bald, B Fendly, JA Lofgren. Monoclonal antibody based ELISAs for measurement of activins in biological fluids. J Immunol Methods 165: 1–10, 1993.

29. M Gassmann, P Thommes, T Weiser, U Hubscher. Efficient production of chicken egg yolk antibodies against a conserved mammalian protein. The Faseb Journal 4: 2528–2532, 1990.

30. TK Woodruff, LA Krummen, S Chen, G DeGuzman, R Lyon, DL Baly, DE Allison, S Garg, WL Wong, N Hebert, JP Mather, P Cossum. Pharmacokinetic profile of recombinant human inhibin A and activin A in the immature rat. I. Serum profile of rh-inhibin A and rh-activin A in the immature female rat. Endocrinol 132: 715–724, 1993.

31. MB Sigel, YN Sinha, WP VanderLaan. Production of antibodies by inoculation into lymph nodes. Methods Enzymol 93: 3–12, 1983.

32. BD Bennett, GL Bennett, RV Vitangcol, JRS Jewett, J Burnier, W Henzel, D Lowe. Extracellular domain-IgG fusion proteins for three human natriuretic peptide receptors. Hormone pharmacology and application to solid phase screening of synthetic peptide antisera. J Biol Chem 266: 23060–23067, 1991.

33. A-J van Zonneveld, H Veerman, H Pannekoek. Autonomous functions of structural domains on human tissue-type plasminogen activator. Proc Natl Acad Sci USA 83: 4670–4674, 1986.

34. C Lucas, LN Bald, MC Martin, RB Jaffe, DW Drolet, M Mora-Worms, G Bennett, AB Chen, PD Johnston. An enzyme-linked immunosorbent assay to study human relaxin in human pregnancy and in pregnant rhesus monkeys. J Endocrinol 120: 449–457, 1989.

35. JP Tam. Synthetic peptide vaccine design: synthesis and properties of a high-density multiple antigenic peptide system. Proc Natl Acad Sci USA 85: 5409–5413 1988.

36. GP Smith. Filamentous fusion phage: novel expression vectors that display cloned antigens on the virion surface. Science 228: 1315–1317, 1985.

37. J McCafferty, AD Griffiths, G Winter, DJ Chiswell. Phage antibodies: filamentous phage displaying antibody variable domains. Nature 348: 552–554, 1990.

38. G Winter, AD Griffiths, RE Hawkins, HR Hoogenboom. Making antibodies by phage display technology. Annu Rev Immunol 12: 433–455. 1994.

39. JW Goding. Use of staphylococcal protein A as an immunological reagent. J Immunol Methods 62: 1–13, 1978.

40. SA Chen, B Reed, T Nguyen, N Gaylord, GB Fuller, J Mordenti. The pharmacokinetics and absorption of recombinant human relaxin in nonpregnant rabbits and rhesus monkeys after intravenous and intravaginal administration. Pharm Res 10: 223–227, 1993.

41. W Prince, KJ Harder, S Saks, BR Reed, AB Chen, AJS Jones. An enzyme-linked immunosorbent assay for the quantitation of tumor necrosis factor-alpha in serum. J Pharm Biomed Anal 5: 793–802, 1987.

42. R Clark, K Olson, G Fuh, M Marian, D Mortenson, G Teshima, S Chang, H Chu, V Mukku, E Canova-Davis, T Somers, M Cronin, M Winkler. Long-acting growth hormones produced by conjugation with polyethylene glycol. J Biol Chem 271: 21969–21977, 1996.

43. NB Afeyan, NF Gordon, FE Regnier. Automated real-time immunoassay of biomolecules. Nature 358: 603–604, 1992

44. M deFrutos, FE Regnier. Tandem chromatographic immunological analysis. Anal Chem 65: 17A–23A, 1993.

45. Pierce Catalog and Handbook. Rockford Illinois, 1994.

46. CH Albini, J Sotos, B Sherman, A Johanson, A Celniker, N Hopwood, T Quattrin, BJ Mills, M MacGillivray. Pediatr Res 29: 619–622, 1991.

47. L Iyer, M Adam, J Amiral, J Fareed, E Bermes. Development and validation of two enzyme-linked immunosorbent assay (ELISA) methods for recombinant hirudin. Seminars in Thrombosis and Hemostasis 21: 184–192, 1995.

48. SL Jones, JC Cox, JM Shepherd, JS Rothel, PR Wood, AJ Radford. Removal of false-positive reactions from plasma in an enzyme immunoassay for bovine interferon-γ. J Immunol Methods 155: 233–240, 1992.

49. RJ Knapp, TP Davis, TF Burks and HI Yamamura. Physiological and Pharmacological Evaluation of Peptide Analogues.

In: D Ward, ed. Peptide Pharmaceuticals. New York: Elsevier, 1991, pp 210–241.

50. CS Randall, TR Malefyt, LA Sternson. Approaches to the Analysis of Peptides. In: VHL Lee, ed. Peptide and Protein Drug Delivery. New York: Marcel Dekker, 1991, pp 247–301.

51. T Chard. An Introduction to Radioimmunoassay and Related Techniques. 4th Edition, Amsterdam: Elsevier, 1990.

52. PJ McConahey, FJ Dixon. Radioiodination of proteins by the use of chloramine-T method. Methods Enzymol 70: 210–213, 1980.

53. PJ Fraker, JC Speck Jr. Protein and cell membrane iodinations with a sparingly soluble chloroamide, 1,3,4,6-tetrachloro-3, 6-diphenylglycouril. Biochem Biophys Res Commun 80: 849–857, 1978.

54. SI Schlager. Radioiodination of cell surface lipids and proteins for use in immunological studies. Methods Enzymol 70: 252–265, 1980.

55. JJ Langone. Radioiodination by use of the Bolton-Hunter and related reagents. Methods Enzymol 70: 221–247, 1980.

56. SA Lieberman, J Bukar, SA Chen, AC Celniker, PG Compton, J Cook, J Albu, AJ Perlman, AR Hoffman. Effects of recombinant human insulin-like growth factor-I (rhIGF-I) on total and free IGF-I concentrations, IGF-binding proteins, and glycemic response in humans. J Clin Endocrinol Metab 75: 30–36, 1992.

57. E Engvall, P Perlman. Enzyme-linked immunosorbent assay, ELISA III. Quantitation of specific antibodies by enzyme-labeled anti-immunoglobulin in antigen-coated tubes. J of Immunology 109: 129–135, 1972.

58. EFG Dickson, A Pollak, EP Diamandis. Time-resolved detection of lanthanide luminescence for ultrasensitive bioanalytical assays. J Photochem Photobiol 27: 3–19, 1995.

59. H Yang, JK Leland, D Yost, RJ Massey. Electrochemiluminescence: A new diagnostic and research tool. Biotechnology 12: 193–194, 1994.

60. H Yu. Enhancing immunoelectrochemiluminescence (IECL) for sensitive bacterial detection. J Immunol Methods 192: 63–71, 1996.

61. EP Diamandis TK Christopoulos, The biotin-(strept)avidin system: principles and applications in biotechnology. Clin Chem 37: 625–636, 1991.

62. A Souillard, M Audran, F Bressolle, R Gareau, A Duvallet, J-L Chanal. Pharmacokinetics and pharmacodynamics of recombinant human erythropoietin in athletes. Blood sampling and doping control. Brit J Clin Pharmacol 42: 355–364, 1996.

63. J-Y le Cotonnec, HC Porchet, V Beltrami, A Khan, S Toon, M Rowland. Clinical pharmacology of recombinant human follicle-stimulating hormone (FSH). 1. Comparative pharmacokinetics with urinary human FSH. Fertility and Sterility 61: 669–678, 1994.

64. RP Ekins. Immunoassay design and optimization. In: CP Price, DJ Newman, eds. Principles and Practice of Immunoassay 2nd Edition. London: Macmillan Reference Ltd., 1997 pp 173–207.

65. N. Sunahara, S. Kurooka, K Kaibe, Y Ohkaru, S Nishimura, K Nakano, Y Sohmura, M Iida. Simple enzyme immunoassay methods for recombinant human tumor necrosis factors α and its antibodies using a bacterial cell wall carrier. J Immunol Methods 109: 203–214, 1988.

66. OA Khan, Q Xia, CT Bever Jr, KP Johnson, HS Panitch, SS Dhib-Jalbut. Interferon beta-1b serum levels in multiple sclerosis patients following subcutaneous administration. Neurology 46: 1639–1643, 1996.

67. T Porstmann, ST Kiessig. Enzyme immunoassay techniques: an overview. J Immunol Methods 150: 5–21, 1992

68. PK Nakane, A Kawaoi. Peroxidaselabeled antibody. A new method of conjugation. J Histochem Cytochem 22: 1084–1091, 1974

69. AKJ Goka, MJG Farthing. The use of 3,3′,5,5′-Tetramethylbenzidine as a peroxidase substrate in microplate enzyme-linked immunosorbent assay. J Immunoassay 8: 29–41, 1987.

70. B-S Kuo, GD Nordblom, RC Dudeck, LS Kirkish, DS Wright. Disposition kinetics of human epidermal growth factor (hEGF1-53) and its truncated fragment (hEG1-48) in rats. Drug Metab Disp 24: 96–104, 1996.

71. LC Eaton. Host cell contaminant protein assay development for recombinant biopharmaceuticals. J Chromatography A, 705: 105–114, 1995

72. W Harrington, J Binz, R Mertz, W Faison, W Oliver, K Brown. Comparison of electrochemiluminescence, scintillation proximity assay and radioimmunoassay in the determination of insulin concentrations in rat serum and perfused pancreas effluent. Pharm Res 14: S680, 1997. Abstract No. 4147.

73. R Paulussen, B Lee, K Pham, M Kelley. The quantification of RWJ 47428-021 in

human serum: comparison of RIA and electrochemiluminescence (ECL) immunoassays. Pharm Res 14: S683, 1997. Abstract No. 4158.

74. D Schmalzing, W Nashabeh, M Fuchs. Solution-phase immunoassay for determination of cortisol in serum by capillary electrophoresis. Clin Chem: 41: 1403–1406, 1995.

75. D Schmalzing, W. Nashabeh. Capillary electrophoresis based immunoassays: a critical review. Electrophoresis 18: 2184–2193, 1997.

76. LB Koutny, D Schmalzing, TA Taylor, M Fuchs. Microchip electrophoretic immunoassay for serum cortisol. Anal Chem 68: 18–22, 1996

77. TM Phillips. Measurement of recombinant interferon levels by high performance immunoaffinity chromatography in body fluids of cancer patients on interferon therapy. Biomedical Chromatography 6: 287–290, 1992.

78. JE Battersby, VR Mukku, RG Clark, WS Hancock. Affinity purification and microcharacterization of recombinant DNA-derived human growth hormone isolated from an in vivo model. Anal Chem 67:447–455, 1995.

79. JW Park, R Stagg, GD Lewis, P Carter, D Maneval, DJ Slamon, H Jaffe, HM Shepard. Anti-p185HER2 monoclonal antibodies: biological properties and potential for immunotherapy. In: RB Dickson, ME Lippman, eds. Genes, Oncogenes, and Hormones. Advances in Cellular and Molecular Biology of Breast Cancer. Norwell MA: Kluwer Academic Publishers, 1992, pp 193–211.

80. BE Fayer, PP Soni, MH Binger, DR Mould, H Satoh. Determination of humanized anti-TAC in human serum by a sandwich enzyme linked immunosorbent assay. J Immunol Methods 186: 47–54, 1995

81. LG Bachas, CD Tsalta, ME Meyerhoff. Binding proteins as reagents in enyme-linked competitive binding assays of biological molecules. Biotechniques 1: 42–55, 1986.

82. NB Modi, SA Baughman, BD Paasch, A Celniker, SY Smith. Pharmacokinetics and pharmacodynamics of TP-9201, a GPIIbIIIa antagonist, in rats and dogs. J Cardiovascular Pharmacol 25: 888–897, 1995.

83. VP Shah, KK Midha, S Dighe, IJ McGilveray, JP Skelly, A Yacobi, T Layloff, CT Viswanathan, CE Cook, RD McDowall, KA Pittman, S Spector. Analytical Methods Validation: Bio-

availability, Bioequivalence and Pharmacokinetic Studies. J Pharm Sci 81: 309–312, 1992.

84. Text on Validation of Analytical Procedures: Definitions and Terminology. ICH-Q2A Federal Register Mar 1995 (60 FR 11260).

85. Text on Validation of Analytical Procedures: Methodology. ICH-Q2B Food and Drug Administraion (Docket No. 96D-0030), May 1997.

86. MA Mohler, CJ Refino, SA Chen, AB Chen, AJ Hotchkiss. D-Phe-Pro-Arg-Chloromethylketone: Its potential use in inhibiting the formation of in vitro artifacts in blood collected during tissue-type plasminogen activator thrombolytic therapy. Thromb Haemostas 56: 160–164, 1986.

87. E Siefried, P Tanswell. Comparison of specific antibody, D-Phe-Pro-Arg-chloromethylketone and aprotinin for prevention of in vitro effects of recombinant tissue-type plasminogen activator on haemostasis parameters. Thromb Haemostas 58: 921–926, 1987.

88. DC Stump, EJ Topol, AB Chen, A Hopkins, D Collen. Monitoring of hemostasis parameters during coronary thrombolysis with recombinant tissue-type plasminogen activator. Thromb Haemostas 59: 133–137, 1988.

89. N Sizemore, RC Dudeck, CM Barksdale, GD Nordbloom, WT Mueller, P McConnell, DS Wright, A Guglietta, B-S Kuo. Development and validation of two solid-phase enzyme immunoassays (ELISA) for quantitation of human epidermal growth factors (hEGFs). Pharm Res 13: 1088, 1996.

90. PJ Lowe, CS Temple. Calcitonin and insulin in isobutylcyanoacrylate nanocapsules: Protection against proteases and effect on intestinal absorption in rats. J Pharm Pharmacol 46: 547–552, 1994.

91. MA Mohler, JE Cook, G Baumann. Binding Proteins of Protein Therapeutics. In: BL Ferraiolo, MA Mohler, CA Gloff, eds. Protein Pharmacokinetics and Metabolism. New York: Plenum Publishing, 1992, pp 35–71.

92. A Corti, C Boiesi, S Merli, G Cassani. Tumor necrosis factor (TNF)α quantitation by ELISA and bioassay: effects of TNFα-soluble TNF receptor (p 55) complex dissociation during assay incubations. J Immunol Methods 177: 191–198, 1994.

93. WF Blum, BH Brier. Radioimmunoassays for IGFs and IGFBPs Growth Regul 4 Suppl: 11-30, 1994.

94. RW Furlanetto, LE Underwood, JJ Van Wyk, AJ D'Ercole. Estimation of Somatomedin-C levels in normals and patients with pituitary disease by radioimmunoassay. J Clin Invest 60: 648–657, 1977.

95. AC Celniker, S Chen, N Spanski, J Pocekay, AJ Perlman. IGF-I assay methods and pharmacokinetics of free IGF-I in the plasma of normal human subjects following intravenous adminstration. Proc US Endocrine Soc Vol 2 Abstract No 1144 p 310.

96. DW Leung, SAS Spencer, G Cachianes, RG Hammonds, C Collins, WJ Henzel, R Barnard, MJ Waters, WI Wood. Growth hormone receptor and serum binding protein: purification, cloning and expression. Nature 330: 537–543, 1987.

97. T Jan, MA Shaw, G Baumann. Effects of growth hormone-binding proteins on serum growth hormone measurements. J Clin Endocrinol Metab 72: 387–391, 1991.

98. AC Celniker, AB Chen, R Wert, B Sherman. Variability in the quantitation of circulating growth hormone using commercial assays. J Clin Endocrinol Metab 68: 469–476, 1989.

99. I Engelberts, S Stephens, GJM Francot, CJ Van Der Linden, WA Buurman. Evidences for different effects of soluble TNF-receptors on various TNF measurements. Lancet 338: 515, 1991.

100. JP Gosling. A decade of development of immunoassay methodology. Clin Chem 36: 1408–1427, 1990.

101. JE Gray, V Peterman, R Newton, S-YP King, HJ Pieniaszek Jr. ELISA determination and preliminary pharmacokinetics of modified human rIL-1β in dogs. Res Com Chem Pathol Pharmacol 81: 233–241, 1993.

102. BL Ferraiolo, MA Mohler. Goals and Analytical Methodologies for Protein Disposition Studies. In: BL Ferraiolo, MA Mohler, CA Gloff, eds. Protein Pharmacokinetics and Metabolism. New York: Plenum Publishing, 1992, pp 1–33.

103. BL Ferraiolo, GB Fuller, B Burnett, E Chan. Pharmacokinetics of recombinant human Interferon gamma in the rhesus monkey after intravenous, intramuscular and subcutatneous administration. J Biol Resp Modifiers 7: 115–122, 1988.

104. E Rinderknecht, BH O'Connor, H Rodriguez. Natural human interferon-gamma complete amino acid sequence and determination of sites of glycosylation. J Biol Chem 259: 6790–6797.

105. A Hotchkiss, CJ Refino, CK Leonard, JV O'Connor, C Crowley, J McCabe, K Tate, G Nakamura, D Powers, A Levinson, M Mohler, MW Spellman. The influence of carbohydrate structure on the clearance of recombinant tissue type plasminogen activator. Thromb Haemostas 59: 480–484, 1988.

106. M Davidian, DM Giltinan. Some simple methods for estimating intra-individual variability in nonlinear mixed effects models. Biometrics 49: 59–73, 1993.

107. RP Ekins. The precision profile: its use in assay design, assessment and quality control. In: WM Hunter, JET Corrie, eds. Immunoassays in Clinical Medicine. Edinburgh, Churchill Livingstone, 1983, pp 76–105.

108. DW Marquardt. An algorithm for least squares estimation of non-linear parameters. J Soc. Indust. Appl Meth 11: 431–441, 1963.

109. RW Gerlach, RJ White, SN Deming, JA Palasota, JM Van Emon. An evaluation of five commercial immunoassay data analysis software systems. Anal Biochem 212: 185–193, 1993.

110. B Tonshoff, WF Blum, M Vickers, S Kurilenko, O Mehls, E Ritz. Quantification of urinary insulin-like growth factors (IGFs) and IGF BP-3 in healthy volunteers before and after stimulation with recombinant human growth hormone. Eur J Endocrinol 132: 433–437, 1995.

111. T Inaba, S Shimada, M Gonda, J Mori, T Tokunaga, M Geshi, R Torii. Enzyme immunoassay of gonadotropin releasing hormone in the canine hypothalamus and plasma using monoclonal antibodies. Br. Vet J 150: 85–92, 1994.

112. A Supersaxo, WR Hein, H Steffen. Effect of molecular weight on the lymphatic absorption of water soluble compounds following subcutaneous administration. Pharm Res 7: 167–169, 1990.

113. V Bocci, GP Pessina, L Paulesu, C Nicoletti. The lymphatic route VI. Distribution of recombinant interferon-$\alpha_2$ in rabbit and pig plasma and lymph. J Biol Resp Modifiers 7: 390–400, 1988.

114. BL Ferraiolo, M Mohler, J Cook, A Chen, B Reed, J O'Connor, R Keck. The metabolism of recombinant human interferon-gamma (IFN-gamma) in rhesus monkeys. Pharm Res 7: S-46, 1990.

115. P Tanswell, M Schluter, J Krause. Pharmacokinetics and isolated liver perfusion of carbohydrate modified recombinant tissue-type plasminogen activator. Fibrinolysis 3: 79–84, 1989.

116. EW Wawrzynczak, AJ Cumber, GD Parnell, PT Jones, G Winter. Blood clearance in the rat of a recombinant mouse monoclonal antibody lacking the N-linked

oligosaccharide side chains of the $C_H2$ domains. Mol Immunol 29: 213–220, 1992.

117. H Tanaka, T Kaneko. Development of a competitive radioimmunoassay and a sandwich enzyme-linked immunosorbent assay for a recombinant human granulocyte colony-stimulating factor. Application to a pharmacokinetic study in rats. J Pharmacobio-Dyn 15: 359–366, 1992.

118. A Meager, H Leung, J Wooley. Assays for tumor necrosis factor and related cytokines. J Immunol Methods 116: 1–17, 1989.

119. JO Westgard, PL Barry, MR Hunt. A multi-rule Shewhart chart for quality control in clinical chemistry. Clin Chem 27: 493–501, 1981.

# 18

# Pharmacogenomics

**Ann K. Daly and Jane Grove**
*University of Newcastle upon Tyne, Newcastle upon Tyne, United Kingdom*

## I. INTRODUCTION

Pharmacogenomics can be defined as the study of the genetic factors that determine drug efficacy and toxicity and is closely allied to the subject of pharmacogenetics, which has been mainly concerned with study of the genetics of drug metabolism. Using sequence data from the Human Genome Project, it is increasingly possible to consider the consequences of interindividual variation in sequences of genes encoding drug targets as well as drug metabolizing enzymes and use this information both to design new drugs and to individualize drug therapy with existing drugs. However, for this approach to be successful, it is necessary to understand the exact genetic defects that give rise to specific diseases and also to understand the basis of interindividual variation in both pharmacokinetics and pharmacodynamics for specific drugs. This can be achieved if both the extent of interindividual variation in DNA sequences and the consequences of this variation is known. While the genetic defects associated with a number of single gene disorders such as cystic fibrosis and Huntington's chorea have been identified (1, 2), progress has been slower with polygenic disorders including diabetes, asthma and many forms of cancer where a number of differ-

ent gene defects and also environmental factors may together contribute to the disease. However, once genetic factors predisposing to these diseases are better understood it is likely that the possibility of individualized drug therapy to target one or more of the gene products whose abnormal expression or function contributes to the disease state will be possible. The application of pharmacogenomics to clinical treatment is still at a very early stage. However, there are already some examples of the use of knowledge of a patient's genotype for specific genes to determine what treatments are appropriate including the genotyping of Alzheimer's disease patients for apolipoprotein E (APOE) alleles to predict whether a response to drug treatment is likely (3), genotyping for deficiency in the cytochrome P450 CYP2D6 to determine dose requirement for antidepressant or antipsychotic drugs (4) and genotyping for angiotensinogen gene variants to determine whether a low sodium diet will be of benefit in mild hypertension (5).

In addition to the possibility that different combinations of gene defects may give rise to the same disease in different individuals, it is also clear that individuals may vary considerably in their disposition of individual drugs. The molecular basis of interindividual variation in drug metabolism is now well understood and it is increasingly possible to predict how an individual will metabolize a drug by simple genetic tests (4, 6). Other parts of the drug disposition process, particularly drug distribution, may also show interindividual variation. Drug receptor sequences including those for several adrenergic and dopaminergic receptors may vary between individuals though studies on the consequences of this variation up to the present have focussed mainly on its contribution to polygenic diseases rather than on drug responses (7–9).

There is also sequence variation in genes encoding biologically active peptides. This is frequently due to base changes in the promoter region that may alter the amount of peptide synthesized, either constitutively or in response to a signal. There is evidence for associations between polymorphisms affecting levels of certain cytokines including those in the tumor necrosis $\alpha$ (TNF $\alpha$) and interleukins-6 and -10 (IL-6 and IL-10) genes and susceptibility to specific polygenic diseases and certain infectious diseases (10–12).

Since there is still only limited information available on genetic factors directly affecting responses to therapy with peptide drugs, we will, in this chapter, use examples of sequence variation in genes encoding drug metabolizing enzymes and cytokines to illustrate how sequence variation can be detected, its functional significance determined, and routine methods for detection of specific alleles developed.

## II. DETECTION OF NOVEL POLYMORPHISMS

### A. Background

With increasing numbers of human gene sequences available, it is now possible to use this information as a basis for further studies to determine the extent of interindividual sequence variation and a number of different techniques are available for the detection of novel functional polymorphisms. A polymorphism is normally defined as an alternative form of a gene (allele) which occurs at a frequency of at least 0.01. Variant alleles seen at frequencies of less than 0.01 are regarded as isolated mutations rather than polymorphisms. Polymorphisms occur every 50 to 1000 bp of DNA with the precise frequency depending particularly on whether the sequence encodes a protein. These polymorphisms are typically single base pair substitutions which are referred to as single nucleotide polymorphisms (SNPs) (13), and the majority will be neutral without functional effect. In addition to SNPs, larger insertions and deletions, and so called "variable number of tandem repeats" (VNTR) sequences where short sequences are repeated a variable number of times may also occur. All polymorphisms are of interest as markers for genetic linkage studies in family studies on disease susceptibility but those of direct functional significance are of most interest in studies on both interindividual variation in drug metabolism and response, and on those polygenic disorders where family studies are not feasible.

The precise approach adopted in the detection and characterization of novel functional polymorphisms will depend on a number of factors. If individuals can be phenotyped for the gene product of interest e.g. levels of enzyme activity measured, it is usually possible to study the gene in detail in a small number of affected subjects. An alternative approach where phenotypic analysis is not feasible is to screen a larger number of healthy individuals for novel polymorphisms in the gene of interest, and to determine whether any polymorphisms detected are functional by either expression of the variant form *in vitro* or in some cases by screening for the polymorphism in a case-control study on a disease relevant to the gene.

## B. The Polymerase Chain Reaction (PCR)

The development of the polymerase chain reaction has revolutionized the study of genetic polymorphisms as well as many other aspects of genetics. The technique involves the use of a thermostable form of DNA polymerase which can resist multiple cycles of heat denaturation together with chemically synthesized sequence-specific oligonucleotide primers to amplify target DNA sequences, facilitating their analysis (14). Although the technique is most easily applied to the amplification of DNA sequences in the range 100 to 1000 bp, it is possible to amplify target sequences of up to 30 kb approximately if appropriate conditions are used including the use of a mixture of several types of thermostable DNA polymerases (15).

PCR-based methods can be used for the cloning of new genes but in the analysis of DNA sequence variation, they are usually used for the amplification of sections from genes of known sequence either to search for novel polymorphisms or to screen for known polymorphisms (genotyping), Both polymorphism scanning and genotyping are mainly performed on genomic DNA. Since genomic DNA does not vary in sequence between tissues, it is convenient to isolate DNA from a blood sample. However, it is also possible to extract DNA from other types of sample including buccal scrapings, hair roots, serum and pathological specimens (16). It is also possible to carry out mutation screening and genotyping on mRNA by synthesis of cDNA by the enzyme reverse transcriptase prior to PCR analysis (RT-PCR) (17). However, analysis of mRNA will only allow the detection of polymorphisms in the coding sequence and polymorphisms in the promoter region or in introns which may also affect gene expression will not be detected. Following initial amplification by PCR, there are a number of approaches that can be used to identify novel polymorphisms.

## C. DNA Sequencing

Currently the most efficient method for the detection of novel polymorphisms involves genome sequencing. This method offers a high degree of accuracy but its feasibility for use in a study of polymorphism detection will depend on the precise size of gene to be screened, the number of samples needing to be analyzed and the resources available which will probably in turn be determined by the importance of the gene and the reason it is being studied. However, currently available DNA sequencing methods do suffer from certain limitations. In particular, detecting heterozygosity and sequencing certain types of DNA sequences can be problematic (18).

In general, most DNA sequencing currently makes use of the Sanger dideoxy sequencing method (19) which can be carried out manually or using an automated system. Studies for the detection of novel polymorphisms are generally carried out on genes whose sequence has already been determined and the location of exons and some information on the upstream regulatory sequences are usually available. Initial sequencing can be carried out on the exons, and 20 to 50 bp of each flanking intron sequence. Generally, selected areas are amplified by PCR and sequenced directly. It is also possible to clone PCR products into suitable vectors and carry out sequencing on plasmid DNA. This approach is useful where the quality of sequence from direct sequencing of PCR products is poor perhaps due to frameshifts altering the reading frame of one allele but it is essential to sequence a number of clones from each PCR product because of the possibility that thermostable *Taq* polymerase will introduce errors during the PCR process.

## D.  Mutation Scanning Methods

Mutation scanning methods enable the screening of samples, usually PCR products, for novel polymorphisms without actually sequencing the DNA. The advantage of these methods is that they are generally more rapid, and allow a higher throughput than DNA sequencing. Once evidence for the existence of a polymorphism has been obtained by a scanning method, it is then necessary to carry sequencing to determine the precise sequence change, and to localize it, though some of the scanning methods also yield information about the location of the polymorphism.

### 1.  Single Strand Conformation Polymorphism (SSCP) Analysis

SSCP is currently the most widely used polymorphism scanning technique because of its simplicity and relatively high sensitivity. Unlike some of the other methods discussed below, it can be carried out using standard molecular biology equipment that will be readily available in most laboratories. The basis of the method is that if a double-stranded PCR product is denatured and applied to nondenaturing polyacrylamide electrophoresis gel, both strands will form intrastrand secondary structure whose electrophoretic mobility will differ in a sequence dependent manner (20). Where a mobility shift is detected, the sample is sequenced to determine the sequence change and its location. The precise sensitivity of SSCP analysis in mutation detection remains controversial with estimates of approx. 80% commonly quoted for a 300 bp PCR product when the analysis is done under at least two different electrophoretic conditions varying by temperature and/or glycerol content of the gel (21). The original description of the technique involved using [32]P-labeled primers and sequencing gel equipment (20) but methods without radiolabeling, and

using smaller polyacrylamide gels have also been developed (22). It is possible to carry out analysis of larger PCR products by cleaving with restriction enzymes prior to electrophoresis or by using a low pH gel buffer system (23).

### 2.  Other Mutation Scanning Methods

Heteroduplex analysis is often used in tandem with SSCP (24). When DNA samples are heterozygous for polymorphisms, heteroduplexes will form and may show slower electrophoretic mobility to homoduplexes. When the polymorphism is a small insertion or duplication, heteroduplexes will appear particularly prominent on the gel but mobility shifts may also occur for point mutations. It is possible to carry out SSCP and heteroduplex analysis on a single gel as sufficient DNA renaturation is likely to occur to make heteroduplexes detectable. By use of several different gel conditions, close to 100% success in detecting polymorphisms by the combined technique has been reported (24).

A variety of more specialized mutation scanning methods have been developed but are beyond the scope of this article. They include denaturing gradient gel electrophoresis, RNase cleavage, chemical cleavage, enzyme mismatch cleavage, mass spectrometry and denaturing high-performance liquid chromatography (18). With all these methods, it is necessary to characterize the mutation by DNA sequencing once the scanning technique has found evidence for its existence.

### E.  Studies on the Functional Significance of Novel Polymorphisms

Once novel polymorphisms have been identified, the next step in their study is usually assessment of their functional significance. In the case of polymorphisms in a protein coding region, the most important consideration is whether they result in any alterations in the protein sequence. If a polymorphism results in a premature stop codon or frameshift, there is a high likelihood of functional significance and many amino acid substitutions will also affect function, particularly it they are non-conservative or occur in an area of the protein critical to structure or function. However, many novel polymorphisms are silent or result in conservative amino acid substitutions. There are reports of some silent polymorphisms altering RNA stability (25) but in general such polymorphisms are unlikely to be of functional significance though the polymorphism may be of interest for studies of linkage to single nucleotide polymorphisms (SNPs). Where an amino acid substitution has occurred, direct assessment of its functional significance will require sufficient of the mutant protein for assessment of its biological properties, usually involving an enzyme or binding assay.

It is more difficult to predict effects of polymorphisms occurring outside protein coding sequences but those in promoter regions may affect gene expression, those in introns can interfere with RNA splicing and those at the 3'-end of a gene might affect RNA stability.

## 1. Studies on Promoter Region Polymorphisms Using Luciferase as Reporter Gene

The functional significance of promoter region polymorphisms can be determined using genetic reporter systems in which the ability of regulatory elements upstream of the coding region of the gene (*cis*-acting elements) to modulate the rate of transcription is measured indirectly by determining their effect on expression of a reporter gene coding for a specific enzyme activity.

Typically, either the entire promoter region of interest is cloned upstream of the reporter or part of the upstream sequence without the actual promoter is cloned upstream of a strong promoter (usually virus-derived) fused to the reporter gene. The sequence of interest can be isolated by PCR of genomic DNA followed by cloning of the PCR product either directly into the vector containing the reporter gene or by initial cloning into one of a number of the vectors designed for cloning PCR products followed by subcloning into the reporter gene vector. Where the effect of a polymorphism is being examined, it will be necessary either to prepare the variant sequence by site-directed mutagenesis or by using genomic DNA positive for the polymorphism as template for a PCR reaction.

The reporter gene construct is normally transfected into an appropriate cell line to achieve transient expression of the reporter and the reporter activity assayed. When measuring the effect of regulatory elements on gene expression, it is essential to include an internal control that will distinguish differences in the level of transcription from differences in the efficiency of transfection or in the preparation of the extracts. This can be achieved by cotransfecting the experimental reporter plasmid with a control vector consisting of a constitutive promoter driving the expression of a second reporter gene which can be independently measured in the same extract. One important consideration when using this dual reporter system is to ensure that expression from the control plasmid is not too high placing a large demand on cellular transcriptional/translational apparatus which may limit expression from the experimental plasmid generating misleading results. Dual reporter systems allow the effectiveness of promoter regions to be easily compared and are a particularly valuable method to demonstrate the effect of promoter region polymorphisms.

Reporter gene activity can be detected by measuring reporter mRNA directly by either Northern blot analysis, ribonuclease protection assays or RT-PCR. However, it is often preferable to either measure protein assays or enzyme activity using one of a range of convenient, versatile and easy assays now available. Antibody-based assays are relatively insensitive but detect the reporter protein when it is both active and inactive and can be used to visualize reporter protein expression in cells by *in situ* staining and immunohistochemistry. Reporters can also be assayed by measuring enzyme activity which is a more sensitive method due to the small amount of enzyme required to generate the reaction products. One limitation of enzyme assays is the high background if there is endogenous enzyme activity in the cell.

The chloramphenicol acetyltransferase (CAT) gene is a widely-used reporter gene used to assay promoter activity in transfected mammalian cell lines (26). The CAT protein is relatively stable in mammalian cells and as the mRNA has a fairly short half-life, it is suitable for transient assays measuring protein accumulation. CAT catalyses the transfer of the acetyl group from acetyl-CoA to chloramphenicol so CAT activity can be assayed by incubating cells with $^{14}C$ labeled chloramphenicol and detecting product formation by thin layer chromatography or organic extraction. The CAT protein can also be quantitated with an enzyme-linked immunosorbant assay (ELISA).

*Escherichia coli lacZ* encoding β-galactosidase, which catalyses the hydrolysis of certain sugars, is commonly used as a control reporter. Specialised substrates such as *o*-nitrophenyl-β-D-galactopyranoside allow the enzyme activity to be measured spectrophotometrically, fluorometrically or luminometrically. One limitation of this system is the presence of endogenous lysosomal β-galactosidase in certain cell lines. Because of the endogenous enzyme activity it is vital to include "no DNA" negative control extracts or untransfected cells as comparisons. Approaches which normalize the experimental reporter activity to β-galactosidase activity have been developed. One method involves the determination of the protein content of the extracts, and subsequently using equal amounts of protein for the reporter assays. Another approach is to use β-galactosidase assay conditions favoring the *E. coli* enzyme such as a pH of 7.3–8, or using extracts pre-heated to 50°C. In some specialized cells (e.g. gut epithelial cells) very high endogenous β-galactosidase expression makes this system unsuitable as a reporter.

Another common reporter system utilizes the firefly (*Photinus pyralis*) luciferase gene as a reporter. Luciferase has a shorter half-life than the CAT reporter making it especially suitable for transient assays assessing inducible, short-lived effects. The luciferase reaction uses D-luciferin and ATP in the presence of $Mg^{2+}$ and $O_2$ to generate light which can be quantitated. The inclusion of coenzyme A in the reaction results in a sustained light reaction with greater sensitivity (27) allowing accurate quantitation of luminesence in a luminometer or scintillation counter. The luciferase assay is approximately 30–1000 times more sensitive than CAT assays (28) and results can be obtained more quickly.

More recently, a dual luciferase reporter assay system using distinct luciferases as the control and experimental reporters has been developed (29). The firefly and sea pansy (*Renilla reniformis*) luciferases have distinct enzyme structures and substrate requirements permitting discrmination between the resulting bioluminesent reactions. The firefly luciferase can therefore be used as the experimental reporter while the sea pansy luciferase is used as the control and the normalized response calculated as the ratio of the experimental firefly luciferase to the control sea pansy luciferase activity. The dual luciferase reporter system provides sequential quantitation of both firefly and sea pansy luciferase in cell lysates in a simple assay procedure. The use of a luciferase reporter in both components of this system makes it one of the most sensitive, robust and rapid assays.

The size of the regulatory element subcloned into reporter vectors often has a significant effect on expression of the reporter. Expression is presumably affected by the presence of additional activator or repressor sites present in longer fragments but short promoter fragments can be transcriptionally ineffective suggesting that other regions are important for expression. Although reporter systems can provide information about the promoter activity, the relevance of systems to the *in vivo* situation is debatable. Expression from different reporter contructs often gives conflicting results. For example, Wilson *et al.* (30) found that a TNFα promoter region polymorphism resulted in significantly higher expression than the wild type sequence in constucts containing a promoter fragment from −585 to +106 linked to CAT. In contrast, Brinkman *et al.* (31) found no differences in expression with and without the polymorphism in constructs carrying the promoter region from −619 to +108 linked to CAT.

4. *Determination of the Effect of Polymorphisms Resulting in Altered Protein Sequences by Expression in vitro*

When novel polymorphisms in the coding region of proteins that appear to be of possible functional significance have been detected, the significance of these polymorphisms is usually assessed by expression of both wild-type and mutant proteins *in vitro* and a comparison of properties (e.g. enzyme activity) between wild-type and mutant forms. Successful expression will require a full length wild-type cDNA encoding the protein of interest. If not already available to the investigator, it is possible to prepare cDNAs by reverse transcriptase PCR of RNA isolated from a cell line or tissue known to express detectable mRNA encoding the required protein. A variety of commercial kits for the isolation of RNA and for reverse transcriptase PCR are available and may be useful for those without previous experience of these techniques. Following PCR, the cDNA will normally be cloned into either a vector designed for the general cloning of PCR products or an appropriate expression vector. In either case, it is advisable to sequence several clones to ensure that a cDNA of the correct sequence has been obtained as the *Taq* polymerase used in the PCR reaction may introduce sequence errors. If the cDNA corresponding to the allelic variant of interest is not already available, the wild-type cDNA can be mutated by site-directed mutagenesis for which a number of methods are now available (32). Alternatively mRNA from an individual positive for the polymorphism can be used as a template for RT-PCR.

The choice of an expression system for the full length cDNA will depend on a number of factors and is discussed in detail by Goeddel (33). Bacterial systems have the advantages of achieving high expression and of also being relatively cheap. However, eukaryotic proteins often undergo posttranslational processing, including phosphorylation, glycosylation, and proteolytic cleavage and in this case, a protein expressed in *E. coli* may not show the normal biological activity. Problems with codon usage may also be encountered when attempting to express eukaryotic genes in bacteria.

Prokaryotic expression systems that facilitate protein purification have been developed. These systems enable a fusion protein consisting of the full length protein of interest with an additional protein sequence tag with properties that will assist with purification by affinity chromatography fused to the 5′-end. Examples of such sequence tags include glutathione *S*-transferase, poly-histidines and maltose binding protein. In some cases, the tag may be removed by digestion of the purified protein with a specific protease (e.g. factor Xa for the maltose binding protein) following purification (34).

Alternative expression systems to *E. coli* include yeast, insect cells and mammalian cells. Discussion of the advantages and disadvantages of these systems are beyond the scope of the present article but are described in detail in a recent review article (35). In the case of peptides, expression in bacterial systems has a good chance of being successful but very high levels of expression of both peptides and peptide receptors in insect cells using baculovirus vectors have recently been reported (36, 37).

## F.  Genotyping Methods

### 1.  *Isolation of DNA*

There is a wide range of methods available for DNA preparation. Choice of method will depend on a number of factors, particularly the amount and type of material available and the type of analysis to be carried out. If a blood sample is to be genotyped only for a single polymorphism by a straightforward PCR reaction, there are a variety of rapid, small scale methods available, including several involving commercial kits. However, if PCR amplification of a long DNA sequence (long PCR) is to be carried out or a DNA sample for long-term storage is needed, high molecular weight DNA free of protein contamination should be prepared and many of the published methods are cumbersome requiring multiple phenol extractions. We routinely prepare DNA by a method involving deproteinization with perchloric acid which is rapid and inexpensive and produces DNA which has an average size of at least 30 kb and is suitable for long term storage at 4°C (38). This method is suitable for blood samples, lymphocytes and a variety of cultured cells but for human liver samples, we have found that it is necessary to carry out proteinase K digestion and phenol extraction (39) to obtain high quality DNA. For DNA extraction from formalin-fixed pathological specimens, proteinase K digestion and phenol extraction should also be used. A useful method is described by Jackson *et al.* (16) though successful analysis of DNA extracted from such samples will depend on the nature and amount of tissue in the sections used.

### 2.  *PCR-based Genotyping Analysis Using the CYP2D6 and IL-10 Polymorphisms as Examples*

Once a polymorphism of interest has been identified and characterized, the next step will often involve genotyping of large numbers of samples, usually involving amplification of the region of interest followed by use of a suitable detection method. The precise detection method will depend on the individual polymorphism being studied. Though automated high throughput detection methods are being developed, for most laboratories use of either PCR followed by digestion of product using a restriction enzyme specific for individual alleles, allele-specific PCR involving two parallel PCR reactions with a single common primer and two different allele-specific primers, PCR using sequence-specific primers or PCR followed by SSCP detection are likely to be the most appropriate methods to use for detection of point mutations or insertions or deletions smaller than 5 bp. Where the polymorphism involves a deletion or insertion of more than 5 bp approx., altered size of product should be detectable directly by gel electrophoresis of the PCR product either on polyacrylamide (for deletions or insertions of less than 20 bp) or agarose (for larger insertions or deletions).

In general, we have obtained the most satisfactory and consistent results by carrying out a PCR reaction followed by detection with either a restriction digestion (PCR–RFLP) or by SSCP analysis. With respect to restriction enzyme digestion, it is important to have an internal control for the enzyme activity in each tube. This can be an additional restriction site in the PCR product which is not subject to polymorphism or an additional PCR product with a site for the enzyme that will yield products of different non-interfering size. Although allele-specific PCR and sequence-specific PCR often work well and involve fewer steps, we have found that the specificity of the PCR reactions which is vital to the successful use of these methods is affected by variations in the laboratory temperature which presumably results in slight changes in the temperature of the annealing step in the thermocycler and we therefore only recommend use of these methods where the laboratory temperature is constant at all times. For many genes, it is important to use the highest annealing temperatures compatible with satisfactory amplification of the required sequence with the chosen primers to ensure that other sequences with homology to the gene of interest such as pseudogenes are not also amplified. A number of other factors important in successful PCR include choice of heat stable DNA polymerase, and whether additional refinements such as "hot start" or long PCR conditions are required (15). We have found that heat stable DNA polymerases from different suppliers vary in their ability to successfully amplify using different primer sets and suggest that enzyme from the same supplier should be used consistently for particular reactions.

## 3. CYP2D6 Genotyping

Genotyping as a method for the identification of individuals deficient in the cytochrome P450 enzyme CYP2D6 which metabolizes a range of commonly prescribed drugs is now well established. This common polymorphism usefully illustrates a number of factors that need to be considered when genotyping for genetic polymorphisms. In particular, design of specific PCR primers is especially important due to the presence within the *CYP2D* gene cluster of two pseudogenes with >90% homology to *CYP2D6* (40). It is essential to ensure that only *CYP2D6*-specific sequences are amplified to avoid interference from the pseudogenes and this can be accomplished by choosing primers specific for *CYP2D6* at the 3'-end and by using the maximum annealing temperature compatible with adequate amplification. Another problem is that there a number of different alleles associated with the absence of CYP2D6 activity and for accurate prediction of an individual's enzyme activity, it is important to genotype for at least the four most common of these alleles, namely *CYP2D6*3 CYP2D6*4*, *CYP2D6*5* and *CYP2D6*6* (see Table 1). Individuals negative for all variant alleles are predicted to show normal CYP2D6 activity, individuals with one variant allele will show lower than normal levels of activity but individuals with two variant alleles will lack CYP2D6 activity. Screening for the *CYP2D6*3*, *CYP2D6*4* and *CYP2D6*6* alleles can be performed by RFLP-PCR using the enzyme *Bst*NI in the case of *CYP2D6*4* where a site for this enzyme present in the wild-type sequence has been lost and the enzymes *Bsa*AI and *Bst*NI in the case of *CYP2D6*3* and *CYP2D6*6*. where sites for these enzymes are introduced by the use of primers with single base pair mismatches which bind adjacent to the polymorphisms of interest. Detailed descriptions of the methods are provided

**Table 1**   Common CYP2D6 Variant Alleles

| Allele | Base change | Effect | Reference |
|--------|-------------|--------|-----------|
| CYP2D6*3 | $A_{2637}$ deletion | Frameshift | 48 |
| CYP2D6*4 | $G_{1934}A$ | RNA splicing altered | 48 |
| CYP2D6*5 | CYP2D6 deleted | No protein | 42 |
| CYP2D6*6 | $T_{1795}$ deleted | Frameshift | 49 |

Some of these alleles may contain additional polymorphisms but only the inactivating mutations are listed.

in Daly *et al.* (38) (*CYP2D6*3* and *CYP2D6*4*) and Sachse *et al.* (41) (*CYP2D6*6*). In the case of *CYP2D6*5*, a large deletion of approx. 12 kb is present. The sequences flanking the deletion show considerable homology to other areas of the *CYP2D* locus and to detect the *CYP2D6*5* allele, it is necessary to use primers specific for sequences some distance from the deletion breakpoints and to use long PCR methods (42). Under these conditions, the presence of the *CYP2D6*5* allele is indicated by a PCR product of 3.5 kb whereas in subjects negative for the deletion, a product of 15 kb is obtained.

## 4. IL-10 Genotyping

Interleukin-10 (IL-10) is an important anti-inflammatory cytokine (43). Three polymorphisms have been identified in the human IL-10 promoter region at positions −1082, −819 and −592 (44, 45). In healthy Caucasians the nucleotides A and G are found approximately equally often at position −1082 while a C-A substitution at −592 which is in complete linkage disequilibrium with a C-T substitution at −819 is found at a frequency of 0.2 (44). The polymorphisms at positions −592 and −1082 have both been reported to affect levels of IL-10 produced following stimulation with a modulator such as interferon and may thus be relevant to disease susceptibility (44, 45). Development of convenient PCR assays for large-scale genotyping studies is therefore important.

PCR-RFLP can be used to detect the $C_{-592}A$ polymorphism due to the substitution resulting in the generation of an *Rsa*I restriction site. Briefly, PCR is used to amplify a 412 bp fragment in 50 $\mu$l buffer (10 mM Tris-HCl, pH 8.8, 1.5 mM $MgCl_2$, 50 mM KCl, 0.1% (v/v) Triton X-100) containing 1 $\mu$g test DNA, 200 $\mu$M dNTPs, 0.25 $\mu$M of each primer (5′-CTTAGGT-CACAGTGACGTGG and 5′-GGTGA-GCACTACCTGACTAGC) and 2 units *Taq* polymerase. The PCR conditions are 35 cycles at 94°C for 1 min, 50°C for 1 min and 70°C for 1 min, followed by 1 cycle at 70°C for 10 min. The resulting products are incubated with 2 units of *Rsa*I and electrophoresed on a 2% agarose gel. Each series of reactions includes a "no DNA" control and three samples of known genotype.

As the $G_{-1082}A$ substitution does not destroy or generate a restriction enzyme recognition sequence, development of a suitable genotyping assay was more difficult. The approach of introducing a new restriction site by using a primer with a single mismatch which binds close to the polymorphism used successfully for *CYP2D6* genotyping was unsuccessful due to a high G-C content in the area. Instead, we found that the most useful approach was the use of SSCP to analyze PCR products. A 321 bp DNA fragment is amplified by PCR using primers 5′-AAGCTTCTGTGGCTGGAGTC and 5′-CCAGAGACTTTCCAGATATCTG-AAGAAG using the same PCR conditions as described above for the −592 polymorphism. The −1082 genotype is then determined by SSCP on a 1× MDE polyacrylamide gel (200 mm × 200 mm × 1.5 mm) in accordance with the manufacturer's recommendations (FMC Bioproducts, USA). Samples are electrophoresed at 200 V and 4°C for 16 h. Following electrophoresis, the separated DNA strands are visualized by staining with ethidium bromide as described by Daly *et al.* (38). PCR products from individuals with known genotypes are included on each gel as standards allowing genotypes to be assigned to other samples on the gel.

## 5. Likely Future Developments in Genotyping-DNA Chips and Other Automated Methods

With the increasing information now available on functionally significant polymorphisms, there is considerable interest in using this information in clinical diagnosis. The usefulness of this approach is as yet largely unproven but it is clear that successful use of genetic information in routine clinical practise will require rapid and accurate screening of patient samples for relevant polymorphisms. There are a number of ways of achieving rapid high throughput genotyping including the automation of existing methodology for both DNA extraction and PCR amplification followed by use of an automated allele detection method. Detection methods with potential for automation include single nucleotide primer extension where a primer adjacent to the polymorphism site is extended by a single base pair using fluorescently tagged bases (46). A more novel detection method involves the use of "DNA chips" consisting of microarrays of oligonucleotides on glass supports and detection of genotype on the basis of hybridization of the initial PCR product to a series of specific oligonucleotides which can cover all the possible sequence variants (47). Analysis of hybridization patterns requires confocal microscopy and the technical complexity of DNA chips may limit the technique to specialized laboratories requiring very high throughputs. However, in many cases rapid analysis of samples may be more important than high throughput and, in these cases, DNA extraction and PCR analysis for well-defined polymorphisms can be carried out very successfully using relatively inexpensive equipment and the wide range of assay kits now available to facilititate these processes.

## REFERENCES

1. JR Riordan, JM Rommens, B Kerem, N Alon, R Rozmahel, Z Grzelczak, J Zielenski, S Lok, N Plavsic, JL Chou, ML Drumm, MC Iannuzzi, FS Collins, LC Tsui. Identification of the cystic fibrosis gene: cloning and characterization of complementary DNA. Science 245: 1066–1073, 1989.
2. The Huntington's Disease Collaborative Research Group. A novel gene containing a trinucleotide repeat that is expanded and unstable on Huntingdon's disease chromosomes. Cell 72: 971–983, 1993.
3. J Poirier, MC Delisle, R Quirion, I Aubert, M Farlow, D Lahiri, S Hui, P Bertrand, J Nalbantoglu, BM Gilfix, S Gauthier. Apolipoprotein E4 allele as a predictor of cholinergic deficits and treatment outcome in Alzheimer's disease. Proc Natl Acad Sci USA 92: 12260–12264, 1995.
4. FJ Gonzalez, JR Idle. Pharmacogenetic phenotyping and genotyping. Present status and future potential, Clin Pharmacokinet 26: 59–70, 1994.
5. V Glaser. Myriad's lifestyle test targets borderline hypertensives. Nature Biotechnology 16: 13, 1998.
6. AK Daly. Molecular basis of polymorphic drug metabolism. J Mol Med 73: 539–553, 1995.
7. P Propping, MM Nothen. Genetic variation of CNS receptors—a new perspective for pharmacogenetics. Pharmacogenetics 5: 318–325, 1995.
8. C Missale, SR Nash, SW Robinson, M Jaber, MG Caron. Dopamine receptors: from structure to function. Physiol Rev 78: 189–225, 1998.
9. BJ Lipworth. Clinical pharmacology of beta(3)-adrenoceptors. Br J Clin Pharmacol 42: 291–300, 1996.

10. W McGuire, AVS Hill, CEM Allsopp, BM Greenwood, D Kwjatkowski. Variation in the TNF-α promoter region association with susceptibility with susceptibility to cerebral malaria. Nature 371: 508–511, 1994.

11. D Fishman, P Woo. A polymorphism in the interleukin-6 (IL-6) gene conferring high IL-6 production is associated with systemic-onset JCA. Arthritis and Rheumatism 40: SS249, 1997.

12. JT van Dissel, P van Langevelde, RGJ Westendorp, K Kwappenberg, M Frolich. Anti-inflammatory cytokine profile and mortality in febrile patients. Lancet 351: 950–953, 1998.

13. FS Collins, MS Guyer, A Chakravarti. Variations of a theme: cataloging human DNA sequence variation. Science 278: 1580–1581, 1997.

14. RK Saiki, S Scharf, F Faloona, KB Mullis, GT Horn, HA Erlich, N Arnheim. Enzymatic amplification of beta-globin genomic sequences and restriction site analysis for diagnosis of sickle-cell anemia. Science 230: 1350–1354, 1985.

15. S Cheng, C Fockler, WM Barnes, R Higuchi. Effective amplification of long targets from cloned inserts and human genomic DNA. Proc Natl Acad Sci USA 91: 5695–5699, 1994.

16. DP Jackson, JD Hayden, P Quirke. Extraction of nucleic acid from fresh and archival material. In: MJ McPherson, P Quirke, GR Taylor, eds. PCR, a practical approach. Oxford: IRL Press, 1991, pp 29–50.

17. JBDM Edwards, P Ravassard, C Icard-Liepkalns, J Mallet. cDNA cloning by RT-PCR. In: MJ McPherson, BD Hames, GR Taylor, eds. PCR 2, a practical approach. Oxford: IRL Press, 1995, pp 89–118.

18. RGH Cotton. Slowly but surely towards better scanning for mutations. Trends in Genetics 13: 43–46, 1997.

19. F Sanger, S Niclen, AR Coulsen. DNA sequencing with chain terminating inhibitors. Proc Natl Acad Sci USA 74: 5463–5467, 1977.

20. M Orita, Y Suzuki, T Sekiya, K Hayashi. Rapid and sensitive detection of point mutations and DNA polymorphisms using the polymerase chain reaction. Genomics 5: 874–877, 1989.

21. M Ravnik-Glavac, D Glavac, M Dean. Sensitivity of single-strand conformation polymorphism and heteroduplex method for mutation detection in the cystic fibrosis gene. Human Mol Genet 3: 801–807, 1994.

22. T Hongyo, GS Buzard, RJ Calvert, CM Weghorst. "Cold SSCP": a simple, rapid and non-radioactive method for optimized single-strand conformation polymorphism analyses. Nucl Ac Res 21: 3637–3642, 1993.

23. Y Kukita, T Tahira, SS Sommer, K Hayashi. SSCP analysis of long DNA fragments in low pH gel. Hum Mutat 10: 400–407, 1997.

24. B Gerrard, M Dean. Single-stranded conformation polymorphism and heteroduplex analysis. In: RGH Cotton, E Edkins, S Forrest, eds. Mutation detection: a practical approach. Oxford: IRL Press, 1998, pp 25–34.

25. J Milland, D Christiansen, BR Thorley, IF McKenzie, BE Loveland. Translation is enhanced after silent nucleotide substitutions in A + T-rich sequences of the coding region of CD46 cDNA. Eur J Biochem 238: 221–230, 1996.

26. CM Gorman, LF Moffat, BH Howard. Recombinant genomes which express chloramphenicol acetyltransferase in mammalian cells. Mol Cell Biol 2: 1044–1051, 1982.

27. KV Wood. Luc genes: introduction of colour into bioluminescence assays. J Biolumin Chemilumin 5: 107–114, 1990.

28. M Pazzagli, JH Devine, DO Peterson, TO Baldwin. Use of bacteria and firefly luciferases as reporter genes in DEAE-dextran mediated transfection of mammalian cells. Anal Biochem 204: 315–323, 1992.

29. K Doyle. Protocols and applications guide. 3rd ed. Madison, WI.: Promega Corporation, 1996, pp 216–219.

30. AG Wilson, JA Symons, TL McDowell, HO McDevitt, GW Duff. Effects of a polymorphism in the tumor necrosis factor α promoter on transcriptional activation. Proc Natl Acad Sci USA 94: 3195–3199, 1997.

31. BMN Brinkman, D Zuijdgeest, EL Kaijzel, FC Breedveld, CL Verweij. Relevance of the tumor necrosis factor α (TNFα) −308 promoter polymorphism in TNFα gene regulation. J Inflamm 46: 32–41, 1996.

32. T Clackson, D Gussow, PT Jones. General applications of PCR to gene cloning and manipulation In: MJ McPherson, P Quirke, GR Taylor eds. PCR, a practical approach. Oxford: IRL Press, 1991, pp 187–214.

33. DV Goeddel. Systems for heterologous gene expression. Meth Enzymol 185: 3–10, 1990.

34. P Riggs. Protein expression. In: FM Ausebel, R Brent, RE Kingston, DD Moore, JG Seidman, JO Smith, K Struhl, eds. Current Protocols in Molecular Biology, Volume 2. New York, Greene Associates/Wiley Interscience, 1994 pp 16.6.1.–16.6.10.

35. TA Kost. Gene expression systems in the genomics era-editorial overview. Cur. Opin. in Biotech. 8: 539–541, 1997.

36. W Wang, L Yum, MC Beinfeld. Expression of rat pro cholecystokinin (CCK) in bacteria and in insect cells infected with recombinant baculovirus. Peptides 18: 1295–1299, 1997.

37. T Kusui, MR Hellmich, LH Wang, RL Evans, RV Benya, JF Battey, RT Jensen. Characterisation of gastrin-releasing peptide receptor expressed in SF9 insect cells by baculovirus. Biochemistry 34: 8061–8075, 1995.

38. AK Daly, VM Steen, KS Fairbrother, JR Idle. *CYP2D6* multiallelism. Meth Enzymol 272: 199–210, 1996.

39. N Blin, DW Stafford. A general method for isolation of high molecular weight DNA from eukaryotes. Nucl Acids Res 3: 2303–2308, 1976.

40. S Kimura, M Umeno, RC Skoda, UA Meyer, FJ Gonzalez. The human debrisoquine 4-hydroxylase (*CYP2D*) locus: sequence and identification of a polymorphic *CYP2D6* gene, a related gene, and a pseudogene. Am J Hum Genet 45: 889–905, 1989.

41. C Sachse, J Brockmoller, S Bauer, I Roots. Cytochrome P450 2D6 variants in a Caucasian population: allele frequencies and phenotypic consequences. Am J Hum Genet 60: 284–295, 1997.

42. VM Steen, OA Andreassen, AK Daly, T Tefre, A-L Borresen, JR Idle, A-K Gulbrandsen. Detection of the poor metabolizer-associated *CYP2D6(D)* gene deletion allele by long-PCR technology. Pharmacogenetics 5: 215–223, 1995.

43. WE Paul, RA Seder. Lymphocyte responses and cytokines. Cell 76: 241–251, 1994.

44. DM Turner, DM Williams, D Sankaran, M Lazarus, PJ Sinnott, IV Hutchinson. An investigation of polymorphism in the interleukin-10 gene promoter. Eur J Immunogenet 24: 1–8, 1997.

45. J Mascali, L Borish, A Aarons, J Rumbyrt, D Klemm. Sequence and characterisation of the human interleukin (IL)-10 promoter. J Allergy Clin Immunol 95: 343.

46. JM Shumaker, A Metspalu, CT Caskey. Mutation detection by solid-phase primer extension. Hum Mutat 7: 346–354, 1996.

47. A Pease, D Solas, EJ Sullivan, MT Cronin, C Holmes, SPA Fodor. Light-generated oligonucleotide arrays for rapid DNA sequence analysis. Proc Natl Acad Sci USA 91: 5022–5026, 1994.

48. M Kagimoto, M Heim, K Kagimoto, T Zeugin, UA Meyer. Multiple mutations of the human cytochrome P450IID6 gene (CYP2D6) in poor metabolisers of debrisoquine. J Biol Chem 265: 17209–17214, 1990.

49. R Saxena, GL Shaw, MV Relling, JN Frame, DT Moir, WE Evans, N Caporaso, B Weiffenbach. Identification of a new variant CYP2D6 single base pair deletion in exon 3 and its association with the poor metabolizer phenotype. Hum Mol Genet 3: 923–926, 1994.

# 19

# Analysis of Polypeptide Bioactivity

**David C. Wood, Q. Khai Huynh, and Joseph B. Monahan**
*Monsanto/Searle Company, St. Louis, Missouri*

## I. INTRODUCTION

The discovery and development of a protein therapeutic is ultimately aimed toward producing a specific *in-vivo* response. In many cases, the drug discovery process takes a mechanism based approach in which a hypothesis exists for the target of the protein drug and the activity which the drug should possess to modify the target. Therefore, a primary goal is to characterize the functional activity of the protein as early and completely as possible in order to maximize the chances for success once a protein begins clinical trials. The driving force behind these studies is the importance of separating the protein's intrinsic specific activity in an appropriate *in-vitro* assay from the fate of the drug *in-vivo*. There are many *in-vivo* processes (proteolysis, non-specific adsorption, clearance, etc) which can negatively affect the apparent potency of a protein and which are difficult to control, therefore the need exists to distinguish protein specific activity from these biological processes.

Another area where accurate and rapid analysis of bioactivity is essential occurs during the studies designed to produce initial quantities of a newly discovered gene product for testing of its utility as a drug. These are often referred to as "proof-of-concept" studies. The use of heterologous expression systems allows the rapid generation of relatively large amounts of the protein, as analyzed by SDS-PAGE or other denaturing methods, but does not ensure that the protein thus made will be in an active conformation. This is especially true for proteins expressed as inclusion bodies in *Escherichia coli* (1). Such proteins must be renatured from their unfolded or partially folded state, where bioactivity assays are the definitive proof that at least a portion of the protein has been renatured to an active conformation. However, bioactivity assays can also be critical for novel proteins secreted from mammalian or insect cell cultures, especially if specific co-factors or post-translational modifications are required for biological activity (or perhaps not yet known to be required!).

This chapter describes general principles for assessing activity of several types of protein therapeutics and gives examples of existing drugs which illustrate these principles. The types of activities fall into two general classes of mechanism: enzyme catalysis and protein-protein interaction, which present very different analytical problems. At different stages of the drug discovery process, assay needs may vary with regard to throughput, accuracy, and precision, and these issues are discussed.

In addition, we discuss emerging technologies which hold exciting promise but also present technical challenges to be overcome before gaining widespread use.

## II. GENERAL ANALYTICAL ISSUES FOR ASSAY OF PROTEIN BIOACTIVITY

The design and execution of meaningful protein function assays depends upon attention to the general principles of analytical chemistry (2). For both enzymes and receptor ligands, it is important to precisely define the conditions of the assay so that a unit of activity will be reproducible between different laboratories. For enzymes, a commonly used unit is the international unit equal to product formation of 1 micromole per minute under optimal and defined conditions of temperature, pH, [substrate], etc. (3). For receptor ligands, the definition of unit activity is less consistent due to the wide array of functional responses which can be measured (4). However, a general unit of activity can be defined in which a reference lot of known concentration of the ligand produces a half maximal response ($EC_{50}$) per unit volume under prescribed conditions.

For any analytical method, the following parameters must be considered in order to obtain meaningful data. Because of the complexity of many bioassay systems, the degree of reproducibility may be lower than in more direct chemical measurements, hence the need for careful attention to the statistical significance of the measured values. There are many good references to the statistics of small data sets, and the correct use of these principles is essential for interpretation of bioactivity data (5). Relative standard errors of 20% or greater are not uncommon in cell proliferation assays, for example. This degree of variability will be sufficient to support some conclusions, but some relatively large differences in bioactivity between samples will not be statistically real owing to the inherent variability (commonly referred to as "error") of the measurement.

Accuracy. This property of an analytical method is related to the closeness of a measured quantity to the true value. For example, the weight of a 10,000 g standard, which was calibrated using a National Bureau of Standards reference mass, may be determined using a balance. The closer the measured value is to 10,000, the more accurate the assay (6).

In the case of new protein drugs, the lack of certifiable reference standards often requires that a somewhat arbitrarily chosen production lot of the protein is set aside for use as a reference lot. Thereafter, the functional activity of the protein can be expressed as a percentage of the reference lot activity. When the protein concentration of the sample is also known, the activity may be expressed as specific activity in units per mg.

Precision. The precision of an assay refers to the degree of similarity between replicate measurements of the same measured quantity. In the example above, a series of weights of the 10,000 g standard of 10,049, 10,050, and 10,049 g would have good precision, but poor accuracy compared to a single determination of 10,001 g. Greater precision is often an advantage in finding systematic errors in a method which cause the measured value to be offset by a constant amount. When an assay method is highly variable (noisy), the accuracy of the method will be more difficult to assess. Relative standard deviation (expressed as a percentage) provides the most useful measure of precision, since it does not depend on the units of measure (7).

Linearity. Linearity refers to the relationship between the concentration of a molecule and its measured property.

For example, protein absorbance at 280 nm (where aromatic amino acids have maximal absorbance) has been found to obey the Lambert-Beer law (8) in which a plot of protein concentration vs. absorbance is linear, as shown in Fig 1a. However, as shown in Fig. 1b, the standard curve of protein concentration vs. A595 for the Bradford assay (9) is generally a quadratic function. The assessment of linearity is also complicated by the range of analyte concentration which is being measured. An assay may have a linear response over a relatively narrow range, and a non-linear, but well-defined response relationship over a wider range. The use of non-linear regression analyses which are commonly available in spreadsheet and graphics software may allow a mathematical transformation of the data to yield a linear plot over a wider range. The Lambert-Beer law represents such a transformation, in which absorbance gives the linear relationship with analyte concentration, but is the logarithm of the measured property of percent transmittance (8).

Sensitivity.   The measured response must be detectable using the practically available amounts of protein drug and substrate. Often this value is limited by the specific radioactivity of the substrate employed or by the extinction coefficient of a colorimetric substrate. Assay sensitivity may be further subdivided into two properties of an assay, the limit of detection, LOD, and limit of quantitation, LOQ (10). In the context of protein bioactivity, LOD represents the lowest concentration of the protein which gives a detectable response. LOQ, on the other hand, is the lowest concentration of the protein which gives a quantitatible response with acceptable precision and accuracy. The degree to which LOQ is greater than LOD depends largely on the precision of the assay.

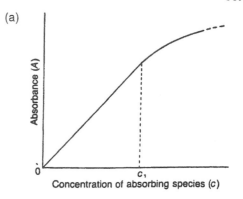

(a)

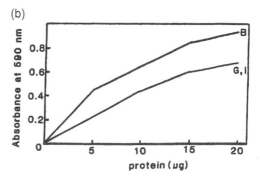

(b)

**Fig. 1**   (a) Relationship between absorbance (A) and concentration (c) of an absorbing substance. Such a graph is often called a *Beer's law plot* or, more generally, a *working curve*. (From Ref. 2, p 617). (b) Coomassie Blue protein assay standard curves for three different proteins. Bovine immunoglobulin G(I), bovine serum albumin (B), and gelatin (G) were weighed and solubilized at 2 mg/ml and 0.1 mg/ml in water. Aliquots of these solutions were used to produce the standard curves shown above using Method 6. (From Ref. 11, p. 62)

Throughput.   The throughput of an assay may be defined as the number of assay results obtained per unit time, and is often the rate limiting factor in the study of protein function. Many bioassay responses take several days or even weeks to occur, which can severely limit the type and number of experiments which can be carried out. On the other hand, many enzyme assays are complete in a matter of minutes. An important distinction can be made between assay throughput and assay turnaround time: In principle, an assay with the relatively long incubation period of one week, hence slow turnaround time, can have the same throughput as an assay with a one day incubation period by initiating a new assay every day and waiting one week for the answer. There is considerable risk, however, that the results of the first assay may invalidate those subsequently started, therefore the development of assays with shorter turnaround remains a priority in most cases.

Protein Quantitation methods. An accurate protein quantitation is key to accurate and reproducible analysis of function in the various lots of protein or protein variants that will be generated, and for successful study of the mechanism of action of a protein drug. This is necessary in order to determine specific activity as measured in activity units/mg total protein. For some purposes, precision rather than accuracy of protein quantitation is sufficient to study function, however, studies of receptor-ligand binding stoichiometry or enzyme active site titration will be greatly affected by inaccurate measures of protein concentration. Although protein quantitation is simple in concept, the great variety of buffer solutions and range of protein concentrations which are encountered at the various stages of study of a protein drug require application of different techniques at different times.

An excellent reference to the various methods which may be applied is given by Stoschek (11). In the early stages of study of a protein prior to purification to homogeneity, the Coomassie dye binding method (9, 12) has proven to be the most generally useful and has largely supplanted the more tedious Lowry assay (13). However, if detergents are present in the buffers, they will often interfere with color formation in Coomassie dye binding, but do not usually interfere with the bicinchoninic acid (BCA) assay method (14). However, the BCA assay is subject to interferences in the presence of common buffer components such as mercaptoethanol, dithiothreitol, ammonium sulfate, glucose, or ampholytes (15). The use of a consistent source of protein standard and a consistent protocol is essential for comparison of various lots of the protein, and with good technique can result in standard errors of less than 10%. The greatest limitation to the accuracy of the above methods is the necessity at the early stages of a protein drug project to use an arbitrary reference protein standard, which may differ in its colorimetric response from the protein of interest. Although use of a previously quantitated BSA or IgG standard is sufficient for many purposes, accuracy confidence will be improved by use of a lot of the protein drug itself to prepare standard curves. This strategy will avoid differences in dye binding and reactivity which do exist between proteins, and becomes practical when larger quantities of purified protein become available. In this case, the reference lot should be accurately quantitated using amino acid composition or a calculated protein extinction coefficient (see below).

After a polypeptide drug as been purified to homogeneity, the use of protein absorbance spectra has proven to be the method of choice for accurate protein quantitation, having significant advantages in throughput and convenience over amino acid composition analysis. In the initial stages of study, a calculated extinction coefficient based on the protein's known amino acid composition is sufficiently accurate and relatively simple to implement (16, 17). The greatest limitation on accuracy is generally the absorbance data, which may be subject to interference by light scattering or to limitations of spectrophotometer linearity. As larger quantities of the purified protein become available, the extinction coefficient may be experimentally determined by correlation of the absorbance spectrum, corrected for light scattering in the 320–360 nm range, with amino acid analysis results from a well-validated protocol (18).

Accurate quantitation of peptide drugs, (arbitrarily defined as having fewer than 30 amino acids) as opposed to the larger protein drugs, is most readily carried out by amino acid composition analysis. A simple weight measurement of the peptide will be subject to variability owing to adsorbed water, and the Bradford assay or absorbance spectra methods lack accuracy for small peptides due to the specifics of their composition, which may affect reactivity with protein quantitation reagents.

An ELISA (see Chapter 17) can provide a measure of immunoreactive protein in a crude sample, but may be difficult to interpret in terms of functionally active molecules. Because of their dependence on epitope-specific antibody interaction, the use of ELISA for comparison of protein point mutation variants for specific activity is difficult to validate. However, for comparison of different lots of the same protein in a crude mixture, ELISA does provide a sensitive method for quantitation of the protein of interest.

## III.   ANALYSIS OF ENZYME BIOACTIVITY

Enzymes such as asparaginase, streptokinase, tissue plasminogen activator and DNase have been used as drugs for many years (19). Recently, with the rapid progress in molecular biology, these enzymes have been cloned, expressed, purified and characterized for therapeutic use (19). Bioactivity of enzymes can be determined by *in vitro* enzyme assay, cell-based assay or assays measuring metabolite changes occurring in organs or whole animals. In this chapter, we concentrate on the analysis of enzymes by an *in vitro* enzyme assay to determine their specific activity and kinetic parameters including catalytic efficiency, turnover number, maximum reaction velocity (Vmax) and Michaelis constant ($K_m$) for substrates.

## A.   *In vitro* Specific Activity

The catalytic activity of an enzyme can be expressed in any appropriate units. However, in order to compare the catalytic activity between preparations of enzymes, a more standardized unit is needed. The most commonly used quantity is the *Unit*, also referred to as the *International Unit* or *Enzyme Unit*. A *Unit* of the enzyme activity is defined as the amount of product (nmoles, $\mu$moles or mmoles) formed from the substrate per unit time (second, minute or hour) (20–22). *Specific activity* is defined as Units per mass of enzyme (mg of protein) (20–22). When the molecular mass of the enzyme is known, it is possible to express the activity as the *molecular activity*, defined as the number of units per mole of enzyme or the number of products formed per mole of enzyme per minute assuming that the enzyme has no more than one active site (20–22). Also, when the number of active sites is known, the activity can be expressed as the *catalytic centre activity* which indicates moles used to express the activity. Catalytic activity of an enzyme can be determined following enzyme kinetic experiments.

## B.   Enzyme Kinetics

Michaelis and Menten (20–22) have proposed a model to determine the kinetic parameters of enzymes. In this model, the rate of catalysis ($V$) varies with the substrate concentration [$S$], defined as the number of moles of product ($P$) formed per second. At a fixed concentration of enzyme ($E$), $V$ is almost linearly proportional to $[S]$ when $[S]$ is small. At a high concentration of $[S]$, $V$ is nearly independent of $[S]$.

$$E + S(\overset{k_1}{\underset{k_2}{\rightleftharpoons}})ES \overset{k_3}{\rightleftharpoons} E + P \qquad (1)$$

As shown in Eq. (1), the substrate $S$ forms an $ES$ complex with enzyme $E$ with a rate constant $k_1$. The $ES$ complex can dissociate to $E$ and $S$ with a rate constant $k_2$ or it can form the product $P$ with a rate constant $k_3$. Equation (1) indicates that the catalytic rate is equal to the product of the $ES$ complex and $k_3$:

$$V = k_3 [ES] \qquad (2)$$

In the steady state, the concentration of the intermediate stays the same while the concentration of starting materials and products are changing [Eq. (3)]:

Rate of formation of

$$ES = k_1[E][S] = (k_2 + k_3) [ES] \qquad (3)$$

or

$$[ES] = [E][S]/(k_2 + k_3)/k_1 \qquad (4)$$

With $K_m = (k_2 + k_3)/k_1$, we have:

$$[ES] = [E][S]/K_m \qquad (5)$$

in which $K_m$ is a new constant called the Michaelis constant. Since the concentration of the enzyme is equal to the total concentration $[E_t]$ minus the concentration of the ES complex, we have:

$$[ES] = ([Et] - [ES])/[S]/K_m \qquad (6)$$

or

$$[ES] = [Et][S]/([S] + K_m) \qquad (7)$$

By substituting this expression for $[ES]$ into Eq. (2), we get

$$V = k_3[Et][S]/([S] + K_m) \qquad (8)$$

The maximum rate, $V_{max}$, will be obtained when the catalytic site on the enzyme is saturated with substrate; that is, when $[S]$ is much greater than $K_m$ so that $[S]/([S] + Km)$ approaches 1. Thus

$$V_{max} = k_3[E_t] \qquad (9)$$

Equation (8) can be changed to:

$$V = V_{max}[S]/(K_m + [S]) \qquad (10)$$

This is the Michaelis-Menten equation (2–4) which indicates that at a fixed concentration of enzyme $(E)$, $V$ is almost linearly proportional to $[S]$ when $[S]$ is small and at high concentrations of $[S]$, $V$ is nearly independent of $[S]$. Equation (10) also shows that when the concentration of the substrate is equal to $K_m$, $V$ will be equal to half of $V_{max}$. Thus, $K_m$ is equal to the substrate concentration at which the reaction rate is half its maximum value.

## C. Determination of $K_m$ and $V_{max}$

The Michaelis constant $K_m$ and the maximum rate of catalysis can be determined from the rate of catalysis of the enzyme at different concentrations of the substrate. After rearranging Eq. (10) we have:

$$1/V = 1/V_{max} + K_m/V_{max} \cdot 1/[S] \quad (11)$$

Equation (11) is the Lineweaver-Burk equation (2–4). When $1/V$ is plotted against $1/[S]$, a straight line is obtained with the intercepts directly indicating reciprocal values of $K_m$ and $V_{max}$. This plot, called the Lineweaver-Burk reciprocal plot, has some disadvantages in that one must adjust to axes scaled as reciprocal values: the lowest points represent the highest rates and substrate concentrations and *vice versa*, and the high concentration points tend to be compressed towards the $1/V$ axis. To avoid any error that may occur when using the Lineweaver-Burk reciprocal plot, Eq. (11) can be changed to:

$$V = (-K_m) \cdot V/[S] + V_{max} \qquad (12)$$

A plot of $V$ against $V/[S]$, called the Eadie-Hofstee plot (2–4), gives $K_m$ and $V_{max}$ values with less experimental errors. It is important to note that the kinetic parameters of enzymes may vary depending on the presence of activator of the enzymes. Details of kinetic experiments in this case have been described (2–4).

## D. Turnover Number and the Catalytic Efficiency $k_{cat}/K_m$ Ratio

The *turnover number* of an enzyme is the number of substrate molecules converted to product by an enzyme molecule per unit time when the enzyme is fully saturated with the substrate. The turnover number is therefore equal to the kinetic constant $k_3$. Turnover number ($k_3$) can be determined from Eq. (9):

$$\text{Turnover number} = V_{\max}/[E_t] \qquad (13)$$

When the concentration of substrate is much smaller than $K_m$ and the concentration of enzyme $[E]$ is nearly equal to the total concentration of enzyme $[E_t]$, from Eqns (2) and (5), we have:

$$V = (k_3/K_m) \cdot [S][E_t] \qquad (14)$$

which means the catalytic rate depends on the value of $k_3/K_m$ ratio and the concentration of substrate $[S]$. Since $K_m$ is equal to $(k_2 + k_3)/k_1$, the ratio of $k_3/K_m$ is equal to $(k_3 \cdot k_1)/(k_2 + k_3)$. When $k_3$ is much greater than $k_2$, the $k_3/K_m$ ratio is nearly equal to $k_1$. This indicates that the value of $k_3/K_m$ depends on $k_1$, the rate of formation of the *ES* complex. In the more complex reaction, the maximal catalytic rate, called $k_{cat}$, depends on several rate constants rather than $k_3$ alone. In these cases, the parameter for these enzymes is the catalytic efficiency $k_{cat}/K_m$ ratio.

## E. Tissue Plasminogen Activator – Bioactivity Characterization of an Enzyme Drug

An example of an enzyme with great utility as a drug is tissue plasminogen activator (tPA), a serine protease which catalyzes the conversion of (glu or lys)-plasminogen to plasmin (23–26). Like urokinase, the other physiologic plasminogen activator, tPA is synthesized in endothelial cells (23–26). Since plasmin proteolytically degrades the fibrin network associated with blood clots (23), tPA has been used as a thrombolytic agent in patients suffering from thromboembolic diseases (26). tPA is a single chain glycoprotein with 530 amino acids (23–26). tPA can be cleaved into two chains consisting of a 38 kDa heavy chain which contains the fibronectin finger-like domain, the EGF-domain, the kringle 1 domain, the kringle 2 domain, and a 31 kDa light chain containing the serine protease domain (24). Glycosylation sites of tPA were identified on Asn-117, Asn-184 and Asn-448 (24). tPA isolated from different sources varies in its carbohydrate content (23–28) and has specific activity in the range from 67,000 to 229,000 IU (International Unit) per mg of protein (27–29). Fibrinogen, fibrin or polylysine (soluble fibrin substitute) stimulates the activation of human but not rat oocyte tPA (30, 31).

tPA activity can be determined by many methods, such as the traditional plate and clot lysis assays, the highly sensitive plasminogen rich [125]I-fibrin coat microtiter well assay, a proteolytic cleavage assay using a chromogenic substrate (for example S-2251, D-valyl-leucyl-lysyl-p-nitroanilide) instead of plasminogen, and radio-immunoassay. Among these assays, the clot lysis assay is widely used to determine the activity of tPA since its conditions are most similar to those of the *in vivo* functions of tPA, while the chromogenic substrate for plasmin has been used primarily for kinetic studies. Neither of these assays distinguished between urokinase and tPA. Recently, the use of antibodies against tPA or urokinase in activity assay allows for functional suppression of one of the plasminogen activators (23–26).

Kinetic studies of plasminogen activation of isolated human tPA in the presence and absence of fibrinogen or fibrin with the chromogenic substrate S-2251 using Lineweaver-Burk plots have been reported (30). For kinetics experiments, different concentrations of plasminogen were incubated at 37°C with

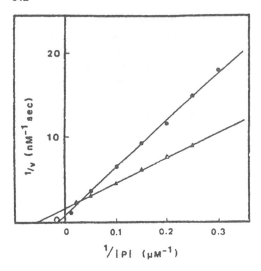

**Fig. 2** Lineweaver-Burk plots of activation of Glu-plasminogen (3.3 to 100 $\mu$M) by 150 IU/ml of uterine tPA (•) and of Lys-plasminogen (4–50 $\mu$M) by 21 IU/ml of uterine tPA (▲) (19). The activation was carried out in phosphate buffered saline, pH 7.4 containing 0.25% gelatine and 0.006% Tween 80 (reproduced from Ref. 30 with permission).

tPA in phosphate buffered saline, pH 7.4, containing 0.25% gelatin and 0.0006% Tween 80. At timed intervals, the plasmin generated was measured spectrophotometrically using S-2251 after 20 fold

dilution of the sample (30). In the absence of an activator, plots of 1/V versus 1/plasminogen (Fig. 2) obeyed Michaelis-Menten kinetics with $K_m = 65 \, \mu$M and $k_{cat} = 0.06 \, s^{-1}$ for glu-plasminogen but $K_m = 19 \, \mu$M and $k_{cat} = 0.2 \, s^{-1}$ for lys-plasminogen. The $k_{cat}/K_m$ ratio were found to be $0.001 \, \mu M^{-1} \, s^{-1}$ for glu-plasminogen and $0.011 \, \mu M^{-1} \, s^{-1}$ for lys-plasminogen (Table I). Addition of fibrinogen increased the activation rate of tPA with a $K_m$ value of 28 $\mu$M and $k_{cat}$ of $0.3 \, s^{-1}$ for glu-plasminogen as compared to 1.8 $\mu$M and $0.3 \, s^{-1}$ for lys-plasminogen. In the presence of fibrin, $K_m$ and $k_{cat}$ values of tPA were significantly decreased with $K_m$ of 0.16 $\mu$M and $k_{cat}$ of $0.1 \, s^{-1}$ for glu-plasminogen as compared to 0.02 $\mu$M and $0.2 \, s^{-1}$ for lys-plasminogen (Table 1).

Fibrin, however, had no effect on the kinetic parameters of rat oocyte tPA. As shown in Table 2, the soluble fibrin substitute polylysine lowered the $K_m$ of human tPA by 30-fold but had no effect on the rat oocyte tPA (31). Polylysine also had no significant effect on the $V_{max}$ value (31). Further studies on the oocyte tPA activity in the conditioned medium of rat insulinoma cell line and analysis of the activity of oocyte tPA after fractionation by SDS polyacrylamide gel elec-

**Table 1** Kinetic Parameters of Plasminogen Activation by Two-Chain Uterine tPA (Reproduced from Ref. 30 with permission)

|  | Glu-plasminogen | Lys-plasminogen |
|---|---|---|
| $K_m (\mu M)$ | 65 | 19 |
| $k_{cat} (s^{-1})$ | 0.06 | 0.2 |
| $k_{cat}/K_m (\mu M \, s^{-1})$ | 0.001 | 0.011 |
| *In the presence of fibrinogen* | | |
| $K_m (\mu M)$ | 28 | 1.8 |
| $k_{cat} (s^{-1})$ | 0.3 | 0.3 |
| $k_{cat}/K_m (\mu M \, s^{-1})$ | 0.011 | 0.17 |
| *In the presence of fibrin* | | |
| $K_m (\mu M)$ | 0.16 | 0.02 |
| $k_{cat} (s^{-1})$ | 0.1 | 0.2 |
| $k_{cat}/K_m (\mu M \, s^{-1})$ | 0.63 | 10.0 |

**Table 2** Kinetic Parameters of Isolated Human and Rat Oocyte in the Presence of Polylysine (13). Initial Rate of Reaction Was Measured Using 0.46–7.36 $\mu$M of Either Lys-Plasminogen (Lys-pmg) or Glu-Plasminogen (Glu-pmg) (Reproduced from Ref. 31 with permission)

| Enzyme | Substrate | Polylysine $\mu$g/ml | $K_m$ $\mu$M | $V_{max}$ nM plasmin/min | $k_{cat}$ s$^{-1}$ |
|---|---|---|---|---|---|
| Human tPA[a] | Lys-pmg | 0 | 24.0 | 0.15 | 0.21 |
| Human tPA[a] | Lys-pmg | 3.3 | 0.78 | 0.089 | 0.13 |
| Oocyte tPA[b] | Lys-pmg | 0 | 2.1 | 0.12 | NA[c] |
| Oocyte tPA[b] | Lys-pmg | 0.33 | 2.3 | 0.19 | NA[c] |
| Oocyte tPA[b] | Lys-pmg | 3.3 | 3.2 | 0.20 | NA[c] |
| Oocyte tPA[b] | Lys-pmg | 3.3 | 2.1 | 0.19 | NA[c] |
| Human tPA[a] | Glu-pmg | 0 | 23.0 | 0.035 | 0.050 |
| Human tPA[a] | Glu-pmg | 3.3 | 0.79 | 0.052 | 0.075 |
| Oocyte tPA[b] | Glu-pmg | 0 | 1.3 | 0.015 | NA[c] |
| Oocyte tPA[b] | Glu-pmg | 3.3 | 1.0 | 0.017 | NA[c] |

[a] Measured using 25 pg (0.36 fmoles) of human tPA.
[b] Measured using tPA from the equivalent 1 oocyte.
[c] NA, not applicable.

trophoresis suggested that the oocyte tPA activity may be already fully stimulated, possibly by an endogenous oocyte component(s) (31).

The *in vitro* $^{125}$I-fibrin clot lysis assay also has been widely used to determine the activity of tPA since its conditions closely mimic those of the *in-vivo* substrates of tPA. The assays were performed in phosphate buffered saline supplemented with plasminogen or in freshly citrated human plasma. Fig. 3 showed the activity of the wild type glycosylated tPA was determined in comparison to those of its non-glycosylated variants (32). The fibrinolytic potency of the variants was comparable to that of the wild type tPA at the activator concentrations of 17–51 nM ($\sim$1–3 mg/ml). At concentrations of 0.5-5.1 nM ($\sim$0.03–0.3 mg/ml), the variant enzymes, however, had lower fibrinolytic potency than the wild type enzyme.

## F.  Active Site Titration

The study of the enzyme activity can follow several approaches. One of them is to analyze the specific activity and kinetic parameters of the reaction catalyzed by the enzyme as described. This approach may give us information about the velocity at which the process takes place or which substrate(s) bind and product(s) leave the enzyme. In some cases, however, information about the mechanism of action of the enzyme and its active site is required to develop a better enzyme, i.e, higher specific activity, greater catalytic efficiency or longer half-life *in vivo* (33, 34).

The active site of the enzyme is the place in which the substrate(s) contact the enzyme. Information about the amino acid residues and their roles at this site is important to understand the catalytic mechanism of the enzyme. Among many methods to study the enzyme's active site, crystallography is the most potent and accurate. However, it has some limitations such as the need of obtaining the enzyme in good crystalline form for analysis, expensive equipment, and a time consuming process. Results from crystallographic study often need to be confirmed by other methods such as site-directed mutagenesis or nuclear magnetic resonance since the enzyme structure in aqueous solution may

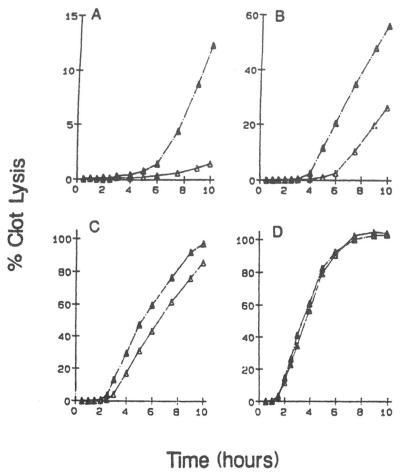

## Time (hours)

**Fig. 3**   Clot lysis of glycosylated and non-glycosylated tPA. The [125]I-fibrin clots was determined at increasing concentrations of glycosylated (△) and non-glcosylated (▲) tPA in phosphate buffered saline supplemented with plasminogen. (150 µg/ml). Panel A–D indicate final tPA concentrations of 17, 51, 170 and 510 pM (∼1,3,10 and 30 µg/ml), respectively (reproduced from Ref. 32 with permission).

differ from that of the crystalline form (33, 34).

The other powerful method to probe the enzyme's active site is chemical modification using group-specific or affinity labeling reagents. In this method the specific reagent will bind covalently to amino acid residue(s) resulting in alteration of the enzyme structure. By monitoring the kinetic parameters of the reaction as well as isolating and analyzing the modified residue-containing peptide,

the identity and the role of the modified amino acid residue(s) will be suggested. Details of kinetic analyses of the chemical modification of enzymes such as irreversible and reversible reactions, formation of dead-end complex (20, 33) and specific reagents for modifying amino acids (33, 34) have been extensively published. Similar to other method(s), chemical modification also has its limitations, since the reagent used must be very specific for certain amino acid residues in the active site and many factors such as pH or buffer may effect the velocity of the reaction process and the results of the modification. Also, results from chemical modification need to be carefully considered, since in some cases, the modified residues may not be located at the active site due to conformation change of the enzyme during modification.

Beside the purpose of monitoring the active site structure, chemical modification is also useful in developing enzymes with better properties as reported for L-asparaginase, the enzyme that hydrolyzes asparagine to aspartic acid and ammonia. L-asparaginase has been used in the treatment of acute lymphoblastic leukemia (35). However, the toxicity (immunoreactivity) of the purified enzyme often limits its repeated use in patients. Modification of the enzyme from *E. coli* with polyethylene glycol results in an enzyme that has less immunogenicity and enhanced pharmacologic properties (36). Modification of L-asparaginase lysine residues with poly-DL-lysine also showed a definite reduction in immunogenicity and longer half life in animals (37). L-asparaginase covalently modified with dextrans also resulted in less immunoreactivity than the unmodified enzyme (38) and crosslinking the enzyme with albumin has been shown to make it less susceptible to proteolytic degradation in the *in-vitro* system (39).

## G.  Conclusion

During enzyme assay, factors such as pH, temperature, buffer, ionic strength, substrates, assay mixture, mixing method, enzyme purity and stability may affect the performance and validity of the assay. These factors need to be carefully considered to obtain meaningful results. Among these factors, from the chemical reaction's point of view, pH and temperature may be the most important (20, 22). Most enzymes have an optimum pH at which their activity is maximal. The pH activity relationship of enzymes depends on the acid-base behavior of the enzyme and substrate as well as many other factors. The pH optimum of enzymes *in vitro* is not necessarily identical to that of enzymes *in vivo*, which may be higher or lower than the optimum pH *in vitro*. This indicates that pH could play an important role in control of the enzyme activity *in vivo*. The rate of catalytic activity of enzymes also depends on the temperature of the reaction. Although the temperature coefficient ($Q_{10}$) varies depending on the energy of activation of the reaction, its activity increases two-fold for each 10°C rise in temperature. Except for a few examples, most enzymes, denature at temperatures above 55°C resulting in a decreased rate of catalytic activity (20–22).

While it is difficult to cover all aspects of analysis of therapeutic enzymes in the space available for this chapter, the references given in the appropriate sections of the text should help fill some of the gaps. The subjects related to enzyme kinetics and problem solving during enzyme assays have also been reported in detail (20–22).

## IV.  ANALYSIS OF MODULATORS OF RECEPTOR-LIGAND INTERACTIONS

The structural complexity inherent in protein ligands results in unique issues with respect to quantitation and functional analysis. The key parameters to address in evaluating the functional activity of a protein drug or drug candidate are potency, efficacy and selectivity. The therapeutic goal of drugs which modulate protein-ligand receptor interactions may be to activate the receptor as in the case of some growth and differentiation factors, or stimulate the immune response as in the case of the interferon family of cytokines and some anti-receptor monoclonal antibodies. Alternatively, blockade of receptor activation through the use of soluble receptors, monoclonal antibodies directed against the ligand of interest or generation of antagonistic ligands via mutational approaches are also being promoted as therapeutic approaches. Modulators of receptor activation may be therapeutically useful in inflammation mediated diseases (eg asthma, rheumatoid arthritis, sepsis) and cancer.

To date protein-ligand drug candidates have either been: (1) isolated or fractionated from biological starting material; (2) cloned into a suitable expression system as native sequences and purified; or (3) altered to enhance some functional property through mutagenesis, chimera formation and/or covalent modification. These molecules may be designed to either activate (functional agonist) or inhibit (receptor antagonist) the hormone/growth factor/cytokine structural and functional classes. Agonistic ligands function through specific interaction with and activation of cell surface receptors in a very selective manner. Receptor activation typically occurs through ligand induced hetero- or homo- multimerization of receptor subunits and subsequent activation of signalling kinases [for review see (40)]. These signalling cascades ultimately result in alteration of gene expression and modulation of cellular activity. *In vitro* functional analysis of these drug candidates takes advantage of the multiple steps involved in the activation pathway from ligand receptor interactions, through early phosphorylation events and mediator release to the longer term cellular functional readouts (eg cellular differentiation, proliferation, viral death). As technology advances towards the requirement to analyze larger numbers of potential drug candidates, rapid, quantitative, higher throughput *in vitro* assays which are reliable measures of activity have become a necessity in many cases. In the paragraphs below we will discuss *in vitro* assays utilized in the discovery of protein based drugs modulating receptor-ligand interactions. These assays may also be utilized in the development of such drugs, coupled with physical measurements, to evaluate functional activity directly and structural integrity and purity indirectly.

## A. Receptor Binding Analysis

The majority of growth factors, cytokines and interleukins function through association with and activation of cell surface receptors. In general the ligand binds to a specific receptor subunit and the receptor is activated and signal transduced following receptor oligomerization. This receptor oligomerization results in higher affinity complexes composed of homodimers as is the case with, for example, granulocyte-colony stimulating factor (G-CSF), erythropoietin, Flt-3 and growth hormone receptors or results in heteromultimers as is the case for granulocyte-macrophage colony stimulating factor (GM-CSF), interleukin(IL)-2, -3, -4, -5, -6 and interferon receptors. Receptor binding analyses have been established for the above, and their interactions quantitated and characterized, leading to the result that interactions of these protein ligands with their receptors proceed with nM-pM affinities.

Ligand-receptor interactions can be described quantitatively through the use of kinetic constants which are a measure of the rate of ligand association ($k_a$) and dissociation ($k_d$) from the receptor. The equilibrium dissociation constant ($Kd$) is generally utilized to describe the affinity of receptor-ligand interactions. The accurate determination of binding constants which describe these interactions is dependent on the use of equations commensurate with experimental conditions [for review see (41)]. In the case of a reversible, bimolecular interaction for a homogenous population of noninteracting sites, the binding of ligand ($L$) to receptor ($R$) is described by Eq. (1).

$$L + R \rightleftharpoons LR \tag{1}$$

At equilibrium:

$$Kd = [L][R]/[LR] \tag{2}$$

Where the $Kd$ is defined as the equilibrium dissociation constant. The total quantity of receptor ($Rt$) can be expressed as:

$$[Rt] = [R] + [LR] \tag{3}$$

and

$$[R] = [Rt] - [LR] \tag{4}$$

Substitution of (4) into Eq. (2) and rearrangement leads to the Scatchard equation (42):

$$[LR]/[L] = -[LR]/Kd + Rt/Kd \tag{5}$$

Data generated from a saturation isotherm where binding is measured as a function of added ligand over several concentrations can be plotted as a Scatchard plot of $[LR]$ vs $[LR]/[L]$. The $Kd$ value can be derived from the slope of this line ($-1/Kd$) and the number of sites determined from the x-intercept ($Rt$). A nonlinear Scatchard plot is indicative of a more complex interaction than that described in Eq. (1) (43). Nonlinear curve fitting is also available for the analysis of dose-response curves (44). The nonlinear analysis is more powerful for the analysis of more complex interactions involving cooperativity or heterogenous binding sites.

Kinetic properties of ligand-receptor interactions are measured utilizing association and dissociation reaction time courses where:

$$Kd = k_d/k_a \tag{6}$$

An association rate constant ($k_a$) can be generated using the following second order rate equation where the total ligand concentration is $[Lt]$ and $[LRe]$ is the concentration of the ligand receptor complex at equilibrium:

$$\ln \{[[LRe]([Lt]$$
$$- [LR][LRe]/[Rt])]/[[Lt]([LRe]$$
$$- [LR])]\} = k_a t \{[Lt][Rt]/[LRe]$$
$$- [LRe]\} \tag{7}$$

The slope of a plot of $\ln \{[[LRe]([Lt] - [LR][LRe]/[Rt])]/[[Lt]([LRe] - [LR])]\}$ vs time is the association constant $k_a$. Under conditions where $[Lt] \gg [LRe]$ equation 6 may be simplified using a pseudo first order relationship:

$$\ln \{[LRe]/([LRe] - [LR]\} = k_a t$$
$$- ([Lt][Rt]/[LRe]) \tag{8}$$

Association kinetic studies should be carried out at several concentrations of both ligand and receptor. The determination of the kinetic constants using either Eq. (7) or (8) require the independent determination of $Rt$. The slope of the pseudo first order plot ($\ln \{[LRe]/([LRe] - [LR]\}$ vs $t$) yields $k_{obs}$, and a plot of $k_{obs}$ vs ligand concentration from experiments utilizing several ligand concentrations results in a linear curve having a slope of $k_a$ and a y-intercept of $k_d$. Using this method, no independent determination of $Rt$ is necessary.

Dissociation studies are typically carried out by elimination of the forward reaction after equilibrium is achieved through extensive dilution or addition of a large amount of excess unlabelled ligand. The dissociation rate equation under these conditions where $[LR_0]$ is the concentration of the ligand receptor complex immediatly after dilution and [LR] is the concentration of complex at time t following the initiation of dissociation is as follows:

$$\ln [LR]/[LR_0] = -k_d t \tag{9}$$

If the reaction is fully reversible, the slope of the first order plot ($\ln [LR]/[LR_0]$ vs $t$) will be equal to $-k_d$.

In cases where ligands of interest are not available in labelled form, competition studies are typically used to indirectly determine ligand affinity. In these experiments a constant concentration of radiolabelled ligand is added to receptor in the presence of varying concentrations of unlabelled competitive ligand. A decrease in labelled ligand associated with the receptor is determined as a function of ligand concentration. The concentration of unlabelled ligand resulting in a 50% decrease in labelled ligand binding is referred to as it's $IC_{50}$. $IC_{50}$ values are dependent on assay conditions but can be converted to an equilibrium constant ($Ki$) for the ligand using the Cheng-Prusoff relationship (45):

$$Ki = IC_{50}/\{1 + ([L]/Kd)\} \qquad (10)$$

In this equation $[L]$ is the concentration of the free labelled ligand and $Kd$ is the equilibrium dissociation constant of the labelled ligand. Equation (10) is valid when $Kd \gg Rt$ and $[L] \gg [LR]$ and is very useful when comparing $IC_{50}$ values from different laboratories or assays carried out under several conditions.

Receptor binding studies have several advantages over cellular functional assays or in vivo analysis. Receptor binding assays are rapid and quantitative and allow for structural alterations in protein ligands to be expressed as differences in association/dissociation rate constants as well as equilibrium binding constants. Binding studies have significant limitations however, in that the analysis is typically restricted to measuring the nature of the interaction of ligand with receptor, providing little information as to the functional consequence of this interaction. Ligand efficacy, in terms of agonist/antagonist activity requires an additional assessment of receptor activation or cellular response. While receptor

binding affinity in and of itself is not a measure of functional activity, it is often well correlated with biological potency when analyzing proteins of similar efficacy (Fig. 4).

Receptor binding studies are often carried out on fractionated cells, cell membranes, cell lines or isolated tissue. Utilizing receptors normally expressed on cells and tissues has the advantage of measuring binding site affinities on biologically functional receptors, potentially on the target cell of interest. These cells and tissues will generally have the entire complement of receptor subunits at physiologically relevent concentrations. Additionally, it might be possible to measure function on these same cells. For example, one can measure receptor binding of IL-3 to the human proerythroleukemic cell line TF-1 and also measure protein kinase phosphorylation (JAK and MAP kinases) along with cell proliferation (46). Some limitations of this approach include a limited signal due to low receptor expression, in that many of these cell lines have only tens to a few hundred receptors per cell. For example, approximately $5 \times 10^6$ cells expressing 300 receptors/cell are required for a 1000–2000 cpm signal using an iodinated ligand (specific radioactivity of 1000 Ci/mmol).

In an effort to circumvent the issue of low numbers of naturally occurring receptors on cells or cell lines, many full length receptors have been cloned and specifically expressed on cell lines which do not typically express the receptor of interest. Using this approach, cell clones may be chosen which express specific numbers of receptors, up to several hundred thousand. Additionally, only the receptor subunit of interest is expressed on these cells, resulting in more straightforward data interpretation. In some cases several receptor subunits have been cloned into cells allowing interactions to be measured on the receptor complex.

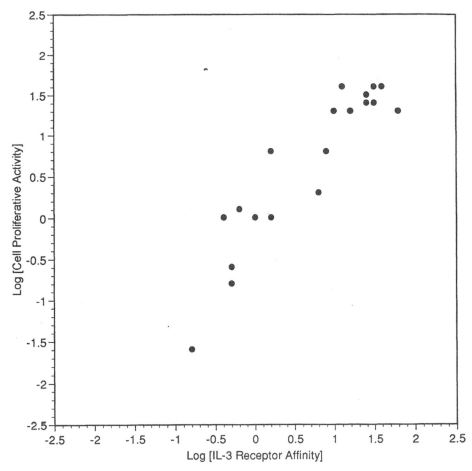

**Fig. 4** Correlation plot of IL-3 receptor affinity vs proliferative activity of synthetic IL-3 mutants. The affinity of IL-3 mutants was determined on the human AML-193 cell line expressing both alpha and beta receptors. Proliferation of AML-193 cells induced by IL-3 mutants was quantitated by [3H]thymidine incorporation as described (46).

The interpretation of binding and functional data derived from cells expressing large numbers of receptors must be done with care. For example, increasing the receptor concentration in the cell membrane may enhance the rate of ligand induced multimerization potentially resulting in higher observed affinity or signal transduction potency. While the cloned, cellular expressed receptor has some limitations, it has been successfully utilized in comparative studies, evaluating various mutants of the natural ligand or to provide evidence of uniformity of structure or ligand preparation.

An alternative receptor source for ligand affinity determination is the cloning and expression of the extracellular domain of cell surface receptors. Many of these cytokine and growth factor receptors are expressed naturally in both soluble (extracellular domain) and membrane associated form. The soluble forms of these receptors have been alternatively demonstrated to function in attenuation or augmentation of the biological response (47). Some soluble receptors including IL-2, IL-4, IL-5, IL-1, and TNF-alpha act as a ligand sink and through binding, prevent the ligand from interacting with the

**Table 3** IL-5 Receptor Characteristics

| Receptor Source | Subunit Composition | [$^{125}$]IL5 | | Antagonist sIL5Rα (IC$_{50}$, nM) |
| | | Kd (pM) | Bmax (sites/cell) | |
|---|---|---|---|---|
| Eosinophils | αβ | $653 \pm 133$ | $1331 \pm 250$ | 3.4 |
| TF-1 Cell Line | αβ | $660 \pm 154$ | $419 \pm 61$ | 15 |
| HL-60 Cell Line | αβ | $388 \pm 132$ | $448 \pm 29$ | ND |
| BHK Transfected Cells (IL5Rα) | α | $442 \pm 173$ | $1.5–5 \times 10^5$ | 4.0 |
| Purified sIL5Rα-Fc Fusion Protein | α | $804 \pm 240$ | $3.5 \times 10^9$/well | 5.5 |

IL5 binding constants (Kd and Bmax) were determined utilizing saturation isotherms and Scatchard transformations. Values reported are mean ± SEM from at least three experiments carried out in triplicate. Cold competition experiments with logit-log analysis were used to determine IC$_{50}$ values for sIL5Rα. Values reported are values from one to three separate experiments carried out in triplicate.

functional cell surface receptor [for review see (48)]. It is this ability to inhibit ligand interaction with the cell associated receptor which make certain soluble receptors such as those for tumor necrosis factor-alpha (TNFα) promising drug candidates. The soluble IL-6 receptor alpha chain on the other hand, may bind ligand in solution and this complex subsequently interacts with the signal transducing GP-130 subunit on cells to form a functional, activated receptor complex (49). The cells expressing the GP-130 subunit to which the IL-6-soluble receptor complex binds may or may not contain the membrane associated full length alpha chain thereby enhancing responses on cells which contain the IL-6 receptor and interestingly providing de novo IL-6 responsiveness to cells devoid of the alpha chain but containing the shared GP-130 signalling subunit.

The cloned soluble receptors may be expressed as a fusion protein to expedite receptor purification and assay development. Various fusion partners have been utilized in this regard including the Fc domain of IgG and poly-His (His$_6$) tags. Solid phase assays are readily established by immobilization of the soluble receptor on microtiter plates precoated with protein-A or Ni to specifically immobilize receptor-Fc or receptor-His$_6$ fusions, respectively. Using the soluble receptor in this format results in a simple, high-throughput screen and eliminates the need for continuous cell culture to provide a receptor source. Binding studies using protein-A immobilized soluble IL-5 receptor alpha chain-Fc fusion protein resulted in Kd values for IL-5 comparable to those obained with BHK cells expressing IL-5 receptor alpha chain and TF-1 cells expressing the functional IL-5 receptor alpha and beta subunits (Table 3). The simplicity of using the soluble receptor also leads to some limitations, in that interaction with a single receptor subunit may not provide the most accurate substitute for the far more complex membrane associated multisubunit functional receptor. This solid phase binding format may also be utilized for evaluation of antibody affinity/avidity through immobilization of antigen.

Most of the above techniques require the labelling of a protein ligand with a tag incorporating radioactivity or fluorophore and the affinity measured using a competitive binding format. A significant advance in protein-protein interaction quantitation has come about with the development of technology which measures these interactions directly in real time. Many of the cytokine/growth factor/interleukin receptor interactions have been measured using the surface plasmon resonance approach of BIACORE (50). The interactions of IL-5 with the the soluble alpha chain, human growth hormone with the soluble domain of it's receptor and erythropoietin with it's receptor have been quantitated using surface plasmon resonance. This technology has the significant advantage of generating real time kinetic and equilibrium analysis rapidly without the need for ligand labelling.

## B. *In vitro* Cellular Functional Assays

While receptor binding analysis provides quantitative information on the characteristics of the molecular interaction between ligand and receptor, it does not provide the critical assessment with respect to receptor activation, the functional consequence of this protein-protein interaction. There are numerous functional readouts which may be exploited to quantitate the level of receptor activation. The choice of assay should be predicated, at least in part, on the nature of the response required to be modulated, the target cell identity, and the necessary throughput. Both potency and efficacy of agonists provide useful metrics for the characterization of molecules. While potency and maximum level of inhibition are useful measurements of antagonistic/blocking molecules, these parameters may be assay and cell type specific, necessitating the inclusion of proper comparator molecules in the analyses and the generation of complete concentration-response curves. Antagonism of cellular response through ligand depletion using monoclonal antibodies, soluble receptors, or alternatively, antagonistic ligands generated through mutagenesis is typically quantitated in the presence of a fixed concentration of agonistic ligand. The $IC_{50}$ value generated is dependent on the concentration of agonist present in the reaction and the nature of the assay, so again it is important to clearly define these parameters in each assay and generate complete concentration-response curves. We will not attempt to review the extensive array of assays which are available to evaluate potential drug candidates, choosing instead to provide representative examples of commonly utilized assay formats such as ligand induced protein phosphorylation, mediator release assays, and growth and proliferation assays.

Phosphorylation is a general signalling mechanism utilized by protein ligand-activated cell surface receptors. These receptors may either possess an intrinsic kinase domain where catalytic activity is enhanced through receptor dimerization or alternatively, noncovalently associated protein kinases are activated following receptor dimerization [for review see (51, 52)]. The resulting signal is transduced through a complex series of protein-protein interactions coupled with phosphorylation, dephosphorylation and translocation events ultimately leading to modulation in gene expression and cell function. Many signalling pathways have been partially or fully delineated over the past several years, and assays have been developed which focus on these events most proximal to receptor activation. Phosphorylation of receptors, Janus kinases (JAKs) and mitogen activated protein (MAP) kinases occur within minutes of receptor activation, can be evaluated using cell lysis, immunoprecipitation and Western blotting. These methods are somewhat tedious, only semiquantitative and not conducive to a high-throughput format.

Mediator release from cells following receptor activation is another functional readout which has been exploited for the evaluation of both agonistic and antagonistic drug candidates. Cytokines, lipid mediators such as leukotrienes and prostaglandins, and metalloproteinases are examples of molecules released from cells following stimulation which have been quantitated using either ELISA, enzyme assay or other functional readout. These analyses are generally rapid, with the response occurring within one hour, sensitive, and quantitative. Choice of the target cell, method of cellular activation, and mediator to be quantitated are key variables which should be researched thoroughly.

Growth and proliferation may be induced in cells by cytokines and growth factors in a receptor specific manner and can be quantitated utilizing several standard techniques. The choice of cell type is a critical variable. The most relevant cells to use are those having characteristics comparable to the target of the drug. Cells isolated from human tissue should be considered, however, the tissue's cellular complexity, tissue availability, and donor to donor variability often preclude the routine use of this approach. Immortalized human cell lines expressing the receptor of interest and dependent on the protein-ligand for growth are very versatile and commonly utilized due to rapid growth, a single stable phenotype, and a consistent response. A third approach is the transfection of the receptor of interest into a stable cell line devoid of the receptor. Using this approach one can control the number of receptors and the ratio of subunits, however the signalling pathway may be somewhat different than that present in the target cell. While the above approaches have their limitations, each has been successfully utilized in a screening paradigm as a predictor of in vivo potency and efficacy and thereby constitute an integral part of the functional analyses scheme.

## C.  Selectivity Assays

Selectivity assays for protein drugs are often considered less critically important than those for medicinal chemistry-generated small organic molecules. Protein drugs have enhanced inherent potential selectivity brought about through evolution. This selectivity is exemplified through the ability of these proteins to efficiently carry out specific functions in a physiologic milieu of thousands of biomolecules. However, complications arise due to the pleiotropic nature of these protein ligand molecules. Many of these molecules may have several distinct activities in vivo in addition to the particular one of interest resulting in potentially negative side effect profiles. This may result from their interaction with multiple receptors or different combinations of receptor subunits. Alternatively, receptors activate distinct second messenger and effector systems in a cell type specific manner. Therefore, selectivity analysis of protein drug candidates remains a critical issue and one which should be addressed early on.

The importance of addressing drug selectivity is illustrated by the properties of the hematopoietic cytokine IL-3. IL-3 is a potent mediator of hematopoiesis and promotes differentiation of pluripotent progenitor stem cells through the granulocytic pathway-enhanced production of neutrophils. IL-3 has been marketed for use in adjunctive therapy for the regeneration of neutrophils depleted during cancer chemotherapy. Although IL-3 is a selective and potent molecule, it is also known to have proinflammatory properties brought about by induction of inflammatory mediator release which may result in dose limiting side effects commonly referred to as cytokine flu. Efforts have been initiated to separate the desired hematopoietic growth factor activity from the inflammatory activity of this molecule.

Toward this end, specific assays to measure IL-3 activity in-vitro were established to support an extensive mutagenesis effort. It was necessary that these assays had sufficiently high throughput to support the analysis of hundreds of mutants. *In-vitro* analysis in 96 well format for evaluation of the potency of these molecules to induce AML-193 and TF-1 cell proliferation as well as leukotriene release proved to be effective predictors of *in-vivo* potency and safety. This approach resulted in the generation of IL-3 receptor agonists with improved potency and therapeutic ratio (46).

The development of reliable *in-vitro* assays to quantitate the potency, efficacy and specificity of drug candidates is critical to the drug discovery process. While *in-vivo* efficacy depends on numerous factors which are not addressed by *in-vitro* studies, the characterization of the functional activity of compounds in isolated systems remains a valuable predictor of *in-vivo* activity.

## V. FUNCTIONAL ANALYSIS IN DESIGN OF VACCINES

Vaccines may be considered a type of protein therapeutic (indeed, the oldest form of polypeptide drug) in that protein sequences are the signal recognized by the immune system, whether an attenuated organism, recombinant viral capsid subunit protein, or chemically synthesized peptide is the immunogen (53, 54). The ultimate test of function of a vaccine is whether it confers protection in human volunteers against a direct challenge from the disease causing organism. However, relatively few of these studies are carried out, owing to ethical concerns, expense, and the long timelines involved. It is the role of *in-vitro* immunological assays used in the testing of new vaccines to provide the best possible prediction of this protective immune response. These assays provide data which allow a decision to be made whether to initiate controlled clinical studies in man. Vaccine functional testing is then carried out by *in-vivo* studies in mammalian species, including primates, and ultimately in man.

The strategies for conferring an active immune response by vaccination involve many factors which depend heavily on the life cycle of the infectious organism as well as it's interactions with the immune system. Depending upon the nature of the pathogen, strategies for the design of an effective vaccine may focus toward eliciting either cell-mediated or humoral responses (55). In this model, the immune response to extracellular bacteria and some parasites is characterized by production of high levels of antibodies generated from the humoral response, which is associated with T-helper CD4$^+$ cells of the Th2 subtype which produce high levels of cytokines IL-4, IL-5, IL-6, IL-11, and IL-13 (56). In a separate arm of the immune response elicited by intracellular pathogens, such as viruses and some bacteria and protozoans, cellular cytolytic activity is associated with a Th1 subset of CD4+ T-cells which produce the inflammatory cytokines IL-2, IFN-gamma, and TNF-beta. Clearly, the levels of cytokines produced in animal testing of a vaccine can be an important marker of the quality of a particular immune response.

Currently, recombinant DNA technology is being exploited to bring forward polypeptide based vaccines with potentially fewer side effects than the attenuated whole cell or "inactivated" virus particles which have been the mainstay of vaccines in this century. This technology for the identification and production of immunogenic polypeptide sequences has lead to assay strategies which test pathogen epitopes alone and in combination. A major goal, therefore, is the identification of the optimal combination of epitopes derived from a pathogen which lead to induction of active immunity.

A general strategy for the discovery of an effective polypeptide based vaccine could involve:

1. Heterologous expression of candidate polypeptides in a recombinant system.

**Table 4**  A Single Injection of the (PAM)$_2$-HTL-CTL
Construct Produces Long-Lasting CTL Immunity. BALB/c
Mice Were Injected Subcutaneously with 10 nmol/Mouse of
the (PAM)$_2$-HTL-CTL Peptide. At the Indicated Time After
Immunization, Five Animals Were Killed and CTL Activity
Determined as Explained in Table I and Methods (From Ref.
59, with permission)

| Time after immunization | Antigen-specific CTL activity* |
|---|---|
| *wk* | *LU/IU splenocytes* |
| 2 | 19 ± 12 |
| 4 | 22 ± 12 |
| 8 | 52 ± 25 |
| 16 | 41 ± 24 |
| 26 | 41 ± 23 |
| 55 | 25 ± 27 |

*Values are means ± SD.

2.  Their purification for use as
    immunogens to prepare anti-
    bodies.
3.  Use of those antibodies to test for
    neutralizing effect on the target
    pathogen.

Some of the strategies employed for analysis of hepatitis B virus (HBV) vaccine candidates can serve to illustrate useful
functional assays. In recognition of the
importance of HLA type in antigen presentation for elicitation of a cytotoxic
T-lymphocyte (CTL) response, methods
for isolation of peptides bound to MHC
complexes have been developed (57, 58).
A peptide comprised of residues 18–27 of
the HBV core antigen was conjugated to
tetanus toxoid peptide and two palmitic
acid molecules to enhance immunogenicity (59). The long-term CTL response
in mice was evaluated out to 55 weeks and
was maintained, as shown in Table 4. The
three-piece, covalently linked molecule
above was then used in a Phase I study
testing safety and tolerability in normal
human volunteers, in which the antigen
specific, primary CTL response was determined as shown in Fig. 5. In this example,
the strong primary CTL response was the
desired result for a therapeutic vaccine
aimed at chronic hepatitis B infection.

A vaccine drug is intended to elicit a
protective immune response over an
extended period of time against challenge
by an infectious organism. While such a
complex biological response represents an
extreme challenge to the *in-vitro* analysis
of function, significant advances have been
made as the workings of the immune system become better understood. The general strategies for functional analysis of
potential new vaccines employ a variety of
methods to quantitate the level and specificity of antibodies generated against
potential vaccine preparations as well as
analyses of the cell-mediated arm of the
immune response. The use of molecular
biology methods to identify critical
epitopes coupled with a more complete
understanding of the immune system may
enable use of vaccines for a wider variety
of diseases in the future.

## VI.  FUNCTIONAL ANALYSIS IN GENE THERAPY

Recent advances in gene therapy is of
interest to polypeptide bioactivity analysis
because many of these strategies have as
their aim the production of a protein or
peptide as the active species. In this sense,

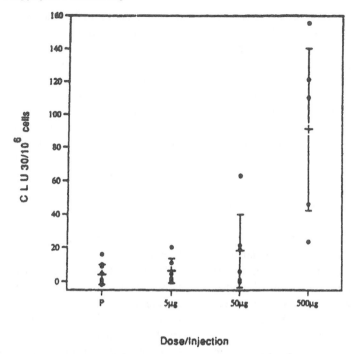

**Fig. 5** Immune response to Theradigm-HBV in normal volunteers. Human volunteers were injected with the indicated preparations. CTL activity was determined as described in Methods in PBMC samples obtained 7, 14 and 28 d after the primary and the booster injection. CLU represents the sum of the lytic activity from all of the time points for each individual subject. (From Ref. 59, with permission.)

gene therapy may be considered as an alternate protein drug delivery technology. In this section, we will discuss the general issues surrounding functional analysis of a protein therapeutic delivered by a gene therapy agent. Most typically, the delivery agent is a viral vector which has been engineered to express a protein *in-vivo* having known properties from prior *in-vitro* studies. The expressed protein may have gene regulatory properties, receptor binding affinity, or may be an enzyme.

While this set of therapeutic strategies is still in the early stages of development, it holds great promise and also presents a unique set of challenges for analysis of function. Gene therapy and more conventional protein delivery systems offer the perceived advantage of the drug substance having the same structure as a native protein, hence minimal issues with regard to immunogenicity of the drug and subsequent attenuation of it's effect. However, the analysis of therapeutic levels of the protein drug can require accurate quantitation of a small increment above a large background.

In general, it is important to distinguish between the determination of protein function in the biological fluid and whether a therapeutic effect is achieved. In deciding whether a gene therapy strategy will be successful in humans, animal model studies may be carried out to test the therapeutic concept. If the desired therapeutic effect is not achieved, determination of the circulating level of the protein's activity can be of great utility. If the activity is not present, then changes in the gene delivery vector may be pursued. However, if the desired level of functional activity is present, then effort may be directed at greater tissue localization or reevaluation of the therapeutic concept.

In gene therapy settings, activity of the protein must often be detected *in vivo* above a background of the same or similar activity. In the case of inherited deficiencies, such as some hemophilias (60), protein activity is completely absent in the patient, while in other cases the needed protein is expressed at insufficient levels. In the former case, the background of activity against which the gene therapy treatment group of animals is being tested will be minimal while in the latter case, the background activity may be considerable. Immunoassays may be inadequate in sensitivity and range to distinguish endogenous from newly expressed activity, unless specific epitopes for detection are incorporated into the viral vector.

The assay sensitivity needed may be very high owing to the relatively low titer of virus and the localized delivery to specific cell populations, as opposed to systemic distribution of a drug. To obtain sufficient expression levels for detection, viral vectors may be cultured *in vitro* and the products detected by standard biochemical assay methods. While this type of assay is useful for many purposes, such as production of an infectious virus, the levels determined of the expressed protein may have little relevance to *in vivo* conditions, both in levels of protein produced, but also in the sites of localization of the viral agent.

Some examples from recent studies involving the clotting factors IX and X will serve to illustrate strategies used to answer the question of functional gene expression in a gene therapy setting. The expression level of Factor IX produced by means of an adenoviral vector was evaluated for the purpose of assessing routes leading to stable, long-term expression of the factor (61). In this study, expression level of Factor IX in plasma was determined by ELISA, but functional activity of the expressed protein from an enzymatic assay was not determined.

In a DNA mediated transfection study of Factor IX into rat, liver functional analysis of Factor IX correlated with Western blot immunoassay of the protein expression level (62). Table 5 shows the Factor IX activity from a standard *in-vitro* blood clotting assay in three groups of rats. Importantly, the reported activity had the average endogenous Factor IX activity subtracted out, which was approximately 50% of the measured activity of the treated rats.

Retrovirus-mediated expression of a human Factor X in rats was carried out (63) to develop a treatment for the severe hemophilia resulting from Factor X deficiency in humans. In this study, an antibody with selectivity for human Factor X instead of rat Factor X was used to immunoprecipitate the human, retrovirally expressed enzyme, which was then assayed for enzymatic activity using a standard colorimetric assay (Table 6). This immunoprecipitation strategy was a useful means to eliminate the endogenous background of rat Factor X activity which would have confounded the analysis. The correlation between human Factor X antigen levels as detected by ELISA and Factor X activity levels suggested that the human Factor X produced by their retroviral vector was all functional.

## VII. FUTURE DIRECTIONS AND CONCLUSIONS

The wide variety of biological responses elicited by protein therapeutics makes it difficult to generalize about the design and execution of bioassays. Each assay will of necessity have components specific to the protein's mechanism of action. In addition, the varying needs of each laboratory regarding sensitivity, selectivity, throughput, and ruggedness will dictate the particular reagents, apparatus, instrumentation, cell lines, and tissues which will be most useful.

In addition to these general principles, the importance of some emerging technologies has become apparent. First, the area of assay automation aimed

**Table 5** hFIX Activity in Plasma from Rats Transfected with the PEPCK-hFIX Gene. The Relative Concentration of hFIX in the Blood of Rats Treated with the DNA Complex Was Evaluated. One Unit of FIX Activity in 1 ml of Normal Human Plasma Is Equivalent to 100% Functional Activity or $\approx 3\,\mu g$ of FIX per ml. Background FIX Activity in the Rat Plasma Was Determined in Each Assay by Using Plasma from Untransfected Controls and Was Subtracted from the FIX Activity of Transfected Rats. The Average Background of FIX Activity for the Control Animals Was 79.5%, 42.6%, and 56.8% of the Experimental Value in Exp. A–C, Respectively. (From Ref. 62)

| Exp. | Rat | Sampling time, days | hFIX activity, unit/ml |
|------|-----|---------------------|------------------------|
| A    | 1   | 2   | 0.040 |
|      | 2   | 2   | 0.045 |
|      | 3   | 4   | 0.045 |
|      | 4   | 4   | 0.025 |
| B    | 5   | 6   | 0.330 |
|      | 6   | 8   | 0.135 |
|      | 7   | 12  | 0.160 |
|      | 8   | 12  | 0.075 |
|      | 9   | 32  | 0.125 |
| C    | 10  | 48  | 0.350 |
|      | 11  | 72  | 0.005 |
|      | 12  | 136 | 0.105 |

**Table 6** Summary of Results in Rats (From Ref. 63)

| Rat No. | hFX Antigen ($\mu g/mL$)* | Functional hFX† | Percentage of hFX That Functions‡ | Transduction Efficiency§ | Normalized Expression ($\mu g$ hFx/mL/1% transduction)¶ |
|---------|------------------|-----------------|-----------------------------------|--------------------------|---------------------------------------------------------|
| 113 | $1.1 \pm 0.5$ (14) | $17.25 \pm 0.75$ | $123 \pm 15$ | $0.3 \pm 0.1$ | $3.67 \pm 1.29$ |
| 114 | $2.3 \pm 0.1$ (29) | $54 \pm 10.7$ | $174 \pm 35$ | $1.5 \pm 0.5$ | $1.53 \pm 0.52$ |
| 115 | $0.88 \pm 0.06$ (11) | $9.5 \pm 0.5$ | $79 \pm 7$ | $1 \pm 0.5$ | $0.88 \pm 0.44$ |
| 119 | $2.71 \pm 0.1$ (34) | $27 \pm 3.2$ | $108 \pm 13$ | $1.1 \pm 0.5$ | $2.46 \pm 1.23$ |

Plasma was obtained at six weeks after transduction. For all columns, the result ± the standard deviation is shown.
* The hFX antigenic activity in micrograms per milliliter is shown, with the percentage of normal human levels shown in parentheses.
† The hFX functional activity was determined and is reported as the percentage of normal human activity.
‡ The percentage of hFX that functions was determined by dividing the hFX functional activity by the hFX antigen.
§ Transduction efficiency was determined by using the PCR-based assay shown in Fig. 3A. The percentage of liver cells that were transduced was calculated by averaging the results obtained from three separate assays.
¶ Normalized expression was obtained by dividing the hFX antigen levels by the transduction efficiency.

at high throughput screening of drug candidates has become a major focus of the drug discovery process in pharmaceutical companies (64–66). Assay automation in this arena can improve reproducibility and throughput while also reducing costs. When coupled with adequate data management systems to allow facile interpretation of the resulting huge volume of information, the automation of functional assays may significantly change the way experiments in the future are designed, as well as how they are executed. However, basic principles of analysis are still the basis of the assay method, and may not be compromised in pursuit of automation. A poorly designed, automated assay will rapidly generate mountains of data which is at best irrelevant and at worst, misleading.

A second emerging technology which may impact functional assays lies in the miniaturization of protein and nucleic acid separations on micromachined silicon chips (67–69). By use of extremely thin channels and electroosmotically driven reagents, both the speed of separation and assay sensitivity have been dramatically improved. In one example, a post-column reactor was microfabricated in combination with capillary electrophoresis and used for separation and derivatization of amino acids with o-phthaldialdehyde to yield a fluorescent derivative (70). Such a tool could be used to separate substrate from product in an enzyme reaction and detect the product with exquisite sensitivity.

In this chapter we have only scratched the surface of what is an expanding field of study posing continual challenges to the biochemist. The impact of current genomics and bioinformatics initiatives which will soon revolutionize the study of biology and medicine has yet to be fully realized, especially in the area of protein functional analysis. It is likely that mining of the human genome database will generate many more ques-

tions regarding biological function of expressed proteins than current bioactivity assay technologies will support. As more information regarding gene expression is correlated with disease states, the data are likely to demonstrate the involvement of several metabolic pathways and receptors simultaneously in many diseases. Hence, there will likely arise the need for analysis of multiple bioactivity responses and their synergies in ever more complex models of disease.

## REFERENCES

1. FAO Marston. The purification of eukaryotic polypeptides synthesized in E. coli. Biochemistry 240: 1–12, 1986.
2. DJ Peters, JM Hayes, GM Hieftje, Chemical Separations and Measurements—Theory and Practice of Analytical Chemistry, Philadelphia: W.B. Saunders, 1974, pp 1–2.
3. D Metzler. Biochemistry: The Chemical Reactions of Living Cells. New York: Academic Press, 1977, p 303.
4. this chapter, pp 615–623.
5. R Ballentine. Treatment of Gaussian Measurement Data. In: JM Brewer, AJ Pesce, RB Ashworth, eds. Experimental Techniques in Biochemistry. Englewood Cliffs, NJ: Prentice-Hall, 1974, pp 10–31.
6. DJ Peters, JM Hayes, GM Hieftje, Chemical Separations and Measurements—Theory and Practice of Analytical Chemistry, Philadelphia: W.B. Saunders, 1974, pp 8–9.
7. R Ballentine. Treatment of Gaussian Measurement Data. In: JM Brewer, AJ Pesce, RB Ashworth, eds. Experimental Techniques in Biochemistry. Englewood Cliffs, NJ: Prentice-Hall, 1974, pp 14.
8. DJ Peters, JM Hayes, GM Hieftje, Chemical Separations and Measurements- Theory and Practice of Analytical Chemistry, Philadelphia: W.B. Saunders, 1974, pp 614–616.
9. MM Bradford. A Rapid and Sensitive Method for the Quantitation of Microgram Quantities of Protein Utilizing the Principle of Protein-Dye Binding. Anal. Biochem 72: 248–254, 1976.
10. C DeSain. Documentation Basics: Master Method Validation Protocols. BioPharm 5: 30–34, 1992.
11. CM Stoschek. Quantitation of Protein. In: Murray P. Deutscher, ed. Methods in Enzymology Vol. 182, Guide to Protein Purification. San Diego: Academic Press, Inc., 1990, pp 50–68.

12. C Stoschek. Increased Uniformity in the Response of the Coomassie Blue G Protein Assay to Different Proteins. Anal. Biochem. 184: 111–116, 1990.

13. OH Lowry, NJ Rosebrough, AL Farr, RJ Randall. Protein Measurement with the Folin Phenol Reagent. J. Biol. Chem. 193: 265–275, 1951.

14. PK Smith, RI Krohn, GT Hermanson, AK Mallia, FH Gartner, MD Provenzano, EK Fujimoto, NM Goeke, BJ Olson, DC Klenk. Measurement of Protein Using Bicinchoninic Acid. Anal. Biochem. 150: 76–85, 1985.

15. RE Brown, KL Jarvis, KJ Hyland. Protein Measurement Using Bicinchoninic Acid: Elimination of Interfering Substances. Anal. Biochem. 180: 136–139, 1989.

16. SC Gill, PH. von Hippel. Calculation of Protein Extinction Coefficients from Amino Acid Sequence Data. Anal Biochem 182: 319–326, 1989.

17. CN Pace, F Vajdos, L Fee, G Grimsley, T Gray. How to measure and predict the molar absorption coefficient of a protein. Protein Science 4: 2411–2423, 1995.

18. J McEntire. Biotechnology Product Validation, Part 5: Selection and Validation of Analytical Techniques. BioPharm 7: 68–79, 1994.

19. VHL Lee. Peptide and Protein Drug Delivery. New York: Marcel Dekker, Inc, 1991, pp 1–56.

20. M Dixon, E C Webb, Enzymes (3rd Edition). Longman Ltd, London, 1979.

21. L Stryer in Biochemistry 4th Edition, New York: WH Freeman and Company, 1995, pp 181–206.

22 KF Tipton. Enzyme Assays. In: R Eisenthal, M J Danson, eds. A Practical Approach. R L Press, 1992, pp 1–58.

23. F Bachmann, EKO Kruithof. Tissue plasminogen activator: Chemical and physiological aspects. Thromb and Haemost 10: 6–17, 1984.

24. J Krause. Catabolism of tissue-type plasminogen activator (t-PA), its variants, mutants and hybrids. Fibrinolysis, 2: 133–142, 1988.

25. V Gurewich. The sequential, complementary and synergistic activation of fibrin-bound plasminogen by tissue plasminogen activator and pro-urokinase. Fibrinolysis, 3: 69–66, 1989.

26. D Collen. Tissue type plasminogen activator (t-PA) and single chain urokinase type plasminogen activator (scu-PA): Potential for fibrin specific thrombolytic therapy. Prog. Hemost. Thromb. 8: 1–18, 1986.

27. N K Harakas, JP Chaumann, DT Connoly, AJ Wittwer, JV Olander, J Feder. Large scale purification of tissue plasminogen activator from cultured human cells. Biotechnology Progress, 4: 149–158, 1988.

28. HR Lijnen, D Gheysen, F Foresta, L Pierard, P Jacob, A Bollen, D Collen. Characterization of a mutant o recombinant human single chain urokinase type plasminogen activator (scu-PA), obtained by substitution of arginine 156 and lysine 158 with threonine. Fibrinolysis 2: 85–93, 1988.

29. M Booyse, J Scheinbuks, PH Lin, M Traylor, R Bruce. Isolation and inter-relationships of the multiple molecular tissue-type and urokinase type plasminogen activator forms produced by cultural human vein endothelial cells. J. Biol. Chem. 263: 15129–15138, 1988.

30. H Hoylaerts, DC Rijken, HR Lijnen, D Collen. Kinetics of the activation of plasminogen by human tissue plasminogen activator. J Biol Chem 257: 2912–2919 1982.

31. TA Bicsak, AJW Hsuch. Rat oocyte tissue plasminogen activator is a catalytically efficient enzyme in the absence of fibrin. Endogenous potentiation of enzyme activity. J Biol Chem 264: 630–634, 1989.

32. L Hansen, Y Blue, K Baroner, D Collen, GR Larsen. Functional effects of asparagine-linked oligosaccharide on natural and variant human type plasminogen activator. J Biol Chem 263: 15713–15719, 1988.

33. AN Glazer, RJ Delange, DS Sigman, Chemical Modification of Proteins— Selected Methods and Analytical Procedures, North Holland and American Elseview Publishing Company, 1975.

34. R L Lundblad, CM Noyes. Chemical Reagents for Proteins Modifications, Volumes I and II, CRC-Press, 1984.

35. CM Haskell. L-Asparaginase: Human toxicity and single agent activity in nonleukemic neoplasms. Cancer Treat Rep 65: 5759, 1981.

36. D Park, A Abuchowski, S Davis. Pharmacology of Escherichia coli L-asparaginase polyethylene glycol adduct. Am Cancer Res 1: 373–376, 1981.

37. R Uren, BJ Hargis, P Beardsley. Immunological and pharmacological characterization of poly DL-alanyl modified Erwinia carotovora L-asparaginase. Cancer Res 42: 4068–4071, 1982.

38. TE Wileman, RL Foster, P. N. Elliot. Soluble asparaginase-dextran conjugates show increase circulatory persistence and lowered antigen reactivity. J Pharm Pharmacol 338: 264–271, 1986.

39. MJ Poznansky, M Shandling, MA Salkie. Advantages in the use of L-aspara-

ginase-albumin polymer as an antitumor agent. Cancer Res. 42: 1202–1205, 1982.

40. A Miyajima, T Kitamura, N Harada, T Yokata, K Arai. Cytokine receptors and signal transduction. In: WE Paul, ed. Annu Rev Immunol 10. Palo Alto: Annual Reviews, Inc., 1992, pp 295–331.

41. GA Weiland, PB Molinoff. Quantitative anlaysis of drug-receptor interactions: 1. Determination of kinetic and equilibrium properties. Life Sciences 29: 313–330, 1981.

42. G Scatchard. The Attractions of Proteins for Small Molecules and Ions. Ann NY Acad Sci 51: 6600–672, 1949.

43. PB Molinoff, BB Wolfe, GA Weiland. Quantitative analysis of drug-receptor interactions: II. Determination of the properties of receptor subtypes, Life Sciences 29: 427–443, 1981.

44. A De Lean, PJ Munson, D Rodbard. Simultaneous analysis of families of sigmoidal curves: application to bioassay, radioligand assay, and physiological dose-response curves. Am J Physiol 235: E97–E102, 1978.

45. Y Cheng, WH Prusoff. Relationship between the inhibition constant (Ki) and the concentration of inhibitor which causes 50 percent inhibition ($I_{50}$) of an enzymatic reaction. Biochem Pharmacol 22: 3099–3108, 1973.

46. JW Thomas, CM Baum, WF Hood, B Klein, JB Monahan, K Paik, M Abrams, A Donnelly, JP McKEarn. Potent interleukin-3 receptor agonist with selectively enhanced hematopoietic activity relative to rhIL-3. Proc Natl Acad Sci USA 92: 3779–3783, 1995.

47. JB Monahan, N Siegel, R Keith, M Caparon, L Christine, R Compton, S Cusik, J Hirsch, M Huynh, C Devine, J Polazzi, S Rangwala, B Tsai, J Portanova. Attenuation of IL-5 mediated signal transduction, eosinophil survival and inflammatory mediator release by a soluble human IL-5 receptor. J Immunol 159: 4024–4034, 1997.

48. ML Heaney, DW Golde. Soluble cytokine receptors. Blood 87: 847–857, 1996.

49. H Yasukawa, K Yasukawa, S Natsuka, M Murakami, K Yamasaki, M Hibi, T Taga, T Kishimoto. Structure-function analysis of human IL-6 receptor: dissociation of amino acid residues required for IL-6 binding and for IL-6 signal transduction through gp130. EMBO J. 12: 1705–1712, 1993.

50. AS Szabo, L Stolz, R Granzow. Surface plasmon resonance and its use in biomolecular interaction analysis (BIA). Curr Opin Struct Biol 5: 699–705, 1995.

51. CJ Bagley, JM Woodcock, FC Stomski, AF Lopez. The structural and functional basis of cytokine receptor activation: lessons

from the common beta subunit of the granulocyte-macrophage colony-stimulating factor, interleukin-3 (IL-3) and IL-5 receptors. Blood 89: 1471–1482, 1997.

52. A Kazlauskas. Receptor tyrosine kinases and their targets. Curr Opin Genet Dev 4: 5–14, 1994.

53. JA Bellanti, M.D. Immunology III, Philadelphia, PA: W.B. Saunders Co.,1985, pp 510–512.

54. GL Ada. Vaccines. In: WE Paul, ed. Fundamental Immunology, 2nd Ed. New York: Raven Press, pp 985–1032, 1989.

55. SL Constant, K Bottomly. Induction of TH1 and TH2 CD4+ T Cell Responses: The Alternative Approaches. In: WE Paul, ed. Ann Rev Immun Palo Alto, CA: Annual Reviews Inc., 1997, pp 297–322.

56. DT Fearon, RM Locksley. The Instructive Role of Innate Immunity in the Acquired Immune Response. Science 272: 50–53, 1996.

57. F Sinigaglia, P Romagnoli, M Guttinger, B Takacs, JR L Pink. Selection of T Cell Epitopes and Vaccine Engineering. In: Murray P. Deutscher, ed. Methods in Enzymology Vol. 203, Antibodies and Antigens, San Diego: Academic Press, Inc., 1991, pp 370–386.

58. WJ Storkus, H.J. Zeh III, RD Salter, and MJ Lotze. Identification of T-Cell Epitopes: Rapid Isolation of Class I-Presented Peptides from Viable Cells by Mild Acid Elution. J Immunotherapy 14: 94–103, 1993.

59. A Vitiello, G Ishioka, HM Grey, R Rose, P Farness, R LaFond, L Yuan, FV Chisari, J Furze, R Bartholomeuz, RW Chesnut. Development of a Lipopeptide-based Therapeutic Vaccine to Treat Chronic HBV Infection. J Clin Invest 95, 341–349, 1995.

60. AR Thompson. Progress Towards Gene Therapy for the Hemophilias. Thromb Hemost 74: 45–51, 1995.

61. Y Dai, M Roman, RK Naviaux, IM Verma. Gene Therapy via primary myoblasts: Long-term expression of factor IX protein following transplantation in vivo. Proc Nat Acad Sci USA 89: 10892–10895, 1992.

62. JC Perales, T Ferkol, H Beegen, OD Ratnoff, RW Hanson. Gene Transfer in-vivo: Sustained expression and regulation of genes introduced into the liver by receptor-targeted uptake. Proc Natl Acad Sci USA 91: 4086–4090, 1994.

63. M Le, T Okuyama, S-R Cai, SC Kennedy, WM Bowling, MW Flye, KP Ponder. Therapeutic Levels of Functional Human Factor X in Rats After Retroviral-Mediated Hepatic Gene Therapy. Blood 89: 1254–1259, 1997.

64. SD Hamilton, JW Armstrong, RA Gerren, AM Janssen, JV Peterson, RA Stanton. An Overview of Automated Biotechnology Screening. Laboratory Robotics and Automation 8: 287–294, 1996.

65. R Pauwels, H Azijn, M-P de Bethune, C Claeys, K Hertogs. Automated Techniques in Biotechnology. Curr Opin Biotech 6: 111–117, 1995.

66. J Peccoud. Automating Molecular Biology: A Question of Communication. Bio/Technology 13: 741–745, 1995.

67. JM Ramsey, SC Jacobson, MR Knapp. Microfabricated chemical measurements systems. Nature Medicine 1: 1093–1096, 1995.

68. A Manz. Ultimate speed and sample volumes in electrophoresis. Biochem Soc Trans 25: 278–281, 1997.

69. DJ Harrison, K Fluri, K Seiler, Z Fan, CS Effenhauser, A Manz. Micromachining a Miniaturized Capillary Electrophoresis-Based Chemical Analysis System on a Chip. Science 261: 895–897, 1993.

70. SC Jacobson, LB Koutny, R Hergenroeder, AW Moore, Jr., JM Ramsey. Microchip capillary electrophoresis with an integrated postcolumn reactor. Anal Chem 66: 3472–3476, 1994.

# 20

# Pharmacokinetics and Pharmacodynamics of Protein Therapeutics

**Rene Braeckman***
*Chiron Corporation, Emeryville, California*

## I.  INTRODUCTION

Pharmacokinetics is the study of the rate processes that are responsible for the time course of the level of an exogenous compound in the body. The processes involved are absorption (A), distribution (D), metabolism (M), and excretion (E) (Fig. 1). The pharmacokinetics of proteins, peptides, peptoids and other biotechnology products are an important factor in their pharmacodynamics, i.e., the time course of their pharmacological effect. Therefore, knowledge of the pharmacokinetics and pharmacodynamics of a pharmaceutical drug in humans and laboratory animals is important in the choice of dose levels and dose regimens. Similarly, the toxicokinetics (pharmacokinetics in toxicology studies, including higher doses than used clinically) and toxicodynamics (time course of undesired effects) are important for the design of toxicology studies (dose levels and dose regimens) as well as in determining safety margins and extrapolating toxicological data to humans.

In this chapter, the pharmacokinetics (PK) of protein therapeutics will be described, followed by the complex relationship with the pharmacodynamic (PD) effect. The immunogenicity of protein therapeutics will be described also, including a discussion on how antibody formation can influence the PK and PD. The PK, PD, and immunogenicity are interrelated, as depicted in Fig. 2. The PK of protein drugs discussed in this chapter includes elimination, biodistribution, plasma binding, modeling, and interspecies scaling. Furthermore, the heterogeneity of proteins caused by their structure, differences in glycosylation, and intended chemical modifications, mainly to improve the PK characteristics will be discussed. Besides the physicochemical characteristics of protein drugs, treatment parameters (such as dose, route, regimen, formulation, and concomitant medications) and the *in vivo* environment (species, physiological and pathophysiological status, target organ, etc.) may also influence the PK, PD, and immunogenicity (Fig. 2). A better understanding of this complex relationship and the processes themselves is essential for the drug development process of protein therapeutics, in particular for the establish-

---

*  *Current affiliation*: Ceptyr, Inc., Bothell, Washington

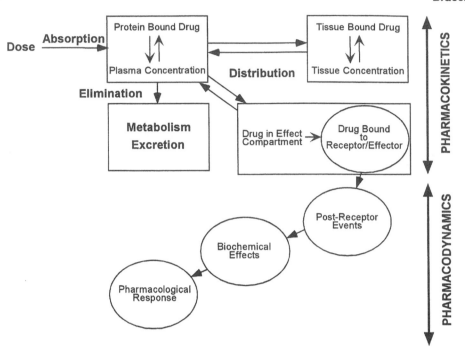

**Fig. 1** Physiological scheme of pharmacokinetic and pharmacodynamic processes.

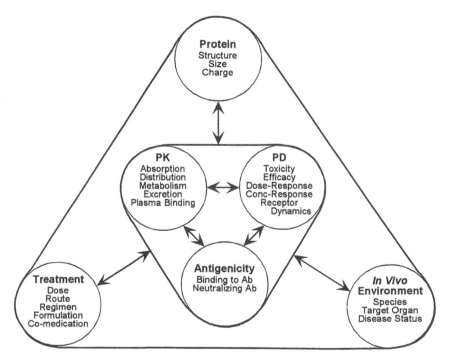

**Fig. 2** Depiction of the complex relationship between the pharmacokinetics (PK), pharmacodynamics (PD), and immunogenicity of protein therapeutics. Protein characteristics, treatment parameters, and the *in vivo* environment may influence all three processes.

**Table 1** Clearance Mechanisms for Peptides and Proteins as a Function of Molecular Weight (MW). Other Determining Factors Are Size, Charge, Lipophilicity, Functional Groups, Sugar Recognition, Vulnerability for Proteases, Aggregation to Particles, Formation Complexes with Opsonization Factors, etc. The Indicated Mechanisms Overlap, and Fluid-phase Endocytosis Can in Principle Occur Across the Entire MW Range. Adapted from Meijer and Ziegler (24).

| MW | Site of Elimination | Dominating Clearance Mechanism | Determinant Factor |
|---|---|---|---|
| <500 | blood liver | extracellular hydrolysis passive nonionic diffusion | |
| 500–1,000 | liver | carrier-mediated uptake passive nonionic diffusion | structure lipophilicity |
| 1,000–50,000 | kidney | glomerular filtration | MW |
| 50,000–200,000 | kidney liver | receptor-mediated endocytosis receptor-mediated endocytosis | sugar, charge |
| 200,000–400,000 | | opsonization | $\alpha_2$-macroglobulin IgG |
| >400,000 | | phagocytosis | particle aggregation |

ment of the most optimal treatment parameters.

## II. PHARMACOKINETICS

### A. Elimination of Protein Therapeutics

It is commonly accepted that peptide and protein drugs are metabolized through identical catabolic pathways as endogenous and dietary proteins. Generally, proteins are broken down into amino acid fragments that can be re-utilized in the synthesis of endogenous proteins. Although history has shown that proteins can be powerful and potentially toxic compounds, their products of metabolism are not considered to be a safety issue. This is in contrast with small organic synthetic drug molecules from which potentially toxic metabolites can be formed. The study of the metabolism of protein drugs is also very complicated because of the great number of fragments that can be produced. The mechanisms for elimination of peptides and proteins are outlined in Table 1.

### 1. Proteolysis

Most if not all proteins are catabolized by proteolysis. Proteolytic enzymes are not only widespread throughout the body, they are also ubiquitous in nature, and

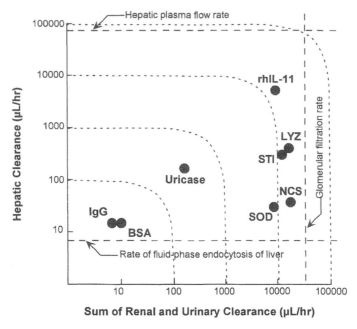

**Fig. 3** Hepatic and renal clearances of proteins in mice. LYZ: lysozyme, STI: soy bean trypsin inhibitor, NCS: neocarzinostatin, IgG: immunoglobulin G, BSA: bovine serum albumin, rhIL-11: recombinant human interleukin-11. [From Takagi *et al.* (123)]

therefore the potential number of catabolism sites on any protein is very large (1–3). It has been shown for interferon-gamma (INF-$\gamma$) that truncated forms are present in the circulation after dosing of rhesus monkeys with rIFN-$\gamma$. The rate and extent of production of these metabolites may be dependent on the route of administration. This, and the cross-reactivity of these degraded forms in the ELISA may be responsible for the observation of a bioavailability of more than 100% after subcutaneous administration of rIFN-$\gamma$ (4). Proteolytic activity in tissue may be responsible for the loss of protein after subcutaneous administration.

## 2. Renal Excretion and Metabolism

Metabolism studies of peptide and protein drugs were performed to identify the organs responsible for metabolism (and/or excretion), and their relative contribution to the total elimination clearance.

The importance of the kidney as an organ of elimination was assessed for rIL-2 (5), M-CSF (6), GH, gonadotropin releasing hormone (GnRH) (7, 8), and rIFN-$\gamma$ (9) in nephrectomized animals. The relative contributions of renal and hepatic clearances to the total plasma clearance of several other proteins are shown in Fig. 3.

The different renal processes that are important for the elimination of proteins are depicted in Fig. 4. Based on the observation that only trace amounts of albumin pass the glomerulus, it is believed that macromolecules have to be smaller than 69 kD to undergo glomerular filtration (10). Glomerular filtration and excretion is most efficient however for proteins smaller than 30 kD (11). Peptides and small proteins (< 5 kD) are filtered very efficiently, and their glomerular filtration clearance approaches the glomerular filtration rate (GFR, ~120 mL/min in humans). For molecular weights exceeding 30 kD, filtration falls off sharply. The

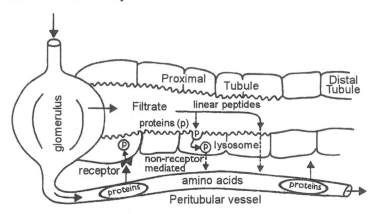

**Fig. 4** Pathways of renal elimination of proteins, including glomerular filtration, catabolism at the luminal membrane, tubular absorption followed by intracellular degradation, and postglomerular peritubular uptake followed by intracellular degradation. [From Rabkin *et al.* (12)].

correlation between molecular weight and Glomerular Sieving Coefficient (GSC) is shown in Table 2. The GSC is the ratio of concentration in the ultrafiltrate to that in plasma. It is actually the effective molecular radius that determines the degree of sieving by the glomerulus (Fig. 5) (12). The glomerular barrier is also charge selective: the clearance of anionic molecules is impaired relative to that of neutral molecules, and the clearance of cationic macromolecules is enhanced. The influence of charge on glomerular filtration is especially important for molecules with a radius greater than 20 nm (13). The charge-selectivity of glomerular

filtration is related to the negative charge of the glomerular filter due to the abundance of glycosaminoglycans. Anionic proteins, such as TNF-$\alpha$, IFN-$\beta$, and IFN-$\gamma$, are therefore repelled (3).

After glomerular filtration, peptides (melanostatin, for example) can be excreted unchanged in the urine. In contrast, complex polypeptides and proteins are actively reabsorbed by the proximal tubules by luminal endocytosis and then hydrolyzed within the intracellular lysosomes to peptide fragments and amino acids (13, 14). Consequently, only small amounts of intact protein are detected in the urine. The kidney appears to be the

**Table 2** Molecular Size and Glomerular Filtration. The Glomerular Sieving Coefficient (GSC) Is Calculated as the Ratio of Concentration in the Ultrafiltrate to the Plasma Concentration. From Rowland and Tozer (121).

| Protein | Molecular Weight (kD) | Glomerular Sieving Coefficient (GSC) |
|---|---|---|
| Insulin | 6 | 0.89 |
| Lysozyme | 14 | 0.80 |
| Myoglobin | 16.9 | 0.75 |
| Growth hormone (GH) | 20 | 0.72 |
| Superoxide dismutase (SOD) | 32 | 0.33 |
| Bence Jones | 44 | 0.08 |
| Albumin | 69 | 0.001 |
| IgG | 160 | 0.000 |

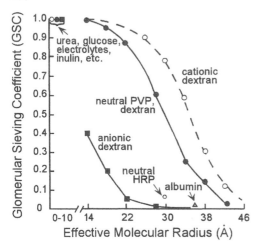

**Fig. 5** Glomerular sieving curves of several macromolecules. The different Glomerular Sieving Coefficients (GSC) reflect the influence of size, charge, and rigidity of molecules. [From Arendshorst and Navar (124)].

most dominant organ for the catabolism of small proteins (15). Examples of proteins undergoing tubular reabsorption are calcitonin, glucagon, insulin, growth hormone, oxytocin, vasopressin, and lysozyme (11). The amino acids are returned to the systemic circulation. Cathepsin D, a major renal protease, is responsible for the hydrolysis of IL-2 in the kidney [16]. Important determinants for tubular reabsorption of proteins are their physicochemical characteristics such as net charge and number of free amino groups (15). Cationic proteins are more susceptible to reabsorption than anionic proteins (17). Renal tubular cells also contain an active transporter for di- and tripeptides (18).

Small linear peptides (<10 amino acids) such as angiotensin I and II, bradykinin, and LHRH are subjected to luminal membrane hydrolysis. They are hydrolyzed by enzymes in the luminal surface of the brush border membrane of the proximal tubules and the small peptide fragments and amino acids are subsequently reabsorbed, further degra-

ded intracellularly and/or transported through the cells into the systemic circulation (19).

Peritubular extraction of proteins from the postglomerular capillaries and intracellular catabolism is another renal mechanism of elimination (20). This route of elimination was demonstrated for IL-2, insulin (21, 22), calcitonin, parathyroid hormone, vasopressin and angiotensin II (15). It is believed that the peritubular pathway exists mainly for the delivery of certain hormones to their site of action, i.e., to the receptors on the contraluminal site of the tubular cells.

The quantitation of glomerular filtration, tubular reabsorption, peritubular extraction, and renal metabolism can be determined by isolated perfused kidney experiments, as has been done for recombinant human interleukin-11 (rhIL-11, 19 kD) (Fig. 6). During a single passage through the kidney vein, 2% of the rhIL-11 dose was filtered in the glomerulus. Most of the filtered rhIL-11 was reabsorbed in the proximal tubules (1.7% of dose), such that only a small fraction of filtered protein appeared in the urine (0.3% of dose). In

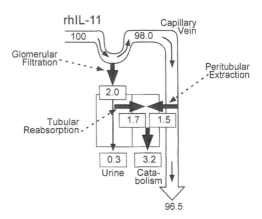

**Fig. 6** Renal disposition profile of recombinant human interleukin-11 (rhIL-11) in kidney perfusion experiments. Values by each process are expressed as percent of dose by a single passage through the renal capillary vein (23).

addition, 1.5% of the dose during a single pass was taken up by the tubules from the capillary site. The fraction of the dose entering the tubular epithelial cells, followed by catabolization (3.2%), had approximately equal contributions from both uptakes (tubular reabsorption and peritubular extraction) (23). If renal disposition is the major pathway, concentrations of the protein are much higher in the cortex than in the medulla.

### 3. Hepatic Metabolism

Besides proteolytic enzymes and renal catabolism, the liver has also been shown to contribute significantly to the metabolism of protein therapeutics. The rate of hepatic catabolism, which determines in part the elimination half-life, is largely dependent on the presence of specific amino acid sequences in the protein (24). Before intracellular hepatic catabolism, proteins and peptides need to be transported from the blood stream to the liver cells. An overview of the different mechanisms of hepatic uptake of proteins is listed in Table 3.

Molecules of relatively small size and with highly hydrophobic characteristics permeate the hepatocyte membrane by simple nonionic passive diffusion. Peptides of this nature include the cyclosporins (cyclic peptides) (25). Other cyclic and linear peptides of small size (<1.4 kD) and hydrophobic nature (containing aromatic amino acids), such as renin and cholecystokinin-8 (CCK-8; 8 amino acids), are cleared by the hepatocytes by a carrier-mediated transport (26). After internalization into the cytosol, these peptides are usually metabolized by microsomal enzymes (cytochrome P-450IIIA for cyclosporin A) or cytosolic peptidases (CCK-8). Substances that enter the liver via carrier-mediated transport are typically excreted into the bile by the multispecific bile-acid transporter. These hepatic clearance pathways are identical to those known for most small organic hydrophobic drug molecules.

For larger peptides and proteins, there is a multitude of energy-dependent carrier-mediated transport processes available for cellular uptake. One of the possibilities is receptor-mediated endocytosis (RME), such as for insulin and EGF (27–29). In RME, circulating proteins are recognized by specific hepatic receptor proteins (11). The receptors are usually integral membrane glycoproteins with an exposed binding domain on the extracellular side of the cell membrane. After binding of the circulating protein to the receptor, the complex is already present in or moves into coated pit regions, and the membrane invaginates and pinches off to form an endocytotic coated vehicle that contains the receptor and ligand (internalization). The vesicle coat consists of proteins (clathrin, adaptin, and others) which are then removed by an uncoating adenosine triphosphatase (ATPase). The vesicle parts, the receptor, and the ligand dissociate and are targeted to various intracellular locations. Some receptors, such as the LDL, asialoglycoprotein and transferrin receptors, are known to undergo recycling. Since sometimes several hundred cycles are part of a single receptor's lifetime, the associated RME is of high capacity. Other receptors, such as the interferon receptor, undergo degradation. This leads to a decrease in the concentration of receptors on the cell surface (receptor downregulation). Others (insulin and EGF receptors, for example) undergo both recycling and degradation (11).

For glycoproteins, if a critical number of exposed sugar groups (mannose, galactose, fucose, N-acetylglucosamine, N-acetylgalactosamine, or glucose) is exceeded, RME through sugar-recognizing receptors is an efficient hepatic uptake mechanism (24). Important carbohydrate receptors in the liver are the asialoglycoprotein receptor in hepatocytes

**Table 3** Hepatic Uptake Mechanisms for Proteins and Protein Complexes. Compiled from Several Sources (see references in the text), Including Reviews by Kompella and Lee (11), by Marks, Gores, and Larusso (26), and by Cumming (33). RME = Receptor-Mediated Endocytosis.

| Cell Type | Uptake Mechanism | Proteins/Peptides Transported |
|---|---|---|
| Hepatocytes | Anionic passive diffusion Carrier-mediated transport | Cyclic and linear hydrophobic peptides (<1.4kD) (cyclosporins, CCK-8) |
| | RME: Gal/GalNAc receptor (Asialoglycoprotein receptor) | N-acetylgalactosamine-terminated glycoproteins Galactose-terminated glycoproteins (e.g.: desialylated EPO) |
| | RME: Low Density Lipoprotein Receptor (LDLR) | LDL, apoE- and apoB-containing lipoproteins |
| | RME: LDLR-Related Protein (LRP receptor) | $\alpha_2$-macroglubulin, apoE-enriched lipoproteins, lipoprotein lipase, (LpL), lactoferrin, t-PA, u-PA, complexes of t-PA and u-PA with plasminogen activator inhibitor type 1 (PAI-1), TFPI, thrombospondin (TSP), TGF-$\beta$ and IL-1$\beta$ bound to $\alpha_2$-macroglubulin |
| | RME: Other receptors | IgA, glycoproteins, lipoproteins, immunoglobulins, Intestinal and pancreatic peptides, Metallo- and hemoproteins, Transferrin, insulin, glucagon, GH, EGF |
| | Nonselective pinocytosis (nonreceptor-mediated) | albumin, antigen-antibody complexes, some pancreatic proteins, some glycoproteins |
| Kupffer cells | endocytosis | Particulates with galactose groups |
| Kupffer and Endothelial cells | RME | IgG N-acetylglucosamine-terminated glycoproteins |
| | RME: Mannose receptor | Mannose-terminated glycoproteins (e.g.: t-PA, renin) |
| | RME: Fucose receptor | Fucose-terminated glycoproteins |
| Endothelial cells | RME: Scavenger receptor | Negatively charged proteins |
| | RME: other receptors | VEGF, FGF (?) |
| Fat-storing cells | RME: Mannose-6-phosphate receptor | Mannose-6-phosphate-terminated proteins (e.g.: IGF-II) |

and the mannose receptor in Kupffer and liver endothelial cells (30–32). The high-mannose glycans in the first kringle domain of rt-PA have been implicated in its clearance, for example (33).

Low density lipoprotein receptor-related protein (LRP) is a member of the low-density lipoprotein (LDL) receptor family responsible for endocytosis of several important lipoproteins, proteases, and protease-inhibitor complexes in the liver and other tissues (34). Examples of proteins and protein complexes for which hepatic uptake is mediated by LRP are listed in Table 3. The list includes many endogenous proteins, including some that are marketed or being developed as drugs, such as t-PA, u-PA, and tissue factor pathway inhibitor (TFPI). There are observations indicating that these proteins bound to the cell-surface proteoglycans are presented to LRP for endocytosis, thus facilitating the LRP-mediated clearance. It seems likely that proteoglycans serve to concentrate LRP ligands on the cell surface, thereby enhancing their interaction with LRP. Interestingly, none of the LRP ligands compete against each other for the LRP receptor, which is very large ($\sim$650 kD) and contains multiple distinct binding sites (35).

Uptake of proteins by liver cells is followed by transport to an intracellular compartment for metabolism. Proteins internalized into vesicles via an endocytotic mechanism such as RME undergo intracellular transport towards the lysosomal compartment near the center of the cell. There, the endocytotic vehicles fuse with or mature into lysosomes, which are specialized acidic vesicles that contain a wide variety of hydrolases capable of degrading all biological macromolecules. Proteolysis is started by endopeptidases (mainly cathepsin D) that act on the middle part of the proteins. Oligopeptides as the result of the first step are further degraded by exopeptidases. The resulting amino acids and dipeptides reenter the metabolic pool of the cell (24). The hepatic metabolism of glycoproteins may occur slower than the naked protein because protecting oligosaccharide chains need to be removed first. Metabolized proteins and peptides in lysosomes from hepatocytes, hepatic sinusoidal cells and Kupffer cells may be released into the blood. Degraded proteins in hepatocyte lysosomes can also be delivered to the bile canaliculus and excreted by exocytosis.

A second intracellular pathway for proteins is the direct shuttle or transcytotic pathway (11). The endocytotic vesicle formed at the cell surface traverses the cell to the peribiliary space, where it fuses with the bile canalicular membrane, releasing its contents by exocytosis into bile. This pathway, described for polymeric immunoglobulin A, bypasses the lysosomal compartment completely.

Receptor-mediated uptake of protein drugs by hepatocytes followed by intracellular metabolism sometimes causes dose-dependent plasma disposition curves due to the saturation of the active uptake mechanism at higher doses. As an example, EGF administered at low doses (50 $\mu$g/kg and lower) to rats showed an elimination clearance proportional to hepatic blood flow, since the systemic supply of drug to the liver is the rate limiting process for elimination. At high doses (>200 $\mu$g/mL), the hepatic clearance is saturated, and extrahepatic clearance by other tissues is the dominant factor in the total plasma clearance. At intermediate doses of EGF, both hepatic blood flow and EGF receptors responsible for the active uptake affect the total plasma clearance (36).

For some proteins, receptor-mediated uptake by the hepatocytes is so extensive that hepatic blood clearance approaches its maximum value, liver blood. As examples, recombinant tissue-type and urokinase-type plasminogen activator (rt-PA and ru-PA, respectively) have been shown to behave as high clearance drugs, and both reductions and increases in liver blood flow affect their clearance in the same direction (37, 38). This may be important for patients with myocardial infarction who may have variations in liver perfusion caused by diminished cardiac function or concomitant vasoactive drug treatment. Also, liver blood flow decreases during exercise, and increases after food intake.

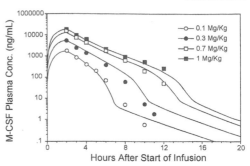

**Fig. 7** Observed and predicted plasma concentration-time profiles of M-CSF after 2-hour intravenous infusions of 0.1–1 mg/kg in cynomolgus monkeys. A two-compartmental pharmacokinetic model with a linear clearance pathway and a parallel Michaelis-Menten elimination pathway was used.

### 4. Receptor-Mediated Elimination by Other Cells

For small synthetic drugs, the fraction of the dose bound to receptors at each moment after administration is usually negligible, and receptor-binding is reversible mostly without internalization of the receptor-drug complex. For protein drugs, however, a substantial part of the dose may be bound to the receptor, and receptor-mediated uptake by specialized cells followed by intracellular catabolism may play an important part in the total elimination of the drug from the body. A derivative of G-CSF, nartograstim, and most likely G-CSF itself, is taken up by bone marrow through a saturable receptor-mediated process (39). It has been demonstrated for M-CSF that besides the linear renal elimination pathway, there is a saturable non-linear elimination pathway that follows Michaelis-Menten kinetics (6, 40). The importance of the nonlinear elimination pathway was demonstrated by a steeper dip in the plasma concentration profile at lower M-CSF concentrations (Fig. 7). At higher levels, linear renal elimination was dominant, and the nonlinear pathway was saturated. The nonlinear pathway could be blocked by coadministration of carrageenan, a macrophage inhibitor, indicating that receptor-mediated uptake by macrophages was likely responsible for the nonlinear elimination (6). This is especially relevant since M-CSF stimulates the proliferation of macrophages. It

is also possible that the receptor-mediated uptake and the effect of M-CSF are closely linked. Indeed, it was observed that after chronic administration of M-CSF the nonlinear elimination was probably induced by autoinduction since M-CSF increases circulating levels of macrophages. Although autoinduction and consequently accelerated metabolism of most drugs is related to a loss of the pharmacological effect, for M-CSF, it may be an indication of sustained pharmacodynamic activity. Similar kinetics were observed for other hematopoietic stimulating factors such as G-CSF (41) and GM-CSF (42). Michaelis-Menten (saturable) elimination was also described for t-PA (43), and recently for a recombinant amino-terminal fragment of bactericidal/permeability-increasing protein (rBPI$_{23}$) (44).

### B. Distribution of Protein Therapeutics

Once a molecule reaches the blood stream, it encounters the following processes for intracellular biodistribution: distribution through the vascular space, transport across the microvascular wall, trans-

**Table 4** Transport Mechanisms for Proteins from the Systemic Circulation Across the Capillary Endothelia. There Are Three Types of Capillary Endothelia with Different Permeabilities for Macromolecules. From Taylor and Granger (122); Reviewed by Kompella and Lee (11).

| Type of Capillary Endothelium | Barrier/Transport Mechanism | Particle Size Subject to Passage | Typical Tissues |
|---|---|---|---|
| Continuous (non-fenestrated) | Basement membrane (Basal lamina) supported by collagen | 50–110 nm | Muscle, CNS, Subcutis, Bone, Skin, Cardiac muscle, Lung arterial capillaries, Connective tissue |
| Fenestrated | Pinocytotic vehicles, Diaphragm fenestrae, Open fenestrae, Intercellular junctions, Basal lamina | 50–800 nm | Renal glomeruli, Peritubular, Intestinal villi, Synovial tissue, Endocrine glands, Choroid plexus (brain), Ciliary body of the eye |
| Discontinuous (sinusoidal) | Large pores (fenestrae), Pinocytotic vesicles | 1,000–10,000 nm | Liver, Spleen, Bone marrow, Postcapillary venules of lymph nodes |

port through the interstitial space, and transport across cell membranes. The biodistribution of macromolecules is determined by the physicochemical properties of the molecule, and by the structural and physicochemical characteristics of the capillaries responsible for transendothelial passage of the molecule from the systemic circulation to the interstitial fluid. In addition, the presence of receptors determines the biodistribution to certain tissues, including extracellular association and/or intracellular uptake. Capillary endothelia are of three types, in increasing order of permeability: continuous (non-fenestrated), fenestrated, and discontinuous (sinusoidal) (Table 4). The most likely dominant mode of transport of macromolecules in nonfenestrated capillaries is through interendothelial junctions (Fig. 8). Through these junctions, there are two modes of transport (45). The convective transport, often the most important for macromolecules, is dependent on a pressure difference between the vascular and interstitial spaces. The diffusive transport is driven by a concentration gradient.

Capillaries selectively sieve macromolecules based on their effective molecular size, shape, and charge. Because of the large size of proteins, their apparent volume of distribution is usually relatively small. The initial volume of distribution after intravenous injection is approximately equal to or slightly higher than the total plasma volume. The total volume of distribution is generally twice or less than twice the initial volume of distribution. Although this is sometimes interpreted as a low tissue penetration, it is difficult to generalize. Indeed, adequate concentrations may be reached in a single target organ because of receptor mediated uptake, but the contribution to the total volume of distribution may be rather small.

Besides size, it appears that the charge-selective nature of continuous

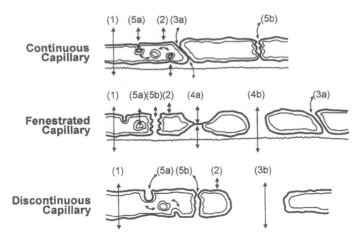

**Fig. 8** Transvascular exchange. Transport pathways in capillary endothelium.
(1) endothelial cell
(2) lateral membrane diffusion;
(3) interendothelial junctions: (a) narrow, (b) wide
(4) endothelial fenestrae: (a) closed, (b) open
(5) vesicular transport: (a) transcytosis, (b) transendothelial channels.
Note that water and lipophilic solutes share pathways (1), (3), and (4). Lipophilic solutes may use pathway (2) as well. Hydrophilic solutes and macromolecules use pathways (3) and (4). Macromolecules may also follow pathway (5). [From Jain (125)].

capillaries and cell membranes may also be important for the biodistribution of proteins. Information for this is available from studies with different types of Cu,Zn-superoxide dismutase (Cu,Zn-SOD), which are similar in molecular weight (33 kD), but have different net surface charges, and are isolated from different species. Tissue equilibration of the positively charged sheep Cu,Zn-SOD was much faster than for the negatively charged bovine Cu,Zn-SOD. In addition, the positively charged Mn-SOD equilibrated much faster than the negatively charged human Cu,Zn-SOD, although Mn-SOD is much bigger (88 kD). A trend towards increasing anti--inflammatory activity, for which interstitial concentrations are important, was observed with increasing isoelectric point. It was suggested that the electrostatic attraction between positively charged proteins and negatively charged cell membranes might increase the rate and

extend of tissue biodistribution. Most cell surfaces are negatively charged because of the abundance of glycosaminoglycans in the extracellular matrix.

Tissue binding is also important for the biodistribution of the heparin-binding proteins, including the fibroblast growth factor family (such as FGF-1 and FGF-2) (46), vascular endothelial growth factor (VEGF) (47), platelet-derived growth factor (PDGF), tissue factor pathway inhibitor (TFPI) (48), amphiregulin (AR), and epidermal growth factor (EGF). Proteins of this group contain a highly positively charged tail which electrostatically binds to low-affinity binding sites consisting of heparin sulfate proteoglycans (acidic glycosaminoglycans) (49, 50). These binding sites are abundant on the vascular endothelium and liver, and are responsible for the majority of cell surface binding of these proteins. The rapid and extensive binding to the vascular endothelium of protein drugs in this class is most likely the explanation for their rapid distribution phase after i.v. injection, and their relatively large volume of distribution. Binding of growth factors to proteoglycans has been proposed to provide a mechanism for growth factor recruitment at the cell surface, presentation to specific receptors, regulation of their action on target cells at short range, and establishment of a growth factor gradient within a tissue.

A major *in vivo* pool of some of the heparin-binding proteins is probably associated with the vascular endothelium, and is released into the circulation quickly after injection of heparin. Since heparin is structurally similar to the cell-surface glycosaminoglycans, the proteins bind to circulating heparin, depleting the intravascular pool. This was demonstrated, for example, for TFPI (51, 52) and basic FGF (FGF-2) (46).

Biodistribution studies with the measurement of the protein drug in tissues are necessary to establish tissue distribution. Because of the difficulty of performing biodistribution studies, the intensity and duration of the pharmacological effects of the drug are sometimes used as an indirect measurement of drug levels in a target organ or tissue. Biodistribution studies are usually performed with radiolabeled compounds. Biodistribution studies are imperative for small organic synthetic drugs since long residence times of the label in certain tissues may be an indication of tissue accumulation of potentially toxic metabolites. Because of the possibility of re-utilization of amino acids from protein drugs in endogenous proteins, such a safety issue does not exist. Therefore, biodistribution studies for protein drugs are usually performed to assess drug targeting to specific tissues, or to detect the major organs of elimination (usually kidneys and liver).

If the protein contains a suitable amino acid such as tyrosine or lysine,

an external label such as $^{125}$I can be chemically coupled to the protein (4). Although this is easily accomplished and a high specific activity can be obtained, the protein is chemically altered. Therefore, it may be better to label proteins and other biotechnology compounds by introducing radioactive isotopes during their synthesis by which an internal atom becomes the radioactive marker. For recombinant proteins, this is accomplished by growing the production cell line in the presence of amino acids labeled with $^3$H, $^{14}$C, $^{35}$S, etc. This method is not routinely used because of the prohibition of radioactive contamination of fermentation equipment. Moreover, internally labeled proteins may be less desirable than iodinated proteins because of the potential for re-utilization of the radiolabeled amino acid fragments in the synthesis of endogenous proteins and cell structures. Irrespective of the labeling method, but more so for external labeling, the labeled product should have demonstrated physicochemical and biological properties identical to the unlabeled molecule (53).

In addition, as for all types of radiolabeled studies, it needs to be established whether the measured radioactivity represents intact labeled protein, or radiolabeled metabolites, or the liberated label. Trichloro-acetic acid- precipitable radioactivity is often used to distinguish intact protein from free label or low-molecular-weight metabolites, which appear in the supernatant. Proteins with re-utilized labeled amino acids and large protein metabolites can only be distinguished from the original protein by techniques such as PAGE, HPLC, immunoassays, or bioassays. This discussion implies also that the results of biodistribution studies with autoradiography can be very misleading. Although autoradiography is becoming more quantitative, one never knows what is being measured qualitatively without specific assays. It is therefore sometimes better to perform biodistribution studies by collection of the tissues, and the specific measurement of the protein drug in the tissue homogenate.

A method was developed to calculate early-phase tissue uptake clearances based on plasma and tissue drug measurements during the first five minutes after intravenous administration (27). The short time interval has the advantage that metabolism and the tissue efflux clearance presumably can be ignored. As an example, with this method, dose-independent (non-saturable) uptake clearance values were observed for a recombinant derivative of hG-CSF, nartograstim, for kidney and liver (39). In contrast, a dose-dependent reduction in the uptake clearance by bone marrow with increasing doses of nartograstim was observed. These findings suggested that receptor mediated endocytosis of the G-CSF receptor in bone marrow may participate in the nonlinear properties of nartograstim. Since G-CSF is one of the growth factors that stimulates the proliferation and differentiation of neutropoietic progenitor cells to granulocytes in bone marrow, the distribution aspects of nartograstim into bone marrow are especially relevant for the pharmacodynamics. In addition, since G-CSF and nartograstim are catabolized in the bone marrow cells after receptor-mediated uptake, the biodistribution into bone marrow is also a pathway for elimination of these molecules. Unlike for classical small synthetic drugs, it is not uncommon for biotechnology derived drugs that biodistribution, pharmacodynamics, and elimination are closely connected.

Besides receptor-mediated uptake into target organs and tissues, other proteins, or macromolecules in general, distribute into tissues in more nonspecific ways. It was demonstrated in at least one study with tumor-bearing mice that high total systemic exposure of target-nonspecific macromolecules was the most important factor determining the extent of tissue uptake (10). Consequently, molecules with physicochemical characteristics that minimize hepatic and renal elimination clearances showed the highest tumoral exposure. Compounds with relatively low molecular weights (approximately 10 kD) or positive charges were rapidly eliminated and showed lower tumor radioactivity accumulation; large (>70 kD) and negatively charged compounds (carboxymethyl dextran, BSA, mouse IgG) showed prolonged retention in the circulation, and high tumoral levels. An example is the murine urokinase (muPA) EGF-like domain peptide of 48 amino acids, muPA(1–48). This peptide is a urokinase receptor antagonist under consideration as an anticancer drug since urokinase has been implicated in invasive biological processes such as tumor metastasis, trophoblast implantation, and angiogenesis. muPA(1–48) was fused to the human IgG constant region. The fused molecule [IgG-muPA(1–48)] retained its activity of inhibition of the murine UPA receptor, but has a much longer *in vivo* elimination half-life (79 versus 0.5 hr; Fig. 9) (54). The half-life increase was due to both a decrease in elimination clearance (4.3 versus 95 mL/hr/kg) and an increase in the peripheral volume of distribution (434 versus 43 mL/kg). Although the fused molecule was substantially larger, tissue distribution increased, possibly because of substantial tissue binding. This is in contrast with some polyethylene glycol-modified (PEGylated) molecules such as PEG IL-2 (polyethylene glycol-modified interleukin-2) for which size increase resulted in a smaller distribution volume compared with the original molecule (see below).

Biodistribution into the lymphatics after subcutaneous (s.c.) injection deserves special attention since it is a rather unique transport pathway for macromolecules. Following s.c. administration, drug can be transported to the systemic circulation by absorption into the blood capillaries or

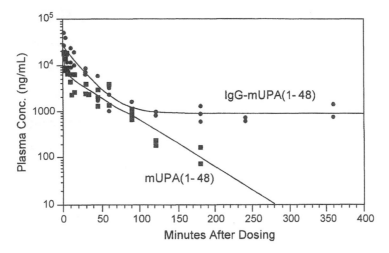

**Fig. 9** Fusion of the murine urokinase EGF-like peptide of 48 amino acids with human IgG [IgG-mUPA(1–48)] resulted in a much longer half-life than the original peptide [mUPA(1–48)]. The data were modeled according to a linear two-compartmental model (54).

by the lymphatics. Since the permeability of macromolecules through the capillary wall is low, they were found to enter blood indirectly through the lymphatic system (55, 56). Compounds with a molecular weight larger than 16 kD are absorbed mainly (>50%) by the lymphatics, while compounds smaller than 1 kD are hardly absorbed by the lymphatics at all (Fig. 10).

Biodistribution into target organs, usually receptor-mediated, is important for the pharmacodynamics of protein

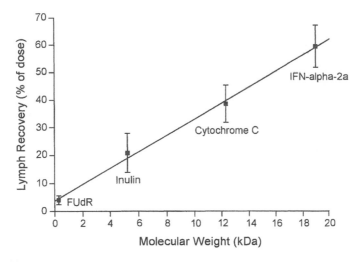

**Fig. 10** Correlation between the molecular weight and the cumulative recovery of rIFNα-2a (19 kD), cytochrome c (12.3 kD), inulin (5.2 kD), and FUdR (256.2 kD) in the efferent lymph from the right popliteal lymph node following s.c. administration into the lower part of the right hind leg of sheep. Each point and bar shows the mean and standard deviation of three experiments performed in different sheep. The line drawn is the best fit by linear regression analysis of the four mean values (r = 0.998; p > 0.01). [From Supersaxo *et al.* (56)].

drugs. For some proteins, saturable receptor-mediated tissue uptake in target organs is responsible for nonlinear kinetics (57). For example, the uptake clearance of rhEPO by bone marrow and spleen exibited clear saturation in rats. Also, a single high dose of rhEPO caused a reduction of uptake clearance by bone marrow and spleen, while repeated injections caused an increase of the tissue uptake clearance, especially by the spleen, in a dose-dependent manner (57). Hematopoietic parameters such as hematocrit and hemoglobin concentration changed accordingly, suggesting that changes in the uptake clearance were caused by down- or upregulation of EPO receptors.

## C. Plasma Binding of Protein Therapeutics

The binding of drugs to circulating plasma proteins can influence both the distribution and elimination of drugs, and consequently their pharmacodynamics. Since it is generally accepted for small drug molecules, including small proteins, that only the unbound drug molecules can pass through membranes, distribution and elimination clearances of total drug are usually smaller than those of free drug. Accordingly, the activity of the drug is more closely related to the unbound drug concentration than to the total plasma concentration. For other protein drugs, however, plasma binding proteins may act as facilitators of cellular uptake processes, especially for drugs that pass membranes by active processes. When a binding protein facilitates the interaction of the protein therapeutic with receptors or other cellular sites of action, the amount of bound drug influences the pharmacodynamics directly.

Numerous examples of binding proteins are reported for proteins: IGF-I and IGF-II, t-PA, growth hormone, DNase (58), nerve growth factor, etc. (59). Some proteins have their own naturally occur-

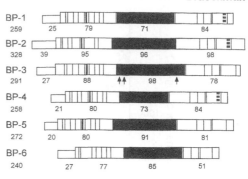

**Fig. 11** Schematic diagram of the structures of the human IGF-I binding proteins, with the number of amino acids. Each protein contains at least 18 cysteines, that are aligned (vertical lines), and each protein has also a cysteine-free region (solid bar). [From Clemmons (61)].

ring binding proteins that bind the protein specifically. As an example, six specific binding proteins were identified for IGF-I (Fig. 11), denoted as IGFBP-1 to IGFBP-6 (60, 61). The IGFBPs are high affinity, soluble carrier proteins that transport IGF-I (and IGF-II) in the circulation (61). IGFBP-3 appears to be the most important binding protein for IGF-I since it is the most abundant in human serum and tissues. At least 95% of the total human serum concentration of IGF-I is bound to IGFBP-3 (62). The remainder is mainly bound to IGFBP-1, -2, and -3. Less than 1% of the total is free under normal conditions (61). IGFBP-3 seems to act as a reservoir for IGF-I, and as such to protect the organism against acute insulin-like hypoglycemic effects. Indeed, the hypoglycemic effect is related to the free IGF-I plasma concentration. In this case, the binding protein limits the accessibility of IGF-I to receptors since all binding proteins have substantially higher affinities for IGF-I than the IGF receptors (63). In contrast, the delayed, indirect effects of IGF-I, such as its anabolic effects, may be related to the bound IGF-I levels. This is supported by evidence that the IGFBPs may play an active role in the interaction with target cells, and may act as facilitators for the delivery of IGF-I to certain receptors (61). One example is the demonstration that the affinity of the binding protein for IGF-I (IGFBP-6) at the cell surface is lower than in solution, which would make it easier for IGF-I to leave its association with the binding protein and to engage in binding with a cell-based receptor. As such, the IGFBPs may act as inhibitors for certain IGF-I effects, and as stimulators for other IGF-I effects.

It is demonstrated that the elimination half-life of bound IGF-I is significantly prolonged relative to that of free IGF-I (59, 64, 65). This suggests that unbound IGF-I only is available for elimination by routes such as glomerular filtration and peritubular extraction. The binding proteins for IGF-I are also responsible for the complicated pharmacokinetic behavior of IGF-I. The IGFBPs can be saturated at high IGF-I plasma concentrations, typically reached after therapeutic administration of IGF-I. At high doses, the binding proteins saturate and leave a larger proportion of free protein available for elimination. Additionally, the nonlinear pharmacokinetics of IGF-I are complicated by the fact that the concentrations and relative ratios of the IGFBPs change with time during chronic dosing. The binding proteins are also very different between species, which makes interspecies scaling of the IGF-I pharmacokinetics for IGF-I impossible.

Another example is growth hormone (GH), for which a specific high-affinity binding protein homologous with the extra cellular domain of the growth hormone receptor is present in human plasma (66, 67). At least two GH-binding proteins (GHBP) have been identified in plasma with respectively high and low binding affinities for GH (59). GHBP binds about 40–50% of circulating GH at low GH concentrations of about 5 ng/mL (68). At higher circulating GH levels, the binding proteins become saturated (Fig. 12). The

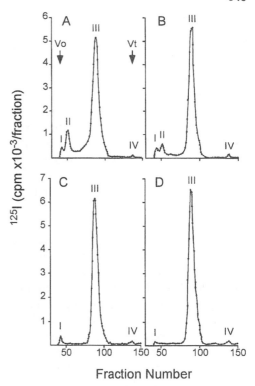

**Fig. 12** Gel filtration profiles of $^{125}$I-hGH in plasma on Sephadex G-100. $V_0$ and $V_t$ are the void and total volumes, respectively. A. Blank plasma with endogenous level of hGH only; B. 126 ng/mL hGH added; C: 10 g/mL hGH added; D. tracer only (no plasma). Peak III corresponds to monomeric hGH; peak II and the plateau region between peaks II and III refer to the plasma-bound hGH; peak IV is free iodide. Higher hGH concentrations saturate the binding proteins as peak II becomes smaller relative to peak III (C versus B versus A). [From Baumann *et al.* (126)]

clearance of bound GH is about ten-fold slower than that of free GH (69). Consequently, the binding proteins prolong the elimination half-life of GH, and as a result, enhance or prolong its activity. On the other hand, plasma binding of GH prevents access of free GH to its receptors, and this could decrease its activity (59).

Other protein therapeutics seem to bind to circulating proteins in a more nonspecific way. As an example, a recombinant derivative of hG-CSF, narto-gastrim, showed 92% binding in rat plasma, presumably to albumin (39). A wide variety of soluble receptors with specific binding characteristics for proteins are also circulating in plasma. Soluble receptors have generally reduced ligand affinity compared with their membrane-bound counterparts, and the circulating receptor concentration may be insufficient to effectively compete with binding to the cell-bound receptor (70). The soluble IL-2 receptor (sIL-2R), for example, consists of the released α-chain of the membrane-bound IL-2 receptor on activated lymphocytes. The affinity of IL-2 for sIL-2R is approximately 10 and 1000-fold less than for the medium-affinity ($\beta\gamma$-subunit) and high-affinity ($\alpha\beta\gamma$-subunit) IL-2 receptors, respectively (71, 72). Soluble receptors for other proteins may bind their ligand more strongly. For example, soluble low density lipoprotein receptor-related protein (sLRP) seems to act as a competitive inhibitor of ligand uptake by cell surface-bound LRP (73). Increased levels of soluble receptors may influence the pharmacodynamics of their ligands, but can also extend the half-life.

## D.   Pharmacokinetic Models

Pharmacokinetic models are widely used to describe and predict the time course of the drug in plasma and tissues. These models include compartmental models and physiological models (74, 75). Compartmental models represent the body as a number of well-stirred compartments. Typically, transfers of drug between the compartments are first-order processes, although this is not a strict requirement. The required number of compartments (usually one to three) is empirically determined from the data. In physiologically based pharmacokinetic (PBPK) models, the body is represented as a physiological system of the various, most relevant body

organs and tissues. The model uses organ and tissue volumes, and distribution throughout the body is determined by blood flow (flow-limited distribution) and tissue partition coefficients (diffusion-limited distribution). Physiological PK models are usually applied in animal studies from which tissue concentration data are available, such as for $\beta$-endorphin in rats (76).

## E.   Interspecies Scaling

Techniques for the prediction of pharmacokinetic parameters in one species from data derived from other species have been applied for many years (77, 78). Such scaling techniques use various allometric equations based on body weight using the following allometric equation:

$$P = a.W^b$$

where $P$ is the pharmacokinetic parameter being scaled, $W$ is the body weight, $a$ is the allometric coefficient, and $b$ is the allometric exponent. Although $a$ and $b$ are specific constants for any compound and for each pharmacokinetic parameter, the exponent $b$ seems to average around 1 for volume terms such as the volume of distribution and 0.75 for rates such as elimination and distribution clearances. Since the elimination half-life of any drug is proportional to the volume of distribution and inversely proportional to the elimination clearance, $b$ is about 0.25 for elimination half-lives. Allometric scaling of pharmacokinetic parameters has been difficult for small synthetic drug molecules, especially for those drugs with a high hepatic clearance and quantitative and/or qualitative interspecies differences in metabolism. In contrast, the biochemical and physiological processes that are responsible for the pharmacokinetic fate of biologics such as peptides and proteins are better conserved across mammalian species. As such, allometric scaling for those compounds has been more reliable

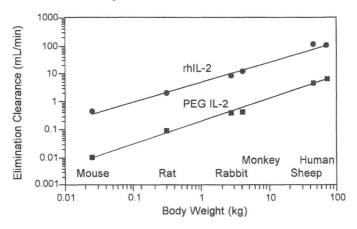

**Fig. 13** Allometric interspecies scaling of the elimination clearance of IL-2 and PEG IL-2.

and accurate (79). It is our experience that the systemic exposure in humans of proteins that follow linear pharmacokinetics can be predicted within a factor of two from pharmacokinetic data obtained from three to four animal species. As a typical example, we could scale the pharmacokinetic parameters for IL-2 and PEG IL-2, as demonstrated in Fig. 13, for the elimination clearance. Notice that the regression lines for both compounds are parallel, which is expected if PEGylation decreases the clearance to the same degree in all species.

A helpful, although potentially less accurate prediction can be made based on pharmacokinetic data from one species to another based on the average allometric exponents for volumes ($b = \sim 1$) and clearances ($b = \sim 0.7$) for other compounds. Interspecies scaling is helpful in the prediction of doses for animal models of disease, toxicology studies, and for the first human studies. Indeed, if the efficacious concentration of a protein drug is known from *in vitro* studies, one might predict the dose needed to reach these levels in an animal efficacy or toxicology model when pharmacokinetic data are known from another species. Similarly, if an estimation of the maximum tolerated exposure can be made, allometric scaling

may be helpful to determine the highest dose in toxicology studies. The dose that results in efficacious concentrations may be taken as the lowest dose in the toxicology studies. Additionally, the efficacious dose in humans can be estimated from the animal pharmacokinetic data, and, as a rule, the starting dose in the first human study (dose-escalation study) is smaller by a factor of two or more, based on safety considerations.

It needs to be emphasized that allometric scaling techniques are useful tools for the prediction of doses used in dose-ranging studies, but they can never replace such studies. The advantage of including such dose prediction in the protocol design of dose-ranging studies is that a smaller number of doses need to be tested before finding the final dose level. Interspecies dose predictions simply narrow the range of doses in the initial pharmacological efficacy studies, the animal toxicology studies, and the human safety and efficacy studies.

## III. PHARMACODYNAMICS

Although the time course of the compound at the receptor or effector site is the desired knowledge to predict or explain the pharmacodynamics (PD), accurate drug level data at that site are difficult to obtain. In most cases, pharmacokinetic (PK) data are limited to plasma concentration data.

During the last decade, the application of PD models for *in vivo* effect data has increased tremendously (80). In addition, the PD models have been linked to PK models, and this approach has made integrated PK/PD analysis possible. PK/PD modeling has been reviewed extensively for small drug molecules (81–86), but relatively few publications are available for proteins, or biotechnological therapeutics in general (87). PD models are based on the law of mass action of drug-receptor interaction, classically called the occupancy theory (88). These models that express the effect as a function of drug concentration are known for a long time from the classical *in vitro* pharmacological experiments wherein receptors in organ baths or tissue strips were exposed to a drug concentration. A similar situation occurs *in vivo* when the effect concentration is the concentration in plasma, an effect compartment or biophase (89).

## A.   Direct Effects

Sometimes, the effect concentration in the PD model equations can be set equal to the plasma concentration when there is a direct relationship between the plasma drug concentration and the pharmacological effect. These are the direct effect PK/PD models. Fig. 14 shows an example of a PK/PD model wherein the effect is directly related to the concentration in the central compartment (the plasma concentration). Any appropriate compartmental model, or other PK model that predicts the plasma concentration-time curve can be used. In the direct effect PK/PD models, the effect-time profile follows the plasma-concentration profile, and the maximum effect occurs at the time of the peak plasma concentration.

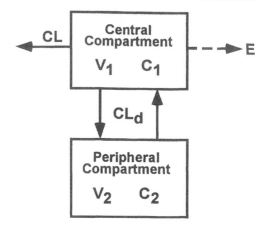

**Fig. 14**   Example of a direct effect PK/PD model. The PK model is a typical two-compartmental model with a linear elimination clearance from the central compartment (CL) and a distributional clearance (CL$_d$). $C_1$ and $C_2$ are the concentrations in the central and peripheral compartments, and $V_1$ and $V_2$ are their respective apparent volumes of distribution. The effect (E) is a function of $C_1$.

A typical example is the thrombolytic effect of tissue factor pathway inhibitor (TFPI). The increase of the prothrombin time (PT) during continuous infusion of *Escherichia Coli* derived recombinant human TFPI in a two-week toxicology study in cynomolgus monkeys was directly related to the TFPI plasma concentrations (Fig. 15) (90). The PD model equation is

$$E = E_0 + S\ C$$

where $E$ is the effect (PT); $E_0$ is the baseline effect (predose PT), $C$ is the TFPI plasma concentration, and $S$ is the slope of the effect-concentration curve. The slope represents the sensitivity of the effect or the potency, i.e., change in PT (in sec) per unit change of concentration ($\mu$g/mL). An integrated PK/PD analysis according to a two-compartmental PK model and a direct effect PD model explained the observed data (Fig. 16) (90).

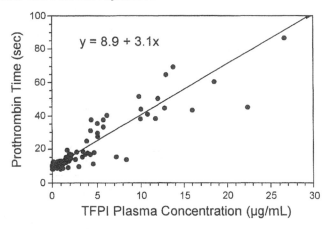

**Fig. 15** Direct relationship of the increase in prothrombin time (PT) and the plasma concentrations of TFPI after continuous i.v. infusion of TFPI in cynomolgus monkeys.

## B. Indirect Effects

For most effects after administration of peptides and proteins, however, no such direct relationship can be observed. In a lot of cases, the maximum effect is reached at times later than the maximum plasma concentration, and sometimes, a considerable effect can still be measured at times where the plasma drug levels have fallen below the limit of detection. The temporal differences between drug exposure and onset/duration of effect has created the idea that for peptides and proteins, there is no relationship between plasma drug levels and effect. The opposite is true: if the effect is drug related, there must be a relationship between plasma drug concentrations and the time course of effect intensity. A plot of the concentration-effect relationship from non-steady state conditions, i.e., when the plasma concentrations rise and fall, such as following an i.v. infusion or an extravascular dose, is a helpful diagnostic of the temporal features of drug effect. In such plots, effect delays manifest themselves as (*counterclockwise*) *hysteresis*. Delays caused between the appearance of drug in plasma and the appearance of the pharmacodynamic response, by processes such as distribution

into the biophase or cascade-type post-receptor events (Fig. 1) may cause counterclockwise hysteresis. The relationship can be described by more complicated combined PK/PD models. Two basic approaches are available. The first one is the family of PK/PD link models, the second approach uses the indirect effect PK/PD models.

### 1. PK/PD Link Models

The temporal delay of the effect appearance and duration in the PK/PD link models is explained by a distributional delay (91). In this case, drug concentrations in a slowly equilibrating tissue compartment with plasma are directly related to the effect intensity. Since the peak level of drug in the biophase is reached later than the time of the peak plasma concentration, the peak effect also occurs later than the plasma peak level. Although theoretically the biophase drug concentration may equal the drug concentration in a peripheral compartment, it rarely happens that a peripheral pharmacokinetic compartment acts as the biophase or effect compartment. More often the biophase is a small part of a pharmacokinetic compartment that from

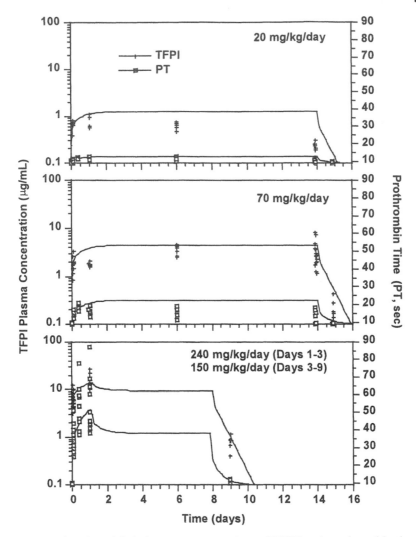

**Fig. 16** Measured and modeled plasma concentrations of TFPI and prothrombin times (PT) after i.v. infusion of 20–240 mg/kg/day TFPI in cynomolgus monkeys as part of a 14-day toxicology study. The modeling was performed according to the PK/PD model outlined in Fig. 14, with increase of PT, above the baseline (predose) PT, that was directly and linearly proportional to circulating TFPI concentrations.

a pharmacokinetic point of view cannot be distinguished from other tissues within that compartment. Compartmental modeling with plasma concentration-time data is just not sensitive enough to isolate the biophase as a separate compartment without the availability of measured drug concentration data in the biophase. The solution to this problem has been to postulate a hypothetical effect com-partment linked to the central com-partment (or to a peripheral compartment in some cases) (Fig. 17). Drug distributes into the effect compartment (this is the link) but since the amount of drug in the effect compartment is rather small, no actual mass transfer is implemented in the pharmacokinetic part of the PK/PD model. The concentration used in the pharmacodynamic part of the PK/PD

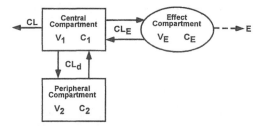

**Fig. 17** Example of a typical PK/PD link model. A hypothetical effect compartment is linked to the central compartment of a two-compartmental pharmacokinetic model. The concentration in the effect compartment ($C_E$) drives the intensity of the pharmacodynamic effect (E). $CL_E$ is the linear clearance for distribution of drug to the effect compartment and elimination from the effect compartment. $V_E$ is the apparent volume of distribution in the effect compartment. All PK parameters are identical as in Fig. 14.

mean serum concentration profile of insulin after a single s.c. injection of 10 U in ten volunteers, and the corresponding effect measured as the glucose infusion rate to maintain an euglycemic state. Fig. 19 shows the hysteresis in the effect-concentration relationship, and how a typical sigmoidal effect-concentration curve is obtained with the hypothetical effect concentration from a one-compartmental PK/PD link model (93).

model is then the drug concentration in the effect compartment. Although this PK/PD model is constructed with tissue distribution as the reason for the delay in effect, the distribution clearance to the effect compartment can be interpreted as including other reasons of delay, such as transduction processes and secondary post receptor events. The hypoglycemic effect of insulin has been modeled by this type of PK/PD model (92, 93). Fig. 18 shows the

## 2. Indirect Effect Models

Another and better approach to include effect delays in PK/PD modeling based on post receptor events has been the indirect effect models (94–97). In this modeling approach, the observed effect is an indirect effect, i.e., is not the primary effect, but rather a consequence of rate-limiting transduction and other post receptor events. In the simplest indirect effect model, a balance of two processes (Fig. 20) maintains the effect, which together form the biosignal flux. The first process is the production of the effect, determined by a zero-order production rate. The second process is the decrease or disappearance of the effect by a first-order dissipation rate. In a normal (predose; no

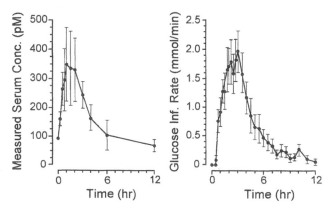

**Fig. 18** Mean measured serum insulin concentrations after a single 10-U subcutaneous dose of regular insulin in ten volunteers (left panel); corresponding glucose infusion rates needed to maintain euglycemia (right panel). [From Woodworth *et al.* (93).]

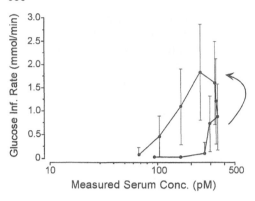

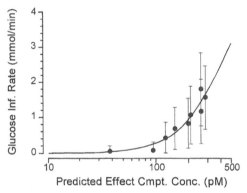

Fig. 19 Relationship between the glucose infusion rate to maintain euglycemia versus serum insulin concentrations after a single sub-cutaneous dose of 10 U regular insulin in ten volunteers (upper panel). The time-dependent hysteresis, which is an indication of the indirect nature of the effect, is indicated by the arrow. The lower panel shows the sigmoidal relationship between the effect and the predicted effect compartment concentration, demonstrating the collapse of the hysteresis loop [From Woodworth et al. (93)].

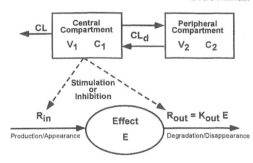

Fig. 20 Pharmacodynamic indirect effect model wherein the effect is maintained by equilibrium between a zero-order appearance rate, $R_{in}$, and a first-order disappearance rate, $R_{out}$. A drug effect is caused by stimulation or inhibition of $R_{in}$ or $R_{out}$. The degree of stimulation or inhibition is dependent on the plasma drug concentration. The PD parameters are $R_{in}$, $K_{out}$ (the first order rate constant for effect disappearance), $EC_{50}$ (the concentration that produces 50% of maximum inhibition or stimulation), and $E_{max}$ (the maximum inhibition or stimulation). The pharmacokinetic model is identical as in Fig. 14.

drug present) situation, both processes are in equilibrium and homeostasis of the effect is maintained (baseline effect). Drug effect is caused by stimulation or inhibition of either the production or disappearance rate. The degree of stimulation or inhibition is determined by the plasma concentration.

This type of model was recently applied to model the effects of IL-2 treatment in HIV patients (98). The PK model for IL-2 included two compartments with a time-varying serum clearance, which was related to concentrations of the soluble IL-2 receptor (sIL2R). Increasing circulating sIL2R levels were used as a surrogate marker for the upregulation of the cell-based IL-2 receptor, which causes probably an increase of the receptor-mediated clearance of IL-2 after chronic dosing. Indirect PK/PD models with IL-2 stimulation of the formation rates were used for sIL2R as well as for the serum levels of TNF-α (tumor necrosis factor-alpha), which were increased by IL-2.

A similar model (Fig. 21) was developed to describe the effects of the antibody response on the pharmacokinetics of the anti-B7.1 murine monoclonal antibody M24 (of the IgG2a class) after chronic administration in rhesus monkeys. B7.1 is a transmembrane glycoprotein on antigen presenting cells that binds to CD28 on T-cells. This ligand-receptor interaction leads to full T-cell activation in acute

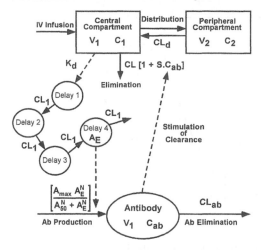

**Fig. 21** Pharmacodynamic indirect effect model of the antibody response after M24 administration in rhesus monkeys. M24 stimulates the antibody production (titer $C_{ab}$) according to an $E_{max}$ model (parameters $A_{max}$, $A_{50}$, and N), driven by the amount $A_E$ in delay compartment 4. The delay between the systemic M24 appearance and the formation of systemic anti-M24 antibodies is modeled with four delay compartments (in/out delay clearance $CL_1$). $CL_{ab}$ is the elimination clearance of the antibody, and the antibody titer is scaled to $V_1$. The pharmacokinetics of M24 are modeled according to a two-compartmental model (identical to Fig. 14). However, the M24 clearance (CL) increases linearly with increasing antibody titers according to a slope S.

graft-versus-host disease (AGVHD) after allogeneic bone marrow transplantations. M24 in combination with cyclosporin A may have benefit in the prevention of transplant rejection. The plasma concentrations of M24 after daily injections for 14 consecutive days in rhesus monkeys were measured by ELISA. During the first six days, plasma levels of M24 increased steadily approaching steady state, but decreased gradually thereafter to almost undetectable levels, although dosing continued until Day 14 of the study. Increasing plasma titers of anti-M24 antibodies were detected on Day 9–16, when most of the M24 molecules were present as a complex with the anti-M24 antibodies (the M24 ELISA measured both free and bound M24). The M24 plasma levels were modeled according to a two-compartmental model. The rhesus-anti-mouse antibody (RAMA) response was modeled according to an indirect effect model wherein the anti-M24 antibody production was stimulated by an increase in the M24 plasma levels ($E_{max}$ model) in a delayed fashion (with four delay compartments). The effect of the antibody response on M24 plasma levels was modeled by an increase of the clearance of M24 according to a linear stimulatory model. The measured and modeled M24 plasma concentrations and antibody response after i.v. administration of 0.05 mg/kg/day is shown in Fig. 22.

## C.  Complex PK/PD Models

Many protein therapeutics have multiple and/or biphasic responses, which are indirect in nature. In some cases, more complicated PK/PD models based on the basic models described above are necessary. An example is the model that was developed to explain the effects of subcutaneous PEG IL-2 (polyethylene glycol-modified interleukin-2) in rats. An early moderate decrease of the number of blood lymphocytes was followed by a pronounced increase of the blood lymphocyte count. The model (Fig. 23) describes the production of lymphocytes in tissues, after which they traffic into the blood pool, followed by trafficking out of blood and subsequent degradation. The indirect, but relatively early decrease of the number of blood lymphocytes after PEG IL-2 administration is caused by a stimulation of the lymphocyte traffic out of blood. The indirect, delayed increase of the blood lymphocytes is modeled by a stimulation of the production rate of lymphocytes. The effect delay relative to the maximum plasma levels of PEG IL-2 is caused by dis-

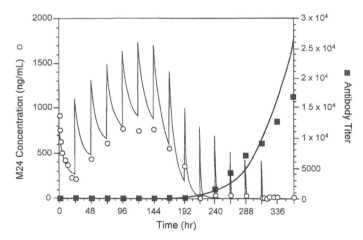

**Fig. 22**  Measured and modeled M24 plasma concentrations and antibody response after daily 30-min i.v. infusions of 0.05 mg/kg M24 for 14 days in rhesus monkeys. The modeling was performed according to the PK/PD model depicted in Fig. 21.

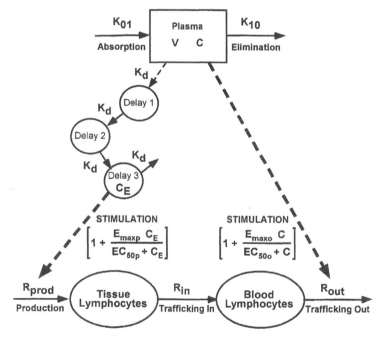

**Fig. 23**  PK/PD model for changes in blood lymphocytes after s.c. administration of PEG IL-2 in rats. The PK model is a one-compartmental model with first-order absorption (rate constant $K_{01}$) and elimination (rate constant $K_{10}$). PEG IL-2 stimulates the trafficking out of blood and/or catabolism of lymphocytes (first-order rate $R_{out}$) according to an $E_{max}$ model (parameters $E_{maxo}$ and $EC_{50o}$), which is a function of the PEG IL-2 plasma concentration (C). The delayed increase of blood lymphocytes is modeled in two consecutive ways: 1. Three delay compartments with first-order input and output rates (rate constant $K_d$) resembling distribution and transduction delays; 2. Stimulation of lymphocyte production in tissues according to an $E_{max}$ model (parameters $E_{maxp}$ and $EC_{50p}$), which is a function of the effect concentration $C_E$. Tissue lymphocytes traffic into the blood pool (first-order rate $R_{in}$).

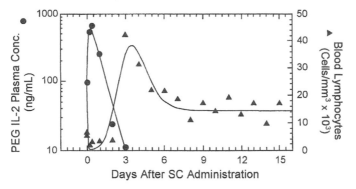

**Fig. 24** Measured and observed PEG IL-2 pharmacokinetics and pharmacodynamics (changes in blood lymphocyte count) after subcutaneous administration of 10 MIU/kg in rats. The modeling was performed according to the PK/PD model depicted in Fig. 23.

tributional and post receptor events through three delay compartments. An additional delay is caused by the fact that the lymphocytes need to travel from their site of production to blood before an increase can be measured. Fig. 24 demonstrates the goodness-of-fit of the modeling, but also shows that the blood lymphocyte increase on Day 3 post dosing occurs after most of the PEG IL-2 is eliminated. PK/PD models like this one link the effect to drug exposure in a quantitative way, which allows extrapolations to other dose levels and routes of administration. It is obvious that a better understanding of the receptor-mediated and post receptor transduction mechanisms of protein drugs may contribute to the creation of representative and useful PK/PD models.

The stimulation of erythropoiesis by recombinant human erythropoietin (EPO) therapy in patients with uremic anemia, was analyzed with a different PD model (99). The model assumes a linear stimulatory effect of EPO on the production rate of red blood cells, as measured by hematocrit. In the first stage (Fig. 25), EPO increases the hematocrit because red cells are produced at an increased rate and none of the newly produced red cells die yet. However, after reaching one life span,

the red cells start dying at the increased rate they were produced and a new steady state is reached. Although the effect of EPO is almost immediate after dosing initiation, it takes time to develop fully.

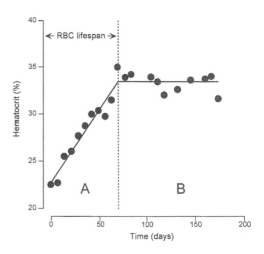

**Fig. 25** Hematocrit (Hct) in an uremic patient on a constant EPO dose of $3 \times 4000$ U/week. Stage A: Hct increases because EPO stimulates erythrocyte production and none of the newly formed red cells are old enough to die; Stage B: EPO maintains increased red cell production, but red cells die at faster rate since newly formed cells from Stage A exceed their average life span. Consequently a new steady state is reached. [From Uehlinger *et al.* (99).]

This is an alternate mechanism responsible for the appearance of hysteresis in concentration-effect curves.

## D. Dose-Response and Concentration-Response Curves

PK/PD modeling is especially useful in Phase II studies wherein the relationship between dose (and/or concentration) and response for new drug candidates needs to be established. The outcome from these studies does not only convince the sponsors and the regulatory agencies of an existing drug effect, but also assists in the selection of the optimal dose for Phase III trials. For many biotechnology-derived drugs, the Phase II trial was nonexistent or too small (one dose level for example) to design a successful or optimal Phase III trial. To complicate this matter even more, it is believed that biotechnology drugs (and potentially all types of drugs) sometimes show bell-shaped dose-response curves, i.e., there is a dose that gives a maximum response, and any increase beyond this dose level results in a further decrease of the response. As an example, lower doses of rINF-$\gamma$ in multiple myeloma patients seemed to induce a greater increase in natural killer activity than higher doses (100). Bell-shaped dose-response curves were also observed for superoxide dismutase (SOD) in four different animal models of myocardial infusion, *in vivo* and *in vitro*, using different SOD preparations (Fig. 26) (101).

## IV. IMMUNOGENICITY

Immunogenicity is the ability to induce the formation of antibodies, a prerequisite for antigenicity, which is the ability to react with specific antibodies. Since most protein therapeutics are obtained from nonhuman sources, administration of heterologous (nonhost) proteins may

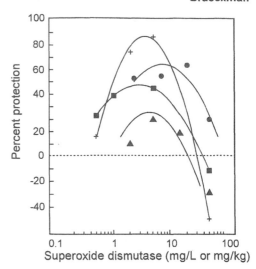

**Fig. 26**  Bell-shaped dose-response curves for superoxide dismutase (SOD) in four animal models of myocardial reperfusion injury: A. Ischemic rabbit hearts treated with hrMn-SOD, monitoring recovery of developed tension; B. hypoxic rat hearts treated with yeast Cu,Zn-SOD, monitoring creatine kinase release; C. hypoxic rabbit hearts treated with hrCu,Zn-SOD, monitoring lactate dehydrogenase release; D. rabbit hearts with ligated coronary arteries in vivo treated with hrCu,Zn-SOD, monitoring infarct size (101).

result in an immunogenic response. This can potentially lead to hypersensitivity reactions, and anaphylactic shock in some cases. As an example, bovine Cu,Zn-SOD (Orgotein) as a treatment various arthritic diseases was withdrawn from several European countries because of hypersensitivity. However, in general, immunogenicity has not been the primary limitation for the clinical use of proteins; poor PK and PD are frequently the major obstacles for efficacy.

Antibody formation is often observed after chronic dosing of human proteins in animal studies, but is also occurring in clinical studies. Also, human-derived proteins are not always truly nonimmunogenic. Immunogenicity can be a problem in the study of protein drugs since the presence of antibodies can complicate the interpretation of preclinical and clinical studies by inactivating (neutralizing) the biological activity of the protein drug. Additionally, protein-antibody complex formation may affect the distribution, metabolism, and elimination of the protein drug. Neutralizing antibodies may inactivate the biological activity of the protein by blocking its active site or by a change of the tertiary structure by steric effects. Antibodies are most likely to be induced when the protein is foreign to the host. Examples of such situations are when mouse-derived monoclonal antibodies are administered to humans, or when human recombinant proteins are tested for safety in animals. Extravascular injections (s.c., i.m., ...) are also more likely to stimulate antibody production than intravenous administrations, presumably because of the higher degree of protein precipitation and aggregation at the injection site. This was demonstrated for IL-2 (102) and INF-$\beta$ (103, 104).

Antibodies may directly neutralize the activity of the protein. As an example, drug failure with luteinizing hormone releasing factor (LHRH) following chronic administration correlated with serum antibody titers (105). If neutralization occurs, it indicates that at least some fraction of the antibody population binds at or near the active site, which blocks activity (106). Irrespective of the neutralizing capabilities of the antibodies formed, they may also indirectly affect the efficacy of a protein drug by changing its pharmacokinetic profile. Elimination clearances of protein drugs may be either increased or decreased by antibody formation and binding. An increase of the clearance is observed if the protein-antibody complex is eliminated more rapidly than the unbound protein (107). This may occur when high levels of the protein-antibody complex stimulate its clearance by the reticuloendothelial system (108). In other

situations, the serum concentration of a protein can be increased if binding to an antibody slows its rate of clearance, because the protein-antibody complex is eliminated slower than the unbound protein (106). In this case, the complex may act as a depot for the protein and, if the antibody is not neutralizing, a longer duration of the pharmacological action may occur.

Whether an increase or a decrease of the clearance is manifested depends on the dose level administered. At low doses, protein-antibody complexes delay clearance because their elimination is slower than the unbound protein. In contrast, at high doses, higher levels of protein-antibody complex result in the formation of aggregates, which are cleared more rapidly than the unbound protein.

Posttranslational changes in cloned proteins that are different from the endogenous human protein (such as the addition of an extra terminal amino acid, or the lack of proper glycosylation) may increase the immunogenicity of the protein.

An approach to minimize immunogenicity is the conjugation of proteins to various hydrophylic polymers (109). This is the same approach used to extend the circulating half-life. Conjugation of proteins with polymers masks determinant sites on the protein surface which are responsible for recognition by target organs for clearance of the protein and for immunogenic recognition. Since this approach may shield the protein from receptor interaction as well, receptor binding affinity is often decreases as well.

## V. HETEROGENEITY OF PROTEIN THERAPEUTICS

### A. Structural Heterogeneity

The identity, purity and potency of small synthetic drugs can be demonstrated analytically, and consequently, they are usually completely defined in terms of their chemical structure. Peptides, proteins, and other biotechnologically derived compounds are usually more complex compounds, and it is generally not possible to define them as discrete chemical entities with unique compositions. The physicochemical and biochemical characteristics of proteins are not only dependent on the amino acid sequence (primary structure), but also on the shape and folding (secondary and tertiary structures), and the relationship between the protein molecules themselves, such as the formation of aggregates (quaternary structure). Biotechnologically derived and endogenous proteins may be heterogeneous at each structural level. For natural IFN-$\gamma$, for example, six naturally occurring C-terminal sequences have been identified (110–112).

## B.   Glycosylation of Protein Therapeutics

Post-translational modification of proteins, such as the degree of glycosylation of amino acid residues, is also a source of heterogeneity. The secreted and membrane-associated proteins of almost all eukariotic cells are glycosylated, and different glycoproteins have also different carbohydrate contents, from $\sim 3\%$ for serum IgG to >40% for erythropoietin (EPO). The degree of glycosylation differs according to the cell line used for production. For example, GM-CSF and M-CSF are nonglycosylated in bacterial cell lines such as *E. Coli*, moderately glycosylated in yeast, and heavily glycosylated in mammalian cell lines. Receptor binding studies with GM-CSF have shown that the receptor affinity decreases with increased glycosylation (113).

Another classical example is recombinant human tissue-plasminogen activator (t-PA). Although the active

enzyme was first derived from *E. coli* cultures, this cell line lacks several desirable biological activities, such as glycosylation ability and the ability to form the correct three-dimensional t-PA structure. Finally, recombinant t-PA was cloned into a Chinese hamster ovary (CHO) cell line. These mammalian cells carried out the glycosylation, disulfide bond formation, and proper folding similar to human cells (114).

Because the large structural diversity of glycans attached to proteins and the fact that each protein is generally associated with a population of different glycan structures, most glycoproteins are produced as a set of glycosylation variants or glycoforms (115). Both pharmacokinetic and pharmacodynamic properties of glycoproteins can be affected by changes in type and degree of glycosylation. A single amino acid mutation in t-PA or the removal of carbohydrate on a single amino acid in t-PA resulted in plasma concentration profiles that were very different from natural t-PA (Fig. 27) (116). In other studies, the plasmin-dependent conversion of single-chain to two-chain t-PA was

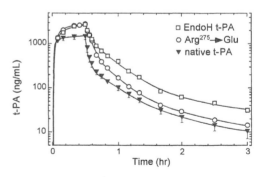

**Fig. 27**   t-PA plasma concentrations after 30-min i.v. infusions of 0.6 mg/kg t-PA in groups of 4 rabbits. The figure shows the marked effect on clearance of a single amino acid mutation (Arg$_{275}$->Glu) or of removal of high mannose carbohydrate at Asn$_{114}$ by the enzyme endoglycosidase H (EndoH t-PA), as compared to native t-PA. [From Tanswell (116).]

influenced by glycosylation at one amino acid (117).

Because of several sugar-recognizing receptors in the liver, the appropriate presentation of carbohydrates on proteins can lead to their rapid elimination from the blood stream (see hepatic metabolism above). In other cases however, sugar substituents (mostly terminal sialic acid moieties) on the protein appeared to protect against hepatic uptake. For example, desialylation of rEPO and rGM-CSF lead to their rapid removal from the circulation and localization to the liver (33), suggesting a protective role of the sugar groups. Removal of the large number of sialic acid residues on EPO exposes the galactose residues that can then be recognized by the hepatic galactose-recognizing asialoglycoprotein receptors.

## C. Chemical Modifications of Protein Therapeutics

Besides the mostly unwanted heterogeneity of protein drugs introduced by the manufacturing process, other chemical modifications of protein and peptide drugs are intentional to obtain molecules with specified characteristics. Variant proteins can be engineered that differ from natural proteins by exchange, deletion, or insertion of single amino acids, or longer sequences up to entire domains. Small changes in the chemical structure of proteins may cause differences in pharmacokinetics and pharmacodynamics. In addition, mutations may affect glycosylation patterns and conformational changes (117), which in turn may affect clearance (Fig. 27) and receptor interactions.

Modification of peptide and protein drugs with the aim of changing the pharmacological activity may at the same time affect the pharmacokinetic behavior of the molecules. In other instances, the increase of duration of response may be

exclusively attributed to a change in the pharmacokinetics such as an increase in residence time. Such modifications include amino acid substitution, deletions and additions, cyclization, drug conjugation, glycosylation or deglycosylation, etc.

The elimination half-life of many peptide and protein drugs is rather small. Consequently, frequent dosing or continuous infusion is necessary to maintain efficacious plasma levels of the drug. Several approaches have been applied to decrease the elimination clearance of biotechnological drugs. One approach is chemical modification such as PEGylation, i.e., the covalent attachment of monomethoxy polyethylene glycol polymer (PEG) to the protein. An example is PEG IL-2, which usually consists of a mixture of rhIL-2 molecules (MW 15 kD) with 1 to 5 or more PEG polymers attached to each molecule at the $\varepsilon$-amino portions of the lysine residues. The production process determines the average number of PEG residues attached, but any process results in a mixture. With each PEG addition, the molecular weight increases with about 7 kD, but because of the attraction of water molecules, the hydrodynamic size increases even more (95–250 kD). Increasing the degree of PEGylation decreases the elimination clearance and the volume of distribution (Fig. 28). Since the elimination clearance usually decreases relatively more than the decrease in volume of distribution, the elimination half-life of PEG IL-2 is longer than for IL-2. Based on the relationship between elimination clearance and effective molecular weight, it is possible to calculate the optimal degree of PEGylation to obtain the desired systemic exposure (118, 119).

The effect of prosthetic sugar groups on elimination and targeting is illustrated by the comparison of the pharmacokinetics of native glucose-oxidase (GO), deglycosylated GO (dGO), and galactosylated GO (gGO) in mice (120). A saturable mechanism was responsible for

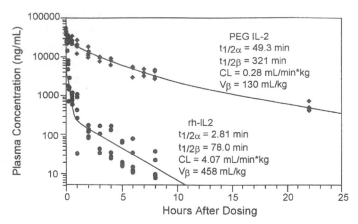

**Fig. 28** Pharmacokinetics of recombinant human interleukin-2 (rhIL-2) and its PEGylation form (PEG IL-2) in rats after i.v. bolus administration of 0.25 mg/kg. The data were described by a linear two-compartmental pharmacokinetic model.

GO and dGO uptake by mononuclear phagocytes, although there was a substantial difference in elimination half-life (10 min. for GO; 100 min. for dGO). In contrast, gGO had a half-life of 4 min. and was taken up preferably by hepatocytes, presumably through hepatic galactose receptors. This is an example where receptor-mediated endocytosis through sugar-recognizing receptors is an efficient hepatic uptake mechanism for glycoproteins. However, when terminal sialic acid residues on the carbohydrate moieties of glycoproteins shield the receptor-binding sugars, hepatic receptor-mediated endocytosis is lower than for the desialylated analogues (24). This has been demonstrated for rEPO and rGM-CSF (33). The protection by sialic residues appears to be a natural mechanism essential for the normal survival of enzymes, acute-phase proteins (such as $\alpha_1$-acid glycoprotein), and most plasma proteins of the immune system.

## REFERENCES

1. VHL Lee. Enzymatic barriers to peptide and protein absorption. CRC Crit Rev Ther Drug Carrier Syst 5: 69–97, 1988.

2. V Bocci. Metabolism of anticancer agents. Pharmacol Ther. 34: 1–49, 1987.

3. V Bocci. Catabolism of therapeutic proteins and peptides with implications for drug delivery. Adv Drug Del Rev 4: 149–169, 1990.

4. BL Ferraiolo, MA Mohler. Goals and analytical methodologies for protein disposition studies. In: BL Ferraiolo, MA Mohler, CA Gloff, eds. Protein pharmacokinetics and metabolism. New York: Plenum Press,1992, pp 1–33.

5. JA Gibbons, Z-P Luo, ER Hannon, RA Braeckman, JD Young. Quantitation of the renal clearance of interleukin-2 using nephrectomized and ureter-ligated rats. J Pharmacol Exp Ther 272: 119–125, 1995.

6. RJ Bauer, JA Gibbons, DP Bell, Z-P Luo, JD Young. Nonlinear pharmacokinetics of recombinant human macrophage colony-stimulating factor (M-CSF) in rats. J Pharmacol Exper Ther. 268(1):152-158, 1994.

7. PJ Collipp, JR Patrick, C Goodheart, SA Kaplan. Distribution of tritium labeled human growth hormone in rats and guinea pigs. Proc Soc Exp Biol Med 121: 173–177, 1966.

8. ALC Wallace, BD Stacy. Horm Metab Res 7: 135–138, 1975.

9. J Mordenti, SC Chen, BL Ferraiolo. Pharmacokinetics of interferon-gamma. In: AHC Kung, RA Baughman, JW Larrick, eds. Therapeutic proteins. Pharmacokinetics and pharmacodynamics. New York: W.H. Freeman and Company, 1992, pp 187–199.

10. Y Takakura, T Fujita, M Hashida, H Sesaki. Disposition characteristics of macromolecules in tumor-bearing mice. Pharm Res 7: 339–346, 1990.

11. UB Kompella, VHL Lee. Pharmacokinetics of peptide and protein drugs. In: VHL Lee, ed. Peptide and protein drug delivery. New York: Marcel Dekker,1991, pp 391–484.

12. R Rabkin, DC Dahl. Renal uptake and disposal of proteins and peptides. In: KL Audus, TJ Raub, eds. Biological barriers to protein delivery. New York: Plenum Press, 1993, pp 299–338.

13. T Maack, CH Park, MJF Camergo. Renal filtration, transport, and metabolism of proteins. In: DW Seldin, G Giebisch, eds. The kidney: Physiology and pathophysiology. New York: Raven Press, 1985, pp 1173–1803.

14. DA Wall, T Maack. Endocytic uptake, transport, and catabolism of proteins by epithelial cells. Am J Physiol 248: C12–C20, 1985.

15. T Maack, V Johnson, ST Kau, J Figueiredo, D Sigulem. Renal filtration, transport, and metabolism of low-molecular weight protein: a review. Kidney Int 16: 251–270, 1979.

16. H Ohnishi, JT Chao, KK Lin, H Lee, TM Chu. Role of the kidney in metabolic change of interleukin-2. Tumor Biol 10: 202–214, 1989.

17. T Maack. Renal handling of low molecular weight proteins. Am J Med 58: 57–64, 1975.

18. V Ganapathy, FH Leibach. Carrier-mediated reabsorption of small peptides in renal proximal tubule. Am J Physiol 25: F945–F953, 1986.

19. FA Carone, DR Peterson. Hydrolysis and transport of small peptides by the proximal tubule. Am J Physiol. 238: F151–F158, 1980.

20. R Rabkin, J Kitaji. Renal metabolism of peptide hormones. Mineral Electrolyte Metab 9: 212–226, 1983.

21. M Hellfritzsch, S Nielsen, EI Christensen, JT Nielsen. Basolateral tubular handling of insulin in the kidney. Contrib Nephrol 68: 86–91, 1988.

22. R Rabkin, MP Ryan, WC Duckworth. The renal metabolism of insulin. Diabetologica 27: 351–357, 1984.

23. A Takagi, Y Yabe, Y Oka, K Sawai, Y Takakura, M Hashida. Renal disposition of recombinant interleukin-11 in the isolated perfused rat kidney. Pharm Res 14(1): 86–90, 1997.

24. DKF Meijer, K Ziegler. Mechanisms for the hepatic clearance of ologopeptides and proteins. In: KL Audus, TJ Raub, eds. Biological barriers to protein delivery. New York: Plenum Press, 1993, pp 339–408.

25. K Ziegler, G Polzin, M Frimmer. Hepatocellular uptake of cyclosporin A by simple diffusion. Biochim Biophys Acta 938: 44–50, 1988.

26. DL Marks, GJ Gores, NF LaRusso. Hepatic processing of peptides. In: MD Taylor, GL Amidon, eds. Peptide-based drug design. Controlling transport and metabolism. Washington, DC: American Chemical Society, 1995, pp 221–248.

27. DC Kim, Y Sugiyama, H Satoh, T Fuwa, T Iga, M Hanano. Kinetic analysis of in vivo receptor-dependent binding of human epidermal growth factor by rat tissues. J Pharm Sci. 77: 200–207, 1988.

28. SJ Burwen, AL Jones. Hepatocellular processing of endocytosed proteins. J Electron Microsc Tech. 14: 140–151, 1990.

29. Y Sugiyama, M Hanano. Receptor-mediated transport of peptide hormones and its importance in the overall hormone disposition in the body. Pharm Res 6: 192–202, 1989.

30. G Ashwell, AG Morell. The role of surface carbohydrates in the hepatic recognition and transport of circulating glycoproteins. Adv Enzymol 41: 99–128, 1974.

31. G Ashwell, J Harford. Carbohydrate-specific receptors of the liver. Ann Rev Biochem 51: 531–554, 1982.

32. RJ Fallon, AL Schwartz. Receptor-mediated delivery of drugs to hepatocytes. Adv Drug Deliv Rev 4: 49–63, 1989.

33. DA Cumming. Glycosylation of recombinant protein therapeutics: control and functional implications. Glycobiology 1(2): 115–130, 1991.

34. DK Strickland, MZ Kounnas, WS Argraves. LDL receptor-related protein: a multiligand receptor for lipoprotein and proteinase catabolism. Faseb J 9(10): 890–898, 1995.

35. MS Nielsen, A Nykjaer, I Warshawsky, AL Schwartz, J Gliemann. Analysis of ligand binding to the $\alpha_2$-macroglobulin receptor / low density lipoprotein receptor-related protein. J Biol Chem 270(40): 23713–223719, 1995.

36. T Murakami, M Misaki, S Masuda, Y Higashi, T Fuwa, N Yata. Dose-dependent plasma clearance of human epidermal growth factor in rats. J Pharm Sci 83(10): 1400–1403, 1994.

37. JMT van Griensven, KJ Burggraaf, J Gerloff, WA Gunzler, H Beier, R Kroon, LGM Huisman, RC Schoemaker, K Kluft, AF Cohen. Effects of changing liver blood flow by exercise and food on kinetics and

dynamics of saruplase. Clin Pharmacol Ther 57: 381–389, 1995.

38. JMT van Griensven, LGM Huisman, T Stuurman, G Dooijewaard, R Kroon, RC Schoemaker, K Kluft, AF Cohen. Effect of increased liver blood flow on the kinetics and dynamics of recombinant tissue-type plasminogen activator. Clin Pharmacol Ther. 60: 504–511, 1996.

39. T Kuwabara, T Uchimura, K Takai, H Kobayashi, S Kaboyashi, Y Sugiyama. Saturable uptake of a recombinant human granulocyte colony-stimulating factor derivative, nartograstim, by the bone marrow and spleen of rats in vivo. J Pharmacol Exper Ther 273: 1114–1122, 1995.

40. A Bartocci, DS Mastrogiannis, G Migliorati, RJ Stockert, AW Wolkoff, ER Stanley. Macrophages specifically regulate the concentration of their own growth factor in the circulation. Proc Natl Acad Sci USA 84: 6179–6183, 1987.

41. H Tanaka, T Kaneko. Pharmacokinetics of recombinant human granulocyte colony-stimulating factor in the rat. Drug Metabol Dispos 19(1): 200–204, 1991.

42. WP Petros, J Rabinowitz, AR Stuart, CJ Gilbert, Y Kanakura, JD Griffin, WP Peters. Disposition of recombinant human granulocite-macrophage colony-stimulating factor in patients receiving high-dose chemotherapy and autologous bone marrow support. Blood 80: 1135–1140, 1992.

43. P Tanswell, G Heinzel, A Greischel, J Krause. Nonlinear pharmacokinetics of tissue-type plasminogen activator in three animal species and isolated perfused rat liver. J Pharmacol Exp Ther 255: 318–324, 1990.

44. RJ Bauer, K Der, N Ottah-Ihejeto, J Barrientos, AHC Kung. The role of liver and kidney on the pharmacokinetics of a recombinant amino terminal fragment of bactericidal/permeability-increasing protein in rats. Pharm Res 14(2): 224–229, 1997.

45. RK Jain, LT Baxter. Extravasation and interstitial transport in tumors. In: KL Audus, TJ Raub, eds. Biological Barriers to Protein Delivery. New York: Plenum Press, 1993, pp 441–465.

46. B Medalion, G Merin, H Aingorn, H-Q Miao, A Nagler, A Elami, R Ishai-Michaeli, I Vlodavski. Endogenous basic fibroblast factor displaced by heparin from the luminal surface of human blood vessels is preferentially sequestered by injured regions of the vessel wall. Circulation 95: 1853–1863, 1997.

47. BQ Shen, GG DeGuzman, TZ Zioncheck. Characterization of vascular endothelial growth factor binding to rat liver sinusoidal cells in vitro. Western Regional Meeting of the American Association of Pharmaceutical Scientists, South San Francisco, 1997.

48. M Narita, G Bu, GM Olins, DA Higuchi, J Herz, GJJ Broze, AL Schwartz. Two receptor systems are involved in the plasma clearance of tissue factor pathway inhibitor in vivo. J Biol Chem 270(42): 24800–24804, 1995.

49. M Yanagishita, VC Hascall. Cell surface heparan sulfate proteoglycans. J Biol Chem 267(9451–9454), 1992.

50. DM Templeton. Proteoglycans in cell regulation. Crit Rev Clin Lab Sci 29(2): 141–184, 1992.

51. J-B Hansen, PM Sandset, K Raanaas Huseby, N-E Huseby, A Nordoy. Depletion of intravascular pools of Tissue Factor Pathway Inhibitor (TFPI) during repeated or continuous intravenous infusion of heparin in man. Thromb Haemost 76(5): 703–709, 1996.

52. PM Sandset, U Abildgaard, ML Larsen. Heparin induces release of extrinsic coagulation pathway inhibitor (EPI). Thrombosis Research 50: 803–813, 1988.

53. HPJ Bennett, C McMartin. Peptide hormones and their analogues: distribution, clearance from the circulation, and inactivation in vivo. Pharm Rev 30: 247–292, 1979.

54. R Braeckman. Pharmacokinetics and pharmacodynamics of peptide and protein drugs. In: DJA Crommelin, RD Sindelar, eds. Pharmaceutical Biotechnology. An Introduction for Pharmacists and Pharmaceutical Scientists. Amsterdam: Harwood Academic Publishers,1997, pp 101–121.

55. A Supersaxo, W Hein, H Gallati, H Steffen. Recombinant human interferon alpha-2a: delivery to lymphoid tissue by selected modes of application. Pharm Res 5(8): 472–476, 1988.

56. A Supersaxo, WR Hein, H Steffen. Effect of molecular weight on the lymphatic absorption of water-soluble compounds following subcutaneous administration. Pharm Res 7(2): 167–169, 1990.

57. M Kato, H Kamiyama, A Okazaki, K Kumaki, Y Kato, Y Sugiyama. Mechanism for the nonlinear pharmacokinetics of erythropoietin in rats. J Pharmacol Exp Ther. 283(2): 520–527, 1997.

58. M Mohler, J Cook, D Lewis, J Moore, D Sinicropi, A Championsmith, B Ferraiolo, J Mordenti. Altered pharmacokinetics of recombinant human deoxyribonuclease in

rats due to the presence of a binding protein. Drug Metabol Dispos. 21(1): 71–75, 1993.

59. MA Mohler, JE Cook, G Baumann. Binding proteins of protein therapeutics. In: BL Ferraiolo, MA Mohler, CA Gloff, eds. Protein pharmacokinetics and metabolism. New York: Plenum Press, 1992, pp 35–71.

60. RC Baxter. Circulating binding proteins for the insulinlike growth factors. Trends Endocrinol Metab 4: 91–96, 1993.

61. DR Clemmons. IGF binding proteins and their functions. Molecular Reproduction and Development 35: 368–375, 1993.

62. RC Baxter, JL Martin. Structure of the Mr 140,000 growth hormone-dependent insulin-like growth factor binding protein complex: determination by reconstitution and affinity labeling. Proc Natl Acad Sci USA 86: 6898–6902, 1989.

63. DR Clemmons, MH Dehoff, WH Busby, ML Bayne, MA Cascieri. Competition for binding to IGFBP-2, 3, 4 and 5 by the insulin-like growth factors and IGF analogs. Endocrinology 132: 890–895, 1992.

64. KL Cohen, SP Nissley. The serum half-life of somatomedin activity: Evidence for growth hormone dependence. Acta Endocrinol 83: 243–258, 1976.

65. J Zapf, C Hauri, M Waldvogel, ER Froesch. Acute metabolic effects and half-lives of intravenous insulin-like growth factor I and II in normal and hypophysectomized rats. J Clin Invest 77: 1768–1775, 1986.

66. DW Leung. Growth hormone receptor and serum binding: Purification, clonong and expression. Nature. 330: 537–543, 1987.

67. AC Herington, S Ymer, J Stevenson. Identification and characterization of specific binding proteins for growth hormone in normal human sera. J Clin Invest 77: 1817–1823, 1986.

68. G Baumann, K Amburn, MA Shaw. The circulating growth hormone (GH)-binding protein complex: A major constituent of plasma GH in man. Endocrinology 122: 976–984, 1988.

69. G Baumann, MA Shaw, TA Buchanan. In vivo kinetics of a covalent growth hormone-binding protein complex. Metabolism 38: 330–333, 1988.

70. S Rose-John, PC Heinrich. Soluble receptors for cytokines and growth factors: generation and biological function. Biochem J. 300: 281–290, 1994.

71. T Taniguchi, Y Minami. The IL-2/IL-2 receptor system: a current overview. Cell 73: 5–8, 1993.

72. SD Voss, JA Hank, CA Nobis, P Fisch, JA Sosman, PM Sondel. Serum levels of the low-affinity interleukin-2 receptor molecule (TAC) during IL-2 therapy reflect systemic lymphoid mass activation. Cancer Immunol Immunother 29: 261–269, 1989.

73. KA Quinn, PG Grimsley, Y-P Dai, M Tapner, CN Chesterman, DO Owensby. Soluble low sensity lipoprotein receptor-related protein (LRP) circulates in human plasma. J Biol Chem 272: 23946–23951, 1997.

74. M Gibaldi, D Perrier. Pharmacokinetics. Second ed. New York: Marcel Dekker, Inc, 1982.

75. DWA Bourne. Mathematical Modeling of Pharmacokinetic Data, ed. Lancaster, PA: Technomic Publishing Company, Inc, 1995.

76. H Sato, Y Sugiyama, Y Sawada, T Iga, M Hanano. Physiologically based pharmacokinetics of radioiodinated human beta-endorphin in rats. An application of the capillary-limited model. Drug Metab Dispos 15(4): 540–550, 1987.

77. RL Dedrick. Animal Scale-Up. J Pharmacokinet Biopharm 1(5): 435–461, 1973.

78. H Boxenbaum. Time Concepts in Physics, Biology and Pharmacokinetics. J Pharm Sci 75(11): 1053–1062, 1986.

79. J Mordenti, SA Chen, JA Moore, BL Ferraiolo, JD Green. Interspecies Scaling of Clearance and Volume of Distribution Data for Five Therapeutic Proteins. Pharm Res 8(11): 1351–1359, 1991.

80. P Girard, P Nony, JP Boissel. The place of simultaneous pharmacokinetic pharmacodynamic modeling in new drug development: trends and perspectives. Fundam Clin Pharmacol 4(Suppl 2): 103s–115s, 1990.

81. WA Colburn, JW Blue. Using pharmacokinetics and pharmacodynamics to direct pharmaceutical research and development. Applied Clinical Trials 1(1): 42–46, 1992.

82. TL Schwinghammer, PD Kroboth. Basic Concepts in Pharmacodynamic Modeling. J Clin Pharmacol 28: 388–394, 1988.

83. NHG Holford. Concepts and usefulness of pharmacokinetic-pharmacodynamic modelling. Fundam Clin Pharmacol 4(2): 93s–101s, 1990.

84. H Derendorf, G Hochhaus. Handbook of Pharmacokinetic/Pharmacodynamic Correlation. London: CRC Press, 1995.

85. PD Kroboth, VD Schmith, RB Smith. Pharmacodynamic Modelling—Application to New Drug Development. Clinical Pharmacokinetics 20(2): 91–98, 1991.

86. J-L Steimer, M-E Ebelin, J Van Bree. Pharmacokinetic and Pharmacodynamic data and models in clinical trials. Euro-

pean Journal of Drug Metabolism and Pharmacokinetics 18(1): 61–76, 1993.

87. WA Colburn. Peptide, peptoid, and protein pharmacokinetics/pharmacodynamics. In: PD Garzone, WA Colburn, M Mokotoff, eds. Peptides, peptoids, and proteins. Pharmacokinetics and pharmacodynamics. Cincinnati: Harvey Whitney Books, 1991, pp 93–115.

88. T Kenakin. Drug-Receptor Theory. In: T Kenakin, ed. Pharmacologic Analysis of Drug-Receptor Interaction. New York: Raven Press, 1993, pp 1–38.

89. NHG Holford, LB Sheiner. Pharmacokinetic and Pharmacodynamic Modeling in Vivo. Crit Rev Bioengin 5: 273–322, 1981.

90. A Childs, J Grevel, DA Baron, DL Reynolds, RD McCabe, EG Burton, DE Johnson, RA Braeckman. Population PK/PD of tissue factor pathway inhibitor (TFPI) in cynomolgus monkeys during a 12-day toxicology study. Fund Appl Toxicol 30(1): 104, 1996.

91. NHG Holford, LB Sheiner. Understanding the Dose-Effect Relationship: Clinical Application of Pharmacokinetic-Pharmacodynamic Models. Clinical Pharmacokinetics 6: 429–453, 1981.

92. S Hooper. Pharmacokinetics and Pharmacodynamics of Intravenous Regular Human Insulin. In: PD Garzone, WA Colburn, M Mokotoff, eds. Pharmacokinetics and Pharmacodynamics. Peptides, Peptoids, and Proteins. Cincinnati, OH: Harvey Whitney Books, 1991, pp 128–137.

93. JR Woodworth, DC Howey, RR Bowsher. Establishment of time-action profiles for regular and NPH insulin using pharmacodynamic modeling. Diabetes Care 17(1): 64–69, 1994.

94. R Nagashima, RA O'Reilly, G Levy. Kinetics of pharmacologic effects in man: The anticoagulant action of warfarin. Clin Pharmacol Ther 10(1): 22–35, 1968.

95. NL Dayneka, V Garg, WJ Jusko. Comparison of Four Basic Models of Indirect Pharmacodynamic Responses. J Pharmacokinet Biopharm 21(4): 457–478, 1993.

96. RA O'Reilly, G Levy. Kinetics of the anticoagulant effect of bishydroxycoumarin in man. Clin Pharmacol Ther 11(3): 378–384, 1970.

97. WJ Jusko, HC Ko. Pharmacodynamics and Drug Action. Physiologic Indirect Response Models Characterize Diverse Types of Pharmacodynamic Effects. Clin Pharmacol Ther 56(4): 406–419, 1994.

98. SC Piscitelli, A Forrest, S Vogel, J Metcalf, M Baseler, R Stevens, JA Kovacs. A novel PK/PD model for infused interleukin-2

(IL-2), in HIV-infected patients. Ninety-Seventh Annual Meeting of the American Society for Clinical Pharmacology and Therapeutics, Florida: Lake Buena Vista, 1996, pp 152.

99. DE Uehlinger, FA Gotch, LB Sheiner. A pharmacodynamic model of erythropoietin therapy for uremic anemia. Clin Pharmacol Ther 51: 76–89, 1992.

100. S Einhorn, A Ahre, H Blomgren, B Johansson, H Mellstedt, H Strander. Interferon and natural killer activity in multiple myeloma. Lack of correlation between interferon-induced enhancement of natural killer activity and clinical response to human interferon-α. Int J Cancer 30(1): 167–172, 1982.

101. BA Omar, SC Flores, JM McCord. Superoxide dismutase. In: AHC Kung, RA Baughman, JW Larrick, eds. Therapeutic Proteins. Pharmacokinetics and Pharmacodynamics. New York: W.H. Freeman and Company, 1992, pp 295–315.

102. RL Krigel, KA Padavic-Shaller, AR Rudolph, S Litwin, M Konrad, EC Bradley, RL Comis. A Phase I study of recombinant interleukin-2 plus recombinant β-interferon. Cancer Res 48: 3875–3881, 1988.

103. AP Larocca, SC Leung, SG Marcus, CB Colby, EC Borden. Evaluation of neutralizing antibodies in patients treated with recombinant interferon-$\beta_{ser}$. J Interferon Res 9(Suppl. 1): S51–S60, 1989.

104. M Konrad, A Childs, T Merigan, E Bordon. Assessment of the antigenic response in humans to a recombinant mutant interferon beta. J Clin Immunol 7: 365–375, 1987.

105. GM Brown, GR Van Loon, CW Hummel, LJ Grota, A Arimura, AV Schally. Characteristics of antiboy produced during chronic treatment with LHRH. J Clin Endocrinol Metab 44: 748–790, 1977.

106. PK Working. Potential effects of antibody induction by protein drugs. In: BL Ferraiolo, MA Mohler, CA Gloff, eds. Protein pharmacokinetics and metabolism. New York: Plenum Press, 1992, pp 73–92.

107. MG Rosenblum, BW Unger, JU Gutterman, EM Hersh, GS David, JM Fincke. Modification of human leucocyte interferon pharmacology with monoclonal antibody. Cancer Res 45: 2421–2424, 1985.

108. S Sell. Immunology, immunopathology and immunity. 4th ed. Amsterdam: Elsevier, 1987.

109. FF Davis, GM Kazo, ML Nucci, A Abuchowski. Reduction of immunogenicity and extension of circulatinh half-life of peptides and proteins. In: VHL Lee, ed.

Peptide and Protein Drug Delivery. New York: Marcel Dekker, Inc., 1991, pp 831–864.

110. E Rinderknecht, BH O'Connor, H Rodriguez. Natural human interferon-gamma: complete amino acid sequence and dtermination of sites of glycosylation. J Biol Chem 259: 6790–6797, 1984.

111. E Rinderknecht, LE Burton. Biochemical characterization of natural and recombinant IFN-gamma. In: H Kirchner, H Schellenkens, eds. The biology of the interferon system 1984. Amsterdam: Elsevier, 1985, pp 397–402.

112. Y-CE Pan, AS Stern, PC Familletti, FR Khan, R Chizzonite. Stuctural characterization of human interferon gamma. Eur J Biochem 166: 145–149, 1987.

113. JB Stoudemire. Pharmacokinetics and metabolism of hematopoietic proteins. In: BL Ferraiolo, MA Mohler, CA Gloff, eds. Protein pharmacokinetics and metabolism. New York: Plenum Press, 1992, pp 189–222.

114. JR Ogez, R van Reis, N Paoni, SE Builder. Recombinant human tissue-plasminogen activator: biochemistry, pharmacology, and process development. In: PD Garzone, WA Colburn, M Mokotoff, eds. Peptides, peptoids, and proteins. Cincinnati: Harvey Whitney Books, 1991, pp 170–188.

115. RB Parekh. Effects of glycosylation on protein function. Curr Opin Struct Biol 1: 750–754, 1991.

116. P Tanswell. Tissue-type plasminogen activator. In: AHC Kung, RA Baughman, JW Larrick, eds. Therapeutic proteins. Pharmacokinetics and pharmacodynamics. New York: W.H. Freeman and Company, 1992, pp 255–281.

117. A Wittwer, SC Howard. Glycosylation at Asn-184 inhibits the conversion of single-chain tissue-type plasminogen activator by plasmin. Biochemistry 29: 4175–4180, 1990.

118. MJ Knauf, DP Bell, P Hirtzer, Z-P Luo, JD Young, NV Katre. Relationship of effective molecular size to systemic clearance in rats of recombinant interleukin-2 chemically modified with water-soluble polymers. J Biol Chem 263(29): 15064–15070, 1988.

119. RJ Bauer, JL Winkelhake, JD Young, RZ Zimmerman. Protein drug delivery by programmed pump infusion: Interleukin-2. In: AHC Kung, RA Baughman, JW Larrick, eds. Therapeutic proteins. Pharmacokinetics and pharmacodynamics. New York: W.H. Freeman and Company, 1992, pp 239–253.

120. S Demignot, D Domurado. Effect of prosthetic sugar groups on the pharmacokinetics of glucose-oxidase. Drug Des Del 1: 333–348, 1987.

121. M Rowland, TN Tozer. Clinical Pharmacokinetics. Concepts and Applications ed. Baltimore: Williams & Wilkins, 1995.

122. AE Taylor, GN Granger. In: EM Renkin, CC Michel, eds. Handbook of physiology. Bethesda, MD: American Physiological Society, 1984, pp 467–520.

123. A Takagi, H Masuda, Y Takakura, M Hashida. Disposition characteristics of recombinant human interleukin-11 after a bolus intravenous administration in mice. J Pharmacol Exper Ther 275(2): 537–543, 1995.

124. WJ Arendshorst, LG Navar. Renal circulation and glomerular hemodynamics. In: RW Schrier, CW Gottschalk, eds. Diseases of the kidney. Boston: Little, Brown, 1988, pp 65–117.

125. RK Jain. Transport of molecules across tumor vasculature. Cancer Metastasis Rev 6: 559–594, 1987.

126. G Baumann, MW Stolar, K Amburn, CP Parsano, BC DeVries. A specific growth hormone-binding protein in human plasma: Initial characterization. J Clin Endocrinol Metab 62: 134–141, 1986.

# 21

# Approaches to Validating the Identity of Peptides and Proteins

**Ian R. Tebbett and Nancy D. Denslow**
*University of Florida, Gainesville, Florida*

## I. INTRODUCTION

Interest in the use of biologically active proteins and peptides as potential therapeutic agents has grown dramatically in recent years. The synthesis of these compounds has now become fairly routine, and many laboratories are able to provide custom made peptides for investigational use. Before any meaningful study can be performed however, it is first necessary to confirm the identity of the peptide or protein, determine its purity, and evaluate its stability. These concerns are often overlooked in the rush to test for pharmacological activity. A standardized approach to method validation however, can result in the saving of an inordinate amount of time in the long run. This chapter presents a discussion of the sources of contamination which can be encountered with synthetic peptides and proteins, the methods used for the detection of such contaminants, and those factors which should be considered when validating such a method.

### A. Synthetic Peptides

The advent of solid phase peptide synthesis (1), made it possible to design instruments to synthesize peptides semi-automatically. The instruments start out with the C-terminal amino acid attached to an insoluble polymeric support and perform subsequent operations on the growing peptide linked to the resin. A variety of supports and chemistries have evolved.

A major challenge these methods confront is dealing with contaminants that accompany a peptide. Before discussing contaminants, it is important to step through the process. The C-terminal amino acid is linked to the insoluble support through a cleavable ester or amide bond. The first step is to remove the amino protecting group so it can react with the carboxy terminus of the next incoming amino acid. In most instances dicyclohexylcarbodiimide or HBTu (Benzotriazol-yl-tetramethylamonium hexafluorophosphate) are used as condensing agents. Excess reagents and by-products are removed by several solvent washes. This procedure continues until all the desired amino acids are added. Finally, the peptide is removed from the resin and the protecting groups are removed from all the side chain functional groups, with an acid treatment.

The most popular chemistries include the use of tert-butyloxycarbonyl- amino acids (t-Boc) and more recently, the 9-fluorenylmethyloxycarbonyl (Fmoc) (2). These methods vary in how they deprotect the amino acids prior to con-densation. The t-Boc method requires a two-step protocol, using TFA to remove the t-Boc protecting group and a strong acid to cleave the peptide from the resin which will also deprotect the benzyl-based blocking groups on the side chains. The Fmoc chemistry, on the other hand, uses base to remove the blocking group. Because the Fmoc chemistry is sensitive to base and resists acid, it allows the use of acid sensitive anchors between the C-terminal amino acid and the resin, and the use of acid-sensitive blocking groups for the side chains of amino acids. Thus, only at the end of the synthesis is TFA used to remove the synthetic peptide from the resin and to deprotect all the R groups. The simplicity of this technique has stimulated improve-ments at an unprecedented pace.

Peptide synthesis is not always per-fect. Undesirable contaminants in the final product may result from a number of problems, including the presence of some sequences missing an internal amino acid, or truncated sequences. These problems arise from situations in which the reactants may not reach all of the functional groups and cause incomplete coupling. The pro-tecting group on the N-terminus may not be fully removed, leaving a portion of the growing peptide unable to form a peptide bond with the incoming amino acid. When this happens, some of the growing strands do not incorporate the next amino acid but can still be activated and coupled in the subsequent steps. These problems may be corrected by increasing the time allowed for the reaction or by applying the reagents in considerable excess. The presence of failure sequences in the synthetic material creates serious problems in purification, since the properties of such contaminants may be quite similar to those of the target compound.

Other problems that arise occasion-ally are racemization of certain amino acids, cyclization of β-benzylaspartyl residues: pyroglutamyl formation of an

N-terminal glutamyl residue, alkylation of tyrosine, tryptophan, and methionine; and oxidation of certain amino acids, such as cysteine or methionine. Most of these side reactions can be controlled by additives and scavengers of one sort or another. For example, alkylation of tyrosine or trypto-phan can be controlled by adding phenol, water, thioanisole and 1,2-ethanedithiol. It is prudent to be aware of these modification possibilities in order to set up adequate analytical methods to detect them.

## 1.   Analytical HPLC

High performance liquid chromatography is a popular analytical method to qualify (or purify) a synthetic peptide because of its exquisite sensitivity, speed and resolving power. By far the most common method is reverse phase, usually using columns of octadecyl (C18)-bonded silica. In some circumstances, however, it is beneficial to use ion exchange chromatography or size-exclusion as a second dimension. Reverse phase chromatography is usually carried out at pH 2–3 with TFA as a counter-ion. The peptide is eluted from the column with an acetonitrile gradient.

Usually the major peak observed in the chromatogram is the peptide of interest. The purity is determined by integrating the main and side peaks observed and calculating the main peak's percentage of the total area. This is accep-table since, in general, the impurities are of peptide origin with a comparable absorb-ance at the detection wavelength. Each of the peaks should be collected for further analysis by other methods. In the case of a severely flawed synthesis, the main peak may not be the desired peptide, but an altered form or failed product. If the target peptide is not resolved from the con-taminants sufficiently by HPLC, several parameters can be altered and optimized to increase the resolution: gradient con-ditions, organic modifier, counter ion, temperature, column pore size and particle size, solvent composition and flow rate. It is also possible to change from a gradient to an isocratic procedure (a constant unchanging concentration of organic modifier) close to the time the target peptide elutes from the column. This tactic is helpful for separating closely eluting contaminants from the product of interest by allowing a gradual separation. Generally, optimizing these parameters results in excellent separations.

Preparative HPLC purification of synthetic peptides can be performed on larger columns, up to 1 inch in diameter, by changing the flow rate in proportion to the cross-sectional area of the larger column. For the best separations, the column should not be overloaded, and multiple fractions should be taken across the peak. These should be analyzed by mass spectrometry to determine which fractions can be pooled.

## 2. Capillary Electrophoresis

Capillary electrophoresis (CE) is emerging as a good orthogonal separation method to reverse phase chromatography for synthetic peptides because it depends on different chemical and physical properties. Contaminants that might not have been resolved by reverse phase may be resolved by CE. CE is also attractive because the separations are relatively rapid and consume very little sample. The technique usually matches HPLC in its quantitative precision and accuracy as well as separation reproducibility. CE is a good method to employ on samples that have already been purified by reverse phase to ensure that the pure peak is indeed free of contaminants.

The separation by CE is based on the net charge of the peptide; the rate of migration is a function of the magnitude of the charge, the mass of the peptide, its Stoke's radius and its intrinsic viscosity (3). CE is typically performed in capillary tubes of 25–100 mm inner diameter and 50–100 cm in length, filled with an electrolyte, the composition of which is dictated by the pH of the separation. The resolution is affected by the applied voltage, capillary length and diameter, buffer and sample introduction. Of these properties, charge is the most amenable to systematic experimental manipulation. This property can be changed over a wide range since the charge of a peptide is a function of the pH of the buffer. For best peptide characterization, two consecutive runs should be carried out, one basic and one acidic. The basic run should be at pH 10.5, because most peptides tend to be soluble at high pH and because all sample components will pass through the detector fairly quickly. The acidic run should be at pH 2.0, since the very different charge selectivity will separate any pH 10.5 comigrations.

CE differs from other electrophoretic techniques in the contribution of electroendosmosis to the separation (4). This is the bulk flow of electrolyte caused by the surface charge on the wall of the capillary . Since the electroosmotic flow rate usually exceeds the electrophoretic mobility, both positively and negatively charged species migrate toward the cathode. Consequently, both positively and negatively charged peptides can be analyzed in a single run. If separation is incomplete, additional parameters can be modified. Adding 10–20% acetonitrile, for example, increases solubility and may change the separation. Additives such as zwitterions, ion-pairing reagents, chaotropic agents and detergents can be used.

## 3. Amino Acid Analysis (AAA) and Sequencing by Edman Chemistry

The best way to confirm that the correct set of amino acids has been incorporated into the target product and to help troubleshoot a synthesis that may contain some molecules with deleted internal amino acids or truncated peptides, is Amino Acid Analysis (AAA). To determine that the correct sequence of amino acids has been incorporated into the peptide calls, on the other hand, for Edman Chemistry sequencing. The proverbial nightmare is a peptide that is built by mistake in reverse order, which can happen if a novice is involved. In this circumstance, all other methods of quality control would show a single peptide product of the correct composition, hydrophobicity and charge. Edman Chemistry sequencing is not advised as a general quality control method, however, because it is expensive and time consuming.

Amino acid composition can be determined at several steps during the synthesis, and, if necessary, can be used to monitor the building of peptides. This is useful if a particular step in the synthesis is in doubt. The protected and support-bound peptide can be hydrolyzed by the normal procedures and analyzed by ion-exchange chromatography using post-column derivatization by ninhydrin. Standards should be incorporated into the analysis to match the expected amino acids.

AAA is best used to check the main peak plus all the side peaks collected from an analytical reverse phase HPLC run. Because the analysis is both rapid and relatively cheap, it is easy to determine with certainty whether the major peak is the target peptide and to understand the nature of the side peaks. Being quantitative, AAA also aids in establishing the correct concentration of a peptide solution.

## 4. Mass Spectrometry

Recent developments in desorption ionization techniques and instrumentation have made mass spectrometry a complementary technology to evaluate the purity and composition of synthetic peptides. The most widely used methods involve ionizing peptides by electrospray (ESI) or by matrix-assisted laser desorption ionization (MALDI), although fast atom bombardment and plasma desorption methods are still in use. The newest techniques require only femtomole to picomole concentrations of peptide and can be performed in minutes with excellent results. It is now possible to directly analyze the effluent from an HPLC or CE separation in real time by LC/MS using electrospray ionization coupled to a quadrupole or ion trap mass analyzer. Alternatively, it is possible to spot 1 ml aliquots from collected peaks in the presence of an organic matrix such as α-cyano-4-hydroxycinnamic acid for analysis by MALDI TOF (Matrix-assisted laser-desorption ionization Time-of- Flight) mass spectrometry. Another problem associated with MS analysis for peptides has been raising the mass limit. Newer machines however are capable of handling molecules exceeding 150,000 mass units.

The application of mass spectrometry to peptides and proteins is well documented and reviewed (5–7). Every analytical laboratory either has a mass spectrometer available within the facility or has access to one, making these determinations routine for quality control. It is now possible to monitor the removal of all blocking groups from the final product and to determine whether the product contains failed or truncated sequences simply by measuring the mass and comparing it to the mass predicted by the amino acid composition. It is also possible to detect oxidation, amino acid cyclizations and alkylations. Mass spectrometry is the method of choice to detect the presence of an impurity that might have copurified with the peptide.

In addition to measuring the mass, it is possible to use mass spectrometry to obtain the sequence of the peptide by post-source decay using MALDI TOF equipped with an ion reflector or by collision induced fragmentation in triple quadrupole or ion trap instruments. These techniques have become rather sophisticated in the last few years, with the possibility of doing MS to the nth power using the new commercially available ion trap instruments. It is now possible to determine the full identity of a contaminant with these techniques.

The art of peptide synthesis continues to develop with new applications for synthetic biologically active peptides and inhibitors. Fortunately, locked in step with the synthesis are new methods to analyze the structure of the products and determine the presence of contaminants.

## B. Proteins

Purifying proteins, especially from animal tissues, has always been difficult, requiring substantial time and resources. New developments, however, in particular the use of affinity columns and gel electrophoresis, eliminate a number of less specific purification steps used in the past, thereby accelerating the process. It is still challenging to purify a protein to complete homogeneity, removing all contaminants, proteinaceous and non-proteinaceous. How clean a protein needs to be depends on how it will be used. Therapeutic uses require exhaustive testing to ensure that no foreign material accompanies the protein. Even a minor contaminant can lead to side reactions that may be fatal. In this case, it is critical to evaluate the protein not only for its physical impurities, but also for its potential to cause immunogenic or allergic responses (8). Exhaustive guidelines for validating the purity of such proteins are in place (9–11). If the protein is to be used strictly for research, absolute purity may not be required.

### 1. Common Sources of Contamination

### a. Contaminating Proteins

It is critical to determine that the sample is free of contaminating proteins. The most difficult to detect and eliminate are those of similar molecular weight, pI and hydrophobicity. The best way to find such

impurities is by HPLC, using at least two complementary chromatographic separations (for example reverse phase on a C4 column and ion-exchange, or using two types of ion pairing agents in reverse phase) with the aid of a diode array detector. Another powerful method involves the use of capillary electrophoresis (CE), a separation method that gives very high resolution because of the high number of theoretical plates. This technique, normally requiring only minute amounts of protein ($<1$ pmol), separates proteins based on their isoelectric properties. Finally, high resolution peptide mapping by HPLC alone or in combination with electrospray mass spectrometry (5, 12) or in combination with MALDI TOF mass spectrometry (13), offers a quick and precise method for characterizing proteins. If authentic pure protein is available for comparison, proteolytic digests of both the authentic and the recombinant protein under the same conditions should generate the same fragmentation pattern. Differences are interpreted as contaminants.

New techniques with mass spectrometry remove the requirement for pure authentic protein, because the mass assignments for each proteolytic fragment of the recombinant protein are accurate enough on their own to determine if the fragments are all derived from the parent protein. Any fragments that do not fit may belong to a contaminant or indicate incorrect primary sequence, or a post-translational modification.

### b. Proteolytically Clipped Proteins

Many proteins become proteolytically cleaved during the purification scheme, even in the presence of protease inhibitor cocktails, due to the ubiquitous distribution of proteolytic agents in cell lysates. In some circumstances, it may be difficult to separate the fragments from the product of interest, especially when the protein is nibbled from one end or the other by exoproteases. The nibbled fragments are almost the same size and have nearly identical physical properties. This task becomes particularly difficult when the contamination is below normal detection levels. For proteins used therapeutically, this problem may have major ramification. Degraded proteins, or proteins that are not folded correctly, may be more antigenic than their intact counterparts, causing the patient to produce antibodies to the protein itself. Assuring purity from this type of contamination may require determining the N- and C-terminal sequences.

### 2. Physical Characterization of Purified Proteins

### a. Checking the Primary Structure of Proteins

Once the purity of the protein has been determined, it is imperative to prove that the isolated protein indeed has the correct amino acid sequence. There are a number of analytical techniques that can be used to characterize the protein, each providing different but overlapping answers. The more exacting techniques generally require expensive instruments such as HPLCs or mass spectrometers. Other techniques are less precise, but provide reasonable data if authentic protein is available as a control.

The first approach is to determine the mass of the protein, especially as it compares to the calculated theoretical mass, if this is available. This can be accomplished in most laboratories by SDS PAGE. The mass determined in this manner is accurate to about ±5% of the real mass. A more precise measure of the mass can be obtained by mass spectrometry, either electrospray or matrix- assisted laser-desorp tion ionization time- of-flight (MALDI TOF). These measurements are generally within 0.01–0.05% of the correct mass.

Next, it is useful to determine the pI of the recombinant protein. This can be accomplished by isoelectric focusing (IEF) gel electrophoresis; using flat bed agarose or acrylamide gels or slab gels. The pI is determined from a standard curve constructed using IEF standards, proteins of known pI. If the native protein is available, it is good practice to include it in an adjacent lane. Differences in pI from the expected value are indicative of differences in charged amino acids.

A sample of the protein (1–5 ug) should be analyzed for its amino acid composition. This indicates whether all the expected amino acids are present. A second sample (25–50 pmol) should be submitted for N-terminal amino acid sequencing. This sample can be provided on a PVDF membrane, electroblotted from an SDS gel, or if free of salts, in solution. This analysis verifies that the protein starts with the expected sequence, an important consideration if the N-terminal Met is cleaved after translation. A blocked N-terminus, as occurs when the terminal amino acid is post-translationally modified, requires special procedures to be identified (14). If enough protein is available (>500 pmol), it may be possible to obtain sequence information from the C-terminus as well. While this method requires more material and only provides information on 4–5 amino acids, it does confirm that the intact protein has been isolated.

The most important consideration in characterizing a protein is to show that its primary amino acid sequence matches the predicted translation of the DNA coding sequence. In theory, one can simply digest the protein with an enzyme or chemical agent that is highly specific and compare the resulting fragments with those expected. There are computer programs that predict fragmentation patterns for proteins with all of the commonly used enzymes or chemical agents. In practice, however, the comparison can be a little tricky because often digestions do not go to completion, or peptide fragments behave differently on HPLC than predicted due to slight variations in technique.

High resolution peptide maps (fingerprinting) provide information on internal sequences of the purified protein. The best diagnostic fingerprints are analyzed by HPLC, where retention times of over a hundred different fragments can often be compared. This method is particularly effective if coupled to mass spectrometry where each HPLC peak can be identified by its mass and compared to the expected theoretical masses. It is highly likely that a difference in amino acid sequence between a newly purified protein and authentic protein would change the mobility of at least one fragment by HPLC and it should certainly be discernible by mass spectrometry.

On occasion, SDS PAGE must be used as the final purification step. In this case, it is possible to digest the protein in a gel slice, extract the fragments and analyze them by HPLC and mass spectrometry. It is now possible to start with 30–50 pmol of protein in a gel slice and get a reasonably good separation by HPLC that is amenable to further analysis by either Edman chemistry sequencing or mass spetrometry.

### b. Post-translational Modifications

Among the most frequent post-translational modifications are phosphorylations, N-methylation, acetylation, glycosylation, farnesylation, and sialic acid capping of oligosaccharides. Pinpointing these modifications using traditional approaches can be time--consuming, requiring specific methods for each modification. The usual method involves digestion of the protein, separation of the fragments by HPLC, and further analysis of the fragments by N-terminal sequencing using Edman Chemistry. A major disadvantage to this approach is that some of the modifications are not stable when exposed to the chemicals used in sequencing, and are missed. For example, phosphorylated Ser and Thr residues are not easily detected because they undergo $\beta$ elimination of the phosphoryl group (15). Instead, a system using radioactive tracers has been developed in which the assignment of the phosphorylation sites is based on the release of $[^{32}P]Pi$ during Edman degradation (16). Among the most problematic modifications are unwanted, non-specific modifications including the oxidation of Met residues, or the chance deamidation of Asn or Gln. These modifications are best detected by careful analysis of the fragments by mass spectrometry. The presence of reducing agents such as 2-mercaptoethanol or dithiothreitol (DTT) in the isolation buffer often prevents oxidation of Met.

### c. Activity

Finally, the biological activity of the isolated protein should be studied carefully. When possible, the activity should be compared to the authentic protein. Lower activity is often an indicator that the protein has not folded correctly, or that it has been improperly post-translationally modified. A low specific activity indicates the presence of other proteinaceous contaminants.

## II. VALIDATION OF ANALYTICAL PROCEDURES: METHODOLOGY

Having decided upon an analytical method to verify the identity and purity of a peptide or protein, that method must in turn also be validated. There are many approaches which may be taken to validate an analytical procedure. This is particularly true of analytical methods for biotechnological products, this being a relatively new and quickly developing field. However the main objective of method validation is to demonstrate that the procedure is adequate for its intended use, and some general guidelines can be described for this purpose. The following is based on the FDA's guidelines for Good

Laboratory Practice as it applies to the analysis for proteins and peptides.

An analytical procedure can generally be considered to include the following components; extraction of the test compound from a matrix, construction of a standard curve, quantitation of the test article, and quality control of the analytical process. The following are key parameters which should be considered when validating such a process:

A.  The purity and identity of the neat test sample.
B.  Homogeneity of solutions.
C.  Extraction recovery.
D.  Range of standard curve and control samples.
E.  Linearity and Precision of standard curves.
F.  Sensitivity.
G.  Specificity.
H.  Accuracy and Precision.
I.  Stability.
J.  Quality control.

## A.  Determination of Identity and Purity of Test Compound

An analytical procedure should be validated using well-characterized reference materials as standards. Although the degree of purity necessary depends on the intended use of the standard, it is imperative that the purity is known since this will have an impact on the determination of sensitivity, specificity and accuracy of the method. The purity and identity of the neat test compound should be determined at the start and end of the project. Chromatographic purity is typically determined by high performance liquid chromatography (HPLC), and analysis of the sample should show the presence of a single major peak in the chromatogram with a retention time corresponding to that of previously injected standards. The identity of the compound is generally verified by means of mass spectrometry. Ideally, the full mass spectrum of the sample should be compared with that of an injected standard.

## B.  Homogeneity of Solutions

When the reference standard consists of a suspension or even a solution of the compound in a vehicle, it is important to check the homogeneity of the preparation before using it for quantitative analysis. An aliquot should be taken from the top, middle, and bottom of the sample container. Each aliquot is then analyzed in triplicate by the analytical method and the mean concentration used for quantitative work. Concentrations throughout the mixture should fall within a previously determined acceptable range for the sample to be considered as homogenous.

## C.  Extraction Recovery

Analytical procedures often require some degree of sample preparation prior to analysis. Examples include extraction from an aqueous matrix, derivatization, and concentration. The efficiency of such procedures must be determined before any quantitative analysis can be performed. Extraction recovery of a drug or metabolite is determined by comparison of the peak area observed for a nonextracted standard solution in an appropriate solvent injected directly into the chromatographic system, to the peak area observed with a solution prepared at the same concentration in the appropriate matrix, extracted and injected into the chromatographic system. The recovery procedure is performed at a minimum of three concentrations with several replicates at each concentration.

## D.  Range of Standard Curve and Control Samples

The acceptable range of the assay is normally derived from linearity studies and depends on the intended application of the procedure. It is established by confirming that the analytical procedure provides an acceptable degree of linearity, accuracy, and precision when applied to samples containing amounts of analyte within or at the extremes of the specified range of the analytical procedure. The standard curve is constructed so that the maximum calibrator is $\geq 20\%$ of the expected maximum concentration and the minimum calibrator is at least $10\% >$ than the minimum quantifiable value. The minimum quantifiable value is supported by chromatographic data and depends largely on the degree of background noise in the chromatogram (see below). The number of calibrators used to construct a standard curve is typically four to seven depending on the range of concentrations to be quantified. Additionally a minimum of three control samples at low, intermediate and high concentration values should be prepared independent of the analyst performing the assay procedure. The high control sample is approximately 10% less than the maximum standard curve calibrator and the low control sample is approximately 10% greater than the minimum standard curve calibrator. Following the initial development of the chromatographic system and extraction procedure, sufficient volumes of the standard curve and control samples are prepared in drug free, pooled serum (or other biological matrix as required) to allow use in the remainder of the study. Each calibrator and control is divided into individual aliquots as determined by the requirements of the assay and stored frozen (at either $-20$ or $-80°C$ dependent on sample stability) until needed. Standard curve calibrators and control samples are prepared for each compound of interest (drug and/or metabolites) in each biological matrix to be analyzed during the study. Dependent on the assay specificity studies, it may be necessary to prepare separate standard curve and control samples for each animal species.

## E. Linearity and Precision of Standard Curves

Having determined the range of the analytical procedure, the linear relationship between concentration and response should be demonstrated across this range. If there is a linear relationship, test results should be evaluated by appropriate statistical methods, for example, by calculation of a least squares regression line. A minimum of five concentrations is recommended to establish linearity. The standard curve is determined from the calibrators by a linear least squares fit to the equation $y = mx + b$, where $x =$ concentration and $y =$ ratio of drug peak area (or whichever response factor is being measured) to internal standard peak area. The regression line is not forced through zero. A correlation coefficient, representing acceptable linearity for this particular assay, should be predetermined and justified.

Precision is assessed from the slope of replicate standard curves run on the same day and over a minimum two week period. The $y$-intercept should typically be less than 25% of the value of the minimum standard curve calibrator. Throughout the study period, changes in curve slope, $y$-intercept and residuals should be assessed to check for occurrence of proportional or determinate error. Having ascertained the range and linearity of the analytical method, selected sets of the test compound are then analyzed by the proposed procedure. These samples should fall within a previously determined acceptable range of their target concentrations.

## F. Sensitivity

The detection limit of an assay can be determined using any one of several approaches. However as with all of the other validation parameters, the choice of such an approach must be justified. The detection limit can be determined visually, by analyzing samples of known concentration and establishing the minimum level at which the analyte can reliably be detected. This approach however introduces an element of operator subjectivity to the assay and is not recommended. Sensitivity is generally reported by reference to signal to noise ratios. It should be remembered however that this approach is only applicable to analytical procedures which do in fact exhibit baseline noise. Determination of the signal-to-noise ratio is performed by comparing measured signals from samples with known low concentrations of analyte with those of blank samples and establishing the minimum concentration at which the analyte can be reliably detected. A signal-to-noise ratio greater than 3:1 or 2:1 is generally considered acceptable for estimating the detection limit, although the ratio used is debatable and basically is assay dependent. The magnitude of the background noise is determined by analyzing an appropriate number of blank samples and calculating the standard deviation of their responses.

The detection limit and the method used for determining the detection limit should be presented in the method validation report, together with suitable chromatograms to justify the determination. If the detection limit has been estimated by extrapolation of a standard curve, this estimate may subsequently be validated by the independent analysis of a suitable number of samples known to be near the detection limit.

*1. Quantitation Limit*

The same considerations and approaches as described for detection limit also apply to the determination of the Limit of Quantitation (LOQ). The LOQ can be determined visually by analysis of samples with known concentrations of analyte and by establishing the minimum level at which

the analyte can be quantified with acceptable accuracy and precision. Again, the usual approach is by reference to signal to noise ratio and establishing the minimum concentration at which the analyte can be reliably quantified. A typical signal-to-noise ratio is 10:1. The quantitation limit and the method used for determining the quantitation limit should be presented in the validation report in the same way as for limit of detection.

**G. Specificity**

Identification tests should be able to discriminate between compounds of closely related structures which are likely to be present. The specificity of an assay may be demonstrated by obtaining positive results from samples containing the analyte, and negative results from samples which do not contain the analyte. In addition, the assay should be applied to materials structurally similar to or closely related to the analyte to confirm that a positive response is not obtained. The choice of such potentially interfering materials should be made after reasonable consideration of the interferences that could be present in the sample. For chromatographic procedures, representative chromatograms should be used to demonstrate specificity. If impurities and/or excipients are available, specificity can be demonstrated by simply spiking a suitable matrix with the analyte in question together with suitable concentrations of potentially interfering compounds. Resolution of the analyte peak from any other compound verifies specificity.

Demonstrating specificity of the assay if impurity or degradation product standards are unavailable is more involved but can be achieved by comparing the test results of samples containing impurities or degradation products to a second validated method. This should include samples stored under relevant conditions of light, heat, humidity, acid/base hydrolysis, and oxidation. Comparison of the assay results and/or impurity profiles of the new procedure with that of the validated method can be used to show specificity. Peak purity tests using chromatographic or spectroscopic techniques may be useful to show that the analyte chromatographic peak is not attributable to more than one component. Samples of the appropriate biological matrix known to be drug free are analyzed both with and without a known quantity of drug and/or metabolites added to the sample to demonstrate lack of interference from endogenous materials.

## H. Accuracy and Precision

Accuracy should be established across the specified range of the analytical procedure. This can be achieved using one of several different approaches. Ideally the analytical procedure is used to assay a reference standard of known concentration. In the absence of a known standard however, accuracy can be verified by comparison of the results of the proposed analytical procedure with those of a second well-characterized procedure, the accuracy of which has been previously defined. Accuracy can also be inferred from precision, linearity, and specificity data although this approach is more difficult to justify. It is recommended that accuracy should be assessed using a minimum of nine determinations over a minimum of three concentration levels covering the specified range. Accuracy is generally reported as the difference between the mean and the accepted true value together with the confidence intervals.

Validation of tests for assay and for quantitative determination of impurities includes an investigation of precision. The standard deviation, relative standard deviation (coefficient of variation), and confidence interval, should be reported for each type of precision investigated. Intraday and interday precision are deter-mined on three control samples (low, medium and high concentration range) prepared independent of the analyst as described above. Replicate analysis of control samples is generally performed on the same day and over a minimum period of two weeks. Precision is expressed as the relative standard deviation of the intraday and interday replicate analysis for each control sample. Minimum acceptable accuracy and precision criteria are prospectively defined based on the specific requirements of each study.

## I. Stability

The stability of the analytical method depends on the type of procedure under study. Robustness of the assay indicates the reliability of an analysis with respect to deliberate variations in method parameters. If it is determined that measurements are susceptible to variations in analytical conditions, those conditions should be suitably controlled or a precautionary statement should be included in the procedure. Such parameters may include the stability of the analyte, pH, temperature and humidity. Stability of standard curve, control and subject samples is prospectively demonstrated for the sample collection (including collection devices and containers) and storage conditions to be used for each study. Stability of extracted biological samples is determined during assay development/validation by replicate injections of the sample over a 24 hour period or longer, as applicable to a particular assay. Variation in peak area and peak area ratio of drug (and/or metabolites) and internal standard should typically be less than 5% and independent of time of injection.

## J. Quality Control

Standard curve and control samples are analyzed with each assay batch of samples.

Replicate control samples are spread evenly throughout each batch so that a control sample is injected at least every 10 samples and each control sample injected a minimum of three times. Control samples are assessed after each assay run to ensure suitable within batch stability and reproducibility. All biologic specimens from a given animal should be assayed in the same batch. In addition, many analytical techniques now incorporate a means of testing system suitability. System suitability is based on the concept that the equipment, electronics, analytical operations, and samples to be analyzed constitute an integral system that can be evaluated as such. System suitability test parameters to be established for a particular procedure depend on the type of procedure being validated.

## III.  CONCLUSION

It should always be remembered that the synthesis of any compound is not infallible. This is particularly true of complex peptides and proteins, and the opportunity for an error in the sequence or the introduction of a contaminant into the sample is a distinct possibility. Before using such samples for any investigative work it is imperative that their identity, purity and stability be verified by the use of a suitably validated analytical method.

## REFERENCES

1.  RB Merrifield. Solid phase peptide synthesis. The synthesis of tetrapeptide. J Amer Chem Soc 85: 2149, 1963.
2.  LA Carpino, BJ Cohen, KE Stephens, SY Sadat-Aalaee, JH Tien, DC Langridge. (Fluoren-9-ylmethoxy) carbonyl (Fmoc) amino acid chlorides. Synthesis, characterization and application to the rapid synthesis of short peptide segments. J Org Chem 51: 3732, 1986.
3.  TE Wheat. Principles and practice of peptide analysis with capillary zone electrophoresis.

In: BM Dunn, MW Pennington, eds. Peptide Analysis Protocols. New Jersey: Humana Press, 1994, pp 65–84.
4.  JW Jorgensen, KD Lukacs. Capillary zone electrophoresis. Science 222: 266–272, 1983.
5.  SA Carr, ME Hemling, MF Bean, GD Roberts. Integration of mass spectrometry in analytical biotechnology. Anal Chem 63: 2802–2824, 1991.
6.  SD Patterson. From electrophoretically separated protein to identification: Strategies for sequence and mass analysis. Anal Biochem 221:1-15, 1994.
7.  RJ Cotter. Time-of-flight mass spectrometry for the structural analysis of biological molecules. Anal Chem 64: 1027–1039,1992
8.  AF Bristow. Purification of proteins for therapeutic use. In: ELV Harris, S Angal, eds. Protein Purification Applications. A Practical Approach. Oxford, England: IRL Press, 1989, pp 29–44.
9.  FDA Handbook. Points to consider in the production and testing of new drugs and biologicals produced by recombinant DNA technology. Bethesda, MD: Office of Biologics Research and Review, Center for Drugs and Biologics.
10.  MJ Geisow. Characterizing recombinant proteins. Biotechnology 9: 921–924, 1991.
11.  RL Garnick, MJ Ross, RA Baffi. Characterization of proteins from recombinant DNA manufacture. In: Y-Y H Chiu, J L Gueriguian, eds. Drug Biotechnology Regulation. Scientific Basis and Practices. Bethesda MD: FDA, 1991, pp 263–313.
12.  HA Scoble, SA Martin. Characterization of recombinant proteins. Methods in Enzymol 193: 519–536, 1990.
13.  DJC Pappin, P Hojrup, AJ Bleasby. Rapid identification of proteins by peptide-mass fingerprinting. Current Biology 3: 327–333, 1993.
14.  N LeGendre, M Mansfield, A Weiss, P Matsudaira. Purification of proteins and peptides by SDS-PAGE. In: P Matsudaira, ed. A Practical Guide to Protein and Peptide Purification for Microsequencing. San Diego, CA: Academic Press, 1993, pp 74–101.
15.  J-C Mercier, F Grosclaude, B Ribadeau-Dumas. Primary structure of bovine S1 casein. Complete Sequence. Eur J Biochem 23: 41–51, 1971.
16.  RE Wettenhall, P Cohen. Isolation and characterization of cyclic AMP-dependent phosphorylation sites from rat liver ribosomal protein S6. FEBS Lett 140: 263–269, 1982.

# 22

# Solution Structure Determination of Proteins by Nuclear Magnetic Resonance Spectroscopy

**Nicholas J. Skelton**
*Genentech Inc., South San Francisco, California*

**Walter J. Chazin**
*The Scripps Research Institute, La Jolla, California*

## I. INTRODUCTION

The ability to interpret biological function in terms of molecular structure is contributing significantly to rapid developments in the fields of biological chemistry, biochemistry, molecular biology, pharmacology, and immunology. Until 10 or so years ago, knowledge of protein three-dimensional structure was based largely on single crystal x-ray and neutron diffraction studies. Since proteins are inherently flexible polymeric molecules that are highly sensitive to their environment, it is important to obtain structural information about them under conditions that more closely approximate the physiological environment. Over the past decade, NMR spectroscopy has become the first technique available for the determination of the complete three-dimensional structures of proteins in solution at a resolution comparable to x-ray and neutron diffraction studies, and it is now well accepted as a viable and complementary approach to crystallography. Furthermore, NMR provides an alternative to single crystal studies for those proteins that fail to form

crystals, that form crystals unsuitable for diffraction experiments, or that crystallize only under very specific and extreme conditions. A second, powerful application of NMR is for detailed characterization of the internal motions and the dynamics of intermolecular processes. Knowledge of protein dynamics is important for understanding the true implications of a protein structure within the context of complex biological systems; there is an increasing appreciation among structural biologists of the fundamental importance of structural dynamics.

NMR technology has now reached the stage where small ($<30$ kDa) protein hormones, receptor fragments, enzymes and the complexes that they form with inhibitors, are amenable to direct structural analysis by solution NMR. Thus, from the pharmacology perspective, NMR-derived structural and dynamic information can now be used not only to study drug candidates, but also to define the details of their interaction with target proteins. The use of such information can play a major role in understanding functional aspects of a protein, such as the effects of site-directed mutation (e.g. to determine if the mutation affects the structure beyond the altered residue), or in the design of proteins with altered functional properties (e.g. in the production of protein antagonists to a receptor based on an agonist protein of known structure). Further, the ability to characterize dynamic aspects of a protein may have implications for the ability of small molecules to bind to proteins, or for protein ligands to be recognized by multiple receptors.

The steps required for the determination of a protein structure from NMR data are depicted schematically in Fig. 1. For any NMR study there is always an initial step of assigning the resonance for each nucleus. The ability to carry out these procedures on peptides and proteins was dependent on the development of two-dimensional NMR spectroscopy because there is very extensive overlap of NMR resonances in the 1D spectrum of these biopolymers (for example, see Fig. 3 below). It has been only 13 years since the first report (1) of the use of a large number of proton-proton distance, dihedral angle and hydrogen-bond constraints derived from NMR spectroscopy to calculate the three-dimensional structure of a protein in solution. The past eight years has seen the development and application of heteronuclear methodologies ($^{1}$H, $^{15}$N and $^{13}$C NMR) for the determination of the structure of proteins that cannot be analyzed by the standard homonuclear $^{1}$H NMR approach, i.e. for small proteins with suboptimal spectral characteristics and for larger proteins in the range 15–30 kDa. These newer techniques involve the detection of $^{13}$C or $^{15}$N, nuclei which have a low natural abundance. Thus, for these experiments, samples must be produced with $^{13}$C and/or $^{15}$N enrichment, either via chemical synthesis or by expression in cells grown on media highly enriched in these isotopes.

Once resonances are assigned, secondary structure can be readily deter-

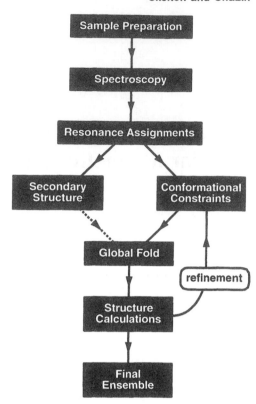

**Fig. 1** Flow diagram of the steps involved in solving a protein structure from NMR data. After initial sample preparation and optimization of NMR spectral characteristics by varying experimental conditions, and experiments are run to assign resonance positions. The location of elements of regular secondary structure can be ascertained from a qualitative analysis of the data. A more detailed analysis leads to the generation of conformational restraints that can in many, but not all, cases be used to define a global fold, or can be used as input to a structure calculation program. Refinement of the conformational constraints usually is time consuming, and requires many rounds of structure calculation before the final ensemble of conformations is determined.

mined and then a series of distance and dihedral angle constraints on the conformation of the polypeptide chain can be generated from a battery of NMR experiments. These distance and dihedral angle constraints are used as input for computer programs that calculate structural coordinates. This process is repeated many times to create an ensemble of structures, each of which satisfy the input constraints equally well. Distance geometry and restrained molecular dynamics are the two main methods for performing structure calculations; a mixture of the two, termed hybrid simulated annealing, is commonly used. Although dynamic aspects of a molecule will have an effect on the number and quality of structural constraints, internal mobility is usually investigated in a separate series of experiments. There is no one single experiment that can report on the wide range of time scales of motion in peptides and proteins, so a variety of experiments must be utilized.

In this chapter, we touch on all aspects of protein structure determination by NMR. In Sec. II we review briefly basic NMR concepts, and go on to discuss the approaches used to assign resonances in Sec. III. Sections IV, V and VI describe how the three-dimensional structures are determined starting from an overview of the global fold through the generation of structural constraints from the NMR data to calculation of structures. Section VII addresses the assessment of the quality of NMR solution structures. Section VIII summarizes NMR characterization of protein internal dynamics, and Sec. IX highlights the important differences between NMR solution and x-ray crystal structures. We conclude with an overview of some of the future directions of solution NMR as a tool to study proteins.

## II. BASIC NMR CONCEPTS

We begin with a primer on NMR terminology, briefly outlining some of the fundamental NMR parameters and concepts and how they manifest themselves in the NMR spectrum. Several of the key observables are depicted schematically in Fig. 2. A more detailed explanation of the fundamentals of NMR spectroscopy can be found in many other reviews and text books [for example see (2–5)].

NMR-Active Nuclei. NMR spectra containing well resolved resonances and narrow lines are usually obtained only for nuclei having a nuclear spin angular momentum quantum number of $1/2$. In the case of proteins, this limits analysis to $^1$H, $^{13}$C and $^{15}$N. The separation of nuclear energy levels vary from one type of nucleus to another as reflected in an efficiency factor termed the gyro-magnetic ratio ($\gamma$). Hence, each of these nuclei have characteristic resonance frequencies: for an applied magnetic field sufficient to make $^1$H resonate at 500 MHz (11.7 T), $^{13}$C will resonate around 125.7 MHz and $^{15}$N will resonate close to 50.68 MHz. It is also important to consider the natural abundance of these nuclei, which is low for $^{13}$C (1.11%) and $^{15}$N (0.36%). Only $^1$H NMR is readily accessible for systematic analysis unless some form of isotopic enrichment can be implemented during the sample preparation process.

Fourier Transform NMR. High-resolution NMR spectra are best obtained by the pulsed-Fourier transform method; the details and advantages of this method are beyond the scope of this chapter and are described in detail elsewhere (3, 5). An initial powerful pulse of radio frequency energy (a rf pulse) is applied close to the resonant frequency of the nuclei of interest (usually one nuclear type will be affected, e.g. all $^1$H or all $^{15}$N), causing changes in the populations of the nuclear spin angular momentum energy levels. After the pulse, the "excited" nuclei resonate with frequencies characteristic of their environment (see next section), and induce an oscillating voltage in the receiver coil of the spectrometer which constitutes the NMR signal. This signal, or free induction decay (fid) is recorded over a period of milliseconds to seconds,

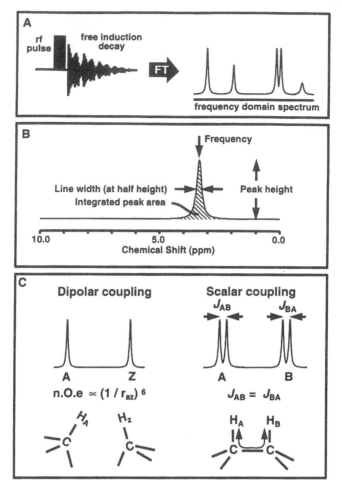

**Fig. 2** Fundamental concepts in NMR spectroscopy. (A) Pulsed Fourier transform NMR. Initial excitation by an rf pulse gives rise to a time domain free induction decay. Fourier transformation of these data produce the frequency domain spectrum. (B) Attributes that define a peak: frequency (chemical shift), peak height, line width at half height and intensity (integrated peak area). (C) Coupling interactions between protons. Dipolar coupling is a through-space effect and is observed between nuclei less than ~6 Å apart; it has an effect on peak intensities and is most readily observed in NOESY spectra. Scalar coupling is a through-bond effect. In peptides and proteins, it is observed between nuclei separated by three or fewer covalent linkages. A scalar coupling leads to splitting of the resonances into two components. The magnitude of the scalar coupling may be obtained by measuring the separation of the components.

digitized and stored. The fid contains a complex mixture of frequencies, one component contributed by each of the magnetically distinct nuclei influenced by the initial rf pulse. The individual frequency components may be extracted from the fid by mathematical manipulation, most commonly the Fourier transformation: the initial time domain data is converted into a frequency domain spectrum. The result is a series of peaks whose positions (frequencies), intensities and line widths (see below) are characteristic of the particular sample. NMR is an inherently insensitive technique, and in order to improve the signal to noise ratio such a single pulse experiment is repeated many times, and the results co-added prior to Fourier transformation.

Chemical Shifts. The frequency of each resonance line depends primarily on the type of nucleus involved, but also depends upon the surrounding nuclei. The resonance frequencies are described in terms of a "chemical shift" relative to a resonance of a standard reference compound. The quantity of chemical shift is defined as absolute frequency divided by applied field strength, with unit of parts per million, or ppm. Thus, the chemical shift is by definition invariant with the applied magnetic field, i.e. each nucleus will have the same chemical shift no matter what instrument is used to record the spectrum or what experiment is recorded. Data are readily compared provided the chemical shifts are given relative to the same reference compound (6). The chemical shift is very sensitive to the covalent attachment of atoms, so that nuclei of similar chemical disposition often have similar chemical shifts: e.g. backbone amide protons ($H^N$) resonate between 7 and 10 ppm, backbone alpha protons ($H^\alpha$) between 3.0 and 6.0 ppm, aliphatic methylene and methine protons between 3.5 and 1.0 ppm and methyl protons between 2.0 and 0.0 ppm (Fig. 3). The non-covalent environment also has an effect on chemical shift, hence each chemically distinct nucleus in a molecule usually has a distinct frequency position.

Peak Intensities. The intensity of a particular resonance in the NMR spectrum varies in a linear fashion with the concentration of that nuclear species. The NMR phenomenon is relatively insensitive [see (2) or (3) for a detailed explanation of this], hence large amounts of sample are required for a detailed analysis (see Sec. 3). The absolute sensitivity of an NMR signal does depend on the nuclear type being investigated, e.g. $^1H$ is intrinsically more sensitive than $^{13}C$ or $^{15}N$. However, for a given nuclear type, the resonances arising from each chemically distinct nucleus should have equivalent peak intensities.

Line Widths. The widths of the NMR resonance lines of a biomolecule (usually termed "line widths") are determined primarily by the rate of reorientation of the molecule in solution, which is characterized by the rotational correlation time, $\tau_c$. Increasing molecular size leads to slower reorientation, a larger $\tau_c$, faster decay of the NMR signal and broader NMR resonance lines. This size-dependent line broadening provides one of the most significant limitations to the general application of NMR: increasing molecular weight is associated with broader resonances, lower signal-to-noise ratios and decreased signal resolution. Self-association or non-specific aggregation increases the apparent molecular weight and such behavior limits the systems that are amenable to study by NMR.

Scalar Coupling. NMR-active nuclei within a biomolecule interact with one another by a variety of mechanisms. The nuclear energy levels of one nucleus are affected by the spin state of neighboring spins. One type of interaction is transmitted via the electrons in the intervening covalent bonds and is referred to as scalar or spin-spin coupling. The magnitude of these interactions depends on the two nuclei involved, the number of bonds between the nuclei, and the chemical nature of the intervening atoms. For scalar coupled protons in proteins, the effect is usually observable only for nuclei separated by two or three bonds and has a magnitude of 0–18 Hz. One-bond scalar couplings are considerably larger, e.g. ~35 Hz for C—C couplings, ~90 Hz for N-H couplings and ~140 Hz for C—H couplings.

The effect of scalar coupling is to split the intrinsic resonance of a nucleus

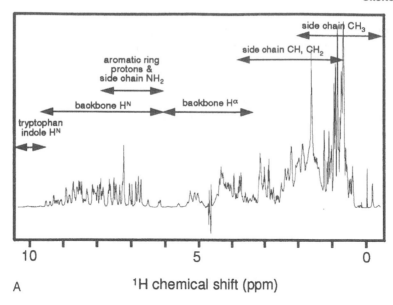

A                    ¹H chemical shift (ppm)

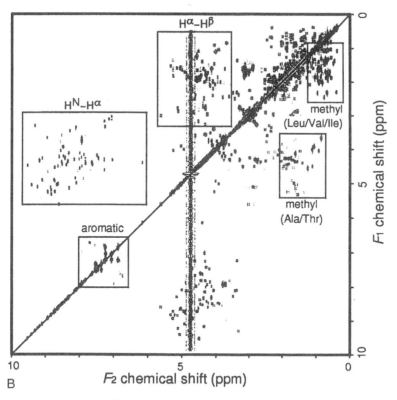

B          F₂ chemical shift (ppm)

**Fig. 3**  (A). One-dimensional ¹H NMR spectrum of a protein labeled with the typical frequencies over which different types of protons resonate. The data were acquired at 500 MHz with a 1.5 mM sample of ubiquitin ($M_w = 8.5$ kDa) at pH 5.8 and 30°C. Chemical shifts are relative to an internal chemical shift reference (DSS) at 0.0 ppm. (B). 2D COSY spectrum of ubiquitin [same sample conditions as (A)], with the characteristic "fingerprint" regions identified.

into several components, giving rise to fine-structure or peak multiplicity. When the scalar coupling is smaller than the linewidth, only peak broadening (no fine structure) is observed. Of greatest importance for multi-dimensional NMR, the scalar coupling between two nuclei allows magnetization to be transferred between them after appropriate manipulation by rf pulses. Such transfer is exploited to give rise to the critical *cross-peaks* in homonuclear and heteronuclear multi-dimensional correlation spectra (2D, 3D or 4D). In addition, the size of three-bond couplings depends on the dihedral angle formed by the four atoms involved (the Karplus relationship—see below). Thus, measurements of the magnitude of scalar couplings can be used to infer structurally useful information on the intervening torsion angle between the scalar coupled spins.

Dipolar Coupling. The other fundamental interaction between nuclei occurs not through bonds, but rather through space, and is referred to as *dipolar* coupling. The strength of the interaction is dependent on the gyromagnetic ratio and is therefore dominated by couplings to protons. The most common measurement of this interaction is via the *nuclear Overhauser effect* (*NOE*) between two protons, which varies in a distance-dependent ($1/r^6$) manner, where $r$ is the inter-proton separation. In typical situations, an observable effect occurs when $r$ is <6 Å. Quantitation of the relative size of the NOEs allows inter-proton distances to be estimated. The NOE is usually detected in multi-dimensional (2D, 3D or 4D) NOESY experiments. The inter-nuclear distances obtained from NOESY spectra are the primary input to the determination of protein structures from NMR data.

Two-dimensional Spectra. Theoretically, every chemically distinct nucleus in a molecule will give rise to a discrete resonance position in the NMR spectrum. Practically, as larger molecules are studied, resonances in the 1D spectrum will start to coincide with one another, complicating the observation of through-bond or through-space interactions between nuclei; the unambiguous observation of such interactions is crucial to making resonance assignments and obtaining structural constraints from NMR data. The effective resolution of the spectrum can be dramatically improved by increasing the number of dimensions in which the resonances are displayed. In the two-dimensional case (Fig. 4A), this is achieved by first detecting the resonance frequencies during time $t_1$ (giving rise to frequency domain $F_1$ after Fourier transformation), then causing scalar or dipolar coupling ("mixing") from one spin to another, and finally detecting the new resonance frequencies during time $t_2$ (giving rise to the frequency domain $F_2$). The technical details of how this is achieved are beyond the scope of this review; interested readers should see in-depth references (2, 3, 5).

The essential element of the 2D experiment is that the history of transfer (or lack thereof) from one coupled spin to another can be tracked throughout. The results of a 2D experiment are typically viewed as a simple 2D matrix, with peak intensities indicated by contours (Fig. 4B). The entries on the matrix diagonal correspond to no transfer between nuclei during the experiment ($F_1 = F_2$). These *diagonal peaks* contain little information beyond what is present in the normal one-dimensional spectrum. The off-diagonal matrix elements ($F_1 \neq F_2$) correspond to events where transfer from one nucleus to the other did occur during the experiment. These *cross peaks* show that there is connectivity between A and B, i.e. they are correlated (scalar or dipolar coupled) with each other (Fig. 4B).

The particular type of mixing sequence present in Fig. 4A is primarily

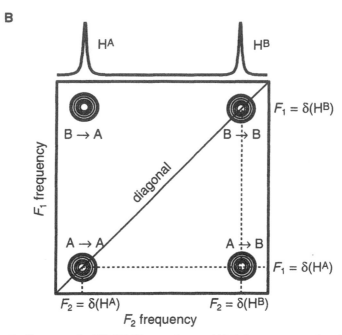

**Fig. 4** Schematic diagram of a 2D NMR experiment. (A) Pulse sequence showing the two periods where frequencies are measured ($t_1$ and $t_2$) with a mixing scheme between them. (B). Schematic two-dimensional frequency domain spectrum for a two spin system obtained from Fourier transformation of the time domain data acquired using the pulse sequence in (A). Peaks along the $F_1 = F_2$ diagonal correspond to nuclei that resonate at the same frequency during $t_1$ and $t_2$. Off diagonal peaks correspond to nuclei that coupled during the mixing scheme. For example, the peak at lower right represents the coupling of spin A [$F_1 = \delta(H^A)$] to spin B [$F_2 = \delta(H^B)$].

what differentiates one 2D NMR experiment from another. For example, by a suitable choice of rf pulses and delays in the mixing scheme, magnetization may be transferred between $^1$H spins that are connected by: a single scalar coupling [correlated spectroscopy or COSY (7, 8)]; two consecutive scalar couplings [relayed-COSY or R-COSY (9)]; three consecutive scalar couplings [double relayed-COSY or DR-COSY (10)]; any number of consecutive scalar couplings [total coher-ence spectroscopy or TOCSY (11, 12)]; a dipolar coupling [nuclear Overhauser enhanced spectroscopy or NOESY (13, 14)]. The uses of these types of experiments are discussed in more detail below.

Magnetization may also be transferred between nuclei of different types. For example, the heteronuclear single quantum coherence (HSQC) or heteronuclear multiple quantum coherence (HMQC) experiments can be used to correlate protons with their directly attached

$^{15}$N or $^{13}$C via the large one bond $^{1}J_{HX}$ scalar coupling (15–18). Such spectra have $^{1}$H in one dimension and $^{15}$N or $^{13}$C chemical shifts in the other dimension, hence do not contain diagonal peaks; a single peak will be present for each H-N (or H-C) group in the sample.

Higher-dimensional spectra. Two-dimensional spectra are now commonplace in the analysis of peptides and small proteins by $^{1}$H NMR. For proteins larger than about 80 residues, 2D $^{1}$H spectra become too overlapped for a complete analysis. 3D homonuclear $^{1}$H spectra may alleviate these problems in some cases but have not found widespread usage (19–23). Further improvements in spectral dispersion are obtained by using $^{13}$C and/or $^{15}$N labeled samples to collect three dimensional data. For example, the schematic experiment in Fig. 4A may be amended for a variety of spectra to measure the frequency of the $^{15}$N or $^{13}$C nucleus attached to the protons whose frequencies are measured in $F_2$ of the $^{1}$H 2D experiment. Magnetization is transferred between the proton and attached heteronucleus ($^{13}$C, $^{15}$N) using HSQC transfer. This is depicted schematically in Fig. 5A. Such three-dimensional spectra are usually visualized as a series of two dimensional sub-spectra containing two $^{1}$H chemical shift axes, with each plane corresponding to a different heteronuclear chemical shift (Fig. 5B). Because the chemical shifts of $^{13}$C and $^{15}$N are not directly correlated to the chemical shift of the attached proton, spreading out the peaks into the third dimension based on the heteronuclear shift is a very powerful way to resolve proton resonances. This approach is readily extended to four dimensions (the chemical shifts of the heteronuclei attached to both interacting protons are measured). The extension to four dimensions comes at a cost of time required to acquire a spectrum so that there is acceptable digital resolution in all four dimensions. With the available experimental and analysis tools, 4D spectra push the limits of current capabilities. While experiments with higher dimensionality can be readily envisaged, there have yet to be any practical applications demonstrated.

## III.  RESONANCE ASSIGNMENTS

NMR provides a powerful tool to study structure and internal dynamics of biomolecules. The power of the technique over other forms of spectroscopy lies in the ability to provide information about specific atoms or small groups of atoms within the molecule of interest. In order to make use of these data, the resonance positions of all of the corresponding nuclei that appear in the spectrum must be known. Thus, the initial stage in any investigation by NMR spectroscopy is the determination of the *sequence-specific assignments* of the resonances in the spectrum. This section describes how these assignments are made both for native proteins where only $^{1}$H nuclei are readily available for NMR experiments, and for recombinant proteins in which heteronuclei ($^{13}$C or $^{15}$N) have been enriched in the sample.

### A.  Sample Requirements and Experimental Conditions

NMR spectroscopy has intrinsically low sensitivity, and consequently there are some very strict sample requirements for recording NMR spectra. The most severe requirement for NMR is the need to work at very high concentrations. Basic 1D spectra can be acquired on samples as low as 20–50 $\mu$M using a standard 500 MHz spectrometer and 5 mm probe, but considerably higher concentrations are required to obtain the multi-dimensional spectra needed for structure determination. Concentrations of 1 mM or greater

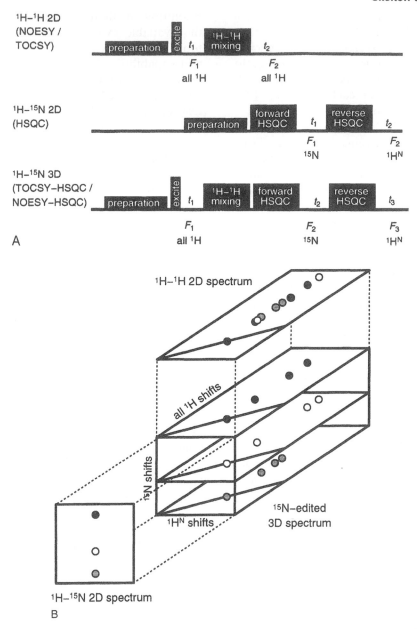

**Fig. 5** Schematic diagrams of a 3D NMR experiment as a combination of two 2D experiments. (A) pulse sequences. The initial $^1$H–$^1$H 2D scheme may be combined with a 2D $^1$H$^N$–$^{15}$N correlation experiment to yield a 3D $^{15}$N-edited experiment. Commonly, a $^1$H–$^1$H TOCSY or NOESY sequence is combined with a $^1$H–$^{15}$N correlation spectrum (HSQC). (B) Resolution of resonance degeneracy in 2D spectra by heteronuclear editing in a third dimension. In the schematic $^1$H 2D spectrum (top), three residues have degenerate amide protons, hence the correlations to other protons cannot be attributed unambiguously to a single amide proton. The $^1$H–$^{15}$N 2D correlation spectrum (lower left) indicates that the $^{15}$N resonances of the three amide groups are not degenerate. The 2D experiment may be extended to measure the chemical shifts of the $^{15}$N atom attached to each amide proton in a third dimension, allowing the three sets of peaks (open circles, filled circles and shaded circles) to be unambiguously resolved on separate planes of the 3D matrix.

are typically sought. In favorable cases, it is possible to collect the requisite NMR spectra on samples of roughly 0.4 mM using new higher field instruments ($^1$H frequencies of 750–800 MHz). The quantity of protein required is dependent on sample volume, which in standard 5 mm diameter probes is either $\sim 0.5$ ml for standard NMR tubes or $\sim 0.25$ ml when using susceptibility-matched NMR tubes. In either case, this corresponds to many milligrams of protein (0.5 ml of a 1 mM solution corresponds to 10 mg of a 20 kDa protein). Larger bore 8 mm probes are becoming increasingly popular, even though they require 1.0–2.5 ml of protein solution, because there can be a substantial improvement in sensitivity over 5 mm probes due to the considerably larger volume of solution within the NMR receiver coil. In theory, 3–5 fold lower concentrations can be used, which can be especially important for proteins with limited aqueous solubility or with a tendency to self-associate (aggregate) at $\approx 1$ mM concentration. Finally, it is recommended that the protein samples used for NMR structure determination be of high purity (>95%) because the protein spectra are so complex that the resonances of impurities are not always readily distinguishable *a-priori*.

Due to the extreme complexity of the NMR spectrum and the sensitivity to experimental conditions, it is necessary to screen parameters such as buffer, ionic strength, pH, temperature and concentration. The objective is to minimize adverse effects on the state of the molecules (e.g. low solubility, poor stability, aggregation) while optimizing the characteristics of the resulting spectrum (e.g. linewidth, sensitivity, spectral overlap). This typically involves other analytical techniques such as gel electrophoresis, UV absorption, analytical ultracentrifugation, light scattering, gel filtration chromatography or CD spectroscopy, in addition to preliminary 1D and 2D NMR spectroscopy.

## B. Homonuclear $^1$H Resonance Assignments of Proteins

The approach to obtaining $^1$H NMR assignments is shaped by fact that $^1$H-$^1$H scalar couplings are rarely observable between atoms separated by more than three bonds. Since scalar coupling interactions will only connect protons contained in the same amino acid residue (there are at least four bonds between protons in adjacent residues), the first step in the $^1$H assignment process involves identifying resonance frequencies for each separate amino acid in the protein. These discrete collections of spins are referred to as the amino acid *spin-systems*. There should be one backbone-based spin-system present for every residue in the protein. For certain side-chains there are four bonds between some of their adjacent side-chain protons (Phe, Tyr, His, Trp, Asn, Gln, Met), hence a discrete side-chain spin system will have to be identified. The second step in the assignment process uses through-space interactions (dipolar coupling) to correlate each spin-system(s) with a specific residue in the polypeptide chain. Assigning $^1$H resonances in this way is known as the *sequential assignment strategy*, and was developed by Wüthrich and co-workers [reviewed in (24)]. The method is generally applicable for well-behaved polypeptides up to $\sim 12,000$ Da, although applications to specific systems approaching 20,000 Da have been reported [e.g. (25, 26)].

Spin-system identification. Initially, the $^1$H resonances are categorized based on their chemical shifts. For the vast majority of residues in a protein, backbone $H^N$ and $H^\alpha$, aromatic side-chain protons, aliphatic side-chain protons and methyl protons resonate in specific regions of the 1D spectrum (Fig. 3A). The corresponding sections of 2D spectra are termed "fingerprint" regions because they can be rapidly assayed to monitor the integrity of a protein sample (Fig. 3B).

Fingerprint correlations involving back-bone $H^N$-$H^\alpha$, aromatic rings protons, Ala/Thr methyl groups and Leu/Ile/Val methyl groups are most useful, while more pronounced resonance overlap of the $H^\alpha$–$H^\beta$ correlations makes this region less useful.

The amide protons have superior chemical shift dispersion (i.e. a given amide resonance is less likely to be coincident with another amide proton) and usually form the basis for identifying spin systems (Fig. 6) (27). The side-chain resonance frequencies associated with a given amide proton are most readily deter-mined by the complementary analysis of

the cross-peaks to the backbone amide proton resonance (28) in the following series of experiments: COSY, in which cross-peaks arise from a single scalar coupling transfer (i.e. $H^\alpha$–$H^N$ cross-peaks); R-COSY, DR-COSY, in which cross-peaks arise from transfer via two or three scalar coupling steps (i.e. $H^\beta$–$H^{N,}$ $H^\gamma$–$H^N$ cross-peaks); TOCSY, in which cross-peaks arise from transfer through few, many or all possible intervening scalar couplings (cross-peaks from $H^N$ to all side-chain resonances). In instances where the identification of side-chain res-onances in this manner is incomplete, additional information can be obtained

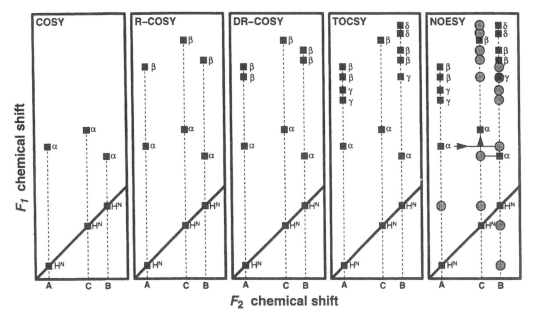

**Fig. 6** Backbone amide proton-based spin system identification from scalar correlation spectra. The schematic spectra show the results for three different spin systems. The square boxes indicate the peak positions expected for a three amino acid fragment of a protein in various correlation experiments. Thus, the COSY identifies only the $-H^\alpha$-$H^N$ correlations, the relayed-COSY adds a single $H^\beta$–$H^N$ correlation to each amide proton, the double relayed-COSY adds the position of both $H^\beta$ for $\beta$-methylene-containing amino acids, whereas the TOCSY connects all side-chain resonances to the backbone amide proton. The number and chemical shifts of the side-chain res-onances suggest that residue A is a "5-spin" spin system (glutamate, glutamine or methionine), residue B is a leucine and residue C is an alanine. The inter residue peaks in the schematic NOESY (circles) indicate the manner in which the three spin systems are positioned within the primary sequence, with sequential NOEs observed at the resonance positions of residue $i$ and the amide proton position of residue $i+1$. In this schematic example, the spin systems can be placed in the order A-B-C (from N to C-terminus) on the basis of these peaks.

from a corresponding analysis of the same experiments, but originating at the side-chain terminus rather than the backbone amide proton (28). The analysis can be further complemented by two-quantum (and occasionally three-quantum) correlation experiments, which have great value for the unambiguous and complete assignment of side-chain resonances (29).

The spin-system topology and constituent chemical shifts identified in the scalar correlation experiments may be used to assign each spin-system to a unique residue type (alanine, glycine, proline, isoleucine, lysine, leucine, arginine, serine, threonine or valine) or to a subset of amino-acid types: "three spin" ($H^\alpha$ and $H_2^\beta$ resonances for cysteine, aspartate, phenylalanine, histidine, asparagine, tyrosine, and tryptophan); "five-spin" ($H^\alpha$, $H_2^\beta$ and $H_2^\gamma$ resonances for glutamate, methionine and glutamine) (Fig. 7). If the coherence transfer experiments do not identify all resonance positions (e.g. in long side-chains where TOCSY transfer from side-chain terminal protons to the amide proton is inefficient), some of the "unique" spin systems may not be readily distinguished at this stage of analysis. For example, the cross-peaks from an amide proton to one methyl resonance and one other aliphatic proton could arise from a leucine, valine or isoleucine residue. In these cases, the aliphatic regions of the scalar correlation spectra need to be carefully inspected to identify other resonances within the spin system.

Aromatic protons of tyrosine, phenylalanine, tryptophan and histidine residues, the methyl group of methionine, and side-chain amide protons of glutamine and asparagine residues are at least four bonds distant from other side-chain protons, hence form discrete side-chain spin-systems that cannot be readily associated with a backbone amide resonance by scalar coupling experiments. Association of the side-chain and backbone resonances of these spin systems has to be made on the basis of intraresidue through-space (NOE) correlations. For example, in the case of a tyrosine or phenylalanine residue, one of the most intense NOEs to $H^\delta$ is generally from $H^\beta$ of the same residue because these protons are always less than 2.8Å apart.

Sequential Connectivities. In the second stage of the $^1H$ resonance assignment process, the previously identified spin systems must be attributed to a specific residue in the protein. Knowledge of the protein primary sequence is essential for this process. If the protein sequence contains only a single copy of any of the spin system types described above, then sequence specific assignment of that spin system is trivial. However, this is relatively rare even in proteins with only 50 amino acids, so at the outset a particular amino acid spin system could result from any one of the residues of this type present in the protein.

The assignment process is completed by analysis of through-space dipolar coupling (NOE) interactions. Statistical analysis of the proton positions inferred from x-ray crystal structures of proteins has shown that the majority of short inter-proton distances between $H^N$, $H^\alpha$ and $H^\beta$ are either intra-residue or between residues adjacent in the primary sequence (30). Thus, observation of intense NOEs from $H^N$, $H^\alpha$ and/or $H^\beta$ of one spin system to $H^N$ of a second spin system almost always indicates that the two spin systems are adjacent in the primary sequence and correspond to residue $i$, and residue $i+1$, respectively. Such sequential NOE cross peaks and the interproton distances corresponding to them are given the shorthand notation $d_{NN}$, $d_{\alpha N}$ and $d_{\beta N}$. Identification of a series of such sequential NOE interactions places several spin systems in the order $i$, $i+1$, $i+2 \ldots i+n$. Since the spin systems have been categorized to one or a small number of possible amino acid types, the sequence of adjacent spin systems

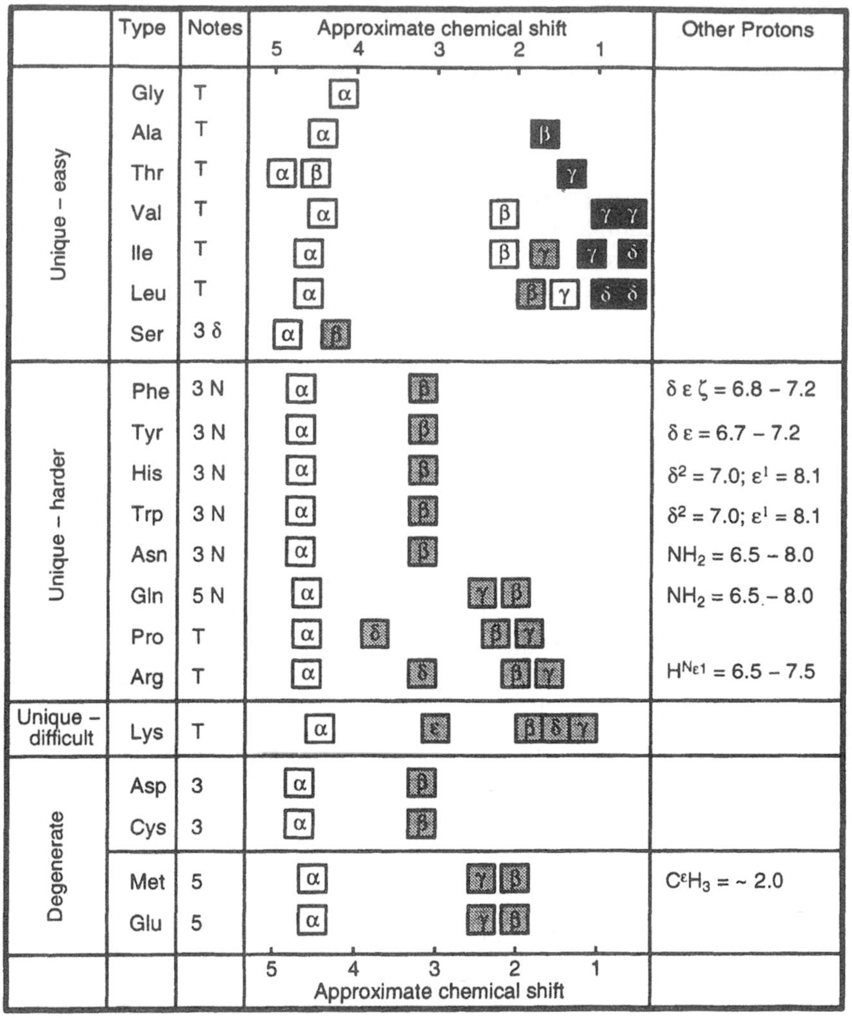

| | Type | Notes | Approximate chemical shift (5 → 1) | Other Protons |
|---|---|---|---|---|
| **Unique – easy** | Gly | T | α (~4) | |
| | Ala | T | α (~4); β (~1.5) | |
| | Thr | T | α, β (~4.3–4.2); γ (~1.2) | |
| | Val | T | α (~4); β (~2); γ γ (~1) | |
| | Ile | T | α (~4); β (~2); γ (~1.4), γ (~1), δ (~0.9) | |
| | Leu | T | α (~4); β γ (~1.7); δ δ (~0.9) | |
| | Ser | 3 δ | α (~4.5), β (~4) | |
| **Unique – harder** | Phe | 3 N | α (~4.5); β (~3.2) | $\delta\,\varepsilon\,\zeta = 6.8 - 7.2$ |
| | Tyr | 3 N | α (~4.5); β (~3) | $\delta\,\varepsilon = 6.7 - 7.2$ |
| | His | 3 N | α (~4.5); β (~3.2) | $\delta^2 = 7.0;\ \varepsilon^1 = 8.1$ |
| | Trp | 3 N | α (~4.5); β (~3.2) | $\delta^2 = 7.0;\ \varepsilon^1 = 8.1$ |
| | Asn | 3 N | α (~4.5); β (~2.8) | $NH_2 = 6.5 - 8.0$ |
| | Gln | 5 N | α (~4.5); γ β (~2.2) | $NH_2 = 6.5 - 8.0$ |
| | Pro | T | α (~4.5); δ (~3.6); β γ (~2) | |
| | Arg | T | α (~4.5); δ (~3.2); β γ (~1.8) | $H^{N\varepsilon1} = 6.5 - 7.5$ |
| **Unique – difficult** | Lys | T | α (~4.5); ε (~3); β δ γ (~1.7) | |
| **Degenerate** | Asp | 3 | α (~4.5); β (~2.8) | |
| | Cys | 3 | α (~4.5); β (~3) | |
| | Met | 5 | α (~4.5); γ β (~2.1) | $C^\varepsilon H_3 = \sim 2.0$ |
| | Glu | 5 | α (~4.5); γ β (~2.1) | |

**Fig. 7** Identification of $^1$H spin system types. Analysis of the position and pattern of side-chain $^1$H resonance positions allows designation to an amino acid type. The approximate location of the side-chain $^1$H resonances are indicated by white boxes for methines, gray boxes for methylenes and black boxes for methyl groups. The box position indicates the approximate chemical shift of the resonance. Notes: "T" indicates that an amino acid type may be identified on the basis of the pattern of peaks only; "3" and "5" indicate a 3-spin or 5-spin spin system (see text), respectively, that requires other information to unambiguously identify amino acid type; "δ" and "N" indicate that chemical shift information or intra residue NOEs to extreme positions of side-chain, respectively, are required to make assignments.

eventually will match a unique section of the primary sequence; a schematic example of this is shown in the right most panel of Fig. 6. For most proteins, this happens with segments of four or five adjacent spin systems, and permits the spin systems to be assigned *sequence specifically*. This process is more easily accomplished if the spin systems are well characterized into amino acid types (*i.e.* the majority of side-chain resonance positions have been identified).

The observation of $d_{NN}$, $d_{\alpha N}$ and $d_{\beta N}$ NOEs may also occur between non-sequential residues as a result of secondary or tertiary structure in the protein (24). The resulting ambiguity in the assignment process can be reduced by identifying the set of $d_{NN}$, $d_{\alpha N}$ and $d_{\beta N}$ NOEs as opposed to basing the assignment on just one (24). The matching of spin system type with the primary sequence then enables the remaining NOEs to be assigned to longer-range contacts (vide supra). In the final analysis, the assignment process should encompass *all* spin systems, and self consistency is perhaps the best measure of the validity of the assignments.

This section has outlined the fundamental basis for obtaining $^1H$ sequential resonance assignments. However, in practical applications, it is not always the case that these steps are followed in sequence or that each step is fully completed before passing on to the next step. For example, there have been very few instances in which spin systems are completely identified before the sequential resonance assignment is started. Furthermore, only rarely has the assignment of nearly all proton resonances been attained [for examples of complete $^1H$ assignments see (31, 32)].

## C.   Heteronuclear Resonance Assignments

The application of the homonuclear $^1H$ approach for resonance assignments becomes increasingly more difficult as the size of the system under study increases. Problems arise due to poor coherence transfer via three bond $^1H$-$^1H$ scalar couplings and severe resonance overlap. All of the approaches to overcome these problems involve inclusion of $^{15}N$ and/or $^{13}C$ nuclei in the analysis.

Homonuclear $^1H$ sequential assignment aided by uniform $^{15}N$ incorporation. The simplest heteronuclear strategy involves utilizing $^{15}N$ nuclei to

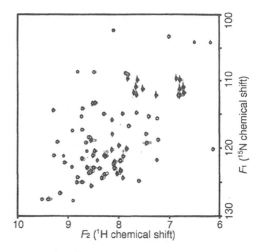

**Fig. 8**   $^1H$–$^{15}N$ correlation spectrum (HSQC) of ubiquitin. The data were acquired at 500 MHz, 27°C, pH 5.8 on a 1 mM sample uniformly labeled with $^{15}N$.

disperse or edit $^1H$ experiments into a third dimension, while essentially following the homonuclear $^1H$ approach to sequential assignment (33). The success of this approach is based on the fact that there is virtually no correlation between the value of the $^{15}N$ and $^1H$ resonance frequencies, as can be seen in the $^{15}N$-$^1H$ correlation spectrum shown in Fig. 8. The objective of these experiments is to overcome any resonance overlap of the key backbone amide protons by separating each amide strip according to the chemical shift of the $^{15}N$ atom to which it is bonded as shown schematically in Fig. 5. For example, the assignment of cross peaks from side-chain protons to an overlapped set of $H^N$ resonances in a 2D $^1H$ TOCSY spectrum may be resolved in the 3D $^{15}N$-edited TOCSY-HSQC because the attached $^{15}N$ nuclei will not usually be coincident (33). Similarly, NOEs involving the overlapped $H^N$ may be resolved in the 3D $^{15}N$-edited NOESY-HSQC. The next logical extension of this is to disperse the side-chain proton resonances involved in the correlations according to the chemical shift of their directly attached $^{13}$C or $^{15}$N atom in a 4D $^{13}$C-edited, $^{15}$N-edited HSQC-NOESY-HSQC experiment.

The applicability of this extended homonuclear $^1$H sequential resonance assignment strategy is limited by the ability to obtain side-chain connectivities through the three-bond vicinal $^1$H scalar couplings in COSY or TOCSY experiments. Thus, the issue of poor coherence transfer is not addressed at all. In practice, this approach is valid for proteins up to $\sim 20$ kDa in favorable cases [e.g. (34, 35)]. The method is also useful for smaller proteins that suffer from limited dispersion of the $H^N$ resonances.

Heteronuclear approaches to identify side-chain resonances. For proteins above about 10 kDa, transfer through three bond $^1$H-$^1$H scalar couplings in homonuclear $^1$H COSY or TOCSY experiments starts to become highly inefficient. To circumvent this problem, protein samples enriched in $^{13}$C can be used to identify side-chain resonances in 3D HCCH-COSY or 3D HCCH-TOCSY experiments in which transfer occurs through one or several C—C scalar couplings, respectively (36, 37). The advantage of this approach is that the one bond $^1$H-$^{13}$C ($\approx 140$ Hz) and $^{13}$C—$^{13}$C ($\approx 35$ Hz) couplings are significantly larger than the homonuclear three bond scalar couplings ($<13$ Hz). 3D-HCCH spectra contain proton chemical shifts in two of the dimensions, and the $^{13}$C chemical shift of the carbon attached to one of the protons in the third dimension. This facilitates the identification of complete side-chain $^1$H spins systems, which in turn enables the application of heteronuclear-edited NOE-based sequential assignment method to proteins with molecular masses up to $\sim 20$ kDa [e.g. see (38)]. This method is equally applicable to the identification of the side-chain resonances for all other heteronuclear assignment strategies.

Triple Resonance Assignment Strategy. The second, and most powerful, heteronuclear approach to obtain sequence specific resonance assignments involves triple resonance ($^1$H, $^{13}$C, $^{15}$N) experiments developed primarily by Bax and coworkers [(39, 40) and reviewed in (2)]. The homonuclear $^1$H based strategies discussed above ultimately relies on the assumption that most NOEs between backbone protons are sequential; the triple resonance approach is conceptually superior since it uses only scalar (through-bond) correlations. Furthermore, the sequential resonance assignment procedure for the heteronuclear strategy does not require having a significant portion of the side-chain assignments; the sequential assignments are obtained via correlations along the backbone and the side-chains are connected to the backbone using separate experiments.

The critical aspect of the heteronuclear triple resonance strategy is the presence of non-zero scalar couplings between $^{13}$C and $^{15}$N atoms across the amide bond, which enables the complete sequential assignment by through-bond coupling only. The method makes use of a series of experiments in which backbone $^1$H, $^{13}$C and $^{15}$N resonances are correlated with each other via the large and relatively uniform one and two bond scalar couplings between them, hence the descriptor "triple resonance method". The magnitude of the scalar coupling constants along the polypeptide backbone that are used commonly in triple resonance experiments are presented in Fig. 9. The assignment process is independent of spin system type and also highly efficient: the basic approach is applicable to proteins up to $\sim 30$ kDa and in favorable cases even larger systems. Side-chain resonances are correlated to the assigned backbone atoms in separate HCCH-type experiments, as described above. Although a myriad of experiments have been developed to assign

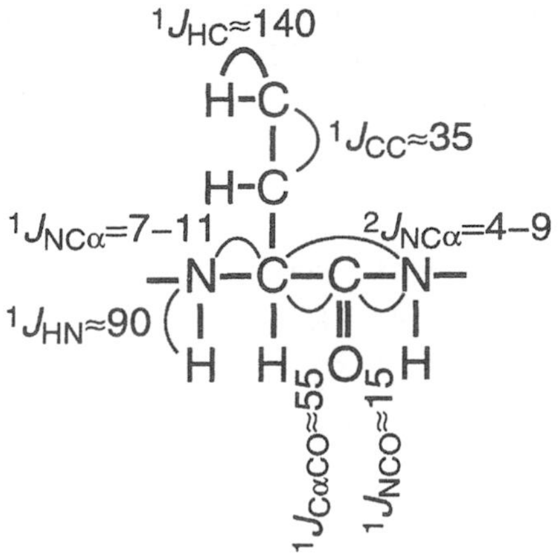

**Fig. 9** $^1$H, $^{13}$C and $^{15}$N scalar coupling constants. The range of values observed in proteins are given in the figure (in Hz). These couplings are used to assign the resonance frequencies via the triple resonance methods described in the text.

$^{13}$C/$^{15}$N labeled proteins in this fashion (see below), it is important to realize that these experiments in and of themselves, do not provide the information essential for structure calculations. While the $^1$H-$^1$H NOEs still provide the principle determinant of protein structure, the triple resonance experiments are critical to the assignment of the peaks in 2D, 3D or 4D homonuclear and heteronuclear NOESY spectra.

Over the last few years, a shorthand notation has arisen for describing the nature of the experiments and the corresponding peaks that are observed. Thus, the symbols HN, N, HA, CA, CO, HB and CB are used to describe the involvement of $H^N$, N, $H^\alpha$, $C^\alpha$, $C'$, $H^\beta$ and $C^\beta$ nuclei, respectively, in the particular pathway of the experiment; the symbols are listed in the order in which they become "excited" during the pulse sequence. Nuclei that are involved in the pathway but whose frequencies are not actually measured in the experiment are included in parentheses. Finally, transfer from a

proton is always assumed to be to the directly bonded heavy atom, hence "HNN" (i.e. transfer from HN to N) may be abbreviated to "HN". More details of this nomenclature can be found in Chapter 7 of Ref. (2).

As an example, the experiment HN(CO)CA contains correlations between the $H^N$ and N resonances of one residue ($i$), and the $C^\alpha$ resonance of the preceding residue ($i-1$). The "(CO)" indicates that the magnetization is transferred from $N_i$ to $C^\alpha_{i-1}$ via the intervening CO; indeed, this is how we know that the HN and CA frequencies measured in the experiment are in adjacent amino-acid residues. The spectrum is acquired in a 3D fashion with the three frequency axes in the spectrum corresponding to the $H^N$, N and $C^\alpha$ chemical shifts. An HNCA experiment is highly complementary because it contains correlations between HN and N resonances of one residue, and the CA resonances of the same *and* the preceding residue. Two sets of correlations arise because the one bond coupling between N and $C^\alpha$ of the same residue and the two bond coupling between N and $C^\alpha$ of the preceding residue, are both large enough for efficient transfer (Fig. 9). Due to the smaller size of the two bond coupling, some of the interresidue correlations may not be observed, especially for larger proteins.

The combination of the information from the HNCA (intraresidue and some sequential correlations from $C^\alpha$ to HN) and the HN(CO)CA (only sequential correlations between HN and $C^\alpha$) experiments provides an obvious route to sequential assignment since any given $C^\alpha$ resonance may be linked to both its intraresidue and sequential $H^N$ and N resonances. This approach breaks down if there is chemical shift degeneracy amongst the $C^\alpha$ resonances (a common problem). Ambiguities can be resolved by using additional experiments that give alternative correlations (Fig. 10), for example between sequential $H^N$ and CO groups (the HNCO experiment) and intraresidue $H^N$ and CO groups [in HN(CA)CO or (HACA)CO (CA)NH experiments (41, 42)]. Alternatively, the $H^\alpha$ spins may be linked together

by combining information from HCACO, HCA(CO)N and H(CA)NH experiments. Combinations of the above mentioned experiments give at least two and often

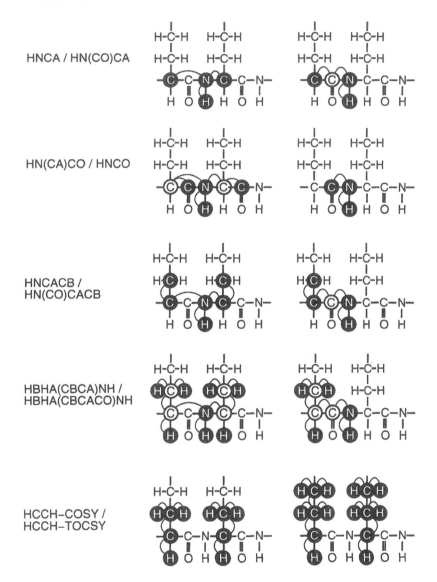

**Fig. 10** Schematic view of some triple resonance experiments used to assign $^1H$, $^{15}N$ and $^{13}C$ resonances of a protein. The thin curved lines indicate the routes by which magnetization transfer takes place. Filled circles indicate nuclei whose resonance positions are actually measured, whilst open circles indicate nuclei through which magnetization is transferred, but whose frequencies are not measured. The left-hand experiments in the top four rows all utilize coupling between $^{15}N$ and $^{13}C^\alpha$; both the intraresidue $^1J$ and the sequential $^2J$ are of sufficient size to permit transfer, hence two peaks will be observed for each amide group. Thus, the top four rows of experiments are used in pairs so that the inter residue correlations can be positively identified by the right-hand experiments. See text for more details.

more independent pathways to link adjacent amino acids, without any previous knowledge of the spin system types. In practice, knowledge of at least some of the side-chain spin system is required, in order to "align" these backbone assignments with the protein amino acid sequence; such information is usually obtained from HCCH-COSY and HCCH-TOCSY experiments (see above), or via the experiments described in the next section.

Triple-resonance experiments incorporating side-chain correlations. A particularly powerful extension of the HNCA/HN(CO)CA strategy involves the acquisition of CBCA(CO)NH and CBCANH experiments to obtain further correlations to the $\beta$ carbon (43–45). These experiments provide all of the information of their HNCA-based predecessors, plus they have peaks for the $C^\beta$ carbons. Thus, adjacent residues may be matched on the basis of $C^\beta$ chemical shift if there is substantial $C^\alpha$ chemical shift degeneracy. In addition to the extra correlation, the $C^\beta$ chemical shifts are particularly powerful because they are highly characteristic of the identity of the side-chain (46). The disadvantage of these experiments is they are more susceptible to signal losses through relaxation, and are therefore not always applicable to larger protein systems. For proteins with rather favorable relaxation properties, it is possible to extend the direct correlations from the backbone even further out along the side-chains. For example, the HBHA(CBCACO)NH and HBHA(CBCA)NH experiments may be used to align backbone amide groups based on the $H^\beta$ and $H^\alpha$ resonance frequencies. The ultimate extension of this concept involves using carbon spin-locking (TOCSY) experiments to obtain complete side-chain $^{13}C$ and $^1H$ assignments in HCC(CO)NH-TOCSY and HCCNH-TOCSY spectra (47, 48).

One additional important consideration regarding the heteronuclear side-chain experiments is that $^{13}C$ chemical shifts are highly characteristic of their chemical environment. Thus, the specific pattern of $^{13}C$ shifts observed is highly dependent on the identity of the amino acid side-chain, very much more so than the pattern of $^1H$ chemical shifts. As a consequence, the heteronuclear approach readily lends itself to automated analysis and considerable effort has been put forward in this area (49–53). However, automated analysis has yet to be shown convincingly as a robust approach to obtaining complete assignments. The strategy invariably breaks down due to the incompleteness of the data and the tendency for severe resonance overlap in certain regions of the $^1H$, $^{13}C$ and $^{15}N$ spectra.

$^2H$ incorporation. Although $^{13}C$ and $^{15}N$ enrichment allows a significant increase in the size of protein that can be analyzed in detail by NMR, the general applicability of the approach is eventually limited by the rapid relaxation during the course of the experiment, particularly for $C^\alpha$. In fact, the intrinsic line widths of proteins are increased by $^{13}C$ enrichment due to scalar and dipolar couplings among the $^{13}C$ and $^1H$ atoms. The rapid relaxation may be circumvented in part by preparing a sample that is highly deuterated. Bacteria can be adapted to grow in media containing only $^{13}C$, $^{15}N$ and $^2H$ labeled nutrients, leading to proteins that have >90% incorporation of these isotopes (54, 55). The reduced rate of relaxation of $^2H^\alpha$–$^{13}C^\alpha$ compared to $^1H^\alpha$–$^{13}C^\alpha$ has been shown to increase the efficiency of triple resonance experiments involving $C^\alpha$ (56). $^2H$ decoupling during the pulse sequence offers further benefits (56). As a caveat to this approach, the only protons remaining in a protein sample that is highly deuterated will be those that are able to exchange with solvent, namely the amide protons. Thus, analysis of other protons, and in particular NOEs involving side-chain protons, are seriously compromised.

With a triple-labeled sample (i.e. $^2$H, $^{15}$N and $^{13}$C), $^{15}$N, $^{13}$C and $^1$H$^N$ resonances can be readily assigned (57), secondary structure may be estimated from $^{13}$C chemical shifts, and in favorable cases, a low resolution view of the global fold may be obtained from NOEs between amide protons (58). However, the lack of side-chain protons severely limits the precision with which such a structure can be determined (from NOEs). The resolution can be improved by utilizing selective biosynthetic labeling patterns so that certain methyl groups are protonated in an otherwise deuterated protein [e.g. (59)]. This strategy can provide very useful distance constraints to increase the definition of the protein core, particularly for helical proteins. The optimal solution to the conundrum of spectral improvement versus decreased structural resolution is still a source of debate, with some groups proposing the use of a single fractionally $^2$H labeled sample for all data collection (50–60% appearing optimal) (60), whilst others recommend data collection on two samples, one with high levels of $^2$H, and one with no $^2$H incorporation (61). The latter "combined" strategy is complicated by secondary isotope shifts, in particular the variation of $^{13}$C chemical shifts with the number of attached protons vs. deuterons; this effect seems to vary significantly with amino acid type and position within the side-chain (62). The fractional labeling approach suffers from sensitivity losses in NOESY data, and peak broadening due to secondary isotope shifts. However, no matter which strategy is followed, the use of $^2$H labeling clearly enables access of NMR structural studies to proteins that are considerably larger than those accessible by other solution methods.

Site-directed Labeling Strategies. In all of the above discussions, we have tacitly assumed that all resonances in a molecule will be assigned and analyzed from a single sample. However, this need not be the case (63). For example, it is possible to build up assignments in a piecemeal fashion by selectively labeling only a certain subset of amino acids (64–66). Although much more expensive in terms of sample preparation, this approach offers a considerable degree of spectral simplification. In other instances, only certain nuclei are of interest (e.g. at a protein's active site) and these can be specifically labeled, identified and used to monitor the system. These goals can be achieved through the use of unique patterns of labeling the protein, wherein only one or a few types of atoms or amino acids are enriched in $^{13}$C and/ or $^{15}$N. Residue-specific reverse labeling, wherein the protein is fully labeled (e.g. $^2$H or $^{13}$C/$^{15}$N) except for specific unlabeled amino acids has also been utilized in a similar manner (67–69). Knowledge of biosynthetic pathways can be also extremely valuable, providing a means to produce samples with unique functional group-specific labeling patterns (59, 70). One important example of this is in the stereo-specific assignment of the pro-chiral methyl groups in valine and leucine residues (71). While these approaches appear to have vast potential, they are currently used only in specialized applications.

## IV.   SECONDARY STRUCTURE AND GLOBAL FOLD

Not long after the homonuclear $^1$H sequential resonance strategy was developed, it was shown that details of the local backbone geometry can be obtained by a straight-forward extension of the analysis. The relative intensities of sequential $d_{NN}$, $d_{\alpha N}$ and $d_{\beta N}$ NOE cross peaks and the

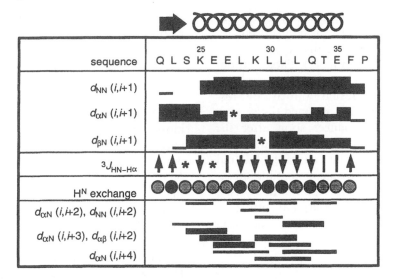

**Fig. 11** Summary of NMR parameters that are used to define elements of regular secondary structure. NOEs observed between two residues are indicated by boxes or bars; the thickness of the bar indicates the relative NOE intensity. Upward and downward pointing arrows indicate $^3J_{HN-H\alpha} > 8.0$ Hz or $< 6.0$ Hz, respectively. Shaded and filled circles indicate slow or very slow rates of amide proton exchange with solvent, respectively. Asterisks indicate that data could not be determined because of resonance degeneracy. The presence of intense $H^N$–$H^N$ NOEs, small values of $^3J_{HN-H\alpha}$ and medium range NOEs indicate the presence of a helix between residues K25 and E35. The intense $H^\alpha$–$H^N$ sequential NOEs and large $^3J_{HN-H\alpha}$, together with other longer-range NOE information (not shown) and slowed $H^N$ exchange with solvent, indicate that residues Q22–S24 are part of a short anti-parallel $\beta$-sheet. The data in this figure are actually derived from the four helix bundle calcium-binding protein calbindin $D_{9k}$ in the apo state [adapted from Ref. (32)].

magnitude of backbone amide scalar couplings ($^3J_{HN-H\alpha}$) are characteristically different for helical and extended conformation. The observation of intense $d_{NN}$ NOEs and small $^3J_{HN-H\alpha}$ coupling constants ($< 6.0$ Hz) are indicative of helical conformation (e.g. see Fig. 11); observation of intense $d_{\alpha N}$, weak $d_{NN}$ and $d_{\beta N}$ NOEs and large $^3J_{HN-H\alpha}$ coupling constants ($> 8.0$ Hz) are indicative of extended conformation (24). Specific elements of secondary structure can be assigned from the observation of these patterns over several consecutive residues, in addition to complementary evidence of hydrogen bonds from slowed backbone amide proton exchange and characteristic patterns of medium range NOEs in helices or long range NOEs across $\beta$-strands. The chemical shifts of C', $C^\alpha$, $C^\beta$ and $H^\alpha$ are

highly sensitive to backbone conformation and it is possible to identify elements of regular secondary structure from the Chemical Shift Index (CSI), a cumulative analysis of these chemical shifts (72). Thus, secondary structure can be identified directly from chemical shifts obtained by the heteronuclear triple resonance assignment approach without recourse to NOE information.

Once the elements of secondary structure are identified, it becomes possible to establish the nature of the global fold of many proteins. For $\beta$-sheet proteins or domains, the supersecondary structure, including the packing order of the strands and their relative (parallel or anti-parallel) orientation, is directly evident from the characteristic cross-strand NOEs between backbone protons. The nature of the packing of $\beta$ supersecondary structures and helical elements can be deduced frequently from a few critically positioned long-range NOEs (Fig. 12). Thus, without recourse to extensive calculations, important structural results (albeit of low resolution) can be obtained in a very straightforward manner soon after the sequential assignments are made. It is important to note that a coarse global fold represents the maximum possible structural resolution at this level of analysis; a very considerable additional effort is required to obtain a higher resolution structure.

## V. THE GENERATION OF EXPERIMENTAL RESTRAINTS FOR STRUCTURE DETERMINATION

The sequence specific assignment of NMR resonances requires many different experiments, weeks of spectrometer time, and substantial input from the spectroscopist. However, a table of chemical shifts does not, by itself, reveal much information about the three-dimensional structure and dynamics of the protein under investigation. They are nonetheless an essential prerequisite to achieving more biochemical goals. The chemical shift, scalar couplings and dipolar couplings are all sensitive to molecular conformation, and quantitation of these parameters provides the basis of all structural analyses using NMR data. Although the chemical shift is the easiest parameter to measure, interpretation of chemical shifts in structural terms is very difficult due to the subtle interplay between internal dynamics and conformation, and the large number of factors contributing to this highly sensitive parameter [e.g. (73)]. Dipolar interactions and scalar couplings form the majority of restraints used for all protein structures calculated to date using NMR data, although other parameters such as certain

subsets of chemical shifts are coming into increasingly greater use.

### A. Distance Restraints Between Protons

The NOE (defined in Sec. 2) is by far the most important parameter in the determination of a protein structure from NMR data. For a rigid molecule, the size of an NOE between two protons depends upon their separation, on the distribution of other protons in the molecule, and on the time scale of global reorientation of the protein (2–4, 74). Once magnetization has been transferred via the NOE from $H^A$ to $H^B$, it may then start to transfer to other protons that are close to $H^B$ (e.g. proton $H^C$). Thus, a cross peak will be observed between $H^A$ and $H^C$ even though the separation of $H^A$ and $H^C$ may too great for direct NOE transfer of magnetization between them; this effect is known *as spin diffusion*. Given that the NOE does have a $1/r^6$ dependence on inter-nuclear separation, one might envision using the size of an NOE ($S_{ref}$) between two protons that are a known distance apart ($r_{ref}$, e.g. fixed by covalent geometry) to determine other inter proton distances ($r_i$) by simply taking the ratio of the NOE intensities:

$$r_i = r_{ref}(S_{ref}/S_i)^{1/6}$$

This will not be accurate, however, because of the dependence of a given NOE intensity on the location of all other protons in the molecule; this will be different for each proton-proton pair. In addition, motions within the protein may also have a differential effect on some NOEs.

For these reasons, the NOEs are not interpreted as specific distances, but rather are used to generate distance ranges with explicit upper and lower bounds. The NOESY cross-peaks are usually grouped into broad categories (strong, medium, weak, etc.) on the basis of their intensity. A distance upper bound for each category

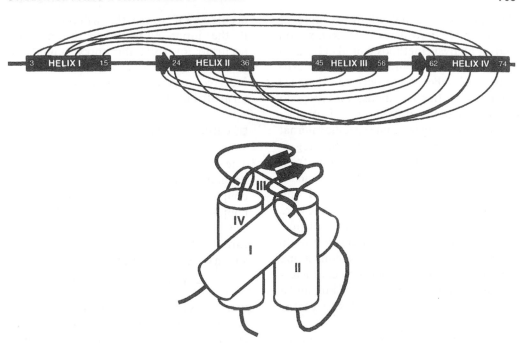

**Fig. 12** The use of long-range NOEs to determine a global fold. Helices and β-strands are represented by bars and arrows, respectively in the top panel. A selection of the NOEs observed between the elements of regular secondary structure are indicated by the thin lines. The NOE data indicate that: adjacent helices are roughly antiparallel, Helix II and Helix IV are parallel, Helix I and Helix III are not in contact and that the β-strands are arranged in an antiparallel fashion. This information allowed the global fold to be deduced (lower view) from a subset of only 17 critically important NOEs [adapted from Ref. (32)].

(e.g. 3.0, 4.0 and 5.0 Å) is estimated empirically for each NOESY spectrum. The distance bound categorization is based on observed NOE intensities between protons of known separation (*e.g.* geminal methylene protons, vicinal protons in aromatic rings) or between protons in regions of regular secondary structure whose separations are known (*e.g.* sequential $H^\alpha$–$H^N$, cross-strand $H^\alpha$–$H^\alpha$ and cross-strand $H^\alpha$–$H^N$ NOEs in β-sheets, or $H^\alpha$–$H^N$ medium range NOEs in helices). The upper bounds determined in this manner need to be set conservatively so that NOE peaks that have contributions from spin diffusion (i.e. the NOE peak is artificially large relative to the actual proton separation) are not restrained too tightly.

Ultimately the inherent inaccuracy of estimating distance bounds should be compensated for by more rigorous relaxation matrix analysis [e.g. see discussion in (75)]. Once the first three-dimensional structural coordinates are available, inter-proton distances can be determined from these atomic coordinates and a list of theoretical NOE intensities can be calculated using a computer program that is able to solve the relaxation matrix equations for the system of interest (76). This process is known as complete relaxation matrix calculation or NOE back-calculation. After suitable scaling, the calculated NOE intensities are compared to the experimentally observed NOE intensities to judge how well the structures reproduce the experimental data. Usually, the agreement is judged by a figure of merit, or NOE $R$-factor (77). Relaxation matrix analysis methods can also be used to obtain more accurate and precise estimates of the NOE-derived distance bounds (78). In all methods, it is typical to assume overall motions are isotropic, while accounting for some specific internal motions such as aromatic ring flipping or methyl group rotation. Deviations from isotropic motion are one source of inconsistency between the experimental and theoretical NOE intensities. Methods are available to incorporate non-isotropic motional properties of proteins, but to do so requires additional time-consuming experimental measurements to characterize the dynamic properties in detail, and a large increase in the complexity of the computational analysis (79).

In practice, the upper bounds for the distance constraint ranges are set on the basis of the NOE and the lower bound is set as the sum of the van der Waal's radii. For a $^1$H-only analysis of a small protein (under 10 kDa), 8–12 useful NOE peaks are commonly identified per residue. The superior resolving power afforded by $^{13}$C or $^{15}$N greatly increases the number of assignable NOEs, and in some cases, as many as 25-30 NOE restraints per residue have been used to determine a structure (80). In this context "useful" indicates that the ascribed upper bound is more restraining than allowed by covalent geometry; certain intraresidue or sequential NOEs though assigned, are not useful because the upper bound set from the NOE-based estimation is less constraining than the limitations based on the covalent geometry. The precision of the final structures will depend not only on the number of NOE restraints identified, but also on the distribution of the NOEs throughout the molecule. Structures calculated with a high proportion of "long-range" NOEs (between residues more than four amino acids apart in the primary sequence) tend to provide the highest precision and

accuracy. Of course, this will only be true if the input NOE upper bounds are accurate; overly tight upper bounds or NOE assignment errors can lead to incorrect structures or ensembles that have an artificially high precision (78, 81).

## B.  Dihedral Angle Restraints

The magnitude of scalar couplings are dependent on the bond angles between the coupled spins. Vicinal three-bond ($^3J$) couplings are particularly useful, in that they are dependent on the intervening three-bond dihedral angle. This relationship has been expressed in the form:

$$J = A\cos^2\theta + B\cos\theta + C$$

which was first described by Karplus using valence bond theory (Fig. 13) (82). The constants $A$, $B$ and $C$ depend upon the particular nuclei involved, and $\theta$ is the dihedral angle formed by the three bonds. Values for $A$, $B$ and $C$ have been derived by correlation of observed $J$ values with the corresponding dihedral angles measured in protein structures determined by x-ray crystallography or NMR methods (83–86). To determine constraints on dihedral angles along the backbone and certain side-chains, measurements of $^1$H-$^1$H scalar couplings can be made from homonuclear $^1$H COSY-type experiments or from corresponding heteronuclear experiments for isotopically labeled samples. For labeled proteins, additional vicinal scalar couplings can be measured for the $^{13}$C and $^{15}$N nuclei ($^{15}$N-$^1$H, $^{13}$C-$^1$H, $^{13}$C-$^{13}$C, $^{15}$N-$^{13}$C) by using E-COSY-type or intensity-modulation strategies. General reviews on the methods for measuring coupling constants are available (87, 88).

Note that given the shape of the Karplus curve (Fig. 13), as many as four different conformations may give rise to the same value of $^3J$. The dihedral angle range enforced during structure calculations must include all of the valid sol-

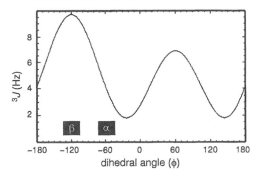

**Fig. 13** A Karplus curve. The variation of $^3J_{HN-H\alpha}$ is plotted as a function of the backbone dihedral angle $\phi$. The dihedral angle formed by the four atoms ($H^N$–N–$C^\alpha$–$H^\alpha$) $= |\phi - 60°|$. The approximate $\phi$ angles found in $\alpha$-helices and $\beta$-sheets are indicated by the black boxes. The curve was generated using the following constants in the Karplus equation: A = 6.4, B = −1.4, C = 1.9.

utions to the Karplus curve, although some of the possible orientations may be ruled out on steric grounds. The various dihedral angle ranges may also be distinguished experimentally by measuring several different couplings across a common central bond, since these usually have different angular dependencies. For example, the $\chi_1$ angle can be determined by measuring $^3J_{H\alpha-H\beta}$ and $^3J_{15N-H\beta}$, which is sufficient to distinguish the three classical $\chi_1$ rotamers (Fig. 14). As with the NOE distance restraints, dihedral angle restraints are set conservatively (usually not restrained to ranges less than 60°) to allow for limited motional averaging within the bottom of the low energy rotamer wells. If substantial motional averaging between rotamers occurs, angle constraints are not meaningful, even if the coupling constants are very accurately measured (89). For example, extensive backbone motion will result in a $^3J_{HN-H\alpha}$ coupling constant of about 7.0 Hz, the weighted average between extended and helical conformations (values of $\sim 7$ Hz are observed in short, unstructured linear peptides). Since very large and very small

values are possible only if there is limited motional averaging, the reliability of the interpretation of $^3J$ is highest for scalar couplings at the extrema expected from the Karplus relationship (89).

$J$ values can also be used in combination with NOEs to define dihedral angle constraints. This kind of analysis enables stereo-specific assignments for the prochiral methylene protons and the methyl groups of valine and leucine. As shown in Fig. 14, in certain cases, the relative magnitude of specific NOEs can be sufficient to distinguish among $\chi_1$ rotamers (89). The combined scalar coupling and NOE analysis is often facilitated by grid-search programs [e.g. HABAS (90), STEREOSEARCH (91), DTAGS (92)]. Such methods can provide much more extensive and systematic analysis of dihedral angle constraints on the local geometry of proteins.

## C. Distance Restraints for Hydrogen Bonds and Integral Waters of Hydration

The amide protons along the polypeptide backbone are acidic and exchange with solvent protons. In folded proteins, slow rates of amide proton exchange are associated with shielding of the amide proton from solvent, most commonly from hydrogen bonding interactions (93). Knowledge of hydrogen bonds can be converted into specific distance and angular constraints between the atoms involved in the interaction. However, the observation of slowed amide proton exchange only indicates the hydrogen bond donor, not the acceptor. Thus, restraints for particular hydrogen bonds are usually only enforced in well defined regions of regular secondary structure, in which only one possible hydrogen bond acceptor is consistent with the NOE data defining the secondary structure (24). Hydrogen bond restraints can have a large impact on the nature and precision of the resulting

| $\chi_1$ | +60° | 180° | −60° |
|---|---|---|---|
| Rotamer | 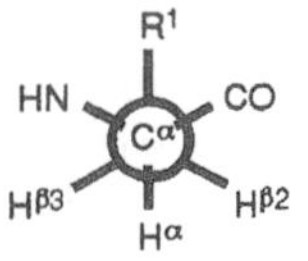 (rotamer diagram: HN, CO, $R^1$, $C^\alpha$, $H^{\beta3}$, $H^{\beta2}$, $H^\alpha$) | 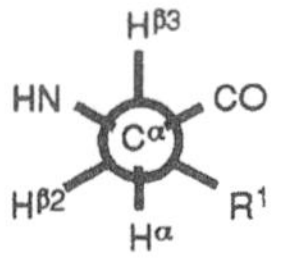 (rotamer diagram: HN, CO, $H^{\beta3}$, $C^\alpha$, $H^{\beta2}$, $R^1$, $H^\alpha$) | 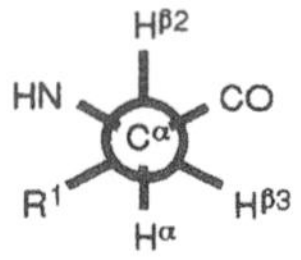 (rotamer diagram: HN, CO, $H^{\beta2}$, $C^\alpha$, $R^1$, $H^{\beta3}$, $H^\alpha$) |
| $^3J_{\alpha\beta2}$ $^3J_{\alpha\beta3}$ | small small | small large | large small |
| $^3J_{N\beta2}$ $^3J_{N\beta3}$ | large small | small small | small large |
| NOE $(\alpha,\beta2)$ NOE $(\alpha,\beta3)$ | intense intense | intense weak | weak intense |
| NOE $(H^N,\beta2)$ NOE $(H^N,\beta3)$ | weak intense | intense intense | intense weak |

**Fig. 14** Analysis of side chain rotamers and stereospecific assignments of $\beta$-methylene groups. Side chains occupying only one of the three staggered rotamer positions depicted at top give rise to characteristic patterns of $^3J$ and intra residue NOEs. Measurements of both $^3J_{\alpha\beta}$ and $^3J_{N\beta}$ lead to unambiguous stereospecific assignment and rotamer determination. For $^1$H-only data, assignments cannot be made unambiguously without analysis of intra residue $H^N$–$H^\beta$ NOEs; this can be complicated by the dependence of the inter proton distances on the backbone $\phi$ angle. Small and large $^3J_{\alpha\beta}$ are $<4$ and $>11$ Hz, respectively, whilst small and large $^3J_{N\beta}$ couplings are $<1.5$ and $>4.5$ Hz, respectively.

structures, particularly in the early stages of refinement, and should be applied only if the acceptor can be identified with a high degree of certainty.

NMR also offers the possibility to identify individual water molecules that form an integral part of the protein structure. The techniques to identify such water molecules involve the measurement of a group of NOEs from the water resonance to specific protons of the protein (94). These are intrinsically low sensitivity experiments and the most efficient methods are gradient versions of homonuclear 3D ROESY-TOCSY (95) and 2D water-selected $^{15}$N-$^1$H ROESY-HSQC and $^{13}$C-$^1$H ROESY-HSQC (96, 97). Careful manipulation of the water sig-

nal is critical to the success of these low sensitivity experiments (98). The integral water molecules are usually identified with the assistance of the three-dimensional structures, and are utilized only in the later stages of refinement of solution structures.

## VI. STRUCTURE CALCULATIONS

The NMR-derived distance restraints are analyzed with the aid of computational methods, to arrive at coordinates for the three-dimensional solution structure. The objective of the calculation is to identify all of the regions of conformational space that are compatible with the observed NMR parameters. There is no unique solution to this problem because only a subset of all possible distance, dihedral and hydrogen bonding restraints are observed, and because the restraints are included as ranges of allowed values due to experimental uncertainties. Consequently, the procedure involves repetition of the calculation many times, until an ensemble correctly representing all of the available conformational space is produced. It is intrinsic to this procedure that the structure is only properly represented by the full ensemble and that each member satisfies the input data equally well. The two most common approaches to generation of structures are distance geometry (DG) and restrained molecular dynamics (rMD). Whichever method is used to generate structures, great care must be taken to ensure that conformational space is adequately sampled. Thus, a "good" ensemble of structures is one in which violations of the input restraints are minimized, while the RMSD between members of the family is at its maximum for the specified region of the conformational space.

## A.   Distance Geometry

Distance geometry is often used to calculate an initial set of structures since the method does not require, and thus cannot be biased by, an initial model of the structure (99). Popular implementations of DG use either the metric matrix algorithm (100, 101) or the variable target function (102, 103). The primary weakness of the distance geometry approaches is the lack of chemical knowledge being incorporated into the output structures. Thus, distance geometry structures need to be further refined in a molecular mechanics force field. A problem with the metric matrix algorithms is that they are very computationally expensive because the size of the matrix is proportional to the square of the number of atoms in the protein. The primary advantage of the distance

geometry approach is the thoroughness of the conformational search.

## B.   Restrained Molecular Dynamics

Restrained molecular dynamics (rMD) simulations provide both an alternative and a complement to DG approaches. The method is based on use of standard molecular mechanics force fields supplemented by flat-bottomed, square-well pseudo-energy restraining potentials for the NMR-derived restraints (104, 105). The basic concept of this approach is that NMR-derived potentials guide the structure towards the desired low-energy conformation. The refinement requires an initial "starting" structure, which is first heated to some high target temperature, then slowly cooled back down to either room temperature or 0 K. This "annealing" process is designed to search a wide area of conformational space in the vicinity of the starting structure, then find the most stable (lowest energy) conformation. In addition to the standard chemical potentials and the experimental restraints, restraints are imposed to enforce proper chirality and the planarity of aromatic rings and peptide units.

The most computationally efficient implementations of the rMD method use a simplified force field in which only bond length, bond angle and repulsive van der Waal's terms are retained (electrostatic and attractive van der Waal's terms are ignored). These approaches are termed restrained dynamical simulated annealing (rSA) (106). Some groups have reported rSA or energy minimized DG structures for the final structural ensemble, whereas others refine such structures further by formal rMD with the complete force-field. A highly computationally efficient hybrid distance geometry-rSA method has been reported (107). In this approach, initial structures are generated in which only a subset of atoms are included, with the remainder added by reference to standard amino acid templates. The resulting structures generally have the correct global fold but poor local geometry. These are then refined (annealed) to remove the local violations of NMR restraints and covalent inconsistencies present in the starting structures using the simplified force-field noted above.

In addition to the standard implementation of distance and dihedral angle restraining potentials in rMD simulations, direct NMR parameter refinement has been developed (NOE intensities (108, 109), chemical shifts [73, 110–113), scalar coupling (114)]. The strategy involves implementing pseudo-energy potentials to force agreement between the NMR parameters calculated from the current structural model and the experimental values. Direct NOE intensity refinement is the most common example. This method proceeds via relaxation matrix calculations on the structure during the refinement process: the pseudo-energy term depends on the difference between observed and back-calculated NOE intensities (108, 109). Thus, the structures are refined so as to minimize the NOE intensity R-factor and maximize the agreement between the structures and the experimental NOE intensities. This process requires a complete relaxation matrix analysis to be performed at every step of the annealing cycle and therefore is very computationally demanding.

For any of the direct NOE refinement methods to be valid, the accuracy of the experimental NOE intensities must be high. This places emphasis on obtaining NOE spectra with very flat base lines (even slight rolls or tilts in the baseline can have dramatic effects on cross-peak volumes) and otherwise free of spectral artifacts. In addition, a dynamical model must be incorporated that is capable of reproducing the effects of molecular motion known to be present in proteins, such as fast rotation of methyl groups and 180° flips of aromatic side-chains (115). Problems with

systematic deficiencies in the experimental data (116) and incomplete accounting of dynamic effects have so far prevented the wide-spread use of back-calculation, even though the procedure does provide one of the few ways to independently cross-check the accuracy of protein structures against the experimental NMR data.

## VII. STRUCTURE REFINEMENT AND ASSESSMENT OF THE QUALITY OF STRUCTURES

Determination of a structure using NMR data is an iterative process involving a series of progressive steps of calculating structures and identifying additional experimental constraints. In the first round of calculations, only NOE restraints that are unambiguous on the basis of chemical shifts are included. While these structures tend to be of relatively low precision, they are completely reliable. These structures are used to assign additional NOEs, for example those with possible contributions from two proton pairs on the basis of chemical shift, but for which only one pair of protons is in close proximity in the structure. Inspection of initial structures may also permit additional stereospecific assignments or hydrogen bonding partners to be identified. The course of NMR refinement involves a series of calculations incorporating more and more data, and yielding structures that are of increasingly higher precision. At each stage of the analysis, the structures are analyzed to see whether they have good covalent geometry, and how well they agree with the input data. These analyses give an indication of how reliable the structure is, which can be important if the results are to form the basis of other experiments. Further, such analysis can be used to rank structures in order of quality, and members of the ensemble of low quality can be ignored from any further analysis. One advantage of this iterative process is that errors in the data can be corrected before the final ensemble of structures is determined.

Discrepancies between the input distance and dihedral angle restraints occur because they are not rigidly enforced during the structure calculations; mutually exclusive restraints (e.g. due to conformational heterogeneity), or in rMD and SA calculations other components of the force field, may prevent every constraint from being satisfied. Tabulations of such restraint violations are an important consideration of the quality of a structure. The violations are usually listed as the number of violations above a specified threshold, or as an RMS average over all violations. Structures of high quality and integrity should not have any violations above 0.1 Å, or have RMS deviations greater than 0.010 Å. Failure to meet these criteria suggests that either NOEs have been misassigned, or else the calibration of NOE intensity to distance upper bounds is incorrect (typically, upper bounds are too tight). Annealing algorithms are very efficient at minimizing the violation of incorrect restraints, especially in areas of low restraint density because the energetic penalty incurred for minimizing the erroneous restraint can be distributed among the other components of the force field. Thus, even though restraints may be violated by only a few tenths of an Ångstrom, the upper bounds may actually be in error by many Ångstroms. Restraints (including NOEs, dihedral angles and hydrogen bond distances) should always be analyzed for cases that are consistently violated (i.e. violated in most if not all members in the ensemble) even if the violation is small, because this is a tell-tale sign of a restraint error.

One commonly-used parameter to monitor structure calculations is the precision to which each coordinate position is defined. As noted in Section 5, due to the imprecise nature of the input NMR restraints, an ensemble of structures must be calculated and each of these are valid representations of the data. An example of an ensemble of NMR solution structures is shown in Fig. 15A. The degree of coordinate uncertainty between members of the ensemble depends upon the number, type and distribution of restraints throughout the molecule (117, 118). The precision is defined by calculating a root mean squared displacement (RMSD) of a subset of atoms either from a mean structure, or as a mean of all pairwise comparisons within the ensemble. For a uniform distribution of coordinates, these two methods give results that are arithmetically related (81) with the former being a factor of $\sqrt{2}$ smaller for large ensembles. The RMSD from the mean is most often reported. A variety of RMS values are often given for different regions and the entire molecule. Since the distribution of restraints is not uniform, some parts of the structure are invariably well defined (e.g. the backbone atoms of secondary structure elements or side-chains in the hydrophobic core), whilst others will be poorly defined (typically, chain termini, solvent exposed side-chains and loops). This relationship is shown numerically in the top two panels of Fig. 16, and is also readily apparent in the corresponding structural ensemble in Fig. 15A. Clearly, the structure of apo calbindin $D_{9k}$ is not uniformly defined by the NMR data, with the termini, the $Ca^{2+}$-binding loops and Helix III being less well defined (RMSD > 1.2 Å).

The ensemble of conformations used to describe the structure should contain enough members to maximize the RMSD, as only then will all conformations consistent with the primary data be represented. Methods have been proposed to determine the minimum number of structures required to accurately represent the complete conformational space (119–121). Although a low RMSD makes analysis of an ensemble easy, minimizing an RMSD

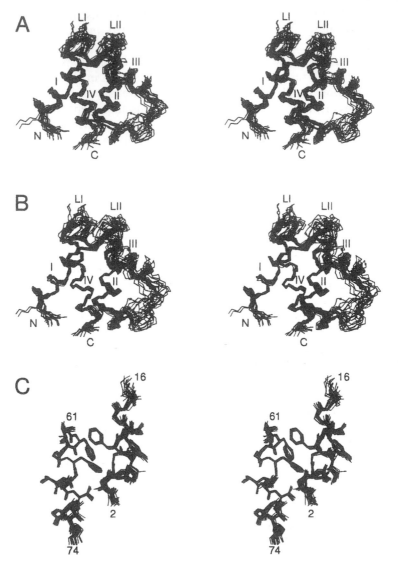

**Fig. 15** The solution structure of apo calbindin $D_{9k}$ (158). (A) Ensemble of structures overlaid on the backbone N, $C^\alpha$ and C atoms of residues 2-74 (mean RMSD from the mean structure $= 1.11 \pm 0.14$ Å). Only backbone N, $C^\alpha$ and C atoms are shown. Helices are labeled I–IV and the $Ca^{2+}$-binding loops as LI and LII. (B) Same ensemble overlaid using the backbone heavy atoms of helices I, II and IV only (mean RMSD from the mean structure over all three helices $= 0.38 \pm 0.08$ Å). (C) Overlay of all heavy atoms of Helices I and IV showing the well defined nature of the side chains packing in the hydrophobic core between these two helices.

to the detriment of structural quality must be avoided. Including overly tight upper bounds for distance restraints (see below), or failure to account for possible motional averaging in the conversion of scalar couplings to dihedral angle restraints can lead to artificially well defined or "pinned" structures (81). The RMSD is an extremely valuable tool to follow the progression of structure refinement, but since the basis

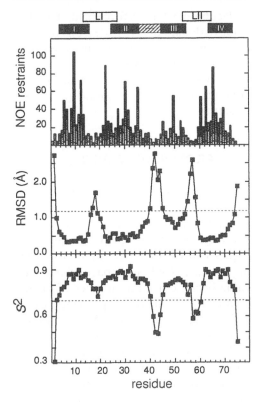

**Fig. 16** Comparison of input restraints, co-ordinate uncertainty and backbone dynamics for apo calbindin $D_{9k}$ (131). The distribution of NOE-derived distance restraints is plotted in the upper panel. The mean RMS displacement of the backbone atoms from the mean structure is plotted in the middle panel, with a dotted line drawn at the value of 1.2 Å. The generalized order parameters ($S^2$) determined from the "model-free" analysis of $^{15}N$ relaxation data are presented in the lower panel, with the dotted line drawn at 0.7. The helices, $Ca^{2+}$-binding loops and the linker are identified by solid, open and shaded bars above the graph, respectively.

for comparisons are highly subjective, it is not a reliable parameter to gauge the absolute quality of a structure.

The average number of restraints per residue is an important parameter to consider when assessing the quality of an NMR structure (80). Structures of low resolution may be obtained with as few as five restraints per residue, while the most precise structures obtained from homonuclear $^1H$ data may have >15 restraints per residue. For the latter cases, the mean RMS deviation of backbone atoms from the mean structure may be <0.5 Å for the well defined regions of secondary structure. The example shown in Fig. 15 for the apo state of calbindin $D_{9k}$ is determined to this level of resolution (13.3 NOE restraints/residue; mean RMSD of backbone atoms from the mean structure = $0.60 \pm 0.10$ Å for regions of regular secondary structure). The use of $^{15}N$ or $^{13}C$ labeling usually allows many more $^1H$-$^1H$ NOEs to be unambiguously identified, increasing the number of restraints per residue to between 20 and 25, and leading to the highest precision structures (backbone RMSD in the range 0.3–0.5 Å). Thus, many hundreds of NOE cross-peaks must be assigned unambiguously for use in these calculations. In addition, the utilization of dihedral angle restraints and the stereospecific assignment of prochiral groups (especially $C^\beta H_2$ groups, and $(CH_3)_2CH$ groups of valine and leucine residues) are critical to obtaining structures of high precision (90).

One measure of the quality of a structure coming into increasingly greater use is the comparison of the structural characteristics to standard values in protein structure databases. Statistical surveys of a large number of high resolution crystal structures have been used to compile mean values and standard deviations of the bond lengths, bond angles, and dihedral angles found in proteins (122). Large deviations from such values are often indicative of structures that have been calculated using erroneous restraints, or are otherwise incorrect. Such deviations are usually presented as RMS deviations from "ideality." The rMD or rSA calculations actually restrain the protein to have bonds and angles close to some equilibrium value and tabulations of the energies and forces associated with the deviation from these values are readily generated; low values for these energy terms provide another indication of high structural quality. Note that there are many force fields in common use today, all of which have slightly different "ideal" values. Thus, analysis of structures using different force fields may lead to different conclusions about deviations from "ideality".

The value and distribution of backbone and side-chain dihedral angles provide another means to assay the quality of a structural ensemble. As noted many years ago, not all protein backbone conformations are energetically equivalent and some combinations of $\phi$ and $\psi$ are distinctly disfavored (123). Computer programs such as PROCHECK or PROCHECK-NMR (124, 125) categorize dihedral angles into favorable, allowed, generously allowed, and unfavorable ranges based on structural databases, and provide a ready means to tabulate and display these angles. Examination of the residues in each category allows a quick identification of potential problems. The well ordered, low RMSD regions of a structure are expected to have dihedral angles within the allowed ranges for all members of the ensemble.

Van der Waal's or Lennard-Jones energy terms are also reported as a means to assay the quality of structures, with a large negative value being indicative of a well packed hydrophobic core. This energy term is actively minimized during rMD refinement, hence is often large and negative for the final structures. A simplified, approximate van der Waal's term is used in the hybrid rSA and related refinement strategies. The full Lennard-Jones energy is then taken as an unbiased view of the quality of the structures, since it is not directly included in the SA refinement. However, the exact choice of atomic radii and the parameters defining the Lennard-Jones potential have a large effect on the sign and magnitude of the resulting energy. As with any term from a molecular mechanics force field, the actual values calculated have little if any meaning except in a relative sense, since there is no scaling or reference state. Thus, the Lennard-Jones energy is not a particularly good measure of the absolute quality of a structure.

## VIII. PROTEIN DYNAMICS

Analysis of not only structure, but also dynamics, can be of utmost importance in understanding a protein biopolymer. Although a number of experimental techniques can characterize internal motions in proteins, NMR is of particularly great value because a large number of nuclear sites can be studied independently over a wide range of time scales. Analysis of protein dynamics can also be important for understanding the consequences of ligand binding or site-specific mutation. The motions of proteins are highly complex, in particular because there are so many degrees of freedom. Correspondingly, motions occur on multiple timescales, spanning a range greater than 10 orders of magnitude! For example, even in a small protein such as calbindin $D_{9k}$ (75 amino acids, $M_r$ 8600), a variety of dynamical processes have been characterized at widely different timescales: backbone amide exchange (minutes to days) (126); *cis-trans* proline isomerism (minutes) (127); aromatic ring flips (ms) (128); $Ca^{2+}$-binding events ($\mu s$ to ms) (129); backbone librations (ps to ns) (130, 131). The types and approximate time scales of motion occurring within a protein, and the NMR parameters that are sensitive to them, are summarized in Fig. 17.

The most obvious indicator of internal dynamics of proteins manifested in NMR spectra is the width of specific resonance lines. Typically, narrow lines can be correlated with regions of high flexibility, and broad lines with slow motional processes, however, quantitative

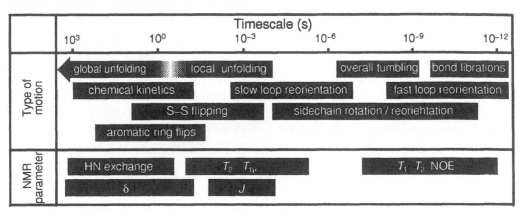

**Fig. 17** NMR analysis of protein dynamics. A selection of internal motions experienced by proteins is shown by the upper bars, while the lower bars indicate NMR parameters whose quantitation can lead to characterization of the motion. Note that the motions listed in the upper panel may occur over a wider range of times than that depicted, but may not be amenable to study by the NMR parameters listed below; e.g. global unfolding may take longer than $10^3$ s whereas aromatic ring flips may occur in a much shorter period than $10^{-1}$ s. Notes: HN exchange, amide proton exchange with solvent; $\delta$, slow "chemical" exchange leading to multiple resonances for a particular nucleus; $T_2$ $T_{1\rho}$, linewidth variations obtained from measurement of transverse relaxation times; $J$, scalar coupling indicative of motional averaging; $T_1$ $T_2$ NOE, measurement of $^{15}$N/$^{13}$C longitudinal, transverse and $\{^1$H$\}$ NOE relaxation parameters and analysis by the model-free or other approach.

analysis is complicated. Measurements of the temperature dependence of line widths and classical dynamic NMR theory can be applied, for example, to examine the flipping of aromatic rings in proteins where there are reasonably high energy barriers to ring rotation (93, 132).

A second frequently measured parameter is the rate of amide proton exchange with solvent. This parameter reports on dynamics only in an indirect manner, i.e. the observed rates can be correlated with certain large-scale molecular motions such as domain movements and global unfolding. It is these events that enable amide proton exchange to occur, typically on the microsecond to days timescale (93, 126). The rates can be measured for specific amide protons over a wide range of conditions by dissolving protonated, lyophilized protein in $^2$H$_2$O and recording successive $^1$H-$^1$H COSY, TOCSY or NOESY spectra to follow the decay of H$^N$ peak intensities, or by transfer of saturation from water to rapidly

exchanging amide protons (133). A substantial increase in both the spectral and temporal resolution of these measurements is obtained through the use of $^1$H-detected 2D $^{15}$N-$^1$H HSQC correlation spectroscopy on $^{15}$N-labeled samples (134).

Characterization of backbone and side-chain dynamics on the picosecond to nanosecond time scale is possible through the measurement of $^{15}$N and $^{13}$C relaxation parameters ($T_1$, $T_2$, *NOE*) in specifically designed $^1$H-detected 2D heteronuclear correlation experiments (135). Specialized approaches for the study of methyl $^{13}$C relaxation are available (136, 137). Characterization of slower $\mu$sec-ms dynamics can also be made, using $T_{1\rho}$-type relaxation experiments [e.g. (138, 139)]. Sensitivity-enhanced pulse sequences are utilized to acquire data from solutions of isotopically enriched protein, thus the sensitivity is excellent and the accuracy and precision of the results are high (140).

Relaxation data are typically analyzed using formalisms such as that of Lipari and Szabo (141, 142), which provide a correlation time for the global reorientation of the molecule ($\tau_c$), a measure of relative mobility (the generalized order parameter, $S^2$) and an internal correlation time ($\tau_e$) for each NH (or CH) bond vector in the molecule. $S^2$ varies between unity for bond vectors that are fixed with respect to the molecular framework, and zero, for completely isotropic reorientation of the bond vector with respect to global motion of the protein. Fig. 16 (lower panel) shows a plot of the $S^2$ values for each residue in the protein apo calbindin $D_{9k}$. This so called "model free" analysis assumes that the internal motions are much faster than the overall tumbling, and that the global reorientation of the molecule is isotropic. The latter is not always valid (especially for proteins that are not spherically symmetric), which can lead to systematic errors in the resulting motional parameters. The use of relaxation data to determine the rotational diffusion tensor for a protein is more readily able to account for anisotropic motion, and may provide a more robust view of molecular dynamics [e.g. see (143)]. The "spectral density mapping" approach provides yet another alternative, in which the relaxation data are analyzed to identify dynamic heterogeneity along the protein backbone, and to classify the approximate frequency of these motions (144, 145).

The application in most common use today is $^{15}$N relaxation analysis to analyze the regions of an NMR structure that have high RMSDs. When relaxation data show that the corresponding region of the protein has higher than average flexibility, it is very likely that the high coordinate uncertainty in that region of the structure is due to motion. The correlation between increases in RMSD and decreases in order parameter is clearly evident for apo calbindin $D_{9k}$ (Fig. 16). Comparison of the corresponding plots indicate that the high coordinate uncertainty of the termini, the linker, and portions of the $Ca^{2+}$-binding loops are likely to be the result of increased flexibility in these regions. In contrast, when the relaxation data show that the corresponding region of the protein is no more flexible than any other region (e.g. for portions of the $Ca^{2+}$-binding loops of apo calbindin $D_{9k}$), then it is likely that the high uncertainties are due to the inability to identify a sufficient number of restraints to precisely define that region of the structure. Typically, this means that additional experiments must be acquired in an attempt to identify more restraints. However, the "lack of data" interpretation is not conclusive because the high uncertainties could conceivably be caused by significant motions that occur outside the particular time-scale to which the relaxation experiments are sensitive.

## IX.  COMPARISON OF NMR AND X-RAY STRUCTURES

Although x-ray crystallography and NMR are both capable of resolving protein structures at atomic detail, differences in the nature of the primary data and method of calculation make direct comparison of the results of the two techniques non-trivial. The most obvious difference is that an ensemble of structures is usually calculated from NMR data, while x-ray analysis results in a single conformation for each molecule in the asymmetric unit. This arises because x-ray diffraction provides a direct measure of the average atom coordinate position, whereas NMR provides the ensemble of all conformations consistent with the experimental data. Uncertainty in the definition of the coordinate position is directly incorporated into the NMR structural ensemble. The average coordinate position is determined by averaging over the full conformational ensemble. The uncertainty in the x-ray structure is superimposed upon the average coordinate position as a B-factor, which reflects both static disorder and molecular motions within the constraints of the crystalline lattice.

An additional point of importance concerns comparison of NMR ensembles to other structures, which normally is carried out by calculating mean coordinate positions of the NMR ensemble. The simple calculation of geometric average coordinates usually produces structures having poor covalent geometry. Consequently, mean coordinates are often subjected to restrained energy minimization to regularize their bond lengths, bond angles, etc. Although this does improve the chemical quality of the mean structure, it necessarily introduces bias relative to comparisons made against the geometric mean. Choosing a single structure from the ensemble that is closest to the mean structure provides an alternative single coordinate set for comparison (146). However, any attempt to represent an ensemble by a single structure inevitably causes information to be lost, hence it is frequently the case that the clarity of a single structure may have to be sacrificed in order to perform detailed and accurate comparisons.

Addressing the accuracy of a structure is generally more reliable for an x-ray structure determination since theoretical calculation of structure factors (and diffraction patterns) are less prone to systematic errors. Analysis of R-factors and free R-factors provide an unbiased and powerful means to assess the integrity of a structure (147). As discussed in Sec. 6.2, such an approach is not so straightforward for the NMR case.

One shortcoming of an NMR structure determination is that all of the restraints are either short distances, or localized dihedral angles. The potential exists for NMR structures to be approximately correct at the local level (5–10 Å),

but grossly in error over longer distances. This is especially true for long extended molecules, or multi-domain proteins with few contacts between the domains. The problem is compounded by the imperfection of force fields in regions of low restraint density since non-bonded, long-range forces may have a dominant effect. X-ray diffraction patterns, by definition, report on all atomic positions, hence long-range structural uncertainties are not an issue. However, the relative orientation of distant parts of elongated molecules or multidomain proteins may be affected by crystal packing artifacts.

In the study of homologous proteins (i.e. proteins related by high sequence similarity or altered by site-specific mutagenesis), x-ray crystallographic analysis of the modified protein can rapidly provide a high-resolution structure based on the known structure of the native form (if the molecular replacement approach can be applied). Such an efficient means to solve related structures from NMR data does not exist. However, NMR provides a powerful means to quickly assess the structural similarity of two proteins, with the level of analysis depending on the extent of structural similarity. For example, the effect of site-specific mutation on the folding (or lack thereof) of the protein is readily apparent from the 1D $^1$H NMR spectrum. Comparisons of characteristic fingerprint regions of 1D and 2D $^1$H spectra provide general insights into the extent of structural differences. This is all possible without having to make sequence-specific assignments, hence can be completed in one day or less. For a more detailed study, resonance assignments are necessary to interpret the observed spectral differences. However, the assignment process is usually greatly expedited by comparison of spectra collected on the modified and native (previously assigned) protein, since the patterns of cross-peak distribution will often be similar, even if there are some chemical shift differences.

Experiments that contain many correlations for each spin system (e.g. TOCSY) are particularly useful in this regard, and analysis is greatly facilitated by special software tools available in NMR analysis computer programs.

X-ray crystallography offers the ability to obtain a set of molecular coordinates to extremely high atomic resolution. This is particularly powerful for relatively static, well-packed regions of proteins. But as the flexibility increases, the diffraction data becomes progressively diffuse and the ability to accurately define the average coordinate position diminishes. In the NMR structure, increased flexibility causes a weakening of NOE interactions, but the NMR resonances if anything become stronger. The NMR characterization of dynamic processes is perhaps the greatest feature of this technique.

One difference frequently observed between NMR and x-ray structures of identical proteins is the orientation of solvent accessible side-chains. Inspection of the primary NMR data indicate that these side-chains often sample many rotamer states in solution, and correspondingly are ill defined in the solution structure. Such ambiguity is not always seen for the equivalent side-chains in an x-ray structure, suggesting that either the crystalline environment is sufficiently anisotropic that only a single conformation is energetically favorable, or else the conformations are sufficiently similar that inspection of the electron density cannot distinguish between the possibilities and only a single orientation is included in the model. Thus, an x-ray structure may give a misleading sense of side-chain localization relative to what is actually present in the protein in solution. Joint SA refinements using both NMR and x-ray data have been reported in an attempt to satisfy both data sets simultaneously (148, 149). As a caveat in NMR structures, due to the low number of experimental NMR

restraints, the orientation of a surface side-chain may easily be biased by other components in the force field. For example, during rMD calculations carried out *in vacuo* instead of in the presence of explicit water molecules, optimization of van der Waal's or electrostatic forces may cause side-chains to pack onto the surface of the protein (150).

## X. OVERVIEW AND PROSPECTS

The determination of the solution structures of proteins by NMR has come a long way in a very short period of time. This has occurred due to a confluence of factors, including advances in magnet and electronics engineering, computer hardware and software, the fundamental understanding of the NMR physics of proteins, and the facility and flexibility of sample preparation by recombinant DNA methods. The methods currently available enable structures to be determined for proteins up to 30–40 kDa. While this is a great limitation, many large proteins are comprised of smaller functional domains that fold independently, hence NMR may still be used to study the individual modules. The structure can then be built-up from structure of these smaller parts, plus studies of the domain interfaces. Such a strategy has proven to be particularly successful for modular proteins such as those involved in cell adhesion, clotting and fibrinolysis [e.g. see (151)].

Fundamental biological processes of pharmacological interest invariably invovle the interaction of one molecule with another, whether it be a drug binding to a protein active site, or the complex formed between two or more biopolymers. Although this chapter has focused on the tools and methods of analysis used to determine the structures of single protein molecules in solution, the methods are readily extended to study the structure and dynamics of multimeric species. Even with the current size limit of 30–40 kDa, NMR is able to provide of insights into many biologically interesting proteins and their interactions with other molecules; this area of research is the subject of a separate chapter in this volume.

A relatively new advance in biomolecular NMR that is already beginning to show signs of promising results is the use of NMR with combinatorial chemistry to discover high affinity ligands for pharmaceutical targets (152, 153). This approach, termed "SAR by NMR", takes advantage of the high sensitivity of the NMR chemical shift to structural perturbations, which enables a highly efficient screening method for protein ligands from a library of compounds. Furthermore, these same experiments allow a qualitative determination of the binding site on the protein for each ligand. The lead compounds are then used in structure-based rational design of new agents with very high affinities and selectivities, which in turn are also very efficiently analyzed using "SAR by NMR" (153).

One particularly exciting area under development in biomolecular NMR is the realization that residual heteronuclear dipolar couplings can be measured even for proteins in solution (154). Dipolar couplings allow highly accurate measurements of the angular orientation between specific inter-nuclear vectors and the applied magnetic field and can be used as a structural refinement tool. This information is particularly powerful because it is completely distance independent and establishes structural relationships at high resolution (155); this type of "long-range" ordering is completely lacking in current NMR structure refinements. Recent experiments by Tijandra and Bax have demonstrated that these residual dipolar couplings can be amplified by causing partial alignment of protein molecules in solution by using combinations of lipophilic compounds known to form liquid crystals (156). These methods hold

considerable promise for providing sub-Å structural resolution at a level comparable to x-ray crystallography, which is required for high-resolution structural biology (157). In effect, the measurement of residual dipolar couplings also provides a complement to specialized labeling strategies (e.g. fractional deuteration) as a means for extending the size of protein for which complete three- dimensional solution structures can be determined; measurements are extremely simple and do not require complicated experimental approaches and therefore can be applied to proteins that do not give rise to triple resonance spectra of adequate quality for structure determination.

And finally, in thinking about the three-dimensional structures of proteins, it is very important to consider that the protein biopolymer has a high degree of intrinsic flexibility. The characterization of motions by NMR requires separate experiments for each of the timescales of interest, so it is extremely difficult to generate a cohesive understanding of these dynamical properties and then incorporate this into the time-averaged structure. The static structures we determine by x-ray crystallography or NMR are but snapshots of these highly dynamic polypeptides. Neither crystal nor solution structures are accurately represented by a single low-energy conformer. One of the major challenges in structural biology today is the development of methods that fuse structural and dynamic information to generate a unified and comprehensible view for analyzing biomolecular function.

## XI. ACKNOWLEDGEMENTS

We thank Drs. Wayne Faibrother, Tahnia Fairbrother, Melissa Starovasnik and Celia Schiffer for encouragement and critical reading of this manuscript. W J C acknowledges financial support from the NIH (GM RO1-40120, GM PO1-48495),

NSF (MCB-9604568) and the American
Cancer Society (FRA-436).

## REFERENCES

1. MP Williamson, TF Havel, K Wüthrich. Solution conformation of the protease inhibitor IIA from bull seminal plasma by $^1$H NMR and distance geometry. J Mol Biol 182: 295–315, 1985.
2. J Cavanagh, WJ Fairbrother, AG Palmer III, NJ Skelton. Protein NMR spectroscopy, principles and practice. San Diego: Academic Press, 1995.
3. RR Ernst, G Bodenhausen, A. Wokaun. Principles of nuclear magnetic resonance in one and two dimensions. Oxford: Clarendon Press, 1987.
4. A Abragam. Principles of Nuclear Magnetism. Oxford: Clarendon Press, 1961.
5. AE Derome. Modern NMR techniques for chemistry research. Oxford: Pergamon, 1987.
6. DS Wishart, et al. $^1$H, $^{13}$C and $^{15}$N chemical shift referencing in biomolecular NMR. J. Biomol. NMR 6: 135–140, 1995.
7. WP Aue, E. Bartholdi, RR Ernst. Two dimensional spectroscopy. Application to NMR. J Chem Phys 64: 2229–2246, 1976.
8. M Rance, et al. Improved spectral resolution in COSY $^1$H NMR spectra of proteins via double quantum filtration. Biochem Biophys Res Commun 117: 479–485, 1983.
9. G Eich, G Bodenhausen, RR Ernst. Exploring nuclear spin systems by relayed magnetization transfer. J Am Chem Soc 104: 3732, 1982.
10. G. Wagner. 2D Relayed coherence transfer spectroscopy of a protein. J. Magn. Reson. 55:151-156, 1983.
11. A Bax, DG Davis. MLEV-17-Based 2D homonuclear magnetization transfer spectroscopy. J Magn Reson 65: 355–360, 1985.
12. L Braunschweiler, RR Ernst. Coherence transfer by isotropic mixing: application to proton correlation spectroscopy. J Magn Reson 53: 521–528, 1983.
13. S Macura, RR Ernst. Elucidation of cross relaxation in liquids by two-dimensional N.M.R. spectroscopy. Mol Phys 41: 95–117, 1980.
14. A Kumar, RR Ernst, K Wüthrich. A 2D NOE experiment for the elucidation of complete $^1$H-$^1$H cross-relaxation networks in biological macromolecules. Biochem Biophys Res Commun 95: 1–6, 1980.
15. L Mueller. Sensitivity enhanced detection of weak nuclei using heteronuclear multiple quantum coherence. J Am Chem Soc 101: 4481–4484, 1979.
16. G Bodenhausen, DJ Ruben. Natural abundance nitrogen-15 NMR by enhanced heteronuclear spectroscopy. Chem Phys Lett 69: 185–189, 1980.
17. A Bax, M Ikura, LE Kay, DA Torchia, R Tschudin. Comparison of different modes of two-dimensional reverse-correlation NMR for the study of proteins. J Magn Reson 86: 304–318, 1990.
18. TJ Norwood, J Boyd, JE Heritage, N Soffe, ID Campbell. Comparison of techniques for $^1$H-detected heteronuclear $^1$H-$^{15}$N spectroscopy. J Magn Reson 87: 488–501, 1990.
19. H Oschkinat, C Cieslar, AM Gronenborn, GM Clore. 3D homonuclear HOHAHA-NOE spectroscopy in $H_2O$ and its application to proteins. J Magn Reson 81: 212–216, 1989.
20. JP Simorre, D Marion. A method aimed at obtaining a complete set of cross peaks in single-scan high-resolution homonuclear 3D NMR. J Magn Reson 94: 426–432, 1991.
21. R Boelens, GW Vuister, TMG Koning, R Kaptein. Observation of spin diffusion in biomolecules by three-dimensional NOE-NOE spectroscopy. J Am Chem Soc 111: 8525–8526, 1989.
22. GW Vuister, R Boelens, A Padilla, R Kaptein. Statistical analysis of double NOE transfer pathways in proteins as measured in 3D NOE-NOE spectroscopy. J Biomol NMR 1: 421–438, 1991.
23. R Bernstein, A Ross, C Cieslar, TA Holak. A practical approach to calculations of biomolecular structures from homonuclear 3D NOE-NOE spectra. J Magn Reson 101B: 185–188, 1993.
24. K Wüthrich. NMR of Proteins and Nucleic Acids. New York: Wiley, 1986.
25. C Redfield, CM Dobson. Sequential $^1$H NMR assignments and secondary structure of hen egg white lysozyme in solution. Biochemistry 27: 122–136, 1988.
26. BCM Potts, G. Carlström, K Okazaki, H Hidaka, WJ Chazin. $^1$H NMR assignments of apo calcyclin and comparative structural analysis with calbindin $D_{9k}$ and S100b. Protein Science 5: 2162–2174, 1996.
27. WJ Chazin, PE Wright. A modified strategy for identification of $^1$H spin systems in proteins. Biopolymers 26: 973–977, 1987.
28. WJ Chazin, M Rance, PE Wright. Complete assignment of the $^1$H NMR spectrum of French bean plastocyanin. Application of an integrated approach to spin system identification in proteins. J Mol Biol 202: 603–622, 1988.

29. M Rance, WJ Chazin, C Dalvit, PE Wright. Multiple-quantum nuclear magnetic resonance. Meth Enzymol 176: 114–134, 1989.

30. M Billeter, W Braun, K Wüthrich. Sequential resonance assignment in protein $^1$H NMR spectra. Computation of sterically allowed $^1$H-$^1$H distances and statistical analysis of $^1$H-$^1$H distances in single crystal protein conformations. J Mol Biol 155: 321–346, 1982.

31. WJ Chazin, PE Wright. Complete assignment of the $^1$H NMR spectrum of French bean plastocyanin. Sequential resonance assignments, secondary structure and global fold. J Mol Biol 202: 623–636, 1988.

32. NJ Skelton, S Forsén, WJ Chazin. $^1$H NMR resonance assignments, secondary structure, and global fold of apo bovine calbindin $D_{9k}$. Biochemistry 29: 5752–5761, 1990.

33. D Marion, et al. Overcoming the overlap problem in the assignment of $^1$H NMR spectra of larger proteins by the use of 3D heteronuclear $^1$H-$^{15}$N Hartman-Hahn-multiple quantum coherence and nuclear Overhauser-multiple quantum coherence spectroscopy: application to interleukin 1$\beta$. Biochemistry 28: 6150–6156, 1989.

34. PC Driscoll, GM Clore, D Marion, PT Wingfield, AM Gronenborn. Complete resonance assignment for the polypeptide backbone of interleukin 1$\beta$ using three-dimensional heteronuclear NMR spectroscopy. Biochemistry 29:3542–3556, 1990.

35. BJ Stockman, et al. Sequence-specific $^1$H and $^{15}$N resonance assignments for human dihydrofolate reductase in solution. Biochemistry 31: 218–229, 1992.

36. A Bax, GM Clore, AM Gronenborn. $^1$H-$^1$H correlation via isotropic mixing of $^{13}$C magnetization, a new 3D approach for assigning $^1$H and $^{13}$C spectra of $^{13}$C-enriched proteins. J Magn Reson 88: 425–431, 1990.

37. LE Kay, M Ikura, A Bax. Proton-proton correlation via carbon-carbon couplings: a three-dimensional NMR approach for the assignment of aliphatic resonances in proteins labeled with carbon-13. J Am Chem Soc 112: 888–889, 1990.

38. RT Clubb, V Thanabal, C Osborne, G Wagner. $^1$H and $^{15}$N resonance assignments of oxidized flavodoxin from *Anacystis nidulans* with 3D NMR. Biochemistry 30: 7718–7730, 1991.

39. M Ikura, LE Kay, A Bax. A novel approach for sequential assignment of $^1$H, $^{13}$C and $^{15}$N spectra of larger proteins: heteronuclear triple-resonance three-di-

40. LE Kay, M Ikura, R Tschudin, A Bax. Three-dimensional triple-resonance NMR spectroscopy of isotopically enriched proteins. J Magn Reson 89: 496–514, 1990.

41. R Bazzo, DO Cicero, G Barbato. A new 3D pulse sequence for correlating intraresidue HN, N and CO chemical shifts in $^{13}$C, $^{15}$N-labeled proteins. J Magn Reson 110B: 65–68, 1996.

42. F Löhr, H Rüterjans. A new triple resonance experiment for the sequential assignment of backbone resonances in proteins. J. Biomol. NMR 6: 189–197, 1995.

43. S Grzesiek, A Bax. An efficient experiment for sequential backbone assignment of medium-sized isotopically enriched proteins. J Magn Reson 99: 201–207, 1992.

44. S Grzesiek, S Bax Correlating backbone amide and side chain resonances in larger proteins by multiple relayed triple resonance NMR. J Am Chem Soc 114: 6291–6293, 1992.

45. M Wittekind, L Mueller. HNCACB, a high-sensitivity 3D NMR experiment to correlate amide-proton and nitrogen resonances with alpha- and beta-carbon resonances in proteins. J. Magn. Reson. 101B: 210–205, 1993.

46. S Grzesiek, A Bax. Amino acid type determination in the sequential assignment procedure of uniformly $^{13}$C/$^{15}$N-enriched proteins. J Biomol NMR 3: 185–204, 1993.

47. S Grzesiek, J Anglister, A Bax. Correlation of backbone amide and aliphatic side-chain resonances in $^{13}$C/$^{15}$N-enriched proteins by isotropic mixing of $^{13}$C magnetization. J Magn Reson B101: 114–119, 1993.

48. RT Clowes, W Boucher, CH Hardman, PJ Domaille, ED Laue. A 4D HCC(CO)NNH experiment for the correlation of aliphatic side-chain and backbone resonances in $^{13}$C/$^{15}$N-labeled proteins. J Biomol NMR 3: 349–354, 1993.

49. R. Powers, et al. $^1$H, $^{15}$N, $^{13}$C, and $^{13}$CO assignments of human interleukin-4 using three-dimensional double- and triple-resonance heteronuclear magnetic resonance spectroscopy. Biochemistry 31: 4334–4346, 1992.

50. MS Friedrichs, L Mueller, M Wittekind. An automated procedure for the assignment of proteins $^1$HN, $^{15}$N, $^{13}$C$^\alpha$, $^1$H$^\alpha$, $^{13}$C$^\beta$ and $^1$H$^\beta$ resonances. J Biomol NMR 4: 703–726, 1994.

51. RP Meadows, ET Olejniczak, SW Fesik. A computer-based protocol for semiautomated assignments and 3D structure deter-

mination of proteins. J Biomol NMR 4: 79–96, 1994.

52. JB Olson Jr., JL Markley. Evaluation of an algorithm for the automated sequential assignment of protein backbone resonances: a demonstration of the connectivity tracing assignment tools (CONTRAST) software package. J Biomol NMR 4: 385–410, 1994.

53. D Zimmerman, C Kulikowski, L Wang, B Lyons, GT Montelione. Automated sequencing of amino acid spin systems in proteins using multidimensional HCC (CO)NH-TOCSY spectroscopy and constraint propagation methods from artificial intelligence. J Biomol NMR 4: 241–256, 1994.

54. DM LeMaster. Deuterium labeling in NMR structural analysis of larger proteins. Quart Rev Biophys 23: 113–174, 1990.

55. M Sattler, SW Fesik. Use of deuterium labeling in NMR: overcoming a sizeable problem. Structure 4: 1245–1249, 1996.

56. S Grzesiek, J Anglister, H Ren, A Bax. $^{13}$C line narrowing by $^{2}$H decoupling in $^{2}$H/$^{13}$C/$^{15}$N-enriched proteins. Applications to triple resonance 4D J connectivity of sequential amides. J Am Chem Soc 115: 4369–4370, 1993.

57. T Yamazaki, W Lee, CH Arrowsmith, DR Muhandiram, LE Kay. A suite of triple resonance NMR experiments for the backbone assignment of $^{15}$N, $^{13}$C, $^{2}$H labeled proteins with high sensitivity. J Am Chem Soc 116: 11655–11666, 1994.

58. RA Venters, WJ Metzler, LD Spicer, L Mueller, BT Farmer III. Use of $^{1}$H$^{N}$-$^{1}$H$^{N}$ NOEs to determine protein global folds in perdeuterated proteins. J Am Chem Soc 117: 9592–9593, 1995.

59. KH Gardiner, MK Rosen, LE Kay. Global folds of highly deuterated, methyl protonated proteins by multidimensional NMR. Biochemistry 36: 1389–1401, 1997.

60. D Nietlispach, et al. An approach to the structure determination of larger proteins using triple resonance NMR experiments in conjunction with random fractional deuteration. J Am Chem Soc 118: 407–415, 1996.

61. DS Garrett, et al. Solution structure of the 30 kDa N-terminal domain of Enzyme I of the E. coli phosphoenolpyruvate:sugar phosphotransferase system by multidimensional NMR. Biochemistry 36: 2517–2530, 1997.

62. RA Venters, BT Farmer III, CA Fierke, LD Spicer. Characterizing the use of perdeuteration in NMR studies of large proteins: $^{13}$C, $^{15}$N and $^{1}$H assignments of human carbonic anhydrase II. J Mol Biol 264: 1101–1116, 1996.

63. LP McIntosh, FW Dahlquist. Biosynthetic incorporation of $^{15}$N and $^{13}$C for assignment and interpretation of nuclear magnetic resonance spectra of proteins. Quart Rev Biophys 23: 1–38, 1990.

64. M Kainosho, T Tsuji. Assignment of the three methionyl carbonyl carbon resonances in Streptomyuces subtilisin inhibitor by a carbon-13 and nitrogen-15 double labeling technique: a new strategy for structural studies of proteins. Biochemistry 21: 6273–6279, 1982.

65. LP McIntosh, AJ Wand, DF Lowry, AG Redfield, FW Dahlquist. Assignment of the backbone $^{1}$H and $^{15}$N resonances of bacteriophage T4 lysozyme. Biochemistry 29: 6341–6362, 1990.

66. T Kigawa, Y Nuto, S Yokoyama. Cell-free synthesis and amino acid-selective isotope labeling of proteins for NMR analysis. J Biomol NMR 6: 129–134, 1996.

67. P Brodin, T Drakenburg, E Thulin, S Forsén, T Gundström. Selective proton labeling of amino acids in deuterated bovine calbindin $D_{9k}$. A way to simplify $^{1}$H NMR spectra. Protein Engineering 2: 353–358, 1989.

68. H Kuboniwa, et al. Solution structure of calcium-free calmodulin. Nature Struct Biol 2: 768–776, 1995.

69. Y Oda, et al. $^{1}$H NMR studies of deuterated ribonuclease HI selectively labeled with protonated amino acids. J Biomol NMR 2: 137–147, 1992.

70. T Szyperski, D Neri, B Leiting, G Otting, K Wüthrich. Support of $^{1}$H NMR assignments in proteins by biosynthetically directed fractional $^{13}$C-labeling. J Biomol NMR 2: 323–334, 1992.

71. D Neri, T Szyperski, G Otting, H Senn, K Wüthrich. Stereospecific NMR assignments of the methyl groups of valine and leucine in the DNA-binding domain of the 434-repressor by biosynthetically directed fractional $^{13}$C labeling. Biochemistry 28: 7510–7516, 1989.

72. DS Wishart, BD Sykes. The $^{13}$C chemical shift index: A simple method for the identification of protein secondary structure using $^{13}$C chemical shift data. J Biomol NMR 4: 171–180, 1994.

73. K Ösapay, DA Case. Analysis of proton chemical shifts in regular secondary structure of proteins. J Biomol NMR 4: 215–230, 1994.

74. JH Noggle, RE Shirmer. The Nuclear Overhauser Effect: Chemical Applications. New York: Academic Press, 1971.

75. BA Borgias, M Gochin, DJ Kerwood, TL James. Relaxation matrix analysis of 2D NMR data. Prog NMR Spect 22: 83–100, 1990.

76. JW Keepers, TL James. A theoretical study of distance determinations from NMR. Two-dimensional nuclear Overhauser effect spectra. J Magn Reson 57: 404–426, 1984.

77. C Gonzalez, JAC Rullmann, AMJJ Bonvin, R Boelens, R Kaptein. Towards an NMR R factor. J Magn Reson 91: 659–664, 1991.

78. PD Thomas, VJ Basus, TL James. Protein solution structure determination using distances from two-dimensional nuclear Overhauser effect experiments: Effect of approximations on the accuracy of derived structures. Proc Natl Acad Sci USA 88: 1237–1241, 1991.

79. TMG Koning, R Boelens, GA ver der Marel, JH van Boom, R Kaptein. Structure determination of a DNA octamer in solution by NMR spectroscopy. Effect of fast local motions. Biochemistry 30: 3787–3797, 1991.

80. GM Clore, AM Gronenborn. Two, three-four- dimensional NMR methods for the obtaining larger and more precise 3D structures of proteins in solution. Ann Rev Biophys Biophys Chem 20: 29–63, 1991.

81. T Havel. The precision of protein structures determined from NMR data: reality or illusion. In V Renugopalakrishnan, PR Carey, ICP Smith, S-G Huans, AL Storer, eds. Proteins: Structure, Dynamics, Design. Leiden, Holland: ESCOM Science Publishers, 1991, pp 110–115.

82. M Karplus. Contact electron-spin coupling of nuclear magnetic moments. J Phys Chem 30: 11–15, 1959.

83. A DeMarco, M Llinás, K Wüthrich. Analysis of the $^1$H-NMR spectra of ferrichrome peptides. I. The non-amide protons. Biopolymers 17: 617–636, 1978.

84. A Pardi, M Billeter, K Wüthrich. Calibration of the angular dependence of the amide proton $C^\alpha$-proton coupling constant, $^3J_{HN\alpha}$, in a globular protein. Use of $^3J_{HN\alpha}$ for identification of helical secondary structure. J Mol Biol 180: 741–751, 1984.

85. LJ Smith, MJ Sutcliffe, C Redfield, CM Dobson. Analysis of $\phi$ and $\chi_1$ torsion angles for hen lysozyme in solution from $^1$H NMR spin-spin coupling constants. Biochemistry 30: 986–996, 1991.

86. GW Vuister, A Bax. Quantitative J correlation: A new approach for measuring homonuclear three bond $J(H^N-H^\alpha)$ coupling constants in $^{15}$N-enriched proteins. J Am Chem Soc 115: 7772–7777, 1993.

87. GT Montelione, SD Emerson, BA Lyons. A general approach for determining scalar coupling constants in polypeptides and proteins. Biopolymers 32: 327–334, 1992.

88. GW Vuister, et al. Measurement of homo- and heteronuclear J couplings from quantitative J correlation. Meth Enzymol 239: 79–105, 1994.

89. SG Hyberts, W Märki, G Wagner. Stereospecific assignment of side-chain protons and characterization of torsion angles in eglin C. Eur J Biochem 164: 625–635, 1987.

90. P Güntert, W Braun, M Billeter, K Wüthrich. Automated stereospecific $^1$H NMR assignments and their impact on the precision of protein structure determination in solution. J Am Chem Soc 111: 3997–4004, 1989.

91. M Nilges, GM Clore, AM Gronenborn. $^1$H-NMR stereospecific assignments by conformational data-base searches. Biopolymers 29: 813–822, 1990.

92. GP Gippert, PE Wright, DA Case. Recursive torsion angle grid search in high dimensions: A systematic approach to NMR structure determination. J Biomol NMR 11: 241–263, 1998.

93. G Wagner. Characterization of the distribution of internal motions in BPTI using a large number of internal NMR probes. Quart Rev Biophys 16: 1–57, 1983.

94. G Otting, E Liepinsh, K Wüthrich. Protein hydration in aqueous solution. Science 254: 974–980, 1991.

95. G Otting, E Liepinsh, K Wüthrich. Proton exchange with internal water molecules in the protein BPTI in aqueous solution. J Am Chem Soc 113: 4363–4364, 1991.

96. S Mori, JM Berg, PCM Zijl. Separation of intramolecular NOE and exchange peaks in water exchange spectroscopy using spin-echo filters. J Biomol NMR 7: 77–82, 1996.

97. T-L Hwang, S Mori, AJ Shaka, PCM Zijl. Application of phase-modulated CLEAN chemical exchange spectroscopy (CLEANEX-PM) to detect water-protein proton exchange and intermolecular NOEs. J Am Chem Soc 119: 6203–6204, 1997.

98. S Grzesiek, A Bax. The importance of not saturating $H_2O$ in protein NMR. Applications to sensitivity enhancement and NOE measurements. J Am Chem Soc 115: 12593–12594, 1993.

99. TF Havel, GM Crippen, ID Kuntz. Effects of distance constraints on macromolecular conformation. II. Simulation of experimental results and theoretical calculations. Biopolymers 18: 73–81, 1979.

100. TF Havel. An evaluation of computational strategies for use in the determination of protein structure from distance constraints obtained by nuclear magnetic resonance. Prog Biophys Mol Biol 56: 43–78, 1991.

101. GM Crippen, TF Havel. Distance Geometry and Molecular Conformation. Taunton, England: Research Studies Press, 1988.

102. W Braun, N Go. Calculation of protein conformations by proton-proton distance constraints. A new efficient algorithm. J Mol Biol 186: 611–626, 1985.

103. P Güntert, W Braun, K Wüthrich. Efficient computation of 3D protein structures in solution from NMR data using the program DIANA and the supporting programs CALIBA, HABAS and GLOMSA. J Mol Biol 217: 517–530, 1991.

104. AT Brünger, GM Clore, AM Gronenborn, M Karplus. Three-dimensional structure of proteins determined by molecular dynamics with interproton distance restraints: application to crambin. Proc Natl Acad Sci USA 83: 3801–3805, 1986.

105. GM Clore, AT Brünger, M Karplus, AM Gronenborn. Application of molecular dynamics with interproton distance restraints to three-dimensional protein structure determination. A model study of crambin. J Mol Biol 191: 523–551, 1986.

106. M Nilges, GM Clore, AM Gronenborn. Determination of three-dimensional structures of proteins from interproton distance data by dynamical simulated annealing from a random array of atoms. FEBS Lett 239: 129–136, 1988.

107. M Nilges, GM Clore, AM Gronenborn. Determination of three-dimensional structures of proteins and from interproton distance data by hybrid distance geometry-dynamical simulated annealing calculations. FEBS Lett 239: 317–324, 1988.

108. P Yip, DA Case. A new method for refinement of macromolecular structures based on nuclear Overhauser effect spectra. J Magn Reson 83: 643–648, 1989.

109. AMJJ Bonovin, R Boelens, R Kaptein. Direct NOE refinement of biomolecular structures using 2D NMR data. J Biomol. NMR 1: 305–309, 1991.

110. J Kuszewski, J Qin, AM Gronenborn, GM Clore. The impact of direct refinement against $^{13}C\alpha$ and $^{13}C\beta$ chemical shifts on protein structure determination by NMR. J Magn Reson 106B: 92–96, 1995.

111. J Kuszewski, AM Gronenborn, GM Clore. The impact of direct refinement against proton chemical shifts in protein structure determination by NMR. J Magn Reson 107B: 293–297, 1995.

112. H-b Le, J Pearson, AC de Bois, E Oldfield. Protein structure refinement and prediction via NMR chemical shifts and quantum chemistry. J Am Chem Soc 117: 3800–3807, 1995.

113. JG Pearson, J-F Wang, JL Markley, H-b Le, E Oldfield. Protein structure refinement using 13-C NMR spectroscopy chemical shifts and quantum chemistry. J Am Chem Soc 117: 8823–8829, 1995.

114. DS Garrett, et al. The impact of direct refinement against three bond $HN-C^{\alpha}H$ coupling constants on protein structure determination by NMR. J Magn Reson 104B: 99–103, 1994.

115. H Liu, PD Thomas, TJ James. Averaging of cross-relaxation rates and distances for methyl, methylene and aromatic ring protons due to motion or overlap. J Magn Reson 98: 163–175, 1992.

116. M.J Dellwo, DM Schneider, AJ Wand. Modifications of the rate matrix required for quantitative analysis of NOESY spectra of proteins. J Magn Reson 103B: 1–9, 1994.

117. GM Clore, MA Robien, AM Gronenborn. Exploring the limits of precision and accuracy of protein structures determined by NMR spectroscopy. J Mol Biol 231: 82–102, 1993.

118. D Zhao, O Jardetsky. An assessment of the precision and accuracy of protein structures determined by NMR. Dependence on distance errors. J Mol Biol 239: 601–607, 1994.

119. H Widmer, A Widmer, W Braun. Extensive distance geometry calculations with different NOE calibrations: New criteria for structure selection applied to sandostatin and BPTI. J Biomol NMR 3: 307–324, 1993.

120. M Akke, S Forsén, WJ Chazin. Solution structure of $(Cd^{2+})_1$ calbindin $D_{9k}$ reveals details of the stepwise structural changes along the apo-$(Ca^{2+})_1$II-$(Ca^{2+})_2$ binding pathway. J Mol Biol 252: 102–121, 1995.

121. JA Smith, L Gomez-Paloma, DA Case, WJ Chazin. MD docking driven by NMR-derived restraints to determine the structure of the calicheamicin $\gamma_1^I$ oligosaccharide domain complexed to duplex DNA. Magn Res Chem 34S: 147–155, 1996.

122. RA Engh, R Huber. Accurate bond and angle parameters for X-ray protein structure refinement. Acta Cryst A47: 392–400, 1991.

123. GN Ramachandran, V Sasisekharan. Conformation of polypeptides and proteins. Adv Protein Chem 23: 283–437, 1968.

124. RA Laskowski, MW MacArthur, DS Moss, JM Thornton. PROCHECK: a program to check the stereochemical quality of protein structures. J Appl Crystallogr 26: 283–291, 1993.

125. RA Laskowski, JAC Rullmann, MW MacArthur, R Kaptein, JM Thornton. AQUA and PROCHECK-NMR: programs for checking the quality of protein structures solved by NMR. J Biomol. NMR 8: 477–486, 1996.

126. NJ Skelton, J Kördel, M Akke, WJ Chazin. NMR studies of the internal dynamics in apo, $(Cd^{2+})_1$, and $(Ca^{2+})_2$ calbindin $D_{9k}$. The rates of amide proton exchange with solvent. J Mol Biol 227: 1100–1117, 1992.

127. J Kördel, S Forsén, T Drakenburg, WJ Chazin. The rate and structural consequences of proline cis-trans isomerization in calbindin $D_{9k}$: NMR studies of the minor (cis-Pro43) isoform and the Pro43Gly mutant. Biochemistry 29: 4000–4009, 1990.

128. NJ Skelton, J Kördel, S Forsén, WJ Chazin. Comparative structural analysis of the calcium free and bound states of the calcium regulatory protein Calbindin $D_{9k}$. J Mol Biol 213: 593–598, 1990.

129. S Forsén, J Kördel, T Grundstöm, WJ Chazin. The molecular anatomy of a calcium-binding protein. Acc Chem Res 26: 7–14, 1993.

130. J Kördel, NJ Skelton, M Akke, AG Palmer, WJ Chazin. Backbone dynamics of calcium-loaded calbindin $D_{9k}$ studied by 2D proton-detected $^{15}N$ NMR spectroscopy. Biochemistry 31: 4856–4866, 1992.

131. M Akke, J Kördel, NJ Skelton, AG Palmer, WJ Chazin. Effects of ion binding on the backbone dynamics of calbindin $D_{9k}$ determined by $^{15}N$ relaxation. Biochemistry 32: 8932–8944, 1993.

132. ID Campbell, CM Dobson, GR Moore, SJ Perkins, RJP Williams. Temperature dependent molecular motions of a tyrosine residue of ferrocytochrome C FEBS Lett 70: 96–100, 1976.

133. S Spera, M Ikura, A Bax. Measurement of the exchange rates of rapidly exchanging amide protons: application to the study of calmodulin and its complex with a myosin light chain kinase fragment. J Biomol NMR 1: 155–165, 1991.

134. D Marion, M Ikura, R Tschudin, A Bax. Rapid recording of 2D NMR spectra without phase cycling. Application to the study of hydrogen exchange in proteins. J Magn Reson 85: 393–399, 1989.

135. LE Kay, DA Torchia, A Bax. Backbone dynamics of proteins as studied by $^{15}N$ inverse detected heteronuclear NMR spectroscopy: application to staphylococcal nuclease. Biochemistry 28: 8972–8979, 1989.

136. LK Nicholson, et al. Dynamics of methyl groups in proteins as studied by proton-detected $^{13}C$ NMR spectroscopy. Application to leucine residues of staphylococcal nuclease. Biochemistry 31: 5253–5256, 1992.

137. AG Palmer, RA Hochstrasser, DM Millar, M Rance, PE Wright. Characterization of amino acid side-chain dynamics in a zinc-finger peptide using $^{13}C$ NMR spectroscopy and time-resolved fluorescence spectroscopy. J Am Chem Soc 115: 6333–6345, 1993.

138. T Szyperski, P Luginbühl, G Otting, P Güntert, K Wüthrich. Protein dynamics studied by rotating frame $^{15}N$ spin relaxation times. J Biomol NMR 3: 151–164, 1993.

139. M Akke, AG Palmer III. Monitoring macromolecular motions on microsecond to millisecond time scales by $R_{1\rho}$-$R_1$ constant relaxation time experiments. J Am Chem Soc 118: 911–912, 1996.

140. NJ Skelton, et al. Practical aspects of two-dimensional proton-detected $^{15}N$ spin relaxation measurements. J Magn Reson B102: 253–264, 1993.

141. G Lipari, A Szabo. Model-free approach to the interpretation of nuclear magnetic resonance relaxation in macromolecules. 1. Theory and range of validity. J Am Chem Soc 104: 4546–4559, 1982.

142. G Lipari, A. Szabo. Model-free approach to the interpretation of nuclear magnetic resonance relaxation in macromolecules. 2. Analysis of experimental results. J Am Chem Soc 104: 4559–4570, 1982.

143. LK Lee, M Rance, WJ Chazin, AG Palmer III. Rotational diffusion anisotropy of proteins from simultaneous analysis of $^{15}N$ and $^{13}C$ nuclear spin relaxation. J. Biomol. NMR 9: 287–298, 1997.

144. JW Peng, G Wagner. Mapping spectral density functions using heteronuclear NMR relaxation measurements. J Magn Reson 98: 308–332, 1992.

145. JW Peng, G Wagner. Frequency spectrum of NH bonds in eglin C from spectral density mapping at multiple fields. Biochemistry 34: 16733–16752, 1995.

146. MJ Sutcliffe. Representing an ensemble of NMR-derived protein structures by a

single structure. Protein Sci. 2: 936–944, 1993.

147.  AT Brünger. The free R value: a novel statistical quantity for assessing the accuracy of crystal structures. Nature 355: 472–474, 1992.

148.  B Shaanan, et al. Combining experimental information from crystal and solution studies. Science 257: 961–964, 1992.

149.  CA Schiffer, R Huber, K Wüthrich, WF van Gunsteren. Simultaneous refinement of the structure of BPTI against NMR data measured in solution and X-ray diffraction data measured in single crystals. J Mol Biol 241: 588–599, 1994.

150.  J Kördel, DA Pearlman, WJ Chazin. Protein solution structure calculations in solution: solvated molecular dynamics refinement of calbindin $D_{9k}$. J Biomol NMR 10: 231–43, 1997.

151.  ID Campbell, AK Downing. Building protein structure and function from modular units. Trends Biotechnology 12: 168–172, 1994.

152.  SB Shuker, PJ Hajduk, RP Meadows, SW Fesik. Discovering high-affinity ligands for proteins: SAR by NMR. Science 274: 1531–1534, 1996.

153.  PJ Hajduk, et al. Discovery of potent nonpeptide inhibitors of stromelysin using SAR by NMR. J Am Chem Soc 119: 5818–5827, 1997.

154.  N. Tjandra, S. Grzesiek, A. Bax. Magnetic field dependence of nitrogen-proton J splittings in $^{15}$N-enriched ubiquitin resulting from relaxation interference and residual dipolar couplings. J. A. Chem. Soc. 118:1996.

155.  N Tjandra, JG Omichinski, AM Gronenborn, GM Clore, A Bax. Use of dipolar $^1$H-$^{15}$N and $^1$H-$^{13}$C couplings in the structure determination of magnetically oriented macromolecules in solution. Nature Struct Biol 4: 732–738, 1997.

156.  N Tjandra, A Bax. Direct measurement of distances and angles in biomolecules by NMR in a dilute liquid crystalline medium. Science 278: 1111–1114, 1997.

157.  GM Clore, AM Gronenborn. NMR structures of proteins and protein complexes beyond 20,000 $M_r$. Nature Struct Biol 4: 849–853, 1997.

158.  NJ Skelton, J Kördel, WJ Chazin. Determination of the solution structure of apo calbindin $D_{9k}$ by NMR spectroscopy. J Mol Biol 249: 441–462, 1995.

# 23

# NMR Studies of Protein-Peptide Complexes: Examples from the Calmodulin System

**A. Joshua Wand, Jeffrey L. Urbauer, Mark R. Ehrhardt, and Andrew L. Lee**
*University of Pennsylvania, Philadelphia, Pennsylvania*

## I. INTRODUCTION

Advances in the field of nuclear magnetic resonance (NMR) since its first observation fifty years ago have been staggering. As the technology has developed there has been a corresponding explosion in the number of studies of proteins by NMR methods (for recent reviews, see 1–5). The continuing interest in the interaction of proteins with small ligands and biopolymers has provided a potent driving force for the development of new NMR techniques to probe the structural and dynamic characteristics of protein complexes. This review focuses on the methodologies and strategies that have been developed to study complexes between proteins and peptide ligands which often act as models of larger protein-protein complexes. The review is structured to first provide an overview of the underlying methodologies and to then illustrate the basic themes with several detailed examples taken from this laboratory's studies of calmodulin-peptide complexes.

## II. METHODOLOGY

A basic issue facing the NMR spectroscopist when approaching a noncovalent complex of a protein with a ligand molecule is the question of the lifetime of the complex. Is the complex in slow, intermediate or fast exchange with its dissociated components on the NMR timescale? Significant structural information may be gained in the fast exchange limit by use of the transferred NOE (e.g., 6) although this and related approaches are inherently limited (for a review, see 7, 8). Notable examples of the use of the fast exchange limit to provide detailed structural information about the ligand in the protein bound state include the characterization of interactions between calmodulin and small peptides (9, 10). However, one must conclude that a complex in slow exchange on the NMR time scale with its dissociated components is the most ideal condition for detailed structural studies. In this situation, the full power of multinuclear, multidimensional NMR methods can be applied to the structural and dynamic characterization of the complex.

The use of heteronuclear NMR brings to the forefront the issue of isotopic enrichment of the ligand and/or the protein. Manipulation of the fact that two separable entities are involved is often extremely advantageous. Basically, the resonance assignment and structure determination problems presented by a stable binary complex of proteins is no different than that of single protein except for the fact that the former may be isotopically manipulated in a more selective manner. Three distinct isotopic enrichment strategies have been used to simplify the resonance assignment problem presented by protein-ligand complexes: 1) Uniform deuteration of one component; 2) uniform $^{13}C$ and/or $^{15}N$ enrichment of one component; and 3) uniform $^{13}C$ and/or $^{15}N$ enrichment of both components of the complex.

Ironically, it appears that the uniform isotopic enrichment of proteins by biosynthetic means is often far easier and more cost effective than uniform or even selective isotopic enrichment of peptides or other small ligands by chemical synthesis. Enrichment of proteins with $^{13}C$, $^{15}N$ and, to a lesser extent, $^{2}H$ is now often routinely achieved by bacterial expression of the structural gene during growth on labeled media. Routes to reasonably cost effective eucaryotic expression of proteins using $^{13}C$, $^{15}N$-containing minimal or rich media are also now available (e.g. 11–13).

Uniform deuteration has been used to simplify the $^{1}H$ spectra of protein-ligand complexes such as that between calmodulin and melittin (14) and between cyclophilin and cyclosporin (15). In these cases, the subsequent analysis of the structure of the bound peptide relied entirely on $^{1}H$-$^{1}H$ interactions. In the context of $^{1}H$ resonance assignments, which must rely on $^{1}H$-$^{1}H$ J-coupling to provide intraresidue side chain resonance correlations and subsequently identification, the size of the complex becomes an issue. Though significant improvements have been made in isotropic mixing sequences employed in TOCSY experiments (e.g., 16), the limitations presented by a long effective correlation time on chemical shift correlation via direct $^{1}H$-$^{1}H$ J-coupling places a severe restriction on the size of the complex that can be efficiently studied by this approach. This is especially true when the spectrum of the bound peptide differs greatly from that of the free peptide and therefore requires highly reliable chemical shift correlation via J-coupling to allow comprehensive resonance assignments to be obtained.

A more flexible strategy is to incorporate $^{15}N$ and/or $^{13}C$ into the peptide ligand itself. This serves to provide a heteronuclear chemical shift for purposes of resolution and a means to distinguish $^{1}H$ resonances of the ligand from those of the receptor protein. In cases where the ligand is made by chemical synthesis, the cost effective availability of suitably isotopically enriched precursors may often be limited. This is especially true for peptides and deoxyoligonucleotides. One example, discussed extensively below, is the use of $\alpha$-$^{15}N$-labeled tBOC-protected amino acids to prepare a peptide corresponding to the calmodulin binding domain of the smooth muscle myosin light chain kinase (17).

In some cases extensively isotopically enriched ligand can be prepared by biosynthetic means. Recent examples include the uniform enrichment of cyclosporin with $^{13}C$ and its subsequent use to study the conformation of the peptide bound to cyclophilin (18, 19). Surprisingly, fusion protein expression vectors, which can be used to rescue small peptides, have not yet been used extensively for isotopic enrichment of small peptides though there have been notable exceptions (20).

Characterization of the interface of a protein-ligand complex was made much more feasible by the general development of heteronuclear multidimensional NMR. The use of HCCH-TOCSY and related experiments (1–5) allows nearly complete resonance assignments for long side chains which are often at the interface between proteins and their bound ligands. Additional experiments have been developed recently to facilitate assignment of specific amino acid side chains or side chain groups. For instance, experiments have been developed that provide a reliable path to the assignment of methionine methyls (21), which are often at the center of hydrophobic interfaces, to the stereospecific assignment of the primary amide $NH_2$ groups of asparagine and glutamine (22), and to the arginine guandino group (23) which often participates in ionic interactions. All three of these assignment experiments rely on isotopic enrichment [e.g., (24)]. A number of strategies using both homonuclear and heteronuclear NMR approaches have also been developed to study bound water molecules which are often critical to the structural integrity of a variety of protein complexes [for reviews see (25,26)].

In cases where isotopic enrichment of the bound peptide becomes problematic, a somewhat more difficult spectroscopic approach may be used to isolate $^1H$ -$^1H$ interactions exclusively involving the unlabeled peptide. This is achieved by uniformly enriching the receptor protein with $^{15}N$ and/or $^{13}C$ and editing the spectrum of the complex on the basis of whether or not a given $^1H$ is J-coupled to a heteronucleus. In this approach, pioneered by Wüthrich and coworkers, a spin echo difference scheme is used to suppress $^1H$-$^{13}C$ or $^1H$-$^{15}N$ pairs on the basis of the difference in sign of magnetization arising from protons that are directly bonded to an NMR active heteronucleus. This standard X-filter selects for those $^1H$ resonances that are scalar coupled to NMR active X-nuclei [for a review see (27)]. Application of a suitable X-filter selects only signals due to protons coupled to labeled X-nuclei. This same approach applied in the opposite sense is termed the reverse X-filter; only signals due to protons coupled to NMR inactive X-nuclei survive. However, as pointed out by Fesik and coworkers (28), reverse filtration is a much more stringent task than simply selecting $^1H$ nuclei that are J-coupled to a

heteronucleus. Reverse X-filtered experiments therefore often employ multiple strategies to suppress resonances arising from protons bonded to $^{15}N$ or $^{13}C$. A schematic illustration is shown in Fig. 1.

Pulse sequences employing so-called adiabatic inversion pulses which are swept at a rate which is tuned to the H-X coupling constant versus carbon chemical shift profile of the labeled molecule have been introduced (Fig. 2) (30). The basic

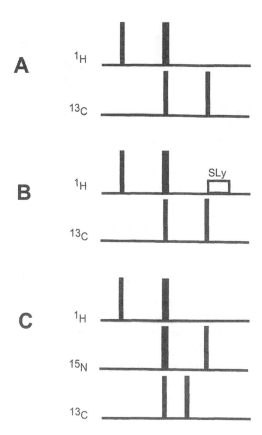

**Fig. 1** Schematic illustration of the reverse X-filter due to Fesik and coworkers (28) and Ikura and Bax (29). Scheme A shows the basic reverse filter which can be followed by either acquisition or evolution (28, 29). In scheme B, the residual $^{13}C$ signal is suppressed by its conversion to multiple quantum coherence and by the spin lock pulse (SLy) (28). Scheme C shows that both $^{15}N$ and $^{13}C$ bonded $^1H$ signals can be suppressed by appropriate choice of the refocusing delays (29).

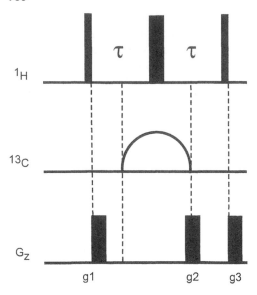

**Fig. 2** Swept adiabatic pulse purging scheme of Kay and coworkers (30). This particular scheme uses a carbon WURST pulse to minimize residual magnetization from protons attached to $^{13}C$. The rate of the linear frequency sweep is chosen so that traverse magnetization arising from protons attached to $^{13}C$ is antiphase while that bonded to $^{12}C$ is in-phase at the time of the final gradient pulse.

purging scheme is deceptively simple as it generally requires relatively extensive customization for optimal results. Nevertheless, this approach overcomes much of the difficulty presented by variable $^{1}H$-$^{13}C$ coupling constants albeit with some increase in experimental complexity (30).

The library of experiments employing X-filtration and reverse X-filtration to simplify the NMR spectra of protein complexes is now relatively stable [for a review, see (31)] with only a few recent additions (e.g., 30, 32). With filtering experiments added to the modern library of multidimensional NMR experiments available to the protein chemist, the study of protein-peptide complexes becomes straightforward. There are now many examples of such work.

Studies of protein-peptide complexes have been commonly used as models of their often much larger parent protein-protein complexes. Early examples include the use of synthetically $^{15}N$-enriched peptide to provide the first direct confirmation of the amphiphilic helix model for the structure of calmodulin-binding domains bound to calmodulin (17) and the first characterization of the fast dynamics of a bound domain (33). The calmodulin-peptide complexes also fueled the development of many of the reverse filtered experiments described above and led to the dramatic characterization of the complex between calmodulin and the calmodulin-binding domain of the myosin light chain kinase (34). The interaction of calmodulin with peptides corresponding to calmodulin-binding domains of target proteins also serves to illustrate the use of the transferred NOE to characterize the structure of a bound peptide (10) and the fast exchange limit to cross assign spectra (9). Early work with cyclosporin illustrated a variety of approaches to the protein-peptide complex including deuteration (15) and $^{13}C$-enrichment coupled with X- and reverse X-filtering (18, 19). The structural characterization of bound peptide ligands with irregular secondary structure is exemplified by the more recent work on the complex between SH2 domains and phosphotyrosine peptide ligands (35, 36), between thioredoxin and its target peptide from the transcription factor NF kappa B (37) and between the SH3 domains and their peptide ligands (38, 39).

While the methods available for solving the resonance assignment problems peculiar to protein complexes are now well established and complete, approaches available to the study of dynamics at the interface are less evolved. Recent developments in the use of both $^{13}C$ (40, 41) and $^{2}H$ relaxation (42) in conjunction with existing $^{15}N$ relaxation methods [for a review, see (43)] improve the ability to use relaxation techniques to probe the effect of complexation on the internal dynamics of both ligand and protein. Deuterium relaxation has recently been employed to probe the effects of complexation by the SH2 domain on its phosphotyrosine-containing peptide ligand (44, 45) and $^{15}$N relaxation has been used to probe the role of protein dynamics in gating ligand binding to the HIV protease (46) . It appears that the effects of complexation on the internal dynamics of proteins may be directly related to changes in the fundamental thermodynamic properties of the system (47, see also 48, 49). This is a most exciting development.

Use of the exchange of backbone amide hydrogens with solvent is now a well established approach to the characterization of less frequent motions than those dominating NMR relaxation phenomena. In favorable cases it has been possible to measure hydrogen exchange behavior directly in a protein-ligand complex. This approach has been especially illuminating in studies of calmodulin-peptide complexes where hydrogen exchange behavior revealed not only the dynamics of a peptide bound to calmodulin but also allowed the sequence of steps in the binding and release of the peptide ligand to be inferred (50). This follows from the view that hydrogen exchange can be used to monitor the breakage of hydrogen bonds arising from largely local motions, motions involving larger but still subglobal fragments of the protein, or from motions leading to global unfolding of the protein (51, 52). Access to binding and dynamic behavior within protein complexes that are too large for direct study can still be obtained by NMR analysis of hydrogen exchange (HX) labeling patterns. In this case, time dependent H-D exchange labeling is performed in the complex and the protein interaction surface, marked by amide sites that are made slow exchanging in the complex, is analyzed later in one of

the separated partners. This approach has been applied to protein-monoclonal antibody interactions (53–56), enzyme-inhibitor complexes (40, 41), the interaction of proteins with micellar systems (57) and interactions within protein-protein complexes (58, 59, 60, 61). Thus it is seen that while high resolution x-ray crystallographic and NMR techniques can provide precise details of equilibrium structures of proteins, the dynamics of the transitions between the potentially numerous transient conformational states can be probed using hydrogen exchange techniques.

## III. EXAMPLES FROM THE CALMODULIN COMPLEXES

To help illustrate the various approaches and strategies outlined above, we will appeal to the work of our laboratory on studies of the biophysical basis for the high affinity interaction between calmodulin and small peptides corresponding to the calmodulin-binding domains of regulated proteins. Calmodulin is a small (148 amino acids), acidic protein that occupies a central place in calcium-mediated signal transduction in eucaryotes. It is a primary player in the activation of a wide range of cellular responses to an increase in intracellular calcium. Calmodulin has been found to regulate a wide range of cellular targets and is consequently central to control of a number of critical cellular events (62). Over 70 potential targets of regulation by calmodulin have been identified. Calmodulin has been the object of extensive biophysical characterization, particularly by NMR methods, and provides a series of examples of the use of NMR to examine protein-peptide complexes.

In the crystal, calcium-saturated calmodulin (CaM) has two globular domains separated by a long alpha helix (63). A ribbon representation of the structure of calcium-saturated calmodulin is shown in Fig. 3.

**Fig. 3**  Ribbon representation of the backbone of calcium-saturated calmodulin as determined by x-ray crystallography (63).

Chemical cross-linking (64), low angle X-ray scattering (64), and NMR spectroscopy (65) have been used to demonstrate the dynamic disorder of the central region of the bridging helix. The amphiphilic helix model (62) for the structure of a peptide corresponding to the calmodulin-binding domain of the myosin light chain kinase (MLCK) when bound to CaM was first directly confirmed by NMR-based studies (17) and the structure of an entire complex subsequently characterized by both NMR (34, 66, 67) and crystallographic (68) methods. The structure of the complex was quite different than originally envisaged (62). The dynamic aspects of this classic interaction between the calcium-saturated calmodulin and the calmodulin-binding domain of the smMLCK have also recently been investigated in our laboratory [e.g., (33, 50); Lee et al., unpublished results].

The calmodulin system provides extensive examples of the various ways of approaching the resonance assignment task in protein complexes. As suggested

above, one need not take a particularly novel approach if complete labeling is available for a complex in slow exchange with its free components. An early example of using selective labeling of a peptide in a complex was the use of chemical synthesis to introduce α-amino $^{15}$N-labeling into the peptide corresponding to the calmodulin-binding domain of the smooth muscle myosin light chain kinase. Here, chemical synthesis was used to label 9 of 19 amide sites in the following peptide sequence (smMLCKp): Acetyl-Ala-Arg-Arg-Lys-Trp-Gln-Lys-Thr-Gly-His-Ala-Val-Arg-Ala-Ile-Gly-Arg-Leu-Ser-NH$_2$. Roth and coworkers used this peptide with filtering to focus in on the issue of whether or not the peptide was helical when bound to calmodulin, an open question at the time (17). An expansion of the amide-amide NOE region of the reverse filtered NOESY spectrum of the complex of the labeled peptide with unlabeled calcium-saturated calmodulin is shown in Fig. 4. The appearance of sequential amide-amide NOEs clearly

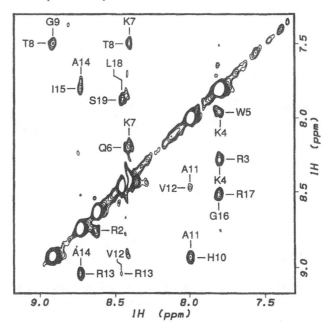

**Fig. 4** Half X-filtered NOESY spectrum of the 1:1 CaM · smMLCKp complex. The mixing time was 100 ms. The resonance assignments correspond to those of the bound peptide. Note the asymmetry in NOEs involving only one labeled $^{15}$N-amide. Excellent selection for NOEs involving at least one proton bonded to $^{15}$N was achieved. Reprinted with permission from Roth et al. (17). Copyright 1991 American Chemical Society.

confirmed the helical nature of the bound peptide. Thus, labeling in conjunction with X-filtering allowed for a rapid and direct determination of the structure of the peptide ligand without having to deal with the spectral problem presented by the protein.

Subsequently, the structure of the entire complex between calcium-loaded calmodulin and a peptide corresponding to the calmodulin-binding domain of the skeletal muscle myosin light chain kinase was determined (34). This *tour de force* by Ikura and coworkers represents a striking step forward by NMR in the area of the structural characterization of protein · peptide complexes. A ribbon representation of the complex is shown in Fig. 5. The striking feature is the fact that calmodulin essentially buries the entire helical region of the bound peptide. This is a highly unusual binding interaction.

These studies relied on the extensive use of reverse filtering to obtain resonance assignments and structural restraints for the bound peptide itself and for its close distance interactions with calmodulin. To illustrate the kind of data available by these techniques, the doubly reverse filtered NOESY spectrum of a complex between one of the calmodulin-binding domains of the phosphorylase kinase and calcium-saturated calmodulin is shown in Fig. 6.

The CaM · smMLCKp complex is in slow exchange on the NMR time scale with its dissociated components. In cases where the complex is in fast exchange on the NMR time scale the trNOE may be used with some success to focus in on the structure of the bound peptide while avoiding the complexities presented by the protein. There are several examples of this

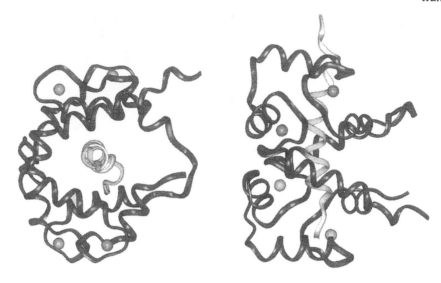

**Fig. 5** Ribbon representation of the complex between calcium-loaded calmodulin and a peptide corresponding to the calmodulin-binding domain of the skeletal muscle myosin light chain kinase determined using NMR methods by Ikura et al. (34).

approach in studies of CaM · peptide complexes, particularly by Vogel and coworkers (10). Our laboratory has also used this approach to characterize the interaction between the D-amino acid analog of the smMLCKp peptide used by Roth et al. (17). Here, the idea was to explore the sensitivity of this particular protein-ligand interaction to chirality (9). This study also provided a hint as to the

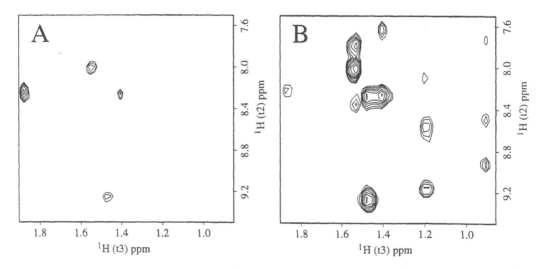

**Fig. 6** Expansion of the reverse filtered $^{13}$C- resolved NOESY spectrum (A) and the corresponding $^{13}$C-resolved NOESY spectrum (B) of the complex between a peptide corresponding to the calmodulin binding domains of the phosphorylase kinase and calcium-saturated calmodulin. The pulse sequence of Pascal et al. (36) was used for the reverse filtered experiment with a mixing time of 100 ms (Urbauer and Wand, unpublished results).

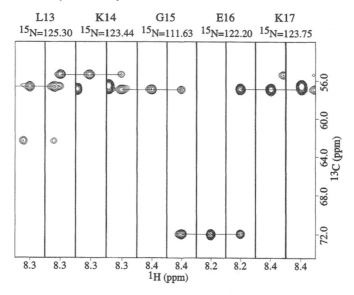

**Fig. 7** Strips from HNCA and HN(CO)CA spectra of a complex between a peptide corresponding to the calmodulin-binding domain of neuromodulin in complex with unlabeled apocalmodulin. Adapted with permission from Urbauer et al. (84). Copyright 1995 American Chemical Society.

free energy path experienced by an unstructured peptide encountering calmodulin and then reorganizing to its final, lowest energy structure (see below). The CaM · D-smMLCKp complex is in fast exchange with its dissociated components and this feature was used to quickly cross assign calmodulin in the complex by reference to the resonance assignments of free calcium-saturated calmodulin (9). Analysis of the chemical shift changes brought about by binding of the peptide indicated that the complexation did not result in collapse to the compact structure seen for the L-amino acid smMLCKp peptide.

Biosynthetic labeling of peptide ligands opens the door to the application of many robust and novel experiments which are not possible without $^{15}N$ and/or $^{13}C$-labeling. We have used fusion peptide systems to successfully prepare uniformly $^{15}N$ and $^{15}N,^{13}C$-enriched peptides using either constructs with glutathione S transferase or thioredoxin His-tag carrier proteins. The thioredoxin His tag system provides many advantages including high

solubility (i.e., avoidance of inclusion bodies) and one step purification by affinity chromatography. We have used the thioredoxin His tag system to prepare peptides corresponding to several different calmodulin-binding domains. Uniform $^{15}N,^{13}C$-labeling allows for rapid assignment of the resonances of the bound peptide without the confusion presented by those of calmodulin. This is illustrated in Fig. 7 which shows triple resonance HNCA and HN(CO)CA spectra of a uniformly $^{15}N,^{13}C$-labeled peptide corresponding to the calmodulin-binding domain of neuromodulin in complex with unlabeled apocalmodulin. This provided a direct and extremely efficient way to probe the role of the peptide-protein interactions in the stabilization of this particular complex (see below).

To illustrate the potential of NMR spectroscopy to probe the dynamics of protein · peptide complexes we turn again to the calmodulin system. Previous to recent studies, very little was known about how calmodulin initially recognizes the domain, how it collapses from an initial encounter complex to the final compact state, or the energetics of these events and associated structural changes. In order to monitor the dynamics of the molecular recognition process one could attempt to follow the kinetic events directly, as has recently been attempted using stopped-flow fluorescence spectroscopy [see the discussion by Török and Whitaker (69)], or by quantifying the equilibrium mani-fold of states accessible to the complex. As suggested above, hydrogen exchange methods in principle allow access to the equilibrium manifold of states, and we have employed them in our recent study of the energetics and dynamics of CaM in complex with a peptide corresponding to the calmodulin-binding domain of the smooth muscle myosin light chain kinase (50).

A variety of NMR-based techniques were employed to measure the hydrogen exchange rates of the backbone amide hydrogens of 18 of the 19 amino acid residues of a peptide analogue of the smooth muscle myosin light chain kinase calmodulin-binding domain (smMLCKp) while bound to calcium saturated calmodulin (CaM) (50). Exchange rates spanning six orders of magnitude were measured using a battery of techniques which allowed determination of the exchange rates of both $^{15}N$- and $^{14}N$-bonded amide hydrogens of the bound smMLCKp peptide. Key to this approach is the ability to filter the $^1H$ spectrum on the basis of the presence of bonded $^{15}N$-nuclei. The heteronuclear $^{15}N$-$^1H$ cor-relation spectrum of the labeled peptide bound to calcium saturated calmodulin was assigned and the structure determined as discussed above (17).

The nine $^{15}N$-labeled amide sites of the smMLCK peptide allow for direct, selective detection of their bonded hydrogens even in the presence of $^1H$ res-onances arising from amide and aromatic hydrogens due to calmodulin. In the case

of relatively slow hydrogen exchange rates ($<0.04$ min$^{-1}$), the exchange rate can be simply measured in "real time" by sequential acquisition of HSQC spectra following initiation of exchange by dilution of protonated sample into $D_2O$ buffer. The rate of exchange at a given amide site is then obtained from a simple single exponential fit of the integrated peak volumes as a function of exchange time. Typical examples of exchange of an amide hydrogen for a deuterium from solvent are shown in Fig. 8. Exchange was initiated by dilution of an $H_2O$ solution of the complex into the $D_2O$ exchange buffer in order to avoid potential refolding artifacts from the lyophilized state. The use of serially acquired HSQC spectra employing spin locks for water suppression allowed the exchange rates at six of the nine labeled sites to be determined under the conditions employed. Essentially identical rates were determined using CaM to smMLCKp ratios of 1.1:1 and 2.2:1. This is important as in order to probe the dynamics of the bound smMLCKp peptide, the measured exchange rate must, of course, arise from events that occur in the bound state and not be contaminated by exchange from the free, random coil peptide. This con-dition can be satisfied by taking advantage of the high affinity (Ka $\sim 10^9$ M$^{-1}$) of the peptide for CaM. The concentration of *free* peptide when the CaM:smMLCKp ratio is 1.1:1 and the CaM concentration is millimolar will be at least an order of magnitude greater than when the CaM:smMLCKp ratio is 2.2:1. Hence, the fraction of time that a given amide NH will spend in the free peptide state will correspondingly decrease. As the observed rate of exchange is the same under both conditions, the observed hydrogen exchange rates are indeed dominated by exchange from the *bound* state(s) of the peptide under the conditions.

The rates of exchange of the amide hydrogens of Ala-1 and Lys-7 are too rapid to measure by the method of serial acqui-

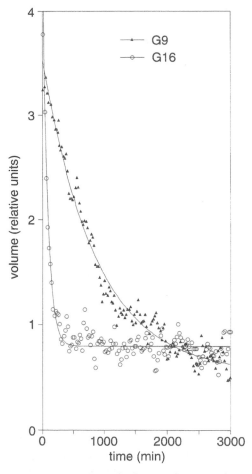

**Fig. 8** Examples of the method used to measure relatively slow amide hydrogen exchange rates at $^{15}$N-labeled sites of the smMLCKp peptide bound to calcium saturated calmodulin using $^{15}$N-$^1$H correlation spectroscopy. Shown are the decays of the intensity of the amide $^{15}$N-$^1$H cross peaks of Gly-9 and Gly-16 of the smMLCKp peptide bound to calmodulin in a set of serially acquired $^{15}$N-HSQC NMR spectra upon dilution of a protonated sample into D$_2$O buffer at pH* 6.98. The observed rates are $1.2 \times 10^{-3}$ min$^{-1}$ and $1.3 \times 10^{-2}$ min$^{-1}$. Reprinted with permission from Ehrhardt et al. (50). Copyright 1995 American Chemical Society.

sition of HSQC spectra. The rate of exchange of the amide hydrogen of Lys-7 falls in the range that could be quantitated by serial acquisition of $^{15}$N-filtered one dimensional $^1$H spectra. A sample of the complex was fully exchanged into D$_2$O, lyophilized, wetted with D$_2$O and diluted into H$_2$O. At pH 5.94, Lys-7 is resolved in the $^{15}$N-filtered $^1$H spectrum from all but the very slowly exchanging amide of Val-12. The rate of exchange could then be determined from the increase in integrated intensity as a function of exchange time. The rate of exchange of the amide hydrogen of Ala-1 is even faster and was determined by saturation transfer experiments undertaken at pH 7.35, 7.79, and 8.07 (see Fig. 9). Volumes of a given peak were measured in the absence ($M_o$) and the presence ($M_{ps}$) of water presaturation. The ratio of these volumes and an assumed average T$_1$ of 125 msec were used to calculate the exchange rate of the amide in question from equation 1 (70).

$$k = \frac{(M_o/M_{ps} - 1)}{T_1} \qquad (1)$$

It should be noted that the generalized order parameters of the nine $^{15}$N-$^1$H bonds of the bound smMLCKp peptide have been determined and found to be uniformly high across the full length of the peptide (33) (See below). Thus it is unlikely that variations in local dynamics would give rise to significant variation in the effective $^1$H spin lattice relaxation making the assumption of a uniform effective T$_1$ reasonable.

Hydrogen exchange rates at slowly exchanging unlabeled amide nitrogen sites of the bound smMLCKp peptide were measured by following the decay of amide-amide NOEs involving one amide hydrogen at an $^{15}$N-labeled site and another amide hydrogen at an unlabeled site in serially acquired $^{15}$N-HSQC-NOESY spectra. Under the so-called EX$_2$ conditions used here, the rate of exchange of the amide hydrogen of the unlabeled amino acid is equal to the rate of change in the NOE crosspeak volume minus the rate of change of the autocorrelation peak volume arising from the labeled amino

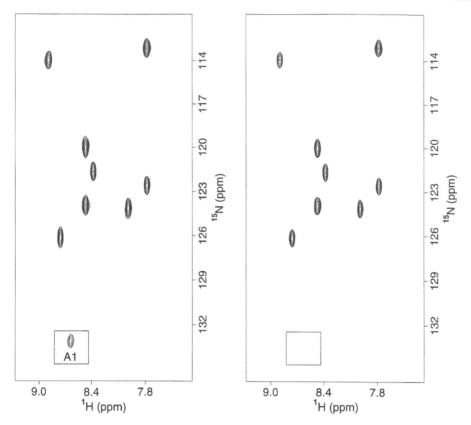

**Fig. 9** Determination of the rate of exchange of the amide hydrogen of Ala-1 by saturation transfer. $^{15}$N-HSQC without saturation of solvent (left panel) and with saturation of solvent (right panel) show the effects of saturation transfer to the amide hydrogen of Ala-1 which is in rapid exchange on the NMR timescale with solvent. These spectra were obtained at pH* 8.07 and 25 C and when combined with similar spectra obtained at other pH values provide an estimate of the slowing factor. Reprinted with permission from Ehrhardt et al. (50). Copyright 1995 American Chemical Society.

acid in the $^{15}$N-HSQC-NOESY spectrum. An example is shown in Fig. 10. The use of NOE cross peaks in $^{15}$N-HSQC-NOESY spectra to obtain hydrogen exchange rates is the least sensitive of the techniques employed here. Comparison of the behavior of two cross peaks to an unlabeled amide proton flanked by two labeled amide protons indicates that rates determined by this method can vary by up to two fold. Finally, the kinetic behavior of the "auto" peaks of the $^{15}$N-HSQC-NOESY spectra compare favorably with the decay rates obtained using serially acquired $^{15}$N-HSQC spectra.

To measure fast exchange rates at unlabeled amide sites, $^{15}$N-HSQC-NOESY spectra were collected at pH values ranging from 6.31 to 8.59. The rate of exchange at the labeled Lys-4 site was also determined in this manner. An NOE mixing time of 75 ms was employed which is in the linear region of the NOE-buildup curves at 500 MHz. A simple two spin relaxation matrix was used to model the effects of saturation transfer over a range of conditions and relaxation time constants. For a two spin case, analytical solutions are available for the eigenvalues and vectors of the relaxation matrix mak-

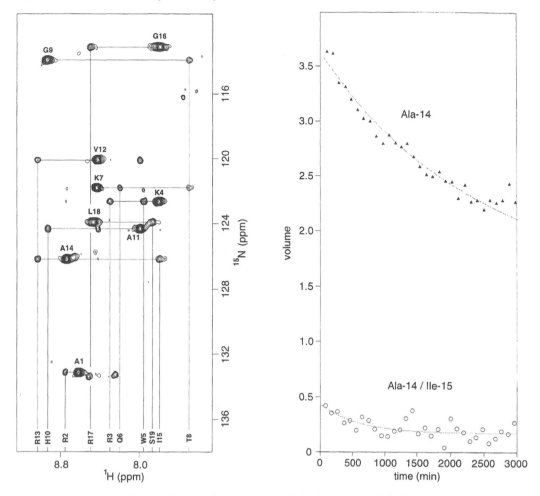

**Fig. 10** Examples of the method used to measure relatively slow amide hydrogen exchange rates at [14]N-sites of the smMLCKp peptide bound to calcium saturated calmodulin using $^{15}$N-$^1$H correlation spectroscopy. The NOE cross peaks arising between adjacent amide hydrogens in the smMLCKp peptide bound to calmodulin are indicated in the left panel and have been assigned by standard methods [Roth et al., 1991 (17)]. The right panel shows the decays of intensity of the Ala-14 $^{15}$N-$^1$H autopeak and the Ile-15 [14]NH to Ala-14 $^{15}$NH NOE crosspeak in serially acquired HSQC-NOESY spectra collected with an NOE mixing time of 75 ms. Reprinted with permission from Ehrhardt et al. (50). Copyright 1995 American Chemical Society.

ing simulations facile. The effect of chemical exchange enters into the diagonal terms of the rate matrix (71). Using the standard expressions relating pH to the rate of hydrogen exchange and the recently recalibrated empirical factors for local sequence dependence (72) reasonably precise estimates of the effective exchange rate can be determined by fitting the

observed cross peak intensities as a function of pH. Very generous allowances for the imprecision of estimates of the intrinsic spin lattice and cross relaxation rates have only modest effects on the accuracy of the obtained hydrogen exchange rates. Examples are shown in Fig. 11.

It should be noted that the rates of exchange of the slowly exchanging amides

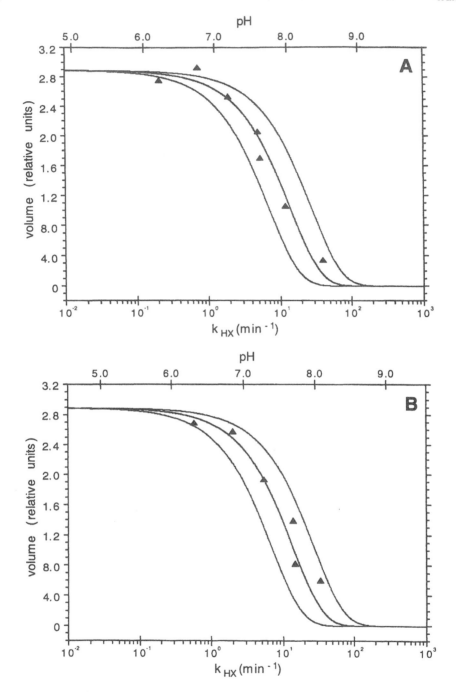

**Fig. 11** Examples of the method used to measure relatively fast amide hydrogen exchange rates at $^{14}$N-sites of the smMLCKp peptide bound to calcium saturated calmodulin using $^{15}$N-$^{1}$H correlation spectroscopy. Shown are the intensities of the Lys-4 $^{15}$N-$^{1}$H autopeak (Panel A) and Lys-4 $^{15}$NH to Trp-5 $^{14}$NH NOE crosspeak as a function of pH. The effective rate for exchange is also shown on the lower abscissa. The center line is the best fit to the data and the outer lines are calculated using the assumed error on the fitted rate and illustrate the conservative nature of the estimate. Reprinted with permission from Ehrhardt et al. (50). Copyright 1995 American Chemical Society.

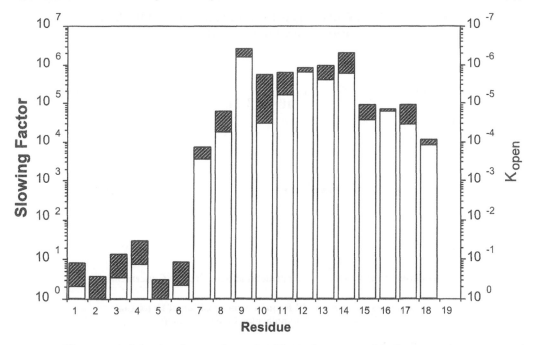

**Fig. 12** Histogram of slowing factors determined from the measured hydrogen exchange rates of the smMLCKp peptide bound to calmodulin and the predicted exchange rates of the free peptide. Adapted with permission from Ehrhardt et al. (50). Copyright 1995 American Chemical Society.

follow the predicted pH dependence over the range of pH used indicating that exchange is occurring under the so-called $EX_2$ condition and, importantly, that the complex is structurally unaffected by the range of solution conditions employed.

Slowing factors for 18 of the 19 amide sites of the bound smMLCKp peptide were calculated using observed hydrogen exchange rates obtained using the various approaches summarized above and the rates predicted for the free, unstructured peptide according to the empirical factors due to Bai et al. (72). Shaded regions of the histogram shown in Fig. 12 correspond to the experimental uncertainty and are determined from up to three independent measurements. In the case of His-10 and Ala-11, the shaded regions of the histogram also include the effects of uncertainty in the charge state of the histidine side chain in the hydrogen exchange competent state(s). It should be

mentioned, however, that the general pH dependence of the hydrogen exchange rates of His-10 and Ala-11 are consistent with a pKa of ∼ 6.5-7 for the histidine side chain in the exchange competent state(s). The exchange rate of Ser-19 could not be accurately determined due to spectral degeneracies that occur at the low pH necessary to accurately estimate the rate.

The general dependence of the slowing factors obtained for the bound smMLCKp peptide as a function of sequence position is characteristic for hydrogen bond breakage arising from helix-coil transitions. Such transitions lead to "end fraying" and result in more stable helical hydrogen bonding towards the center of the peptide. The helix-coil transition for isolated helices is often treated in a statistical mechanical way utilizing various models (e.g., 73, 74). With such treatments, one can calculate the fraction of time a given amide NH is

## Interpretation of Hydrogen Exchange in the smMLCKp
## Peptide Bound to Calmodulin

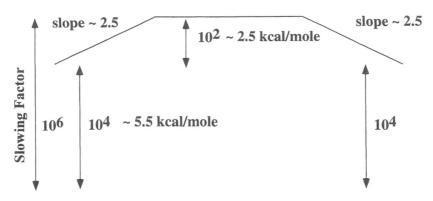

**Fig. 13** Schematic illustration of the origin of the observed hydrogen exchange rates of the smMLCKp peptide bound to calcium-saturated calmodulin.

hydrogen bonded given residue specific parameters of the model (S and s for Zimm-Bragg). For model peptides in water, most individual amino acids generally have S values between 0.5 and 2.0 [see, for example, Lyu et al. (75) and Rohl and Baldwin (76)]. The apparent S-value is a reflection of the effective hydrogen bond stability which in turn is manifested in the hydrogen exchange slowing factors. For proteins, such as cytochrome c, observed patterns of hydrogen exchange slowing factors of helical segments indicate effective S parameters on the order of 10 or larger (77). Such large effective S values are indicative of extensive interactions between the helix and surrounding protein.

It is important to note then that the observed end-fraying of the bound smMLCKp peptide results in a relatively *small* range of slowing factors indicating

low effective S values. This is most consistent with exchange arising from helix-coil transitions in a relatively unhindered state, perhaps suggesting that exchange occurs in the free peptide state. However, this is clearly not the case, as shown above. In addition, the slowing factors at the ends of the helical segment are very large, on the order of $10^4$. Such large values for the ends of helical segments of free peptides are not predicted by small S parameters. The physical explanation of these observations is outlined schematically in Fig. 13. Given the range of effective S values reflected in the slope of the end fraying, one can model the helix-coil transition such that a range of $10^2$ in slowing factors is predicted using nucleation (s) parameters on the order of $10^{-3}$, a value which is seen in model peptides. If the helix-coil manifold is considered the only set of states from which hydrogen exchange can occur (i.e., they provide the dominant structural reorganizations that lead to transient hydrogen bond breakage and exposure of the amide NH to solvent), then initial slowing factors of $10^4$ require that there be a major reorganization that moves the complex from one that suppresses the helix-coil transitions to one that allows them to occur with parameters characteristic of a relatively unhindered peptide (i.e., extensively solvated but bound).

This picture leads to a model where this equilibrium results in an opening of the complex such that the peptide is brought to the surface of calmodulin and extensively exposed to solvent. The observed hydrogen exchange slowing factors require that this equilibrium be characterized by an equilibrium constant of $10^4$ ($\Delta G \sim 5.5$ kcal/mole). The pattern of slowing factors indicates that helix-coil transitions in the peptide occur in the reorganized state of the complex. In the reorganized state, the peptide is extensively solvated and the helix-coil transitions create a manifold of states leading to transient breakage of intrahelical hydrogen bonding, allowing for exchange

of amide hydrogens with solvent. This manifold is characterized by an effective equilibrium constant spanning the least and most protected amide NH of $\sim 10^2$ (corresponding to a $\Delta G$ of $\sim 2.5$ kcal/mole). The reorganized complex is in direct equilibrium with the dissociated complex. These dynamic processes are schematically summarized in Fig. 14.

When this process is viewed in reverse, a free energy scale is revealed for the binding of the unstructured free peptide to calmodulin followed by the induction of helical structure and collapse to the compact complex. This view is consistent with the results for the binding of a D-amino acid analogue of the smMLCKp peptide to CaM in which collapse to the compact complex is frustrated but induction of helical structure on the surface of calmodulin is essentially complete (9). In a kinetic sense, the collapse to the compact complex would presumably be associated with a relatively large barrier and may in fact be the source of biphasic behavior described in a preliminary report of a stopped flow fluorescence study of the binding kinetics of a related peptide (69).

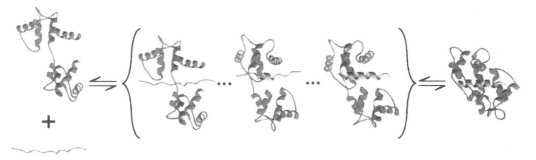

**Fig. 14** Schematic illustration of the physical model used to interpret the observed hydrogen exchange of the smMLCKp peptide bound to calmodulin. The random coil peptide forms an initial encounter complex with an open dumbbell-like calmodulin structure. The general surface interactions promote helix-coil transitions which lead to the formation of the helix on the surface of the protein. This manifold of states is characterized by an effective free energy difference between the least and most stable hydrogen bond of about 2.5 kcal/mole. The complete helix provides properly oriented contacts to allow for the collapse of the complex. This latter reorganization is characterized by a free energy change of about 5.5 kcal/mole. Reprinted with permission from Ehrhardt et al. (50). Copyright 1995 American Chemical Society.

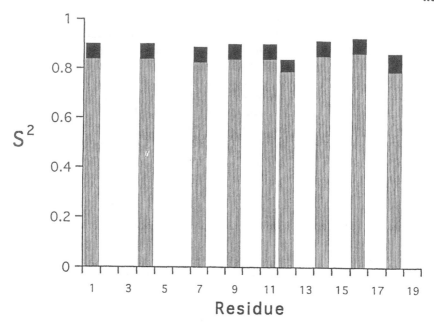

**Fig. 15** Generalized order parameters obtained at selectively [15]N-labeled amide sites of the smMLCKp peptide bound to calcium-saturated calmodulin. Reprinted with permission from Chen et al. (33). Copyright 1993 Academic Press.

The foregoing illustrations have concentrated on the use of hydrogen exchange to probe the free energy manifold of protein-peptide complexes via the structural fluctuations that occur over it. Of additional interest are dynamics which are much more frequent and are presumably much smaller in scope. Very fast, subnanosecond dynamics can be approached by a variety of NMR relaxation techniques. We have used both [15]N and [2]H relaxation to probe the dynamical consequences of the high affinity interaction between calmodulin and the smMLCKp peptide. [15]N relaxation studies unequivocally showed that the bound peptide displays main chain dynamics that are quite similar to integral units of secondary structure of native-state proteins (33). This is summarized in Fig. 15 where the generalized order parameters obtained for labeled sites in the peptide (see above) where generally $\sim 0.85$, which is typical of stable units of secondary structure. That is, from

the point of view of the dynamics of the main chain, the helical peptide of the ligand domain is no different than another helical element within calmodulin itself.

More recently we have applied deuterium relaxation techniques to probe the dynamics of the interface between calmodulin and the smMLCKp peptide. Introduction of deuterium is easily achieved by expression of the protein on a minimal media containing 50% $D_2O$ as well as [15]N nitrogen and [13]C carbon sources. The resulting [13]C-HSQC spectrum of the 1:1 complex between 50% [2]H,[15]N,[13]C-calmodulin and the unlabeled smMLCKp peptide is shown in Fig. 16. Transverse and longitudinal deuterium relaxation of the methyl groups of calmodulin were measured with high sensitivity using the methods of Kay and coworkers (42). Example decay curves are shown in Fig. 16.

NMR offers the unique opportunity to comprehensively probe the structural

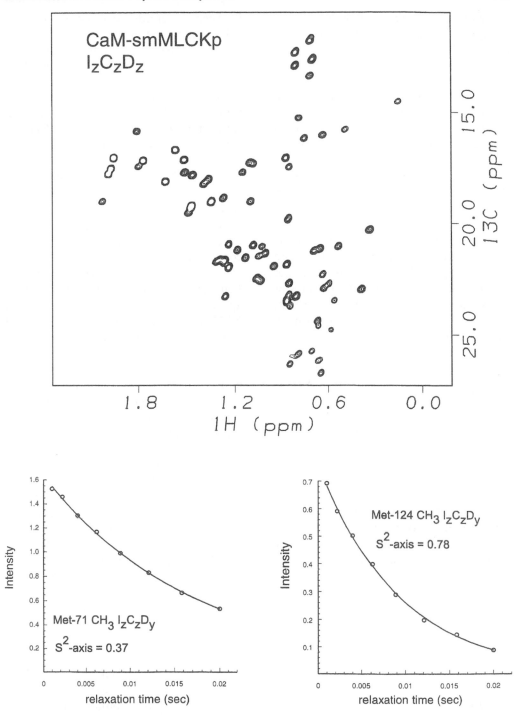

**Fig. 16** Panel A, $^{13}$C-HSQC spectrum of the 1:1 complex between 50% $^{2}$H,$^{15}$N,$^{13}$C-calmodulin and the unlabeled smMLCKp peptide. Panel B, sample decay curves of methyl carbons of two methionines of calmodulin. Generalized order parameters for the axis of the methyl groups were obtained from data collected at 17.6 and 14.1 Tesla. Lee et al., unpublished results.

consequences of a range of environmental changes. Obvious environmental variables include temperature, pH, and ionic strength. A perhaps less obvious environmental variable is hydrostatic pressure. We have recently shown that hydrostatic pressure apparently offers one of the few routes to the explicit separation of the contributions of hydrophobic and ionic interactions to the formation of protein complexes (20). For example, protein interfaces are frequently composed of both intermolecular salt links as well as extensive hydrophobic surface contacts. Ionic interactions often have significant negative $\Delta V$ of dissociation and are therefore usually exquisitely sensitive to pressure, having midpoint dissociation pressures of a few hundred bar (79, 80). In contrast, van der Waals interactions are usually characterized by small $\Delta V$ of dissociation and are relatively pressure insensitive often having midpoint dissociation pressures of a few thousand bar (78–80). This difference perhaps offers the only means to isolate and define the relative contributions of each type of interaction to the affinity and specificity of protein complexes. We have recently demonstrated the first instance of high pressure triple resonance NMR spectroscopy and used this capability to trace the origin of the high pressure reorganization of a novel complex involving calmodulin (20). As discussed above, many aspects of the structure-function relationships of calmodulin have been illuminated by the structures of apocalmodulin (81, 82), calcium-saturated calmodulin (63), and calcium-saturated calmodulin in complex with the calmodulin-binding domains of a few target proteins (34, 67, 68, 83). Most target proteins are bound by calcium-saturated calmodulin only. In contrast, neurogranin and neuromodulin bind to calmodulin in the absence of calcium and not in the presence of bound calcium. We have recently reported NMR (84) and fluorescence (85) studies of the interaction

between a peptide corresponding to the calmodulin-binding domain of neuromodulin (Neuro-p) and calcium-free apocalmodulin. The acidity of calmodulin and the seven basic amino acids of this domain suggest that electrostatic interactions contribute strongly to complex formation. The apocalmodulin · Neuro-p complex is 50% dissociated in 40 mM KCl. Application of high hydrostatic pressure red shifts the Neuro-p tryptophan fluorescence emission spectrum indicating a regional change in polarity (85). Importantly, this event is *unimolecular* and does not involve dissociation of the complex (85).

Since the pioneering work of Benedek and Purcell (86), two types of approaches to achieving high pressure NMR capability have evolved. The high pressure *probe* technique takes the design principle that the entire RF coil and sample are to be pressurized. This original design strategy has been adopted and extended by Jonas and coworkers (87). The second general approach is the high pressure *cell* technique which employs thick walled glass, Vespel or sapphire tubes or reinforced quartz capillaries. We have developed a high pressure cell that allows state-of-the-art NMR spectroscopy to be safely performed at kbar pressures without modification of the spectrometer. We have employed a sapphire tube fitted to a BeCu charging valve and have used liquid pressurization to increase the reliability and safety of the device. A detailed description of the design and operation of the pressure cell will be presented elsewhere. The NMR spectra of both free apocalmodulin and in the 1:1 complex with Neuro-p have recently been assigned using triple resonance spectroscopy (20). Upon binding to apocalmodulin, the $^{15}$N-HSQC spectrum of the peptide changes significantly (compare Panels A & B, Fig. 17). Especially noteworthy are the two resonances associated with the Arg side chain guanidino moieties

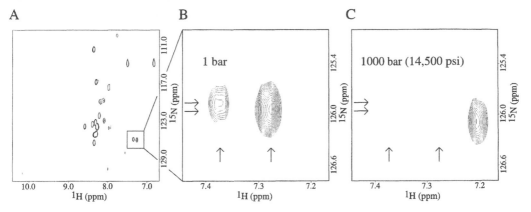

**Fig. 17** Characterization of the pressure-induced reorganization of the apocalmodulin · Neuro-p complex by high resolution NMR. The peptide used has the sequence GSQASWRGHITRK KLKGEKG. Panel A shows the $^1$H-$^{15}$N HSQC spectrum of the 1:1 apocalmodulin · Neuro-p complex (4 mM) in the sapphire high pressure tube [5 mm (o.d.)/1.5 mm (i.d.), ~80 mL active volume] at ambient pressure (~1 bar). Expansions showing the $\varepsilon$ $^1$H -$^{15}$N correlations of R5 and R10 of the bound peptide (aliased from ~85 ppm) at ambient pressure (~1 bar) and 1 kbar are shown in Panels B and C, respectively. Arrows indicate peak centers at ambient pressure. Main chain assignments of the Neuro-p peptide bound to apocalmodulin were obtained using triple resonance HNCA and HN(CO)CA spectra (see Fig. 7). All spectra were acquired with a Varian Unity Inova 600 MHz spectrometer using a standard z-gradient triple resonance probe and a custom built high pressure sapphire NMR cell. Adapted with permission from Urbauer et al. (84). Copyright 1996 American Chemical Society.

of the peptide. In the free peptide, hydrogen exchange with solvent is efficient enough to cause total loss of cross peaks from these groups. Upon binding to apocalmodulin, strong cross peaks arising from these groups appear in the $^{15}$N-HSQC indicating their participation in hydrogen bonding interactions with apocalmodulin. Application of high pressure results in significant but localized changes in the spectrum of the bound peptide (Fig. 17). At 1.0 kbar there is a marked reduction in the intensity of the two guanidino cross peaks indicating increased hydrogen exchange with solvent. Under the conditions used here, the effect of a 1 kbar change in pressure on the underlying chemistry of hydrogen exchange is expected to increase the intrinsic exchange rate by only ~3 fold [see Carter et al. (88) and references therein]. This cannot explain the observed changes in cross peak intensities even if significant hydrogen

exchange can occur from the salt linked (hydrogen bonded) state which, as Fig. 17 indicates, is not the case. The only remaining interpretation is also the simplest: that pressure disrupts the ionic interactions between the peptide and protein resulting in exposure of the peptide's Arg guanidino hydrogens to solvent. As noted above, the non-salt linked state has intrinsic hydrogen exchange rates resulting in total loss of cross peak intensity. Thus, under the conditions used here, the residual intensities of the guanindo resonances are a direct measure of the salt linked state of the Arg side chain where hydrogen exchange is significantly retarded.

Cross peak intensities therefore provide a direct measure of the equilibrium between the salt linked and separated ion states and indicate, through the usual expression $K(p) = K° \exp(\Delta V/RT)$, an apparent free energy change of ~2 kcal mol$^{-1}$ and a standard volume change of

$\sim 60 \, \text{mL mol}^{-1}$. These values are consistent with those determined from the pressure-sensitivity of the fluorescence of the peptide's Trp side chain (85). Comparison of triple resonance HNCA-J spectra of the bound peptide obtained at 1 bar and at 1.0 kbar provide access to changes in the main chain torsion angle phi via $^3J_{HNHa}$. These data show that the main chain of the bound peptide is only slightly perturbed during this reorganization (20). The absence of significant perturbations of the chemical shifts of backbone amide NH resonances of the bound peptide also reinforces the conclusion that the secondary structure of the peptide remains largely unchanged through the pressure induced transition.

Though the central elements embodied in the basic amphiphilic helix were identified some time ago as critical to the maintenance of a high affinity complex with calmodulin (62), almost nothing is known about the relative importance of ionic and hydrophobic interactions in the generation of a high affinity complex and in the selection of a unique structure. The contrast between specificity and affinity is of critical importance for calmodulin which must bind a wide array of proteins. Here we have demonstrated that the disruption of two ionic interactions between calmodulin and the Neuro-p domain is the origin of the pressure sensitivity of the complex, as suspected previously (85), and that these interactions contribute $\sim 2 \, \text{kcal mole}^{-1}$ towards the total binding free energy. It is has also been shown that the main ionic interactions between the peptide and calmodulin are not required for binding nor are they required for the maintenance of helical peptide structure. Taken together, these results also support a recently proposed model for the molecular recognition events leading from the initial encounter complex to the final low energy complex structure where general hydrophobic interactions dominate affinity while ionic interactions dominate in

maintaining the uniqueness of the complex (9, 50).

## IV. CONCLUSIONS

The last few years have seen a significant maturation of the techniques available to characterize the structure and dynamics of not only proteins but also nucleic acids, carbohydrates and lipids. The interaction of proteins with other small molecules and biopolymers can now be studied by NMR in great detail under a range of experimental conditions. The characterization of protein complexes by NMR can be expected to be a crowning achievement of the NMR technique.

## ACKNOWLEDGEMENTS

We are grateful to Ms. Marilyn Stevens for assistance in the preparation of the manuscript. The authors gratefully acknowledge the support provided by National Institutes of Health grant DK-39806. ALL is the recipient of an NIH postdoctoral fellowship (GM18114).

## REFERENCES

1. MF Leopold, JL Urbauer, AJ Wand. Resonance assignment strategies for the analysis of NMR spectra of proteins. Mol Biotech 2: 61–93, 1994.
2. GM Clore, AM Gronenborn. Structures of larger proteins, protein-ligand and protein-DNA complexes by multi-dimensional heteronuclear NMR. Prog Biophys Mol Biol 62: 153–184, 1994.
3. LE Kay. Pulsed field-gradient multi-dimensional NMR methods for the study of protein structure and dynamics. Prog Biophys Mol Biol 63: 277–299, 1995.
4. G Wagner. An account of NMR in structural biology. Nature Struct Biol 4(suppl s): 841–844, 1997.
5. GM Clore, A. Gronenborn. NMR structures of proteins and protein complexes beyond 20,000 Mr. Nature Struct Biol 4(suppl s): 849–853, 1997.
6. GK Jarori, N Murali, RL Switzer, BD Rao. Conformation of MgATP bound to

5-phospho-alpha-D-ribose 1-diphosphate synthetase by two-dimensional transferred nuclear Overhauser effect spectroscopy. Eur J Biochem 230: 517–524, 1995.

7. LY Lian, IL Barsukov, MJ Sutcliffe, KH Sze, GCK Roberts. Protein-ligand interactions: Exchange processes and determination of ligand conformation and protein-ligand contacts. Methods Enzymology 239C: 657–700, 1994.

8. F Ni. Recent developments in transferred NOE methods. Prog Nucl Magn Reson Spectrosc 26: 517–606, 1994.

9. PJ Fisher, FG Prendergast, MR Ehrhardt, JL Urbauer, AJ Wand, SS Sedarous, DJ McCormick, PJ Buckley. Calmodulin interacts with amphiphilic peptides composed of all D-amino acids. Nature 368: 651–653, 1994.

10. M Zhang, HJ Vogel, H Zwiers. Nuclear magnetic resonance studies of the structure of B50/neuromodulin and its interaction with calmodulin. Biochem Cell Biol 72: 109–116, 1994.

11. AP Hansen, AM Petros, AP Mazar, TM Pederson, A Rueter, SW Fesik. A practical method for uniform isotopic labeling of recombinant proteins in mammalian cells. Biochemistry 31: 12713–12718, 1992.

12. R Powers, DS Garrett, CJ March, EA Frieden, AM Gronenborn, GM Clore. $^1$H, $^{15}$N, $^{13}$C, and $^{13}$CO assignments of human interleukin-4 using three-dimensional double- and triple-resonance heteronuclear magnetic resonance spectroscopy. Biochemistry 31: 4334–4346, 1992.

13. M Kainosho. Isotope labelling of macromolecules for structural determinations. Nature Struct Biol. 4: 858–861, 1997.

14. SH Seeholzer, M Cohn, JA Putkey, AR Means, HL Crespi. NMR studies of a complex of deuterated calmodulin with melittin. Proc Natl Acad Sci USA 83: 3634–3638, 1986.

15. VL Hsu, IM Armitage. Solution structure of cyclosporin A and a nonimmunosuppressive analog bound to fully deuterated cyclophilin. Biochemistry 31: 12778–12784, 1992.

16. A Mohebbi, AJ Shaka. Improvements in carbon-13 broadband homonuclear cross-polarization for 2D and 3D NMR. Chem Phys Lett 178: 374–378, 1991.

17. SM Roth, DM Schneider, LA Strobel, MF VanBerkum, AR Means, AJ Wand. Structure of the smooth muscle myosin light-chain kinase calmodulin-binding domain peptide bound to calmodulin. Biochemistry 30: 10078–10084, 1991.

18. Y Theriault, TM Logan, R Meadows, L Yu, ET Olejniczak, TF Holzman, RL Simmer, SW Fesik. Solution structure of the cyclosporin A/cyclophilin complex by NMR. Nature 361: 88–91, 1993.

19. C Spitzfaden, W Braun, G Wider, H Widmer, K Wüthrich. Determination of the NMR solution structure of the cyclophilin A-cyclosporin A complex. J Biomol NMR 4: 463–482, 1994.

20. JL Urbauer, JH Short, L Dow, AJ Wand. Structural analysis of a novel interaction by calmodulin: high-affinity binding of a peptide in the absence of calcium. Biochemistry 34: 8099–8109, 1995.

21. A Bax, F Delaglio, S Grzesiek, GW Vuister. Resonance assignment of methionine methyl groups and chi 3 angular information from long-range proton-carbon and carbon-carbon J correlation in a calmodulin-peptide complex. J Biomol NMR 4: 787–797, 1994.

22. LP McIntosh, E Brun, LE Kay. Stereospecific assignment of the NH$_2$ resonances from the primary amides of asparagine and glutamine side chains in isotopically labeled proteins. J Biomol NMR 9: 306–312, 1997.

23. T Yamazaki, SM Pascal, AU Singer, JD Forman-Kay, LE Kay. NMR pulse schemes for the sequence-specific assignment of arginine guandino $^{15}$N and $^1$H chemical shifts in proteins. J Am Chem Soc 117: 3556–3564, 1995.

24. SM Pascal, T Yamazaki, AU Singer, LE Kay, JD Forman-Kay. Structural and dynamic characterization of the phosphotyrosine binding region of a Src homology 2 domain–phosphopeptide complex by NMR relaxation, proton exchange, and chemical shift approaches. Biochemistry 34: 11353–11362, 1995.

25. G Otting, E Liepinsh, K Wüthrich. Protein hydration in aqueous solution. Science 254: 974–980, 1991.

26. K Wüthrich, G Otting, E Liepinsh. Protein hydration in acqueous solution. Faraday Disc 93: 35–45, 1992.

27. G Otting, K Wüthrich. Heteronuclear filters in two-dimensional [$^1$H,$^1$H]-NMR spectroscopy: combined use with isotope labelling for studies of macromolecular conformation and intermolecular interactions. Quart Rev Biophys 23: 39–96, 1990.

28. G Gemmecker, ET Olejniczak, SW Fesik. An improved method for selectively observing protons attached to $^{12}$C in the presence of $^1$H-$^{13}$C spin pairs. J Magn Reson 96: 199–204, 1992.

29. M Ikura, A Bax. Isotope-filtered 2D NMR of a protein-peptide complex: Study of a skeletal muscle myosin light chain kinase fragment bound to calmodulin. J Am Chem Soc 114: 2433–2440, 1992.

30. C Zswahlen, P Legault, SJF Vincent, J Greenblatt, R Konrat, LE Kay. Methods for measurement of intermolecular NOEs by multinuclear NMR spectroscopy: Application to a bacteriophage 1 N-peptide/box B RNA complex. J Am Chem Soc 119: 6711–6721, 1997.

31. AJ Wand, JH Short. Nuclear magnetic resonance studies of protein-peptide complexes. Methods Enzymology 239: 700–717, 1994.

32. W Lee, MJ Revington, C Arrowsmith, LE Kay. A pulsed field gradient isotope-filtered 3D $^{13}$C HMQC-NOESY experiment for extracting intermolecular NOE contacts in molecular complexes. FEBS Lett 350: 87–90, 1994.

33. C Chen, Y Feng, JH Short, AJ Wand. The main chain dynamics of a peptide bound to calmodulin. Arch Biochem Biophys 306: 510–514, 1993.

34. M Ikura, GM Clore, AM Gronenborn, G Zhu, CB Klee, A Bax. Solution structure of a calmodulin-target peptide complex by multidimensional NMR. Science 256: 632–658, 1992.

35. MM Zhou, RP Meadows, TM Logan, HS Yoon, WS Wade, KS Ravichandran, SJ Burakoff, SW Fesik. Solution structure of the Shc SH2 domain complexed with a tyrosine-phosphorylated peptide from the T-cell receptor. Proc Natl Acad Sci USA, 92: 7784–7788, 1995.

36. SM Pascal, AU Singer, G Gish, T Yamazaki, SE Shoelson, T Pawson, LE Kay, JD Forman-Kay. Nuclear magnetic resonance structure of an SH2 domain of phospholipase C-gamma 1 complexed with a high affinity binding peptide. Cell 77: 461–472, 1994.

37. J Qin, GM Clore, WM Kennedy, JR Huth, AM Gronenborn. Solution structure of human thioredoxin in a mixed disulfide intermediate complex with its target peptide from the transcription factor NF kappa B. Structure 3: 289–297, 1995.

38. S Feng, TM Kapoor, F Shirai, AP Combs, SL Schreiber. Molecular basis for the binding of SH3 ligands with non-peptide elements identified by combinatorial synthesis. Chem Biol 3: 661–670, 1996.

39. S Feng, SL Schreiber. Enanantiometric binding elements interacting at the same site of an SH3 protein receptor. J Am Chem Soc 119: 10873–10874, 1990.

40. AJ Wand, RA Bieber, JL Urbauer, RP McEvoy, Z Gan. Carbon relaxation in fractionally randomly $^{13}$C-enriched proteins. J Magn Reson Series B 108: 173–175, 1995.

41. AJ Wand, JL Urbauer, RP McEvoy, RA Bieber. Internal dynamics of human ubiquitin by $^{13}$C relaxation studies of randomly fracitionally labeled protein. Biochemistry 35: 6116–6125, 1996.

42. DR Muhandiram, T Yamazaki, BD Sykes, LE Kay. Measurement of $^{2}$H $T_1$ and $T_{1r}$ relaxation times in uniformly $C^{13}$ labeled and fractionally $^{2}$H labeled proteins in solution. J Am Chem Soc 117: 11536–11544, 1995.

43. JW Peng, G Wager. Investigation of protein motions via relaxation measurements. Methods Enzymology 239: 563–596, 1994.

44. LE Kay, DR Muhandiram, NA Farrow, Y Aubin, JD Forman-Kay. Correlation between dynamics and high affinity binding in an SH2 domain interaction. Biochemistry 35: 361–368, 1996.

45. NA Farrow, R Muhandiram, AU Singer, SM Pascal, CM Kay, G Gish, SE Shoelson, T Pawson, JD Forman-Kay, LE Kay. Backbone dynamics of a free and phospho-peptide-complexed Src homology 2 domain studied by $^{15}$N NMR relaxation. Biochemistry 33: 5984–6003, 1994.

46. LK Nicholson, T Yamazaki, DA Torchia, S Grzesiek, A Bax, SJ Stahl, JD Kaufman, PT Wingfield, PY Lam, PK Jadhav, CN Hodge, PJ Domaille, CH Chang. Flexibility and function in HIV-1 protease. Nat Struct Biol 2: 274–280, 1995.

47. M Akke, R Bruschweiler, AG Palmer III. NMR order parameters and free energy: an analytical approach and its application to cooperative $Ca^{2+}$ binding by calbindin $D_{9k}$. J Am Chem Soc 115: 9832–9833, 1993.

48. DW Yang, YK Mok, JD Forman-Kay, NA Farrow, LE Kay. Contributions to protein entropy and heat capacity from bond vector motions measured by NMR spin relaxation. J Mol Biol 272: 790–804, 1997.

49. Z Li, S Raychaudhuri, AJ Wand. Insights into the local residual entropy of proteins provided by NMR relaxation. Protein Science 5: 2647–2650, 1996.

50. MR Ehrhardt, JL Urbauer, AJ Wand. The energetics and dynamics of molecular recognition by calmodulin. Biochemistry 34: 2731–2738, 1995.

51. Y Bai, TR Sosnick, L Mayne, SW Englander. Protein folding intermediates: native-state hydrogen exchange. Science 269: 192–197, 1995.

52. Y Bai, SW Englander. Future directions in folding: the multi-state nature of protein structure. Proteins 24: 145–151, 1996.

53. Y Paterson, SW Englander, H Roder. An antibody binding site on a protein antigen defined by hydrogen exchange and two-

dimensional NMR. Science 249: 755–759, 1990.

54. J Orban, P Alexander, P Bryan. Hydrogen-deuterium exchange in the free and immunoglobin G-bound protein GB-domain. Biochemistry 33: 5702–5710, 1994.

55. L Mayne, Y Paterson, D Cerasoli, SW Englander. Effect of antibody binding on protein motions studied by hydrogen exchange labeling and two-dimensional NMR. Biochemistry 31: 10678–10685, 1992.

56. DC Benjamin, DC Williams, SJ Smith-Gill, GS Rule. Long range changes in protein antigen-antibody interaction. Biochemistry 31: 9539–9545, 1992.

57. K Thornton, DG Gorenstein. Structure of glucagon-like peptide (7-36) amide in a dodecylphosphocholine micelle as determined by 2D NMR. Biochemistry 33: 3532–3539, 1994.

58. MH Werner, DE Wemmer. Identification of a protein binding surface by differential amide hydrogen exchange measurements. J Mol Biol 225: 873–889, 1992.

59. DN Jones, M Bycroft, MJ Lubienski, AR Fersht. Identification of the barstar binding site of barnase by NMR spectroscopy and hydrogen-deuterium exchange. FEBS Lett 331: 165–172, 1993.

60. Q Yi, JR Erman, JD Satterlee. Studies of protein-protein association between yeast cytochrome c peroxidase and yeast iso—1 ferricytochrome c by hydrogen-deuterium exchange and two-dimensional NMR. Biochemistry 33: 12032–12041, 1994.

61. M Jeng, SW Englander, K Pardue, JS Rogalsky, G McLendon. Structural dynamics in an electron transfer complex. Nat Struct Biol 1: 234–238, 1994.

62. KT O'Neil, WF De Grado. How calmodulin binds its targets: sequence independent recognition of amphiphilic alpha-helices. Trends Biochem Sci 15: 59–64, 1990.

63. YS Babu, CE Bugg, WJ Cook. Structure of calmodulin refined at 2.2 A resolution. J Mol Biol 204: 191–204, 1988.

64. A Persechini, RH Kretsinger. The central helix of calmodulin functions as a flexible tether. J Biol Chem 263: 12175–12178, 1988.

65. DB Heidorn, J Trewhella. Comparison of the crystal and solution structures of calmodulin and troponin C. Biochemistry 27: 909–915, 1988.

66. M Ikura, S Spera, G Barbato, LE Kay, M Krinks, A Bax. Secondary structure and side-chain 1H and 13C resonance assignments of calmodulin in solution by heteronuclear multidimensional NMR spectroscopy. Biochemistry 30: 9216–9228, 1991.

67. SM Roth, DM Schneider, LA Strobel, MFA VanBerkum, AR Means, AJ Wand. Characterization of the secondary structure of calmodulin in complex with a calmodulin-binding domain peptide. Biochemistry 31: 1443–1451, 1992.

68. WE Meador, AR Means, FA Quiocho. Target enzyme recognition by calmodulin: 2.4 A structure of a calmodulin-peptide complex. Science 257: 1251–1255, 1992.

69. K Török, M Whitaker. Taking a long, hard look at calmodulin's warm embrace. BioEssays 16: 221–224, 1994.

70. S Spera, M Ikura, A Bax. Measurement of the exchange rates of rapidly exchanging amide protons: application to the study of calmodulin and its complex with a myosin light chain kinase fragment. J Biomol NMR 1: 155–165, 1991.

71. S Macura, RP Ernst. Elucidation of cross relaxation in liquids by two-dimensional NMR spectroscopy. Mol Phys 41: 95–117, 1980.

72. Y Bai, JS Milne, L Mayne, SW Englander. Primary structure effects on peptide group hydrogen exchange. Proteins 17: 75–86, 1993.

73. H Qian, JA Schellman. Helix-coil theories: A comparative study for finite length polypeptides. J Chem Phys 96: 3987–3994, 1992.

74. S Lifson, A Roig. On the theory of helix-coil transition in polypeptides. J Chem Phys 34:1963-1974, 1961.

75. PC Lyu, MI Liff, LA Marky, NR Kallenbach. Side chain contributions to the stability of alpha-helical structure in peptides. Science 250: 669–673, 1990.

76. CA Rohl, JM Scholtz, EJ York, JM Stewart, RL Baldwin. Kinetics of amide proton exchange in helical peptides of varying chain lengths. Interpretation by the Lifson-Roig equation. Biochemistry 31: 1263–1269, 1992.

77. AJ Wand, H Roder, SW Englander. Two-dimensional 1H NMR studies of cytochrome c: hydrogen exchange in the N-terminal helix. Biochemistry 25: 1107–1114, 1986.

78. JL Silva, G Weber. Pressure stability of proteins. Annu Rev Phys Chem 44: 89–113, 1993.

79. G Weber, HG Drickamer. The effect of high pressure upon proteins and other biomolecules. Quart Rev Biophys 16: 89–112, 1983.

80. CR Robinson, SG Sligar. Hydrostatic and osmotic pressure as tools to study macromolecular recognition. Methods Enzymology 259: 395–427, 1995.

81. M Zhang, T Tanaka, M Ikura. Calcium-induced conformational transition revealed by the solution structure of apo calmodulin. Nat Struct Biol 2: 758–767, 1995.

82. H Kuboniwa, N Tjandra, S Grzesiek, H Ren, CB Klee, A Bax. Solution structure of calcium-free calmodulin. Nat Struct Biol 2: 768–776, 1995.

83. WE Meador, AR Means, FA Quiocho. Modulation of calmodulin plasticity in molecular recognition on the basis of x-ray structures. Science 262: 1718–1721, 1993.

84. JL Urbauer, MR Ehrhardt, RJ Bieber, PF Flynn, AJ Wand. High resolution triple resonance NMR spectroscopy of a novel calmodulin • peptide complex at kilobar pressures. J Am Chem Soc 118: 11329–11330, 1996.

85. MR Ehrhardt, L Erijman, G Weber, AJ Wand. Molecular recognition by calmodulin: pressure-induced reorganization of a novel calmodulin-peptide complex. Biochemistry 35: 1599–1605, 1995.

86. GB Benedek, EM Purcell. Nuclear magnetic resonance in liquids under high pressure. J Chem Phys 22:2003-2012, 1954.

87. J Jonas, A Jonas. High-pressure NMR spectroscopy of proteins and membranes. Annu Rev Biophys Biomol Struct 23: 287–318, 1994.

88. JV Carter, DG Knox, AJ Rosenberg, Pressure effects on folded proteins in solution. Hydrogen exchange at elevated pressures. Biol Chem 253: 1947–1953, 1978.

# 24

# Circular Dichroism and Fourier Transform Infra-Red Analysis of Polypeptide Conformation

**Graeme J. Anderson**
*Manchester Metropolitan University, Manchester, United Kingdom*

## I.  INTRODUCTION

Circular Dichroism (CD) and Fourier transform Infra-Red (FTIR) are two distinct spectroscopic techniques which are used to give information concerning the secondary structure of a variety of biological molecules including polysaccharides, nucleotides, peptides and proteins. In the latter two categories, to be discussed here, these techniques can be used to give information concerning the secondary structure(s) present, and can be applied to a range of molecules from small peptides to large proteins.

CD and FTIR can detect and resolve the multiple conformations present for small to medium sized peptides (such as hormones and neurotransmitters) in solution, the approximate percentages of each component and changes resulting from alterations in medium, pH, salt concentration, temperature and so forth. In addition, differences in conformation resulting from single or multiple amino acid substitutions are rapidly distinguished, which may help to resolve conformational preferences for different receptor subtypes.

CD and FTIR spectroscopies are commonly applied to the study of large proteins and again will help to identify the various secondary structures present. This is useful for identification of the structural class of the protein; any gross changes resulting from, for example, ligand binding can be easily followed. The advent of genetic engineering and the large amount of protein which can be readily produced, means that a rapid analysis of whether the protein has folded correctly relative to the native form of the protein, which is of the utmost importance to both industry and academia, is readily ascertained.

The application of CD and FTIR techniques gives only a global snapshot of the molecular structure. The techniques to determine structures at the atomic level will continue to be multi-dimensional NMR and x-ray diffraction. However, CD and FTIR techniques described in greater detail within this chapter are best thought of as complimentary to x-ray/NMR; the presence of secondary structural elements under specified conditions of temperature, pH and solvent are best determined prior to lengthy and intensive structural determination to atomic resolution.

## II.  CIRCULAR DICHROISM STUDIES OF POLYPEPTIDE CONFORMATION

### A.  Theory of Circular Dichroism

Normal light, as with all electromagnetic waves, consists of randomly orientated electronic and magnetic vectors which are perpendicular to the direction of travel of the wave. The commonly encountered plane polarized light [Fig. 1(a)] differs in that the direction of the electronic vector is constant (confined to a plane) while the magnitude is varied. This form of light is used in the technique of polarimetry, when the rotation of light by optically active compounds is measured.

Circular Dichroism spectroscopy makes use of circularly polarized light, one form of which is illustrated in Fig. 1(b). Here, the magnitude of the electric vector is held constant while the direction is varied; right circularly polarized (rcp) light traces the outline of a right-handed helix, while left circularly polarized (lcp)

light rotates in the opposite direction and follows a left-handed helical path.

Circular Dichroism is the differential absorbance, by a chiral molecule, of left or right circularly polarized light (1). This is normally expressed as the differences in molar extinction coefficients, $\Delta\varepsilon$.

$$\Delta\varepsilon = \varepsilon_l - \varepsilon_r \qquad (1)$$

Occasionally, units of CD are expressed in terms of the molar ellipticity, $[\Phi]$, particularly in older papers and is still prevalent in some of the biochemical literature. This is easily converted to the more proper $\Delta\varepsilon$, using the formula

$$\Delta\varepsilon = [\Phi]/3298 \qquad (2)$$

In the CD of polymers, the observed spectrum is no longer due to the absorbance of the single individual chromophores. The secondary structural elements present combine to form an array which gives rise to a new set of excited transitions. Each electronic transition can therefore have several CD bands, depending upon the interaction of the circularly polarized light with the orientated chromophores (known as exciton splitting/coupling). Thus each type of pep-

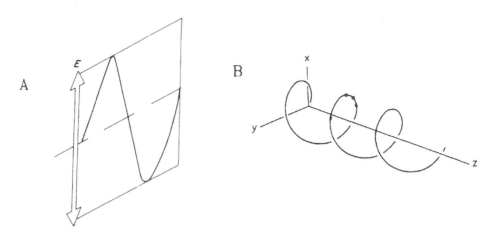

**Fig. 1**  Representation of (A) plane polarized light in which the direction of the electric vector is constant (to within a sign) while the magnitude is varied and (B) circularly polarized light where the direction of the electric vector is varied. Right circularly polarized light is shown. [Adapted from (A) Ref. (1) and (B) Ref. (2)]

tide structure (α-helix, β-sheet, β-turn etc.) can be considered to have a unique and distinct CD spectrum, within the constraints of flexibility of the backbone. Where several conformational components are present (either from globular structures or conformational equilibria of small peptides), the resultant CD spectrum is a weighted average of each of these standard spectra (2).

In peptides and proteins there are a number of chromophores which can absorb within the spectrum, the most important of which is the peptide (amide) bond. The other absorbances, arising from aromatic side-chains and disulphide bonds, can give very useful information (see later) but are generally interferences within the main backbone region (3). There are two main absorbances associated with the peptide chromophore, the low energy nπ* transition centred around 220 nm and the higher energy ππ* transition at 190 nm (3). The low energy transition is very sensitive to the solvent in which the spectrum is recorded, varying by up to 10 ± nm. The ππ* transition in contrast moves only by ~2 nm to higher wavelength on moving to more hydrophobic solvents.

The peptide chromophore gives rise to the characteristic CD spectra; in certain cases additional information can be obtained by observing bands at higher wavelengths arising from aromatic (Phe, His, Tyr, Trp) or disulphide linkages (Cys-Cys). These lower energy transitions are not associated with any secondary structure components, but can still be used to give useful information when observing changes in protein structure, ligand binding etc. (see Sec. II.B.1.)

### B. CD Spectra and Secondary Structure

*a. α-Helix*

The best characterized and most easily recognized CD spectra of peptide and protein secondary structure is that of the α-helix (Fig. 2). This is characterized by a high wavelength minima at 222 nm (nδ*

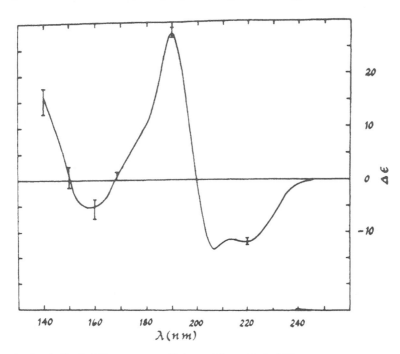

**Fig. 2**  Standard α-helical CD spectrum. [Adapted from Ref. (4)]

transition) and a minima and maxima at 208 nm and 190 nm respectively, from exciton splitting of the $\pi\pi^*$ transition (4). A large number of peptides and homopolymers show this characteristic CD spectrum and this secondary structure is a common component of protein structures. The CD spectrum is a weighted average of all the components present in solution; because the $\alpha$-helical spectrum is of such strong intensity, the CD spectra of many proteins look very similar in appearance to this spectrum, with only small changes apparent.

### b. β-Sheet

$\beta$-sheets are common components of protein secondary structure and can exist in two distinct forms: the parallel $\beta$-sheet, where the two chains run in the same direction (NC or CN) and the anti-parallel $\beta$-sheet with the two chains running in opposite directions.

The CD spectra of idealized $\beta$-sheets are very similar, characterized by a negative band at $\sim215$ nm ($n\pi^*$) and $\pi\pi^*$ exciton bands at $\sim198$ nm (positive) and $\sim175$ nm (negative) (4); this is shown in Fig. 3. In contrast to the $\alpha$-helix, the $\beta$-sheet CD spectrum shows much greater variation, varying with side-chain, solvent and salt concentration. It has been suggested that the best method for distinguishing between the parallel, and anti-parallel forms is by measurement of the difference in $\lambda_{max}$ between CD and UV absorbance. This is $\sim13$ nm for parallel $\beta$-sheets, and $\sim5$ nm for anti-parallel.

Theoretical studies have indicated that the differences between parallel and anti-parallel forms are negligible in relation to the twisting of the sheets which can occur in large globular protein structures. Sheets can occur in parallel, anti-parallel or mixed forms with varying degrees of twisting and thus give rise to CD spectra with a greater degree of variation than that observed for the $\alpha$-helix.

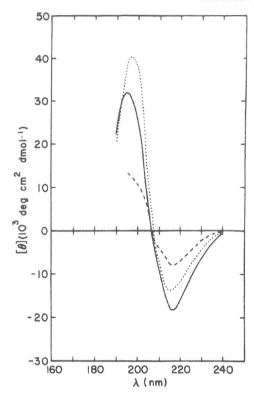

**Fig. 3** Standard $\beta$-sheet cd spectra derived from a number of peptides. [From Ref. (2)]

### c. β Turns

There is a significant degree of variation within turn structures and therefore a different representative spectrum for each type (5). The type II $\beta$-turn CD spectrum (Fig. 4) has the negative $n\pi^*$ band at 220–225 nm, with a positive $\pi\pi^*$ band at 200–205 nm (Class C spectrum in the nomenclature of Woody, 1974) (6). Qualitatively this spectrum resembles that of the $\beta$-sheet, but is moved to higher wavelength. The other common type of turn, the type I (and the closely related type III) (class B spectra in the nomenclature of Woody, 1974) has a radically different spectrum to that of the type II (Fig. 4), with a weak negative band at 225–230 nm and a positive band at 205-210 nm (7); these bands are generally weaker than in the type II turn.

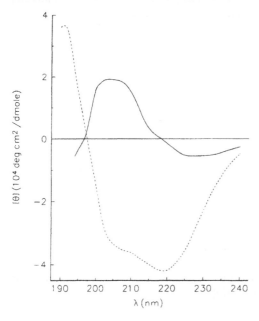

**Fig. 4** β-turn CD spectra representing the type II (dotted line) and type I (solid line) turns. [From Ref. (2)]

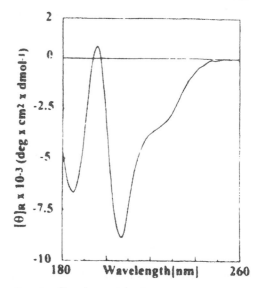

**Fig. 5** Circular Dichroism spectrum of a model $3_{10}$ helical peptide. [From Ref. (9)]

### d. $3_{10}$ Helix

The $3_{10}$ helix is a relatively common structural feature in proteins (alongside the α-helix and the β-sheet), but has until recently received little attention in terms of identification by CD means, previous studies suggesting these would be indistinguishable from α-helical structures (8). The structure is similar to the α-helix, but is more elongated with 3.0 residues per turn and has different hydrogen bonding patterns. A recent paper (9) describes the synthesis and CD spectrum of an α-Me octapeptide which adopts a right-handed $3_{10}$ helix (see Fig. 5). Although similar to the α-helical CD spectrum (negative bands at 207 and 222 nm), there are significant differences: a slight positive shoulder at 195 nm, and another negative band at 184 nm. Most significantly, the ratio R, $[\Phi]_{222}/[\Phi]_{207}$ is ~0.4 (R ~ 1 in α-helical polypeptides). The authors suggest that several published conformational analyses may have

to be re-interpreted in light of these findings.

### e. Extended and 'Random' Structures

The CD spectrum attributed to a "random coil" or "unordered" conformation has been the subject of much debate and controversy in recent years. Initial models for such peptide structure were obtained from studies of homopolymers such as poly(Lys) and poly (Glu)at pH 7 (10, 11). These spectra are similar in appearance to the CD spectrum of poly (Pro) II (12), which exists in water and other polar media as a left-handed extended helical structure with 3 residues per turn (Fig. 6). Several researchers have thus suggested that peptides exhibiting this CD spectrum must contain a significant amount of poly(Pro) II type structure, which has been supported by low temperature CD (13), globular protein analysis (14), and vibrational CD measurements (15). Spectra exhibiting a negative component at 200–205 nm and a weak positive band

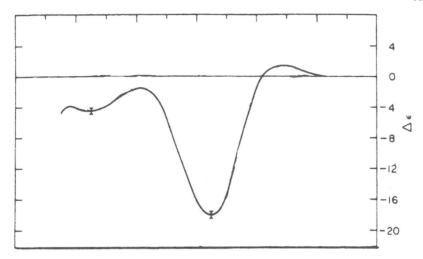

**Fig. 6**  Poly(Pro) II standard CD spectrum, which has previously been interpreted (incorrectly) as from an unordered conformation. [Adapted from Ref. (2)]

centred around 222 nm, can thus be said to contain a substantial quantity of poly(Pro) II type structures.

What, therefore, can be deduced about the nature and CD spectrum of fully "unordered" peptides? At present, such peptide models as exist show a negative band at ~200 nm, as with poly(Pro) II discussed above, but with a weak *negative* shoulder around 220 nm (Fig. 7) (16). Much work remains to be done towards the prediction of CD spectra of truly ran-

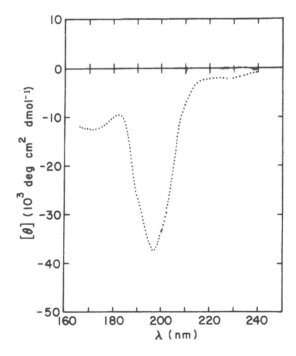

**Fig. 7**  CD spectrum resulting from an unordered peptide conformation. [Adapted from Ref. (2)]

dom peptide structures and in obtaining reliable model peptides.

### 1. Effects of Aromatic and Disulphide Bonds on the CD Spectrum

Contributions from aromatic residues and disulphide bonds to the CD spectrum of peptides and proteins can be significant (17) and can, in favourable circumstances, give information concerning tertiary structural changes resulting form a number of factors such as ligand binding, folding/unfolding and metal binding. The contributions from these two groups will be treated separately, but researchers should be aware that the situation can become highly complex should both aromatic and disulphide bonds be present within the same molecule.

Table I lists the bands arising from aromatic side-chains from the amino acids phenylalanine (Phe), tyrosine (Tyr), tryptophan (Trp) and histidine (His) in both the near and far-UV. There is no contribution from the peptide backbone to the near-UV CD spectra (250–320 nm) of peptides and proteins and in many cases only a limited number of aromatic residues are present. The vibronic fine structure observed in the near-UV can occasionally be assigned to a particular type of residue of even a particular amino acid within the primary sequence (18). As aromatic residues are hydrophobic and often buried inside protein structures at the catalytic/binding site, they can provide a subtle probe of the changes which occur upon ligand binding, and thus the tertiary structure of the molecule being studied. A recent example of peptide conformational analysis using near-UV CD spectroscopy has been carried out by Cebrat and co-workers (19), studying analogues of cyclolinopeptide A.

As illustrated in Table 1, aromatic residues contribute to the far-UV CD spectra of peptides and proteins, with transitions which are an order of

**Table 1** Aromatic Residues and Far-UV CD Spectra

| Residue | Wavelength (nm) |
|---------|-----------------|
| Phe | 257; 206/188 |
| Tyr | 274; 222/193 |
| Trp | 280; 219 |
| His | 211 |

magnitude greater than those in the near-UV. In molecules with a significant number of aromatic residues, it may be expected that CD spectra are obtained which do not correspond precisely to those arising from amide secondary structural elements alone. Indeed, many reports attribute anomalous spectra to significant aromatic effects (20–23), particularly since model compounds have shown that "frozen" residues give rise to an intensification in the CD of 10–1000 times (24).

Disulphide bonds absorb within the region 260–300 nm; there are two transitions which are degenerate when the $\chi_{ss}$ angle is 90 (gauche conformer) absorbing at ~260 nm. Deviation from this ideal angle leads to a splitting of the n$\sigma$* transitions to higher and lower wavelengths; application of a quadrant rule for significant deviations from the 90° dihedral angle allows the chiral sense of the disulphide bond to be determined (25). In general, the disulphide contributions to the near-UV can be distinguished from the aromatic contributions by their rather broad appearance and their lack of fine structure.

### C. Circular Dichroism Studies of Peptides and Polypeptides

The following section (and Sec. 2.4) are not intended to be an extensive survey of the CD study of peptide (and polypeptide conformation), but rather to illustrate, with the use of recent relevant examples, the types of studies being carried out at present.

Small peptides ($\geq 20$ amino acids) do not adopt a single conformation in solution under normal conditions of temperature, salt concentration etc. but rather exist as a mixture of rapidly interconverting conformers in which several folded conformations may predominate. Circular Dichroism is thus an excellent technique for investigation of such conformational equilibria, as the spectrum obtained is a weighted average of all the structures present in solution. Perturbation of the equilibria by modification of factors such as solvent polarity, temperature and salt concentration (26) can allow the predominant conformers to be determined in conditions which may approach those found in the relevant biological medium. Such studies are also important in designing conformationally restricted analogues which can mimic the 'bioactive conformation'.

An exciting development in peptide chemistry is the use of "template" structures with defined conformations (*de novo* synthesis) which have the ability to recognize specific receptors or target molecules, and thus have great potential as therapeutic agents. Maynard and Searle (27) designed a 16-mer peptide to mimic the anti-parallel $\beta$-sheet structure of the DNA binding motif of the *met* repressor protein dimer. Circular Dichroism was used to monitor the folding and conformations of this peptide in water and water/methanol solutions and indicated a mixture of $\beta$-sheet and $\beta$-turn structures (as anticipated) in equilibrium with a "random coil." CD studies have also been used to monitor the successful synthesis of peptides which form amphipathic $\alpha$-helices (21–28 residues) as anti-microbial agents, and their subsequent self-association to form coiled-coils (28, 29). A two-state equilibrium was identified for several analogues between "random coil" and alpha-helical structures, which tended towards the latter conformation with increasing concentration of micelles. The effects of double D-amino acid substitutions on a $\beta$-sheet model peptide has recently been investigated (30). The model peptide existed as (see Fig. 8) predominantly helix in TFE/water (1:1), an unordered conformation in aqueous solution and a $\beta$-sheet between residues 9 and 18 in SDS micelles and water/apolar interfaces by CD measurements.

The correlation between peptide conformation and biological activity has been, and remains, one of the central goals of peptide chemistry; the "bioactive" conformation and its exact relationship to the solution conformation remains elusive. Studies of peptide conformation in media closely related to the biological environment are thus important in helping to resolve the conformation-activity problem. Castiglione Morelli and colleagues (31) studied the conformation of a number of chemoattractant pentapeptides in water, ethanol, and trifluoroethanol, whilst performing assays for biological activities. They found that although all peptides were a mixture of rapidly interconverting forms (folded and unfolded), the chemotactic activity was related to the peptide's propensity towards a folded conformation. In a similar vein, studies have been carried out on cyclized 7-mer and 14-mer peptides derived from the central loop of fasciculin (32). This study attempted to relate the contribution of this loop region to the acetylcholinesterase inhibitory activity of fasciculin. CD investigations (Fig. 9) were used to identify the presence of $\beta$-sheet and $\beta$-turns in both peptides and to calculate the percentages of each component; the conformationally constrained 14-mer peptide was partially active albeit with a relatively high inhibitory constant, suggesting that a different structure is required in the "bioactive" conformation.

Recording the CD spectra as a function solvent polarity, pH and ionic strength can be used to investigate structural transitions of peptides from one con-

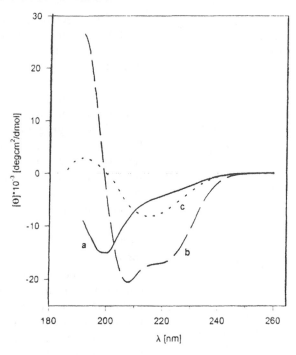

**Fig. 8** Conformation of a $\beta$-sheet model peptide investigated as a function of solvent polarity by CD. The spectrum is recorded in (a) water, (b) TFE:H$_2$O (1:1) and (c) 15 mM SDS. [From Ref. (30)]

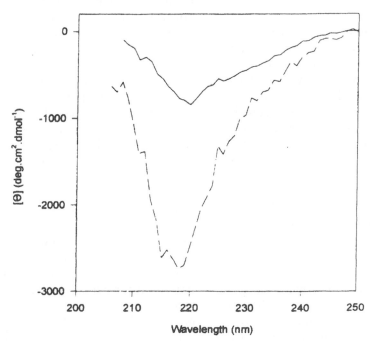

**Fig. 9** CD spectra of cyclised 7-mer (solid line) and 14-mer (dashed line) peptides in water. [From Ref. (32)]

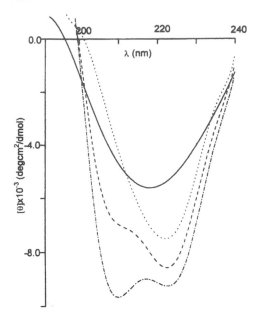

**Fig. 10** CD spectra of N-terminal trichosanthin derivatives in water pH 10.2 (—), MeOH:H₂O (1:1) (·····), TFE:H₂O (1:1) (---) and HFP:H2O (1:1) (-•-). [From Ref. (33)]

formation to another, an important consideration in the process of protein folding, in addition to helping to resolve conformational equilibria. Hu et al (33) have studied three analogues from the N-terminus of trichosanthin small domain in a variety of solvents and salt concentrations. Figure 10 shows the effect of solvent polarity on the CD spectrum, which the authors attribute to an equilibrium between α-helical and β-sheet conformations. Conformational equilibria were also investigated in studies on the 44 residue peptide derived from the HIV glycoprotein gp120 (34). In aqueous solution the peptide was found to exist as a mixture of three conformations, the α-helix, β-strand and random-coil; the percentages of each of these components were calculated using CD simulations, based upon CD spectra of known secondary structures.

CD spectroscopy also readily lends itself to the study of the problem of aggregation in peptide and protein systems; although generally an undesired effect, such aggregation is important in specific biological systems as G-protein activation which may involve helix-helix aggregation. The aggregation of a C-terminal fragment peptide from the Angiotensin II receptor (35) was monitored by recording the CD spectrum as a function of solvent composition and indicated an increasing helical content as a function of TFE. The structure of the peptide was further refined using two-dimensional NMR measurements.

In order to study peptide and protein conformation in biologically relevant media, it is often desirable to analyze conformational parameters in lipidic environments to stimulate cell membranes. Such investigations are difficult to perform by NMR and x-ray techniques, lending an additional importance to such CD and FTIR studies. The most easily and commonly studied media is that of sodium dodecylsulphate (SDS) which forms micelles into which the peptide can readily insert. The M13 coat protein from *Escherichia Coli* (28 residues) was investigated by CD and solution/solid-state NMR techniques (36). CD results indicated $42 \pm 10\%$ α-helical content upon addition of SDS, in good agreement with subsequent NMR studies. More relevant perhaps are studies carried out in lipid vesicles, a situation more akin to the condition within the cell itself. Liu et al (37) studied the effect of substituting different amino acids within a 25-mer template on the peptide insertion into lipid vesicles. They found that the electrostatic interactions provided by anionic lipids are important for peptide insertion into such membranes, a result which may be important for understanding how signal peptides evoke their function. In a similar vein, Zhang et al (38) studied the conformations of two distinct signal peptides (the

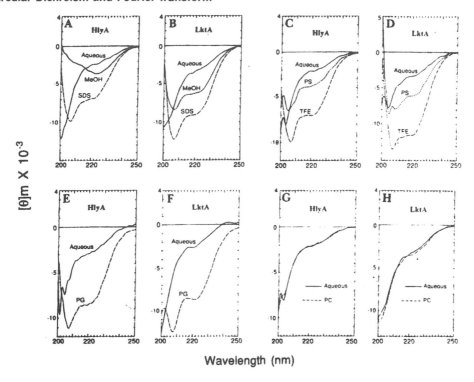

$[\theta]m \times 10^{-3}$

Wavelength (nm)

**Fig. 11** CD spectra of signal peptides HlyA and LktA in various membrane mimetic environments and aqueous solution. [From Ref. (38)]

C-terminal sequences of HlyA and LktA) which have similar transport properties, but have different primary sequences. Studies were carried out in phosphatidylserine (PS), dipalmitoylphosphatidylglycerol (PG) and dipalmitoylphosphatidylcholine (PC) lipids (Fig. 11). Both peptides exhibited a change from "random coil" to alpha helical conformation upon addition of negatively charged lipid vesicles; in contrast, addition of neutral lipids (such as PC) did not produce structure in either peptide, suggesting an interaction (specific or non-specific) between the lipid head groups and positively charged residues from the signal peptides, which induces α-helical arrangement. NMR studies are reported to be underway to further refine the conformations of these signal peptides.

Peptide mimetics (pseudopeptides) are commonly synthesised once biologi-

cally active peptides or protein fragments are identified and characterized, in order to obtain analogues with enhanced *in vivo* stability, enzyme inhibition, receptor specificity, conformation, and biological response (39). The analogues can encompass either modification of the amino acid side-chains, the peptide backbone or both. For example, Cebrat et al (19) synthesized analogues of the immunosuppresive peptide cyclolinopeptide A (CLA) in their search for water soluble derivatives, in which Phe was replaced by p-sulfophenylalanine. Monitoring the CD of the near UV allowed a direct comparison between the conformations of the side-chains in the sulfonated and native derivatives (Fig. 12). Derivatives containing the [Phe*9]CLA replacement showed equivalent CD spectra to CLA itself, indicating an identical side-chain conformation in these two peptides. The

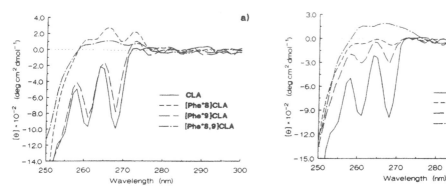

**Fig. 12** CD spectra of cyclolinopeptide A (CLA) derivatives in the aromatic region. The (a) cyclic and (b) linear analogues contain a sulphonated phenylalanine residue at the positions indicated. [From Ref. (19)]

[Phe*8]CLA and [Phe*8,9]CLA derivatives on the other hand showed distinctly different near-UV CD spectra. These observations correlated precisely with the observed biological activity in all cases.

A recent study (40) details an interesting CD analysis of a pseudopeptide in which the side-chain is restricted via an $\alpha,\beta$-dehydro derivative. The pentapeptide contains three consecutive dehydrophenylalanine ($\Delta$-Phe) residues which were shown to individually induce a $\beta$-bend in previous studies of short peptides. CD and x-ray studies identified a left-handed $3_{10}$-helical structure for this peptide, despite the presence of two Val residues. Such studies may help in the design of non-peptidic molecules of defined structure (see also *de novo* synthesis above).

The advent of molecular biological techniques has resulted in the rapid production of proteins of defined primary structure in which one or more amino acids are systematically altered, to assess the affect upon biological function, ligand binding conformation etc. Folding of genetically engineered or chemically synthesised proteins is a process which lends itself readily to CD investigations, as each intermediate state can, at least in theory, be analyzed. For instance, Tevelev

and co-workers expressed and purified wild-type and mutant tumour supressor protein p16[INK4A] and verified their structural similarities by CD measurements (41). They were also able to follow the guanidinium chloride induced denaturation of the native and mutant proteins, and relate different conformational behaviour to specific amino acid residues within each mutant. Likewise, chemically synthesized proteins can be examined for structural integrity and compared to the native protein. The convergent synthesis of the 124-mer protein secretory phospholipase A2 from human has been carried out and folded to produce an enzymatically active molecule containing seven disulphide linkages (42). Circular Dichroism rapidly confirmed the that the structures of synthetic and native proteins were indistinguishable (Fig. 13).

Structural studies of purified proteins by CD and structural prediction methods are rather commonplace and readers are directed towards the chemical/biochemical literature for recent studies (see for example Refs 43 and 44). CD spectroscopy is also often used to study the changes in protein structure resulting from complexation with ligands and/or ions. See Refs (45) and (46) for current instances of these types of analyses.

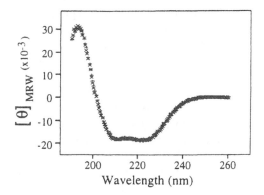

**Fig. 13** CD spectra of synthetic ($\times$) and recombinant ($+$) Phospholipase $A_2$. [From Ref. (42)]

## III. FOURIER TRANSFORM INFRA-RED STUDIES OF POLYPEPTIDE CONFORMATION

### A. Theory of Fourier Transform Infra-Red

Infra-Red spectroscopy is an important technique in the analysis of a vast range of molecules from gas phase to solid phase, encompassing applications in the environment, manufacturing industry, pharmaceuticals, medicine, construction and astronomy.

In organic chemistry, each molecule gives rise to a unique spectrum which can be used in characterization as well as providing a wealth of information concerning the various functional groups present. Infra-Red radiation is of sufficient energy to cause an increase in the vibrational motion; the precise frequency of vibration is dependant on the masses of the bonded atoms and also the bond strengths. (Each vibration appears as a band, rather than a single line, because of associated rotational energy changes) Each type of functional group present in a molecule will thus give rise to a characteristic vibrational band in the IR spectrum, pro-

viding the basis for the use of IR as a investigative tool.

The complexity of Infra-Red spectra arise from the variety of atoms which can contribute to a specific vibration, and the number of stretching and bending modes associated with each type of functional group. In large molecules such as proteins and peptides, spectra are complex and it is impossible to assign every band present; polyglycine for example contains approximately fifty Infra-Red active bands (47). Fortunately, in peptides and proteins it is possible to observe particular regions of the spectrum which have characteristic vibrational bands associated with the amide linkage.

Conformational analysis of peptides and proteins (see Sec. III.B.) using a dispersive spectrometer is not possible, due to interferences from solvent signals and low signal to noise ratios. The use of Fourier transform methods has transformed the use of IR as a spectroscopic technique and allows for the rapid gathering of data on peptide and protein systems. The Fourier Transform Infra-Red (FTIR) spectrometer uses an interferometer to scan over the total range ($4000-600\ cm^{-1}$); this is in contrast to the well known FT-NMR technique where a brief excitation is followed by a Fourier Transform of the subsequent time-domain decay. The lack of such analogue-to-digital converters operating at IR frequencies necessitates the use of a different approach (48). The application of FT techniques to IR spectroscopy results in spectra with greater signal-to-noise ratios or a spectrum with given signal-to-noise ratios for a reduced acquisition period.

Recording FTIR spectra of peptides and proteins is generally (but not always) carried out in deuterated solvents, in order to remove strong signals from the OH bending frequency ($\sim 1640\ cm^{-1}$) which lies in the middle of the conformation sensitive amide I mode. Irrespective of the solvent used, this spectrum must be digitally subtracted from the peptide/protein solvent spectrum to yield a spectrum free from solvent interferences. It is also worth noting that peptides and proteins purified by HPLC (particularly by reverse-phase chromatography) may contain a significant amount of trifluoroacetic acid (TFA, $CF_3CO_2H$) or other highly absorbing trace contaminants which can lead to false assignment of amide I bands. TFA for example absorbs strongly at $\sim 1673 \, cm^{-1}$ but is readily removed by ion-exchange chromatography.

It is necessary to separate the overlapped amide I bands into their components by the application of line-narrowing techniques such as derivation or deconvolution. Deconvolution techniques were the first to be applied and consist of multiplying the cosine Fourier transformed band with an exponential function to yield a new function. This is subsequently reverse transformed to yield a spectrum in which the band is narrowed (49, 50). Derivative techniques employ similar methods to that of deconvolution, weighting the Fourier domain data with quadratic (or quartic) functions to give second (or fourth) derivative spectra, with line narrowed bands (49). Both cases of resolution enhancement require good signal-to-noise ratios (of the order of 500:1) to ensure that artefacts are not introduced via magnification of random noise and water vapour signals.

## B.  FTIR Spectra and Secondary Structure

There are two main regions within the FTIR spectrum from which structural information related to peptides and proteins can be extracted:

$3460–3300 \, cm^{-1}$, from the NH stretch of amide bonds

$1620–1700 \, cm^{-1}$, from carbonyl stretch of amide bonds

Within the NH stretch region, there are generally two distinct bands, between $3460–3400 \, cm^{-1}$ from non hydrogen-bonded stretching modes and $\sim 3300 \, cm^{-1}$ from hydrogen-bonded amide stretches. Measurement of the pertinent areas of the two regions can give an indication of the amount of hydrogen-bonding within the molecule, particularly in globular protein molecules. Very often the band at $3300 \, cm^{-1}$ (Amide A) is accompanied by a second band at $\sim 3100 \, cm^{-1}$ (Amide B); the latter band arises from a coupled interaction with the amide II band. As result, it is not straightforward to use these bands for secondary structural analysis and is further complicated by the strong aqueous OH stretching band (at $3400 \, cm^{-1}$) which tends to interfere with the amide A band.

The amide I band ($1620–1700 \, cm^{-1}$) has been the most important source of information for peptide and protein secondary structural analysis; other bands from amide II to IX have been identified but provide limited information on secondary structure. The amide I band arises primarily from the carbonyl stretching mode of the amide bond; the frequency has been found to be sensitive to the presence of hydrogen bonding and mirrors differences in the secondary structure (51–53). Table 2 lists the main frequencies of the amide I band associated with secondary structural elements, in $D_2O$.

In general the spectra obtained are broad featureless overlapped bands, par-

**Table 2**  Frequencies and Secondary Structural Elements

| 2° Structure | Wavenumber ($cm^{-1}$) |
|---|---|
| α-helix | 1650-1658 |
| β-sheet (parallel + anti-parallel) | 1624-1641 |
| β-sheet (anti-parallel) | 1670-1680 |
| β-turn | 1666-1688 |
| "random" | ~1644 |

ticularly in proteins where a number of absorbances are present, which must be resolved into their component bands using second derivative or deconvolution line-narrowing techniques (see Sec. III.A.). It is possible in some instances to quantitate the secondary structural bands, using a wide variety of techniques (52 and references therein), but I will concentrate on studies which use resolution enhanced FTIR spectroscopy in a qualitative sense only.

The $\alpha$-helix gives rise to an intense band between 1650–1658 cm$^{-1}$. This band is considered to be relatively constant in frequency, although some instances have been reported (polyLys and polyGlu) where the helical frequency is moved to 1635–1646 cm$^{-1}$, due to interactions of the charged side chains with the helical backbone (54). $\beta$-sheets do not give rise to band assignments which are as unique as those of the $\alpha$-helix, reflecting the heterogeneity in structural types associated with the $\beta$-sheet. Both the parallel and anti-parallel $\beta$-sheet structures give rise to a strong band centred around 1632 cm$^{-1}$ (but can vary between 1624–1641 cm$^{-1}$) with the anti-parallel structure showing a weak band at higher wavenumber between 1670–1680 cm$^{-1}$. It is thus theoretically possible to distinguish between the parallel and anti-parallel forms of the $\beta$-sheet, but this is complicated by the possible overlap with $\beta$-turn absorbances, and great care must be exercised when attempting to distinguish between the two forms (55).

Turn conformations have been extensively studied by Krimm and Bandekar (51), who predicted the amide I band positions for the standard type I, II and III turns. However, turns are rarely present in peptides or proteins with standard dihedral angles, and the large variations result in a corresponding spread of frequencies of the bands. It has thus not been possible to distinguish between the various subclasses of turns experimentally; turn conformations are thus to

be found within the range 1666–1688 cm$^{-1}$, overlapping with the high wavenumber band from anti-parallel $\beta$-sheet. The $3_{10}$ helix, equivalent to the poly(type III turn), should also be classed amongst the turn conformations and is best distinguished using CD spectroscopy (see Sec. II.B.).

FTIR spectroscopy has not been used to distinguish between the true "unordered" conformation and the conformation containing a significant amount of poly(Pro) II structure. These are generally classed together as "random" structures and exhibit a broad amide I band around 1644 cm$^{-1}$. The extent of hydrogen bonding from solvent molecules has been studied (56) and found to significantly effect the position of the random coil band. In $H_2O$ solutions, the random coil and helical bands overlap; the general consensus at present is that these two conformations cannot be distinguished in this solvent.

## C. Fourier Transform Infra-Red Studies of Peptides and Polypeptides

Nisin is a positively charged 34-mer peptide with antimicrobial activity, and is widely used in the food industry as a preservative for milk and meat products. Little is known about the antimicrobial action of this peptide other than that it exerts its activity by interaction with membranes. El-Jastimi and Lafleur (57) studied the conformations of nisin in membranous environments containing both neutral and negatively charged lipids by FTIR spectroscopy. Nicin showed much greater affinity for the negatively charged phosphatidylglycerol (PG) than the zwitterionic phosphatidylcholine (PC), due to the electrostatic interactions between peptide and lipid. FTIR studies concluded that membrane bound nisin contained a higher proportion of $\beta$-turns than the free form in water, due to an increase in the intensity of the band at

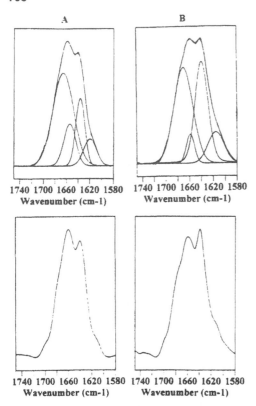

**Fig. 14** Deconvolved FTIR spectra of nicin bound to (A) DPPC and (B) DPPG lipids (lower spectra) and the curve fitted bands (upper spectra). [From Ref. (57)]

$1673\,\mathrm{cm}^{-1}$. In spite of the large differences in affinity between neutral and charged lipids, the FTIR spectra of nicin bound to PG and PC lipids is similar (Fig. 14), suggesting that the nature of the membrane interface is not important.

In a similar vein, FTIR has been used to study a 27-mer peptide corresponding to the cytoplasmic tail from HIV-1 envelope glycoprotein in lipid vesicles (phosphatidylcholine (PC)/phosphatidylserine (PS)/cholesterol and PC/cholesterol) (58). The leucine-zipper peptide aggregated in aqueous solution but interacted in an identical manner with both neutral and charged vesicles, giving rise to $\sim 60\%$ α-helical structure in a membrane environment.

The monomeric form of insulin has been little studied in solution, mainly due to its tendency to aggregate at the pH and concentrations required for such studies. FTIR has been used to study the conformations adopted by insulin in sodium dodecylsulphate (SDS) micelles (59) in which insulin exists as a monomer. The secondary structural elements identified by FTIR were subjected to a quantitative analysis and compared to the dimeric form of insulin in tris- buffered solution. The SDS-insulin spectrum was notable for the absence of a band at $1637\,\mathrm{cm}^{-1}$ (β-sheet) which was correlated with the lack of monomer-dimer equilibrium.

Lamthanh and co-workers (60) used both CD and FTIR to characterize a disulphide cyclized 18-mer peptide derived from snake curaremimetic toxin. This sequence elicits a neutralising response to the toxin, and contains epitopes for both B- and T-cells. FTIR spectra were recorded under a variety of experimental conditions of pH, temperature, solvent and in SDS micelles. In each case the peptide showed a preference for both β-turn and β-sheet conformations; the position of the β-turn has been predicted and is thought to be important in the immunogenic properties of the peptide.

FTIR spectroscopy has been used to study the conformational behaviour of unusual di- and tri-peptides containing an amino acid, fullero-3,4-proline, in which proline is joined to a buckminsterfullerene moiety (61). Spectra were recorded in deutero-chloroform and deutero-dimethylsulphoxide over a range of concentrations. The N-H stretching frequencies were consistent with folded conformations; the carbonyl regions showed complex patterns, similar to those previously assigned to β-turn conformations. The structure was further refined by [1]H NMR studies and confirmed as a type II β-turn. The authors suggest that it may thus be feasible to use this highly unusual, conformationally constrained amino acid to induce secondary structures in small peptides with a defined conformation.

NMR and x-ray crystallography suffer from their inability to study proteins in lipidic environments; these molecules cannot be readily studied in a condition approximating to the natural state, lending importance to CD (see Sec. III.C.) and FTIR investigations. Phospholipase $A_2$ (PLA$_2$) for example, which catalyses the conversion of glycerophospholipids to fatty acid and lysophospholipid, is a $Ca^{2+}$ dependent enzyme which becomes activated upon binding to phospholipid membranes. PLA$_2$ has been investigated by NMR and x-ray crystallography in micelles, but its structure in phospholipid bilayers (its natural substrate) is not known. Tatulian and co-workers (62) studied PLA$_2$ in phospholipid bilayers (phosphatidylglycerol and phosphatidylcholine vesicles) using FTIR spectroscopy, to investigate structural changes upon complexation with lipids. Free PLA$_2$ shows the presence of a single $\alpha$-helical band at $1651 \, cm^{-1}$, whereas the presence of PG/PC vesicles induce a structural change and the band is split into two components at $1658 \, cm^{-1}$ and $1650 \, cm^{-1}$ (Fig. 15). This higher frequency component was identified as resulting from the presence of a more flexible helical region. Likewise, the structure of adipocyte (fat cell) fatty acid-binding protein (63) has recently been studied in a variety of phospholipid vesicles. The protein was shown by FTIR to interact with the lipid bilayer; the predominant $\beta$-sheet in this case remained essentially unchanged, with minor differences possibly resulting from changes within the short $\alpha$-helical "caps".

Another situation where FTIR gives rapid information with minimum sample preparation relative to NMR and x-ray studies, is the study of the complexes formed between proteins during a number of biochemical events. A recent paper by Nishimura et al (64) highlights the use of FTIR in the study of the complex formed

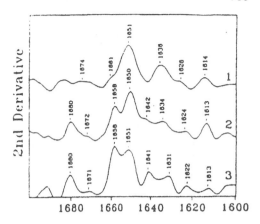

**Fig. 15** FTIR spectra of phospholipase $A_2$ (PLA$_2$) in (1) $D_2O$ buffer, (2) PC/PG vesicles and (3) PC/PC:PG bilayers. [From Ref. (62)]

between transducin and activated rhodopsin, which is important in visual transduction processes. The spectral contributions from metarhodopsin I and uncomplexed metarhodopsin II were subtracted, giving the FTIR spectrum of the complex only. Small changes (to lower frequency) in the carbonyl band and intensities were detected, consistent with an increase in the strength of hydrogen-bonding of one or a small number of peptide bonds during complexation. FTIR was thus able to detect small local interactions in the absence of global conformational changes. The complex formed in the photosystem II reaction centre consists of five proteins with a total molecular weight of $\sim 95 \, kDa$, and is therefore beyond the range for study by other conventional spectroscopic techniques. The structures of the individual proteins within the complex have been studied; FTIR has recently been used to study the secondary structural elements of the intact complex (65). Decomposition of the amide I band has been carried out to identify the various components present and the thermal stability of the complex investigated by FTIR temperature studies.

A novel and exceptionally useful method for determining the conformation

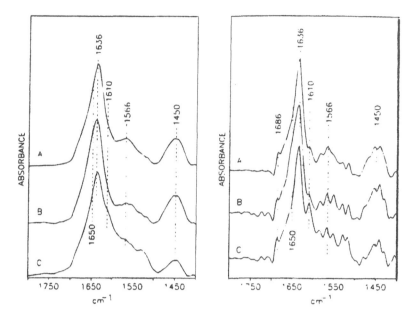

**Fig. 16** Isotope labelling experiment followed by FTIR spectroscopy. (A) free receptor, (B) G-CSF/receptor complex and (C) $^{13}C/^{15}N$ labeled G-CSF/receptor complex. The original spectra are shown on the left and the resolution enhanced on the right. [From Ref. (67)]

of peptides and/or proteins during complexation is via the method of isotope labeling. Although rather expensive and time-consuming, this technique can provide unambiguous answers to changes resultant upon receptor/ligand binding. In this process the peptide is prepared using $^{13}C$ and $^{15}N$ isotopes and by these means the amide I' (and NH stretching band) is moved to $\sim 30\text{--}40\,cm^{-1}$ lower wavenumbers, (66) separating the protein/protein complex bands away from those of the bound receptor or ligand and allowing each to be analyzed independently. This method has recently been applied to the study of the cytokine-receptor interactions by investigation of the complex between isotopically labeled granulocyte colony-stimulating factor (G-CSF) and its receptor (67). Fig. 16 shows the deconvolved FTIR spectra of the receptor, unlabeled G-CSF/receptor and labeled G-CSF/receptor respectively. In the latter case (C), the peak from the receptor at $1636\,cm^{-1}$ is well

separated from that of G-CSF at $1610\,cm^{-1}$, in contrast to the situation in spectrum (B) (unlabeled G-CSF/receptor). A thorough examination of the complex reveals that there is a slight increase in the $\alpha$-helical content of the receptor and a corresponding conformational change to $\alpha$-helix at the expense of $3_{10}$ helix in the G-CSF protein structure.

As in the case of CD spectroscopy, FTIR has commonly been used to study recombinant proteins, either to confirm the secondary structural features are identical to the native protein or probe the conformation and stability of proteins not previously studied. Jiang and co-workers, for example, studied the secondary structure and the effects of disulphide reduction and temperature variation on recombinant granulocyte-macrophage colony-stimulating factor (GM-CSF) in aqueous solution (68). Similarly, the effects of $Fe^{2+}$, $Fe^{2+}$/dopamine and phosphorylation upon the secondary structure of recombinant tyrosine hydroxylase (TH) was studied by FTIR spectroscopy (69); the secondary structure was unaffected by the addition of $Fe^{2+}$ or $Fe^{2+}$/dopamine, but gave rise to an increase in protein stability, determined in temperature studies, and a heightened resistance to proteolysis. Phosphorylation, followed by iron (II) binding was found to significantly alter the protein conformation via a decrease in unordered structure and a concomitant increase in $\alpha$-helical structure. FTIR spectroscopy has also been used to study the dynamics of recombinant ribonuclease A, ribonuclease T1 and several variants, by monitoring of H-D exchange rates of several sensitive amide bands (70). Application of liner temperature-gradients and temperature-jump techniques were used to predict a "transient global unfolding" mechanism for the H-D exchange process.

## IV. CONCLUSIONS

The two previous sections have focused upon the applications of Circular Dichroism and Fourier Transform Infra-Red to the study of the secondary structural parameters of peptides and proteins, in isolation. In the majority of cases, structural analysis is best carried out using a combination of these two techniques (a number of the examples cited above do indeed contain both CD and FTIR experiments, but have tended to concentrate on one particular technique). At the very least, such studies will provide the best and most relevant experimental conditions under which the conformations are further investigated (by techniques such as NMR and crystallography) and a global view of the expected secondary structures. Additionally, a number of examples have been cited where large proteins are studied in conditions which are not, at present, amenable to investigation by NMR and x-ray crystallography

(membranous environments, substrate binding, metal binding, protein complexes etc.) and where CD and FTIR provide rapid and instructive answers. In spite of recent rapid advances (such as 3- and 4-D NMR studies of fully $^{13}C$ and $^{15}N$ labeled proteins), CD and FTIR will remain the techniques of choice for the foreseeable future for the speedy analysis of (changes in) secondary structure in peptides and proteins.

## REFERENCES

1. P Bayley. Introduction to Spectroscopy for Biochemists. SB Brown, ed. New York: Academic Press, 1981, pp 150–234.
2. RW Woody. Circular Dichroism and the Conformational Analysis of Biomolecules. GD Fasman, ed. New York: Plenum Press, 1996, pp 25–67.
3. S Brahms, J Brahms. Determination of protein secondary structure in solution by vacuum ultraviolet circular dichroism. J Mol Biol 138: 149–178, 1980.
4. WC Johnson, I Tinoco. Circular dichroism of polypeptide solutionsin the vacuum ultraviolet. JACS 94: 4389–4390, 1972.
5. A Perczel, M Hollosi. Circular Dichroism and the Conformational Analysis of Biomolecules. GD Fasman, ed. New York: Plenum Press,1996, pp 285–380.
6. RW Woody. Peptides, Polypeptides and Proteins. ER Blout, FA Bover, M Goodman, N Loton, eds. New York: John Wiley and Sons, 1974, pp 338–348.
7. GD Rose, LM Gierasch, JA Smith. Turns in peptides and proteins. Adv Prot Chem 37: 1–109, 1985.
8. TS Sudha, EKS Vijayakumar, P Balaram. Circular dichroism studies of helical oligopeptides- can $3_{10}$ and alpha-helical conformations be chiroptically distinguished? Int J Peptide Protein Res 22: 464–468, 1983.
9. C Tonolio, A Polese, F Formaggio, M Crisma, J Kamphuis. Circular dichroism spectrum of a peptide $3_{10}$ helix. JACS 118: 2744–2745, 1997.
10. ML Tiffany, S Krimm. New chain conformations of poly(glutamic acid) and polylysine. Biopolymers 6: 1379–1382, 1968.
11. RW Woody. Circular dichroism. Methods Enzymol 246: 47–48, 1995.
12. R Mandel, Holzworth G. Circular dichroism of orientated helical polypeptides. J Chem Phys 12: 655–674, 1973.

13. AF Drake, G Siligardi, WA Gibbons. Reassessment of the electronic circular dichroism criteria for random coil conformations of poly(L-lysine) and the implications for protein folding. Biophys Chem 31: 143–146, 1988

14. AA Adzhubei, MJE Sternberg. Left-handed polyproline II helices commonly occur in globular proteins. J Mol Biol 229: 472–493, 1993.

15. MG Paterlini, TB Freedman, LA Nafie. Vibrational circular dichroism spectra of three conformationally distinct states and an unordered state of poly(L-lysine) indeuterated aqueous solution. Biopolymers 25: 1751–1765, 1986.

16. S Brahms, J Brahms, G Spach, A Brack. Identification of $\beta$, $\beta$-turns and unordered conformations in polypeptide chains by vacuum ultraviolrt circular dichroism. PNAS 74: 3208–3212, 1977.

17. RW Woody, AK Dunker. Circular Dichroism and the Conformational Analysis of Biomolecules GD Fasman, ed. New York: Plenum Press, 1996, pp 110–157.

18. RW Woody. Aromatic side-chain contributions to the far ultraviolet circular dichroism of peptides and proteins. Biopolymers 17: 1451–1467, 1978.

19. MA Cebrat, M Lisowski, IZ Siemion, M Zimecki, Z Wieczorek. Sulfonated analogues of cyclolinopeptide a: synthesis, immunosuppressive activity and CD studies. J Pep Res 49: 415–420, 1997.

20. RW Woody. Contribution of tryptophan side-chains to the far-ultraviolet circular dichroism of proteins. Eur Biophys J, 23: 34–71, 1994.

21. AC Dong, V Kery, J Matsuura, MC Manning, JP Kraus, JF Carpenter. Secondary structure of recombinant human cystathionine beta-synthase in aqueous solution: effect of ligand binding and proteolytic truncation. Arch Biochem Biophys 344: 125–132, 1997.

22. GJ Anderson, PI Haris, D Chapman, JT Romer, GK Toth, I Toth, WA Gibbons. Synthesis and spectroscopy of membrane receptor proteins: the $\gamma$ subunit of the IgE receptor. Eur J Biochem 207: 51–54, 1994.

23. GJ Anderson, PI Haris, D Chapman, AF Drake. The conformational equilibria of a renin inhibitor peptide in solution. Biophys Chem 52: 173–181, 1994.

24. P Bayley. Introduction to Spectroscopy for Biochemists. SB Brown, ed. London: Academic Press, 1980, pp 185–191.

25. RW Woody. Application of the Bergson model to the optical properties of chiral sulphides. Tetrahedron 29: 1273–1283, 1973.

26. G Siligardi, AF Drake, P Mascagni, P Neri, L Lozzi, N Niccolai, WA Gibbons. Resolution of the conformational equilibria in linear peptides by circular dichroism in cryogenic solvents. Biochem Biophys Res Comm 143: 1005–1011, 1987.

27. AJ Maynard, MS Searle. NMR structural analysis of a $\beta$-hairpin peptide designed for DNA binding. Chem Comm 1297–1298, 1997.

28. AD McLachlan, M Stewart. Tropomycin coiled-coil interactions: evidence for an unstaggered structure. J Mol Biol 98: 293–304, 1975.

29. MM Javadpour, MD Barkley. Self-assembly of designed anti-microbial peptides in solution and micelles. Biochemistry 36: 9540–9549, 1997.

30. E Krause, M Beyermann, H Fabian, M Dathe, S Rothermund, M Bienert. Conformation of a water soluble $\beta$-sheet model peptide: a circular dichroism and Fourier-transform infra-red study of double D-amino acid replacements. Int J Peptide Protein Res 48: 559–568, 1996.

31. MA Castaglione Morelli, F Bisaccia, S Spisani, S Traniello, AM Tamburro. Structure-activity for some elastin-derived peptide chemoattractants. J Pept Res 49: 492–499, 1997.

32. RJ Falkenstein, C Pena. Synthetic peptides derived from the central loop of fasciculin: structural analysis and evaluation as inhibitors of acetylcholinesterase. Biochimica et Biophysica Acta 1340: 143–151, 1997.

33. HY Hu, ZX Lu, YC Du. Solution conformation of N-terminal fragments of trichosanthin small domain (TCS 182-200). J Pept Res 49: 113–119, 1997.

34. D-K Chang, W-JChien, S-F Cheng, S-T Chen. NMR and circular dichroism studies on the conformation of a 44-mer peptide from a CD4-binding domain of human immunodeficiency virus envelope glcoprotein. J Pept Res 49: 432–443, 1997.

35. L Franzoni, G Nicastro, TA Pertinhez, M Tato, CR Nakaie, ACM Paiva, S Schreier, A Spisni. Structure of the C-terminal fragment 300–320 of the rat angiotensin II $AT_{1A}$ receptor and its relevance with respect to G-protein coupling. J Biol Chem. 272: 9734–9741, 1997.

36. B Bechinger. Structure and dynamics of the M13 coat signal sequence in membranes by multidimensional high-resolution NMR and solid-state NMR. Prot: Struct Funct Genet 27: 481–492, 1997

37. L-P Liu, CM Deber. Anionic phospholipids modulate peptide insertion into membranes. Biochemistry 36: 5476–5482, 1997.

38. F Zhang, Y Yin, CH Arrowsmith, V Ling. Secretion and circular dichroism analysis of the C-terminal signal peptides of HlyA and LktA. Biochemistry 34: 4193–4201, 1995.

39. BA Morgan, JA Gainor. Approaches to the discovery of non-peptide ligands for peptide receptors and peptidases. Ann Rep Med Chem 24: 243–252, 1989.

40. RM Jain, KR Rajashankar, S Ramakumar, VS Chauhan. First observation of left-handed conformation in a dehydro peptide containing two L-Val residues. Crystal and solution structure of Boc-L-Val-ΔPhe-ΔPhe-ΔPhe-L-Val-OMe. J Am Chem Soc 119: 3205–3211, 1997.

41. A Tevelev, I-JL Byeon, T Selby, K Ericson, H-J Kim, V Kraynov, M-D Tsai. Tumor suppressor p16$^{INK4A}$: structural characterisation of wild-type and mutant proteins by NMR and circular dichroism. Biochemistry 35: 9475–9487, 1996.

42. TM Hackeng, CM Mounier, C Bon, PE Dawson, JH Griffin, SBH Kent. Total chemical synthesis of enzymatically active human type II secretory phospholipase $A_2$. Proc Natl Acad Sci USA 94: 7845–7850, 1997.

43. A Boffi, JB Wittenberg, E Chiancone. Circular dichroism spectroscopy of Lucina I hemoglobin. FEBS Lett 411: 335–338, 1997.

44. PN Farnsworth, B GrothVasselli, NJ Greenfield, K Singh. Effects of temperature and concentration on bovine lens alpha-crystallin secondary structure: a circular dichroism spectroscopic study. Int J Biol Macromol 20: 283–291, 1997.

45. L Mauricio, TR Lima, G dePratGay. Conformational changes and stabilisation induced by ligand binding in the DNA-binding domain of the E2 protein from human papillomavirus. J Biol Chem 272: 19295–19303, 1997.

46. G Kassam, A Manro, CE Braat, P Louie, SL Fitzpatrick, DM Waisman. Characterisation of the heparin binding properties of annexin II tetramer. J Biol Chem 272: 15093–15100, 1997.

47. EA Cooper, K Knutson. Physical Methods to Characterise Pharmaceutical Proteins. JA Herron et al, eds. New York: Plenum Press, 1995, pp 101–143.

48. AG Marshall, FR Verdun. Fourier Transforms in NMR, Optical and Mass Spectrometry. Amsterdam: Elsevier Science Publishers, 1990, pp 331–367.

49. DG Cameron, DJ Moffatt. Deconvolution, derivation and smoothing of spectra using Fourier transforms. J Test Eval 12: 78–85, 1984.

50. RJ Markovich, C Pidgeon. Introduction to infrared spectroscopy and applications in the pharmaceutical sciences. Pharm Res 8: 663–675, 1991.

51. S Krimm, J Bandekar J. Vibrational spectroscopy and conformation of peptides, polypeptides and proteins. Adv Prot Chem 38: 181–364, 1986.

52. PI Haris, D Chapman. Does Fourier transform infrared spectroscopy provide useful information on protein structures? Trends Biochem Sci 17: 328–333, 1992.

53. EA Cooper, K Knutson. Physical Methods to Characterise Pharmaceutical Proteins. JN Herron et al, eds. New York: Plenum Press, 1995, pp 101–143.

54. NA Nevskaya, YN Chirgadze. Infrared spectra and resonance interactions of amide-I and II vibrations of α-helix. Biopolymers 15: 637–648, 1976.

55. WK Surewicz, MA Moscarello, HH Mantsch J. Secondary structure of the hydrophobic myelin protein in a lipid environment as determined by Fourier-transform infrared spectroscopy. Biol Chem 262: 8598–8602, 1987.

56. YN Chirgadze, BV Shestapalov, SY Venyaminov. Intensities and other spectral parameters of infrared amide bands of polypeptides in the β- and random forms. Biopolymers 12: 1337–1351, 1973.

57. R El-Jastimi, M Lafleur. Structural characterisation of free and membrane-bound nisin by infrared spectroscopy. Biochim Biophys Acta 1324: 151–158, 1997.

58. Y Kliger, Y Shai. A leucine zipper-like sequence from the cytoplasmic tail of the HIV-1 envelope glycoprotein binds and perturbs lipid bilayers. Biochemistry 36: 5157–5169, 1997.

59. G Veccio, A Bossi, P Pasta, G Carrea. Fourier-transform infrared conformational study of bovine insulin in surfactant solution. Int J Peptide Protein Res 48: 113–117, 1996.

60. H Lamthanh, M Leonetti, E Nabedryk, A Menez. CD and FTIR studies of an immunogenic disulphide cyclised octapeptide, a fragment of a snake curaremimetic toxin. Biochim Biophys Acta 1203: 191–198, 1993.

61. A Bianco, T Bertolini, M Crisma, G Valle, C Tonolio, M Maggini, G Scorrano, M Prato. β-turn induction by a $C_{60}$-based fulleroproline: synthesis and conformational characterisation of Fpr/Pro small peptides. J Pept Res 50: 158–170, 1997.

62. SA Tatulian, RL Biltonen, LK Tamm. Structural changes in a secretory phospholipase $A_2$ induced by membrane

binding: a clue to interfacial activation? J Mol Biol 268: 809–815, 1997.

63. A Gericke, ER Smith, DJ Moore, R Mendelsohn, J Storch. Adipocyte fatty acid-binding protein: interaction with phospholipid membranes and thermal stability studied by FTIR spectroscopy. Biochemistry 36: 8311–8317, 1997.

64. S Nishimura, J Sasaki, H Kandori, T Matsuda, Y Fukada, A Maeda. Structural changes in the peptide backbone in complex formation between activated rhodopsin and transducin studied by FTIR spectroscopy. Biochemistry 35: 13267–13271, 1996.

65. J De Las Rivas, J Barber. Structure and thermal stability of photosystem II reaction centres studied by infrared spectroscopy. Biochemistry 36: 8897–8903, 1997.

66. PI Haris, GT Robillard, AA van Dijk, D Chapman. Biochemistry, 36, 6279–6284, 1992.

67. T Li, T Horan, T Osslund, G Stearns, T Arakawa. Conformational changes in G-CSF complex as investigated bi isotope-edited FTIR spectroscopy. Biochemistry 36: 8849–8857, 1997.

68. H Jiang, Z Song, M Ling, S Yang, Z Du. FTIR studies of recombinant human granulocyte-macrophage colony-stimulating factor in aqueous solutions: secondary structure, disulphide reduction and thermal behaviour. Biochim. Biophys. Acta 1294: 121–128, 1996.

69. A Martinez, J Haavik, T Flatmark, JLR Arrondo, A Muga. Conformational properties and stability of tyrosine hydroxylase studied by infrared spectroscopy. J Am Chem Soc 271: 19737–19742, 1996.

70. J Backmann, C Schultz, H Fabian, U Hahn, W Saenger. Thermally induced hydrogen exchange processes in small proteins as seen by FTIR spectroscopy. Prot: Struct Funct Genet 24: 379–387, 1996.

# 25

# Electron Paramagnetic Resonance of Peptides and Proteins

**Derek Marsh**
*Max-Planck-Institut für biophysikalische Chemie, Göttingen, Germany*

## I. INTRODUCTION

Peptides and proteins do not in general contain paramagnetic centers and therefore in the unlabeled form do not posses an int.insic EPR spectrum. The exceptions to this are certain metalloproteins and a relatively small, but growing, number of proteins that have been found to contain stable free radicals. The spin labeling technique consists of attaching a stable nitroxide free radical to the molecule of interest as an environmentally sensitive reporter. The labeling method is generally applicable, also in the case of systems with intrinsic paramagnetic centers.

The sensitivity to structure and environment arise principally from the anisotropy of nitroxide EPR spectra with respect to angular orientation of the label (1). Rotational motion modulates this anisotropy and results in splittings and linewidths in the EPR spectrum that are dependent on the amplitude and rate of motion. This spectral dependence on the local or global mobility of the spin-labeled group is one of the primary indicators relating to structure and molecular dynamics. Conventional spin label EPR spectra are sensitive to rotational correlation times down to approximately 10 ns, which are characteristic of local motions or the global rotation of peptides and small proteins in solution (2). Saturation transfer EPR is sensitive to rotational correlation times in the microsecond regime, which are characteristic of the global rotation of supramolecular assemblies or proteins and peptides in membranes (2).

A further class of spin label experiments involves the interaction of the spin label with other paramagnetic species (3–5). This may either be with fast-relaxing paramagnetic ions or molecular oxygen, which are preferentially soluble in hydrophilic or hydrophobic environments, respectively, or with the same or other spin-labelled species (in double-labelling experiments). Such experiments can give information on translational mobility and collisional accessibility, and also can yield direct distance measurements. These also are aspects that are related directly to both molecular structure and dynamics.

All these classes of experiments are capable of giving most information, if combined with site-directed spin-labeling. In this case the point of spin label attachment is scanned systematically throughout the amino acid sequence, normally by introducing single cysteine residues with site-directed mutagenesis. Comparison of relative mobilities and accessibilities of the spin-labeled groups then reveals periodicities that are characteristic of the local secondary structure, tertiary fold, and/or transmembrane topology in the case of integral membrane proteins (3).

In addition to fundamental molecular structural and dynamic features, the EPR spectra may also be used to study binding of spin-labeled ligands (6, 7), the kinetics of covalent spin-labelling (8), and the conformational changes of spin-labeled proteins (1). Such measurements rely directly on the sensitivity of the spectra to changes in local environment.

Reviews on various aspects of spin labeling methodology may be found in references (1, 2, 4, 5, 8–10). The series of volumes on "Spin Labeling" edited by Berliner (11–14) additionally contain most valuable information.

## II.  SPIN LABELS FOR PEPTIDES AND PROTEINS

Nitroxide free radicals that are stabilized by flanking methyl groups are used as spin labels. Those common for protein or peptide labeling are based on the 5-membered pyrrolidine or pyrrolinyl ring, or on the 6-membered piperidine ring. The pyrrolidine/pyrrolinyl ring is relatively rigid, whereas the piperidine ring can undergo boat-chair conformational transitions. It should also be noted that the 5-membered ring nitroxides are chemically much more stable, e.g. against reduction, than are the 6-membered ring nitroxides (15). The spin labels most used for studying peptides and proteins are nitroxide free radical analogues of the standard reagents in protein chemistry that are used for covalent modification reactions (see Fig. 1). Amongst these are derivatives of maleimide (e.g. 3-maleimido-2,2,5,5-tetramethyl-1-pyrrolidinyloxyl), of iodoacetamide [e.g. 3-(2-

iodoacetamido)-2,2,5,5-tetramethyl-1-pyrrolidinyloxyl], and of isothiocyanate (e.g. 4-isothiocyanato-2,2,6,6-tetramethylpiperidinoxyl). Several of these derivatives, with differing lengths of the linkage between the active center and the nitroxide ring, are available commercially from Sigma Chemical Co. (St. Louis, MO). Synthesis of these labels and many others is described by Gaffney (16). At neutral pH, the reaction of these labels is primarily with sulphydryl groups, except for the isothiocyanate derivative that has a specificity for amine groups, including the N-terminal amine. Of considerable current interest is the methanethiosulphonate (MTS) spin label derivative [(1-oxyl-2,2,5,5-tetramethylpyrroline-3-methyl)methanethiosulphonate; (17)] that has found considerable application in site-directed spin-labeling (3), in much the same way as MTS itself is used for labeling in cysteine scanning mutagenesis. The MTS spin label is available commercially from Reanal (Budapest, Hungary). Other spin labels of interest are the chloromercuri reagents [e.g. N-(1-oxyl-2,2,6,6-tetramethylpiperid-4-yl) p-chloromercuribenzamide; (18)] because of their high reactivity and lack of the reversibility that is potentially involved in the Michael addition reaction (19). A spin-labeled derivative of carbodiimide [N-(2,2,6,6-tetramethylpiperidine-1-oxy)-N'-cyclohexylcarbodiimide] can be used for labeling also carboxylic acid side chains (20).

The possibility to use different protein spin-labeling reagents allows versatility in probing active-site geometries (6, 21–23), side-chain mobility (24), segmental flexibility, conformational changes, and overall molecular rotational diffusion. This possibility is particularly crucial in finding labels with some flexibility, for detecting protein conformational changes or, at the opposite extreme, in obtaining labeling without independent segmental flexibility to probe overall protein rotation by using satu-

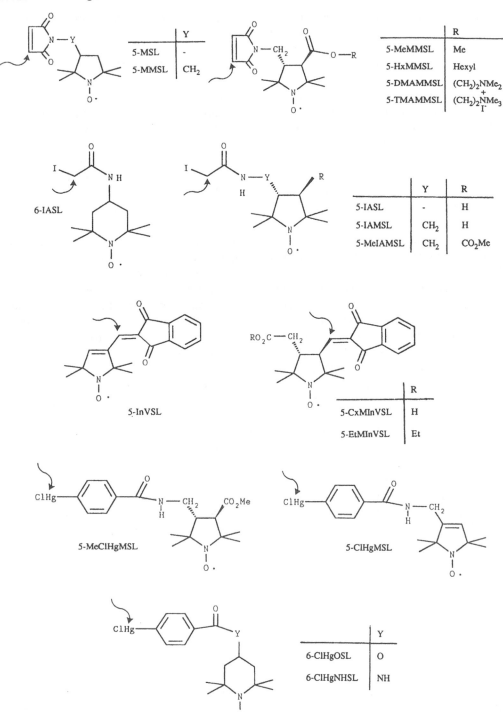

**Fig. 1** Chemical structures of nitroxide spin-label derivatives used in the covalent modification of proteins. From *top* to *bottom*: maleimides, iodoacetamides, vinyl ketones (here the indanedione derivative), and chloromercuri-derivatives. The arrows show the points of nucleophilic attack in the reaction with sulphydryl- or amino group-containing protein side chains (19).

ration transfer EPR (19). In the latter case, the β-vinyl ketone spin labels have proved to be particularly useful (25). Of these, the indanedione derivative (see Fig. 1) has proved to be almost optimal with respect to its reactivity, that results from the electron withdrawing capacity of the indanedione moiety, and its lack of mobility that results from the bulk of the indandione substituent and the shortest possible link from the vinyl active centre to the nitroxide ring. In addition, a second substituent at the nitroxide ring may be used for selective targeting of the spin-label, and also further to enhance its steric immobilisation [see Fig. 1 and Refs (15, 19)].

Specifically spin-labeled amino acids that contain the nitroxide ring as a side chain, or contain the α-carbon as a constituent of the nitroxide ring, are discussed by Hideg and Hankovszky (26). Nitroxide amino acids have been acylated to a suppressor t-RNA and introduced into the protein (bacteriophage T4 lysozyme) by *in vitro* translation-transcription with a DNA containing the corresponding nonsense codon at the desired position in the gene (27).

A most useful review of peptide and protein spin labels, including syntheses, has been given by Hideg and Hankovszky (26). A list of structures and molecular weights of commonly used protein spin labels is given in the Appendix to this chapter.

## III. EPR METHODOLOGY

The general instrumental and practical aspects of EPR are given in the textbook by Knowles et al. (9). Specific considerations for spin label EPR are given in Refs (1, 8, 28). Practical spin-label EPR directed towards ion channel studies is dealt with in Refs (29, 30).

Some typical spectrometer settings for spin label applications are given in Table 1. These are intended merely as a guide. In practice, optimisation must be performed to check that the modulation amplitude is not so high as to cause linebroadening, and that the scan time is not so short (relative to the time constant) as to cause spectral distortion. The combinations of scan time and filter time constant that are given in Table 1 are representative for a single scan. Normally signal averaging would be used, in which case scan times can be shorter, with corresponding reduction of the filter time constant. Formally, signal averaging and analogue filtering are essentially equivalent, but signal averaging has the advantage of removing the effects of long-term drifts. The latter can be important for very weak samples.

In addition, the microwave power setting is important, especially for

**Table 1** Typical EPR Spectrometer Settings for Spin Label Measurements on Peptides or Proteins. The Values Are Given as a Guide for Spin Labels of Different Rotational Mobilities. Field Scan Width (100 G), Modulation Frequency (100 kHz), and Field Centre and Microwave Frequency (typically 3240 G for 9.12 GHz) Are Constant.

| Immobilisation: | Weak | Moderate | Strong | ST-EPR[a] |
|---|---|---|---|---|
| Modulation amplitude (G) | 0.2–0.4 | 0.5–1.0 | 1.0–2.0 | 5.0 |
| Microwave power (mW) | 1–10 | 10–20 | 20–30 | 25–40 |
| Time constant (s)[b] | 0.075–0.125 | 0.125–0.25 | 0.25–0.5 | 0.5–2.0 |
| Scan time (min)[b] | 4 | 4 | 4 | 16 |

[a] Saturation transfer EPR: modulation frequency 50 kHz, with second-harmonic detection 90°-out-of-phase at 100 kHz. The microwave power is chosen to give $H_1 = 0.25$ G at the sample (31, 34).

[b] These values are coupled, and here are typical for a single scan without signal averaging.

quantitating spin concentrations. For the latter it is important to avoid saturation (i.e., the EPR signal height should increase linearly with the square root of the microwave power), or at least to make sure that spectra are compared under identical degrees of saturation. When only spectral lineshapes are of interest the requirements for the microwave power setting are less stringent. Then it is sufficient to ensure that the microwave power is sufficiently low as not to cause saturation broadening of the spectrum. In the case of multi-component spectra, this criterion applies to the component that is most easily saturated.

For saturation transfer EPR studies of protein rotational diffusion, it is important that the spectra be recorded under standard conditions of microwave power, modulation amplitude and modulation frequency. This is because calibrations of the spectral lineshapes in terms of the rotational correlation time of the spin-labeled species refer to these standard conditions (31, 32). Saturation transfer EPR experiments apply only to samples that are strongly immobilized on the conventional EPR timescale. The ST-EPR spectra are recorded under moderate saturation, with overmodulation and detection at the second harmonic, 90-out-of-phase with respect to the field modulation (33). The degree of saturation is specified by the value of the $H_1$-microwave field at the sample: $H_1 = 0.25$ G. This is related to the microwave power, which must be adjusted to yield this standard condition for each sample, and possibly also for each temperature [see Ref. (34)]. This is because different samples have different microwave losses. Also small samples, centred in the cavity, are essential to avoid the effects of inhomogeneities in both the $H_1$- and modulation fields (31, 34). Setting of the modulation phase, which is not very crucial for conventional EPR, is, however, critical for saturation transfer EPR. The phase setting must be done with high accuracy by minimizing the out-of-phase signal at low (i.e., non-saturating) microwave power. A protocol for this so-called self-null method is given in Ref. (8).

Peptides in non-lossy apolar organic solvents cause little problems and can be examined in a standard 4 mm-diameter quartz EPR-sample tube. Aqueous solutions of peptides or proteins are troublesome because of the high dielectric loss of water at microwave frequencies. The optimum sample cell in these cases is a quartz flat cell with an internal spacing of 0.3 mm. This must be carefully oriented in the centre of the rectangular resonant microwave cavity, transverse to its long axis. For a cylindrical microwave cavity, a cylindrical sample capillary oriented along the cavity axis is optimum. For reasons of convenience, glass capillaries of ≤1 mm internal diameter may also be used with rectangular cavities. The 100 $\mu$l glass capillaries used with positive delivery pipettes are useful disposable sample cells that may be easily flame-sealed. For conventional EPR the capillaries may be filled, although the active sample length is ≤3 cm. For saturation transfer EPR, however, the sample length must be restricted to 5 mm. For membranous samples, part of the effect of the dielectric loss from water can be overcome by packing the membranes, using low-speed centrifugation, and then removing the excess supernatant.

EPR is relatively insensitive compared with optical spectroscopy and the spin label concentrations needed are relatively high. The actual requirements depend on the linewidths of the spectra. For very mobile spectra, concentrations in the range 5–10 $\mu$M are adequate; for intermediate states of immobilisation, this increases to approximately 50 $\mu$M; and for very strongly immobilised spectra and ST-EPR applications, concentrations of the order of 300 $\mu$M are required. Care must be taken, however, to ensure that the spin label concentrations are not so high as to induce spectral broadening by (collisional) spin-spin interactions. These values of the spin-label concentration required are just a rough guide. Better sensitivity, when the quantity of sample is limited, can be achieved with a loop-gap resonator combined with a microwave preamplifier (35).

## IV. QUANTITATIVE AND SPECTRAL ANALYSIS

Spin-label EPR is a versatile form of spectroscopy and many types of experiments may be performed, yielding a wide range of different information on the spin-labeled system and its molecular interactions. Those experiments that are applicable to peptides and proteins are described here, along with their methods of analysis.

### A. Determination of Concentrations

Quantitative analysis of concentrations is in many ways technically more demanding than the analysis of spectral lineshapes and splittings. This is because sample size, positioning and solvent composition can quite critically affect the spectrometer sensitivity—principally as a result of microwave losses and $H_1$- and modulation field inhomogeneities that were referred to in the previous section. For samples that are not limited in size, it is advisable to fill the sample cell to beyond its active length in the microwave cavity, and to have a holder that positions the cell reproducibly in the cavity. Saturation of the EPR signal should be avoided, unless sample and standard have closely similar saturation properties. Ideally, the same solvent would be used for sample and standard, because microwave losses degrade the cavity Q-factor which directly affects sensitivity. A small, non-saturable, solid sample can

be used as an internal secondary standard. DPPH ($\alpha,\alpha$-diphenyl-$\beta$-picrylhydrazyl) is suitable for such purposes (Aldrich, Milwaukee, WI). In this case, the microwave power must be adjusted to give the same signal height for the secondary standard under conditions that apply to sample and primary standard, respectively. See Ref. (36) for considerations related to secondary standards. Quantitation is performed by evaluating the second integral of the EPR spectrum, because the conventional spectral display (after phase-sensitive detection) is the first derivative of the absorption spectrum. Spectral lineheights may be compared directly only if the linewidths of both spectra are the same.

### B. Rotational Mobility

Rotational correlation times, $\tau_R$, or the corresponding rotational diffusion coefficients ($D_R = 1/6\tau_R$), are parameters that are related both to the size and to the anisotropic shape of the rotating species [see e.g. Refs (2, 37)]. In addition, the anisotropy of the rotational diffusion relative to the principal magnetic axes of the spin label is also a structurally relevant parameter.

The rotational correlation times of relatively small, rapidly tumbling, spin-labeled species in isotropic solvents may be determined from the relative linebroadening of the three $^{14}$N-hyperfine lines in the EPR spectrum. The linewidths of the different hyperfine lines are given by:

$$\Delta H(m_I) = A + Bm_I + Cm_I^2 \qquad (1)$$

where $m_I = +1$, 0, $-1$ for the low-field, central and high-field lines, respectively. Both the coefficients $B$ and $C$ depend on the rotational correlation times, but in different ways for anisotropic rotational diffusion. The relative values of these coefficients therefore may be used to diagnose different axial modes of anisotropic rotation diffusion.

**Table 2** Numerical Constants Relating Line-Width Coefficients B and C to Rotational Correlation Times, in Eqns 5 and 6, and Range of Validity of the Ratio C/B for Axial Rotation (38).

| Rotation axis | $|C/B|$ | $c_1$ $(s \cdot G^{-1})$ | $c_2$ | $b_1$ $(s \cdot G^{-1})$ | $b_2$ |
|---|---|---|---|---|---|
| isotropic | $\approx 1$ | $1.19 \cdot 10^{-9}$ | 0 | $-1.22 \cdot 10^{-9}$ | 0 |
| z-axis | $< 0.89$ | $1.28 \cdot 10^{-9}$ | $-0.0115$ | $-0.335 \cdot 10^{-9}$ | 1.130 |
| y-axis | 0.8–8.8 | $4.95 \cdot 10^{-9}$ | 0.775 | $-1.38 \cdot 10^{-9}$ | 0.113 |
| x-axis | 0.5–1.5 | $-2.85 \cdot 10^{-9}$ | 1.447 | $1.28 \cdot 10^{-9}$ | 1.912 |

The linewidth coefficients $B$ and $C$ can be measured most readily from the relative lineheights of the low-field, central and high-field lines ($h_{+1}$, $h_0$ and $h_{-1}$, respectively), together with the linewidth, $\Delta H_o$, of the central line:

$$B = \frac{\Delta H_o}{2} \left( \sqrt{\frac{h_0}{h_{+1}}} - \sqrt{\frac{h_0}{h_{-1}}} \right) \qquad (2)$$

$$C = \frac{\Delta H_o}{2} \left( \sqrt{\frac{h_0}{h_{+1}}} - \sqrt{\frac{h_0}{h_{-1}}} - 2 \right) \qquad (3)$$

For isotropic rotation, the ratio $|C/B|$ is approximately unity. For preferential anisotropic rotation about one of the nitroxide principal axes, however, this ratio can depart considerably from unity, in a diagnostic manner as is indicated in Table 2. The spin label principal axes are defined with the x-axis along the N—O bond, and the z-axis parallel to the nitroxide 2p-$\pi$ orbital (i.e. perpendicular to the plane of the nitroxide ring). The anisotropy in the rotational diffusion is specified by the effective rotational correlation times, $\tau_{R//}$ and $\tau_{R\perp}$, for axial rotation which are related by (2):

$$\tau_{R//} = \frac{2\tau_{R\perp}\tau_{22}}{3\tau_{R\perp} - \tau_{22}} \qquad (4)$$

where $\tau_{R\perp} \equiv \tau_{20}$. These correlation times may be determined from the experimental linewidth parameters, according to the relations (38):

$$\tau_{R\perp} = c_1(C + c_2 B) \qquad (5)$$

$$\tau_{22} = b_1(B + b_2 C) \qquad (6)$$

where the coefficients $c_1$, $c_2$ and $b_1$, $b_2$ depend on the mode of anisotropic rotation and are given in Table 2.

At the other extreme, for slow rotational motion that is approaching the limits of motional sensitivity of conventional spin label EPR spectroscopy (correlation times $\geq 10$ ns), calibrations have been given for the rotational correlation time, $\tau_R$, in terms of the maximum outer splitting, $A'_{zz}$, and the linewidths, $\Delta H_{m_I}$, of the outer hyperfine peaks (39). These are defined in terms of the corresponding values, $A^R_{zz}$ and $\Delta H^R_{m_I}$, at the rigid limit by the following relations for isotropic rotation:

$$\tau_R = a \left( 1 - \frac{A'_{zz}}{A^R_{zz}} \right)^b \qquad (7)$$

$$\tau_R = a'_{m_I} \left( \frac{\Delta H_{m_I}}{\Delta H^R_{m_I}} - 1 \right)^{b'_{m_I}} \qquad (8)$$

where the calibration parameters $a$, $b$ and $a'_{m_I}$, $b'_{m_I}$ are given in Table 3, with $m_I = +1$ for the low-field line and $m_I = -1$ for the high-field line. The choice of the rigid limit parameters, $A^R_{zz}$ and $\Delta H^R_{m_I}$, is critical at the extremes of the calibration, and the values given in Table 3 are to serve only as a guide.

For intermediate rates of rotational motion ($\leq 10$ ns), the only real choice for obtaining correlation times is by spectral simulation. Programmes that are generally applicable in all rotational regimes for both isotropic and anisotropic rotation are described in Ref. (41). Calibrations for anisotropic rotation about the nitroxide

**Table 3**  Parameters for Fitting the Correlation Time Calibrations for Slow Rotational Motion Given by Eqns 7 and 8 (39, 40).

| Diffusion model | $A_{zz}^R$ (G) | $a$ ($10^{-9}$ s) | $b$ | $m_I$ | $\Delta H_{m_I}^R$ (G) | $a'_{m_I}$ ($10^{-8}$ s) | $b'_{m_I}$ |
|---|---|---|---|---|---|---|---|
| Isotropic | $\sim 32$ | 0.54 | $-1.36$ | $+1$ | 2.4 | 1.15 | $-0.943$ |
|  |  |  |  | $-1$ | 2.7 | 2.12 | $-0.778$ |
| $y$-axis | $\sim 32$ | 0.2596 | $-1.396$ |  |  |  |  |

$y$-axis which are applicable in both the slow and intermediate transitional regimes are given in Table 3 (40). The slow motional calibrations for isotropic rotation that are given in Table 3, refer principally to motion of the nitroxide $z$-axis, in the case that the rotation is anisotropic.

## C.  Saturation Transfer EPR (ST-EPR)

For rotational motions that are beyond the limits of sensitivity of conventional spin label EPR (correlation times $\sim 1$–$100\,\mu s$), saturation transfer EPR methods (33) must be used to study the rotational diffusion. The rotational correlation times, $\tau_R^{eff}$, are obtained from diagnostic lineheight ratios, $L''/L$, $C'/C$ and $H''/H$, defined for motionally sensitive intermediate positions in the low-field, central and high-field regions, respectively, relative to rotationally insensitive peaks at corresponding turning points in the spectrum (32, 33). Alternatively, the integrated intensity, $I_{ST}$, of the ST-EPR spectrum, relative to the double integrated intensity of the conventional EPR spectra, may be used (32). The latter has the advantage that it is additive in multicomponent systems. Calibrations, which are established from isotropically rotating spin-labeled haemoglobin in solutions of known viscosity, may be expressed in the following form (42):

$$\tau_R^{eff} = k_{ST}/(R_o - R) + b_{ST} \qquad (9)$$

where $R$ is the diagnostic ST-EPR parameter ($L''/L$, $C'/C$, $H''/H$ or $I_{ST}$), $R_o$ is the limiting value of $R$ for no motion, and $k_{ST}$ and $b_{ST}$ are the ST-EPR calibration constants. The values of these calibration coefficients are given in Table 4, for each of the diagnostic ST-EPR parameters. For isotropic rotational motion, $\tau_R^{eff}$ is the true correlation time by definition. For axially anisotropic rotational diffusion, it has been shown by spectral simulation that the isotropic calibration spectra may be used, but the values of $\tau_R^{eff}$ that are obtained are related to the true correlation times approximately by (44):

$$\tau_R^{eff} = \frac{2\tau_{R//}\tau_{R\perp}}{\tau_{R\perp}\sin^2\theta + \tau_{R//}(1+\cos^2\theta)} \qquad (10)$$

where $\tau_{R//}$, $\tau_{R\perp}$ are correlation times for rotation about and perpendicular to the principal rotational axis ($\tau_{R//} \ll \tau_{R\perp}$), respectively, and $\theta$ is the angle between the nitroxide $z$-axis and the principal axis of rotation. For an integral transmembrane peptide or protein, it may be assumed that $\tau_{R\perp}$ is so long that $\tau_R^{eff}$ is determined solely by $\tau_{R//}$. The latter rotational correlation time is then given by $\tau_{R//} = \frac{1}{2}\tau_R^{eff}\sin^2\theta$, the maximum value of which is $\frac{1}{2}\tau_R^{eff}$, which also corresponds to the statistically most likely orientation of the nitroxide axes relative to the rotation axis.

**Table 4** Parameters for Rotational Correlation Time Calibrations of the Diagnostic Lineheight Ratios ($R \equiv L''/L$, $H''/H$ or $C'/C$) and Normalized Integral Intensities ($R \equiv I_{ST}$) of Second-Harmonic Saturation Transfer EPR Spectra, According to Eq. 9. [Adapted from Ref. (43)]

| $R$ | Range fitted | | $R_0$ | $k_{ST}$ ($\mu$s) | $b_{ST}$ ($\mu$s) |
|---|---|---|---|---|---|
| $L''/L$ | 0.2 | 2.0 | 1.825 | 105.6 | 63.8 |
| $H''/H$ | 0.2 | 2.0 | 2.17 | 407 | 210 |
| $C'/C$ | 0.2 | 1.0 | 1.01 | 21.3 | 21.1 |
|  | −0.4 | 1.0 | 0.976 | 11.9 | 7.82 |
| $I_{ST}$ | $0.15 \cdot 10^{-2}$ | $1.0 \cdot 10^{-2}$ | $1.07 \cdot 10^{-2}$ | 0.400 | 43.6 |

## D. Spin-Spin Interactions with Paramagnetic Species

Spin-spin interactions arising from Heisenberg exchange interactions between spin labels, or between a spin label and another paramagnetic species, give rise to a line broadening of the conventional spin-label EPR spectrum that is given by (10):

$$\Delta H = \Delta H_o + (\hbar/g\beta_e)\tau_{ex}^{-1} \qquad (11)$$

where $\Delta H_o$ is the linewidth in the absence of exchange, and $\tau_{ex}^{-1}$ is the spin exchange frequency; $g$ is the electron $g$-factor, $\beta_e$ is the Bohr magneton and $\hbar$ is Planck's constant divided by $2\pi$. Heisenberg exchange requires electron orbital overlap. It is therefore of relatively short range. For different spin-labeled molecules, or for a spin label and another paramagnetic species, it can be used as a measure of their mutual accessibility. The exchange frequency is directly proportional to the concentration of the paramagnetic species, and, for mutually accessible species, it is a measure of the collision frequency between the two.

Effects on the conventional EPR linewidths are seen only for relatively strong exchange. For weak exchange interactions, or correspondingly for low concentrations of paramagnetic species, the saturation properties of the spin label EPR spectrum at high microwave powers are much more sensitive to the spin exchange. Saturation of the spectrum is

determined by the spin-lattice relaxation rate which is given by (4, 43):

$$\frac{1}{T_1} = \frac{1}{T_1^o} + k_{RL}[c_R] \qquad (12)$$

where $T_1^o$ is the value of the spin-lattice relaxation time (i.e., of $T_1$) in the absence of paramagnetic relaxant, and $[c_R]$ is the concentration of the paramagnetic relaxant. The factor $k_{RL}$ depends on the translational diffusion coefficient and the cross-section for collision in the case of Heisenberg exchange ($\tau_{ex}^{-1} = k_{RL}[c_R]$), or on the distance of closest approach in the case of magnetic dipole-dipole interactions. Saturation measurements are much more sensitive to spin-spin interactions than are linewidth measurements because $1/T_1^o \ll \Delta H_o$ (cf. Eqns 11 and 12).

## E. Progressive Saturation and Non-Linear EPR

The degree of saturation can be determined either from the variation in the double-integrated intensity of the conventional EPR spectra with progressively increasing microwave power, $P$, or from the integrated intensity, $I_{ST}$, of the ST-EPR spectrum normalized to the double-integrated intensity of the conventional first-derivative EPR spectrum (i.e., to the spin-label concentration). The second, non-linear EPR method is less conventional, but is more sensitive to saturation (4, 43). Saturation has opposite effects on the linear and non-linear EPR spectra, respectively: the intensity of the conventional EPR spectrum is reduced, relative to that in the absence of saturation at the same power, whereas the intensity of the ST-EPR spectrum is increased by saturation.

For progressive saturation experiments, the dependence of the double-integrated intensity of the conventional EPR spectrum is given by (45):

$$S(P) = \frac{S_o \sqrt{P}}{\sqrt{1 + 3P/P_{1/2}}} \qquad (13)$$

where $S_o$ is a scaling factor, and the half-saturation power, $P_{1/2}$, is determined by the condition: $S(P_{1/2}) = \frac{1}{2} S_o \sqrt{P_{1/2}}$. The value of $P_{1/2}$ that is determined from the saturation curve (see Eq. 13) is inversely proportional to the spin-lattice relaxation time, $T_1$. Therefore, the value of $\Delta P = P_{1/2} - P_{1/2}^o$, where $P_{1/2}^o$ is the value of $P_{1/2}$ in the absence of paramagnetic relaxants, may be used as an accessibility parameter or more generally as a measure of the enhancement in relaxation rate.

For non-linear EPR experiments (46), the normalized integrated intensity of the ST-EPR spectrum is approximately linearly proportional to $T_1$ (47). Therefore an accessibility parameter or relaxation enhancement can be defined for the ST-EPR experiment by $\Delta(1/I_{ST}) = 1/I_{ST} - 1/I_{ST}^o$, where $I_{ST}^o$ is the value of $I_{ST}$ in the absence of relaxation enhancement. Non-linear EPR is more sensitive to relaxation enhancement than is conventional ESR. This is because the intensity of the non-linear spectrum is directly proportional to $T_1$, whereas that of the linear spectrum (cf. Eq. 13) is approximately proportional to $T_1^{-1/2}$ (48).

## V. SPIN LABEL MEASUREMENTS ON PEPTIDES AND PROTEINS

The different types of EPR measurements that can be made on spin-labelled peptides

and proteins fall principally into three groups. The first is spectrometric quantitation of relative intensities or concentrations. The second is determination of rotational mobility, and the third is detection and quantitation of spin-spin interactions. Each of these allow characterization of several crucial properties of the spin-labeled system, as will be seen below.

### A. Binding and Labeling Kinetics

The binding of spin-labeled peptides to proteins, receptors or membranes can be studied from the mobility differences of the bound or free spin label. A good example of the method is illustrated by the binding of a spin-labeled hapten to an antibody (6). In this case, the free hapten has very sharp EPR lines in solution. This makes quantitation, by assaying depletion of the free spin label signal on binding, relatively straightforward. The free signal provides an internal standard. In more complicated cases, where the mobility difference between bound and free labels is not so large, spectral subtraction may be necessary to separate and quantitate the individual components. For all cases, absolute spin intensities may be determined by double integration of the first-derivative spectrum, relative to a standard. A specific example of the binding of spin-labeled peptides to proteins is given by the binding of insulin B-chains and of melittin to α-crystallin (49). In addition to measuring binding constants, the EPR spectra of spin labels at different sites on the peptide, in the bound form, were used to obtain information on the mode of binding. In addition, the tetrameric oligomerisation of the 26-residue amphiphilic peptide melittin at high ionic strength, in solution, has been demonstrated by the change in spin-label mobility (50). The rotational correlation times in water and higher-viscosity sucrose solutions were also correlated with the aggregate size. For isotropic rotation, the correlation time is given by (2):

$$\tau_R = \frac{\eta V}{k_B T} \tag{14}$$

where $\eta$ is the viscosity of the medium, $V$ is the volume of the rotating species, $k_B$ is Boltzmann's constant and $T$ is the absolute temperature. The rotational correlation time is therefore directly proportional to the aggregate size.

A similar method can be used for studying the kinetics of covalent binding of a spin-labeled reagent to a peptide or protein. In these cases, the free spin label will have sharp lines because the labeling reagents are of relatively small molecular size. An example of such techniques is offered by the covalent labeling of the Ca-ATPase by spin-labelled maleimide and other spin-label reagents (8, 51, 52). Stoichiometries of labeling can be determined by calibration with standard solutions of the labeling reagent. In this way, the protection of specific labeling sites on the Ca-ATPase by the binding of ATP could be demonstrated.

### B. Conformational Sensitivity

Conformational changes may be detected and quantitated (e.g. on substrate binding) from the difference in mobility of a covalently-attached spin label, in the two different conformations of the protein. This method has met with considerable success, particularly with allosteric proteins. The key to the method is to label at a site where the changes in the spin label EPR spectrum are readily detectable, without inactivation by the chemical modification. An excellent example is the glycolytic enzyme, phosphofructokinase labeled at the most reactive thiol with the iodoacetamide spin label (53–55). In this case, the two different conformations exhibit spin label spectra that are very well resolved from one another, and are exquisitely sensitive, in their relative intensities, to the binding of allosteric effectors. A further example is given by the mitochondrial enzyme aspartate aminotransferase (56).

The antibody-hapten system also affords an illustration of the use of spin-labelling to map out active-site geometries, in this case the antibody combining site (22, 23). By using spin-labeled haptens with different conformationally predictable molecular geometries and different separations between the binding epitope and the nitroxide ring it was possible to probe both the width and depth contours of the binding pocket. These can be deduced from the restriction in spin label mobility by steric interaction with the walls of the binding site. For analogous studies using covalent labels, those available with variable distances between the reactive centre and spin label ring are appropriate (24). In the antibody combining site, it is also possible to detect the presence of charged residues located in the binding crevice from the electrostatic effects on the electron spin densities that are reflected in the overall size of the hyperfine splittings.

### C. Secondary Structure in Solution

In addition to probing structural folds, spin label mobility may also reflect the secondary structure of peptides or proteins. For this, systematic spin labeling throughout the primary sequence (i.e., nitroxide scanning) is necessary. As a result of the tertiary folding, the rotational mobility of the side chains may display the periodicity that is characteristic of the particular secondary structure. For instance, an approximate 3.6-residue repeat frequency is exhibited by the spin labels attached to residues in the amphiphilic $\alpha$-helical C-terminal domain of colicin E1 (57). Similarly, a 2-residue periodicity is found in the mobility of spin-labeled side chains in the $\beta$-sheet regions of the cellular retinal binding protein (58). This method

**Fig. 2** Relative strengths of $i \rightarrow i+3$ (i.e., $\alpha$-helical) and $i \rightarrow i+4$ (i.e., $3_{10}$-helix) interactions in the sequence of alanine-based peptides that are double spin-labeled in the regions shown. The thickness of the bars indicates the relative strengths of the interactions (61).

has subsequently been refined with spin-labeled cysteine mutants of T4 lysozyme to differentiate between buried residues, tertiary interactions, helix surfaces, and loops (59). A sterically hindered MTS spin label with a 4-methyl substituent in the pyrrolinyl nitroxide ring was introduced in this latter study to obtain spin-label spectra that were sensitive principally to protein backbone dynamics.

Secondary structure of proteins/peptides in solution may also be reflected by the exposure of spin-labeled residues to paramagnetic relaxation reagents. Accessibility to paramagnetic ions, their complexes, or molecular oxygen, can be used to detect solvent-exposed spin-labeled side chains. This method is complementary to that involving side chain mobility and has been applied successfully to both T4 lysozyme and the cellular retinal binding protein (58).

Many of the above methods for determining secondary structure rely on tertiary interactions, differential burying of side chains within the tertiary fold, or binding to other complexes, for their success. For small peptides in solution, they may be less effective. In this case, a method based on Heisenberg spin-spin exchange interactions in doubly-labeled peptides is more generally applicable. This method has been introduced by Millhauser and coworkers (60, 61). The approach is relatively simple in principle and has analogies with the use of nuclear Overhauser effect enhancements in the determination of secondary structure by NMR, but has the advantage that the length scale for interaction is larger. Rankings between $i \rightarrow i+3$ and $i \rightarrow i+4$ interactions are established by measuring the spectral broadening of the conventional EPR spectra for double labeling at the two pairs of relative positions. A convenient measure of the linebroadening is the distance from the middle of the spectrum to the outside width at half-height, relative to that for the corresponding single label (61). This parameter has been calibrated in terms of the strength of the exchange interaction, which is assumed to have a steeply exponential decay with distance apart of the spins. In this way, it has been found possible to distinguish between $\alpha$-helical and $3_{10}$-helix regions of alanine-based peptides in aqueous solution (see Fig. 2).

An extension of the double-spin label method may be used to delineate conformational changes between domains in the tertiary fold of larger peptides or proteins. In this way, the hinge-bending conformational change that is induced by substrate binding in T4 lysozyme has been demonstrated in solution (62).

## VI. APPLICATION TO MEMBRANE PROTEINS AND PEPTIDES

Spin label EPR methods are especially powerful when applied to membrane systems. From a practical point of view, this is not least because they are equally applicable to opaque or highly scattering samples. Much incidental information can also be obtained from spin-labeled lipids that are not considered in detail here, but see e.g. Refs (1, 63, 64).

### A. Binding to Membranes

Membrane binding can be assayed readily for spin-labeled peptides and proteins. Some examples of soluble peptides and proteins in which there are large differences in the EPR spectra of bound and free forms are: the 25-residue basic peptide corresponding to the protein kinase C and calmodulin binding domain of the MARCKS protein (65), the 25-residue presequence of cytochrome $c$ oxidase subunit IV (66), cytochrome $c$ and apocytochrome $c$ (67), the C-terminal fragment of colicin E1 (68), melittin (50), and the diphtheria toxin T-domain (69).

### B. Conformational Changes

Conformational changes associated with the transport cycle of the Ca-ATPase have been assayed by the changes in population of the coexisting spectral components from the iodoacetamide spin-labeled protein, in response to $Ca^{2+}$ and MgATP (52, 70). Subsequently, improved spectral resolution of the components has been obtained by using a deuterated iodoacetamide spin label to remove the broadening from unresolved proton hyperfine splittings (71). The studies of the spin-labeled Ca-ATPase have also

been extended to transient measurements of the conformational changes by using laser photolysis of caged ATP with EPR detection (72). Differential accessibility of a maleimide spin label on the Ca-ATPase from opposite sides of the membrane has been probed at different points in the transport cycle by using aqueous ascorbate as a spin-label reducing agent (73).

### C. Rotational Diffusion and Oligomeric Assembly: ST-EPR

The rotational correlation times, $\tau_{R//}$, of transmembrane integral protein and peptide assemblies are in the microsecond regime because of the high effective intramembranous viscosity. The rotational diffusion therefore requires study by saturation transfer EPR. From hydrodynamic theory, the rotational correlation time of the protein in the membrane is given by (74):

$$\tau_{R//} = \frac{2}{3k_B T} \left( \frac{\eta_m V_m}{F_{R//,m}} + \frac{\eta_o V_o}{F_{R//,o}} \right) \quad (15)$$

where $V_m$ and $V_o$ are the volumes of intramembranous and extramembranous parts of the protein, respectively; $\eta_m$ and $\eta_o$ are the intramembranous and extramembranous viscosities, respectively; $k_B$ is Boltzmann's constant and $T$ is the absolute temperature. The volume terms are modified by a shape parameter, $F_{R//,i}$ ($i \equiv m,o$), which for an elliptical cross-section is given by (75):

$$F_{R//,i} = \frac{2(a_i/b_i)}{1 + (a_i/b_i)^2} \quad (16)$$

where $(a_i/b_i)$ is the ratio of the lengths of the axes of the ellipse. For a circular cylinder: $F_{R//,i} = 1$. Normally, the term in Eq. (15) that involves the aqueous viscosity ($\eta_o \approx 1$ cP) may be neglected compared with that containing the membrane viscosity ($\eta_m \sim 2.5$–5 P). The rotational correlation time measured by ST-EPR is therefore determined by the size (and shape) of the intramembranous part of the protein alone. Neglecting changes in shape, the correlation time is directly proportional to the degree of oligomerisation of the protein in the membrane: $\tau_{R//}$ $(n_{agg}) = n_{agg} \times \tau_{R//}$ $(n_{agg} = 1)$, where $n_{agg}$ is the aggregation number.

By comparing with estimates of the monomer dimensions, it was concluded that the Na,K-ATPase is probably present as a dimer or higher oligomer in the membrane (76), and cytochrome $c$ oxidase (77) and the ADP-ATP carrier (78) are probably both dimers of approximate cross-sections $10 \times 6$ nm and $4.6 \times 4$ nm, respectively, whereas rhodopsin is a monomer (79). Aggregation in the membrane was demonstrated to be induced by high concentrations of polyethyleneglycol for the Na,K-ATPase (74), and for rhodopsin in membranes of long- or short-chain phospholipids by the hydrophobic mismatch with the protein (79). Aggregates of the IsK K-channel-related peptide in the $\beta$-sheet form were also found to have long rotational correlation times in lipid membranes (80). ST-EPR rotational diffusion measurements have also been used to demonstrate that the intramembranous assembly of the Na,K-ATPase is preserved on removing the extramembranous parts of the protein by trypsinisation (81).

A complementary method for studying the state of intramembranous assembly of integral proteins and peptides is from the conventional EPR spectra of spin-labeled lipid chains (64, 82). The lipids directly associated at the intramembranous surface of the protein are resolved in the EPR spectrum, and may be quantitated by difference spectroscopy to give the stoichiometry of the protein-associated lipids. This is a direct measure of the intramembranous perimeter of the protein, which then may be compared with various possible packing motifs. The lipid stoichiometry for transmembrane $\beta$-strands is much less

than that for $\alpha$-helices and is determined by the degree of tilt of the peptide monomers (80). This lipid spin label method has also been used, in conjunction with ST-EPR of the labelled protein, to study trypsinised Na,K-ATPase and hydrophobic matching with rhodopsin that were discussed above (79, 81).

If the aqueous viscosity is increased to become more comparable to that of the membrane viscosity, e.g., by adding sucrose or glycerol, then the viscosity dependence of the rotational correlation time that is given by the second term on the right of Eq. (15) may be used to determine the size of the extramembranous portions of integral proteins. This has been done both for the Na,K-ATPase (74) and for the Ca-ATPase (83). In both cases, the results obtained from ST-EPR were consistent with those from low-resolution electron crystallography.

## D.  Transmembrane Topology and Secondary Structure: Paramagnetic Relaxants

Saturation EPR experiments with paramagnetic relaxation enhancement agents apply *par excellence* to membrane systems. This is because there are now three "compartments": aqueous, lipid, and protein, as opposed to just two for aqueous solution, which allows better spatial discrimination. Charged paramagnetic ion relaxants are confined to the aqueous phase, molecular oxygen is preferentially concentrated in the hydrophobic lipid phase of the membrane, and electroneutral paramagnetic ion complexes have a distribution between the two that possibly emphasises the interface. In addition, using spin-labelled lipids as relaxants can take advantage of the structure of the membrane to give further spatial resolution by locating the spin label at different positions in the lipid molecule.

The most direct structural information has been obtained by combining saturation studies with site-directed spin labeling, i.e. nitroxide/cysteine scanning mutagenesis (3, 84). Charged paramagnetic ion complexes (e.g. chromium oxalate) then identify those parts of the protein that are external to the membrane. Molecular oxygen identifies those side chains buried in the membrane that are on the hydrophobic surface of the protein facing the lipid. In the latter case, the paramagnetic relaxation enhancement by oxygen exhibits a periodicity characteristic of the secondary structure, as illustrated by helix E of bacteriorhodopsin (85). Spin labels facing the interior of the protein are much less affected by oxygen than are those facing the lipid. Oxygen has an increasing profile of concentration into the bilayer (86), whereas neutral paramagnetic ion complexes (e.g., Ni(II)ethylenediamine diacetate) have a decreasing concentration profile in this direction. Therefore, the ratio of the relaxation enhancements by the two, $\Delta P_{1/2}(O_2)/\Delta P_{1/2}(NiEDDA)$, is a sensitive measure of the depth that a lipid-exposed spin label is buried in the bilayer (87). This method was calibrated by using the lipid-exposed residues on helix D of bacteriorhodopsin and also spin labels on the lipid chains. The logarithm of $\Delta P_{1/2}(O_2)/\Delta P_{1/2}(NiEDDA)$ was found to have a linear dependence on the distance from the membrane surface to the centre of the membrane (87).

## E. Membrane Insertion: Paramagnetic Relaxants and Spin-Spin Interactions

Differential interactions with the variously located paramagnetic relaxants may also be used to study the degree of membrane insertion of proteins and peptides, particularly those involved in protein translocation across the membrane. This has been done, principally by using paramagnetic oxygen and ions, for apocytochrome $c$, i.e., the precursor of

cytochrome $c$ (88); for prePhoE, i.e., the precursor of the phosphate transporter, including its signal sequence (89); and for the 25-residue peptide corresponding to the presequence of cytochrome $c$ oxidase subunit IV (66). For these studies, apocytochrome $c$ was spin-labeled at Cys-14 or Cys-17, at which the haem group is subsequently attached in the holoprotein. It was further demonstrated that this N-terminal part of the protein penetrates deeply up to the centre of the membrane. This was done by using interactions with spin labels located at different positions in the lipid molecules in order to alleviate saturation of the spin label on the protein (90). Spin-labeled cytochrome $c$ was used as a control for the experiments with apocytochrome $c$, and was found to be located at the membrane surface, as expected for a peripheral membrane protein. In all cases, complementary experiments with spin-labeled lipids also gave supporting evidence for membrane penetration by the above proteins. This was detected from the specific restriction in mobility of the spin-labeled lipid chains, in a manner analogous to that found with integral transmembrane proteins (91). Spin-spin interactions between labeled peptides can also be used for studying their oligomeric assembly in the membrane, e.g. for alamethicin channel formation (92).

## F. Surface Association

Melittin, a 26-residue basic amphiphilic peptide, was spin-labeled specifically on different lysine residues, and oxygen and chromium oxalate were used as paramagnetic relaxants to determine its mode of binding to lipid membranes (93). It was found that the peptide was bound as an amphiphilic helix oriented along the membrane surface, with the hydrophobic surface pointing towards the membrane and most of the basic side chains exposed to water. Significantly, binding of this peptide does not induce specific motional restriction of the spin-labelled lipid chains (94). Another example of surface binding that has been investigated by paramagnetic relaxants and site-directed spin-labeling is the 25-residue basic MARCKS peptide which binds in an extended conformation on the membrane surface (65).

### G.  Depth Measurements: Dipole-Dipole Interactions

Finally, the depth in the membrane of a spin-labelled group can be determined from the distance-dependent magnetic dipole-dipole interactions with a fast-relaxing paramagnetic ion species in the aqueous phase. This method has been tested and calibrated with lipid chains spin-labelled at different positions in gel-phase bilayer membranes (95). For relaxation enhancement by aqueous $Ni^{2+}$ ions, the saturation transfer EPR enhancement parameter, $\Delta(1/I_{ST})$, was found to be linearly dependent on $Ni^{2+}$ ion concentration. The gradient of this concentration dependence was found to be related to the depth, $R_m$ nm, of the spin label in the membrane by:

$$\frac{d\Delta(1/I_{ST})}{d[Ni^{2+}]} =$$
$$14.8 \text{nm}^3 \cdot \left[\frac{1}{R_m^3} + \frac{1}{(R_m - d_l)^3}\right] mM^{-1}$$
$$(17)$$

where $d_l = 4.0$–$4.5$ nm is the membrane thickness. Equation (17) provides a calibration for estimating positional information of an unknown system with a single site of labeling. In general, however, it is better to establish independent calibrations with spin-labeled lipids in the membrane system of interest (5). This has been done in performing analogous progressive saturation EPR experiments on the $Ni^{2+}$-induced relaxation enhancement of spin-labeled cysteine mutants of

the 50-residue bacteriophage M13 coat protein in membranes (96).

### VII.  CONCLUSION

This chapter has dealt with the different types of EPR measurements that may be made on spin-labeled peptides and proteins, and the information that can be obtained from them. This information ranges from the structural to the dynamic. Spin-label EPR has unique strengths in the latter area, both rotational and translational mobility being amenable to measurement, over a wide dynamic range when conventional EPR is combined with saturation measurements. The spin-label EPR spectra can also be very sensitive to both intermolecular interactions and conformational changes. The very powerful combination of site-directed spin-labeling with paramagnetic relaxants and saturation measurements rounds off the repertoire of possibilities.

### REFERENCES

1.  D Marsh. Electron spin resonance: spin labels. In: E Grell, ed. Membrane Spectroscopy. Molecular Biology, Biochemistry and Biophysics. Berlin, Heidelberg, New York: Springer-Verlag, 1981, pp 51–142.
2.  D Marsh, LI Horváth. Spin-label studies of the structure and dynamics of lipids and proteins in membranes. In: AJ Hoff, ed. Advanced EPR. Applications in Biology and Biochemistry. Amsterdam: Elsevier, 1989, pp 707–752.
3.  WL Hubbell, C Altenbach. Site-directed spin-labeling of membrane proteins. In: SH White, ed. Membrane protein structure: experimental approaches. New York: Oxford University Press, 1994, pp 224–248.
4.  D Marsh. Progressive saturation and saturation transfer ESR for measuring exchange processes of spin-labelled lipids and proteins in membranes. Chem Soc Rev 22: 329–335, 1993.
5.  D Marsh, T Páli, LI Horváth. Progressive Saturation and Saturation Transfer EPR for Measuring Exchange Processes and Proximity Relations in Membranes. In: LJ

Berliner, ed. Spin Labeling: The Next Millennium. Biological Magnetic Resonance, Vol. 14. New York: Plenum Publishing Corporation, 1998, pp 23–82.

6. RA Dwek, R Jones, D Marsh, AC McLaughlin, EM Press, NC Price, AI White. Antibody-hapten interactions in solution. Phil Trans Roy Soc Lond B272: 53–74, 1975.

7. D Marsh, A Watts. ESR spin label studies. In: CG Knight, ed. Liposomes: from Physical Structure to Therapeutic Applications. Amsterdam: Elsevier/North-Holland Biomedical Press, 1981, pp 139–188.

8. D Marsh. Electron spin resonance: spin label probes. In: JC Metcalfe, TR Hesketh, eds. Techniques in Life Sciences. Techniques in Lipid and Membrane Biochemistry. Vol. B4/II, B426. Amsterdam: Elsevier, 1982, pp 1–44.

9. PF Knowles, D Marsh, HWE Rattle. Magnetic Resonance of Biomolecules. London: Wiley-Interscience, 1976.

10. D Marsh. Experimental methods in spin-label spectral analysis. In: LJ Berliner, J Reuben, eds. Spin-Labeling. Theory and Applications. Biological Magnetic Resonance, Vol. 8. New York: Plenum Publishing Corp., 1989, pp 255–303.

11. LJ Berliner, ed. Spin Labeling. Theory and Applications. Vol. I. New York: Academic Press. 1976.

12. LJ Berliner, ed. Spin Labeling. Theory and Applications. Vol. II. New York: Academic Press. 1979.

13. LJ Berliner, J Reuben, eds. Spin Labeling. Theory and Applications. Biological Magnetic Resonance, Vol. 8. New York: Plenum Press. 1989.

14. LJ Berliner, ed. Spin Labeling: The Next Millennium. Biological Magnetic Resonance, Vol. 14. New York: Plenum Publishing Corporation. 1998.

15. M Esmann, K Hideg, D Marsh. Analysis of thiol-topography in Na,K-ATPase using labelling with different maleimide nitroxide derivatives. Biochim Biophys Acta 1112: 215–225, 1992.

16. BJ Gaffney. The chemistry of spin labels. In: LJ Berliner, ed. Spin Labeling. Theory and Applications. New York: Academic Press, 1976, pp 184–238.

17. LJ Berliner, J Grunwald, HO Hankovsky, K Hideg. A novel reversible thiol-specific spin label: papan active site labeling and inhibition. Anal Biochem 119: 450–455, 1982.

18. JC Boeyens, HM McConnell. Spin-labeled hemoglobin. Proc Natl Acad Sci USA 56: 22–25, 1966.

19. M Esmann, PC Sar, K Hideg, D Marsh. Maleimide, iodoacetamide, indanedione,

and chloromercuric spin label reagents with derivatized nitroxide rings as ESR reporter groups for protein conformation and dynamics. Anal Biochem 213: 336–348, 1993.

20. A Azzi, MA Bragadin, AM Tamburro, M Santato. Site-directed spin-labeling of the mitochondrial membrane. Synthesis and utilization of the adenosine triphosphatase inhibitor (N-(2,2,6,6-tetramethyl-piperidyl-1-oxyl)-N′-(cyclohexyl)-carbodiimide). J Biol Chem 248: 5520–5526, 1973.

21. RA Dwek, JCA Knott, D Marsh, AC McLaughlin, EM Press, NC Price, AI White. Structural studies on the combining site of the myeloma protein MOPC 315. Eur J Biochem 53: 25–39, 1975.

22. BJ Sutton, P Gettins, D Givol, D Marsh, S Wain-Hobson, KJ Willan, RA Dwek. The gross architecture of an antibody-combining site as determined by spin-label mapping. Biochem J 165: 177–197, 1977.

23. KJ Willan, D Marsh, CA Sunderland, BJ Sutton, S Wain-Hobson, RA Dwek, D Givol. Comparison of the dimensions of the combining sites of the dinitrophenyl-binding immunoglobulin A myeloma proteins MOPC 315, MOPC 460 and XRPC 25 by spin-label mapping. Biochem J 165: 199–206, 1977.

24. M Esmann, K Hideg, D Marsh. Conventional and saturation transfer EPR spectroscopy of Na,K-ATPase modified with different maleimide-nitroxide derivatives. Biochim Biophys Acta 1159: 51–59, 1992.

25. M Esmann, HO Hankovszky, K Hideg, JA Pedersen, D Marsh. Vinyl ketone reagents for covalent protein modification. Nitroxide derivatives suited to rotational diffusion studies by saturation transfer electron spin resonance, using membrane-bound Na,K-ATPase as an example. Anal Biochem 189: 274–282, 1990.

26. K Hideg, OH Hankovszky. Chemistry of spin-labeled amino acids and peptides: some new mono- and bifunctionalized nitroxide free radicals. In: LJ Berliner, J Reuben, eds. Spin-Labeling. Theory and Applications. Biological Magnetic Resonance, Vol. 8. New York: Plenum Publishing Corp., 1989, pp 427–488.

27. VW Cornish, DR Benson, CA Altenbach, K Hideg, WL Hubbell, PG Schultz. Site-specific incorporation of biophysical probes into proteins. Proc Natl Acad Sci USA 91: 2910–2914, 1994.

28. OH Griffith, PC Jost. Instrumental aspects of spin labeling. In: LJ Berliner, ed. Spin Labeling. Theory and Applications. New York: Academic Press, 1976, pp 251–272.

29. JBC Findlay, D Marsh. Channel structure. In: R Ashley, ed. Ion Channels—A Practical Approach. Oxford: IRL Press, 1995, pp 241–267.

30. D Marsh. Spin label ESR spectroscopy and FTIR spectroscopy for structural/dynamic measurements on ion channels. Meth Enzymol, 294 C: 59–92, 1999.

31. MA Hemminga, PA de Jager, D Marsh, P Fajer. Standard conditions for the measurement of saturation transfer ESR spectra. J Magn Reson 59: 160–163, 1984.

32. LI Horváth, D Marsh. Analysis of multicomponent saturation transfer ESR spectra using the integral method: application to membrane systems. J Magn Reson 54: 363–373, 1983.

33. DD Thomas, LR Dalton, JS Hyde. Rotational diffusion studied by passage saturation transfer electron paramagnetic resonance. J Chem Phys 65: 3006–3024, 1976.

34. P Fajer, D Marsh. Microwave and modulation field inhomogeneities and the effect of cavity Q in saturation transfer ESR spectra. Dependence on sample size. J Magn Reson 49: 212–224, 1982.

35. W Froncisz, JS Hyde. The loop-gap resonator: a new microwave lumped circuit ESR sample structure. J Magn Reson 47: 515–521, 1982.

36. J-H Sachse, MD King, D Marsh. ESR determination of lipid diffusion coefficients at low spin-label concentrations in biological membranes, using exchange broadening, exchange narrowing, and dipole-dipole interactions. J Magn Reson 71: 385–404, 1987.

37. CR Cantor, PR Schimmel. Biophysical Chemistry. Vol. 2. New York: Freeman, 1980.

38. D Marsh. ESR probes for structure and dynamics of membranes. In: PM Bayley, RE Dale, eds. Spectroscopy and the Dynamics of Molecular Biological Systems. London: Academic Press, 1985, pp 209–238.

39. JH Freed. Theory of slowly tumbling ESR spectra for nitroxides. In: LJ Berliner, ed. Spin Labeling. Theory and Applications. New York: Academic Press Inc., 1976, pp 53–132.

40. CF Polnaszek, D Marsh, ICP Smith. Simulation of the ESR spectra of the cholestane spin probe under conditions of slow axial rotation: application to gel phase dipalmitoyl phosphatidylcholine. J Magn Reson 43: 54–64, 1981.

41. DJ Schneider, JH Freed. Calculating slow motional magnetic resonance spectra: a user's guide. In: LJ Berliner, J Reuben, eds. Spin-Labeling. Theory and Applications. Biological Magnetic Resonance, Vol. 8. New York: Plenum Publishing Corp., 1989, pp 1–76.

42. D Marsh, LI Horváth. A simple analytical treatment of the sensitivity of saturation transfer EPR spectra to slow rotational diffusion. J Magn Reson 99: 323–331, 1992.

43. D Marsh. Exchange and dipolar spin-spin interactions and rotational diffusion in saturation transfer EPR spectroscopy. Appl Magn Reson 3: 53–65, 1992.

44. BH Robinson, RL Dalton. Anisotropic rotational diffusion studied by passage saturation transfer electron paramagnetic resonance. J Chem Phys 72: 1312–1324, 1980.

45. T Páli, LI Horváth, D Marsh. Continuous-wave saturation of two-component, inhomogeneously broadened, anisotropic EPR spectra. J Magn Reson A 101: 215–219, 1993.

46. D Marsh, VA Livshits, T Páli. Non-linear, continuous-wave EPR spectroscopy and spin-lattice relaxation: spin-label EPR methods for structure and dynamics. J Chem Soc, Perkin Trans 2, 2545–2548, 1997.

47. T Páli, VA Livshits, D Marsh. Dependence of saturation-transfer EPR intensities on spin-lattice relaxation. J Magn Reson B 113: 151–159, 1996.

48. D Marsh, LI Horváth. Influence of Heisenberg spin exchange on conventional and phase-quadrature EPR lineshapes and intensities under saturation. J Magn Reson 97: 13–26, 1992.

49. ZT Farahbakhsh, QL Huang, LL Ding, C Altenbach, HJ Steinhoff, J Horwitz, WL Hubbell. Interaction of α-crystallin with spin-labeled peptides. Biochemistry 34: 509–516, 1995.

50. C Altenbach, WL Hubbell. The aggregation state of spin-labeled melittin in solution and bound to phospholipid membranes: evidence that membrane-bound melittin is monomeric. Proteins 3: 230–242, 1988.

51. P Champeil, S Buschlen-Boucly, F. Bastide, C. Gary-Bobo. Sarcoplasmic reticulum ATPase. Spin labeling detection of ligand-induced changes in the relative reactivities of certain sulfhydryl groups. J Biol Chem 253: 1179–1186, 1978.

52. C Coan, S Verjovski-Almeida, G Inesi. $Ca^{2+}$ regulation of conformational states in the transport cycle of spin-labeled sarcoplasmic reticulum ATPase. J Biol Chem 254: 2968–2974, 1979.

53. R Jones, RA Dwek, IO Walker. Conformational states of rabbit muscle phosphofructokinase investigated by a spin label probe. FEBS Lett 26: 92–96, 1972.

54. R Jones, RA Dwek, IO Walker. Spin-labelled phosphofructokinase and its

interactions with ATP and metal-ATP complexes as studied by magnetic-resonance methods. Eur J Biochem 34: 28–40, 1973.

55. R Jones, RA Dwek, IO Walker. Spin-labelled phosphofructokinase. Eur J Biochem 60: 187–198, 1975.

56. M Sterk, H Hauser, D Marsh, H Gehring. Probing conformational states of spin-labeled aspartate aminotransferase by ESR. Eur J Biochem 219: 993–1000, 1994.

57. AP Todd, J Cong, F. Levinthal, C Levinthal, WL Hubbell. Site-directed mutagenesis of colicin E1 provides specific attachment sites for spin labels whose spectra are sensitive to local conformation. Proteins 6: 294–305, 1989.

58. WL Hubbell, HS Mchaourab, C Altenbach, MA Lietzow. Watching proteins move using site-directed spin labeling. Structure 4: 779–783, 1996.

59. HS Mchaourab, MA Lietzow, K Hideg, WL Hubbell. Motion of spin-labeled side chains in T4 lysozyme. Correlation with protein structure and dynamics. Biochemistry 35: 7692–7704, 1996.

60. SM Miick, AP Todd, G Millhauser. Position-dependent local motions in spin-labeled analogues of a short α-helical peptide determined by electron spin resonance. Biochemistry 30: 9498–9503, 1991.

61. WR Fiori, SM Miick, GL Millhauser. Increasing sequence length favors α-helix over $3_{10}$-helix in alanine-based peptides: evidence for a length-dependent structural transition. Biochemistry 32: 11957–11962, 1993.

62. HS Mchaourab, KJ Oh, CJ Fang, WL Hubbell. Conformation of T4 lysozyme in solution. Hinge-bending motion and the substrate-induced conformational transition studied by site-directed spin labeling. Biochemistry 36: 307–316, 1997.

63. D. Marsh. ESR spin label studies of lipid-protein interactions. In: A Watts, JJHHM de Pont, eds. Progress in Protein-Lipid Interactions, Vol. 1, Ch. 4. Amsterdam: Elsevier, 1985, pp 143–172.

64. D Marsh, LI Horváth. Structure, dynamics and composition of the lipid-protein interface. Biochim Biophys Acta 1376: 267–296, 1998.

65. Z Qin, DS Cafiso. Membrane structure of protein kinase C and calmodulin binding domain of myristoylated alanine-rich C-kinase substrate determined by site-directed spin labeling. Biochemistry 35: 2917–2925, 1996.

66. MME Snel, AIPM de Kroon, D Marsh. Mitochondrial presequence inserts differently into membranes containing cardiolipin and phosphatidylglycerol. Biochemistry 34: 3605–3613, 1995.

67. MME Snel, B de Kruijff, D Marsh. Interaction of spin-labeled apocytochrome c and spin-labeled cytochrome c with negatively charged lipids studied by electron spin resonance. Biochemistry 33: 7146–7156, 1994.

68. Y-K Shin, C Levinthal, F Levinthal, WL Hubbell. Colicin E1 binding to membranes: time-resolved studies of spin-labeled mutants. Science 259: 960–963, 1993.

69. H Zhan, KJ Oh, Y-K Shin, WL Hubbell, RJ Collier. Interaction of the isolated transmembrane domain of diphtheria toxin with membranes. Biochemistry 34: 4856–4863, 1995.

70. C Coan, G Inesi. $Ca^{2+}$-dependent effect of ATP on spin-labeled sarcoplasmic reticulum. J Biol Chem 252: 3044–3049, 1977.

71. SM Lewis, DD Thomas. Resolved conformational states of spin-labeled Ca-ATPase during the enzymatic cycle. Biochemistry 31: 7381–7389, 1992.

72. DD Thomas, EM Ostap, CL Berger, SM Lewis, PG Fajer, JE Mahaney. Transient EPR of spin-labeled proteins. In: LJ Berliner, J Reuben, eds. EMR of Paramagnetic Molecules. Biological Magnetic Resonance, Vol. 13. New York: Plenum Press, 1993, pp 323–351.

73. Y Tonomura, MF Morales. Change in state of spin labels bound to sarcoplasmic reticulum with change in enzymic state, as deduced from ascorbate-quenching studies. Proc Natl Acad Sci USA 71: 3687–3691, 1974.

74. M Esmann, K Hideg, D Marsh. Influence of poly(ethylene glycol) and aqueous viscosity on the rotational diffusion of membranous Na,K-ATPase. Biochemistry 33: 3693–3697 1994.

75. F Jähnig. The shape of membrane protein derived from rotational diffusion. Eur Biophys J 14: 63–64, 1986.

76. M Esmann, LI Horváth, D Marsh. Saturation-transfer electron spin resonance studies on the mobility of spin-labeled sodium and potassium ion activated adenosinetriphosphatase in membranes from *Squalus acanthias*. Biochemistry 26: 8675–8683, 1987.

77. P Fajer, PF Knowles, D Marsh. Rotational motion of yeast cytochrome oxidase in phosphatidylcholine complexes studied by saturation-transfer electron spin resonance. Biochemistry 28: 5634–5643, 1989.

78. LI Horváth, A Munding, K Beyer, M Klingenberg, D Marsh. Rotational diffusion of mitochondrial ADP/ATP carrier studied

by saturation-transfer electron spin resonance. Biochemistry 28: 407–414, 1989.

79. NJP Ryba, D Marsh. Protein rotational diffusion and lipid/protein interactions in recombinants of bovine rhodopsin with saturated diacylphosphatidylcholines of different chain lengths studied by conventional and saturation transfer electron spin resonance. Biochemistry 31: 7511–7518, 1992.

80. LI Horváth, T Heimburg, P Kovachev, JBC Findlay, K Hideg, D Marsh. Integration of a K$^+$ channel-associated peptide in a lipid bilayer: conformation, lipid-protein interactions, and rotational diffusion. Biochemistry 34: 3893–3898, 1995.

81. M Esmann, SJD Karlish, L Sottrup-Jensen, D Marsh. Structural integrity of the membrane domains in extensively trypsinized Na,K-ATPase from shark rectal glands. Biochemistry 33: 8044–8050, 1994.

82. D Marsh. Stoichiometry of lipid-protein interaction and integral membrane protein structure. Eur Biophys J 26: 203–208, 1997.

83. M Török, G Jakab, A Bérczi, L. Dux, LI Horváth. Rotational mobility of Ca$^{2+}$-ATPase of sarcoplasmic reticulum in viscous media. Biochim Biophys Acta 1326: 193–200, 1997.

84. WL Hubbell, C Altenbach. Investigation of structure and dynamics in membrane proteins using site-directed spin labeling. Curr Opin Struct Biol 4: 566–573, 1994.

85. C Altenbach, T. Marti, H. Gobind Khorana, WL Hubbell. Transmembrane protein structure: spin labeling of bacteriorhodopsin mutants. Science 248: 1088–1092, 1990.

86. MME Snel, D Marsh. Accessibility of spin-labeled phospholipids in anionic and zwitterionic bilayer membranes to paramagnetic relaxation agents. Continuous wave power saturation EPR studies. Biochim Biophys Acta 1150: 155–161, 1993.

87. C Altenbach, DA Greenhalgh, H Gobind Khorana, WL Hubbell. A collision gradient method to determine the immersion depth of nitroxides in lipid bilayers: application to spin-labeled mutants of bacteriorhodopsin. Proc Natl Acad Sci USA 91: 1667–1671, 1994.

88. MME Snel, B de Kruijff, D Marsh. Membrane location of spin-labeled apocytochrome c and cytochrome c determined by paramagnetic relaxation agents. Biochemistry 33: 11150–11157, 1994.

89. RCA Keller, D ten Berge, N Nouwen, MME Snel, J Tommassen, D Marsh, B de Kruijff. Mode of insertion of the signal sequence of a bacterial precursor protein into phospholipid bilayers as revealed by cysteine-based site-directed spectroscopy. Biochemistry 35: 3063–3071, 1996.

90. MME Snel, D Marsh. Membrane location of apocytochrome c and cytochrome c determined from lipid-protein spin exchange interactions by continuous wave saturation electron spin resonance. Biophys J 67: 737–745, 1994.

91. H Görrissen, D Marsh, A Rietveld, B de Kruijff. Apocytochrome c binding to negatively charged lipid dispersions studied by spin-label electron spin resonance. Biochemistry 25: 2904–2910, 1986.

92. M Barranger-Mathys, DS Cafiso. Membrane structure of voltage-gated channel forming peptides by site-directed spin-labeling. Biochemistry 35: 498–505, 1996.

93. C Altenbach, W Froncisz, JS Hyde, WL Hubbell. Conformation of spin-labeled melittin at membrane surfaces investigated by pulse saturation recovery and continuous wave power saturation electron paramagnetic resonance. Biophys J 56: 1183–1191, 1989.

94. JH Kleinschmidt, JE Mahaney, DD Thomas, D Marsh. Interaction of bee venom melittin with zwitterionic and negatively charged phospholipid bilayers: a spin-label electron spin resonance study. Biophys J 72: 767–778, 1997.

95. T Páli, R Bartucci, LI Horváth, D Marsh. Distance measurements using paramagnetic ion-induced relaxation in the saturation transfer electron spin resonance of spin-labeled biomolecules. Application to phospholipid bilayers and interdigitated gel phases. Biophys J 61: 1595–1602, 1992.

96. D Stopar, KAJ Jansen, T Páli, D Marsh, MA Hemminga. Membrane location of spin-labeled M13 major coat protein mutants determined by paramagnetic relaxation agents. Biochemistry 36: 8261–8268, 1997.

## APPENDIX

Molecular weights of common nitroxides used in protein spin labeling.

| Label | Structure | Formula | Mol. wt. |
|---|---|---|---|
| $N$-(1-Oxyl-2,2,6,6-tetramethylpiperid-4-yl)maleimide | | $C_{13}H_{19}N_2O_3$ | 251.31 |
| $N$-[(1-Oxyl-2,2,5,5-tetramethylpyrroline-3-yl)-methyl]maleimide | | $C_{13}H_{17}N_2O_3$ | 249.29 |
| $N$-(1-Oxyl-2,2,6,6-tetramethylpiperid-4-yl)iodoacetamide | | $C_{11}H_{20}IN_2O_2$ | 339.19 |
| 1-Oxyl-2,2,6,6-tetramethyl-4-isothiocyanato-piperidine | | $C_{10}H_{17}N_2OS$ | 213.32 |
| 1-Oxyl-2,2,5,5-tetramethyl-3-isothiocyanato-methylpyrrolidine | | $C_{10}H_{17}N_2OS$ | 213.32 |
| $N$-(1-Oxyl-2,2,6,6-tetramethylpiperid-4-yl) $p$-chloromercuribenzamide | | $C_{16}H_{22}ClN_2O_2Hg$ | 510.41 |

| Label | Structure | Formula | Mol. wt. |
|-------|-----------|---------|----------|
| (1-Oxyl-2,2,5,5-tetramethylpyrroline-3-methyl)methanethio sulfonate | | $C_{10}H_{18}NO_3S_2$ | 264.40 |
| 2-[(1-oxyl-2,2,5,5-tetramethyl-3-pyrrolin-3-yl)methenyl]indane-1,3-dione | | $C_{18}H_{18}O_3N$ | 296.35 |
| 1-Oxyl-2,2,5,5-tetramethylpyrrolidine-3-carboxylic acid, *N*-Succinimide ester | | $C_{13}H_{19}N_2O_5$ | 283.30 |
| 1-Oxyl-2,2,5,5-tetramethylpyrroline-3-carboxylic acid, *N*-Succinimide ester | | $C_{13}H_{17}N_2O_5$ | 281.29 |

# 26

# UV-Visible and Fluorescence Spectroscopy of Polypeptides and Proteins

**G. M. Anantharamaiah\*, Vinod K. Mishra, Geeta Datta, and Herbert C. Cheung**
*University of Alabama at Birmingham, Birmingham, Alabama*

## I. INTRODUCTION

The physical measurement most frequently performed on polypeptides and proteins probably is the absorption of light in the ultraviolet and visible region of the spectrum. This measurement is readily performed in any biochemistry laboratory, and this technique is used for many different purposes ranging from concentration determinations to resolution of complex structural problems. Proteins and peptides show a strong absorption maximum at 190 nm, which is a $\pi$-$\pi^*$ transition involving the amide bond and is red-shifted when the solvent becomes more polar. This solvent-induced spectral shift arises largely from solvent polarizability (dielectric constant). Investigation of the absorption in this region is useful in structural studies relating to helix-coil transitions of $\alpha$-helical polypeptides. Associated with the amide bond is a much weaker, symmetry-forbidden,

n-$\pi^*$ transition observed at 210–220 nm which is not very useful since it may not be resolvable. Several amino acid side chains have electronic transitions in the region of the strong amide absorption, but these transitions are not observable in proteins because they are very weak compared to the amide $\pi$-$\pi^*$ transition. However, the transitions of the side chains of the three aromatic amino acid residues, tryptophan (Trp), tyrosine (Tyr) and phenylalanine (Phe), can be observed between 230 nm to 300 nm in the near uv region. The phenylalanine absorption observed in the 250 nm region arises from a symmetry-forbidden n-$\pi^*$ transition and has a relatively low molar absorption. Of particular interest to peptide/protein studies are the n-$\pi^*$ transitions observed at 274 nm for tyrosine and in the region of 294 nm for tryptophan. The absorption of these two residues is an order of magnitude higher than that of phenylalanine because the local symmetry of the phenol and indole side chains is much lower than the phenylalanine side chain. In contrast to the $\pi$-$\pi^*$ transition, the n-$\pi^*$ transition is blue shifted when the solvent polarity is increased. These spectral properties of the absorption bands of tyrosine and tryptophan have been used extensively to charac-

*\*Address Correspondance to: Dr. G. M. Anantharamaiah, Department of Medicine, UAB Medical Center, Birmingham, Al 35294. Tel. No. 205-934-1884. Fax 205-975-8079. Email: Ananth@uab.edu*

terize protein conformational changes, determine concentration of the peptide/protein solution, membrane-associating properties of peptides/proteins, the depth of membrane penetration of peptides/proteins and lipid-associated conformation of peptides/proteins.

Generally uv spectral characteristics of the aromatic amino acids are used to monitor changes in a protein. However, some proteins, which possess metals (metalloproteins) also, show characteristic spectra in the visible range. Thus both uv and visible spectroscopy are used to explore changes in proteins.

When a chromophore absorbs light, it is excited to its first singlet electronic state, and returns to the ground state *via* non radiative and/or radiative decay, which involves the emission of a photon. The emission (fluorescence) process is the reverse of light absorption, but this reversal of light absorption lends itself to the investigation of a number of chemical processes that are not accessible by absorption. Light absorption occurs in less than $10^{-15}$ sec, a time scale over which the molecules and their environments are virtually static. On the other hand, the lifetime of a singlet excited state is $10^{-9}$ to $10^{-8}$ sec. Over this nanosecond time window, many molecular processes can occur, including local conformational changes, solvent-cage relaxation, protonation and deprotonation, translational and rotational motions, etc. The excited singlet state has a larger dipole moment than the ground state, and this leads to a stronger dipole interaction of the excited state with aqueous solvent molecules. This solvent relaxation is radiationless and takes about $10^{-12}$ sec to complete. Thus, an equilibrium excited state of a lower energy is reached before photons are emitted. A consequence of this solvent relaxation is that the emitted photon is of lower energy than the absorbed photon and the fluorescence spectrum is red-shifted relative to the absorption

spectrum (Stokes' shift). This dipolar interaction of the excited state is also responsible for the observed red shift of the fluorescence spectrum when the fluorophore is moved from a non-polar medium to a polar medium, and the observed blue shift when it is moved from a polar solvent to a nonpolar solvent or to a lipid environment. These spectral changes are in the opposite direction to those predicted for the absorption spectrum. The red shift of the fluorescence spectrum is accompanied by a decrease in the quantum yield (intensity) due to collisional quenching of the excited state by polar solvent molecules. This quenching provides a further measure of environmental changes. The three aromatic amino acids are fluorescent, but the fluorescence of tryptophan and tyrosine are more readily detected because of their relatively large absorption coefficients and quantum yields when compared to phenylalanine. Peptides and proteins that do not contain tryptophan and tyrosine can be converted into fluorescent species either by covalent modification with fluorophores or by introducing noncovalently bound fluorescent ligands. Both the steady-state and time-resolved fluorescence measurements have been used successfully to follow the binding of ligand to proteins with either unpolarized or polarized emission and to provide information regarding global conformational changes.

A unique characteristic of fluorescence is the frequently observed quenching of the singlet state of one chromophore by the ground state of another chromophore via a mechanism other than collisional quenching. This quenching occurs when the two molecules are in close proximity and the emission dipole of the excited chromophore and the absorption dipole of the non excited chromophore are favorably oriented. This type of quenching is the origin of fluorescence resonance energy transfer (FRET) and is the result of a tightly coupled dipolar interaction between the two chromophores (1, 2). This radiationless interaction transfers part of the excited-state energy of the donor molecule to the acceptor molecule, and this transfer in effect partially deactivates the excited donor and indirectly excites the acceptor. If the acceptor is also fluorescent, its fluorescence becomes enhanced (sensitized fluorescence). Since the efficiency of a dipolar interaction is governed by an inverse sixth power of the distance between the two dipoles as well as the angle between the dipoles, FRET is very sensitive to changes in the separation between two chromophores. With many common donor-acceptor pairs, the effective transfer distance can be as large as 80 Å. This distance allows investigations of a variety of structural perturbations, including the movements of structural elements within a protein and the kinetics of protein folding or unfolding, and sensing of molecular interactions in living cells. FRET data can now be analyzed in terms of a distribution of the intersite distances (3–5) which contains additional structural information for assessment of protein dynamics. Recent developments in the studies of green fluorescent protein (GFP) have demonstrated the feasibility of introducing into the living cell a fusion protein prepared from two different mutants of GFP linked by a short calmodulin-binding peptide (6) or by calmodulin and its target peptide in tandem (7). The two GFP proteins form a donor-acceptor pair for FRET measurements in the cell and the fusion protein can serve as an intracellular calcium indicator. This type of GFP fusion protein is likely to have wide applications as FRET-based sensitive sensors of dynamically modulated protein-polypeptide and protein-protein interactions in live cells.

Some compounds like acrylamide and iodide quench the fluorescence of proteins and peptides. Fluorescence intensity is dependent on quencher concentration. The observed kinetics of the quenching assumes a bimolecular collision (hence this type of quenching is called collisional quenching) which is represented by the following equation

$$F_0/F = 1 + K_{SV} [Q] \qquad (1)$$

Where $F_0$ is the initial fluorescence, $F$ is the fluorescence in presence of the quencher, $Q$, $K_{SV}$ represents a constant, the Stern-Volmer constant, which takes into account the various rate constants of the excitation and deexcitation processes during fluorescence lifetime and $[Q]$ is the concentration of the quencher. The $K_{SV}$ value has been used to express the degree of exposure of Trp in many proteins.

The present chapter reviews the interrelationship between absorption and fluorescence and the complementary nature of the two methods in detecting environmental and global conformational changes in peptides and proteins. Several studies will be discussed to illustrate the applicability of some of the spectroscopic techniques in protein folding, peptide/protein-membrane interaction, conformation of peptides/proteins in membranes, and in peptide/protein-calmodulin interactions.

## II. APPLICATIONS

### A. UV Spectroscopy as an Analytical Tool

#### 1. Determination of Purity

The amino acids tryptophan, tyrosine and phenylalanine possess characteristic uv spectra as shown in Fig. 1. The spectral properties are retained when the amino acids are a part of a protein or a polypeptide. In denaturing solvents, the spectral properties of an amino acid in a protein/polypeptide are similar to those of the free amino acids. If a protein or poly-

**A.**

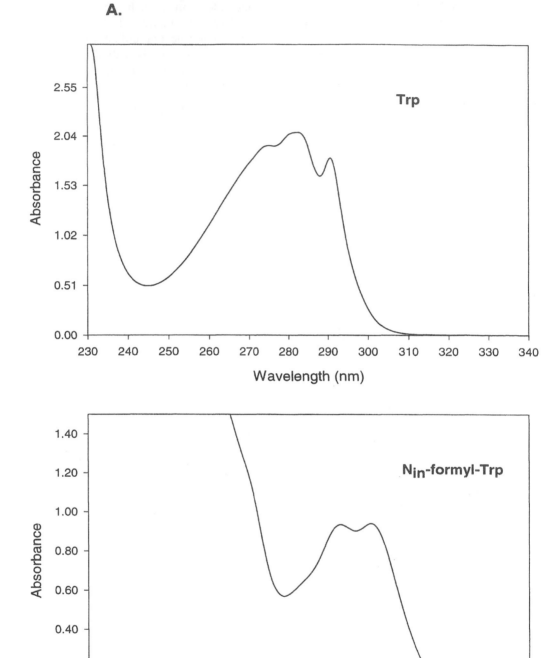

**Fig. 1**  (See legend on p. 802)

**B.**

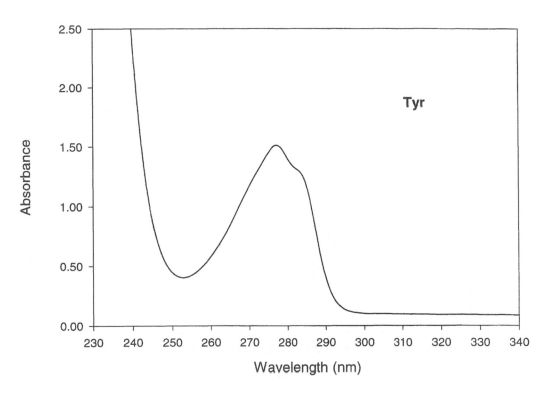

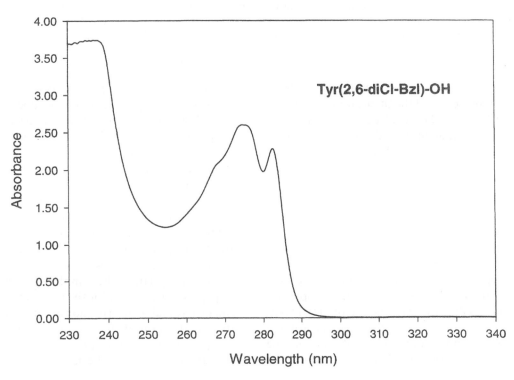

## C.

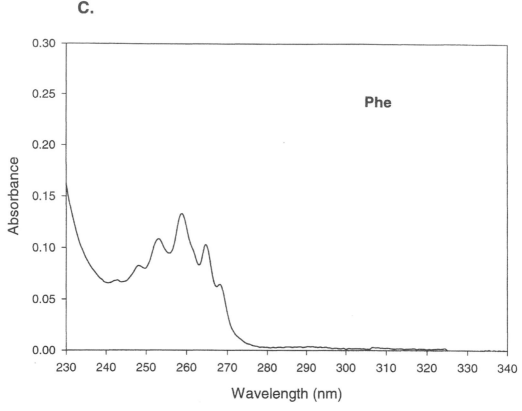

**Fig. 1** uv Absorption spectra of aromatic amino acids and their derivatives. (A) Spectra of 0.25 mM Trp (upper panel) and 0.25 mM Nin formyl Trp (lower panel). (B) Spectra of 1 mM Tyr (upper panel) and 1 mM Tyr(2,6,diCl-Bzl)-OH (lower panel) and (C) Spectrum of 1 mM Phe.

peptide contains more than one of these amino acids, all the three amino acids contribute to the spectrum of the protein, although, as discussed earlier, the contribution from phenylalanine in a protein that possesses tyrosine and/or tryptophan is not very significant. The uv spectrum of tyrosine is pH dependent. At pH 6.0 it has a maximum absorption at 275 nm whereas at pH 12.0 it absorbs maximally at 288 nm, the latter representing the ionized phenolic group. A covalent modification of either Trp or Tyr residues of a protein dramatically affects its spectral properties. This is illustrated in Fig. 1A and 1B. Modification of the indole ring of tryptophan by a formyl group, drastically changes the spectrum of Trp. Similarly, when the

hydroxyl group of Tyr is protected by a protecting group such as dichlorobenzyl (diCl-Bzl), a drastic change in the spectral property is seen (Fig. 1B). These two are common derivatives used during peptide synthesis. Similarly, any change in the phenylalanine structure caused by hydrogenation of the aromatic ring to cyclohexane etc. will be reflected in a change in the spectrum. Fig. 1C shows the spectrum of phenylalanine. Loss of the aromatic ring will result in a loss of spectral properties. Thus, in the synthesis of peptides, uv spectra of the final peptide can give valuable information about either the modification of the Tyr or Trp or the complete removal of the protecting groups that are used in the synthesis.

Tyrosine is also sensitive to the presence of halogens. For example, tyrosine can be easily brominated or iodinated to form 2,6-di-bromo (or 2,6-di-iodo) tyrosine. This also drastically changes the spectral property of the peptide or protein that possesses this amino acid. Similarly, tryptophan is susceptible to oxidizing agents. By obtaining the uv spectra of the peptides and proteins, one can assess the purity of the peptide (protein). Since microenvironments can alter the spectral properties of tyrosine (tryptophan)-containing peptides, the purity of a peptide/protein is determined by obtaining the uv spectrum in either a detergent or under denaturating conditions, so that the spectral properties remain similar to those of the individual amino acids.

## 2.  Determination of Concentration

Most proteins contain at least one of the aromatic amino acids and hence absorb in the uv range. Advantage is taken of this fact to determine the concentration of the protein/peptide. Absorption is dependent on the concentration of the protein/peptide and the Beer-Lambert law is followed.

$$A = \epsilon\, cl \qquad (2)$$

where $A$ is the absorbance at a particular wavelength, $\epsilon$ is the extiction coefficient at that wavelength, $c$ is the molar concentration of the sample and $l$ is the pathlength in cm. The extinction coefficient of a protein depends on the number of Trp or Tyr residues it has. Every aromatic residue contributes to the absorption at a particular wavelength. Thus, knowing the extinction coefficient of the protein/peptide, one can determine the concentration of the protein.

## B.  Protein Folding/Conformation

A polypeptide/protein has to fold into a native tertiary structure that is functionally active. Initially, it had been hypothesized that a protein assumes the lowest free energy state, which it reaches, by random conformational fluctuations. However, a protein folds into its native tertiary structure in a very short time in spite of the fact that there are an astronomically large number of possible conformational states (Levinthal's paradox). The tertiary structure of a protein is held together by many forces, chief of which are hydrogen bonds, salt bridges, disulfide bonds, and hydrophobic interactions. It is assumed that the $\alpha$ helices and the $\beta$ strands form first and then these intermediary structures fold into the native structure. The intermediary structure is also called the molten globule state. The folding process can be represented schematically as follows:

Random coil $\rightarrow$ molten globule $\rightarrow$ folded

Recent reports (8) have shown another "state," called the pre-molten state which comprises of native as well as non-native secondary structure.

## 1.  Characterization of the Molten Globule State

The folding of proteins has been studied by many spectroscopic techniques, including fluorescence and uv-vis spectroscopy. Although the folding of many proteins has been studied, that of $\alpha$-lactalbumin has been investigated in great detail. Fluorescence spectroscopic studies on the thermal unfolding of $\alpha$-lactalbumin (9–11) contributed to the overall characterization of the molten globule state of this protein. It has been shown to be highly heterogeneous with the $\alpha$-helical domains appearing structured while the $\beta$ sheets are not. The most significant finding about this is that it appears to have a native-like tertiary fold even in the molten globule state.

## 2. *Folding Intermediates*

Barstar, a small bacterial RNase inhibitor is another protein whose folding characteristics have been well-studied using fluorescence [intrinsic, as well as using the hydrophobic dye, 8-anilino-1-naphthalene-sulfonic acid (ANS)] and uv spectroscopy. It has been shown by Agashe et al. (12) that the polypeptide chain collapses within 4 ms to a compact globule wherein the hydrophobic core is accessible to the solvent. However, no secondary structure formation is seen at this stage. Shastry and Udgaonkar (13) have shown that the folding pathway of this protein involves three intermediate structures. Using fluorescence spectroscopy and energy transfer experiments Nath and Udgaonkar (14) have also shown that a burst phase change within 4 ms occurs in both wild type barstar and in its Trp mutants. Their kinetic data also show that the burst phase intermediate is sufficiently compact to prevent the hydration of Trp 53, a buried residue, even in the intermediate state. Such studies, using mutant proteins have contributed significantly to the knowledge of folding and folding intermediates of various proteins.

Although uv spectroscopy can be used for most proteins, the use of visible spectroscopy is limited to colored proteins. Cytochrome $c$ (cyt $c$) a protein playing an important role in the electron transport chain, has been studied using uv, visible as well as fluorescence spectroscopy. Cyt $c$ possesses a heme molecule that contributes an absorption maximum at 600 nm and one Trp residue (Trp 59) whose fluorescence has proved to be a useful parameter in studying the overall chain dimensions. The heme is ligated by coordinate bonds to different domains in the protein. Thus, changes observed in the visible region provide a sensitive measure of changes in heme coordination during folding. An important feature of the folding mechanism of cyt $c$ involves a complex interplay of various folding pathways. A unique feature of the folding process of this peripheral membrane protein is the ligation of a methionine at the 80th position to heme iron (Fe-Met80). The single tryptophan residue present at the 59th position becomes quenched during folding through energy transfer with the heme. This is a convenient parameter to monitor overall chain dimensions and heme absorbance which is a function of the oxidation, ligation and spin states of the heme iron. The fluorescence of Trp 59 in native cyt $c$ in solution is completely quenched through energy transfer with heme. However, cyt $c$, when bound to dioleoyl phosphatidylserine (DOPS) vesicles shows a substantial increase in the fluorescence with an emission maximum ($\lambda_{max}$) at 330 nm. Denaturation of this complex with urea results in a further increase in fluorescence intensity and a shift in the $\lambda_{max}$ to 345 nm. The blue shift and reduced spectral intensity of the fluorescence emission for lipid-bound protein with respect to the denatured cyt $c$ shows that Trp 59 is in a more hydrophobic environment when in lipid membranes and that there is still some energy transfer to the heme molecule. The interaction of cyt $c$ with anionic lipid vesicles of DOPS induces a disruption of the native structure of the protein. A kinetic description of cyt $c$ unfolding induced by the interaction with lipid vesicles has been studied by a series of fluorescence- and absorbance-detected stopped-flow measurements. The results show that the tightly packed native structure of cyt $c$ is disrupted at a rate of $1.5 \, s^{-1}$ independent of protein and lipid concentration. This leads to the formation of a lipid-inserted denatured state of the protein. Based on these and other related studies, a kinetic mechanism for cyt $c$ unfolding on the membrane surface has been proposed (15).

### 3. Structural Studies of Synthetic Peptide Analogs of Troponin C

Fluorescence spectroscopy gives information on conformational changes in peptides/proteins. However, it is not amenable to determining the exact conformation of a protein/peptide. Nevertheless, extrapolation of these results are helpful in understanding the conformation. An example of using this technique for determining conformational changes is the effect of $Ca^{2+}$ on troponin C. In the skeletal muscle, binding of $Ca^{2+}$ to troponin C, one of the three subunits of troponin, over the $Ca^{2+}$ concentration range of $10^{-7}$ to $10^{-8}$ M results in the activation of actinomysin ATPase. Chelation of this group of proteins to calcium has been extensively studied using fluorescence spectroscopy. These results, coupled with other methods have shown that the protein is an unusual dumb-bell shaped molecule with the binding pockets for $Ca^{2+}$ at the amino- and carboxyl termini (16). The amino acid sequences of the $Ca^{2+}$ binding loops of the four $Ca^{2+}$ sites in rabbit skeletal troponin C are shown in Fig. 2a. The amino acids are aligned to show the conservation of Gly at position 4 of the loop for sites I and II, and at position 6 for all four sites. The conserved Gly residues are shown in bold italics (underlined). The residues at positions 1, 3, 5, 7, 9, and 12 (bold) are involved in hexa-coordination with $Ca^{2+}$. The segment of sites II from residues 52 to 63 corresponds to helix C in the troponin C structure (17). The sequences of peptides are aligned with the sequence of site II of troponin C. The amino acid residue substitutions made are underlined. The majority of the amino acids here are acidic in nature and can interact with calcium. However, the role of residues like the conserved Gly was not clear. Therefore, peptides with substitutions in Gly were synthesized (Fig. 2b).

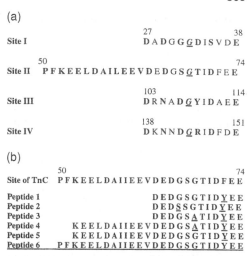

**Fig. 2** (a). Sequence alignment of four $Ca^{2+}$ binding loops of rabbit skeletal TnC (b). Synthetic peptide analogs studied to understand the mechanism of binding of cations by calcium binding site II of TnC. [Taken from Ref. (18)]

The intrinsic fluorescence of these peptides arises from Tyr at position 10 within the loop region. Fluorescence properties of peptides 3 to 6 (18) are given in Table 1. Upon excitation at 276 nm, peptides 3 and 4 showed an increase in fluorescence intensity by about 10 to 18% in the presence of a large excess of $Ca^{2+}$. $Tb^{3+}$ has been used to quench fluorescence due to Tyr. This quenching can be recovered by using $Ca^{2+}$, which displaces $Tb^{3+}$. These studies give information about complexation of peptides with $Ca^{2+}$ or $Tb^{3+}$ that takes place with change in the conformation of peptides. The quenching depends on the ability of peptide to bind to $Tb^{3+}$. The intrinsic fluorescence of peptide 3 was quenched by about 30% when a large excess of $Tb^{3+}$ was added. This suggests transfer of excitation energy from Tyr to bound $Tb^{3+}$. However, such a quenching was not detected with the other peptides. The quenched intrinsic fluorescence of peptide 3 was recovered upon subsequent addition of a large excess of

**Table 1**   Fluorescence Spectral Properties* of Synthetic Peptides

| Peptide intrinsic emission | | Normalized intensity | | | Enhancement of $Tb^{3+}$ emission by peptides |
|---|---|---|---|---|---|
| | $\lambda_{em}$ (nm) | $-Ca^{2+}$ | $+Ca^{2+}$ | $+Tb^{3+}$ | |
| Peptide 3 | 306 | 1.0 | 1.10 | 0.7 | 0.5 |
| Peptide 4 | 308 | 1.0 | 1.18 | 1.0 | 2.9 |
| Peptide 5 | 305 | 1.0 | 1.15 | 1.19 | 16.7 |
| Peptide 6 | 305 | 1.0 | 1.16 | 1.22 | 20.1 |

* The intrinsic emission intensity of each peptide was measured at its peak ($\lambda_{em}$) in the absence of added cations ($Ca^{2+}$). The intensity was taken as 1.0. The intensity determined in the presence of large excess of cations was measured at the same wavelength and normalized to the same value of apo peptide. The emission of free $Tb^{3+}$ measured at 546 nm peak with 276 nm excitation was taken as 1.0, and enhancement induced by peptides was referred to the value of $Tb^{3+}$[18].

$Ca^{2+}$. Recovery of the fluorescence suggests that bound $Tb^{3+}$ was displaced by $Ca^{2+}$. The 10% enhancement induced by $Ca^{2+}$ indicates interaction between the peptide and $Ca^{2+}$.

The possibility that these peptides interacted with $Tb^{3+}$ could be demonstrated by the changes observed in the emission spectrum of $Tb^{3+}$. The emission spectrum of $Tb^{3+}$ ($\lambda_{ex} = 276$ nm) shows four characteristic peaks at 490, 546, 586 and 621 nm. Addition of peptide 3 results in a reduction of all of the four peaks by a factor of two. In contrast, the $Tb^{3+}$ peaks were enhanced in the presence of the other peptides 4, 5 and 6. The absorption spectrum of free $Tb^{3+}$ showed a peak at 262 nm which is red shifted to 277 nm in the presence of peptide 3. A similar red shift was also observed with longer peptides. This spectral shift, which overlaps with the absorption spectrum of Tyr, and the increase in fluorescence observed in the presence of other peptides (in contrast, peptide 3 caused a reduction in emission) provide evidence for energy transfer from the Tyr residue of the longer peptides to $Tb^{3+}$. The quenching of the intrinsic fluorescence of peptide 3 by $Tb^{3+}$ and the quenching of $Tb^{3+}$ emission by peptide 3 are in contrast to the properties observed by peptides 1 and 2. $Tb^{3+}$ enhanced the intrinsic fluorescence of dodecapeptides 1 and 2. The $Tb^{3+}$ emi-

ssion was enhanced by a factor of about 10 in the presence of peptides 1 and 2. Peptide 3 differs from the other two peptides in position 6, where Gly was replaced by Ala. This suggests that the apparent lack of sensitized $Tb^{3+}$ emission by peptide 3 is due to the presence of Ala, the methyl group in Ala probably quenches the $Tb^{3+}$ emission. The sensitization through energy transfer from the Tyr residue was insufficient to overcome internal quenching. When the peptide 3 is lengthened to obtain peptide 4, spectral enhancements were observed in presence of both $Tb^{3+}$ and $Ca^{2+}$. Intrinsic fluorescence is enhanced by a factor of 3. This enhancement, however, is considerably less than the enhancement observed by peptides 1 and 2. The lower enhancement is consistent with the internal quenching and sensitized emission.

Peptides 5 and 6 produced intrinsic emission spectra similar to that of peptide 4 with respect to binding of $Ca^{2+}$. $Tb^{3+}$ induced a 20% increase in intrinsic fluorescence of peptides 5 and 6. The enhancement is consistent with the results obtained by peptides 1 and 2. The presence of Gly at position 6 in the 12-residue loop provides a 17-fold sensitization of $Tb^{3+}$ emission of peptide 5. This is considerably larger than the 3-fold sensitization induced by peptide 4 and 60% larger than that observed with peptides 1 and 2. These results reflect the absence of internal quenching of sensitized $Tb^{3+}$ emission in peptides 5 and 6, and are consistent with the expectation that interaction between the 12-residue loop and the cation is promoted by a short helix on the amino end of the binding loop. The peptide analog studies described here are an example of the use of fluorescence spectroscopy in understanding $Ca^{2+}$ induced conformational changes.

Native cardiac muscle troponin C (cTnC) contains two cysteine residues, Cys-35 and Cys-84, both located in the regulatory N-terminal domain. Cys-35 is within the 12-residue $Ca^{2+}$ binding loop of site 1. This loop differs from that of the skeletal isoform because of amino acid substitutions in critical positions and cannot chelate $Ca^{2+}$. To understand the role of this inactive loop in the activation of cardiac muscle, two monocysteine mutants, cTnC(Cys35Ser) and cTnC(Cys8-Ser), were used to probe the regulatory domain conformations of isolated cTnC and cTnC reconstituted into the ternary troponin complex with the other two subunits, cTnI and cTnT (19). The environmentally sensitive probe, IAANS (2-[4'-(iodoacetamido)anilino]-naphtahlene-6-sulfonic acid) individually attached to Cys-35 and Cys-84 were used in a variety of fluorescence studies to probe the local conformations of the two regions. With isolated cTnC, the IAANS properties suggest that the Cys-84 region of helix D is relatively restricted when the regulatory domain is in the apo state, and becomes relatively open in the $Ca^{2+}$-loaded state. In reconstituted troponin, the Cys-84 region is already in the open conformation even in the apo state. $Ca^{2+}$ binding does not appear to lead to a significant change in the Cys-84 environment. These conformational differences are shown to be due to the interaction between cTnC and cTnI within the ternary complex. In isolated cTnC, IAANS attached to Cys-35 is insenstive to $Ca^{2+}$ binding to the regulatory domain, but in the presence of cTnI the probe senses a $Ca^{2+}$-induced conformational change. This response provides evidence that the inactive loop 1 maybe involved in the $Ca^{2+}$-dependent cTnC-cTnI interaction. Since the latter interaction is obligatory in muscle activation, these results strongly suggest participation of the inactive loop 1 in the trigger mechanism of cardiac muscle.

## C.  Membrane Interactions

### 1.  *Amphipathic Helical Domains Present in Serum Apolipoproteins*

#### a.  *Fluorescence Spectroscopy in Understanding Apo A-I:lipid Interaction*

Lipoproteins are molecular complexes of lipids and apolipoproteins. The major function of lipoproteins is to transport otherwise water-insoluble substances like cholesterol, lipids, and lipid-soluble materials throughout the body. The interest in the structure and function studies of lipoproteins emanates from their intimate involvement in cardiovascular diseases, including atherosclerosis, which result from imbalances in cholesterol metabolism.

Apolipoproteins, in addition to maintaining the structural integrity of lipoprotein particles (20), also function as ligands for lipoprotein receptors and as activators or inhibitors of the enzymes involved in cholesterol metabolism (21). A detailed structural characterization of lipoprotein particles is a formidable challenge because of their structural complexity and heterogeneity. A thorough knowledge of the lipid associating properties of the apolipoproteins is a prerequisite to meet this challenge. Towards this, lipid-associating properties of isolated apolipoproteins, their fragments, and model apolipoprotein-mimetic synthetic peptides have been studied. Fluorescence spectroscopy has been widely used to understand some structural features of these very complex proteins.

Human apo A-I, the major protein component of high density lipoproteins (HDL), has been studied extensively because of its antiatherosclerotic properties (22). For a long time, structural information on this protein was difficult to obtain because of the difficulty in crystallizing this protein. However, recently, a deletion mutant has been crystallized whose x-ray structure has been determined (23). It has been shown that deletion of residues 1–43 had little effect on the secondary structure of apo A-I (24). To support the concept that the major lipid-associating domain in apo A-I lies in the C-terminal of this protein (25) the C terminal deletion mutant, $\Delta(187–243)$ apo A-I was studied using the fluorescent dye ANS. ANS has a low quantum yield in aqueous solution. When it interacts with apo A-I, it produces a 70 fold increase in quantum yield with a concomitant blue shift from 515 nm to 469 nm. When the deletion mutants $\Delta(1–43)$ apo A-I and $\Delta(187–243)$ apo A-I were made to interact with ANS, the fluorescence intensity increase was 200 fold and 40 fold over the solution intensity with a blue shift from 515 nm to 475 nm and 464 nm, respectively. These results suggest that ANS binds to a hydrophobic surface or cavity in the C terminal region of human apo A-I and $\Delta(1–43)$apo A-I. This region is absent in the C-terminal deletion mutant. These results thus support earlier studies that the C-terminal amphipathic helical domain of apo A-I possesses the major lipid-associating domain.

Apo A-I possesses multiple 22mer amphipathic helical domains, most of them punctuated by a Pro. When mixed with phospholipids, apo A-I forms discoidal complexes. The mechanism by which apo A-I forms discoidal complexes with lipids has been the subject of research in many laboratories. Two popular models

are (i) the belt model in which apo A-I is presumed to wrap around the discoidal complexes in the form of a belt and (ii) the picket fence model where the helices of apo A-I are arranged in an up and down fashion (Fig. 3). While many studies exist to indicate that the latter is the structure of apo A-I around discoidal particles, recent x-ray studies of $\Delta(1–43)$ apo A-I (23) provide a convincing argument for the arrangement of apo A-I in the belt model conformation.

There are four tryptophan residues in the apo A-I molecule (at positions 8, 50, 70 and 108). The helicity of apo A-I increases from 50% in aqueous solution to 75% in the presence of lipids such as POPC (26). This conformational change is accompanied by a blue shift in the fluorescence emission maximum from 335 nm to 332 nm. Depending on the peptide:lipid ratio, four types of apo A-I:POPC complexes are formed with varying number of apo A-I molecules: One with two apo A-I, the second set with two and three apo A-I molecules and the third set with four apo

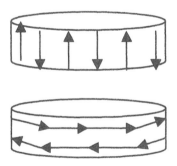

**Fig. 3** A schematic representation of a model describing the discoidal structure of apo A-I:phospholipid complex. Amphipathic helical domains of apo A-I are arranged with their helical axes parallel (upper panel) to the lipid alkyl chains. The arrows show the direction of the helices going up and down the bilayer (fence model). The lower panel shows the helical axes of the helical domains perpendicular to the alkyl chains wrapping round the disc (belt model). The arrows indicate the direction of the helices.

**Table 2** Properties of Apo A-I and Isolated Discoidal Particles of Apo A-I:POPC

| Properties | Apo A-I (in aqueous solution) | Complexes with different amounts of apo A-I | | |
|---|---|---|---|---|
| | | 2 apo A-I | 2/3 apo A-I | 4 apo A-I |
| % α-Helix | 50 | 61 | 72 | 58 |
| Trp fluorescence Polarization | 0.175 | 0.136 | 0.128 | 0.134 |
| $\lambda_{Max}$ (nm) | 335 | 332 | 334 | 332 |
| Trp accessibility to quenchers* | 0.56 | 0.88 | 0.69 | 0.84 |
| Stern-Volmer Constant* | 0.52 | 2.0 | 4.32 | 1.91 |

* Trp accessibility to quenchers was obtained by Stern-Volmer plot of quenching with potassium iodide and $Ksv$ calculated according to equation 1 according to procedure of Jonas et al. (26).

A-I molecules (27). The spectral properties of these particles are given in Table 2.

From Table 2, it is apparent that the four Trp residues are in different environments in different recombinant protein:lipid complexes. Trp residues in complexes 2/3 have an increased mobility compared to the other two particles. The fluorescence quenching values obtained using potassium iodide and the shift in fluorescence maxima indicate that the Trp residues are in the same environment in complexes 2 and 4 but significantly different in complexes 2/3. Trp residues in recombinant complexes 2 and 4 appear to be more protected from the solvent, as the quencher KI was not able to quench the fluorescence of Trp. These results thus show that the apo A-I conformation in complexes 2 and 4 is similar. Based on the differences between 2/3 particles from those of 2 and 4, the authors postulated that each class of particles possesses a unique apo A-I conformation.

*b. Class A Amphipathic α Helix: Use of Fluorescence Spectroscopy in Understanding the Role of Charged Residue Positions for Lipid Association*

The chief secondary structural motif responsible for the lipid-association of exchangeable apolipoproteins is the amphipathic α helix (28). An amphipathic α helix possesses a distinct segregation of polar and nonpolar amino acid residues on the opposite faces of the helix. In addition to the exchangeable apolipoproteins, amphipathic helices are also present in other biologically active peptides and proteins. Based upon a detailed analysis of their physical-chemical and structural properties, amphipathic helices were grouped into seven different classes (29). The amphipathic helices present in exchangeable apolipoproteins were grouped as class A (29).

In order to study the role of interfacial basic amino acid residues in lipid association two peptides were designed. The model peptide 18A possesses the amino acid sequence as shown in Fig. 4. The wheel representation of the helix shows that it has positively charged residues at the polar-nonpolar interface. Another peptide 18R possesses negatively charged residues at the polar-nonpolar interface and positively charged residues at the center of the polar face. Both the peptides have tryptophan at the same position. Thus, the two peptides possess the same hydrophobic face with tryptophan in position 2 from the N-terminus in both the peptides and the charged residue positions are reversed. In the presence of dimyristoyl phosphatidylcholine (DMPC), peptide 18A showed a

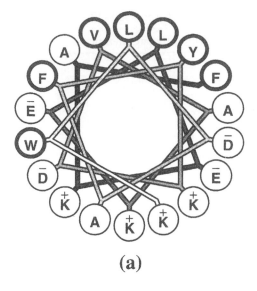

**(a)**

**(b)**

**Fig. 4** Helical wheel representation of 18R and 18A. (a) 18R has negatively charged residues at the polar-nonpolar interface, (b) 18A has positively charged residues at the polar-nonpolar interface.

blue shift in its emission maximum from 361 nm in solution to 352 nm in the presence of lipid. Peptide 18R did not show any shift in its emission maximum. The blue shift observed when the peptide 18A associates with phospholipid is an indication of tryptophan moving from an

aqueous phase to a nonpolar one. Thus, these studies show that the presence of positively charged amino acid residues at the interface play a role in determining the lipid affinity of the peptides (30–33).

To explain the increased lipid affinity of the class A amphipathic helix (34), the "snorkel model" (29) was proposed (Fig. 5). According to this model, the amphipathic basic residues located at the polar-nonpolar interface of the amphipathic helix, when associated with phospholipid, extend toward the polar face of the helix to insert their charged side chain termini into the aqueous medium for solvation (29).

Three peptide analogs were studied to experimentally test the snorkel model (35). The peptides were acetylated at the N-terminal end and amidated at the C-terminal end to remove unfavorable interaction of the helix macrodipole with the end terminal charges (36). In addition to Ac-18A-$NH_2$ and Ac-18R-$NH_2$, a peptide analog, Ac-18A(Lys > Haa)-$NH_2$, in which all the four interfacial lysine (Lys) residues in Ac-18A-$NH_2$ were replaced by homoaminoalanine (Haa) was also studied. The Haa side chain has only two methylene groups compared to four in the Lys side chain and, thus, would be useful to estimate the hydrophobic contribution of Lys side chains to the lipid affinity. All three peptide analogs have a single tryptophan residue located at the polar-nonpolar interface of the helix (Fig. 5) to serve as an intrinsic fluorophore.

A characteristic property of the amphipathic helical peptides is their ability to transform large multilamellar vesicles (MLV) and small unilamellar vesicles (SUV) into smaller discoidal peptide-lipid complexes. To this extent, these peptides mimic the properties of exchangeable apolipoproteins. These discoidal peptide-lipid complexes have a single bilayer of lipid molecules whose acyl chains are shielded from the aqueous environment by the amphipathic helical

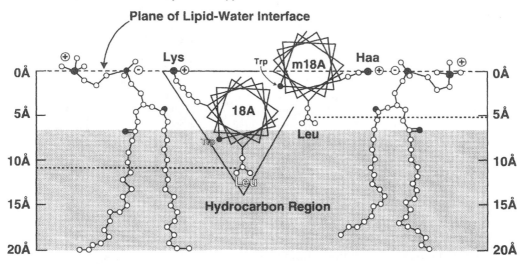

**Fig. 5** A model (Snorkel model) to explain the experimental results describing the depth of burial of peptides 18A and modified 18A (m18A, possessing homoamino alanine instead of Lys at the polar-nonpolar interface) in the lipid bilayer. 18A with Lys (four methylene carbon chain) at the polar-nonpolar interface of the amphipathic helix, is involved not only in exposing the ε-amino group of Lys to the aqueous medium, but also in imparting hydrophobicity to the hydrophobic face. Thus, the entire 18A amphipathic helix is shown as buried in the lipid bilayer. Note also, the cross section of 18A (solid inverted triangle) which is shown as a wedge shape. This wedge shape is responsible for micellizing the lipid bilayer, thus the class A amphipathic helix is involved in detergent action. The position of Trp is shown in 18A and m18A. While 18A is buried in the lipid bilayer up to 11 Å, m18A is buried in the lipid bilayer only upto 5 Å, because of the shorter homoaminoalanine side chain (two methylene carbons), thus, explaining the Trp fluorescence quenching results. The shaded area indicates the non polar environment of the lipid bilayer. The Trp in 18A is in a greater hydrophobic environment compared to Trp in m18A, which is closer to the plane of the lipid-water interface.

peptides. All the three peptides, when mixed with MLV of DMPC at a lipid to peptide molar ratio of 20:1 formed discoidal complexes. The emission spectra of these discoidal complexes showed a further blue shift in the emission maxima of the three peptides, indicating partitioning of the Trp residue in the more hydrophobic environment of the lipid bilayer in the discoidal complexes. The extent of the blue shift was different for the three peptides, with Ac-18A-NH$_2$ exhibiting the largest (emission maximum 329 nm) and Ac-18R-NH$_2$ the smallest blue shift (emission maximum 336 nm). The emission maximum of Ac-18A (Lys > Haa)-NH$_2$ in the discoidal complex was 331 nm. These results indicate that

Trp in Ac-18A-NH$_2$ is buried deeper in the lipid bilayer compared to the Trp in Ac-18R-NH$_2$. The depth of penetration of Trp in Ac-18A(Lys > Haa)-NH$_2$ is intermediate between Ac-18A-NH$_2$ and Ac-18R-NH$_2$.

The depth of penetration of the Trp in the lipid bilayer for the three peptides was further probed using the aqueous phase quenchers, iodide and acrylamide (37, 38). Iodide is negatively charged, whereas acrylamide is uncharged but polar. Acrylamide, because of the absence of a net charge, is expected to penetrate deeper in the nonpolar environment of a lipid bilayer compared to iodide. These aqueous phase quenchers can be used to probe shielding of the Trp from the aque-

ous environment as a result of its partitioning into the lipid bilayers. Analysis of the
collisional quenching using the
Stern-Volmer equation [Eq. (1)] showed
that shielding of Trp is maximum for
Ac-18A-NH$_2$ and minimum for
Ac-18R-NH$_2$ (35). These studies confirmed the "snorkel model" according to
which the presence of Lys residues at
the polar-nonpolar interface in an amphipathic helical peptide increases its lipid
affinity. The results of this study further
indicated that the larger number of
hydrophobic methylene groups in the
lysine side chain compared to that in
Haa side chain impart higher lipid affinity
to Ac-18A-NH$_2$ compared to Ac-18A
(Lys > Haa)-NH$_2$.

c.   *Optimal Length of the Helix*
     *Required for Lipid Association*
     *Studied Using Fluorescence*
     *Spectroscopy*

Exchangeable apolipoproteins possess tandem amphipathic helices, which
are often punctuated by proline (Pro) (39).
To understand the optimal arrangement
of the amphipathic helices and the role of
Pro punctuation in the lipid association,
three model class A peptides, namely 36A,
37aA, and 37pA were studied (40). The
peptide 36A is a dimer of 18A (18A-18A)
arranged in a head-to-tail manner. The
peptide 37aA has an alanine (Ala) residue
(18A-Ala-18A), whereas 37pA has a Pro
residue (18A-Pro-18A) inserted between
the two 18A helices. The helical wheel and
helical net representations of the three peptide sequences are shown in Fig. 6. It is
noteworthy that whereas in 36A both the
Trp residues are located at the polar-
nonpolar interface (Fig. 6a), in 37aA and
37pA, one Trp (Trp 2) is in the polar face
whereas the second Trp (Trp 21) is in the
nonpolar face of the helix (Figs 6b and 6c).
This differential positioning of the two
Trp residues in 37aA and 37pA results
from the fact that the hydrophobic faces

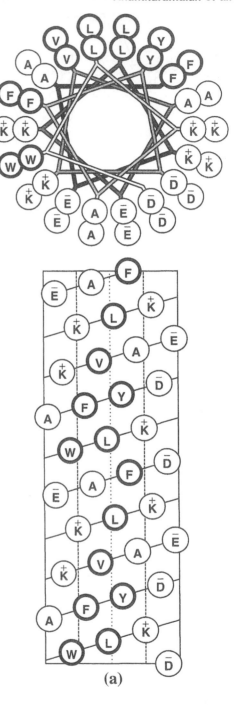

(a)

in the two amphipathic helical segments
are out of register by 100° (40).

The emission spectra of these three
peptides in PBS indicated that compared
to N-acetyltryptophanamide (NATA)

**Fig. 6** Helical wheel (top) and helical net (bottom) representations of the amino acid sequences of 36A (a), 37aA (b), and 37pA (c). Note that while in 36A hydrophobic faces of the two amphipathic helical segments are in register (a), they are out of register by 100° in 37aA (b) and 37pA (c). Also note the differences in the location of the two Trp residues in 36A (a) and in 37aA (b) and 37pA (c).

which has an emission maximum of 350nm, wavelengths of their emission maxima are blue shifted by 9 nm (36A) to 12 nm (37aA) (40). The maximum emission wavelength of the parent peptide 18A was 349 nm. The emission maxima of all the four peptides in the discoidal DMPC complexes were further blue shifted. In the discodal complex, 37aA and 37pA exhibited similar emission maximum (331 nm and 332 nm, respectively), whereas emission maximum of 36A was 337 nm. The emission maximum of 18A was blue shifted (335 nm) compared to that of 36A. The fluorescence quenching experiments with iodide and acrylamide indicated that in the 36A, 37aA, and 37pA discoidal complexes, Trp residues in 37pA are buried most whereas those in 36A are buried least. Interestingly, in the discoidal complex, the Trp residue in 18A is buried deeper than the Trp residues in 36A. The results of this study indicated that Pro punctuation in 37pA imparts a structure to the molecule, which is well suited to interact with the highly curved lipid surface in the discoidal complexes; a long straight helix like 36A is less suitable for interaction with the highly curved lipid surfaces (40).

### d. Fluorescence Properties of Model Class $A_1$, $A_2$, and Y Amphipathic α Helices

The lipid affinity of the different exchangeable apolipoproteins varies appreciably. For example, apo A-I can be displaced by apo A-II from HDL because of the higher lipid affinity of apo A-II compared to apo A-I (41). Different amphipathic helical domains in an exchangeable apolipoprotein also have different lipid affinities (42, 25). A detailed computer analysis (43) of all the class A amphipathic helices present in the exchangeable apolipoproteins revealed that the distribution of charged amino acids in the polar face is not ident-

ical (39). Even in a given exchangeable apolipoprotein, amphipathic helices differ in the distribution of charged amino acid residues in the polar face of the helix (39, 44). The computer analysis of the amphipathic helices present in exchangeable apolipoproteins revealed the presence of subclasses among the class A amphipathic helices (class $A_1$, class $A_2$, and class $A_4$) as well as other classes, namely, class Y, class G*, and a class representing amphipathic helices present in insect apolipoprotein, apolipophorin III (39). Class $A_1$, class $A_2$, and class Y amphipathic helices differ in the clustering of positive and negatively charged amino acid residues in the polar face of the helix. In class $A_1$ helices, the positive residue clusters are distributed at ±90° from the center of the nonpolar face and the separation of positive and negative residue clusters is not as exact as in the class $A_2$ helix (39). In class $A_2$ helices, the midpoints of the positive residue clusters are symmetrically distributed at ±100° from the center of the nonpolar face of the helix, and the separation between the positive and negative residue charge clusters is exact (39). Class Y helices are characterized by the presence of two negative amino acid residue clusters separating the two arms and the base of the "Y" motif formed by three positive amino acid residue clusters in the polar face of the helix (39). The distribution of class $A_1$, class $A_2$, and class Y helices among the different exchangeable apolipoproteins is as follows: class $A_1$ (apolipoproteins A-I and E); class $A_2$ (apolipoproteins A-II, C-I, C-II, and C-III); class Y (apolipoproteins A-I and A-IV) (21). The number of different classes of amphipathic helices present varies among different apolipoproteins (21). The presence of different classes of amphipathic helices has been postulated to modulate the lipid affinity and the functional properties of the different apolipoproteins (39).

To test the hypothesis that differences in the lipid affinity of exchangeable apolipoproteins are due to the presence of different classes of amphipathic helices, model amphipathic helical peptides mimicking the distribution of charged amino acid residues in the polar faces of the class $A_1$, class $A_2$, and class Y helices have been studied (45). The three model peptides studied are: Ac-18$A_1$-NH$_2$ (class $A_1$), Ac-18$A_2$-NH$_2$ (class $A_2$), and Ac-18Y-NH$_2$ (class Y). The helical wheel and helical net representations of the three peptide sequences are shown in Fig. 7. It is important to note that all the three peptides contain a single Trp residue located in the middle of the nonpolar face of the helix (Fig. 7). Such a location of Trp is better suited to study the depth of penetration of amphipathic helices in the lipid bilayer.

In PBS, the quantum yield of Ac-18$A_2$-NH$_2$ is higher than Ac-18$A_1$-NH$_2$ and Ac-18Y-NH$_2$. All the three peptides showed a blue shifted emission maximum and an increase in their quantum yield in the presence of palmitoyloleoyl phosphatidylcholine (POPC) SUV (45). Among the three peptides, Ac-18$A_2$-NH$_2$ exhibited the largest blue shift in the emission maximum and highest quantum yield in the presence of POPC SUV. The partition coefficients of the three peptides into POPC SUV were determined using fluorescence titration. The partition coefficient of Ac-18$A_2$-NH$_2$ was the highest and that of Ac-18$A_1$-NH$_2$ was the lowest. The results of acrylamide quenching in the presence of POPC SUV indicated that Trp residues in Ac-18$A_2$-NH$_2$ and Ac-18Y-NH$_2$ are more shielded from the aqueous environment than in Ac-18$A_1$-NH$_2$ (45). To determine the location of the Trp residue in the lipid bilayer, fluorescence quenching experiments using dibrominated lipids were carried out (46, 47). The results of these experiments indicated that, among the three peptides, Trp in Ac-18$A_2$-NH$_2$ is buried most while the Trp in Ac-18$A_1$-NH$_2$ is buried least in the POPC bilayer (43).

The results of this study indicated that the class $A_2$ amphipathic helix has a higher lipid affinity than class Y amphipathic helix whose lipid affinity, in turn, is higher than that of class $A_1$ amphipathic helix. It is interesting to note that this rank order of lipid affinity is identical to the rank order of the lipid affinity of the exchangeable apolipoproteins containing these classes of amphipathic helix (43). It should be noted that in the natural sequences, there is no strict adherence to the topography of these subclasses of amphipathic helices. However, the original class A motif appears to be present in most of the exchangeable apolipoproteins.

### e. Interaction of Serum Proteins with Liposomes

Liposomes have been used as a vehicle for drug delivery. A number of serum proteins are capable of associating with liposomes. Some of the examples are albumin, immunoglobulins, and fibronectin. These proteins coat the surfaces of multilamellar liposomes of phosphatidylcholine and cholesterol. Small unilamellar liposomes (SUV) composed of dioleoylphosphatidylethanolamine (DOPE) and oleic acid (OA) are stabilized by HDL, apo A-I and synthetic peptide mimics of apo A-I. The concept of vesicle-entrapped dye leakage has been used to determine the ability of peptides to associate with phospholipids.

To demonstrate the ability of apolipoproteins and synthetic peptide mimics of apo A-I to stabilize liposomes, the stability of calcein-entrapped liposomes to BSA-induced lysis has been studied using fluorescent dye leakage at an excitation wavelength of 490 nm and an emission wavelength of 520 nm. The total fluorescence intensity was obtained by adding deoxycholate. This will lyse liposomes and release the entrapped dye.

Percent (%) release of the entrapped dye is calculated using Eq. (1)

$$\% \text{ release} = F_x - F_0/F_t - F_0 \qquad (3)$$

Where $F_0$ is the fluorescence intensity of the liposomes at 1 min. i.e. before the addition of serum proteins. $F_t$ is the total fluorescence intensity of liposomes after the addition of deoxycholate. $F_x$ is the fluorescence intensity at different times before the addition of deoxycholate in the presence of stabilizer proteins (peptides). For studying the stabilization of liposomes by apolipoproteins, HDL and peptides,

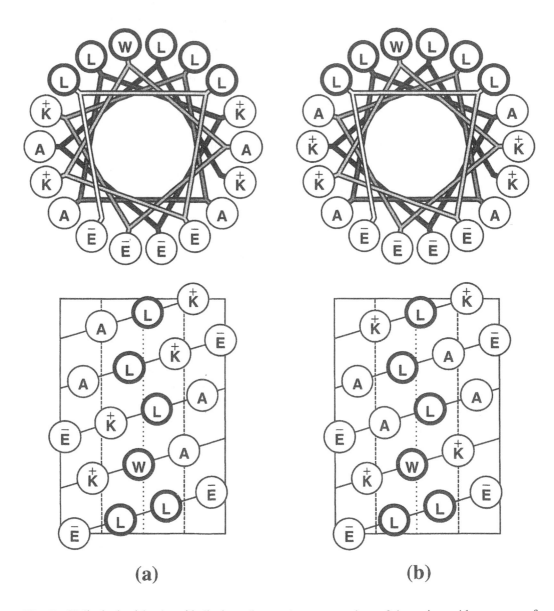

**(a)**                                                    **(b)**

**Fig. 7**   Helical wheel (top) and helical net (bottom) representations of the amino acid sequences of Ac-18A$_1$-NH$_2$ (a), Ac-18A$_2$-NH$_2$ (b), and Ac-18Y-NH$_2$ (c). Note the differences in the distribution of charged amino acids in the polar face, and the location of Trp residue in the nonpolar face, of the helices.

BSA was used to cause the lysis. Stabilizer proteins were added to the cuvette 1 min. after the addition of liposomes and before the addition of BSA. Table 3A shows that addition of peptides alone does not cause any leakage of dye. However, as shown in Table 3B, addition of BSA causes more than 50% leakage of the dye compared to deoxycholate. Table 3B also shows that addition of apo A-I or different peptides inhibit BSA-induced lysis of calcein.

The results indicate that the peptide 18A-Pro-18A is as effective as apo A-I in inhibiting the BSA-induced leakage of calcein. Table 3A shows that HDL is also inhibitory. It has been shown that the protein components of HDL and not the lipid components are inhibitory to BSA-induced vesicle lysis. Examination of the liposomes after the experiments indicated that apo A-I and the inhibitory peptides had transferred to the liposomes. Since it has been postulated that class A amphipathic helical peptides have a wedge shaped cross section, it was postulated that any wedge shaped molecule such as lyso PC could inhibit the destabilization of liposomes. Thus, these studies provided a clue to molecular design that would stabilize cell surfaces from cytotoxic agents (48).

## 2. Lytic Peptides-Membrane Interactions

Lytic peptides are cationic in nature with the positively charged residues present at the center of the polar face. They also possess a bulky hydrophobic face. This is in contrast to class A amphipathic helix described earlier, which is zwitterionic in nature. Based on the snorkel hypothesis one would expect the class L peptides to have a higher lipid binding ability as compared to class A peptides. However, the effect of the peptide on the membrane depends upon the cross sectional shape of the molecule (49). Thus, while class A peptides have a wedge cross sectional shape, class L peptides possess an inverted wedge cross sectional shape. Wedge shaped molecules are able to penetrate and stabilize vesicles. At higher peptide:lipid ratios they also act as detergents and form discoidal structures with lipids. The inverted wedge shaped molecules are able to fuse

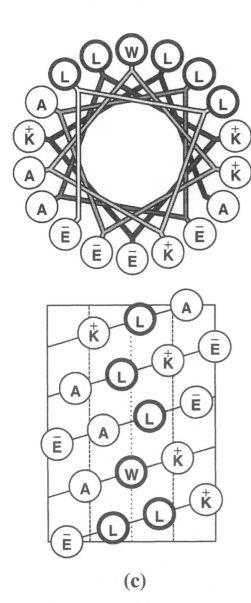

**(c)**

**Table 3A**  Effect of Serum Proteins on Liposome Stability

| Lytic agent | % leakage of calcein at different time (in min.) points | | | | | | | |
|---|---|---|---|---|---|---|---|---|
|  | 0 | 1 | 2 | 4 | 5 | 8 | 9 | 10 |
| BSA | 0 | 5 | 65 | 100 | 100 | 100 | 100 | 100 |
| HDL | 0 | 0 | 0 | 0 | 0 | 0 | 0 | 0 |
| BSA + HDL* | 0 | 0 | 2 | 5 | 8 | 10 | 11 | 12 |

*HDL added was only 1.5 $\mu$g/ml compared to BSA (50 $\mu$g/ml).

**Table 3B**  Effect of Human Apo A-I and Peptide Mimics on BSA-Induced Lysis of Liposomes*

| Peptide (mg) | % leakage with increasing amount of inhibitory peptides† | | | | | | |
|---|---|---|---|---|---|---|---|
|  | 0.025 | 0.05 | 0.075 | 0.1 | 0.15 | 0.2 | 0.3 |
| Apo A-I |  | 80 | 50 | -10 | -15 |  |  |
| 18A | 90 | 85 | 60 | 50 | 40 | 5 | -15 |
| 37pA | 80 | 20 | 10 | -15 | -15 |  |  |
| 37aA | 95 | 92 | 70 | 55 | 40 | 5 | -15 |

*Calcein-entrapped liposomes (containing 15 $\mu$g of lipid) were treated with 50 $\mu$g/ml of BSA.
  Total reaction mixture volume was maintained at 2 ml.
†0.025 to 0.3 mg of inhibitory peptides (apo A-I) are added to the reaction mixture and the dye leakage was
  followed.

membranes. An illustration of such behavior by two model peptides is given below. Kinetics and equilibrium membrane binding of a class L peptide has been compared to a class A peptide (50). A model class L peptide designed had the sequence GIKKFLGSIWKFIKAFV. The helical wheel analysis of this peptide is shown in Fig. 8. Compared to the class A peptide, this peptide possesses positively charged residues at the center of the polar face. The peptide possesses a bulky hydrophobic face. Thus, one would expect that compared to class A amphipathic helical peptide, the class L peptide would be much more effective in binding to phospholipids.

To determine the binding constant, peptides 18L and Ac-18A-NH$_2$ were allowed to interact with different types of lipid vesicles. Kinetics of binding was monitored using a stopped flow spectrophotometer. The lipids used were DOPC, DOPE, DMPG and DOPG. Changes in tryptophan fluorescence were used to monitor association of peptides with membranes.

Binding of the peptides to membranes is accompanied by a blue shift of tryptophan fluorescence emission. The spectra of membrane-bound peptides were found to be shifted compared to the spectra in n-octanol or ethyl acetate. Fluorescence spectra of 18L and Ac-18A-NH$_2$ coincided in organic solvents. However, the peptides behaved differently in the presence of DMPC, DOPC:DOPE (1:1 M/M) or DOPG. The binding isotherm for 18L with all of the lipids appeared to be linear until saturation. For Ac-18A-NH$_2$, such a pattern was observed only for acidic lipids. With zwitterionic lipids, no apparent saturation was observed. Binding constants observed from the linear part of the curves are given in Table 4.

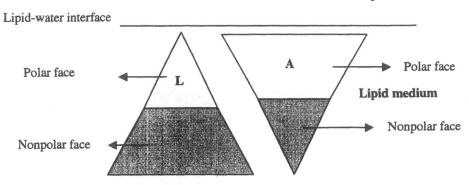

**Fig. 8** A cartoon depicting the wedge shaped Class A (A) amphipathic helix and the inverted wedge shaped Class L (L) amphipathic helix and the way they insert into the lipid medium when they interact with lipids (see Fig. 5 also). The inverted wedge shape of class L comes from the presence of Lys residues at the center of the polar face and bulky hydrophobic residues at the nonpolar face. Such structures can form pores in lipids that can form a hexaagonal phase. This effect is nullified by the wedge shaped class A molecules, as described in the figure.

*a.  Pore Formation and Vesicle Rupture*

As described above, one of the widely used fluorescent assays is the determination of peptide/protein interaction with vesicles. In the usual leakage assay, vesicles with entrapped dye (e.g. carboxyfluorescein, calcein) or a dye and quencher (ANTS/DPX) are placed in a cuvette and leakage is initiated by the injection of a peptide/protein solution into the cell with rapid stirring. Lytic peptides that form

**Table 4**  Binding Parameters for 18L and Ac-18A-NH$_2$

| Peptide | Lipid | Association constant* | Free energy** |
|---------|-------|----------------------|---------------|
| 18L | DOPC:DOPE | $1.35 \times 10^6$ | 10.7 |
|     | DOPC | $1.2 \times 10^6$ | 10.6 |
|     | DOPG | $4.5 \times 10^7$ | 12.8 |
| 18A | DOPC:DOPE | $3.2 \times 10^4$ | 8.5 |
|     | DOPC | $4.0 \times 10^4$ | 8.6 |
|     | DOPG | $3.5 \times 10^6$ | 11.3 |

* K,M$^{-1}$.

** Free energy of peptide lipid association is calculated from the association constant k, as $\Delta G =$ $-RT.L_n$ (k·55.5) $\Delta G_b$, kcal/mol.

The results obtained from these experiments indicate that the binding of 18L is higher than the binding of many of the naturally occurring lytic peptides such as mastoparan, magainin. Binding of both the peptides to anionic lipids are higher than for zwitterionic peptides. Peptide Ac-18A-NH$_2$ is able to solubilize the vesicles to form discoidal structures and thus form clear solutions with MLVs. However, 18L at higher peptide:lipid ratio, increases the trubidity of both zwitterionic and anionic LUV, which is indicative of aggregation of vesicles. Thus, although the class L peptide is able to snorkel deeper into the membrane, because of the cross sectional shape given in Figure 8, compared to class A peptide, has entirely different effect on the membrane. Fluorescence studies give a good handle in understanding of the difference in the behaviour of the two peptides in membranes.

pores of amphipathic helices were studied using the interaction of *Bacillus thuringiensis israelensis* cytotoxic toxin, a gram-positive soil bacteria. These toxins are larvicidal and cytotoxic for a wide range of cells. These toxins are composed of two major domains. One is a receptor binding domain (helix 1) and the other is a pore forming domain (helix 2). Both are amphipathic helical domains and become active only after the toxin binds to receptors. That helix 2 can penetrate phospholipid membranes and reorganize by self-assembly into the formation of a pore is shown by fluorescence energy transfer measurements between donor/acceptor helix 2 in PC vesicles. Resonance energy transfer measurements are done using NBD-labeled helix 2 (NBD-helix 2), which served as an energy donor. Rhodamine-labeled helix 2 (Rho-helix 2) is used as an energy acceptor. Addition of Rho-helix 2 to NBD helix-2 in the presence of vesicles markedly quenches the donor emission, which is due to energy transfer. Resonance energy transfer does not take place when an irrelevant Rho-labeled peptide is used as an acceptor molecule. The percentage of energy transfer could be determined at various Rho-peptide concentrations and using the binding isotherms of NBD-helix 2. A high percentage of energy transfer was obtained only when NBD-helix 2/Rho-helix 2 (1:1M/M) was made to interact. The values obtained are considerably higher than those obtained with random distribution of monomers. Such results therefore suggest that helix-2 is specifically self-associated rather than randomly distributed on the membrane. This mode of pore formation includes the following steps: 1) a fast binding step, as reflected by the rapid increase of the fluorescence of the NBD moiety covalently attached to helix 2 in the presence of vesicles. 2) insertion of helix 2 into the lipid bilayers as reflected by the blue shift of the maximum emission wavelength of NBD moiety when bound to

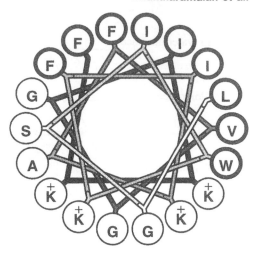

**Fig. 9** A wheel representation of a pore forming peptide, 18L. This peptide has an inverted wedge shape (see Fig. 8 for details) which interacts with lipids in such a way that it can form pores.

vesicles 3) the monomers aggregate into a barrel-like structure, in which a central aqueous pore is formed, which is surrounded by the protein moiety, which is reflected in the pattern of the isotherm and in the resonance energy transfer studies. The increase in the pore size with the further accumulation of helix 2, is again reflected in further energy transfer measurements. A typical structure of a peptide that would form such a pore is given in Fig. 9. Thus, fluorescence studies have been used to understand the interaction of class A amphipathic helical peptides or helical domains of plasma apolipoproteins which bind to phospholipid membranes in a fashion that would stabilize membranes and channel forming peptides and proteins which form self-aggregates at a particular peptide/protein concentrations that causes leakage of contents of cell membranes.

### 3.  Calmodulin Binding Peptides

Calmodulin is a major intracellular receptor for $Ca^{2+}$ and is involved in the regulation of diverse cellular functions. Normal calmodulin function is required for cell activation, proliferation and proper transmembrane signaling. Positively charged amphipathic helical sequences have been shown to be important structural motifs in the recognition of calmodulin by different calmodulin activated enzymes and peptides. Phosphodiesterase catalyzes the hydrolysis of cyclic nucleotides and it shows a basal level of activity. This activity can be increased by several fold in the presence of $Ca^{2+}$ and calmodulin. To determine whether several peptides, for example, peptides derived from gp160, the HIV-I envelope glycoprotein or mellitin, exert any effect on calmodulin-dependent phosphodiesterase activity, dansylated calmodulin has been used (51). It has been shown that dansylation of calmodulin does not alter the ability of calmodulin to activate phosphodiesterase.

Calmodulin binding to peptides is determined by studying the fluorescence emitted by dansyl calmodulin in the presence and in the absence of calmodulin binding peptides. The emission scans are usually run from 400 nm to 600 nm after excitation at 345 nm using spectrofluorometers. Dansyl-calmodulin undergoes a $Ca^{2+}$ dependent increase in base-line fluorescence. Addition of calmodulin binding peptides or mellitin causes an increase in fluorescence intensity and a blue shift in the fluorescence maxima. The gp160 peptides 768–788 and 826–854 also cause a large increase in fluorescence intensity and a blue shift. Using such studies, the binding constant, Kd, of the binding of mellitin, gp160, and the gp160 peptides, 768–788 and 826–854, has been determined to be 3 nM, 31 nM and 41 nM, respectively. In these studies, EGTA can be used to inhibit the binding of peptides to calmodulin. Although the binding constants for the two gp160 peptides are lower than mellitin, the values are similar to other calmodulin binding peptides that are noted in the review of Segrest et al. (39)

The effect of calmodulins and the peptides on phosphodiesterase activity, was studied using mant-cGMP as a fluorescent substrate. Activator deficient bovine brain phosphodiesterase and mant-cGMP are incubated alone and in the presence of 14 nm of calmodulin or calmodulin plus increasing amounts of the inhibitor peptides such as gp160 peptides 768–788 and 826–854. The fluorescence intensity of mant-cGMP is monitored in each case, at 450 nm after excitation at 280 nm. In the presence of calmodulin binding peptides or mellitin, the hydrolysis of mant-cGMP by calmodulin-stimulated phosphodiesterase is inhibited. The concentration of mellitin to inhibit 50% of phosphodiesterase activity has been determined to be 20 nM. For the gp160 peptides the concentrations for the inhibition of 50% of phosphodietserase activity have been determined to be 15 nM and 62 nM for peptides 768–788 and 826–854, respectively. Thus, unlike the binding studies with dansyl calmodulin, the gp160 peptides exhibited a comparable calmodulin binding affinity in the phosphodiesterase inhibition assay. These studies gave a clue, for the first time, to the possible connection between HIV infection and alteration of major intracellular receptors for calcium. The presence of these two domains in the cytoplasmic domain of HIVgp160, putative amphipathic helical segments that resemble calmodulin-binding domains, have been proposed to be important for calmodulin-dependent cell functions.

## III.  CONCLUSIONS

The number of publications utilizing uv, visible and fluorescence spectroscopy is vastly increasing. In this chapter, we have given examples of many applications of these methods. The reader should be aware that we have only touched the tip of the iceberg in this vast subject. With instrumentation becoming more sensitive and precise, the answers to the structural/functional questions utilizing uv-visible and fluorescence spectroscopy will be more accurate. For several proteins and peptides that do not crystallize, x-ray structure is difficult to obtain. In such cases, uv-visible spectroscopy gives a clue to the structural information. As described in this chapter, the type of applications of this technique for studying biological molecules is growing. The use of new generations of fluorescence quenchers covalently attached to different alkyl positions of lipids of different alkyl lengths will provide information on the depth of burial of peptides and proteins on membrane surface and orientation of molecules in the membranes. While the information obtained from these techniques has to be carefully interpreted, in most of the cases, the interpretations are quite straightforward. Studies of proteins with mutations placing Trp residues in various positions will provide information on the overall shape and size of the biologically active peptides and proteins. In the field of protein (peptide-)-membrane interactions, fluorescence spectroscopy has already provided precise answers with respect to the orientation of the molecule on a membrane surface, point of interaction of a molecule in the membrane, and information on the stabilization (or destabilization) effect a molecule may have on the membrane structure. It should be noted that information obtained by fluorescence spectroscopy, when used in conjunction with other physical measurements such as circular dichroism and NMR, described in other chapters of this book, will give a better overall answer to the questions asked such as peptide/protein interaction with membrane, orientation in the membrane bilayer, folding of proteins, interaction with metal ions, and protein-protein interactions. Using the information obtained from these studies, combined with the knowledge gained by other methods described in this book, one can engineer specific peptides and proteins to perform a desired physiological function, thus facilitating the development of potentially important therapeutic agents.

## REFERENCES

1. T Förster. Intermolecular energy migration and fluorescence. Ann. Physik. (Leipzig) 2: 55–75, 1947.
2. HC Cheung. Resonance energy transfer. In JR Lakowicz, ed. Topics in Fluorescence spectroscopy, 2: Principles. New York: Plenum Press, 1991, pp 127–175.
3. E Haas, M Wilchek, E Katchalski-Katzir, IZ Steinberg. Distribution of end-to-end distances of oligopeptides in solution as estimated by energy transfer. Proc Natl Acad Sci USA 72: 1807–1811, 1975.
4. D Amir, E. Haas. Estimation of intramolecular distance distributions in bovine pancreatic trypsin inhibitor by site-specific labeling and nonradiative excitation energy transfer measurements, Biochemistry 26: 2162–2175, 1987.
5. JR Lakowicz, I Gryczynski, HC Cheung, CK Wang, ML Johnson. Distance distrivbutions in native and random coil troponin I from frequency-domain measurement of fluorescence energy transfer. Biopolymers, 27: 821–930, 1988.
6. VA Romoser, PM Hinkle, A Persechini. Detection in living cells of $Ca^{2+}$-dependent changes in the fluorescence emission of an indicator composed of two green fluorescent protein varients linked by a calmodulin-binding sequence. A new class of fluorescent indicators. J Biol Chem 272: 13207–13274, 1987.
7. A Miyawaki, J Llopis, R Heim, JM McCaffrey, JA Adams, M Ikura, RY Tsien. Fluorescent indicators for $Ca^{2+}$ based on green fluorescent proteins and calmodulin. Nature 388: 822–887, 1997.
8. AF Chaffolte, JI Guijarro, YGuillou, M Delepierre, ME Goldberg. The pre-molten globule, a new intermediate in protein folding. J Protein Chem 16: 433–439, 1997.
9. AK Lala, P Kaul. Increased exposure of hydrophobic surface in the molten globule

state of α-lactalbumin. Fluorescence and hydrophobic photolabelling studies. J Biol Chem 267: 19914–18, 1992.

10. G Vanderheeren, I Hanssens. Thermal unfolding of bovine α-lactalabumin. Comparison of circular dichroism with hydrophobicity measurements. J Biol Chem 269: 7090–94, 1994.

11. K Kuwajima. The molten globule state of α-lactalbumin. FASEB J 10: 102–9, 1996.

12. VR Agashe, MC Shastry, JB Udgaonkar. Initial hydrophobic collapse in the folding of barstar. Nature 377: 754–7, 1995.

13. MC Shastry, JB Udgaonkar. The folding mechanism of barstar:evidence for multiple pathways and multiple intermediates. J Mol Biol 247: 1013–1027, 1995.

14. U Nath, JB Udgaonkar. Folding of tryptophan mutants of barstar: evidence for an initial folding pathway. Biochemistry 36: 8602–10, 1997.

15. TJT Pinheiro, GA Elove, A Watts, H Roder. Structural and kinetic description of cytochrome c unfloding induced by the interaction with lipid vesicles. Biochemistry 36: 13122–13132, 1997.

16. M Sundarlingam, R Bergstrom, G Strasburg, ST Rao. Molecular structure of troponin C from chicken skeletal muscle at 3-angstrom resolution. Science 227: 945–948, 1985.

17. O Herzberg, MNG James. Stucture of the calcium regulatory muscle protein troponin-C at 2.8 Å resolution. Nature 213: 653–659, 1985

18. NA Malik, GM Anantharamaiah, A Gawish, H-C Cheung. Structural and biological studies on synthetic peptide analogues of a low affinity calcium-binding site of skeletal troponin C. Biochim Biophys Acta 911: 221–230, 1987.

19. W-J Dong, C-K Wang, AM Gordon, HC Cheung. Disparate Fluorescence Properties of 2-[4'-(Iodoacetamido)anilino]-Naphtahlene-6-Sulfonic Acid Attached to Cys-84 and Cys-35 of troponin C in Cardiac Muscle Troponin. Biophysical J 72: 850–57, 1997.

20. P Alaupovic. Significance of apolipoproteins for structure, function, and classification of plasma lipoproteins. Method Enzymol 263: 32–60, 1996.

21. JP Segrest, DW Garber, CG Brouillette, SC Harvey, GM Anantharamaiah. The amphipathic α helix: a multifunctional structural motif in plasma apolipoproteins. Adv Prot Chem 45: 303–369, 1994.

22. DW Garber, G Datta, KR Kulkarni, P Anderson, S Yamanaka, TL Innerarity, GM Anantharamaiah. Anti-Atherogenic properties of a model amphipathic helical

peptide: Studies in Transgenic mice. Circulation 96, I490, 1997.

23. DW Borhani, DP Rogers, JF Engler, CG Brouillette. Crystal structure of truncated human apolipoprotein A-I suggests a lipid bound conformation. Proc Natl Acad Sci USA 94: 12291–12296, 1997.

24. DP Rogers, LM Roberts, JA Lebowitz, JA Engler, CG Brouillette. Structural analysis of apolipoprotein A-I. Biochemistry 37: 945–955, 1998.

25. MN Palgunachari VK Mishra, S Lund Katz, MC Phillips, SO Adeyeye, S Alluri, GM Anantharamaiah, JP Segrest. Only the two end helixes of eight tandem amphipathic helical domains of human apo A-I have significant lipid affinity. Implications for HDL assembly. Arterisclerosis, Thrombosis & Vascular Biol 16: 328–338, 1996.

26. A Jonas, KE Kezdy, JH Wald. Defined apolipoprotein A-I conformation in reconstituted high density lipoproteins. J Biol Chem 264: 4818–4824, 1989.

27. CG Brouillette, JL Jones, H Kercert, TC Ng, JP Segrest. Structure of the high density lipoproteins: Studies of apo A-I:PC recombinants by high field proton NMR, gradient gel electrophoresis and electron microscopy. Biochemistry 23: 359–367, 1984.

28. JP Segrest, RL Jackson, JD Morrisett, AM Gotto, Jr. A molecular theory of lipid-protein interactions in the plasma lipoproteins. FEBS Lett 38: 247–253, 1974.

29. JP Segrest, H De Loof, JG Dohlman, C. G. Brouillette, GM Anantharamaiah. Amphipathic helix motif: classes and properties. Proteins: Struct Funct Genet 8: 103–117, 1990.

30. GM Anantharamaiah, JL Jones, CG Brouillette, CF Schmidt, BH Chung, TA Hughes, AS Bhown, JP Segrest. Studies of synthetic peptide analogs of the amphipathic helix. Structure of complexes with dimyristoyl phosphatidylcholine. J Biol Chem 260: 10248–10255, 1985.

31. BH Chung, GM Anantharamaiah, CG Brouillette, T Nishida, JP Segrest. Studies of synthetic peptide analogs of the amphipathic helix. Correlation of structure with function. J Biol Chem 260: 10256–10262, 1985.

32. GM Anantharamaiah. Synthetic peptide analogs of apolipoproteins. Method Enzymol 128: 628–648, 1986.

33. RM Epand, A Gawish, M Iqbal, KB Gupta, CH Chen, JP Segrest, GM Anantharamaiah. Studies of synthetic peptide analogs of the amphipathic helix. J Biol Chem 162: 9389–9396, 1987.

34. YV Venkatachalapathi, KB Gupta, H De Loof, JP Segrest, GM Anantharamaiah. Positively charged residues, because of their amphipathic nature, can increase the lipid affinity of the amphipathic helix. In: J Rivier, ed. Peptides: Chemistry and Biology. Leiden: ESCOM Press, 1990, pp 672–673.

35. VK Mishra, MN Palgunachari, JP Segrest, GM Anantharamaiah. Interactions of synthetic peptide analogs of the class A amphipathic helix with lipids. Evidence for the snorkel hypothesis. J Biol Chem 269: 7185–7191, 1994.

36. YV Venkatachalapathi, MC Phillips, RM Epand, RF Epand, EM Tytler, JP Segrest, GM Anantharamaiah. Effect of end group blockage on the properties of a class A amphipathic helical peptide. Proteins: Struct Funct Genet 15: 349–359, 1993.

37. SS Lehrer, PC Leavis. Solute quenching of protein fluorescence. Method Enzymol 49: 22–236, 1978.

38. MR Eftink, CA Ghiron. Fluorescence quenching studies with proteins. Anal Biochem 114: 199–227, 1981.

39. JP Segrest, MK Jones, H De Loof, CG Brouillette, YV Venkatachalapathi, GM Anantharamaiah. The amphipathic helix in the exchangeable apolipoproteins: a review of secondary structure and function. J Lipid Res 33: 141–166, 1992.

40. VK Mishra, MN Palgunachari, S Lund-Katz, MC Phillips, JP Segrest, GM Anantharamaiah. Effect of the arrangement of tandem repeating units of class A amphipathic a-helixes on lipid interaction. J Biol Chem 270: 1602–1611, 1995.

41. A Jonas. Lipid-binding properties of apolipoproteins. In: M Rosseneu, ed. Structure and Function of Apolipoproteins. Boca Raton, FL: CRC Press, 1992, pp 217–250.

42. JT Sparrow, AM Gotto, Jr. Phospholipid binding studies with synthetic apolipoprotein fragments. Ann NY Acad Sci 348: 187–211, 1980.

43. MK Jones, GM Anantharamaiah, JP Segrest. Computer programs to identify and classify amphipathic $\alpha$ helical domains. J Lipid Res 33: 287–296, 1992.

44. GM Anantharamaiah, MK Jones, JP Segrest. An atlas of the amphipathic helical domains of human exchangeable plasma apolipoproteins. In: RM Epand, ed. The Amphipathic Helix. Boca Raton, FL: CRC Press, 1993, pp 109–142.

45. VK Mishra, MN Palgunachari. Interaction of model class $A_1$, class $A_2$, and class Y amphipathic helical peptides with membranes. Biochemistry 35: 11210–11220, 1996.

46. T Markello, A Zlotnick, J Everett, J Tennyson, PW Holloway. Determination of the topography of cytochrome b5 in lipid vesicles by fluorescence quenching. Biochemistry 24: 2895–2901, 1985.

47. FS Abrams, E London. Calibration of the parallax fluorescence quenching method for determiantion of membrane penetration depth: refinement and comparison of quenching by spin-labeled and brominated lipids. Biochemistry 31: 5312–5322, 1992.

48. D Liu, L Huang, M Moore, GM Anantharamaiah, JP Segrest. Interactions of serum proteins with small unilamellar liposomes composed of dioleoylphosphatidylethanolamine and oleic acid: High density lipoprotein, apolipoprotein AI, and amphipathic peptides stabilize liposomes. Biochemistry 29: 3637–3643, 1990.

49. EM Tytler, GM Anantharamaiah, DE Walker, VK Mishra, MN Palgunachari, JP Segrest. Molecular basis for prokaryotic specificity of magainin-induced lysis. Biochemistry 34: 4393–4401, 1995.

50. IV Polozov, AI Polozova, VK Mishra, GM Anantharamaiah, JP Segrest, GM Anantharamaiah, RM Epand. Studies of kinetics and equilibrium membrane binding of class A and class L model amphipathic helical peptides. Biochim Biophys Acta 1368: 343–354, 1998.

51. S Srinivas, RV Srinivas, GM Anantharamaiah, RW Compans, JP Segrest, Cytosolic Domain of the Human Immunodeficiency Virus Envelope Glycoproteins Binds to Calmodulin and Inhibits Calmodulin-regulated Proteins. J Biol Chem 268: 22895–22899, 1993.

# 27

# Ultracentrifugation, Light-, X-ray- and Neutron-Scattering of Peptides, Proteins, and Nucleic Acids in Solution

**Henryk Eisenberg***
*National Institutes of Health, Bethesda, Maryland*

## I. INTRODUCTION

Science in the 20th century, including the biological sciences, has moved at a tremendous pace. Peptides are relatively low-mass molecular weight chains, however they clearly exceed the mass of the amino-acid biochemical building blocks and manifold low-molecular weight chemical compounds playing roles in the biological domain. Proteins and their complexes may assume significant proportions, the nature of which had not been unequivocally recognized into the third decade of our century, slowing down the universal acceptance of the existence of large covalently bonded macromoleculer structures earlier proposed by protein chemists (1). Osmotic pressure is a colligative process useful for the determination, as will be described below, of the number average molar mass of

moderate size particles which do not penetrate through the pores of semipermeable membranes, permeable to solvent and small cosolvent molecules. It is a result (2, 3) of the thermodynamic equilibrium of pressure $P$, temperature $T$ and the chemical potentials $\mu_i$ of the permeable components and constitutes the basis of the methods to be described in this work. However osmotic pressure determination becomes insensitive with increasing molar mass even though progress has been achieved in the design of membranes and the precise determination of small pressure differences.

For the study of size distribution in colloidal systems The Svedberg proceeded to study by optical means the variation with height of the concentration of colloidal particles in a sedimenting system in a centrifugal field (4). Solute particles of large size, heavier than the medium in which they are suspended, would sink to the bottom of the sector-shaped optical cell were it not for the disordering motion due to their thermal energy, discovered by the botanist Robert Brown in 1827. Depending on the buoyant weight of the particles (and the rotational velocity of

*Current affiliation*: Structural Biology Department, Weizmann Institute of Science, Rehovot 76100 Israel. E-mail: BPEISENB @WEIZMANN.WEIZMANN.AC.IL

the ultracentrifuge rotor) they settle with a velocity measurable by the optical system. Alternatively, at lower rotational velocities, conditions could be such that an equilibrium is established, distributing particles in the field, somewhat like the decreasing concentration of air in the atmosphere with increasing height. Svedberg developed suitable equations to derive the correct sedimentation velocity and diffusion coefficients by considering frictionally impeded sedimentation and diffusion, as well as obtaining the molar mass from the ratio of these coefficients, the well-known Svedberg equation. We shall see below that basic equations for sedimentation equilibrium of multi-component systems can be derived on thermodynamic principles relating to the osmotic pressure, not involving hydrodynamic friction considerations (2). Svedberg and Fähreus (5) successfully sedimented haemoglobin to equilibrium determining a correct molar mass of 67,000, which was four times the value of the peptide subunit, 16,700, expected from chemical analysis of the iron content. Moreover, they reached the surprising conclusion that the protein was uniformly sized, in contra-distinction to man-made gold colloids. For a full understanding of mono- or poly-dispersity or complex interactions between species or with solvents, additional theoretical insight and considerable experimental upgrading were required. Of the many interesting studies of Svedberg and collaborators one stood out in particular, namely the study of haemocyanin from the vineyard snail *Helix pomatia* which, according to its copper content, should have had a minimum particle mass of 15,000–17,000. However it sedimented quickly with knife-sharp boundary, disclosing an absolutely uniformly-sized particle with a molar mass in the millions (6). Interestingly, when dynamic light scattering came in use much later (7), haemocyanin was used as a first demonstration because of its huge

and uniform size. The development and early use of the analytical ultracentrifuge have been described in a classical text (8).

Intense activity in analytical ultracentrifugation and progress in the experimental and theoretical aspects of the field were summarized in due time (9, 10). Major experimental aspects included the design of electrically driven and magnetically suspended ultracentrifuge rotors, design of rotors and cells, optical—Schlieren, Lamm scale, interference and light absorption—methods, analysis of boundaries and systems of reversibly interacting components. Gilbert (11) showed that even from a qualitative analysis of sedimenting boundaries of concentration gradients in mixed interacting systems, it was possible to derive conclusions with respect to interactions. Yet the complex nature of the flow, the necessity to disregard diffusion, the imperfect knowledge of the frictional coefficients, and other simplifying assumptions, made a quantitative analysis extremely difficult (12). Vinograd (13) pioneered the analysis of sedimentation equilibrium in a buoyant density gradient, an approach which convincingly led (14) to the establishment of the rules of semiconservative DNA replication in *Escherichia coli* by $^{14}N/^{15}N$ isotope substitution.

Analytical ultracentrifugation reached a peak in activity and usefulness in the seventies, in the determination of molar masses, subunit structures and the study of interacting systems. Its usefulness declined when sequencing and cloning methods became available and the primary structure of both proteins and nucleic acids could be precisely determined on a molecular level. Gel electrophoresis (15) in a matrix of polyacrylamide or agarose, became the standard biochemical and molecular biological procedure for determining the molar mass of minute amounts of proteins and nucleic acids, and for separating complex mixtures. Gel electrophoresis procedures, however useful, rely on empirical calibrations and anomalous results defying standard calibrations may be unwittingly obtained for materials of unknown structure. Equilibrium sedimentation on the other hand is an absolute method based on classical Gibbs thermodynamics. At sedimentation equilibrium the chemical potentials of all components in a multicomponent system must be constant all along the sedimentation radius. This allows exact treatment of multicomponent systems in ultracentrifugation and in the evaluation of the scattering of light, x-rays and neutrons, to be discussed below, resulting from fluctuations under strict thermodynamic constraints (2, 3, 16, 17). These works should be consulted for details and references not given in the present contribution.

New technologies in the construction of analytical ultracentrifuges and refined computer analysis, as well as use of precisely known molar masses, whenever obtained from molecular sequencing or mass spectrometry, presently allow the use of analytical ultracentrifugation and/or the appropriate scattering methods, yielding complementary results, in the study of hydration and complex cosolvent interactions (18), high level interactions between proteins (19), protein-carbohydrate, and protein nucleic acid complexes (20), as well as other systems of biological relevance.

## II. THERMODYNAMICS OF MULTICOMPONENT SYSTEMS

### A. Osmotic Pressure (OP)

Osmotic pressure is a basic manifestation of biological cellular structure and function. In the laboratory it involves solutions separated by "semipermeable" organic membranes, i.e. permeable to the "solvent," component 1, to the "cosolvent," component 3, but impermeable to the "macromolecular" component 2. In practice the number of components is not restricted, however we limit ourselves here to three components to maintain a simpler discussion. In biological systems comp. 1 is usually water, comp. 2 may be a polypeptide, a protein or a nucleic acid, comp. 3 a low molecular weight salt or sugar. An equally satisfactory semipermeable "membrane," however, is afforded by the air space (or vacuum) in a closed desiccator in an "isopiestic" experiment. This membrane is permeable only to the volatile solvent, which is allowed to equilibrate between a container holding all three components and another container enclosing a suitable reference solution containing comp. 1 and the nonvolatile comp. 3 only. Distillation of the solvent, at constant temperature T of the two containers, proceeds until the vapor pressures of both solutions are identical (distillation would continue to dryness if the reference solution would be pure water). A more general statement for the equilibrium condition in osmotic pressure analysis, following the seminal studies of George Scatchard, requires that the chemical potentials $\mu$ of all components (taken in neutral combinations in the case of ionic species) permeable through the membrane, be identical in contiguous phases. A simple osmometer is shown in Fig. 1. In the simplest case of a neutral polymeric solute comp. 2, which cannot pass through the pores of the semipermeable membrane, there is only solvent, comp. 1, on the outside of the compartment enclosed by the membrane, but both components are present inside this compartment. Comp. 1 is free to move across the membrane. What happens when the osmometer compartment is introduced into the solvent beaker is easily predictable on thermodynamic grounds. The chemical potential $\mu_1$ of the solvent is lowered in the inner compartment because of the presence of the dissolved comp. 2. Therefore comp. 1 tends to flow from the outer into the inner compartment, from a higher

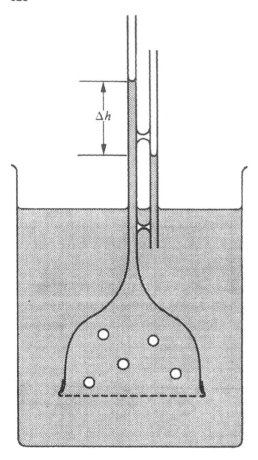

**Fig. 1** Schematic representation of simple osmometer. [From Ref. (63)]

pressure corresponding to this equilibrium state is the OP $\Pi$ equal to

$$\Pi = \Delta h \rho g \quad (\text{dyne/cm}^3), \qquad (2)$$

where $\Delta h$ is the liquid level difference, $\rho$ is the density and $g$ is the gravitational constant. In analogy to dilute gaseous systems it is possible to expand the OP in a virial series in powers of the number density $\rho_2$ of the "nondiffusible" comp. 2 per unit volume of solution

$$\Pi/\rho_2 k T = 1 + \mathcal{B}_2 \rho_2 + \mathcal{B}_3 \rho_2^2 + \dots, \quad (3)$$

where $k$ is Boltzmann's constant, and $\mathcal{B}_2$ and $\mathcal{B}_3$ are "virial" coefficients. An important point to note is that neither OP, nor the other major techniques which we plan to discuss, equilibrium sedimentation, light-, x-ray- and neutron scattering, determine molar masses in a direct way. All of these methods count particles and determine weighted averages of the particle count. Molecular "weight," or rather molar masses $M$ in correct terminology, are obtained because we express particle density in terms of an experimentally measurable concentration unit related to particle mass. Thus $c_2 = \rho_2 M_2/N_A$ (g/ml), where $N_A$ is Avogadro's number. Eq. (3) is now rewritten in the familiar form

$$\Pi/c_2 RT = M_2^{-1} + A_2 c_2 + \dots, \qquad (4)$$

where $R = N_A k$ is the gas constant and $A_2 = \mathcal{B}_2 N_A/M_2^2$ (ml mole/g$^2$) is the second virial coefficient in the units usually used (Avoid being confused by the conventional fact that subscripts 2 and 3 refer to either component denomination, or to the order of the virial coefficients). The intercept in the plot $\Pi/c_2 RT$ against $c_2$ yields the reciprocal of $M_2$ and the initial linear slope of the plot is equal to $A_2$ and expresses particle-particle interactions. When $c \to 0$ the osmotic pressure $\Pi$ is proportional to $c/M$ (mole/ml), and $M$ (g/mole) is expressed in terms of g of dry material weighed into the solvent (or in terms of some secondary experimental parameter which relates an optical absorp-

to a lower chemical potential. This process would go on indefinitely were it not for the fact that the influx of component 1 raises the liquid level in the measuring capillary and therefore raises the pressure $P$ in the inner compartment (a reference capillary identical to the measuring capillary dips into the solvent to allow for corrections due to capillary rise). Increase in pressure raises $\mu_1$, and solvent influx ceases as soon as

$$d\mu_1 = d\mu_1' = 0. \qquad (1)$$

Outer polymer-free solution components in the dialysis-equilibrium experiment are designated by primes. The difference in

tion coefficient for instance or any other analytical elementary determination to a total dry weight). It is therefore completely independent of processes, such as solvation for instance, which may occur in solution without a change in number of particles per unit volume. Incomplete understanding of this point has led, and is still leading, to confusion in many instances particularly when more complicated systems are involved.

For a given value of $c$, $\Pi$ decreases as $M$ increases (the number of particles decreases with increasing value of $M$) demonstrating that the utility of OP determination is restricted to the low molecular weight range, peptides rather than proteins in the context of the present work. The usefulness of both equilibrium sedimentation and the scattering methods though increases with increasing molar mass, as will be shown below.

An ionized polymer in a single solvent (Fig. 2), complicates matters significantly. Because of the long-range electrostatic interactions between the macro-ions and the small counter-ions, as well as between the macro-ions themselves, the OP cannot be expanded into a virial series in powers of the concentration. If, by some artifice, we could remove the charge from the macro- and counter-ions and still constrain all particles to remain on one side of the membrane (although no longer naturally constrained to it by the requirement of electroneutrality) then we would (in the limit of vanishing concentration $c$) measure the number-average molar mass of the dissolved particles- both large and small, hardly the aim of the experiment. To avoid this situation, an additional low-molecular weight electrolyte comp. 3 is added, one of its charged ions being identical to the polyelectrolyte counterion, to maintain a total of three components.

A typical example of a three-component system is, for instance: comp. 1, water; comp. 2, the sodium salt of deoxyribonucleic acid, DNA; and comp. 3, sodium chloride at some suitable concentration, 0.1 M or 1.0 M, for instance. The complications arising by having a more complex buffer made up of two or more components diffusible through a semipermeable membrane are not too difficult to handle. All macromolecules (comp. 2) may be identical (a pure protein or an intact nucleic acid) or they may be polydisperse with respect to molar mass but part of a homologous series in which all members are chemically identical, a polypeptide for instance, or closely similar to each other. More complicated situations arise for instance as a result of density heterogeneity in DNA particles in equilibrium sedimentation in a density gradient or when mixtures of chemically dissimilar polymers are considered. The three component systems discussed in this work present the basis for the study of the more complex interacting systems of current research interest, and a reliable guide to the choice of the most appropriate method of analysis, or combination of complementary methods to be used.

We return now to the analysis of the osmotic system comprising three components (Fig. 3), picked in electroneutral combinations from the ionic species. This

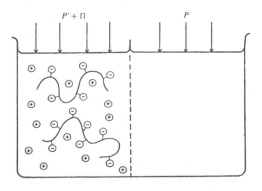

**Fig. 2** Schematic representation of osmotic membrane experiment; charged polyelectrolyte molecules are restricted to one side of semipermeable membrane and no low-molecular salt is present. [From Ref. (63)]

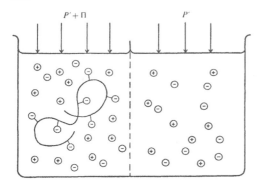

$P' + \Pi$                     $P'$

**Fig. 3** Same as Fig. 2, but low-molecular weight uni-univalent salt has been added to system. [From Ref. (63)]

can be done in ways that are different from the present procedure, which actually corresponds to the way components may be stored in bottles and weighed into the solutions. The fact that an ionized low-molecular weight component is now present produces a profound change in the nature of the solution. The screening of the fixed charges of the macro-ion by oppositely charged low-molecular weight ions drastically reduces the range of the electrostatic forces. The addition of the low-molecular weight salt again allows the OP to be expanded in a virial series *cf* Eq. (4). The reciprocal of the molar mass $M_2^{-1}$ is indeed obtained in the extrapolation of $\Pi/c_2RT$ to $c_2 = 0$. Considering however that as molecules or moles are "counted," as mentioned above, this is only one useful way to express the results of the experiment. Thus, the question sometimes asked, whether the mass of the counter-ions (or the degree of "binding" of counter-ions to, or their exclusion by the Donnan effect from, the macro-ion) is included in the molar mass determination is irrelevant. If $c$, for instance, is expressed as g/ml of NaDNA in solution, then $M_2$ will correspond to g/mole of NaDNA (including the stoichiometric number of $Na^+$ counterions), even if the experiment is performed in a solution in which CsCl concentration dominates, and not NaCl.

In the fundamental thermodynamic derivations concentrations are usually given as molalities $m_J$, moles of comp. $J$ per kg of comp. 1; weight molalities $w_j = m_j M_j/10^3$ are given in g of comp. $J$ per g of comp. 1. Concentrations on a volume basis are the molarity, $C_J$, the number of moles of comp. $J$ per liter of solution (nota bene!), and $c_j$, in g of comp. $J$ per ml of solution. The two schemes are related by

$$C_J = 10^3 m_J / V_m,  \qquad (5)$$

where $V_m$ is the volume of the solution in ml (of density $\rho$) containing 1 kg of comp. 1. For three components

$$V_m = (10^3 + m_2 M_2 + m_3 M_3)/\rho$$
$$= 10^3(\bar{v}_1 + \bar{v}_2 w_2 + \bar{v}_3 w_3),  \qquad (6)$$

where $\bar{v}_J = \bar{V}_J/M_J$ is the partial specific and

$$\bar{V}_J = (\partial V/\partial n_J)_{P,T,m} = (\partial V_m/\partial m_J)_{P,T,m}$$

the partial molal volume—the change in volume $V$ following addition of $n_J$ molecules of component $J$, at constant pressure, temperature and molality $m$ of all components, except that indicated in the differentiation.

Advantages or disadvantages in the use of the various concentration units depend upon the character of the experiment performed. The molality units are independent of temperature and pressure and, in a closed system, addition of one component does not change the concentrations of the others. Preparation of solutions by weight ensures, whenever feasible, the highest accuracy. Molarity scales are more convenient when volumetric manipulations are involved. Redistribution of diffusible components in a dialysis process complicates the use of the molality scale. For proteins and nucleic acids concentrations are often conveniently determined on a volume-based scale without interference from other components, by elementary (nitrogen, sulphur or phosphorus) analysis or ultraviolet absorption, provided the necessary conversion factors are available. If, for example, the amino-acid composition of a protein is known, the amount of nitrogen or sulphur per mole (or as a submultiple of a mole) is thereby known, and the molar mass can be determined for the case of no side groups ionized, or for the sodium or other salt corresponding to ionization of carboxylic or phosphate groups (in the case of a nucleic acid), and so on. Ambiguities associated with "dry weight," not easily determinable, are in this way avoided, as well as conceptual difficulties connected with unknown degrees of association, "condensation" of counter-ions on charged polyelectrolyte chains, or contributions from other "bound" low-molecular-weight materials (salts or neutral molecules) to the molecular complex.

The activity $a_J$ of any solute component $J$ is given by

$$\mu_J = \mu_J^0 + RT \ln a_J = \mu_J^0$$
$$+ RT \sum_i v_{iJ} \ln m_i + RT\beta_J, \qquad (7)$$

in which $\mu_J^0$ is the chemical potential in the standard state at 1 atm. and $RT\beta_J$ is the excess chemical potential or RT times the logarithm of the activity coefficient $\gamma_J (a_J = \gamma_J m_J)$ for component $J$. In the reference state with all $m_J$ at infinite dilution, $\gamma_J$ approaches unity; the $\beta_J$ and $\gamma_J$ are, in general, functions of pressure, temperature, and all the concentrations. The number of moles of species $i$ included in a mole of component $J$ is $v_{iJ}$, hence $m_i = \sum_i v_{iJ} m_J$. The $v_{iJ}$ are taken in electrically neutral combinations for the concentrations of the components to be independent thermodynamic variables. The derivatives $a_{JK}$ (or $\mu_{JK} \equiv RT\, a_{JK}$) defined by

$$a_{JK} \equiv [\partial \ln a_J / \partial m_K]_{P,T,m}$$
$$= [\partial \ln a_K / \partial m_J]_{P,T,m}$$
$$= \sum_i (v_{iJ} v_{iK}/m_i) + \beta_{JK}, \qquad (8)$$

will be encountered.

If, for example, comp. 2 is a $Z$-valent ion $P$ along with its complement of $Z$ univalent counter-ions $X$ and comp. 3 is a 1–1 electrolyte $XY$ (for example NaCl), then we can calculate from Eq. (8)

$$a_{22} = 1/m_2 + [Z^2/(m_3 + Z\, m_2)] + \beta_{22},$$
$$(9)$$

$$a_{23} = [Z/(m_3 + Z\, m_2)] + \beta_{23}, \qquad (10)$$

and

$$a_{23} = 1/m_3 + [1/(m_3 + Z\, m_2)] + \beta_{33}.$$
$$(11)$$

Alternate definitions of components lead to different equivalent expressions.

A thermodynamic derivation of the OP equation follows. Consider the outer polymer-free solution, the dialysate, to be at fixed atmospheric pressure $P'$ with the equilibrium pressure acting on the polymer solution equal to $P = P' + \Pi$. A three-component system will be considered (constant temperature $T$ is maintained throughout this work, and will not be indicated in subscripts). The condition for osmotic equilibrium is that, in addition to Eq. (1),

$$d\mu_3 = d\mu_3' = 0 \qquad (12)$$

for the diffusible components, both pressure and composition being maintained constant in the outer solution. The Gibbs-Duhem equation (2, 3) then reduces, at constant temperature and on the molality scale, to

$$V_m dP = m_2\, d\mu_2.$$

For the change of pressure with $m_2$ of the non-diffusible species

$$V_m(d\Pi/dm_2) = m_2(\partial \mu_2/\partial m_2)_\mu, \qquad (13)$$

where $d\Pi/dm_2 \equiv (\partial P/\partial m_2)_\mu$, and the subscript $\mu$ signifies constancy of the chemical potentials of the diffusible components 1 and 3.

The total variation $(\partial\mu_2/\partial m_2)_\mu$ is next expressed in terms of the derivatives $(\partial\mu_i/\partial m_j)_{P,m} = \mu_{ij}$, $\mu_{ij} = \mu_{ji}$ and $(\partial\mu_i/\partial P)_m$, which equals the partial molal volume $\overline{V}_J$. From

$$d\mu_2 = \mu_{22}dm_2 + \mu_{23}\,dm_3 + \overline{V}_2 dP, \quad (14)$$

$$(\partial\mu_2/\partial m_2)_\mu = \mu_{22} + \mu_{23}(\partial m_3/\partial m_2)_\mu + \overline{V}_2\,d\Pi/dm_2 \quad (15)$$

and from

$$d\mu_3 = \mu_{23}dm_2 + \mu_{33}dm_3 + \overline{V}_3 dP, \quad (16)$$

$$(\partial m_3/\partial m_2)_\mu = -\mu_{23}/\mu_{33} - (\overline{V}_3/\mu_{33})(d\Pi/dm_2) \quad (17)$$

Substitution of Eq. (17) into (15) and the latter into (13) yields

$$[(V_m - \{\overline{V}_2 - (\mu_{23}/\mu_{33})\overline{V}_3\}m_2)/m_2]$$

$$d\Pi/dm_2 = \mu_{22} - (\mu_{22}^2/\mu_{33}). \quad (18)$$

Eq. (18) can be approximated by dropping terms which are inconsequential in relation to the best experimental accuracy which can be attained, to yield (the approximate-equality symbol will be used to indicate this action)

$$V_m^0(d\Pi/dm_2) \approx [\mu_{22} - (\mu_{23}^2/\mu_{33})]m_2 \quad (19)$$

or, with the use of Eq. (8)

$$(V_m^0/RT)(d\Pi/dm_2) \approx 1$$

$$+ \left[\sum_i(\nu_{i2^2}/m_i) + \beta_{22} - (a_{23}^2/a_{33})\right]^0 m_2 + \ldots, \quad (20)$$

which may be compared with the virial expansion Eqns (3) and (4), in the differential form. The term in square brackets on the r.h.s. of Eq. (20) equals twice the second virial coefficient $B_2(m)$, in molality units. It is related to $A_2$ of Eq. (4) by

$$B_2(m) = (M_2^2 A_2/V_m^0)$$

$$- (1/V_m^0)(\partial V_m/\partial m_2)_\mu^0. \quad (21)$$

The second term on the r.h.s. of Eq. (21) may not be neglected with respect to the first term. Use of Eqns (9) to (11) in Eq. (20) and neglect of the $\beta_{JK}$ terms leads to an idealized value $B_2(m) = Z^2/4m_3$ for the second virial coefficient. In some applications the ideal value of the charge $Z$ is substituted by an effective value $iZ$ with $i$ ranging from 0 to unity, however this is of no concern to the present discussion. The value of the second virial coefficient is very large at low salt concentrations $m_3$ for highly charged macromolecules such as nucleic acids, for instance, however it may be appreciable at very low salt concentrations in the case of polypeptides and proteins as well. With increasing concentrations $m_3$ the contribution of the charge term to the virial coefficients diminishes and contributions due to size and shape of the macromolecule and hydrophilic and hydrophobic interactions preponderate. It should be noted that for uniunivalent electrolyte components 3 the charge contribution to the third virial coefficient equals zero.

## B. Distribution of Diffusible Components

When, as seen above, the ionized macromolecular comp. 2, $PX_Z$, is added to a solution of a low-molecular-weight salt comp. 3, $XY$ (having an ion in common with comp. 2) in a solvent, comp. 1, the chemical potential $\mu_3$ is raised if no other additional molecular mechanism such as ion "binding," for instance, occurs. If the solution containing all the three components is in equilibrium across a semipermeable membrane with a solution containing the diffusible comps 1 and 3, designated by primes, only, then the chemical potentials of the diffusible components will reach identity on both sides of the membrane. This corresponds to the classical Donnan equilibrium and the unequal distribution of ions across the membrane is known as the Donnan distribution. We generally require electroneutrality in a membrane equilibrium

$$m_u + m_Y = m_X, \quad \text{and} \quad m'_Y = m'_X \tag{22}$$

where $Z m_2 = m_u$, and equality of the activities of diffusible components, rather than of the ionic species, in the inner and outer phases.

$$a_X a_Y = a'_X a'_Y = (a'_3)^2 \tag{23}$$

The "ideal" Donnan distribution is first calculated by identifying activities with concentrations and use of Eq. (22) to obtain

$$m_Y(m_Y + m_u) = (m'_Y)^2. \tag{24}$$

This quadratic equation is solved for $m_Y$

$$m_Y = -(1/2)m_u \\ + m'_Y[1 + (m_u^2/4m_Y'^2)]^{1/2}, \tag{25}$$

and the square root expanded at low concentrations $m_u$, to obtain an expression for $\Gamma$

$$\Gamma \equiv (m_Y - m'_Y)/m_u \\ = -0.5 + (m_u/8\,m'_Y) \\ - (m_u^3/128\,m_Y'^3) + \ldots, \tag{26}$$

The limiting value $-0.5$ of the "ideal" Donnan distribution coefficient $\Gamma$ when $m_u \to 0$, defines salt-rejection as a negative and salt-binding as a positive contribution; comp. 3 is therefore "rejected" in this ideal calculation from the solution compartment containing comp. 2. In the non-ideal case the value of $\Gamma$ will not be equal to $-0.5$; it may be larger or smaller or even positive if "binding" of component 3 is involved. It has been customary, as mentioned before in the case of the virial expansion, to account for nonideality by assuming that the polyelectrolyte macromolecule carries an "effective" charge $iZ$, that $1-i$ counter-ions are

"bound" to the macromolecule, and that $i$ counter-ions are "free" and behave "ideally," or subject to the classical Debye-Hückel laws for simple electrolytes. The value of $\Gamma$ in the limit $m_u \to 0$ is then $-0.5i$, and it varies between zero (for $i = 0$, all counter-ions bound) and $-0.5$ (for $i = 1$, all counter-ions "free"). Theoretical models for the formulation of this concept have been advanced and are still being proposed. Within the more thermodynamic context of this article the value of $\Gamma$ is obtained as an experimental quantity. It will be seen in the following that it is usually not possible to associate unique molecular mechanisms with the distribution parameters and the measured value usually reflects, in addition to the Donnan mechanism discussed above, salt "binding" (of comp. 3) and/or solvation (binding of comp. 1). The case when comp. 3 is not an electrolyte will also be discussed.

From the expansion, at low concentrations $m_u(m_3^0 \equiv m'_3)$

$$m_3 = m_3^0 + (\partial m_3/\partial m_u)_\mu^0 m_u + \ldots, \tag{27}$$

we find $\Gamma^0 = (\partial m_3/\partial m_u)_\mu^0 = Z^{-1}(\partial m_3/\partial m_u)_\mu^0$. The limiting value of the second term on the r.h.s of Eq. (17) can be neglected with respect to the first term and, to a good approximation

$$\Gamma = (\partial m_3/\partial m_u)_\mu \approx -Z^{-1} a_{23}/a_{33}. \tag{28}$$

The distribution parameter $\Gamma$ thus reflects the change in activity of comp. 3 with change in concentration of comp. 2, as well as with change in its own concentration.

In an experiment in which components are allowed to equilibrate it is necessary to determine concentrations of diffusible components precisely by analytical methods, for instance halide-ion concentrations on both sides of the membrane. As volume concentrations are usually determined a distribution coefficient $(\partial C_3/\partial C_u)_\mu$ is likely obtained rather than $\Gamma$. The connection between the two quantities, at vanishing concentration $C_u$, can be derived by transforming the concentration units to yield

$$(\partial C_3/\partial C_u)_\mu = \Gamma - 10^{-3}C_3(\overline{V}_u + \Gamma\overline{V}_3), \tag{29}$$

or

$$(\partial c_3/\partial c_u)_\mu = \xi_3 - c_3(\bar{v}_2 + \xi_3\bar{v}_3), \tag{30}$$

where $\overline{V}_u = \overline{V}_2/Z$ is the partial molal volume of equivalent units, $\xi_3 \equiv (\partial w_3/\partial w_2)_\mu = (M_3/M_u)\Gamma$ and $M_u = M_2/Z$ is the mass of equivalent units. The evaluation of distribution coefficients by volume concentration differences is not easy and we shall show below that they may be obtained conveniently in molal units from density increments and analytical ultracentrifugation.

Isopiestic distillation, mentioned in the Introduction, yields a distribution coefficient $(\partial m_3/\partial m_u)_{P,\mu}$ which, for all practical purposes, equals $(\partial m_3/\partial m_u)_\mu$.

## C.  Partial Volumes and Density Increments

Density and refractive index increments are of many-fold interest in the study of multi-component biopolymer solutions. A basic application is in the obtainment of molar masses by analytical ultracentrifugation and light-scattering. If density and refractive index changes are determined both at constant composition and at constant potential of diffusible solutes, interaction parameters reflecting the distribution of diffusible solutes may be conveniently derived. This is practically achievable in view of the availability of the Kratky/Leopold/Stabinger digital precision densitometer and the Rayleigh refractometer.

The experimentally accessible apparent specific volume $\phi_J \equiv \Phi_J/M_J = \Delta V/g_J$, where $\Delta V$ is the volume change upon addition of $g_J$g of comp. J to a solvent mixture of volume $V_s$ (at constant

$P$, $T$ and composition). In the solution process ions may bind and hydration may occur, however this process in itself does not lead to a volume change of the system, which may occur nevertheless, because hydrated water undergoes a volume change due to electrostriction. It is worth noting that in protein and nucleic acid solutions, and at low concentrations of the macromolecular comp. 2 in particular, $\phi_J$ is found to be independent of concentration of comp. 2 and may therefore be identified with the partial specific volume $\bar{v}_2$. Differentiation of Eq. (6) yields

$$(\partial\rho/\partial m_2)_{P,m} = (M_2/V_m)(1-\rho\bar{v}_2). \tag{31}$$

From $c_2 = M_2 m_2/V_m$

$$(\partial c_2/\partial m_2)_{P,m} = (M_2/V_m)(1 - c_2\bar{v}_2). \tag{32}$$

and therefore

$$(\partial\rho/\partial c_2)_{P,m} = (1-\rho\bar{v}_2)/(1-c_2\bar{v}_2). \tag{33}$$

At low concentrations $c_2$, $(\partial\rho/\partial c_2)_{P,m}$ equals $\Delta\rho/c_2$, where $\Delta\rho = \rho - \rho^0$ is the increase in solvent-mixture density $\rho^0$ upon addition of comp. 2, at constant composition of the solvent mixture. At low concentrations $c_2$ we may also write

$$\rho = \rho^0 + (\partial\rho/\partial c_2)_{P,m}c_2, \tag{34}$$

and substitute into Eq. (33) to obtain

$$(\partial\rho/\partial c_2)_{P,m} = (1 - \rho^0\bar{v}_2^0), \tag{35}$$

which is a simple expression to obtain $\bar{v}_2^0$ from the measured values of $(\Delta\rho/c_2)_{P,m}$. Experimental methods have been worked out for the precise evaluation of this quantity (3).

In distinction to the density increment determination at constant solvent composition a quantity of major interest is the density increment at constant chemical potential of diffusible solutes. Experimentally we determine here the density difference between dialyzate and dialysis solvent in a dialysis experiment. In a three component system

$$(\partial\rho/\partial m_2)_\mu = (\partial\rho/\partial m_2)_{P,m}$$

$$+ (\partial\rho/\partial m_3)_{P,m}(\partial m_3/\partial m_2)_\mu$$

$$+ \kappa\rho(d\Pi/dm_2), \qquad (36)$$

where $\kappa = (\partial\ln\rho/\partial P)_m$ is the isothermal compressibility coefficient, which can be neglected in calculations in aqueous solutions. Using procedures described above Eq. (36) transforms to a good approximation into

$$(\partial\rho/\partial m_2)_\mu = (\partial\rho/\partial m_2)_{P,m}$$

$$- (\mu_{23}/\mu_{33})(\partial\rho/\partial m_3)_{P,m}, \qquad (37)$$

and further, in g molarity units, into

$$(\partial\rho/\partial c_2)_\mu = (1 - \rho^0\bar{v}_2^0) + \xi_3(1 - \rho^0\bar{v}_3) \qquad (38)$$

$$= (1 + \xi_3) - \rho^0(\bar{v}_2^0 + \xi_3\bar{v}_3) \qquad (39)$$

Here $\bar{v}_2^0$ the partial specific volume at vanishing comp. 2 concentration and $\bar{v}_3$ the corresponding quantity for comp. 3 at its finite, specified concentration, in the absence of comp. 2. From the experimental values of $(\partial\rho/\partial c_2)_\mu$, $\bar{v}_2^0$ and $\bar{v}_3$ it is possible to calculate the interaction coefficient $\xi_3$.

The value of $(\partial\rho/\partial c_2)_\mu$ may be expressed by introducing an apparent quantity $\phi'$, defined by $(\partial\rho/\partial c_2)_\mu \equiv 1 - \phi'\rho^0$, which, in contradistinction to $\phi_2$ is not a specific apparent volume, because it includes contributions due to the redistribution of components. It has some practical use even though it is changing due to redistribution of comp. 3 Its continued use is due to the custom, which is hard to eradicate, that Archimedes buoyancy terms in the form $1 - \rho\bar{v}_2$, appearing in two-component systems in sedimentation equations for instance, should also be expressed in this form in systems containing more than two components. Adequate care in expressing the results from multicomponent systems is advised.

Whereas the basis for the molality in the definition of $\xi_3$ was comp. 1, the sym-

metry of the thermodynamic system requires that the reduced density increments (due to the addition of comp. 2) may be expressed to within the same assumptions as embodied in Eqns (38) and (39), in the form

$$(\partial\rho/\partial c_2)_\mu = (1 - \rho^0\bar{v}_2^0) + \xi_1(1 - \rho^0\bar{v}_1)$$
$$(40)$$

$$= (1 + \xi_1) - \rho^0(\bar{v}_2^0 + \xi_1 v_1),$$
$$(41)$$

where $\xi_1 = (\partial w_1/\partial w_2)_\mu$ and the weight-molalities $w_J$ are now per gram of comp. 3; due to the ratio $(\partial w_1/\partial w_2)_\mu$, $\xi_1$ is expressed in terms of g of comp. 1 per g of component 2, whereas $\xi_3$ was expressed in g of comp. 3 per g of comp. 1. The two interaction parameters are related in the following way. At vanishing concentration of comp. 2, we use the basic relations $c_1 + c_3 = \rho^0$, $c_1\bar{v}_1 + c_3\bar{v}_3 = 1$ and $w_3 = c_3/c_1$ to obtain, by comparing Eqns (38) and (40)

$$\xi_1 = -\xi_3/w_3. \qquad (42)$$

In a popular, but sometimes misleading, interpretation positive $\xi_1$ is associated with "preferential" binding of component 1, negative $\xi_3$ with "preferential" exclusion of component 3, and vice-versa. However, as will be noticed in Fig. 4 for bovine serum albumin, BSA, in the denaturant guanidinium chloride, GdmCl, and for halophilic malate dehydrogenase, hMDH, in NaCl, $\xi_3$ may be negative at some concentration $w_3$, and becomes positive with increasing $w_3$, vanishing at one stage. Does this mean that hydration, or rather the mysterious "preferential" hydration occurs at high concentrations of comp. 3 only, it vanishes with decreasing $w_3$ and transforms into "preferential exclusion" of water with further decrease in $w_3$? No basic molecular mechanisms should be attached to these parameters without further thought. To do this it is necessary to leave the formally

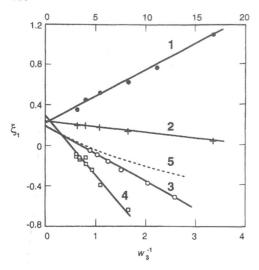

**Fig. 4** Interactoin parameter $\xi_1$, as a function of reciprocal cosolvent weight molality $w_3^{-1}$ (g of water/g of cosolvent), (1) DNA in NaCl, (2) BSA in NaCl, (3) BSA in GdmCl, (4) hMDH in NaCl, (5) BSA in GdmCl, curve calculated according to Ref. (28). Upper scale, NaCl, lower scale, GdmCl. Curves (1) to (3) from Ref. (64), data for curve (4) from Ref. (54). [From Ref. (18)]

correct but information-poor thermodynamic formalism and construct simple or more complex models to represent the experimental data.

An essential feature of a particular model to be considered is that either one of the thermodynamic interaction parameters $\xi_1$ or $\xi_3$ should indicate interactions of comp. 2 with both comps 1 and 3. The model we have used stipulates that $B_1$ g of comp. 1 per g of comp. 2 are "bound" to it. "Bound" is used for lack of a better term. It indeed signifies hydration in the case of water and exclusion of the space occupied by this "bound" material to the other components. In the same way $B_3$ g of comp. 3 per g of comp. 2 refers to space excluded by "bound" comp. 3. Binding of comp. 3 to comp. 2 may be reduced at low concentrations of component 3, however experiments are usually performed at high concentrations of comp. 3 and saturation

on available sites is presumed. In the case of highly charged macromolecules, such as nucleic acids for instance, $E_3$ g of electrolyte comp. 3 per g of comp. 2 are excluded by the Donnan mechanism. In this case we replace $B_3$ by $B_3' = B_3 - E_3$. In this analysis $B_3$ and $E_3$ cannot be separated, however the latter may be estimated by electrostatic calculations. The model described has been called the "invariant" particle model (21) and presents an important operational concept in the structural analysis from the angular dependence of scattering experiments. It is possible to calculate. the total volume of the particles from the relation $V_{\text{tot}} = (\bar{v}_2 + B_1\bar{v}_1 + B_3'\bar{v}_3)$ and confirm this calculation by radius of gyration measurements by light, x-ray or neutron scattering. In an alternate model Schellman (22–24) proposes a "weak" interaction model, in distinction to the "strong" interaction model described above; comp. 3 binds to comp. 2 with increasing concentration, however it additionally replaces "bound" comp. 1, in this process. Timasheff (25) proposes a combined model containing both exchangeable and non-exchangeable solute/solvent and solute/cosolvent complexes, yet it seems experimentally difficult to distinguish between these groups.

The analysis of the "invariant" particle system yields

$$(\partial\rho/\partial c_2)_\mu = (1 - \rho^0\,\bar{v}_2^0) + B_1(1 - \rho^0\bar{v}_1)$$
$$+ B_3'(1 - \rho^0\,\bar{v}_3) \qquad (43)$$

$$= (1 + B_1 + B_3')$$
$$- \rho^0(\bar{v}_2 + B_1\bar{v}_1 + B_3'\bar{v}_3). \qquad (44)$$

Furthermore

$$\xi_3 = B_3' - B_1 w_3, \qquad (45)$$

and

$$\xi_1 = B_1 - B_3'/w_3. \qquad (46)$$

An experimental observation supporting the "invariant" particle concept is that a "contrast variation" plot of $(\partial\rho/\partial c_2)_\mu$ versus $\rho^0$ is usually found to be linear, and it is thus possible to calculate the constants $B_1$ and $B'_3$ from the slope and the intercept of the plot. A serious warning concerns the analysis of this plot by Eq. (39) or (41); $\xi_3$ and $\xi_1$ are functions of $w_3$, and therefore of $\rho^0$, and neglect of this fact has in the past led to erroneous interpretations.

A plot of $\xi_1$ versus $1/w_3$ is linear according to Eq. (46), the slope yielding $-B'_3$ and the intercept $B_1$. A summary of data is presented in Fig. 4 and Table 1, which we will describe here briefly, as they have been described before (18). Curve 1 in Fig. 4 represents DNA in NaCl. The positive slope in this case is due to the preponderance of the Donnan coefficient $E_3$ over the salt binding term $B_3$. The positive intercept yields the correct hydration term $B_3$, $5 \pm 1$ molecules of water per nucleotide, a value confirmed in continuation by crystallographic studies of Kopka et al. (26) and further works. Curve 2 represents BSA in NaCl, again yielding the correct hydration (Table 1) and weak binding of NaCl, in agreement with early observations of Scatchad et al. (27). Curve 3 represents BSA in the denaturating cosolvent GdmCl. The figure and Table 1 indicate significant "binding" of the denaturant to the denatured protein and, from the intercept of the curve at high denaturant concentrations, practically unchanged hydration, within experimental error, upon denaturation. Further investigation of this important conclusion is indicated. Curve 5, disagreeing with our results is simulated from the protein-GdmCl association constants calculated by Makhatadze and Privalov (28) from calorimetric data. Measurements should be extended to the BSA-urea system, claimed to have a lower dissociation constant (28), which should yield a more pronounced curve. Finally, Curve 4 is a summary of halophilic malate dehydrogenase hMDH in NaCl (18), indicating (Table 1) strong salt binding and hydration. These results, critical for an understanding of the adaptation of extreme halophiles to supersaturated saline environments, are receiving support form recently reported x-ray structure reports (29, 30) and high-resolution studies of halophilic ferredoxin (F Frolow, Microbiology and Biogeochemistry of Hypersaline Environments, Jerusalem, June 22–26, 1997). In summary, it may be said that the interpretation of density increments in a number of nucleic acids and protein systems leads to reliable insights into the behavior and interactions of biological macromolecules with water and cosolvents in solution.

In the presentation above of solvent, comp. 1, and electrolyte cosolvent, comp. 3, interactions with a macromolecular (polyelectrolyte) comp. 2, we have shown that by varying the concentration of comp. 3 and subsequent solution density change we can probe volume excluded in the surface and in the interior of component 2, associated with either comps 1 or 3. If we substitute the electrolyte comp. 3 by a neutral water-soluble compound, such as a sugar for instance, its interaction with comp. 2 may decrease and $B_3$ may be close to zero. This is not always true, because sugars may interact with proteins, and interesting results may be obtained (31). Still, sugars of varying size can be used to

**Table 1** Interaction Parameters $B_1$ and $B'_3 = B_3 - E_3$ of Proteins and DNA in Salt Cosolvents, in Units g Water/g Protein (DNA) and g Cosolvent/g Protein (DNA). [From Ref. (18)]

| Substance | $B_1$ | $B'_3 = B_3 - E_3$ |
|---|---|---|
| | g/g | |
| hMDH, NaCl | 0.35–0.45 | 0.08–0.14 |
| BSA, NaCl | 0.23 | 0.012 |
| BSA, GdmCL | 0.18 | 0.27 |
| DNA, NaCl | 0.20 | −0.054 |
| DNA, CsCl | 0.24 | −0.070 |

fractally probe exterior surfaces and interior domains of biological macromolecules. The nucleosome core particle is composed of 146 DNA base pairs winding in 1.75 helical turns around eight histone particles. The gross shape of the nucleosome core particle derived from neutron scattering contrast variation experiments and by low angle x-ray diffraction is that of a flat cylinder having a diameter of 11 nm and a height of 5.7 nm. The total calculated volume ($542\,nm^3$) is considerably in excess of the approximate volume ($216\,nm^3$) of its nucleic acid and protein components. Thus, in addition to these components, and volumes of hydration, the nucleosome core particle must contain interior spaces, contributing to the total volume of the particle. It could thus be argued that density contrast experiments could be undertaken with small sugars, such as sucrose, raffinose or glycerol, capable of penetrating inside the nucleosome core particle, which would probe the volume of the particle including volume excluded by hydration. Indeed it was found (32, 33) that for these small sugars $B_3 \approx 0$ and hydration corresponds to normal DNA and protein values

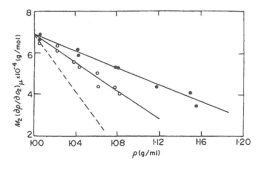

**Fig. 5** The slopes $M_2(\partial\rho/\partial c_2)_\mu$ for nucleosome core particles, from equilibrium sedimentation against solvent mass-density $\rho^0$. Density contrast variation solutes: (•), sucrose ($\xi_1 = 0.318$); (○), $\gamma$-cyclodextrin ($\xi_1 = 1.005$). Broken curve is the dextran curve ($\xi_1 = 2.32$) from Fig. 14 of Ref. (32). [From Ref. (33)]

(Fig. 5). However, when $\gamma$-cyclodextrin, a large sugar which was believed not to penetrate the nucleosome core particle, was used in the density contrast experiments (Fig. 5), the value of $B_1$ increased considerably, whereas $B_3$ stayed close to zero. It was now possible, following what corresponded in essence to fractal surface probing, to calculate the correct volume of the nucleosome core particle, which has since been confirmed by x-ray diffraction structure analysis (34). The discussion given here is an example to indicate constructive uses of density increment results in the evaluation of the properties of biological interactive systems. Refractive index increments will be discussed in the light scattering section.

## III. EXPERIMENTAL METHODOLOGIES

### A. Analytical Ultracentrifugation

#### 1. Equilibrium Sedimentation (ES)

In the analysis of analytical ultracentrifugation I will first analyze equilibrium sedimentation (ES) of a three-component system closely following an approach previously used (3). Extension to more than three components is straightforward and an exact treatment has been given (2). For the derivation of the equation for ES, we note that in a gravitational potential $\phi$, transfer of mass $M$, from a phase $\alpha$, at a potential $\phi^{(\alpha)}$ to a different position at potential $\phi^{(\beta)}$, involves an amount of work $M(\phi^{(\beta)} - \phi^{(\alpha)})$, independent of the chemical nature of the matter. The total potential $\bar{\mu}_i$ of ionic species $i$, is given by the sum of chemical, gravitational and electrostatic terms

$$\bar{\mu}_i = \mu_i + M_i\phi + v_i\mathcal{F}\psi, \qquad (47)$$

where $v_i$ is the valency (with the appropriate sign) of species $i$, $\mathcal{F}$ is the Faraday and $\psi$ is the electrostatic potential of the phase. The theory of heterogeneous equilibrium requires that $\bar{\mu}_i$ be uniform for each species throughout the system so that

$$\mathrm{d}\mu_i + M_i \mathrm{d}\phi + v\mathcal{F}\,\mathrm{d}\psi = 0. \qquad (48)$$

Final equations are evaluated for the equilibrium position of electro-neutral components, as defined previously, as long as the electroneutrality condition is satisfied locally. No complications arise as a consequence of gradients of electric charge resulting from inequality of gravitational forces exerted on ions of unlike mass, as long as supporting electrolyte is present to sufficient extent. The electrostatic terms cancel exactly in the summation of contributions from ionic species forming a neutral macromolecular component, and we obtain for comp. 2 the differential expression

$$\mathrm{d}\mu_2 + M_2\,\mathrm{d}\phi = 0. \qquad (49)$$

The gravitational field potential $\phi$ equals $\phi = -(1/2)\omega^2 r^2$, where $\omega$ is the angular velocity (rad/sec) and $r$ is the distance (cm) from the center of rotation (Fig. 6). From Eqns (49) and (14) one obtains

$$M_2\omega^2 r\,\mathrm{d}r = \mu_{22}\,\mathrm{d}m_2 + \mu_{23}\,\mathrm{d}m_3$$
$$+ \overline{V}_2\,\mathrm{d}P, \qquad (50)$$

and a similar equation can be written for comp. 3

$$M_3\omega^2 r\,\mathrm{d}r = \mu_{23}\,\mathrm{d}m_2 + \mu_{33}\,\mathrm{d}m_3$$
$$+ \overline{V}_3\,\mathrm{d}P. \qquad (51)$$

The Gibbs-Duhem equation for a phase of fixed volume $V$ containing $n_J$ moles of comp. $J$ at constant temperature $T$, yields $V\,\mathrm{d}P = \sum_J n_J\,\mathrm{d}\mu_J$ for change in pressure in the vicinity of the phase. Combination with Eq. (49) and the expression for $\phi$ yields the hydrostatic pressure condition $\mathrm{d}P = \rho\omega^2 r\mathrm{d}r$, where $\rho = \sum_J(n_J M_J/V)$ is the density of the phase. Substitution of the hydrostatic pressure condition into Eqns (50) and (51) and use of Eq. (31), and a similar expression for comp. 3, yield

$$\omega^2 r\,\mathrm{d}r\,V_m(\partial\rho/\partial m_2)_{P,m} = \mu_{22}\,\mathrm{d}m_2$$
$$+ \mu_{23}\,\mathrm{d}m_3, \qquad (52)$$

and

$$\omega^2 r\,\mathrm{d}r\,V_m(\partial\rho/\partial m_3)_{P,m} = \mu_{23}\,\mathrm{d}m_2$$
$$+ \mu_{33}\,\mathrm{d}m_3. \qquad (53)$$

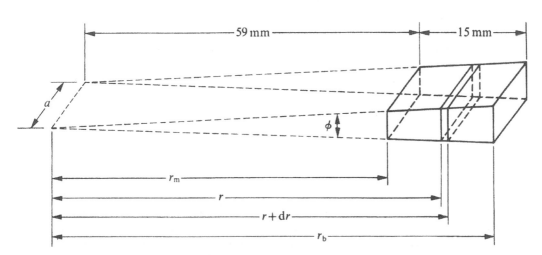

**Fig. 6** Schematic diagram of a sector-shaped ultracentrifuge cell, at a distance from the axis of rotation corresponding to its position in the ultracentrifuge rotor. The angle $\phi$ is generally 4°, but smaller angles are often used. The distance $a$ represents the optical path in the cell, that is the thickness of the liquid column contained between the quartz windows. [From Ref. (10)]

Subscript $m$ refers to the constancy of the molality of the components not appearing in the differentiation. It should be noted that the derivatives $\partial\rho/\partial m_J$ are used instead of the conventional buoyancy expressions $M_J(1 - \rho\bar{v}_j)$, a procedure leading to considerable simplification in the equations for ES and transport in multicomponent systems. To calculate the distribution of comp. 2 $dm_3$ is eliminated between Eqns (52) and (53), Eq. (37) introduces $(\partial\rho/\partial m_2)_\mu$, Eq. (19) introduces the derivative of the OP, and after some transformations (2, 3) to the gram molarity units useful in analytical ultracentrifugation, the correct differential expression for the distribution of comp. 2 is obtained.

$$(\omega^2/2)(\mathrm{d}\ln c_2/\mathrm{d}r^2)^{-1}(\partial\rho/\partial c_2)_\mu$$
$$= \mathrm{d}\Pi/\mathrm{d}c_2, \quad (54)$$

where we have used $\mathrm{d}r^2$ for $2r\,\mathrm{d}r$, and the variables refer to position $r$ in the centrifugal field. Eq. (54) holds for every electroneutral component in the system and subscript 2 may be replaced by subscript $J$. It is exact and completely general without restrictions to thermodynamic ideality or to incompressibility. The introduction of the OP has been done in a formally exact way and establishes a connection with the thermodynamic derivations presented earlier. The validity of the relations is completely independent though as to whether the OP can actually be determined experimentally. The symmetrical form of the equation is, in particular due to the use of the density increments (at constant chemical potentials of components diffusible through a semipermeable membrane), which are also conveniently measureable in many important practical situations. Pressure effects may be manifested in the dependence of the variables, and in particular the density increments, on pressure arising from the velocity of rotation of the ultracentrifuge rotor.

In the case of a homogenous comp. 2 the OP derivative in Eq. (54) can be expanded in a virial series, yielding

$$\mathrm{d}\ln c_2/\mathrm{d}r^2 = (\omega^2/2RT)(\partial\rho/\partial c_2)_\mu$$
$$\times (M_2^{-1} + 2A_2c_2 + \cdots)^{-1}, \quad (55)$$

and, in the limit of vanishing concentration of comp. 2

$$\mathrm{d}\ln c_2/\mathrm{d}r^2 = (\omega^2/2RT)(\partial\rho/\partial c_2)_\mu M_2. \quad (56)$$

If $\rho^0$ is constant throughout the liquid column then, by Eqns (39) or (41), $(\partial\rho/\partial c_2)_\mu$ is also constant throughout, because all the quantities on the right-hand side of these equations are independent of $c_2$ at low macromolecular concentrations. Therefore, if we plot the experimental values of $\ln c_2$ against $r^2$ (Fig. 7), then the slope of this linear plot is proportional to $M_2$. If the density increment and $\omega$ are known, then $M_2$ may be evaluated. This can be considered the classical use of the analytical ultracentrifuge as stated above. Presently $M_2$ is usually known and more extended uses of ES in the analytical ultracentrifuge will be presented but not discussed in detail. The reader is referred to three pertinent sets of recent publications (35–37), covering broad aspects of up-to-date ultracentrifugation analysis.

In general terms the use of the differential Eqns (55) or (56) in volume concentrations units is convenient because concentrations derived by the optical probes used in the ultracentrifuge (interference, Schlieren and light-absorption methods) naturally relate to volume concentration units. As long as we are only interested in the slopes of Fig. 7 it is sufficient to determine any quantity proportional to $c_2(r)$ rather than the absolute value itself. Other features not peculiar to the analysis of multicomponent systems, but arising identically in

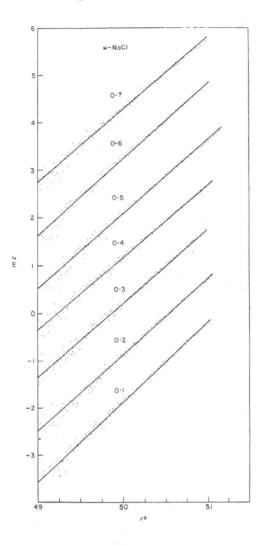

**Fig. 7** Equilibrium sedimentation profiles ln$c$ (in arbitrary concentration units $c$), derived from the measurement of $A_{260}$ against $r^2$; $r$ is the distance, in cm, from center of rotation; measurements of nucleosome core particles at 0.1 to 0.7M NaCl. Each successive representation above 0.1M NaCl is displaced by one unit of ln$c$ for display purposes, to avoid overlap of curves. Angular velocity 8000 revs/min, temperature 15° to 18°C; initial DNA concentration in nucleosome cores $A_{260} = 0.35$. Points represent computer sampling of photoelectric scanner of Model E Beckman ultracentrifuge. Straight lines are computer-generated least-square best fits. [From Ref. (32)]

two-component systems, involve the case of positive virial coefficients which may be derived from a more refined analysis of the experimental results. Downward curvature would then be observed in the plot of Fig. 7. Molecular weight polydispersity, on the other hand, would lead to upward curvature and the slope at each level $r$ is proportional, in the limit of negligible virial coefficients, to the weight-average molar mass $M_w(r)$ at the position $r$ in the centrifuge cell. The integration of the differential equation for ES also proceeds by standard methods applicable to simpler two-component systems. The experimental analysis is based on a combination of measurements over a range of velocities and concentrations for various liquid column lengths, whereby it usually becomes possible to separate nonideality from polydispersity contributions. Additional complications arise from macromolecular systems in chemical associative systems and, in general, both lower ($M_n$) and higher ($M_z$) moments of the molecular weight may be obtained. In case an association between proteins and nucleic acids is studied, advantage may be gained by scanning at additional absorption wavelengths, emphasizing the behavior of this or that component. Edelstein and Schachman (38) have developed a method for the simultaneous evaluation of apparent partial volumes $\phi'$ and molar masses by ES in a medium consisting in one case of light, and in the other case of heavy water. Interaction coefficients with solvents, which determine $(\partial\rho/\partial c_2)_\mu$ are presumed not to be significantly affected by the substitution of hydrogen by deuterium. We shall return to the powerful analysis of H/D substitution in neutron scattering experiments. We emphasize here that molar masses in ES are obtained in whatever units we choose to express concentrations, recalling the discussion presented earlier with respect to molar mass determinations by OP, considerations applicable as well to the analysis of scattering phenomena, to be described below.

As already mentioned before in equilibrium studies of biological macromolecules in aqueous solutions centrifuge speeds are reasonably low, as are concentrations of comp. 2, and water is a rather incompressible liquid. Thus compressibility effects and deviations from van't Hoff's law are usually rather small, and redistribution of diffusible solutes in the absence of comp. 2 is practically negligible. These considerations had to be considered however with care in a recent study (C Ebel, H Eisenberg, R Ghirlando Biophys. J., submitted), in which we aim to examine the interactions of a variety of sugars, varying in size and in nature, at relatively high cosolvent concentrations, with a well defined protein of known crystallographic structure. The use of rabbit muscle aldolase, a crystalline protein tetramer with a molecular weight of 156,841 g/mole, allowed equilibrium centrifugation at 8000 r.p.m., a speed at which the sugar cosolvent redistribution is negligible. An example shown in Fig. 8 represents data of sedimentation equilibrium of aldolase at various sucrose concentrations, at various temperatures, In a presentation of $(\partial\rho/\partial c_2)_\mu$ against the solvent density $\rho^0$ we encounter linear plots which can be analyzed by Eq. (44). As the interaction parameter $B_3$ for raffinose, for instance, also could be considered negligible, we determined *inter alia* $B_1$ and $\bar{v}_2^0$ as a function of temperature, the partial volume temperature dependent values matching well published temperature dependent results (39). This is given as an example of the modern use of the analytical ultracentrifuge for the study of molecular interactions.

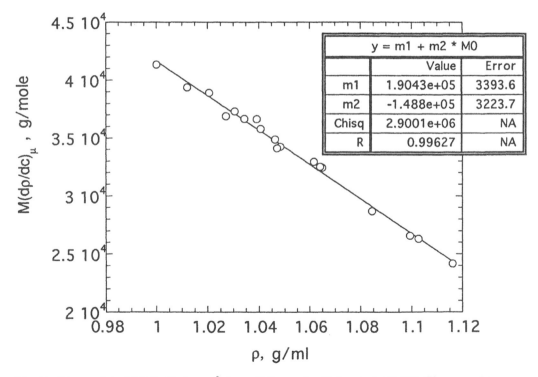

**Fig. 8**    Linear plot of $M_2(\partial\rho/\partial c_2)_\mu$ vs $\rho^0$, for rabbit muscle aldolase in 0.15M NaCl over a wide range of sucrose concentrations, at 20°C (C Ebel, H Eisenberg, R. Ghirlando, Biophys. J., submitted).

Sedimentation equilibrium in a density gradient and self-association by this procedure have been discussed in detail before (3) and will not be analyzed in this work. This concludes the presentation of equilibrium sedimentation in multicomponent systems.

## 2. Velocity Sedimentation (VS)

The discussion of transport methods in this chapter is not exhaustive and refers mainly to problems arising because the sedimenting macromolecules carry charges and the solvent is composed of more than one component. The sedimentation coefficient $s$ is the flow, per unit concentration, per unit centrifugal field, in the absence of concentration gradients. We consider only the numerical value of $s$ and its physical meaning, and disregard the practical reality of the existence of boundaries, although from the shape of the boundaries some interesting bits of information which the ultracentrifuge can provide, are derived (8–10, 40–42). As already mentioned an interesting aspect concerns the analysis in the sedimentation process of protein interactions and associating systems (12). Even though from a qualitative analysis of sedimenting boundaries of concentration gradients in mixed interacting systems it is possible to derive conclusions with respect to the interaction, yet the complex nature of the flow, the necessity to disregard diffusion, the imperfect knowledge of frictional coefficients, and other simplifying assumptions make a quantitative analysis extremely difficult (11). Charge effects add additional complexity (3, 43). In view of the emphasis on equilibrium phenomena in this contribution and the elaborate discussion required for a complete presentation of VS and diffusion in the ultracentrifuge by the methods of non-equilibrium thermodynamics, a qualitative description only will be presented, and a set of final equations, considered most suitable for the interpretation of these nonequilibria phenomena will be given. A detailed discussion was presented by Fujita (40, 44) and Eisenberg (16, 43) showed that in multicomponent systems comprising a macromolecular solute and a number of low-molecular weight components, the reasonable simplifying assumption that the total chemical potential of the low-molecular weight components is uniform throughout the cell leads to less involved and more tractable expressions. For an acquaintance with non-equilibrium thermodynamics consider Katchalsky and Curran (45), based on the pioneering contributions of Onsager, Prigogine, de Groot and Kirkwood.

A simple way to write down sedimentation equations is to consider first the simpler case of a non-ionic three-component system, consisting of one solvent, comp. 1, one macromolecular solute, comp. 2, and a non-ionic cosolvent, comp. 3, the nomenclature we have used in the discussion of the equilibrium phenomena. Complicated expressions can be reduced to simpler form considering $c_2 \to 0$. Diffusion coefficients $D_2$ are derived from the flow of comp. 2 in the absence of an external field and the sedimentation coefficient $s$ is defined as the velocity per unit field in homogeneous solution, that is the $dc_J/dr$ are equal to zero. In the three-component system considered, $C_3$ (on a molar basis) is fairly large and $C_2$ is usually much smaller than $C_3$; also $D_3$ is usually much larger than $D_2$ because of the large disparity in size between the macromolecular comp. 2 and the low-molecular weight comp. 3 (the flows are referred to the solvent comp. 1). Under these conditions it may be assumed that in the VS experiment, while the solution just ahead of the boundary is homogeneous with respect to comp. 2, comp. 3 (and all other low-molecular weight components which might be present in the solution) is essentially in equilibrium, that is $\bar{\mu}_3$ is constant and the gradient $-d\bar{\mu}_3/dr$ is equal to zero (similar considerations apply to the diffusion equation). This then leads to the simple expressions for the sedimentation and diffusion of the macromolecular comp. 2:

$$s_2 = (\partial\rho/\partial m_2)_{P,\mu}(V_m/M_2 f_2), \quad (57)$$

$$D_2 = (d\Pi/dm_2)(V_m/M_2 f_2), \quad (58)$$

where $f_2$ is the frictional coefficient of 1 g of component 2. Division of Eq. (57) by Eq. (58) yields the simple expression, valid at finite concentrations of comp. 2,

$$s_2/D_2 = (\partial\rho/\partial m_2)_{P,\mu}/(d\Pi/dm_2). \quad (59)$$

Transforming to the $c$-concentration scale, and at vanishing concentrations $c_2$, following expansion of the virial series Eq. (4),

$$s_2/D_2 = (\partial\rho/\partial c_2)_{P,\mu} M_2/RT, \quad (60)$$

which is analogous to the Svedberg equation (8) here extended to multicomponent systems by formulation of the density increments earlier described in this work.

It was possible to show that absolute information on molar masses, intermolecular interactions, and distribution of mass in charge-carrying macromolecules can be obtained by equilibrium studies without explicit consideration of the ionic polyelectrolyte nature of the macromolecular and cosolvent components. To be sure, this ionic nature influences the molecular conformation, interaction with other macromolecules, and interaction with small ions in solution. Yet, for many intents and purposes, and in particular for the applications discussed earlier, a thermodynamic framework constructed on the basis of choosing components in electro-neutral combinations from the ionized species, is adequate to solve the aforementioned problems. However when we presently turn to the investigation of transport phenomena it becomes

necessary to restate the problem and to reconsider the question of the extent to which the above conclusions still hold with respect to VS and diffusion experiments, which involve motion of the macromolecules in a nonequilibrium process. Such typical ionic processes as electrophoresis and conduction involve properties of the single constituent ions, and (at finite concentrations) interactions between them. We limit ourselves here however to the study of phenomena in which, as before, electro-neutrality is preserved, even though deviations may be extant. Our query is restricted to the significance of the so called "charge effects" which may arise in the transport of electroneutral components after a microscopic separation of the species in the process of flow. If charged macromolecules are sedimented in the ultracentrifuge, the higher sedimentation tendency of the macromolecules, as contrasted to the slower sedimenting counterions, results in a microscopic separation of charge between the bottom of the cell and the macromolecular boundary (46). This separation establishes an electric field in the intervening column of solution which, in principle, slows down the macromolecular ion and speeds up the counterions. Both ions in fact move with equal intermediate velocity. This is called the *primary charge effect*. In diffusion, on the other hand, the small counterions diffuse faster than the large macroions and pull the latter after them. Thus, whereas in sedimentation the charge effect slows down the large ions (as opposed to an equivalent non-charged particle), in diffusion the contrary is observed. Charge neutrality must be maintained in all situations and a much debated point is whether or not the primary charge effect vanishes at infinite dilution of macromolecules.

With increasing concentration of low-molecular weight salt comp. 3 the charges on the macromolecules are screened, the sedimentation coefficient increases and the diffusion coefficient decreases. This known as the *primary salt effect*. Do these coefficients ever approach the values for uncharged macromolecules when simple salt concentration increases "to infinity"? Distinction must be made between almost rigid globular macromolecules, such as proteins for instance, and chain-like charged macromolecules, such as nucleic acids for instance, whose conformations are sensibly affected by ionic strength. In the first case one might indeed approach (by extrapolation to high ionic strength) the sedimentation and diffusion coefficients of the uncharged species, in the second case one faces an inextricable mixture of charge and conformation effects.

The *secondary salt effect* arises from unequal mobilities and sedimentation coefficients of the ions of comp. 3 in mixed polyelectrolyte-electrolyte solutions. It is less fundamental and can be minimized or eliminated by choosing a supporting electrolyte, such as NaCl for instance, in which both cation and anion have nearly identical sedimentation coefficients.

Although the charge effects in the sedimentation and diffusion of multicomponent polyelectrolyte systems have been extensively investigated both on theoretical and experimental grounds, the problem has not been solved in a satisfactory way. A simple question which has not been answered in unequivocal terms, refers to the problem whether combination of the sedimentation and diffusion coefficients (as is done for non-ionic systems in the Svedberg equation) leads to a well defined value for the molecular weight in the limit of infinite dilution. Another difficulty concerns the evaluation of the hydrodynamic properties of macromolecular coils with excluded volume and the extent to which these coils are freely draining or shield the motion of small ions or solvent molecules. For space filling globular macromolecules hydrodynamic uncertainties may arise from the odd shape of the protein, for instance, or ill defined hydration layers.

The difficulties enumerated above need not lead to an abandonment of the use of sedimentation and diffusion for the characterization of charged species. At high enough concentrations of salt the radius of the Debye-Hückel ionic atmosphere is small when compared to the dimensions of the bulky macromolecules, and with judicious choice of simple salt systems (to avoid secondary charge phenomena) the effects of charge in sedimentation and diffusion may be reduced to a small contribution. Equations to use for sedimentation and diffusion derived by the methods of irreversible thermodynamics (3, 43), essentially reduce to Eqns (57) and (58) derived for non-ionic systems. In principle it is possible to proceed to an experimental verification of a chosen system by performance of ES, VS and dynamic light scattering (47, 48), to be discussed below, for the determination of diffusion coefficients. The latter can also be obtained to lesser accuracy from the broadening of sedimentation boundaries in the ultracentrifuge. The reader interested in the topic of this. section in a deeper way is referred to the references quoted, and further primary quotations stated therein.

## B. Scattering of Radiation

### 1. Light Scattering (LS)

Light scattering will now be discussed as a first example in which time averaged fluctuation theory is used in the interpretation of the forward scattering phenomenon. Forward scattering refers to scattering extrapolated to zero scattering angle $2\theta$ (Fig. 9), which is equivalent to angle-independent scattering if the macromolecular scattering particle is less than about $1/20$ in size of the wavelength $\lambda$ of the light *in vacuo*. Angle-dependent scattering will be referred to below (In

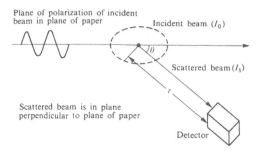

Plane of polarization of incident
beam in plane of paper

Incident beam ($I_0$)

Scattered beam ($I_s$)

Scattered beam is in plane
perpendicular to plane of paper

Detector

**Fig. 9** Geometry of light-scattering experiment. [From Ref. (3)]

light scattering the scattering angle is usually given as $\theta$ but $2\theta$ is used here as is customary in small angle x-ray and neutron scattering). Limitations in the size of this article prevent detailed derivations of the thermodynamics of the scattering phenomena, which were presented for OP derivations in the opening sections of this work. Again, the reader interested in a complete presentation is referred to the material quoted (2, 3, 7, 47–50).

Light incident on a particle from which scattering is to be observed subjects the sample to an electromagnetic alternating field with a frequency in the range of $10^{14}$–$10^{16}$ Hz. In response to this field, the electrons in the sample become a source of secondary dipole radiation. The induced electric dipole moment $p$ may be expanded in terms of the electric field. In somewhat simplified fashion

$$p = \alpha E_0, \qquad (61)$$

where $E_0$ is the field strength of the incident field and $\alpha$ is the polarizability of the particle. Elementary considerations applied to slowly moving scatterers, evaluation of the scattered intensity from the time-averaged square of the amplitude of the scattered light, and summation over $N$ particles per unit volume, lead to Rayleigh's equation for scattering of a dilute gas:

$$R = 16\pi^4 \alpha^2 N / \lambda^4, \qquad (62)$$

where $R$ is known as the Rayleigh factor, and is equal to $(I_s/I_0)r^2$ ($I_s$ is the light scattered from unit volume, $I_0$ is the intensity of the incident beam and $r$ is the distance from the sample to the radiation detector).

For the analysis of strongly interacting condensed systems, the solution is divided into a large number of volume elements containing many molecules, but with linear dimensions small with respect to $\lambda$. The polarizability $\alpha$ of each volume element $V$ (of which there are $V^{-1}$ per ml) fluctuates at each moment around its equilibrium value. As the latter is equal for all scattering volumes, only the mean-square-average fluctuations of $\alpha$ contribute to the scattering. For vertically polarized incident radiation

$$R = (16\pi^4 / \lambda^4 V)\langle (\Delta\alpha)^2 \rangle, \qquad (63)$$

and the remaining problem becomes the evaluation of the ensemble average $\langle (\Delta\alpha)^2 \rangle$. Fluctuations in polarizability, for light, are related to fluctuations in refractive index $n$, and these in turn to fluctuations of concentration of the macromolecular component, linked to osmotic work. Appropriate choice of thermodynamic variables for the case of multicomponent solutions in the statistical-mechanical analysis leads to expressions which, for the three-component system under consideration, are represented in simple form by an equation (2, 3)

$$(\partial n/\partial c_2)^2_{P,\mu}[K\,c_2/\Delta R(0)]$$
$$= (1/RT)(\mathrm{d}\Pi/\mathrm{d}c_2), \quad (64)$$

which bears strong resemblance to the equilibrium sedimentation Eq. (54). The refractive index increment, under the appropriate restrictions, in analogy to the density increment Eq. (38), is given by (2, 3)

$$(\partial n/\partial c_2)_{P,\mu} = (\partial n/\partial c_2)_{P,m} + (\partial n/\partial c_3)_{P,m}$$
$$\times (1 - c_3\bar{v}_3)\xi_3, \qquad (65)$$

an experimentally accessible quantity by refractometry of solutions at dialysis equilibrium, determined at the scattering wavelength (The difference between $(\partial n/\partial c_2)_{P,\mu}$ and $(\partial n/\partial c_2)_\mu$ is inconsequential); $K = (2\pi^2/N_{Av})(n_0^2/\lambda^4)$ is an optical constant, $N_{Av}$ is Avogadro's number; $\Delta R(0)$ refers to the forward (at zero scattering angle) scattering in excess of solvent scattering, $(d\Pi/dc_2)$ may be expanded in a virial series as before [cf Eq. (4)]:

$$(\partial n/\partial c_2)^2_{P,\mu}[K\,c_2/\Delta R(0)]$$
$$= (1/M_2) + 2A_2c_2 + \ldots , \quad (66)$$

Again the choice of the units of $c_2$ determines the value of $M_2$, and the same considerations apply as before.

The considerations presented above refer to a homogeneous macromolecular comp. 2. If the macromolecular particles are heterodisperse in size, but uniform with respect to $(\partial n/\partial c_2)_\mu$, then Eq. (66) yields $M_w$, the weight average molar mass. Effects of heterogeneities in composition on the molar masses determined by light, x-ray and neutron scattering will not be further considered here (17).

The added value in light scattering studies is in the determination of the angular dependence of scattering (49), permitting the evaluation of the size and shape of macromolecular particles exceeding in size $1/20$ the value of $\lambda$. Particle sizes range from small to medium in the case of proteins, for which small-angle x-ray scattering may be the method of choice, to large in the case of nucleic acids, which have been extensively studied by LS. The Rayleigh factor is now redefined as $\Delta R(h) = r^2\Delta/(h))/I_0(1 + \cos^2 2\theta)$, where $h = 4\pi \sin\theta/\lambda$, and the scattering Eq. (66) modifies to

$$(\partial n/\partial c_2)^2_{P,\mu}[K\,c_2/\Delta R(h)]$$
$$= [1/i_n(h)M_2] + 2A_2c_2 + \ldots , \quad (67)$$

where $i_n(h)$ the particle scattering factor, often designated $P(\theta)$ in polymer solution light scattering practice. In LS it is usual to use the Zimm plot (49) for the double extrapolation to zero angle and zero concentration The method is equivalent to Guinier's method (51), used in small-angle x-ray scattering, if the exponential in

$$\Delta R(h)/\Delta R(0) = \exp(-R_g^2 h^2/3), \quad (68)$$

is expanded for small enough values of $h$; $R_g$ is the radius of gyration of the particle, representing a measure of its size (49). For rigid particles, or for a given fixed conformation of a flexible particle, $R_g^2 = \sum m_i r_i^2 / \sum m_i$, where $m_i$ mass elements are considered, each located at a distance $r_i$ from the center of the mass; for spherical particles of radius $R$, for instance, $R_g^2 = (3/5)R^2$ and for long straight rods of length $L$, $R_g^2 = L^2/12$. Within the limits of the applicability of the Zimm plot, a straight-line plot remnant from the days before computers made their appearance in science, a plot of $(\partial n/\partial c_2)^2_{P,\mu}[K\,c_2/\Delta R(h)]$ vs $\sin^2\theta$ should yield a straight line, the slope of which is proportional to $R_g^2$. Equations have been derived for $P(\theta)$ for spheres (native proteins), Gaussian random coils (nucleic acids) and rods (collagen, nucleic acids, virus particles), and Fig. 10 is a plot of $P(\theta)^{-1}$ vs $R_g^2 h^2$ for these various models (49). It is seen that the extent of the validity of the Zimm linear plot is limited and the applicability of the equation of either the Gaussian coil, or the rigid rod, extends to significantly larger values of $R_g^2 h^2$. Reference (49) is a good representation of the evaluation of light scattering data of biological particles, however the classical techniques used at the time of publication have now been replaced by the present use of coherent laser radiation, significant reduction in sample size, and sophisticated computer evaluation.

An additional welcome aspect of light scattering, made possible by the advent of coherent laser radiation, is dynamic light scattering already mentioned (3, 7, 47, 48). Again, it will be discussed very briefly in the present context.

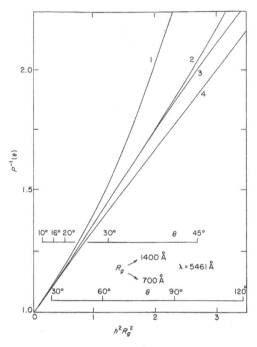

**Fig. 10** The function $P^{-1}(\theta)$ at low values of $h^2 R_g^2$ for spheres (1), Gaussian coils (2), and rigid rods (3); curve (4) is the Zimm limiting slope, independent of the shape of the particles. Horizontal bars indicate range covered at various scattering angles $\theta$ at 5461 Å for two particles with $R_g = 700$ Å, respectively. [From Ref. (49)]

It can be used for the determination of diffusion coefficients, complementing the determination of sedimentation coefficients in the analytical ultracentrifuge, as well as electrophoretic mobility superimposed on Brownian motion, rotational diffusion of asymmetric molecules, internal motion, chemical kinetics for binding of ethidium bromide to DNA by determining the temporal correlation of thermodynamic concentration fluctuations as followed by fluorescent correlation spectroscopy of the fluorescent reaction product (52). The much more sophisticated technique of fluorescence correlation spectroscopy (53), allowing the study of the dynamics of single macromolecules, will not be elaborated here.

In current dynamic light scattering analysis the fluctuations of the photo-current are determined directly. This is known as intensity-fluctuation spectroscopy and the second-order correlation function which leads to the evaluation of the diffusion coefficient, may be determined directly.

In Fig. 11 the angular-independent concentration-dependent static LS of halophilic malate dehydrogenase (hMDH) in 4M NaCl is shown (54). Upon lowering the simple salt concentration hMDH undergoes time-dependent denaturation from the native tetramer to denatured monomeric structure (54). In Fig. 12 the time dependent decrease of the intensity of scattering is shown, as well as the dynamic LS Contin distribution of particle sizes after 1, 24 and 118 hours of denaturation. The interesting result is that the size and size distribution is not affected by time, leading to the conclusion that upon dissociation to monomers and denaturation, the size of the denaturing

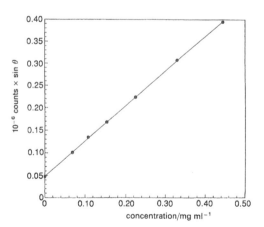

**Fig. 11** Relative light scattering intensities of hMDH without coenzyme NADH in 4 M NaCl, 5 mM Tris, pH 8, as a function of enzyme concentration mg/ml. Wavelength $\lambda = 5145$ Å. Counts were taken in the angular scattering range $\theta = 45 - 135°$ and averaged after multiplying by $\sin \theta$, as no angular dependence of scattering was observed. The point at zero concentration is solvent. [From Ref. (54)]

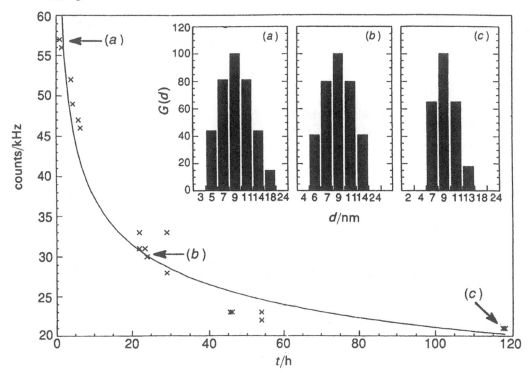

**Fig. 12** Relative light-scattering intensities *vs.* time for hMDH complexed with the coenzyme NADH in 0.8 M NaCl, 5 mM Tris, pH 8, 20°C. Insert: dynamic light-scattering Contin distribution of particle sizes after (a) 1, (b) 24 and (c) 118 hrs. [From Ref. (54)]

enzyme is unchanged. This is confirmed by radius of gyration determination by LS intensity, x-rays and neutron scattering experiments (55, 56). The velocity sedimentation behavior in the ultracentrifuge (54) is shown in Fig. 13, indicating two peaks for the native and the denatured enzyme (in the VS Eq. (57), it is seen that the sedimentation coefficient is determined by both the molar mass and the frictional coefficient). The change in molar mass by denaturation from the native tetramer (M = 113,000 g/mol) to the monomer value [M = 32,638 g/mol, this is the value derived from the sequence (57)] is confirmed by the LS intensity measurements (54). Recent confirmation of the enzyme structure and predicted water and salt interactions are provided by the medium resolution hMDH x-ray dif-

fraction study (29) and by the higher resolution halophilic ferredoxin x-ray structure (30) and still unpublished high resolution work (F Frolow, Microbiology and Biogeochemistry of Hypersaline Environments, Jerusalem, June 22–26, 1997).

## 2. Small Angle X-ray Scattering (SAXS)

The forward small-angle scattering of x-rays (58) is given by

$$I(0)/I_0 = (16\pi^4/r^2\lambda^4 V)\langle(\Delta\alpha)^2\rangle. \quad (69)$$

Because of the much smaller wavelength of x-rays as compared to light, scattering angles in SAXS must perforce be much smaller than in LS to enable proper extrapolation to forward scattering conditions.

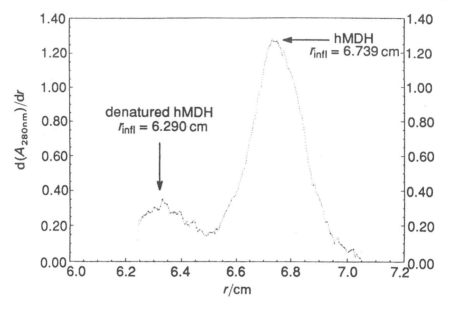

**Fig. 13** Unsmoothed derivative plot $dA_{280}/dr$ vs. $r$ of hMDH complexed with the coenzyme NADH after 120 min VS, in 1.25 M NaCl, 5 mM Tris, pH 8, 42,000 r.p.m., 20°C. [From Ref. (54)]

For further evaluation of SAXS the scattering $I_{el}$ of an electron is introduced

$$I_{el}/I_0 = (16\pi^4/r^2\lambda^4)\alpha_{el}^2, \tag{70}$$

where $\alpha_{el}$ is the polarizability of an electron. Combination of Eqns (70) and (69) yields

$$I(0) = (I_{el}/V\alpha_{el}^2)\langle(\Delta\alpha)^2\rangle. \tag{71}$$

Temperature and pressure fluctuations are disregarded and concentration fluctuations only considered

$$d\alpha = \sum_i (\partial\alpha/\partial m_i)_{P,m}\,dm_i, \tag{72}$$

where the summation is over components 2 and 3 and the $m_i$ are molalities. Subscript $m$, as before, signifies constant molality, besides the component being varied. Also

$$(\partial\alpha/\partial m_i)_{P,m} = (\partial\rho_{el}/\partial m_i)_{P,m}\alpha_{el}V_m, \tag{73}$$

where $\rho_{el}$ is the density in electrons per ml and scattering is considered from a volume $V_m$ containing a fixed number of electrons of component 1.

Substitution of Eq. (73) into Eq. (72), squaring and taking the average $\langle(\Delta\alpha)^2\rangle$, and introducing into Eq. (71), leads to

$$I(0) = I_{el}V_m\sum\sum(\partial\rho_{el}/\partial m_i)_{P,m}$$
$$(\partial\rho_{el}/\partial m_j)_{P,m}\langle\Delta m_i\,\Delta m_j\rangle. \tag{74}$$

Similarly to LS, the problem now reduces to the evaluation of the concentration fluctuations $\langle\Delta m_i\,\Delta m_j\rangle$. These are here multiplied by electron densities rather than refractive index increments. Further steps involve evaluation of the concentration fluctuations and transformation from molalities to volume concentrations, to give the final expression for the scattering of x-rays

$$(\partial\rho_{el}/\partial c_i)_{P,\mu}^2[I_{el}c_i/\Delta I(0)]$$
$$= (1/kT)(d\Pi/dc_i). \tag{75}$$

Eq. (75) is entirely analogous to the corresponding quantity Eq. (64) for LS, the OP is expanded as before in a virial series [Eq. (4)] and the choice of concentration units is again arbitrary, as for ES and LS.

The practical difference between this and the previous two methods discussed is that $(\partial \rho_{el}/\partial c_i)_{P,\mu}$, the increment of electron density at constant $P$ and $\mu_3$ is not a directly measurable quantity in an auxiliary experiment. We can measure $(\partial \rho/\partial c_i)_\mu$, and then transform it into proper electron density; the difference in the restrictions $P$, $\mu_3$ on the one hand, and $\mu$ on the other, is inconsequential as seen before, and will be disregarded.

For a two-component system

$$(\partial \rho_{el}/\partial c_i)_P = I_2 - I_1 \bar{v}_2 \rho^0, \qquad (76)$$

where $I_1$ and $I_2$ are electrons per g of the respective component. All other quantities are in units of mass, as defined above. For the three-component system

$$(\partial \rho_{el}/\partial c_i)_\mu = (I_2 + \xi_3 I_3) - \rho_{el}^0$$
$$\times (\bar{v}_2 + \xi_3 \bar{v}_3). \qquad (77)$$

$\rho^0$ as before is the density (g/ml) of the solvent mixture, in absence of comp. 2, and $\rho_{el}^0$ is the corresponding quantity in electrons/ml. Combination of Eq. (77) with the measurable mass density increment Eq. (39) yields

$$(\partial \rho_{el}/\partial c_2)_\mu = I_2$$
$$- [(I_1 - I_3)/(1 + w_3)]\xi_3$$
$$- [(I_1 + w_3 I_3)/(1 + w_3)]$$
$$\times [1 - (\partial \rho/\partial c_2)_\mu] \qquad (78)$$

where we have, as before, used $w_3 = c_3/c_1$ and (for $c_2 \to 0$), $c_1 + c_3 = \rho^0$ and $c_1 I_1 + c_3 I_3 = \rho_{el}^0$. Eq. (78) shows that although $(\partial \rho_{el}/\partial c_2)_\mu$ is not directly measurable, it may be calculated from quantities which are experimentally accessible. For the evaluation of the $I_i$, only the basic chemical composition of the components is required. Thus $I_2 = N_{Av} \sum n_i I_i / M_2$, where $M_2$ is the molar mass containing $n_i$ atoms with $I_i$ electrons each; $(\partial \rho/\partial c_i)_\mu$ is available from density measurements or ES; $\xi_3$ need not be known

**Table 2** Neutron Scattering Length $b_i$, per g in $H_2O$ Buffers and $I_i$, Electrons per g. [Refs. (55, 56)].

| Neutrons (H2O) (cm/g) | x-rays (e/g) |
|---|---|
| $b_1 = -5.62 \times 10^9$ | $I_1 = 3.343 \times 10^{23}$ |
| $b_2 = 14.8 \ \times 10^9$ | $I_2 = 3.23 \ \times 10^{23}$ |
| $b_3 = 13.59 \times 10^9$ | $I_3 = 2.885 \times 10^{23}$ |

1 electron/gram is equivalent to $2.81 \times 10^{-13}$ cm g$^{-1}$.

precisely as the appropriate term in Eq. (78) contains the difference $(I_1 - I_3)$ as a multiplier. Values of the $I_i$ electron per gram parameters for the system water/hMDH/NaCl are listed in Table 2, in conjunction with the neutron scattering length parameters, to be discussed in the following (55, 56). Details of SAXS evaluation have been given (17) and the limitations of the fluctuation theory should be kept in mind (3).

Problems relating to the angular dependence of scattering are beyond the fundamental subject matter of this chapter and will not be treated extensively. The problem though is so basic to the methods of LS and SAXS that we will briefly compare here the range of $h$ values in which these two methods are applicable, as well as the information which can be obtained by either method. Whereas in LS the radius of gyration is the only parameter which can be obtained from the angular dependence of scattering, SAXS is applicable over a broader range of angular scattering and additional information with respect to size and shape can be obtained, in particular due to the x-ray wavelength, which is in the Angstrom range (59). The range of $h$ in which the two methods are applicable, as well as the information which can be obtained by either method is best seen by examination of Fig. 14 taken from Finch and Holmes (60). In the forward direction of scattering ($n = 0$) interference effects are eliminated [$P(h)$ is equal to unity]. In scattering at all other angles interference occurs for a solution composed of "large" particles. In a

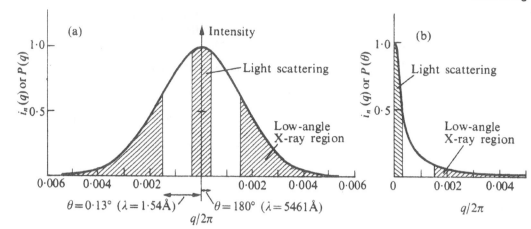

**Fig. 14** Comparison of LS and SAXS. Particle scattering factor *vs* $h/2\pi$; (a) Small spherical particle (radius 100 Å); (b) Rod particle of length 3000 Å and diameter 150 Å. This corresponds to tobacco mosaic virus particle. [From Ref. (60)]. With modern techniques the range of SAXS may be extended lower, to overlap with the LS range.

diffraction experiment on randomly oriented dissolved particles one records, in the limit of vanishing concentration $c_2$ the spherical average of the square of the Fourier transform per particle. Figure 14 shows the origin peak of the particle transform (actually the normalized $P(h)$ function for two particles): (1) a small spherical particle of radius 100 Å; and (2) a rod-like particle of length 3000 Å and diameter 150 Å, corresponding, respectively, to a globular protein and a molecule similar to tobacco mosaic virus. The regions explored by either LS or SAXS are shaded.

The wavelength most commonly used in LS were the mercury lines at 4358 Å and 5461 Å, however presently the use of laser beams has extended this range over almost all the visible spectrum, and the scattering angles explore a range from about 10° to 150°. The corresponding wavelength in SAXS is 1.54 Å for the Cu Kα radiation, and scattering angles as low as 0.0015–0.00015 rad. In both instances, upon the extrapolation of the scattering intensity to $h = 0$ and $c = 0$, the true molar mass is obtained. For molar mass

determinations it is necessary to have an absolute calibration (possibly in terms of a secondary standard) of the scattered intensity, and the values of the refractive index and electron density increments respectively. As already mentioned the shape of the origin peak in Fig. 14 around $h = 0$ and $c_2 = 0$ is independent of the shape of the particle and depends on the distribution of the mass in the particle only, as expressed in the value of $R_g$, the radius of gyration. Neither absolute scattering values nor the values of the constants or the values of the refractive index or electron density increments are required for the determination of $R_g$. In Table 3 are given $R_g$ values for hMDH at various concentrations of NaCl (55, 56) from both SAXS and SANS experiments, to be discussed below. The agreement is quite good and corresponds to the value expected for the solvent and cosolvent bound hMDH molecule (54). Additional information is available for the shape of hMDH from SAXS and SANS at higher scattering angles (3, 56), and the hMDH data have been confirmed by x-ray diffraction (29).

**Table 3** Radius of Gyration Values $R_g$ of hMDH, for Different Salt Conditions, from SAXS and from SANS. [Refs. (55, 56)]

| Salt conditions (NaCl, M) | $R_g$ (neutrons $H_2O$) (Å) | $R_g$ (neutrons $^2H_2O$) (Å) | $R_g$ (x-rays) (Å) |
|---|---|---|---|
| 1.0 | $31 \pm 0.5$ | | $31.4 \pm 0.9$ |
| 1.5 | | | $32.7 \pm 1.3$ |
| 2.0 | | | $31.2 \pm 0.7$ |
| 2.5 | $30 \pm 0.5$ | $29 \pm 0.5$ | |
| 3.0 | $30 \pm 0.5$ | $29 \pm 0.5$ | $31.5 \pm 1.0$ |
| 4.0 | $29.5 \pm 0.5$ | $29 \pm 0.5$ | $32.4 \pm 0.8$ |

### 3. Small Angle Neutron Scattering (SANS)

Having discussed LS and SAXS the extension to SANS does not introduce new principles, either with respect to the thermodynamics or to the angular dependence of scattering. There are however interesting differences which lead to complementarities adding major interest to SANS studies. Scattering here is caused by the molecular nuclei and a scattering length $b_i$ (cm) is associated with the coherent scattering cross-section of each atom $i$ in a given molecule (61). The wavelength in neutron scattering is variable in the Ångstrom and nanometer range. A scattering length $b_J = N_{Av} \sum n_i b_i / M_J$ can now be defined for each component. Similarly to the previously (for x-rays) derived electron densities per g of component, the calculation is based on the chemical composition of the component and in particular, for macromolecular components, only the ratio $n_i/M_j$ need to be known. The forward neutron scattering can now be expressed as

$$(\partial \rho_n / \partial c_2)^2_{P,\mu} [c_2/I_s(0)]$$
$$= (1/kT)(d\Pi/dc_2) \quad (79)$$

where $\rho_n$ (cm$^{-2}$) is the scattering length density (per ml of solution), with proper corrections for geometrical factors and detector configuration (17). As before, the choice of concentration units is left to the experimentor. Upon use of the virial

expansion Eq. (4), and with the concentration units $c_2$(g/ml) the molar mass (g/mole), is obtained

$$(1/N_{Av})(\partial \rho_n / \partial c_2)^2_{P,\mu} [c_2/I_s(0)]$$
$$= (1/M_2) + 2A_2 c_2 + \dots \quad (80)$$

The scattering length density increments $(\partial \rho_n / \partial c_2)_{P,\mu}$, which are in practice indistinguishable from $(\partial \rho_n / \partial c_2)_\mu$, can be obtained from the mass density increment $(\partial \rho / \partial c_2)_\mu$ in similar fashion to the x-ray case [Eqns (77) and (78)], with the neutron scattering lengths $b_i$ substituting for the electrons per gram $I_i$.

$$(\partial \rho_n / \partial c_2)_\mu = (b_2 + \xi_3 b_3)$$
$$- \rho_n^0 (\bar{v}_2 + \xi_3 \bar{v}_3) \quad (81)$$

The interesting aspect of neutron scattering can be appreciated by an examination of Table 2. Whereas the $I_J$ values for x-rays are nearly identical, $b_1$ (for water) is negative, in distinction to all the other positive values. Indeed, the contribution of hydrogen provides the negative sign, and replacement of hydrogen by deuterium leads to a positive value for $D_2O$. More about H/D substitution later. The way in which the $(\partial \rho_n / \partial c_2)_\mu$ values affect the neutron scattering results provides complementary information, as can be seen in the following.

In the invariant particle hypothesis in which the macromolecule associates with $B_1$ and $B_3$ g of solvent and cosolvent respectively, per g of protein, the mass density increment can be written as Eq. (44). Similarly (17) the electron and neutron-scattering length density increments can then be written:

$$(\partial \rho_{el}/\partial c_2)_\mu = (I_2 + B_1 I_1 + B'_3 I_3)$$
$$- \rho_{el}^0 (\bar{v}_2 + B_1 \bar{v}_1 + B'_3 \bar{v}_3) \quad (82)$$

and

$$(\partial \rho_n/\partial c_2)_\mu = (b_2 + B_1 b_1 + B'_3 b_3)$$
$$- \rho_n^0 (\bar{v}_2 + B_1 \bar{v}_1 + B'_3 \bar{v}_3) \quad (83)$$

if the "particle" volume $V_T = (\bar{v}_2 + B_1 \bar{v}_1 + B'_3 \bar{v}_3)$ is constant, as well as $B_1$ and $B_3$, then $(\partial \rho/\partial c_2)_\mu$, $(\partial \rho_{el}/\partial c_2)_\mu$ and $(\partial \rho_n/\partial c_2)_\mu$ vs. $\rho^0$, $\rho_{el}^0$ and $\rho_n^0$, respectively, are straight lines with the same slope which is equal to $V_T$. It is then possible to determine three parameters $B_1$, $B'_3$ and $\bar{v}_2$, for instance, by solving the three equations obtained from Eqns (44) and (83): (1) slope $= V_T = (\bar{v}_2 + B_1 \bar{v}_1 + B'_3 \bar{v}_3)$; (2) "mass" intercept at $\rho^0 = 0 = (1 + B_1 + B'_3)$ and (3) ("neutron" intercept at $\rho_n^0 = 0 = (b_2 + B_1 b_1 + B'_3 b_3)$. Eqns (82) and (83) could also be used in similar fashion, but a combination of Eqns (44) and (82) is not useful because of the similarity of $I_1$, $I_2$ and $I_3$.

A novel plot was proposed (54), which allows a joint analysis of all the scattering and mass density increments. Eqns (44), (82) and (83) are each divided through by their respective intercept at zero solvent density. The resulting "reduced" density increments $(\partial \rho/\partial c_2)_\mu^*$ are dimensionless and can be plotted together as a function of solvent density also divided by the appropriate intercept. All points should fall on the same "straight" line, within the invariant particle hypothesis, with an intercept of 1 and a slope of $V_T$.

$$(\partial \rho/\partial c_2)_\mu^* = (\partial \rho_x/\partial c_2)_\mu /$$
$$(\bar{v}_2 + B_1 x_1 + B'_3 x_3)$$

$$= 1 - [\rho_x^0/(\bar{v}_2 + B_1 x_1$$
$$+ B'_3 x_3)] V_T$$
$$= 1 - (\rho_x^0)^* V_T \quad (84)$$

where $x$ is blank for mass, $el$ and $n$ for x-rays and neutrons, $x_2$ and $x_i$ are unity for mass, $I_2$ and $I_i$ for x-rays and $b_2$ and $b_i$ for neutrons. By using Eq. (84) the mass density, x-rays and NS density increments were plotted together (54) in Fig. 15. It is seen that the same straight line goes through the neutron, x-ray and mass points with an intercept of unity, showing the data to be self-consistent. From these values can be calculated the interaction parameters in g per g of protein, $B_1$ and $B_3$, and the partial specific volume of the protein. The interaction parameters are found to be 0.41 and 0.40 g of water, 0.08 and 0.10 g of salt per g of protein in NaCl and KCl respectively. The partial specific volume is calculated to be $0.73 \text{ cm}^3\text{g}^{-1}$, reasonably close to the value calculated from the amino-acid composition including counterions to the charged amino-acids (B Kernel, thesis. Grenoble, 1997).

The complementarity between the neutron data, on one hand, and the x-ray and mass data, on the other, is well demonstrated in Fig. 15. Neutron data, being close to null solvent scattering do not define a precise slope; they are therefore not sensitive to the volume of the particle, however they are very sensitive to its composition through the value of the intercept $(b_2 + B_1 b_1 + b'_3 b_3)$, which is well defined. On the other hand, the mass and x-ray values are quite far from the intercept at zero solvent density, but they define the slope of the line and, therefore, are sensitive to the total volume of the particle.

A comparison of the angular dependence of scattering of hMDH by SAXS and NS has been made (56), and leads to comparable results as seen in Table 3. Sedimentation, scattering of light, x-rays and neutrons for DNA solutions, have also been summarized (62).

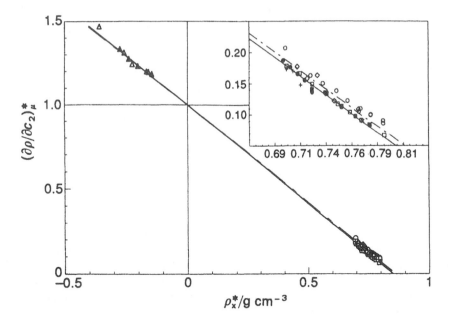

**Fig. 15** Complementary superposition of normalized data from ES, SAXS and SANS of hMDH solutions: $M_2(\partial\rho/\partial c_2)^*_\mu$ vs $\rho^*_x$ (see text). In NaCl ($\bigcirc$), from densimetry; $\square$, $+$, from $s/D$; $\times$, from ES; $\diamond$, from SAXS; $\triangle$, from SANS. In KCl: $\blacksquare$, from ES, $\bullet$, from $s/D$; $\blacktriangle$, from SANS. The two straight lines correspond to a total specific volume of $1.165\ cm^3/g$: [($\_._\_$), NaCl] or $1.178\ cm^3/g$ [($\_$), KCl]. Insert: enlargement of the lower part of the plot (mass and x-ray data). [From Ref. (54)]

As already mentioned, valuable results are obtained by the exchange of H by D in both the solvent and in the macromolecule (17), but will not be discussed in detail here. The present treatment frees the experimentor from the restrictive requirement that for the interpretation of NS experiments it is necessary to achieve scattering length matching conditions. It is the good fortune of neutron scattering that $D_2O/H_2O$ exchange makes the realization of such conditions rather convenient. This is in distinction to x-ray scattering in which the electron density matching is a rather difficult goal. To realize its full potential NS will be applied to systems in which, in addition to the standard $D_2O/H_2O$ scattering length variation technique, other contrast reagents will find increasing use. Thus, biological macromolecules of complex shape and function may associate with water in various ways. All water associated with the particle in one way or another will exchange with $D_2O$ until the composition of all water surrounding the particle is the same as that of the bulk mixture. No information on solvation and particle volume is obtained from forward scattering unless we turn the tables around and calculate a molar volume, *a priori*, assuming a value for the molar mass, which is now easily obtained from sequence or mass spectrometry sources. This a result which can always be obtained if the correct partial specific volume $\bar{v}_2$ is known. From it derives the statement that neutron forward scattering with $D_2O$, $H_2O$ mixtures determines a "dry" particle volume. If one now uses additional components such as glycerol, sugars or salts, which are excluded to the particle hydration areas, then from the values of $(\partial\rho/\partial c_2)_\mu$, $(\partial\rho_{el}/\partial c_2)_\mu$ and $(\partial\rho_n/\partial c_2)_\mu$ we can extract the interaction parameter $\xi_1$, which, when properly interpreted, may yield the volume of regions excluded to these reagents. Further distinction of "water-filled volumes associated with biological particles" can be achieved if other low or higher molecular weight particles are found which are selectively excluded from protected regions in the macromolecular domain. This approach, already documented earlier (Fig 5), considerably enlarges the usefulness of the complementary scattering methods. Recent remarkable developments in the technology of light-scattering intensity have been reviewed (65), as well as an overview has been given of current use of complementary solution study methodologies including light-scattering and novel ultracentrifugation procedures (66).

## IV.  CONCLUSIONS

The central points of this chapter can now be summarized. Low particle concentrations in three-component systems, a macromolecule, a solvent and a cosolvent were considered for simplicity of presentation only. Osmotic pressure and its virial expansion, constitutes the basis for equilibrium sedimentation and fluctuation scattering analysis [Eq. (4)]. The transition to ionized systems is presented, leading to the distribution of diffusible components [Eq. (28)]. Partial volumes and density increments are defined and subjected to precise evaluation and connectivity [(Eqns (38)–(42)]. Analytical ultracentrifugation leading to equilibrium sedimentation [Eqns (54) and (55)] is discussed in detail providing the basis connecting the concentration derivative of the osmotic pressure to the analysis of zero angle, forward scattering. From mass density increments at constant chemical potential of diffusible components interaction parameters may be derived by precision density measurements or equilibrium sedimentation (Fig.

4). Velocity sedimentation is discussed briefly emphasizing basic equations for sedimentation velocity and diffusion [Eqns (57) and (58)] deriving from nonequilibrium thermodynamics. Basic equations for total intensity (static) light scattering are derived for the three-component system [Eqns (64), (66) and (67)]. Dynamic light scattering is introduced but not discussed in detail. The basic equation for small angle X-ray scattering, relating it to the osmotic pressure derivative is derived [Eq. (75)]. The angular dependence of x-ray scattering is discussed and compared to light scattering (Fig. 14). Neutron scattering is evaluated [Eqns (79) and (80)] and compared to the previously discussed methods and the complementarity of the three scattering methods and equilibrium sedimentation is established [Eq. (84) and Fig. 15] in terms of the invariant particle hypothesis. Mass density, electron density and neutron-scattering length density increments are compared in this context [Eqns (44), (82) and (83)]. The angular dependence of small-angle x-rays and neutron scattering is compared (Table 3). Hydrogen/deuterium contrast variation and exchange and its role in neutron scattering is referred to in relation to earlier probing of macromolecular surfaces and interior spaces and volumes excluded by hydration (Fig. 5).

## REFERENCES

1.   C Tanford, J Reynolds. Protein chemists bypass the colloid/macromolecule debate. Ambix 46: 33–51, 1999.
2.   EF Casassa, H Eisenberg. Thermodynamic analysis of multicomponent systems. Adv Protein Chem 19: 287–395, 1964.
3.   H Eisenberg. Biological Macromolecules and Polyelectrolytes in Solution. Oxford: Clarendon Press, 1976.
4.   T Svedberg, H Rinde. The determination of the distribution of size of particles in disperse systems. J Am Chem Soc 45: 943–954, 1923.

5.  T Svedberg, R Fähreus. A new method for the determination of the molecular weight of the proteins. J Am Chem Soc 48: 430–438, 1926.

6.  T Svedberg, E Chirnoaga. The molecular weight of haemocyanin. J Am Chem Soc 50: 1399–1411, 1928.

7.  R Foord, E Jakeman, CJ Oliver, ER Pike, RJ Blagrove, E Wood, AR Peacocke. Determination of diffusion coefficients of haemocyanin at low concentration by intensity fluctuation of scattered laser light. Nature 227: 242–245, 1970.

8.  T Svedberg, K Pedersen. The Ultracentrifuge. London: Oxford University Press, 1940.

9.  JW Williams, KE van Holde, RL Baldwin, H Fujita. The theory of sedimentation analysis. Chem Rev 58: 7115–806, 1958.

10. HK Schachman. Ultracentrifugation in Biochemistry. New York: Academic Press, 1959.

11. GA Gilbert. General Discussion. Disc Faraday Soc 20: 68–71, 1955.

12. LM Gilbert, GA Gilbert, Sedimentation velocity measurements protein association. Meth Enzym 27D: 273–296, 1973.

13. J Vinograd. Sedimentation equilibrium in a buoyant dansity gradient. Meth Enzym 6: 854–870, 1963.

14. M Meselson, FW Stahl. The replication of DNA in Escherichia coli. Proc Natl Acad Sci USA 44: 671–682, 1958.

15. KE van Holde, WC Johnson, PS Hó. Principles of Physical Biochemistry. Englewood Cliffs, NY: Prentice Hall, 1998.

16. H Eisenberg. Multicomponent polyelectrolyte solutions. Part 1. Thermodynamic equations for light scattering and sedimentation. J Chem Phys 36: 1837–1843, 1962.

17. H Eisenberg. Forward scattering of light, x-rays and neutrons. Q Rev Biophys 14: 14–172, 1981.

18. H Eisenberg. Protein and nucleic acid hydration and cosolvent interactions: Establishment of reliable baseline values at high cosolvent concentrations. Biophys Chem 53: 57–68, 1994.

19  R Ghirlando, MB Keown, GA Mackay, MS Lewis, JC Unkeless, HJ Gould. Stoichimetry and thermodynamics of interaction between the Fc fragment of human IgG, and its low affinity receptor FcγRIII. Biochemistry 34: 13320–13327, 1995.

20. S-J Kim, T Tsukyiama, MS Lewis, C Wu. Interaction of the DNA binding domain of Drosophila heat shock factor with its cognate DNA site; A thermodynamic analysis using analytical ultracentrifugation. Protein Science 3: 1040–1051, 1994.

21. A Tardieu, P Vachette, A Gulik, M Le Maire. Biological macromolecules in solvents of variable density: Characterization by sedimentation equilibrium, densimetry and x-ray forward scattering and an application to the 50S ribosomal subunit of Escherichia coli. Biochemistry 20: 4399–4406,1981.

22. JA Schellman. Selective binding and solvent denaturation. Biopolymers 26: 549–550, 1987.

23. JA Schellman. A simple model for solvation in mixed solvents: Applications to the stabilization and destabilization of macromolecular structures. Biophys Chem 37: 121–140, 1990.

24. JA Schellman. The relation between the free energy of interaction and binding. Biophys Chem 45: 273–279, 1993.

25. SN Timasheff. The control of protein stability and association by weak interactions with water. How do solvents affect these processes? Ann Rev Biophys Biomol Struct 22: 63–97, 1993.

26. MI Kopka, AV Fratini, HR Drew, RE Dickerson. Ordered water structure around a B- DNA dodecamer. A quantitative study. J Mol Biol 163: 129–146, 1983.

27. G Scatchard. Physical chemistry of protein solutions. J Am Chem Soc 68: 2315–2320, 1946.

28. G.I. Makhatadze, PL Privalov. Protein interactions with urea and guanidinium chloride: A calorimetric study. J Mol Biol 226: 491–505, 1992.

29. O Dym, M Mevarech, JL Sussman. Structural features that stabilize halophilic malate dehydrogenase from an archaebacterium. Science 267: 1344–1346, 1995.

30. F Frolow, M Harel, JL Sussman, M Mevarech, M Shoham. Insights into protein adaptation to a saturated salt environment from the crystal structure of a halophilic 2Fe-2S ferredoxin. Nature Struct Biol 3: 452–457, 996: *erratum ibid.* 3: 1055, 1996.

31. T Arakawa, SN Timasheff. Stabilization of protein structure by sugars. Biochemistry 21: 6536–6544, 1982.

32. H Eisenberg, G Felsenfeld. Hydrodynamic studies of the interaction between nucleosome core particles and core histones. J Mol Biol 150: 537–555, 1981.

33. KO Greulich, J Ausio, H Eisenberg. Nucleosome core particle structure and structural changes in solution. J Mol Biol 186: 167–173, 1985.

34. TJ Richmond, JT Finch, B Rushton, D Rhodes, A. Klug. Structure of the nucleosome core particle at 7 Å resolution. Nature 311: 532–537, 1984.

35. SE Harding, AJ Rowe, JC Horton, eds. Analytical Ultracentrifugation in Biochemistry and Polymer Science. Cambridge: Royal Soc Chem, 1992.

36. TM Schuster, TM Laue, eds. Modern Analytical Ultracentrifugation. Boston: Birkhäuser, 1994.

37. SE Harding, O Bryan, eds. New developments in analytical ultracentrifugation and related macromolecular modelling techniques. European Biophys J 25: 305–487, 1997.

38. SJ Edelstein, HK Schachman. Measurement of partial specific volume by sedimentation equilibrium in $H_2O$-$D_2O$ solutions. Meth Enzym 27D: 82–98, 1973.

39. TV Chalikian, M Totrov, R Abagyan, KJ Breslauer. The hydration of globular proteins as derived from volume and compressibility measurements: Cross correlating thermodynamic and structural data. J Mol Biol 260: 588–603, 1995.

40. H Fujita. Mathematical Theory of Sedimentation Analysis. New York: Academic Press, 1962.

41. JW Williams, ed. Ultracentrifugal Analysis in Theory and Experiment. New York: Academic Press, 1963.

42. JM Creeth, RH Pain. The determination of molecular weights of biological macromolecules by ultracentrifuge methods. Prog Biophys Mol Biol 17: 217–287, 1967.

43. H Eisenberg, Sedimentation in the ultracentrifuge and diffusion of macromolecules carrying electrical charges. Bioph Chem 5: 243–251, 1976.

44. H Fujita. Foundations of Ultracentrifugal Analysis. New York: Wiley, 1975.

45. A Katchalsky, PF Curran. Nonequilibrium Thermodynamics in Biophysics. Cambridge: Harvard University Press, 1965.

46. KO Pedersen. On charge and specific ion effects on sedimentation in the ultracentrifuge. J Phys Chem 62: 1282–1290, 1958.

47. B Chu. Laser Light Scattering 2nd ed. San Diego: Academic Press, 1990.

48. KS Schmitz. Dynamic Light Scattering by Macromolecules. San Diego: Academic Press, 1990.

49. H Eisenberg. Light scattering and some aspects of small angle x-ray scattering. In: GL Cantoni, DR Davies, eds. Procedures in Nucleic Acid Research Vol 2. New York: Harper and Row, 1971. pp 137–175.

50. P Lindner, Th Zemb. Neutron, X-ray and Light Scattering: Introduction to an Investigative Tool for Colloidal and Polymeric Systems. Amsterdam: North-Holland, 1991.

51. A Guinier, G Fournet. Small Angle Scattering of X-rays. New York: Wiley, 1955.

52. EL Elson, D Magde. Fluorescence correlation spectroscopy. I. Conceptual basis and theory. Biopolymers 13: 1–xx, 1974.

53. M Eigen, R Rigler. Sorting single molecules: Application to diagnostics and evolutionary biotechnology. Proc Natl Acad Sci USA 91: 5740–5747, 1994.

54. F Bonneté, C EbeI, G Zaccai, H Eisenberg. Biophysical study of halophilic malate dehydrogenase in solution: Revised subunit structure and solvent interactions of native and recombinant enzyme. J Chem Soc Faraday Trans 89: 2659–2666, 1993.

55. M Reich, Z Kam, H Eisenberg. A small angle x-ray scattering study of halophilic malate dehydrogenase. Biochemistry 21: 5189–5195, 1982

56. G Zaccai, E Wachtel, H Eisenberg. Solution structure of halophilic malate dehydrogenase from small-angle neutron and x-ray scattering and ultracentrifugation. J Mol Biol 190: 97–106, 1986.

57. F Cendrin, J Chroboczek, G Zaccai, H Eisenberg, M Mevarech. Cloning, sequencing, and expression in Escherichia coli of the gene coding for malate dehydrogenase of the extremely halophilic archaebacterium Haloarcula marismortui. Biochemistry 32: 4308–4313, 1993.

58. H Eisenberg, G Cohen. An interpretation of the low-angle X-ray scattering of DNA solutions. J Mol Biol 37: 355–362. Erratum J Mol Biol 42: 607, 1969.

59. O Glatter, O Kratky, eds. Small Angle X-ray Scattering. London: Academic Press, 1982.

60. JT Finch KC Holmes. Structural studies of viruses. Methods in Virology, vol. 3. In: K Maramorosch, H Koprowski, eds. New York: Academic Press, 1967. pp 351–474.

61. B Jacrot, G Zaccai. Determination of molecular weight by neutron scattering. Biopolymers 20: 2413–2426, 1981.

62. H Eisenberg. Solution properties of DNA: sedimentation, scattering of light, x-rays and neutrons, and viscometry. Chapter 4.3 Landolt-Börnstein New Series Group VII: Biophysics vol 1/c Nucleic Acids, W Saenger, ed. Heidelberg: Springer, 1990. pp 257–276.

63. H Eisenberg. Light scattering intensity studies in multicomponent solutions of bio-

logical macromolecules. In: HZ Cummins, ER Pike, eds. Photon Correlation and Light Beating Spectroscopy. New York: Plenum, 1974. pp 551–567.

64. H Eisenberg, Y Haik, JB Ifft, W Leicht, M Mevarech, S Pundak. S. R. Interactions of proteins and nucleic acids with solutes in concentrated solutions of monovalent salts, relating to hydration, spectral transitions and inactivation of halophilic malate and glutamate dehydrogenases. In: SR Kaplan, M Ginzburg, eds. Ener-getics and Structure of Halophilic Microorganisms. Amsterdam: Elsevier, 1978, pp 13–32.

65. PJ Wyatt. Light scattering and the absolute characterization of macromolecules. Anal Chim Acta 272: 1–40, 1993.

66. P Hensley. Defining the structure and stability of macromolecular assemblies in solution: The re-emergence of analytical ultracentrifugation as a practical tool. Structure 4: 367–373, 1996.

# Index